D0085147

Thermodynamics and Heat Power

Thermodynamics and Heat Power

Sixth Edition

Kurt C. Rolle
University of Wisconsin—Platteville

Upper Saddle River, New Jersey
Columbus, Ohio

Library of Congress Cataloging-in-Publication Data
Rolle, Kurt C.
Thermodynamics and heat power/Kurt C. Rolle. — 6th ed.
p. cm.
ISBN 0-13-113928-2 (alk. paper)
1. Thermodynamics. 2. Heat engineering. I. Title.
TJ265.R84 2005
621.402′1—dc22 2004044598

Editor: Debbie Yarnell
Associate Editor: Kim Yahle
Production Editor: Kevin Happell
Production Coordination: Preparé, Inc.
Design Coordinator: Diane Ernsberger
Cover Designer: Keith Van Norman
Production Manager: Matthew Ottenweller, Deidra Schwartz
Marketing Manager: Jimmy Stephens

This book was set in Times Roman by Preparé, Inc. It was printed and bound by Courier Kendallville, Inc. The cover was printed by Phoenix Color Corp.

Pearson Education Ltd.
Pearson Education Singapore, Pte. Ltd.
Pearson Education, Canada, Ltd.
Pearson Education—Japan
Pearson Education Australia Pty. Limited
Pearson Education North Asia Ltd.
Pearson Educación de Mexico, S.A. de C.V.
Pearson Education Malaysia, Pte. Ltd.

10 9 8 7 6 5 4 3 2 1
ISBN 0-13-113928-2

To Kurt, Loreli, Timothy, Heidi, Charity,
Sunshine, and my grandchildren

Preface

This edition has been prepared with the same goals and intentions of the past five editions: to present, in a clear, correct, and complete manner, the fundamental concepts of thermodynamics and heat transfer and to demonstrate their applications. I have held closely to the presentation of the fifth edition and emphasized design and real-world applications. The understanding or use of calculus is not mandatory for studying the text or doing most of the practice problems. I still feel that the subject matter is too important and practical to be inaccessible to those who do not have a strong background in mathematics.

The major addition to this volume is the inclusion of discussions on using *Engineering Equation Solver (EES)* as a commercial software tool for solving many of the problems encountered in thermodynamics and heat power. *EES*, available from F-Chart Software (ww.fchart.com), is a powerful package for obtaining and using thermodynamic properties and for solving sets of simultaneous equations, but if the reader does not have access to *EES* then this presentation can be ignored without detracting from the text.

The nine software programs that I have made available in previous editions have been revised into a Windows or event-driven format to be more comfortable and user friendly. I hope that they can aid the student in understanding the subject matter and solving many of the problems. These programs are available on the CD packaged with the Instructor's Solutions Manual. It includes programs that can be used to analyze some of the processes and cycles discussed in the text, that can perform the necessary computations for parametric studies, and that connotate *design* for many professionals.

With the increased applications of refrigerant blends in mechanical refrigeration, I have added two refrigerant blends, R-407c and R-502, to the discussion of refrigerants in chapter 12. Further, I have included a discussion of the phase change of mixtures in chapter 13, with particular emphasis on the refrigerant blends, including the observed temperature glide during a phase change for specific blends.

Because of the increased interest in *fuel cells*, I have expanded the presentation and discussion of these devices.

Finally, I have rearranged the appendix table of properties to be more logical and convenient. A complete listing of the appendix tables is now included in the Table of Contents.

Both SI (Système International) and English units are used in this edition. There still is a need for today's students to be fluent in both systems, so the text is split about evenly between the two systems. Practice problems are keyed for both systems, and example problems show the conversions between the systems. The book contains adequate material and enough practice problems to permit emphasis on one of the two units.

The sequence of presentation roughly follows the order of definitions, statements of laws or principles, and applications. The importance of understanding the definitions cannot be overstated. The vocabulary of thermodynamics contains many words of common usage (such as *temperature*, *heat*, and *work*) that are given precise meaning through definitions. Without this precision, most technical problem solving would be vague or impossible. Laws or principles are stated as truths that have no observed contradictions in nature. Applications of the laws are then presented to give the reader a sampling of the types of problems clarified by the thermodynamic approach.

In proceeding from the basic laws to specific applications, the reader is presented with a common methodology for all problems of a thermodynamics nature. From the statements of laws, a few precise equations are developed from which students can proceed with an analysis. The reader is shown how to make statements regarding the physical characteristics of the material involved and how to make simplifying but realistic assumptions that allow for reducing general equations to specific ones. The reader is then shown how to proceed with calculations to obtain quantitative answers. Included are some derivations of specific equations from the general relationships and compilations or tables displaying specific equations. It should be emphasized that an understanding of the underlying assumptions that allow the use of specific relationships is most important.

A book for engineering technology that covers such a popular subject as thermodynamics cannot be expected to present much new or original work. What is presented is well known in the scientific community but is, I believe, presented here in a manner that is especially clear and easily accessible to students.

In chapter 1, material is presented that is expected to set the stage for the sequence of study in thermodynamics. Readers are strongly urged to study the sections on thermodynamic calculations and the method of problem solving. Particular attention should be directed to the subject of computing areas under curves and the use of the computer for these efforts. The program used to facilitate these calculations with the trapezoid rule method is contained on the CD included in the Instructor's Solutions Manual, but it is short enough to be readily copied or entered onto a separate disk.

In chapter 2, the idea of a system is introduced and emphasis is placed on identifying important properties of the systems. Some treatment is given to the instrumentation used in measuring those properties, such as pressure gages and thermometers.

Chapter 3 contains the definitions of work and heat as well as a discussion of how energy changes from one form to another. Also, the idea of a reversible process is introduced and then used throughout the book. The first law of thermodynamics is presented in chapter 4 as a conservation principle applicable to a system. It was decided to consider systems as open, closed, or isolated rather than using the term *control volume*. Some arguments can be made against this terminology, but I felt that this was the best way to reduce the language to the fewest words.

In chapter 5, the reader is shown how to describe the state of a system. First, the three common phases of substances—solid, liquid, and vapor—are discussed with the assistance of phase diagrams. Pressure–volume–temperature relations are introduced with explanations given for when a system can be assumed to be a perfect gas or an incompressible solid or liquid. Compressible gases and liquids are considered and other models are also mentioned. Equations that are used to predict internal energy and enthalpy from temperature are then introduced. Perfect gases and incompressible liquids or solids are given the most attention. Some of the experimental methods for measuring internal energy or enthalpy changes are presented, and then pure substances are discussed. Extensive references are made to the appendix tables of properties (appendix B).

Chapter 6 is somewhat of a watershed for the conservation of energy. The material has been extensively revised and expanded from my previous work, attention being given to processes of pure substances other than ideal gases. Mastery of the material in this chapter would be a good indicator of understanding the material in the first five chapters.

In chapter 7, entropy is presented through the ideas of a cyclic device and heat engines. One never knows how best to introduce an abstraction such as entropy, but its usefulness has been demonstrated and it is used abundantly in the literature. If students are to use thermodynamics in their professional endeavors, they must have an understanding of entropy.

The concepts of available energy are presented based on the useful work idea and the definitions of irreversibility. A consideration of energy by itself, even with some idea of the second law of thermodynamics, can lead to some apparent misconceptions regarding the capabilities of heat engines, batteries, and other power devices. If there are time constraints, the material on available energy given in chapter 8 can be eliminated without loss in continuity.

The material in chapters 9 through 14 represents applications of thermodynamics frequently called *heat power*. Each chapter is reasonably self-contained and distinct, including such topics as existing technological devices and thermodynamic analysis of those devices. In chapters 13 and 14, mixtures are considered more fully than in the fifth edition. In chapter 13, nonreacting mixtures of ideal gases and gases and vapors are analyzed. Psychrometrics, air–water mixtures in particular, are given most attention. In chapter 14, combustion of fuel–air mixtures is treated. This area is one that surely will be emphasized by engineers and technologists in coming years, beyond simply knowing the heating value of fuels. The concepts of air pollution and waste disposal are but two areas involving combustion processes that will require special attention in the future.

In chapter 15, the elements of heat transfer are considered. Although the subject of heat transfer cannot be covered adequately in a single chapter, it is hoped that the more important and straightforward problems addressed by this subject will be appreciated. Many engineering technology curricula do not have a course in heat transfer, or students are not required to take a course in the thermosciences beyond introductory thermodynamics. This chapter is intended expressly for those students. It might also serve as a refresher for individual preparations for professional licensing examinations.

Heating, ventilating, and air conditioning (HVAC) are considered in chapter 16. Here the emphasis is on how thermodynamics and heat transfer can be applied to a very practical, service-oriented field. If one wishes to cover this treatment of HVAC, it would be appropriate to cover the material in chapter 15 immediately before that in chapter 16.

Chapter 17 contains some nontraditional applications of thermodynamics with the intention of showing that the concepts of thermodynamics can be used to analyze any system.

In using this book, the following order of chapters is suggested for a three-semester-hour course in thermodynamics: 1, 2, 3, 4, 5, 6, 7, and at least 9, 11, and 12. An Instructor's Solutions Manual is available to instructors who desire further suggestions of possible sequences of study.

I am pleased that so many users have been loyal to this textbook in its previous editions. I thank each of them. As with all of my past efforts at authorship, I have had much assistance and guidance by others. Any errors are all mine, and I ask that you communicate any errors to me or the publisher. In addition, if you want to communicate with me regarding anything about the text, you may e-mail me at ROLLE@UWPLATT.EDU.

ACKNOWLEDGMENTS

I want to thank the reviewers of this edition, S. Kant Vajpayee, University of Southern Mississippi, and Abulkhair M. Masoom, University of Wisconsin–Platteville.

Contents

1

INTRODUCTION

1–1 SOME REASONS FOR STUDYING THERMODYNAMICS

The most significant contribution to the development and maintenance of our modern technological society has been our ability to extract large amounts of energy from nature through the natural resources. These extractions of energy allow us to control or use work, power, and heat to meet society's demands. The science that explains and predicts how much energy we may extract and how efficiently we may do it for a particular situation is called **thermodynamics**. It is the science that studies energy in its various forms or types and helps to explain why some types of energy are easier to use than others. Because of its subject matter, the science of thermodynamics is used often by engineers and technologists in very practical design problems and in problems of the operation of large or complicated systems.

The measurement of temperature and humidity in the air around us is an application of the ideas of thermodynamics, and the questions of how to reduce heat losses in a building in cold weather and heat gains in hot weather can be answered through a knowledge of thermodynamics. The proper design and selection of heating or air-conditioning systems can be accomplished only with an understanding of the concepts of thermodynamics. A modern room air-conditioning unit, shown in figure 1–1, is one example of a system that has been designed and developed with the ideas of thermodynamics.

The new gas and diesel engines used to provide power for the vehicles of transportation have all been developed for improved performance only because of the applications of thermodynamics, whether that performance is defined as greater power per weight ratio, extended reliability, decreased noise, or fewer identified pollutants. A modern, large diesel engine used in ships, railroad locomotives, or electrical power generation is shown in figure 1–2. The engine is shown on a test bed where it can be operated to measure its efficiency and output accurately. Thermodynamics is used to better determine and understand these results.

Gas turbines and jet engines are analyzed by using the principles of thermodynamics. The modern gas turbine shown in figure 1–3 is used as a jet engine to provide power for aircraft, and another gas turbine shown in figure 1–4 is used to provide power for driving electrical generators. Both engines are examples of devices that have been designed and developed through the principles of thermodynamics; larger, more powerful engines cannot be developed without an understanding of those principles.

Electrical energy, or electrical power, is provided primarily through steam power generating stations. These facilities, an example of which is shown in figure 11–1, produce the majority of all the electrical power worldwide, as well as in the United States. Even modern nuclear power facilities, such as that shown in figure 1–5, are steam power generating stations, and the science of thermodynamics allows us to understand how these inventions can convert coal, oil, natural gas, wood, or nuclear energy into electricity.

FIGURE 1–1 A modern room air conditioner (courtesy of Carrier Corporation, Syracuse, NY)

FIGURE 1–2 Modern diesel engine on a test stand for measuring performance characteristics (courtesy of Krupp MaK Maschinenbau GmbH, Kiel, Germany)

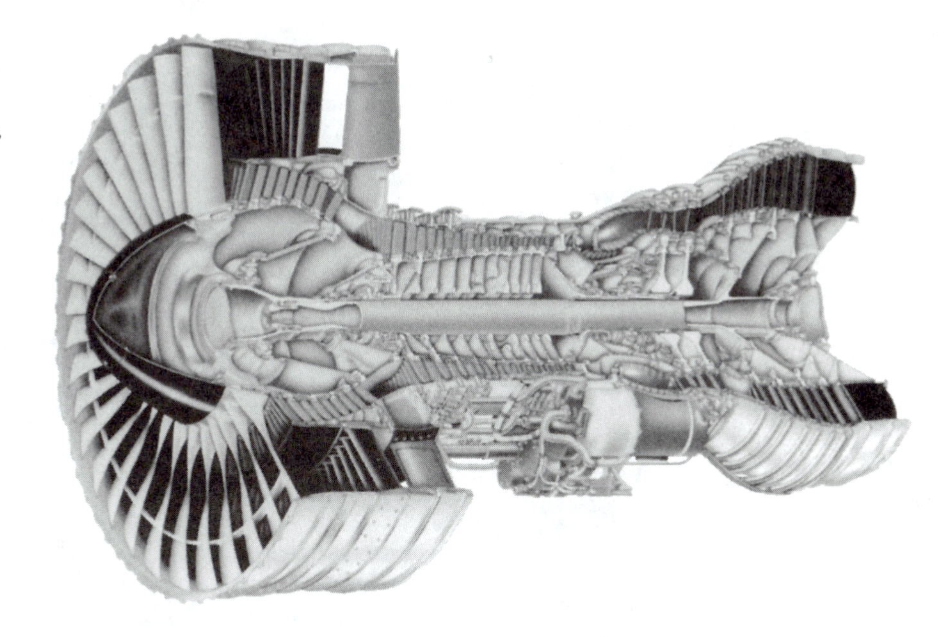

FIGURE 1–3 Cutaway of a modern turbofan jet engine for aircraft propulsion (courtesy of Pratt & Whitney, East Hartford, CT)

FIGURE 1–4 A modern gas turbine used for electrical power generation (courtesy of General Motors Corporation, Industrial Gas Turbine Division, Indianapolis, IN)

Uses for thermodynamic ideas can be found in nearly everything we do. The electronic age has expanded due to the miniaturization of the electronic control circuitry or chips. One of the major problems associated with such miniaturization of components is the difficulty of properly cooling these parts, and the array of modern computers, video systems, camcorders, and other devices function properly only because the chips are adequately cooled. This cooling of parts can be accomplished only in systems designed and developed by engineers and technologists having an understanding of the concepts of heat transfer—an application of thermodynamics. Figure 1–6 shows a camcorder with miniaturized circuitry that provides the user with all of the conveniences of sound/visual recording. If the

FIGURE 1–5 A modern nuclear power plant (courtesy of Commonwealth Edison Company, Byron, IL)

FIGURE 1–6 A modern camcorder (courtesy of Sony Electronics, Inc., Park Ridge, NJ)

goal of continuing miniaturization of electronic circuitry is to be realized, a better understanding of thermodynamics and heat transfer must be developed by engineers and scientists. Even the common items such as the toaster oven and domestic refrigerator, shown in figure 1–7, have been developed only after the concepts of thermodynamics and heat transfer and an appreciation of heat flow have been developed.

Further applications of heat flow can be found in construction. Frost damage to foundations of buildings and roads is due to heat flows and the freezing of water, both phenomena that are studied in thermodynamics. Burying a water line to prevent freezing or using underground water as a heat source for a heat pump system should be done only after thorough consideration of thermodynamic concepts.

FIGURE 1–7 Modern domestic refrigerator (reproduced with permission of the copyright owner, General Electric Company, Louisville, KY)

The design and analysis of all refrigeration systems can be accomplished completely only with thermodynamics. Even the servicing and troubleshooting of large cooling or heating systems is best done by someone who understands and can apply thermodynamic ideas to the system.

Finally, an understanding of the concepts of thermodynamics is mandatory if we want to have a clear appreciation of the effects of environmental pollution. Combustion, an application of thermodynamics to chemically reacting systems, creates products that go into the atmosphere or are buried in the ground. A good background in thermodynamics is necessary for an understanding of the full effects of these products on the environment. The concerted efforts of many engineers and technologists are required to reduce the adverse effects of combustion. Combustion is most obvious as a process occurring in the automobile engine and the coal-burning power plant but also occurs in waste incineration plants and even (as a pyrolytic process) in the burying of waste and garbage, and in the corrosion and rusting of metals.

The use of solar and wind energy will become economical only if the concepts of thermodynamics are better used and applied. The solar hot water heater shown in figure 1–8, the Heliodyne Gobi flat plate solar water collectors, is an example of a device that uses many of the ideas of thermodynamics and heat transfer advantageously. Passive solar energy in architectural designs of buildings, photovoltaics for direct electrical energy generation, and solar air conditioners are three more examples of free, or sustainable, energy supplies that can be used now and in the future. All of these devices and systems, however, can be developed to their full potential only if their designers have an understanding of thermodynamics.

It is clear, then, that an understanding of thermodynamics can be a useful tool in technology. By studying this book you will be exposed to the fundamental ideas of thermodynamics and applications of those ideas to engineering problems.

FIGURE 1–8 This flat plate solar collector system is installed on a research laboratory building at the Biological Reserve, Jasper Ridge, belonging to Stanford University. The system is designed to feed the laboratory's hydronic space heating system, with tubes embedded in the floor of the building circulating heated water. The Heliodyne Gobi collectors are treated as prominent architectural features of the low one-story building. The system consists of 26 Heliodyne Gobi 410 collectors in a closed loop glycol design, heating a storage tank by way of a heat exchanger external to the tank. The heat exchanger is a two-pass shell and tube counterflow device, and the heat transfer fluid is propylene glycol mixed 50/50 with water. A differential temperature thermostat (DTT) reads two 10K-ohm sensors, one on the collector and one on the cold bottom of the tank, and is programmed for an *on* differential of 18°F and an *off* at 4°F. Additional equipment includes an expansion tank sized for the length of the lines and equipment, pressure gages, and pressure relief valves rated at 150 psig. (Courtesy of Heliodyne, Inc. Richmond, California)

New Terms

a	Acceleration	Δ (delta)	Change in a variable
F	Force	δ (delta)	Very small change in a variable
m	Mass		

1–2 HISTORICAL BACKGROUND OF THERMODYNAMICS

The development of thermodynamics can be traced back to the earliest recorded dates in human history. Central to the theme of this development is the human desire to ease or replace manual efforts with additional animate or inanimate sources of power. What follows in this section is a brief sketch of the history of the present science of thermodynamics. It is obviously not a complete exposition; it is meant only to give you some historical insight into how the ideas of thermodynamics originated and expanded.

FIGURE 1–9 Hero's turbine or aeolipile [revised from A. Sinclair, *Development of the locomotive engine* (Cambridge, Mass.: MIT Press, 1970), p. 2; with permission of MIT Press]

Human use of animate power, such as horses and oxen, began around 4000 B.C. and represented the major source of energy through the nineteenth century A.D. By 3500 B.C., wheeled vehicles were used in Mesopotamia to ease the burden of man and beast. Watermills, steam jets, and various mechanical devices were in use during Christ's lifetime, and around A.D. 150 Hero's turbine was invented. This turbine was a globe containing water from which hot steam could escape through two nozzles, as shown in figure 1–9. A fire placed under the device boiled water in the flask, and the steam traveled up the vertical tubes and into the globe. Once in the globe, the steam was expelled through nozzles, thus causing the globe to rotate. It was really nothing but a novelty toy at the time, but it represents a thermodynamic concept of converting inanimate energy from a fuel into an effect (motion).

The science of thermodynamics probably began around 1592 when Galileo used a thermometer to make the first measurement of temperature. The inaccurate and fickle human sense of touch was thus circumvented and replaced by quantitatively describing the hotness or coldness of objects. Meanwhile, the first use of steam for furnishing significant amounts of power for social needs occurred in the late seventeenth century. In 1698, Thomas Savery devised an arrangement of tanks and hand-operated valves to utilize steam and its energy to pump water from a well. (See figure 1–10.) In the pump, steam was produced in a boiler (*a*) and conducted to the two reservoirs (*b*) through hand-operated valves (*c*). The steam was furnished to the reservoirs alternately, in turn pushing water in the reservoir out through pipe (*d*) and to the top. Valve (*c*) would then be closed and a trickle of cold water would condense the steam in the reservoir, thus causing a vacuum to be created. This vacuum would allow water to flow up pipe (*e*) from a low water supply (*f*) into the reservoir. Valve (*c*) would then be opened again to repeat the cycle. By alternating between the right and left reservoirs, a continuous flow of water was created at the top. While representing a historical first, this device was little improvement over animate power. Thomas Newcomen, in 1712, developed a steam-piston engine that was a logical replacement of animate power for pumping water. This arrangement, shown in figure 1–11,

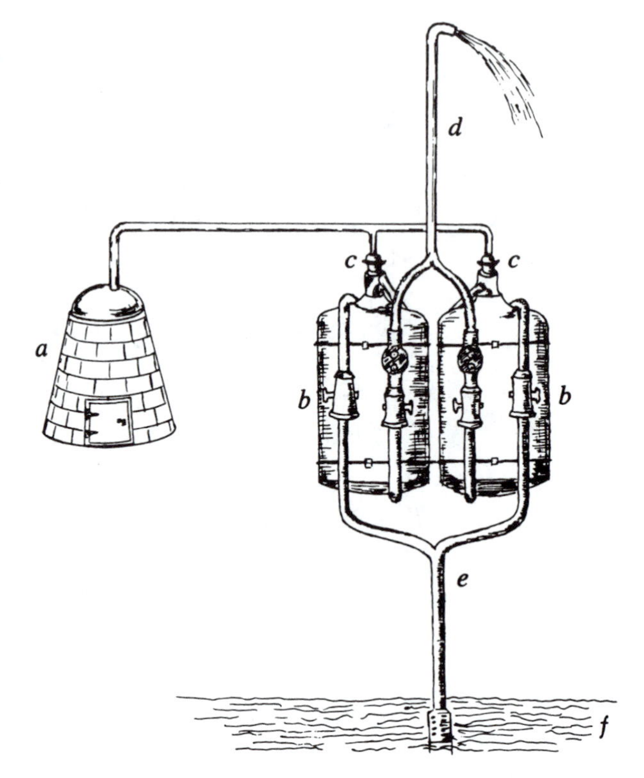

FIGURE 1–10 Savery's pistonless steam-vacuum pump (c. 1698) (revised from figure of photo, Science Museum, London; with permission of The Science Museum, London, England)

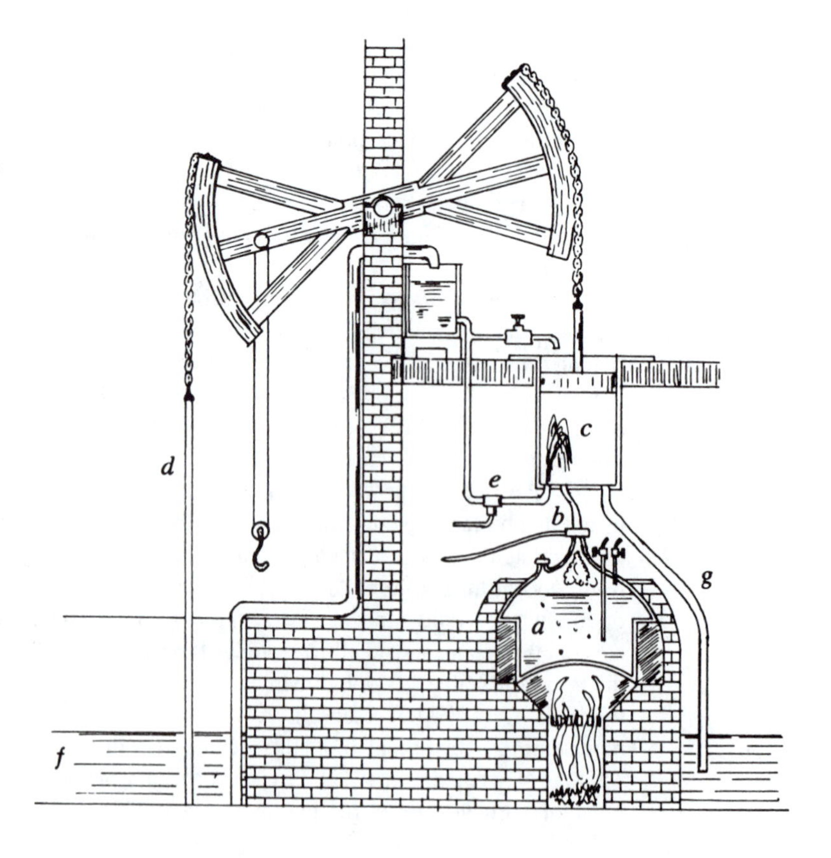

FIGURE 1–11 Newcomen's steam engine (1712) [revised from A. Sinclair, *Development of the locomotive engine* (Cambridge, Mass.: MIT Press, 1970), p. 7; with permission of MIT Press]

provided a cycling motion that we call a **heat engine**. Known as the "atmospheric engine" because a vacuum and atmospheric pressure combined to provide the power stroke, this engine was used for pumping water. Steam produced in the boiler (*a*) was conducted through a hand valve (*b*) to the piston-cylinder (*c*). The steam would push the piston up to the position shown, allowing the pump rod (*d*) to descend into a water supply. The valve (*e*) was then opened to allow a spray of water to condense the steam in the cylinder, causing a vacuum to be created. The piston was then pushed down by atmospheric pressure, the pump rod raised, and water pumped up out of the water supply (*f*). Valve (*e*) was closed, valve (*b*) opened, and the process repeated. Line (*g*) was opened intermittently to allow the condensed steam to flow out of the cylinder.

With improved machining techniques available for manufacturing parts, James Watt developed a steam-piston engine that represented a significant improvement over Newcomen's. Watt's engine, first operated in 1775 to pump water, represents the forerunner of the steam engines used for railroads, ships, and numerous other applications (figure 1–12). Unlike Savery's or Newcomen's devices, which condensed steam inside the working cylinder chamber, Watt's steam engine condensed the spent steam external to the cylinder (*a*). Steam was furnished from a boiler through a pipe (*b*). Valve (*c*), controlled from a tappet rod (*d*), allowed steam to enter the top side of the piston (*e*). This pushed the piston down and, through the walking beam (*f*), raised the pump rods (*g*) and (*h*). This motion drew water out of the reservoir (*i*), through the pipe (*j*), and from reservoir (*k*) to reservoir (*i*). Valve (*l*) was then

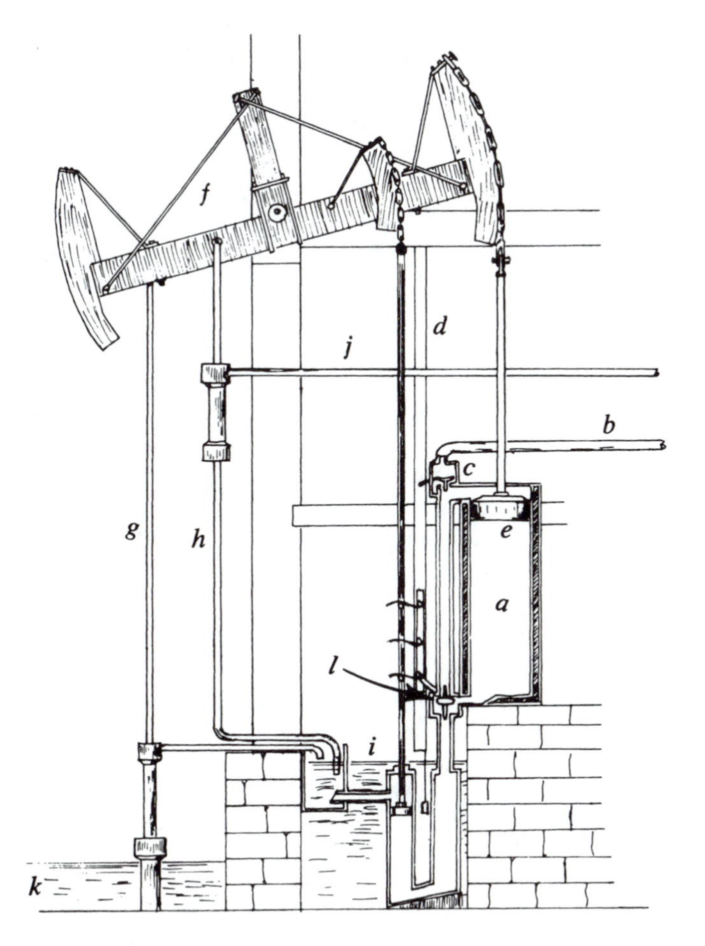

FIGURE 1–12 Watt's steam engine [revised from A. Sinclair, *Development of the locomotive engine* (Cambridge, Mass.: MIT Press, 1970), p. 10; with permission of MIT Press]

shifted to allow steam to enter the bottom of the piston; thus equilibrated, the piston moved to the top to begin a new cycle.

Thermodynamic theory began to take form in 1693, when G. W. Liebnitz pronounced the conservation of mechanical energy (kinetic and potential). Sadi Carnot published a treatise in 1824 that described cycling devices or heat engines and that alluded to the first and second laws of thermodynamics. Twenty-six years later, in 1850, Rudolph Clausius formally stated these two laws of thermodynamics, and in 1854 he identified and defined the property now called **entropy**.

From 1840 to 1848, James Joule experimentally proved the equivalence of heat and work, thus making thermodynamics a quantitative science in the best tradition of Galileo. The internal combustion gasoline engine, later used to provide power for automobiles, trucks, and numerous other devices, was developed around 1860 by Lenoir. This type of engine was first used in vehicles around 1876 by Otto and Benz.

Around 1884, Parson introduced a steam turbine capable of developing significant amounts of power. This type of device, utilizing the popular medium steam, has been a most durable power generator and seems more popular today than ever. Early in the twentieth century, Nernst and Planck separately enunciated the earliest definition of the third law of thermodynamics. These statements have since been refined and revised by various theoreticians.

These advancements in both theory and technology reflect the application of thermodynamics to practical efforts; this usefulness has stimulated much of the interest in further expanding that body of knowledge known as the **science of thermodynamics**.

We see that thermodynamics developed through theory, experiments, and practice. Theoretical advancements came from the giants of thought; many persons, such as Joseph Black, Lord Kelvin, J. W. Gibbs, James Maxwell, L. Boltzmann, H. L. F. Helmholtz, and Albert Einstein, contributed to thermodynamics at least as significantly as those mentioned previously. However, without the experimentation, design, creativity, and artisan ability to machine and fabricate parts precisely, useful engines providing significant amounts of power and devices to use this power would be nonexistent.

Our concern in this book is to understand and use the concepts of thermodynamics, clarified by theoreticians both past and present, for the solution of engineering and technological problems.

1–3 BASIC DIMENSIONS AND UNIT SYSTEMS

Thermodynamics is one of the fundamental engineering sciences, and it has been developed through methods of empirical observation or experimentation. These methods involve watching a physical happening, recording the events, and measuring some of the important changes that may have occurred during the experiment. The basic **dimensions** that you can measure are length (L), mass (m), and time (t), and you can also measure force (F). Length, mass, time, and force are all related through Newton's second law of motion, which is usually written

$$F = ma \qquad \textbf{(1–1)}$$

where F is a force that imparts an acceleration a to a mass m.

For our purposes, we will consider force to be a basic dimension that we can measure directly. When dimensions are measured, a number is determined. For instance, if you want to know the length of your index finger, you measure it with a ruler and determine a numerical value representing the length of your index finger. That number has a label or unit associated with it so that you can be more precise in describing the finger's length. In thermodynamics we use two systems of units to associate with the basic dimensions and

TABLE 1–1 Basic dimensions and units

Quantity	Système International	English
Length	meter (m)	foot (ft)
Mass	kilogram (kg)	pound-mass (lbm)
Force	newton (N)	pound-force (lbf)
Time	second (s)	second (s)
Temperature	K or °C	°R or °F
Energy	joule (J)	foot-pound (ft · lbf)

other terms: the Système International (SI) and the English system. The units used for the basic dimensions in these two systems are given in table 1–1. Temperature and energy, quantities derived from the basic dimensions, are included in table 1–1 for reference purposes.

The SI system utilizes prefixes to make the units more flexible over a wide range of values for dimensions and other quantities. For instance, the prefix "milli" represents $^1/_{1000}$ or 10^{-3}; 1 millimeter (mm) is equal to $^1/_{1000}$ meter. Similarly, 1 kilogram is equal to 1000 grams and can be converted to 1000 grams if that is a more convenient unit. Other prefixes and their conversions are given in table 1–2. Sometimes you will need to convert from SI units to English units or from English to SI. Conversions of the units for the basic dimensions between these two systems are given in table 1–3, and more conversions are given on the front inside cover.

TABLE 1–2 Système International (SI) prefixes

Amount	Multiple	Prefix	Symbol
1 000 000 000	10^9	giga	G
1 000 000	10^6	mega	M
1 000	10^3	kilo	k
100	10^2	hecto	h
10	10	deka	da
0.1	10^{-1}	deci	d
0.01	10^{-2}	centi	c
0.001	10^{-3}	milli	m
0.000 001	10^{-6}	micro	μ
0.000 000 001	10^{-9}	nano	n

TABLE 1–3 Conversion factors between SI and English units

Unit	Multiply by:	To Convert to:
meter (m)	3.2808	feet (ft)
ft	0.3048	m
kilogram (kg)	2.2046	pound-mass (lbm)
lbm	0.45359	kg
newton (N)	0.2248	pound-force (lbf)
lbf	4.4484	N
joule (J)	0.737	ft · lbf
ft · lbf	1.356	J

1–4 THERMODYNAMIC CALCULATIONS AND UNIT CANCELLATIONS

This book contains many examples and practice problems that are characteristic of those encountered in engineering applications. The solutions to these problems are very often numerical answers calculated from mathematical equations. Frequently, these equations are arranged algebraically in such a way that a specific term or variable may be solved. Thus, the student must be able to perform arithmetic and algebraic operations to gain a clear understanding of the principles of thermodynamics. The discussion and examples that follow show the types of problems often used in this book.

As you study these examples, notice that the calculations involve two ideas: first, the actual arithmetic computation, which gives a numerical answer; and second, the fact that almost every term or number inserted into the equations has associated with it a unit of measure. Sometimes the number has no units (and is therefore said to be *unitless*), but more often the number does have a unit. The units should always be included in a calculation and used to determine the units of the answer. Like numbers, units can be multiplied, divided, added, or subtracted. Always remember that in adding or subtracting, the numbers must have the same units. As an example, if you add 2.0 kilograms to 200 grams, either the kilograms must be changed (or converted) to grams or the grams to kilograms. Such conversions are made with **conversion factors**, some of which are given in table 1–3 and on the front inside cover. The answer to this example is 2.2 kg or 2200 g. When units (or numbers having those units) are multiplied or divided, the resulting unit would be the product of the two original units for multiplications, or the quotient of the dividend/divisor. As an example, suppose that 2.0 kilograms is divided by 200 grams. Here the result is just 0.01 kilogram/gram, but there are 1000 grams per kilogram, so that the answer would be expressed as

$$0.01 \text{ kilogram/gram} \times 1000 \text{ grams/kilogram} = 10 \text{ (unitless)}$$

Notice here that the grams cancel each other, as do the kilograms. The answer 10 is the simplest form, although the answer 0.01 kilogram/gram is also correct. **Unit cancellation** is the name of the operations of including units in the calculations and performing the arithmetic and algebraic manipulations of those units. You will find that the unit cancellation method saves time and effort, even though it appears to involve trivial, unnecessary extra effort as you do simple problems. Nevertheless, use the method for very easy problems, and thereby develop efficient habits so that you will be able to handle more difficult problems when they appear.

The following examples show the types of problems you will encounter in later chapters:

EXAMPLE 1–1 A perfect gas satisfies the relationship $pV = mRT$. If $p = 1.01 \times 10^5$ N/m², $m = 3$ kg, $R = 0.287$ N·m/kg·K, and $T = 300$ K, determine the value of V.

Solution We may algebraically solve for the term V, so that

$$V = \frac{mRT}{p}$$

Substituting the values into the proper terms, we obtain

$$V = \frac{(3\text{ kg})(0.287\text{ N}\cdot\text{m/kg}\cdot\text{K})(300\text{ K})}{1.01 \times 10^5\text{ N/m}^2}$$

Answer

$$= 0.00256\text{ m}^3$$

EXAMPLE 1–2 The amount of work achieved or expended during a particular action or process is

$$Wk = \frac{1}{1-n}(p_2V_2 - p_1V_1)$$

If $n = 1.4$, $p_2 = 220 \times 10^5\ \text{N/m}^2$, $V_2 = 0.01\ \text{m}^3$, $p_1 = 16 \times 10^5\ \text{N/m}^2$, and $V_1 = 0.09\ \text{m}^3$, determine the amount of work Wk.

Solution Here we may readily substitute numerical values into the relationship to obtain

$$Wk = \frac{1}{1-1.4}(220 \times 10^5\ \text{N/m}^2 \times 0.01\ \text{m}^3 - 16 \times 10^5\ \text{N/m}^2 \times 0.09\ \text{m}^3)$$

Answer

$$= -190{,}000\ \text{N}\cdot\text{m}$$

EXAMPLE 1–3 Air is commonly assumed to behave as a perfect gas, describable by the equation $pv = RT$. The gas constant for air R can be taken to be 53.3 ft·lbf/lbm·°R. If air is at a pressure p of 2100 lbf/ft^2 and at a temperature T of 600°R, what is the specific volume v?

Solution We may observe that the specific volume can be solved from the perfect gas relationship. Thus,

$$v = \frac{RT}{p}$$

and we can then obtain

$$v = \frac{(53.3\ \text{ft}\cdot\text{lbf/lbm}\cdot{}^\circ\text{R})(600^\circ\text{R})}{2100\ \text{lbf/ft}^2}$$

Answer

$$= 15.2\ \text{ft}^3/\text{lbm}$$

EXAMPLE 1–4 During a particular process the relationship between the variables p and V has been found to be $pV^n = \text{constant} = C$, where $n = 1.29$. For two conditions, we know the value of p, say, p_1 and p_2. We also know the value for V_2, and we need to know the value for V_1. The known values are $p_1 = 14$ psi, $p_2 = 280$ psi, and $V_2 = 0.02\ \text{ft}^3$.

Solution We may write the given relationship as

$$p_1V_1^n = C = p_2V_2^n$$

or

$$V_1^n = V_2^n\frac{p_2}{p_1}$$

and

$$V_1 = V_2\left(\frac{p_2}{p_1}\right)^{1/n}$$

Substituting the values into this equation, we obtain

$$V_1 = (0.02\ \text{ft}^3)\left(\frac{280\ \text{psi}}{14\ \text{psi}}\right)^{1/1.29}$$

Answer

$$= 0.204\ \text{ft}^3$$

In these four examples, notice that the equation used to solve for the required answer must often be obtained after an algebraic manipulation of a given relationship. For instance, in example 1–1 we determined the volume of a perfect gas after rearranging the common form of the equation describing a perfect gas, $pV = mRT$. Other examples of algebraic manipulations are given in examples 1–3 and 1–4. In example 1–2 the necessary equation to solve for the amount of work was furnished; only the requisite care in handling the numbers and their units was required. You will find it helpful to try some of the practice problems at the end of this chapter to review and gain more confidence in handling algebra and computations.

1–5 FURTHER THERMODYNAMIC CALCULATIONS

In section 1–4 we reviewed some of the mathematics that you will use later in the book. Here we review additional ideas and mathematics that you have seen and used before but with some definitions and notation that you may not have seen. First, let's consider the term **variable**. A variable is a quantity that can have different numbers assigned to it. For instance, in algebra, x and y are often used as symbols for variables. Here, x (or y) may have any number assigned, such as 2, 3.14, 780, or $^1/_4$. There are an infinite number of values that x or y could take. In thermodynamics these variables can become physical quantities, such as distance, area, pressure, temperature, volume, or heat. In the chapters that follow, we will see in what many of those variables or parameters are. Let us consider two different kinds of variables. The first is the one we have just identified: a variable that can have any value given to it. This is called an **independent variable** because it can take any value—it is independent of other variables or unknown quantities. The second type of variable is a **dependent variable**, a variable whose value depends on the value of an independent variable. This means that, for instance, if x is an independent variable and y is a dependent variable of x, we know the value of y automatically if x is given. In algebra y is said to be a *function of* x or is just written

$$y = f(x) \tag{1–2}$$

where $f(x)$ means "a function of x." Equation (1–2) does not say what the function is; it says only that y is a dependent variable and is a function of x. Also, by using algebraic manipulations you could rearrange the exact equation indicated in equation (1–2) to find that x is dependent on y, or $x = f(y)$. So nothing prevents designating any variable as an independent variable, as long as it does not depend on another variable.

There are three convenient ways to describe how, say, y varies with x as in equation (1–2): we can list in a table the values of y as they correspond with x, plot on an $x-y$ graph (or $x-y$ coordinates) the values of y as functions of x, or use an algebraic equation. In example 1–5 these three methods of showing the relationship between y and x are used; table 1–4 lists x (the independent variable) and y (the dependent variable), the function of x is shown in figure 1–13, and the equation $y = 1.6x$ is the algebraic relationship. Some-

TABLE 1–4

x	$y = f(x)$
0	0
1	1.6
2	3.2
3	4.8
4	6.4
5	8.0
6	9.6

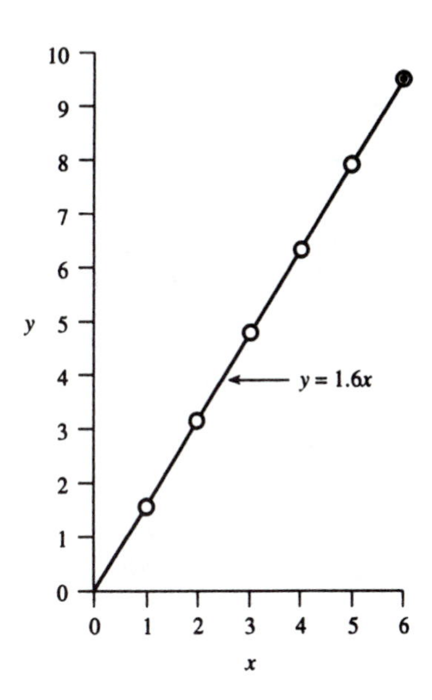

FIGURE 1–13 Plot of $y = 1.6x$ in example 1–5

times the algebraic equation is difficult to determine from the given table data or graph, but if that equation is known, you should easily be able to make up a table and plot a graph of that particular function, $y = f(x)$. Also, remember that thermodynamics has many other independent variables, and very seldom will x and y be the symbols used for them. You will see and use P, V, T, and many other symbols for variables, but it is important to see what we do with those variables in the subsequent discussion and examples.

EXAMPLE 1–5 For the function $y = 1.6x$, plot the graph and tabulate the function between $x = 0$ and $x = 6$ in increments of one digit.

Solution The table of the function $y = f(x) = 1.6x$ is given in table 1–4, and the graph of the function between $x = 0$ and $x = 6$ is shown in figure 1–13.

In thermodynamics we often need to calculate the area under a curve or line on a graph of y versus x. By the "area under the curve" we mean that geometric area enclosed by the x-axis, the line connecting the plotted points on the graph, and vertical lines connecting these two. We call the line connecting the points "a curve" even if the line may be straight. Thus, the general phrase "area under a curve" can be an area under a straight line, under a smoothly curving line, or under an irregular line. Figure 1–14 shows five examples of areas under curves. The first example (figure 1–14a) is for an area of a triangle, which is one-half the base (10 m^3) times the height (5 kN/m^2); the second is a rectangle; and the third is a rectangle with a base given by two different x's. In the fourth and fifth examples, the area is calculated by adding the areas of a triangle and a rectangle. Notice that these areas could be found from the trapezoid area formula. (See appendix A–1.) The trapezoid area is given by the product of one-half the base times the sum of the two heights. In figure 1–14d and e the base of the trapezoid was determined from the difference between two values, x_1 and x_2. The common way of writing the difference is

$$\Delta x = x_2 - x_1 \tag{1–3}$$

and the symbol Δ is just an easy way of writing "the **difference between** x_2 and x_1." Notice that we insist that it be $x_2 - x_1$ and not $x_1 - x_2$. It is for this reason that the area is

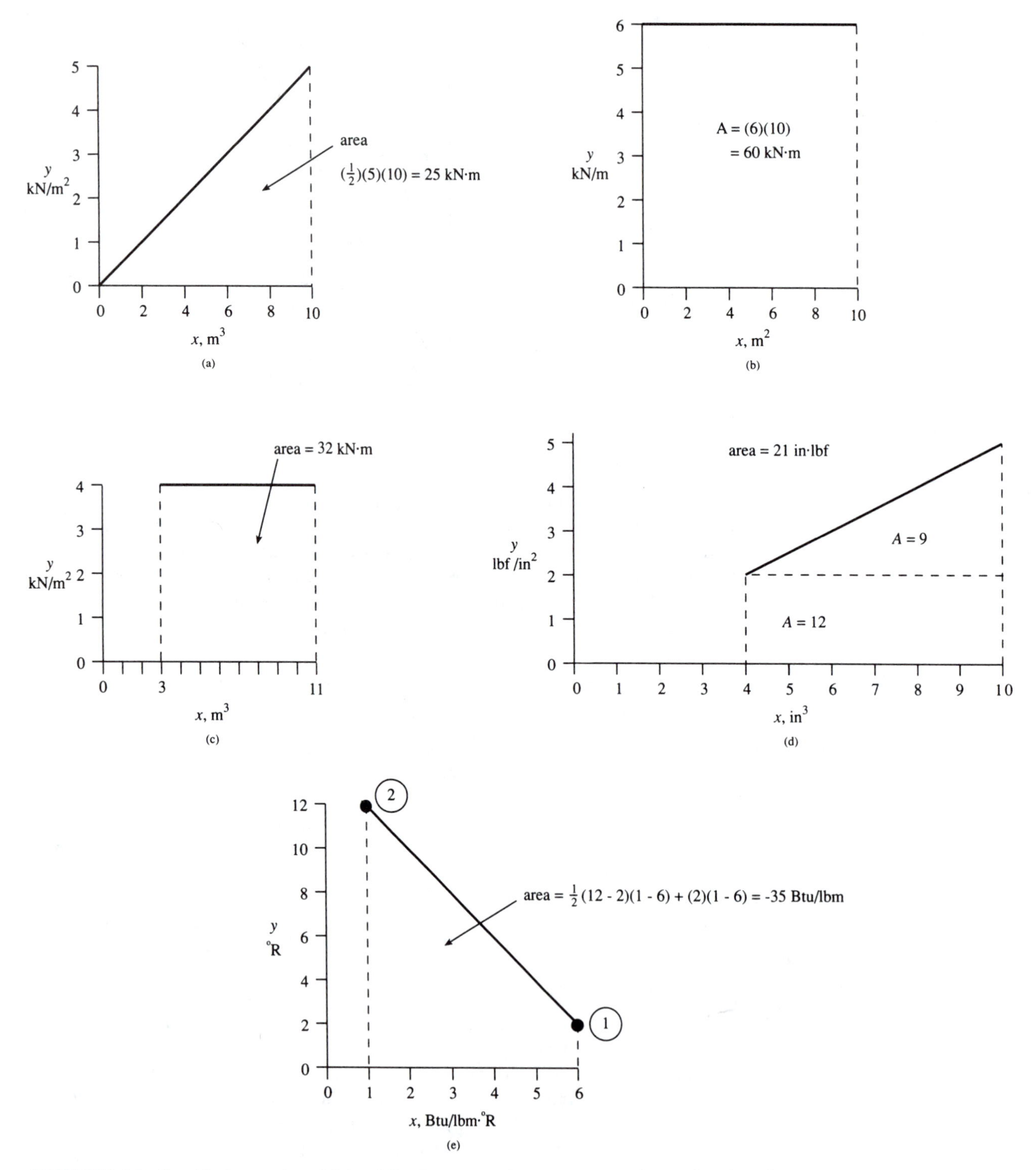

FIGURE 1–14 Graphical examples of determining the areas under some curves of y as a function of x

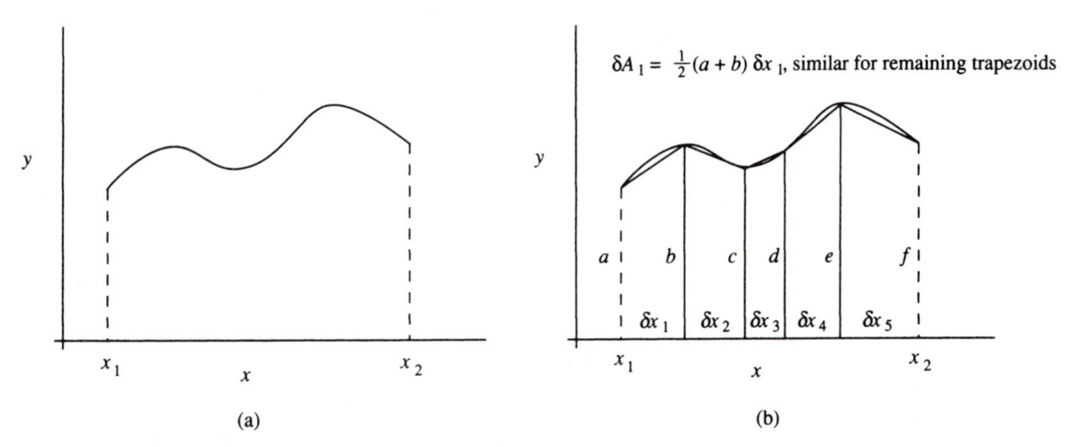

FIGURE 1–15 Example of method for determining approximate area under a curve: (a) graph of a function $y = f(x)$; (b) approximating the area with trapezoids

given a negative sign in the fifth example of figure 1–14; the base, or Δx, is negative. In thermodynamics we always call x_1 the first or initial value and x_2 the value after some time has elapsed or something has happened. It is always assumed that time proceeds from "1 to 2."

Sometimes we will need to determine the area under a curve that is not a straight line or even a regular curve. Figure 1–15a is a curve of $y = f(x)$ which is very irregular. The area under this curve from x_1 to x_2 would be difficult to determine exactly, but a value that is close to the exact area can be found by dividing the area up into small trapezoids having small bases, δx, as is done in figure 1–15b. Notice that δx is a change in x, but such a small change that a straight line is able to be drawn between the two points on the curve. Then each of the small trapezoids has an area δA. If all the δA's are added up or summed, the total area under the curve between x_1 and x_2 is found. This is written

$$A = \sum_{i=1}^{n} \delta A_i \tag{1–4}$$

where the symbol Σ represents the sum of a number of different small areas δA_i and the i is just an index number, or the "i th" area. On the summation sign the "$i = 1$" and the "n" mean that the sum of the small areas, $\Sigma\ \delta A_i$, will include the first (or number 1), the second, the third, and so on, through the "nth" δA_i. Often, the $i = 1$ and the n will not be written; you are then to assume that the summation will be done from the first through the nth terms. Equation (1–4) would then be written

$$A = \sum \delta A_i$$

We can also see, again looking at figure 1–15b, that each δA_i is equal to the product of its base δx_i and its average height y_i. The average height y_i is equal to one-half the sum of the two heights, or y values, at x and at $x + \delta x$:

$$\delta A_i = y_i(\delta x_i) \tag{1–5}$$

Substitution into equation (1–4) gives us

$$A = \sum y_i(\delta x_i) \tag{1–6}$$

Using equation (1–6) to find the area under a curve requires much arithmetic if there are many points on a curve or if the curve is very irregular, but routine arithmetic can be done very quickly and accurately on a computer. First, let's do a problem demonstrating the use of equation (1–6).

EXAMPLE 1–6 In figure 1–16a, the function $y = f(x)$ is shown. Estimate the area under the curve by using the idea of equation (1–6).

Solution First the area is divided into smaller areas by identifying a number of points on the curve and the x and y values for those points. This procedure is shown in the graph of figure 1–16b, and the x and y values are given in table 1–5. In this example there are five small trapezoids of δA, each having a base δx determined by a difference between two x values. These δx's are listed in table 1–5 between the entries for the six x and y points. The values for y_i were calculated from $\frac{1}{2}$ times the sum of the y values at the two adjacent x rows, or at x and at $x + \delta x$. In the column farthest to the right is the value of each of the areas δA calculated from the equation

$$\delta A = y_i\,\delta x$$

Finally, the sum of the right column gives the total area A, or the sum of the δA's. In this example the total area is approximately 299.75 N · cm, or 2.9975 N · m (joules). For engineering purposes, this answer could be given as 3 J.

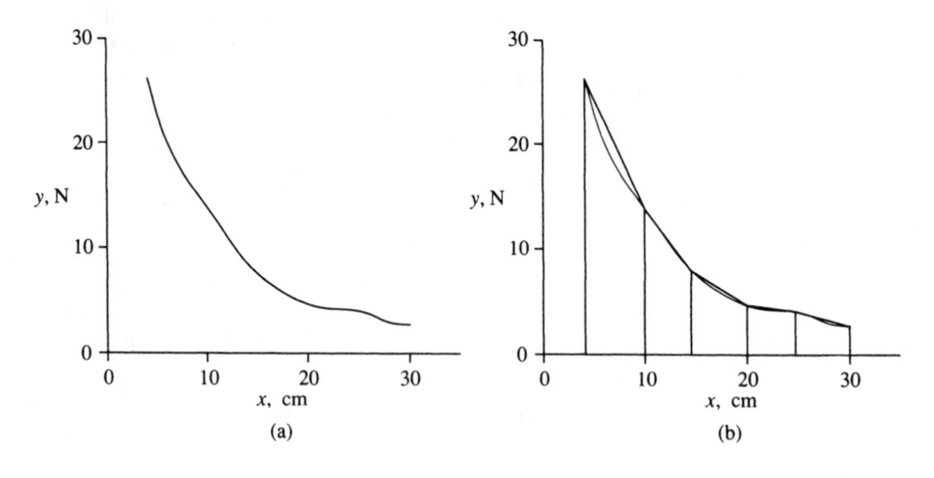

FIGURE 1–16 (a) Function $y = f(x)$ for example 1–6; (b) dividing the area under the curve into small trapezoids

TABLE 1–5 Computation of the approximate area under a curve for example 1–6

x cm	y N	δx cm	y_i N	δA N · cm
3.0	27.0			
		7.0	20.75	145.25
10.0	14.5			
		5.0	11.95	50.75
15.0	9.4			
		5.0	8.05	40.25
20.0	6.7			
		5.0	6.0	30.0
25.0	5.3			
		5.0	4.9	24.5
30.0	4.5			
				$\Sigma\,\delta A = 299.75$ N · cm

A computer can assist in many ways with the calculations relevant to table 1–5. For instance, a computer program can be written that asks for the number of x and y values (in example 1–5 there would be six pairs) and the values for each of the x and y pairs, and then computes the approximate area. Appendix A–5 lists a program called AREA that does this, and the student is encouraged to use this program on a digital computer to determine the areas under curves.

Often a graph will not be given or easily constructed from a set of data points, yet an "area under the curve" can still be found. The next example demonstrates this.

EXAMPLE 1–7

Table 1–6 gives data for the specific heat (c_v) as it changes with temperature (T) for a perfect gas. The internal energy or thermal energy (u) for a perfect gas is defined as the product of c_v and T, or

$$u = c_v T \tag{1–7}$$

Determine the change in internal energy from 50°F to 150°F. Notice that the units for specific heat are Btu/lbm · °F. As we shall see later, the Btu, or British thermal unit, is an energy unit equal to 778 ft · lbf. From equation (1–7) it can be seen that the internal energy will have units of Btu/lbm. Also, the change in the internal energy will then have units of Btu/lbm.

TABLE 1–6

T °F	c_v Btu/lbm · °F
25	0.118
50	0.120
75	0.123
100	0.125
125	0.128
150	0.131
175	0.129
200	0.128

Solution

The change in the internal energy is $\Delta u = u_2 - u_1$. The change is also equal to the change in the product of c_v and T, $\Delta c_v T$. But in this case you should notice that *both* c_v and T are changing, so we must write that $\Delta c_v T$ is equal to a sum of small changes in $c_v T$; that is,

$$\Delta c_v T = \sum \delta c_v T = \sum c_{v_i}\, \delta T \tag{1–8}$$

where c_{v_i} is the average c_v during the small temperature change. These three equalities are also equal to the change in the internal energy, so we can write

$$\Delta u = \sum c_{v_i}\, \delta T \tag{1–9}$$

and can recognize that this is the same as an area under a curve if x is now T and y is c_v. Table 1–7 presents the data of $c_v = f(T)$ and the average c_v between small temperature changes δT. The table also lists the small changes, $\delta u = c_{v_i}\, \delta T$, and at the bottom the sum of the δu's or Δu is given. This example could be done using the AREA program to determine the area under curves.

Therefore, the internal energy change between 50°F and 150°F is 12.5375 Btu/lbm or 12.5 Btu/lbm if you round the answer to three significant figures.

TABLE 1–7 Results of the change in internal energy for example 1–7

T °F	c_v Btu/lbm·°F	δT °F	c_{v_i} Btu/lbm·°F	$c_{v_i}\,\delta T$ Btu/lbm
50	0.120			
		25	0.1215	3.0375
75	0.123			
		25	0.1240	3.1000
100	0.125			
		25	0.1265	3.1625
125	0.128			
		25	0.1295	3.2375
150	0.131			
				$\Delta u = \Sigma\, c_{v_i}\,\delta T = 12.5375$

Notice that in this section we have determined a way of finding a particular quantity; in example 1–6 the area A was found to be a sum of small areas, each of which was equal to a base times an average height. In example 1–7 the internal energy (equal to specific heat times temperature) changed with temperature, and this change was found from a calculation involving a sum of small internal energy changes. In both examples there was an independent variable (x or T) and a dependent variable (y or c_v). The area A (in example 1–6) and the internal energy change Δu (in example 1–7) were also dependent variables because they were functions of the other variables. Throughout this book the idea of finding a quantity by summing many small parts will be used. You should also be aware that we defined the area A by equation (1–6) and found the area and not a change in area; the internal energy was defined as $u = c_v T$, so that the change in internal energy was then found from equation (1–9). A special case of equation (1–9) should be considered: if c_v is a constant and does not change, the average value of c_v is just the same c_v, and the change in the internal energy is given by

$$\Delta u = c_v\,\Delta T \tag{1–10}$$

As we shall see in later sections of this book, two other quantities, work and heat, will be defined by equations in the same way that area was defined in example 1–6. We will calculate the heat and the work, but, by *definition*, the change in work or the change in heat will be meaningless terms that you should not even consider.

Also, we will find that other quantities (properties) will be defined so that a change in those quantities will be determined in the same way that the internal energy change was found in example 1–7. In these cases the absolute value of, say, energy will have meaning only as a change from an arbitrary zero point.

CALCULUS FOR CLARITY 1–1

It can be seen from the previous discussions that the idea of a small element or area, such as the small area $\delta A_1 = (1/2)(a + b)\,\delta x_1$ shown in figure 1–15b, requires an arbitrary decision about how small or large δx can be. From calculus, we know that an area A, as defined by equation (1–4),

$$A = \sum_{i=1}^{n} \delta A_i$$

is exact only if the small areas δA_i are as small as possible, that is, in the limit as δA_i approaches zero (0). If these small areas are made to approach zero, then it follows that the number of these small areas must be large to be able to obtain the area A. As δA_i approaches zero, the number of these areas must approach an infinitely large number. In this example of an area under a curve, the area is determined by the product of an

CALCULUS FOR CLARITY 1–1, continued

independent variable x (or δx_i) and a dependent variable y_i. Here, y_i is called dependent because it depends on x_i or is dependent on x. We write this as $y(x)$ to denote the dependency of y on x. Since the small areas are made to approach zero, and their number to approach infinity, then the changes in the independent variable δx_i will be made to approach zero, and their number to approach infinity. These are the ideas of the fundamental theorem of integral calculus and may be expressed as

$$A = \lim_{n\to\infty} \sum_{i=1}^{n} \delta A_i = \lim_{n\to\infty} \sum_{i=1}^{n} y_i\, \delta x_i = \int_{x_1}^{x_2} y(x)\, dx \tag{1–11}$$

where A is the area evaluated between the values of $x = x_1$ and $x = x_2$, as indicated in figure 1–15b. The last quantity is read, "the integral of $y\,dx$ between the limits of $x = x_1$ to $x = x_2$." If the function $y(x)$ is known, then the area or integral of $y\,dx$ can be evaluated *exactly*. That is, there is a unique solution for equation (1–11) if the relationship $y(x)$ is known. In appendix A–4, some integrals of the more common algebraic relationships are given. If the function $y(x)$ has no known or convenient analytic expression, then the student may still want to evaluate such an area or sum of small elements by the approximation methods.

EXAMPLE 1–8 Determine the area under a curve defined by the analytic expression

$$pV = 12(\text{kPa}\cdot\text{m}^3)$$

between the limits of $V = V_1 = 0.1\ \text{m}^3$ and $V = V_2 = 1.0\ \text{m}^3$ as shown in figure 1–17.

Solution We see that the area under the curve $pV = 12$ may be determined from the defining equation (1–11),

$$A = \int_{V_2}^{V_1} p(V)\, dV$$

where the independent variable x is V and the dependent variable is $y(x) = p(V)$. Using some algebra, we obtain

$$p(V) = \frac{12}{V}$$

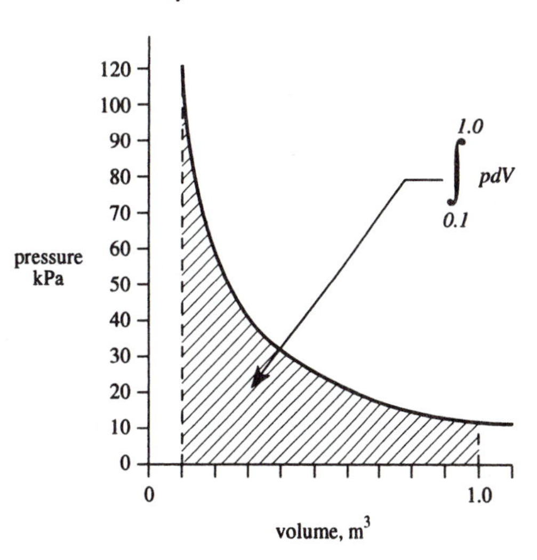

FIGURE 1–17 Determining area under a curve using integral calculus

CALCULUS FOR CLARITY 1–1, continued

and the area is then

$$A = \int_{0.1\ m^3}^{1.0\ m3} \frac{12}{V}\, dV = 12[\ln V]_{0.1\ m^3}^{1.0\ m^3}$$

where the integral $\int dV/V = \ln V$ (see appendix A–4d). Continuing, we have

$$A = 12\ \text{kPa}\cdot\text{m}^3[\ln 1.0 - \ln 0.1] = 12\ \text{kPa}\cdot\text{m}^3\left[\ln \frac{1.0}{0.1}\right]$$

Answer

$$= 27.63\ \text{kPa}\cdot\text{m}^3 = 27.63\ \text{kN}\cdot\text{m}$$

Notice in this example that transposing the limits of integration to $V_1 = 1.0\ \text{m}^3$ to $V_2 = 0.1\ \text{m}^3$ gives the result

$$A = 12\ \text{kPa}\cdot\text{m}^3[\ln 0.1 - \ln 1.0] = 12\ \text{kPa}\cdot\text{m}^3\left[\ln \frac{0.1}{1.0}\right]$$

$$= -27.63\ \text{kN}\cdot\text{m}$$

which is the magnitude obtained by using the original limits, but with a negative sign. The sign will be a significant result, as we will see in chapter 3.

Often the integral form of equation (1–11) can be used to determine the change in a new variable rather than an area under a curve. For instance, in example 1–7 the change in the internal energy of a perfect gas was determined from equation (1–9), by using the definition of internal energy for a perfect gas given by equation (1–7). The following example demonstrates how the integral forms may also be used for determining changes in internal energy, or changes in other variables that may be defined similarly to equation (1–7).

EXAMPLE 1–9

The specific heat of a certain perfect gas is given by the relationship

$$c_v = 0.247 + 5.3 \times 10^{-5}T - 8.1 \times 10^{-9}T^2\ (\text{Btu/lbm°R})$$

Determine the change in the internal energy per pound-mass, Btu/lbm, if the temperature increases from 500°R to 700°R.

Solution

The change in the internal energy of a perfect gas is given by equation (1–9), where the dependent variable $y(x)$ is the specific heat term $c_v(T)$ and the independent variable is temperature T. The integral in this situation gives the change in the internal energy of the gas, Δu, instead of the area under a curve. That is,

$$\Delta u = \int c_v(T)\, dT$$

In this example, the integration limits are from $T_1 = 500°\text{R}$ to $T_2 = 700°\text{R}$. Integrating the general integral $\int x^n\, dx = [1/(n+1)]x^{n+1}$ (given in appendix A–4b) results in

$$\Delta u = \int 0.247\, dT + \int (5.3 \times 10^{-5})T\, dT - \int (8.1 \times 10^{-9})T^2\, dT$$

$$= 0.247(T_2 - T_1) + (5.3 \times 10^{-5})(1/2)(T_2^2 - T_1^2) - (8.1 \times 10^{-9})(1/3)(T_2^3 - T_1^3)$$

$$= 0.247(700 - 500) + (5.3 \times 10^{-5})(1/2)(700^2 - 500^2)$$

$$- (8.1 \times 10^{-9})(1/3)(700^3 - 500^3)$$

Answer

$$= 55.17\ \text{Btu/lbm}$$

1–6 METHOD OF PROBLEM SOLVING

Students of thermodynamics expect to "understand" the subject matter after a reasonable amount of study. For a student to "understand" means that the student will be able to solve problems that would be met in professional work. In this book we emphasize the solving of problems that practicing engineers face, which requires knowledge of the principles of thermodynamics. We stress that the best way to find a solution to a problem is first to use and develop those principles before rushing headlong toward a quick solution. With the continuing development and availability of computer workstations, PCs, and more powerful hand-held calculators, students may want to utilize one of the many mathematical software packages, in conjunction with the appropriate hardware, that can be used to aid in setting down equations and obtaining solutions. Whether you choose one of these options or whether you use the simple "pencil and paper" method, you are encouraged to approach all the problems in this book by using the following steps, in order:

1. Use a separate sheet of paper for each problem. Number or otherwise identify each problem, and include the date.
2. State the problem or situation in your own words. Use suitable sketches if needed to help describe the problem.
3. Identify the system involved in the problem. In chapter 2 the idea of a system is developed and described so that you can know better how to identify one.
4. Carefully list the known values and the unknowns to be found. In chapter 2 the idea of a *state* is given, and we will see that the knowns and unknowns can often be listed as properties of a particular state.
5. List those assumptions that will make the problem easier to solve without making a solution wrong. For instance, if a problem involves air at room temperature and pressure, you can assume that the air is a perfect gas, and an answer from that assumption could be correct. If the problem involves using liquid air at $-150°C$ or $-239°F$, you should not assume a perfect gas because it is not even approximately a perfect gas. Here you might assume an incompressible liquid, and then an answer could be correct. Other assumptions that you should try to consider are whether work is involved, whether heat is involved, and whether the changes occur steadily (called steady state) or under physical restraints. This is an important step in all problem solving and involves judgment and experience, both of which you will develop as you study the material.
6. Identify the process or processes involved. At this time you should see a process as a change that is occurring in a system. The particular way in which the changes are coming about should be known (a much more detailed description of some of those processes is given in chapter 6).
7. Apply one or more of the conservation principles (conservation of mass or conservation of energy) or the second law of thermodynamics. If it has not been solved at one of the earlier steps, the problem will be solved after completing this step.
8. Always be neat and thorough, erasing errors and writing clearly and legibly. Check your answer to see if it is reasonable. Here you must have some experience and judgment, which come with time. But often an answer is clearly wrong, such as determining a negative volume for a gas, say, or heat flowing from cold to hot regions. If the answer appears to be wrong, check it; go back to steps 3 through 8 and see if you have missed something—maybe an answer that seems wrong is actually right. Those cases where an answer seems right but is actually wrong are much harder to find and are found only through considerable study.

1–7 COMPUTER METHODS FOR THERMODYNAMIC PROBLEMS

There are a number of commercial software packages available for routine and sophisticated technical calculations. A package that has been designed for use in thermodynamic analyses is *Engineering Equation Solver* (*EES*), which will be referenced at various times in this textbook. One of this program's capabilities is the simultaneous solution of a set of equations, which may be linear or nonlinear. *EES* also has a selection of thermodynamic properties making it very useful for thermodynamic analysis. *EES* can be obtained from F-Chart Software, P.O. Box 628013, Middleton, Wisconsin 53562, and an operating manual is normally supplied with the software. After loading the program, which is formatted for Windows operating systems, you should have a window with a tool bar across the top and a general description of *EES*. There should be an *OK* box or some box to indicate continuation of the program. By clicking *New* in the *File* menu, the *Equation* window shown in figure 1–18 should appear. Since *EES* is continually being revised with new versions, you may find a slightly different format, but it should contain all of the functions shown in figure 1–18.

Let us familiarize ourselves with *EES* by solving some equations.

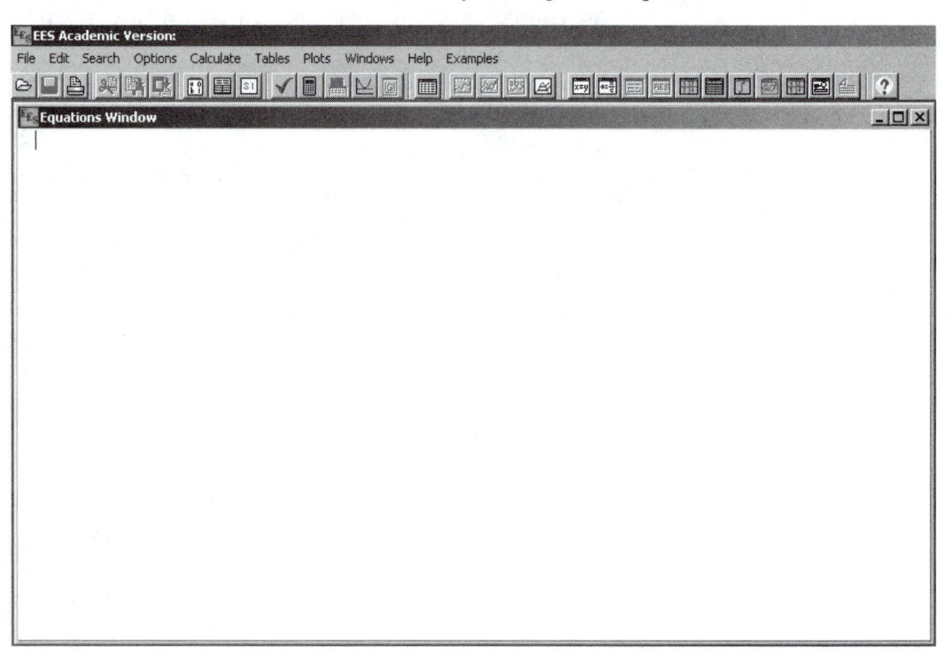

FIGURE 1–18 The *Equation* window *EES*. With permission of F-Chart Software.

EXAMPLE 1–10 Determine the value for p when V is 0.1 for the relationship

$$pV^{1.2} = 20$$

Solution Using *EES*, we will write in the *Equation* window

$$p*V**1.2 = 20 \text{ or } p*V\wedge 1.2 = 20$$
$$V = 0.1$$

The equation box should then look like that in figure 1–19.

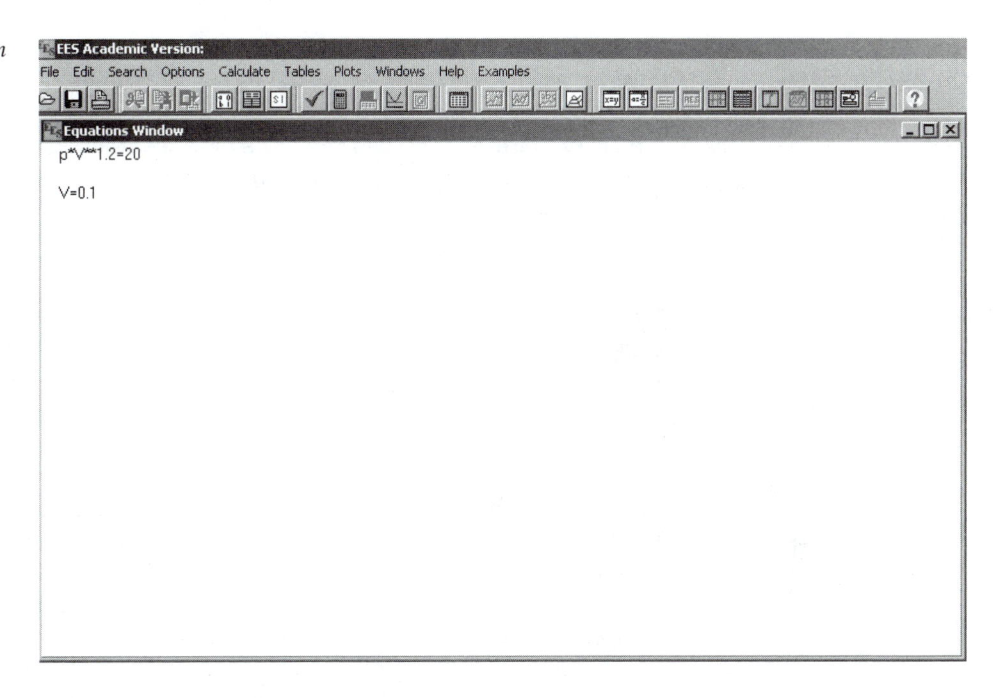

FIGURE 1–19 The *Equation* box for *EES* after entering equations for example 1–10. With permission of F-Chart Software.

If we now select the *Solve* command from the *Calculate* menu, we see a window with a solution as shown in figure 1–20. The solution for p when V is 0.1 is 317. You could check this with your hand calculator or other means. Notice also that the window dialogue indicates that no time was used to do the calculation. In reality, a few nanoseconds were used.

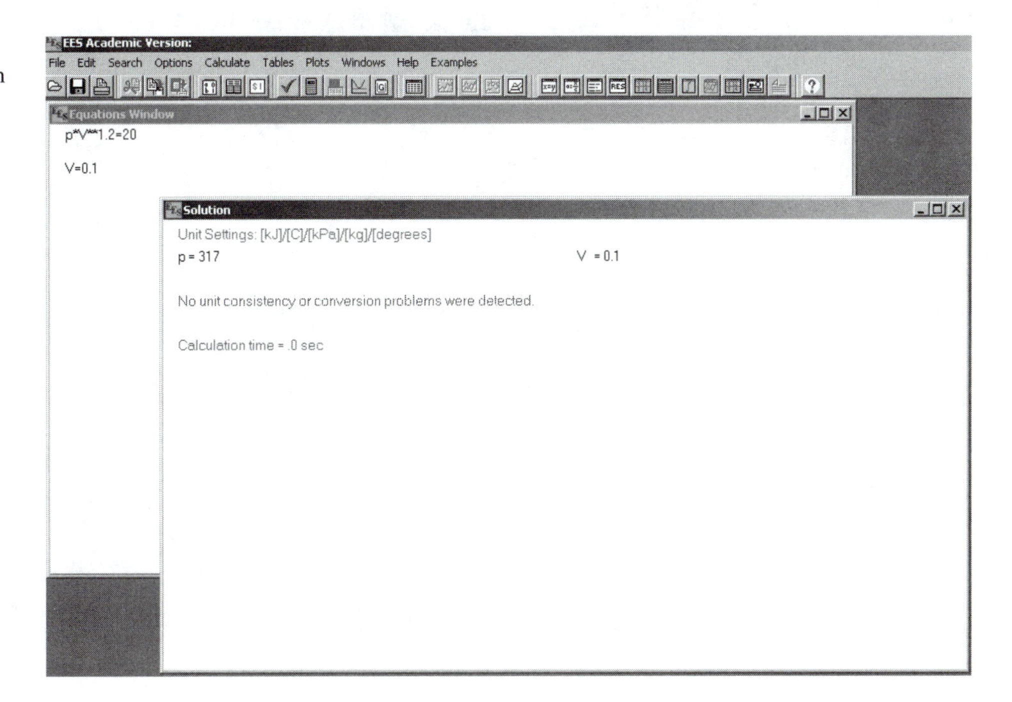

FIGURE 1–20 Solution for example 1–10. With permission of F-Chart Software.

EXAMPLE 1–11 Plot p-V for the relationship of example 1–10 between values of 0.1 and 1.1 for V in increments of 0.1.

Solution From example 1–10 we can determine all of the values of p. *EES* does this automatically with the *tables* menu, as shown on the top of the windows box in figure 1–18. First, enter only the equation

$$p*V**1.2 = 20$$

Then, click *New Parametric Table* in the *Tables* menu, and the window shown in figure 1–21 should appear. In the window, set *Number of Runs* to 11 and click on p. A highlight background should appear. Click *ADD* → and p should then appear in the right box. Do the same for V, so that both p and V are in the right box. Click *OK*. The *New Parametric* table should appear as shown in figure 1–22

Now, click *OK* again. This will result in a spreadsheet table with two open columns with 11 rows as shown in figure 1–23. Click in the box for V and *Run 1* and enter 0.1. Drop down to the next row and enter 0.2 in the box for V and *Run 2*. Continue down for all 11 values of V. The window should then look like that in figure 1–24.

Now click *Solve Table* (or *F3*) under the *Calculate* menu. An intermediate window may appear as shown in figure 1–25.

Click *OK* and the completed table should appear as shown in figure 1–26.

To plot these results, click *New Plot* under the *Plots* menu. Then select *X · Y Plot*, which displays a *New Plot Setup* window. The y-axis should be p and the x-axis should be V. Click *OK*. The plot shown in figure 1–27 should result. You could reverse the plot easily by selecting p for the x-axis and V for the y-axis.

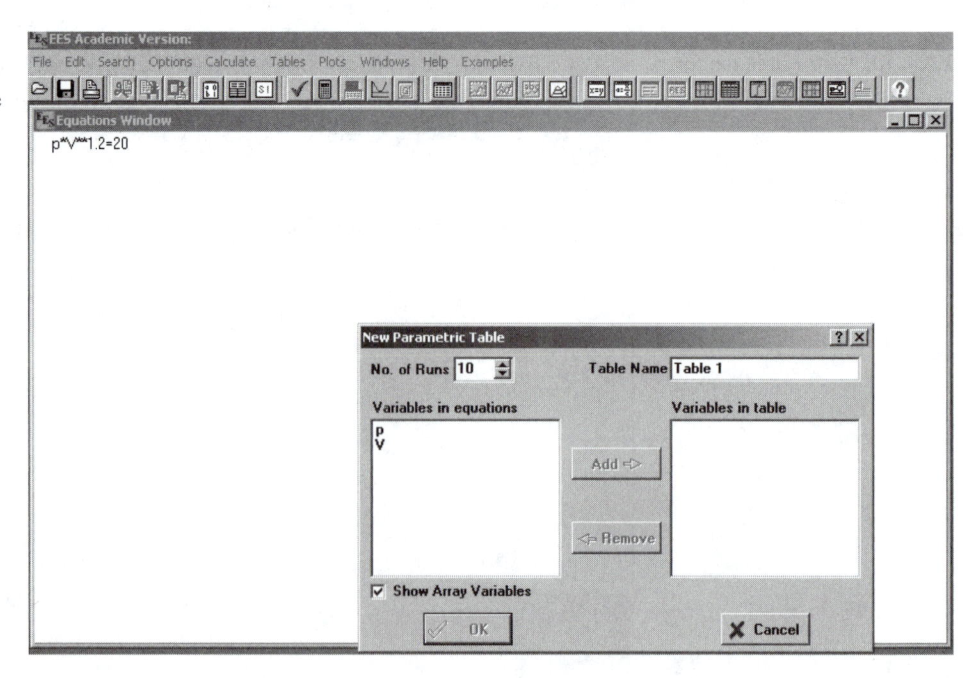

FIGURE 1–21 *New Parametric* table setting for creating values of p for example 1–11. With permission of F-Chart Software.

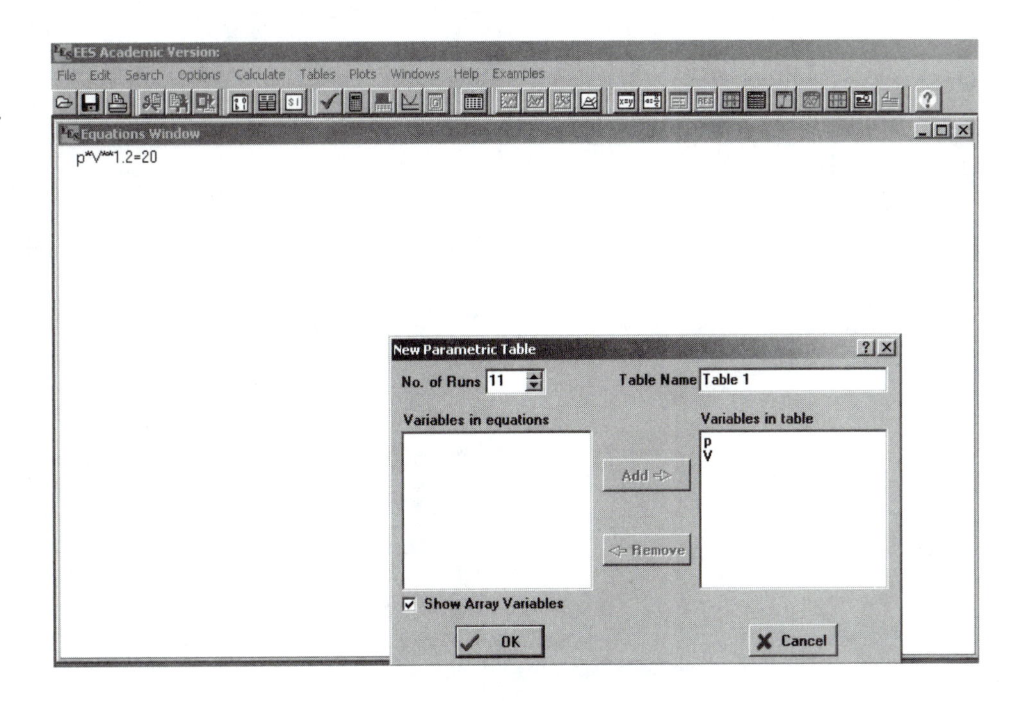

FIGURE 1–22 *New Parametric* table for creating values of p for example 1–11. With permission of F-Chart Software.

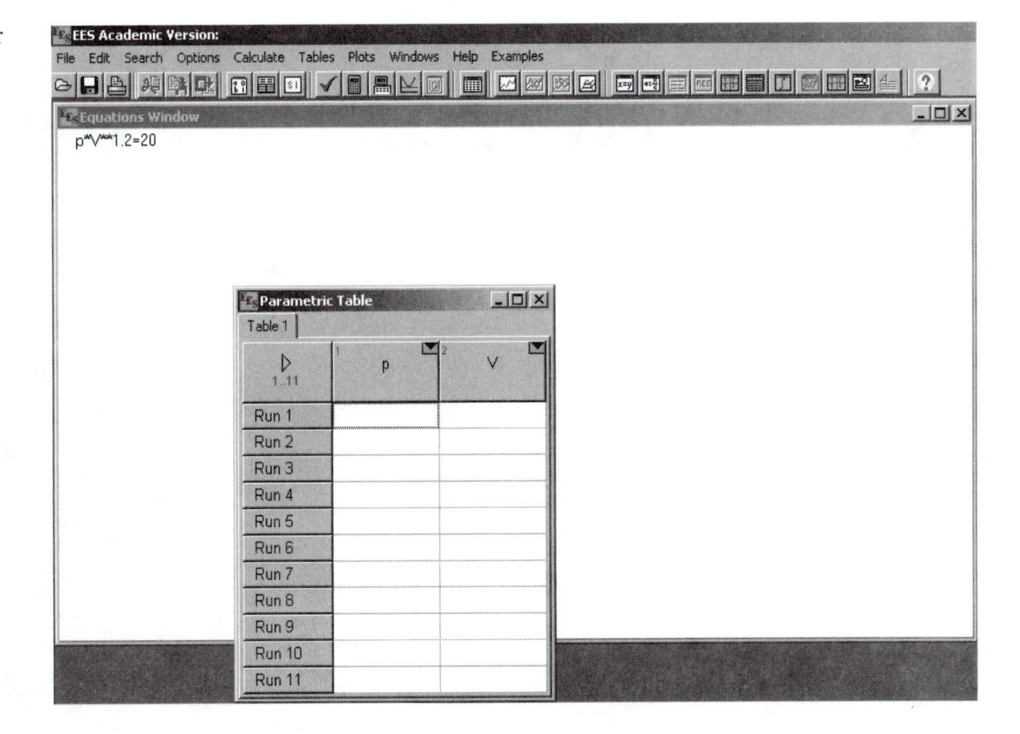

FIGURE 1–23 Window for new table of example 1–11. With permission of F-Chart Software.

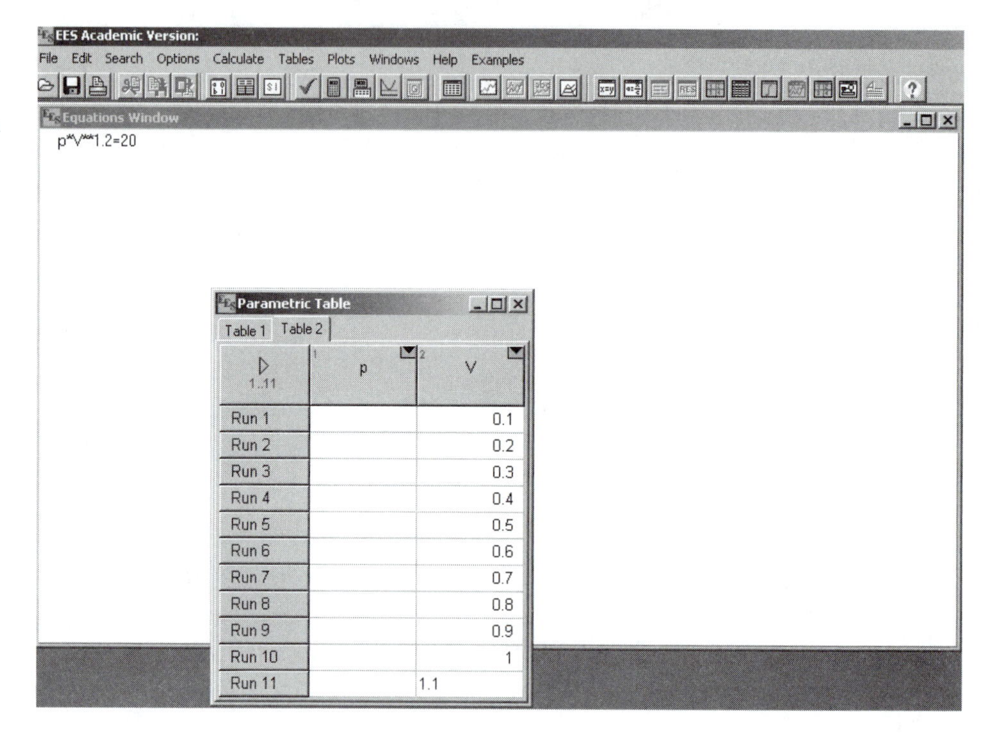

FIGURE 1–24 *New Parametric* table ready to determine the values of p in example 1–11. With permission of F-Chart Software.

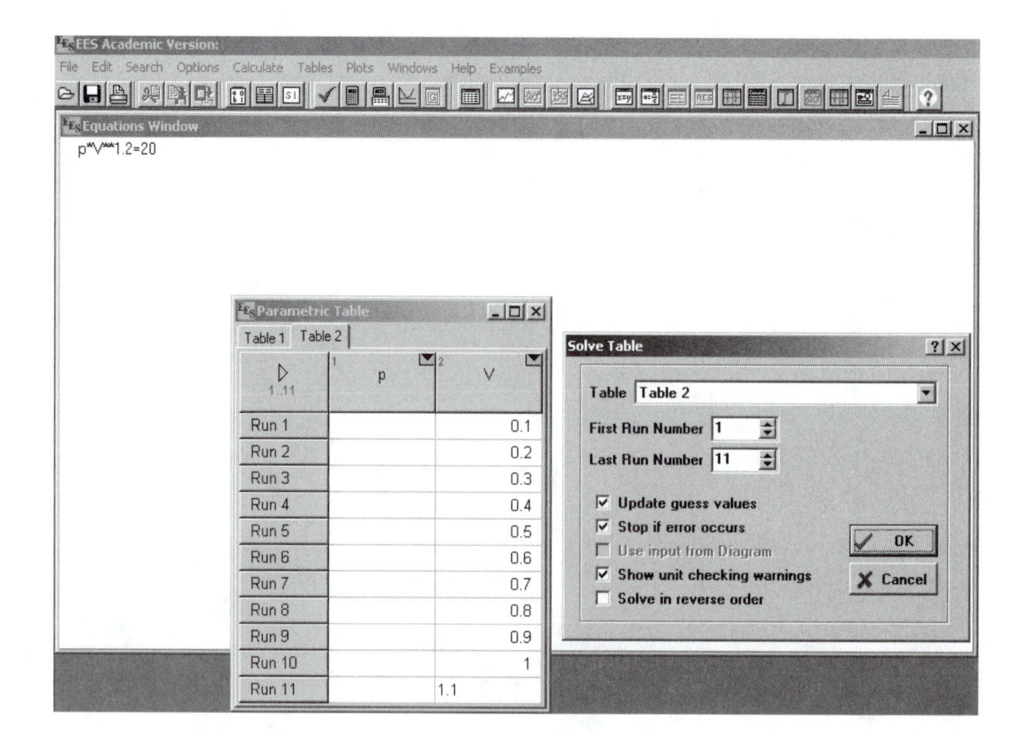

FIGURE 1–25 *Solve* table for example 1–11. With permission of F-Chart Software.

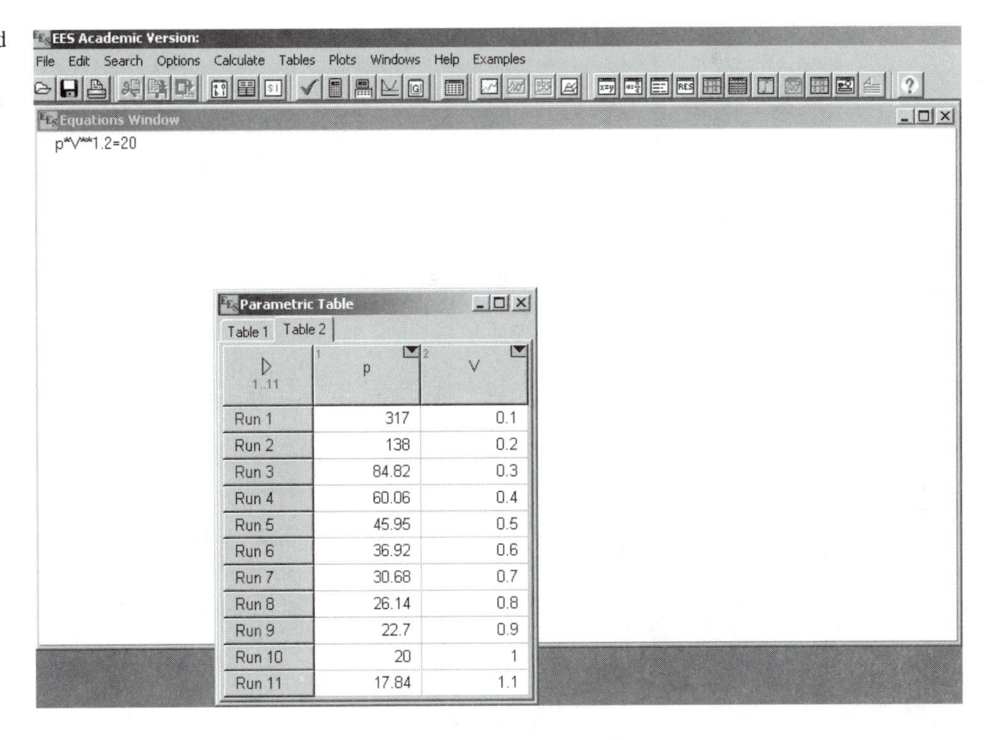

FIGURE 1–26 The completed table for example 1–11. With permission of F-Chart Software.

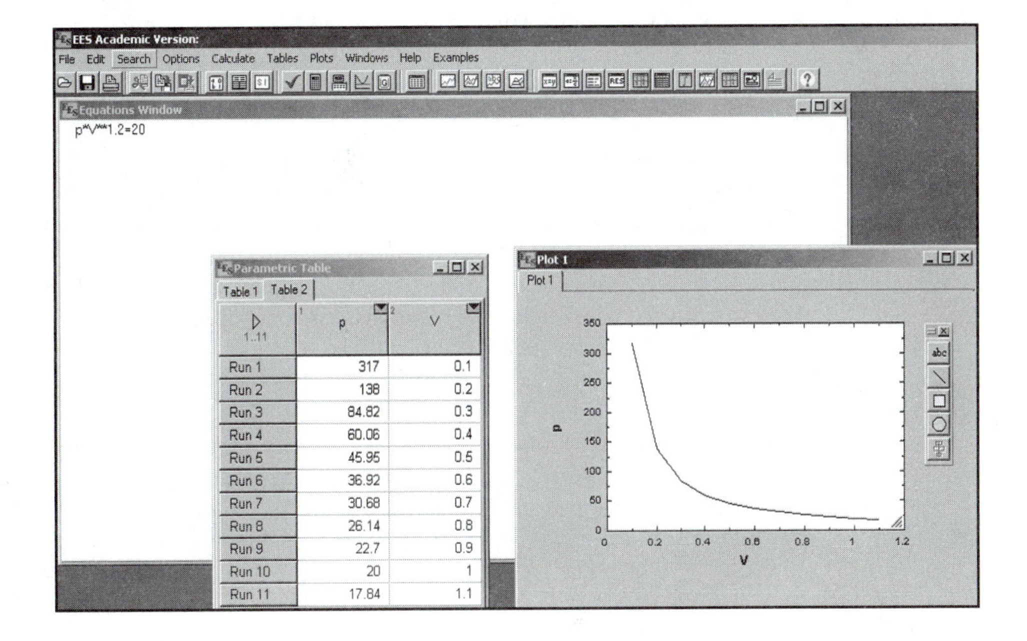

FIGURE 1–27 Plot of *p-V* for example 1–11. With permission of F-Chart Software.

EXAMPLE 1–12 Determine the solution to the following set of equations:

$$Q = Wk - 0.32T$$
$$Wk = 4.5x - 0.005x^3$$
$$T = 5.6x$$
$$Q = \frac{378}{T}$$

Solution We write this set of equations in the equation box as

```
Q = Wk - .32*T
Wk = 4.5*x - .005*x ^ 3
T = 5.6*x
Q = 378/T
```

Clicking the *Solve* option gives the result for the four unknowns:

$$Q = 39$$
$$Wk = 42$$
$$x = 133$$
$$T = 23$$

The *EES* program uses an iterative solution, beginning with a value of 1.0 for all variables and continuing until it converges on the answer. Often, if the same set of equations is solved more than once, the answers may have slight differences in the third or fourth significant figure. If the equations are such that the iteration diverges, the program will tell you by indicating no solution and either you need to change the default initial estimate of 1.0 to another value, or check the equations for correctness. Finally, computers are unable to make the determination of correct equations; only you can do that.

1–8 SUMMARY

In this chapter, we introduced thermodynamics and discussed how it developed over the years. The fundamental dimensions used in thermodynamics, from which all other dimensions are derived, are mass, length, time, and force. The two unit systems used in this book are the Système International (SI) and the English system.

In section 1–5 the ideas of independent variables, dependent variables, and functions were presented. We wrote that, for a dependent variable y of the independent variable x, y is a function of x:

$$y = f(x) \tag{1–2}$$

This function can be plotted on an $x-y$ graph; the result of this plot is a line, called a curve. The idea of calculating the area under that curve over some change in x, Δx, was introduced. The area A was found from the definition

$$A = \sum y_i \, \delta x \tag{1–6}$$

where y_i is the average y over the very small change in x, δx, computed from the relation

$$y_i = (1/2)[(y \text{ at } x) + (y \text{ at } x + \delta x)]$$

Internal energy for a perfect gas is defined as

$$u = c_v T \tag{1–7}$$

Then the change in the internal energy over some temperature change is found from the equation

$$\Delta u = \sum c_{v_i}\, \delta T \tag{1–9}$$

where c_{v_i} is the average specific heat during a small temperature change, δT. The average specific heat was calculated from the equation

$$c_{v_i} = (1/2)[(c_v \text{ at } T) + (c_v \text{ at } T + \delta T)]$$

If c_v is found to be a constant (exactly), the equation used to calculate Δu becomes

$$\Delta u = c_v\, \Delta T \tag{1–10}$$

Finally, a method of solving thermodynamics problems was presented in section 1–6. In succeeding chapters we explain that method in more detail and expose the reader to the principles of thermodynamics.

PRACTICE PROBLEMS

Section 1–4

Evaluate each quantity in problems 1–1 through 1–10.

1–1 $[(3.70)(40.1)]/[(136)(270)(3)]$

1–2 $(1870)(26.0)(9.80)$

1–3 $(260)^2$

1–4 $(260)^{1/4}$

1–5 $(62.1)(35.1/26.1)^{1.6}$

1–6 $(333)[1/(1 - 1.2)]$

1–7 (a) $1.3 \sin(25°)$
(b) $3.7 \sin(2\pi/9)$

1–8 (a) $(5.6 \text{ kJ}) \cos(160°)$
(b) $(9.1 \text{ Btu/lbm}) \cos(\pi/16)$

1–9 (a) $6.48 \log(37.6)$
(b) $(0.2 \text{ kN} \cdot \text{m}) \ln(37{,}000)$

1–10 $e^{1.7}$, $e^{-20.0}$, $e^{\pi/2}$

1–11 Solve for P:

$$3P + 17 \text{ (psi)} = (22 \text{ psi})(\cos 28°)$$

1–12 Solve for x:

$$x^3 = 324 \text{ ft}^3$$

1–13 Solve for V:

$$V^2 + 2V = 265 \text{ m}^6$$

1–14 Solve for T:

$$27.315°\text{C} = 27.600°\text{C} - 0.003T$$

1–15 Solve the equation of a perfect gas $pV = mRT$ for T.

1–16 For the equation $xy^{1.6} = 2.3$, solve for x in terms of y, then for y in terms of x.

Section 1–5

1–17 Determine the area under the curve shown in figure 1–28 between $V = 0.06$ and 1.5 m^3.

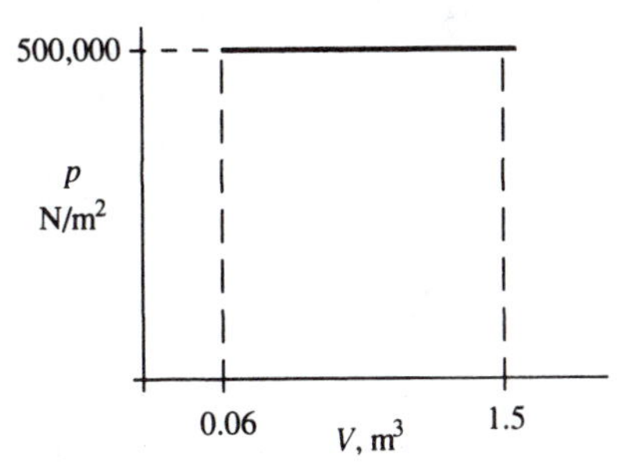

FIGURE 1–28

1–18 Determine the area under the curve shown in figure 1–29 between $T = 10°\text{C}$ and $100°\text{C}$.

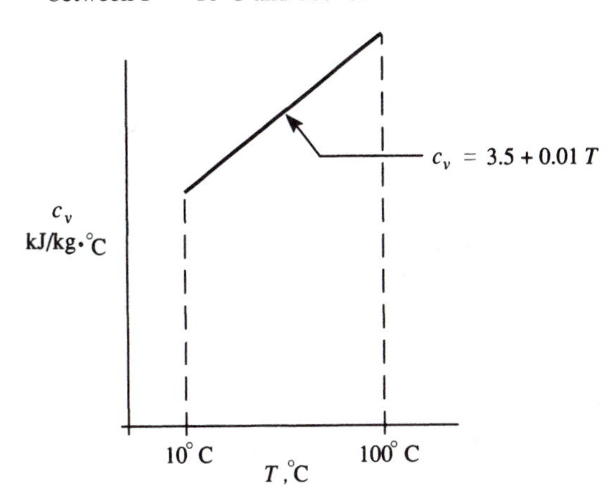

FIGURE 1–29

1–19 Determine the area under the curve shown in figure 1–30 between $V = 1.0$ and $4.0\ \text{ft}^3$.

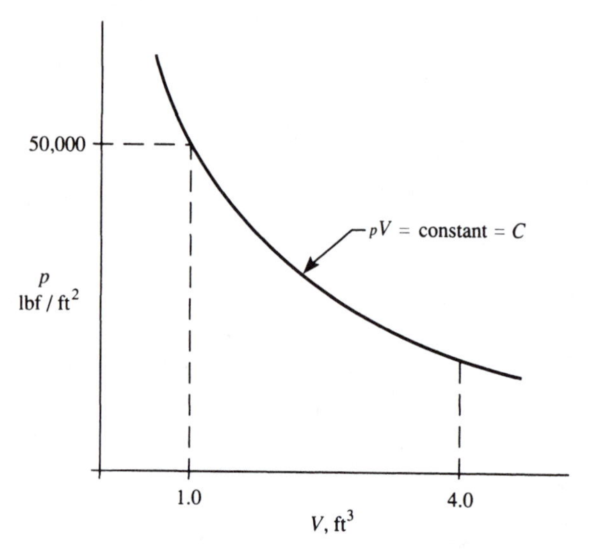

FIGURE 1–30

1–20 Determine the area under the curve shown in figure 1–31 between $V = 15.0$ and $100.0\ \text{in}^3$.

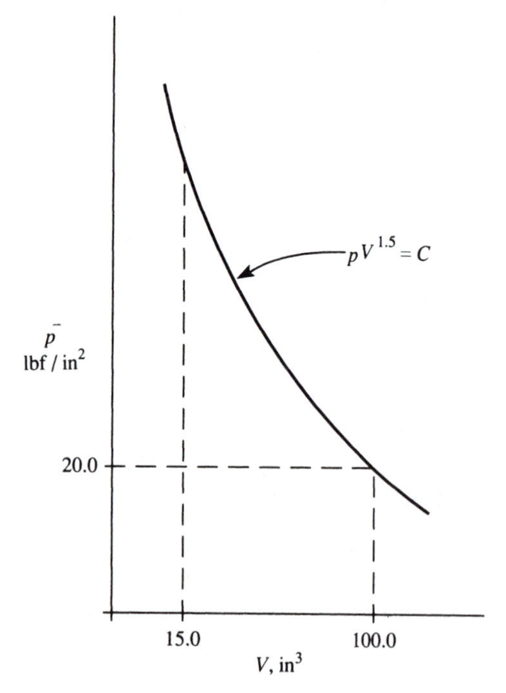

FIGURE 1–31

1–21 Given the following data, determine the area under a curve between $V = 0.010$ and $0.020\ \text{ft}^3$ of a curve of $p = f(V)$.

p, lbf/in^2	V, ft^3
1000	0.010
900	0.0108
800	0.0117
700	0.0130
600	0.0145
500	0.0160
400	0.020

1–22 Given the following data, determine the area under the curve of a T–s diagram between $s = 6.78$ and 6.960 kJ/kg · K.

T, K	s, kJ/kg · K
3400	6.78
3500	6.81
3600	6.831
3700	6.873
3800	6.904
3900	6.942
4000	6.960

1–23 Given the following data, determine the area under the curve in a T–s graph between 500°R and 800°R.

T, °R	S, Btu/°R
500	3.456
600	3.789
700	3.954
800	4.002
900	4.011

1–24 For the function $p = 20.5v$, determine the area under the curve of $p = f(v)$ between $v = 1$ and 10.

1–25 For the function $c_v = 3.56 + 0.0346\,T$ (kJ/kg · K) of a perfect gas, determine the change in the internal energy between 100°C and 500°C.

1–26 For the function $pv^{1/2} = 2700$, determine the area under the curve between $v = 10$ and 300, by using $\delta v = 10$ and then by using the result of appendix A–4e, with $B = 2700$ and $n = {}^1/_2$.

Section 1–7

1–27 Solve the following set of equations:

$$x + 2y = 3.4$$
$$x^2 + y^2 = 4.5$$

1–28 Solve the following set of equations:

$$s = 3.458$$
$$Ts^{1.4} = 4456$$

1–29 For the relationship

$$pV^{1.4} = 280$$

determine the values of p for values of V from 0.1 to 2.0, in increments of 0.1.

1–30 Plot p–V for the result of problem 1–29.

1–31 Solve the following set of equations:

$$Wk = pv^{1.4}$$
$$pv = 4.56T$$
$$Wk = Q - 0.234T$$
$$Q = \frac{456}{T}$$
$$T = 23p$$

THE THERMODYNAMIC SYSTEM 2

In this chapter the reader is instructed on what thermodynamic systems are and how they may be identified. Then some of the properties that describe thermodynamic systems are defined and ways of measuring them are discussed. After finishing this chapter the reader should be familiar with pressure, temperature, density, specific volume, some of the devices used to measure pressure and temperature, and the various types of energy.

New Terms

A	Area	T	Temperature
g	Local gravitational acceleration	U	Internal energy
g_c	32.17 ft · lbm/lbf · s^2, constant in English units	u	Specific internal energy
		V	Volume
G_u	Universal gravitational constant	v	Specific volume
KE	Kinetic energy	$\overline{V}$	Velocity
ke	Specific kinetic energy	W	Weight
L, l	Length	x, y	Length
p_a	Atmospheric pressure	z	Reference elevation above zero potential energy
p_g	Gage pressure		
p_{gv}	Vacuum gage pressure	γ (gamma)	Specific weight
p	Pressure	pe	Specific potential energy
PE	Potential energy	R	Gas constant
r	Radius	η (eta)	Efficiency
SG	Specific gravity	ρ (rho)	Density

2–1 THE SYSTEM

When you wish to solve a technical problem, the first step is to identify what is most important. You need to focus on what the problem really is, what is being affected, or what is affecting something else. In thermodynamics we find this to be crucial to nearly all the problems. In chapter 1, a method of solving problems was presented, and there, actual problem solving began with the third step: identifying the system. Let's now define what we mean by *system or thermodynamic system.*

Thermodynamic system: ***Any region in space that occupies a volume and has a boundary.***

In solving a technical problem through thermodynamics, you must identify the system and its boundaries. For instance, suppose you wish to know the power required to operate a refrigerator. In this circumstance the system boundary would be the outside surface of the refrigerator; everything included inside this surface would be the system. On the other hand, if you are concerned only with the operation of a compressor within the refrigerator, the compressor itself is the system.

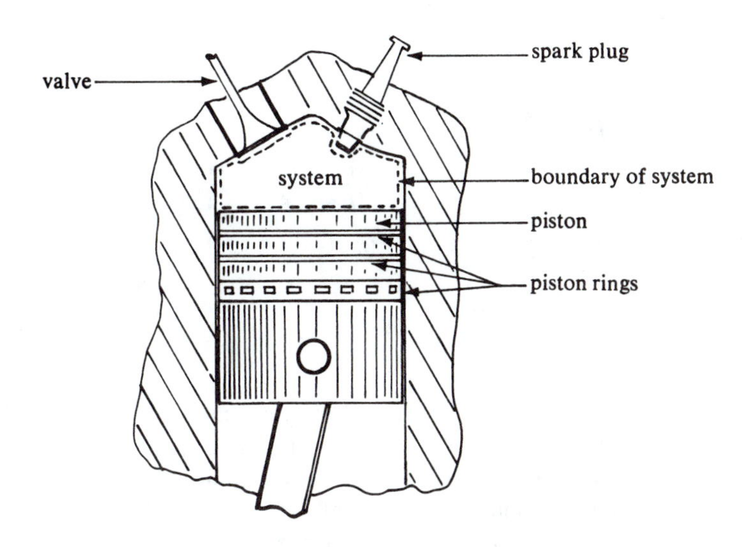

FIGURE 2–1 Piston-cylinder system

As another example, let's consider the internal combustion reciprocating engine of an automobile. If you are interested in the total operation of the automobile, your system might contain the entire vehicle, including the engine, fuel tank, battery, controls, and maybe even the passengers. However, if you wish to study the detailed manner in which power is extracted from the fuel and converted into mechanical energy, the system might be only one cylinder of the engine itself and not even the actual surfaces of the cylinder. This one-cylinder system, or **piston-cylinder** for our purposes, is shown in figure 2–1, with the boundary indicated by a dashed line. This figure illustrates some important characteristics that a boundary has, and as you look at it, you should recognize that it represents a dynamic system in which the piston is in motion at all times. In addition, the valves open and close at opportune instances, either allowing fuel and air to enter the system or exhausting the burned gases. Now, obviously the boundary can move (because the piston and valves move), and we can even shuffle fuel, air, and exhaust gases across the boundary; but not all system boundaries have this capability, and we will see in section 3–6 that the difference between two types of systems—open and closed—will be determined by whether or not matter crosses the boundary. Open and closed systems will then be an important part of the thermodynamic analysis, each handled in a slightly different manner.

Another system is identified in figure 2–2. This is a balloon made of thin rubber that can stretch and contract. The system boundary of the right-side configuration is drawn outside the balloon surface and envelopes two separate materials: air in the balloon and the rubber balloon itself. In analyzing this system, you must be careful. In figure 2–1 the system is homogeneous at any given time, but the balloon in figure 2–2 is not. By **homogeneous**, we mean that there is one and only one distinct uniform material throughout the system at any instant in time. In figure 2–2 we have two different materials, the balloon and the air, each of which has distinct ways of handling energy, thereby affecting the analysis differently.

In defining or establishing a system, whenever possible, select a boundary that will enclose homogeneous materials. In the balloon problem we were concerned with rubber as well as the air, so that ignoring all but the air would be undesirable. If we wished, however, to specify a homogeneous system in the balloon, we could do so by denoting a boundary inside the balloon, thus defining a system composed of air alone, as indicated in the left-side configuration of figure 2–2.

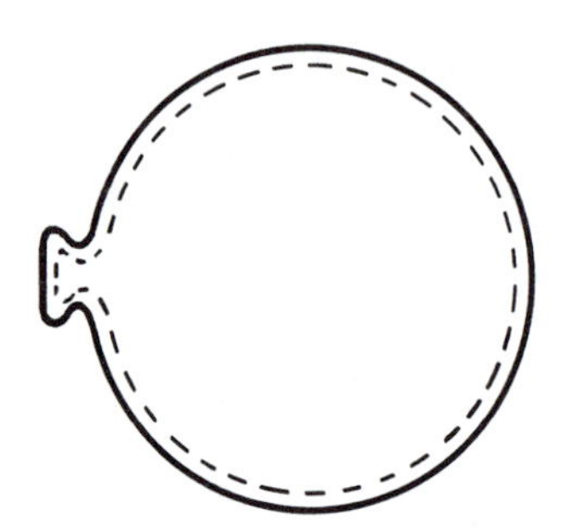

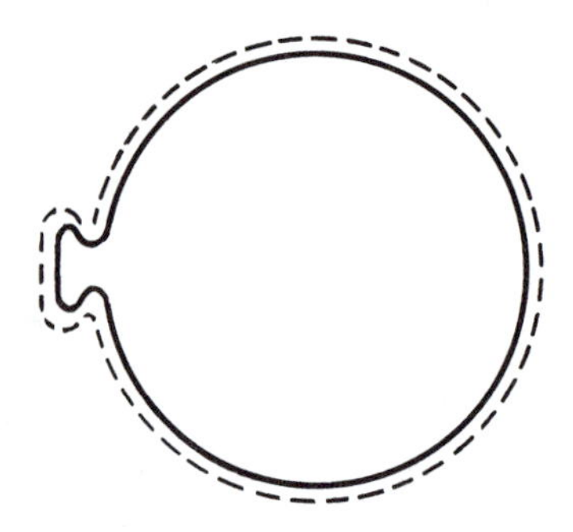

FIGURE 2–2 Inflated balloon with system defined by boundary inside or outside balloon material

As we have seen, the boundary and the enclosed volume of a system can be arbitrary, and the possible variations are infinite. Let's allude to the physical world. The volume outside the boundary we will call the **surroundings** of the system, and the sum of the system and its surroundings we will call the **universe**. The universe, if we are completely rigorous and correct, is infinite, but we need not concern ourselves with that aspect in this book. We will consider the surroundings as only that physical part of the universe which is capable of affecting our system during the period with which we are concerned.

In addition, once we have assured ourselves of either a homogeneous system or a simple composition of homogeneous systems (like the balloon problem), the internal details of the dynamics or design of the system are not needed. That is, for the piston-cylinder system of figure 2–1, the materials, dimensional sizes, or velocities and accelerations of the valves, pistons, spark plugs, or cylinder wall are not necessarily important in the thermodynamic analysis, provided we know that the system is a homogeneous gas inside the cylinder.

2–2 ELEMENTARY THEORY OF MATTER

Thermodynamic systems are volumes in space, and these volumes usually contain matter or some material. If the volume contains no matter, it is a perfect vacuum, and in chapter 8, it will be shown that such a volume or system can be used to obtain power or work. However, for now let us consider only those systems that do contain matter. It is matter that contains, collects, sorts, transfers, and gives off energy, and a brief description of its structure should be beneficial.

Matter is composed of **atoms**, each of which in turn is composed of a cluster (one or more) of **protons** and **neutrons** very tightly compressed into a **nucleus**. The nucleus is surrounded by a cloud of one or more very small particles, called **electrons**, found in different concentric levels, or **shells**, encircling the nucleus. Protons have a positive charge; electrons, a negative charge; and neutrons, as the name implies, have no charge, that is, they are neutral. The electrons, protons, and neutrons of different atoms vary in number and arrangement, and each unique arrangement is called an **element**. There are naturally occurring and fabricated elements. (See table B–1). A simplistic diagram of an atom is shown in figure 2–3, a sketch of a hydrogen atom.

Each shell of an atom can hold a certain maximum number of electrons; for example, the first shell can hold two electrons, and the second, eight. When the outer shell, that is, the last shell that contains electrons, contains the maximum number it can hold, that atom is in a stable state and does not readily combine with other atoms. For example, neon atoms have two electrons in their first shell and eight in their second and are thus stable. However, if an atom has an incomplete outer shell and does not contain the maximum number of electrons it can hold, it is unstable and will readily combine with other atoms to make its outer shell complete. When two atoms combine, one gains the electrons that the

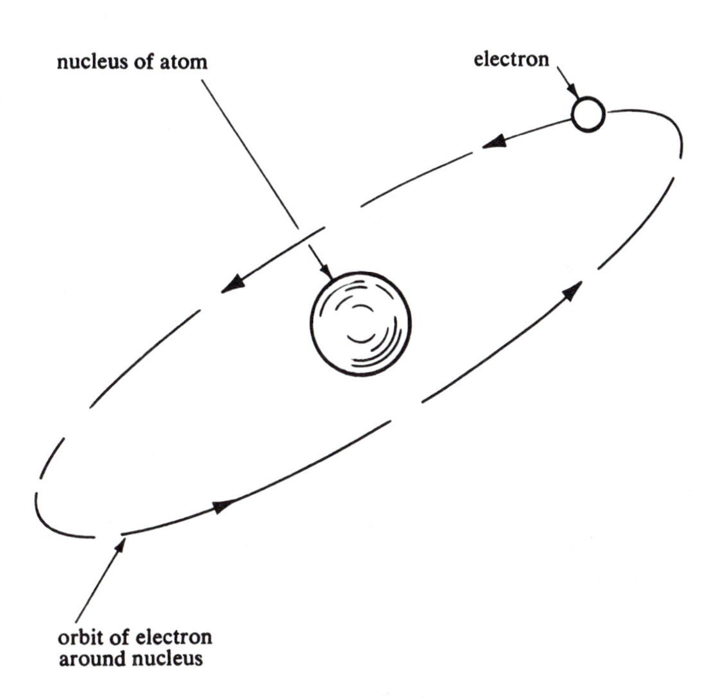

FIGURE 2–3 Simplified visualization of hydrogen atom

other loses; in combining, atoms form different types of chemical bonds, such as ionic or metallic (figure 2–4). When atoms combine, they form what are called **molecules**.

In addition to gaining or losing electrons, atoms can form bonds by "sharing" one or more pairs of electrons; this type of bond is called **covalent**. In this sharing of electron pairs, each electron in a pair comes from a different atom. A hydrogen molecule is a good example of covalent bonding, or electron sharing (figure 2–5). Each hydrogen molecule contains two atoms, each of which, when it exists singly without the other, has only one electron in its outer shell and needs one more to make it complete. However, when the atoms combine to form a molecule, each shares the other's electron, so that each has two electrons and thus a completed outer shell.

Atoms and molecules have mass and occupy volumes. The atomic weights of the first 103 elements are listed in table B–1. Each element is denoted by a symbol; for instance, carbon is C and has an atomic weight of 12.0, hydrogen is denoted as H with an atomic weight of 1.008 (or just 1.0). The atomic weights for all other elements are found directly below the symbol for the elements. The atomic weights are to be interpreted as the mass of that element when it occupies 1 mole of space at atmospheric pressure and temperature. The mole is a unit of volume; the volume occupied by an amount of an element whose mass is equal to its atomic weight in grams is called the gram-mole. Thus we say that carbon has an atomic weight of 12.0 grams/gram-mole, hydrogen 1.0 gram/gram-mole, and so on. Also the atomic weight could be interpreted as the mass in kilogram units or pound-mass (lbm) units, so that the mass of carbon is 12.0 kg/kg · mol or 12.0 lbm/lbm · mol. The kg · mol has a volume 1000 times larger than the g · mol, and the lbm · mol unit is approximately 454 times larger than the g · mol (because 454 g = 1 lbm). Clearly, the

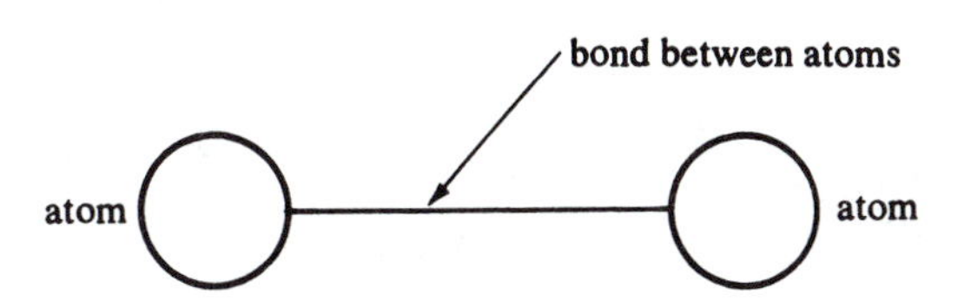

FIGURE 2–4 Simplified view of metallic or ionic bond in molecule composed of two atoms

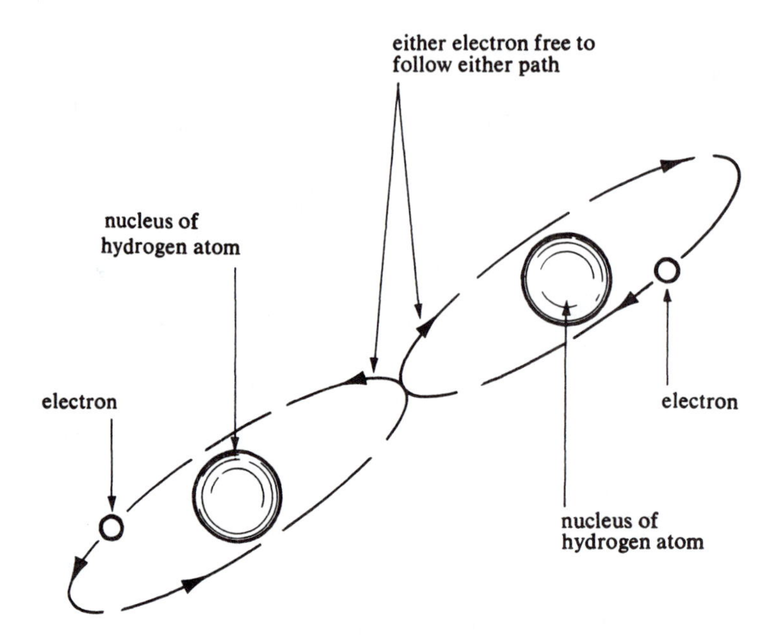

FIGURE 2–5 Hydrogen molecule

mole is a relative unit and is made specific by the interpretation of the mass unit for the atomic weight. The amount of matter is also sometimes visualized through Avogadro's number, which is equal to 6.022×10^{23} atoms or molecules per gram-mole. That is, experimentally it has been determined that there are 6.022×10^{23} atoms or molecules in a volume of 1 gram-mole of any element at standard atmospheric conditions.

In our approach to thermodynamics we will not be concerned with counting or otherwise describing individual atoms or molecules, but we will consider very large amounts of atoms or molecules mixed together, called a **substance**. If the substance is composed of only one type of molecule, we will call this a **pure substance**. Water, composed of molecules made up of two hydrogen (H) atoms and one oxygen (O) atom and written H_2O, is an example of a pure substance.

The atoms and molecules of substances behave differently, depending on the conditions of the surroundings in which the substance is immersed. Consequently, we will want to know the arrangement and interaction of these particles, the **phase** of the element or atoms. We will concern ourselves with three phases of a substance: solid, liquid, and gaseous. The solid phase of a substance is characterized by rigid bonds between atoms or molecules, bonds that can generally stretch, bend, and break (figure 2–6). Also, the solid phase will, under many conditions, be characterized by a crystal structure.

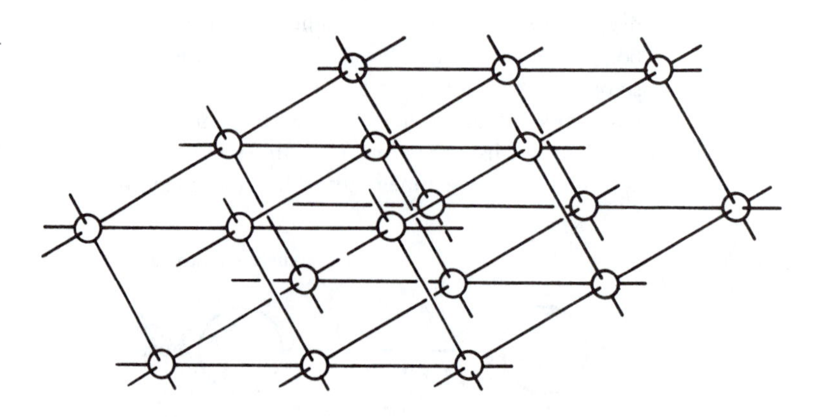

FIGURE 2–6 Solid phase—typical "lattice" composed of metallic or ionic bonds and atoms in a rigid configuration

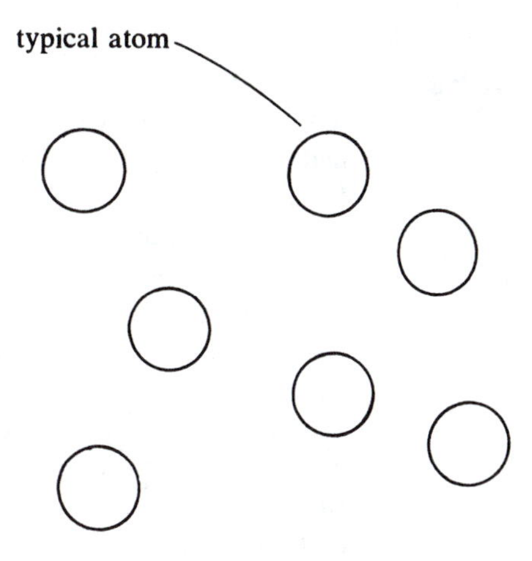

FIGURE 2–7 Gaseous phase—little or no evidence of bonds between atoms; atoms free to move independently in any direction

The gaseous phase has free-floating molecules with some, although rather insignificant, force between them (figure 2–7). We will be discussing a substance called a **perfect gas**—a gas that has no forces, other than collision forces, between the individual atoms or molecules. The collision for a perfect gas is assumed to be "perfect" in the sense that no permanent distortion occurs among the particles after colliding.

The liquid phase is behaviorally like a mixture of the solid and gaseous phases. The molecules are more closely packed than in a gas, but the bonds between the molecules are passive. That is, the molecules will adhere to one another under very light forces but can easily slide apart and shift arrangements without great forces, as shown in figure 2–8. We could characterize liquids as being able to retain some volume (one gallon, for instance), but not shape. If left alone in space, gaseous substances cannot even retain their volume, let alone their shape, whereas solids retain volume and shape.

At low temperatures, materials generally are in a solid or a liquid phase; as the temperature rises, they will pass to a gaseous phase. We will see that "high" or "low" temperature has no absolute value thermodynamically, so that some materials, such as air, do indeed have liquid and solid phases at their "low" temperatures, and some solids have liquid and gaseous phases at their "high" temperatures. In this book we are concerned with the solid, liquid, and gaseous phases of pure substances; mixtures of pure substances; and substances that will be approximately pure substances.

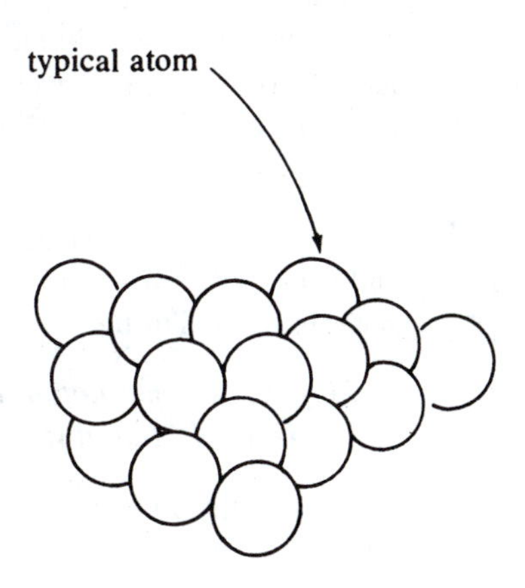

FIGURE 2–8 Liquid phase—atoms attracted by bonds that allow sliding between atoms but prevent separation of individual atoms

2–3 PROPERTY

To describe and analyze a system, we must know some of the quantities that are characteristic of it. These quantities, called **properties**, include volume, mass, weight, pressure, temperature, density, shape, position in space, velocity, energy, specific heat, color, taste, and odor. The list could go on and on, and the longer the list, the better description we could give of our system.

We shall separate properties into two general classifications: **intensive properties** and **extensive properties**. An intensive property is independent of the mass or total amount of the system: for example, color, taste, odor, velocity, density, temperature, and pressure. Extensive properties are all properties that are dependent on the total amount of the system, for example, mass, weight, energy, and volume. It is worth noting that any extensive property can be made intensive by dividing by the mass. In this book we generally denote extensive properties by capital letters and intensive properties by lowercase letters. Exceptions to this notation are mass m, which is an extensive property, and velocity $\overline{V}$, which is intensive. Also, we frequently use the term **specific** to denote an intensive property. Thus we use the term **specific energy** to describe the energy per unit of mass of some material. The term **total** is used to describe extensive properties, such as **total energy** to denote the energy in a given amount of mass.

2–4 STATE OF A SYSTEM

A complete list of the properties of a system describes its state. In order that a list not get too long and confusing, we will generally assume that the system is composed of pure substances of one phase, or of simple inert mixtures of the three phases, gaseous, liquid, and solid. We will see that only a few properties need to be known under these conditions, and these are generally as follows:

Type of substance (element)
Volume
Weight or mass
Pressure
Density or specific volume
Temperature
Energy

2–5 PROCESS

The primary reason for describing the state of a system is to analyze the system as a power-producing or power-consuming quantity. **Process** is a change of state, which can occur in a number of ways. For example, if we change one or more of the properties, such as energy, pressure, temperature, or volume, we will have gone through a process. Our concern will be evaluating the state immediately before and after this change. An important point to remember is that work, power, and heat can occur only during processes and only across the boundary of the system. We will return to this later.

2–6 CYCLES AND CYCLIC DEVICES

Having a system that changes its state, thereby producing or using work or heat, is all well and good, but if we want an engine that runs continuously without seemingly changing its state further and further, we must occasionally return to our stabilizing point or initial state.

> **Cycle:** ***A combination of two or more processes that, when completed, return the system to its initial state; a system operating on a cycle is called a cyclic device.***

Notice that there is no change in energy or property over a complete cycle. There can, however, be work and heat added or extracted, which is the essence of the power-producing and power-consuming devices of our society—we return the engine (or cyclic device) to its initial condition periodically (for example, 300 times per minute if we have an engine running at 300 revolutions per minute), and it still continues to produce work.

The special form of the cyclic device that transfers heat into work is called a **heat engine** and is of major concern to us in the study of thermodynamics. Examples of heat engines are the internal combustion engine of an automobile, the jet engine used to power aircraft, and the steam turbine electric power generating system.

2–7 WEIGHT AND MASS

The **mass** of a system or object is a measure of its amount and is a fundamental dimension that we determine experimentally. The normal units for mass are kilograms (kg) in the SI system and the pound-mass (lbm) in the English system. The "slug" is sometimes used in the English system but will not be used in this book. The symbol used for mass in this book is m.

The **weight** of an object or system is the force of attraction of the Earth to that object due to gravity and will be denoted by the symbol W. *Newton's law of gravity* is written as

$$F = \frac{G_u m_1 m_2}{r^2} \tag{2–1}$$

where F is the **force** of attraction between two bodies having masses m_1 and m_2. The distance between the centers of the two bodies is r, and G_u is **the universal gravitational constant**, determined as

$$G_u = 6.67 \times 10^{-11}\ \text{m}^3/\text{kg} \cdot \text{s}^2$$

or

$$G_u = 106.84 \times 10^{-11}\ \text{ft}^3/\text{lbm} \cdot \text{s}^2$$

Two bodies that are separated but attracted to each other due to the force of gravity F are shown in figure 2–9a. Notice that the force F acts as an attraction for both bodies.

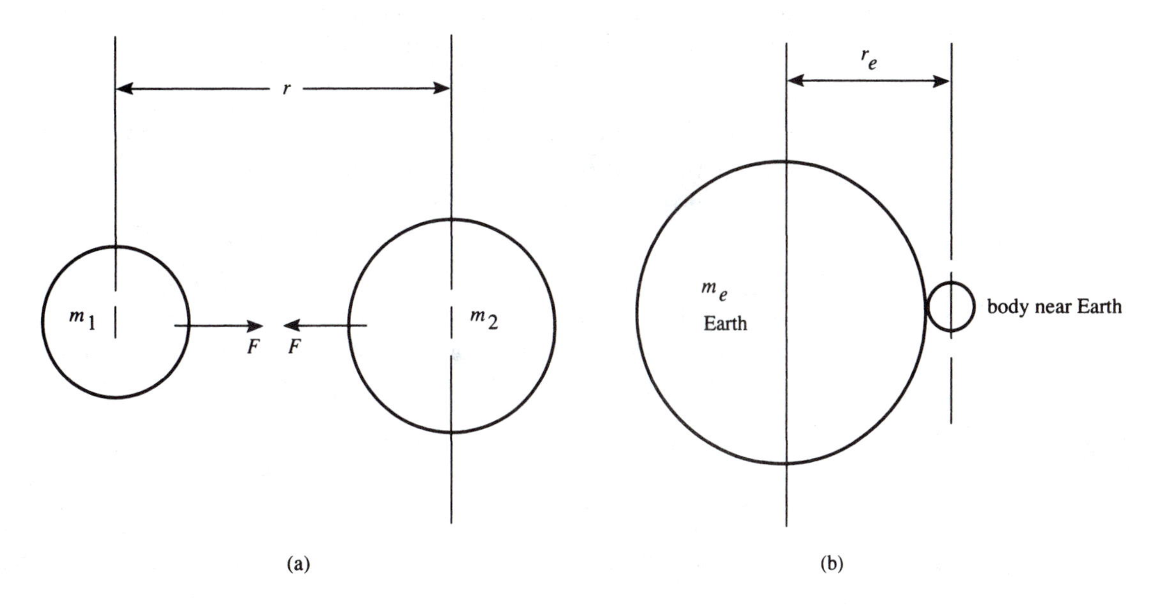

FIGURE 2–9 Action of gravity

Figure 2–9b illustrates the situation where one of the masses is much larger than the other, such as occurs for a body near Earth. In this case the mass m_1 is Earth's mass, m_e, which has been calculated by astronomers and found to be approximately 5.983×10^{24} kg. The distance between the center of Earth and the center of a body near Earth is nearly the radius of Earth. Since Earth is not perfectly round, the radius varies, but it is approximately equal to 3963 miles or 6378 kilometers (km). Therefore, in equation (2–1) the term $G_u m_1/r^2$ is a constant and is approximately equal to

$$\frac{(6.67 \times 10^{-11}\ \text{m}^3/\text{kg}\cdot\text{s}^2)(5.983 \times 10^{24}\ \text{kg})}{(6.378 \times 10^6\ \text{m})^2} = 9.81\ \text{m/s}^2$$

This is called the **gravitational acceleration**, or g*. Then we can write Newton's law of gravity for a body near Earth as $F = mg$, where m is the mass of the body and F is the weight W, or

$$W = mg \tag{2–2}$$

Since Earth's radius changes at various locations on Earth, the gravitational acceleration g varies at the different locations. In table B–2 are listed the local gravitational acceleration in both English and SI units. From equation (2–2) it should be noted that the mass of a body or system does not change but the weight does, depending on g. In the English unit system the basic unit for force has been the pound-force (lbf). The lbf is also the unit used for the weight because, as we have just seen, weight is a force due to gravity. As a consequence of having a unit of pounds used to describe two fundamentally different properties, mass and force, a reexamination of equation (2–2) is necessary. Also, from Newton's laws of motion, it is well known that

$$F = ma \tag{2–3}$$

where F is force acting on a mass m which is experiencing an acceleration a. **Acceleration**, a, is the rate of change of speed or velocity per unit of time; the gravitational acceleration, g, is a special type of acceleration. If an object were to fall toward Earth and if the wind resistance were zero, the object would accelerate at an amount equal to g. So equations (2–2) and (2–3) are statements relating the same terms, and they can be rewritten with a constant of proportionality, C. Thus, equations (2–2) and (2–3) are then

$$F = Cma \quad \text{and} \quad W = Cmg \tag{2–4}$$

where C is to be determined. In SI units F has units of newtons (N), mass has units of kilograms (kg), and acceleration has units of meters per second per second (m/s^2). The newton is defined as

$$1\ \text{N} = 1\ \text{kg}\cdot\text{m/s}^2 \tag{2–5}$$

so that C in equation (2–4) is just $1\ \text{N}\cdot\text{s}^2/\text{kg}\cdot\text{m}$. Since the newton is defined in such a convenient way, the constant C is really unnecessary and is deleted. In the English system, however, C must have the units of $\text{lbf}\cdot\text{s}^2/\text{lbm}\cdot\text{ft}$ in order that equations (2–3) and (2–4) be dimensionally correct. That is, from equation (2–4), we have

$$C = \frac{F}{ma} = \frac{W}{mg} \tag{2–6}$$

* Since the Earth is not a perfect sphere, but has its largest diameter at the equator and least between the north and south poles, the gravitational acceleration on the Earth's surface varies slightly with latitude on the Earth. This variation is tabulated in table B–2 of the appendix. By agreement, the average gravitational acceleration on the Earth is taken as 9.81 m/s^2 or 32.174 ft/s^2.

and this equation checks the units required of C. Also, if the definition of 1 lbf is taken to be equal to the force required to accelerate 1 lbm at 32.174 ft/s^2, then C must have a value of 1/32.174. The purpose of choosing the value of 32.174 for the acceleration in this definition is to allow for the weight and the mass to have the same numerical value on Earth's surface, when mass is in pounds-mass and weight in pounds-force, and the gravitational acceleration, g, is 32.174 ft/s^2. The constant C is more often written as $1/g_c$, where $g_c = 32.174\ \text{lbm}\cdot\text{ft/lbf}\cdot\text{s}^2$, and then equation (2–3) becomes

$$W = \frac{1}{g_c}mg \quad \text{(English units, } m \text{ in lbm)} \tag{2–7a}$$

and (2–4) becomes

$$F = \frac{1}{g_c}ma \tag{2–7b}$$

The slug is a mass unit that is occasionally used in engineering. A slug is defined as 32.174 lbm, so that pounds-force is equal to the force required to accelerate 1 slug at 1 ft/s^2. Therefore, if the slug is used for mass, the constant C is again equal to 1, and equations (2–3) and (2–2) can be used for the relationships among force, weight, mass, and acceleration.

EXAMPLE 2–1 A person has a mass of 100 kg. Determine his or her weight at 40° latitude and 1500 m above sea level.

Solution From table B–2, we find that g has a value of 9.7976 m/s^2 at 40° latitude and 1500-m elevation. Thus, from equation (2–2), we have

$$W = mg$$
$$= (100\ \text{kg})(9.7976\ \text{m/s}^2)$$

Answer
$$= 979.76\ \text{kg}\cdot\text{m/s}^2 = 979.76\ \text{N}$$

or

Answer
$$W = 980\ \text{N}$$

EXAMPLE 2–2 An automobile engine weighs 300 lbf at 1000-ft elevation and 20° latitude. What is the mass of the engine, and what would it weigh if it were located at 20° latitude and 4000 ft?

Solution From table B–2 we find the value of g to be 32.105 ft/s^2. Then, because the weight was expressed in pound-force units, we use equation (2–7) and obtain

$$m = \frac{W}{g}g_c$$
$$= \frac{(300\ \text{lbf})(32.17\ \text{ft}\cdot\text{lbm/s}^2\cdot\text{lbf})}{32.105\ \text{ft/s}^2}$$

Answer
$$= 300.6\ \text{lbm}$$

If the engine were located at 4000-ft elevation, the value for g would be 32.096 ft/s^2, and the weight would be found from equation (2–7) to be

$$W = \frac{(300.6\ \text{lbm})(32.096\ \text{ft/s}^2)}{32.17\ \text{ft}\cdot\text{lbm/lbf}\cdot\text{s}^2}$$

Answer
$$= 299.9\ \text{lbf}$$

EXAMPLE 2–3 What is the weight W of a suitcase having a mass of 1 slug when the system is
(a) At sea level and 40° latitude?
(b) At 1000 ft above sea level and 40° latitude?

Solution (a) The value of g is 32.158 ft/s^2 at sea level and 40° latitude. (See table B–2.) Then, from equation (2–2),

$$W = mg$$
$$= 1 \text{ slug} \times 32.158 \text{ ft/s}^2$$

Answer
$$= 32.158 \text{ lbf}$$

We see that

$$1 \text{ slug} = 1 \text{ lbf} \cdot \text{s}^2/\text{ft}$$

(b) At 1000 ft, we see from table B–2 that g is 32.155 ft/s^2, so

$$W = 1 \text{ slug} \times 32.155 \text{ ft/s}^2$$

Answer
$$= 32.155 \text{ lbf}$$

Alternatively, the mass of the system in pound-mass units is

$$m = 1 \text{ slug} \times 32.174 \text{ lbm/slug}$$

and from equation (2–7),

$$W = \frac{1}{32.174 \text{ ft/s}^2 \times \text{lbm/lbf}} \times 1 \text{ slug} \times 32.174 \text{ lbm/slug} \times 32.155 \text{ ft/s}^2$$
$$= 32.155 \text{ lbf}$$

Although the mass unit of slugs provides somewhat simplified equations by eliminating a factor of $1/g_c$ from the force–mass relationships as shown by equations (2–2) and (2–3) as compared to (2–7), we will generally use pound-mass units unless otherwise specified. The slug is used in some reference literature, but the pound-mass is generally used for reporting thermodynamic data.

2–8 VOLUME, DENSITY, AND PRESSURE

Volume and mass are the primary measures of the quantity of matter in a system.

Volume: ***An extensive and geometric property having a value characterized by a length times a height times a width, simply "length cubed."***

Volume

Volume will be denoted by the letter V and expressed in the units cubic meters (m^3). Other units that may be used to express volume are liters (L) and cubic centimeters (cm^3 or cc). In the English system the common units for volume are cubic feet (ft^3), gallons (gal), and cubic inches (in^3). Various conversion factors for volume are given in the front inside cover conversion table.

Specific Volume

Many times we will want to know the volume occupied by a unit mass of the system. This is called the **specific volume** and is an intensive property denoted by the letter v. If the matter is homogeneous, then

$$v = \frac{V}{m} \tag{2–8}$$

Specific volume will be described with the units of cubic meters per kilogram (m^3/kg) or liters per kilogram (L/kg). In English units specific volume will be described by cubic feet per pound-mass (ft^3/lbm), cubic feet per slug (ft^3/slug), or gallons per pound-mass (gal/lbm).

Density

Density is a property used frequently in describing a system. It is denoted in this book by the lowercase Greek letter ρ, and for homogeneous materials we have

$$\rho = \frac{m}{V} \tag{2–9}$$

Note that density is an intensive property and is the inverse of specific volume. Thus, we have

$$\rho = \frac{1}{v} \tag{2–10}$$

expressed in units of kilograms per cubic meter (kg/m^3), kilograms per liter (kg/L), and grams per cubic centimeter (g/cm^3). The English units used to describe density are pounds-mass per cubic foot (lbm/ft^3), slugs per cubic foot (slugs/ft^3), and pounds-mass per cubic inch (lbm/in^3).

Specific Weight

Occasionally, a property given by the weight per unit volume is desired. This we call the **specific weight**, denoted by the lowercase Greek letter γ. The specific weight for homogeneous materials is calculated from

$$\gamma = \frac{W}{V} \tag{2–11}$$

The units of specific weight are newtons per cubic meter (N/m^3), newtons per liter (N/L), and newtons per cubic centimeter (N/cm^3). In the English unit system the common units for specific weight are pounds-force per cubic foot (lbf/ft^3), pounds-force per cubic inch (lbf/in^3), and pounds-force per gallon (lbf/gal^3). When the density is given in lbm/ft^3, the specific weight is defined as

$$\gamma = \frac{g}{g_c}\rho \tag{2–12}$$

In SI units or in the English system, if density is given in slugs/ft^3, the specific weight can be found from the equation

$$\gamma = \rho g \tag{2–13}$$

Specific Gravity

Liquids are sometimes described by their **specific gravity**. This quantity, denoted by SG, is the ratio of the density of the fluid described to the density of water where the water is at 4°C (39.2°F). The density of water at this temperature has been found to be $1000\ \text{kg/m}^3$ $(62.43\ \text{lbm/ft}^3)$, so the specific gravity may be computed from the following equations:

$$\text{SG} = \begin{cases} \dfrac{\rho}{1000} & (\rho \text{ in kg/m}^3) \quad \text{(2–14)} \\[2ex] \dfrac{\rho}{62.43} & (\rho \text{ in lbm/ft}^3) \quad \text{(2–15)} \end{cases}$$

Note that specific gravity is unitless and should have the same value in all unit systems. Table 2–1 lists values of specific gravity for some common liquids.

Pressure

Pressure is a property that we will use often in this book. It can be defined by the equation

$$p = \frac{F}{A} \tag{2–16}$$

TABLE 2–1 Density and specific gravity of some liquids at 20°C (68°F) and 101 kPa (14.7 psi)

Liquid	Density, ρ kg/m^3	Density, ρ lbm/ft^3	Specific Gravity, SG
Benzene	895	55.9	0.8950
Carbon tetrachloride	1,587	99.1	1.5950
Ethyl alcohol	788	49.2	0.7893
Glycerine	1,259	78.6	1.2600
Mercury	13,536	845.0	13.5460
Methyl alcohol	793	49.5	0.7928
Water	998	62.3	0.9982

where p is the pressure and F is a force distributed uniformly over an area A which is perpendicular to that force. A graphical view of this definition is given in figure 2–10.

From the definition we can see that the units of pressure must be force per unit of area or newtons per square meter (N/m^2). This unit is called the pascal (Pa), so

$$1\ \text{Pa} = 1\ \text{N/m}^2$$

The pascal is such a small unit that in many engineering problems the value is very large and unhandy to write. For instance, atmospheric pressure is about 101,000 Pa. It is more convenient to use the kilopascal (kPa), and thus in this book the kilopascal will most often be used for pressure. At high pressures, the megapascal (MPa) will be used, the megapascal being 1000 kPa. Also, the bar is used occasionally as a pressure unit; the bar is defined as 10^5 Pa or 100 kPa. The bar is a convenient unit (the atmospheric pressure is 1.01 bar), but we will generally not use it in this book. In English units, the units most often used for pressure are the pound per square inch (psi), the pound per square foot (psf), and the inch of mercury (in Hg). For very small pressure, the inch of water (in H_2O) is frequently used. All of these various English pressure units are used in this book. On the front inside cover are listed conversions among the various pressure units.

Pressure is most often observed in liquids or gases or fluids. It can occur in solids, but for our purposes, when we speak of pressure, we mean a force acting on an area of a fluid or on the surface of a solid. When it acts on a fluid, pressure propagates in all directions; that is, if a pressure is applied to a fluid in a downward direction at a point and if the fluid is in a static condition where it cannot move, there is pressure acting on the fluid at that point in all directions. This is true for liquids or gases. A graphical two-dimensional sketch

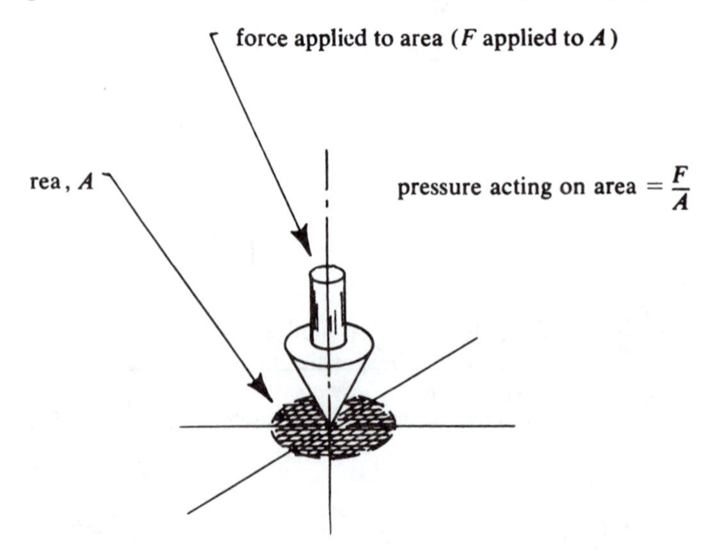

FIGURE 2–10 Pictorial representation of pressure

FIGURE 2–11 Pascal's law for liquids and gases

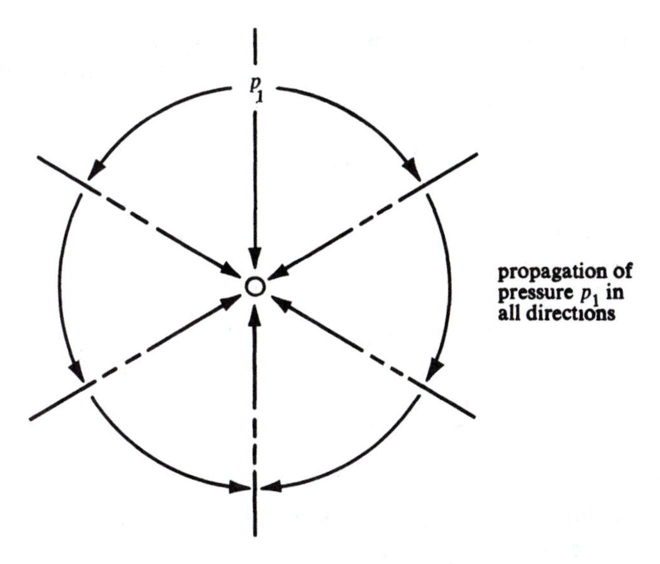

of this phenomenon is shown in figure 2–11, and this particular action is called *Pascal's law*. A number of useful relations result from it, some of which we develop next.

Pressure can vary within a fluid, and in most engineering applications the variation is dependent on the vertical position within the fluid. To understand how such variations can occur within a fluid, consider the vertical tube of cross-sectional area A and height l shown in figure 2–12. The tube is filled with water at 4°C (39.2°F), and the pressure at the top of the tube is observed to be the weight of a volume of air (the atmosphere) divided by A.

FIGURE 2–12 Pressures acting on water in a tube

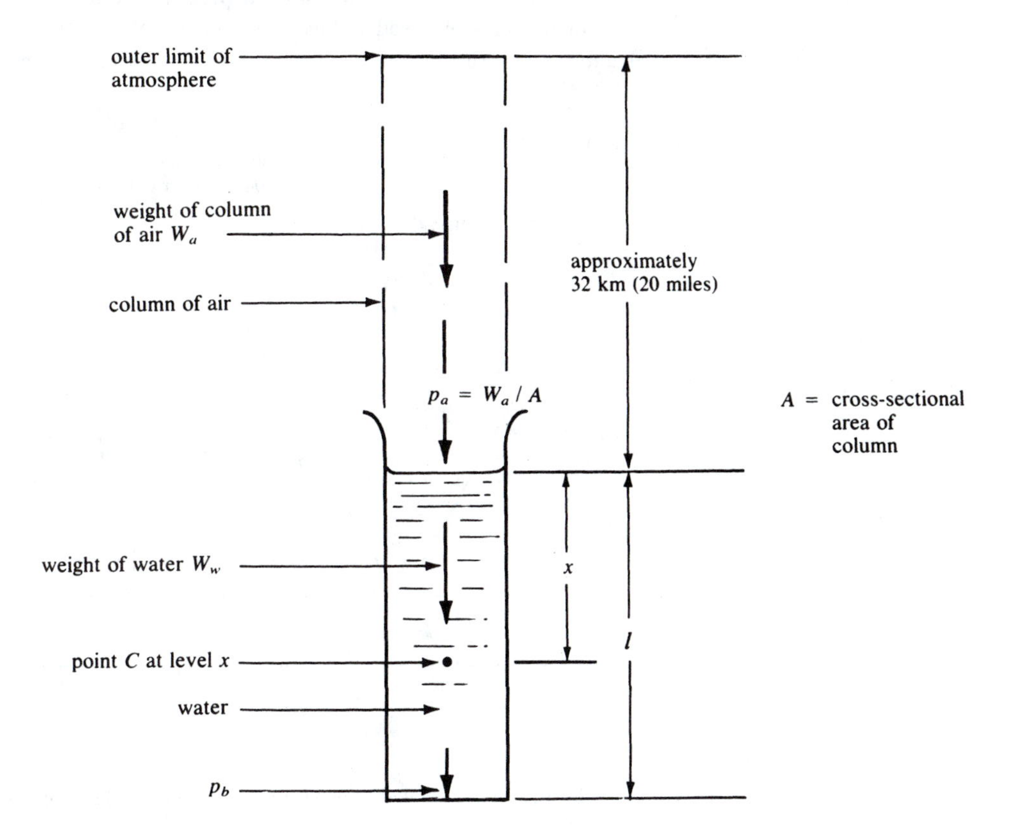

We call this pressure **atmospheric pressure** p_a, and its value is near 101 kPa or 14.7 psi on the surface of the Earth. Table B–3 gives the U.S. standard day values of atmospheric pressure for various locations on or near the Earth. Standard day is defined as annual averages at 45° latitude within the United States (probably never realized exactly).

At the bottom of the water column, we see that the pressure p_b is the pressure due to the weight of the water column W_w plus p_a:

$$p_b = \frac{W_w}{A} + p_a$$

We could also substitute for W_w and obtain

$$W_w = \gamma_w \times l \times A$$

where γ_w is the specific weight of water, yielding

$$p_b = \gamma_w l + p_a$$

Additionally, if we look at some other point in the water level, say, point C at x in figure 2–12, the pressure will be

$$p_c = \gamma_w x + p_a$$

Notice that the pressure horizontally through the water is the same but that it changes vertically and is increasing at lower depths. This is a result of Pascal's law, which allows for the pressure to be transmitted in all directions. It is the weight of the fluid above the water that is causing the pressure increase, and the water farther down has more water above it. Referring again to figure 2–12, the vertical depth x is called the **pressure head**, and the pressure p_a due to the atmosphere is called the **atmospheric pressure**. As another illustration of the importance of knowing the pressure head and the action of Pascal's law in fluids, in figure 2–13 it is shown that the pressure at a given head (say, x_1) is the same in columns A, B, C, and D. That is,

$$p_1 = \gamma_w x_1 + p_a$$

Similarly, at level 2,

$$p_2 = \gamma_w x_2 + p_a$$

for all four columns.

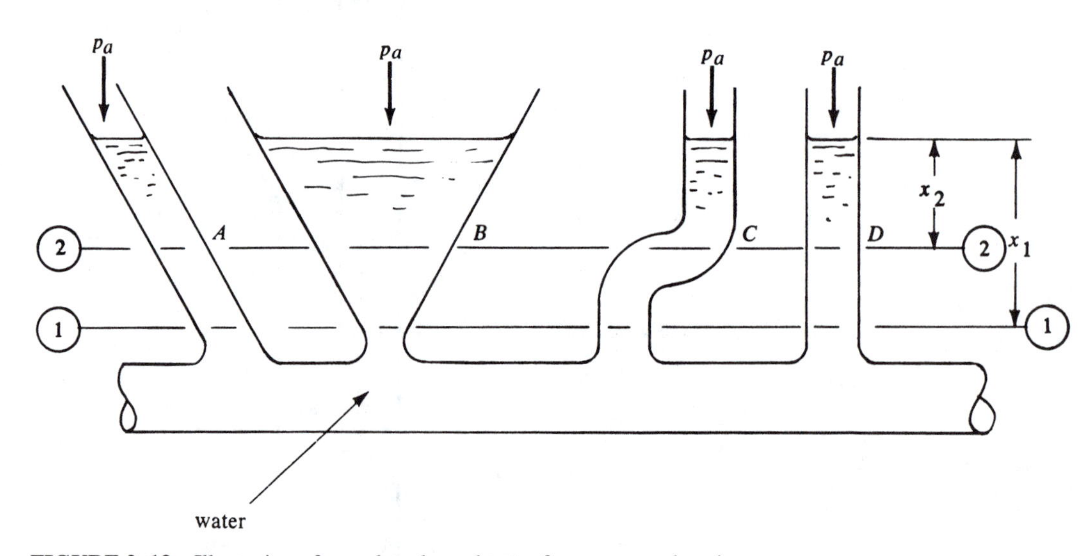

FIGURE 2–13 Illustration of complete dependence of pressure to elevation

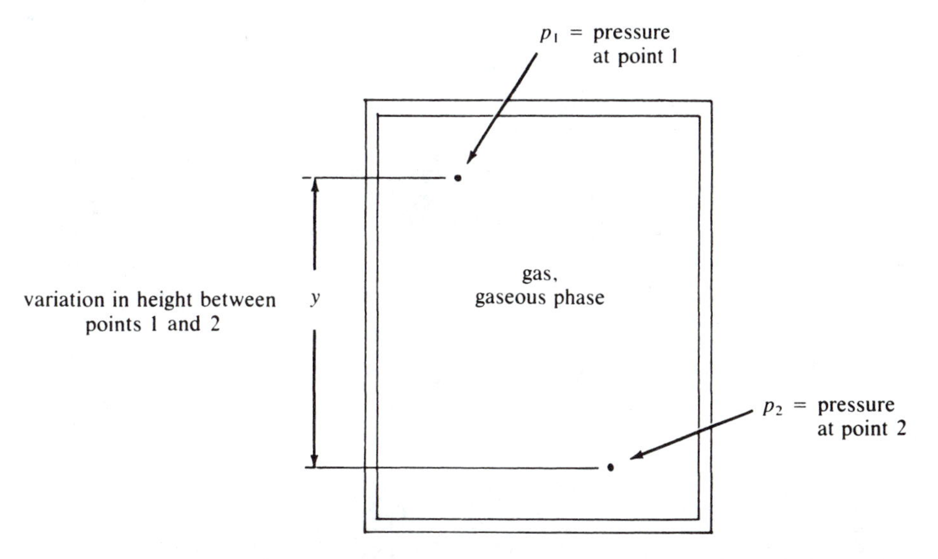

FIGURE 2–14 Illustration of independence of pressure with elevation in gaseous systems

For systems that contain gases, we will generally neglect the pressure changes due to elevation differences because the specific weight times the elevation change ($\gamma \times y$ term) will be several orders of magnitude less than the system pressure at the top; that is, we will say that

$$p_1 = p_2 = \text{constant} \quad \text{as identified in the system of figure 2–14}$$

Pressure is measured with devices called **pressure gages**. Manometers are probably the simplest pressure gages, and they are widely used. An example of a manometer is shown in figure 2–15. These devices use a fluid to measure a pressure head in a tube and

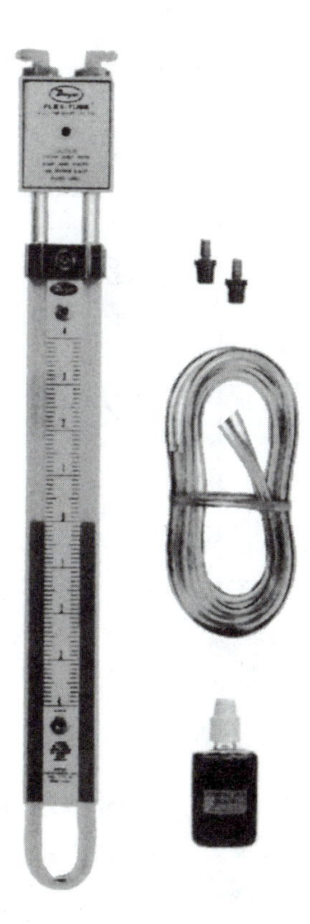

FIGURE 2–15 Commercial manometer

are thus a direct application of Pascal's law. The fluids most often used are water and manometer fluids. Most manometer fluids are slightly lighter than water, having specific gravities of around 0.8 to 0.95. Mercury is also used in manometers and in barometers. Figure 2–16 shows schematics of five common types of manometers; the U-tube type, the well type, the inclined-tube type, the micromanometer type, and the barometer. In the U-tube type, the pressure difference between p_1 and p_2 is sensed as a difference in height between the two fluid levels in the vertical tubes. Figure 2–17a shows a system B at pressure p_B, and a U-tube manometer is used to determine that pressure. The left leg of the

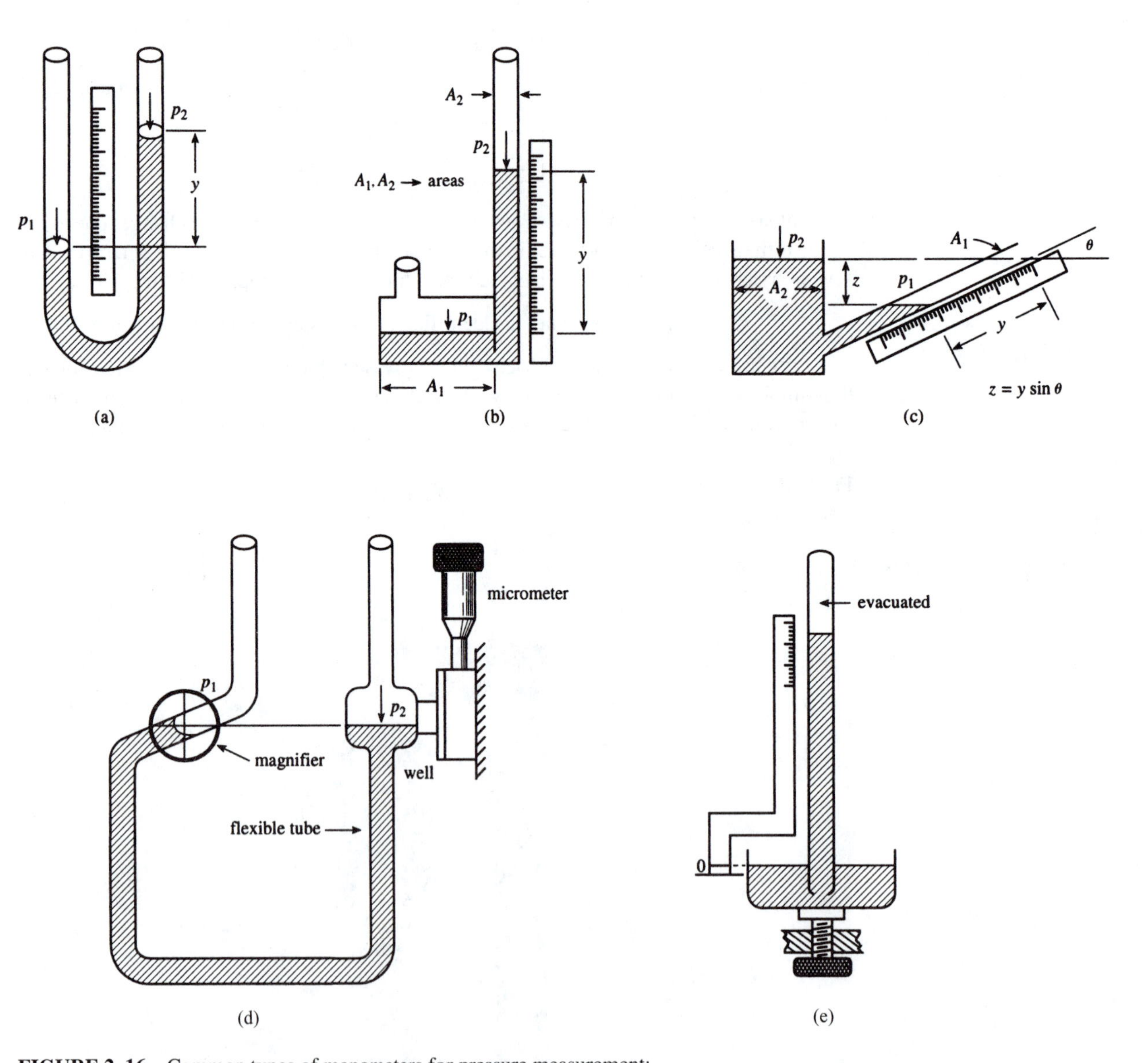

FIGURE 2–16 Common types of manometers for pressure measurement:
(a) U-tube manometer; (b) well-type (single-leg) manometer; (c) inclined manometer; (d) micromanometer; (e) barometer

manometer is open to the atmosphere, and if the pressure in the container, p_B, is greater than the atmospheric pressure p_a, we see from figure 2–17b that

$$p_B A = p_a A + \gamma y A$$

Dividing through by the area A, we obtain

$$p_B = p_a + \gamma y$$

Notice that we are physically determining the term y in this measurement, but we will call the term γy the **gage pressure**, p_g, so that

$$p = \text{pressure of a system}$$

or

$$p = p_g + p_a \tag{2–17}$$

The term γ used to find the gage pressure is the specific weight of the manometer fluid, and y is the pressure head in length units. Many manometers include a scale (as shown in figure 2–16) that gives y as the length equivalent to what y would be if mercury or water were the manometer fluid. In such cases γ would then be the specific weight of that fluid (mercury or water).

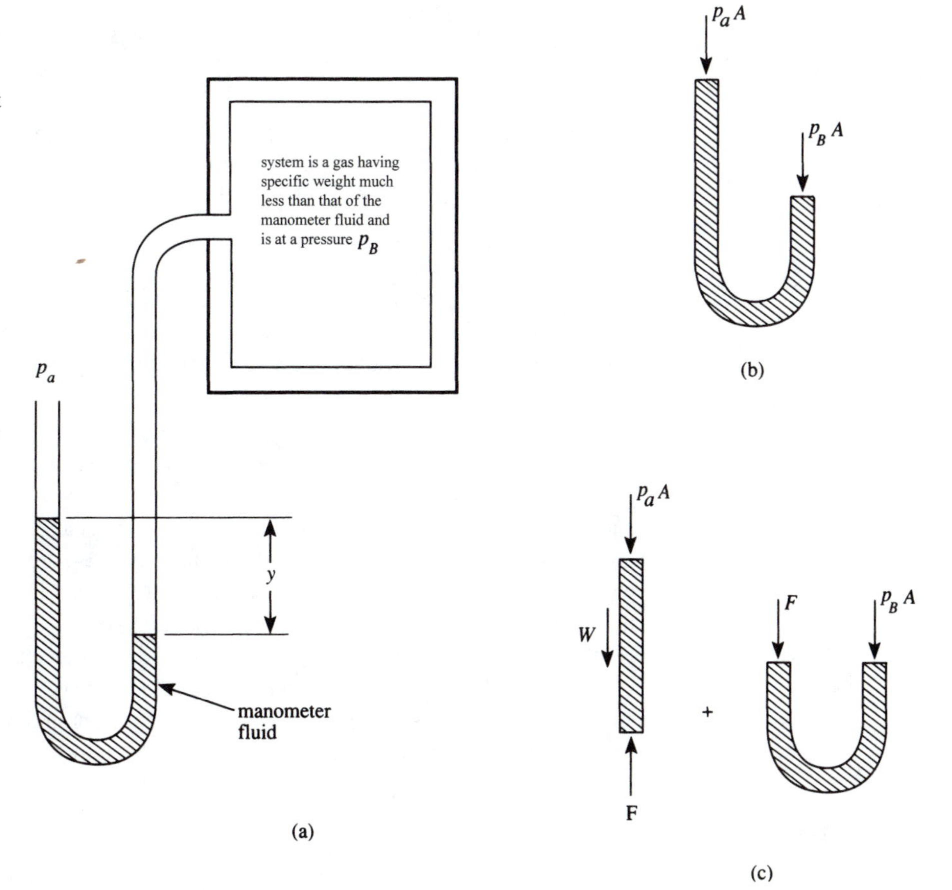

FIGURE 2–17
(a) Manometer for pressure measurement; (b) free body of manometer fluid; (c) equivalent free body of manometer fluid

Since the height of a column of fluid is frequently used to describe a pressure measurement, some discussion of the conversion should be made. Consider a 1-mm column of mercury. The specific weight of mercury can be computed from equation (2–13) by using the density of mercury from table 2–1.
Thus,

$$\begin{aligned}\gamma &= \rho g \\ &= (13{,}536\ \text{kg/m}^3)(9.8\ \text{m/s}^2) \\ &= 132{,}653\ \text{N/m}^3\end{aligned}$$

and, using

$$p_g = \gamma y \tag{2–18}$$

for the gage pressure, we find that

$$\begin{aligned}p &= (132{,}653\ \text{N/m}^3)(1 \times 10^{-3}\ \text{m}) \\ &= 132.65\ \text{N/m}^2 \\ &\simeq 133\ \text{Pa}\end{aligned}$$

Therefore, the pressure generated by a 1-mm column of mercury is equal to 133 Pa, or

$$1\ \text{mm Hg} = 0.133\ \text{kPa} \tag{2–19}$$

Similarly, for a 1-in column of mercury, the gage pressure can be shown to be 0.491 psi. Thus,

$$1\ \text{in Hg} = 0.491\ \text{psi} \tag{2–20}$$

Both equations (2–19) and (2–20) are actually conversion factors, but they are important enough for special notice. If the pressure p_B in figure 2–17 were less than atmospheric pressure, the pressure in system B would be

$$p_B = p_a - \gamma y$$

Also, if p_B is less than atmospheric pressure, the system is at a partial vacuum and the term γy in the preceding equation is called the **vacuum gage pressure**, p_{gv}. The pressure gage is then usually called a **vacuum gage**. In general, the pressure of a vacuum system is

$$p = p_a - p_{gv} \tag{2–21}$$

where, again, p_{gv} is the vacuum pressure actually measured by a vacuum gage.

The well-type manometer (shown in figure 2–16b) is a variation of the U-tube type and is popular because only one column of fluid needs to be measured. The "well" or other leg of the manometer is large enough that any pressure change does not appreciably change its level. The gage pressure of a well-type manometer is just the column height times the specific weight of the manometer fluid, similar to the U-tube. The inclined tube-type manometer (shown in figure 2–16c) is used most to sense small pressure differences, such as air drafts, fan pressures, and the like. Since one of the legs is inclined, a small pressure difference results in a much larger fluid movement in that leg, and this distance, y, can be conveniently and accurately determined. As indicated in the figure, inclined tube manometers use a well for the second leg of the device, so the distance y directly determines the pressure head, just as with the well-type manometer.

The micromanometer is a device used to determine extremely small pressure differences. Figure 2–16d shows a schematic of such a gage and may be used to explain its operation. The device uses a well-type arrangement, which allows for fluid movement under the magnifier when the pressure p_1 or p_2 changes. The micrometer adjustment is then

FIGURE 2–18 Principle of mercury barometer

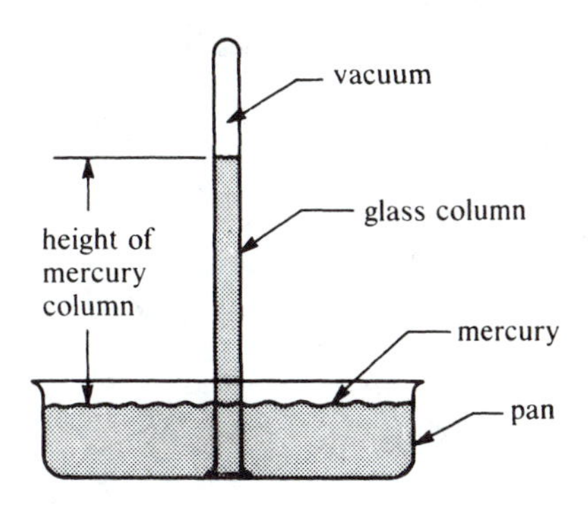

made to restore the fluid to its original level before the pressure change, and the difference in micrometer readings is then equal to the pressure head.

Figure 2–16e shows the barometer, a special form of the manometer used to measure atmospheric pressure. It accomplishes the measurement of atmospheric pressure because the vertical leg is evacuated above the manometer fluid level. Figure 2–18 shows the principle of a simple barometer having a vacuum above a vertical column of fluid. Of course, there is not a perfect vacuum in the barometer, but it is very nearly so, and for most engineering purposes the pressure in the leg can be said to be zero. Barometers utilize a well to take advantage of the convenience of that arrangement. Mercury is nearly always used in barometers, so that the atmospheric pressure is frequently given as a mercury pressure head, such as mm Hg or in Hg. Also, for precision determinations of the atmospheric or barometric pressure, an adjustment is sometimes included in the design (as shown in figure 2–16e) so that the well level can be zeroed or adjusted to a standard level. Such an adjustment assures that the vertical pressure head measurement is accurate. The height of mercury columns in barometers on the Earth are commonly around 760 mm Hg (100.8 kPa) or 29.92 in Hg. Table B–3 lists standard atmospheric pressures at various locations on and near the Earth.

EXAMPLE 2–4 If a mercury barometer reads 720 mm Hg, determine the atmospheric pressure in kPa and psi.

Solution The atmospheric pressure is given as 720 mm Hg, so that we may convert to kPa through equation (2–19):

$$p = (720 \text{ mm Hg})(0.133 \text{ kPa/mm Hg})$$

Answer
$$= 95.76 \text{ kPa}$$

In pressure units of psi we may use the conversion factor listed in the table on the inside front cover, 1 psi = 6.895 kPa. Thus,

$$p = (95.76 \text{ kPa})\left(\frac{1 \text{ psi}}{6.895 \text{ kPa}}\right)$$

Answer
$$= 13.9 \text{ psi}$$

Perhaps the most commonly used pressure gage is the bourdon gage. These devices utilize a thin hollow tube bent in a circular pattern that has a cross section that is oval or elliptic. Fluid fills the tube, and if the fluid experiences pressure, the tube will tend to straighten from its circular pattern. The operating elements of a bourdon gage are shown in figure 2–19, where you can see that a small straightening or movement of the tube is magnified mechanically into a rotation of a gage needle. Bourdon gages are nearly always

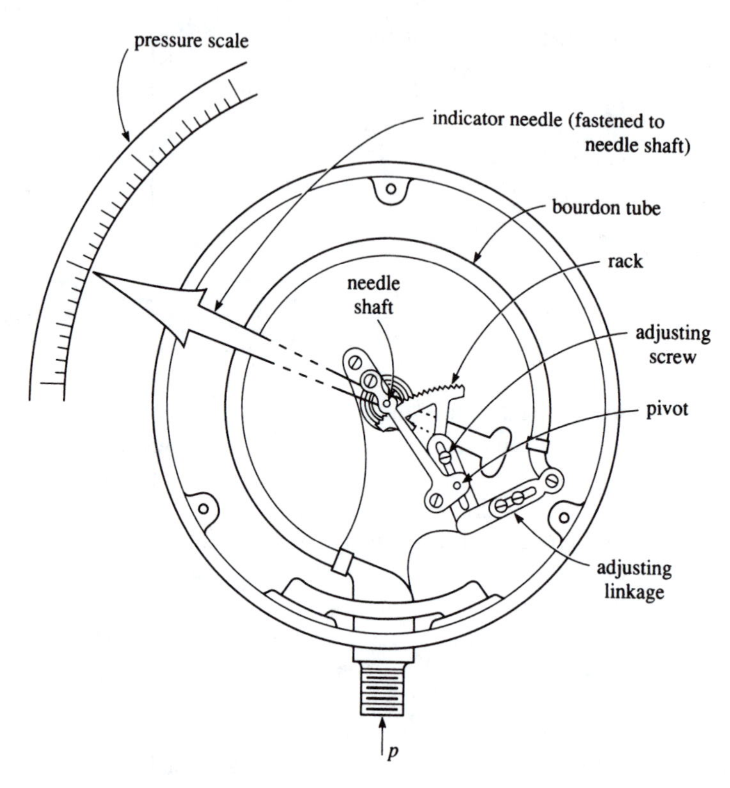

FIGURE 2–19 Schematic of a bourdon-tube pressure gage

encased in a housing to protect the gage itself from dust and contaminates. Again, as with manometers, bourdon gages measure a difference in pressure, or a gage pressure. Equation (2–17) or (2–21) is used to determine a system pressure, depending on whether the bourdon gage is designed as a pressure or a vacuum gage.

EXAMPLE 2–5 The water tank shown in figure 2–20 has a bourdon pressure gage that reads 100 mm Hg. What is the pressure in the tank in kilopascals if the atmospheric pressure is 101 kPa?

FIGURE 2–20 Water pressure tank

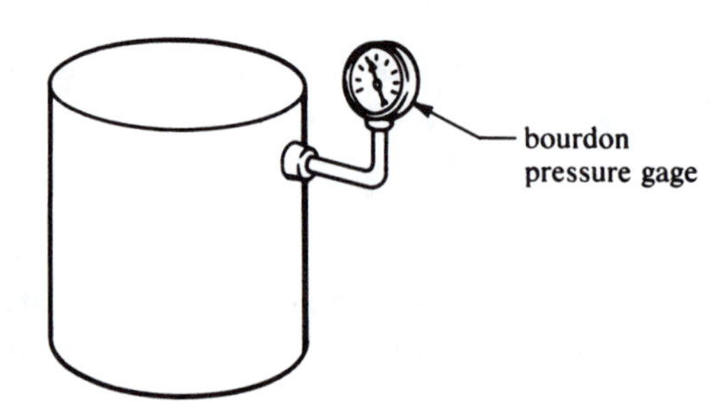

Solution The pressure indicated by the gage is the gage pressure, 100 mm Hg. We may readily convert to the units of kilopascals by using equation (2–19). Thus,

$$p_g = (100 \text{ mm Hg})(0.133 \text{ kPa/mm Hg})$$
$$= 13.3 \text{ kPa}$$

Using equation (2–17), we then obtain

$$p = 13.3 \text{ kPa} + 101 \text{ kPa}$$

Answer
$$= 114.3 \text{ kPa}$$

EXAMPLE 2–6 The steam leaving a turbine is at a pressure of 70 kPa vacuum. Determine the absolute pressure if the atmospheric pressure is 101 kPa.

Solution From equation (2–21) we have

$$p = p_a - p_{gv}$$

and

$$p = 101 \text{ kPa} - 70 \text{ kPa}$$

Answer

$$= 31 \text{ kPa}$$

EXAMPLE 2–7 What is the vacuum pressure inside the combustion chamber of a gasoline engine at a position when the pressure is 14.2 psia? The atmospheric pressure is 14.7 psi.

Solution We may assume the combustion chamber to be a vacuum chamber during the time that the pressure is less than atmospheric. Then, using equation (2–21), we obtain

$$p_{gv} = p_a - p$$

$$= 14.7 \text{ psi} - 14.2 \text{ psia}$$

Answer

$$= 0.5 \text{ psi vacuum}$$

There are other devices that are designed to measure pressures. In many arrangements, use is made of mechanical movement of some component and with various methods of sensing that motion. A common device used to sense pressure mechanically is a diaphragm, shown in figure 2–21. Here the diaphragm can move because of a pressure difference between the two surfaces, and this movement can then be measured with a strain gage (electrical resistance device), an optical sensor, an electro-optical sensor, or an electrical capacitance device. Other pressure gages, or transducers as they are described when electrical or electronic signals are used, use piezo-electric devices (which generate an electric charge under motion or pressure change), linear variable differential transformers (LVDTs), and direct change in electrical resistance with pressure. There are other devices and concepts developed and developing for pressure measurements, which are beyond the scope of this book.

FIGURE 2–21 Diaphragm pressure gage

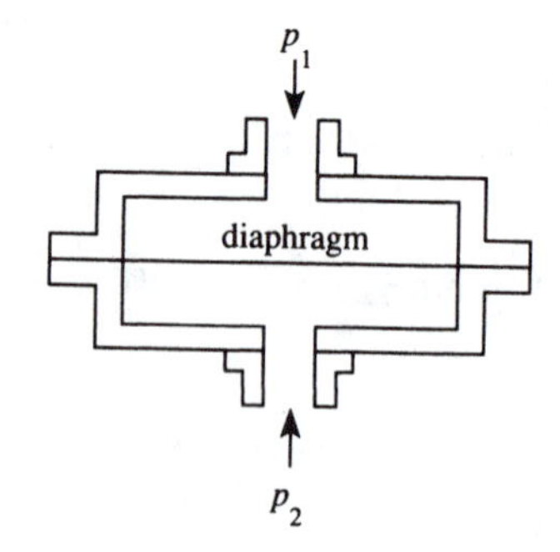

As a final example in this section, let us consider the following manometer problem.

EXAMPLE 2–8 Two pipes *A* and *B* are used to convey water and air, respectively, as shown in figure 2–22. Pressure gage *A* indicates a pressure of 1.01 psig (pounds per square inch gage) in the water-filled pipe. If the atmospheric pressure is 14.69 psi, find:

(a) Absolute pressure at center of pipe *A*.
(b) Absolute pressure in air-filled pipe *B*.
(c) Gage pressure measured by pressure gage (*B*).

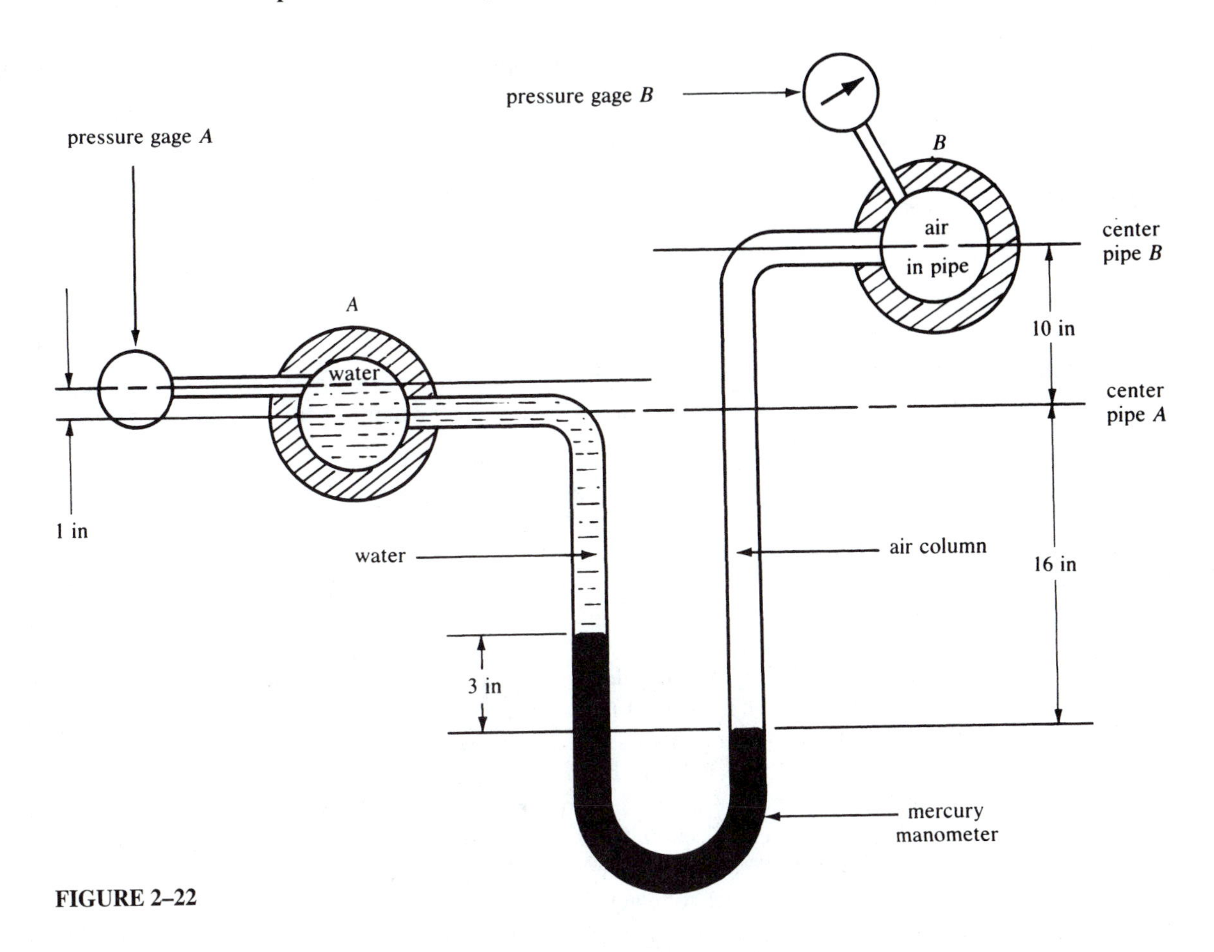

FIGURE 2–22

Solution (a) Pressure gage A (which may be a mercury U-tube manometer) is located 1 in above the center of the pipe. The absolute pressure at the gage location is, from equation (2–17),

$$p = p_g + 14.69 \text{ psi} = 1.01 \text{ psig} + 14.69 \text{ psi}$$
$$= 15.70 \text{ psia}$$

Corrected to the center of the pipe A, the pressure is

$$p_A = 15.70 \text{ psia} + (\gamma_w \times 1 \text{ in})$$

where γ_w is the specific weight of the water in pipe A. We have

$$\gamma_w = 62.4 \text{ lbf/ft}^3 \simeq 0.036 \text{ lbf/in}^3$$

so

$$p_A = 15.70 + (0.036 \times 1)$$

Answer
$$= 15.736 \text{ psia}$$

(b) To determine the pressure of the air in pipe B, we analyze the mercury manometer. The pressure p_B, sensed by gage B, although not exactly the pressure at the center of the pipe B due to the height variation, is close enough for engineering or most scientific work because air is a gas with very low specific weight. Thus,

$$p_B = \text{constant in pipe } B$$

Also, the pressure acting on the right side of the U-tube mercury manometer, p'_B, is given by

$$p'_B = p_B + (\gamma_{\text{air}} \times 26 \text{ in})$$

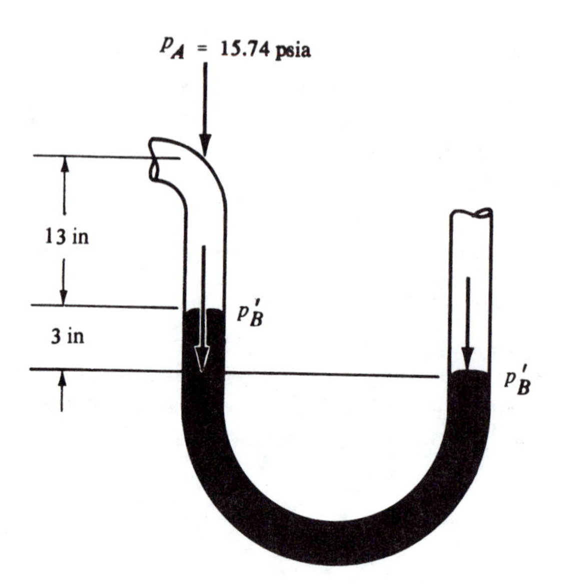

FIGURE 2–23 Pressures acting on mercury manometer in example 2–8

For air at 68°F and low pressure,

$$\gamma_{air} \approx 0.00004 \text{ lbf/in}^3$$

so that, for our case,

$$p_B' \approx p_B$$

From figure 2–23, showing pressures acting on the mercury manometer, we have

$$\begin{aligned} p_B' &= p_A + (\gamma_w \times 13 \text{ in water}) + (\gamma_{Hg} \times 3 \text{ in Hg}) \\ &= 15.74 \text{ psia} + (0.036 \text{ lbf/in}^3 \times 13 \text{ in}) + (0.491 \text{ lbf/in}^3 \times 3 \text{ in}) \\ &= 15.74 + 0.47 + 1.47 \\ &= 17.68 \text{ psia} \end{aligned}$$

The pressure in pipe B is, therefore,

Answer

$$p_B = 17.68 \text{ psia}$$

(c) The gage pressure in pipe B must be, from equation (2–17),

Answer

$$p = p_B - 14.69 = 2.99 \text{ psig}$$

2–9 EQUILIBRIUM AND THE ZEROTH LAW OF THERMODYNAMICS

In mechanics, we utilize equilibrium to determine forces acting on bodies. This type of equilibrium we call **mechanical equilibrium**, and it is indeed an important part of our scientific efforts. In thermodynamics we are concerned with **thermal equilibrium**. The property that determines this type of equilibrium we call **temperature**, and we postulate the following:

> **Zeroth law of thermodynamics:** ***Two separate bodies that are in thermal equilibrium with a third body are also in thermal equilibrium with each other.***

This law tells us that we can measure temperature through thermal equilibrium of bodies and be assured that it is independent of the material involved.

Remember that if two separate bodies at different temperatures are brought into contact with each other, thermal equilibrium will be reached and retained when the temperature is the same in both bodies. Interestingly, temperature equalizes in this condition, but the energies of the two bodies do not necessarily equalize.

2–10 TEMPERATURE AND THERMOMETERS

Temperature is the property that describes the hotness of a system. The higher the temperature, the hotter the substance, and the lower the temperature, the less hot or more cold the substance is. Although these statements appear so obvious as to be unnecessary, it should be noted that hotness is the only thing measured with temperature. Temperature does not measure the heat or the thermal energy of a substance directly, but you could define a scale based on one over the temperature and call it the *coldness*; you would then be able to have a number describing the coldness or lack of hotness in a substance.

Temperature is often recorded in degrees Fahrenheit (°F), which is the English unit, or in degrees Celsius (°C), the SI unit. These two scales are defined by the boiling and freezing (or melting) points of pure water at 101-kPa pressure. These are as follows: the boiling point is 100°C or 212°F and the freezing point is 0°C or 32°F.

The setting T_F is the temperature in degrees Fahrenheit, and T_C is the temperature in degrees Celsius. By plotting T_F versus T_C as shown in figure 2–24, we see that

$$T_F = \frac{180°\text{F}}{100°\text{F}} T_C + 32°$$
$$= {}^9\!/_5 T_C + 32°$$

or

$$T_C = {}^5\!/_9 (T_F - 32°) \tag{2–22}$$

Temperature has been measured via a number of methods, all of which involve the zeroth law of thermodynamics, where a temperature meter (thermometer) comes to thermal equilibrium with a system. When the thermometer reaches thermal equilibrium with the system being measured, some property of the thermometer has changed, and that property is then measured. For instance, the common *mercury-in-glass thermometer* is able to sense temperatures because the volume of a quantity of mercury changes with temperature (or hotness). The sensitivity of the mercury-in-glass thermometer is obtained by restricting the mercury to a very small column (or capillary tube), and a scale is put on that column to indicate degrees of temperature as shown in figure 2–25. The volume of the mercury is called a **thermometric property** of the thermometer. Other common temperature-measuring devices are the bimetallic strip, the thermocouple, and the thermistor. The *constant-value gas* or *perfect gas thermometer* is a theoretically practical thermometer as well, and we will discuss it here. All of these devices have thermometric properties, properties that serve

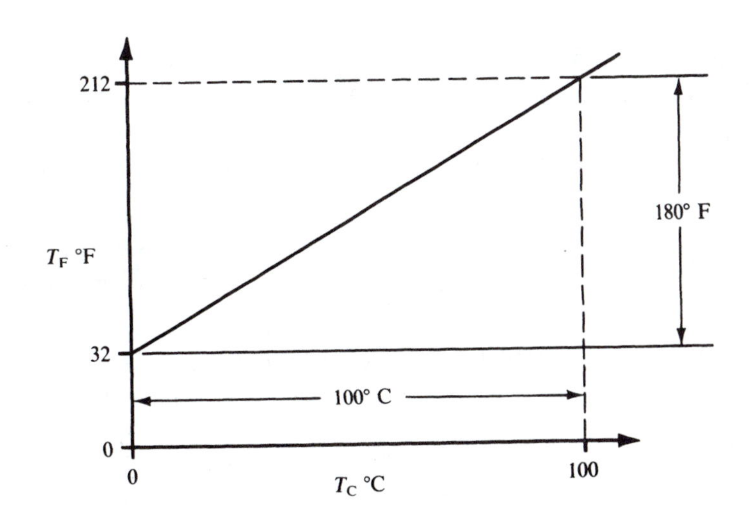

FIGURE 2–24 Relation of Celsius and Fahrenheit temperature scales

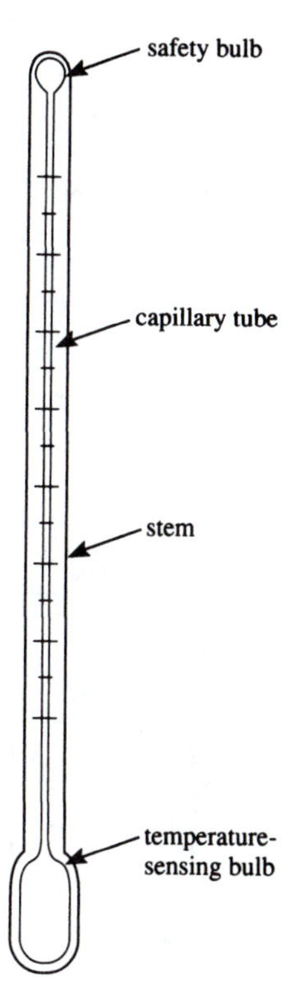

FIGURE 2–25 Mercury-in-glass thermometer

to indicate the hotness of a measured system. The thermometric properties of various thermometers are listed in table 2–2.

The *bimetallic strip* is a temperature-sensing device that uses the fact that two metals, bonded together as shown in figure 2–26, will expand lengthwise at different rates to bend the bimetallic strip. By suitable calibration a scale can be used to record the bending caused by a change in temperature.

TABLE 2–2

Type of Thermometer	Thermometric Property
Mercury-in-glass	Volume
Bimetallic strip	Linear thermal expansion difference of two metals
Thermocouple	Induced electrical voltage
Thermistor	Electrical resistance
Pyrometer	Optical radiation or thermistor
Constant-volume gas	Pressure of gas

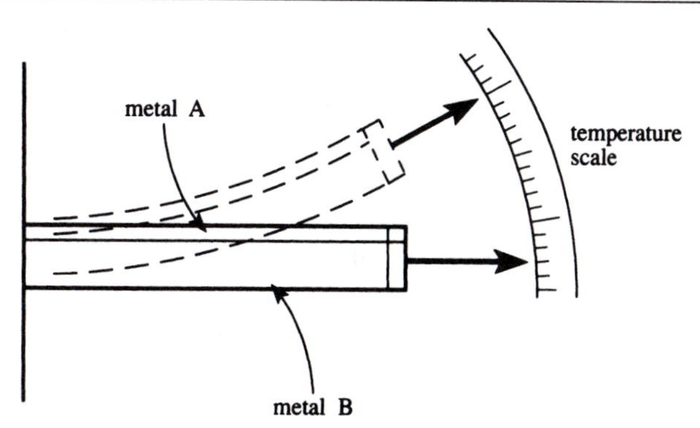

FIGURE 2–26 Bimetallic strip temperature measurement

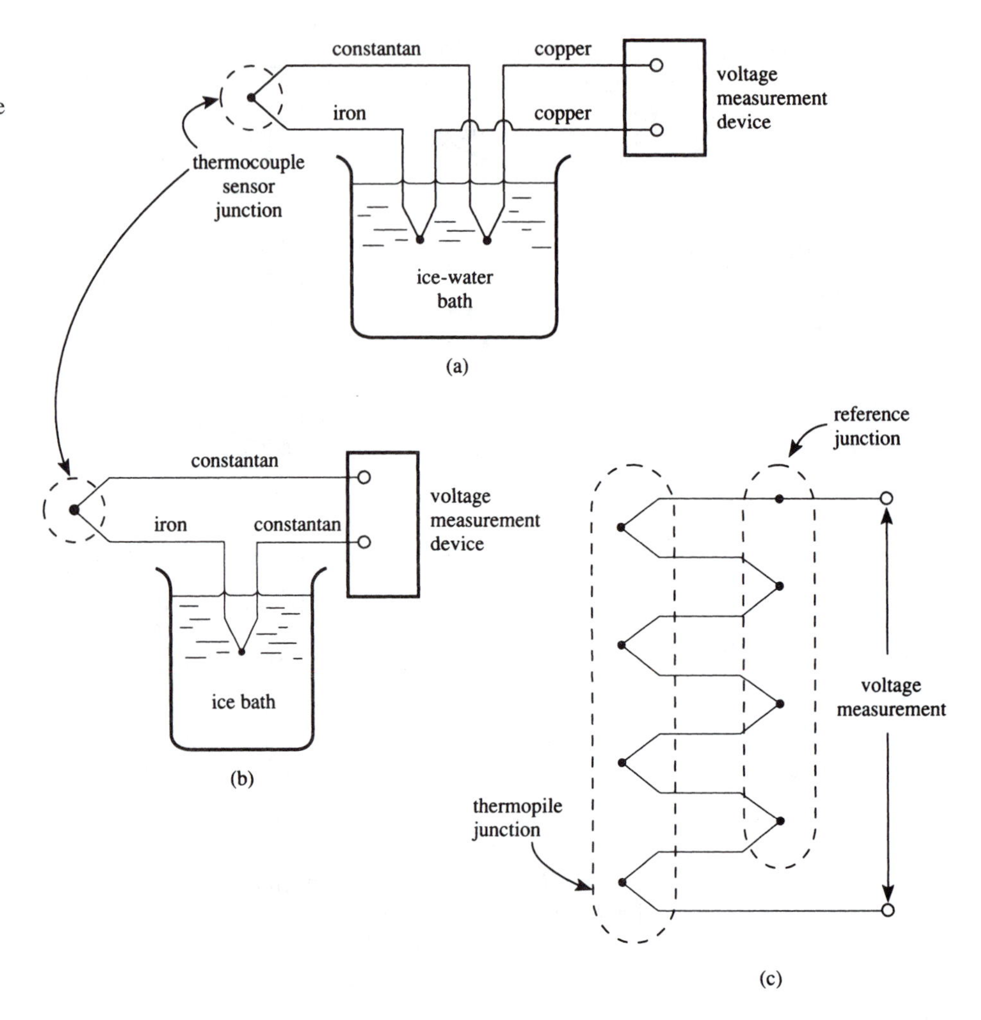

FIGURE 2–27 (a) Simple thermocouple—copper wire leads; (b) simple thermocouple — constantan wire leads; (c) thermopile (four-junction)

The *thermocouple* is a commonly used device for measuring large temperature differences or high temperatures. The device is made of two different electrical (thermocouple) wires welded together at one junction point, as shown in figure 2–27a. When the temperature of the junction changes, an electromotive force (EMF) or electrical voltage is created at the junction, and this voltage can be measured by a voltmeter or other electronic device. Some of the commonly used thermocouple wire combinations are: copper–constantan (designated as type T), iron–constantan (type J), and chromel–alumel (type K). There are many other types of thermocouples that are not considered here. The junctions where the lead wire from the voltmeter attaches to the thermocouple wire must be at the same temperature to avoid unnecessary complications. If the two junctions are put in an ice bath (at 0°C), the two junctions will be at the same temperature, and a conversion chart such as that shown in table 2–3 could be used to convert voltage measurements directly into temperature readings. If a thermocouple reading is between table entries, a linear interpolation should provide an acceptable temperature reading. If the lead wires are different from either of the thermocouple wires, an arrangement like that shown in figure 2–27a must be used, with the additional precaution that the terminals at the voltmeter must be of the same metal as the lead wires. Many of the new electronic thermocouple voltmeters have built-in compensation to take the place of the ice bath. Also, if greater sensitivity is required, a series of thermocouples may be connected as shown in figure 2–27c. Such an arrangement of thermocouples is called a *thermopile*.

TABLE 2–3 Thermal EMF (mV) for some common thermocouples as functions of temperature (reference junction at 0°C)

Temperature		Copper vs. Constantan	Iron vs. Constantan	Chromel vs. Alumel
°F	°C	T	J	K
−300	−184.4	−5.341	−7.519	−5.632
−250		−4.745	−6.637	−5.005
−200	−128.9	−4.419	−5.760	−4.381
−150		−3.365	−4.623	−3.538
−100	−73.3	−2.581	−3.492	−2.699
−50		−1.626	−2.186	−1.693
0	−17.8	−0.674	−0.885	−0.692
50		0.422	0.526	0.412
100	37.8	1.518	1.942	1.520
150		2.743	3.423	2.667
200	93.3	3.967	4.906	3.819
250		5.307	6.425	4.952
300	148.9	6.647	7.947	6.092
350		8.085	9.483	7.200
400	204.4	9.523	11.023	8.314
450		11.046	12.564	9.435
500	260.0	12.572	14.108	10.560
600		15.834	17.178	12.865
700	371.1	19.095	20.253	15.178
800			23.338	17.532
1000	537.8		29.515	22.251
1200				26.911
1500	815.6			33.913
1700				38.287
2000	1093.3			44.856
2500	1371.1			54.845
3000	1648.9			

EXAMPLE 2–9 Convert a chromel–alumel thermocouple measurement of 3.100 millivolts (mV) to degrees Celsius and degrees Fahrenheit.

Solution For a chromel–alumel type K thermocouple, 3.100 mV is between the entries of 2.667 (at 150°F) and 3.819 (at 200°F) of table 2–3. The linear interpolation can be written

$$\frac{T - 200°\text{F}}{150 - 200°\text{F}} = \frac{3.100\text{ mV} - 3.819\text{ mV}}{2.667\text{ mV} - 3.819\text{ mV}} = 0.624$$

Then

$$T - 200 = (150 - 200)(0.624) = -31.2$$

and

Answer

$$T = 168.8°\text{F}$$

From equation (2–22), the temperature in degrees Celsius is

Answer

$$T_C = (5/9)(T_F - 32) = 76.0°\text{C}$$

Thermistors are temperature-measuring devices whose electrical resistance varies with a change in temperature so that for a given electrical resistance there is a specific temperature associated with that resistance value. A thermistor is essentially a precise length of electrical conducting wire that is exposed and able to sense its surrounding temperature. Thus, the thermistor can be made as small as the smallest practical electrical conducting wire diameter, and it can then be used to accurately measure rapid temperature changes or very small temperature changes. Thermistors are often preferred over thermocouples for these two reasons. However, they tend to be useful over smaller ranges of temperature than thermocouples. Thermistors can also have very nonlinear temperature–electrical resistance relationships. This nonlinearity requires that more experimental data be obtained to establish an accurate calibration than are needed for thermocouples, which tend to be linear in their temperature–EMF relationships.

Pyrometers are temperature measuring devices that use thermal radiation to sense temperatures. These devices have the advantage of being able to measure temperatures of surfaces or objects without contact and even far away from the pyrometer. The pyrometer operates best when there is a vacuum or air between it and the object. Thermal radiation is composed of visible light, infrared light, and ultraviolet light, and its color and intensity are indicators of the surface of the object's temperature. As a result of this fact, a class of pyrometers called *optical pyrometers* are used by comparing the light from a fine light filament to the field of the surface to be measured. This procedure, which is indicated in figure 2–28a, involves adjusting or changing the light filament brightness (which is calibrated to a tempera-

FIGURE 2–28 Schematics and cross-sections of (a) optical pyrometer and (b) total radiation pyrometer

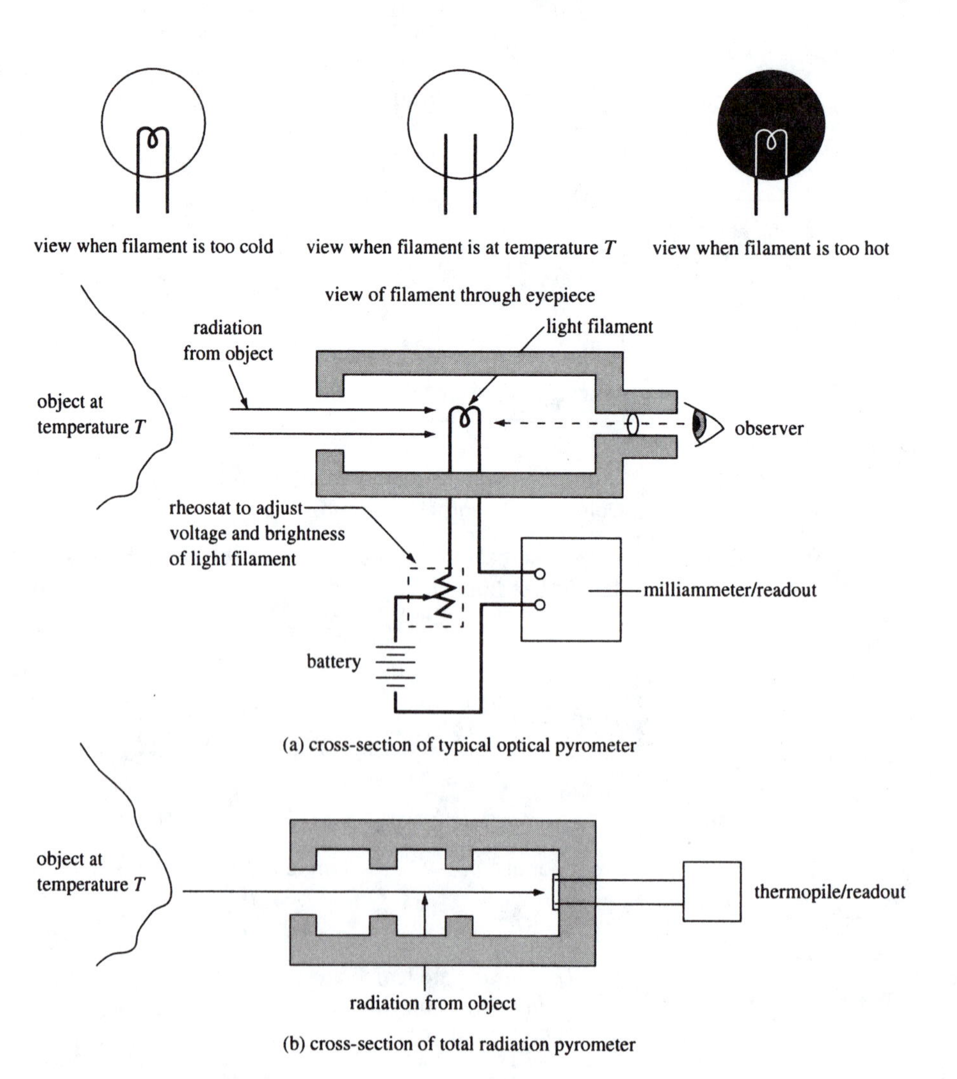

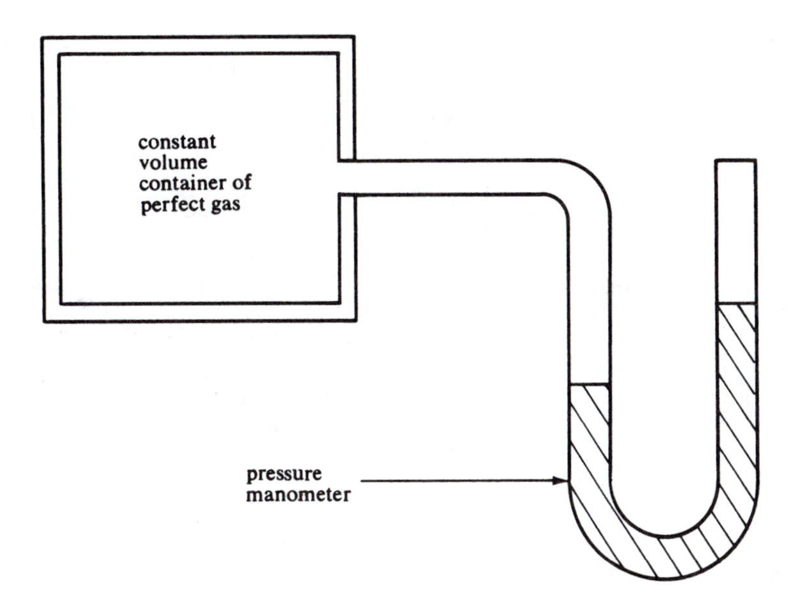

FIGURE 2–29 Gas thermometer

ture scale) until the filament appears to disappear in the field. For this reason the *optical pyrometer* is sometimes referred to as the *disappearing filament pyrometer*.

Another pyrometer, sometimes referred to as the *total radiation pyrometer*, uses a thermopile to sense the thermal radiation reaching it from an object whose temperature is to be measured. A schematic of this device is shown in figure 2–28b. These devices are advantageous for automatic data acquisition. They are often used to measure the sun's surface temperature and other applications when continuous temperature monitoring is wanted.

Finally, if a rigid container is filled with a perfect gas as shown in figure 2–29 and if a pressure gage (such as a manometer) is attached to the chamber, the pressure measured by the gage will be proportional to the temperature of the gas inside the container. The equation for a perfect gas is

$$pV = mRT$$

where R is called the **gas constant**. We will discuss this property further later in the book. The gas constant is experimentally determined and serves to make the perfect gas equation more nearly correct.

Solving for the pressure, we obtain

$$p = \frac{m}{V}RT = \rho RT \tag{2–23}$$

where ρ is the density defined in equation (2–9). But the density and the gas constant R are constant, so that we may write

$$p = (\text{constant})T \tag{2–24}$$

Suppose now that we use the apparatus of figure 2–29 to measure temperature–pressure relationships of the three gases, A, B, and C. We could easily generate pressure–temperature data points as shown in figure 2–30. Some of the data points fit straight lines, and the point where these lines intersect with the vertical axis (point Z in figure 2–30) we define as the *absolute zero point* of temperature. If the temperature scale in figure 2–30 had been in degrees Fahrenheit, we would have found that

$$Z = -459.4°\text{F}$$

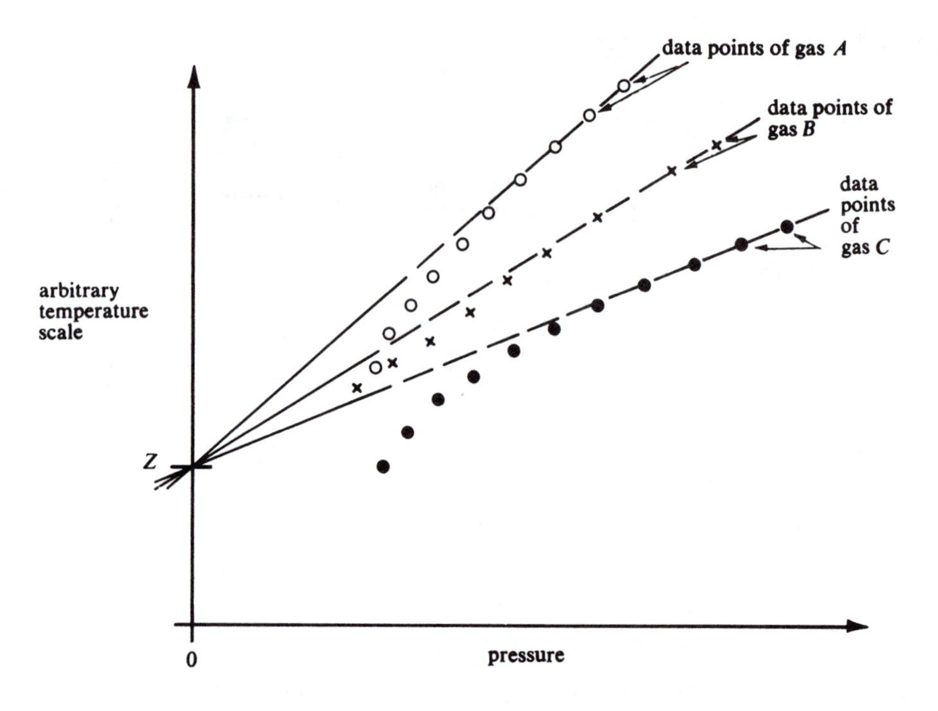

FIGURE 2–30 Method for determining absolute-zero temperature

Similarly, if we had used degrees Celsius, then

$$Z = -273°\text{C}$$

From these values of absolute zero, we then say that the temperature of a body is

$$T = (T_F + 459.4) \text{ degrees Rankine (°R)} \quad \textbf{(2–25)}$$

or

$$T = (T_C + 273) \text{ kelvin (K)} \quad \textbf{(2–26)}$$

EXAMPLE 2–10 A thermodynamic system composed of liquid nitrogen is at 72 kelvin. Convert this to degrees Celsius, Fahrenheit, and Rankine.

Solution From equation (2–26), we have

$$T_C = T - 273 \text{ K}$$
$$= 72 - 273$$

so

Answer
$$T_C = -201°C$$

To convert this temperature to degrees Fahrenheit, we use equation (2–22), rearranged:

$$T_F = (9/5)T_C + 32$$

Answer
$$= (9/5)(-201°\text{C}) + 32 = -330°\text{F}$$

The temperature in degrees Rankine is from equation (2–25),

Answer
$$T = T_F + 459.4 = 129.4°\text{R}$$

2–11 ENERGY

Energy is a property of a body or system that can be defined as follows:

Energy: ***The capacity of a given body to produce physical effects external to that body.***

In this definition the word "effects" implies that physical changes can occur, such as movement or changes in size, color, temperature, or numerous other changes in the physical character of objects. A better grasp of energy as an idea may be obtained by considering the various types of energy: kinetic, potential, and internal.

Kinetic Energy

A body or particle has energy by virtue of its movement. As an example, a stone lying on the ground is motionless and incapable of effecting change. Pick the stone up, however, and throw it against a window, and this will produce a change directly attributable to the motion of the stone. This energy is called ***kinetic energy*** (KE) and is given by

$$\mathrm{KE} = \frac{1}{2} m \overline{V}^2 \tag{2–27}$$

where $\overline{V}$ is the velocity of the stone. If we substitute equation (2–2) into this result, we obtain

$$\mathrm{KE} = \frac{1}{2} \frac{W}{g} \overline{V}^2 \tag{2–28}$$

from which we can see that the units are newton-meters $(\mathrm{N \cdot m})$. A newton-meter is defined as a joule (J). In the English system the units for energy are foot-pounds $(\mathrm{ft \cdot lbf})$ or British thermal units (Btu). The **specific kinetic energy**, or kinetic energy per unit mass, is

$$\mathrm{ke} = \frac{\mathrm{KE}}{m} \tag{2–29}$$

and has the units $\mathrm{N \cdot m/kg}$ or J/kg. Equation (2–29) may be combined with equation (2–27) to give

$$\mathrm{ke} = \frac{1}{2} \overline{V}^2 \qquad (\mathrm{J/kg}) \tag{2–30}$$

In the English system one must use equation (2–7) in equation (2–28) to obtain

$$\mathrm{KE} = \frac{m}{2g_c} \overline{V}^2 \qquad (\mathrm{ft \cdot lbf}) \tag{2–31}$$

Further, the specific kinetic energy is then given by the relationship

$$\mathrm{ke} = \frac{1}{2g_c} \overline{V}^2 \qquad (\mathrm{ft \cdot lbf/lbm}) \tag{2–32}$$

EXAMPLE 2–11

A stream of water coming from a hose is found to have a velocity of 30 m/s. Determine the kinetic energy per kilogram of this water.

Solution

The kinetic energy per kilogram is computed from equation (2–30) as

$$\mathrm{ke} = {}^1\!/_2 \overline{V}^2$$

where the velocity $\overline{V}$ is 30 m/s for the water. Then

$$\mathrm{ke} = {}^1\!/_2 (30\ \mathrm{m/s})^2$$

Answer

$$= 450\ \mathrm{J/kg}$$

EXAMPLE 2–12 A piston of an internal combustion engine is found to have a velocity of 180 ft/s at a particular instant in its travel. If the piston weighs 2 lbf, determine its KE and ke. Use the value of 32.2 ft · lbm/lbf · s^2 for g_c and 32.2 ft/s^2 for g.

Solution From equation (2–32), the kinetic energy per pound-mass is readily computed:

$$\text{ke} = \frac{(180 \text{ ft/s})^2}{(2)(32.2 \text{ ft} \cdot \text{lbm/lbf} \cdot \text{s}^2)}$$

Answer
$$= 503.1 \text{ ft} \cdot \text{lbf/lbm}$$

From equation (2–28), the kinetic energy is then easily found to be

$$\text{KE} = \frac{(2 \text{ lbf})(180 \text{ ft/s})^2}{(2)(32.2 \text{ ft/s}^2)}$$

Answer
$$= 1006 \text{ ft} \cdot \text{lbf}$$

Potential Energy

In section 2–7 it was shown that the weight of an object is the same as a force of attraction between the Earth and that object. This force of attraction represents a "potential" for motion where the object would tend to move closer to the Earth's center. We call this capability ***total potential energy***, denote it by PE, and calculate it from

$$\text{PE} = Wz \tag{2–33}$$

where z is a vertical reference distance between the center of the Earth and the center of the object that we say has the potential energy. The units of potential energy are easily seen to be foot-pounds and joules. The potential energy per unit of mass, or the **specific potential energy**, is given by the relationship

$$\text{pe} = \frac{\text{PE}}{m} = \frac{Wz}{m} \tag{2–34}$$

The units of the specific potential energy are joules per kilogram (J/kg) or foot-pounds per pound-mass (ft · lbf/lbm). Additionally, equation (2–34) can be simplified by substituting equation (2–2). Thus,

$$\text{pe} = gz \qquad \text{J/kg} \tag{2–35}$$

Similarly, for English units we can substitute equation (2–7) into (2–34) to obtain

$$\text{pe} = \frac{gz}{g_c} \qquad \text{ft} \cdot \text{lbf/lbm} \tag{2–36}$$

EXAMPLE 2–13 Water dropping over a 300-m fall loses potential energy. If we say that the potential energy is zero at the bottom of the fall, what is the water's potential energy per kilogram at the top?

Solution From equation (2–35), we obtain

$$\text{pe} = (9.8 \text{ m/s}^2)(300 \text{ m})$$

Answer
$$= 2940 \text{ J/kg}$$

EXAMPLE 2–14 A 5-ton elevator rises 10 stories, where each story is 10 ft high. What increases in potential energy does the elevator have in this motion upward?

Solution A 5-ton elevator weighs 5 tons × 2000 lbf/ton, or 10,000 lbf. The increase in potential energy is found from equation (2–33) to be

$$\text{PE} = (10{,}000 \text{ lbf})(10 \text{ stories} \times 10 \text{ ft})$$

Answer
$$= 1 \times 10^6 \text{ ft} \cdot \text{lbf}$$

Internal Energy

Internal energy can be defined as the energy of a system that cannot be associated with kinetic or potential energies, and it will be denoted by the symbol U. It exists as two distinct types: sensible and latent. *Sensible internal energy* is associated with the hotness or coldness of the object and so can be measured by temperature. *Latent internal energy* is associated with the melting, freezing, boiling, and condensing of substances and cannot be measured by temperature of thermometers. We shall see later how internal energy is determined. We will not use the term "heat" to describe this property, but sometimes thermal energy will be equated with internal energy. *Heat* will be assigned a much different meaning in chapter 3.

Internal energy, being a form of energy, has the same units associated with it as the other forms of energy, kinetic and potential. Thus, we will use joules, foot-pounds, and British thermal units to describe internal energy. The specific internal energy u is given by the equation

$$u = \frac{U}{m} \tag{2–37}$$

and has the units of joules per kilogram, foot-pounds per pound-mass, and British thermal units per pound-mass.

Internal energy is frequently described as the property that reflects mechanical energy of the molecules and atoms of the material. Generally, the contributions to internal energy are as follows:

1. Translational or kinetic energy of the atoms or molecules as indicated in figure 2–31a
2. Vibrational energy of the individual molecules due to straining of the atomic bonds at increasing temperatures (figure 2–31b)
3. Rotational energy of those molecules that spin about an axis as shown in figure 2–31c

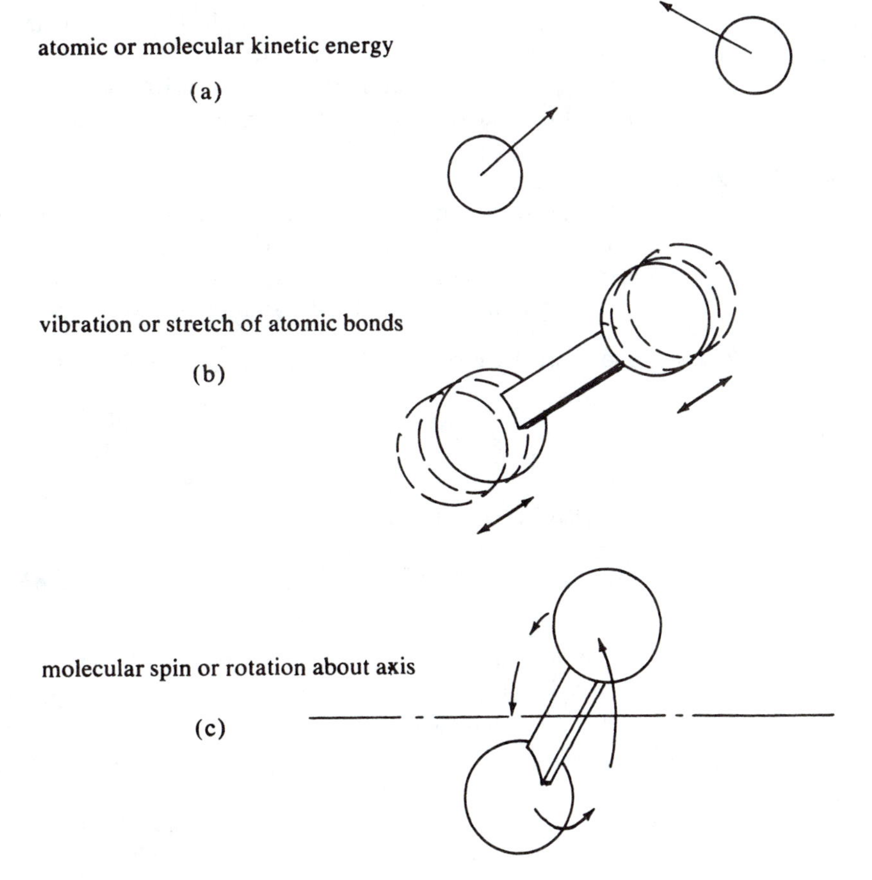

FIGURE 2–31 Molecular motions contributing to internal energy

There are other forms of energy, such as electromagnetic energy, chemical energy, and strain energy (caused by the stretching of solid materials), which must be included in any complete analysis of a thermodynamic problem; but in this book we consider the total energy E of a system to be given by

$$E = \text{KE} + \text{PE} + U \tag{2–38}$$

and the energy e to be given by

$$e = \text{ke} + \text{pe} + u \tag{2–39}$$

If the previous equations are substituted into the appropriate terms of equation (2–39), we have

$$e = \begin{cases} \frac{1}{2}\overline{V}^2 + gz + u & \text{J/kg} \quad (2\text{–}40) \\ \dfrac{\overline{V}^2}{2g_c} + \dfrac{g}{g_c}z + u & \text{ft}\cdot\text{lbf/lbm} \quad (2\text{–}41) \end{cases}$$

2–12 EFFICIENCY

The idea of efficiency is used frequently to describe systems and engines and other devices. **Efficiency** η can be defined as

$$\eta = \frac{\text{output}}{\text{input}} \tag{2–42}$$

In later sections we will define other types of efficiency, but they all involve an input and an output. Efficiency is usually given as a percentage, so that the value calculated from equation (2–42) would need to be multiplied by 100 to obtain percentage values. Efficiencies cannot be greater than 1 or 100% and are always something less than 100%. For instance, automobile engines are usually about 10% to 25% efficient, electrical power plants about 40%, and hydroelectric power plants about 85%.

EXAMPLE 2–15

A windmill is used to pump water. It was found that 4 million (4×10^6) ft · lbf of energy was actually used during a 10-minute period when the wind was blowing at 45 ft/s. This amount of energy was the increase in potential energy of the water in raising it from the well. Assuming that 680,000 lbm of air passed through the windmill during this period, calculate the efficiency of converting the kinetic energy in the wind to potential energy in the water.

Solution

The input to the windmill is the kinetic energy of the wind and can be found from equation (2–31):

$$\text{KE} = \frac{m}{2g_c}\overline{V}^2 = \frac{680{,}000 \text{ lbm}}{(2)(32.174 \text{ ft}\cdot\text{lbm/lbf}\cdot\text{s}^2)}(45 \text{ ft/s})^2$$
$$= 21{,}399{,}000 \text{ ft}\cdot\text{lbf} \approx 21.4 \times 10^6 \text{ ft}\cdot\text{lbf}$$

Then the efficiency is just found from equation (2–42):

Answer

$$\eta = \frac{4 \times 10^6 \text{ ft}\cdot\text{lbf}}{21.4 \times 10^6 \text{ ft}\cdot\text{lbf}} = 18.7\%$$

EXAMPLE 2–16

An electrical power plant burns coal to produce 10 MW of power. If the plant efficiency is 38% and each kilogram of coal releases 30,000 kJ of energy, determine the amount of coal required to operate the plant for 1 hour.

Solution

In this example, the input is required to be found from the efficiency equation (2–42), in the form

$$\text{input} = \frac{\text{output}}{\eta} = \frac{1}{0.38}(\text{output})$$

The output is 10 MW for 1 hour, so

$$\text{output} = (10 \text{ MW})(1 \text{ h}) = (10 \text{ MW})(3600 \text{ s}) = 36{,}000 \text{ MJ}$$

and because 1 kg of coal releases 30,000 kJ of energy, the amount of coal required in kilograms is the input (by the coal) divided by the energy per kilogram of coal. The input required by the coal is

$$\text{input} = \frac{36{,}000 \text{ MJ}}{0.38} = 94{,}736{,}842 \text{ kJ}$$

The required amount of coal is then

$$\frac{94{,}736{,}842 \text{ kJ}}{30{,}000 \text{ kJ/kg}}$$

Answer

$$= 3157.9 \text{ kg for 1 hour}$$

2–13 UNITS REVISITED

In section 1–3 and 1–4 we presented the fundamental dimensions and the unit systems (the SI and English) used in this book. There it was emphasized that units should always be included in problem solving, and here we want to reemphasize that point. In this chapter many terms or properties have been defined and then used in determining other dimensions or properties. These quantities (or derived properties) have units that can also be derived from the fundamental dimension units. A number of the more important thermodynamic quantities are given in the List of Symbols and Thermodynamic Notation of appendix D. Also, table 2–4 gives some derived base units that are equivalent to the normal units to aid you in problem solving. For instance, in an equation containing, say, pressure, it would be more helpful to use the derived SI unit $\text{kg/m} \cdot \text{s}^2$ instead of the pascal (Pa). Then the kg, m, and s units would more likely occur in units of other terms of the equation (compared to the pascal), allowing you to make better use of the unit cancellation method of section 1–4.

TABLE 2–4 Some derived thermodynamic units

	Système International		English System	
Quantity	**Unit**	**Derivation**	**Unit**	**Derivation**
Area	m^2	m^2	ft^2	ft^2
Volume	m^3	m^3	ft^3	ft^3
Density	kg/m^3	kg/m^3	lbm/ft^3	lbm/ft^3
Force	N	$\text{kg} \cdot \text{m/s}^2$	lbf	$\frac{1}{32.17}\ \text{lbm} \cdot \text{ft/s}^2$
Pressure	Pa	$\text{kg} \cdot \text{m/m}^2 \cdot \text{s}^2$	psi	$\frac{1}{144}\ \text{lbf/ft}^2$
Energy	J	$\text{kg} \cdot \text{m}^2/\text{s}^2$	ft · lbf	$\frac{1}{32.17}\ \text{lbm} \cdot \text{ft}^2/\text{s}^2$
Specific energy	J/kg	m^2/s^2	ft · lbf/lbm	
Entropy	J/K	$\text{kg} \cdot \text{m}^2/\text{s}^2 \cdot \text{K}$	ft · lbf/°R	$\frac{1}{32.17}\ \text{lbm} \cdot \text{ft}^2/\text{s}^2 \cdot {}^\circ\text{R}$
Specific entropy	J/kg · K	$\text{m}^2/\text{s}^2 \cdot \text{K}$	ft · lbf/lbm · °R	
Power	W	$\text{kg} \cdot \text{m}^2/\text{s}^2$	horsepower (hp)	550 ft · lbf/s

2–14 SUMMARY

In this chapter the concept of the system has been introduced, and identifying the boundaries of this system has been stressed. Once the system has been identified (not too easy a task in many cases), its quantitative description is determined through the system properties, the total list of which describes the state of the system. The important properties for a thermodynamic analysis are as follows: mass, weight, density, volume, specific volume, pressure, temperature, and energy. Energy has been typified as potential, kinetic, or internal.

The following list of formulas describes the important points of this chapter.

Mass/weight:

$$F\ (\text{newtons}) = ma \qquad m \text{ in kg} \tag{2–3}$$

$$W\ (\text{newtons}) = mg \qquad m \text{ in kg} \tag{2–2}$$

$$F\ (\text{lbf}) = \frac{ma}{g_c} \qquad m \text{ in lbm} \tag{2–7b}$$

$$W\ (\text{lbf}) = \frac{mg}{g_c} \qquad m \text{ in lbm} \tag{2–7a}$$

Density, volume/pressure:

$$v = \frac{1}{\rho} = \frac{V}{m} \tag{2–8), (2–10}$$

$$p = p_g + p_a \tag{2–17}$$

$$p = p_a - p_{gv} \tag{2–21}$$

Temperature/energy:

$$T\ (°\text{R}) = T\ (°\text{F}) + 459.4° \tag{2–25}$$

$$T\ (\text{K}) = T\ (°\text{C}) + 273° \tag{2–26}$$

$$T\ (°\text{R}) = \frac{9}{5} T\ (\text{K})$$

$$\text{KE}\ (\text{J}) = \frac{1}{2} m\overline{V}^2 \qquad m \text{ in kg} \tag{2–27}$$

$$\text{ke}\ (\text{J/kg}) = \frac{1}{2} \overline{V}^2 \tag{2–30}$$

$$\text{KE}\ (\text{ft} \cdot \text{lbf}) = \frac{m\overline{V}^2}{2g_c} \qquad m \text{ in lbm} \tag{2–31}$$

$$\text{ke}\ (\text{ft} \cdot \text{lbf/lbm}) = \frac{\overline{V}^2}{2g_c} \tag{2–32}$$

$$\text{PE}\ (\text{J}) = Wz = mgz \qquad m \text{ in kg} \tag{2–33}$$

$$\text{pe}\ (\text{J/kg}) = gz \tag{2–35}$$

$$\text{PE}\ (\text{ft} \cdot \text{lbf}) = \frac{mgz}{g_c} \qquad m \text{ in lbm}$$

$$\text{pe}\ (\text{ft} \cdot \text{lbf/lbm}) = \frac{gz}{g_c} \tag{2–36}$$

$$u = \frac{U}{m} \tag{2–37}$$

DISCUSSION QUESTIONS

Section 2–1

2–1 What is meant by a system?

2–2 Why does a system need a boundary?

Section 2–2

2–3 What is meant by the term mole (or mol)?

2–4 Is a gram-mol different than a lbm-mol?

Section 2–3

2–5 What is meant by the term property?

2–6 Why are extensive properties different than intensive properties?

2–7 What is specific energy?

Section 2–4

2–8 What is meant by a system's state?

Section 2–5

2–9 What is a process?

Section 2–6

2–10 What is meant by the term cycle?

Section 2–7

2–11 How are mass and weight of system related?

2–12 What is g_c?

Section 2–8

2–13 What is meant by the term specific volume?

2–14 What is meant by the term specific weight?

2–15 What is meant by the term specific gravity?

2–16 What is density?

2–17 Why is gage pressure different than absolute pressure? Which of these pressures is "felt" by the system?

Section 2–9

2–18 What purpose is served by the Zeroth law of thermodynamics?

Section 2–10

2–19 What is temperature?

2–20 What is a thermopile?

Section 2–11

2–21 What is energy?

2–22 What is internal energy?

Section 2–12

2–23 In terms of efficiency, what are some "outputs" from a system?

2–24 In terms of efficiency, what are some "inputs" to a system?

Section 2–13

2–25 What is a derived unit?

PRACTICE PROBLEMS

Sections 2–7 and 2–8

Problems that use SI units are indicated by an (M) under the problem number; those that use English units are indicated by an (E). Mixed unit problems are indicated by a (C).

2–1 (M) A 2-kg cube of gold is weighed on a spring scale in two locations. The first location has a local gravitational acceleration of 9.80 m/s^2, and the second has a gravitational acceleration of 9.78 m/s^2. At which location will the gold's weight W in newtons be greater?

2–2 (M) Three kilograms of water are vaporized in a boiler. What is the weight of this water in newtons if the local acceleration of gravity is 9.79 m/s^2?

2–3 (E) A weight scale located at Earth's sea level indicates that a gallon of water weighs 8.333 lbf. Determine the mass of 1 gallon in units of slugs and pounds-mass.

2–4 (E) A battery weighing 32 lbf at Earth's sea level is transported to the moon. Determine the weight of the battery on the moon, where $g = 5.47$ ft/s^2.

2–5 (C) Referring to necessary tables, convert the following masses:
(a) 1 lbm to grams
(b) 2 lbm to kilograms
(c) 20 slugs to pounds-force at Earth's sea level
(d) 100 g to dynes at Earth's sea level
(e) 200 kg to pounds-force at Earth's sea level

2–6 (M) A cylindrical tank filled with a fluid has a diameter of 1 m and a length of 1.5 m. It weighs 6000 N, where the gravitational acceleration is 9.82 m/s^2. Determine:
(a) Volume occupied by the fluid
(b) Specific weight of the fluid
(c) Density of the fluid
(d) Specific gravity of the fluid

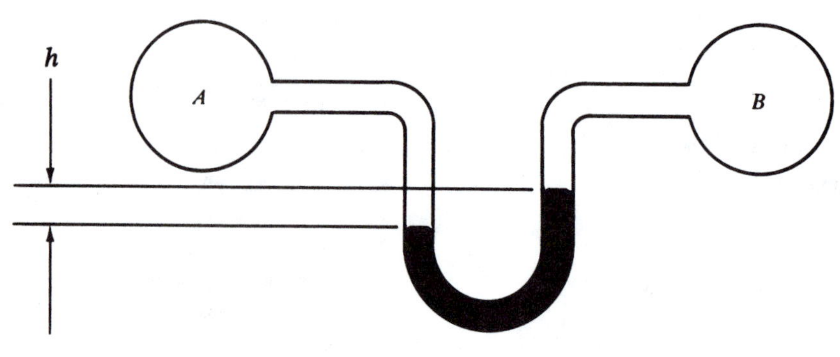

FIGURE 2–32

2–7 **(M)** A balloon is filled with a gas having a specific volume of 0.9 m^3/kg. If the inside volume of the balloon is 4800 cm^3, what is the weight of the gas-filled balloon? Ignore the weight of the balloon itself and assume that $g = 9.78\ m/s^2$.

2–8 **(M)** A tank is filled with hydrogen gas, and a gage indicates that the pressure inside the tank is 1.0 kPa.

(a) If the atmospheric pressure is 101 kPa, what is the pressure inside the tank?

(b) What is the pressure in the tank if the atmospheric pressure is 768 mm Hg?

2–9 **(M)** A tank 5 m high is half full of water, and air at 13-kPa gage pressure is occupying the remaining volume. The tank is 1.5 m in diameter and the contents are at 20°C.

(a) What is the gage pressure at the top of the water?

(b) What is the gage pressure at the bottom of the tank?

(c) If the atmospheric pressure is 101 kPa, find the absolute pressure of (a) and (b).

2–10 **(E)** Steam has a specific volume of 10.07 ft^3/lbm at a pressure of 80 psia and 900°F. At this state, assuming that $g = 32.1\ ft/s^2$, determine:

(a) Density

(b) Specific weight

(c) Specific gravity

2–11 **(E)** Two tanks, A and B, contain air (figure 2–32). Tank A is at a pressure of 20 psig, and tank B is at a pressure of 18 psig. If the two tanks are connected by a U-tube manometer, as shown, what is the difference in height of the mercury column, h? (*Note:* The specific weight of mercury is 845 lbf/ft^3, and the specific weight of air is 0.076 lbf/ft^3.)

2–12 **(E)** A hydraulic pump produces a pressure of 250 psig in an oil line as shown in figure 2–33. What force in pounds-force will be applied to the 1-ft-diameter piston?

2–13 **(E)** Determine the maximum vacuum pressure possible inside a tank with the following surrounding conditions:

(a) An atmospheric pressure of 14.8 psia

(b) An atmosphere at a pressure of 14 in Hg

2–14 **(C)** Convert the following pressures:

(a) 14.7 psi to inches of mercury

(b) 460 mm Hg to kPa

(c) 300 in Hg to pounds per square inch

(d) +50 psi to kPa

(e) 20 kPa to pounds per square inch

(f) 20 in. water gage pressure (WG) to psig

(g) 50 cm water to kPa

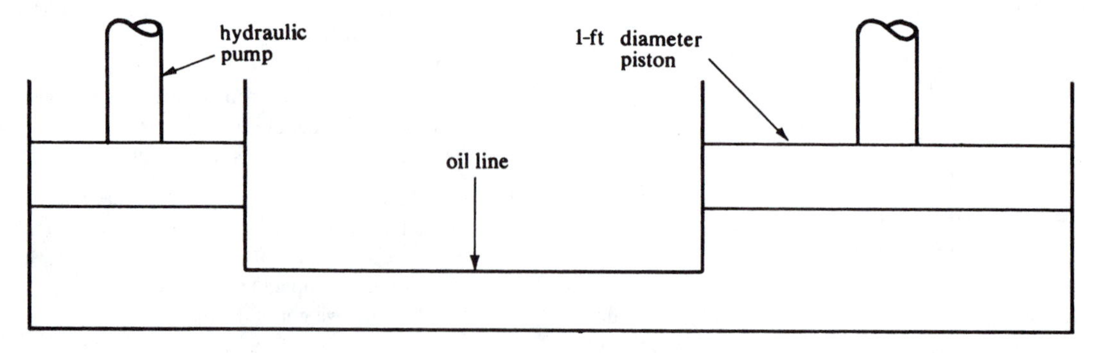

FIGURE 2–33

2–15 (E) The pressure in a steam line is read as 955 psi from a bourdon gage, as shown in figure 2–34. If the atmospheric pressure is known to be 14.4 psi, what is the absolute pressure in the steam line?

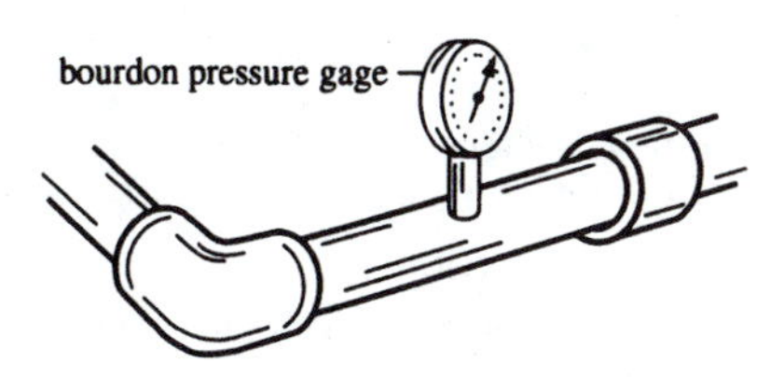

FIGURE 2–34 Pressure in a steam line

2–16 (M) A vacuum pressure gage is used to monitor the pressure in a vacuum chamber. When the gage indicates a pressure of 80 kPa and the atmospheric pressure is known to be 100.4 kPa, what is the absolute pressure in the chamber?

Sections 2–9 and 2–10

2–17 (M) A thermometer indicates 70°C when it is in thermal equilibrium with block *A* and 100°C when in thermal equilibrium with block *B*. Are blocks *A* and *B* in thermal equilibrium? State a condition when blocks *A* and *B* are in thermal equilibrium (figure 2–35).

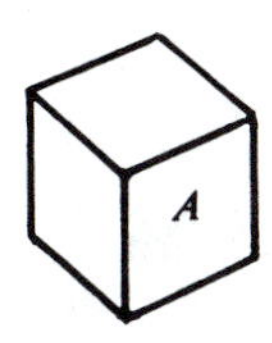

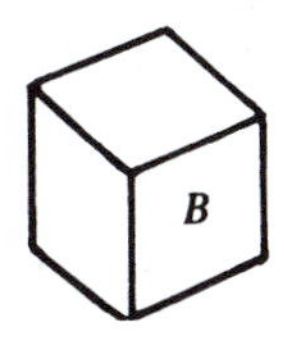

FIGURE 2–35

2–18 (E) A copper–constantan thermocouple needs to operate from 50°F to 400°F. What maximum voltage will be measured?

2–19 (M) An iron–constantan thermocouple has a reading of 8.700 mV when in an oven. What temperature is the oven in degrees Celsius?

2–20 (M) A new temperature scale, the N scale, based on the melting and freezing of the element cesium is defined as

$$0°\text{N} = \text{melting point of cesium} = 28.5°\text{C}$$
$$100°\text{N} = \text{boiling point of cesium} = 690°\text{C}$$

What temperature is absolute zero in this new scale?

2–21 (E) A temperature scale is proposed which has values of inverse Rankine temperature. If we call this scale the "down scale" and identify it as T_D, then $T_D = 1/T$ (°R) as proposed. Plot T_D versus T on regular graph paper between values of $1/10$°R and 10°R. What is the slope of the curve at a point where $T = 1°\text{R} = 1°T_D$?

2–22 (C) A temperature scale called an L scale is suggested, which has the values of $T_L = \log T$ °R. Determine the value of T_L in terms of the Kelvin scale (K).

2–23 (C) Convert the following temperatures:
(a) 140°F to degrees Rankine
(b) 88°F to degrees Rankine
(c) 230°F to degrees Celsius
(d) 87 K to degrees Rankine

2–24 (C) Convert the following temperature values to both degrees Rankine and Kelvin:
(a) 412°F
(b) 32°F
(c) 117°C
(d) 72°C

2–25 A thermopile is a number of thermocouples connected in series, as shown in figure 2–36. If the thermopile is made up of eight copper–constantan thermocouples, determine the expected signal from the device in millivolts (mV) when the temperature is 200°F.

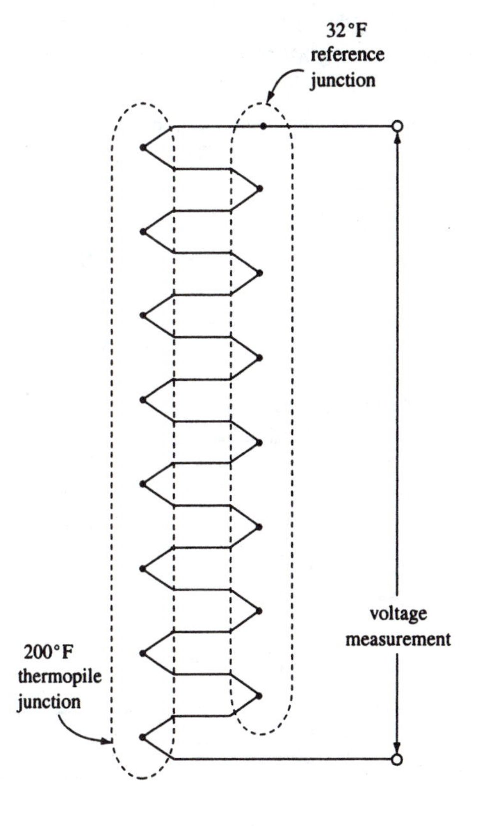

FIGURE 2–36

Section 2–11

2–26 (M) A 45,000-kg aircraft travels at 1000 km/h at an altitude of 3000 m. The local gravitational acceleration at this altitude is 9.81 m/s^2. Determine
(a) the kinetic energy of the aircraft
(b) the potential energy of the aircraft if the sea level is considered as the plane of zero potential energy

2–27 (M) Assume that you are taking a trip on a train. You carry a filled suitcase that weighs 170 N, and during the course of your trip you are told that the train is moving at 140 km/h.
(a) What do you observe the kinetic energy of the suitcase to be, before you are told the velocity of the train?
(b) What do you interpret the kinetic energy of the suitcase to be after knowing the train's velocity?

2–28 (M) A 1-kg piece of wood and a 1-kg piece of steel are dropped simultaneously into 20-m-deep water from a bridge that is 40 m high. Find the change in potential energy or the available potential energy of the:
(a) Wood
(b) Steel

2–29 (M) A pump is used to remove water from a well into a tank. If the well is 75 m deep, what energy per kilogram of water must be supplied by the pump in this process? Assume that $g = 9.8\ m/s^2$.

2–30 (M) Steam flows through a 5-cm-diameter pipe with a velocity of 24 m/s. What is the kinetic energy (ke) of the steam per unit mass?

2–31 (M) A 1-kg piston in an internal combustion engine travels at 60 m/s at a specific instant. What kinetic energy does the piston possess?

2–32 (M) Angular kinetic energy (AKE) is a form of mechanical energy that is usually not accounted for in the kinetic energy. A system has the following amounts of energy:

$$\begin{aligned} \text{AKE} &= 150\ \text{kJ} \\ \text{KE} &= 100\ \text{kJ} \\ \text{PE} &= 20\ \text{kJ} \\ U &= 35\ \text{kJ} \end{aligned}$$

(a) Determine the total energy of the system.
(b) Determine the total mechanical energy of the system.

2–33 (E) A balloon weighing 10 oz at a location where $g = 31.7\ ft/s^2$ is released at 20° latitude and floats to an altitude of 6000 ft above sea level. If we released the balloon from 1000 ft above sea level and defined this elevation as where potential energy is zero, determine
(a) The total potential energy of balloon when floating
(b) The total potential energy of balloon if it had been at sea level
(c) Total potential energy of balloon at release
(*Note:* 16 ounces = 1 lbm.)

2–34 (E) One pound-mass of mercury at 426°F has 150 Btu of internal energy under a certain condition, while it also has 28 Btu of KE and 2 Btu of PE. Determine the total energy of the mercury.

2–35 (E) During a windstorm, the velocity of the wind is measured at 70 mph. What is the specific kinetic energy of the air under this condition?

2–36 (E) Ten pounds-mass of steam flowing through a pipe are found to have 15,000 Btu of internal energy U. If the kinetic energy ke is 500 Btu/lbm, and the potential energy pe is 100 Btu/lbm, determine the total energy of the 10-lbm steam, E, and the energy e.

2–37 (E) Freon flows through a circuit of pipes and components shown in figure 2–37. If the freon has a velocity of 2 ft/s at

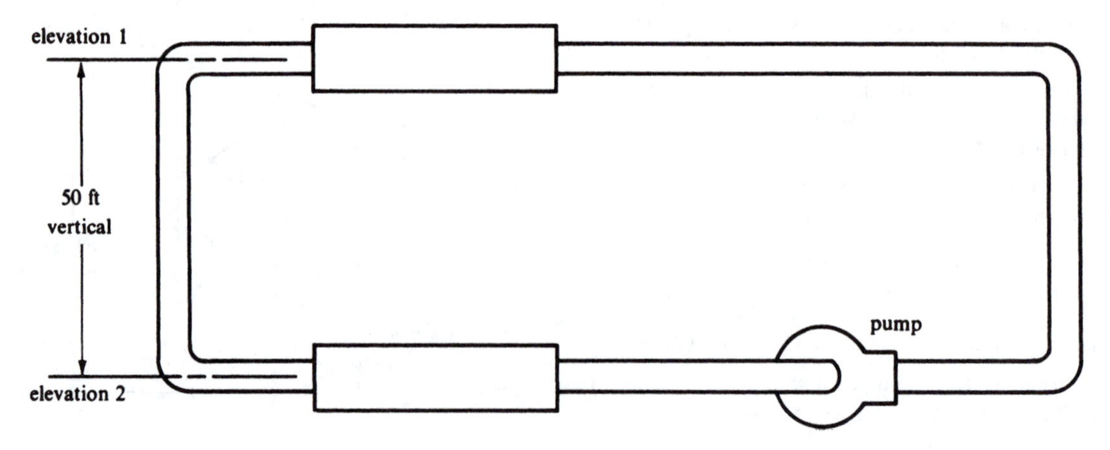

FIGURE 2–37

both points 1 and 2 in the circuit, what is the difference in energy per pound-mass of the freon between 1 and 2, if internal or thermal energy is neglected? Assume that $g = g_c$.

2–38 (E) For problem 2–37, determine the kinetic energy at points 1 and 2 (figure 2–37).

2–39 (E) For problem 2–37, determine the specific potential energy at points 1 and 2, if the plane of zero potential energy is assumed to be 100 ft below point 2.

Section 2–12

2–40 (E) A heating furnace burns fuel oil that can release 140,000 Btu/gal of thermal energy. If the furnace has a combustion efficiency of 92%, how much thermal energy can be expected to be obtained from 100 gal of fuel oil?

2–41 (M) Solar energy reaching the Earth is about 1000 W/m^2, and photovoltaic devices can convert light into electricity with an efficiency of about 8%. How large a photovoltaic panel must one have to obtain 2.5 kW of power under ideal conditions? Express your answer in square meters (m^2).

2–42 (E) A hydroelectric power plant has a dam 100 ft high, and water can flow through the dam and water turbines at 1 million lbm per minute over a 60-ft drop. If the turbines can convert the loss in potential energy of the water into electric power at an efficiency of 70%, how much power can the facility produce in megawatts? (*Note:* 550 ft · lbf/s = 1 hp = 0.746 kW.)

2–43 (M) A 200-MW power plant is known to require 1.6×10^6 kg of coal per day. If the coal is assumed to provide 30,000 kJ per kilogram of coal of thermal energy, determine the overall plant efficiency.

2–44 (E) A remote 5-kW electric power unit is driven by a diesel engine. The engine uses 0.4 gal of diesel fuel per hour. If the fuel has 180,000 Btu of thermal energy in each gallon, determine the efficiency of the unit.

2–45 (M) A rechargeable battery can supply 3600 W · s of energy before it needs to be recharged again. If it is found that 3800 J is needed to bring the battery back to a recharged state, what is the battery's efficiency?

2–46 (M) A wind farm consisting of a number of windmills, or wind generators, is proposed to produce 25 megawatts (MW) of electrical power. If there are 100 wind generator machines installed, each capable of producing 250 kW of power in a 30 kilometer per hour (kph) wind and with a conversion efficiency of 38%, how much wind power needs to be available to produce the rated 25 MW of power?

Section 2–13

2–47 (C) The Reynolds number, Re, is a dimensionless (unitless) number frequently used in fluid mechanics, thermodynamics, and physics. This number is found by the defining equation

$$\text{Re} = \frac{\rho \bar{V} D}{\mu_k}$$

where D is the characteristic length having units of meters or feet and μ_k is the viscosity. In SI and English units, what dimensions does viscosity have?

2–48 (C) In terms of SI and English units, determine the dimension x in the following equations, using table 1–1, table 2–4, or the list of symbols when necessary:

(a) $x = T\,\Delta s/\Delta t$
(b) $x = \Delta h/\Delta T$
(c) $x = T\,\Delta S$
(d) $x = RT/(v - c)^2$
(e) $x = pv/T$

2–49 (C) Show that both sides of the following equation agree dimensionally:

$$\frac{\Delta p}{L} = \frac{\rho \bar{V}^2}{D g_c}\left(C \frac{\mu_k}{D \bar{V} p}\right)$$

C is a dimensionless constant and L has units of length. See problem 2–47 for definitions of μ_k and D.

2–50 (C) What are the dimensions of C in the following equations?

(a) $C = pv^{1.7}$
(b) $C = pv^{1.3}$
(c) $C = pv/v^{2.3}$
(d) $C = p$
(e) $C = T$

3 WORK, HEAT, AND REVERSIBILITY

In this chapter, we define work and derive some useful relationships for calculating its magnitude or amount. The methods of computing areas under curves are used to find the work, particularly when the force varies. The concept of heat is introduced and defined. Then the concept of reversibility is introduced, along with the items or causes for irreversibilities in processes. Following this, the concepts of the equivalence and differences of work and heat are discussed. The thermodynamic system, introduced in chapter 2, is classified into three general types: open, closed, and isolated. This classification is introduced now to allow for better understanding of the motivation behind typifying systems. The chapter ends with a tabulation of the forms of energy, providing the foundation for the conservation of energy principle presented in subsequent chapters.

New Terms

b	Spring constant or spring modulus	t	Time
D	Diameter	Wk	Work
F_f	Friction force	Wk_{cs}	Closed-system work
F_N	Normal force	Wk_{shaft}	Shaft work
J	778 ft · lbf/Btu, mechanothermal conversion factor	wk	Work per unit of mass
		$\dot{Wk}$	Power
n	Polytropic exponent	θ (theta)	Angular displacement or rotation
N	Angular speed of rotation		
Q	Heat	T (tau)	Torque
$\dot{Q}$	Heat transfer		
q	Heat per unit mass or specific heat transfer		

3–1 WORK

Work connotes an active or dynamic state during which some mechanical effort has been exerted. This visualization is embodied in our definition of work:

Work: ***Force times distance through which the force acts constantly.***

$$\delta Wk = F\,\delta x \tag{3–1}$$

where δWk is a very small amount of work and δx is a very small distance through which the constant force F is acting. Further, as discussed in section 1–5, the force F may be considered as an average force over the distance δx. Now, if a number of these small work terms are added together, a significant amount of work, Wk, is obtained. Thus, just as in chapter 1 [equation (1–4)], where very small areas were summed to obtain an area under a curve, work can be determined from a sum of very small work terms, δWk. If the force F is a function of x (or dependent on x) and if F is then plotted with x on a graph as shown in figure 3–1, the area under that curve resulting from the plot of $F = f(x)$ is just the work, Wk.

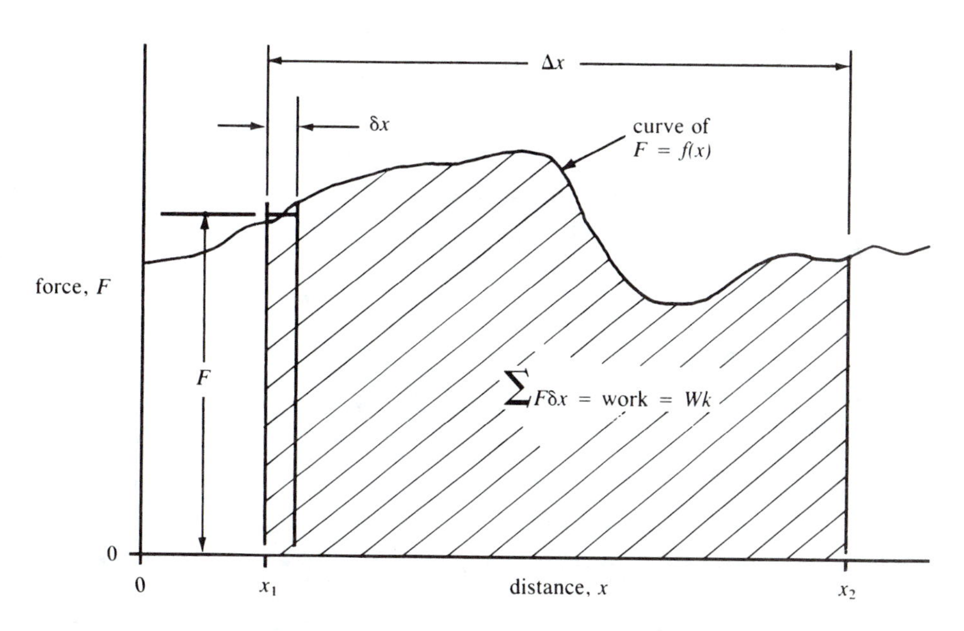

FIGURE 3–1 Typical force–distance relationship for a process involving work

This is so because the definition of work [equation (3–1)] is the same as the definition for a very small area δA under the curve, and the sum of those small areas is the total area, or the work. So we now have an important principle for calculating work:

$$Wk = \sum F\,\delta x \qquad \textbf{3–2}$$

Notice also that the work depends on the function $F = f(x)$ and the values of x_1 and x_2. If, for instance, the work is to be counted from $x_1 = 0$, then, in figure 3–1, the work would be more than the shaded area. Also, if the sum of the small work terms (or small areas under the curve) were added from right to left so that x_2 is smaller than x_1, the work would be negative $(-)$, as we discussed in section 1–5 about areas being negative if the sum proceeded from right to left. The reason that the work can be positive or negative is that both force and distance (called **displacement**) are vectors; that is, they both have a magnitude and a direction, and if the direction of the force is opposite to the displacement, the product is negative. If force and displacement are in the same direction, the work will be positive. Also, if you recall *Newton's third law of motion* (every force or action has an opposing reaction), we must be clear as to what force F is implied in equations (3–1) and (3–2). For our purposes, the force F is the force or action of the system, as opposed or opposing the agent external to the system. In many cases, the idea of a free-body diagram from mechanics will be useful in deciding which force should be involved in the calculation of work. Finally, the significance of the positive and negative signs associated with work may best be seen by considering another definition for work:

Work: ***Energy in transition across the boundary of a system, which can always be identified with a mechanical force acting through a distance.***

We see, then, that work is the boundary effect on a system as opposed to the internal property of energy. It is an effect that transfers energy to or from a system due to the mechanical action of force acting through a distance. Therefore, if the work is positive, we say that energy (in the form of work) is flowing out of the system being studied, and if work is negative, the energy is flowing into the system. That is,

positive work = work out of a system
negative work = work into a system

CALCULUS FOR CLARITY 3–1

Work may be defined rigorously from equation (3–1) as

$$dWk = F\,dx \qquad \textbf{(3–3)}$$

where dWk is a differential amount of work done by a force F acting through a differential displacement dx. A finite amount of work is then found from the result

$$Wk = \int dWk$$

Using equation (3–3) results in the following:

$$Wk = \int F\,dx \qquad \textbf{(3–4)}$$

The relationship between F and x, where $F = f(x)$, must be known to perform the integration of equation (3–4).

The calculation of work can become a bewildering exercise if the student does not first understand the foregoing definitions for work, particularly equation (3–2). We will see many specialized equations in this book that can be used to compute work for special cases, but it is crucial that you understand that all of them come from equation (3–2). Let us consider the special case when the force acting through a displacement is constant. This is not the same as equation (3–1), where the force is constant over a very small displacement δx, but rather, the situation where the force is constant over the full range from, say, x_1 to x_2. A simple demonstration of how this can occur is shown in figure 3–2, where a system is lifted vertically through a distance or displaced upward.

FIGURE 3–2

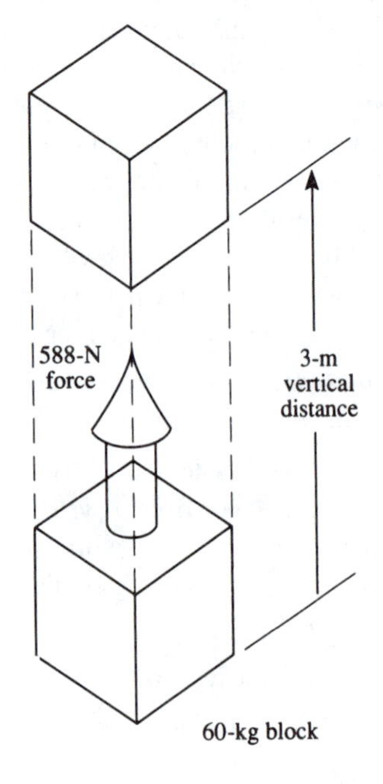

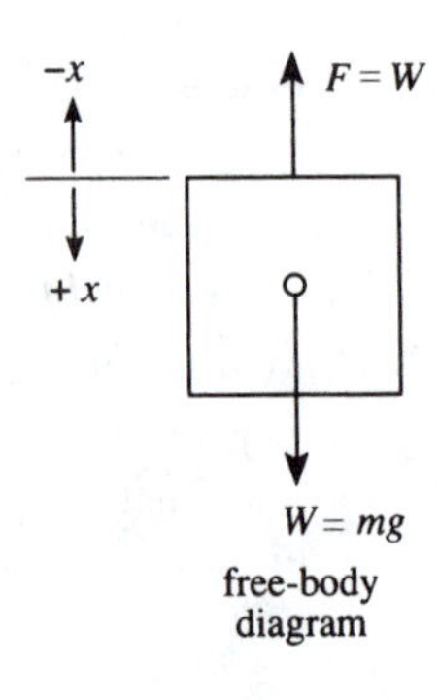

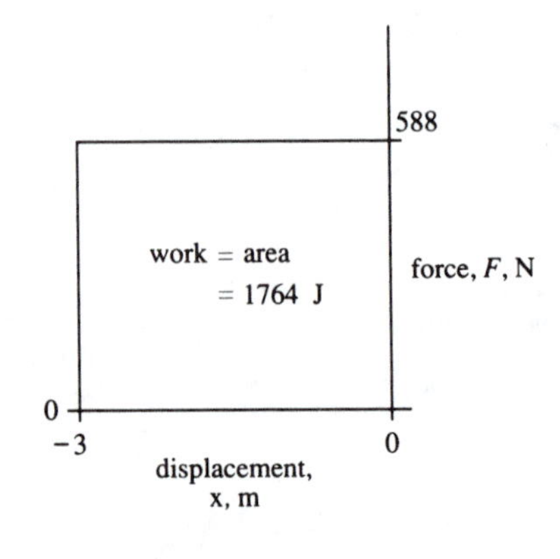

EXAMPLE 3–1 Determine the work done in lifting a 60-kg block vertically 3 m.

Solution The force required to lift the block is just equal to the weight of the block, as given by equation (2–2):

$$\begin{aligned} W &= mg \\ &= (60\ \text{kg})(9.8\ \text{m/s}^2) \\ &= 588\ \text{N} \end{aligned}$$

The force is opposite in direction, as indicated in the free-body diagram of figure 3–2. As demonstrated in the sketch, the weight (which is the force of the system) is directed oppositely to the displacement, which is upward, so that if we assign down as positive, we must assign up as negative. The result will be the same if you assign up as positive and down as negative. Using our convention of down as positive, the total displacement of the system will be -3.0 m, as indicated in the graph of force versus displacement in figure 3–2. Now, using equation (3–2), we can factor the F out of the sum because F is going to be a constant all the time anyway. This gives us

$$Wk = F \sum \delta x = F\,\Delta x$$

or

$$\begin{aligned} Wk &= (588\ \text{N})(-3\ \text{m}) = -1764\ \text{N}\cdot\text{m} \\ &= -1.764\ \text{kJ} \end{aligned}$$

Answer

Notice in example 3–1 that the units of work are the same as for energy. Also, the negative sign indicates that work is into the system or block, and this is converted into an increased potential energy. We will see later that work and the various forms of energy can be equated in the conservation of energy. Also, notice that in example 3–1 the work was finally calculated from the equation

$$Wk = F\,\Delta x \tag{3–5}$$

or work equals force times distance. This relationship is true only if the force is constant.

Let us now consider a common situation: when force varies with the displacement. Springs are mechanical devices that require more and more force to stretch them (if they are extension springs) or compress them (if they are compression springs). In figure 3–3, we show an extension spring that requires an increasing force as it gets longer. If we call the length of the unstretched spring when no force is applied the free length, l_o, and the actual spring length during loading l, then the displacement of the force of the spring (and the opposing force causing the spring to stretch) is just $l - l_o$ (call it x). But, in figure 3–3 it can be seen that the displacement and the spring force, F_s, are in opposite directions again, as in example 3–1. Many springs are such that the force, F_s, is linearly related to the spring displacement, x, or $F_s = -bx$. The negative sign here takes care of the fact that the force and displacement are in opposite directions, and b is called the spring constant or spring modulus. The spring constant must have units of force per length to satisfy the equation. If this relationship is plotted on a graph, a curve like that shown in figure 3–3 results. The area under that curve will be the work done on or by the spring.

EXAMPLE 3–2 An extension spring is extended to 3 in more than its free length by an external force. If the spring constant is 30 lbf/in and if it then shortens by 2 in, determine the work of the spring.

Solution The spring begins at a state where it is 3 in longer than its free length, the spring displacement x_1 is -3 in, and the force of the spring is

$$F_s = -bx = -(30\ \text{lbf/in})(-3\ \text{in}) = +90\ \text{lbf}$$

FIGURE 3–3

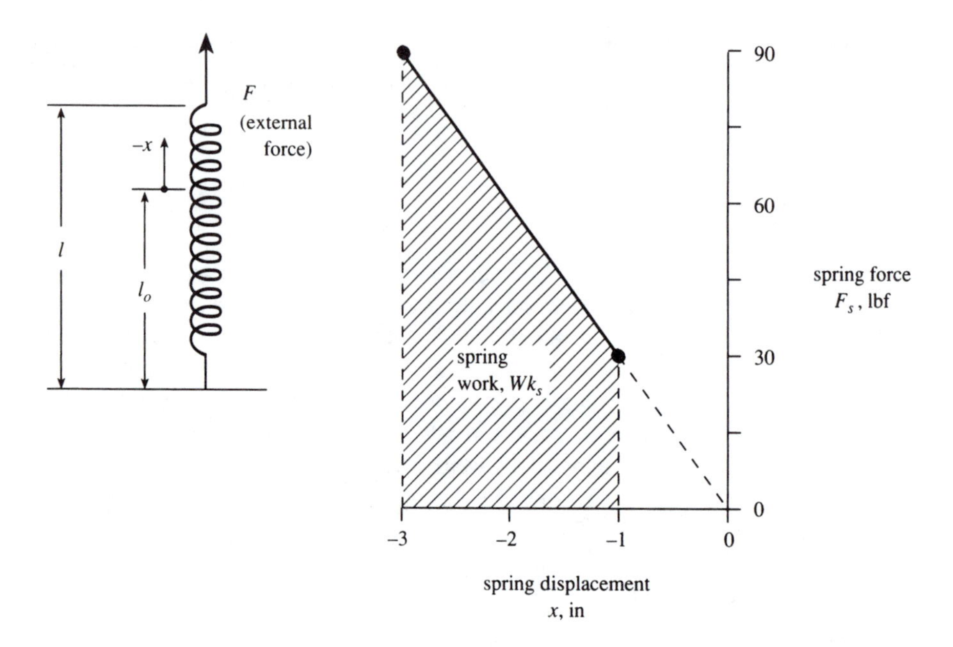

The spring then shortens by 2 in so that the spring displacement x_2 is −1 in (−3 in + 2 in) and the force is

$$F_s = -(30 \text{ lbf/in})(-1 \text{ in}) = 30 \text{ lbf}$$

In figure 3–3 is shown the graph of F_s against x, and the work of the spring is the area under the curve from x_1 to x_2. We could calculate the work from equation (3–2), or just calculate the geometric area under the curve directly, since that area is a trapezoid. Also, the area could be separated into two areas, a rectangle and a triangle. Let us calculate the area from the trapezoidal rule. Then the area (or work) is

$$Wk = \left(\frac{1}{2}\right)(F_{s_1} + F_{s_2})(x_2 - x_1)$$

$$= \left(\frac{1}{2}\right)(90 \text{ lbf} + 30 \text{ lbf})[-1 \text{ in} - (-3 \text{ in})]$$

Answer

$$= +120 \text{ in} \cdot \text{lbf}$$

Notice in example 3–2 that the work is positive, so that the spring is doing work. We say that the "spring energy" was used to do the spring work. Also, if the spring forces are replaced by the term bx in the preceding equation, the spring work is then found to be

$$Wk = \left(\frac{1}{2}\right)(b)(x_2^2 - x_1^2) = Wk_s \qquad \textbf{(3–6)}$$

We will use equation (3–6) to calculate the work of a spring when its displacement changes from x_1 to x_2 and whether the spring is an extension or a compression spring.

CALCULUS FOR CLARITY 3–2

The use of calculus often provides a more straightforward means of determining the work of a system. In example 3–1, the work done by the block was determined from the product of the force (the system weight) times the distance through which the block was lifted.

CALCULUS FOR CLARITY 3–2, continued

If the rigorous definition for work, equation (3–3), is used, then we have, for **constant force** F,

$$Wk = \int F\,dx = F\int dx = F(x_2 - x_1) = F\,\Delta x$$

In example 3–2, the spring force varied linearly with the displacement such that $F = -bx$. Using equation (3–3) again, we would have

$$Wk_s = \int -bx\,dx = -b\int x\,dx = -b\left[\frac{1}{2}\right](x_2^2 - x_1^2)$$

which is the same equation we use without using calculus. The integration of $\int x^n\,dx$, however, gives the result more conveniently if the student has been exposed to calculus. Notice that the way in which the system force varies with the displacement is crucial in determining the work.

Many mechanical devices provide or use work through a rotating shaft. These rotating machines, or cyclic devices, have work occurring because of a torque acting through an angular distance or displacement. This is equivalent to the definition of work of equation (3–2) and can be written

$$Wk_{\text{shaft}} = \sum \mathrm{T}\,\delta\theta \qquad \textbf{(3–7)}$$

where the "shaft" subscript indicates that it is **shaft work**; T is **torque** or moment; and θ is an angular distance or **displacement** through which T is acting, in radians. Torque and angular displacement are terms well used in mechanics, and from the relationships

$$\mathrm{T} = F \times r \quad \text{and} \quad \delta\theta = \frac{\delta x}{r}$$

you can see that equations (3–7) and (3–2) are equal to each other. The direction of the torque relative to the angular displacement must be accounted for. For positive work, the torque is applied with the rotation, and for negative work, the torque opposes rotation. Angular displacement is often known or given in revolutions rather than radians. Using the conversion of 1 revolution $= 2\pi$ radians or 2π radians/rev gives the correct units for equation (3–7).

EXAMPLE 3–3 A gas engine supplies a torque of 3 kN · m for 300 revolutions. Determine the work done by the engine.

Solution Using equation (3–7), we see that the torque is constant, so that it can be factored out of the summation, and the summation of $\delta\theta$ is just the total angular displacement and is in the same direction as the torque. Thus,

$$Wk_{\text{shaft}} = \mathrm{T}\sum \delta\theta = \mathrm{T}\theta$$

$$= (3\ \text{kN}\cdot\text{m})(300\ \text{rev})(2\pi\ \text{rad/rev})$$

Answer

$$= 5655\ \text{kN}\cdot\text{m}$$

CALCULUS FOR CLARITY 3–3

Shaft work is rigorously defined through the equation

$$Wk_{shaft} = \int T \, d\theta \tag{3–8}$$

and may be computed if the relationship between the shaft torque T and the angular displacement θ is known.

EXAMPLE 3–4 A pump is driven by an electric motor through a rotating shaft. The torque required to rotate a pump from standstill is usually greater when the pump is first started, called the start-up torque. If the pump shaft start-up torque varies with angular displacement for the first 20 revolutions of start-up according to the relationship

$$T = -30 \text{ (in} \cdot \text{lbf)} + 0.5 \text{ (in} \cdot \text{lbf/rev)} \, \theta$$

where θ is the angular displacement of the pump shaft from start-up in revolutions, determine the shaft work required to rotate the pump through the first three revolutions of start-up.

Solution Since the start-up torque of the pump is known, we may use equation (3–8) to determine the work. We integrate between $\theta_1 = 0$ to $\theta_2 = 3$ (revolutions) to obtain

$$Wk_s = \int T \, d\theta = \int (-30 + 0.5\theta) \, d\theta = -30 \int d\theta + (0.5) \int \theta \, d\theta$$

$$= -(30 \text{ in} \cdot \text{lbf})(\theta_2 - \theta_1) + (0.5 \text{ in} \cdot \text{lbf/rev})\left[\frac{1}{2}\right](\theta_2^2 - \theta_1^2)$$

$$= -87.75 \text{ in} \cdot \text{lbf/rev}$$

Using the conversion 2π rad/rev, we have

Answer

$$Wk_s = (-87.75)\left(\frac{1}{2\pi}\right) = -13.97 \text{ in} \cdot \text{lbf} = -1.16 \text{ ft} \cdot \text{lbf}$$

Let us now look at a type of apparatus that we will see often in the remainder of this book. The apparatus, shown in figure 3–4, includes a freely sliding piston of radius r inside a cylindrical chamber with suitable guides and linkages to convert the piston

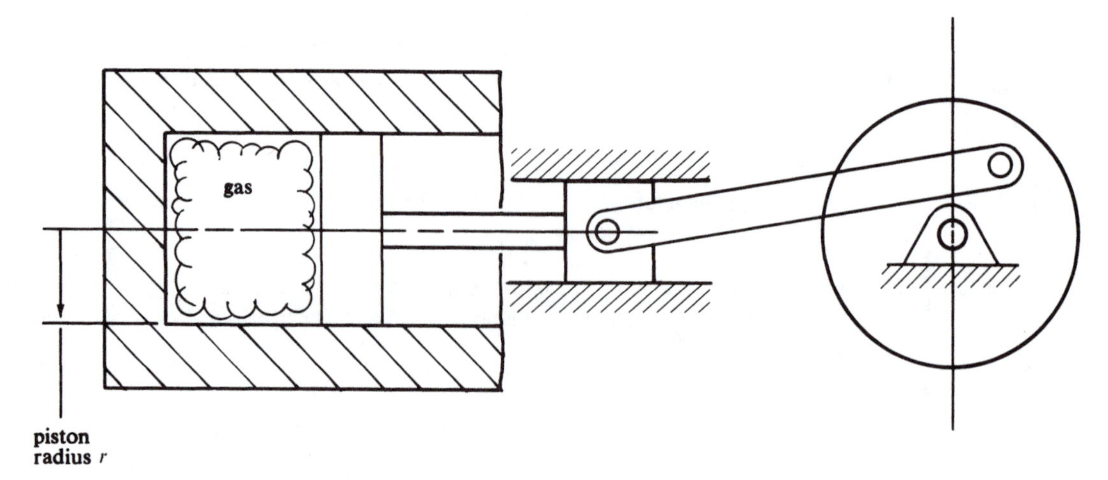

FIGURE 3–4 Axial cross section of piston-cylinder device

motion into a rotation of a flywheel or shaft. We call the apparatus a piston-cylinder, and normally it will contain a gas of some sort behind the piston as indicated in figure 3–4. Unless it is otherwise assumed, the gas will not be allowed to leak out past the piston, and the gas will then be called a closed system. Now, we notice that as the gas in the piston-cylinder expands and pushes the piston out, the pressure (which is pushing the piston out) will probably drop. This is not necessarily so, but would be a natural happening if the gas were left alone to expand. The force acting on the piston by the system (the gas) is just the pressure times the area, $F = pA$, and if this is substituted into equation (3–2) for work, we have

$$Wk = \sum pA\,\delta x$$

But the term $A\,\delta x$ is a very small change in the volume of the gas or system. If we call this change δV, the work is

$$Wk_{cs} = \sum p\,\delta V \tag{3–9}$$

The work determined from equation (3–9) is the work that results from a closed system when it changes volume. Sometimes this is called boundary work, but we will call it **closed-system work,** Wk_{cs}. If the pressure of the system or gas were plotted as the volume changes, a graph much like that in figure 3–5 would probably result. The area under the curve is equal to the work given by equation (3–9). Also, the **work per unit of system mass** can be calculated by dividing the Wk_{cs} by the system mass, m. We will write this as wk_{cs} and notice that it is equal to the right side of equation (3–9) if it were divided by m. Then

$$wk_{cs} = \sum p\frac{\delta V}{m} = \sum p\,\delta v \tag{3–10}$$

Many special forms of equations (3–9) and (3–10) will be found in thermodynamics. Let us now consider some of them.

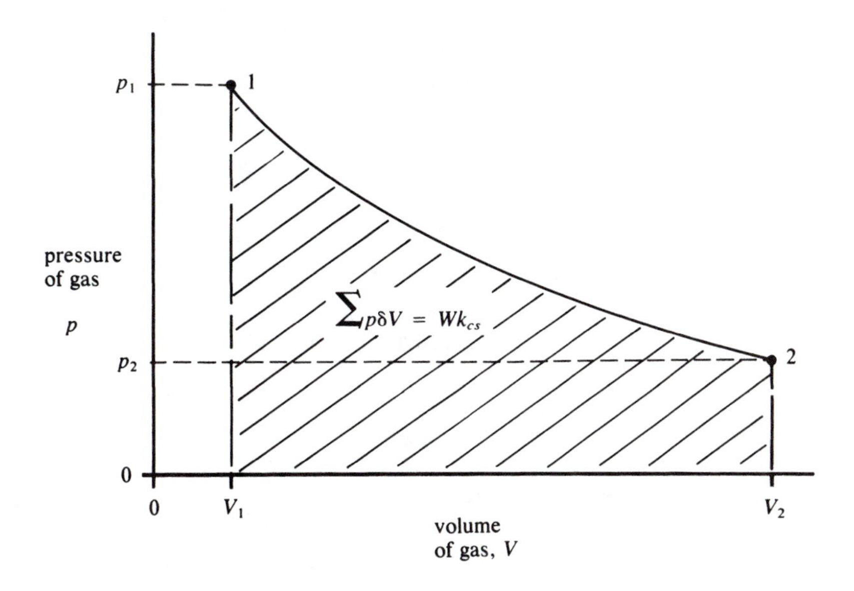

FIGURE 3–5
Pressure–volume, p–V, diagram of closed thermodynamic system involving a work process

CALCULUS FOR CLARITY 3–4

The general equation for the boundary work or closed-system work of a system can be derived from equation (3–3) by noting that the system force is the pressure times the area over which that pressure acts, or $F = pA$. Then the closed-system work is

$$Wk_{cs} = \int pA\,dx = \int p\,dV \tag{3–11}$$

and the work per unit of mass m is

$$wk_{cs} = \frac{Wk_{cs}}{m} = \int p\,d\frac{V}{m} = \int p\,dv \tag{3–12}$$

EXAMPLE 3–5 An incompressible fluid is one that does not change its density or volume as the pressure changes on that fluid. Consider water as an incompressible fluid, with density of 1000 kg/m^3, compressed or pressurized from 100 kPa to 1000 kPa. Determine the work of this compression process.

Solution The work is determined from equation (3–9). Since the water is assumed to be incompressible, its volume remains constant and δV is zero. Therefore, the work must be zero.

Example 3–5 shows that the boundary work for a closed system is zero for the case when the volume remains constant. If the pressure remains constant during a particular process and the volume changes, equation (3–9) can be written

$$Wk_{cs} = p\sum \delta V = p\,\Delta V \tag{3–13}$$

and equation (3–10) becomes

$$wk_{cs} = p\,\Delta v \tag{3–14}$$

These two equations are correct for constant-pressure processes only and should not be used for other types of processes.

If a piston-cylinder device contains a gas, the pressure of the gas often decreases with increasing volume and increases with decreasing volume of the gas. We say that the pressure varies inversely with volume, and a general form of this relationship is plotted in figure 3–5. A general mathematical equation for this relationship is

$$p = \frac{C}{V^n}$$

or

$$pV^n = C \tag{3–15}$$

and this is called the **polytropic equation**, where n is the polytropic exponent and C is a constant of proportionality. In thermodynamics, the value of n is usually between 1 and 2, except for the constant-pressure process $(n = 0)$ and the constant-volume process $(n = \infty)$. If $n = 1$, equation (3–15) is

$$pV = C$$

and the area under the curve for this relationship, the boundary work given by equation (3–9), can then be written

$$Wk_{cs} = C\sum \frac{\delta V}{V}$$

and this can be shown to be equal to

$$Wk_{cs} = C \ln \frac{V_2}{V_1} \tag{3–16}$$

where V_2 and V_1 are the final and initial volumes of the gas and C is a constant equal to p_1V_1 or p_2V_2.

If n is not equal to 1 but is a positive value such as 1.4 or 1.5, substituting equation (3–15) into (3–9) gives

$$Wk_{cs} = C \sum \frac{\delta V}{V^n}$$

and it can be shown that this is equal to the equation

$$Wk_{cs} = \frac{1}{1-n}(p_2V_2 - p_1V_1) \tag{3–17}$$

Equations (3–13), (3–16), and (3–17) will be used extensively in this book for those processes where they are applicable.

CALCULUS FOR CLARITY 3–5

The closed-system work for the general polytropic equation for pressure–volume, $pV^n = C$, can be derived by using calculus:

$$Wk_{cs} = \int p\,dV = \int C\left[\frac{1}{V^n}\right]dV = C\int \frac{dV}{V^n} = C\int V^{-n}\,dV$$

If n is equal to one (1), then the work between V_1 and V_2 is

$$Wk_{cs} = C\int \frac{dV}{V} = C[\ln V_2 - \ln V_1] = C \ln \frac{V_2}{V_1}$$

which agrees with equation (3–16). If n is not equal to one (1), then the work is

$$Wk_{cs} = C\left[\frac{1}{1-n}\right](V_2^{1-n} - V_1^{1-n})$$

But we also know that $C = p_1V_1^n = p_2V_2^n$ for the polytropic equation (3–15), so that the work is

$$Wk_{cs} = \left[\frac{1}{1-n}\right](p_2V_2^nV_2^{1-n} - p_1V_1^nV_1^{1-n})$$

The V^n's cancel and the result is

$$Wk_{cs} = \left[\frac{1}{1-n}\right](p_2V_2 - p_1V_1)$$

which is equation (3–17). Notice that this result holds for any relationship of $pV^n = C$, as long as n is not one (1). It applies if, for instance, n is less than zero, such as in the examples where $p = CV$ when $n = -1$.

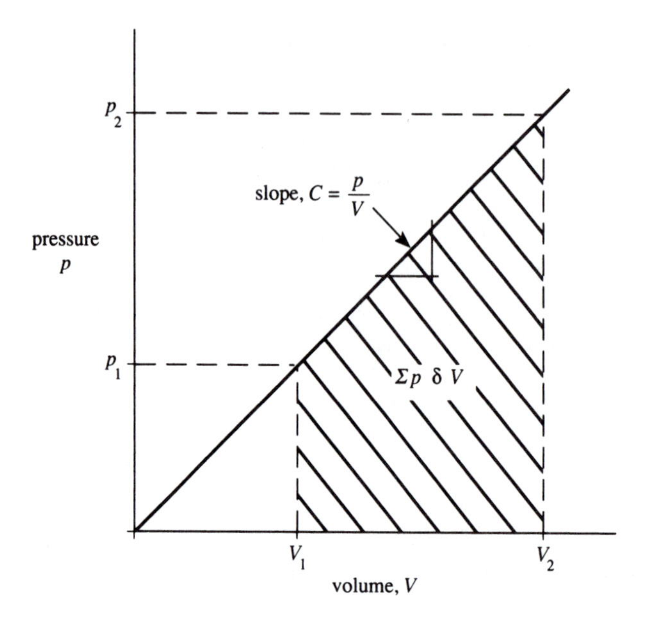

FIGURE 3–6 Example of physical situation where pressure is directly related to volume

Now let us consider a situation where the pressure increases or decreases directly with volume; that is,

$$p = CV \tag{3–18}$$

An example of when such a condition might occur is when a balloon or inflatable container is filled with a gas and the gas is then heated or cooled to change its pressure. Here the balloon acts somewhat like a spring to resist increased volume by requiring increased gas pressure. For the case when gas pressure increases linearly with the volume, as in equation (3–18), the resulting graph of p of V is shown in figure 3–6. The work done by the gas would be equal to the area under the curve and is determined by equation (3–9), revised by equation (3–18) to be

$$Wk_{cs} = \sum CV\,\delta V = \text{area under curve}$$

For the area under the curve of figure 3–6 between V_1 and V_2, at pressures p_1 and p_2, the work is just the trapezoidal area

$$Wk_{cs} = \left(\frac{1}{2}\right)(p_2 + p_1)(V_2 - V_1)$$

and, since $p_2 = CV_2$ and $p_1 = CV_1$,

$$\begin{aligned} Wk_{cs} &= \left(\frac{1}{2}\right)(CV_2 + CV_1)(V_2 - V_1) \\ &= \left(\frac{1}{2}\right)(C)(V_2 + V_1)(V_2 - V_1) \\ &= \left(\frac{1}{2}\right)(C)(V_2^2 - V_1^2) \end{aligned} \tag{3–19}$$

where now $C = p_1/V_1 = p_2/V_2 =$ slope of curve $p = f(V)$. Notice the similarity between equation (3–19) and equation (3–6) for a spring.

EXAMPLE 3–6 A hot-air balloon 10 ft in diameter is filled with air at 14.8 psia. If the balloon volume is proportional to the air pressure as given by equation (3–18), determine the work done by the air to increase the balloon's diameter to 11 ft.

Solution The balloon's volumes are first determined from the equation for a sphere (see appendix A–1):

$$V_1 = \frac{1}{6}\pi D_1^3 = \frac{1}{6}(\pi)(10\text{ ft})^3 = 523.6\text{ ft}^3$$

$$V_2 = \frac{1}{6}\pi D_2^3 = \frac{1}{6}(\pi)(11\text{ ft})^3 = 696.9\text{ ft}^3$$

and these are also the volumes of the air in the balloon. Since the pressure $p_1 = 14.8$ psia, the constant C is

$$C = \frac{p_1}{V_1} = \frac{(14.8\text{ lbf/in}^2)(144\text{ in}^2/\text{ft}^2)}{523\text{ ft}^3} = 4.07\text{ lbf/ft}^5$$

and from equation (3–19), we get

$$Wk_{cs} = \frac{1}{2}C(V_2^2 - V_1^2) = \frac{1}{2}(4.07\text{ lbf/ft}^5)[(696.9\text{ ft}^3)^2 - (523.6\text{ ft}^3)^2]$$

Answer

$$= 430{,}428\text{ ft}\cdot\text{lbf}$$

The air in the balloon does work on the balloon in an amount of 430,428 ft · lbf to expand it from 10 ft to 11 ft in diameter.

As a final example in this section, consider the case when the pressure varies irregularly with the volume during a process. Equation (3–9) must then be used to calculate the work, and the AREA program for calculating areas under curves given in appendix A–5 can be used to aid in the computation. Of course, the area under a curve can often be separated into convenient, well-known areas and the calculation then done with hand calculators, as in the examples of chapter 1.

EXAMPLE 3–7 Consider the p–V diagram of figure 3–7a. Determine the work if the pressure is that of a fluid in a volume that begins at 0.04 m^3 and expands to 0.26 m^3.

Solution The work is the area under the curve and can be found from equation (3–9):

$$Wk_{cs} = \sum p\,\delta V$$

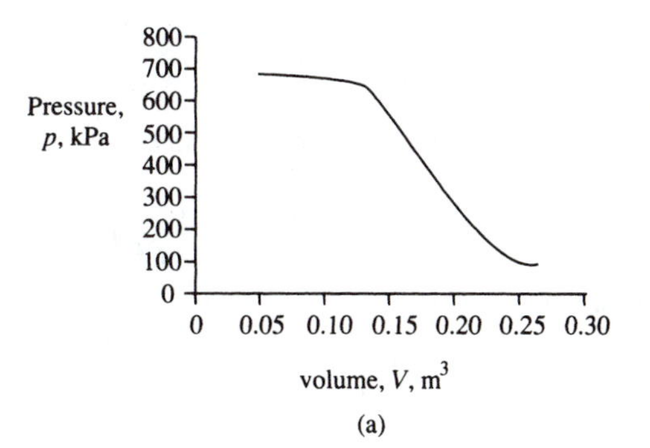

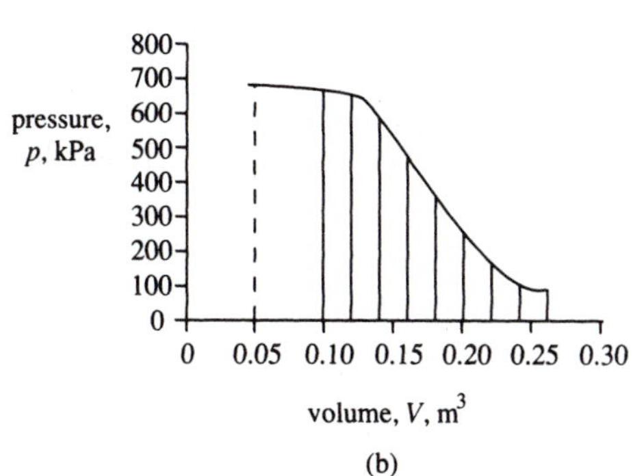

FIGURE 3–7 (a) p–V diagram for fluid of example 3–7; (b) area separated into small elements for determination of Wk

TABLE 3–1 Results of determining work in example 3–7

p kPa	V m^3	p_i kPa	δV m^3	$p\,\delta V$ kN·m
680	0.04	680	0.06	40.80
680	0.10	675	0.02	13.50
670	0.12	600	0.02	12.00
530	0.14	465	0.02	9.30
400	0.16	335	0.02	6.70
270	0.18	235	0.02	4.70
200	0.20	165	0.02	3.30
130	0.22	110	0.02	2.20
90	0.24	85	0.02	1.70
80	0.26			
			$\sum p\,\delta V = 94.20$ kN·m	

If the AREA program in appendix A–5 is used, values for p and V need to be determined at various points where the area is separated into small elements as shown in figure 3–7b. Here the area is arbitrarily broken into nine small areas, and 10 points need to be determined. The data of p and V are read from the graph and are recorded in table 3–1.

Using the AREA program and entering the p's and V's from table 3–1 gives, for the area, or the work,

Answer

$$Wk_{cs} = +94.20 \text{ kN}\cdot\text{m (kJ)}$$

The same result can be obtained by hand calculating the values, as displayed in the last three columns of table 3–1, and by using the technique of computing small trapezoids as introduced in chapter 1.

3–2 POWER

The rate of doing work or work per unit of time is called **power** and is defined as

$$\dot{W}k = \frac{\delta Wk}{\delta t} \tag{3–20}$$

where δWk is a small amount of work done during a small time period δt. The "dot" above the work symbol represents the rate of doing work. Typically, the units for power are energy per unit of time: kJ/s (= kilowatts, kW), Btu/s, ft·lbf/s, or horsepower, hp (1 hp = 550 ft·lbf/s). If the work is done uniformly so that power is constant over a time period, $\Delta t = \sum \delta t$, equation (3–20) can be written

$$\dot{W}k = \frac{Wk}{\Delta t} \tag{3–21}$$

Also, the work can then be found from equation (3–21) if a constant power is known for a given period Δt:

$$Wk = \dot{W}k\,\Delta t \tag{3–22}$$

Using the definition of work [equation (3–1)] in equation (3–20) gives

$$\dot{W}k = F\left(\frac{\delta x}{\delta t}\right)$$

but $\delta x/\delta t$ is just the speed or velocity of a body or the speed at which the force F is applied, so that we may write, in general,

$$\dot{W}k = F\overline{V} \tag{3–23}$$

EXAMPLE 3–8 Determine the average power required in kilowatts and horsepower to move an elevator weighing 2000 lbf vertically through 40 ft in 10 s.

Solution From equation (3–23), the average power can be found if the force and the velocity are known. The force, *F*, is just the weight of the elevator, or 2000 lbf, and the average velocity is the 40 ft divided by 10 s, 4 ft/s. Then

$$\dot{W}k = (2000\ \text{lbf})(4\ \text{ft/s}) = 8000\ \text{ft}\cdot\text{lbf/s}$$
$$= \frac{8000\ \text{ft}\cdot\text{lbf/s}}{550\ \text{ft}\cdot\text{lbf/s}\cdot\text{hp}}$$
$$= 14.5\ \text{hp}$$

From the table on the inside front cover, 1.34 hp = 1 kW, so

$$\dot{W}k = \frac{14.5\ \text{hp}}{1.34\ \text{hp/kW}}$$

Answer
$$= 10.8\ \text{kW}$$

CALCULUS FOR CLARITY 3–6

From calculus, *power* is defined as the first derivative of work with respect to time:

$$\dot{W}k = \frac{dWk}{dt} \quad \textbf{(3–24)}$$

The "dot" is used in the same manner as in equation (3–20).) Work is given by equation (3–3), so that power is

$$\dot{W}k = \frac{F\,dx}{dt} = F\frac{dx}{dt}$$

The first derivative of displacement with respect to time is velocity, so that equation (3–23) results directly from this relationship.

It is sometimes important to know when power is greatest or maximum. Using calculus and assuming that force depends upon velocity, we find that the maximum (or minimum) power occurs when the first derivative of power with respect to the velocity is set equal to zero.

EXAMPLE 3–9 The force required to slide a particular block over a surface is found to depend on the speed or velocity at which the block slides according to the relationship

$$F = 30 - 2\overline{V}\ (\text{kN})$$

where the velocity $\overline{V}$ is in m/s. Determine the speed at which maximum power is required to slide the block over the surface.

Solution The power is given by equation (3–23) as

$$\dot{W}k = (30 - 2\,\overline{V})(\overline{V}) = 30\overline{V} - 2\overline{V}^2$$

and the maximum power will occur when the first derivative of power with respect to velocity is set equal to zero:

$$\frac{dWk}{dt} = 0 = \frac{d}{dt}[30\overline{V} - 2\overline{V}^2] = 30 - 4\overline{V}$$

Solving for the velocity gives

Answer
$$\overline{V} = \frac{30}{4} = 7.5\ \text{m/s}$$

CALCULUS FOR CLARITY 3–6, continued

The power required at this velocity is

$$\dot{W}k = [30 - (2)(7.5)](7.5) = 112.5 \text{ kW}$$

Work can be determined from the integration of power over a time interval:

$$Wk = \int \dot{W}k \, dt \qquad \textbf{(3–25)}$$

If the power is constant over the interval equation, then equation (3–22) results.

Previously, we found that the shaft work done by a rotating machine or cyclic device was given by equation (3–7):

$$\delta Wk_{\text{shaft}} = \text{T}\,\delta\theta$$

Substituting this relationship into the definition of power, we find that the power transmitted through a rotating shaft is

$$\dot{W}k_{\text{shaft}} = \text{T}\frac{\delta\theta}{\delta t}$$

But $\delta\theta/\delta t$ is the angular speed or velocity of a rotating shaft in radians per time unit. If we say that N is the angular speed in revolutions per minute (rpm), the angular speed is

$$\frac{\delta\theta}{\delta t} = 2\pi N \quad (\text{rad/min})$$

or

$$\frac{\delta\theta}{\delta t} = \frac{2\pi N}{60} \quad (\text{rad/s})$$

so that

$$\dot{W}k_{\text{shaft}} = \frac{2\pi}{60}\text{T}N \qquad \textbf{(3–26)}$$

In this equation, the torque, T, is in N · m or ft · lbf units and N is in revolutions per minute (rpm).

EXAMPLE 3–10 A large gas turbine develops 2000 kW at 8000 rpm. Determine the torque developed by the engine in kN · m.

Solution We can make use of equation (3–26), where power is known and the angular speed in rpms, N, is also known. Thus, rearranging this equation, we find the torque to be

$$\text{T} = \frac{60}{2\pi}\frac{\dot{W}k_{\text{shaft}}}{N} = \left(\frac{30}{\pi}\right)\left(\frac{2000 \text{ kW}}{8000 \text{ rpm}}\right)$$

Answer $$= 2.387 \text{ kN} \cdot \text{m}$$

EXAMPLE 3–11 A small air compressor requires a torque of 24 in · lbf to rotate its power shaft. If the air compressor is to run at 1800 rpm, determine the power required in horsepower.

Solution From equation (3–26), the shaft power is

$$\dot{W}k_{\text{shaft}} = (24 \text{ in} \cdot \text{lbf})\left(\frac{2\pi}{60}\right)(1800 \text{ rpm})\left(\frac{1}{12} \text{ ft/in}\right)$$

Answer $$= 376.99 \text{ ft} \cdot \text{lbf/s} = 0.685 \text{ hp}$$

3–3 HEAT

Heat is a word that probably has been more mistreated in technological language than any other single word. Following the manner of defining work in section 3–1, we will define heat:

Heat: ***Energy in transition across the boundary of a system that cannot be identified with a mechanical force acting through a distance.***

Heat occurs in a process when there is some temperature difference between the system and its surroundings. The direction of energy transition is always toward the area of lesser temperature. Heat will leave a system if it is hotter than its surroundings; if it is cooler, heat will enter the system. This energy transition will continue in the same direction until the system and its surroundings are thermally insulated from each other or until thermal equilibrium is reached.

Heat will be identified by the symbol Q, and the heat per unit of mass by q. In the old metric system of units, the term *calorie* was used to describe heat. This unit is defined as follows:

1 Calorie: ***The amount of heat required to raise the temperature of 1 gram of water 1°C when the water is at 4°C.***

Frequently, the kilocalorie, equal to 1000 calories, is used and called the "large calorie." The kilocalorie is often used to describe the energy associated with food. Thus one speaks of consuming "so many" calories, that is, kilocalories.

The calorie is related to the customary unit of energy, the joule, by the conversion listed on the front inside cover: 4.1868 J = 1 calorie. Since the SI system does not use the calorie as a proper unit, in this book we use only joules or kilojoules to describe heat.

In the English system, the common unit used to describe heat is the British thermal unit (Btu), given by the following definition:

1 Btu: ***The amount of heat required to raise the temperature of 1 lbm of water 1°F when the water is at 39°F.***

The rate at which heat flows from hot to cold bodies is called **heat transfer** and is defined as

$$\dot{Q} = \frac{\delta Q}{\delta t} \tag{3–27}$$

where δQ is a small amount of heat flowing during a small time period δt. The units of heat transfer are energy per unit of time: kJ/s (kW), Btu/s, or Btu/h. Heat transfer is usually considered to occur through one of three distinct modes or methods—conduction, convection, or radiation—and in later portions of this book we will consider these modes.

3–4 REVERSIBILITY

In thermodynamics, it becomes convenient to consider processes as being reversible. What this idea means is that the process may be reversed or allowed to proceed backward from the direction of the actual situation—it is reversible. When this is done, the work and heat of the process must be reversed, and the direction of motion of the system, if it involves motion, must be reversed. It happens that any real process is not reversible, and we should consider why this is so and what is ignored when we assume that a process is reversible. Let us consider a piston-cylinder with gas pushing against the piston, as in section 3–1. The work done by the gas is given by the expression $\sum p\, \delta V$ or the area under a curve in a p–V diagram. If the work is completely transferred through the piston-cylinder device and is used to compress or alter the same gas to bring it back to its original state, this would be

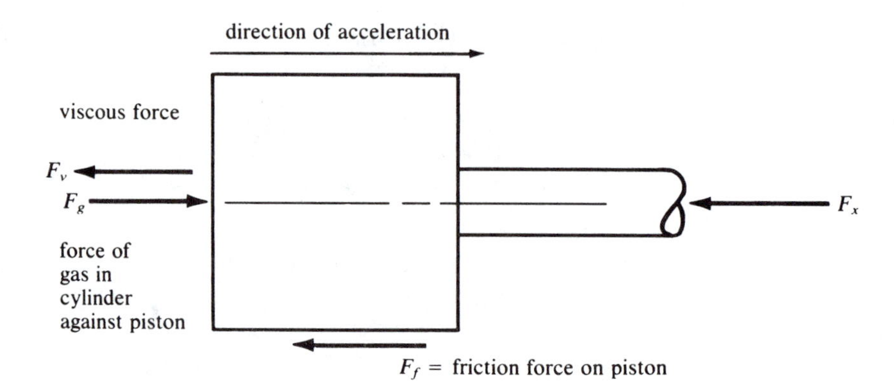

FIGURE 3–8 Forces acting on massless piston in actual piston-cylinder work process with friction and acceleration

called **reversible work**. But work is never completely reversible, and some of the causes for irreversibilities can be considered by looking at the forces acting on a piston. The gas pushing the piston applies a force, F_g, as shown in figure 3–8. Here we shall assume that the piston is going to accelerate or move to the right. The resisting forces are: F_x, an **external force**; F_f, a **friction force**; and F_v, a **viscous force**. The external force, F_x, represents the force that could be used as work, and if F_x were equal to F_g, the process would be reversible, at least as far as the piston-cylinder and gas (the system) would be concerned. But we write for the external force

$$F_x = F_g - F_f - F_v$$

Notice here that the force of the gas needs to overcome friction in order to move the piston. Thus, the friction force F_f is indicated as opposing the gas force. Also, the piston has a mass and will exhibit an inertial force ($F = ma$) opposing its change in motion. Even if the piston is considered to have no mass (and therefore no inertial force), the gas itself does have mass, and it also has viscosity. These two effects contribute to the viscous force in the following way. The gas will tend not to want to expand and follow directly with the piston if there is a finite change in motion involved—it has its own inertia, which detracts from the gas pressure. In addition, the viscosity of the gas tends to act as a resistance to prevent it from moving outward with the piston—an internal friction, so to speak. As the piston moves out (or in), the viscosity acts to dissipate energy as friction dissipates the kinetic energy of the gas as its motion is changing. This is sometimes called **viscous dissipation** and is always present when a real gas or liquid moves or is stirred.

The work transmitted outside the piston-cylinder from the gas is then just

$$Wk_x = Wk_{cs} - \text{friction work} - \text{viscous work} \tag{3–28}$$

where now the friction work is associated with the friction force, and viscous work with the viscous dissipative work. If the process is now reversed so that the gas is being worked on instead of doing work, the last two terms of equation (3–28) are reversed because they all act against motion as shown in figure 3–8, and these effects mean that less of the external work is available for compressing the gas. We say that the friction work and the viscous dissipative work are irreversible work (although notice that they are the terms that have been reversed in the analysis; that is, they always oppose the input effort). The only way to have a reversible process in which no irreversible work is present is to have the process proceed very slowly, with no friction, and with no viscous effects. All of these methods are impractical or impossible (such as eliminating friction and viscous effects), but serve to show the directions one may take to reduce irreversibilities (and therefore reduce inefficiencies in processes).

Another cause of irreversibilities in a process is the transfer of heat through a finite temperature difference with no work obtained from the heat transfer. Of course, we have noticed that heat is the flow of energy due to a temperature difference and that the flow is always from hot to cold. There have been no observed cases where heat flows from cold to hot without external input. Later we will see that refrigeration cycles, or refrigerators, "pump" heat from cold to hot but only by requiring input power and then only through a cyclic device. Therefore, unless one obtains a sufficient amount of work from a heat transfer process from hot to cold, an actual heat transfer, or a spontaneous heat transfer, is irreversible. The only way to say that heat transfer or heat occurs in a reversible manner is to say that it occurs over an infinitesimal (very, very small) temperature difference, or no temperature difference, and then (as it turns out) it will require an infinitely long time to transfer any actual heat. We will consider reversible heat to be that which occurs over no temperature difference between the system and its surroundings. Even then, if the inside of the system is at a different temperature than its boundary, the process will be irreversible if heat is transferred internally. Clearly, reversible heat or heat transfer requires that the system and its surroundings be at the same temperature, and if the system temperature changes during a process, the surroundings must change correspondingly.

Sometimes the term **internally reversible** will be used to describe a system that does not have viscosity or internal temperature differences which heat flows down. Such a description can be used under certain cases for perfect gas, which we consider later.

Some other causes for irreversibility that you will encounter in actual processes are

Electric resistance
Hysteresis effects in magnets and electric motors
Shock waves in air or other fluids
Inelastic deformation of solids (deformation beyond the yield strength)
Internal damping (like a shock absorber)
Combustion of gases and other spontaneous chemical reactions
Mixing of different substances
Mixing of the same two substances when both are first at different temperatures or pressures
Osmosis
Flow of a viscous fluid or gas along a solid surface

We have discussed the following causes for irreversibilities:

Friction between two solid surfaces
Expansion or compression of a fluid or gas at a finite rate
Spontaneous heat transfer

In this book, many of the processes and problems will be assumed to be reversible because of the difficulty of determining the irreversible work terms. In fact, the major problems of thermodynamics today involve the analytic determination of irreversibilities, and these are beyond the scope of this book.

Let us now consider two examples of irreversible work.

EXAMPLE 3–12 A sandpaper block is rubbed on a walnut wood surface. If the coefficient of friction (friction force/normal force) between sandpaper and wood is 0.2, determine the work done in rubbing the wood a distance of 300 m (back and forth) with a downward force (or normal force) of 40 N. (See figure 3–9.)

FIGURE 3–9

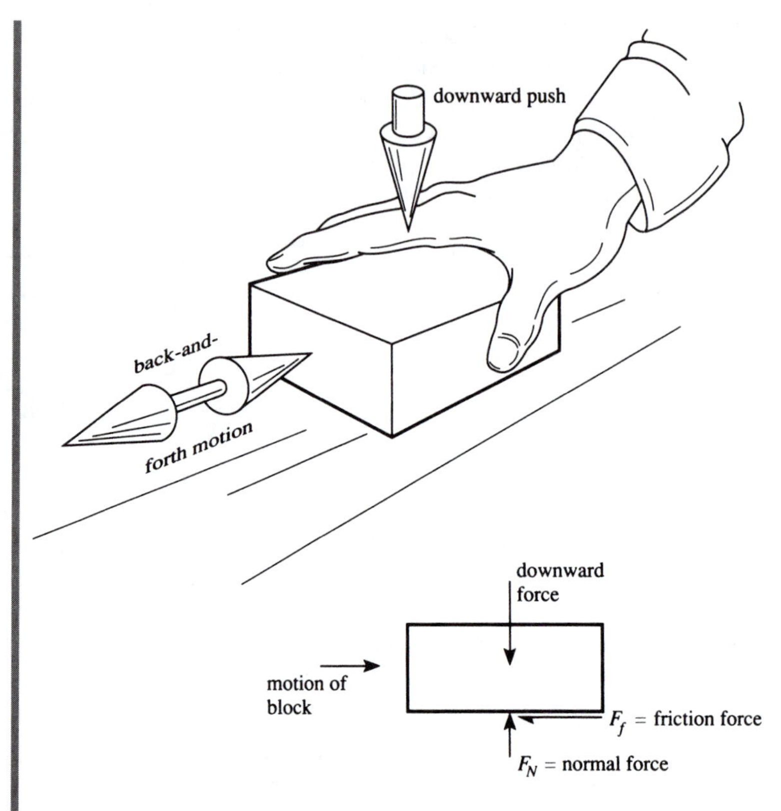

Solution

From mechanics, the coefficient of friction is given as the ratio of the friction force to the **normal force**. Therefore,

$$0.2 = \frac{F_f}{F_N}$$

and the friction force is

$$F_f = 0.2(F_N)$$
$$= (0.2)(40\ \text{N}) = 8\ \text{N}$$

The work done against friction, or the irreversible work, is just the friction force times the distance through which that force acts, 300 m. Then

$$Wk_{\text{friction}} = (F_f)(\text{distance})$$
$$= (8\ \text{N})(300\ \text{m}) = 2400\ \text{N}\cdot\text{m}$$

Answer

$$= 2.4\ \text{kJ}$$

EXAMPLE 3–13

An electric mixer is used to combine the ingredients for making a cake. The mixer has two beaters, each of which rotates at 100 rpm and has a torque of 15 N · cm due to the viscous resistance of mixing. Determine the irreversible power consumed by the mixer and the irreversible work consumed in mixing for 4 min.

Solution

A sketch of the mixer is shown in figure 3–10. As this figure indicates, the work is a shaft work done by two rotating shafts through a torque or moment. Using equation (3–26), we can compute the shaft power; and that is also the irreversible power. It is power that cannot

FIGURE 3–10 Viscous dissipation in a mixer

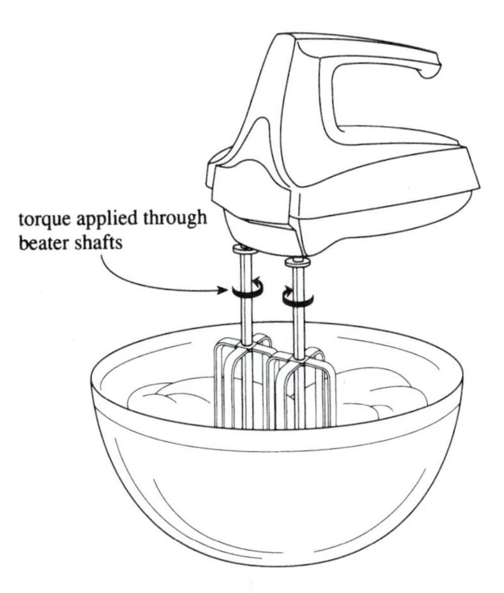

be taken back out and is completely dissipated through viscous dissipation of the cake ingredients. Then

$$\dot{W}k_{\text{shaft}} = (15\text{ N}\cdot\text{cm})\left(\frac{1\quad\text{m}}{100\text{ cm}}\right)\left(\frac{2\pi}{60}\right)(100\text{ rpm})(2\text{ beaters})$$

Answer

$$= 3.14159\text{ J/s} = 0.00314159\text{ kW}$$

Solution

The irreversible work, or shaft work for this example, is just that from equation (3–22):

$$\begin{aligned} Wk_{\text{shaft}} &= (\dot{W}k_{\text{shaft}})(\Delta t) \\ &= (0.00314159\text{ kJ/s})(4\text{ min})(60\text{ s/min}) \\ &= 0.754\text{ kJ} \end{aligned}$$

Answer

3–5 THE EQUIVALENCE OF WORK AND HEAT

Work and heat have been defined as mutually exclusive phenomena; that is, if the energy transfer is heat, then it cannot be work, and vice versa. But, heat and work are both energies in transition (or energy being transferred), so work could ultimately affect a system exactly as if the process had involved heat instead of work. The reverse of this statement is not always true, as the second law of thermodynamics will later demonstrate; heat cannot always affect a system exactly like work. There is, however, an equivalence between the common unit of heat, the joule (J), and work, the newton-meter (N · m), or

$$1\text{ J} = 1\text{ N}\cdot\text{m}$$

In the English system, the common unit for describing heat, the British thermal unit (Btu), and the unit for work, the foot-pound (ft · lbf), are related by the conversion

$$778.16\text{ ft}\cdot\text{lbf} = 1\text{ Btu}$$

or

$$778.16\text{ ft}\cdot\text{lbf/Btu} = 1$$

Some authors treat the conversion factor 778.16 ft · lbf/Btu as an algebraic term and assign to it the letter *J*. In this book, we refrain from this practice and consider the conversion from common heat units to common work units as equivalent to converting from feet to inches or any other unit conversion.

Note that when the conversion factor above is recalled, we can place heat, work, and energy all in the same units of British thermal units or foot-pounds, as we choose. Let us look at an example of this conversion.

EXAMPLE 3–14 The internal energy of 3 lbm of air is 60 Btu. How much energy is this in ft · lbf?

Solution This is strictly a unit conversion, so

$$U = 60 \text{ Btu} \times 778 \text{ ft} \cdot \text{lbf/Btu}$$

Answer

$$= 46{,}680 \text{ ft} \cdot \text{lbf}$$

Obviously, 1 Btu represents a much greater *amount* of energy than 1 ft · lbf, and it is for this reason that the equivalence of the units of heat and work was not readily accepted by the scientific community when first proposed by James Joule in 1842.

3–6 TYPES OF SYSTEMS

When the concept of the system was introduced in chapter 2, it was emphasized that the identification of the system was a first step in the thermodynamic method of solving real problems. Here we classify the system into one of three types, *open*, *closed*, or *isolated*, indicating that the *second step* in solving a problem is determining what type the identified system is. So that the learner can make the distinction, we identify the three types of systems:

Open system: ***A system whose boundaries allow for mass transfer, heat transfer, and work. That is, the amount of mass and energy in an open system can change.***

Closed system: ***A system whose boundaries allow for heat transfer and work, but not mass transfer. That is, the amount of mass of a closed system always remains the same, but the amount of energy can change.***

Isolated system: ***A system whose boundaries prevent mass transfer, heat transfer, and work. That is, the amount of mass and energy of an isolated system remains the same.***

Certain peculiarities exist for each of the three systems and will be notated with appropriate subscripts (such as Wk_{cs} for work of a closed system) when needed.

3–7 THE FORMS OF ENERGY

During the past two chapters, *energy* has been a term that entered the discussion frequently. We will see in chapter 4 that conserving energy is the major task of thermodynamics, called the *conservation of energy* or the *first law of thermodynamics*. It involves, naturally, that ubiquitous property of the system, energy; in fact, other properties of the system (such as mass, pressure, volume, and temperature) will be measured so that the amount of energy can be subsequently determined. We know that in open or closed systems, the amount of energy can be increased or decreased; thus, energy can be *static* (stationary) or *dynamic* (moving from one place to another).

Table 3–2 is a concise list of the forms in which energy exists and makes clear the distinction between static and dynamic energy. In studying the table, note that work and heat exist only when energy is dynamic, that is, only when energy is in a state of motion or transition. (See section 2–5.) As soon as energy becomes static, it changes form, and the forms of heat and work cease to exist. In chapter 4, we will see the importance of this change.

TABLE 3–2 Forms of energy

Form	Type	Condition
Static energy	Potential Kinetic Internal (thermal) Electromagnetic Strain Chemical	System property
Dynamic energy (i.e., energy in transition)	Work Heat Heat transfer	Not a system property Dependent on process Occurs only during a process

3–8 SUMMARY

In this chapter, work was defined as a force acting through a distance. In particular, a small amount of work was defined as

$$\delta Wk = F\,\delta x \tag{3–1}$$

and a finite amount of work as

$$Wk = \sum F\,\delta x \tag{3–2}$$

For the special case when force is constant throughout a process, this equation is

$$Wk = F\,\Delta x \tag{3–5}$$

Other forms of this equation were given for work done through springs, namely,

$$Wk = \left(\frac{1}{2}\right)(b)(x_2^2 - x_1^2) = Wk_s \tag{3–6}$$

and for rotating shafts at constant torque T:

$$Wk_{\text{shaft}} = \sum \mathrm{T}\,\delta\theta \tag{3–7}$$

For piston-cylinders or any system whose volume changes, the boundary work is

$$Wk_{cs} = \sum p\,\delta V \tag{3–9}$$

which, per unit mass, is

$$wk_{cs} = \sum p\,\delta v \tag{3–10}$$

Special forms of the boundary work for unique processes are, for cases when pressure is constant,

$$Wk_{cs} = p\Delta V \tag{3–13}$$

for polytropic processes, when $pV^n = C$,

$$Wk_{cs} = \frac{1}{1-n}(p_2V_2 - p_1V_1) \tag{3–17}$$

and, if $n = 1$, so that $pV = C$,

$$Wk_{cs} = C \ln \frac{V_2}{V_1} \tag{3–16}$$

Power was defined as the time rate of doing work and, if the power is constant over a period Δt; can be written as

$$\dot{W}k = \frac{Wk}{\Delta t} \tag{3–21}$$

and

$$Wk = \dot{W}k\,\Delta t \tag{3–22}$$

The power can also be computed from the equation

$$\dot{W}k = F\overline{V} \tag{3–23}$$

and for rotating shafts

$$\dot{Wk}_{\text{shaft}} = \frac{2\pi}{60}\text{T}N \qquad \textbf{(3–26)}$$

Heat was defined as the energy crossing the boundary of a system that could not be described through a force acting through a distance. Heat transfer was defined as the time rate of heat crossing the system boundary.

Heat was visualized as flowing from hot to cold, but we also discussed that reversible heat can flow only through a zero-temperature difference. Thus, an actual heat transfer from hot to cold is an irreversible process, unless reversible work is obtained from the process simultaneously. Also, a discussion was given about the other causes of irreversibilities in processes. Friction and viscous dissipation were given as two examples of irreversible effects and irreversible work. Three types of systems were defined:

Closed systems, having the same mass but able to have heat and work

Open systems, having mass flow, heat, and work capabilities

Isolated systems, always having the same mass and energy

The forms of energy were discussed, and an equivalence between heat and work was made.

DISCUSSION QUESTIONS

Section 3–1

3–1 What is meant by *work*?

3–2 Why is the work predicted by equation 3–9 *boundary work*?

Section 3–2

3–3 What is meant by the term *power*?

3–4 Why are the units of energy sometimes written as *kilowatt-hours*?

Section 3–3

3–5 What is meant by the term *heat*?

3–6 What is the *calorie*?

3–7 What is *heat transfer*?

Section 3–4

3–8 Why do friction and viscosity make work irreversible?

Section 3–5

3–9 What is "equivalent" about *heat* and *work*?

Section 3–6

3–10 What are the three of systems?

Section 3–7

3–11 Why are heat and work not properties of a system?

PRACTICE PROBLEMS

Section 3–1

Problems that use SI units are indicated by an (M) under the problem number; those that use English units are indicated by an (E). Mixed unit problems are indicated by a (C). Problems marked with an asterisk (*) after the problem number are often more difficult and best analyzed with the use of calculus.

3–1 (M) A force of 20 N is required to slide a 30-kg box horizontally across a platform 20 m long. What work is required?

3–2 (M) In problem 3–1, what work is required to lift the 30-kg box vertically 20 m?

3–3 (E) A 30-lbm container is picked up and put on a shelf 3 ft above the floor. If local gravitational acceleration is 31.8 ft/s^2, determine the work done in lifting the container from the floor to the shelf.

3–4 (E) A sled is pushed up a 70-ft incline of 45° which is frictionless. If the sled weighs 80 lbf, determine the work done.

3–5 (E) A compression spring is used to close valves in an auto engine. If the spring constant is 100 lbf/in, determine the force required to deflect the spring (shorten it) $^3/_8$ in from its free length, and find the work done.

3–6 (M) A 15-cm-long spring having a modulus of 180 N/cm is deflected an amount that requires 1.8 J of work. Determine the deflection of the spring.

3–7 (E) A spring deflected 1 in from its free length is deflected 1 in farther. If the modulus of the spring is 140 lbf/in, what work is required to deflect the spring the second inch?

3–8 (M) An extension spring has a constant of 6.4 kN/m and is given a pretension by extending it 2 cm. It is then extended 8 cm more. Determine the force at its first and second states and the work required to extend the spring during this change from the first to the second states.

3–9 (E) A 3000-lbm automobile accelerated at a constant rate from rest to 60 mph in 10 s on a flat, straight highway.

(a) How much work was done by the auto's engine in this process? (*Hint:* force = mass × acceleration; distance = $^1/_2 at^2$.)

(b) If the acceleration to 60 mph required 15 s, what was the engine's work output?

3–10 (M) Water is usually considered to be incompressible in its liquid phase. How much work is required to compress 3 kg of water from 100 kPa to 500 kPa in a piston-cylinder?

3–11 (E) If a brass rod is considered to be incompressible, determine the amount of boundary work done in stretching the rod 1 in.

3–12 (M) For certain reversible processes of a closed system, the pressure–volume relationships are as given by the solid lines in figure 3–11. Find the work for these processes.

3–13 (M) In figure 3–11, the dashed lines represent a p–V relationship for a process that involves work. Find the work done.

3–14 (M) A hydraulic cylinder is extended by means of hydraulic oil at 14,000 kPa. The cylinder bore (diameter) is 8 cm and its stroke is 20 cm. Determine the work done by the piston, neglecting friction between the piston and the cylinder.

3–15 (M) A piston-cylinder retracts due to a partial vacuum, which allows the atmosphere to push the piston into the cylinder. If the atmospheric pressure is 100 kPa and the volume of the cylinder decreases by 3 m^3, determine the work done by the atmosphere on the piston.

3–16 (E) For the balloon example 3–6, determine the work done on the atmosphere by the balloon if the atmospheric pressure is 14.7 psia.

3–17 (E) During the intake stroke of an auto engine, 6 in^3 of air is taken in at 14.6 psia. Determine the work done by (or on) the engine during this process.

3–18 (M) A pump shaft requires a torque of 75 N · m to overcome internal friction. Determine the work done in rotating the pump shaft through 100 revolutions. (*Note:* 1 rev = 2π radians.)

3–19 (E) A torque wrench is used to tighten bolts. If 120 ft · lbf is applied to the wrench through 25°, determine the work done. (*Note:* $2\pi/360$ radians = 1°.)

3–20* (M) For a process where $pV = C$, $p_1 = 200$ kPa, $V_1 = 0.5\ m^3$, and $p_2 = 1600$ kPa, determine the final volume. If the work can be described by $\sum p\,\delta V$, determine its value for this process from 200 to 1600 kPa.

3–21* A piston-cylinder device contains a gas that expands by the relationship $pV = C$. If $p_1 = 500$ psia, $V_1 = 1.4$ in^3, and $V_2 = 15$ in^3, determine the final pressure and the boundary work.

FIGURE 3–11

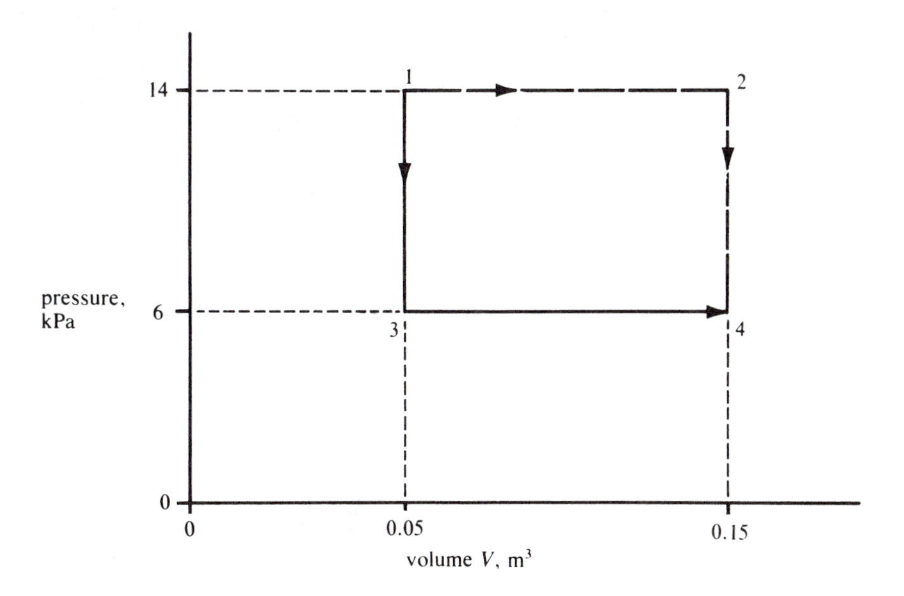

3–22* **(M)** A polytropic process is defined by the relationship $pV^n = C$. For a gas expanding in a piston-cylinder in a polytropic manner with $n = 1.35$ and going from $p_1 = 6$ MPa, $V_1 = 0.02$ to $1.0\ \text{m}^3$, determine the final pressure. Then determine the boundary work.

3–23* **(E)** A certain gas in a piston-cylinder is compressed from 14.6 psia, 0.33 ft³/lbm, to 120 psia, 0.057 ft³/lbm. If the compression process is to be approximated by the polytropic one, where $pv^n = C$, determine the exponent, n, and the boundary work of the gas.

3–24 **(M)** For the process displayed on the p–V diagram shown in figure 3–12, approximate the work between states 1 and 2.

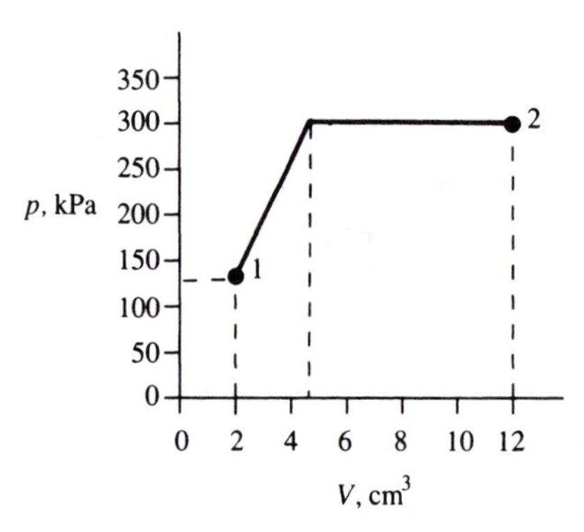

FIGURE 3–12

3–25 **(E)** For the process given by the p–V diagram shown in figure 3–13, determine the work between states 1 and 2.

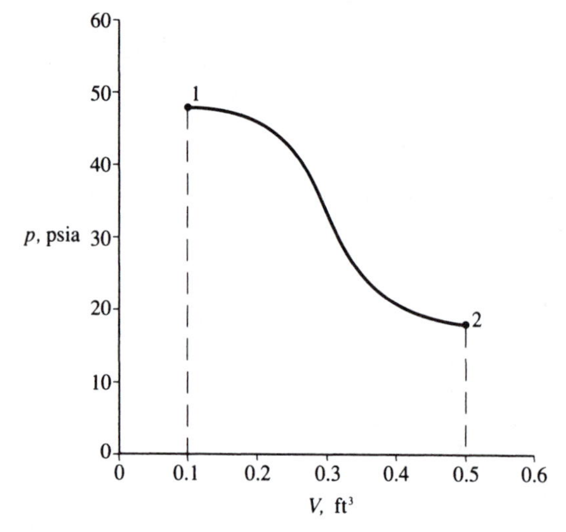

FIGURE 3–13

3–26 **(M)** For the process indicated by the force–displacement $(F - x)$ diagram shown in figure 3–14, determine the work between the displacements from x_1 and x_2.

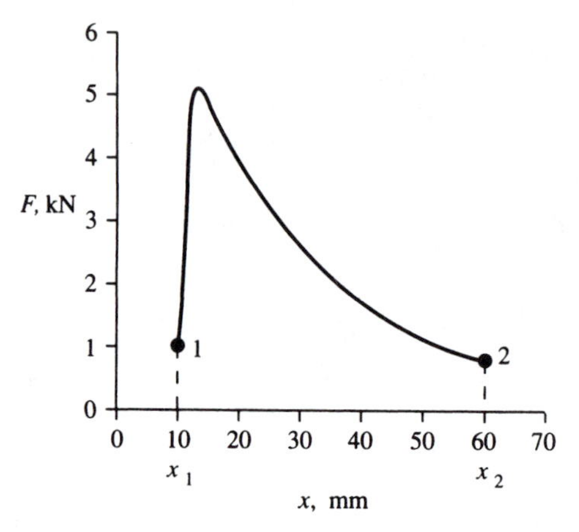

FIGURE 3–14

3–27 **(E)** For the process indicated on the p–V diagram shown in figure 3–15, determine the work between the change in volume, $V_2 - V_1$.

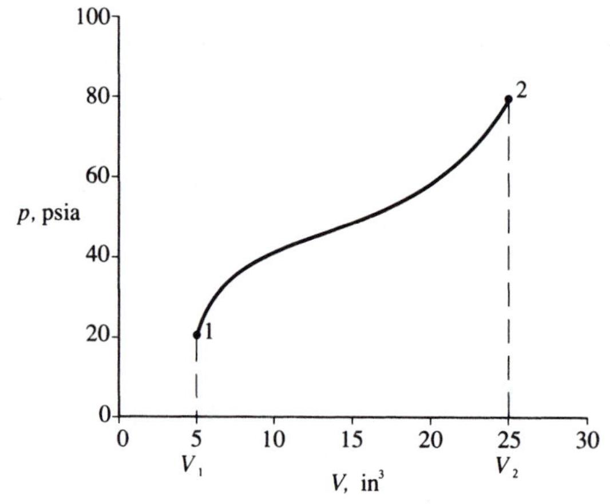

FIGURE 3–15

3–28 **(M)** The torque varies with angular position on a large turntable as shown in figure 3–16. Determine the work done in rotating the turntable through one revolution.

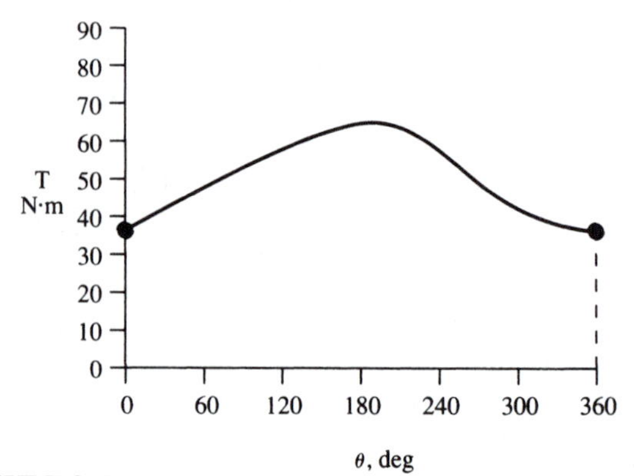

FIGURE 3–16

3–29 (E) For the torque–angular position relationship shown in figure 3–17, determine the work involved in rotating 120°.

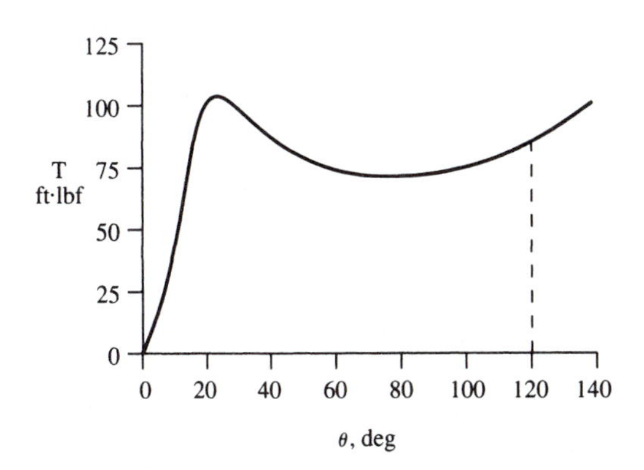

FIGURE 3–17

3–30* (E) A nonlinear compression spring has a force–displacement relationship that is described by the equation

$$F = bx^2$$

where the spring constant $b = 50\ \text{lbf/in}^2$. Determine the work done by the spring as it expands from a compression of $x_1 = 3$ in to $x_2 = 1$ in.

3–31* (M) For a certain process, it is found that $pV^{1.5} = 18.5$, where p is in kPa and V is in m^3. Determine the work done if the volume increases from 2 to 3 m^3.

3–32* (M) A gaseous system is in a container such that the system pressure is described by the equation

$$p = 3V^3 - 0.3V\ (\text{bars})$$

for V, the volume, greater than 1.5 m^3 and less than 3 m^3. Determine the boundary or closed-system work of the gaseous system as the volume decreases from 2.6 to 1.6 m^3. Notice that the pressure is given in bars.

3–33* (M) The start-up torque supplied by a certain electric motor is given by the relationships

$$T = 10\theta - 0.05\theta^2\ (\text{N}\cdot\text{m}) \qquad \text{if } 0 \le \theta \le 100 \text{ rev}$$
$$T = 500\ \text{N}\cdot\text{m} \qquad \text{if } \theta > 100 \text{ rev}$$

where θ is in revolutions. Determine the start-up work done by the electric motor during the first 200 revolutions of the motor.

3–34* (M) It is essential to understand Newton's law of gravitation, as expressed by equation 2–1, in the launching of rockets for projecting satellites into orbits around the Earth or for space travel. Determine the work required to launch a 1-kg mass into space if friction and wind resistance is neglected. Notice that the weight of the 1-kg mass changes, becoming small as the mass moves farther from the Earth. Also, the launch occurs on the Earth's surface where the radius is just the Earth's radius and the final distance or radius approaches infinity (∞) as the mass is in space. Then determine the distance out from the Earth's surface where 99.5% of the work of launching the mass into space has been accomplished.

Section 3–2

3–35 (M) Determine the average power produced in 7 s if the work is 750 J.

3–36 (E) Determine the average power consumed if 80,000 ft · lbf is used in 2.3 s.

3–37 (M) There is 380 W of power furnished for 2 h. Determine the work done.

3–38 (E) A total of 125 hp is produced by an engine which runs at this performance level for 30 min. Determine the work done by the engine.

3–39 (M) A force of 628 N is applied to pull a wagon at 20 m/s. Determine the power required.

3–40 (E) A rocket is launched through a thrust force of 7,000,000 lbf. What power is produced when the rocket's velocity is 100 ft/s?

3–41 (M) Cardboard boxes move along a conveyor at 1.5 m/s. If the air resistance of each box is 2 N, determine the power required to keep the conveyor moving.

3–42 (E) Bales of hay of 40 lbf move up a conveyor set at a 30° angle to the ground. If the hay bales are moving at 1.5 ft/s, determine the power required to move each bale, neglecting any wind or air resistance.

3–43 (E) An electric motor is running at 1200 rpm and can produce $^1/_2$ hp. Determine the torque that the motor could apply.

3–44 (M) A two-stage air compressor runs at 600 rpm and requires a torque of 70 N · m to drive it. What is the power requirement?

3–45 (M) A generator produces 160 kW at 1800 rpm. What torque is required to drive the generator?

3–46 (E) A 3.5-hp lawn mower motor runs at 3200 rpm when it produces its rated power. Determine the torque it produces at this condition.

3–47 (M) The daily electrical power usage in kilowatts (kW) for a particular building is shown in figure 3–18. Determine the total energy usage of the building for one day or 24 hours.

3–48* (M) A certain system is able to exert an external force F according to the relationship

$$F = 15{,}000 - 500\overline{V}\ (\text{N}) \quad \text{for } \overline{V} < 30\ \text{m/s}$$

where the velocity $\overline{V}$ is in m/s. Determine the maximum power that can be supplied and the velocity at which this will occur.

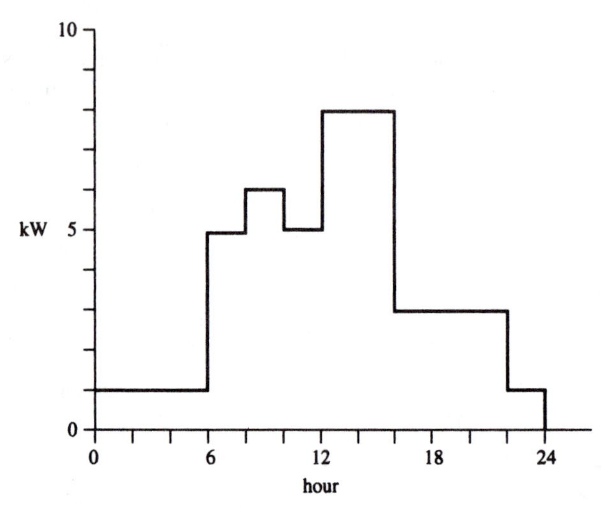

FIGURE 3–18 Daily electrical power usage for a particular building

3–49* **(E)** A generator system has a start-up torque that can be described by the equation

$$T = 5.15 - (2 \times 10^{-4})N^2 \text{ (ft·lbf)} \quad \text{for } N < 160 \text{ rad/s}$$

where N is in radian/s. Determine the maximum power required at start-up and the angular speed at which this occurs.

Section 3–3

3–50 **(E)** There is 7000 Btu of heat to be added to hot water. If the heat can be added at a rate of 80 Btu/s, determine the time required to do this.

3–51 **(M)** A window of a building loses 1.2 J/s of heat during a 24-h period. Determine the heat loss during this time in kJ.

3–52 **(E)** A solar collector receives 670 Btu/s of radiant heat. How long will it take to receive 20,000 Btu?

3–53 **(M)** Solar energy is received at a certain location on the Earth in the amount of 700 W/m^2. Determine the amount of heat (in kJ) that a 1-m^2 area will receive in 1 h.

3–54 **(E)** An office building is found to gain 200,000 Btu/h of heat on a hot summer day over the period from 6 A.M. to 6 P.M. The same building loses 20,000 Btu/h over the first 6 hours and the last 6 hours of the same day. What is the total heat gain or loss of the building for the day in Btu. Then determine the average heat gain or loss for the day in Btu/h.

Section 3–4

3–55 **(M)** A roller bearing needs 1.6 N·m of torque to rotate the inner race while holding the outer race fixed. Determine the amount of irreversible work done in rotating the bearing through 360°.

3–56 **(E)** A journal bearing is found to have a resistance of 7 in·oz when rotating. Determine the irreversible power used when the shaft being supported by the journal bearing is rotating at 800 rpm (see figure 3–19).

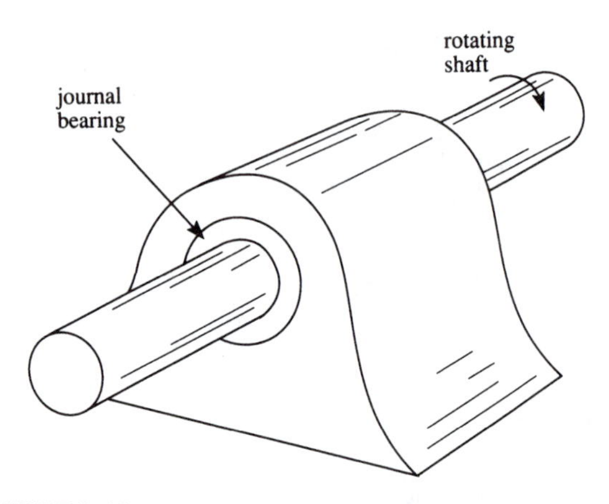

FIGURE 3–19

3–57 **(M)** A blender is used to make carrot juice. If a torque of 30 N·m is required at 720 rpm, determine the irreversible power of the blender.

3–58 **(E)** An auto tire must convey 32 hp during freeway driving at 55 mph. If slippage (and therefore tire wear) requires 2% of the power transmitted through the tires, determine the slippage force between tire and road.

3–59 **(E)** A tractor slippage in field work is equal to a 100-lbf loss in the drawbar pull (see figure 3–20) at 4.4 ft/s. Determine the power loss attributable to irreversible work of slippage.

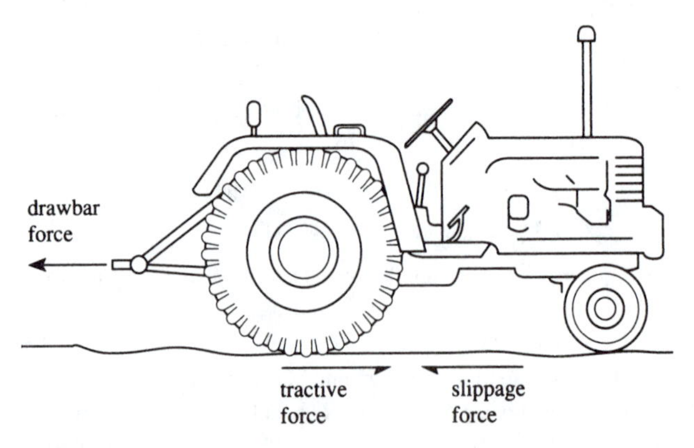

FIGURE 3–20

3–60 **(M)** A crawler tractor has a drawbar loss due to slippage of 30 kN (see figure 3–20) at 7 m/s. Determine the power loss due to slippage.

3–61 (E) Sandpaper has a coefficient of friction of 0.3 when used in a particular sanding operation. Determine the power lost as irreversible power when a force of 6 lbf is applied down against the surface to be sanded and the sanding is done at a speed of 5 ft/s.

3–62 (M) Worm gear friction is quoted as 0.2% of transmitted power at 10 kW, 1200 rpm. Determine the expected dynamic friction torque of the worm gear.

3–63 (E) A large circulating air fan is driven by a motor furnishing 150 hp. If the fan is 85% efficient, what is the reversible and irreversible power of the fan?

3–64 (M) A 4-newton (N) billiard ball at a particular location on a billiard table (no pockets) is struck so that its velocity is 0.4 m/s just as it starts to roll away from the cue. The ball is the only one on the table, the side boards are perfectly elastic, and the ball is found to roll a distance of 20 meters before coming to a stop in the center of the table. What is the rolling resistance of the ball/table in millinewtons (mN)?

Section 3–5

3–65 (C) Convert the following:

(a) 17 Btu/lbm to ft · lbf/lbm
(b) 3350 ft · lbf to British thermal units
(c) 2,000,000 in · oz to British thermal units
(d) 27.8 kJ to newton-meters
(e) 3000 MW to basic SI units

CONSERVATION OF MASS AND THE FIRST LAW OF THERMODYNAMICS

4

In this chapter, the conservation principles of mass and energy are introduced as the general vehicles for solving thermodynamics problems. The method of solving many thermodynamics problems proceeds with the application of these two principles after identifying the system, identifying the necessary equilibrium states by a listing of the properties, and recognizing the process or processes through an identification of the work and heat, as we saw in chapter 3. The conservation of mass is presented for the general system and for the steady-flow, steady-state conditions. Uniform flow is also introduced, with emphasis on the filling and emptying processes of open systems.

The conservation of energy, or first law of thermodynamics, is stated, and the equations representing this concept are formulated for the system, with particular emphasis placed on the closed system. The isolated system is discussed to illustrate the conversion of energy from one form to another when no work or heat is present. Flow work and enthalpy are introduced and used to formulate the equations of the first law of thermodynamics applied to the open system. Particular attention is given to the steady-flow energy equation for steady-state open systems. Some treatment is also given to the non-steady-state situations for filling and emptying tanks and other open systems where uniform flow occurs.

New Terms

E	Energy	$\dot{m}$	Mass flow rate
e	Specific energy	pV	Flow work
H	Enthalpy	$\dot{V}$	Volume flow rate
h	Specific enthalpy	Wk_{os}	Open-system work

4–1 CONSERVATION OF MASS

One of the most fundamental concepts of science is that mass is indestructible; that is, mass can be neither created nor destroyed.* This is the principle known as the **conservation of mass**, and for a closed or an isolated system, we write

$$\text{mass} = \text{constant} \tag{4–1}$$

If the system is open, so that mass can be transferred into or out of it, the statement of the conservation of mass is written

$$m_{\text{in}} - m_{\text{out}} = \Delta m_{\text{system}} \tag{4–2}$$

where m_{in} is the mass entering the system, m_{out} is the mass leaving the system, and Δm_{system} is the change in mass of the system. (See figure 4–1.) The term Δm_{system} is positive

* An exception to this principle is the theory of relativity, which relates mass and energy, or rest energy, of the mass by the equation $E = mc^2$, where c is the velocity of light. Thus, for certain processes such as nuclear reactions, mass and energy are conserved together but not individually.

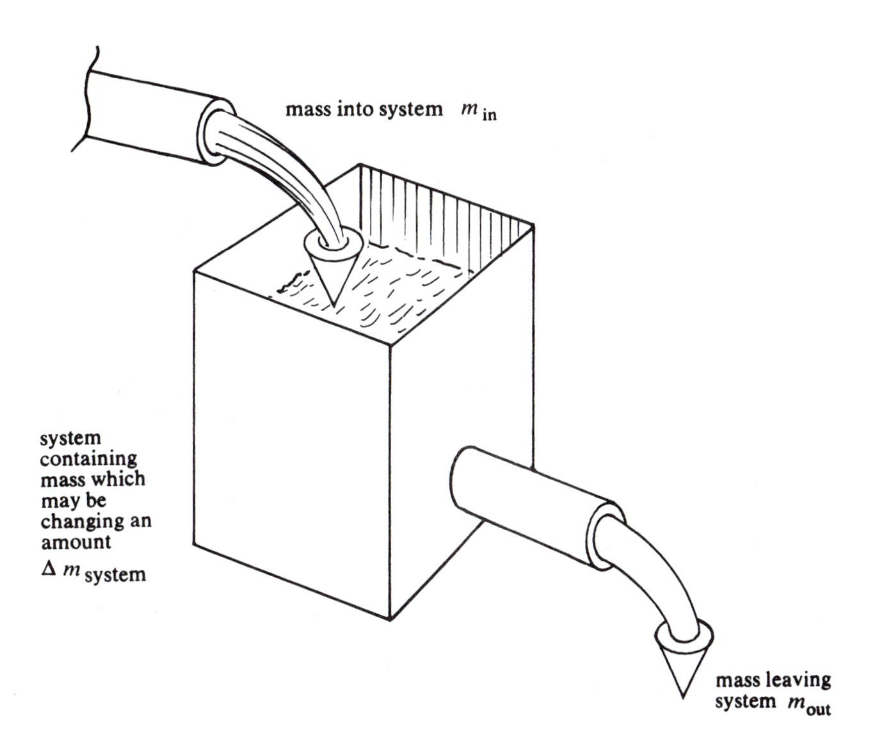

FIGURE 4–1 Conservation of mass

if the system is gaining mass and negative if it is losing mass. Equation (4–2) implies that all terms are related to a common period, Δt, which can be divided into each of the terms in the equation, giving a new form of the mass balance, written

$$\frac{m_{in}}{\Delta t} - \frac{m_{out}}{\Delta t} = \frac{\Delta m_{system}}{\Delta t} \tag{4–3}$$

Now, if the period is very small, we write it as δt, and the change in the system's mass may also be very small, written δm_{system}. If a very short period δt is considered, the amounts of mass entering and leaving the system are very small, and we then write these as δm_{in} and δm_{out}, so that equation (4–3) becomes

$$\frac{\delta m_{in}}{\delta t} - \frac{\delta m_{out}}{\delta t} = \frac{\delta m_{system}}{\delta t} \tag{4–4}$$

In this book, we write $\delta/\delta t$ by the shortcut notation of putting a dot above the term as we did for power and heat transfer in chapter 3. Thus, we write equation (4–4) as

$$\dot{m}_{in} - \dot{m}_{out} = \dot{m}_{system} \tag{4–5}$$

The two terms $\dot{m}_{in}$ and $\dot{m}_{out}$ are called the **mass flow rate into the system**, or inflows, and the **mass flow rate out of the system**, or outflows, respectively. The term $\dot{m}_{system}$ is the rate at which the system mass changes with respect to time. The mass flow rate is commonly found from the equation

$$\dot{m} = \rho A \overline{V} \tag{4–6}$$

where A is the cross-sectional area across which mass is moving with average velocity $\overline{V}$ and with density ρ.

CALCULUS FOR CLARITY 4–1

The mass flow equation (4–6) may be derived by noting that the mass flowing across an area A is the density ρ times its volume V. The volume is the area A times the distance through which the mass has traveled in a time interval δt as shown in figure 4–2. We then write that this mass m is

$$m = \rho V = \rho A\,\delta x \tag{4–7}$$

The average mass flow rate for the time interval δt is given by

$$\frac{m}{\delta t} = \frac{\rho A\,\delta x}{\delta t} \tag{4–8}$$

If we take δt very short, or in the limit as δt approaches zero, then this equation becomes

$$\lim_{\delta t \to 0}\left[\frac{m}{\delta t}\right] = \lim_{\delta t \to 0}\left[\frac{\rho A\,\delta x}{\delta t}\right]$$

but

$$\lim_{\delta t \to 0}\left[\frac{\rho A\,\delta x}{\delta t}\right] = \rho A \lim_{\delta t \to 0}\left[\frac{\delta x}{\delta t}\right]$$

so that

$$\lim_{\delta t \to 0}\left[\frac{m}{\delta t}\right] = \rho A \lim_{\delta t \to 0}\left[\frac{\delta x}{\delta t}\right] \tag{4–9}$$

The left side is the mass flow rate, $\dot{m}$, and the right side is $\rho A\bar{V}$ because $\text{Lim}\ \delta t \to 0[\delta x/\delta t] =$ velocity $(\bar{V})$. Thus, the mass flow rate is given by equation (4–6).

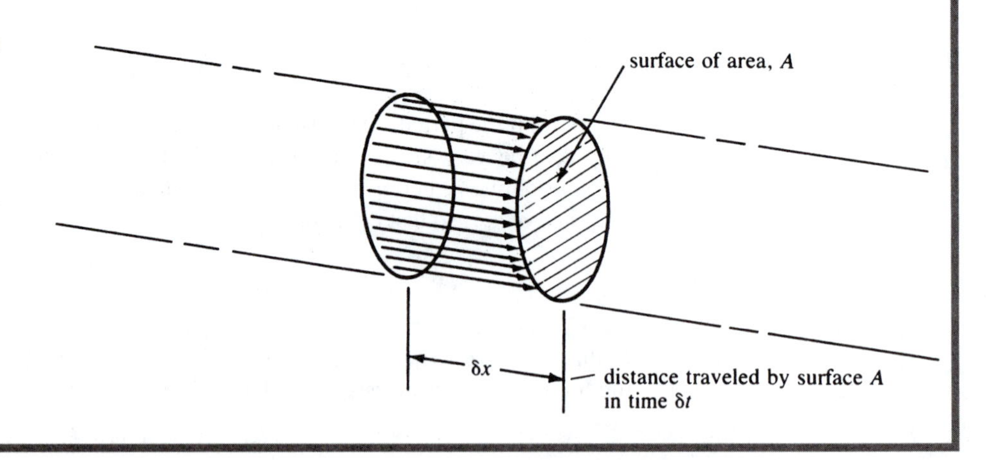

FIGURE 4–2 Illustration of mass flow

EXAMPLE 4–1 Kerosene is pumped into an aircraft fuel tank through a hose that has an inside diameter of 4 cm. If the velocity of the kerosene is 8 m/s through the hose, determine the mass flow rate. Assume that kerosene has a density of 800 kg/m^3.

Solution The mass flow rate is obtained from equation (4–6). The area of the hose is given by a circular section of 2-cm radius; thus,

$$A = \pi(2\ \text{cm})^2 = 12.6\ \text{cm}^2$$

or

$$A = 0.00126\ \text{m}^2$$

The mass flow rate is then

Answer

$$\dot{m} = (800\ \text{kg/m}^3)(0.00126\ \text{m}^2)(8\ \text{m/s}) = 8.06\ \text{kg/s}$$

EXAMPLE 4–2 Water flows from a 1-in-diameter faucet with a velocity of 8.7 ft/s. Determine the mass flow rate of the water leaving the faucet.

Solution The water is crossing the circular area of the faucet outlet, equal to π times the radius squared. Then the area A is calculated by

$$A = \pi\left(\frac{1}{2}\right)^2 \text{in}^2$$

or

$$A = 0.785\ \text{in}^2$$

The water is here assumed to be at 78°F, at which temperature the density is approximately 62.4 lbm/ft^3. We then determine the mass flow rate from equation (4–6):

$$\dot{m} = \rho A\overline{V}$$

Substituting values into this equation gives us

$$\dot{m} = (62.4\ \text{lbm/ft}^3)(0.785\ \text{in}^2)(8.7\ \text{ft/s})$$

To convert to consistent units, we must multiply by the factor 1/144 ft^2/in^2, so

$$\dot{m} = (62.4\ \text{lbm/ft}^3)(0.785\ \text{in}^2)(8.7\ \text{ft/s})\left(\frac{1\ \text{ft}^2}{144\ \text{in}^2}\right)$$

Thus,

Answer

$$\dot{m} = 2.96\ \text{lbm/s}$$

The volume flow rate, defined as the volume of material crossing an area per unit of time and written

$$\dot{V} = \frac{\delta V}{\delta V} \qquad \textbf{(4–10)}$$

is another term for describing flow rate. Since the volume δV can be described by $A\ \delta x$, we have, from equation (4–10),

$$\dot{V} = A\overline{V} \qquad \textbf{(4–11)}$$

The volume flow rate is described by units of cubic meters per second (m^3/s), cubic feet per minute (ft^3/min), gallons per hour (gal/h), gallons per minute (commonly written gpm), or any other compatible combination of volume per unit time.

EXAMPLE 4–3 Determine the volume flow rate for the kerosene in example 4–1.

Solution Using equation (4–11), we obtain

Answer

$$\dot{V} = A\overline{V} = (0.00126\ \text{m}^2)(8\ \text{m/s}) = 0.010\ \text{m}^3/\text{s}$$

EXAMPLE 4–4 Determine the volume flow rate of the water in example 4–2.

Solution We may use equation (4–11) to obtain the volume flow rate. Then

$$\dot{V} = (0.785\ \text{in}^2)(8.7\ \text{ft/s})\left(\frac{1\ \text{ft}^2}{144\ \text{in}^2}\right)$$

Answer
$$= 0.0474\ \text{ft}^3/\text{s}$$

Further, using the conversion of 7.48 gal = 1 ft^3, we find that

$$\dot{V} = (0.0474\ \text{ft}^3/\text{s})(7.48\ \text{gal/ft}^3)$$

Answer
$$= 0.355\ \text{gal/s}$$

Recall the preceding discussions of the conservation of mass as described by equation (4–1) or (4–5) for closed and open systems, respectively; there is no principle of conservation of weight or volume, and caution must be used whenever reference is made to volume flow rate of weight flow rate—a conversion to mass flow rate is a safe approach.

4–2 STEADY FLOW

As mass flows through a system, there is often no loss or gain of mass in the system itself. This tells us that, since $\dot{m}_{\text{system}} = 0$,

$$\dot{m}_{\text{in}} - \dot{m}_{\text{out}} = 0$$

or

$$\dot{m}_{\text{in}} = \dot{m}_{\text{out}} \quad \textbf{(4–12)}$$

which indicates that all the mass flows into the system must equal (exactly) the mass flows leaving the system. This condition is called **steady flow** or **steady state** and is frequently encountered in engineering and technological applications. Any engine producing power, refrigerator cooling foods, generator producing electric energy, or any device that is intended to perform for extended periods of time is in steady flow, or some of the components are in steady flow.

EXAMPLE 4–5 A nozzle is commonly used to change the velocity of liquids or gases, by changing the cross-sectional area of the flow line. (See section 10–4 for a more complete description of nozzles.) Suppose we have a nozzle with air passing through so that, within the nozzle, no loss or accumulation of air occurs. The air is entering the nozzle with a velocity of 24 m/s and a density of 1.28 kg/m^3. The density of the air leaving is 1.10 kg/m^3. The nozzle is circular in cross-sectional area and reduces evenly from an entrance diameter of 60 cm to an exit diameter of 30 cm. Determine the velocity of the air leaving the nozzle. (See figure 4–3.)

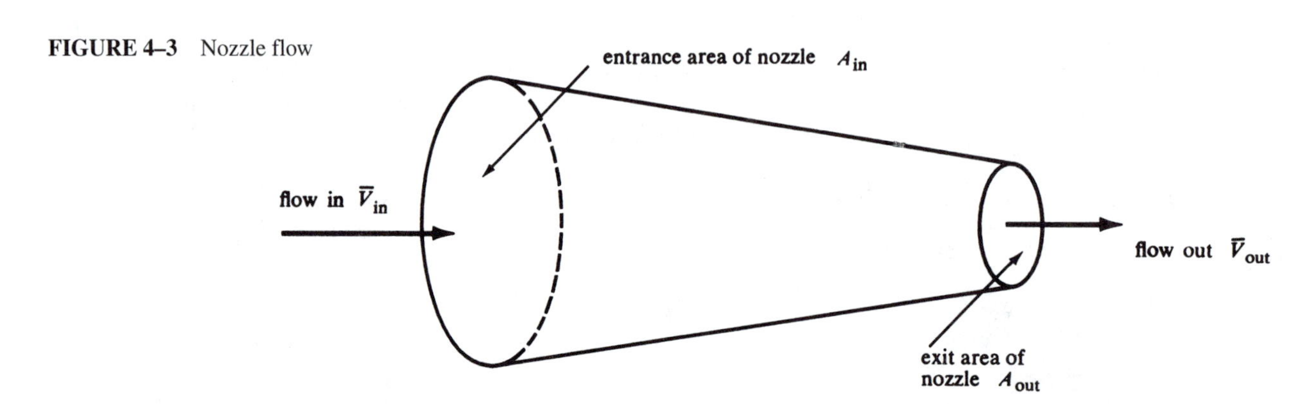

FIGURE 4–3 Nozzle flow

Solution

This problem is an example of steady flow; that is,

$$\dot{m}_{\text{system}} = 0$$

so that we may use equation (4–12):

$$\dot{m}_{\text{in}} - \dot{m}_{\text{out}} = 0$$

If we replace these two terms by using equation (4–6), we obtain

$$\rho_{\text{in}} A_{\text{in}} \overline{V}_{\text{in}} - \rho_{\text{out}} A_{\text{out}} \overline{V}_{\text{out}} = 0 \quad \textbf{(4–13)}$$

We can then substitute numbers into this equation, an important steady-flow relation that we will use often. Substituting values into equation (4–13) yields

$$(1.28\ \text{kg/m}^3)[(\pi)(0.3\ \text{m})^2](24\ \text{m/s}) - (1.10\ \text{kg/m}^3)[(\pi)(0.15\ \text{m})^2](\overline{V}_{\text{out}}) = 0$$

which can be solved for $\overline{V}_{\text{out}}$:

Answer

$$\overline{V}_{\text{out}} = 111.7\ \text{m/s}$$

EXAMPLE 4–6

A carburetor, shown schematically in figure 4–4, mixes air with fuel to provide a combustible mixture for an internal combustion engine. Determine the amount of mixture of fuel and air flowing through the carburetor if 0.01 lbm/s of fuel is consumed and the amount of air per pound-mass of fuel is 20 lbm.

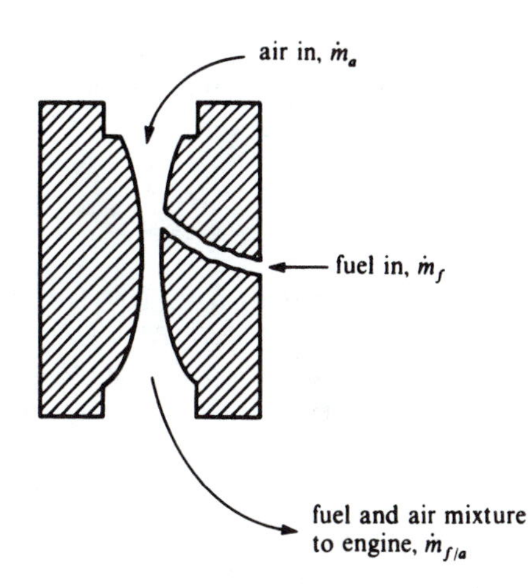

FIGURE 4–4 Schematic diagram of carburetor of internal combustion engine

Solution

Assuming steady flow through the carburetor, we have

$$\dot{m}_a + \dot{m}_f = \dot{m}_{f/a}$$

where

$$\dot{m}_f = 0.01\ \text{lbm/s}$$

and

$$\dot{m}_a = (20\ \text{lbm air/lbm})(\dot{m}_f) = (20)(0.01) = 0.2\ \text{lbm/s}$$

Thus,

Answer

$$\dot{m}_{f/a} = 0.2\ \text{lbm/s} + 0.01\ \text{lbm/s} = 0.21\ \text{lbm/s}$$

4–3 UNIFORM FLOW

In section 4–2, steady flow was identified as the condition of an open system where its mass remains constant and the inflows are exactly equal to the outflows. In this section, we consider situations where the inflows and outflows are not equal to each other and the control volume, while changing its amount of mass, will have a uniform density and state at any given instant in time. We call this condition **uniform flow**, and conservation of mass will be given by equation (4–5), which indicates that the difference between the two terms (inflows and outflows) is equal to the change of mass of the open system:

$$\dot{m}_{\text{in}} - \dot{m}_{\text{out}} = \dot{m}_{\text{system}} \tag{4–5}$$

If the open system does not have any mass flowing out (i.e., it has no outflows), the change in the system's mass is just equal to the inflows (i.e., the mass flow in), written

$$\dot{m}_{\text{in}} = \dot{m}_{\text{system}} \tag{4–14}$$

and this process is called the **filling process**.

Recalling the definition for the rate of change of the system mass, we can write equation (4–14) as

$$\dot{m}_{\text{in}} = \frac{\delta m_{\text{system}}}{\delta t}$$

and the change in the system mass for a period δt is just

$$(\dot{m}_{\text{in}})\,\delta t = \delta m_{\text{system}} \tag{4–15}$$

Now, if we want to know the change of the system mass for a finite time period Δt, equation (4–15) becomes

$$\sum (\dot{m}_{\text{in}})\,\delta t = m_2 - m_1 \tag{4–16}$$

where m_2 is the mass of the system after the filling process ends and m_1 is that just before it began. In many engineering design problems or for estimations, it is assumed that the system is empty to start, so that m_1 is then zero, the mass of the system is m_2, or just m, and equation (4–16) is

$$\sum (\dot{m}_{\text{in}})\,\delta t = m_{\text{system}} \tag{4–17}$$

For the special situations where the mass flow rate is constant for a given period, equation (4–17) reduces to

$$\dot{m}_{\text{in}}\Delta t = m_{\text{system}} \tag{4–18}$$

The case where there is no inflow to the system, but only mass flowing out, is called the **emptying process**, and the conservation of mass for the system becomes

$$-(\dot{m}_{\text{out}}) = \dot{m}_{\text{system}} \tag{4–19}$$

For uniform flow, the change in mass of the system is

$$\sum \dot{m}_{\text{out}}\delta t = m_1 - m_2 \tag{4–20}$$

For the case where the mass flow out is constant, equation (4–20) becomes

$$\dot{m}_{\text{out}}\Delta t = m_1 - m_2 \tag{4–21}$$

EXAMPLE 4–7 A railroad tank car is to be filled with liquid ammonia at a rate of 10 kg/s. If the tank car is 25 m long and is 4 m in diameter, determine the time required to fill the car if it is empty and the ammonia has a density of 715 kg/m^3.

Solution The tank car is the system that is to be filled, and because the flow is constant, equation (4–18) may be used for uniform flow conditions. The filling time can then be solved from this equation after determining the mass of the ammonia in the car after filling. We recognize that the mass is equal to the density times the volume from equation (2–9), and the volume is just that for a cylinder of diameter 4 m and length 25 m. Hence,

$$V = \pi(\text{radius})^2(\text{length}) = \pi(2\text{ m})^2(25\text{ m})$$
$$= 314.16\text{ m}^3$$

Then the mass is

$$m = \rho V = (314.16\text{ m}^3)(715\text{ kg/m}^3) = 224{,}624\text{ kg}$$

It follows that the filling time Δt is, from equation (4–18),

$$\Delta t = \frac{m}{\dot{m}_{\text{in}}}$$
$$= \frac{224{,}624\text{ kg}}{10\text{ kg/s}} = 22{,}462.4\text{ s}$$

Answer
$$= 6.2\text{ h}$$

Let us now consider a problem involving mass flowing in from two sources and one flow out, all flows of which are constant.

EXAMPLE 4–8 A cylindrical mixing tank having a diameter of 2 ft and containing 620 lbm of water is being filled from two water lines, one line delivering hot water at a rate of 0.7 lbm/s and a second line of $^5/_8$-in diameter delivering cold water at 8 ft/s. If we assume that the tank has an exit port of $^3/_4$-in diameter from which mixed water discharges at 12 ft/s, determine the rate of change of water level in the tank and the mass of the water in the tank 10 s after flow begins. (See figure 4–5).

Solution For determining the rate of change of water level, the rate of change of mass in the mixing tank must be found. Using the principles of conservation of mass, we write equation (4–5):

$$\dot{m}_{\text{in}} - \dot{m}_{\text{out}} = \dot{m}_{\text{system}}$$

Here, the system is the mixing tank. Now,

$$\dot{m}_{\text{in}} = \text{mass flow from line } A + \text{mass flow from line } B$$

which gives us

$$\dot{m}_{\text{in}} = 0.7\text{ lbm/s} + \left[\underset{(\text{density})}{\rho_B}\underset{(\text{area})}{A_B}\underset{(\text{velocity})}{\overline{V}_B}\right]$$

The velocity of line B, $\overline{V}_B$, is 8 ft/s; the area of B is $\pi(5/16)^2$ in^2. We assume the density to be 62.4 lbm/ft^3. Then the mass flow, from line B, is

$$\dot{m}_B = (62.4\text{ lbm/ft}^3)\left[\pi\left(\frac{5}{16}\right)^2\left(\frac{1}{144}\text{ft}^2\right)(8\text{ ft/s})\right]$$

and it follows that

$$\dot{m}_B = 1.064\text{ lbm/s}$$

FIGURE 4–5 Mass flow balance

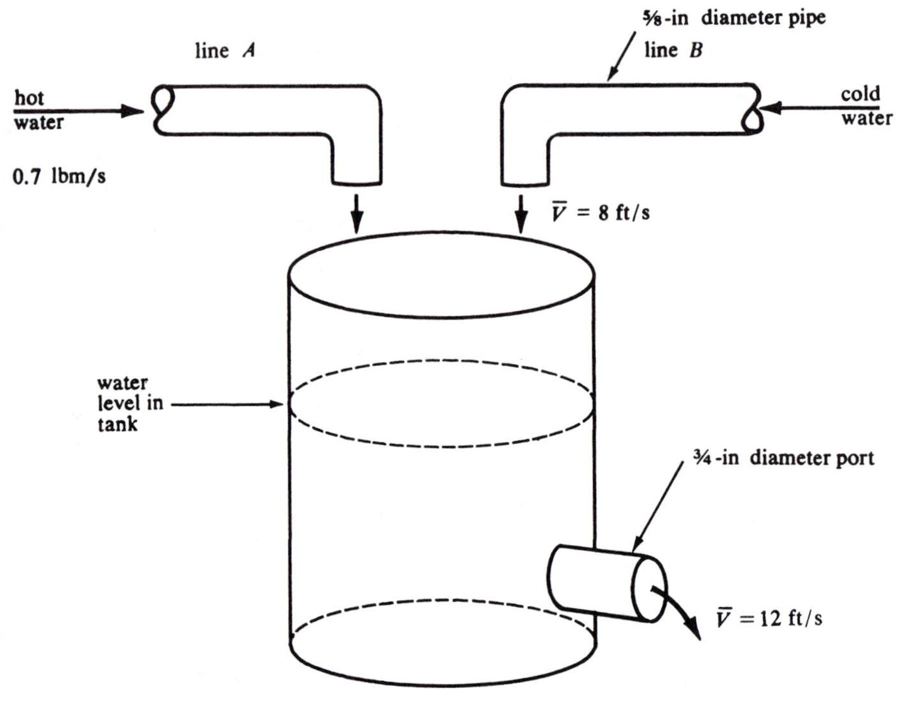

So

$$\dot{m}_{\text{in}} = 0.700 \text{ lbm/s} + 1.064 \text{ lbm/s}$$
$$= 1.764 \text{ lbm/s}$$

The mass flow out of the system is just that leaving the port of area $\pi(3/8)^2 \text{ in}^2$. Assume that the density of the mixed water is 62.0 lbm/ft^3. We then have

$$\dot{m}_{\text{out}} = (62.0 \text{ lbm/ft}^3)\left[\pi\left(\frac{3}{8}\right)^2 \text{in}^2\right]\left(\frac{1}{144}\text{ft}^2/\text{in}^2\right)(12 \text{ ft/s})$$
$$= 2.28 \text{ lbm/s}$$

The mass rate of change for the system, $\dot{m}_{\text{system}}$, is then $1.764 - 2.28$ lbm/s or -0.516 lbm/s, from equation (4–5).

The rate of change of the water in the mixing tank is $\dot{m}_{\text{system}}/\rho A_{\text{tank}}$,

So

$$\overline{V}_{\text{level}} = \frac{(1.764 - 2.28) \text{ lbm/s}}{(62.0 \text{ lbm/ft}^3)(\pi)(1 \text{ ft}^2)}$$

or

Answer

$$\overline{V}_{\text{level}} = \frac{-0.516}{62\pi(1 \text{ ft}^2)} = -0.00265 \text{ ft/s}$$

The negative sign here means that the level is dropping and the mass of the system is decreasing.

Ten seconds after flow has commenced, the mass is determined from

$$\dot{m}_{\text{system}} = \frac{\Delta m_{\text{system}}}{\Delta t}$$

or, more clearly, with a little algebra,

$$\Delta t\, \dot{m}_{\text{system}} + m_{\text{system, initial}} = m_{\text{system}} \text{ at 10 s}$$

Then, since $\dot{m}_{\text{system}} = -0.516$ lbm/s initially from the preceding calculation, we have

$$-(0.516 \text{ lbm/s})(10 \text{ s}) + 620 \text{ lbm} = m_{\text{system}} \text{ at } 10 \text{ s}$$

or

Answer

$$m_{\text{system}} \text{ at } 10 \text{ s} = 614.84 \text{ lbm}$$

From example 4–8, you can see that the mass balance equation (4–5) can involve more than two terms on the left side. In equation (4–5), you should notice that m_{in} and m_{out} are, in general, sums of all the masses flowing in and out, respectively; thus,

$$\sum \dot{m}_{\text{in}} - \sum \dot{m}_{\text{out}} = \dot{m}_{\text{system}} \qquad \textbf{(4–5) (revised)}$$

Now let us look at the situation where the mass flow is not constant.

EXAMPLE 4–9 A 2000-liter tank is full of milk. It is known that if milk is drained from the tank (see figure 4–6), the mass flow rate will decrease from the maximum when the tank is full according to the relationship indicated in figure 4–7. Determine the amount of milk drained in 35 min, assuming that the tank has 1500 liters at the start. Assume that milk has a density of 900 kg/m^3.

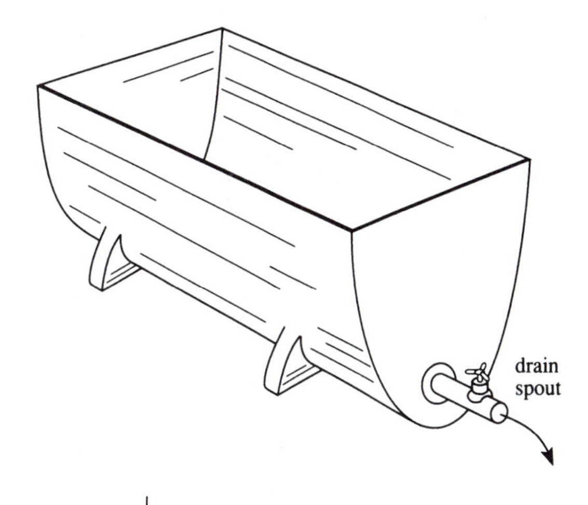

FIGURE 4–6 Emptying process of example 4–9

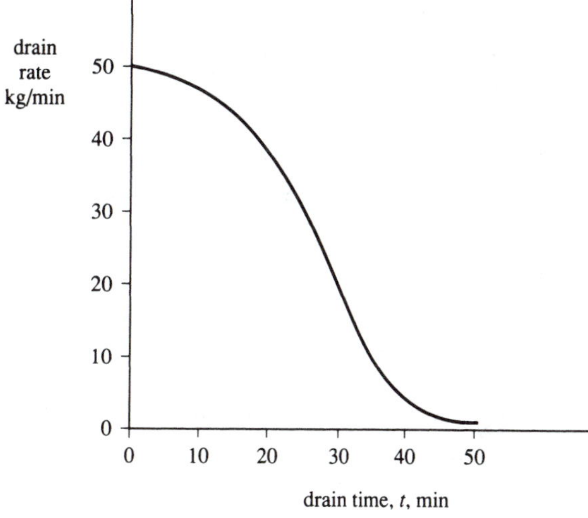

FIGURE 4–7

Solution

First we recognize that the tank is an open system and is subject to an emptying process, the mass balance of which is described by equation (4–20). We cannot use equation (4–21) here because the mass flow is changing during the process. The term $m_1 - m_2$ in equation (4–20) is the amount of milk that will be drained in 35 min, and the term $\Sigma\, \dot{m}_{out}\, \delta t$ will be calculated. We can use the same procedure that was used to calculate areas under curves for work and the change in the internal energy. Here, however, the computation requires mass flow rate and small time changes. Table 4–1 lists the mass flows at times from figure 4–7. By using the AREA program listed in appendix A–5, we can readily calculate the amount of milk drained in 35 min. Table 4–1 includes the listings for a hand calculation of this result for reference. Either way, the answer is 1266.25 kg of milk.

TABLE 4–1 Results of uniform flow emptying of milk tank of example 4–9

Mass Flow Rate kg/min	Time min	δt min	$\dot{m}_{out}$ kg/min	$\dot{m}_{out}\, \delta t$ kg
50	0			
		5	49.125	245.625
48.25	5			
		5	47.125	235.625
46	10			
		5	44.25	221.25
42.5	15			
		5	40.000	200
37.5	20			
		5	34	170
30.5	25			
		5	24.75	123.75
19	30			
		5	14.00	70
9	35			
				$\Sigma\, \dot{m}_{out}\, \delta t = 1266.25$ kg

CALCULUS FOR CLARITY 4–2

The uniform flow processes are often best analyzed with the use of calculus. In particular, the emptying process usually involves an outflow of mass which is not constant, and equation (4–20) is then best used in the form

$$\dot{m}_{out}\, dt = -dm_{system} \qquad \textbf{(4–20) (revised)}$$

This is a differential equation of first order and can often be solved directly, particularly if the outflow can be described as a function of the system mass. Example 4–10 considers this situation.

EXAMPLE 4–10

If the drain rate or outflow of the water tank shown in figure 4–8 is given by the equation

$$\dot{m}_{out} = \rho A_{out}\sqrt{2gy}$$

determine the time for half of the water to drain out of the tank. The water density may be taken as 1000 kg/m^3.

Solution

Since there is only an outflow of water from the tank, we use the revised equation (4–20) and note that the water height y can be written

$$y = \frac{m_{system}}{\rho A_s}$$

Then equation (4–20) becomes

$$\rho A_{out}\sqrt{2gm_{system}/\rho A_s}\, dt = -dm_{system}$$

CALCULUS FOR CLARITY 4–2, continued

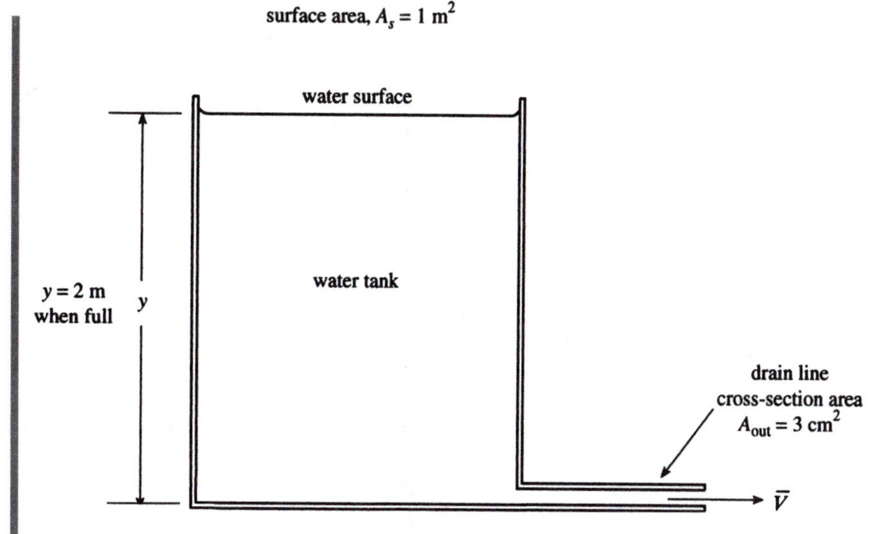

FIGURE 4–8 Cross-sectional view of water tank and drain line

Separating the variables so that the system mass is on the right side gives

$$dt = -\sqrt{\frac{A_s}{2g\rho A_{out}^2}}\frac{dm_{system}}{\sqrt{m_{system}}}$$

Substituting the values for the terms results in the equation

$$dt = -[23.8\ \text{s/kg}^{1/2}]\frac{dm_{system}}{\sqrt{m_{system}}}$$

Integrating both sides gives, for $t = 0$ initially,

$$t = [47.6\ \text{s/kg}^{1/2}]\left[\sqrt{m_{system,\,i}} - \sqrt{m_{system,\,f}}\right]$$

Since we want to know the time when the tank is half full, the final system mass, $m_{system,\,f}$, is one-half of the initial system mass, $m_{system,\,i}/2$, and the time is then

$$t = [47.6]\left[1 - \frac{1}{\sqrt{2}}\right]\sqrt{m_{system,\,i}} = 13.94\sqrt{m_{system,\,i}}$$

The initial mass of the system was

$$m_{system,\,i} = \rho V = (1000\ \text{kg/m}^3)(2\ \text{m}^3) = 2000\ \text{kg}$$

and the time is then

Answer

$$t = 623.4\ \text{s} = 10.39\ \text{min}$$

4–4 CONSERVATION OF ENERGY

As a system goes through a process, some properties of the system are altered, and we have postulated that the mass is conserved. If the system is closed or isolated, the system mass remains unchanged for any process; but if the system is open, the mass will change according to equation (4–5). Similarly, we now postulate that energy is conserved in any process of a system, and we write

$$E_{in} - E_{out} = \Delta E_{system} \tag{4–22}$$

where a positive energy change of the system implies an accumulation of energy in the system and a negative value implies loss of system energy. In terms of rates of change, we say that

$$\dot{E}_{\text{in}} - \dot{E}_{\text{out}} = \dot{E}_{\text{system}} \quad \textbf{(4–23)}$$

in a manner similar to the conservation of mass in section 4–1.

In equation (4–22), E_{in} and E_{out} represent energy crossing the boundary of the system under scrutiny—they are terms of energy in transition. From chapter 3, we recall the definitions of heat and work, which are precisely the energy terms under discussion. We quite arbitrarily assign the following sign conventions:

+*heat*—implies energy *into* the system
+*work*—implies energy *out of* the system
−*heat*—implies energy *out of* the system
−*work*—implies energy *into* the system

Using Q for heat and Wk for work, we then obtain the following from equation (4–22):

$$Q - Wk = \Delta E_{\text{system}} \quad \textbf{(4–24)}$$

Notice that we arrived at equation (4–24) by saying that

$$E_{\text{in}} = +Q \quad \text{or} \quad E_{\text{in}} = -Wk$$

and

$$E_{\text{out}} = -Q \quad \text{or} \quad E_{\text{out}} = +Wk$$

With a positive heat transfer, we "heat up" the system, and with negative heat, we "cool down" the system. Similarly, work gained from a system is positive; work put into the system is negative.

Recalling from chapter 3 that a rate of heat transfer $\dot{Q}$ is a form of the rate of energy transition and defining here $\dot{W}k$ as the rate of work or **power**, we then obtain the following from equation (4–23):

$$\dot{Q} - \dot{W}k = \dot{E}_{\text{system}} \quad \textbf{(4–25)}$$

The remainder of this book is devoted to clarification and examples of equations (4–24) and (4–25).

Remember that the energy being used in these equations is any of the forms of energy in the static condition referred to in chapter 3, that is, kinetic, potential, or internal energy. Other forms or adjectives are equally acceptable, such as *strain*, *electromagnetic*, or *chemical*, but only the three common ones will generally be considered in this book.

EXAMPLE 4–11 A fuel tank is filled with propane gas, and heat is transferred from the surroundings at a rate of 30 Btu/h. Determine the increase in energy of the propane gas over a 24-h period. (See figure 4–9.)

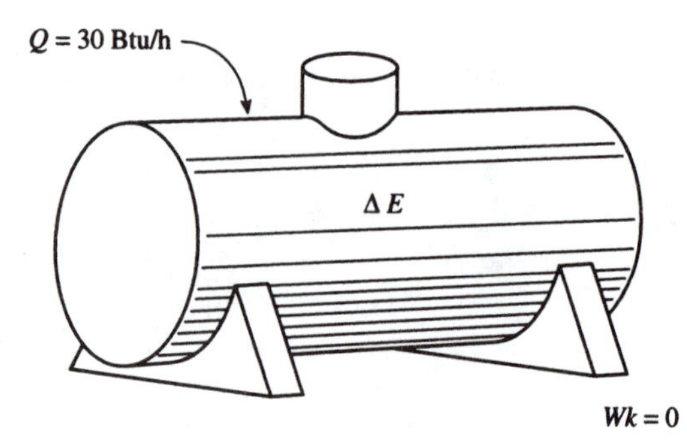

FIGURE 4–9 Propane tank subject to heat transfer

Solution

We first identify the system to be analyzed—the propane gas inside the fuel tank. Assuming that the tank is rigid and no gas or air is entering or leaving the tank, we have no mass change of the system and no reversible work since there is no change in volume. We then write the first law as

$$\Delta E = Q - Wk$$

or

$$\Delta E = Q$$

since no work is present, or

$$Wk = 0$$

Because the rate of heat transfer was given, we can write

$$\dot{E} = \dot{Q}$$

so that, since $\dot{Q}$ is 30 Btu/h,

$$\dot{E} = 30 \text{ Btu/h}$$

The definition of the rate of energy change is

$$\dot{E} = \frac{\delta E}{\delta t} \qquad \textbf{(4–26)}$$

where δt is the time during which the energy changes an amount δE. From this equation, the change in energy can be found if the rate of energy change and the time period are known. Thus, we can write

$$\sum \dot{E}\,\delta t = \Delta E$$

and if the rate of change of energy is constant,

$$\dot{E}\,\Delta t = \Delta E$$

and then

$$\Delta E = (30 \text{ Btu/h})(24 \text{ h})$$

Answer

$$= 720 \text{ Btu}$$

EXAMPLE 4–12

A piston-cylinder device does 7800 ft · lbf of work while losing 3.7 Btu of heat. What is the change of energy of the contents of the piston-cylinder? Give the answer in SI units as well as English units.

Solution

Here we have a closed system where energy is conserved, so we write equation (4–24) as

$$\Delta E = Q - Wk$$

Since 7800 ft · lbf of work is done, work is a positive quantity in the equation. The amount of heat lost is a negative quantity. We then obtain

$$\Delta E = -3.7 \text{ Btu} - 7800 \text{ ft} \cdot \text{lbf}$$

Since 778 ft · lbf is equal to 1 Btu, we have

$$\Delta E = -3.7 \text{ Btu} - \frac{7800}{778} \text{ Btu}$$

Answer

$$= -13.7 \text{ Btu}$$

In SI units we have, from the table on the inside front cover, the conversion 1054 J = 1 Btu, so

$$\Delta E = -13.7 \times 1054 \text{ J}$$

Answer

$$= -14{,}400 \text{ J} = -14.4 \text{ kJ}$$

EXAMPLE 4–13 An electric power generating station, sketched in figure 4–10, produces 100 MW of power. If the coal releases 900×10^6 kJ/h of energy, determine the rate at which heat is rejected from the power plant.

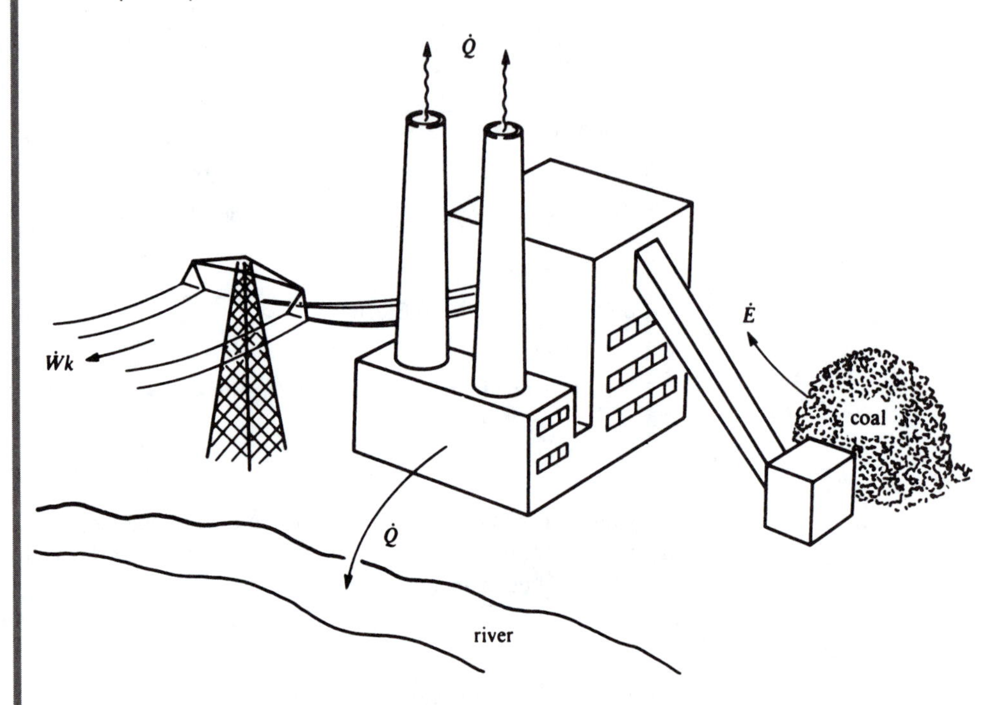

FIGURE 4–10 Power station as a system

Solution The system is the power station and the coal pile adjacent to the facility. From conservation of energy, we get

$$\dot{Q} = \dot{E} + \dot{W}k$$

and

$$\begin{aligned} Q &= -900 \times 10^6 \text{ kJ/h} + 100 \text{ MW} \\ &= -900 \times 10^3 \text{ MJ/h} + 100 \text{ MW} \\ &= -250 \text{ MJ/s} + 100 \text{ MW} \\ &= -250 \text{ MW} + 100 \text{ MW} = -150 \; MW \end{aligned}$$

Answer

The negative sign given the rate of energy release $\dot{E}$ indicates the condition of decreasing energy, whereas the negative sign for the heat release $\dot{Q}$ indicates that the heat was rejected by the system to the river or atmosphere.

4–5 THE FIRST LAW OF THERMODYNAMICS FOR A CLOSED SYSTEM

First law of thermodynamics: *Energy can be neither created nor destroyed but can only be converted to its various forms.*

When considering the conservation of energy, or first law of thermodynamics, as it applies to a closed system, we can write the energy balance as

$$\Delta E = Q - Wk_{cs} \tag{4–27}$$

or

$$m\,\Delta e = mq - mwk_{cs}$$

or

$$\Delta e = q - wk_{cs} \quad \textbf{(4–28)}$$

where q and wk_{cs} are heat per unit mass of the system and work per unit mass of the system, respectively. The system mass m is usually expressed in kilograms or pounds-mass. If the process involves boundary work, we can see that the Wk_{cs} is given by equation (3–9):

$$Wk_{cs} = \sum p\,\delta V$$

We saw in chapter 3 that the boundary work can be found from a number of different special equations, depending on the type of process involved, and it would be good for the reader to go back to section 3–1 and note those versions.

We can also consider time as a variable, so equation (4–27) becomes

$$\dot{E} = \dot{Q} - \dot{W}k_{cs} \quad \textbf{(4–29)}$$

or

$$\dot{e} = \dot{q} - \dot{w}k_{cs} \quad \textbf{(4–30)}$$

where $\dot{e}$ is the rate of change of the system's energy per unit of the system's mass, $\dot{q}$ the heat transfer per unit system mass, and $\dot{w}k_{cs}$ the power per unit mass. Clearly, the first law of thermodynamics for a closed system is a direct application of the principle of the conservation of energy applied to a given mass.

4–6 THE FIRST LAW OF THERMODYNAMICS FOR AN ISOLATED SYSTEM

Considerations of isolated systems eliminate any heat or work or mass transfer. All we can say about an isolated system then is

$$\Delta E = 0$$

Since $\Delta m = 0$ for isolated systems and $E = me$, it follows that

$$m\,\Delta e = 0 \quad \text{or} \quad \Delta e = 0 \quad \textbf{(4–31)}$$

for a system having mass m and energy e. All that happens here is that energy is converted from one form to another, and interestingly, there is no indication outside the isolated system of what is transpiring inside. Some consider our universe as an isolated system, although it has not been established as such and no actual isolated systems have been identified in the strictest sense.

EXAMPLE 4–14

Figure 4–11 shows a closed chamber whose walls do not allow heat transfer (called adiabatic walls) and which contains 9 kg of dust uniformly distributed. The kinetic energy of the dust is 100 J as the dust is moving about in the chamber. Determine the changes in energy as the dust settles to the bottom.

Solution

Since the chamber does not allow heat transfer ($Q = 0$) and since it is closed off, it cannot convey mechanical work ($Wk = 0$); thus, we write

$$\Delta E = 0$$

or

$$\Delta \text{KE} + \Delta \text{PE} + \Delta U = 0$$

The initial kinetic energy is 100 J and the final is zero. Then

$$\begin{aligned}\Delta \text{KE} &= \text{KE}_{\text{final}} - \text{KE}_{\text{initial}} \\ &= 0 - 100\ \text{J}\end{aligned}$$

Answer

$$= -100\ \text{J}$$

FIGURE 4–11 Adiabatic chamber containing dust

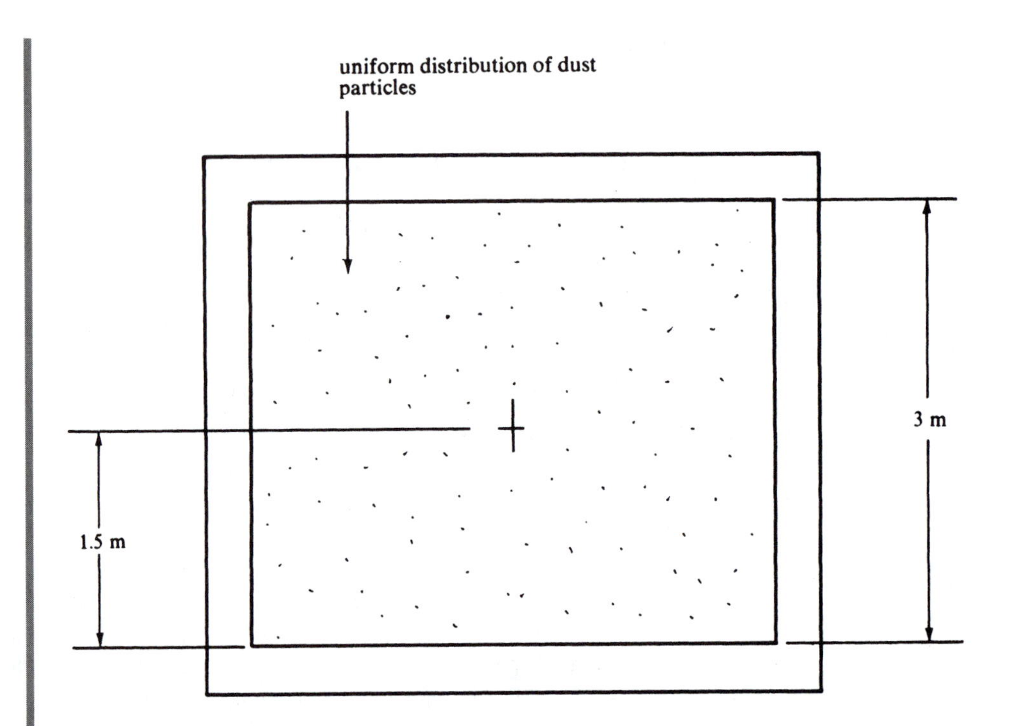

Since the dust is uniformly distributed, the initial potential energy of the dust is simply the total weight of the dust particles times the average height. Thus,

$$\begin{aligned} PE_{initial} &= mgz = (9\ kg)(9.8\ m/s^2)(1.5\ m) \\ &= 132\ kg \cdot m^2/s^2 = 132\ N \cdot m \\ &= 132\ J \end{aligned}$$

where the gravitational acceleration is assumed to be 9.8 m/s². The final potential energy is zero if the bottom of the chamber is assigned as the level of no potential energy. Then

Answer

$$\begin{aligned} \Delta PE &= PE_{final} - PE_{initial} = 0 - 132\ J \\ &= -132\ J \end{aligned}$$

and

$$\Delta U = -\Delta KE - \Delta PE$$

or

$$\begin{aligned} \Delta U = U_{final} - U_{initial} &= -(-100\ J) - (-132\ J) \\ &= 232\ J \end{aligned}$$

If we arbitrarily say that

$$U_{initial} = 0$$

then

Answer

$$U_{final} = 232\ J$$

In example 4–14, we used the term **adiabatic wall** to denote no heat transfer. Throughout the text we will refer quite often to an **adiabatic process**, which is a process where there is no heat transfer (Q or $\dot{Q}$ is zero), whether the process applies to open, closed, or isolated systems. Example 4–14 is an adiabatic process.

4–7 FLOW ENERGY AND ENTHALPY

Thermodynamics has been shown to apply quite generally to closed and isolated systems. No mass transfer is involved, and only energy transfer in the form of work and heat is of concern. But most of the systems encountered in technology are open systems, and we should hope to use the thermodynamic approach for these as well. Let us then look carefully at the unique mechanism of the open system—mass moving across the system boundary. In figure 4–12, we have defined an open system and for simplistic purposes provided only two pipes, A and B, through which mass may enter or leave the system. Now consider some mass moving into the system on the left boundary of the system in figure 4–12. In order that the mass fully enter the system, it must move the distance x_A along pipe A. But for it to move this distance *someone or something external must exert a force* F_A through the distance x_A. We then say that there is flow energy required to "flow the mass" into the system:

$$\text{flow energy} = F_A x_A$$

Also, F_A is equal to the pressure p_A of the system times the cross-sectional area of the pipe, A. Then

$$\text{flow energy at } A = p_A A x_A = p_A V_A \tag{4–32}$$

since

$$\begin{aligned} A x_A &= \text{volume of the mass entering at } A \\ &= V_A \end{aligned}$$

Frequently, the term *flow work* is used for flow energy, but for this development, let us use the term *flow energy*.

The flow energy per unit mass at A is

$$p_A \frac{V_A}{m_A} = p_A v_A \tag{4–33}$$

At pipe B, we consider that mass is leaving the system, so the mass will move out on its own, propelled by a back pressure p_B, and, associated with it, the flow energy out of the system. At B, the flow energy per unit mass is $p_B v_B$ in a completely analogous development as for pipe A. Keep in mind, though, that while work (or flow energy) was required external to this system at A to shove the mass in, work is done by the system in ejecting mass at B.

FIGURE 4–12 Open system

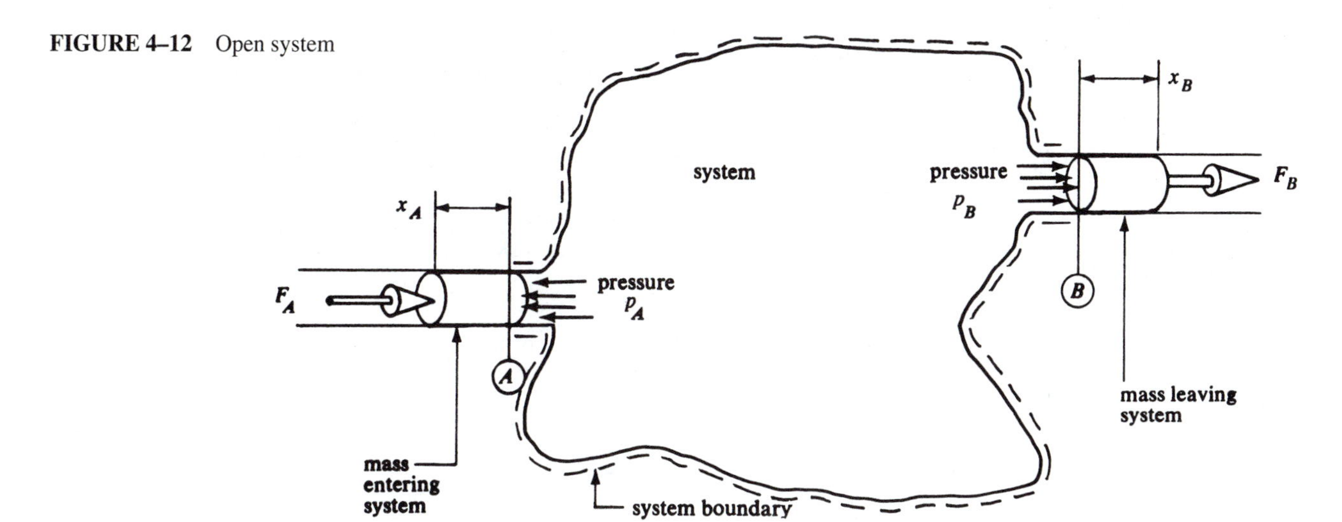

FIGURE 4–13 Open system

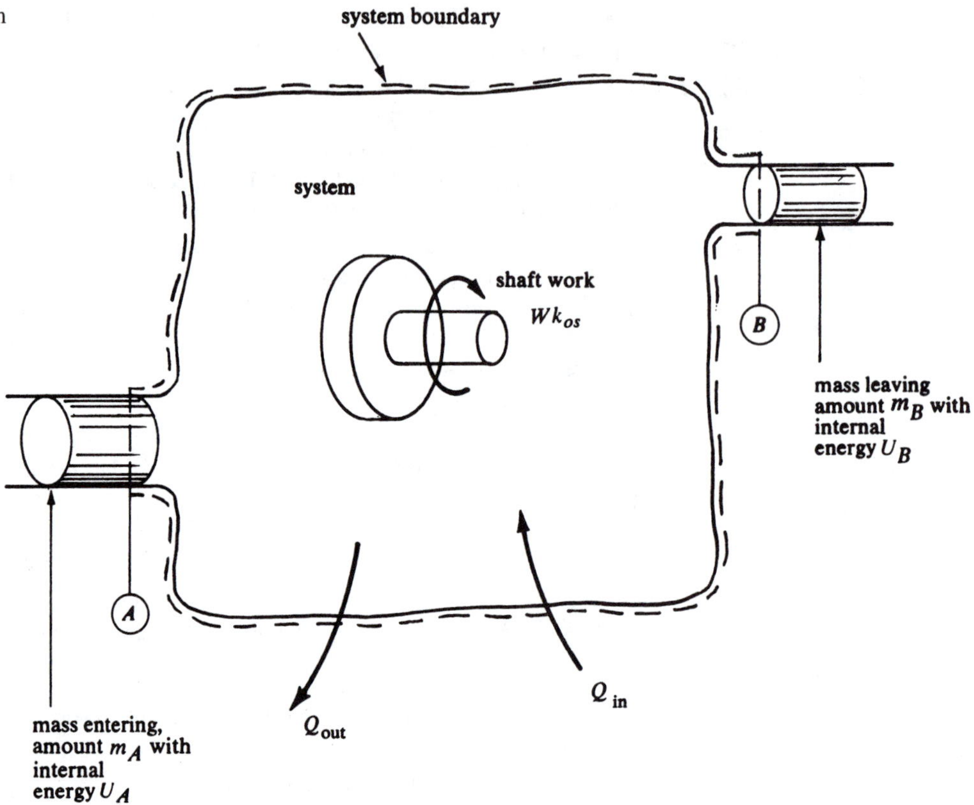

Now let us look at the energy balance or first law for this system. In figure 4–13 is our open system with some mechanical additions to account for the fact that work and heat may occur. We write equation (4–24) as

$$\Delta E = Q - Wk$$

or, since, in this illustration,

$$Wk = \text{shaft work} - \text{flow energy at } A + \text{flow energy at } B$$

we have, for our energy balance,

$$\Delta E = Q - Wk_{os} - \text{flow energy at } B + \text{flow energy at } A \tag{4–34}$$

The flow energy at A was *into* the system, or done *on* the system, and therefore negative. The flow energy at B was *out* and hence positive. Upon substitution into the energy balance, both signs change, and equation (4–34) results.

The term Wk_{os} is the open-system work, here given as the shaft work. Using equation (4–33) for flow energy, we then substitute these into equation (4–34) to obtain

$$\Delta E = Q - Wk_{os} - p_B v_B m_B + p_A v_A m_A \tag{4–35}$$

If we break up the energy terms on the left of this equation into their identifiable types, we have

$$\Delta E = \Delta \text{KE} + \Delta \text{PE} + \Delta U \tag{4–36}$$

but

$$\Delta E = \Delta E_{\text{system}} - E_A + E_B$$

where E_A and E_B are KE + PE + U evaluated at A and B, respectively. Then equation (4–35) becomes

$$\Delta E_{\text{system}} - \text{KE}_A - \text{PE}_A - U_A + \text{KE}_B + \text{PE}_B + U_B = Q - Wk_{os} - m_B p_B v_B + m_A p_A v_A \qquad \textbf{(4–37)}$$

We can write U_B as $m_B u_B$ and U_A as $m_A u_A$. Then, if we take the flow energy terms to the left side of the equation, we get

$$\Delta E_{\text{system}} - \text{KE}_A - \text{PE}_A - m_A u_A - m_A p_A v_A + \text{KE}_B + \text{PE}_B + m_B u_B + m_B p_B v_B = Q - Wk_{os}$$

In this equation, we could algebraically combine the terms $u + pv$, namely,

$$m_A(u_A + p_A v_A) \quad \text{and} \quad m_B(u_B + p_B v_B)$$

The quantities inside the parentheses here are defined as the **specific enthalpy** and are denoted by h. Then $h_A = u_A + p_A v_A$ and $h_B = u_B + p_B v_B$, or, generally,

$$h = u + pv \qquad \textbf{(4–38)}$$

Specific enthalpy is specified by units of energy per unit mass, and if it is multiplied by the mass, the resulting property is called **enthalpy**, H. We have

$$\begin{aligned} H &= mh \\ &= mu + mpv \\ &= U + pV \end{aligned} \qquad \textbf{(4–39)}$$

Total enthalpy is specified by energy units. Remember that enthalpy and total enthalpy are merely mathematical combinations that may or may not have physical significance in a given problem.

EXAMPLE 4–15 An air pump shown in figure 4–14 takes in air at 14.7 psia, 78°F, and with a density of 0.075 lbm/ft^3 at station 1. The pump compresses the air flowing to the tank so that, at station 2, the density is found to be 0.050 lbm/ft^3 and the pressure is 100 psia. Determine the flow energy and specific enthalpy of the air at stations 1 and 2, if the internal energy of the air is 60 Btu/lbm at (1) and 180 Btu/lbm at (2).

Solution The flow energy at station 1 is equal to $p_1 v_1$ or p_1/ρ_1, so that flow work at (1) = 14.7 lb/in^2 × 1 ft^3/0.075 lbm. Converting to compatible units, we get

FIGURE 4–14 Air pump

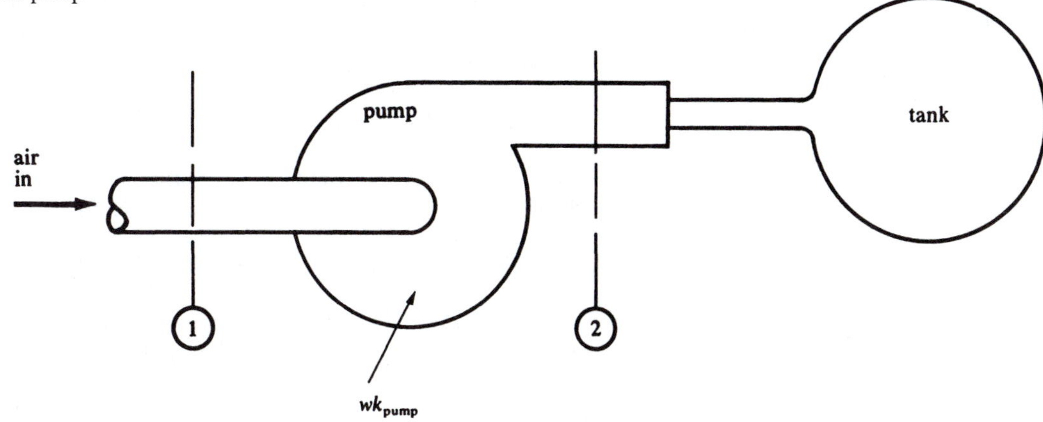

Answer

$$14.7 \text{ lb/in}^2 \times 144 \text{ in}^2/\text{ft}^2 \times 1 \text{ ft}^3/0.075 \text{ lbm} = 28{,}224 \text{ ft}\cdot\text{lbf/lbm}$$

At station 2,

$$\text{flow energy} = +100 \text{ lb/in}^2 \times 144 \text{ in}^2/\text{ft}^2 \times 1 \text{ ft}^3/0.050 \text{ lbm}$$

Answer

$$= 288{,}000 \text{ ft}\cdot\text{lbf/lbm}$$

To calculate the specific enthalpy, we use equation (4–38), and at station 1 we obtain

$$h_1 = u_1 + p_1 v_1$$

or

$$h_1 = 60 \text{ Btu/lbm} + 28{,}224 \text{ ft}\cdot\text{lbf/lbm}$$

Answer

$$= 96.3 \text{ Btu/lbm}$$

Then, at station 2, we have

$$h_2 = 180 \text{ Btu/lbm} + 288{,}000 \text{ ft}\cdot\text{lbf/lbm}$$

Answer

$$= 550 \text{ Btu/lbm}$$

EXAMPLE 4–16 Steam at 6 MPa pressure and 400°C has a specific volume of 0.047 m^3/kg and a specific enthalpy of 3174 kJ/kg. Determine the internal energy per kilogram of steam.

Solution

From equation (4–38), we may write

$$u = h - pv$$

or

$$\begin{aligned} u &= 3174 \text{ kJ/kg} - (60 \times 10^5 \text{ N/m}^2)(0.047 \text{ m}^3/\text{kg}) \\ &= 3174 \text{ kJ/kg} - 2.82 \times 10^5 \text{ N}\cdot\text{m/kg} \\ &= 3174 \text{ kJ/kg} - 282 \text{ kJ/kg} \\ &= 2892 \text{ kJ/kg} \end{aligned}$$

Answer

4–8 THE FIRST LAW OF THERMODYNAMICS FOR AN OPEN SYSTEM

One way to consider the energy balance or first law of thermodynamics is as the bookkeeping rule of balancing the "energy budget" of a system. All energy must be accounted for in a given process as is done in the closed and isolated systems. Here let us take an accounting of the energy of an open-system process—very generally and intuitively—in the same manner as in section 4–7. We could generalize the system as anything, such as a pump, fan, radiator, cooling tower, boiler, turbine, or a biological system. Figure 4–15 shows a general open system that allows heat transfer, work, and mass flow into and out of the region. Assuming that mass enters at station 1 and leaves at station 2, then, from $\Delta E = Q - Wk$, we write, per unit of time,

$$\dot{\text{KE}}_2 + \dot{\text{PE}}_2 + \dot{H}_2 - \dot{\text{KE}}_1 - \dot{\text{PE}}_1 - \dot{H}_1 + \dot{E}_s = \dot{Q}_{\text{in}} - \dot{Q}_{\text{out}} - \dot{W}k - \dot{W}k_s \quad \textbf{(4–40)}$$

Combining $\dot{W}k$ and $\dot{W}k_s$ into the term called $\dot{W}k_{os}$, combining $\dot{Q}_{\text{in}}$ and $\dot{Q}_{\text{out}}$ into $\dot{Q}$, and merely recalling the convention of signs for heat and work (section 4–4), we have

$$\dot{\text{KE}}_2 + \dot{\text{PE}}_2 + \dot{H}_2 - \dot{\text{KE}}_1 - \dot{\text{PE}}_1 - \dot{H}_1 + \dot{E}_s = \dot{Q} - \dot{W}k_{os} \quad \textbf{(4–41)}$$

This equation is generally too cumbersome to use, so if we neglect kinetic and potential energy changes of the system, we have $\dot{E}_s = \dot{U}_s$, so that

$$\dot{\text{KE}}_2 + \dot{\text{PE}}_2 + \dot{H}_2 - \dot{\text{KE}}_1 - \dot{\text{PE}}_1 - \dot{H}_1 + \dot{U}_s = \dot{Q} - \dot{W}k_{os} \quad \textbf{(4–42)}$$

FIGURE 4–15 General open system

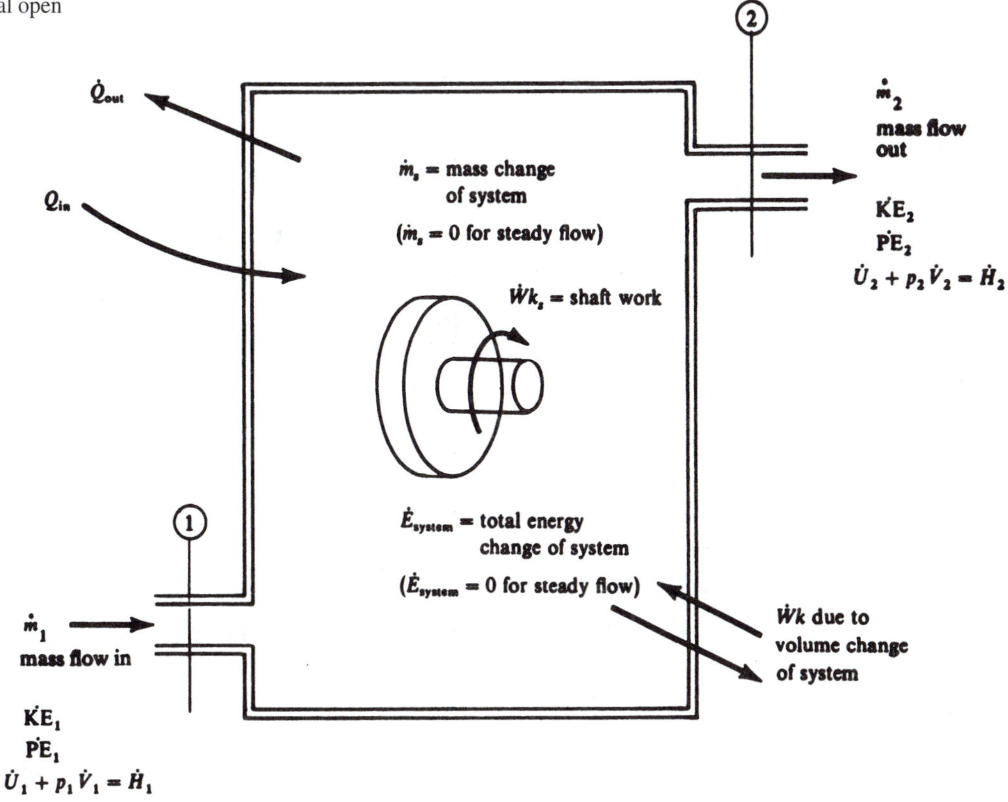

Even this equation can be extremely difficult to use under completely general conditions of variable flow, mass change of system, or variations in heat or work. For most common engineering problems, the assumption of steady flow is sufficiently general. Steady flow means that the mass flow rates in and out are constant in time and that there is no change in system mass with time. This is equivalent, then, to writing

$$\text{system mass} = \text{constant} \tag{4–43}$$

and

$$\dot{m}_2 = \dot{m}_1 = \text{constant} \tag{4–44}$$

The mass flow rates in and out are the same from conservation of mass. If steady state is assumed (the state of the system remains constant in time), then

$$\dot{W}k_{os} = \text{shaft power} = \dot{W}k_s, \qquad \dot{E}_s = \dot{U}_s = 0$$

and the energies evaluated at stations 1 and 2 are constant in time. The steady-flow steady-state energy equation (hereafter referred to as the *steady-flow energy equation*) is then written

$$\dot{\text{KE}}_2 + \dot{\text{PE}}_2 + \dot{H}_2 - \dot{\text{KE}}_1 - \dot{\text{PE}}_1 - \dot{H}_1 = \dot{Q} - \dot{W}k_{os} \tag{4–45}$$

or

$$\dot{m}_2(\text{ke}_2 + \text{pe}_2 + h_2) - \dot{m}_1(\text{ke}_1 + \text{pe}_1 + h_1) = \dot{m}q - \dot{m}wk_{os} \tag{4–46}$$

Since $\dot{m} = \dot{m}_1 = \dot{m}_2$, equation (4–46) can be reduced to

$$\Delta\text{ke} + \Delta\text{pe} + \Delta h = q - wk_{os} \tag{4–47}$$

Expanding this equation gives us a well-known form of the first law of thermodynamics applied to the open system under steady-flow conditions:

$$\left.\begin{aligned} &(\text{kJ/kg}):\quad \frac{1}{2}(\overline{V}_2^2 - \overline{V}_1^2) + g(z_2 - z_1) + h_2 - h_1 = q - wk_{os} \\ &(\text{Btu/lbm}):\quad \frac{\overline{V}_2^2 - \overline{V}_1^2}{2g_c} + \frac{g(z_2 - z_1)}{g_c} + h_2 - h_1 = q - wk_{os} \end{aligned}\right\} \tag{4–48}$$

Quite frequently, power or rates of heat transfer are specified. The steady-flow equation for rates, equivalent to equation (4–48), is equation (4–46), here expanded and rewritten as

$$\left.\begin{aligned} &(\text{kJ/kg}):\quad \dot{m}\left(\frac{1}{2}(\overline{V}_2^2 - \overline{V}_1^2) + g(z_2 - z_1) + h_2 - h_1\right) = \dot{m}q - \dot{m}wk_{os} \\ &(\text{Btu/lbm}):\quad \dot{m}\left(\frac{\overline{V}_2^2 - \overline{V}_1^2}{2g_c} + \frac{g(z_2 - z_1)}{g_c} + h_2 - h_1\right) = \dot{m}q - \dot{m}wk_{os} \end{aligned}\right\} \tag{4–49}$$

All terms on the left in equation (4–49) should be easily identifiable with some physical quantity. On the right side of the equation, heat or heat transfer rates can be calculated from those equations discussed in chapter 3, but Wk_{os} is without a defining equation. It can be shown that the open-system work, Wk_{os}, is given by the equation

$$Wk_{os} = -\sum V\,\delta p - \Delta\text{PE} - \Delta\text{KE} \tag{4–50}$$

and if the kinetic and potential energy changes are negligible, then

$$Wk_{os} = -\sum V\,\delta p \tag{4–51}$$

Equation (4–51) is true for a reversible steady-flow process without kinetic or potential energy changes and is shown geometrically in figure 4–16. We have stated that $Wk_{cs\,\text{rev}}$ is equivalent to the area under a curve in a p–V diagram; the $Wk_{os\,\text{rev}}$ is equal to the area to the left of the curve in the p–V diagram, as can be seen in figure 4–16. We will use equation (4–51) as the general definition for open-system work when kinetic, potential, and chemical energy changes are neglected.

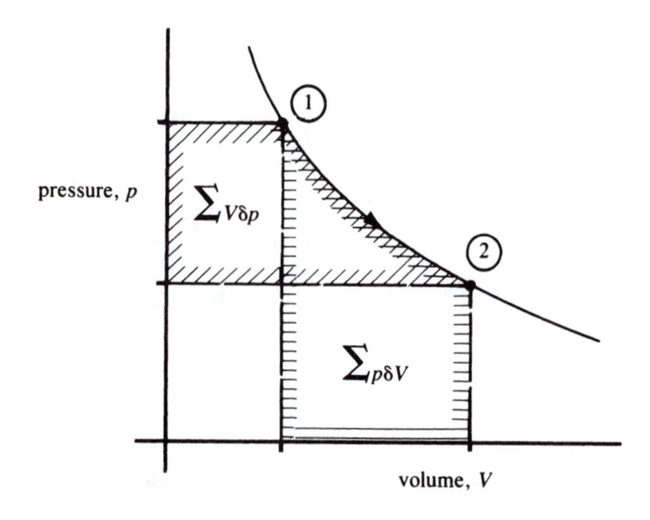

FIGURE 4–16 Graphical comparison of $p\,\delta V$ and $V\,\delta p$ terms

Per unit mass, the reversible work of the open system is

$$wk_{os} = -\sum v\,\delta p \tag{4–52}$$

and the power for an open system under steady-flow conditions is

$$\dot{W}k_{os} = -\dot{m}\sum v\,\delta p \tag{4–53}$$

If kinetic or potential energy changes are not negligible, the work must be obtained from

$$wk_{os} = -\sum v\,\delta p - \Delta\text{ke} - \Delta\text{pe} \tag{4–54}$$

CALCULUS FOR CLARITY 4–3

Equation (4–47) may be written in differential form as

$$d\text{ke} + d\text{pe} + dh = dq - dwk_{os} \tag{4–55}$$

Also, since $h = u + pv$, we have $dh = du + dpv$, and then

$$d\text{ke} + d\text{pe} + du + dpv = dq - dwk_{os} \tag{4–56}$$

or

$$dwk_{os} = dq - du - dpv - d\text{ke} - d\text{pe} \tag{4–57}$$

From the energy equation for a closed system, we have

$$dq - de = dwk_{cs}$$

Neglecting kinetic energy and potential energy changes gives $de = du$. Then equation (4–57) becomes

$$dwk_{os} = dwk_{cs} - dpv - d\text{ke} - d\text{pe} \tag{4–58}$$

For a reversible process, the closed-system work $dwk_{cs} = p\,dv$, which gives

$$dwk_{os} = p\,dv - dpv - d\text{ke} - d\text{pe} \tag{4–59}$$

Using the chain rule of differential calculus, we have

$$dpv = v\,dp + p\,dv \tag{4–60}$$

Substituting this result into equation (4–59) gives, for a reversible steady-flow process,

$$dwk_{os} = p\,dv - v\,dp - p\,dv - d\text{ke} - d\text{pe} = -v\,dp - d\text{ke} - d\text{pe} \tag{4–61}$$

If this equation is integrated from the inlet to the outlet, we have

$$wk_{os} = -\int v\,dp - \Delta\text{ke} - \Delta\text{pe} \tag{4–62}$$

which is equation (4–54) in calculus form. If a given mass is considered, then the open-system work is

$$Wk_{os} = mwk_{os} = -\int V\,dp - \Delta\text{KE} - \Delta\text{PE} \tag{4–63}$$

and this is the calculus form of equation (4–51) if the kinetic and potential energies are accounted for.

EXAMPLE 4–17 A steam turbine running under reversible steady-flow conditions takes in steam at 200 psia and exhausts it at 15 psia. Assuming that the steam has a specific volume of 4.0 ft^3/lbm at the inlet and the pressure–volume relation is

$$p = 228.48 - 7.12v$$

where p is in psia units and v is in ft^3/lbm units, determine the work done per pound-mass of steam flowing through the turbine. Neglect kinetic and potential energy changes of the steam.

Solution If we construct the p–v diagram for the expanding steam, we have the curve shown in the graph of figure 4–17. The work done per unit mass is the shaded area of the figure, or $wk_{os} = -\Sigma\, v\, \delta p$. This area is a rectangle and a triangle; that is,

$$wk_{os} = (200 - 15)\ \text{lbf/in}^2 \times (144\ \text{in}^2/\text{ft}^2)(4\ \text{ft}^3/\text{lbm})$$
$$+ (200 - 15)(144)\ \text{lbf/ft}^2\left(\frac{1}{2}\right)(29.98 - 4.00)\ \text{ft}^3/\text{lbm}$$
$$= 106{,}560\ \text{ft}\cdot\text{lbf/lbm} + 346{,}054\ \text{ft}\cdot\text{lbf/lbm}$$

or

$$wk_{os} = 452{,}614\ \text{ft}\cdot\text{lbf/lbm}$$

or

Answer

$$wk_{os} = 581.8\ \text{Btu/lbm}$$

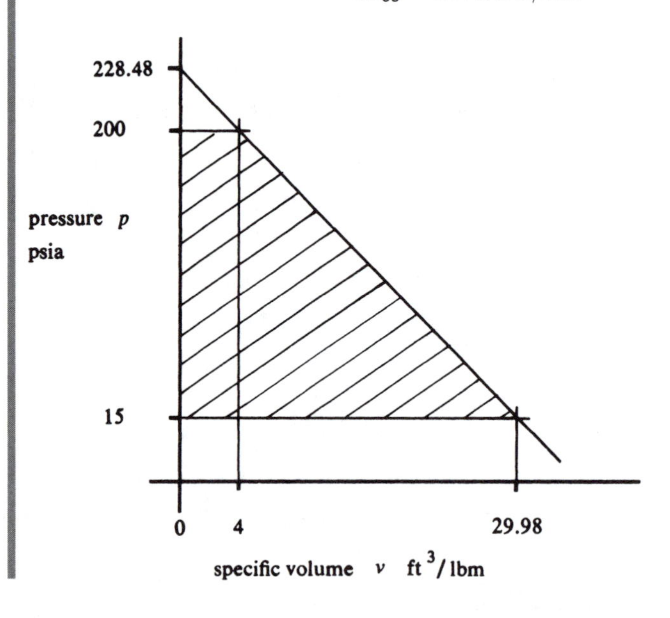

FIGURE 4–17 p–v diagram for process of example 4–17

EXAMPLE 4–18 A water pump, shown schematically in figure 4–18, delivers 1.0 m^3/min at 50-kPa gage pressure through a 2.5-cm-diameter pipe. If the water is at 101-kPa pressure and 27°C and has negligible velocity at inlet station 1, and if there is no heat transfer or friction in the system, determine the power required by the pump, $\dot{W}k_{pump}$.

Solution The pump is an open system and is assumed to have steady flow. If the boundary is then drawn around the pump as shown in figure 4–18, we can apply equation (4–49) to the system:

$$\dot{m}\left[\frac{1}{2}(\overline{V}_2^2 - \overline{V}_2^2) + g(z_2 - z_1) + h_2 - h_1\right] = \dot{Q} - \dot{W}k_{pump}$$

FIGURE 4–18

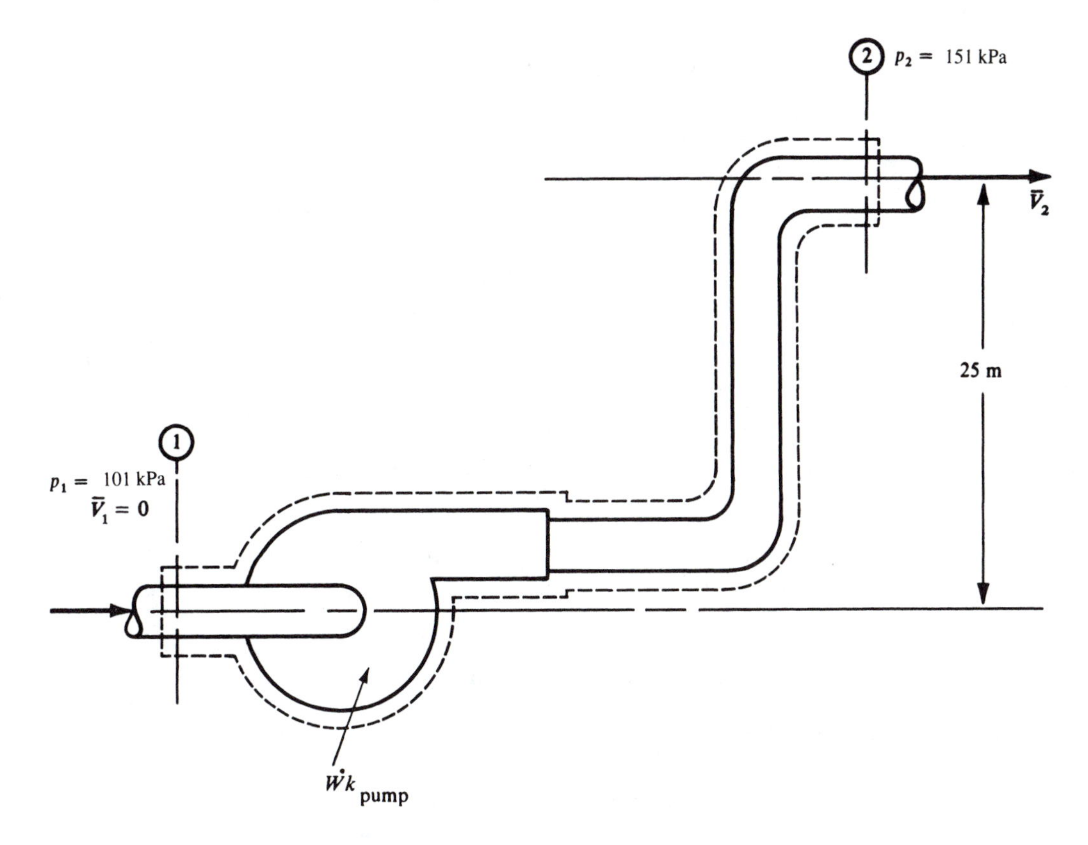

Since no heat is transferred, $\dot{Q}$ is zero and $\bar{V}_1$ is nearly zero. The elevation of the water is increased by 25 m, so that $(z_2 - z_1) = 25$ m. If the internal energy change of the water is neglected, then $u_2 = u_1$, and the enthalpy change becomes

$$h_2 - h_1 = p_2 v_2 - p_1 v_1$$

For water, we may use a value of 1000 kg/m^3 for the density, so the specific volume v becomes 0.001 m^3/kg. The volume flow through the pump is 1.0 m^3/mim, or 0.0167 m^3/s, and the mass flow is simply

$$\dot{m} = \dot{V}\rho$$

or

$$\dot{m} = (0.0167 \text{ m}^3/\text{s})(1000 \text{ kg/m}^3)$$
$$= 16.7 \text{ kg/s}$$

The velocity at station 2 is, from equation (4–6),

$$\bar{V}_2 = \frac{\dot{m}}{A_2 \rho_2}$$

where A_2 is the cross-sectional area at (2). We find that

$$A_2 = \pi(1.25 \text{ cm})^2 = 4.91 \text{ cm}^2 = 4.91 \times 10^{-4} \text{ m}^2$$

Then the velocity is

$$\bar{V}_2 = \frac{16.7 \text{ kg/s}}{(4.91 \times 10^{-4} \text{ m}^2)(1000 \text{ kg/m}^3)}$$
$$= 34.0 \text{ m/s}$$

Substituting the foregoing values into the energy equation yields

$$(16.7\ \text{kg/s})\left[\frac{1}{2}(34.0\ \text{m/s})^2 + (9.8\ \text{m/s}^2)(25\ \text{m}) + (151\ \text{kN/m}^2 \times 0.001\ \text{m}^3/\text{kg}) - (101\ \text{kN/m}^2 \times 0.001\ \text{m}^3/\text{kg})\right] = -\dot{W}k_{\text{pump}}$$

or

$$(16.7\ \text{kg/s})(578\ \text{m}^2/\text{s}^2 + 245\ \text{m}^2\text{s}^2 + 151\ \text{N}\cdot\text{m/kg} - 101\ \text{N}\cdot\text{m/kg}) = -\dot{W}k_{\text{pump}}$$

Notice that the units of m^2/s^2 are equivalent to the units of $\text{N}\cdot\text{m/kg}$. This can be seen by recalling that $1\ \text{N} = 1\ \text{kg}\cdot\text{m/s}^2$, and by substituting this into the $\text{N}\cdot\text{m/kg}$ units, we obtain $\text{kg}\cdot\text{m}\cdot\text{m/s}^2\cdot\text{kg}$ or m^2/s^2. Continuing now, we find that

$$\dot{W}k_{\text{pump}} = -14{,}600\ \text{kg}\cdot\text{m}^2/\text{s}^3$$

or

$$\dot{W}k_{\text{pump}} = -14{,}600\ \text{N}\cdot\text{m/s} = -14{,}600\ \text{J/s}$$

Answer

$$= -14{,}600\ \text{W} = -14.6\ \text{kW}$$

Many pumps are still rated in horsepower (hp) units, and from the inside front cover table, we find that 1.34 hp = 1 kW, so

Answer

$$\dot{W}k_{\text{pump}} = -14.6 \times 1.34 = -19.6\ \text{hp}$$

4–9 SUMMARY

In thermodynamics, the conservation principles are at the foundation of all applications. The first conservation law considered is that of mass:

$$m_{\text{in}} - m_{\text{out}} = \Delta m_{\text{system}} \tag{4–2}$$

Alternatively, we may write

$$\dot{m}_{\text{in}} - \dot{m}_{\text{out}} = \dot{m}_{\text{system}} \tag{4–5}$$

For steady flow, we have

$$\dot{m}_{\text{in}} = \dot{m}_{\text{out}} \tag{4–12}$$

or

$$\rho A\overline{V} = \text{constant} \tag{4–13}$$

For uniform flow filling processes in which no mass is leaving the open system, we have

$$\dot{m}_{\text{in}} = \dot{m}_{\text{system}} \tag{4–14}$$

and for constant mass flow rate in, $\dot{m}_{\text{in}}$,

$$\dot{m}_{\text{in}}\,\Delta t = \Delta m_{\text{system}} \tag{4–18}$$

For uniform flow emptying processes,

$$\dot{m}_{\text{out}} = -\dot{m}_{\text{system}} \tag{4–19}$$

and the constant mass flow rate,

$$\dot{m}_{\text{out}}\,\Delta t = -\Delta m_{\text{system}} \tag{4–21}$$

The conservation of energy, here considered the first law of thermodynamics, is the principle upon which this book is primarily built. For the system, we have

$$\Delta E = Q - Wk \tag{4–24}$$

or

$$\dot{E} = \dot{Q} - \dot{W}k \tag{4–25}$$

and for isolated systems, these reduce to

$$\Delta E = 0 = \dot{E}$$

For the closed system,

$$\Delta E = Q - Wk_{cs} \tag{4–27}$$

where, for reversible processes involving boundary work,

$$Wk_{cs} = \sum p\,\delta V$$

The closed-system energy balance per unit mass is

$$\Delta e = q - wk_{cs}$$

For the open system, flow energy is identified as the energy associated with motion of mass across system boundaries. This term is evaluated from the product pV, or pv per unit mass, and together with internal energy is called *enthalpy* H; that is,

$$H = U + pV \tag{4–39}$$

or

$$h = u + pv \tag{4–38}$$

The general equation of the conservation of energy for an open system is

$$\dot{\text{KE}}_2 - \dot{\text{KE}}_1 + \dot{\text{PE}}_2 - \dot{\text{PE}}_1 + \dot{H}_2 - \dot{H}_1 + \dot{E}_{\text{system}} = \dot{Q} - \dot{W}k_{os} \tag{4–41}$$

The open-system conservation of energy for steady-flow condition is, per unit mass,

$$\frac{1}{2}(\overline{V}_2^2 - \overline{V}_1^2) + g(z_2 - z_1) + h_2 - h_1 = q - wk_{os}$$

where wk_{os} is the work of the open system and is evaluated from the equation

$$wk_{os} = -\sum v\,\delta p - \Delta\text{ke} - \Delta\text{pe} \tag{4–54}$$

DISCUSSION QUESTIONS

Section 4–1

4–1 What is meant by *mass flow rate*?

4–2 What is meant by *volume flow rate*?

Section 4–2

4–3 What is meant by the term *steady flow*?

Section 4–3

4–4 What is meant by the term *uniform flow*?

4–5 What is the *filling process*?

4–6 What is the *emptying process*?

Section 4–4

4–7 Why is work out of a system designated as positive while work in is negative?

Section 4–5

4–8 What is the *first law of thermodynamics*?

Section 4–6

4–9 What is an *isolated system*?

4–10 What is meant by the term *adiabatic*?

Section 4–7

4–11 What is meant by *flow energy*?

4–12 What is *enthalpy*?

Section 4–8

4–13 What is meant by *open system*?

4–14 What is meant by *shaft work*?

4–15 Why is open system work, given by equation (4–52), different than closed system work, given by equation (3–9)?

PRACTICE PROBLEMS

Sections 4–1 and 4–2

Problems that use SI units are indicated by an (M) under the problem number; those that use English units are indicated by an (E). Mixed unit problems are indicated by a (C). Problems marked with an asterisk (*) after the problem number are often more difficult and best analyzed with the use of calculus.

4–1 (M) Water having a density of 1000 kg/m^3 is flowing with a velocity of 3 m/s through a round pipe. There is a restriction within the pipe where the diameter is one-half the normal diameter. Determine the water velocity at the restriction.

4–2 (M) Methyl alcohol at 20°C flows through a plastic tube, and 1 kg/s is required to be delivered. If the plastic tube limits the velocity to 5 m/s or less, what is the minimum acceptable diameter of the tube?

4–3 (M) Air flows through the converging nozzle shown in figure 4–19 with a velocity of 240 m/s at station *A*. The density of the air is found to be 0.48 kg/m^3 at *A* and 1.12 kg/m^3 at *B*. If the area of the nozzle at *A* is 0.1 m^2 and at *B* it is 0.05 m^2, determine

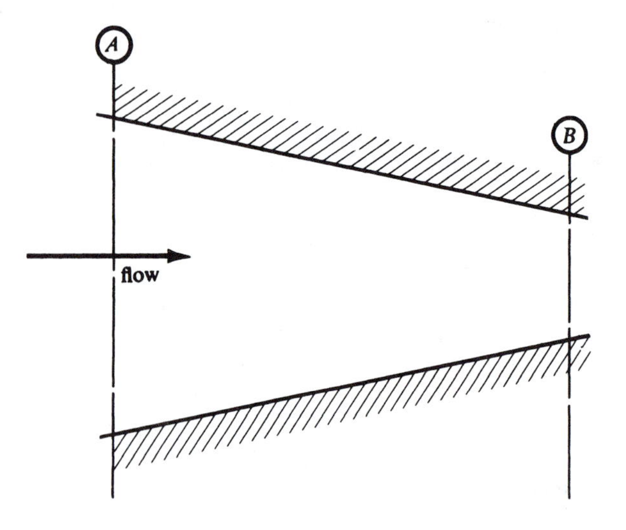

FIGURE 4–19

(a) the mass flow rate of the air through the nozzle.
(b) the velocity of the air at station *B*.

4–4 (M) Water at 20°C is delivered through a pipe at 1.0 m^3/min. If the pipe has an inside diameter of 4 cm, determine the specific kinetic energy of the water.

4–5 (M) Pipes *A* and *B* are joined together to supply pipe *C* as shown in figure 4–20. Pipe *A* is required to convey 1.5 m^3/min of water, and pipe *B* 2.5 m^3/min. If the pipes restrict the maximum velocities to 6 m/s, determine the diameters of pipes *A*, *B*, and *C*.

4–6 (M) In figure 4–21, air flows through a 2.5-cm-diameter duct that is being heated from below. If the entering air has a density of 1.2 kg/m^3 at station *A* and 0.64 kg/m^3 at *B*, find the entering velocity of the air if it is 30 m/s at *B*.

4–7 (M) Forty thousand kg/h of steam is exhausted from a steam turbine at a nuclear power station. The steam, flowing through a 5-cm-diameter pipe with a specific volume of 15 m^3/kg, enters a closed condenser where the steam cools and becomes liquid water with a density of 1000 kg/m^3. This water then flows at 24 m/s through a pipe back to the nuclear reactor, where it is boiled and becomes steam for yet another passage through the steam turbine. What pipe diameter would you select for conveying the water from the condenser?

4–8 (M) A sprinkler consisting of an open square tank 60 cm on a side discharges 1.0 kg/s of water through holes in the tank bottom. If the tank is half full at a certain time, how great a flow must be provided from an external faucet to fill the tank to the top with water in 2 min if the discharge is constant at 1.0 kg/s? The tank is 60 cm tall.

4–9 (M) Into a mixing chamber flows 13 kg/s of water at 12°C, 9 kg/s of water at 85°C, and 20 kg/min of methyl alcohol as shown in figure 4–22. If 23 kg/s of mixed solution is flowing out, determine the rate of accumulation or decline of mass in the mixing chamber.

4–10 (E) A 6-in-diameter pipe is being used to convey 600 lb of water per minute. Assuming that the water has a density of 62.4 lbm/ft^3, determine the velocity of the water.

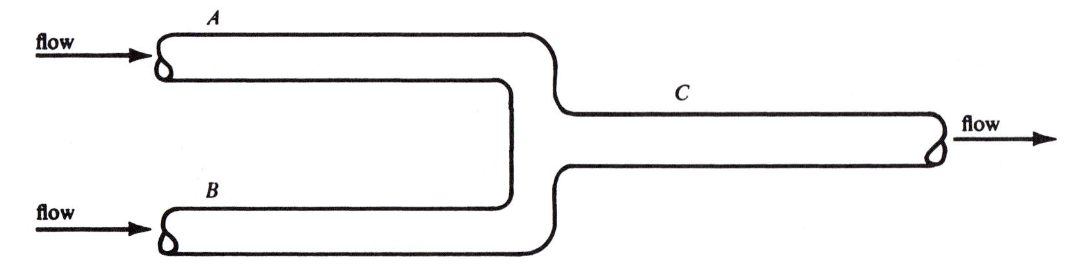

FIGURE 4–20

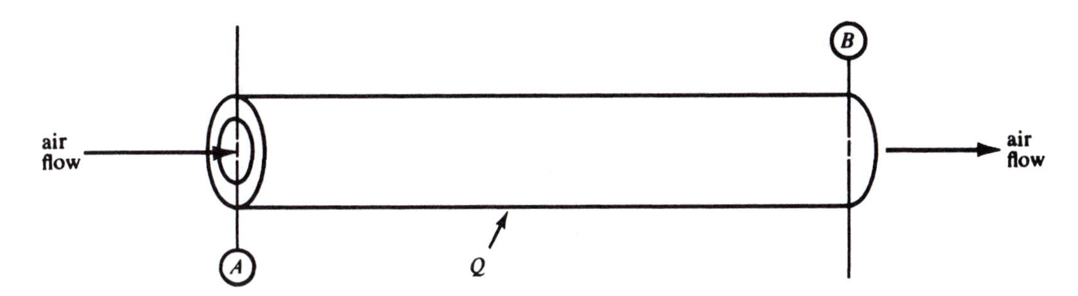

FIGURE 4–21

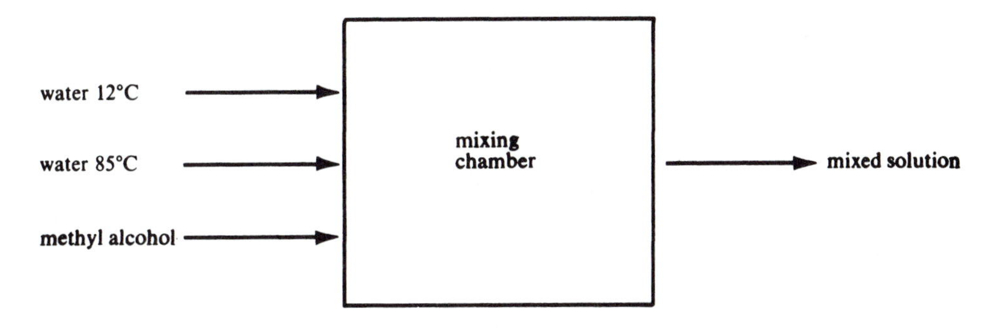

FIGURE 4–22

4–11 (E) Shown in figure 4–23 is a converging-diverging nozzle through which benzene is flowing. Assume that the temperature of the benzene remains at 68°F as it passes through the nozzle. If 60 lbm/s of benzene flows through the nozzle, determine
(a) the mass flow at the throat, station B, and at the exit, station C.
(b) the velocity of benzene at station B.
(c) the velocity at station C.

4–12 (E) Air flows through the converging-diverging nozzle of figure 4–23. The density of the air is found to be 0.045 lbm/ft^3 at station A, 0.060 lbm/ft^3 at station B, and 0.050 lbm/ft^3 at station C. If the velocity of air entering at station A is 400 ft/s, determine
(a) the mass flow rate of air.
(b) the velocity of air at station B.
(c) the velocity of air at station C.

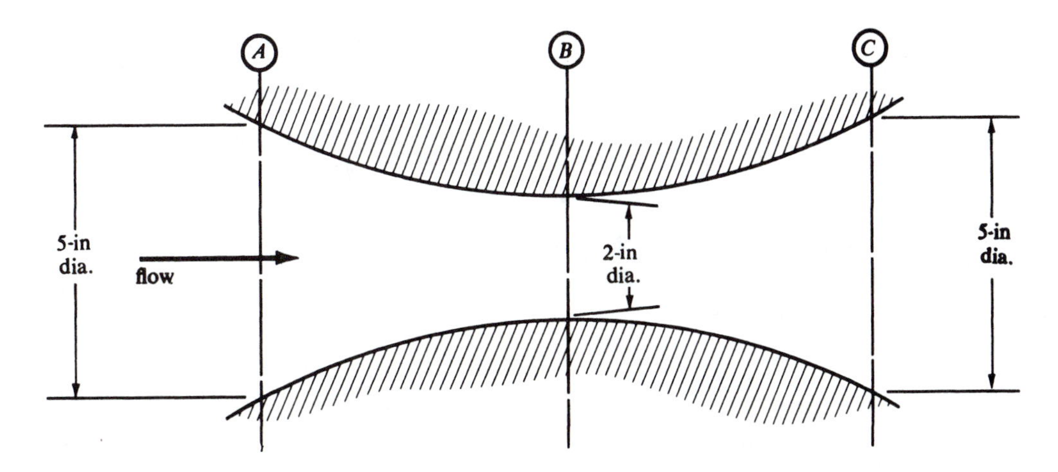

FIGURE 4–23

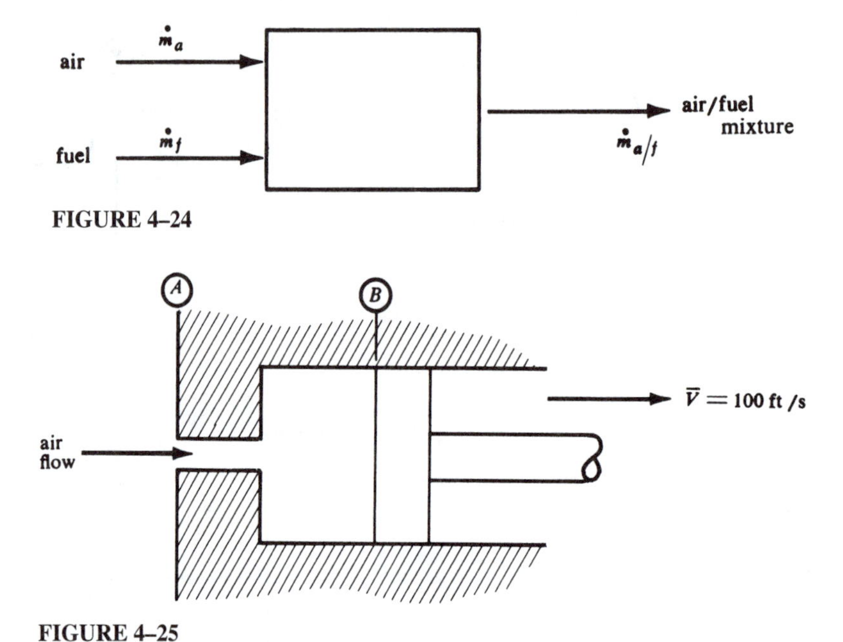

FIGURE 4–24

FIGURE 4–25

4–13 (E) Shown in figure 4–24 is a system diagram for a simple carburetor used to mix air and fuel. Under ideal conditions assume that the carburetor mixes 0.04 lbm of fuel for every lbm of air where the density of air is 0.08 lbm/ft^3. Assume that the fuel has a density of 60 lbm/ft^3 and the required air–fuel mixture is 2 lbm/min. Determine the mass flow rates of fuel and air.

4–14 (E) Water flows through $^3/_8$-in-ID boiler tubes. Before being heated, the water in the pipes has a density of 62.5 lbm/ft^3 and a velocity of 10 ft/s. What would you expect the velocity of the water to be downstream after it has been heated such that the water density is 61.8 lbm/ft^3?

4–15 (E) In a combustor of a jet engine, 30,000 $\text{ft}^3\text{/min}$ of air with a density of 0.06 lbm/ft^3 enters through a cross-sectional area of 1 ft^2. In the combustor, 0.02 lbm of fuel is mixed and burns with every 1 lbm of air. If the burned gases exit through a 1-ft^2 area, determine their velocities. Assume that the burned gases have a density of 0.01 lbm/ft^3.

4–16 (E) A refrigeration system rated at 60 tons uses 260 lbm/min of freon-12. During the flow cycle of Freon through the refrigerator, it is required to lose pressure at a certain point called the *expansion valve*. If the velocity through this valve is restricted to values of 100 ft/s or less, what tube diameter should be used if the freon™ has a density of 78 lbm/ft^3?

4–17 (E) Balloons are filled with air having a specific volume of 12 $\text{ft}^3\text{/lbm}$. If the filled balloons have a volume of 0.5 ft^3 and the air is available at 0.01 lbm/s, determine the time required to fill each balloon.

4–18 (E) In figure 4–25, a 3-in-diameter piston travels outward at 100 ft/s at a given instant. If air at 14.7 psia and 78°F is flowing through a 1-in^2 port at *A*, find the velocity of air through the part required to keep the cylinder at a uniform density.

Section 4–3

4–19 (M) A 5-kL fuel tank is to be filled in 45 min. If fuel oil has a density of 920 kg/m^3, determine the mass flow rate required to fill the tank in this time.

4–20 (M) Liquid oxygen is pumped into a large tank at a rate of 500 lbm/s. Determine how much oxygen is in the tank after 30 s of pumping if the tank was empty to start.

4–21 (E) A transfer pump has start-up characteristics as shown in figure 4–26 and a fluid specific gravity of 1.00. Determine the mass of fluid that the pump can transfer in the first 3 s of operation.

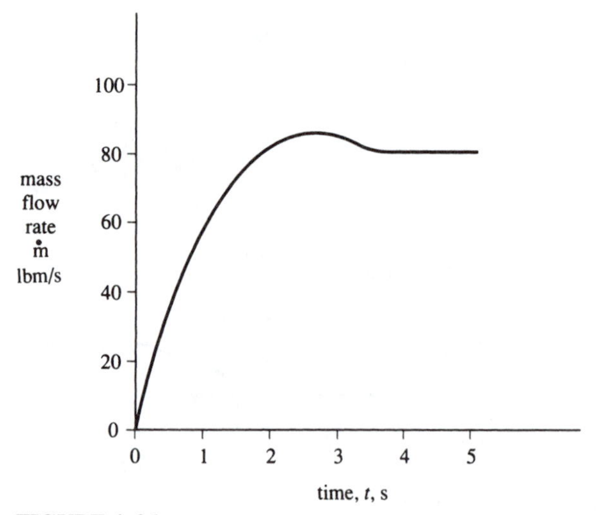

FIGURE 4–26

4–22 (M) Steam is exhausted through a pressure relief valve of a steam line at a rate of 25 kg/s. Determine the amount of steam escaping during a 20-min period.

4–23 (M) Referring to example 4–9 concerning the milk tank, determine the amount of milk drained from the tank during the 20-min period starting 25 min after the beginning of the draining.

4–24 (E) An oil refinery uses a cracking tower, as shown schematically in figure 4–27, to produce various petroleum products from crude oil. Determine the amount of crude oil required, in gallons per minute, to keep the tower in steady state.

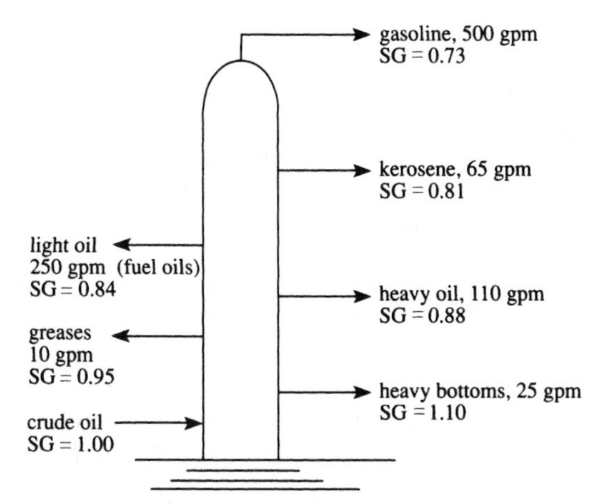

FIGURE 4–27

4–25 (M) A 2 m × 2 m × 1.5 m high tank is one-third full of water when water at 20°C is supplied to the tank at a rate of 3 kg/s. Simultaneously 80 kg/min of water is drained out of the tank as shown in figure 4–28.

(a) Is the tank filling or emptying?

(b) How long before the tank is full or empty?

4–26* (M) Determine how the amount of water in the tank of example 4–10 varies with the drain time. That is, plot the mass of water remaining in the tank against the drain time.

4–27* (E) Consider a 20-gallon ethyl glycol tank with a drain valve that allows a drain rate that is proportional to the amount of ethyl glycol in the tank by the relationship

$$\dot{m}_{\text{out}} = 0.5 m_{\text{system}}$$

where m_{system} is in lbm and $\dot{m}_{\text{out}}$ is in lbm/min. Determine the time required to drain 50 lbm of ethyl glycol from a full tank. Assume that the density of ethyl glycol is 70 lbm/ft^3. Also determine how the amount of ethyl glycol in the tank varies with the drain time.

4–28* (M) A tank contains 300 kg of ammonia vapor at a high pressure when it is considered to be "full." The emptying valve on the tank is found to allow a mass flow of ammonia out of the tank such that

$$\dot{m}_{\text{out}} = 0.1 m_{\text{system}}^{3/4}$$

where m_{system} is in kilograms and $\dot{m}_{\text{out}}$ is in kg/min. Determine the time for half the ammonia to be drained from a full tank.

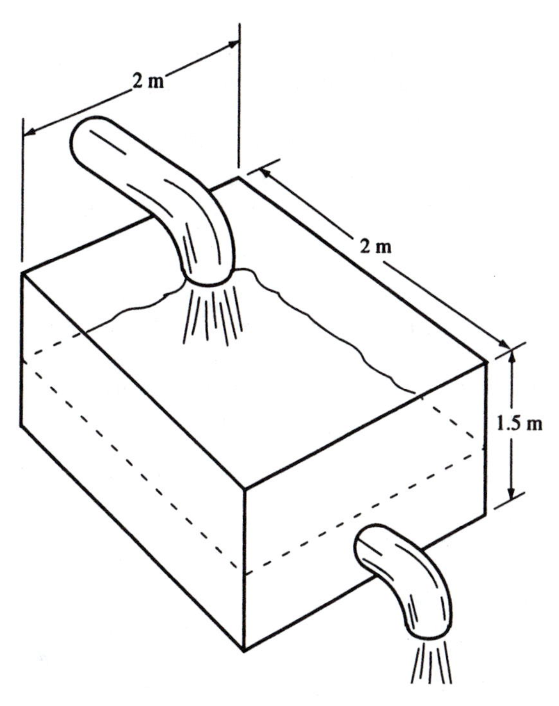

FIGURE 4–28

Sections 4–4 and 4–5

4–29 (M) Gases enclosed in a frictionless piston-cylinder increase their internal energy by 30 kJ when 40 kJ of heat is added. Determine the work of the piston-cylinder and indicate whether it is output or input to the device. Assume that there are no kinetic or potential energy changes.

4–30 (M) For the process where $Wk_{cs} = -200\ \text{N}\cdot\text{m}$ and $Q = 0$, find the change in internal energy.

4–31 (M) For the process where $q = 62.5$ kJ/kg and $wk_{cs} = 60$ kJ/kg, determine the energy change if the system mass is 2 kg.

4–32 (M) For the process involving 0.01 kg of air, find the internal energy change per kilogram if $Wk_{cs} = 20$ kJ and $Q = -10$ kJ.

4–33 (M) Into a closed boiler containing 100 kg of water, 150 kJ/s of heat are transferred. Determine the rate of change of energy of the water and the specific energy change; that is, find $\dot{U}$ and $\dot{u}$.

4–34 (E) For the process where $\Delta U = -20$ Btu and $Q = -20$ Btu, find Wk_{cs}.

4–35 (E) For the process where $wk_{cs} = 778\ \text{ft}\cdot\text{lbf/lbm}$ and $\Delta u = 0.75$ Btu/lbm, determine the heat transferred per pound-mass.

4–36 (E) For a process involving no heat transfer, find the output if the internal energy change is −16.8 Btu.

4–37 (E) For the process where $q = 8$ Btu/lbm and $wk_{cs} = 6224\ \text{ft}\cdot\text{lbf/lbm}$, find the energy change.

4–38 (E) During a reversible process, 10 hp is being delivered external to the system. If the system energy is changing by -10 Btu/s, determine the rate of heat transfer.

4–39 A battery supplies 100 W · h of electric energy. If 10 Btu is lost by the battery during this process, what was the total decline in energy of the battery?

Section 4–5

4–40 (M) Three kilograms of methanol are contained in a closed, perfectly insulated jar (no heat transfers allowed). A paddle is inserted through the top as shown in figure 4–29 and, driven by an electric motor, agitates the methanol so that its internal energy increases by 24 kJ. Determine
(a) Wk_{cs}
(b) the irreversible work Wk_{irr} and the reversible work $Wk_{cs\text{ rev}}$.
(c) Q
(d) wk_{cs}, wk_{irr}, and q

4–41 (M) In figure 4–29, the electric motor supplies 20 kJ of paddle work (or irreversible work) to stir up 3 kg of methanol contained in the insulated jar. Assuming that the jar is not perfectly insulated so that 1 kJ of heat is transferred out during the stirring, find the change in energy of the methanol.

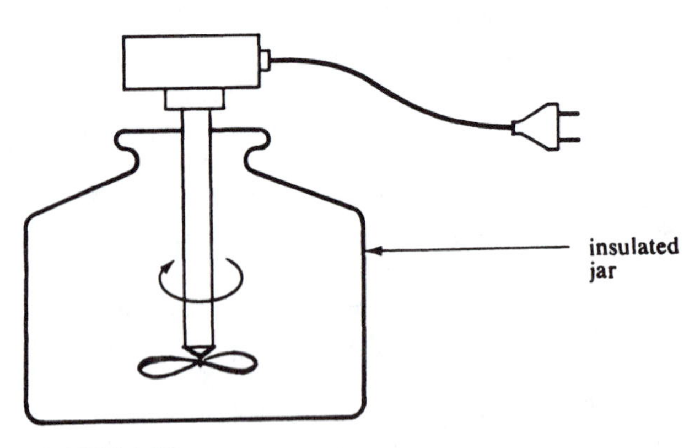

FIGURE 4–29

4–42 (E) One pound-mass of air at 78°F and 75 psia is contained in a piston-cylinder device. During a reversible process where the volume of the cylinder increases by 2 ft^3 due to the piston moving out, the pressure of the air remains constant. Determine the work produced during this process.

4–43 A heat engine drives an electric generator, thereby producing 100,000 W of power. If this process is completely reversible and the heat engine has no change in its internal energy, determine the rate of heat transfer required to the heat engine.

Section 4–6

4–44 (M) Two 1-kg balls, each with 20 J of kinetic energy in the positions shown in figure 4–30, constitute an isolated system along with the insulated container. The balls bounce around and come to rest at the bottom. By what amount has the internal energy increased in the system due to this dissipation process?

4–45 (E) An isolated system composed of 30 lbm and with a total energy of 3000 Btu exists. What are its energy and mass:
(a) 2 h after observing it?
(b) 2 years after observing it?
When do you observe a change in its mass or energy?

4–46 (E) An isolated system is composed of grain in a box as shown in figure 4–31. By what amount is the internal energy capable of increasing if the grain density is 38 lbm/ft^3?

Section 4–7

4–47 (M) Twenty kilograms per second of air with a pressure of 1000 kPa, temperature of 250°C, and a specific volume of 0.232 m^3/kg flows through a pipe. Determine its flow energy or flow work per kilogram.

4–48 (E) Water at 78°F flows through a 2-in-diameter pipe with a velocity of 30 ft/s. If the pressure is 60 psia, find the flow work of the water per pound-mass.

4–49 (E) A fuel pump is used to convey gasoline from a tank to a mixing chamber. At the entrance to the pump, the gasoline has a pressure of 14 psia and a density of 42 lbm/ft^3. As the gasoline leaves the pump, it has a pressure of 14.8

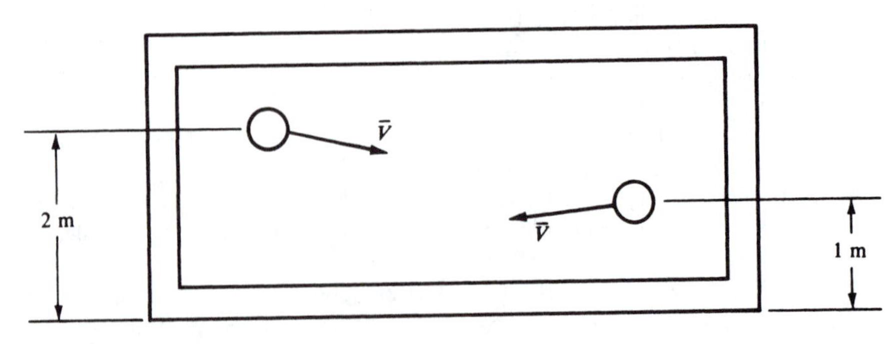

FIGURE 4–30

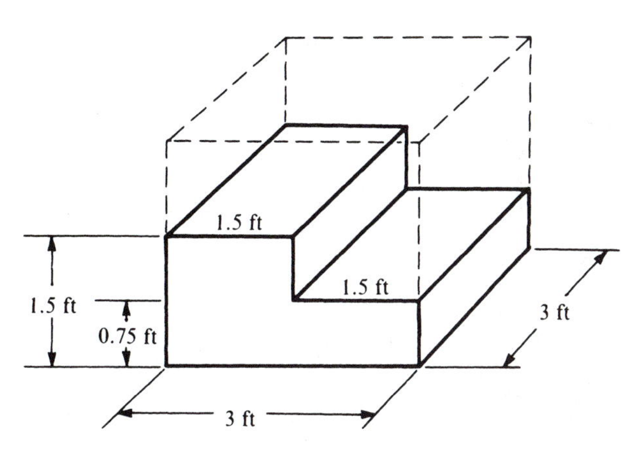

FIGURE 4–31

psia, and its density is 42 lbm/ft^3. If 0.1 lbm/s of fuel is pumped, determine the rate of change in total flow work of the gasoline per unit of time due to the pump.

4–50 (C) Calculate the rate of flow energy $(p\dot{V})$ for
(a) Problem 4–47.
(b) Problem 4–48.

For problems 4–51 through 4–56, calculate flow energy per pound-mass, enthalpy, and total enthalpy.

4–51 (M) Two kilograms of ammonia at 200 kPa, 0.755 m^3/kg, and internal energy of 1405.6 kJ/kg.

4–52 (M) Seven kilograms of steam having a pressure of 200 kPa, specific volume of 1.36 m^3/kg, and specific internal energy of 2839 kJ/kg.

4–53 (M) Seventy-five kilograms of air at 101 kPa density of 1.3 kg/m^3, and internal energy of 180 kJ/kg.

4–54 (E) One pound-mass of nitrogen at 20 psia, 120 Btu/lbm of internal energy, and density of 0.1 lbm/ft^3.

4–55 (E) One pound-mass of nitrogen at 200 psia, 1000 Btu/lbm of internal energy, and 0.1 lbm/ft^3 density.

4–56 (E) Ten pounds-mass of Freon-22, used as a refrigeration medium, under pressure of 112 psia with a specific volume of 0.7 ft^3/lbm and internal energy of 122.93 Btu/lbm.

Section 4–8

4–57 (M) Steam having a specific enthalpy of 160 kJ/kg and pressure of 1 kPa enters an adiabatic pump. Upon leaving the pump, the steam is at a pressure of 1200 kPa and has a specific enthalpy of 170 kJ/kg. No kinetic or potential energy changes occur through the pump. Determine
(a) the pump work, wk_{os}, per kilogram of steam.
(b) the average density of steam, assuming that the pump is reversible in nature.

4–58 (M) Steam with a specific enthalpy of 3278 kJ/kg enters a nozzle at station A shown in figure 4–32 with a velocity of 15 m/s. The exit area of the nozzle at station B is one-

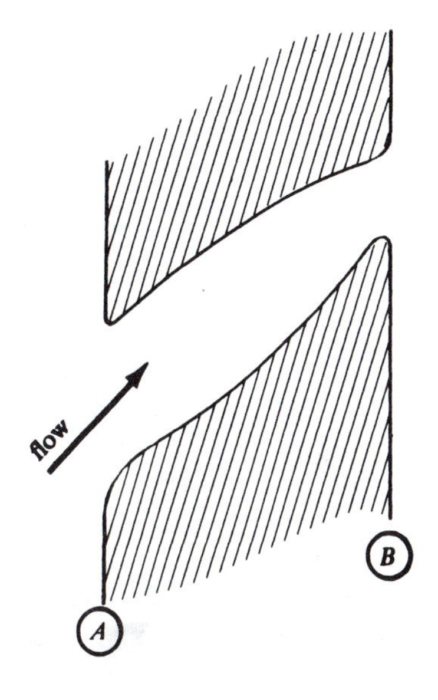

FIGURE 4–32

third the size of the inlet area at A, and the sides are assumed to be adiabatic, allowing no heat transfer. Determine the enthalpy per kilogram of steam leaving the nozzle if the steam is incompressible.

4–59 (M) A simplified single-cylinder internal combustion engine that produces 20 hp is shown in figure 4–33. It loses 40 kJ/min in radiated and conducted heat while using 0.73 kg/min of a fuel–air mixture which is assumed to have an enthalpy of 2600 kJ/kg of mixture. Determine the specific enthalpy of the exhaust gases, assuming that the engine is in a steady state.

4–60 (E) The steam turbine shown in figure 4–34 produces 290 Btu/lbm of work while it loses 8 Btu/lbm of heat. If the entering steam has an enthalpy value of 1530 Btu/lbm and there is negligible kinetic energy loss through the turbine, determine the enthalpy of exiting steam.

4–61 (E) A compressor used in a gas turbine engine increases the pressure of 3000 lbm/min of air from 15 psia to 150 psia without changing the kinetic or potential energy of the air. If the enthalpy of entering air is 118 Btu/lbm and that of the compressed air is 230 Btu/lbm, determine the power required in horsepower to drive the compressor if it is assumed to be reversible and adiabatic.

4–62 (E) In figure 4–35, a fan driven by a $^1/_4$-hp electric motor moves 40 lbm/min of air. Assuming that this process is done without heat transfer and the fan is able to divert the flow of air in a uniform direction, determine the average velocity $\overline{V}_{av}$ of air leaving the fan if it has negligible velocity before passing around the fan. (*Note:* Assume that the air has no enthalpy change.)

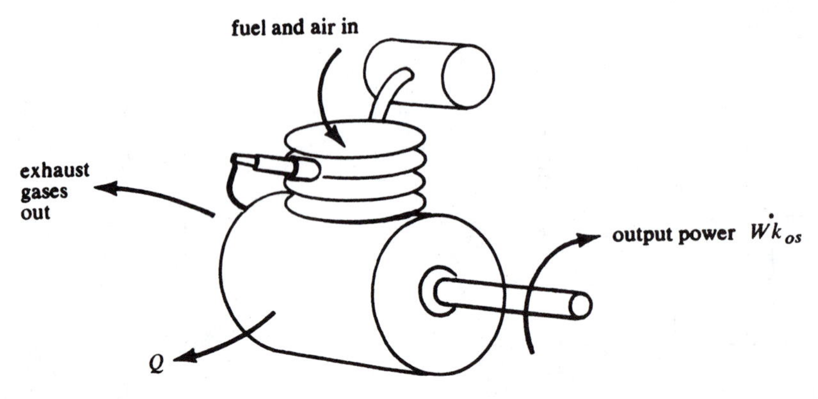

FIGURE 4–33

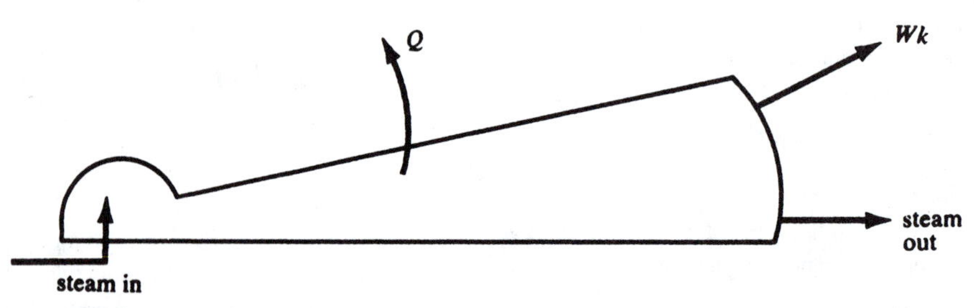

FIGURE 4–34

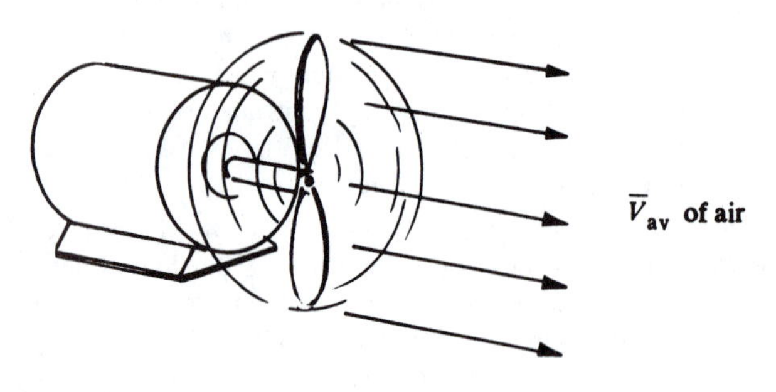

FIGURE 4–35

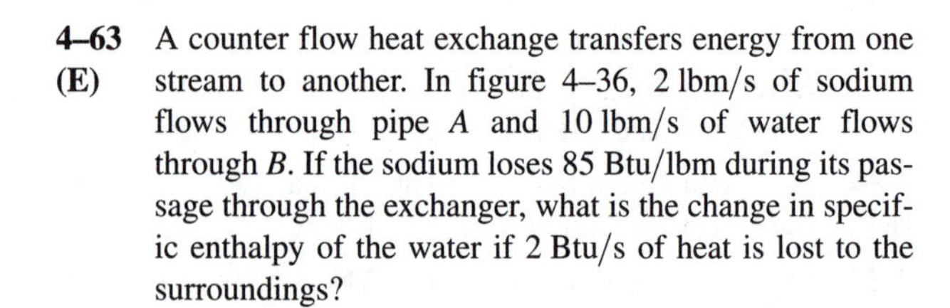

4–63 (E) A counter flow heat exchange transfers energy from one stream to another. In figure 4–36, 2 lbm/s of sodium flows through pipe *A* and 10 lbm/s of water flows through *B*. If the sodium loses 85 Btu/lbm during its passage through the exchanger, what is the change in specific enthalpy of the water if 2 Btu/s of heat is lost to the surroundings?

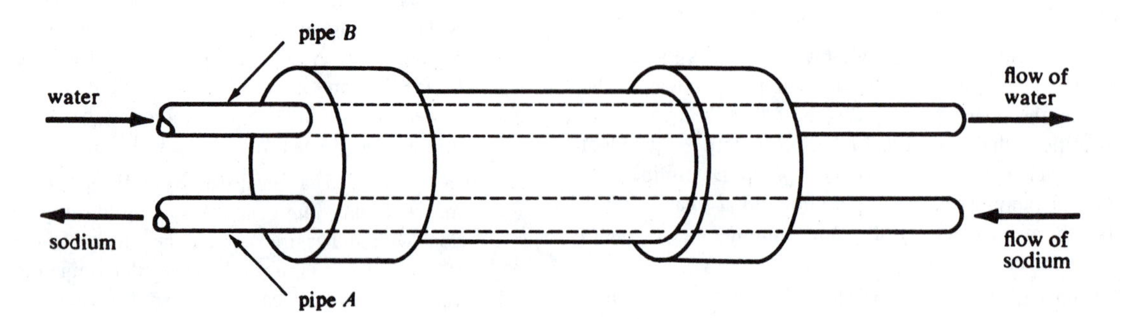

FIGURE 4–36

EQUATIONS OF STATE AND CALORIMETRY

The concepts of energy exchanges (heat and work) and the conservation of energy during such exchanges or processes were presented in the preceding two chapters. In the discussions of these concepts, it was inferred that the description of the system or its state was sufficiently complete. In this chapter, elementary attempts at satisfactorily describing the state of the system are presented. These involve the use of *equations of state* to relate known system properties to the unknown properties at a given location and time.

The idea of a pure substance is presented as a means to simplify the descriptions of thermodynamic systems or the manner in which materials behave. The phase diagram as a graphic means of showing the behavior of a material at different pressure–temperature conditions is presented, and then a discussion of pressure–volume (or density)—temperature relationships is given. Use of the p–v diagram and the three-dimensional p–v–T space is made. Some of the special terminology (particularly regarding phase-change phenomena) is introduced. The perfect gas is presented, and its defining equation, $pV = mRT$, is given attention. Then the superheated vapors are discussed, and the principle of corresponding states or compressibility factors is introduced. Solids and liquids are treated briefly and only as incompressible materials.

Caloric equations of state, which relate thermal energy to temperature and pressure, are discussed, and the perfect gas relationships are introduced through the presentation of Joule's experiment. Superheated vapors are discussed, and the use of the superheat tables for determining their internal energies and enthalpies is demonstrated. Methods of determining internal energies and enthalpy of incompressible substances are shown. Calorimetry or the methods for measuring internal energy and enthalpy are presented, and descriptions are given of some standard testing apparatuses of calorimetry.

5–1 EQUATIONS OF STATE AND PURE SUBSTANCES

Equations of state are equations or mathematical relationships that are used to determine properties of a system or a material at a given state (or a given position and time) from known values or other properties. As we saw in chapter 1, the known properties are the independent variables, and the unknown properties are the dependent variables. Therefore, if we want to know the state of a system, we should know the equations of state relating the various properties of that system at a given condition or state.

New Terms

c	Specific heat of an incompressible material	c_v	Specific heat at constant volume
C_p	Total specific heat at constant pressure	h_f	Enthalpy of saturated liquid
c_p	Specific heat at constant pressure	h_{fg}	Heat of vaporization
C_v	Total specific heat at constant volume	h_g	Enthalpy of saturated vapor
		k	c_p/c_v
		MW	Molecular weight

N_m	Number of moles	u_g	Internal energy of a saturated vapor
p_c	Critical pressure	v_c	Critical specific volume
p_R	Reduced pressure	v_f	Specific volume of a saturated liquid
R_u	Universal gas constant		
s	Specific entropy	v_g	Specific volume of a saturated vapor
s_f	Specific entropy of a saturated liquid		
s_g	Specific entropy of a saturated vapor	v_R	Reduced volume
T_c	Critical temperature	$v_{R'}$	Pseudoreduced volume
T_R	Reduced temperature	Z	Compressibility factor
u_f	Internal energy of a saturated liquid	χ (chi)	Quality

To reduce the complications of determining equations of state, in this book we consider particular materials called **pure substances**. These are substances that are uniform in chemical composition and stable so that they will not change as long as they remain in the state at which they are observed. Examples of pure substances are oxygen gas, water, pure copper, carbon dioxide gas, and air. You should be able to recognize many more pure substances. Air is given as an example of a pure substance even though it is a mixture of many different gases. We are able to do this because we can identify air as a particular combination or mixture of oxygen, nitrogen, argon, and some other gases, each of which remains separate from the other types of gases and will not react with them chemically. Other mixtures of pure substances will react chemically to produce a different chemical species, and these would not be pure substances. An example of a mixture that is not a pure substance is oxygen and hydrogen gas. This mixture reacts chemically to produce water, which is then, finally, a pure substance.

Experimental observations of pure substances have given us the following principle:

> **State principle:** ***The number of independent variables or system properties required to fully know the state of a system composed of a pure substance is one (1) plus the number of different reversible work modes.***

In chapter 4, we saw that reversible work can be done through springs, boundary movement, or any situation where a force acts through a distance. In this book, boundary work, $\sum p\,\delta v$, and spring work will be considered most often. Irreversible work modes do not contribute to those reversible work modes. Therefore, for a system having only one work mode, such as boundary work, and irreversible work, the number of properties required to determine the state of that system would be two, by the state principle. In such a case, if the temperature and the volume of the system were known, all the other thermodynamic properties could be found from these two properties: properties such as pressure, density, specific weight, internal energy, and enthalpy. In this chapter, we demonstrate how we can find those dependent properties.

Pure substances are substances occurring in one particular phase. If we construct a diagram or graph of pressure and temperature (p–T diagram) of a substance, we can qualitatively determine whether the three phases (solid, liquid, and vapor) are most likely to exist. Such a diagram, called a **phase diagram**, is shown in figure 5–1, which indicates that the solid phase occurs at low temperatures and at most pressures. At moderate pressures and temperatures one would find the liquid phase, at high temperatures the vapor or gaseous phase. The three phase regions are separated by three lines, indicated in figure 5–1 as the sublimation curve between solid and vapor, fusion curve (or freezing–melting curve) between solid and liquid phases, and the boiling–condensing curve between the liquid- and vapor-phase regions. These lines can be referred to as the transition curves of the phase diagrams. You may be able to recognize from your past experiences that a substance (water,

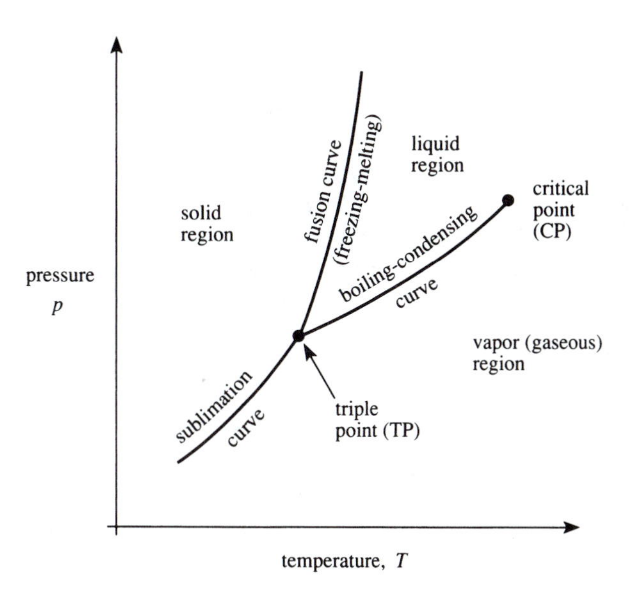

FIGURE 5–1 Phase diagram of a typical pure substance

for instance) goes from solid to liquid to vapor when its temperature is increased at a given constant pressure and crosses the transition curves.

But figure 5–1 also shows that, at constant temperature, you can increase the pressure on a pure substance and change a gas to a liquid or to a solid, or to a liquid and then a solid, depending on the temperature. Unfortunately, although most pure substances do behave as displayed in figure 5–1, water has a very special behavior in that it (almost alone among known materials) has a fusion curve like that shown in figure 5–2. In section 5–2, we shall see that this is true because water, upon freezing, expands or has increased volume.

Notice in figures 5–1 and 5–2 that the three transition curves (sublimation, fusion, and boiling) meet at a point called the **triple point**. This is a condition where all three phases could coexist in a mixture. Thus, water at its triple point could have ice, liquid, and steam all mixed together and be in equilibrium. Also, the boiling curve is shown to extend to a point called the **critical point**, CP. For a substance existing at a pressure and temperature

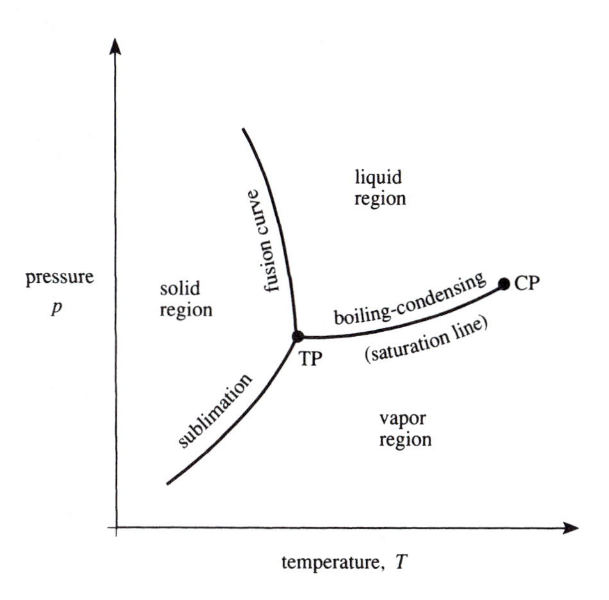

FIGURE 5–2 Phase diagram for water (not to scale)

beyond the critical point, the transition between liquid and vapor is not noticeable. We call this phase the **fluid phase** as a generic term for liquid or vapor. States beyond the critical state are called **supercritical states**.

5–2 PRESSURE–VOLUME–TEMPERATURE RELATIONSHIPS

In section 5–1, we saw that if two properties of a pure substance involved in boundary work (and maybe irreversible work) are known for a given condition, the state or all the other thermodynamic properties can be determined. In figures 5–1 and 5–2, it was shown that for temperature and pressure as the independent properties, the phase of the substance could be predicted. Let us now consider how a third property, volume (or specific volume), can be determined from the two properties. Figure 5–3 shows (in isometric projection, three dimensions) the relationship between specific volume and the temperature and pressure of a pure substance. Notice in the figure that there are surfaces that occur between the phase region when three properties are considered or when a property such as volume is dependent on two separate properties (temperature and pressure). If volume were dependent on only one property, say, pressure, we could construct a graph of volume as a function of p as we did in earlier chapters, where we obtained a line or curve of the function $v = f(p)$. But, now we have that a property requires two known dependent variables, so v is a function of T and p, or $v = f(T, p)$, and a surface results. Notice also in figure 5–3 that there are different distinct surfaces, which are labeled as specific regions: the super-

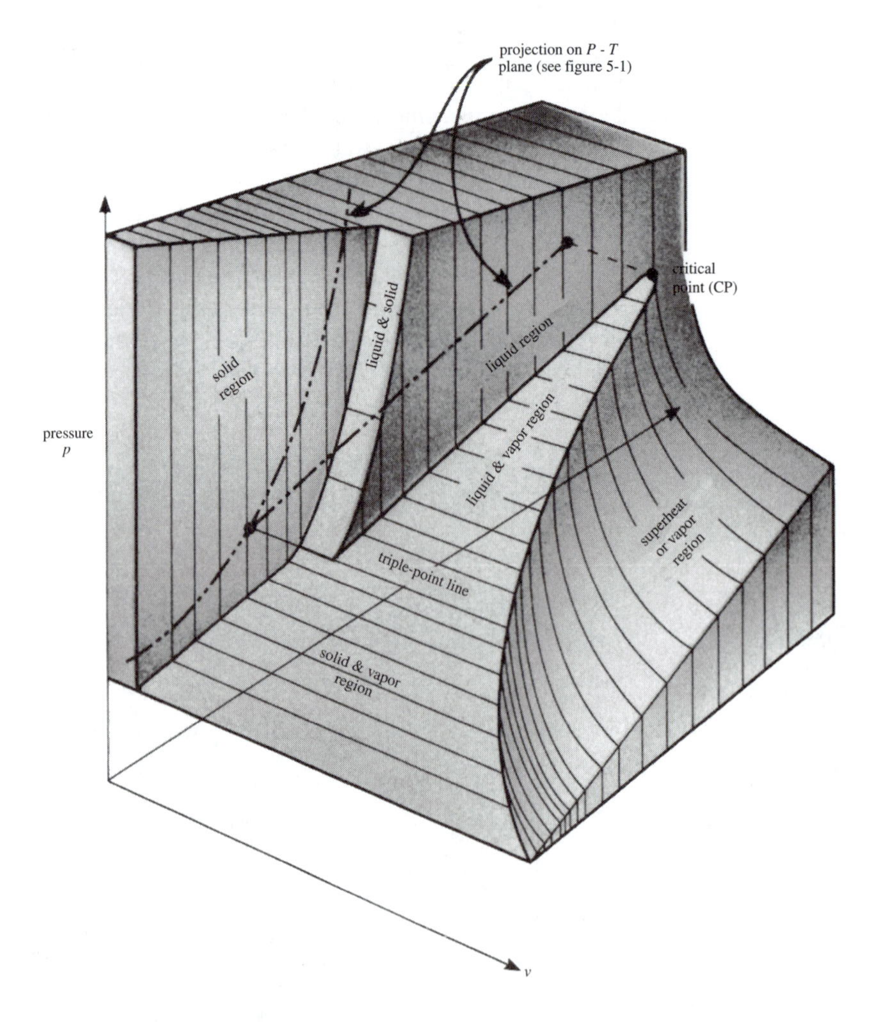

FIGURE 5–3
Pressure–volume–temperature projection of a typical pure substance

heat or vapor region, the solid region, the liquid region, the solid + vapor region, and so on. You can identify six distinct regions on figure 5–3, each of which represents a pure substance's behavior of volume to temperature and pressure.

Now, restudy figure 5–3 and notice when you look along the volume direction back toward the temperature–pressure plane that three surfaces (the solid + vapor, solid + liquid, and liquid + vapor regions) are exactly the transition lines given in the phase diagrams of figures 5–1 and 5–2. This fact means that if the material has a state that is on one of these three surfaces, the temperature and pressure are not independent of each other, and in fact temperature is dependent on pressure or pressure on temperature, whichever you prefer. On the other hand, the three other regions (vapor, liquid, and solid) are oblique to all three axes, T, p, and v, so that temperature and pressure are here independent properties, which is also indicated in figures 5–1 and 5–2. Also notice that the critical point is the topmost point of the liquid + vapor region, and the triple point in the phase diagram is actually a triple line in the three-dimensional condition.

If you have difficulty seeing the three dimensions of figure 5–3 and the various conclusions drawn from it, perhaps a two-dimensional figure would be more acceptable. To construct the two-dimensional relationship of $v = f(T, p)$ of figure 5–3, we could look down the temperature axis or on the p–v diagram, given in figure 5–4. Notice that the oblique surfaces are now flat and seem more useful for analysis. All the curves, surfaces, and points identified from figures 5–1 and 5–3 are included, and there are also included two curves called the saturated liquid and saturated vapor curves. These two lines, the saturated liquid and saturated vapor lines, are unique states of a material where, for saturated

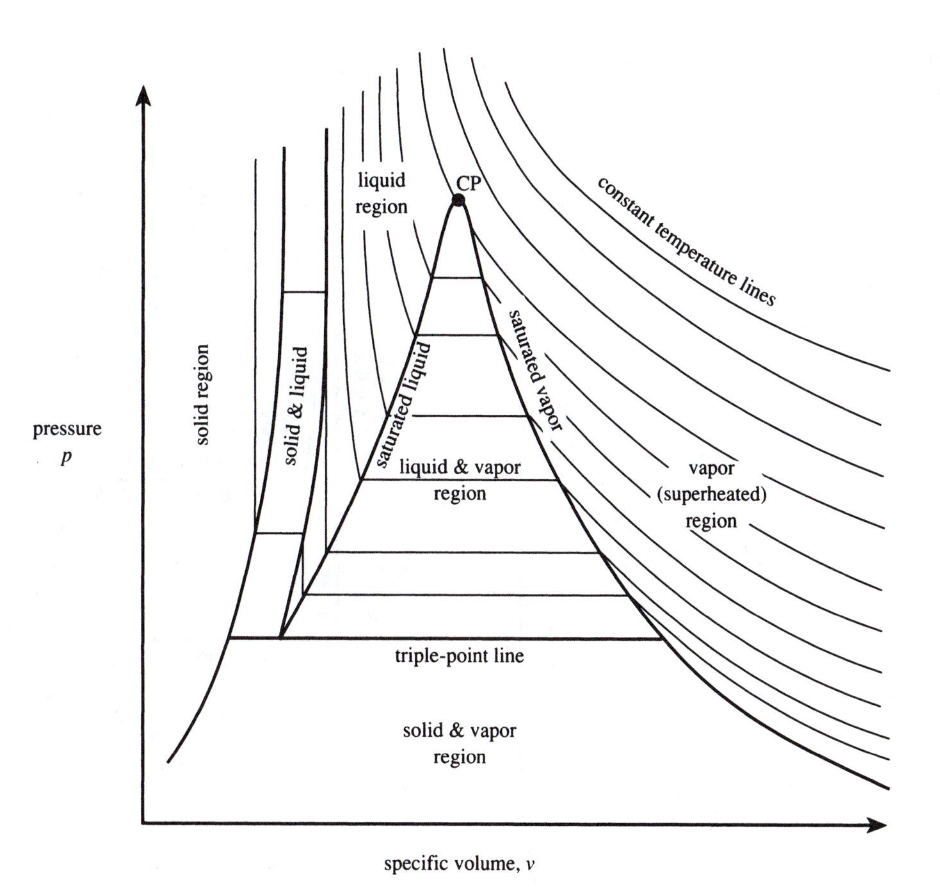

FIGURE 5–4
Pressure–volume projection of a typical pure substance

liquid, the change of phase from liquid to vapor is just beginning, and the saturated vapor line is that where the material has just completed the change to a vapor phase, or (in reverse) the beginning of the phase transition from a vapor to a liquid.

At this time, you may want to refer to the steam table B–11 and notice that it is constructed in such a way that the first column is temperature (or pressure) and the second column is pressure (or temperature). Thus, by knowing one or the other and expecting the substance to be in the liquid–vapor region, the pressure (or temperature) is fixed, as we have just discussed. The next three columns in table B–11 are entries for specific volume; the first column is for the specific volume of the saturated liquid (denoted v_f), the third column is for saturated vapor (denoted v_g), and the middle column is for the difference between v_f and v_g, or $v_g - v_f = v_{fg}$. The subscript f will always be used to denote the condition of saturated liquid so that the enthalpy of a saturated liquid is h_f, the internal energy u_f, and so on. Similarly, the enthalpy for a saturated vapor is denoted h_g, the internal energy u_g, and in general, the subscript g designates saturated vapor. We shall return later to the steam tables and other tables to discuss the details more fully.

It has been observed that water, a very common substance as well as an important engineering material, freezes or melts in a way that is not typical of most materials. It has been mentioned that its volume (specific volume) increases upon freezing or its density decreases. This fact allows ice to float on water and it makes the p–v diagram somewhat more difficult to interpret. Shown in figure 5–5 is the p–v diagram of water, not drawn to scale but accurately describing the situation of the phase regions and the phase transition regions. Notice that the solid + liquid region overlaps the liquid + vapor region, which

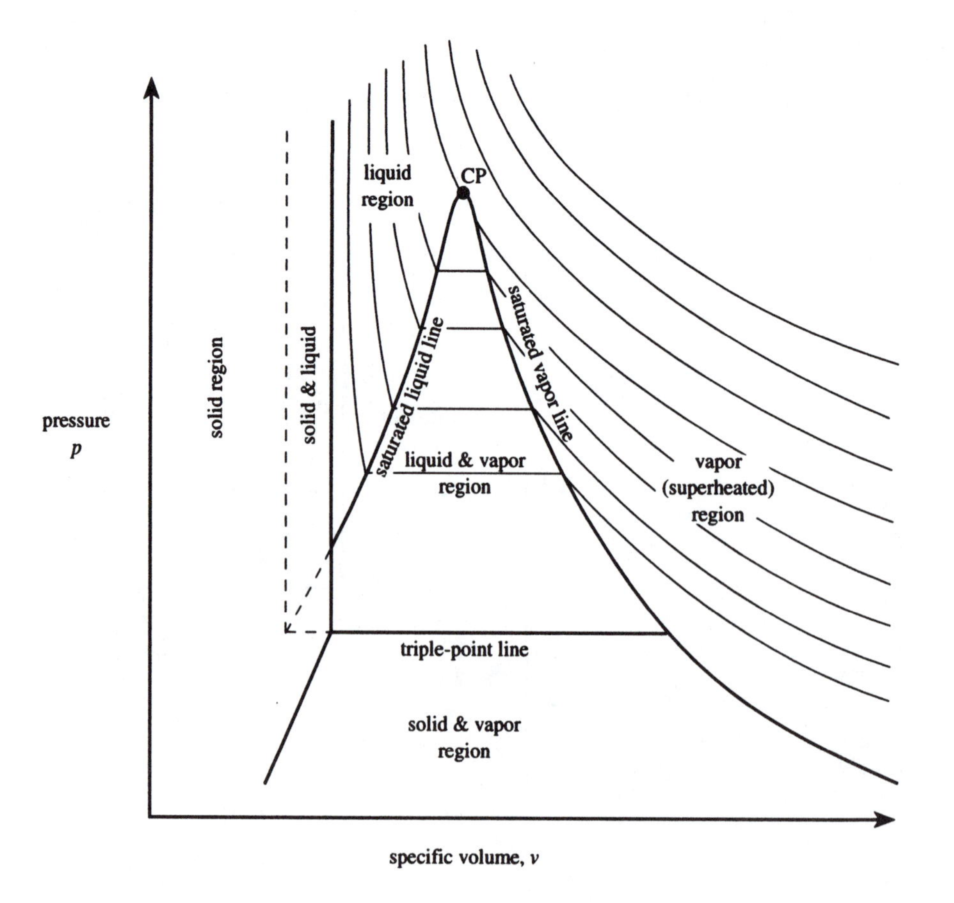

FIGURE 5–5
Pressure–volume projection of water (not to scale)

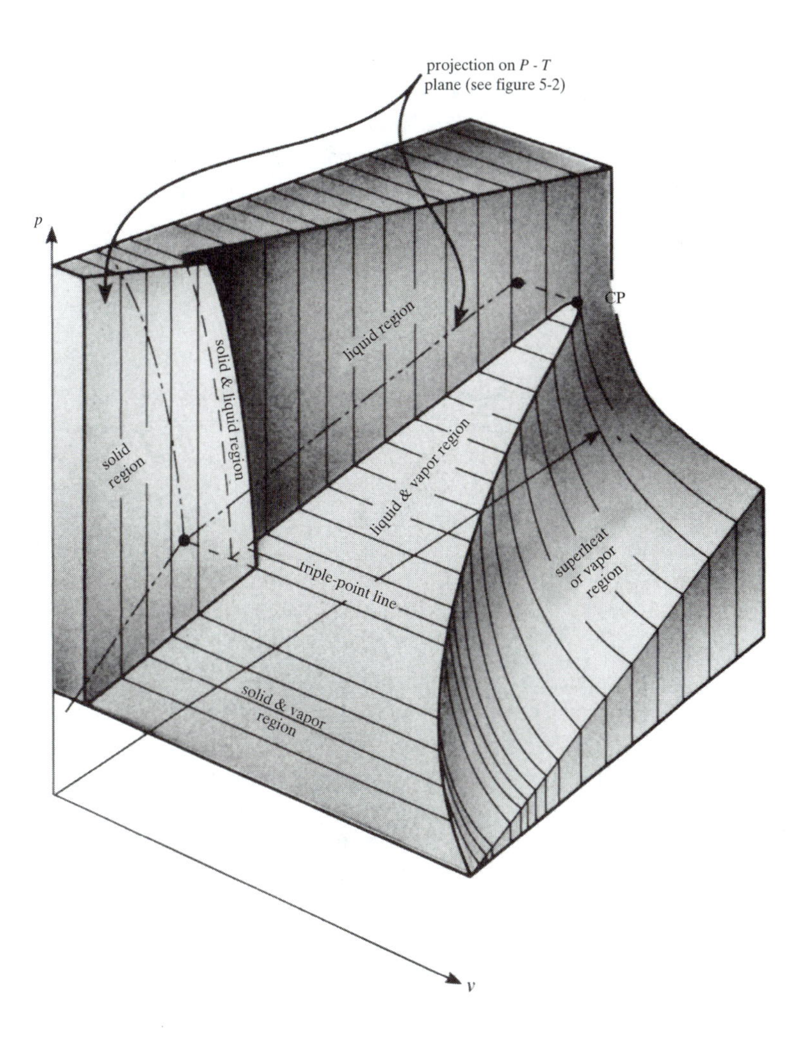

FIGURE 5–6
Pressure–volume–temperature projection of water (not to scale)

again is caused by the expansion of water upon freezing. Figure 5–6 shows the p–v–T projection of water, to illustrate its behavior in the three-dimensional space.

Now, in figures 5–4 and 5–5, notice that lines of constant temperature are included in the diagram. Particularly notice at large volumes or low densities and moderate pressures that constant-temperature lines followed up toward the pressure axis (the ordinate) do not go through the transition regions but are smooth, upward-bending curves. At low enough pressure and density (so that the volume is large), these constant-temperature lines are given by

$$T = \frac{pv}{R}$$

where R is a constant. In fact, this relationship is the same as that for what is defined as a perfect gas:

$$pv = RT \tag{5–1}$$

You have probably seen equation (5–1) before. Perfect gases are vapors which are visualized as being made up of atoms or molecules that do not attract one another in any way. It is also expected of perfect gas molecules that they be so small that they occupy no volume, and that when they hit each other or a container in which they are held, they rebound in a perfect way, with no irreversible work being done. No gases or vapors are perfect gases, but many materials seem to behave in a manner that, for our purposes, can be assumed to exhibit perfect gas behavior. Other forms of the perfect gas equation (5–1) can be written, such as

$$pV = mRT$$
$$pV = N_m R_u T$$
$$\frac{p}{\rho} = RT \tag{5–1}$$

where N_m is the number of moles of perfect gas, in kg · mol, g · mol, or lbm · mol.

The constant R is designated as the gas constant, and values for various real gases are listed in table B–4. The constant R_u is the universal gas constant and has a value of 8.31 J/g · mol · K or 1544 ft · lbf/lbm · mol · °R. The gas constant and the universal gas constant are related through the molecular mass, MW, of the particular substance by the equation

$$R = \frac{R_u}{\text{MW}} \tag{5–2}$$

Values for the atomic mass of the elements are listed in table B–1, and the molecular masses may be obtained with the chemical formula of the particular material involved. It is clear that the various forms of the perfect gas equation (5–1) allow us to determine one of the properties of temperature, pressure, and volume if the other two are known.

The perfect gas relationship is the most common and easily used equation of state, but caution must be exercised in its application. Materials that obey the perfect gas equations have the following qualitative characteristics:

1. Sufficiently rarefied so that no attractive or repulsive forces exist between any of the atoms
2. Sufficiently dense so that the material can be considered a continuous, uniform medium without voids or vacuums
3. Atoms with perfectly elastic collisions (reversible collisions) between themselves and the container enclosing the material

These assumptions are quite difficult to check, but it is generally helpful to "have a feel for" these characteristics. Those materials that obey the perfect gas law include air, most light gases, and free electrons in a solid, electrically conducting material. Obviously, liquids, solids, and dense gases such as moist steam do not obey the perfect gas law, and relationship (5–1) should not be used for defining these materials.

A good rule of thumb for determining when one may assume that a substance is a perfect gas is the following:

1. If the temperature is greater than or equal to the critical temperature and the pressure is less than one-fourth ($^1/_4$) the critical pressure, *or*
2. If the temperature is greater than or equal to about twice the critical temperature and the pressure is less than five times the critical pressure.

The critical temperature and critical pressure are those corresponding to the critical point identified in the phase diagram. Values for the critical temperature and pressure can be found for various materials in table B–5. Using the perfect gas assumption for a substance under conditions other than these two should be done with caution, and only after recognizing that a large error in answers may result.

EXAMPLE 5–1 Oxygen gas is in a 3-m^3 tank at 20°C and 0.6-MPa gage pressure. The atmospheric pressure is 101 kPa. How much gas is in the tank in kilograms?

Solution We first check to see if we can assume a perfect gas. From table B–5, the critical properties of oxygen are

$$p_c = 50.14 \text{ atm} = 50.14 \times 101.325 = 5080.4 \text{ kPa}$$
$$T_c = 154.78 \text{ K}$$

Since the oxygen is at 20°C or 293 K and its pressure is 0.6 MPa or 600 kPa plus the atmospheric pressure of 101 kPa (701 kPa), we can assume a perfect gas by the first of the two conditions listed for a perfect gas. Then, using either equation (5–2) and the molecular mass of oxygen from table B–1 (MW = 16.0 × 2 = 32.0) to find

$$R = \frac{R_u}{\text{MW}} = \frac{8.3143 \text{ kJ/kg} \cdot \text{mol} \cdot \text{K}}{32 \text{ kg/kg} \cdot \text{mol}} = 0.26 \text{ kJ/kg} \cdot \text{K}$$

or reading $R = 260 \text{ J/kg} \cdot \text{K} = 0.26 \text{ kJ/kg} \cdot \text{K}$ from table B–4, we can use equation (5–1) to find the mass. We obtain

$$m = \frac{pV}{RT}$$

or

$$m = \frac{(701 \text{ kPa})(3 \text{ m}^3)}{(0.260 \text{ kJ/kg} \cdot \text{K})(293 \text{ K})}$$

Converting the units of kilopascals and kilojoules to their base units, we have

$$m = \frac{(701 \times 10^3 \text{ N/m}^2)(3 \text{ m}^3)}{(260 \text{ N} \cdot \text{m/kg} \cdot \text{K})(293 \text{ K})}$$

and

Answer
$$m = 27.6 \text{ kg}$$

EXAMPLE 5–2 A perfect gas at 10 psia and 40°F is enclosed in a rigid tank of 3 ft^3. This gas is heated to 540°F and 20 psia, at which point its density is 0.1 lbm/ft^3. Determine the mass and the gas constant for this material.

Solution Since the gas is contained in a container of constant volume (3 ft^3), we can use the relation between density, volume, and mass to determine the mass. Thus,

$$\rho = \frac{m}{V}$$

from which we have

$$m = \rho V$$

Substituting values into this equation yields

$$m = (0.1 \text{ lbm/ft}^3)(3 \text{ ft}^3)$$

or

Answer
$$m = 0.3 \text{ lbm}$$

The gas constant can now be found from the perfect gas equation (5–1):

$$pV = mRT$$

The gas constant can be determined at either state. Let us first calculate it at the state with a pressure of 10 psia and 40°F. We can rearrange equation (5–1) to read

$$R = \frac{pV}{mT}$$

and substituting values into this equation we obtain

$$R = \frac{(10\ \text{lbf/in}^2)(3\ \text{ft}^3)(144\ \text{in}^2/\text{ft}^2)}{(0.3\ \text{lbm})(40 + 460°\text{R})}$$

or

Answer

$$R = 28.8\ \text{ft} \cdot \text{lbf/lbm} \cdot °\text{R}$$

At the second state, as a check on this last answer, we get

$$R = \frac{(20)(3)(144)}{(0.3)(540 + 460)}$$
$$= 28.8\ \text{ft} \cdot \text{lbf/lbm} \cdot °\text{R}$$

and this agrees with the result calculated at the first state, as it should.

The perfect gas equation is a straightforward equation of state that is useful for a large number of engineering applications. Notice that the pressure must be the absolute pressure and not the gage pressure, as indicated in example 5–1. Also, the temperature must be absolute temperature, either Kelvin or Rankine, and not the customary Celsius or Fahrenheit temperatures. A reasonable method for extending the usefulness of the form of the equation is through the principle of corresponding states. This principle states that all pure substances in the fluid region (liquids or vapors) can be described by the equation

$$pv = ZRT \qquad \textbf{(5–3)}$$

where Z is called the **compressibility factor**. The principle of corresponding states further contends that the compressibility factor Z is a function of two variables, the **reduced pressure** (p_R) and the **reduced temperature** (T_R), which are, respectively, defined as

$$p_R = \frac{p}{p_c} \qquad \textbf{(5–4)}$$

and

$$T_R = \frac{T}{T_c} \qquad \textbf{(5–5)}$$

The **critical pressure**, p_c, and the **critical temperature**, T_c, are the same as those given in table B–5.

The function of Z to p_R and T_R is given in charts B–1 and B–2, chart B–1 covering a pressure range from $p_R = 0$ to $p_R = 10.0$, and chart B–2, in more detail, the relation of Z to p_R and T_R between $p_R = 0$ and 1.0. Notice on charts B–1 and B–2 that there are dashed lines directed from the lower left to the upper right. These are labeled as $v_{R'}$, or reduced volume. But the term $v_{R'}$ is called the **pseudoreduced volume** because it is defined as

$$v_{R'} = \frac{vp_c}{RT_c} \qquad \textbf{(5–6)}$$

and a **reduced volume**, consistent with the definitions of equations (5–4) and (5–5), would be

$$v_R = \frac{v}{v_c} \tag{5–7}$$

The **critical volume**, v_c, listed in table B–5, is the actual volume of the materials listed at the critical point.

EXAMPLE 5–3 Consider nitrogen at a gage pressure of 600 psi and at a temperature of $-100°$F. Determine the specific volume and density of the nitrogen if the atmospheric pressure is 1 atm.

Solution: First, again, we check to see if the perfect gas assumption will be appropriate. The pressure is 600 psi/14.696 psi/atm = 40.83 atm gage. The absolute pressure is 41.83 atm. The temperature is $-100 + 460$ or 360°R = $360 \times {}^5\!/_9$K = 200 K. From table B–5, the critical values for nitrogen are 33.54 atm and 126.2 K. Clearly, from the two criteria for perfect gases, the nitrogen will not behave as a perfect gas at the given state of 600 psig, $-100°$F. We then use the principle of corresponding states: $p_R = p/p_c$ = 41.83 atm/33.54 atm = 1.247 and $T_R = T/T_c$ = 200 K/126.2 K = 1.584. From chart B–1, we find Z to be approximately 0.92 and $v_{R'}$ to be nearly 1.17. Using equation (5–3) and a gas constant value from table B–4 of 55.15 ft · lbf/lbm · °R, we have

$$v = \frac{ZRT}{p} = \frac{(0.92)(55.15\ \text{ft}\cdot\text{lbf/lbm}\cdot°\text{R})(360°\text{R})}{(614.696\ \text{psi})(144\ \text{in}^2/\text{ft}^2)}$$

Answer: $$= 0.206\ \text{ft}^3/\text{lbm}$$

An alternative method of finding v is to use equation (5–6):

$$v = \frac{RT_c v_{R'}}{p_c} = \frac{(55.15\ \text{ft}\cdot\text{lbf/lbm}\cdot°\text{R})(126.2 \times 9/5°\text{R})(1.17)}{33.53\ \text{atm} \times 14.696\ \text{psi} \times 144\ \text{in}^2/\text{ft}^2}$$

Answer: $$= 0.206\ \text{ft}^3/\text{lbm}$$

The density of the nitrogen would be found from the inverse or reciprocal of the specific volume,

Answer: $$\rho = \frac{1}{v} = 4.85\ \text{lbm/ft}^3$$

EXAMPLE 5–4 Acetylene is to be stored in a container at 27°C and 3.8 MPa. How large a container is required to store 3000 kg of the acetylene?

Solution: First we determine the absolute temperature and pressure. These are 300 K and 3.8 MPa or 3800 kPa, assuming that the given pressure has already accounted for atmospheric pressure. From table B–5, the critical values are 309.5 K and 61.6 atm or 61.6×101.325 kPa = 6241.62 kPa = 6.24 MPa. Clearly, the state is not one that allows us to assume a perfect gas. The reduced temperature and pressure are

$$T_R = \frac{T}{T_c} = \frac{300\ \text{K}}{309.5\ \text{K}} = 0.969$$

$$p_R = \frac{p}{p_c} = 0.6$$

and from chart B–2, Z is approximately 0.71. The gas constant for acetylene is 320 J/kg·K or 320 N·m/kg·K, from table B–4. Then, using equation (5–3), we find the specific volume:

$$v = \frac{ZRT}{p} = \frac{(0.71)(320\ \text{N}\cdot\text{m/kg}\cdot\text{K})(300\ \text{K})}{3{,}800{,}000\ \text{N/m}^2} = 0.0179\ \text{m}^3/\text{kg}$$

The volume of a container needed to hold 3000 kg of acetylene at this condition is then

$$V = vm = (0.0179\ \text{m}^3/\text{kg})(3000\ \text{kg})$$

Answer $$= 53.8\ \text{m}^3$$

Notice in examples 5–3 and 5–4 that the compressibility factor is actually a correction to the perfect gas equation. We could, in fact, write equation (5–3) as

$$v = Zv_{\text{perfect}} \tag{5–8}$$

where v_{perfect} is the volume predicted from the perfect gas equation. From charts B–1 and B–2 it can also be seen that the factor, Z, tends to make the predictions from the principle of corresponding states for volume smaller than from the perfect gas equation. On the other hand, at pressures greater than about eight times the critical pressure, the principle of corresponding states gives volumes that are greater than that predicted by the perfect gas law. For temperatures greater than about 2.5 times the critical temperature, the compressibility factor is always more than 1.0 and thus gives volumes greater than the perfect gas law.

There are many other equations relating pressure, volume, and temperature which have been suggested to better predict the state of substances from known properties. These are given in the scientific literature and are beyond the scope of this book.

5–3 CALORIC EQUATIONS OF STATE

We would now like to determine equations that predict the thermal properties, internal energy and enthalpy, from the temperature, pressure, or volume. We call these equations the **caloric equations of state**. There is a great temptation to say that the hotness, as measured by the temperature, is the same as the internal energy or maybe even the enthalpy. It turns out that there is a good agreement between temperature and internal energy for a gas, for a liquid, or for a solid, but in the regions of the phase changes, indicated in figures 5–3 and 5–4, there is no direct relationship between the temperature and the internal energy or the enthalpy.

Before we proceed with a detailed investigation of how internal energy is related to the common properties of temperature, pressure, and volume, let us here see how a simple experiment can indicate some powerful conclusions regarding internal energy and its relation to other properties. This experiment, named after James Joule, who first performed it in 1843, involves the use of the apparatus shown in figure 5–7. Tanks A and B are submerged in a water bath at room temperature. Tank A is filled with air at some high pressure, say, 300 psia, while tank B is empty or near minus (−) 14.7 psig. The two tanks are connected by a valve that is opened quickly. The temperature is recorded before and after the valve is opened and is found to be the same. During this process, the air expanded to fill both tanks, the pressure reached an equilibrium in A and B, and the volume of air increased from that in tank A initially to that in both tanks after expanding. The temperature remained constant, however, and since no work or heat was done across the boundary of the two tanks A and

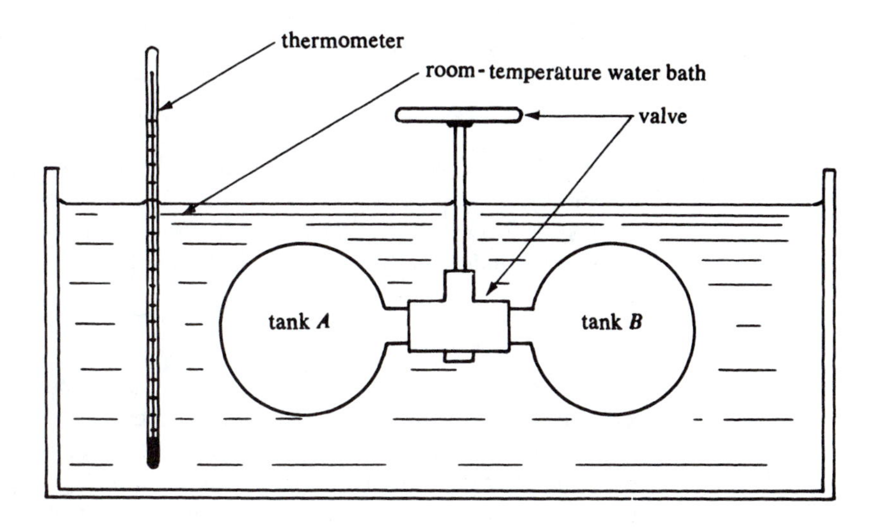

FIGURE 5–7 Apparatus for Joule's experiment

B, from the first law of thermodynamics, the internal energy remained fixed or constant; that is,

$$\Delta U = Q - Wk \tag{5–9}$$

But

$$Q = 0 \quad \text{and} \quad Wk = 0$$

so

$$\Delta U = 0$$

From these observations it is seen that the internal energy of the air cannot be a function of the pressure or volume of the air, and consequently, it must be a function of temperature only. This is true because pressure and volume changed, while temperature and internal energy remained unchanged during the expansion. We then write, for air,

$$U = f(T) \tag{5–10}$$

and this has been found true for all gases that can be considered as perfect gases. The significance here is that the internal energy of a perfect gas is *only* a function of temperature, and this result is a helpful simplification in studying gases.

Now we define a function that satisfies equation (5–10):

$$\delta U = mc_v \, \delta T \tag{5–11}$$

Here, m is the mass, δU is a small amount of internal energy change during a small temperature change δT, and c_v is called the specific heat at constant volume. This is an unfortunate name for c_v, but it is used by nearly everyone studying thermodynamics. Specific heat, c_v, is not related in any direct way to heat (which is an energy flow across a system boundary), and equation (5–11) is not limited to processes of constant volume. That is, equation (5–11) is good for any process of a perfect gas. Sometimes the term mc_v is written C_v and called the **total specific heat** of a perfect gas. Total specific heat depends on how much gas there is; that is, it depends on the amount of mass. Specific heat, c_v, is independent of the amount of mass and is therefore one of the intensive properties. For a unit mass of a perfect gas, we may write

$$\delta u = c_v \, \delta T \tag{5–12}$$

and for a finite amount of internal energy change, we have

$$\Delta u = \sum c_v \, \delta T \tag{5–13}$$

Example 1–7 is an application of equation (5–13) where c_v changes with temperature in a known way.

If c_v is a constant over a wide range of temperatures, particularly over the range of temperatures indicated by equation (5–13), then

$$\Delta u = c_v \, \Delta T \tag{5–14}$$

Also, equation (5–13) or (5–14) can be used for pure substances even if they are not perfect gases but provided that the volume is constant during the temperature change. To repeat, equation (5–13) can always be used for any process of a perfect gas.

Enthalpy, $h = u + pv$, of a perfect gas is a function of temperature only, just as Joule found for internal energy. This can be seen by substituting into the enthalpy definition the relation $pv = RT$. Then $h = u + RT$, and this shows that h is determined by temperature alone. For a small change in enthalpy, δh, we have

$$\delta h = \delta u + \delta RT \tag{5–15}$$

or, using equation (5–12), $\delta h = c_v \, \delta T + R \, \delta T$, which can be factored to yield $\delta h = (c_v + R) \, \delta T$. We then write

$$\delta h = c_p \, \delta T \tag{5–16}$$

where

$$c_p = c_v + R \tag{5–17}$$

and c_p is called the specific heat at constant pressure. Again, this is another unfortunate name for a term that has no direct relationship to heat. It also does not relate directly to constant pressures. For a finite change in enthalpy, equation (5–16) is

$$\Delta h = \sum c_p \, \delta T \tag{5–18}$$

and this equation is good for a perfect gas and for any process of the perfect gas. It may also be used for pure substances that are not perfect gases if the pressure is constant during the temperature change. The total specific heat at constant pressure, C_p, is given by mc_p, and the total enthalpy change is

$$\Delta H = \sum mc_p \, \delta T \tag{5–19}$$

for perfect gases. It is convenient to define the ratio of the two specific heats by

$$k = \frac{c_p}{c_v} \tag{5–20}$$

Values for c_v, c_p, and k for various materials at 25°C or 77°F are given in table B–4. These values can be used with reasonable confidence over a wide range of temperatures, and if the specific heats are constant, equation (5–18) is

$$\Delta h = c_p \, \Delta T \tag{5–21}$$

table B–7 gives some suggested equations used to predict c_p for some perfect gases when c_p does vary with temperature. The scientific literature contains many other detailed descriptions of specific heat variations with temperature, but it is beyond the purposes of this book to present those. Use of equations such as those in table B–7 will give more nearly correct answers for enthalpy and internal energy changes than if the constant values for the specific heats are used from table B–4. In example 5–7, this comparison will be made. Also, many data have been published for internal energy and enthalpy as functions of the temperature of materials acting as perfect gases. Table B–6 lists internal energy and enthalpy of air at various temperatures and at pressures low enough that the air responds as a perfect gas.

EXAMPLE 5–5 Determine the internal energy and enthalpy of air at 300 K (27°C) and at 800 K (527°C). Then predict the average specific heats (c_p and c_v) for air between 27°C and 527°C, and compare your values with the values from table B–4 and by using *EES*.

Solution From table B–6,

$$\left.\begin{aligned} u &= 214.1 \text{ kJ/kg} \\ h &= 300.2 \text{ kJ/kg} \end{aligned}\right\} \quad \text{at } T = 300 \text{ K}$$

and

$$\left.\begin{aligned} u &= 592.3 \text{ kJ/kg} \\ h &= 821.9 \text{ kJ/kg} \end{aligned}\right\} \quad \text{at } T = 800 \text{ K}$$

The average specific heats may be found by using equations (5–14) and (5–21):

$$c_{v_{\text{ave}}} = \frac{\Delta u}{\Delta T} = \frac{592.3 - 214.1 \text{ kJ/kg}}{800 - 300 \text{ K}}$$

Answer

$$= 0.7564 \text{ kJ/kg} \cdot \text{K}$$

$$c_{p_{\text{ave}}} = \frac{\Delta h}{\Delta T} = \frac{821.9 - 300.2 \text{ kJ/kg}}{800 - 300 \text{ K}}$$

Answer

$$= 1.043 \text{ kJ/kg} \cdot \text{K}$$

The specific heats, from table B–4, are, for air,

$$c_v = 0.719 \text{ kJ/kg} \cdot \text{K} \quad \text{and} \quad c_p = 1.007 \text{ kJ/kg} \cdot \text{K}$$

These values, when compared with one another, give

$$\text{percent difference in } c_v = \frac{0.7564 - 0.7190}{0.7564} = 5.2\%$$

$$\text{percent difference in } c_p = \frac{1.043 - 1.007}{1.043} = 3.5\%$$

and using constant values for the specific heats from table B–4 would give reasonable accuracy for air.

To use *EES*, we need to open to the *Equation* window as we discussed in chapter 1, section 1–6. Click *SI Units* in the *Unit System* option of the *Option* menu. Also, check in this menu which unit is used for temperature, whether kelvin or Celsius. The values for the internal energy, enthalpy, and specific heat are determined by writing the following equations in the *Equation* window specifying temperature in degrees Celsius:

u_1 = intEnergy (AIR, T = 27) For air at 300 K or 27 Celsius
h_1 = enthalpy (AIR, T = 27)
cp_1 = specheat (AIR, T = 27)

u_2 = intEnergy (AIR, T = 527) For air at 800 K or 527 Celsius
h_2 = enthalpy (AIR, T = 527)
cp_2 = specheat (AIR, T = 527)

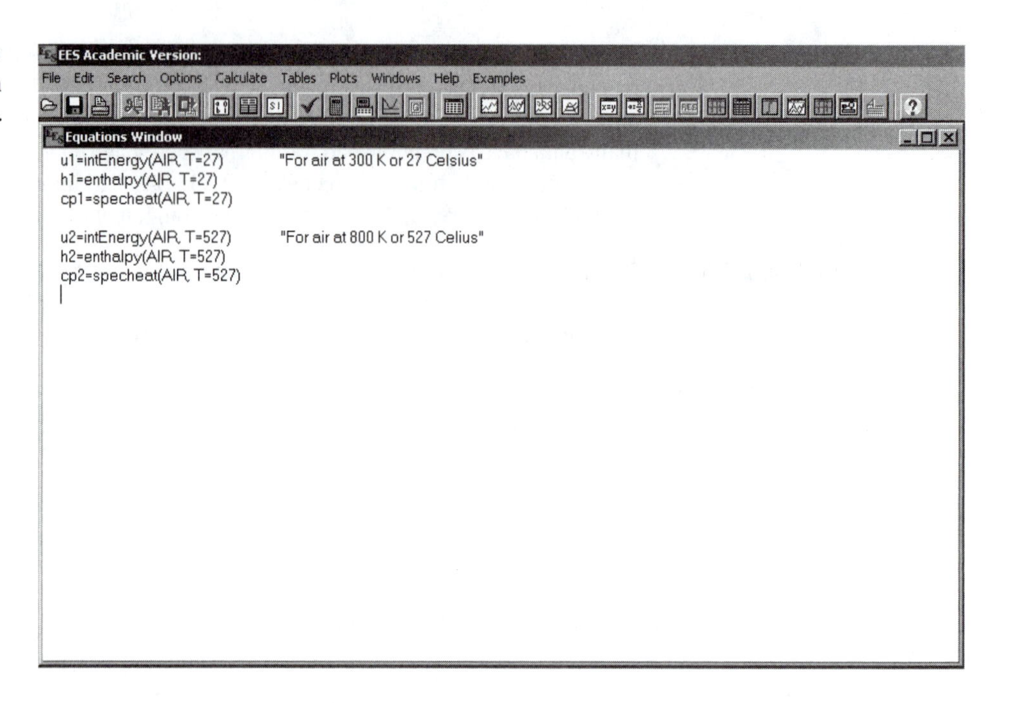

FIGURE 5–8 *EES Equation* window for example 5–5. With permission of F-Chart Software.

The *Equation* window should then look like that in figure 5–8 and if *Solve* in the *Calculate* menu is clicked, *EES* returns the values. These values should be displayed in a window much like that shown in figure 5–9. As you can see by comparing the values obtained from table B–6, the results are in very good agreement.

This example serves to show how the computer can assist in calculations for thermodynamic problems. The program *EES* models air as a perfect gas with variable specific heats, and the properties were determined by specifying the temperature alone. Pure substances that do not behave as perfect gases or incompressible substances will require two independent properties (such as temperature T and specific volume v) to determine the remaining properties. Mixtures of two pure substances will require three independent properties for the determination of the remaining properties.

EXAMPLE 5–6 Determine the internal energy change and enthalpy change of methane if the methane is cooled by 40°F. Assume that methane acts as a perfect gas and has constant specific heats.

Solution From equations (5–14) and (5–21), we have

$$\Delta u = c_v \, \Delta T \quad \text{and} \quad \Delta h = c_p \, \Delta T$$

and the specific heats are found from table B–4:

$$c_v = 0.4079 \text{ Btu/lbm} \cdot {}^\circ\text{R} \quad \text{and} \quad c_p = 0.5318 \text{ Btu/lbm} \cdot {}^\circ\text{R}$$

The temperature change $\Delta T = -40°\text{F} = -40°\text{R}$. Notice that, for temperature differences,

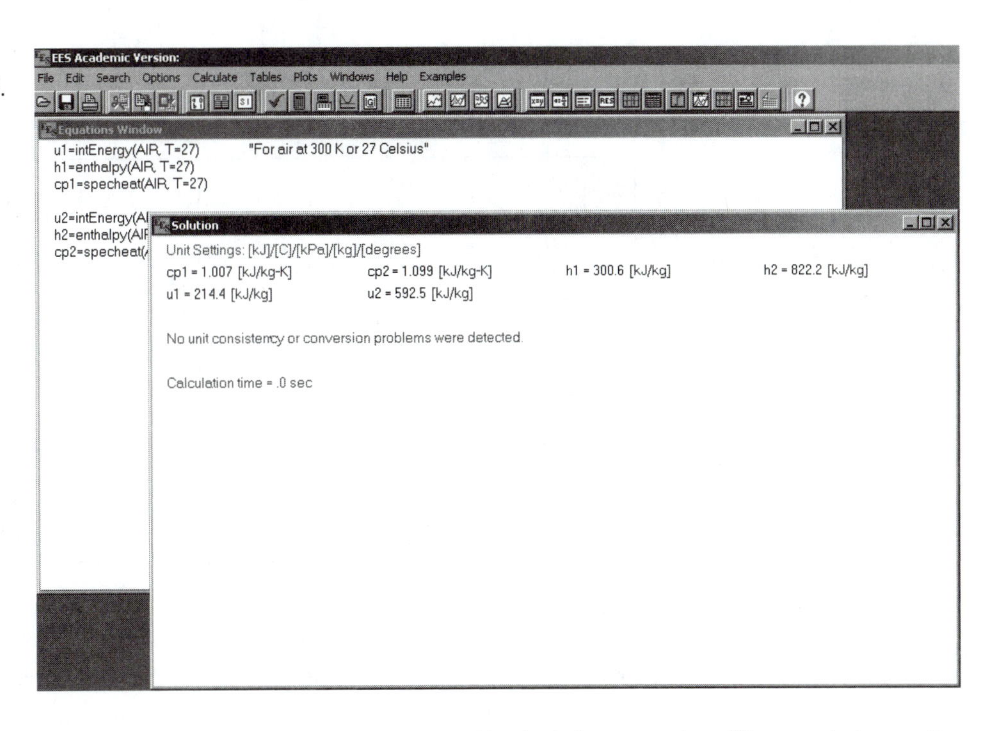

FIGURE 5–9 *Solution* window of *EES* for example 5–5. With permission of F-Chart Software.

the Rankine and Fahrenheit values are identical, because the difference between the scales, a factor of 460 (approximately), cancels out of the difference calculation. Then

$$\Delta u = (0.4079\ \text{Btu/lbm}\cdot{}^\circ\text{R})(-40^\circ\text{R})$$
$$= -16.316\ \text{Btu/lbm}$$

Answer

$$\Delta h = (0.5318\ \text{Btu/lbm}\cdot{}^\circ\text{R})(-40^\circ\text{R})$$
$$= -21.272\ \text{Btu/lbm}$$

Answer

EXAMPLE 5–7 Determine the change in enthalpy per pound-mass for carbon monoxide (CO) as the gas is cooled from 1500°F to 500°F, assuming that the gas does not have a constant specific heat. Compare this to the answer obtained from *EES*.

Solution We will assume that the gas, CO, is a perfect gas, so that

$$\Delta h = \sum c_p\, \delta T \tag{5–18}$$

and we must find the relation c_p has to temperature. From table B–7, we can see that there is a choice of relations for CO, namely,

$$c_p = \left[9.46 - \frac{3.29(10^3)}{T} + \frac{1.07(10^6)}{T^2}\right]\ \text{Btu/lbm}\cdot\text{mol}\cdot{}^\circ\text{R}$$

or

$$c_p = [a + b(10^{-3})T + c(10^{-6})T^2 + d(10^{-9})T^3]\ \text{cal/g}\cdot\text{mol}\cdot\text{K}$$

We shall use the first of these two relationships to compute values for c_p at various temperatures from 1500°F (1960°R) to 500°F (960°R). These values are listed in table 5–2 in the

TABLE 5–1 Results of example 5–7

T °R	c_p Btu/lbm · mol · °R	δT °R	c_p	$c_p\ \delta T$
1960	8.06			
1860	8.00	−100	8.03	−803
1760	7.94	−100	7.97	−797
1660	7.87	−100	7.905	−790.5
1560	7.79	−100	7.83	−783
1460	7.71	−100	7.75	−775
1360	7.62	−100	7.665	−766.5
1260	7.52	−100	7.57	−757
1160	7.42	−100	7.47	−747
1060	7.31	−100	7.365	−736.5
960	7.19	−100	7.25	−725
				$\sum c_p\ \delta T = -7680.5$ Btu/lbm · mol

first two columns for 100°F steps. Then, using the AREA program of appendix A–5, the answer is found to be

Answer

$$\Delta h = -7680.5 \text{ Btu/lbm} \cdot \text{mol}$$

This answer can be given per unit mass instead of per unit mole by dividing the answer by the molecular weight MW of carbon monoxide. The molecular weight is given as 28.011 lbm/lbm · mol (approximately 28 lbm/lbm · mol) from table B–4. Then

Answer

$$\Delta h = -274.2 \text{ Btu/lbm}$$

Table 5–1 also gives the results of a hand calculation of the same data. The answer is the same. Using equation (5–21) and a constant specific heat value $c_p = 0.2485$ Btu/lbm · °R from table B–4 yields

Answer

$$\Delta h = (0.2485 \text{ Btu/lbm} \cdot \text{°R})(-1000 \text{ °R}) = -248.5 \text{ Btu/lbm}$$

and this answer is in substantial agreement with the one obtained by using variable specific heats. The percent difference between the two is about 9.4%.

Open the *Equation* window of *EES* and input the following equation:

delh = enthalpy (CO, T = 500) − enthalpy (CO, T = 1500)

Check that the temperature is to be specified in Fahrenheit (F) units in the English Unit System of the *Unit System* options of the *Options* menu. Then, click *Solve* in the *Calculate* menu and the answer should be displayed in a manner as shown in figure 5–10. As you can see in the window, the answers agree well between the variable specific heat calculation.

EES has the capability of determining the thermodynamic properties of those substances listed in table 5–2. Notice that there appears to be some repetition. (Both methane and CH$ are listed, for instance.) However, those substances listed by their chemical sym-

TABLE 5–2 *EES* functions for thermodynamic properties of the following substances

Perfect Gases	Pure Substances
AIR	Ammonia
AIR_HA	Ammonia_MH
$AIRH_2O$	CarbonDioxide
C_2H_6	Helium
C_3H_8	*n*-Butane
C_4H_{10}	Nitrogen
CH_4	Oxygen
CO	Propane
CO_2	Propane_MH
CH_2	R11
H_2O	R12
N_2	R13
NO	R14
NO_2	R22
O_2	R22_MH
SO_2	R32
	R113
	R114
	R123
	R134a
	R134a_MH
	R141b
	R500
	R600
	Steam (or Water)
	Steam_NBS

*Steam_NBS is a more accurate program for water or steam and extends to very high pressures and also into the liquid water region. The entries "Ammonia_MH, Propane_MH, R22_MH, and R134a_MH" are models for the respective substances that provide property values in agreement with values in the appendix tables. The entry "AIR_HA" is a model for air that provides property values that are more precise than those obtained with AIR.

bol are treated as perfect gases while those with the full name or, for instance R12, R22, etc., are treated as pure substances. An exception to this notation is that AIR and $AIRH_2O$ are treated as perfect gases. Steam or Water are the same program and treat water in the liquid and vapor region and the superheated region as displayed in figures 5–3, 5–4, and 5–5. Steam (or Water) is not accurate at very high pressures and is not recommended for the liquid water region.

Also, one could use the computer to assist even more in a problem of this type and obtain a more precise answer. The way to accomplish this added precision is to write a program that could compute c_p by using the given equation and use small temperature steps, say, 5° steps, instead of the 100° steps in the illustration. Such a program, incorporating the VISUAL BASIC program used to compute areas under curves, is shown in appendix A–6, called DELHCO, and the algorithm for that program could be adapted to many other specialized equations. From that program in appendix A–6 and using 20° steps, the enthalpy change was found to be -255.9 Btu/lbm.

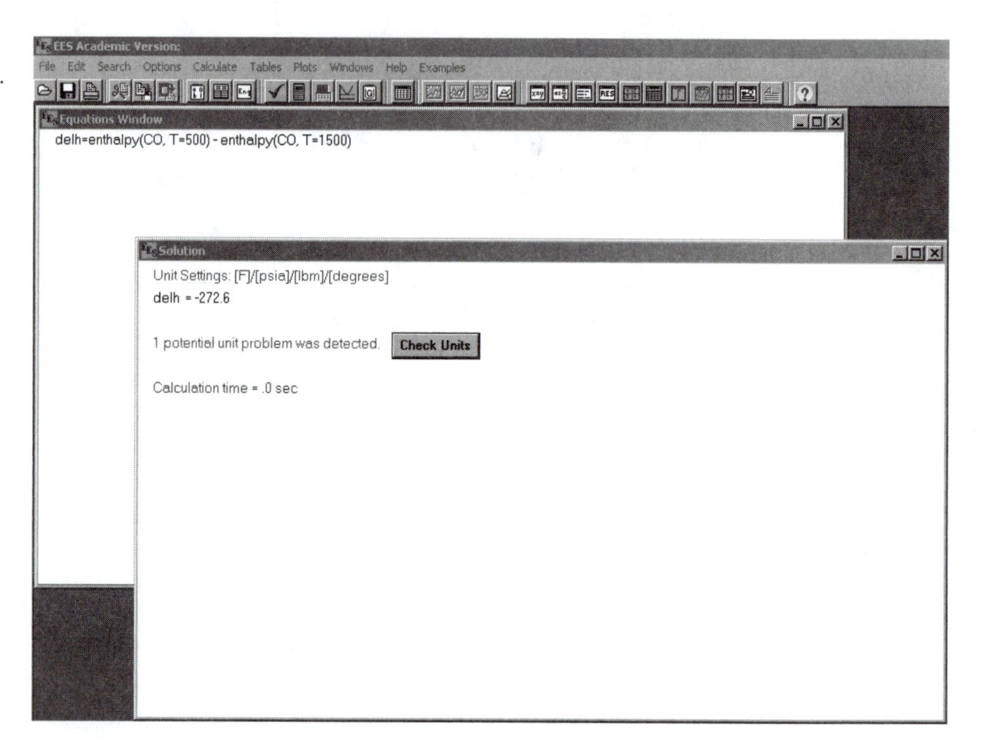

FIGURE 5–10 *Solution* window of *EES* for example 5–7. With permission of F-Chart Software.

There are charts similar to the compressibility charts B–1 and B–2 that relate u and h to temperature and pressure for substances that can be analyzed by the principle of corresponding states. We will not treat those results here. If the reader is interested, a book covering this material is Van Wylen and Sonntag's *Fundamentals of Classical Thermodynamics*, among others.

Caloric equations of state are frequently used for incompressible liquids and solids. *Incompressible* means that the substance will not change density or volume whenever any of the other properties are changed, particularly pressure and temperature. Since volume remains constant for incompressible substances, the internal energy change is just that of equation (5–13), $\Delta u = \sum c_v \, \delta T$. The enthalpy change is given by

$$\Delta h = \Delta u + \Delta p \, v$$

But $\Delta p \, v$ is just $v \, \Delta p$, since v is a constant. Then

$$\Delta h = \sum c_v \, \delta T + v \, \Delta p \tag{5–22}$$

for an incompressible substance. The second term on the right side of this equation, $v \, \Delta p$, is usually very small, so that $\Delta h \simeq \sum c_v \, \delta T$. But we usually associate c_p with an enthalpy change as in equation (5–18), so it has been customary to write equation (5–22) as

$$\Delta h = \sum c \, \delta T + v \, \Delta P \tag{5–23}$$

where c is the specific heat of an incompressible substance. Table B–8 gives specific heats (tabulated as c_p) of some solids and liquids, which are assumed to be incompressible substances. Table B–8 also lists specific heats of some gases, and these are correctly specific heat at constant pressure, c_p.

CALCULUS FOR CLARITY 5–1

The internal energy of a pure substance depends on its state as defined by its temperature, pressure, and specific volume. Since these three properties are interrelated, the internal energy of a pure substance depends on only two of these properties. If we assume that the internal energy is a function of temperature and specific volume, the derivative of internal energy is given by

$$du = \frac{\partial u}{\partial T} dT + \frac{\partial u}{\partial v} dv \tag{5–24}$$

where $\partial u/\partial T$ is the partial derivative of the internal energy with respect to temperature and $\partial u/\partial v$ is the partial derivative with respect to specific volume. The specific heat at constant volume, c_v, is defined as $\partial u/\partial T$, or

$$c_v = \frac{\partial u}{\partial T} \tag{5–25}$$

Then equation (5–24) becomes

$$du = c_v\, dT + \frac{\partial u}{\partial v} dv \tag{5–26}$$

The case of a perfect gas, where the internal energy depends only on the temperature, gives $\partial u/\partial v = 0$, and a differential form of equation (5–12) can be written:

$$du = c_v\, dT \tag{5–27}$$

Incompressible substances experience no change in specific volume, so $dv = 0$ and equation (5–24) reduces again to equation (5–27). A finite change in internal energy is then determined for **perfect gases** and **incompressible substances**:

$$\int du = \int c_v\, dT \quad \text{or} \quad u_2 - u_1 = \int c_v\, dT \tag{5–28}$$

If the specific heat is constant, then equation (5–28) takes on the form of equation (5–14).

The enthalpy of a pure substance, being a property similar to internal energy, depends on temperature, pressure, and specific volume. It has been convenient to consider enthalpy to depend on temperature and pressure, so that the derivative of enthalpy for a pure substance is

$$dh = \frac{\partial h}{\partial T} dT + \frac{\partial h}{\partial p} dp \tag{5–29}$$

The derivative of enthalpy with respect to temperature is defined as the specific heat at constant pressure, c_p:

$$c_p = \frac{\partial h}{\partial T} \tag{5–30}$$

Equation (5–29) then becomes

$$dh = c_p\, dT + \frac{\partial h}{\delta p} dp \tag{5–31}$$

CALCULUS FOR CLARITY 5–1, continued

For a perfect gas, the enthalpy, $h = u + pv$, becomes $h = u + RT$, and

$$\frac{\partial h}{\partial p} = 0$$

since internal energy depends only on temperature and not pressure. Thus, for a **perfect gas**,

$$dh = c_p\, dT \tag{5–32}$$

Also, again using $h = u + RT$ for a perfect gas,

$$dh = du + dRT = du + R\, dT \tag{5–33}$$

Comparing this equation with equations (5–27) and (5–32) gives equation (5–17), $c_p = c_v + R$.

The exact solution to determining the internal energy or enthalpy change for variable-specific-heat perfect gases or incompressible substances can be obtained from equation (5–28) if the relationship between c_v or c_p and T is known. For instance, in example 5–7, the enthalpy change for carbon monoxide was determined by using a numerical approximation. Using the equation for c_p taken from table B–7 for carbon monoxide, we may obtain an **exact solution** from equation (5–32):

$$\begin{aligned}\Delta h &= \int c_p\, dT = \int \left(9.46 - \frac{3.29 \times 10^3}{T} + \frac{1.07 \times 10^6}{T^2}\right) dT \\ &= 9.46(960°\text{F} - 1960°\text{F}) - (3290)\left[\ln\left(\frac{960}{1960}\right)\right] \\ &\quad - (1.07 \times 10^6)\left[\frac{1}{960} - \frac{1}{1960}\right] \\ &= -7657.6\ \text{Btu/lbm} \cdot \text{mol}\end{aligned}$$

The answer using the BASIC program, -7680.5 Btu/lbm · mol, compares favorably with this result.

The derivative of enthalpy may be written, in general, as

$$dh = du + dpv = du + v\, dp + p\, dv \tag{5–34}$$

For an incompressible substance, this equation becomes

$$dh = c_v\, dT + v\, dp$$

and the differential form of equation (5–23), the finite change of enthalpy for an incompressible substance, may be written

Answer

$$\Delta h = \int c_v\, dT + v\, \Delta p \tag{5–35}$$

The second term is usually very small, and the specific heat at constant volume is usually referred to as specific heat, c. In table B–8, it is referred to as c_p for those entries for incompressible fluids or solids.

5–4 CALORIMETRY

We have discussed equations of state and caloric equations of state. In later sections of this book, we consider other thermal properties of pure substances, data that have been accumulated through experimentation and analysis with equations of state. The branch of science that concerns itself with investigations of the thermal properties of substances is called calorimetry, which we define as follows:

> **Calorimetry:** ***The science and technology concerned with precisely measuring energy and enthalpy.***

From calorimetric studies have come the precise measurements of the triple points, the critical points, the saturation curves, the specific heats, and other thermal properties. In this section, we discuss some of the basic methods and equipment that have been used in calorimetry. The purpose of the discussion is to show to the reader some of the concerns, precautions, and design techniques required to have precision test equipment and to develop careful procedures so that accurate measurements and observations of thermal equilibrium states may be made or that thermal changes may be accurately monitored. You may encounter in your professional experiences situations in which precise determinations of specific heat or some other thermal properties will need to be made, or perhaps an accurate calibration of a temperature sensor will be needed. Let us then consider some of the techniques you may want to incorporate into a thermodynamic experiment and also show how some of the data compiled in thermodynamic tables have been experimentally determined.

The **triple point of water** has been found to be a precise reference point that is conveniently reproducible and observable. It is a data point that has been used for basic calibration of thermometers and other temperature sensors. A method used by the U.S. National Bureau of Standards (NBS) for measuring and observing the triple point of water has been the triple-point cell, shown in figure 5–11. In the figure, the cell is shown in vertical cross section as well as in an isometric view, to better visualize its operation and

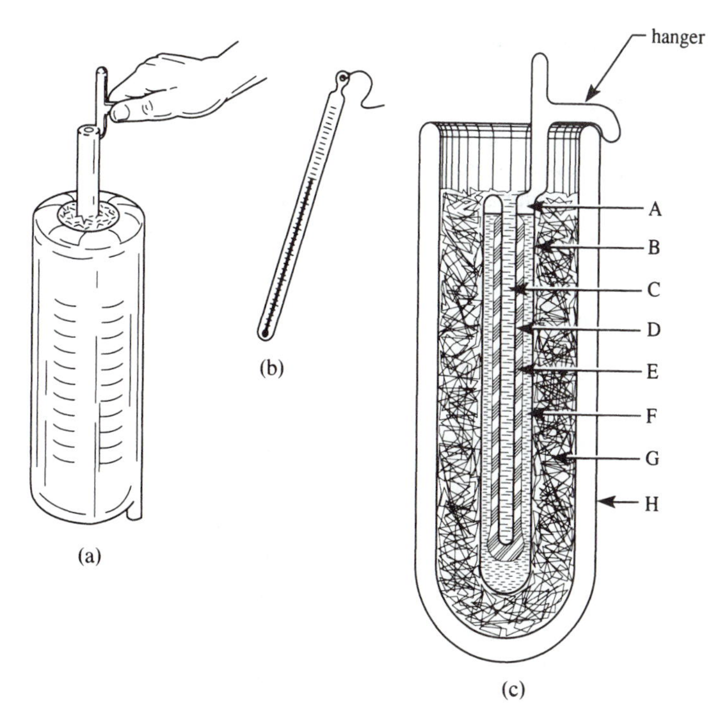

FIGURE 5–11 (a) Isometric view of triple-point cell with partially removed Pyrex cell; (b) thermometer; (c) triple-point cell: A, water vapor; B, Pyrex cell; C, water from ice bath; D, thermometer well; E, ice mantle; F, air-free water; G, flaked ice and water; H, insulated container

design concepts. The triple-point cell is an insulating container surrounding a special Pyrex glass cell. The cell is put in the center of the container and is prevented from moving down or from tipping by a hanger that is part of the cell. The cell, during its final construction, is filled to within about 2 cm ($^3/_4$ in) of the top of the cell with pure air-free water. The cell is then sealed at the top by heating, and it then contains (as nearly as possible) only pure water placed in it. The cell is then placed in an insulating container and an ice bath. The ice bath needs to be supplied continually with ice flakes so that the cell is surrounded by a 0°C (32°F) environment.

The center well of the cell is then filled with dry ice (solid carbon dioxide), and the water against the inside surface of the cell freezes to form an ice mantle. When the ice mantle is thick enough, the dry ice is removed from the well, and ice water is allowed to fill the well. The ice mantle tends to purify the remaining liquid and vapor water, and this is one of the main reasons for causing the water to freeze on the inner surface.

A thermometer or temperature sensor is then placed in the well, and the device is allowed to reach equilibrium; that is, the pure water inside the cell will equilibrate with ice, liquid, and vapor phases, or the triple point. The time for equilibrium to be reached may extend to several hours, and the cells are observed for more than two years. With proper maintenance and observance, the triple point has been shown to be reproducible within 0.00008°C. The observed triple point of water is 0.01°C (32.018°F) and 0.061 kPa (0.08865 psia).

Notice that the essential ingredients for observing thermal equilibrium are long test times, adequate insulation to prevent excessive heat transfer, and concern for purity of the substance. Comparison of results from other cells assures better accuracy and precision.

The methods used to determine the values of specific heats, internal energy, and enthalpy are based on straightforward concepts but involve many complications in design and procedure. Figure 5–12 shows the schematic of nearly all methods used to measure specific heats; that is, a measured amount of heat is transferred to the test specimen, the temperature change is measured, the heat loss is accounted for, and the specific heat is calculated from a form of equation (5–12) or (5–16). Unfortunately, accounting for the heat loss is a major problem in calorimetry, and in figure 5–13 a schematic is shown of an apparatus used to reduce those losses and accurately determine the specific heats of liquids and solids. In this device, the test specimen holder is a cylinder of copper or other heat-conducting material. It has vertical holes drilled in it, and heater elements are embedded in the remaining holder material. The test substance is then put into all of the drilled holes so that when the heating elements are turned on, the test substance will increase temperature

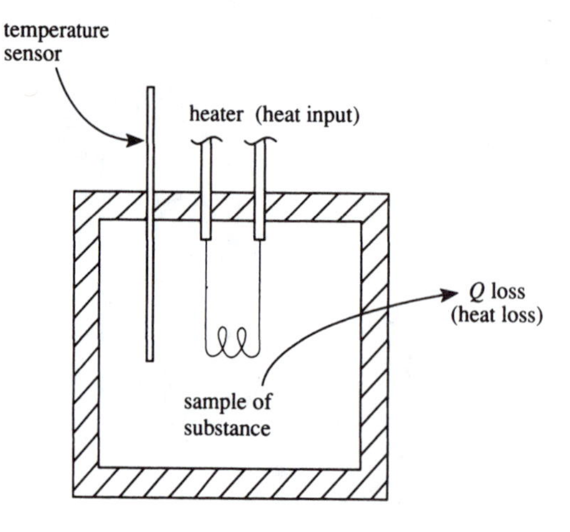

FIGURE 5–12 Schematic of basic device to measure specific heats of substances

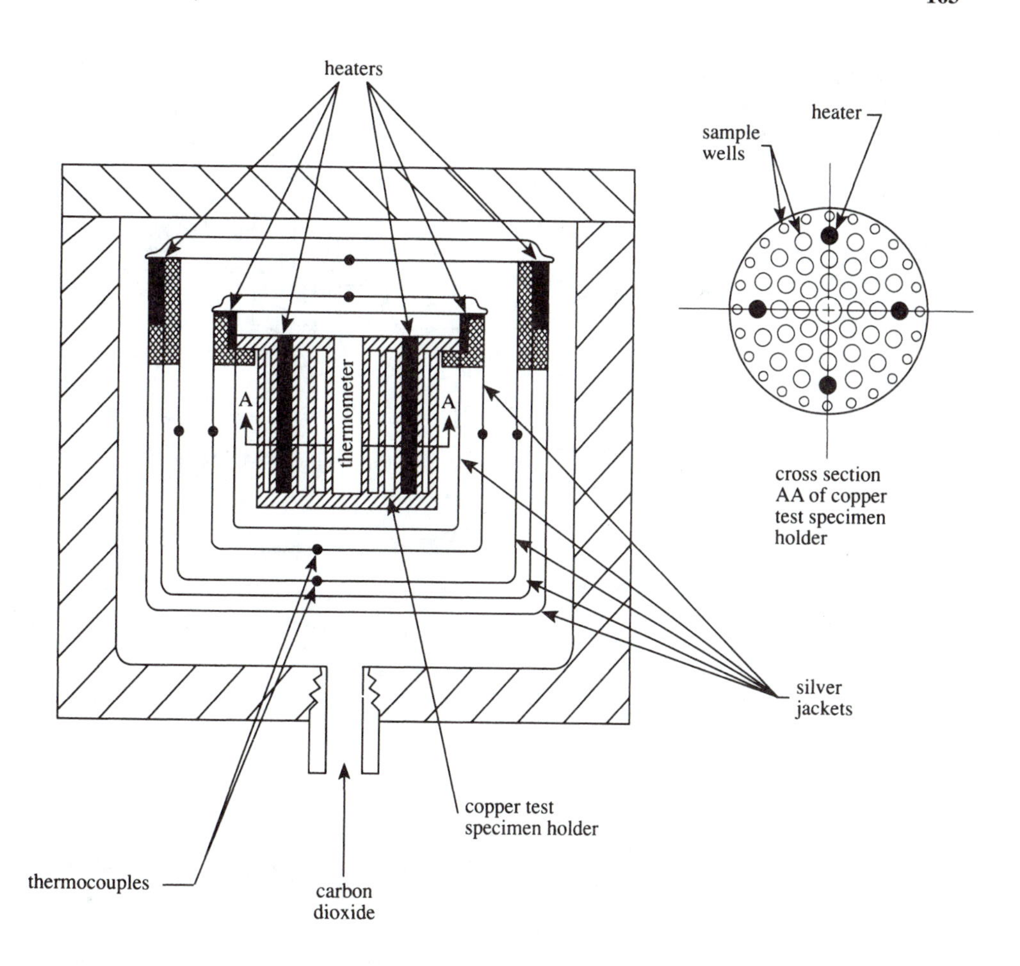

FIGURE 5–13 Calorimeter schematic

in a uniform manner. The holder then has a lip placed on top, and the cell (holder and test substance) is placed in another container. This container, or inner jacket, is made of concentric cylinders of silver or other inert and highly reflective material, and a silver lid is placed on top of this assembly. Finally, the assembly is placed inside another outer jacket of concentric silver cylinders, and this is put into an insulated chamber containing carbon dioxide (CO_2) or other gas. This outer jacket and inner jacket have heating elements so that their temperatures may be kept at the temperature of the test substance.

The method of testing generally uses the following procedure: (1) heat the test substance by a precisely measured amount of heat, simultaneously heating the inner and outer jackets to the same temperature, (2) measure the temperature difference of the test substance after equilibrating, and (3) calculate the specific heat from a form of equation (5–12) or (5–16). The purpose in heating the jackets at the same rate as the test substance is to reduce to zero (or as close as possible to zero) the heat loss from the test substance. The silver jackets also reflect radiant energy back into the test specimen and further reduce heat losses. The carbon dioxide acts to reduce any connective heat losses that might occur if room air were in the chamber.

Finally, we consider an apparatus and procedures used to determine the specific heats or liquid and vapor water and the latent heat or heat required to boil or condense water. The apparatus is shown schematically in figure 5–14. The metal calorimeter shell contains the water sample and is connected by tubing and valves to a water container and a steam receiver. An electric heater and a paddle wheel are inside the container, submersed in liquid

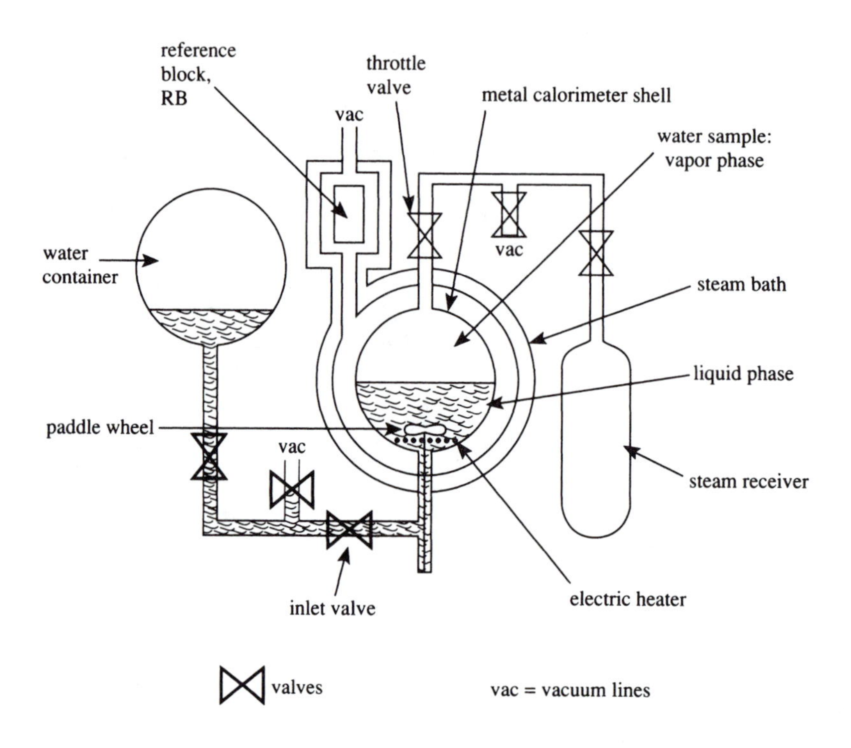

FIGURE 5–14 Schematic diagram of calorimetric apparatus

water. Surrounding the shell is a vacuum chamber which includes a passage and reference block. Surrounding the vacuum chamber is a steam bath at the same temperature as the test water sample. The apparatus allows for at least two fundamental tests: (1) a constant-volume test in which the liquid–vapor mixture is heated and its temperature change is recorded, and (2) a constant-temperature test in which a measured amount of vapor is produced from the liquid phase through a measured amount of heat input to the water. The first of these two tests proceeds with the two valves (inlet and throttle valve) closed so that the amount of water within the shell is constant. Heat is added to the water through the electric heater, and uniform heating is assured with the paddle wheel driven by a small motor. The irreversible work of the paddle wheel must be accounted for in the calculations. During the heating, the steam bath is kept at a constant temperature, but temperature changes of the water sample are so slight as to make negligible the temperature difference between the steam bath and the sample. The vacuum reduces to a minimum any heat losses, and the reference block, RB, is used to monitor the sample temperature. Thermocouples or other temperature sensors are placed around the shell to measure nonuniformities in the sample temperature during the course of testing.

The second test, for latent heat, involves using the apparatus with the throttle valve open and connections so that the vaporized steam may enter the steam receiver. Again, as in the first method, the sample is uniformly heated a measured amount and temperatures are monitored. In this method, the temperatures should remain constant, and the amount of steam in the steam receiver is increased by some amount. The latent heat is then the measured heat input (plus the paddle wheel work plus or minus other heat gains or losses) divided by the mass of vaporized steam.

There are many other methods that can be used to determine thermal properties of substances, such as spectroscopy and optical techniques. Also, use of the p–v–T relationships can be used to predict the specific heats under certain conditions. These methods and others are beyond our purposes in this short discussion of calorimetry.

5–5 PROPERTIES OF PURE SUBSTANCES

The caloric equations of state, which are obtained from experimental data of the substances, are often so complicated that they are beyond the scope of this book. The results from these equations and the experimental data are usually presented in thermodynamic tables. For gases or vapors (usually called superheated vapor), the tables of properties are frequently constructed with pressure and temperature as the independent properties or variables. Superheat tables are given in this book for steam (water) in table B–12, for Refrigerant-12 in table B–16, for Refrigerant-22 in table B–18, for Refrigerant-134a in table B–22, for Refrigerant-123 in table B–21, and for ammonia in table B–20. The use of these tables will now be demonstrated.

EXAMPLE 5–8

Determine the properties *v, u, h,* and *s* (entropy) for the following substances and at the given states:
(a) Water at 1000°F and 300 psia
(b) Ammonia at 1.0 MPa and 100°C
(c) Refrigerant-22 at 90 psia and 120°F
Use the appendix tables and compare these results to those you obtained with *EES*.

Solution

(a) From the steam superheat table B–12 at 1000°F and 300 psia, the following values are read:

$$v = 2.8585\ \text{ft}^3/\text{lbm}$$
$$h = 1526.2\ \text{Btu/lbm}$$
$$s = 1.7964\ \text{Btu/lbm}\cdot{}^\circ\text{R}$$

The internal energy is computed from the equation

$$\begin{aligned} u &= h - pv \\ &= (1526.2\ \text{Btu/lbm}) - \frac{(300\ \text{lbf/in}^2)(2.8585\ \text{ft}^3/\text{lbm})(144\ \text{in}^2/\text{ft}^2)}{(778\ \text{ft}\cdot\text{lbf/Btu})} \\ &= 1367.5\ \text{Btu/lbm} \end{aligned}$$

Open the *EES Equation* window and check the unit system to be sure that English units are used, temperature is in degrees Fahrenheit, and pressure is psia. Then enter the equations

$$v = \text{volume (steam, } T = 1000, p = 300)$$
$$u = \text{intEnergy (steam, } T = 1000, p = 300)$$
$$h = \text{enthalpy (steam, } T = 1000, p = 300)$$
$$s = \text{entropy (steam, } T = 1000, p = 300)$$

Clicking *Solve* in the *Calculate* menu gives

$$h = 1526\ \text{Btu/lbm} \qquad s = 1.796\ \text{Btu/lbm}^\circ\text{R} \qquad u = 1368\ \text{Btu/lbm}$$
$$v = 2.86\ \text{ft}^3/\text{lbm}$$

These values are consistent with those obtained from the appendix tables.
(b) From table B–20, the properties of ammonia at 1.0 MPa and 100°C are

$$v = 0.174\ \text{m}^3/\text{kg}$$
$$h = 1652.9\ \text{kJ/kg}$$
$$s = 5.644\ \text{kJ/kg}\cdot\text{K}$$

and

$$\begin{aligned} u = h - pv &= 1652.9\ \text{kJ/kg} - (1000\ \text{kN/m}^2)(0.174\ \text{m}^3/\text{kg}) \\ &= 1478.9\ \text{kJ/kg} \end{aligned}$$

Open the *EES Equation* window and check the unit system to be sure that SI units are used, temperature is in degrees Celsius, and pressure is kPa. Then, enter the equations

$$v = \text{volume (ammonia\_MH}, T = 100, p = 1000)$$
$$u = \text{intEnergy (ammonia\_MH}, T = 100, p = 1000)$$
$$h = \text{enthalpy (ammonia\_MH}, T = 100, p = 1000)$$
$$s = \text{entropy (ammonia\_MH}, T = 100, p = 1000)$$

Notice that the substance model is ammonia_MH, which should give values in agreement with the appendix tables. Clicking *Solve* in the *Calculate* menu will give the property values, and these results should be

$h = 1667$ kJ/kg $\quad s = 5.643$ kJ/kg K $\quad u = 1493$ kJ/kg $\quad v = 0.1739\ m^3/kg$

These values are consistent with those obtained from the appendix tables.

(c) From table B–18, the properties of Refrigerant-22 at 90 psia and 120°F are

$$v = 0.74120\ ft^3/lbm$$
$$h = 121.894\ Btu/lbm$$
$$s = 0.24376\ Btu/lbm \cdot {}^\circ R$$

and

$$u = h - pv = 121.894\ Btu/lbm - \frac{(90 \times 144\ lbf/ft^2)(0.74120\ ft^3/lbm)}{778\ ft \cdot lbf/Btu}$$
$$= 109.547\ Btu/lbm$$

Open the *EES Equation* window and check the unit system to be sure that English units are used, temperature is in degrees Fahrenheit, and pressure is psia. Then, enter the equations

$$v = \text{volume (R22\_MH}, T = 120, p = 90)$$
$$u = \text{intEnergy (R22\_MH}, T = 120, p = 90)$$
$$h = \text{enthalpy (R22\_MH}, T = 120, p = 90)$$
$$s = \text{entropy (R22\_MH}, T = 120, p = 90)$$

Notice that here, as in part b, the substance model is R22_MH to provide values in agreement with the appendix tables. Clicking *Solve* in the *Calculate* menu will give the following property values:

$h = 121.9$ Btu/lbm $\quad s = 0.2437$ Btu/bm °R $\quad u = 109.5$ Btu/lbm $\quad v = 0.7412\ ft^3/lbm$

These values are consistent with those obtained from the appendix tables.

The sublimation, fusion (freezing–melting), and saturation (boiling–condensing) regions of pure substances can be visualized as mixtures of two pure substances. For instance, the saturation region can be looked at as a mixture of pure saturated liquid and pure saturated vapor. The proportion of saturated vapor, or the ratio of the mass of the saturated vapor to the total mass of a given mixture in the saturated region, is called the **quality**, χ We write this as

$$\chi = \frac{m_g}{m_T} = \frac{m_g}{m_g + m_f} \tag{5–36}$$

where m_g is the mass of saturated vapor, m_f is the mass of saturated liquid, and m_T the total mass or the sum of m_g and m_f. The total volume of a mixture in the saturated region, V_T, is the sum of the volume of the two components, vapor and liquid, so that

$$V_T = V_g + V_f \tag{5–37}$$

But $V_g = (v_g)(m_g)$, $V_f = (v_f)(m_f)$, and $V_T = (v)(m_T)$. Equation (5–37) then becomes

$$vm_T = v_g m_g + v_f m_f$$

and if we substitute for m_f the term $m_T - m_g$, then

$$vm_T = v_g m_g + v_f(m_T - m_g)$$

Now, if this equation is divided by m_T, we find, for the specific volume of a saturation mixture having quality χ,

$$v = \chi v_g + (1 - \chi)v_f \tag{5–38}$$

where $\chi = m_g/m_T$. This equation can also be expressed slightly differently if it is expanded, $v = \chi v_g + v_f - \chi v_f = v_f + \chi(v_g - v_f)$, so that

$$v = v_f + \chi v_{fg} \tag{5–39}$$

where $v_{fg} = v_g - v_f$. Values of v_f, v_g, and v_{fg} are given in the saturation tables of substances. In this book these values can be found for steam in table B–11, for Refrigerant-12 in table B–15, for Refrigerant-22 in table B–17, for Refrigerant-134a in table B–22, for Refrigerant-123 in table B–21, for ammonia in table B–19 for Refrigerant-407c in table B–23, for Refrigerant-502 in table B–24, and for mercury in table B–14. By using similar reasoning for enthalpy, internal energy, and entropy (s) that we have just used for volume, we can show that, for a saturation mixture having quality χ,

$$\begin{aligned} h &= \chi h_{fg} + h_f \\ u &= \chi u_{fg} + u_f \\ s &= \chi s_{fg} + s_f \end{aligned} \tag{5–40}$$

In these equations, the fg terms are

$$h_{fg} = h_g - h_f, \quad u_{fg} = u_g - u_f, \quad \text{and} \quad s_{fg} = s_g - s_f$$

The enthalpy, h_{fg}, is frequently referred to as the heat of vaporization, the heat of condensation, or the latent heat. Values for h_{fg} usually vary from a maximum at near the triple point to zero at the critical point. These properties of the saturated liquid and vapor are given in the saturation tables, where the specific volume values are listed.

EXAMPLE 5–9 Determine the enthalpy of saturated mercury vapor, h_g, at 450-kPa pressure.

Solution From table B–14, which lists the properties of mercury vapor, we see that the enthalpy of the saturated vapor h_g is 345.4 kJ/kg at 400 kPa and 346.3 kJ/kg at 500-kPa pressure. The difference between the two values is 0.9 kJ/kg over a pressure difference of 100 kPa. We may then use a method of proportions or a pressure–enthalpy plot as shown in figure 5–15. From the figure, we can see that the value for the enthalpy at 450 kPa must be exactly midway between the values at 400- and 500-kPa pressure. Thus,

Answer

$$h = 345.4 + 0.45 = 345.9 \text{ kJ/kg}$$

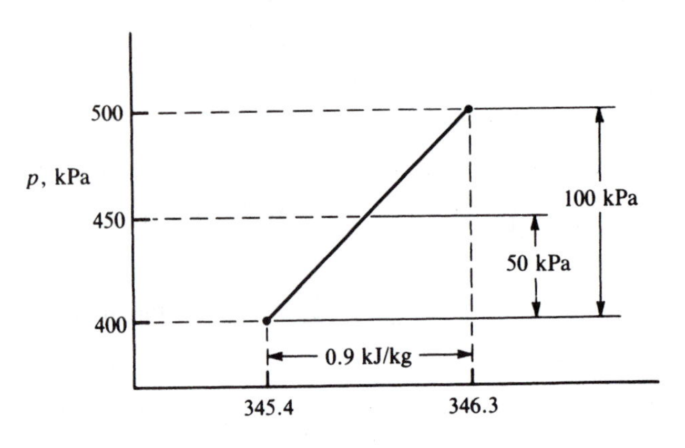

FIGURE 5–15 Graphical illustration of interpolation in example 5–9

Using the method of proportions, we write

$$\frac{450\text{ kPa} - 400\text{ kPa}}{500\text{ kPa} - 400\text{ kPa}} = \frac{h - 345.4\text{ kJ/kg}}{346.3\text{ kJ/kg} - 345.4\text{ kJ/kg}}$$

and then

$$\frac{50\text{ kPa}}{100\text{ kPa}} = \frac{h - 345.4\text{ kJ/kg}}{0.9\text{ kJ/kg}}$$

$$(0.9)(0.5) = h - 345.4\text{ kJ/kg}$$

$$345.4\text{ kJ/kg} + 045\text{ kJ/kg} = h$$

Answer

$$h = 345.9\text{ kJ/kg}$$

EXAMPLE 5–10 Determine the enthalpy and specific volume of steam at 560°C and 3000 kPa. Also, determine the enthalpy and specific volume of ammonia at 100°C and 1400-kPa pressure.

Solution The properties of steam can be found from table B–12. We read

Answer $$h = 3593\text{ kJ/kg}$$

Answer $$v = 0.126\text{ m}^3\text{/kg}$$

The properties of ammonia are found from table B–20. We read

Answer $$h = 1653.8\text{ kJ/kg}$$

Answer $$v = 0.1217\text{ m}^3\text{/kg}$$

EXAMPLE 5–11 Determine the specific volume of saturated steam in the vapor state, v_g, at 50 psia.

Solution From table B–11, we find that values are listed for 49.2 psia but not for 50 psia. If we then interpolate between the values at 40 and 60 psia, we can find v_g at 50 psia. Here we have 50 psia being midway between 40 and 60, so

$$\frac{50\text{ psia} - 40\text{ psia}}{60\text{ psia} - 40\text{ psia}} = \frac{v_g - 10.497\text{ ft}^3\text{/lbm}}{7.174\text{ ft}^3\text{/lbm} - 10.497\text{ ft}^3\text{/lbm}}$$

$$\frac{10}{20} = \frac{v_g - 10.497\text{ ft}^3\text{/lbm}}{-3.323\text{ ft}^3\text{/lbm}}$$

$$(-3.323\text{ ft}^3\text{/lbm})\left(\frac{1}{2}\right) = v_g - 10.497\text{ ft}^3\text{/lbm}$$

Answer

$$8.8355\text{ ft}^3\text{/lbm} = v_g$$

EXAMPLE 5–12 A saturation mixed of steam having 85% quality (85% of the steam is vapor by mass) has a temperature of 200°C. Determine the pressure, specific volume v, and enthalpy h.

Solution Since the steam is a saturation mixture, we use table B–11, "Saturated steam table: temperature." At 200°C, the pressure, or the saturation pressure p_{sat}, is

Answer
$$p_{sat} = 1.555 \text{ MPa} = 1555 \text{ kPa}$$

From equation (5–38), the specific volume is

$$v = \chi v_g + (1 - \chi)v_f$$
$$= (0.085)(0.127 \text{ m}^3/\text{kg}) + (1 - 0.85)(0.00115 \text{ m}^3/\text{kg})$$

Answer
$$= 0.10812 \text{ m}^3/\text{kg}$$

From equation (5–40), the enthalpy is

$$h = \chi h_{fg} + h_f$$
$$= (0.85)(1941 \text{ kJ/kg}) + 852 \text{ kJ/kg}$$

Answer
$$= 2501.85 \text{ kJ/kg}$$

EXAMPLE 5–13 Determine the saturation temperature and the specific volume of steam at 90 psia and having a quality of 20%. Also determine the volume of the vapor as a fraction of the total volume.

Solution Again, since the steam is a saturated mixture of vapor and liquid, we use the saturation table B–11, "Saturated steam table: pressure." Notice that either temperature or pressure can be used, but for convenience, we will use the table that lists pressure in this problem. At 90 psia, the saturation temperature is

Answer
$$T_{sat} = 320.28°\text{F}$$

The specific volume is given by equation (5–38):

$$v = \chi v_g + (1 - \chi)v_f$$
$$= (0.20)(4.895 \text{ ft}^3/\text{lbm}) + (1 - 0.20)(0.0177 \text{ ft}^3/\text{lbm})$$

Answer
$$= 0.99316 \text{ ft}^3/\text{lbm}$$

The volume of the saturated vapor for 1 lbm of mixture is given by χv_g or $(0.20)(4.895) = 0.979 \text{ ft}^3$. The total volume of 1 lbm of mixture is v (m = 1 lbm), or 0.99316 ft^3, so the volume of the vapor as a fraction of the total is

Answer
$$\frac{V_g}{V_T} = \frac{0.979 \text{ ft}^3}{0.99316 \text{ ft}^3} = 98.5\%$$

Notice in this example that for a mixture of 20% vapor by mass, the volume is 98.5% of the total. The reason is the very large difference in density or specific volume between the liquid and vapor at the saturation state.

EXAMPLE 5–14 Determine the specific volume, enthalpy, and entropy of a saturation mixture of mercury at 180 kPa with a quality of 75%. Also, determine the temperature.

Solution Since the mercury is a saturated mixture, we use table B–14. From this table, properties must be interpolated between 160 and 200 kPa. Since the pressure is exactly midway between 160 and 200 kPa, we may use the "average" of the values at 160 and 200 kPa. The temperature, or saturation temperature, is 390.15°C. This result is one-half the sum of the saturation temperatures at the two listed pressures $(383.2 + 397.1°\text{C})/2 = 390.15°\text{C}$. The volume, enthalpy, and entropy are determined as follows:

$$v = \chi v_g + (1 - \chi)v_f$$

The volume $v_g = (0.164 + 0.133\ \text{m}^3/\text{kg})/2 = 0.1485\ \text{m}^3/\text{kg}$. The volume of saturated liquid is not listed, so we shall assume that it is approximately the same as the inverse of the density of mercury in its normal liquid state. The density of mercury is listed in table 2–1 as 13,536 kg/m^3. Then v_f is approximately equal to $1/13{,}536\ \text{kg/m}^3 = 0.0000739\ \text{m}^3/\text{kg}$. The volume is then

$$v = (0.75)(0.1485\ \text{m}^3/\text{kg}) + (1 - 0.75)(0.0000739\ \text{m}^3/\text{kg})$$

Answer

$$= 0.11139\ \text{m}^3/\text{kg}$$

The enthalpy is

$$h = \chi h_g + (1 - \chi)h_f$$
$$= (0.75)(341.85\ \text{kJ/kg}) + (1 - 0.75)(60.93\ \text{kJ/kg})$$

Answer

$$= 271.62\ \text{kJ/kg}$$

The entropy is

$$s = \chi s_g + (1 - \chi)s_f$$
$$= (0.75)(0.5625\ \text{kJ/kg}\cdot\text{K}) + (1 - 0.75)(0.1385\ \text{kJ/kg}\cdot\text{K})$$

Answer

$$= 0.4565\ \text{kJ/kg}\cdot\text{K}$$

The values of h_f, h_g, s_f, and s_g were determined as indicated before: one-half the sum of the values at 160 and 200 kPa.

Notice in these examples that the pressure and temperature are directly related to each other. If one of them is known, the saturation table gives the other property. The saturation line shown in figure 5–1 could be plotted exactly on a pressure–temperature diagram by using the data obtained from the appropriate saturation table.

There may be some concern by the student as to which property table one should refer to if pressure and temperature are known. The following criteria will help in that decision process:

If given $T > T_{\text{sat}}$ at a given pressure, use superheat tables.

If given $T < T_{\text{sat}}$ at a given pressure, use incompressible liquid or compressed liquid tables.

If given $p < p_{\text{sat}}$ at a given temperature, use superheat tables.

If given $p > p_{\text{sat}}$ at a given temperature, use incompressible liquid or compressed liquid tables.

Let us now see how the compressed liquids may be analyzed. Table B–13, "Steam table of compressed liquid (English units)," lists properties of liquid water for given pressures and temperatures in a manner similar to the superheat tables. Here, however, the properties are given as differences between the experimentally determined exact values and the property values at saturated liquid conditions. Notice that the specific volume differences are nearly zero. At the maximum of 3000 psia and 600°F, the difference is only 0.00088 ft^3/lbm, which is, for most engineering purposes, negligible. The enthalpy differences are approximately the same as those one would find by using the equation for incompressible liquid [equation (5–23)], written as

$$h - h_f = c(T - T_{\text{sat}}) + v_f(p - p_{\text{sat}}) \qquad \textbf{(5–41)}$$

and which we now consider.

EXAMPLE 5–15 Determine the enthalpy of water at 1000 psia and 300°F from table B–13, and compare it to that obtained by the approximate equation (5–41).

Solution From table B–13, we find that $h = h_f + 1.74$ Btu/lbm. From table B–11, $h_f = 269.7$ Btu/lbm at 300°F. The saturation pressure here is 67.01 psia. Then

Answer

$$h = 269.7 \text{ Btu/lbm} + 1.74 \text{ Btu/lbm} = 271.44 \text{ Btu/lbm}$$

From equation (5–41), we get

$$h = h_f + v_f(p - p_{\text{sat}})$$

since $T = T_{\text{sat}}$ for h_f. Then

$$h = 269.7 \text{ Btu/lbm} + \frac{(0.0175 \text{ ft}^3/\text{lbm})(1000 - 67.01 \text{ lbf/in}^2)(144 \text{ in}^2/\text{ft}^2)}{778 \text{ ft}\cdot\text{lbf/Btu}}$$

Answer

$$= 269.7 + 3.02 = 272.72 \text{ Btu/lbm}$$

These two results for enthalpy agree within 0.3%. Table B–13 for SI units lists the ideal work done in compressing water from its saturated liquid state to a compressed liquid state. The entries on that table, found at given temperature and pressure values, are essentially the same values as $v\,\Delta p$, which was used in example 5–15 for the approximation equation. In this book, we usually assume that all liquids and solids are incompressible.

5–6 SUMMARY

In this chapter, the idea of the pure substance was introduced and the state principle was presented. From the state principle, it was shown that the state of a pure substance may be defined by two independent properties, such as volume and temperature. The phase diagram or pressure–temperature diagram was discussed, and the pressure–volume diagram of pure substances was given. The pressure–volume–temperature relationships were then considered. The perfect gas was introduced, and its defining equations,

$$pv = RT$$
$$pV = mRT \tag{5–1}$$
$$pV = N_m R_u T$$

were analyzed. Then the principle of corresponding states was used to show that, for many materials, the p–V–T relationship

$$pv = ZRT \tag{5–3}$$

is a better approximation than the perfect gas equation. The compressibility factor Z is a function of p_R and T_R, where

$$p_R = \frac{p}{p_c} \tag{5–4}$$

$$T_R = \frac{T}{T_c} \tag{5–5}$$

Caloric equations of state were introduced to show how temperature and pressure or volume affect the internal energy and enthalpy of pure substances. For perfect gases, it was shown, by the use of Joule's experiment, that

$$\Delta u = \sum c_v\,\delta T \tag{5–13}$$

$$\Delta h = \sum c_p\,\delta T \tag{5–18}$$

and for constant specific heats of a perfect gas,

$$\Delta u = c_v \, \Delta T \tag{5–14}$$

$$\Delta h = c_p \, \Delta T \tag{5–21}$$

For a perfect gas, it was shown that

$$c_p - c_v = R \tag{5–17}$$

and the ratio of the specific heats was defined as

$$k = \frac{c_p}{c_v} \tag{5–20}$$

For incompressible materials, it was shown that the caloric equations of state may be written

$$\Delta u = \sum c \, \delta T$$

and

$$\Delta h = \sum c \, \delta T + v \, \Delta P \tag{5–23}$$

It was then shown that properties for pure substances not behaving as perfect gases or as incompressible materials could be obtained in thermodynamic property tables. The use of the property tables in this book was demonstrated. It was shown that in the saturation region, temperature and pressure are not independent properties, but are directly related to each other. The following guidelines showed us which tables to use:

If given $T > T_{\text{sat}}$ at given pressure, use superheat.
If given $T < T_{\text{sat}}$ at given pressure, use compressed liquid.
If given $p < p_{\text{sat}}$ at given temperature, use superheat.
If given $p > p_{\text{sat}}$ at given temperature, use compressed liquid.

Within the saturation region, the quality is defined as the mass fraction of saturated vapor to the total mass, $\chi = m_g/m_T$. Also, the properties in the saturation region, as functions of the quality, are

$$v = \chi v_g + (1 - \chi)v_f = \chi v_{fg} + v_f \tag{5–38}$$

and

$$\begin{aligned} u &= \chi u_g + (1 - \chi)u_f = \chi u_{fg} + u_f \\ h &= \chi h_g + (1 - \chi)h_f = \chi h_{fg} + h_f \\ s &= \chi s_g + (1 - \chi)s_f = \chi s_{fg} + s_f \end{aligned} \tag{5–40}$$

From the compressed liquid table of water, we saw that, for many engineering problems, we may use

$$v = v_f \text{ at given temperature}$$

$$h = h_f + v_f(p - p_{\text{sat}}) \text{ at given temperature}$$

$$u = u_f \text{ at given temperature}$$

DISCUSSION QUESTIONS

Section 5–1

5–1 What is meant by a *pure substance*?

5–2 What is meant by *equation of state*?

5–3 What purpose does a *phase diagram* serve?

Section 5–2

5–4 Why is the *perfect gas equation* used so much?

5–5 What is the *gas constant*?

5–6 Why is the compressibility factor used?

Section 5–3

5–7 What is the specific heat at constant volume?

5–8 What is the specific heat at constant pressure?

Section 5–4

5–9 What is the *triple point*?

5–10 What is a calorimeter?

Section 5–5

5–11 What is meant by *saturated vapor*?

5–12 What is meant by *saturated liquid*?

5–13 What is *superheat*?

5–14 What is meant by *sub-cooled liquid*?

5–15 What is meant by *quality*?

PRACTICE PROBLEMS

Problems that use SI units are indicated by an (M) under the problem number; those that use English units are indicated by an (E).

Section 5–1

5–1 For a pure substance in the liquid phase at a temperature between the critical point and the triple point, to what phase does it change if the pressure is reduced enough while holding temperature constant?

5–2 If a pure substance is in the liquid phase and at a temperature between the critical and triple points, to what phase will it eventually change if the pressure is increased at constant temperature?

5–3 If a pure solid substance is at a pressure below the triple point and its temperature is increased, to what phase will it eventually change?

5–4 If a pure vapor substance has a temperature below the triple point and its pressure is then increased while holding temperature constant, to what phase will it change?

5–5 A pure solid substance increases temperature at constant pressure which is greater than that of the triple point. To what phase will it change?

5–6 A pure vapor substance is cooled at constant pressure which is greater than the triple point but less than the critical pressure. To what phase does it eventually change?

Section 5–2

Given the following conditions, determine the unknown property of a perfect gas, for problems 5–7 through 5–18.

5–7 (M) Pressure is 140 kPa gage, volume is 0.085 m^3, mass is 0.7 kg, and the gas constant is 300 J/kg · K.

5–8 (M) Density is 1.5 kg/m^3, the gas constant is 600 J/kg · K, and the temperature is 80°C.

5–9 (M) Oxygen gas at 1200 kPa and 400°C.

5–10 (M) A gas whose molecular weight is 13.5 kg/kg · mol and that is at a gage pressure of 200 kPa. The temperature is 800 K.

5–11 (M) Two hundred seventy grams of argon at a pressure of 160 kPa and a volume of 1.3 m^3.

5–12 (M) A gas in a 3-L container at a pressure of 300 kPa and a temperature of 700°C, and with a mass of 0.66 g.

5–13 (E) A gas having a specific volume of 7 ft^3/lbm, a pressure of 120 psia, and a temperature of 1000°F.

5–14 (E) Determine the gage pressure if the atmospheric pressure is 14.7 psia, the gas constant is 96 ft · lbf/lbm · °R, the temperature is 700°R, and the specific volume is 10 ft^3/lbm.

5–15 (E) Carbon dioxide gas (CO_2) at a pressure of 15 psia and 90°F.

5–16 (E) Seventy pounds-mass of a gas are contained in a rigid container at 200 psia and 80°F. The gas is then expanded to fill a 2000-ft^3 volume at a pressure of 20 psia and a temperature of 70°F. Determine the volume of the rigid container.

5–17 (E) During a constant-pressure process, 0.05 lbm of hydrogen gas increases in temperature from 70°F to 200°F. Determine the final volume of the gas if its density is initially 0.09 lbm/ft^3.

5–18 (M) A constant-temperature process is executed during which the pressure of an ideal gas increases by a ratio of 10 to 1. If the gas was initially enclosed in 2 L, determine the final volume.

5–19 (E) Use compressibility factors to determine the specific volume for nitrogen gas at 45 atm and 80°F. Then determine the percent error one would have if the perfect gas equation were used.

5–20 (M) Nitrogen gas (N_2) is at 42 atm pressure and 130 K in a bottle of 0.02 m^3 volume.
(a) Does the nitrogen behave as a perfect gas?
(b) What is the mass of the nitrogen in the bottle?

5–21 (M) Carbon dioxide, CO_2, exists at 0.01 m^3/kg and 310 K. What is its pressure if the principle of corresponding states is used?

5–22 (E) Consider steam at 1600 psia and 1200°F. Determine the specific volume by using the principle of corresponding states, and compare the result to that from the steam table B–12.

5–23 (M) Determine the density of Refrigerant-12 at 3 MPa and 200°C by using the principle of corresponding states.

5–24 (E) Determine the temperature of methane at 780 psia and 0.7 ft^3/lbm by using the principle of corresponding states.

5–25 (M) Determine the density of air at −20°F and 1000 psia by using the principle of corresponding states.

5–26 (E) One of the earliest attempts at deriving a more general equation of state was the van der Waals relationship

$$\left(p + \frac{a}{v^2}\right)(v - b) = RT$$

where a and b are constants. This relationship was introduced to describe the possible inelastic collisions of gas molecules and the volume that they occupy. It seems to be somewhat better than the perfect gas equation but is not often used today. Determine the temperature of carbon monoxide gas at 20 psia and 6 ft^3/lbm by using the van der Waals equation of state with the values

$$a = 375 \text{ atm} \cdot \text{ft}^6/\text{mol}^2$$
$$b = 0.63 \text{ ft}^3/\text{mol}$$

Compare your answer with that derived from using the perfect gas law.

Section 5–3

Use table B–4 for gas constants when needed.

5–27 (M) Determine the increase in internal energy of 2 kg of argon if its temperature increases from 30°C to 130°C.

5–28 (M) Six hundred thirty grams of propane gas is cooled from 38°C to 13°C. Determine the total internal energy change and the specific internal energy change.
Compare your answer with one you obtained by using *EES*.

5–29 (M) Helium gas is cooled from 80°C to 20°C in 60 min. Determine the internal energy change and the rate of energy change per gram.

5–30 (M) Neon gas exhibits a change in temperature of 1000°C. Determine its change in internal energy per gram.

5–31 (M) Determine the change in internal energy of ammonia (NH_3) as its temperature increases from 700 K to 800 K. (*Hint:* Use table B–7 and the equation $c_v = c_p - R$.) Compare your answer with one you obtained by using *EES*.

5–32 (E) Calculate the change in internal energy per pound-mass of sulfur dioxide as its temperature increases by 50°R. Compare your answer with one you obtained by using *EES*.

5–33 (E) Three pounds-mass of a perfect gas, having constant specific heat, can absorb 70 Btu of heat while increasing temperature by 80°R. Determine C_v and c_v of the gas.

5–34 (E) Determine, within 1.1% error, the change of internal energy of oxygen gas between 1000°R and 2000°R. (*Hint:* Use table B–7 and the equation $c_v = c_p - R$.) Compare your answer with one you obtained by using *EES*.

5–35 (E) The specific heat, c_v, of a certain gas obeys the relation

$$c_v = [(0.25) + 0.01\,T] \text{ Btu/lbm} \cdot \text{°R}$$

Determine the change in internal energy of this gas as its temperature increases from 200°F to 300°F.

5–36 (M) A perfect gas increases temperature from 70°C to 130°C. Determine the change in specific enthalpy if the specific heat is 1.0 kJ/kg · K.

5–37 (M) Determine the increase in the enthalpy of 100 kg of sulfur dioxide gas (SO_2) between 30°C and 110°C. Compare your answer with one you obtained by using *EES*.

5–38 (M) A perfect gas is known to have a gas constant of 14.3 kJ/kg · K and a ratio of specific heats, k, of 1.405. Determine c_p and c_v.

5–39 (M) Assuming variable specific heats, determine the enthalpy change of propane gas (C_3H_8) as it is cooled from 98°C to 20°C. Check your answer against the answer you get by using constant specific heat values. Compare your answer with one you obtained by using *EES*.

5–40 (E) Determine the enthalpy change of helium as it is cooled from 120°F to 60°F.

5–41 (E) Determine the values of k and c_p of a perfect gas that has c_v of 0.225 Btu/lbm · °R and a gas constant of 66 ft · lbf/lbm · °R.

5–42 (E) A perfect gas is known to have a gas constant value of 78.5 ft · lbf/lbm · °R and a ratio of specific heats, k, of 1.28. Determine c_p and c_v.

5–43 (E) *n*-Butane, C_4H_{10}, is subjected to a temperature change from 100°F to 1000°F. Use the equation for specific heat, c_p, from table B–7 to determine the enthalpy change. This problem may best be solved with the computer. Compare your answer with one you obtained by using *EES*.

5–44 (M) Methyl alcohol, CH_3OH, is cooled from 200°C to 10°C. Use an equation from table B–7 for specific heat values and the computer to determine the enthalpy change of the methyl alcohol.

5–45 (E) Common bricks are heated from 70°F to 300°F. Determine the enthalpy change of one brick if each brick weighs 5 lbm. (Use table B–8.)

5–46 (M) Common plywood is cooled by 20°C. What is the enthalpy change per unit mass? (Use table B–8.)

5–47 (E) Cast iron is heated to 500°F from room temperature of 65°F. What is the enthalpy change per unit mass? (Use table B–8.)

5–48 (M) Ethylene glycol (a permanent antifreeze) changes temperature by as much as 125°C. How much does its enthalpy change? (Use table B–8.)

5–49 (E) Engine oil exhibits a temperature change of 250°F. What enthalpy change per unit mass would this represent? (Use table B–8.)

5–50 (M) Determine the internal energy and enthalpy of air at 7°C. (Use table B–6.) Compare your answer with one you obtained by using *EES*.

5–51 (E) Determine the internal energy and enthalpy of air at 165°F from table B–6. Compare your answer with one you obtained by using *EES*.

5–52 (M) Determine the enthalpy change of water vapor as its temperature changes from 250°C to 450°C. Use the perfect gas model and compare with the value you obtained by using *EES*.

5–53 (M) Propane at 30°C escapes from a pressurized tank and cools to −40°C. Use the perfect gas model to estimate the enthalpy change. Then compare your answer with the result you obtained from using *EES*.

5–54 (E) Determine the enthalpy change of water vapor as its temperature changes from 800°F to 500°F. Use the perfect gas model, and then compare your answer with the value you obtain from using *EES*.

5–55 (M) Benzene (C_6H_6) is heated from 200°C to 400°C. Use the perfect gas model to estimate the internal energy change.

5–56 (E) Estimate the enthalpy change for propane gas heated from 100°F to 350°F. Compare your result with one obtained from *EES*.

5–57 (M) A certain perfect gas is found to have a specific heat at constant pressure that can be described by the equation

$$c_p = 2.25 - 0.0006T + 0.000003T^2$$

where the specific heat units are kJ/kg K and the temperature is in kelvin. Determine the enthalpy change of 1 kg of the gas when the gas temperature changes from 27°C to 127°C. (*Note:* This problem can be solved most conveniently with calculus; however, an approximate answer can be obtained by using finite differences.)

Section 5–5

5–58 (E) Determine the saturation pressure of ammonia at 75°F.

5–59 (M) Determine the saturation temperature of steam at 750 kPa.

5–60 (M) Determine the saturation pressure of R-22 at − 20°C.

5–61 (E) Determine the saturation temperature of mercury at 33 psia.

5–62 (M) Determine the specific volume and enthalpy of mercury at 0.4 MPa and 55% quality.

5–63 (E) Determine the internal energy of R-12 at −45°F and 32% quality.

5–64 (E) Determine the specific volume and enthalpy of steam at 1.0 psia and 92% quality. Also determine the percent by volume of saturated vapor in the mixture. Compare your answer with one you obtained by using *EES*.

5–65 (M) Determine the percent by volume of saturated vapor in R-22 at −25°C and 25% quality. Compare your answer with one you obtained by using *EES*.

5–66 (M) Determine u_g for steam at 200°C.

5–67 (E) Determine u_g for steam at 400 psia. What is its temperature?

5–68 (M) Determine u_f for steam at 2.0 kPa. What is its temperature? Compare your answer with one you obtained by using *EES*.

5–69 (E) Determine u_f for steam at 0.5 psia. What is its temperature?

5–70 (E) Steam at 65% quality is at 260°F. Determine its pressure and h, u, v, and s.

5–71 (M) Steam at 80% quality is at 150°C. Determine its pressure and h, u, v, and s.

5–72 (M) Determine h, u, v, and s for steam at 240°C and 500 kPa. What will be the steam's enthalpy if the pressure is increased to 5000 kPa while the temperature remains at 240°C? Compare your answer with one you obtained by using *EES*.

5–73 (E) Determine the enthalpy of steam at 300°F if the pressure is 200 psia.

5–74 (E) Determine the enthalpy, specific volume, and internal energy of R-134a at 120°F if its quality is 90%. Compare your answer with one you obtained by using *EES*.

5–75 (M) Determine the enthalpy, specific volume, and internal energy of R-134a at 80°C if its quality is 15%. Compare your answer with one you obtained by using *EES*.

5–76 (M) Water is boiling at 120°C with a quality of 20%. Determine its density and fraction of liquid by volume.

5–77 (M) Saturated liquid water at 30°C flows through an ideal pump, where its pressure increases to 4 MPa as it leaves the pump. Determine the enthalpy of the water leaving the pump and the ideal pump work per kilogram of water.

5–78 (E) Water at 66°F enters an ideal pump as a saturated liquid and leaves at 1000 psia. Determine the ideal pump work and the enthalpy per unit mass of the water leaving the pump.

PROCESSES

In chapter 5, equations relating system properties at one particular state were introduced. Here we introduce equations called *process equations*, which determine and relate a system's properties at various states during a process. The advantage of an accurately descriptive process equation is that it gives additional information necessary to determine the amount of work done during a reversible process. The reversible process is given heaviest emphasis in this text, but some attempt is made to show how irreversibilities might affect predicted properties and work.

The chapter topics are arranged by type of substances. First the perfect gas process is considered. The constant-pressure (isobaric) process, constant-volume (isometric), constant-temperature (isothermal), polytropic, and reversible adiabatic processes are all analyzed. The reversible cases will be given most attention; however, the adiabatic efficiency of turbines, compressors, and nozzles is defined and demonstrated. Compressible fluids are treated briefly, and incompressible liquids and solids are both considered. Strain energy in an elastic, incompressible solid and Bernoulli's equation for incompressible, constant-temperature fluids are just two of the topics considered. Finally, the processes of pure substances are analyzed. Steam, the refrigerants (R-12, R-22, R-123, R-134a, R-407c, and R-502), and ammonia are all treated for those situations most likely to be used in the various applications of the remainder of the book, including the throttling process of pure substances.

Tables that list the various forms of the relationships resulting from process equations are presented at the end of the chapter. These tabulations can circumvent many of the mathematical manipulations that students may find too difficult, and they should be useful references for subsequent problem solving.

New Terms

SE	Strain energy	η_s	Adiabatic efficiency
Y	Young's modulus	ξ (xi)	Strain $= \Delta l/l$
Δl	Length change of elastic material		

6–1 PROCESSES OF PERFECT GASES

We consider next the various thermodynamic processes that involve perfect gases, beginning with the constant-pressure processes.

Constant-Pressure Processes

The most common process occurring in our mechanized, technological society is probably the constant-pressure process—any closed system executing a change in volume against the atmosphere is involved in a constant-pressure process. There are, of course, many other situations that produce this type of process; we investigate a general one.

For a closed reversible system, the work is calculated from $\sum p\,\delta V$. In particular, if the pressure is constant, we have

$$Wk_{cs} = p \sum \delta V = p(V_2 - V_1)$$

or

$$Wk_{cs} = p(\Delta V) = p(V_2 - V_1) \quad \textbf{(6–1)}$$

Here V_2 is the final volume of the closed system, and V_1 is the initial volume. This result has been reached previously but is repeated here for completeness. It is correct for any material—liquid, imperfect gas, or perfect gas.

CALCULUS FOR CLARITY 6–1

The work of a reversible process for a closed system is

$$Wk_{cs} = \int p\,dV$$

and if the pressure is constant,

$$Wk_{cs} = p \int dV = p\,\Delta V$$

which is equation (6–1).

For perfect gases, we have other results that are quite useful. For the constant-pressure process, we have

$$p_1 = p_2 = \text{constant} \quad \textbf{(6–2)}$$

where the subscripts 1 and 2 represent initial and final states. For the perfect gas, this implies that

$$\frac{V_1}{T_1} = \frac{V_2}{T_2}$$

or

$$\frac{V_1}{V_2} = \frac{T_1}{T_2} \quad \textbf{(6–3)}$$

Using the relationship $v = V/m$, we also have

$$\frac{v_1}{v_2} = \frac{T_1}{T_2} \quad \textbf{(6–4)}$$

and from the relationship $\rho = 1/v$, we obtain

$$\frac{\rho_2}{\rho_1} = \frac{T_1}{T_2} \quad \textbf{(6–5)}$$

From these results, and given enough data, we can calculate properties of perfect gases at different states. These relationships hold for either open or closed systems.

For the open system, we have indicated that reversible work can be found from the equation

$$Wk_{os} = -\sum V\,\delta p$$

During the constant-pressure process, this sum vanishes; that is,

$$\delta p = 0$$

so that

$$Wk_{os} = 0$$

and for an open system, the only contributions to work when pressure remains constant must be in the form of kinetic or potential energy changes. Frequently, the constant-pressure process is referred to as the **isobaric process**.

The heat q is obtained from the first-law energy equation for the closed or the open system, depending on the type of system. For a closed system, the heat is given by $q = \Delta u + wk_{cs}$, but the wk_{cs} term is just $p\,\Delta v$, so that, for the constant-pressure process,

$$q = \Delta u + p\,\Delta v = \Delta h \qquad \textbf{(6–6)}$$

For the open system and applying the steady-flow energy equation, we find, for no kinetic or potential energy changes, that $q = \Delta h$ again. Equation (6–6) is good for the constant-pressure process for both open and closed systems.

In chapter 1, we listed the suggested method of problem solving, the critical steps being: (1) to identify the system as open, closed, or isolated with a boundary; (2) to identify the known states by the properties at those states and the states not fully known (not all properties known); (3) to identify the type of substance: perfect gas, incompressible substance, or pure substance; (4) to make any assumptions necessary to simplify the problem (such as reversible conditions) without creating an unrealistic situation; (5) to identify the type of process involved (such as isobaric or, as we shall see, others); and (6) to proceed with solving the problem through the conservation principles (conservations of mass and energy). Let us now look at two examples of isobaric processes or constant-pressure processes of perfect gases.

EXAMPLE 6–1 In figure 6–1, air flows through a chamber at 2-atm pressure, with no change in velocity. If the density of the gas increases by a factor of 2.5, and the inlet temperature is 650°C, determine the work, the exhaust temperature, and the heat per kilogram of air.

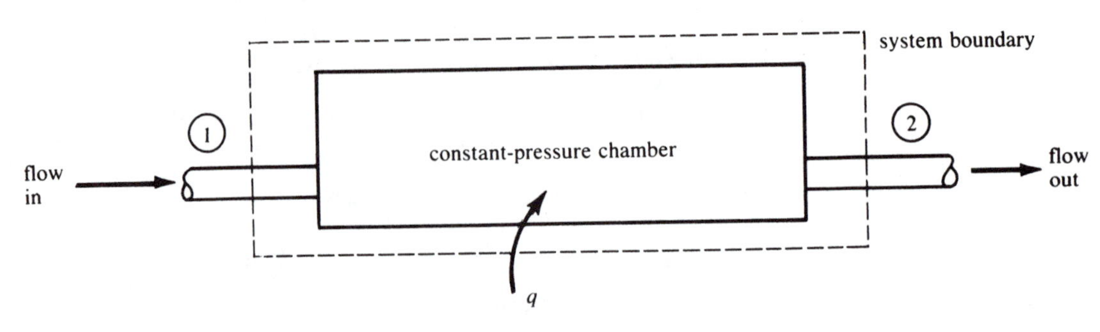

FIGURE 6–1

Solution

Since here we are concerned with a system through which gas flows, we see that the system is open. The boundary is shown in figure 6–1 with heat indicated and states 1 and 2 at the inlet and outlet. The states are as follows:

$$\text{State 1:}\quad p_1 \qquad\qquad \text{State 2:}\quad p_2 = p_1$$
$$T_1 = 650°\text{C (923 K)} \qquad\qquad T_2$$

Air is the working substance, and because the temperatures are well above the critical temperature (132.41 K, table B–5) and the pressure is not near the critical pressure (37.25 atm, table B–5), we shall assume it to be a perfect gas. Also, we assume constant specific heats. The process is one of constant pressure so that the work is just zero. We also assume that the kinetic and potential energies remain the same.

We seek the exhaust temperature from equation (6–5). We then have

$$\frac{\rho_2}{\rho_1} = 2.5 = \frac{T_1}{T_2} = \frac{(650 + 273)\text{ K}}{T_2}$$

yielding, for the exhaust temperature,

Answer

$$T_2 = \frac{923\text{ K}}{2.5} = 369\text{ K}$$

After eliminating the appropriate terms, the heat is found from the energy equation to be

$$q = h_2 - h_1 = c_p(T_2 - T_1)$$

The specific heat c_p is 1.007 kJ/kg · K from table B–4, and

Answer

$$q = (1.007\text{ kJ/kg}\cdot\text{K})(369\text{ K} - 923\text{ K})$$
$$= -557.9\text{ kJ/kg}$$

The negative sign indicates that the gas is cooled or the heat is moving out of the system instead of in, as was indicated in figure 6–1.

EXAMPLE 6–2

Five pounds-mass of nitrogen is contained in a piston-cylinder at 100 psig (gage pressure) and 80°F. The atmospheric pressure is 14.5 psia, and the nitrogen temperature is increased to 90°F through heat and irreversible work. Determine the final volume of the nitrogen and the boundary work if the pressure is to remain at 100 psig at all times during the process.

Solution

In this example, the system is the nitrogen inside the piston-cylinder, and the boundary is drawn just at the cylinder wall and the piston face. In figure 6–2 is shown the diagram of this situation, and we see that the system is closed. The states are as follows:

$$\text{State 1:}\quad p_1 = 100\text{ psig} + 14.5\text{ psia} = 114.5\text{ psia}, \quad T_1 = 80 + 460 = 540°\text{R}$$
$$\text{State 2:}\quad p_2 = p_1 = 114.5\text{ psia}, \quad T_2 = 90 + 460 = 550°\text{R}$$

The nitrogen gas has a pressure below one-fourth of its critical pressure (33.54 atm, table B–5) and greater than the critical temperature (126 K = 226.8°R, table B–5). This satisfies the criteria from chapter 5 for the assumption of a perfect gas. We shall also assume that the specific heats are constant and those listed in table B–4, $c_v = 0.1774$ Btu/lbm · °R and $c_p = 0.2483$ Btu/lbm · °R. Also, the gas constant R is 55.15 ft · lbf/lbm · °R from table B–4. Since the pressure is constant, we have, from equation (6–3), that $V_2 = V_1(T_2/T_1)$. The volume at state 1 is calculated from the perfect gas law:

$$V_1 = \frac{mRT_1}{p_1}$$
$$= \frac{(5\text{ lbm})(55.15\text{ ft}\cdot\text{lbf/lbm}\cdot°\text{R})(540°\text{R})}{(114.5\text{ lbf/in}^2)(144\text{ in}^2/\text{ft}^2)}$$
$$= 9.03\text{ ft}^3$$

FIGURE 6–2

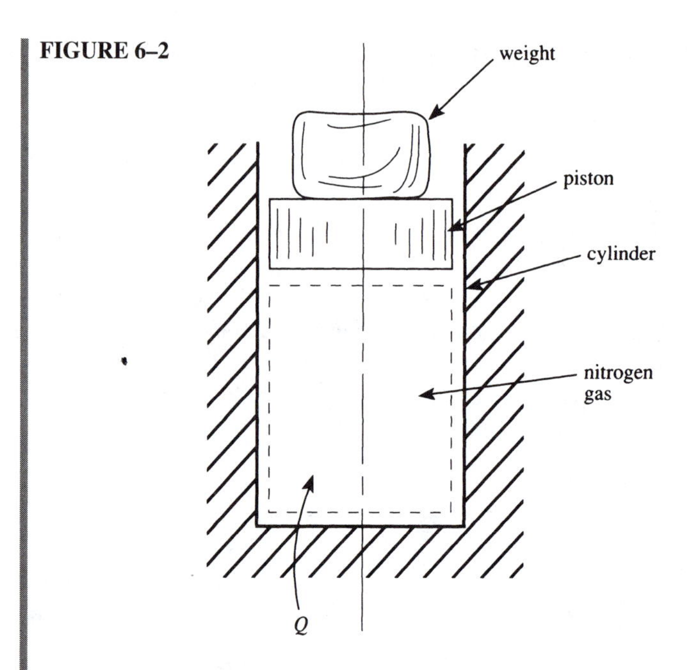

Then

Answer

$$V_2 = (9.03 \text{ ft}^3)\left(\frac{550°\text{R}}{540°\text{R}}\right) = 9.20 \text{ ft}^3$$

The boundary work is just $p\,\Delta V$ from equation (6–1). Then

Answer

$$\begin{aligned} Wk_{cs} &= (114.5 \text{ lbf/in}^2)(144 \text{ in}^2/\text{ft}^2)(9.20 \text{ ft}^3 - 9.03 \text{ ft}^3) \\ &= 2802.96 \text{ ft} \cdot \text{lbf} \end{aligned}$$

or

Answer

$$Wk_{cs} = 3.60 \text{ Btu}$$

Notice that even though this process was irreversible, we could still compute the boundary work of the nitrogen gas.

Constant-Volume Processes

The processes of systems where the volume remains constant are sometimes referred to as **isometric processes**. When perfect gases are considered, the following relationship holds between states 1 and 2:

$$\frac{p_1}{p_2} = \frac{T_1}{T_2} \qquad \textbf{(6–7)}$$

For a closed system the boundary work, Wk_{cs}, is zero because no volume change occurs, and the heat is then given by

$$Q = \Delta U = m \sum c_v \, \delta T \qquad \textbf{(6–8)}$$

For open systems, the shaft work, wk_{os}, is given by $wk_{os} = -V\,\Delta p$, or per unit mass, $wk_{os} = -v\,\Delta p$ if kinetic and potential energies are neglected. The heat for a steady-flow condition of an open system is then $q = \Delta h - v\,\Delta p$.

EXAMPLE 6–3 Heat is added to a piston-cylinder containing air at 1800 kPa and 650 K while the volume is held constant. The amount of heat is such that the final pressure is 3200 kPa. Determine the final temperature and the amount of heat added per kilogram of air.

Solution Here the system is closed and is the air inside the piston-cylinder. The states are as follows:

Initial state (1):	Final state (2):
$p_1 = 1800$ kPa	$p_2 = 3200$ kPa
$T_1 = 650$ K	

The air pressure is below the critical pressure, and the temperatures are more than twice the critical temperature (132.41 K, table B–5), so we shall assume air to be a perfect gas having constant specific heats. Then the final temperature may be found from

$$T_2 = T_1\left(\frac{p_2}{p_1}\right)$$

$$= (650\text{ K})\left(\frac{3200\text{ kPa}}{1800\text{ kPa}}\right)$$

Answer

$$= 1156\text{ K}$$

The amount of heat added per unit mass is

$$q = \Delta u = \sum c_v\, \delta T$$

and for constant specific heats from table B–4,

$$q = c_v\, \Delta T$$

$$= (0.719\text{ kJ/kg}\cdot\text{K})(1156\text{ K} - 650\text{ K})$$

Answer

$$= 363.8\text{ kJ/kg}$$

EXAMPLE 6–4 Inside the closed chamber of a piston-cylinder device, fuel and air burn at constant volume, releasing 3.5 kJ of energy. If the volume is 1000 cm^3, the temperature is 280°C before burning, and the pressure is 600 kPa, determine the temperature and pressure of the gases after burning if the gases behave like air.

Solution The process is one of a closed system, so we may write the energy equation as

$$\Delta U = Q \qquad (Wk_{cs} = 0)$$

for constant-volume closed systems. Using specific heats and temperatures, we have

$$\Delta U = mc_v\, \Delta T = Q = 3.5\text{ kJ}$$

For air, $c_v = 0.719$ kJ/kg·K and $R = 0.287$ kJ/kg·K from table B–4. We may solve for the mass from the perfect gas equation. We find that

$$m = \frac{pV}{RT}$$

$$= \frac{(600{,}000\text{ N/m}^2)(1000\text{ cm}^3)}{(0.287\text{ kJ/kg}\cdot\text{K})(553\text{ K})(10^6\text{ cm}^3/\text{m}^3)}$$

$$= 0.00378\text{ kg}$$

Then, from the preceding,

$$\Delta T = \frac{Q}{mc_v} = \frac{3.5 \text{ kJ}}{(0.00378 \text{ kg})(0.719 \text{ kJ/kg} \cdot \text{K})}$$

or

Answer

$$\Delta T = 1288 \text{ K} = T_2 - T_1$$

Since $T_1 = 280°\text{C} = 553 \text{ K}$, we find that

$$T_2 = 1841 \text{ K}$$

The pressure after burning is found from equation (6–7):

$$p_2 = p_1 \frac{T_2}{T_1} = (600 \text{ kPa})\left(\frac{1841 \text{ K}}{553 \text{ K}}\right)$$

Answer

$$= 2000 \text{ kPa}$$

Constant-Temperature Processes

Constant-temperature processes are often called **isothermal processes**; in this book, both terms are used. For isothermal processes of perfect gases, and from the perfect gas law, we find that

$$\frac{p_1}{p_2} = \frac{V_2}{V_1} = \frac{v_2}{v_1} = \frac{\rho_1}{\rho_2} \tag{6–9}$$

and the relation between p and v for a process is $pv = C$, where C is a constant. We call this a process equation, to indicate that it helps describe the process. As we saw in chapter 3, when the process equation is $pv = C$, the boundary work is given by

$$Wk_{cs} = C \ln \frac{V_2}{V_1} \tag{3–16}$$

The same quantity (boundary work) could be found from

$$Wk_{cs} = C \ln \frac{p_1}{p_2} \tag{6–10}$$

Also, for closed systems, the internal energy change will always be zero for an isothermal process of a perfect gas because the internal energy depends on temperature alone. The heat is then just equal to the boundary work.

For open systems, the shaft work per unit mass, $-\sum v \, \delta p$, is

$$wk_{os} = -C \ln \frac{p_2}{p_1} = C \ln \frac{p_1}{p_2} \tag{6–11}$$

Again (as with the boundary work), since $pv = C$ or $p_1/p_2 = v_2/v_1$, equation (6–11) could be written

$$wk_{os} = C \ln \frac{v_2}{v_1} \tag{6–12}$$

For perfect gases, the constant C is pv, but it is also equal to RT because $pv = RT$. Therefore, we have, for perfect gases,

$$wk_{cs} = RT \ln \frac{V_2}{V_1} = RT \ln \frac{v_2}{v_1}$$

$$wk_{os} = RT \ln \frac{v_2}{v_1}$$

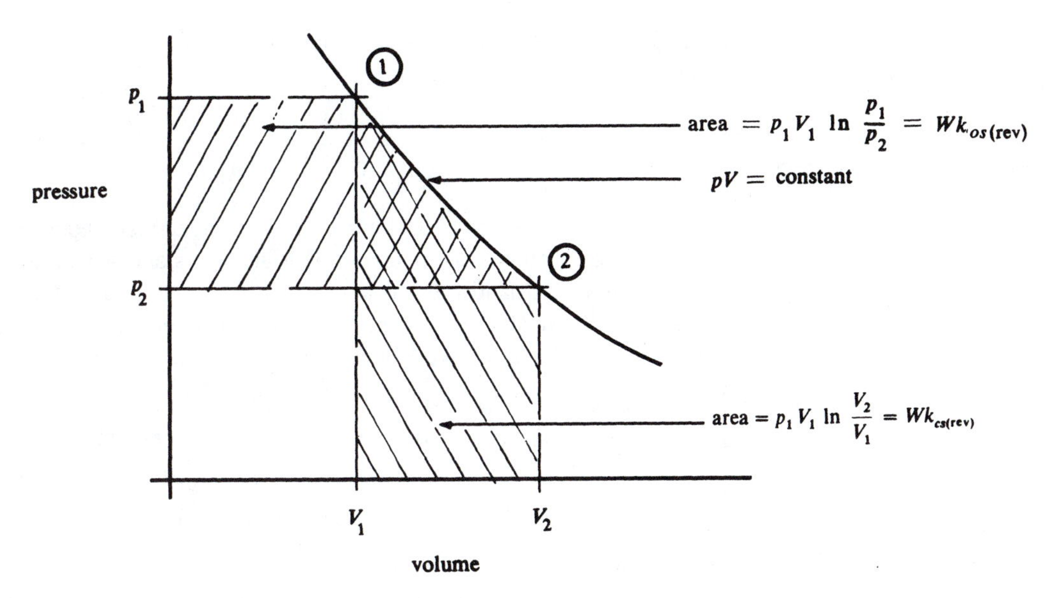

FIGURE 6–3 p–V diagram for isothermal process with a perfect gas

or the shaft work and the boundary work are the same, neglecting kinetic and potential energies. We shall see later that compressors are sometimes designed as piston-cylinder devices and sometimes as steady-flow or rotating centrifugal devices. If the compression is done isothermally, the amount of work will be the same for either device, as just shown. Also, in figure 6–3, the pressure–volume diagram is shown for a constant-temperature process, including the areas representing the boundary work, Wk_{cs}, and the shaft work, Wk_{os}.

For the open system through which a perfect gas flows, the enthalpy will remain constant if the process is isothermal. Then the heat for a steady-flow situation will be equal to the shaft work.

EXAMPLE 6–5 During the expansion of a perfect gas from 1200-kPa pressure and a specific volume of 1.1 m^3/kg to 101-kPa pressure, the temperature remains constant. Determine the constant C in the process equation $pv = C$ describing the process, and determine the final specific volume.

Solution The constant C is equal to the product of the pressure and specific volume at any state during the process. Thus, at the initial condition, we have

$$C = (1200 \text{ kPa})(1.1 \text{ m}^3/\text{kg})$$
$$= (1200 \text{ kN/m}^2)(1.1 \text{ m}^3/\text{kg})$$

Answer
$$= 1320 \text{ kJ/kg}$$

Also, we may write the process equation as

$$p_1 v_1 = p_2 v_2$$

yielding

$$v_2 = v_1 \frac{p_1}{p_2}$$

Substituting values into this relationship gives

$$v_2 = (1.1\ \text{m}^3/\text{kg})\left(\frac{1200\ \text{kPa}}{101\ \text{kPa}}\right)$$

Answer

$$= 13.1\ \text{m}^3/\text{kg}$$

EXAMPLE 6–6 During the compression of 0.01 lbm of air in a cylinder (see figure 6–4), heat is transferred through the cylinder walls to keep the air at a constant temperature. The air pressure increases from 15 psia to 150 psia after the air is fully compressed. The initial specific volume of the air is 7.4 ft^3/lbm. Determine the operating air temperature, the change in internal energy and in enthalpy, the work done, and the heat transferred during this process.

Solution This is an isothermal process, and we will assume it to be reversible as well. If the air is behaving like a perfect gas, which we assume, the operating temperature can be found from

$$T = \frac{pV}{mR} = \frac{pv}{R}$$

or, initially,

$$T_1 = \frac{p_1 v_1}{R} = \frac{(15\ \text{lbf/in}^2)(7.4\ \text{ft}^3/\text{lbm})(144\ \text{in}^2/\text{ft}^2)}{53.3\ \text{ft}\cdot\text{lbf/lbm}\cdot{}^\circ\text{R}}$$

so that

$$T_1 = 300^\circ\text{R}$$

and then

$$T_2 = 300^\circ\text{R}$$

The change in internal energy is

$$\Delta U = mc_v\,\Delta T = 0$$

and for the enthalpy change, we have

$$\Delta H = mc_p\,\Delta T = 0$$

The work done is reversible; thus, we obtain, from equation (3–16),

$$Wk_{cs} = C \ln \frac{V_2}{V_1}$$

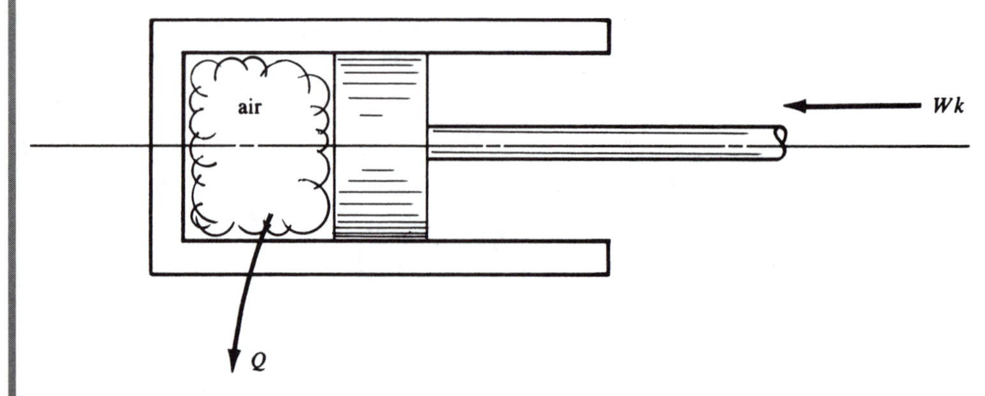

FIGURE 6–4 Isothermal process of a piston-cylinder device

or, more conveniently, from equation (6–10),

$$Wk_{cs} = C \ln \frac{p_1}{p_2}$$

The constant is determined first:

$$\begin{aligned} C &= p_1V_1 = p_1mv_1 \\ &= (15\ \text{lbf/in}^2)(0.01\ \text{lbm})(7.4\ \text{ft}^3/\text{lbm})(144\ \text{in}^2/\text{ft}^2) \\ &= 159.8\ \text{ft} \cdot \text{lbf} \end{aligned}$$

Then

$$\begin{aligned} Wk_{cs} &= (159.8\ \text{ft} \cdot \text{lbf})\left(\ln \frac{15}{150}\right) \\ &= (159.8)\left(-\ln \frac{150}{15}\right) \\ &= (159.8)(-\ln 10) \end{aligned}$$

Answer $$= -368\ \text{ft} \cdot \text{lbf}$$

The heat transferred is equal to the work done, so

Answer $$Q = -368\ \text{ft} \cdot \text{lbf}$$

and Q is, as the sign indicates, removed from the system. For the irreversible isothermal process, the internal energy change can still be zero, but the work and heat increase in absolute values; that is, more work is required and more heat transfer is demanded to retain constant temperature.

EXAMPLE 6–7 Refrigerant-12 (Freon-12) is compressed from 100 psig to 200 psig at 1000°F in a steady-flow device. Determine the minimum work required and the heat transferred during the compression per unit mass. Assume that the atmospheric pressure is 14.6 psia.

Solution The system is identified as open steady flow, and the states are as follows:

Inlet (state 1):	Outlet (state 2):
$p_1 = 114.6$ psia	$p_2 = 214.6$ psia
$T_1 = 1000°\text{F}$	$T_2 = T_1 = 1000°\text{F}$

The Freon has critical values from table B–5 of $p_c = 40.6$ atm, $T_c = 385.16$ K (=693.3°R), so we shall assume that the Freon acts as a perfect gas. The work, neglecting kinetic and potential energies, is

$$\begin{aligned} wk_{os} &= RT \ln \frac{p_1}{p_2} \\ &= (12.78\ \text{ft} \cdot \text{lbf/lbm} \cdot °\text{R})(1460\ \text{R})\left(\ln \frac{114.6\ \text{psia}}{214.6\ \text{psia}}\right) \end{aligned}$$

Answer $$= -11705.2\ \text{ft} \cdot \text{lbf/lbm}$$

or

Answer $$wk_{os} = -15.05\ \text{Btu/lbm}$$

Here the value for R was found in table B–4. The heat is just the same as the work, $q = -15.15$ Btu/lbm, so the heat is removed from the system and work is put in.

Polytropic Processes

In chapter 3, we considered situations involving work where the pressure–volume relationship was given by the equation $pV^n = C$ [equation (3–15)], called the **polytropic equation**, where n is a constant called a **polytropic exponent**. Those processes where the pressure and volume are described by equation (3–15) are called **polytropic processes**, and for a process from state 1 to state 2, this equation can be written

$$\frac{p_1}{p_2} = \left(\frac{V_2}{V_1}\right)^n = \left(\frac{v_2}{v_1}\right)^n \tag{6–13}$$

Also, for perfect gases, the following relationships hold between the temperatures:

$$\frac{T_1}{T_2} = \left(\frac{V_2}{V_1}\right)^{n-1} = \left(\frac{v_2}{v_1}\right)^{n-1} \quad \text{and} \quad \frac{p_1}{p_2} = \left(\frac{T_1}{T_2}\right)^{n/(n-1)} \tag{6–14}$$

The boundary work done in a polytropic process is given by equation (3–17):

$$Wk_{cs} = \frac{1}{1-n}(p_2V_2 - p_1V_1)$$

For a perfect gas, the pV terms can be replaced by mRT terms:

$$Wk_{cs} = \frac{mR(T_2 - T_1)}{1-n} \tag{6–15}$$

For open systems, the shaft work of a polytropic reversible process is given by $wk_{os} = \sum v\,\delta p$, which becomes, through the process equation $pv^n = C$,

$$wk_{os} = \frac{n}{1-n}(p_2v_2 - p_1v_1) \tag{6–16}$$

For perfect gases, equation (6–16) can be written

$$wk_{os} = \frac{nR}{1-n}(T_2 - T_1) \tag{6–17}$$

The values for n typically fall between 1.1 and 1.5. If you look carefully at equation (3–15) ($pV^n = C$), you can see that is the process equation for the polytropic process and can be interpreted as a general process equation. That is, if n is set equal to 1, we have the process equation for the isothermal process of a perfect gas; if it is set equal to zero, it identifies the isobaric process ($p = C$); and so on. Table 6–1 gives a listing of the values of n if $pV^n = C$ is considered as a general process equation of perfect gases. In table 6–1, the reversible adiabatic process is listed for the case where $n = k$ ($=c_p/c_v$), and this process is discussed in section 6–2. In figure 6–5, the pressure–volume diagram shows the representation of the five particular processes listed in table 6–1.

TABLE 6–1 Polytropic exponent

Process	Value of n in Polytropic Equation $pV^n = C$
Isobaric	0
Isometric	∞
Isothermal	1
Reversible adiabatic	k
Polytropic	n

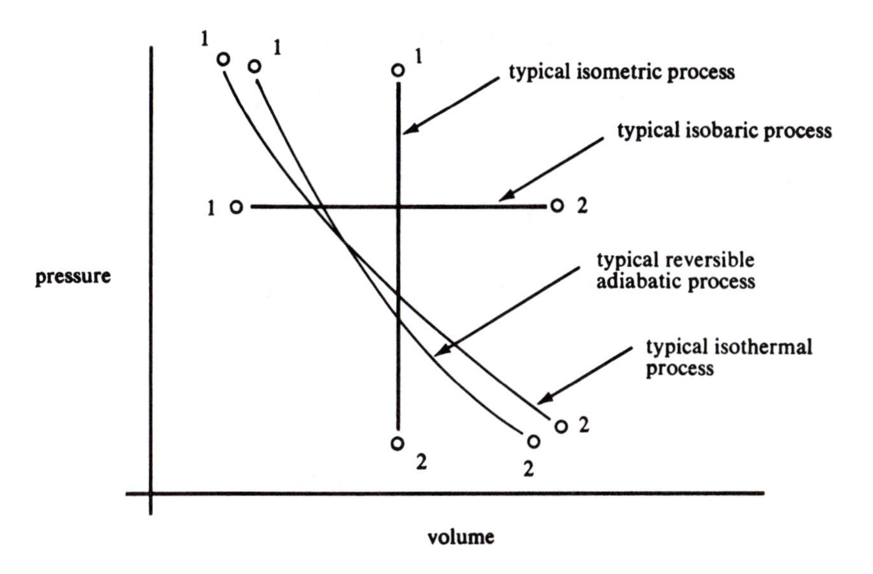

FIGURE 6–5 Various polytropic processes shown on the p–V plane for perfect gases

EXAMPLE 6–8 Air is polytropically expanded through a reversible turbine from 1500-kPa to 100-kPa pressure. The inlet air temperature is 1200 K, the mass flow rate is 1.0 kg/s, and the polytropic exponent n is 1.5. Determine the power produced and the rate of heat transfer.

Solution The system is an open one with air the working medium. The conditions at the inlet and outlet are as follows:

$$\text{State 1: } p_1 = 1500 \text{ kPa} \qquad \text{State 2: } p_2 = 100 \text{ kPa}$$
$$T_1 = 1200 \text{ K}$$

Since the critical temperature is well below 1200 K and the critical pressure well above 1500 kPa, we assume air to be a perfect gas. Also, we assume constant specific heats. The work is the shaft work because the process is reversible and given by equation (6–16) or (6–17). Using either of these two equations requires that the temperature or volume of the second state be found. Using equation (6–14), we rearrange to find the temperature T_2:

$$T_2 = T_1\left(\frac{p_2}{p_1}\right)^{(n-1)/n}$$

Thus,

$$T_2 = (1200 \text{ K})\left(\frac{100 \text{ kPa}}{1500 \text{ kPa}}\right)^{0.5/1.5}$$
$$= 487 \text{ K}$$

The work is then obtained from the computation

$$wk_{os} = \frac{1.5}{1 - 1.5}(0.287 \text{ kJ/kg}\cdot\text{K})(487 - 1200)\text{K}$$
$$= 614 \text{ kJ/kg}$$

The power can be found from its relationship with the mass flow rate:

$$\dot{W}k_{os} = \dot{m}(wk_{os})$$

"Plugging in" the numbers gives

$$\dot{W}k_{os} = (1.0 \text{ kg/s})(614 \text{ kJ/kg})$$

Answer

$$= 614 \text{ kJ/s} = 614 \text{ kW}$$

Using the conversion to horsepower units, we get

Answer

$$\dot{W}k_{os} = 823 \text{ hp}$$

Since we neglect kinetic and potential energy changes through the turbine, we may write the energy equation for the turbine as

$$\dot{m}\,\Delta h = \dot{Q} - \dot{W}k_{os}$$

or

$$\dot{Q} = \dot{m}\,\Delta h + \dot{W}k_{os}$$

from which we may now compute the heat transfer. Thus,

$$\dot{Q} = \dot{m}c_p(T_2 - T_1) + \dot{W}k_{os}$$

Using a value for the specific heat of 1.007 kJ/kg · K, we have

$$\dot{Q} = (1.0 \text{ kg/s})(1.007 \text{ kJ/kg}\cdot\text{K})(-713 \text{ K}) + 614 \text{ kW}$$

Answer

$$= -104 \text{ kW}$$

EXAMPLE 6–9 During a reversible polytropic compression of an ideal gas, the following data are found:

$$p_1 = 14.7 \text{ psia}$$
$$v_1 = 14.5 \text{ ft}^3/\text{lbm}$$
$$p_2 = 165 \text{ psia}$$
$$v_2 = 2.42 \text{ ft}^3/\text{lbm}$$

Determine the polytropic exponent n.

Solution

Since the process is reversible polytropic, for a perfect or ideal gas we write the process equation

$$p_1v_1^n = p_2v_2^n$$

or

$$\left(\frac{v_1}{v_2}\right)^n = \frac{p_2}{p_1}$$

and

$$(n)\ln\frac{v_1}{v_2} = \ln\frac{p_2}{p_1}$$

Then

$$n = \frac{\ln(p_2/p_1)}{\ln(v_1/v_2)}$$

and we compute n to obtain

$$n = \frac{\ln(165 \text{ psia}/14.7 \text{ psia})}{\ln(14.5 \text{ ft}^3/\text{lbm}/2.42 \text{ ft}^3/\text{lbm})}$$

Answer

$$= 1.35$$

6–2 ADIABATIC PROCESSES OF PERFECT GASES

Processes that do not have any heat or heat transfer are called **adiabatic processes**. Therefore, for any substance going through an adiabatic process, from the first law of thermodynamics for a closed system, we find that

$$Wk = \Delta E \tag{6–18}$$

For processes involving reversible boundary work only, and neglecting kinetic and potential energies, equation (6–18) becomes

$$Wk_{cs} = \Delta U$$

or

$$wk_{cs} = \Delta u \tag{6–19}$$

for unit mass of the substance. For adiabatic processes of perfect gases, we have

$$wk_{cs} = \sum c_v \, \delta T \tag{6–20}$$

and we already know that the wk_{cs} is $\sum p \, \delta v$. It can be shown that, for reversible adiabatic processes of perfect gases having constant specific heats (see section 7–8) the relationship between p and v is

$$pv^k = c \tag{6–21}$$

where $k = c_p/c_v$ and c is a constant. This is the process equation for the reversible adiabatic process involving perfect gases with constant specific heats. In table 6–1, it is shown that equation (6–21) is identical to the polytropic equation $pv^n = C$ when $n = k$. Equation (6–21) may be rewritten as

$$\frac{p_1}{p_2} = \left(\frac{V_2}{V_1}\right)^k \tag{6–22}$$

Using equation (6–21) and the perfect gas equation, $pv = RT$ or $pV = mRT$, we can show that the following are also true:

$$\frac{V_2}{V_1} = \left(\frac{T_1}{T_2}\right)^{1/k-1} \quad \text{and} \quad \frac{p_1}{p_2} = \left(\frac{T_1}{T_2}\right)^{k/k-1} \tag{6–23}$$

In addition, it can be shown that the boundary work, $\sum p \, \delta V$, is given by

$$Wk_{cs} = \frac{1}{1-k}(p_2V_2 - p_1V_1) \tag{6–24}$$

Since this is an equation for perfect gases, it could easily be written as

$$Wk_{cs} = \frac{mR}{1-k}(T_2 - T_1) \tag{6–25}$$

Also, for the open system, the energy equation for the adiabatic process is given by equation (4–42) with Q set to zero:

$$\dot{KE}_2 + \dot{PE}_2 + \dot{H}_2 - \dot{KE}_1 - \dot{PE}_1 - \dot{H}_1 + \dot{U}_s = \dot{\not{Q}} - \dot{Wk}_{os} \tag{6–26}$$

Note that the kinetic and potential energies of the system are neglected. For steady-flow conditions, this equation reduces to equations (4–49) with q set to zero; that is,

$$\left.\begin{aligned} &(\text{kJ/kg}): \quad \dot{m}\left(\frac{1}{2}(\overline{V}_2^2 - \overline{V}_1^2) + g(z_2 - z_1) + h_2 - h_1\right) = \dot{m}\not{q} - \dot{m}wk_{os} \\ &(\text{Btu/lbm}): \quad \dot{m}\left(\frac{\overline{V}_2^2 - \overline{V}_1^2}{2g_c} + \frac{g(z_2 - z_1)}{g_c} + h_2 - h_1\right) = \dot{m}\not{q} - \dot{m}wk_{os} \end{aligned}\right\} \quad \text{(6–27)}$$

and for cases where the kinetic and potential energies are neglected, the steady-flow equation is $\dot{m}wk_{os} = \dot{m}(h_1 - h_2)$ for one inlet and one outlet. This can also be written

$$wk_{os} = h_1 - h_2 \quad \text{(6–28)}$$

for the shaft work per unit mass flowing through an open system. The shaft work may also be found from the relationship $-\sum v\,\delta p$ if kinetic and potential energies are neglected. For the reversible adiabatic case of a perfect gas with constant specific heats, this becomes

$$wk_{os} = \frac{k}{1-k}(p_2v_2 - p_1v_1) \quad \text{(6–29)}$$

The mechanical efficiency of devices is sometimes used to describe their irreversibilities. By this term, we shall mean the actual work obtained as a fraction of the work done by a reversible adiabatic one and call it the **adiabatic efficiency**, η_s, given by

$$\eta_s = \begin{cases} \dfrac{wk_{\text{act}}}{wk_{cs}} & \text{(closed system involving work done by the system)} \\ \dfrac{wk_{\text{act}}}{wk_{os}} & \text{(open system involving work done by the system)} \end{cases} \quad \text{(6–30)}$$

where wk_{act} means the actual work obtained from a device. From equation (6–30), it can be seen that the actual work is always less than the work done reversibly and adiabatically because the adiabatic efficiency will always be less than 100% (1.0). For power-consuming devices such as pumps and compressors, the adiabatic efficiency is defined as the ideal or reversible adiabatic work as a fraction of the actual work required. Thus,

$$\eta_s = \begin{cases} \dfrac{wk_{cs}}{wk_{\text{act}}} & \text{(closed system involving work done by the system)} \\ \dfrac{wk_{os}}{wk_{\text{act}}} & \text{(open system involving work done by the system)} \end{cases} \quad \text{(6–31)}$$

From these definitions, you can see that the actual work will always be more than the reversible adiabatic work because the efficiency will be less than 100%, but now the actual work is the ideal work divided by the efficiency. Adiabatic efficiencies of 65% to 95% are common for present-day devices, whether they are power consumers or power producers.

EXAMPLE 6–10 A perfect gas is allowed to expand reversibly and adiabatically in a piston-cylinder from 1500 kPa and 0.05 m^3/kg to 101-kPa pressure. If the gas has a specific heat ratio, k, of 1.38, determine the work done during this process.

Solution The system in this example is a closed one and is a perfect gas. For a reversible adiabatic process, we may then use for the process equation, $pv^k = C$, from which the work is found from the equation

$$wk_{cs} = \frac{1}{1-k}(p_2v_2 - p_1v_1)$$

The states are as follows:

State 1: $p_1 = 1500$ kPa State 2: $p_2 = 101$ kPa
$v_1 = 0.05\ \text{m}^3/\text{kg}$

To find the final specific volume v_2, we write the process equation in the form

$$p_1 v_1^k = p_2 v_2^k$$

Then

$$v_2 = v_1\left(\frac{p_1}{p_2}\right)^{1/k}$$

Substituting values into this relationship gives

$$v_2 = (0.05\ \text{m}^3/\text{kg})\left(\frac{1500\ \text{kPa}}{101\ \text{kPa}}\right)^{1/1.38}$$
$$= 0.353\ \text{m}^3/\text{kg}$$

Using this result, we can now compute the work:

$$wk_{cs} = \frac{1}{1 - 1.38}[(101 \times 10^3\ \text{N/m}^2)(0.353\ \text{m}^3/\text{kg}) - (1500 \times 10^3\ \text{N/m}^2)(0.05\ \text{m}^3/\text{kg})]$$
$$= (-2.63)(-0.393 \times 10^5\ \text{N}\cdot\text{m/kg})$$

Answer
$$= 103\ \text{kJ/kg}$$

EXAMPLE 6–11 A gas turbine receives air at 2000°R, allows reversible adiabatic expansion of the air past the turbine blades, and exhausts this air at 2 lbm/s to the atmosphere at 1000°R. Determine the heat transferred, the work produced per pound-mass of air, and the power produced if there are negligible kinetic and potential energy changes for the air.

Solution Our system is here specified as the gas turbine, which is a device converting energy from a flowing stream (such as air) to a rotational energy of a wheel or shaft. The system is an open one, as indicated in figure 6–6, so we write the first law as

$$\Delta H = Q - Wk_{os}$$

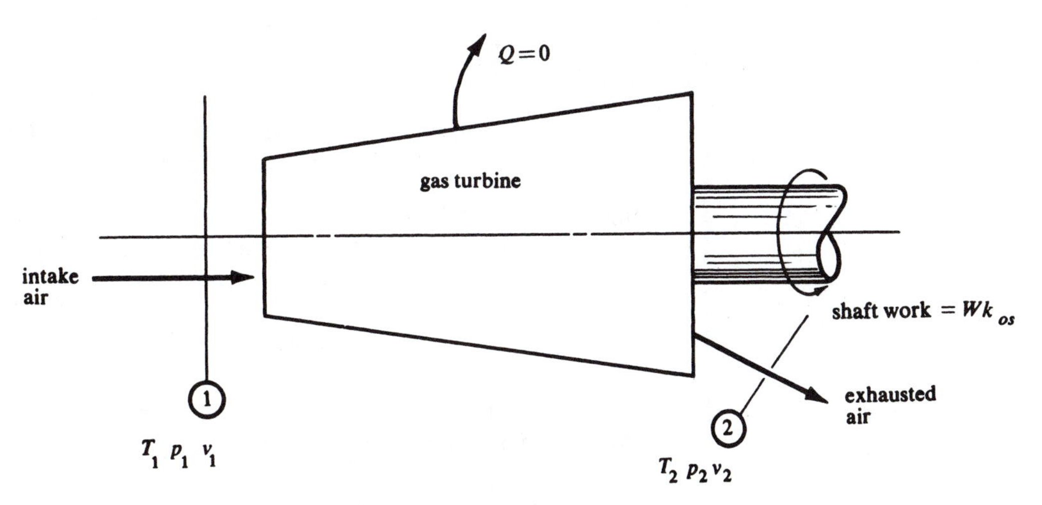

FIGURE 6–6 Sketch of system for adiabatic expansion of air in a gas turbine

In terms of rates per unit time, this becomes

$$\dot{H} = \dot{Q} - \dot{W}k_{os}$$

which can be rewritten as

$$\dot{m}\,\Delta h = \dot{Q} - \dot{m}wk_{os}$$

The enthalpy change, Δh, is obtained from

$$\Delta h = c_p\,\Delta T$$

or

$$\Delta h = h_2 - h_1 = c_p(T_2 - T_1)$$

after assuming that the air is a perfect gas with constant specific heat. From table B–4, we find c_p to be 0.24 Btu/lbm · °R. Then

$$\Delta h = 0.24\ \text{Btu/lbm}\cdot{}^\circ\text{R} \times (1000^\circ\text{R} - 2000^\circ\text{R}) = -240\ \text{Btu/lbm}$$

and because the process is adiabatic, $\dot{Q} = 0$. Then

Answer

$$wk_{os} = 240\ \text{Btu/lbm air}$$

The mass flow rate, $\dot{m}$, is given as 2 lbm/s, so that

$$\dot{W}k_{os} = 2(240)\ \text{Btu/s} = 480\ \text{Btu/s}$$

Converting to horsepower units gives the solution:

Answer

$$\dot{W}k_{os} = 480\ \text{Btu/s} \times 1.41\ \text{hp}\cdot\text{s/Btu} = 676.8\ \text{hp}$$

Irreversibilities included in this process will reduce the actual output power by reducing the enthalpy change or producing strain energy of the system. In the irreversible adiabatic case, we expect deformation or wear in the system.

EXAMPLE 6–12

An adiabatic compressor is used to compress argon gas from 15 psia to 160 psia, and its adiabatic efficiency is 85% under these conditions. If the argon is at 70°F at the low-pressure state, determine the work done by the compressor per unit mass of argon flowing through it. Also determine the temperature of the argon at the high pressure.

Solution

The system in this example is an adiabatic compressor with argon as the substance. We shall assume that the compressor is a steady-flow device, so that the system is an open one. The states are as follows:

State 1: $p_1 = 15$ psia State 2: $p_2 = 160$ psia
$T_1 = 70^\circ\text{F} = 530^\circ\text{R}$

Argon has a critical pressure of 47.996 atm and a critical temperature of 150.72 K. Comparing the conditions of the argon at states 1 and 2, we assume that the argon will behave as a perfect gas. The work done by the compressor is given by equation (6–31):

$$wk_{act} = \frac{wk_{os}}{\eta_s}$$

Here, wk_{os} is the shaft work of a reversible adiabatic compressor operating between the same two pressures. For a reversible adiabatic compressor,

$$wk_{os} = h_{2s} - h_1$$

where h_{2s} is the enthalpy at state 2 identified with the reversible adiabatic compression between states 1 and 2. The subscript s identifies a condition that means "isentropic" or constant entropy. Entropy is an important property that we define and discuss in chapter 7. For our purposes here, we assume constant specific heats for the argon, so

$$wk_{os} = -c_p(T_{2s} - T_1)$$

The temperature, T_{2s}, is given by

$$T_{2s} = (T_1)\left(\frac{p_2}{p_1}\right)^{(k-1)/k}$$

from equation (6–23). Substituting into this equation, we find that

$$\begin{aligned} T_{2s} &= (530°\text{R})(160\text{ psia}/15\text{ psia})^{(1.668-1)/1.668} \\ &= 1367.7°\text{R} \end{aligned}$$

The value for k was taken from table B–4. The reversible work is then found with the value of c_p from table B–4, 0.1244 Btu/lbm · °R:

$$\begin{aligned} wk_{os} &= -(0.1244\text{ Btu/lbm}\cdot°\text{R})(1367.7°\text{R} - 530°\text{R}) \\ &= -104.2\text{ Btu/lbm} \end{aligned}$$

The compressor work is then

$$\begin{aligned} wk_{act} &= \frac{-104.2\text{ Btu/lbm}}{0.85} \\ &= -122.6\text{ Btu/lbm} \end{aligned}$$

Answer

The final temperature of the argon at state 2 is found from the actual steady-flow equation for the compressor:

$$-wk_{act} = h_2 - h_1$$

where the kinetic and potential energies are neglected. For the perfect gas with constant specific heats, this is

$$-wk_{act} = c_p(T_2 - T_1)$$

and

$$\begin{aligned} T_2 &= T_1 - \frac{wk_{act}}{c_p} \\ &= 530°\text{R} - \frac{-122.6\text{ Btu/lbm}}{0.1244\text{ Btu/lbm}\cdot°\text{R}} \\ &= 1515.5°\text{R} \end{aligned}$$

Answer

Notice that the argon has a higher temperature after compression through the actual compressor than if it had been compressed in a reversible adiabatic one. A problem involving turbines or engines where power or work is produced would show that the final temperature of a perfect gas expanding through such a device having an adiabatic efficiency of less than 100% would be more than the temperature of the same gas expanding through a reversible adiabatic device.

Let us now look at a filling process of an adiabatic open system.

EXAMPLE 6–13 A well-insulated tank is connected to an oxygen line, which is also well insulated. There is a shutoff valve in the connecting line as shown in figure 6–7. The oxygen is at 65 psig and 25°F in the line. The tank has a capacity of 5 ft^3 and is completely empty. The valve is then opened and oxygen is allowed to fill the tank. Determine the temperature of the oxygen in

FIGURE 6–7 Schematic for example 6–13

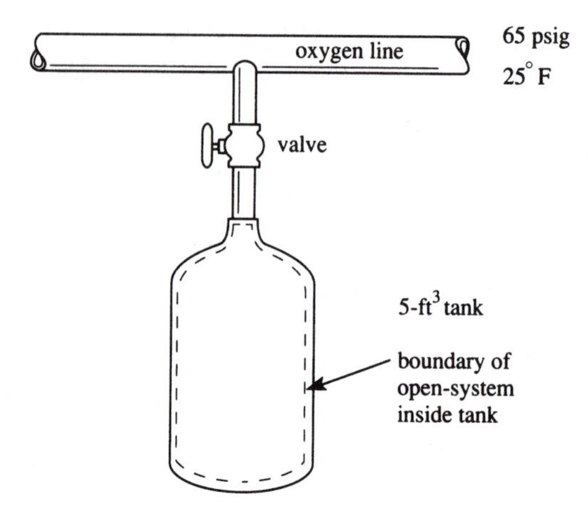

the tank after filling and the amount of oxygen that is then in the tank, in pound-mass units. Assume that atmospheric pressure is 14.6 psia.

Solution

The system is an open one, and its boundary is shown as a dashed line in figure 6–7. The substance is oxygen, and the two states, beginning and ending, are as follows:

$$\text{State 1: } p_1 = 0 \qquad \text{State 2: } p_2 = 65 \text{ psig} = 79.6 \text{ psia}$$
$$V_1 = 5 \text{ ft}^3 \qquad V_2 = 5 \text{ ft}^3$$

Notice that the tank pressure is the same as the line pressure after the tank has filled the 5 ft^3. We should also see that this is not a steady-flow problem but rather one involving a change in the system's energy, so we use equation (6–26), after neglecting the kinetic and potential energies and setting the work term to zero (there is no shaft work). That is,

$$-\dot{H}_1 + \dot{U}_s = 0$$

The first term accounts for the oxygen entering the system through the valve. We may write this as

$$\dot{H}_1 = \dot{m}h_1$$

Also, the internal energy change of the system may be written

$$\dot{U}_s = \frac{m_f u_f - m_i u_i}{\Delta t} = \frac{m_f u_f}{\Delta t}$$

where the subscripts i and f represent the initial and final states of the system. Then the energy balance is

$$-\dot{m}h_1 + \frac{m_f u_f}{\Delta t} = 0$$

Multiplying through by Δt, setting $\dot{m}\,\Delta t = m$, and transposing gives

$$mh_1 = m_f u_f = mu_f$$

because $m_f = m$. If we assume that oxygen is a perfect gas with constant specific heats, we have

$$mc_p T_1 = mc_v T_f$$

or

$$T_f = T_1\left(\frac{c_p}{c_v}\right) = kT_1$$

The temperature of the oxygen in the tank after filling is

$$T_f = (1.396)(25°F + 460)$$

Answer

$$= 677°R = 217°F$$

The amount of oxygen in the tank can be determined from the perfect gas equation. Solving for the mass yields

$$m = \frac{pV}{RT}$$

$$= \frac{(79.6\ \text{lbf/in}^2)(5\ \text{ft}^3)(144\ \text{in}^2/\text{ft}^2)}{(48.29\ \text{ft}\cdot\text{lbf/lbm}\cdot°R)(677°R)}$$

Answer

$$= 1.753\ \text{lbm}$$

Notice that the oxygen increases the temperature in the filling process, and the tank cannot be filled to the amount that one would predict if the temperature were assumed constant. That is, if we said that T_f was 25°F or 485°R, the mass would have been predicted to be 2.447 lbm—nearly 40% too high. If the oxygen were now cooled to a lower temperature in the tank, more oxygen could be put into the tank, until finally we might be able to have nearly 2.4 lbm of oxygen in the tank.

CALCULUS FOR CLARITY 6–2

If the filling problems, such as example 6–13, are analyzed with the use of calculus, then the energy balance is written

$$\frac{-dH_1}{dt} + \frac{dU_s}{dt} = 0 \tag{6–32}$$

The rate of enthalpy inflow into the system is

$$\frac{dH_1}{dt} = \frac{d(m_1h_1)}{dt} = m_1\frac{dh_1}{dt} + h_1\frac{dm_1}{dt}$$

and the state of the fluid entering is steady because the oxygen from the line is at a steady-state pressure and temperature. Then $dh_1/dt = 0$, and $dH_1/dt = h_1\,dm_1/dt = \dot{m}_1h_1$. The energy balance is then

$$\dot{m}_1h_1 = \frac{dU_s}{dt} \tag{6–33}$$

and if we integrate both sides over a time interval t, we have

$$\int \dot{m}_1h_1\,dt = \int \frac{dU_s}{dt}\,dt \tag{6–34}$$

The right side is the change of internal energy of the tank:

$$\int \frac{dU_s}{dt}\,dt = \Delta U = m_{s,f}u_{s,f} - m_{s,i}u_{s,i} \tag{6–35}$$

CALCULUS FOR CLARITY 6–2, continued

Using the mass balance equation (4–16) in the form $\int \dot{m}_1\,dt = m_{s,f} - m_{s,i}$, we find that the left side of equation (6–34) is

$$h_1 \int \dot{m}_1\,dt = h_1(m_{s,f} - m_{s,i}) \tag{6–36}$$

Substituting equations (6–35) and (6–36) into equation (6–34) gives the general equation for steady filling:

$$(m_{s,f} - m_{s,i})h_1 = m_{s,f}u_{s,f} - m_{s,i}u_{s,i} \tag{6–37}$$

For the case where the tank is empty at the start, $m_{s,i} = 0$, and equation (6–37) becomes

$$m_{s,f}h_1 = m_{s,f}u_1$$

or

$$h_1 = u_{s,f} \tag{6–38}$$

6–3 PROCESSES OF COMPRESSIBLE GASES

Compressible gases can behave quite differently from perfect gases, as we saw in chapter 5 in the analysis of gases. Here we look at some gases that do not satisfy the criteria of perfect gases because of too low temperatures or too high pressures. We limit ourselves to pressure–volume–temperature relationships of processes, as given by the principle of corresponding states. The principle of corresponding states may be extended to caloric relationships between the internal energy and temperature and pressure, as well as the relationship between enthalpy and temperature and pressure, but it is beyond our purposes in this book. Here we will look at three processes (the isobaric, or constant-pressure; the constant-volume; and the isothermal) and see some applications where the perfect gas model gives erroneous results.

EXAMPLE 6–14 Methane gas at 5000 kPa and −20°C is heated to 10°C at constant pressure. Determine the density of the methane after being heated.

Solution The methane has the critical properties $p_c = 45.8$ atm and $T_c = 190.7$ K. Since the methane is subjected to 5000-kPa, or 49.3-atm, pressure, and 253-K temperature, we use the principle of corresponding states rather than the perfect gas assumption. We calculate the reduced pressure and temperature:

$$p_R = \frac{p}{p_c} = \frac{49.3}{45.8} = 1.076$$

$$T_{R_1} = \frac{T_1}{T_c} = \frac{253}{190.7} = 1.33 \quad \text{(at state 1)}$$

$$T_{R_2} = \frac{T_2}{T_c} = \frac{283}{190.7} = 1.48 \quad \text{(at state 2)}$$

From chart B–1, we find the compressibility factor at state 2. It is approximately 0.90, so we may then use equation (5–3) for specific volume (inverse density):

$$v = \frac{ZRT}{p} \quad \text{and} \quad \rho = \frac{p}{ZRT}$$

We then find that

$$\rho = \frac{5000 \text{ kN/m}^2}{(0.90)(0.519 \text{ kN}\cdot\text{m/kg}\cdot\text{K})(283 \text{ K})}$$

Answer

$$= 37.8 \text{ kg/m}^3$$

EXAMPLE 6–15 Heat is added to benzene in a constant-volume process such that the final temperature is 1400°R. The initial pressure is 600 psia and the temperature is 1000°R. Assuming that the benzene remains chemically inert, determine the final pressure.

Solution Again, using the principle of corresponding states, we first determine the reduced properties:

$$\text{State 1:} \quad p_{R_1} = \frac{p_1}{p_c} = \frac{600}{(48.6)(14.696)} = 0.84$$

$$T_{R_1} = \frac{T_1}{T_c} = \frac{1000°\text{R}}{(562.6 \text{ K})\left(\frac{9}{5}°\text{R/K}\right)} = 0.987$$

$$\text{State 2:} \quad T_{R_2} = \frac{T_2}{T_c} = \frac{1400}{(562.6)\left(\frac{9}{5}\right)} = 1.38$$

Next, from chart B–1, we find state 1 (approximately) as shown in figure 6–8. The compressibility factor is approximately 0.54, and $v_{R'}$ is 0.59 or 0.6. Then state 2 is found by moving along the $v_{R'}$ constant line to $T_R = 1.38$. Here, the reduced pressure is found as 1.8. The pressure at state 2 is then

$$p_2 = (p_R)(p_c) = (1.8)(48.6 \text{ atm})$$

Answer

$$= 87.48 \text{ atm} = 1286 \text{ psia}$$

If the perfect gas model were assumed for this problem, the final pressure would be computed directly from

$$p_2 = p_1 \frac{T_2}{T_1} = 840 \text{ psia}$$

which is a significant difference from the more nearly correct answer of 1286 psia.

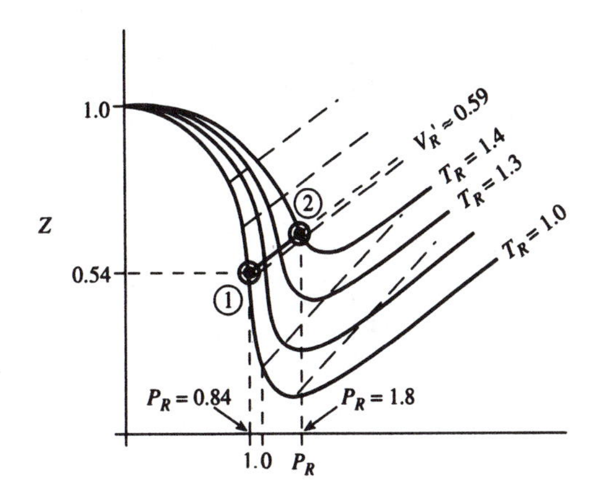

FIGURE 6–8 Method of using compressibility chart to determine final pressure in the constant-volume process of example 6–15

As a final example in this section, we consider an isothermal process of a compressible gas.

EXAMPLE 6–16

Ethane is compressed in a piston-cylinder from 15 atm to 100 atm. If the compression is reversible and done at 100°F, determine the work.

Solution

The system is a piston-cylinder and the substance is ethane. From table B–1, we find the critical properties:

$$p_c = 48.20 \text{ atm} \qquad \text{and} \qquad T_c = 305.43 \text{ K} = 549.8°\text{R}$$

The reduced properties at states 1 and 2 are as follows:

$$\text{State 1:} \quad p_{R_1} = \frac{p_1}{p_c} = \frac{15 \text{ atm}}{48.2 \text{ atm}} = 0.311$$

$$T_{R_1} = \frac{T_1}{T_c} = \frac{560°\text{R}}{549.8°\text{R}} = 1.018$$

$$\text{State 2:} \quad p_{R_2} = \frac{p_2}{p_c} = \frac{100}{48.2} = 2.07$$

Since the process is isothermal, the reduced temperature is constant and equal to 1.018. We then use charts B–1 and B–2 to determine the relationship between volume and pressure so that we can then compute the work from

$$wk_{cs} = \sum p\,\delta v$$

We read, on charts B–1 and B–2, the values of $v_{R'}$ at various states. These are listed in table 6–2. The values for the pressure p are found from the equation $p = (p_R)(p_c)$, and the volume v is found from $v = (v_{R'})(RT_c/p_c)$.

The answer found by using the hand method is shown on the right side of table 6–2, and the final result for the work is -343.21 ft$^3\cdot$lbf/lbm$\cdot$in^2 or $-49{,}422.24$ ft$\cdot$lbf/lbm. The same result can be obtained by using the VISUAL BASIC program AREA in appendix A–5 for computing areas under curves. One would get the proper units of the pressure terms in column (3) of table 6–2 if they are all multiplied by 144 in^2/ft^2 (converting psia to lbf/ft^2 units) before being entered into the AREA program. Otherwise, a simple conversion of the final answer will need to be done, as shown with the hand computation. The work for com-

TABLE 6–2 Example 6–16 results

p_R	$v_{R'}$	p, psia	v, ft^3/lbm	$\bar{p}$	δv	$p\,\delta v$
0.311	3.00	220.3	0.832	287.25	−0.372	−106.86
0.5	1.66	354.2	0.460	425.00	−0.174	−73.95
0.7	1.03	495.8	0.286	566.65	−0.120	−68.00
0.9	0.60	637.5	0.166	708.35	−0.099	−70.13
1.1	0.24	779.2	0.067	850.05	−0.014	−11.90
1.3	0.19	920.9	0.053	991.70	−0.006	−5.95
1.5	0.17	1062.5	0.047	1204.20	−0.003	−3.61
1.9	0.16	1345.9	0.044	1406.10	−0.002	−2.81
2.07	0.15	1466.3	0.042			

$\sum p\,\delta v = -343.21$ ft$^3\cdot$lbf/lbm$\cdot$in^2

pressing a perfect gas isothermally is given by equation (6–10), with the constant equal to RT. Then the work would have been

$$wk_{cs} = (51.43 \text{ ft}\cdot\text{lbf/lbm}\cdot{}^\circ\text{R})(560^\circ\text{R})\left(\ln\frac{15}{100}\right)$$
$$= -54{,}638.6 \text{ ft}\cdot\text{lbf/lbm}$$

and this answer is in some disagreement with the more nearly correct answer we obtained by using the principle of corresponding states.

6–4 PROCESSES OF INCOMPRESSIBLE LIQUIDS

Incompressible liquids are liquids that have densities or volumes that remain constant at all times. Thus, for any change in pressure or temperature, the volume of an incompressible liquid will not change. The boundary work of an incompressible liquid will always be zero, and the shaft work of an open system, neglecting kinetic and potential energies, will be given by

$$wk_{os} = -v\,\Delta p \qquad \textbf{(6–39)}$$

For a closed system of an incompressible substance, the first law of thermodynamics is

$$Q = \Delta U \qquad \textbf{(6–40)}$$

or, because $\Delta U = m\sum c\,\delta T$ for an incompressible substance, from chapter 5 we have

$$Q = m\sum c\,\delta T \qquad \textbf{(6–41)}$$

This equation gives you the first instance in this book where you can see how the term *specific heat* was originally suggested. Here we can see that the heat is related to the temperature change through the specific heat. Of course, the equation is true only for incompressible substances, but there are many applications of those substances. For constant specific heat values, equation (6–41) becomes

$$Q = mc\,\Delta T \qquad \textbf{(6–42)}$$

Values for the specific heats of various liquids are given in table B–8, listed as c_p.

EXAMPLE 6–17

Determine the heat required to increase the temperature of 10 kg of engine oil from 50°C to 100°C.

Solution

We assume that engine oil is incompressible with constant specific heat so that the heat is just given by equation (6–42):

$$Q = mc(T_2 - T_1)$$

From table B–20, we find that $c_p = c = 1.89$ kJ/kg·K = 1.89 kJ/kg·°C and

$$Q = (10 \text{ kg})(1.89 \text{ kJ/kg}\cdot{}^\circ\text{C})(100^\circ\text{C} - 50^\circ\text{C})$$

Answer

$$= 945 \text{ kJ}$$

The steady flow of an incompressible liquid through an open system is a process having many engineering applications. We describe such situations by the steady-flow energy equation:

$$\dot{Q} - \dot{W}k_{os} = \dot{m}c(T_2 - T_1) + \dot{m}(p_2v - p_1v) + \frac{\dot{m}(\overline{V}_2^2 - \overline{V}_1^2)}{2g_c} + \frac{\dot{m}g(z_2 - z_1)}{g_c} \qquad \textbf{(6–43)}$$

For the SI system of units, $g_c = 1\ \text{kg}\cdot\text{s}^2/\text{N}\cdot\text{m}$, and for English engineering units, it is $32.17\ \text{lbm}\cdot\text{s}^2/\text{lbf}\cdot\text{ft}$. The terms $\dot{m}c(T_2 - T_1)$ and $\dot{m}(p_2 v - p_1 v)$ represent the enthalpy change between one inlet and one outlet where the specific heat c is presumed to be constant. Frequently, for incompressible liquids, the specific volume is replaced with the density ρ or the specific weight γ, where $\gamma = \rho g/g_c$. Since $v = 1/\rho$, we have

$$\dot{Q} - \dot{W}k_{os} = \dot{m}c(T_2 - T_1) + \frac{\dot{m}g(p_2 - p_1)}{\gamma g_c} + \frac{\dot{m}(\bar{V}_2^2 - \bar{V}_1^2)}{2g_c} + \frac{\dot{m}g(z_2 - z_1)}{g_c} \tag{6–44}$$

For isothermal processes and those where heat and work are not involved, equation (6–44) reduces to a form called Bernoulli's equation for fluid flow (after dividing through by the mass flow rate):

$$\frac{g(p_2 - p_1)}{\gamma g_c} + \frac{(\bar{V}_2^2 - \bar{V}_1^2)}{2g_c} + \frac{g(z_2 - z_1)}{g_c} = 0 \tag{6–45}$$

EXAMPLE 6–18 A water pump, shown in figure 6–9, operating under steady-flow conditions, moves water at 78°F from a region of low pressure (15 psia) to a region of high pressure (250 psia). Determine the work done per pound-mass of water and the power required if the water is supplied at 20 lbm/s.

Solution The system involved here is a water pump whose boundaries cross an inlet and an exhaust water pipe. Since we are treating water, we assume that it is incompressible and therefore of constant volume. In this circumstance,

$$wk_{os} = -v\,\Delta p = -v(p_2 - p_1)$$

so that, using a value of 0.016 ft³/lbm for the specific volume of water at 78°F, we obtain

$$wk_{os} = -(0.016\ \text{ft}^3/\text{lbm})(250 - 15\ \text{lbf/in}^2)(144\ \text{in}^2/\text{ft}^2)$$

Answer

$$= -541.4\ \text{ft}\cdot\text{lbf/lbm}$$

Now, the power is determined from

$$\dot{W}k_{os} = \dot{m}wk_{os}$$

and

$$\dot{W}k_{os} = (20\ \text{lbm/s})(-541.4\ \text{ft}\cdot\text{lbf/lbm})$$

Answer

$$= -10{,}829\ \text{ft}\cdot\text{lbf/s} = -19.7\ \text{hp}$$

For the irreversible case, this power requirement must be increased.

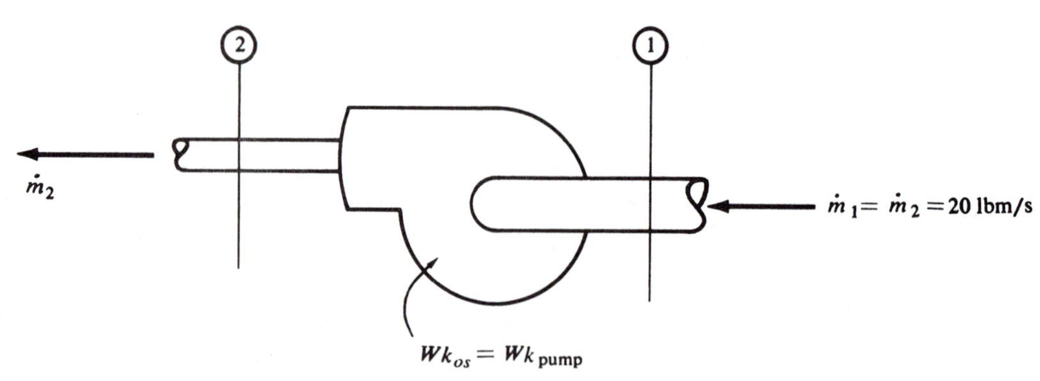

FIGURE 6–9 Water pump of example 6–18

EXAMPLE 6–19 Ethylene glycol flows through a radiator at 10 kg/s while its temperature changes from 92°C to 60°C. The pressure of the radiator is 20 kPa gage, or 120 kPa, because the atmospheric pressure is 100 kPa. Determine the heat transferred from the ethylene glycol in the radiator in kilowatts.

Solution The radiator is an open system, and the ethylene glycol is the working fluid. We assume steady-state conditions and that the ethylene glycol is incompressible and has constant specific heat. The states are as follows:

State 1: $p_1 = 120$ kPa, $T_1 = 92°C$ State 2: $p_2 = 120$ kPa, $T_2 = 60°C$

The work is set to zero, and the kinetic and potential energies are neglected. The specific heat is $c = 2.385$ kJ/kg · K from table B–8, and the steady-flow equation (6–42) becomes

$$Q = mc(T_2 - T_1)$$
$$= (10 \text{ kg/s})(2.385 \text{ kJ/kg}\cdot\text{K})(60°\text{C} - 92°\text{C})$$

Answer
$$= -763.2 \text{ kJ/s} = -763.2 \text{ kW}$$

6–5 PROCESSES OF SOLIDS

Solid substances are nearly always considered to be incompressible for engineering purposes. We assume that all the solids here considered are incompressible, so the equations developed in section 6–4 will apply here as well. The heat transfer involved with a solid is given by the equation

$$Q = m \sum c\, \delta T$$

or, for constant specific heats,

$$Q = mc\,\Delta T \tag{6–46}$$

Values for specific heat are given in table B–8 for various solids.

EXAMPLE 6–20 A 20-kg copper block is heated while it is in a container of water. There is 15 kg of water, and the copper and water are at 10°C before heating and 80°C after heating. Determine the amount of heat transfer to the water bath and to the copper.

Solution The system is a closed one, composed of the copper block and the water. Assuming constant specific heats for the copper and the water, the heat is given by

$$Q = m_w c_w (T_2 - T_1) + m_{cu} c_{cu} (T_2 - T_1)$$

where $m_w = 15$ kg water
$c_w = 4.18$ kJ/kg · K (from table B–8, for water)
$m_{cu} = 20$ kg copper
$c_{cu} = 0.385$ kJ/kg · K (from table B–8, for copper)

Then

$$Q = (15 \text{ kg})(4.18 \text{ kJ/kg}\cdot\text{K})(70 \text{ K}) + (20 \text{ kg})(0.385 \text{ kJ/kg}\cdot\text{K})(70 \text{ K})$$

Answer
$$= 4928 \text{ kJ}$$

The heat transferred to the copper is

$$Q = (20 \text{ kg})(0.385 \text{ kJ/kg}\cdot\text{K})(70 \text{ K})$$

Answer
$$= 539 \text{ kJ}$$

Solids are sometimes capable of supporting forces that tend to stretch or pull them. If a solid is pulled, it is said to be in **tension**, and it is said to be strained, or stretched. In its stretched position, it has strain energy, which is much like the spring energy of a stretched mechanical spring. If the solid is an elastic solid, then while it is an elastic solid, it satisfies the relationship

$$Y = \frac{\text{stress}}{\text{strain}} \tag{6–47}$$

where Y is called the **modulus of elasticity** or **Young's modulus** and the stress and strain are given by the equations

$$\text{stress} = \frac{\text{force}}{\text{area}} = \frac{F}{A}$$

$$\text{strain} = \xi = \frac{\Delta l}{l}$$

The term Δl is the change in length of the elastic solid along the direction in which it is being pulled by the force F. Its original length is l, and the cross-sectional area is A. An example of such a situation is given in figure 6–10. The **strain energy**, SE is another form of energy, which is given by the equation

$$\text{SE} = \frac{1}{2}VY\xi^2 \tag{6–48}$$

and if a solid elastic material is stretched an amount without any heat involved (adiabatic), the work would be the change in the strain energy of the solid. The following example should help clarify this concept.

EXAMPLE 6–21 A cylindrical bar 10 inches long and 1 inch in diameter is subjected to an axial load of 3000 lbf, as shown in figure 6–10. The bar is composed of steel having a modulus of elasticity of 30×10^6 lbf/in^2. Determine the work done for this process if the bar does not change volume when stretched and if no heat is transferred.

Solution We have here a constant-volume process for a closed system, but the system is not simple. We are assuming that the material is perfectly elastic (which means that it will return to its initial shape when external forces are removed) and, therefore, reversible. However, we have interactions between the constituent atoms, and our first law becomes

$$\Delta E = Q - Wk_{cs}$$

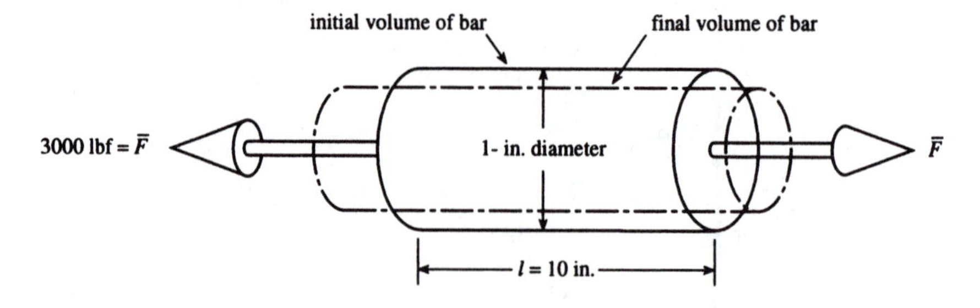

FIGURE 6–10 Strain energy application

We assume no heat transfer or changes in kinetic and potential energy or enthalpy, but we must account for strain energy, SE. Strain energy results from the interaction of the atoms or molecules.

From our first-law equation to the bar, because $Q = 0$,

$$\Delta SE = -Wk_{cs}$$

Since we seek the work required to stress the bar under a 3000-lbf load, we set $\xi_1 = 0$ and, from the definition of the modulus of elasticity, $Y = (F/A)/\xi$, obtain

$$\begin{aligned} \xi_2 &= \frac{3000 \text{ lbf}}{YA} \\ &= \frac{3000 \text{ lbf}}{\left(\pi \times \frac{1}{4} \text{ in}^2\right)(30 \times 10^6 \text{ lbf/in}^2)} \\ &= 1.27 \times 10^{-4} \text{ in/in} \\ &= 1.27 \times 10^{-4} = \xi_2 \end{aligned}$$

Then

$$Wk_{cs} = -\Delta SE = -VY\left(\frac{1}{2}\right)(\xi_2^2 - \xi_1^2)$$

We now substitute numbers into this equation. First, the volume is just the volume of the cylindrical bar; that is,

$$V = (10 \text{ in})\left(\pi \times \frac{1}{4} \text{ in}^2\right)$$

Then

$$Wk_{cs} = -\left(\frac{10\pi}{4} \text{ in}^3\right)(30 \times 10^6 \text{ lbf/in}^2)\left(\frac{1}{2}\right)[(1.27 \times 10^{-4})^2 - 0]$$

or

Answer

$$Wk_{cs} = -1.93 \text{ in} \cdot \text{lbf}$$

This represents the energy expanded in applying 3000 lbf to the bar described and allowing equilibrium to be slowly achieved. In equilibrium, the bar will have been stretched 1.27×10^{-4} in/in of bar length, or a total of $1.27 \times 10 \times 10^{-4}$ or 1.27×10^{-3} in.

Notice that for a modulus of elasticity, Y, that remains constant during a stress process, the work can also be written

$$Wk_{cs} = \frac{1}{2}VY(\xi_2^2 - \xi_1^2) \tag{6–49}$$

If the initial strain, ξ_1, is zero, then

$$Wk_{cs} = \frac{1}{2}VY\xi_2^2 \tag{6–50}$$

This relationship holds true for those materials having no prestress.

CALCULUS FOR CLARITY 6–3

A differential amount of strain is defined as

$$d\xi = \frac{dl}{l} \tag{6–51}$$

The work of an elastic solid may be considered from the definition for work, equation (3–4), using $x = l$ and $dl = dl$:

$$Wk = \int F\, dl$$

The force, F, is given by the stress times the normal area, A, and the stress is Young's modulus, Y, times strain, ξ. The work is then

$$Wk = \int Y\xi A\, dl$$

However, the volume $V = Al$, so introducing l/l into this equation gives

$$Wk = \int Y\xi A\left[\frac{l}{l}\right] dl = \int \frac{YV\xi\, dl}{l} = \int VY\xi\, d\xi \tag{6–52}$$

Integrating equation (6–52) from ξ_1 to some other elastically strained state ξ_2 gives equation (6–49).

6–6 PROCESSES OF PURE SUBSTANCES

Pure substances other than perfect gases are treated in this section. Water (steam), ammonia, R-12, R-22, mercury, R-134a, R-123, R-407c, and R-502 will be treated because they have their property tables included in this book. Other pure substances can also be analyzed with the methods described here if suitable tables of properties are available. In chapter 5, we described the use of thermodynamic tables for purposes of identifying equilibrium states of pure substances. Here we use these same tables to consider processes involving those substances.

For the various types of processes considered in the following examples, it is important to remember to determine the properties by means of the property tables. The use of specific heats to relate the temperatures to the internal energy or enthalpy should be avoided, and the use of the perfect gas relation is not at all applicable. First, let us consider the isobaric or constant-pressure process of closed and open systems.

EXAMPLE 6–22 A steam generator (boiler) produces superheated vapor steam at 8 MPa and 640°C. If the water enters at 8 MPa and 30°C, determine the amount of heat added to the water in the steam generator per kilogram of steam produced. Compare the answer to one obtained by using *EES*.

Solution The steam generator is an open system, and we assume steady-flow conditions for the water. After neglecting kinetic and potential energies and seeing that the shaft work is zero, the steady-flow energy equation becomes

$$q_{add} = h_2 - h_1$$

where state 2 is the superheated state and state 1 the inlet water state. From table B–12, we find that

$$h_2 = 3736 \text{ kJ/kg} \quad \text{at } 640°\text{C} \quad \text{and} \quad 8 \text{ MPa}$$

From table B–13, the pump work is 7.7 kJ/kg at 8 MPa and 30°C. This value is the amount of the increase in enthalpy of water when compressed reversibly and adiabatically from a saturated liquid at 30°C to 8 MPa, or $h - h_f$. From table B–11, h_f is found to be 125.7 kJ/kg at 30°C, so that

$$h_1 = h_f + 7.7 \text{ kJ/kg} = 133.4 \text{ kJ/kg}$$

Answer

$$q_{add} = 3736 - 133.4 = 3602.6 \text{ kJ/kg}$$

Use the *EES* window to check that you are using SI units with temperature in degrees Celsius and pressure in kPa, and then insert the following equations:

$$h_2 = \text{enthalpy}(\text{steam}, T = 640, P = 8000)$$
$$h_1 = \text{enthalpy}\ (\text{steam}, T = 30, P = 8000)$$
$$q = h_2 - h_1$$

Clicking the *calculate* option and then the *solve* option gives the answer:

$$q = 3605$$

This answer has units of kJ/kg and agrees with the answer obtained by using the appendix tables.

EXAMPLE 6–23 A piston-cylinder contains 1 lbm of saturated ammonia at 70°F and 85% quality. The ammonia is then cooled so that it is a saturated liquid at 70°F. Determine the heat transfer and the work during this process.

Solution The system is a closed one bounded by the piston-cylinder with ammonia as the substance. The states are as follows:

State 1: $T_1 = 70°\text{F}$ State 2: $T_2 = 70°\text{F}$

$p_1 = 128.8$ psia $p_2 = p_1$

(saturation pressure)

The heat is determined from the first law,

$$Q = \Delta U + Wk$$

and because the process is at constant pressure,

$$Wk = Wk_{cs} = p\,\Delta V$$

Also,

$$Q = \Delta U + p\,\Delta V = \Delta H = m(h_2 - h_1)$$

Then, using h_f and h_g from table B–19 at 70°F, we obtain

$$\begin{aligned} h_1 &= x_1(h_{fg}) + h_f \\ &= (0.85)(629.1 \text{ Btu/lbm} - 120.5 \text{ Btu/lbm}) + 120.5 \text{ Btu/lbm} \\ &= 552.81 \text{ Btu/lbm} \end{aligned}$$

and

$$h_2 = h_f = 120.5 \text{ Btu/lbm}$$

Then the heat is

$$\begin{aligned} Q &= (1 \text{ lbm})(120.5 \text{ Btu/lbm} - 552.81 \text{ Btu/lbm}) \\ &= -432.31 \text{ Btu/lbm} \end{aligned}$$

The work is obtained from the calculation

$$Wk_{cs} = p\,\Delta V = mp(v_2 - v_1)$$

and the specific volumes are

$$\begin{aligned} v_1 &= \chi_1(v_{fg}) + v_f \\ &= \chi_1(v_g) + (1 - \chi_1)(v_f) \\ &= (0.85)(2.312\ \text{ft}^3/\text{lbm}) + (0.15)(0.026\ \text{ft}^3/\text{lbm}) \\ &= 1.9691\ \text{ft}^3/\text{lbm} \end{aligned}$$

and

$$v_2 = v_f = 0.026\ \text{ft}^3/\text{lbm}$$

The work is then

$$\begin{aligned} Wk_{cs} &= (1\ \text{lbm})(128.80\ \text{lbf/in}^2)(144\ \text{in}^2/\text{ft}^2)(0.026\ \text{ft}^3/\text{lbm} - 1.9691\ \text{ft}^3/\text{lbm}) \\ &= -36{,}039.06\ \text{ft}\cdot\text{lbf} = -46.32\ \text{Btu} \end{aligned}$$

Answer

Notice how the volume decreased during the condensation: $\Delta V = 0.026\ \text{ft}^3 - 1.9691\ \text{ft}^3 = -1.9431\ \text{ft}^3$. If the volume had been held constant during the condensation by locking the piston in place, there would have been a large decrease in the pressure. Let us now look at a constant-volume chamber containing a pure substance as that substance goes through changes of state.

EXAMPLE 6–24 A $1 - \text{m}^3$ rigid container holds 2 kg of water at 120°C. It is then heated to 240°C and finally cooled to 80°C. Determine the conditions of the three states and the heat and work involved in the two processes.

Solution The system is a closed one and the steam is the working substance. At state 1, we have

$$V_1 = 1\ \text{m}^3 \qquad m = 2\ \text{kg}$$

$$v_1 = \frac{V_1}{m} = 0.5\ \text{m}^3/\text{kg}$$

$$T_1 = 120°\text{C}$$

Using table B–11, we notice that v_1 is greater than $v_f(=0.00106)$ but less than $v_g(0.892\ \text{m}^3/\text{kg})$ at 120°C, so that state 1 is a saturation mixture. The pressure $p_1 = 198.5$ kPa, and the quality is found from the calculation

$$\begin{aligned} \chi_1 &= \frac{v_1 - v_f}{v_g - v_f} \\ &= \frac{0.5 - 0.00106\ \text{m}^3/\text{kg}}{0.892 - 0.00106} \\ &= 0.56 = 56\% \end{aligned}$$

The enthalpy at state 1 is

$$\begin{aligned} h_1 &= \chi_1(h_{fg}) + h_f \\ &= (0.56)(2202\ \text{kJ/kg}) + 504\ \text{kJ/kg} \\ &= 1737.12\ \text{kJ/kg} \end{aligned}$$

At state 2, we have

$$v_2 = v_1 = 0.5\ \text{m}^3/\text{kg}$$

$$T_2 = 240°\text{C}$$

From table B–11, it can be seen that v_2 is greater than v_g at 240°C (= 0.0597 m^3/kg), so the steam has become superheated at state 2. From table B–12 at 240°C and at specific volume $v = 0.5\ \text{m}^3/\text{kg}$, we may find p_2, h_2, and the other properties. By interpolation between 400 and 500 kPa (at 240°C), we find p_2:

$$\frac{p_2 - 500}{400 - 500} = \frac{0.50 - 0.464}{0.583 - 0.464} = 0.30252$$

Hence,

$$p_2 = 500 + (400 - 500)(0.30252) = 469.7\ \text{kPa}$$

Also, the enthalpy is

$$\begin{aligned} h_2 &= 2937\ \text{kJ/kg} + (2941 - 2937\ \text{kJ/kg})(0.30252) \\ &= 2935.8\ \text{kJ/kg} \end{aligned}$$

At state 3, we have

$$v_3 = 0.5\ \text{m}^3/\text{kg} \qquad T_3 = 80°\text{C}$$

so that, from table B–11, we find v_3 less than v_g but more than v_f at 80°C. The steam at state 3 is therefore a saturation mixture, and the pressure is the saturation pressure, 47.4 kPa. The quality may be found from

$$\begin{aligned} \chi_3 &= \frac{v_3 - v_f}{v_g - v_f} = \frac{0.5 - 0.00103}{3.41 - 0.00103} \\ &= 0.146 = 14.6\% \end{aligned}$$

The enthalpy is

$$\begin{aligned} h_3 &= \chi_3(h_{fg}) + h_f \\ &= (0.146)(2308\ \text{kJ/kg}) + 335\ \text{kJ/kg} \\ &= 672\ \text{kJ/kg} \end{aligned}$$

The work is zero for both processes because the volume is constant and the heats for the two processes are equal to the changes in internal energy. For process 1–2 (state 1 to state 2),

$$\begin{aligned} Q_{12} &= m(u_2 - u_1) \\ &= m(h_2 - p_2v_2 - h_1 + p_1v_1) \\ &= (2\ \text{kg})[2935.8\ \text{kJ/kg} - (469.7\ \text{kN/m}^2)(0.5\ \text{m}^3) \\ &\quad - 1737.12\ \text{kJ/kg} + (198.5\ \text{kN/m}^2)(0.5\ \text{m}^3)] \\ &= 2126.16\ \text{kJ/kg} \end{aligned}$$

Answer

The heat involved in the process from state 2 to state 3 is

$$\begin{aligned} Q_{23} &= m(u_3 - u_2) \\ &= m(h_3 - p_3v_3 - h_2 + p_2v_2) \\ &= (2\ \text{kg})[672\ \text{kJ/kg} - (47.4\ \text{kN/m}^2)(0.5\ \text{m}^3) \\ &\quad - 2935.8\ \text{kJ/kg} + (469.7\ \text{kN/m}^2)(0.5\ \text{m}^3)] \\ &= -4105.3\ \text{kJ/kg} \end{aligned}$$

Answer

Notice how the pressure increased significantly when the steam was heated and then how the pressure decreased to a partial vacuum condition at state 3 when the steam was cooled.

EXAMPLE 6–25 An evaporation coil of a refrigerator allows 30 g/s of Refrigerant-12 to flow through it from a saturation mixture at −23°C (250 K) and 25% quality to saturated vapor at the exit. The tube has an inside diameter of 1.0 cm, and it is uniform throughout the evaporator. Determine the rate of heat addition to the R-12 and the velocity change of the R-12 in the evaporator.

Solution The system is open and the R-12 is the working fluid. From table B–15, the states are as follows:

State 1: $T_1 = 250$ K State 2: $T_2 = 250$ K
$p_1 = 0.13334$ MPa $p_2 = p_1$ (saturation pressure)
$\chi_1 = 0.25$ (25%) $\chi_2 = 100\%$ (saturated vapor)

The conditions at states 1 and 2 can be further identified:

$$\begin{aligned} v_1 &= \chi_1(v_g) + (1 - \chi_1)(v_f) \\ &= (0.25)(0.12278 \text{ m}^3/\text{kg}) + (0.75)(0.000682 \text{ m}^3/\text{kg}) \\ &= 0.0312 \text{ m}^3/\text{kg} \end{aligned}$$

$$\begin{aligned} h_1 &= \chi_1(h_g) + (1 - \chi_1)h_f \\ &= (0.25)(560.42 \text{ kJ/kg}) + (0.75)(397.25 \text{ kJ/kg}) \\ &= 438.04 \text{ kJ/kg} \end{aligned}$$

$$v_2 = v_g = 0.12278 \text{ m}^3/\text{kg} \qquad \text{and} \qquad h_2 = h_g = 560.42 \text{ kJ/kg}$$

Since the specific volume changes, we shall determine the change in the velocity between the two states caused by the specific volume (or density) change. The change in velocity will then give a kinetic energy change, which we will include in the steady-flow equation. At state 1, the velocity is found from the mass flow rate equation:

$$\overline{V}_1 = \frac{\dot{m}_1 v_1}{A_1} = \frac{(0.33 \text{ kg/s})(0.0312 \text{ m}^3/\text{kg})}{A_1}$$

The area is $\pi D^2/4 = (3.14159)(0.01 \text{ m})^2/4 = 0.0000785 \text{ m}^2$. Then

$$\overline{V}_1 = 11.9 \text{ m/s}$$

At state 2,

$$\begin{aligned} \overline{V}_2 &= \frac{\dot{m} v_2}{A_2} = \frac{(0.03 \text{ kg/s})(0.12278 \text{ m}^3/\text{kg})}{0.0000785 \text{ m}^2} \\ &= 46.9 \text{ m/s} \end{aligned}$$

The velocity change is therefore $\overline{V}_2 - \overline{V}_1 = 35$ m/s. The steady-flow energy equation for the evaporator is

$$\dot{Q} = \dot{m}(h_2 - h_1) + \frac{\dot{m}(\overline{V}_2^2 - \overline{V}_1^2)}{2000}$$

where V is in m/s and h in kJ/kg. Then

$$\begin{aligned} \dot{Q} &= (0.03 \text{ kg/s})(560.42 \text{ kJ/kg} - 438.04 \text{ kJ/kg}) \\ &\quad + \frac{(0.03 \text{ kg/s})[(46.9 \text{ m/s})^2 - (11.9 \text{ m/s})^2]}{2000} \end{aligned}$$

Answer
$$= 3.67 \text{ kJ/s} + 0.03 \text{ kJ/s} = 3.70 \text{ kJ/s}$$

Notice that the kinetic energy change, even with a large-velocity change of the refrigerant, is negligible compared to the enthalpy change. As a result, nearly correct results can usually be obtained by neglecting kinetic energy for these types of problems.

EXAMPLE 6–26 A piston-cylinder is used to compress 30 g of R-22 from 96 kPa at 10°C to 514 kPa while holding the temperature constant. Determine the work and heat transferred in the process. Compare the answer to the one obtained through *EES*.

Solution The system is closed and R-22 is the working substance. The initial and final states are as follows:

$$\text{State 1:}\quad p_1 = 96\text{ kPa}\qquad \text{State 2:}\quad p_2 = 514\text{ kPa}$$
$$T_1 = 10°\text{C}\qquad T_2 = 10°\text{C}$$

The work can be computed from the boundary work:

$$Wk_{cs} = \sum p\,\delta V = m \sum p\,\delta v$$

We can evaluate the summation at constant temperature and find the specific volumes at 10°C and various pressures. From table B–18, we find the following:

$$\text{State 1:}\quad v_1 = 0.2802\text{ m}^3/\text{kg}\qquad \text{State 2:}\quad v_2 = 0.04786\text{ m}^3/\text{kg}$$
$$h_1 = 419.8\text{ kJ/kg}\qquad h_2 = 412.2\text{ kJ/kg}$$

We also find the specific volumes at various pressures between the initial and final pressures. In table 6–3, these data are tabulated in the first two columns.

The work per kilogram of R-22 is 246.21167 kJ/kg from the VISUAL BASIC program AREA used for computing areas under curves. By using long-hand computation as shown in the last columns of table 6–3, we obtain the same result. The work for 30 g is

$$Wk_{cs} = (0.03\text{ kg})(-46.2\text{ kJ/kg})$$

Answer
$$= -1.386\text{ kJ}$$

The heat transferred in the process is

$$Q = \Delta U + Wk_{cs}$$
$$= m(u_2 - u_1) + Wk_{cs}$$

The internal energies are found from the equation $h = u + pv$ or $u = h - pv$. Then

$$u_1 = h_1 - p_1 v_1 = 419.8\text{ kJ/kg} - (96\text{ kN/m}^2)(0.2802\text{ m}^3)$$
$$= 392.9\text{ kJ/kg}$$

TABLE 6–3

Pressure p kPa	Specific Volume v m³/kg	p_{ave} kPa	δv m³/kg	$p\,\delta v$ kJ/kg
96	0.2802			
		103	−0.0373	−3.8419
110	0.2429			
		121	−0.0409	−4.9489
132	0.2020			
		144.5	−0.033	−4.7685
157	0.1690			
		171	−0.0269	−4.5999
185	0.1421			
		201.5	−0.0219	−4.41285
218	0.1202			
		236	−0.01807	−4.26452
254	0.10213			
		275	−0.01497	−4.11675
296	0.08716			
		344.5	−0.02296	−7.90972
393	0.06420			
		422	−0.00884	−3.73048
451	0.05536			
		482.5	−0.0075	−3.61875
514	0.04786			
				$\Sigma\, p\,\delta v = -46.21167$ kJ/kg

and

$$u_2 = h_2 - p_2 v_2 = 412.2\ \text{kJ/kg} - (514\ \text{kN/m}^2)(0.04786\ \text{m}^3)$$
$$= 387.6\ \text{kJ/kg}$$

So

$$Q = (0.03\ \text{kg})(387.6\ \text{kJ/kg} - 392.9\ \text{kJ/kg}) + (-1.386\ \text{kJ})$$

Answer
$$= -1.545\ \text{kJ}$$

Use the *EES* window to check that you are using SI units with temperature in degrees Celsius and pressure in kPa. Then insert the following equations and use the same pressure increments as that for the appendix table solution:

$$v_1 = \text{volume}\ (\text{R22}, T = 10, P = 96)$$
$$v_2 = \text{volume}\ (\text{R22}, T = 10, P = 110)$$
$$v_3 = \text{volume}\ (\text{R22}, T = 10, P = 132)$$
$$v_4 = \text{volume}\ (\text{R22}, T = 10, P = 157)$$
$$v_5 = \text{volume}\ (\text{R22}, T = 10, P = 185)$$
$$v_6 = \text{volume}\ (\text{R22}, T = 10, P = 218)$$
$$v_7 = \text{volume}\ (\text{R22}, T = 10, P = 254)$$
$$v_8 = \text{volume}\ (\text{R22}, T = 10, P = 296)$$
$$v_9 = \text{volume}\ (\text{R22}, T = 10, P = 393)$$
$$v_{10} = \text{volume}\ (\text{R22}, T = 10, P = 451)$$
$$v_{11} = \text{volume}\ (\text{R22}, T = 10, P = 514)$$

$$wk_1 = 103*\ (v_2 - v_1)$$
$$wk_2 = 121*\ (v_3 - v_2)$$
$$wk_3 = 144.5*\ (v_4 - v_3)$$
$$wk_4 = 171*\ (v_5 - v_4)$$
$$wk_5 = 201.5*\ (v_6 - v_5)$$
$$wk_6 = 236*\ (v_7 - v_6)$$
$$wk_7 = 275*\ (v_8 - v_7)$$
$$wk_8 = 344.5*\ (v_9 - v_8)$$
$$wk_9 = 422*\ (v_{10} - v_9)$$
$$wk_{10} = 485.5*\ (v_{11} - v_{10})$$

$$wk = wk_1 + wk_2 + wk_3 + wk_4 + wk_5 + wk_6 + wk_7 + wk_8 + wk_9 + wk_{10}$$

$$u_1 = \text{IntEnergy}\ (\text{R22}, T = 10, P = 96)$$
$$u_{11} = \text{IntEnergy}\ (\text{R22}, T = 10, P = 514)$$

$$q = u_{11} - u_1 + wk$$
$$m = 0.03$$
$$W = m*wk$$
$$QA = q*m$$

Clicking the *Calculate* option and then the *Solve* option gives the answers:

$$wk = -46.08$$
$$q = -51.45$$
$$W = -1.383$$
$$QA = -1.543$$

The resulting values agree with those obtained from the appendix table data.

Notice that the heat is lost or is going out of the system and the work is going into the system. Also, if the R-22 had been approximated by a perfect gas, we would have had $\Delta U = 0$ and $Wk_{cs} = Q = mRT\ln(p_1/p_2)$. The gas constant for R-22 is 0.0955 kJ/kg·K, so that

$$Wk_{cs} = (0.03\ \text{kg})(0.0955\ \text{kJ/kg}\cdot\text{K})(283\ \text{K})\left(\ln\frac{96}{514}\right)$$
$$= -1.36\ \text{kJ}$$

Therefore, by assuming a perfect gas, we would have obtained answers for heat and work that would have been close to the more nearly correct ones obtained by pure substance analysis.

EXAMPLE 6–27 A reversible adiabatic compressor compresses ammonia from a saturated vapor at −30°C to 1200 kPa. Determine the work required and the power required if the flow rate is 5 kg/s.

Solution The system is one of open steady flow with ammonia as the working fluid. The states are as follows:

State 1: $p_1 = 120.0$ kPa (saturated pressure)
$T_1 = -30°\text{C}$
$h_1 = h_g = 1405.6$ kJ/kg (from table B–19)
State 2: $p_2 = 1200$ kPa

We now assume negligible kinetic and potential energies, and because the process is adiabatic, the heat is zero. The steady-flow energy equation is then

$$-wk_{os} = h_2 - h_1$$

We show in chapter 7 that a reversible adiabatic process is one in which the entropy (s) remains constant. Therefore, for this problem, we have that the entropy at 1 equals the entropy at 2, $s_1 = s_2$. But $s_1 = s_g$ at $-30°C = 5.792$ kJ/kg·K from table B–19, and from table B–20, at $s_2 = 5.792$ kJ/kg·K and $p_2 = 1200$ kPa, the corresponding temperature is nearly 140°C. Therefore, $h_2 = 1760.1$ kJ/kg, and the work is

$$-wk_{os} = 1760.1\ \text{kJ/kg} - 1405.6\ \text{kJ/kg}$$

or

Answer
$$wk_{os} = -354.5\ \text{kJ/kg}$$

The power is just the mass flow rate times the work, or

$$\dot{W}k_{os} = \dot{m}wk_{os} = (5\ \text{kg/s})(-354.5\ \text{kJ/kg})$$

Answer
$$= -1772.5\ \text{kW}$$

The minus sign indicates that the work and power are into the compressor.

EXAMPLE 6–28 An adiabatic steam turbine operates with steam at 900 psia and 1200°F. The steam exhausts at 5 psia, and the turbine has an adiabatic efficiency of 92%. Determine the work of the turbine.

Solution The system is an open one and steam is the working medium. At state 1, we have

$p_1 = 900$ psia $T_1 = 1200°\text{F}$
$v_1 = 1.0720\ \text{ft}^3/\text{lbm}$
$h_1 = 1620.6$ Btu/lbm (from table B–12)
$s_1 = 1.7382$ Btu/lbm·°R

At state 2, we have $p_2 = 5$ psia. Also, because the adiabatic efficiency is 92%, we may write

$$wk_{turb} = (0.92)(wk_{os})$$

where wk_{os} is the work of a reversible adiabatic turbine. The reversible adiabatic turbine would operate between state 1 and a state 2*s* given by 5 psia pressure and an entropy $s_{2s} = s_1 = 1.7382$ Btu/lbm · °R. If we look at table B–11 at 5 psia, we find that $s_g = 1.8443$ Btu/lbm · °R and $s_f = 0.2349$ Btu/lbm · °R. Since s_{2s} falls between these two values, we see that the state 2*s* is a saturation mixture. The quality of final state 2*s* after an ideal (reversible adiabatic) process is

$$\begin{aligned} \chi_{2s} &= \frac{s_{2s} - s_f}{s_g - s_f} = \frac{s_{2s} - s_f}{s_{fg}} \\ &= \frac{1.7382 - 0.2349 \text{ Btu/lbm} \cdot °\text{R}}{1.6094 \text{ Btu/lbm} \cdot °\text{R}} \\ &= 93.4\% \end{aligned}$$

The enthalpy h_{2s} at state 2*s* is

$$\begin{aligned} h_{2s} &= \chi_{2s}(h_{fg}) + h_f \\ &= (0.934)(1000.9 \text{ Btu/lbm}) + 130.2 \text{ Btu/lbm} \\ &= 1065.0 \text{ Btu/lbm} \end{aligned}$$

The reversible adiabatic work would then be equal to $h_1 - h_{2s}$, and the actual work is

$$\begin{aligned} wk_{turb} &= (0.92)(1620.6 \text{ Btu/lbm} - 1065.0 \text{ Btu/lbm}) \\ &= 511.15 \text{ Btu/lbm} \end{aligned}$$

Answer

The throttling process is one that we will use in later chapters on applications. A throttle is a steady-flow device that decreases the pressure of a flowing fluid without work and with nearly no heat. The pressure drop occurs because of friction and viscous dissipation. The steady-flow energy equation for a throttle, after neglecting kinetic and potential energy changes, is

$$h_2 = h_1 \quad \text{(constant enthalpy)} \qquad \textbf{(6–53)}$$

All valves, orifices, or tubes or pipes behave as throttles to some extent. The following example demonstrates one important use of the throttle in mechanical refrigerators:

EXAMPLE 6–29 Ammonia flows through a throttle from a saturated liquid state at 100°F to 4 psia. Determine the temperature of the ammonia leaving the throttle. Compare the answer with the one obtained with *EES*.

Solution Using equation (6–53) for a throttle, we find that

$$h_1 = h_2 = h_f \quad \text{at } 100°\text{F} = 155.2 \text{ Btu/lbm} \quad \text{(table B–19)}$$

The outlet pressure is 4 psia, so we may find the outlet temperature because we know the enthalpy. From table B–19, we interpolate between 3.94 psia (at −70°F) and 5.55 psia (at −60°F). Then

$$\begin{aligned} T_2 &= [-60°\text{F} - (-70°\text{F})]\left(\frac{4 \text{ psia} - 3.94 \text{ psia}}{5.55 \text{ psia} - 3.94 \text{ psia}}\right) + (-70°\text{F}) \\ &= (10°\text{F})(0.0373) - 70°\text{F} \\ &= -69.6°\text{F} \end{aligned}$$

Answer

Use *EES* window to check that you are using English units with temperature in degrees Fahrenheit and pressure in psia, and then insert the following equations:

$$h_1 = \text{enthalpy}(\text{ammonia}, T = 100, x = 0.0)$$
$$T = \text{temperature}(\text{ammonia}, h = h_1, p = 4)$$

Clicking the *calculate* option and then the *solve* option gives the answer:

$$T_2 = -69.54$$

This answer has units of degrees Fahrenheit and agrees with the answer obtained by using the appendix tables.

We could proceed to determine other properties at the outlet state if they were needed. The reason we know that the ammonia is at $-69.6°F$ (saturation temperature at 4 psia) is that the enthalpy at the outlet state lies between h_f and h_g at 4 psia, a fact easily seen by studying table B–19. Notice also that, for perfect gases, the throttle does not change the temperature but merely decreases pressure. The reason is that enthalpy is determined only by temperature, so if the enthalpy is constant, so must be the temperature (for perfect gases).

6–7 SUMMARY

We have considered the most common processes encountered in thermodynamics or energy transfer problems. The numerous equations accumulated during these last two chapters are listed in tables 6–4 and 6–5 to provide ready references. Note that table 6–5 lists the

TABLE 6–4 Perfect gas process (constant specific heats)

Term	Reversible Isobaric	Reversible Isometric	Reversible Isothermal	Reversible Adiabatic	Reversible Polytropic
p–v relation	$p = c$	$V = c$	$T = c$	$pV^k = c$	$pV^n = c$
p–v–T relations	$\frac{V_1}{V_2} = \frac{T_1}{T_2}$	$\frac{p_1}{p_2} = \frac{T_1}{T_2}$	$\frac{p_1}{p_2} = \frac{V_2}{V_1}$	$\frac{p_1}{p_2} = \left(\frac{V_2}{V_1}\right)^k$ $\frac{p_1}{p_2} = \left(\frac{T_1}{T_2}\right)^{k/k-1}$	$\frac{p_1}{p_2} = \left(\frac{V_2}{V_1}\right)^n$ $\frac{p_1}{p_2} = \left(\frac{T_1}{T_2}\right)^{n/n-1}$
Wk_{cs}*	$p(V_2 - V_1)$	0	$p_1V_1 \ln \frac{V_2}{V_1}$	$\frac{1}{1-k}(p_2V_2 - p_1V_1)$	$\frac{1}{1-n}(p_2V_2 - p_1V_1)$
Wk_{os}*	0	$V(p_1 - p_2)$	$p_1V_1 \ln \frac{V_2}{V_1}$	$\frac{k}{1-k}(p_2V_2 - p_1V_1)$	$\frac{n}{1-n}(p_2V_2 - p_1V_1)$
Q_{os}	$m\,\Delta h$	$m(\Delta h - v\,\Delta p)$	$p_1V_1 \ln \frac{V_2}{V_1}$	0	$m\,\Delta h + Wk_{os}$
Q_{cs}	$m\,\Delta h$	$m\,\Delta u$	$p_1V_1 \ln \frac{V_2}{V_1}$	0	$m\,\Delta u + Wk_{cs}$
Δh	$c_p\,\Delta T$	$c_p\,\Delta T$	0	$c_p\,\Delta T$	$c_p\,\Delta T$
Δu	$c_v\,\Delta T$	$c_v\,\Delta T$	0	$c_v\,\Delta T$	$c_v\,\Delta T$
Δs	$c_p \ln \frac{T_2}{T_1}$	$c_v \ln \frac{T_2}{T_1}$	$R \ln \frac{V_2}{V_1}$	0	$c_v \ln \frac{T_2}{T_1} + R \ln \frac{V_2}{V_1}$

*Assuming that there are no kinetic, potential, or strain energy changes.

TABLE 6–5 General process (excluding processes involving phase change)

Term	Reversible Isobaric	Reversible Isometric	Reversible Isothermal	Reversible Adiabatic	Reversible Polytropic
p–V relation	$p = c$	$V = c$	$T = c$	$s = c$	$pV^n = C$
Wk_{cs}*	$p(V_2 - V_1)$	0	$\Sigma\, p\, \delta V$	$\Sigma\, p\, \delta V$	$\frac{1}{1-n}(p_2V_2 - p_1V_1)$
Wk_{os}*	0	$V(p_1 - p_2)$	$-\Sigma\, V\, \delta p$	$-\Sigma\, V\, \delta p$	$\frac{n}{1-n}(p_2V_2 - p_1V_1)$
Q_{os}	$m\,\Delta h$	$m(\Delta h - v\,\Delta p)$	$-\Sigma\, V\, \delta p$	0	$m\,\Delta h + Wk_{os}$
Q_{cs}	$m\,\Delta h$	$m\,\Delta u$	$\Sigma\, p\, \delta V$	0	$m\,\Delta U + Wk_{cs}$
Δs	$\Sigma \frac{\delta h}{T}$	$\Sigma \frac{\delta u}{T}$	$\frac{q}{T}$	0	$\Sigma \frac{\delta q}{T}$

Note: The terms are all exact if the summation sign, Σ, is replaced with the integral sign, $\int$, and the notation for a very small amount, δ, is replaced with the differential notation, d. Thus, for instance, the heat of an open-system Reversible Isothermal process is

$$Q_{os} = -\int V\, dp$$

*Assuming that there are no kinetic, potential, or strain energy changes.

general equations that are applicable to any material but which are frequently too cumbersome; table 6–4 lists the specific equations pertinent to perfect gases only. In table 6–4 are listed only the reversible cases, and, in addition, a new property, *entropy*, is tabulated as a change in its value, Δs. We will speak of entropy more in chapter 7, but its formulas are presented here for completeness.

The adiabatic efficiency was defined for any substance involving an adiabatic process. For power producers, the adiabatic efficiency is

$$\eta_s = \frac{wk_{act}}{wk_{os}} \quad \text{or} \quad \frac{wk_{act}}{wk_{cs}} \tag{6–30}$$

For power consumers,

$$\eta_s = \frac{wk_{os}}{wk_{act}} \quad \text{or} \quad \frac{wk_{cs}}{wk_{act}} \tag{6–31}$$

We also saw that a throttle is a steady-flow device through which a pure substance experiences a decrease in pressure and temperature. For a throttle, the enthalpy is constant throughout.

DISCUSSION QUESTIONS

Section 6–1

6–1 What is meant by a *process*?

6–2 How is heat and enthalpy related in a constant pressure process?

6–3 What is meant by a *polytropic process*?

Section 6–2

6–4 What is meant by the term *adiabatic efficiency*?

6–5 Why is the adiabatic efficiency different for power producers (like turbines) than for power consumers (like pumps)?

Section 6–3

6–6 What is the *compressibility factor*?

6–7 Why is the compressibility factor sometimes greater than one? What would this mean?

Section 6–4

6–8 When do you think air can be assumed to be incompressible?

6–9 When do you think liquid water cannot be assumed to be incompressible?

Section 6–5

6–10 What is *tension* with reference to an elastic material subjected to a force?

6–11 What is meant by *stress*?

6–12 What is *strain energy*?

Section 6–6

6–13 What is a *throttling process*?

6–14 Can a *rigid container* be an open system?

PRACTICE PROBLEMS

Section 6–1

Problems that use SI units are indicated by an (M) under the problem number; those that use English units are indicated by an (E).

6–1 (M) Determine the work of a process when the volume changes from 3 m^3 to 5 m^3. The pressure has a constant value of 300 kPa.

6–2 (M) A piston-cylinder contains 0.003 g of air at 120 kPa and 40°C. The piston is then compressed at constant pressure so that the volume of air is half the original value. Determine the final temperature, final volume, and work required for this process.

6–3 (E) Air contained in a frictionless piston-cylinder expands from 1 in^3 to 10 in^3. The mass of air is 0.02 lbm at a temperature of 140°F, and the air has a volume of 1 in^3. If the pressure remains constant during the process, calculate the final air temperature and the work.

6–4 (E) Oxygen gas at 80 psia is heated from 60°F to 180°F. Determine the amount of heat required per pound-mass of oxygen.

6–5 (E) In a rigid chamber, 30 lbm of helium is heated from 60°F and 30 psig to 90°F. If atmospheric pressure is 14.7 psia, determine the final pressure and the heat added to the helium.

6–6 (M) A rigid container is filled with a perfect gas having a gas constant of 200 $J/kg \cdot K$. The volume is 1.0 m^3, the gas pressure is 150 kPa, and the temperature is 30°C. If the gas is heated to 200°C, determine the final pressure and the work.

6–7 (M) For the container and contents described in problem 6–6, assuming the specific heat at constant volume is 1.0 $kJ/kg \cdot K$. determine the reversible work and the irreversible work if the temperature of the gas is increased from 30°C to 200°C by a paddle stirring the gas instead of by heat transfer.

6–8 (E) In a piston-cylinder device, 0.002 lbm of air and a fuel burn, releasing 0.45 Btu of heat at constant volume. If the pressure is 80 psia and the temperature is 300°F before burning, determine the final pressure and temperature. Assume air properties throughout and neglect fuel mass.

6–9 (M) Air in a piston-cylinder is subjected to a reversible isothermal process where the pressure increases from 102 kPa to 306 kPa. Determine the work per unit mass of air for this process if the air temperature is 32°C.

6–10 (E) One pound-mass of air is compressed in a piston-cylinder from 15 psia to 100 psia. If the air temperature remains constant at 85°F, determine the work required.

6–11 (E) Determine the heat transfer during an isothermal expansion of 2 lbm of air from 15 ft^3 to 30 ft^3 if the temperature is 110°F and the process is carried out in a closed container.

6–12 (E) Two cubic feet of air is expanded in a piston-cylinder from 600 psia and 3400°F to a final volume of 6 ft^3. Assuming that the process is isothermal, determine the final

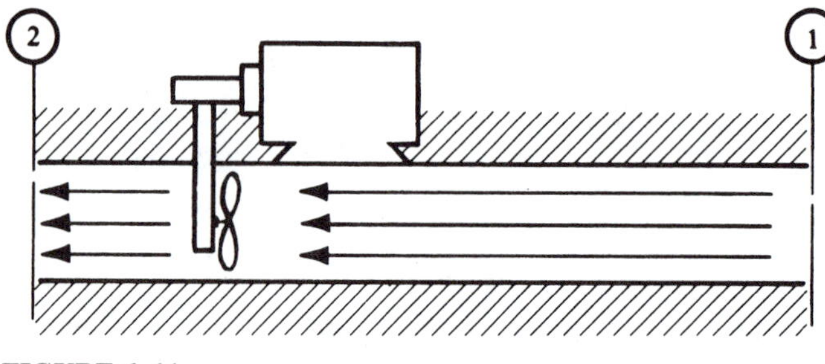

FIGURE 6–11

pressure, the work done, the heat transferred, and the internal energy change. Also, sketch the p–V diagram of this process.

6–13 (E) A fan draws 2 lbm/s of air through the wind tunnel shown in figure 6–11. The following data are given:

$$T_1 = T_2 = 85°\text{F}$$
$$p_1 = 14.7 \text{ psia}$$
$$p_2 = 15.2 \text{ psia}$$
$$\bar{V}_1 = \bar{V}_2$$

Determine the fan motor power requirement and the heat transfer through the wind tunnel walls.

6–14 (M) One kilogram of carbon dioxide (CO_2) is expanded reversibly from 30°C and 200 kPa to 100-kPa pressure. If the expansion is polytropic with $n = 1.27$, determine the work, heat transferred, and energy change. Assume a closed system with no kinetic or potential energy changes.

6–15 (M) A gas at 1000 kPa and 150°C expands to 200 kPa in a piston-cylinder. If the final temperature is 50°C, determine the value for n and the heat transferred if $c_v = 0.7$ kJ/kg · K and $R = 0.28$ kJ/kg · K.

6–16 (M) Seventy grams per second of air flows through a gas turbine, expanding from 1500 kPa to 101 kPa in a reversible polytropic manner. If the inlet air temperature is 2400 K and $n = 1.5$, determine the power generated, the rate of heat transfer, and the rate of enthalpy change. Assume that kinetic and potential energy changes are negligible.

6–17 (E) One pound-mass of air at 82°F and 14.7 psia is compressed reversibly and polytropically ($n = 1.25$) to 6.5 atm in a piston-cylinder. Determine the work required.

6–18 (E) Ammonia is compressed polytropically in a piston-cylinder from 15 psia and 3 ft^3 to 150 psia and 0.4 ft^3. Determine the value for n and the heat transferred.

6–19 (E) A compressor handles 3500 lbm/min of air in a steady-flow polytropic process ($n = 1.36$). The entering air is at 11.5 psia and 85°F and has negligible velocity, while the compressed air is at 115 psia and has a velocity of 30 ft/s. Determine the power required of the compressor, the heat transfer rate, and the rate of enthalpy change of the air.

6–20 (E) Three pounds-mass per second of air is reversibly and polytropically ($n = 1.48$) expanded in a gas turbine from 2100°F to 900°F. If the exhaust pressure is 15 psia, determine the work produced per pound-mass of air, the power generated, and the rate of heat transfer.

6–21 (E) Air is contained in a rigid tank of 20-ft^3 volume. The temperature and pressure of the air are 80°F and 14.7 psia. It is then cooled to -120°F in a first-stage process for liquefaction of air. Determine the pressure of the air after reaching -120°F and the amount of heat lost during the cooling process.

6–22 (M) A well-insulated box contains 5 kg of air at 20°C and 100 kPa. If 200 kJ of work is used to drive a fan in the box, what is the final temperature and pressure.

Section 6–2

6–23 (M) A rotary compressor handles 400 g/s of air, increasing the pressure, reversibly and adiabatically, from 100 kPa to 1000 kPa. If the air temperature initially is 10°C and $k = 1.4$, determine the power required by the compressor. Also, plot the p–v diagram and indicate the work per unit mass on this diagram.

6–24 (M) Two kilograms per second of helium at 300 K are expanded through an adiabatic nozzle and turbine so that the pressure drops to half the initial value. Assuming that the kinetic energy decreases by 30 kJ/s during this expansion, determine the power produced if the process is irreversible and the following process equation holds:

$$pv^k = C$$

Also, determine the change in internal energy per unit time.

6–25 (E) During a reversible adiabatic process of 1 lbm of an ideal gas in a piston-cylinder, the pressure decreases from 183 psig to 2 psig. The atmosphere pressure is 13.8 psia, the gas constant is 37.7 ft · lbf/lbm · °R, and the initial temperature of the gas is 863°F. Determine the initial and

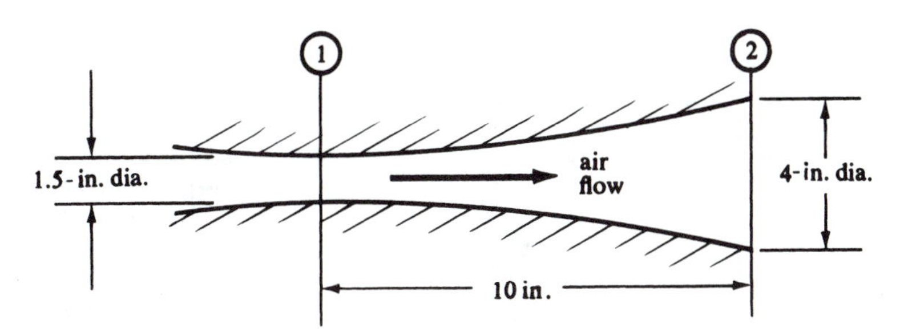

FIGURE 6–12

final gas density, the heat transferred, the work produced, and the change in internal energy. Assume that the specific heat at constant pressure is 0.22 Btu/lbm · °R.

6–26 (E) In a gas turbine, 3000 lbm/min of air are expanded in a reversible adiabatic manner. If we neglect kinetic and potential energy changes of the air flow, and if we find the entering air to have a temperature of 2100°R and a pressure of 230 psia, determine the power developed by the turbine. Assume that the exhaust air pressure is 15 psia.

6–27 (E) One pound-mass per second of oxygen gas (O_2) is expanded through a nozzle reversibly and adiabatically. The dimensions of the nozzle are shown in figure 6–12, and the temperature and pressure of the O_2 are 200°F and 89 psia entering the nozzle. Calculate the following:

(a) Final temperature
(b) Final pressure
(c) Heat transfer rate
(d) Final density of O_2

6–28 (M) In a piston-cylinder device, 10 g of air are compressed reversibly and adiabatically. The air is initially at 27°C and 110 kPa. After being compressed, the air is at 450°C. Determine the final pressure, heat transferred, increase in internal energy, and work required.

6–29 Prove the identity

$$c_p = \frac{kR}{k - 1}$$

for a perfect gas.

6–30 Prove the identity

$$c_v = \frac{-R}{1 - k}$$

for a perfect gas.

6–31 (E) Nitrogen gas is used in a turbine having an adiabatic efficiency of 91%. The inlet conditions are 400 psia and 400°F. If the outlet pressure is 15 psia, determine the turbine work and the outlet temperature.

6–32 (M) An adiabatic air turbine has an efficiency of 85% when it operates between 1 MPa, 227°C, and 100 kPa. Determine the turbine work and the final temperature.

6–33 (E) Expanded adiabatically from 130 psia to 20 psia is 1.4 lbm of carbon dioxide gas. If the efficiency is 72% and the initial temperature was 300°F, determine the work and the final temperature.

6–34 (M) For the compressor of problem 6–23, if the adiabatic efficiency is 86%, determine the work and the outlet temperature.

6–35 (E) An adiabatic compressor having 90% efficiency compresses hydrogen gas from 14.7 psia and 70°F to 100 psia. Determine the work and the final temperature.

6–36 (M) An adiabatic piston-cylinder compressor has an efficiency of 81%. If argon gas is compressed from 100 kPa and 23°C to 1500 kPa, determine the work and the final temperature.

6–37 (E) Air at 102 psia and 62°F is supplied from an air line to fill an empty 5-ft^3 tank that is well insulated. What is the temperature of the air in the tank after filling, and how much air is in the tank, assuming that the tank was completely empty to start? (See Figure 6–13.)

6–38 (M) Nitrogen gas at 850 kPa and 17°C is in a line that is connected to an empty insulated 1.2-m^3 tank through a valve, as shown in figure 6–13. How much nitrogen can be put in the tank, and what will the temperature of the gas in the tank be immediately upon filling?

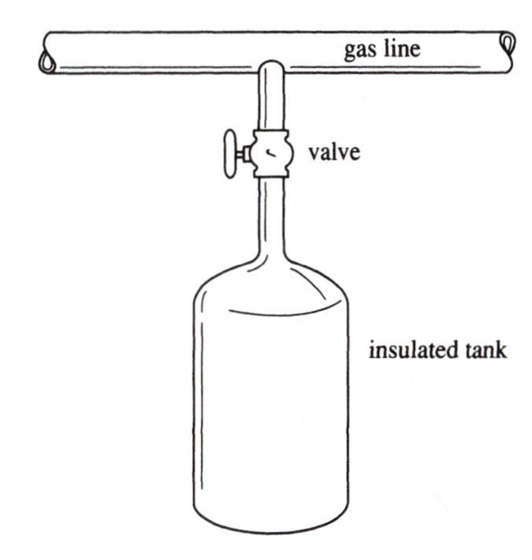

FIGURE 6–13

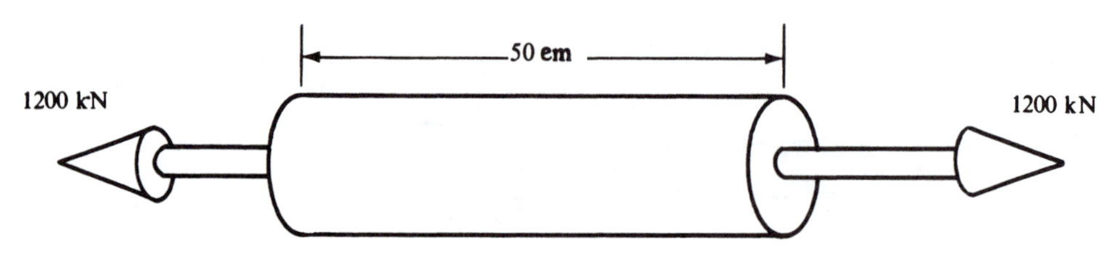

FIGURE 6–14

6–39 (M) An adiabatic gas turbine uses hot air to supply power. The air is supplied to the turbine at 3000 K and 2000 kPa. If the air expands reversibly through the turbine to the atmosphere at 100 kPa, determine the work produced by the turbine for each kilogram of air.

Section 6–3

6–40 (E) Air at 60 atm and -160°F is heated in a rigid tank to 80 atm. What will the temperature be?

6–41 (M) Methane is compressed isothermally in a piston-cylinder device from 15 atm to 100 atm at -45°C. Determine the boundary work done in compressing the methane. (You may want to use the BASIC program in appendix A–5.)

6–42 (E) Ammonia gas flows through a chamber at constant pressure. The specific volume of the ammonia is 0.8 ft^3/lbm entering and 0.95 ft^3/lbm leaving. Assuming no kinetic or potential energy changes, determine the temperature of the ammonia leaving if the initial temperature is 200°F.

6–43 (M) Ammonia is compressed in a piston-cylinder at constant temperature of 133°C from 1000 kPa to 11,000 kPa. Determine the work done on the ammonia. (You may want to use the BASIC program in appendix A–5.)

6–44 (M) Refrigerant-12 (R-12) is heated from 115°C to 220°C in a piston-cylinder and at constant pressure of 35 atm. Determine the work done per kilogram of R-12.

6–45 (E) Ethane at 40 atm is contained in a piston-cylinder. If the ethane is heated from 80°F to 300°F at constant pressure, determine the volume change and the work per pound-mass of ethane.

Section 6–4

6–46 A water pump transfers an incompressible liquid without kinetic or potential energy changes. If the pressure also remains constant, what work is required to drive the pump? Assume the process to be reversible.

6–47 (M) Methyl alcohol having a density of 0.80 g/cm^3 is pumped with no kinetic or potential energy changes from a tank at 100 kPa to a pressurized container at 200 kPa. Determine the work required for a reversible process.

6–48 (E) Kerosene having a density of 51.0 lbm/ft^3 is pumped from a tank at 14.7 psia to a reservoir 30 ft above the tank where the pressure is 20 psia. If 8 lbm/s is transferred, determine the power required by the pump if we assume complete reversibility.

6–49 (M) Ethyl alcohol is heated from -10°C to 25°C. Determine the amount of heat required per kilogram of the alcohol.

6–50 (E) Engine oil, 1 lbm/s, is to be cooled from 130°F to 85°F. Determine the rate of heat removal required to accomplish this cooling.

6–51 (M) Glycerine, 2 kg/s, is to be heated from 15°C to 30°C. Determine the rate of heat addition required.

Section 6–5

6–52 (M) An aluminum pan weighs 1.2 kg. How much heat is required to increase the pan's temperature from 20°C to 120°C?

6–53 (E) Determine the amount of heat required to kiln dry a 5-lbm brick by heating the brick from 65°F to 300°F.

6–54 (M) A round rod 50 cm long and 5 cm in diameter is subjected to a load of 1200 kN in an axial direction as shown in Figure 6–14. If the rod is made of aluminum having a modulus of elasticity of 1.2×10^9 kN/m^2, determine the work or energy absorbed by the rod.

6–55 (E) A block of wood is subjected to a uniform load of 800 lbf, as shown in Figure 6–15. If the modulus of elasticity of the wood is 6×10^5 lbf/in^2, determine the energy absorbed by the wood for this loading.

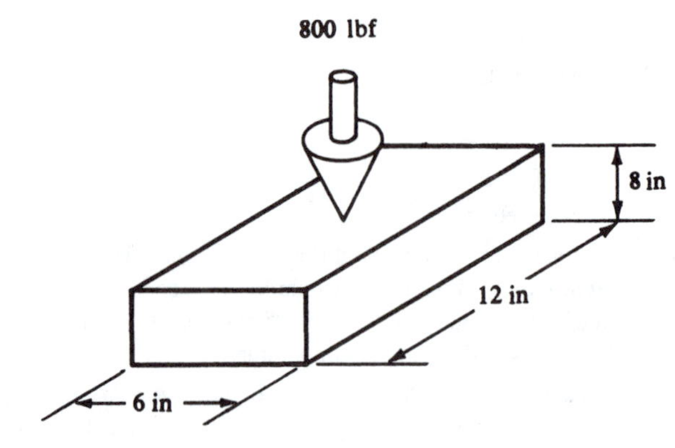

FIGURE 6–15

Section 6–6

6–56 (M) A steam generator receives water at 12 MPa and 30°C and supplies steam at 12 MPa and 720°C at a rate of 100 kg/s. Determine the rate of heat added to the water in the steam generator.

6–57 (E) A rigid container holds 10 lbm of water at 65% quality and 30 psia. The water is then heated to 300°F. Determine the final pressure and the heat added to the water.

6–58 (E) Three pounds-mass of R-12 is in a constant-pressure piston-cylinder chamber at −80°F and 15% quality. If it is then heated to a saturated vapor, determine the change in volume of the R-12 and the heat added.

6–59 (E) A total of 1 lbm/s of R-22 is compressed at 90°F from 10 psia to 150 psia. Determine the work done on the R-22 and the heat transferred. Use of the BASIC program AREA is suggested to find the boundary work.

6–60 (M) An ammonia condenser operates in a large food refrigerator. The ammonia enters as a vapor at 1000 kPa and 60°C and leaves as saturated liquid at 1000 kPa. Determine the heat lost by the ammonia per kilogram.

6–61 (M) Ammonia is compressed in a piston-cylinder at 40°C from 100 kPa to 1000 kPa. Determine the work done on the ammonia and the heat transfer per kilogram of ammonia. Use of the BASIC program for determining boundary work is suggested.

6–62 (E) Five pounds-mass of R-12 in a rigid container is cooled from 40 psia, 90°F, to 10°F. Determine the final pressure and the heat lost by the R-12.

6–63 (M) Two kilograms of ammonia are heated at constant pressure from 18°C, 50% quality to 60°C. Determine the heat.

6–64 (E) Ammonia is compressed reversibly and adiabatically from a saturated vapor at 10°F to 80 psia. Determine the work and the final temperature.

6–65 (M) Steam at 13 MPa and 640°C expands in a piston-cylinder to 8 MPa in a reversible and adiabatic manner. Determine the work done by the steam.

6–66 (E) A reversible adiabatic water pump receives water as saturated liquid at 60°F and increases the pressure to 900 psia. Determine the work done per pound-mass of water.

6–67 (M) Determine the reversible adiabatic pump work involved in pumping saturated liquid water from 10 kPa to 14 MPa.

6–68 (E) Saturated liquid ammonia at 100°F is throttled through an expansion valve to 12 psia. Determine the following:
(a) Ammonia temperature leaving the expansion valve.
(b) Quality of ammonia leaving the expansion valve.

6–69 (M) A total of 2 kg/s of R-22 is condensed from 2.7 MPa and 85°C to a saturated liquid at 2.7 MPa. Determine the rate of heat transferred from the R-22.

6–70 (E) Two pounds-mass of mercury vapor is evaporated in a constant-pressure chamber from 15 psia and 10% quality to a saturated vapor. Determine the volume change and heat added.

6–71 (M) An adiabatic turbine has an efficiency of 85% when steam expands through it from 11 MPa, 800°C, to 60°C. Determine the turbine work per kilogram of steam.

6–72 (E) An adiabatic steam turbine receives steam at 1000 psia and 1200°F. The steam exhausts to 40°F, and the turbine efficiency is 82%. Determine the work of the turbine per pound-mass of steam.

6–73 (M) A well-insulated R-12 compressor receives saturated vapor at −33°C and, with an adiabatic efficiency of 76%, supplies the R-12 at 800 kPa. Determine the work and final temperature of the R-12.

6–74 (E) An adiabatic compressor is used for compressing ammonia from 93.1% quality at −20°F to 160 psia. Determine the work and final temperature of the ammonia if the efficiency is 90%.

6–75 An insulated compressor operates between 60 kPa and 800 kPa when compressing 2 kg/s of R-407c from a saturated vapor to a superheated vapor. Estimate the power required of the compressor, assuming it is reversible.

6–76 (M) A well-insulated water pump having 85% efficiency receives 90°C saturated liquid and increases the pressure to 15 MPa. Determine the work and final enthalpy of the water.

6–77 (E) Refrigerant-22 is compressed adiabatically and with 68% efficiency from a saturated vapor at −5°F to 180 psia. Determine the work per pound-mass of R-22 and the final temperature.

6–78 R-502 is throttled from a saturated liquid at 140°F to 30 psia. Determine the final temperature of the R-502 and its quality at 30 psia.

6–79 (M) A well-insulated compressor is used to compress ammonia from a saturated vapor at −15°C to 400 kPa. If the efficiency is 70%, determine the work required per kilogram of ammonia and the final temperature.

6–80 (E) An adiabatic steam turbine has an efficiency of 90%. Steam expands through the device from 700 psia and 900°F to 1.0 psia. Determine the work per pound-mass of steam.

6–81 (E) A steam turbine receives 6 lbm/s of 1000 psia, 1000°F steam. The steam exhausts as a saturated vapor at 7 psia. If the turbine is producing 2800 hp, what is the rate of heat loss of the turbine.

6–82 (M) A throttling calorimeter is used to determine the quality of a saturation mixture of steam. If steam expands through a throttling calorimeter from 4 MPa to 100 kPa and 160°C, determine the quality of the steam at 4 MPa.

6–83 (E) R-22 is throttled from a saturated liquid at 120°F to 0°F. Determine the final pressure and quality of the R-22.

6–84 (M) Ammonia at saturation conditions of 10% quality at 30°C is throttled to 72 kPa. Determine the temperature and quality of the ammonia at 72 kPa.

6–85 (E) Refrigerant-134a is throttled from a saturated liquid at 20 psig to −40°F. If the atmospheric pressure is 14.570 psia, determine the pressure, enthalpy, and quality at the throttled state.

6–86 (E) Refrigerant-134a is compressed in a well-insulated compressor from −15°F, 85% quality, to 130°F. If the compressor has 100% adiabatic efficiency, determine the quality of the refrigerant leaving the compressor and the work per pound-mass of R-134a.

6–87 (M) An adiabatic compressor that has an adiabatic efficiency of 100% compresses 2 kg/s of Refrigerant-134a from 107 kPa, 75% quality, to 500 kPa. Determine the quality of the R-134a at the high pressure and the power required of the compressor.

6–88 (M) Refrigerant-134a is throttled from a saturated liquid at 80°C to −20°C. Determine the enthalpy and quality of the throttled R-134a.

6–89 (M) Refrigerant-123 is throttled from a saturated liquid at 80°C to 20 kPa. If the low-pressure refrigerant is then evaporated to a saturated vapor, determine the heat added per kilogram of the refrigerant.

6–90 (E) Saturated Refrigerant-123 at −10°F and 98% quality is compressed isentropically to a saturated vapor. Determine the temperature and pressure after compression.

7 HEAT ENGINES AND THE SECOND LAW OF THERMODYNAMICS

In this chapter, cyclic devices called heat engines, which have heat exchanges with their surroundings, are described. The idea is presented that all cyclic devices may have heat and work but do not have net changes in energy over one complete cycle. Some examples of heat engines are discussed, and the Carnot engine with its operating cycle is given.

Entropy is defined by using Carnot's assumption that the ratio of the heat added to the heat rejected in a heat engine is equal to the ratio of the temperatures of the heat reservoirs. The relation between entropy, temperature, and heat is developed by using the temperature–entropy diagrams and the concept of the reversible process.

The thermal efficiency of heat engines is defined, and the maximum efficiency between two constant-temperature reservoirs, that for a Carnot engine, is determined.

The reversed heat engine is considered, and in particular, the two applications of the device, the mechanical refrigerator and the heat pump, are discussed. Examples of the Carnot refrigerator and Carnot heat pump are given.

The second law of thermodynamics is stated, and a qualitative discussion regarding the implications of that law is presented. Primary emphasis in this discussion is on the fact that a thermal efficiency of 100% or more is impossible, that perpetual-motion machines are impossible as we know them, and that a result of operating all real heat engines and heat pumps is always an increase in entropy of the universe.

A qualitative discussion is then given of the meaning of entropy. Along with this discussion, some examples are given to show that entropy may increase in the universe, but that in reversible processes it remains constant. Also, the equations for computing entropy changes are developed for perfect gases, for incompressible liquids and solids, and for pure substances. Finally, the isentropic process is described, and its relationship to the reversible adiabatic process is developed. The perfect gas equation for the reversible adiabatic process, $pv^k = C$, is derived from the isentropic relationship.

The third law of thermodynamics is considered briefly, and then the chapter closes with some long problems involving the Carnot heat engine and heat pump. The analysis of these systems includes some computer assistance.

New Terms

COP	Coefficient of performance	ΔS	Entropy change
COR	Coefficient of refrigeration	η_T	Thermal efficiency
S	Entropy		
$\dot{S}$	Entropy generation	$\Sigma \frac{c_p\,\delta T}{T}$	Encrety

7–1 HEAT ENGINES AND CYCLIC DEVICES

We have defined a thermodynamic system that progresses through a set of processes and periodically returns to its initial or beginning state as a *cyclic device*. We call one complete set of processes that allows the first return of a system to its beginning state a *cycle*. There are many examples of cycling devices, some of which are electrical power plants that use coal or nuclear fuel, electric motors, and mechanical refrigeration devices such as refrigerators and air conditioners. There are other devices that can be treated as cyclic devices if the atmosphere is included as a part of the system: the internal combustion engine that uses gasoline or diesel fuel, gas turbines or jet engines, and steam engines exhausting to the atmosphere.

We call a particular type of cyclic device that transfers heat to its surroundings a *heat engine*. A heat engine and its important thermodynamic characteristics are visualized concisely in a *system diagram* in figure 7–1. Although this diagram contributes no information regarding the details of how the heat engine mechanically functions, it does provide a complete and clear picture of the thermodynamic operation of the devices. Thus, a system diagram can be a worthwhile exercise in the analysis or synthesis of a cycling heat engine. Keep in mind that the term *heat engine* does not necessarily require mechanical components such as gears, shafts, or piston-cylinders, but refers to any device that involves heat and work. In fact, the term *heat engine* or *cyclic device* will mean the particular system that is going through a cycle, and the term *cycle* will mean a set of processes of the working medium of the system (heat engine) that describes one complete cycle. Thus, we shall see that a working material such as air, water, or ammonia will be progressing through a cycle (set of processes) while simultaneously flowing through the actual cyclic device or heat engine.

Notice in figure 7–1 that heat is transferred from some high-temperature source (which may physically be within the system boundary or may be more than one source) to the heat engine. The heat engine then converts the transferred energy to mechanical work and dumps the excess into a low-temperature sink. This is the general operation of all heat engines, and where these phenomena occur in a particular device can be found from a study of a system diagram. Note also that in the cycling device no energy is accumulating or being drained from the engine itself—all energy is taken from some foreign source and deposited in another place. That is, the first law of thermodynamics for a cyclic device,

$$Q - Wk = \Delta E$$

becomes

$$Q - Wk = 0$$

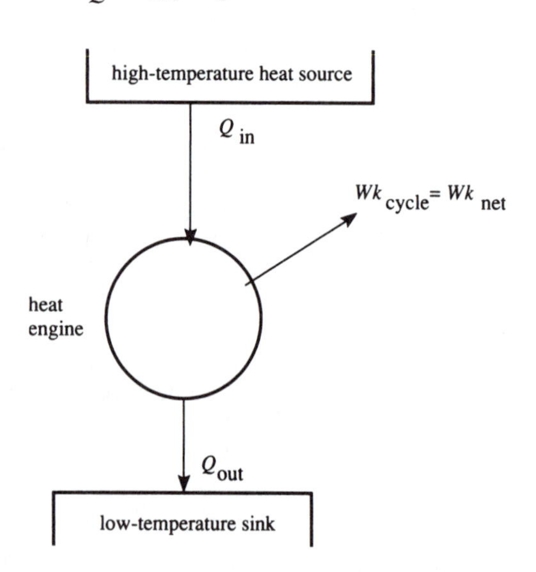

FIGURE 7–1 System diagram

because the energy change is zero over one cycle of the device. From the system diagram of figure 7–1, it can be seen that the heat, Q, is made up of at least two Q's, Q_{in} and Q_{out}. Therefore, we write, for heat engines,

$$Q_{net} = Q_{in} + Q_{out} = Wk_{net} \tag{7–1}$$

where Wk_{net} is the net work done by the heat engine or the amount of work one could expect out of the engine. In equation (7–1), the heat added (Q_{in}) and the heat rejected (Q_{out}) are added together, but in the actual calculations the Q_{out} will always have a negative value attached to it. So, in fact, the Q_{out} will subtract from the Q_{in}, making the net work less than the heat added. Also, notice that all the properties of the heat engine (or the working substance of the heat engine) do not change over one complete cycle. That is, the temperature change is zero, the pressure change is zero, the energy change is zero, the enthalpy change is zero, and so on. As we shall see later, this fact will allow us to define a new property called entropy.

EXAMPLE 7–1 A heat engine operates on a cycle made up of four processes. These four processes and the known values for the heat, work, and energy changes are given in table 7–1. Determine the values for the unknown quantities in this table.

TABLE 7–1

Process	Q, kJ/kg	Wk, kJ/kg	ΔE, kJ/kg
1–2	−5	−106	
2–3		0	485
3–4	0	276	
4–1	−302		

Solution In table 7–1, it can be seen that there are five terms that are not known: the heat of process 2–3; the work of process 4–1; and three energy changes of the processes 1–2, 3–4, and 4–1. For each of the four processes, we may write

$$Q - Wk = \Delta E$$

From this equation, we find that $Q = 485$ kJ/kg for process 2–3. Also, the energy changes are $\Delta E = 101$ kJ/kg for process 1–2 and $\Delta E = -276$ kJ/kg for process 3–4. For the complete cycle, we must have that the sum of the ΔE's is zero, or $\sum \Delta E = 0$. From this equation, we may write

$$\Delta E_{12} + \Delta E_{23} + \Delta E_{34} + \Delta E_{41} = 0$$

and

$$\Delta E_{41} = -\Delta E_{12} - \Delta E_{23} - \Delta E_{34} = -101 - 485 - (-276)$$
$$= -310 \text{ kJ/kg}$$

The final values for all the terms in table 7–1 are given in table 7–2.

TABLE 7–2 Results of example 7–1

Process	Q, kJ/kg	Wk, kJ/kg	ΔE, kJ/kg
1–2	−5	−106	101
2–3	485	0	485
3–4	0	276	−276
4–1	−302	8	−310

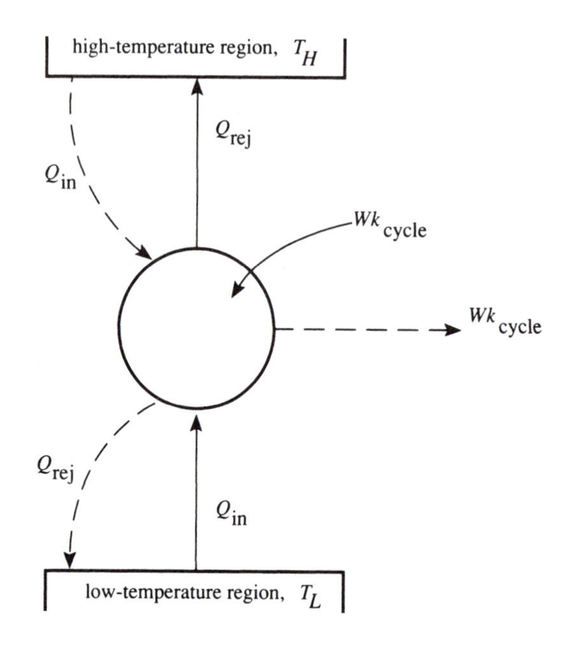

FIGURE 7–2 Reversible heat engine

If we consider the heat engine to be reversed, then we can visualize the energy transfers of heat and work to take either direction. In figure 7–2 is shown such a reversed heat engine where the discontinuous arrows represent the set of transfers that would occur reversed from the set represented by solid arrows. It can be seen that the arrangement of energy transfers diagrammed with the solid arrows is just a device that will provide refrigerating capability; that is, it pumps energy from a lower-temperature into a higher-temperature region by expending work. This type of heat engine is generally referred to as a *heat pump*, or *refrigerator*.

7–2 THE CARNOT ENGINE AND ENTROPY

In 1824, Sadi Carnot published a treatise on thermodynamics in which he invented a cycle composed of four special processes. The heat engine that would run on this cycle (but which no one has yet built) has since been called the **Carnot engine**, and the cycle the **Carnot cycle**. The four processes, in order, that constitute this cycle are as follows:

1–2 Reversible isothermal compression at temperature T_L

2–3 Reversible adiabatic compression from the low temperature T_L to a higher temperature T_H

3–4 Reversible isothermal expansion at temperature T_H

4–1 Reversible adiabatic expansion from temperature T_H to T_L

Notice that the foregoing descriptions of the processes indicate that the cycle is completely reversible because all four processes are reversible. The *p–V* diagram of the Carnot cycle is shown in figure 7–3, and from this diagram we can see that the net work is the sum of four separate work terms. The arrows on the process lines from the states indicate that the direction of the cycle is "clockwise," and the net work will be the shaded area between the four process lines. We may write this as

$$Wk_{net} = \sum_{1-2} p\,\delta V + \sum_{2-3} p\,\delta V + \sum_{3-4} p\,\delta V + \sum_{4-1} p\,\delta V \qquad \textbf{(7–2)}$$

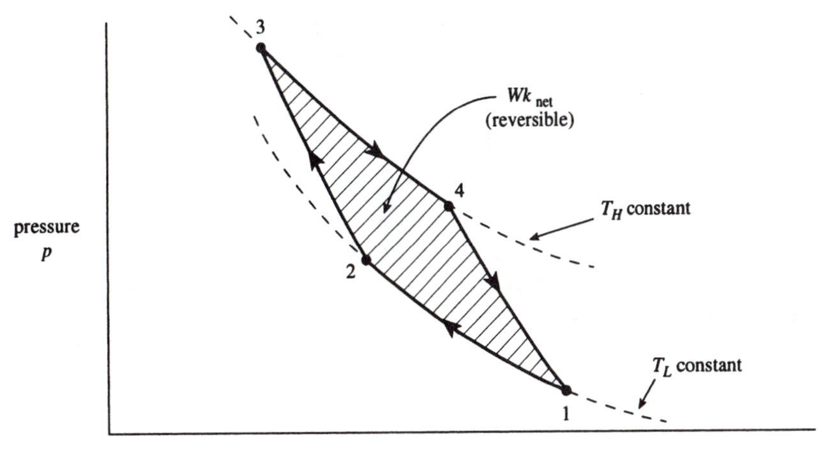

FIGURE 7–3 *p–V* diagram for Carnot cycle (perfect gas)

If the working medium of the Carnot heat engine is a perfect gas with constant specific heats, each of these four terms can be directly computed from the following equations:

$$\sum_{1-2} p\,\delta V = mRT_1 \ln \frac{V_2}{V_1}$$

$$\sum_{2-3} p\,\delta V = \frac{1}{1-k}(p_3V_3 - p_2V_2) = \frac{mR}{1-k}(T_3 - T_2)$$

$$\sum_{3-4} p\,\delta V = mRT_3 \ln \frac{V_4}{V_3}$$

$$\sum_{4-1} p\,\delta V = \frac{1}{1-k}(p_1V_1 - p_4V_4) = \frac{mR}{1-k}(T_1 - T_4)$$

For working substances other than perfect gases, you may want to use something like the VISUAL BASIC program AREA for computing areas under curves (appendix A–5), all the time assuming that states 1, 2, 3, and 4 are known.

CALCULUS FOR CLARITY 7–1

The net work of a reversible heat engine is expressed as the sum of the work terms resulting from each of the reversible processes of the cycle, as indicated by equation (7–2). Precisely,

$$Wk_{\text{net, rev}} = \int_1^2 p\,dV + \int_2^3 p\,dV + \int_3^4 p\,dV + \int_4^1 p\,dV$$

and the right side of this equation is often abbreviated as the cyclic integral, $\oint p\,dV$, so that

$$Wk_{\text{net, rev}} = \oint p\,dV \qquad \textbf{(7–3)}$$

The cyclic integral is the integral for a complete cycle. If, for instance, the cycle were composed of five individual processes, then the cyclic integral would have five separate integral terms. The Carnot cycle has four processes, so its cyclic integral requires four separate integral terms. The cyclic integral also represents the area enclosed by the four *p–V* curves shown in figure 7–3. For a reversible heat engine, the area is the net work.

Now, the Carnot cycle as defined by the four individual processes represents a particular circumstance in which the engine is operating between two temperature regions, one at a high temperature and one at a low temperature. In figure 7–1 and 7–2, the high- and low-temperature regions were not required to be constant; that is, the heat transfers could change the temperature of the two regions. For instance, it would be reasonable to expect the heat removal from the source in figure 7–1 to lower the high-temperature source temperature and the low-temperature sink to increase temperature because of heat transfer to that region. In the description of the Carnot cycle for the Carnot engine, it is required that the heat source be at a constant temperature, T_H, and that the sink be at a different constant temperature, T_L. Also, there is a requirement that only two heat transfers occur in one cycle, Q_H from T_H and Q_L from T_L. Sadi Carnot suggested that the ratio of the two heats, Q_H/Q_L, be equal to the ratio of the corresponding temperature ratio, T_H/T_L. That is,

$$\frac{Q_H}{-Q_L} = \frac{T_H}{T_L} \tag{7–4}$$

where a negative sign is given Q_L (the heat rejected) because Q_L will have a negative value attached to it with respect to the Carnot engine. By including the negative sign in equation (7–4), we have a positive value for the temperature ratio and the heat ratio. The restriction of equation (7–4) is that it applies only to reversible, ideal heat engines operating between two constant-temperature regions, or Carnot engines. Now, if equation (7–4) is rearranged to read $Q_H/T_H = -Q_L/T_L$, or

$$\frac{Q_H}{T_H} + \frac{Q_L}{T_L} = 0$$

we can see that the left side of this equation is a sum of Q/T values for the Carnot engine. We have seen that, for all cyclic devices, the sum of the changes in energy, pressure, temperature, and all other system properties is zero. We therefore say that Q/T can be represented as a system property, because it obeys the necessary rule regarding properties of cyclic devices. To allow for conditions when temperatures may change during heat transfers, we stipulate that Q be very small and that it be done reversibly. We write this term as δQ_{rev}, so $\delta Q_{\text{rev}}/T$ is equal to a very small change in a property. We call this property **entropy**, S, and a finite change in entropy is then

$$\Delta S = \sum \frac{\delta Q_{\text{rev}}}{T} \tag{7–5}$$

The units associated with entropy and its changes are energy per unit temperature, or kJ/K or Btu/°R. Also, in later discussions of entropy, we will often use the entropy per unit mass or the entropy change per unit mass, given by

$$\Delta s = \frac{\Delta S}{m} = \sum \frac{\delta q_{\text{rev}}}{T} \tag{7–6}$$

where δq_{rev} is a very small amount of heat per unit mass of working substance. The units of s or Δs are energy per unit mass per unit temperature: kJ/kg · K or Btu/lbm · °R.

Notice that the Carnot heat engine is a reversible engine, and both of the heat terms are reversible. Since the heats also occur at constant temperature, equation (7–5) is just

$$\Delta S = \frac{Q_{\text{rev}}}{T}$$

For the entropy change due to the high-temperature heat transfer,

$$\Delta S_H = \frac{Q_H}{T_H}$$

and for the low-temperature heat transfer,

$$\Delta S_L = \frac{Q_L}{T_L}$$

Notice also that there is no entropy change during the two reversible adiabatic processes since Q is zero for these processes.

Let us now return to equation (7–5) and consider a very small change in entropy, δS. This would be written

$$\delta S = \frac{\Delta Q_{\text{rev}}}{T} \tag{7–7}$$

and by rearranging, we have

$$\delta Q_{\text{rev}} = T\,\delta S \tag{7–8}$$

Now, to determine the amount of heat transfer during some process, we may sum the right side of equation (7–8) as temperature varies with entropy:

$$Q_{\text{rev}} = \sum T\,\delta S \tag{7–9}$$

The mathematics of such a problem as this one is just like that for computing the boundary work, $\sum p\,\delta V$, but with different variables. We will see later how to determine values for the entropy at given states, but right now notice that a graph in which temperature would be plotted against entropy, called a temperature–entropy or T–S diagram, can represent heat as the area under the curve. Such a diagram is shown in figure 7–4, where an example process curve is drawn and the area under that curve is just the reversible heat transfer. It is important to note that the area under the curve in the T–S diagram is a reversible heat quantity and not that for an irreversible process. We consider circumstances when processes are irreversible in later sections.

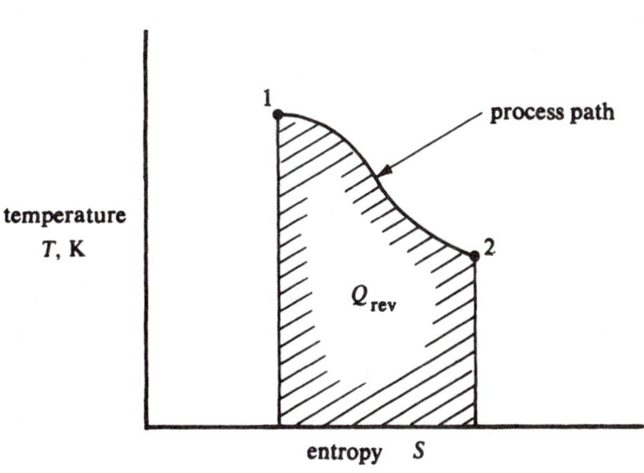

FIGURE 7–4 Characteristics of a process on a T–S plane

CALCULUS FOR CLARITY 7–2

A differential change in entropy is defined as

$$dS = \frac{1}{T}\,dQ_{\text{rev}} \tag{7–10}$$

and a differential amount of reversible heat is then

$$dQ_{\text{rev}} = T\,dS \tag{7–11}$$

CALCULUS FOR CLARITY 7–2, continued

From this equation, we see that the heat

$$Q_{\text{rev}} = \int T\,dS \tag{7–12}$$

and there is a mathematical analogy between this result and the expression for the reversible work, equation (3–11). There is also a geometric analogy: the area under the curve of temperature as entropy changes is reversible heat, and the area under the curve of pressure as volume changes is reversible work. The geometric area under the *T–S* curve equated to reversible heat is indicated in figure 7–4. Also, the net heat of a cycle is the sum of the heat from each of the processes of the cycle, and this is called the cyclic heat

$$Q_{\text{net}} = \oint T\,dS \tag{7–13}$$

where the cyclic integral is the abbreviation for the sum of individual integrals of all of the processes. The net heat from a reversible heat engine can then be described as the area enclosed by the *T–S* curve for one complete cycle. An example of a Carnot heat engine cycle is shown in figure 7–5, where the net heat is the shaped area.

In calculus for clarity 7–1, we described the net work as the cyclic integral of the work. For the heat engine, the energy balance is

$$\oint dQ - \oint dWk = \oint dU \tag{7–14}$$

But the cyclic integral of the internal energy is zero because it is a property of the engine and cannot change over one cycle. Thus,

$$\oint dQ - \oint dWk = 0$$

or

$$\oint dQ = \oint dWk \tag{7–15}$$

This result holds for all heat engines and cycles. For the reversible heat engine,

$$\oint T\,dS = \oint p\,dV \tag{7–16}$$

Hence, the enclosed area of the *p–V* diagram is equal to the enclosed area of the *T–S* diagram for a reversible heat engine. It is not necessarily so for irreversible, or real, heat engines.

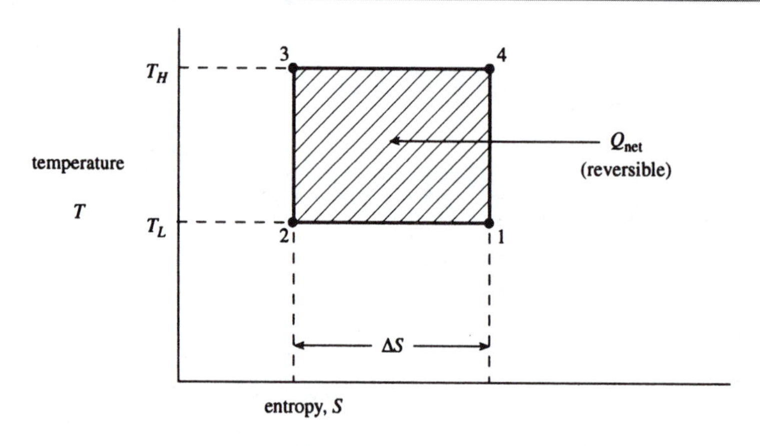

FIGURE 7–5 *T–S* diagram for Carnot cycle

The T–S diagram for the Carnot cycle is shown in figure 7–5, where the states 1, 2, 3, and 4 correspond to those in figure 7–3. In this figure, the term ΔS indicates the magnitude of the entropy changes occurring during the Carnot cycle operation. For instance, in processing through process 1–2, a constant-temperature heat addition, the entropy change is

$$\Delta S = \Delta S_L = \frac{Q_L}{T_L}$$

which will have a negative value, since Q_L is negative. The magnitude of ΔS_L will be the ΔS indicated in figure 7–5. Also, the entropy change occurring over process 3–4 will be

$$\Delta S = \Delta S_H = \frac{Q_H}{T_H}$$

and it will have the same magnitude as ΔS_L, but now have a positive value. The net heat transfer or the sum of the two heat transfers, $Q_H + Q_L$, is indicated as the shaded area surrounded by the four process curves. Again, notice the similarity in concepts between work and heat, as the work was shown in figure 7–3 to be an area surrounded by process curves in a p–V diagram.

EXAMPLE 7–2 A Carnot engine operates between 640°F and 40°F, with 700 Btu added per cycle. Determine the entropy changes occurring in the cycle, the heat rejected, and the net work done by the engine as it operates on the Carnot cycle.

Solution The entropy change occurring during the heat addition process is

$$\Delta S_H = \frac{Q_H}{T_H} = \frac{700\text{ Btu}}{640 + 460°\text{R}}$$

Answer

$$= 0.636\text{ Btu/°R}$$

and for the heat rejection process,

Answer

$$\Delta S_L = \frac{Q_L}{T_L} = \frac{-Q_H}{T_H} = -0.636\text{ Btu/°R}$$

The amount of heat rejected, Q_L, is found from the calculation

$$Q_L = \frac{-Q_H(T_L)}{T_H}$$

$$= \frac{-(700\text{ Btu/°R})(40 + 460°\text{R})}{1100°\text{R}}$$

Answer

$$= -318.18\text{ Btu/cycle}$$

and the net work must be equal to the net heat, so that

$$Wk_{net} = Q_H + Q_L$$

$$= 700\text{ Btu} - 318.1\text{ Btu}$$

Answer

$$= 381.82\text{ Btu/cycle}$$

Notice that all entropy change calculations require absolute temperatures. The foregoing discussion and analysis of the Carnot engine and the entropy applies as well to those situations where rates are to be considered. That is, if the power produced or the rate of heat transfer to or from a heat engine is involved in a problem, we can write equation (7–1) as

$$\dot{Q}_{in} + \dot{Q}_{out} = \dot{W}k_{net} \tag{7–17}$$

and for the Carnot heat engine, equation (7–4) is

$$\frac{\dot{Q}_H}{-\dot{Q}_L} = \frac{T_H}{T_L} \tag{7–18}$$

As a result of these relationships we can find a rate of change of entropy, $\dot{S}$, in units of power per unit degree temperature (such as kW/K or Btu/s · °R). We call this **entropy generation**, and it is defined as

$$\dot{S} = \frac{\dot{Q}_{\text{rev}}}{T} \tag{7–19}$$

Also, the rate of heat transfer may then be written

$$\dot{Q}_{\text{rev}} = T\dot{S} \tag{7–20}$$

For the Carnot engine, entropy generation becomes, for the high-temperature heat transfer,

$$\dot{S}_H = \frac{\dot{Q}_H}{T_H}$$

and for the low-temperature heat rejection,

$$\dot{S}_L = \frac{\dot{Q}_L}{T_L}$$

Since the Carnot engine is reversible and $\dot{S}_H = -\dot{S}_L$, the total entropy generation of the device and its surroundings is zero, but for real (irreversible) heat engines, the entropy generation will always be positive.

EXAMPLE 7–3 Assume that an internal combustion engine operates on a reversible Carnot cycle between 20°C and 2000°C. If this engine generates 100 kW of power, determine the amount of heat added and the heat rejected by the engine.

Solution From equation (7–8), the cyclic work equals the cyclic heat, so

$$\dot{Q}_{\text{cycle}} = \dot{W}k_{\text{cycle}} = 100 \text{ kW}$$

Also,

$$\dot{Q}_{\text{cycle}} = (T_H - T_L)(\dot{S})$$

where $\dot{S}$ is the entropy generation. The heat added is given by the equation

$$\dot{Q}_{\text{add}} = T_H\dot{S}$$

so we have

$$\dot{S} = \frac{\dot{Q}_{\text{cycle}}}{T_H - T_L}$$

Since the temperatures are given, we may compute the entropy generation as

$$\dot{S} = \frac{100 \text{ kW}}{2273 \text{ K} - 293 \text{ K}}$$
$$= 0.0505 \text{ kW/K}$$

The heat added and heat rejected are then found to be, respectively,

$$\dot{Q}_{\text{add}} = (2273 \text{ K})(0.0505 \text{ kW/K})$$

Answer
$$= 114.8 \text{ kW}$$

and

Answer

$$\dot{Q}_{rej} = (293\text{ K})(-0.0505\text{ kW/K}) = -14.8\text{ kW}$$

7–3 THERMAL EFFICIENCY

Efficiency is a term used often to describe the manner in which a heat engine or other cyclic device operates. We have already seen that efficiency can be defined as output divided by input. For a heat engine, which converts heat to work, the input can be identified as heat input, or heat added. The output will be the net work, so that the **thermal efficiency** of a heat engine is

$$\eta_T = \frac{Wk_{net}}{Q_{add}} \tag{7–21}$$

The adjective "thermal" is used to indicate that the efficiency for a heat engine is a measure of the conversion of thermal energy (as heat) to mechanical energy (as work). For any heat engine, we also have, from equation (7–1), that Q_{in} is the Q_{add} term in equation (7–21) for all engines. Therefore, equation (7–21) becomes

$$\eta_T = \frac{Q_{add} + Q_{out}}{Q_{add}}$$

or

$$\eta_T = 1 + \frac{Q_{out}}{Q_{add}} \tag{7–22}$$

Sometimes we will call Q_{out} the rejected heat or heat rejection, Q_{rej}. Notice that equation (7–22) seems to give an efficiency that is greater than 1 (or 100%). This will never be the case since the Q_{out} or heat rejected is always negative for the heat engine, so that the ratio Q_{out}/Q_{add} will be negative, making the efficiency less than 100%. For the Carnot heat engine, the ratio of the heat out to the heat in is given by equation (7–4), so that

$$\eta_T = 1 - \frac{T_L}{T_H} \quad \text{(Carnot engine)} \tag{7–23}$$

Other heat engines will have thermal efficiencies that are found from equation (7–21) or (7–22), but not necessarily equation (7–23). In later chapters, we will see other equations for thermal efficiency, derived from the definition of equation (7–21).

EXAMPLE 7–4 A Carnot engine operates in an atmosphere at 343 K, with an addition of heat from combusting gases burning at 1473 K. Determine the thermodynamic efficiency of the engine.

Solution

We are concerned here with a reversible heat engine that operates between high- and low-temperature regions. We can visualize this Carnot engine better from a system diagram as shown in figure 7–6. In this figure, we see that

$$Q_{add} + Q_{rej} = Wk_{cycle}$$

which is a restatement of equation (7–1). The term Q_{add} is equal to $T_H\,\Delta S$, where T_H is the high-temperature region at 1473 K. Similarly, Q_{rej} is $-T_L\,\Delta S$, where T_L is 343 K. Then

$$(T_H - T_L)\,\Delta S = Wk_{cycle}$$

and from the equation

$$\eta_T = \frac{Wk_{cycle}}{Q_{add}} \times 100$$

FIGURE 7–6 Carnot system diagram for example 7–4

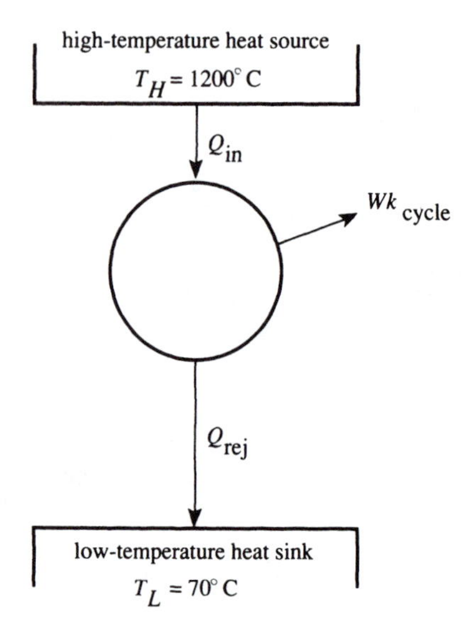

we have, for the general Carnot engine,

$$\eta_T = \frac{(T_H - T_L)\,\Delta S}{T_H\,\Delta S} \times 100$$

or

$$\eta_T = \left(1 - \frac{T_L}{T_H}\right) \times 100$$

Specifically for this problem, we obtain

Answer

$$\eta_T = \left(1 - \frac{343}{1473}\right) \times 100 = 76.7\%$$

7–4 REFRIGERATION AND HEAT PUMP CYCLES

The reversible heat engine or cyclic device is an imaginary system that allows us more easily to study the performance of actual heat engines. The assumption of reversibility allows us to eliminate any friction or inertial effects that would tend to complicate the behavior of these engines, but reversibility also allows us to visualize the engine as being one that we can reverse. If a heat engine is reversed, as in figure 7–2, the net work becomes an input to the device, the heat added comes from a low-temperature region, and the heat is rejected to a high-temperature region. The device shown in figure 7–2 looks very much like a pump that moves heat from low- to high-temperature regions while requiring an input work. In fact, a device such as that in figure 7–2 is sometimes called a heat pump; they are also called *refrigeration devices*. For the heat pump, the first law becomes

$$Wk_{\text{cycle}} = Q_{\text{in}} + Q_{\text{rej}}$$

or

$$Wk_{\text{net}} = Q_{\text{in}} + Q_{\text{rej}} \tag{7–24}$$

In this equation, you should notice that the Q_{rej} term will always have a negative value attached to it. Also, the work will have a negative value because it is put into the system or cyclic device.

The **coefficient of performance** (COP) is sometimes used to describe the performance of heat pump devices. The COP is used in place of thermal efficiency because work is not an output of the heat pumps but is an input. The COP is defined as

$$\text{COP} = \frac{Q_{\text{rej}}}{Wk_{\text{net}}} \tag{7–25}$$

and you can see that it is the inverse or reciprocal of the thermal efficiency if the Q_{rej} is recognized as heat transfer to a high-temperature region. Thus, for a heat pump, the output is visualized as a heat, Q_{rej}, and the input is work. If the device operates as a refrigerator or other such device that is required to cool a region, the **coefficient of refrigeration** (COR) is used. The COR is defined as

$$\text{COR} = \frac{Q_{\text{in}}}{-Wk_{\text{net}}} \tag{7–26}$$

Now, if we use equation (7–24) and substitute for Q_{in} the terms $Wk_{\text{net}} - Q_{\text{rej}}$, equation (7–26) becomes

$$\begin{aligned} \text{COR} &= \frac{Wk_{\text{net}}}{-Wk_{\text{net}}} - \frac{Q_{\text{rej}}}{-Wk_{\text{net}}} \\ &= -1 - \frac{Q_{\text{rej}}}{-Wk_{\text{net}}} \end{aligned}$$

but $-Q_{\text{rej}}/(-Wk_{\text{net}})$ is the COP, so we have

$$\text{COR} = -1 + \text{COP} = \text{COP} - 1 \tag{7–27}$$

From this equation, you can see that the COP is always greater than the COR by exactly 1.

If a heat pump or refrigerator were assumed to follow the reversed Carnot cycle, it would be called a Carnot heat pump or Carnot refrigerator. In this situation, the ratios of the two heat terms would be related to the two temperatures of the low and high regions as given by equation (7–4). It can then be shown that for the Carnot heat pump, the COP is

$$\text{COP} = \frac{T_H}{T_H - T_L} \tag{7–28}$$

and for the Carnot refrigerator, the COR is

$$\text{COR} = \frac{T_L}{T_H - T_L} \tag{7–29}$$

The entropy changes that we considered earlier for Carnot heat engines are equally true for the heat pump cycles; that is, the entropy change occurring through the heat rejection process is

$$\Delta S_H = \frac{Q_H}{T_H}$$

and through the heat addition process is

$$\Delta S_L = \frac{Q_L}{T_L}$$

The term ΔS_H will have a negative value, and the ΔS_L will be positive but with the same magnitude as ΔS_H (call it ΔS). From these two results, we may write $Q_H(Q_{\text{rej}}) = T_H(-\Delta S)$ and $Q_L(Q_{\text{in}}) = T_L(\Delta S)$. The net work for a Carnot heat pump can then be written as $(T_L - T_H)\,\Delta S$. Other reversed cycles that perform either as heat pumps or as refrigerators are considered in later chapters.

EXAMPLE 7–5 A refrigerator operates on a reversed Carnot cycle. If the heat is taken in at −13°C and rejected to 27°C, and if the entropy change is 0.1 J/K, determine the work required to drive the refrigerator, the amount of cooling it will have per cycle, and the COR.

Solution For a reversed Carnot cycle, we refer to figure 7–2 and see that we add heat at the low temperature T_L and reject heat as T_H. The cyclic work is

$$
\begin{aligned}
Wk_{\text{cycle}} &= Q_{\text{add}} + Q_{\text{rej}} \\
&= T_L \, \Delta S - T_H \, \Delta S \\
&= (T_L - T_H) \, \Delta S
\end{aligned}
$$

Converting the temperatures to absolute values, we obtain

$$
\begin{aligned}
Wk_{\text{cycle}} &= (260 \text{ K} - 300 \text{ K})(0.1 \text{ J/K}) \\
&= -4.0 \text{ J/cycle}
\end{aligned}
$$

Answer

The negative sign indicates that work is put into the system. The cooling or refrigerating effect is the heat added, or

$$
\begin{aligned}
Q_{\text{add}} &= T_L \, \Delta S = (260 \text{ K})(0.1 \text{ J/K}) \\
&= 26.0 \text{ J/cycle}
\end{aligned}
$$

Answer

The coefficient of refrigeration, COR, can now be found from either of two equations: equation (7–26) or (7–29). Using equation (7–26), we find that

$$
\begin{aligned}
\text{COR} &= \frac{Q_{\text{add}}}{-Wk_{\text{cycle}}} = \frac{26.0 \text{ J/cycle}}{-(-4.0 \text{ J/cycle})} \\
&= 6.5
\end{aligned}
$$

Answer

The same answer could be found by using equation (7–29):

$$
\begin{aligned}
\text{COR} &= \frac{T_L}{T_H - T_L} \\
&= \frac{260 \text{ K}}{300 \text{ K} - 260 \text{ KL}} \\
&= 6.5
\end{aligned}
$$

Answer

EXAMPLE 7–6 If the refrigerator of example 7–5 operates at 300 cycles per minute, determine the power required and the rate of cooling.

Solution The power is computed from the relationship

$$
\begin{aligned}
\dot{W}k &= (Wk_{\text{cycle}})(300 \text{ cycles/min}) \\
&= -1200 \text{ J/min} = -20 \text{ J/s} \\
&= -0.02 \text{ kW}
\end{aligned}
$$

Answer

The rate of cooling is computed from

$$
\begin{aligned}
\dot{Q}_{\text{add}} &= (Q_{\text{add}})(300 \text{ cycles/min}) \\
&= (26 \text{ J/cycle})(300 \text{ cycles/min}) \\
&= 7800 \text{ J/min} = 130 \text{ J/s} \\
&= 0.13 \text{ kW}
\end{aligned}
$$

Answer

EXAMPLE 7–7 A Carnot heat pump operates between the limits of 10°F and 120°F. Its mechanical aspects are diagrammed in figure 7–7, and it operates with ammonia as the working fluid. At point 1 in the closed cycle, the ammonia is completely saturated liquid at 120°F; at point 4, the ammonia is saturated vapor at 120°F. The ammonia expands adiabatically from (1) to (2), furnishing

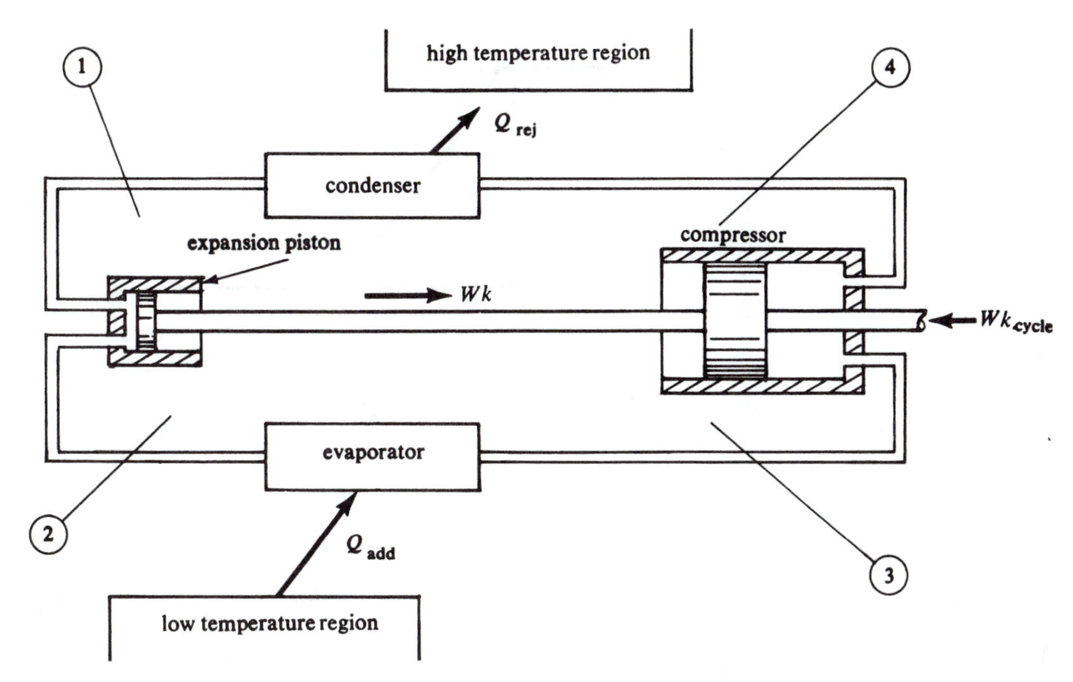

FIGURE 7–7 Arrangement for a Carnot heat pump

some work to help drive the compressor. The operating temperature at point 2 is 10°F. Heat is added to the working fluid in the evaporator and rejected in the condenser. The compressor supplies the energy to the ammonia, which increases in pressure and temperature, making the heat rejection to the high-temperature (120°F or less) region possible. Determine q_{add}, q_{rej}, wk_{cycle}, and the COP. Compare the answers obtained from the appendix tables with those obtained by using *EES*.

Solution

The system is a heat pump system made up of four steady-state components: a piston-cylinder or expansion piston, an evaporator, a compressor, and a condenser. Together these operate to allow ammonia to progress through a Carnot cycle. The *T–S* and *p–v* diagrams for the ammonia are shown in figures 7–8 and 7–9. Notice that the *T–S* diagram includes the saturation line, below which lies the saturation region. We discussed previously the saturation region of pure substances as they occur in *p–V* and *p–T* diagrams, and here we now see how the saturation region appears in the *T–S* diagram. Notice also that points 1, 2, 3, and 4 corresponding to those in figure 7–7 define a Carnot cycle.

FIGURE 7–8 *T–S* diagram for Carnot heat pump using ammonia (not to scale)

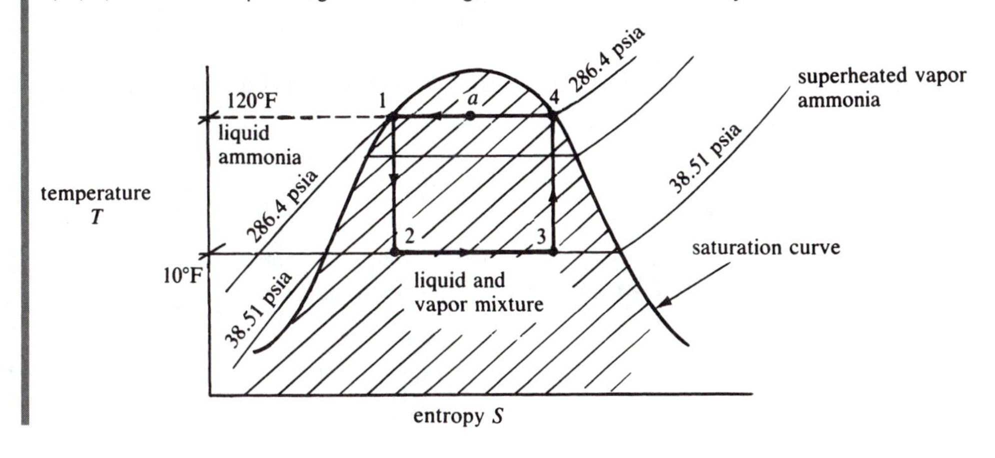

FIGURE 7–9 p–v diagram for Carnot heat pump using ammonia

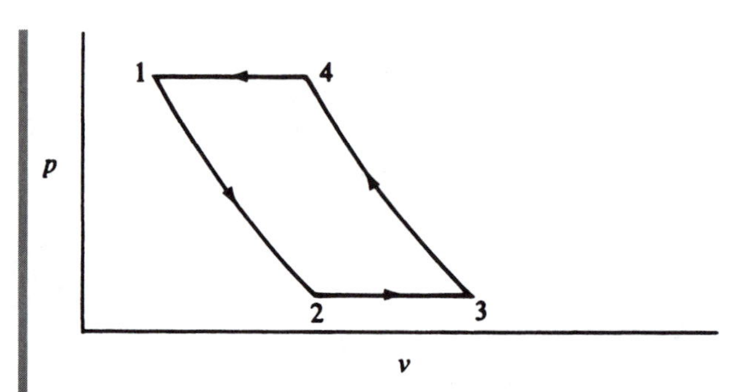

The heat added, q_{add}, is determined by considering the steady-flow energy equation for the evaporator. After neglecting the kinetic and potential energies and the shaft work, we have

$$q_{add} = (h_3 - h_2)$$

In a similar manner, for the rejected heat and the work, we obtain, respectively,

$$q_{rej} = (h_1 - h_4)$$
$$Wk_{cycle} = (h_3 - h_4) + (h_1 - h_2)$$

Thus, we need to find each of the enthalpy values of the ammonia at the four states. At state 1, a saturated liquid at 120°F, from table B–19,

$$h_1 = h_f(\text{at } 120°\text{F}) = 179.0 \text{ Btu/lbm}$$

and

$$s_1 = s_f = 0.3576 \text{ Btu/lbm} \cdot °\text{R}$$

At state 2,

$$s_2 = s_1$$
$$T_2 = 10°\text{F}$$

At state 4, a saturated vapor at 120°F, from table B–19 again,

$$h_4 = h_g = 634.0 \text{ Btu/lbm}$$
$$s_4 = s_g = 1.1427 \text{ Btu/lbm} \cdot °\text{R}$$

At state 3,

$$s_3 = s_4$$
$$T_3 = 10°\text{F}$$

The heat rejected can now be found from the calculation

$$q_{rej} = h_1 - h_4$$
$$= 179.0 \text{ Btu/lbm} - 634 \text{ Btu/lbm}$$

Answer

$$= -455 \text{ Btu/lbm}$$

The heat added is $h_3 - h_2$, but q_{add} is also equal to $T_L(s_3 - s_2)$, since the heat is added at constant temperature. Then

$$q_{add} = T_L(s_3 - s_2)$$
$$= (10 + 460°\text{R})(1.1427 \text{ Btu/lbm} \cdot °\text{R} - 0.3576 \text{ Btu/lbm} \cdot °\text{R})$$

Answer

$$= 368.997 \text{ Btu/lbm}$$

The net work, wk_{cycle}, is equal to the net heat, or

$$\begin{aligned} wk_{\text{cycle}} &= q_{\text{add}} + q_{\text{rej}} \\ &= 368.997 \text{ Btu/lbm} - 455 \text{ Btu/lbm} \\ &= -86.003 \text{ Btu/lbm} \end{aligned}$$

Answer

Finally, the coefficient of performance is determined by one of two methods, either equation (7–25) or (7–28). Using equation (7–25), we find that

$$\text{COP} = \frac{q_{\text{rej}}}{wk_{\text{cycle}}} = \frac{-455 \text{ Btu/lbm}}{-86.003 \text{ Btu/lbm}} = 5.29$$

Answer

Also, using equation (7–28), since the heat pump operates according to a Carnot cycle, we have

$$\text{COP} = \frac{T_H}{T_H - T_L} = \frac{120 + 460°\text{R}}{120 - 10} = 5.27$$

Answer

The slight difference in the two answers is attributable to round-off in the values for entropy and enthalpy as given in table B–19. The following equations are entered into the *EES Equations* window in English units, with temperature in degrees Fahrenheit and pressure in psia:

$$\begin{aligned} h_1 &= \text{enthalpy (ammonia, } T = 120, x = 0.0) \\ s_1 &= \text{entropy (ammonia, } T = 120, x = 0.0) \\ s_2 &= s_1 \\ h_2 &= \text{enthalpy (ammonia, } T = 10, s = s_2) \\ h_4 &= \text{enthalpy (ammonia, } T = 120, x = 1.0) \\ s_4 &= \text{entropy (ammonia, } T = 120, x = 1.0) \\ s_3 &= s_4 \\ h_3 &= \text{enthalpy (ammonia, } T = 10, s = s_3) \end{aligned}$$

$$\begin{aligned} q_a &= h_3 - h_2 \\ q_r &= h_1 - h_4 \\ w_t &= h_1 - h_2 \\ w_c &= h_3 - h_4 \\ w &= w_t + w_c \\ \text{cop} &= q_r/w \end{aligned}$$

Then, clicking the *Calculate* option and *Solve* gives the answers:

$$\begin{aligned} q_a &= 367.9 \text{ (Btu/lbm)} \\ q_r &= -454.1 \text{ (Btu/lbm)} \\ w &= -86.14 \text{ (Btu/lbm)} \\ \text{cop} &= 5.271 \end{aligned}$$

These answers agree well with those obtained by using the appendix table data.

7–5 THE SECOND LAW OF THERMODYNAMICS

We have considered the first law of thermodynamics, and from it came an elementary restriction on any device that we wish to analyze or design—energy cannot be created. It would be a tremendous feat to have an engine that would operate as diagrammed in figure 7–10, but this device is creating energy from nothing, and the first law tells us this is impossible. Engines such as that shown in figure 7–10 are called **perpetual-motion machines of the first kind** (PPM1), and occasionally we hear pronouncements that someone has invented one. They are impossible to achieve, however, because of the natural restrictions stated in the first law, and a simple system diagram can usually pinpoint the fallacy in any such supposed device.

There is, however, a more subtle restriction that we must contend with above and beyond the first law, contained in the second law:

> **Second law of thermodynamics:** ***No heat engine can produce a net work output by exchanging heat with a single fixed-temperature region.***

The effect of this categoric statement is that a restriction is placed on all heat engines, a restriction which requires that both Q_{add} and Q_{rej} be nonzero. You may have considered that to achieve 100% efficiency for the Carnot engine all that was required was for the heat rejected to be zero, or the heat sink to be at 0 K. That is,

$$\eta_T = \left(1 - \frac{T_L}{T_H}\right) \times 100 = 100\% \quad \textbf{(7–30)}$$

when $T_L = 0$. The second law now tells us that it is impossible to reach 100% efficiency for a heat engine because the heat rejected must be greater than zero. For the bulk of technological applications, we must be further restricted to a low-temperature sink that is at the atmospheric state. The term T_L in the efficiency equation (7–30) has a value of between 270 K and 320 K or between 460°R and 550°R for many common cycles.

The second law will not provide us with any new computational or bookkeeping techniques as the first law did, but its underlying usefulness cannot be overestimated. Suppose, for instance, that we have a reversible heat engine that is capable of producing net work from only one temperature region. This machine, called a **perpetual-motion machine of the second kind** (PMM2), would be converting 100% of the heat into work. In addition, because we need be concerned with only one temperature region, our heat engine can run at any lower temperature, and the efficiency of the engine (100%) is independent of everything. This type of device does not in any way violate the first law, yet it has never been achieved. It has a certain advantage over a PMM1 in that it can provide

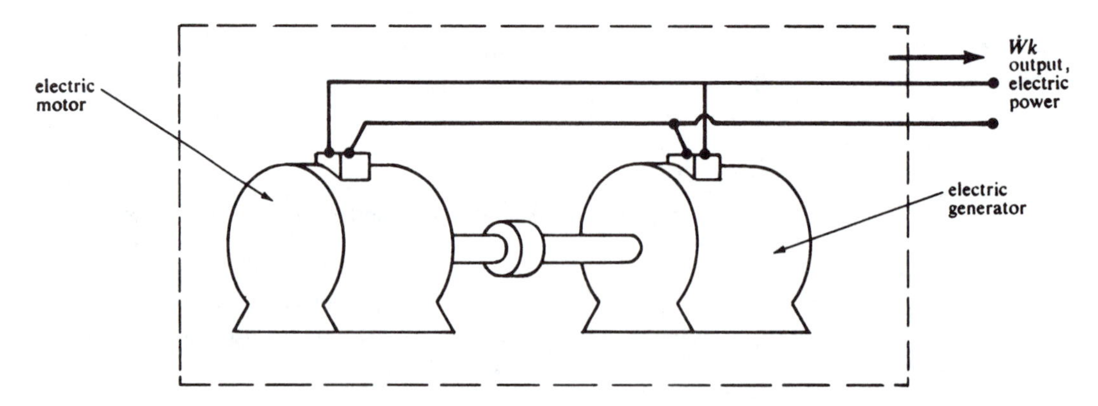

FIGURE 7–10 Example of a perpetual-motion machine

work and refrigeration by extracting heat from any temperature region. The PMM1 merely provided work from nothing. We thus are led to the conclusion that for a heat engine to operate within the laws of nature, it must be somehow dependent on the surroundings. The second law provides the answer.

Other statements of the second law are equivalent to the one given here. One of these alternatives is attributed to Clausius, who stated that it is impossible for a heat pump to have as its only effect the transfer of heat from a low- to a high-temperature region. (Thus, a refrigerator cannot operate without work supplied to it.) The equivalence of the two second-law statements can easily be shown. Suppose, as indicated in figure 7–11a, that we have a reversible heat pump requiring no work input and a reversible heat engine. The arrangement of figure 7–11a is then the same as that of figure 7–11b, and in this second system diagram there is a transfer of heat with only one temperature region. This is an obvious violation of the initial statement of the second law, so we see that Clausius's statement is its equivalent.

Another concept used to explain the second law of thermodynamics is the principle of entropy increase. If a heat engine is studied, the entropy change of everything involved in the operation of that heat engine is called the total entropy change, and it is equal to the entropy changes of the heat engine, the heat source, and the heat sink (or low-temperature region). We say that

$$\Delta S_{\text{total}} = \Delta S_{\text{cycle}} + \Delta S_{\text{source}} + \Delta S_{\text{sink}} \tag{7–31}$$

FIGURE 7–11 Hypothetical heat pump and engine combination

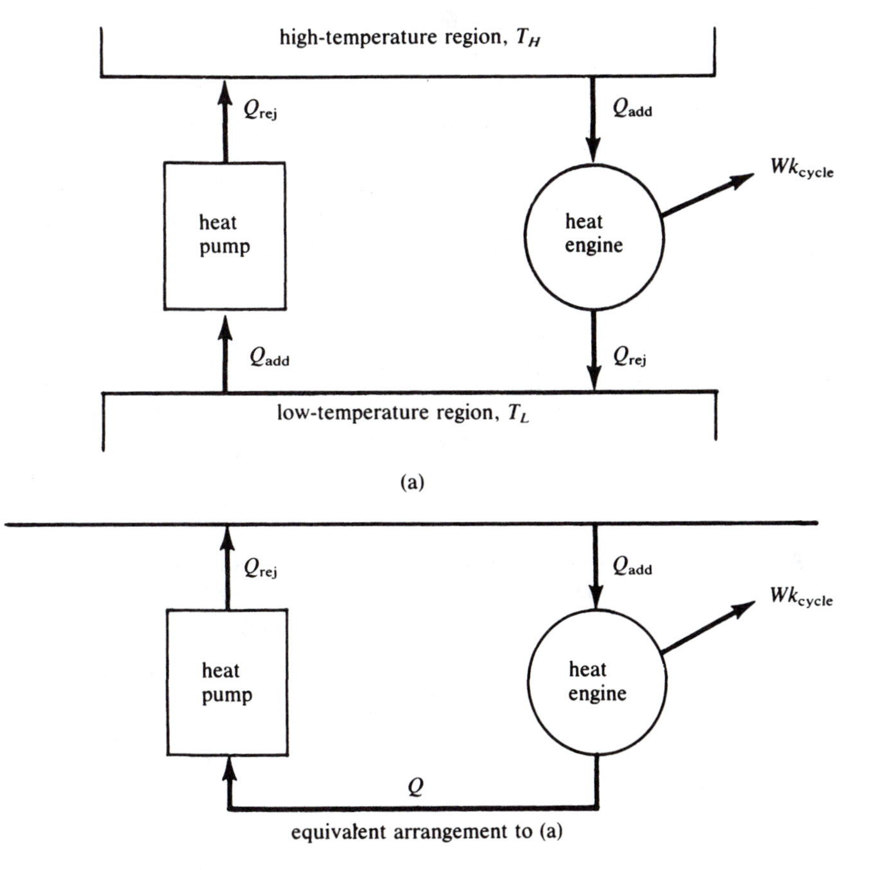

But for the heat engine, whether it is a Carnot heat engine or some real heat engine, $\Delta S_{\text{cycle}} = 0$. If the source and sink are at constant temperatures T_H and T_L, the entropy changes are

$$\Delta S_{\text{source}} = \frac{-Q_H}{T_H} \quad \text{and} \quad \Delta S_{\text{sink}} = \frac{-Q_L}{T_L}$$

but these are just the negative values of the entropy changes for the engine (particularly if the engine is a Carnot engine); that is, ΔS_{source} will be negative (because Q_H is taken positive to the engine), and ΔS_{sink} will be positive (because Q_L is taken negative to the engine and two negatives give a positive). Of course, if the engine is a reversible one such as a Carnot engine, the total entropy change is still zero because $\Delta S_{\text{source}} = -\Delta S_{\text{sink}}$. However, if the engine is a real engine having some efficiency less than a Carnot engine operating between the same two temperatures, the new Q_L will be larger than that for the Carnot, and the equality $Q_H/T_H = -Q_L/T_L$ is now an inequality with $-Q_L$ greater than $Q_H(T_L/T_H)$, or

$$-Q_L > Q_H \frac{T_L}{T_H}$$

From this result, we can see that the entropy change of the sink (which is positive) is larger than it would be if the engine were a Carnot or reversible heat engine. Therefore, equation (7–31) tells us that the total entropy change will be positive for real heat engine operations. How much the entropy change will be (in a positive sense) will be considered in many of the problems and discussion for the remainder of this book. The principle of the increase in entropy can then be written, from equation (7–31), as

$$\Delta S_{\text{total}} \geq 0 \tag{7–32}$$

where the equality holds if the engine is reversible and the inequality holds for real engines. The principle of the increase in entropy is sometimes used as a statement of the second law of thermodynamics. It is a fact that agrees with either of the two previous statements.

Finally, the principle of the increase in entropy can also be considered with a system involved in a process. Suppose that a system is at some constant temperature (T_a) and has a heat transfer (Q) with a region at some other constant temperature (T_b). Now the temperature T_a must be greater than T_b if the heat flows from the system. Then, the total entropy change is $-Q/T_a + Q/T_b$. The negative sign is put on the first term (which is the entropy change of the system) because the heat is flowing out of the system, and the result of this sum is a value greater than zero (positive). This occurs because the second term (which is positive) is larger in magnitude than the first, since T_b is less than T_a.

If you consider the situation where the heat flows into the system, then T_b must be larger than T_a, Q is positive to the system, and again the total entropy change will be positive.

EXAMPLE 7–8 A heat engine operates between 1000°R and 500°R and produces 5000 ft · lbf of energy. If the thermal efficiency is 35%, find the total entropy change and state whether the engine violates the second law of thermodynamics.

Solution Since the net work and thermal efficiency are given, we use equation (7–21) to find the heat added:

$$Q_{\text{add}} = \frac{Wk_{\text{net}}}{\eta_T} = \frac{5000 \text{ ft} \cdot \text{lbf}}{0.35}$$

$$= 14{,}285.7 \text{ ft} \cdot \text{lbf} = \frac{14{,}285.7}{778} \text{ Btu}$$

$$= 18.36 \text{ Btu}$$

Using equation (7–1) for the heat engine, we find the rejected heat:

$$Q_{rej} = Wk_{net} - Q_{add}$$
$$= \frac{5000}{778\ \text{Btu}} - 18.36\ \text{Btu}$$
$$= -11.93\ \text{Btu}$$

The total entropy change is then

$$\Delta S_{total} = \Delta S_{source} + \Delta S_{sink}$$
$$= \frac{-18.36\ \text{Btu}}{1000°\text{R}} + \frac{11.93\ \text{Btu}}{500°\text{R}}$$

Answer

$$= 0.0055\ \text{Btu}/°\text{R}$$

The entropy change is positive, and the heat engine interacts with more than one temperature region, so the engine does not violate the second law of thermodynamics. Notice that, in this example, a Carnot heat engine would have had an efficiency given by the equation

$$\eta_T = 1 - \frac{T_L}{T_H} = 1 - \frac{500°R}{1000°R} = 50\%$$

The fact that the efficiency of the engine was less than that for a Carnot engine is another indication that the engine does not violate the second law.

EXAMPLE 7–9 A piston-cylinder is used to compress a gas at constant temperature. If the gas is at 450 K and its surroundings are at 300 K, determine the total rate of entropy change if 240 W of heat transfer occurs such that the second law is not violated.

Solution The system, which is a gas enclosed in a piston-cylinder, transfers heat with a surrounding. The total rate of entropy change can be written

$$\dot{S}_{total} = \dot{S}_{system} + \dot{S}_{surroundings} \geq 0$$

and if we assume that the heat is flowing from hot to cold,

$$\dot{S}_{total} = \frac{-240\ \text{W}}{450\ \text{K}} + \frac{240\ \text{W}}{300\ \text{K}}$$
$$= +0.267\ \text{W/K} > 0$$

If we had assumed that heat flowed from cold to hot, the rate of entropy change would have been

$$\dot{S}_{total} = \frac{+240\ \text{W}}{450\ \text{K}} - \frac{240\ \text{W}}{300\ \text{K}}$$
$$= -0.267\ \text{W/K}$$

and this is a violation of the second law of thermodynamics. It also violates our intuition, which insists that heat flow from hot to cold.

7–6 ENTROPY AND REVERSIBILITY

Entropy is a property that seems to appear suddenly in thermodynamics, and although it is helpful in explaining the second law, it does not seem to relate to any physical property. I am sure that most of the properties we have been considering are better understood and accepted after you have associated them with some personal experiences: temperature measures hotness, pressure measures an intensity of force, energy measures the potential effects, and so on. But entropy is abstract because it seems not to relate to a basic experience. Perhaps the following discussion will help in your understanding of entropy.

It is based on other persons' observations and interpretations of the concept of entropy and so is limited. Perhaps you will get other insights into this very important property.

Entropy is a measure of unavailable energy in a system; the greater the entropy of a system, the less available is that system's energy for doing work or transferring heat. In mechanical systems, we have seen that as a system increases in volume, it can perform work, but thereby also has a reduced capability to perform other work. It seeks an equilibrium in pressure with its surroundings, and when it expands in volume so that the pressures are equal, the system cannot perform additional work. Thus, a system at high pressure (with respect to its surroundings) has a lower entropy than does the same system upon equilibrating its pressure with the surroundings. Also, if we have a system at a high temperature, it will tend to reach a temperature equal to the surroundings. Here the volume may not change, but entropy will. The high-temperature system can perform work or heat processes and has a low entropy; but after reducing its temperature, it does not have the same capability—it has greater entropy.

In more general terms, entropy is a measure of disorder or randomness. Suppose that we have 10 red marbles and 10 green marbles in separate boxes. All the red are in one box and all the green in another, so we have an ordered system of two boxes. We place the boxes on a high shelf so that we can withdraw the marbles but cannot look in the boxes. Thus, if we wish to get a red marble, we reach up and remove a marble from one of the boxes. If it is green, we are assured of a red marble by selecting a second from the other box. Here we have a system that is ordered and has a low entropy. But we now mix the red and green marbles in one box and return the box to the shelf. The system is well mixed and disordered and has a high entropy. If we want to get a red marble, we are never assured that we will pick one. Perhaps, if we are lucky on the first try, we will get a red marble, but we could also get 10 green ones before we would get one red one. In a less abstract manner, suppose that we have a system composed of pure hydrogen gas (H_2) and pure oxygen gas (O_2), each in a separate container. The system has a low entropy, it is well ordered or well arranged, and it can easily be mixed to produce water. Also, we could get work or electrical energy from it if we wished. However, if we mix the hydrogen and oxygen and produce water (H_2O), we immediately have a higher entropy. The system is less ordered or arranged (we no longer know exactly where the oxygen atoms are), and it is only water. We cannot get work or electrical energy from it anymore. We can thus say that a low entropy implies a broader capability for energy systems, and if we perform reversible processes only, the world will retain this low entropy. The world, as we are reminded by environmentalists, is increasing its entropy, and the choices of power supplies are indeed not as many today as some persons would have us believe.

At a more applied level, we have seen that the total entropy change must be either zero or some positive value for a heat engine or for a process. If the entropy change is zero, the process is reversible, and it can be returned to its original state in a spontaneous manner, that is, without any effort. Also, if a heat engine or cyclic device has no total entropy change, we can assume that the device is reversible—ideal, so to speak. For irreversible processes or irreversible heat engines, the entropy change will be positive for all cases. Clearly, then, entropy change is a good indicator as to whether or not a process is reversible.

7–7 ENTROPY CHANGES

Entropy has been introduced and defined through the change of entropy, ΔS. In fact, the absolute value of entropy cannot be found. Only a relative value can be determined, such as by arbitrarily defining entropy to be zero at some convenient temperature or state and then computing the change in entropy from that zero value. Therefore, the change in entropy will be the important quantity that we will want to determine. Also, for processes or for heat engines or heat pumps, the change in entropy during processes or a process will be more useful than some absolute value at a given state. If you think back to energy, it was the change in energy that allowed for work and heat and not the absolute value of energy at some state. So entropy and energy seem to have much in common, especially regarding analysis. The change in entropy for isothermal processes has been discussed before and is repeated here. The equation used to compute such a change is

$$\Delta S = \frac{Q_{\text{rev}}}{T} \tag{7–33}$$

Most engineering problems involve processes during which the temperature does not remain constant, and to consider entropy changes in those cases, we recall the definition of entropy change:

$$\Delta S = \sum \frac{\delta Q_{\text{rev}}}{T} \tag{7–34}$$

A very small entropy change is

$$\delta S = \frac{\delta Q_{\text{rev}}}{T} \tag{7–35}$$

and from the first law of thermodynamics, the small heat change δQ is given by

$$\delta Q = \delta U + \delta Wk$$

But the term δWk is the boundary work, $p\,\delta V$, so equation (7–35) becomes

$$\delta S = \frac{\delta U}{T} + \frac{p}{T}\delta V \tag{7–36}$$

and a finite change in the entropy is then

$$\Delta S = \sum \frac{\delta U}{T} + \sum \frac{p}{T}\delta V \tag{7–37}$$

It can be shown that, for perfect gases with constant specific heat, c_v, equation (7–37) becomes

$$\Delta S = mc_v \ln \frac{T_2}{T_1} + \text{mR} \ln \frac{V_2}{V_1} \tag{7–38}$$

CALCULUS FOR CLARITY 7–3

A differential amount of entropy change was defined by equation (7–10) as

$$dS = \frac{dQ_{\text{rev}}}{T}$$

and from an energy balance, the differential heat is

$$dQ_{\text{rev}} = dU + dWk_{cs} \tag{7–39}$$

Using equations (7–10) and (3–11), we find that equation (7–39) becomes

$$T\,dS = dU + p\,dV \tag{7–40}$$

This equation relates the properties of a system to each other, and, although it arises from the use of reversible processes for heat and work, it applies to all processes. The differential entropy change is, from equation (7–40),

$$dS = \frac{1}{T}dU + \frac{p}{T}dV$$

CALCULUS FOR CLARITY 7–3, continued

For a finite change,

$$\int dS = \int \frac{1}{T}\, dU + \int \frac{p}{T}\, dV \qquad \textbf{(7–41)}$$

For a perfect gas, $p/T = mR/V$ and $dU = mc_v\, dT$, so

$$\int dS = \int \frac{1}{T} mc_v\, dT + \int \frac{mR}{V}\, dV$$

or

$$\int dS = \int \frac{1}{T} mc_v\, dT + mR \ln \frac{V_2}{V_1} \qquad \textbf{(7–42)}$$

If the specific heat c_v is constant, the change in entropy becomes

$$\int dS = \Delta S = m\left[c_v \ln \frac{T_2}{T_1} + R \ln \frac{V_2}{V_1}\right]$$

which is equation (7–38). Equation (7–40) is sometimes called a ***T–dS* equation**.

Equation (7–38) is a general equation good for perfect gases with constant specific heat values. It is an equation that can be used to determine entropy changes for any process of a perfect gas with constant specific heat. Even for the isothermal process, you may use equation (7–38). In this situation, where temperature is constant ($T_1 = T_2$), we have $\ln(T_2/T_1) = \ln 1 = 0$, and then equation (7–38) becomes

$$\Delta S = mR \ln \frac{V_2}{V_1} \qquad \text{isothermal process} \qquad \textbf{(7–43)}$$

If you will recall, for isothermal processes of a perfect gas, the heat, Q, is $mrT \ln(V_2/V_1)$, and from equation (7–33), we see that, dividing Q by temperature, we obtain equation (7–43) again.

Another equation that gives the entropy change of perfect gases with constant specific heats, one good for any process, is

$$\Delta S = mc_p \ln \frac{T_2}{T_1} - mR \ln \frac{p_2}{p_1} \qquad \textbf{(7–44)}$$

If the specific heat c_p is not constant, we use, not equation (7–44), but rather equation

$$\Delta S = m \sum \frac{c_p\, \delta T}{T} - mR \ln \frac{p_2}{p_1} \qquad \textbf{(7–45)}$$

where the first term on the right, $m \sum c_p\, \delta T/T$, is defined as $m(\phi_2 - \phi_1)$, and then

$$\Delta S = m(\phi_1 - \phi_1) - mR \ln \frac{p_2}{p_1} \qquad \textbf{(7–46)}$$

CALCULUS FOR CLARITY 7–4

Equation (7–40) may be revised by noting that the differential internal energy is

$$dU = dH - dpV = dH - p\, dV - V\, dp$$

Then equation (7–40) becomes

$$T\, dS = dH - p\, dV - V\, dp + p\, dV$$

CALCULUS FOR CLARITY 7–4, continued

or

$$T\,dS = dH - V\,dp \tag{7–47}$$

This equation is often referred to in thermodynamic discussions as the **second *T–dS* equation**. From equation (7–47), the differential entropy change is

$$dS = \frac{1}{T}\,dH - \frac{V}{T}\,dp$$

and a finite change is

$$\int dS = \Delta S = \int \frac{1}{T}\,dH - \int \frac{V}{T}\,dp \tag{7–48}$$

For a perfect gas, this becomes equation (7–46),

$$\Delta S = \int \frac{1}{T} mc_p\,dT - mR \ln \frac{p_2}{p_1}$$

where the integral $\int [1/T] c_p\,dT$ is the exact definition for **encrety**. If the specific heat c_p is constant, then equation (7–46) becomes equation (7–44).

Thermodynamic tables frequently include the values for ϕ, and in table B–6 the last column presents the data for ϕ for air at low pressures such that it is a perfect gas. Notice in that table that only the temperatures need to be known to determine values for ϕ. The ϕ function is sometimes called **encrety**.

If entropy changes per unit mass are needed, then, for perfect gases with constant specific heats, we have

$$\Delta s = c_v \ln \frac{T_2}{T_1} + R \ln \frac{v_2}{v_1} \tag{7–49}$$

and

$$\Delta s = c_p \ln \frac{T_2}{T_1} - R \ln \frac{p_2}{p_1} \tag{7–50}$$

For perfect gases with variable specific heats,

$$\Delta s = \phi_2 - \phi_1 - R \ln \frac{p_2}{p_1} \tag{7–51}$$

These equations were introduced earlier in table 6–4.

EXAMPLE 7–10 At constant pressure, 0.8 lbm of carbon dioxide is expanded from 10 ft^3 to 30 ft^3. Determine the entropy change and the specific entropy change.

Solution The system of this problem is carbon dioxide, and the process is occurring at constant pressure. The entropy change can be determined from equation (7–44) if the carbon dioxide is assumed to be a perfect gas with constant specific heats. Since $p_1 = p_2$, equation (7–44) becomes

$$\Delta S = mc_p \ln \frac{T_2}{T_1}$$

The specific heat, c_p, is 0.2015 Btu/lbm · °R from table B–4. Also, since the process was isobaric (constant pressure) involving a perfect gas, the temperature ratio is equal to the volume ratio; that is, $T_2/T_1 = V_2/V_1$, and the volumes are given in the problem statement. The entropy change is then

$$\Delta S = (0.8\ \text{lbm})(0.2015\ \text{Btu/lbm}\cdot{}^\circ\text{R})\left(\ln \frac{30\ \text{ft}^3}{10\ \text{ft}^3}\right)$$

Answer

$$= 0.177\ \text{Btu/}^\circ\text{R}$$

The specific entropy change (or change per unit mass) is

$$\Delta s = \frac{\Delta S}{m} = \frac{(0.177\ \text{Btu/}^\circ\text{R})}{0.8\ \text{lbm}}$$

Answer

$$= 0.221\ \text{Btu/lbm}\cdot{}^\circ\text{R}$$

EXAMPLE 7–11 Air expands polytropically through a nozzle such that the exponent n is 1.45. The exhaust pressure of the air is 102 kPa, and the temperature is 90°C. If the inlet pressure is 400 kPa gage, determine the change in specific entropy of the air as it passes through the nozzle.

Solution We assume that air is a perfect gas with constant specific heats. Then, for any polytropic process, we can use equation (7–50):

$$\Delta s = c_p \ln \frac{T_2}{T_1} - R \ln \frac{p_2}{p_1}$$

From table B–4, we then find that

$$c_p = 1.007\ \text{kJ/kg}\cdot\text{K}$$

The gas constant R is found to be

$$R = 287\ \text{J/kg}\cdot\text{K} = 0.287\ \text{kJ/kg}\cdot\text{K}$$

Also, the pressure ratio p_2/p_1 is $102\ \text{kPa}/(400 + 101\ \text{kPa})$, assuming an atmospheric pressure of 101 kPa. Then

$$\frac{T_2}{T_1} = \left(\frac{p_2}{p_1}\right)^{(n-1)/n}$$

from equation (6–14). This gives us

$$\frac{T_2}{T_1} = \left(\frac{102}{501}\right)^{0.45/1.45} = 0.610$$

Consequently, the specific entropy change can now be determined from equation (7–50):

$$\Delta s = (1.007\ \text{kJ/kg}\cdot\text{K}) \ln(0.610) - (0.287\ \text{kJ/kg}\cdot\text{K}) \ln(0.204)$$
$$= -0.498\ \text{kJ/kg}\cdot\text{K} + 0.456\ \text{kJ/kg}\cdot\text{K}$$

Answer

$$= -0.042\ \text{kJ/kg}\cdot\text{K}$$

EXAMPLE 7–12 Consider example 7–11, where the variation in the specific heat values are accounted for.

Solution The entropy change for the air is now given by equation (7–51):

$$\Delta s = \phi_2 - \phi_1 - \ln \frac{p_2}{p_1}$$

We will find the two ϕ values from table B–6, but first we must determine the temperatures. At state 1 the temperature is 90°C or 363 K, and at state 2 the temperature will be

$$T_2 = T_1(0.610)$$

where the polytropic equation, $pv^n = c$, is assumed to hold. Then $T_2 = 221.43$ K, and from table B–6,

$$\phi_1 = 2.7071 \qquad \text{(interpolated between 360 K and 380 K)}$$
$$\phi_2 = 2.2103 \qquad \text{(interpolated between 220 K and 240 K)}$$

Then the entropy change is

Answer

$$\Delta s = 2.2103 - 2.7071 - (0.287 \text{ kJ/kg}\cdot\text{K}) \ln(0.204)$$
$$= -0.04082 \text{ kJ/kg}\cdot\text{K}$$

In examples 7–11 and 7–12, notice how close is the agreement between the result assuming constant specific heat values and that for variable specific heat. Also, notice that the entropy change of the air has a negative value. In order that the second law not be violated, there must be an entropy increase of the environment at least as great as the magnitude of the entropy change for the air passing through the nozzle.

Let us now consider incompressible materials and the entropy changes associated with those substances. The entropy changes can be determined from equation (7–37). The internal energy change of an incompressible substance, δU, is equal to $mc\,\delta T$, and the volume change, δV, is zero, so that equation (7–37) can be shown to reduce to

$$\Delta S = mc \ln \frac{T_2}{T_1} \tag{7–52}$$

because the term $mc \sum \delta T/T = mc \ln(T_2/T_1)$. The entropy change per unit mass for incompressible materials is the entropy change divided by the mass:

$$\Delta s = c \ln \frac{T_2}{T_1} \tag{7–53}$$

EXAMPLE 7–13 Ten pounds-mass of water at 70°F is mixed with 15 lbm of water at 160°F. If the mixing is done adiabatically, determine the entropy change.

Solution

The system for this problem is 10 lbm of water and 15 lbm of water. Since the mixing is adiabatic, the first law of thermodynamics or conservation of energy is

$$\Delta U = 0$$

Assuming that the water is incompressible and has constant specific heat $c = 1.0$ Btu/lbm · °R, the energy change is

$$mc(T_2 - 70°\text{F})_{\text{cold}} + mc(T_2 - 160°\text{F})_{\text{hot}} = 0$$
$$(10 \text{ lbm})(1.0 \text{ Btu/lbm}\cdot°\text{R})(T_2 - 70°\text{F}) + (15 \text{ lbm})(1.0 \text{ Btu/lbm}\cdot°\text{R})(T_2 - 160°\text{F}) = 0$$

Solving for T_2, we have

$$10T_2 - 700 + 15T_2 - 2400 - = 0$$
$$25T_2 = 3100$$
$$T_2 = 124°\text{F} = 584°\text{R}$$

The entropy change is the sum of the entropy changes for the cold and hot water:

$$\Delta S_{cold} = (10 \text{ lbm})(1.0 \text{ Btu/lbm} \cdot {}^\circ\text{R}) \ln \frac{584^\circ\text{R}}{530^\circ\text{R}}$$
$$= 0.970 \text{ Btu/}^\circ\text{R}$$

$$\Delta S_{hot} = (15 \text{ lbm})(1.0 \text{ Btu/lbm} \cdot {}^\circ\text{R}) \ln \frac{584^\circ\text{R}}{620^\circ\text{R}}$$
$$= -0.8973 \text{ Btu/}^\circ\text{R}$$

The total entropy change is then

$$\Delta S = +0.970 \text{ Btu/}^\circ\text{R} - 0.8973 \text{ Btu/}^\circ\text{R}$$

Answer
$$= 0.0727 \text{ Btu/}^\circ\text{R}$$

EXAMPLE 7–14 A 2-kg copper block is heated to 110°C from 20°C. The heat transfer occurs with a large oil bath at 150°C. Determine the entropy change of the copper block and of the oil bath, and find the total entropy change.

Solution Assuming that the copper block is an incompressible substance with constant specific heat $c = 0.385$ kJ/kg·K (table B–8), the entropy change of the copper is then given by equation (7–52):

$$\Delta S_{copper} = mc \ln \frac{T_2}{T_1} = (2 \text{ kg})(0.385 \text{ kJ/kg} \cdot \text{K}) \ln \left(\frac{383 \text{ K}}{293 \text{ K}}\right)$$

Answer
$$= 0.2063 \text{ kJ/K}$$

Since the oil bath is large, we assume that its temperature will be constant, and the entropy change of the oil bath is then

$$\Delta S_{oil} = \frac{Q}{T}$$

The heat, Q, is equal to the heat into the copper but with a negative sign because it is leaving the oil bath. Then

$$Q = -mc(T_2 - T_1) = -(2 \text{ kg})(0.385 \text{ kJ/kg} \cdot \text{K})(90 \text{ K})$$
$$= -69.3 \text{ kJ}$$

and the entropy change is

$$\Delta S_{oil} = \frac{-69.3 \text{ kJ}}{423 \text{ K}}$$

Answer
$$= -0.1638 \text{ kJ/K}$$

The total entropy change is

$$\Delta S_{total} = \Delta S_{copper} + \Delta S_{oil}$$

Answer
$$= 0.0425 \text{ kJ/K}$$

Entropy changes of pure substances are determined from the equation

$$\Delta S = S_2 - S_1 = m(s_2 - s_1) \quad \textbf{(7–54)}$$

where s_2 and s_1 are found from tables of the pure substances at known states.

7–8 THE ISENTROPIC PROCESS

For the isentropic process of substances, the entropy change is zero, or the entropy is constant for the process. As we saw in earlier examples, the entropy is also constant for reversible adiabatic processes. Therefore, the isentropic process and the reversible adiabatic process may be considered to be the same. There are some instances when an isentropic process may not be reversible adiabatic, but all reversible adiabatic processes are isentropic. In this book, we use the two terms to mean the same sort of process—one where entropy is a constant.

For an isentropic process involving a perfect gas having constant specific heats, the entropy change equation (7–50) becomes

$$0 = c_p \ln \frac{T_2}{T_1} - R \ln \frac{p_2}{p_1}$$

and using the relationships $R = c_p - c_v$ and $k = c_p/c_v$, we can show that

$$\frac{T_2}{T_1} = \left(\frac{p_2}{p_1}\right)^{(k-1)/k} \tag{7–55}$$

After substituting for the two temperatures $T = pv/R$, we also find that

$$p_2 v_2^k = p_1 v_1^k \quad \text{or} \quad pv^k = C \tag{7–56}$$

Equation (7–56) is used often to describe isentropic or reversible adiabatic equations. It is applicable only for perfect gases with constant specific heats.

EXAMPLE 7–15 A reversible adiabatic nozzle allows air to expand from 600 kPa, 700 K, to 100 kPa. If the air is entering the nozzle at 100 m/s, determine the temperature and the air velocity at the exit.

Solution We use as a system the nozzle and assume steady-flow conditions for the air. Figure 7–12 shows the nozzle and the inlet state (1) and outlet (2). We assume the air to be a perfect gas with constant specific heats. Then the exit temperature may be found directly from equation (7–55), with a value of 1.4 for k:

$$T_2 = T_1\left(\frac{p_2}{p_1}\right)^{(k-1)/k}$$

$$= (700\ \text{K})\left(\frac{100\ \text{kPa}}{600\ \text{kPa}}\right)^{(1.4-1)/1.4}$$

Answer

$$= 419.5\ \text{K}$$

FIGURE 7–12 Converging nozzle

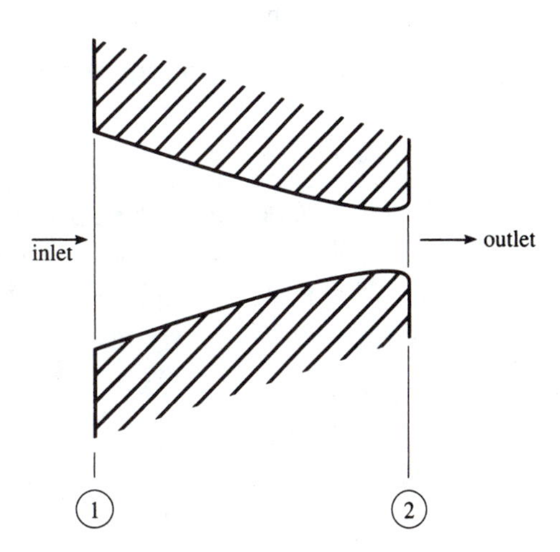

Using the steady-flow energy equation, we have

$$h_1 + \frac{\bar{V}_1^2}{2} = h_2 + \frac{\bar{V}_2^2}{2}$$

and because air is assumed to be a perfect gas with constant specific heats, the enthalpies may be replaced by c_pT, so

$$c_pT_1 + \frac{\bar{V}_1^2}{2} = c_pT_2 + \frac{\bar{V}_2^2}{2}$$

The velocity at the exit, $\bar{V}_2$, is then, after some algebra,

$$\bar{V}_2 = \sqrt{c_p(T_1 - T_2)(2) + \bar{V}_1^2}$$
$$= \sqrt{(1.007\ \text{kJ/kg}\cdot\text{K})(700 - 419.5\ \text{K})(2) + (100\ \text{m/s})^2}$$

The units of the velocity-squared term are m^2/s^2, which is equivalent to J/kg (you may want to check this conversion). We convert the first term, the enthalpy change term, to J/kg by multiplying by 1000 J/kJ. The result is

Answer

$$\bar{V}_2 = \sqrt{564{,}927\ \text{m}^2/\text{s}^2 + 10{,}000\ \text{m}^2/\text{s}^2} = 758.2\ \text{m/s}$$

You may want to notice that the velocity at the inlet does not affect the answer very much. If the inlet velocity were assumed to be zero, the outlet velocity would have been $\sqrt{564{,}927\ \text{m}^2/\text{s}^2}$ or 751.6 m/s, which is within 1% (0.8%) of the correct answer. Therefore, for most nozzle problems, you can solve for the exit velocity by neglecting the inlet velocity and using the equation

$$\bar{V}_2 = \sqrt{(h_1 - h_2)(2)(1000\ \text{J/kJ})}$$

where h is in kJ/kg.

Isentropic processes involving pure substances are important processes that are frequently encountered in thermodynamic analysis. We have already seen in chapter 6 that for the reversible adiabatic process of pure substances, setting the entropy at the end states of the process represented an important step in the solution of the problems. Here we consider further examples of those types of problems.

EXAMPLE 7–16 Steam at 2000 kPa and 560°C is expanded reversibly and adiabatically through a steam turbine until the pressure is 200kPa. Determine the final steam temperature.

Solution The entropy change for this process is zero, so

$$s_2 = s_1$$

and from table B–12, we find that steam at 2000 kPa and 560°C has a specific entropy, s, of 7.596 kJ/kg · K. Therefore, the final specific entropy has this same value, and at 200 kPa, the temperature must be between 160°C and 240°C, as indicated in table B–12. We will interpolate to determine the precise value:

$$s = 7.324\ \text{kJ/kg}\cdot\text{K} \quad \text{at } 160°\text{C}$$
$$= 7.663\ \text{kJ/kg}\cdot\text{K} \quad \text{at } 240°\text{C}$$

Then

$$\Delta S = 0.339\ \text{kJ/kg}\cdot\text{K} \quad \text{with } \Delta t = 80°\text{C}$$

Since $s_2 = 7.596\ \text{kJ/kg}\cdot\text{K}$, we write

$$\frac{7.596\ \text{kJ/kg}\cdot\text{K} - 7.324\ \text{kJ/kg}\cdot\text{K}}{0.339\ \text{kJ/kg}\cdot\text{K}} = \frac{t_2 - 160°\text{C}}{80°\text{C}}$$

$$\frac{0.272}{0.339} \times 80°\text{C} + 160°\text{C} = t_2$$

yielding the result

Answer

$$t_2 = 224°\text{C}$$

EXAMPLE 7–17 A steam nozzle directs steam against turbine blades. The steam enters the nozzle (shown in figure 7–13) at 350 psia and 500°F. The pressure at the nozzle outlet is 300 psia. Neglecting the inlet velocity of the steam, determine the outlet velocity.

Solution

The open system is assumed to be in steady-flow conditions, and we solve, from the steady-flow energy equation, for the outlet velocity, $\bar{V}_2$ (using the English unit form). We have

$$h_1 = h_2 + \frac{\bar{V}_2^2}{2g_c}$$

or

$$\bar{V}_2 = \sqrt{(h_1 - h_2)(2)(g_c)}$$

From table B–12, we find $h_1 = 1251.5\ \text{Btu/lbm}$ and $s_1 = 1.5483\ \text{Btu/lbm}\cdot°\text{R}$. We assume that the nozzle flow is reversible and adiabatic or isentropic, so that $s_2 = s_1 = 1.5483\ \text{Btu/lbm}\cdot°\text{R}$. From table B–12, we see that at 300 psia the entropy of saturated vapor is 1.5105 Btu/lbm · °R and is 1.5703 at 500°F. Therefore, the steam leaves the nozzle between saturated vapor and 500°F (superheated). It is still superheated as it leaves the nozzle. By a linear interpolation between 417.35°F (saturated vapor), where h_g is 1202.9 Btu/lbm, and 500°F, where h is 1257.7 Btu/lbm, we find that $h_2 = 1237.5\ \text{Btu/lbm}$. We then solve for the velocity:

$$\bar{V}_2 = \sqrt{(1251.5 - 1237.5\ \text{Btu/lbm})(2)(32.17\ \text{ft}\cdot\text{lbm/lbf}\cdot s^2)}$$

The Btu units need to be converted to ft · lbf to give a final unit of ft/s. We use the conversion 778 ft · lbf/Btu and obtain

Answer

$$\bar{V}_2 = \sqrt{(900.76\ \text{Btu}\cdot\text{ft/lbf}\cdot\text{s}^2)(778\ \text{ft}\cdot\text{lbf/Btu})}$$
$$= 837.1\ \text{ft/s}$$

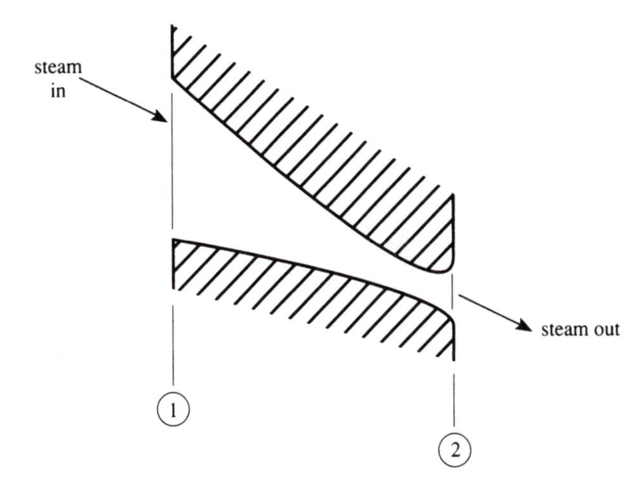

FIGURE 7–13 Steam nozzle for example 7–16

7–9 THE THIRD LAW OF THERMODYNAMICS

In calculating entropy changes due to processes, temperature directly affects the entropy function. A decrease in temperature will induce a decrease in entropy, and an increase in temperature will induce increased entropy. We might ask: How long can entropy be? The answer is given by the third law of thermodynamics:

> **Third law of thermodynamics:** ***Entropy tends to a minimum constant value as temperature tends to absolute zero. For a pure element this minimum value is zero, but for all other substances it is not less than zero, but possibly more.***

This third statement or law is a result of experimentation in the temperature regime near absolute zero and has not been violated; therefore, it is considered a "law." From a practical viewpoint, it tells us that it is impossible to reach an absolute zero temperature by other than a reversible process, because, near the zero point (as illustrated in figure 7–14), the change in entropy is zero and the only irreversible means of lowering entropy further is to have a surrounding that is cooler than absolute zero. This is impossible, so the final approach to absolute zero for cooling any material must be reversible and adiabatic (isentropic). Another manner of observing this is to take the definition, equation (7–5), and set ΔS equal to zero:

$$\Delta S = \sum \frac{\delta Q}{T} = 0$$

In figure 7–14, for substance A near absolute zero, point 1, the entropy s_1 is equal to the entropy when the temperature will be zero, s_0. Then

$$\Delta s = s_0 - s_1 = 0$$

and for this equation to be satisfied, δQ must be zero (adiabatic), or else the term $\delta Q/T$ will approach infinity, which is impossible. The process from 1 to 0 must therefore be adiabatic and reversible.

The point of this discussion is that the state of absolute zero temperature is a highly desirable one for making energy available, but it is impossible to achieve. Materials close to zero temperature are, however, attractive as sources of heat or work.

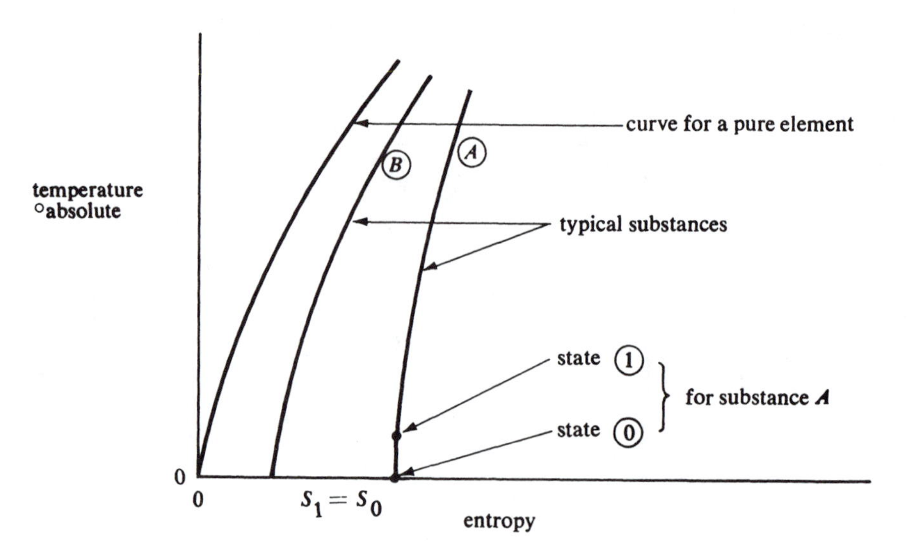

FIGURE 7–14 Behavior of entropy near absolute zero of temperature

7–10 CARNOT CYCLE ANALYSIS

The Carnot heat engine and Carnot heat pump are idealized mechanical devices operating through the Carnot cycle, which we have been using to help explain the operation of actual heat engines and refrigerators. In this section, we show how you may go about analyzing the Carnot cycle so that you can predict the thermal efficiency, net work, input and output heat, and other details of the Carnot heat engine. If the type of working substance is known, you can also determine such properties as the pressures, temperatures, volumes, and entropies at various times in the cycle. We will assume that the heat engine or heat pump uses a perfect gas as the working medium and that the gas has constant specific heats.

Let us consider, as a possible heat engine, a piston-cylinder device as shown in figure 7–15. In this engine, a perfect gas is contained in the cylinder, and heat is added (position 3) so that the gas expands, does work, and pushes the piston down. As the piston reaches the bottom and starts to move up (position 1), heat is removed from the gas, and this lowers the pressure of the gas so that the piston can more easily return to the top position. In the figure, air is assumed to be the working medium, although any gas would work. Figure 7–16 shows the system diagram for the engine, which is the same as that shown in figure 7–1. Also, the Carnot cycle defined in section 7–2 is repeated here:

1–2 Reversible isothermal compression at temperature T_L

2–3 Reversible adiabatic compression from the low temperature T_L to a higher temperature T_H

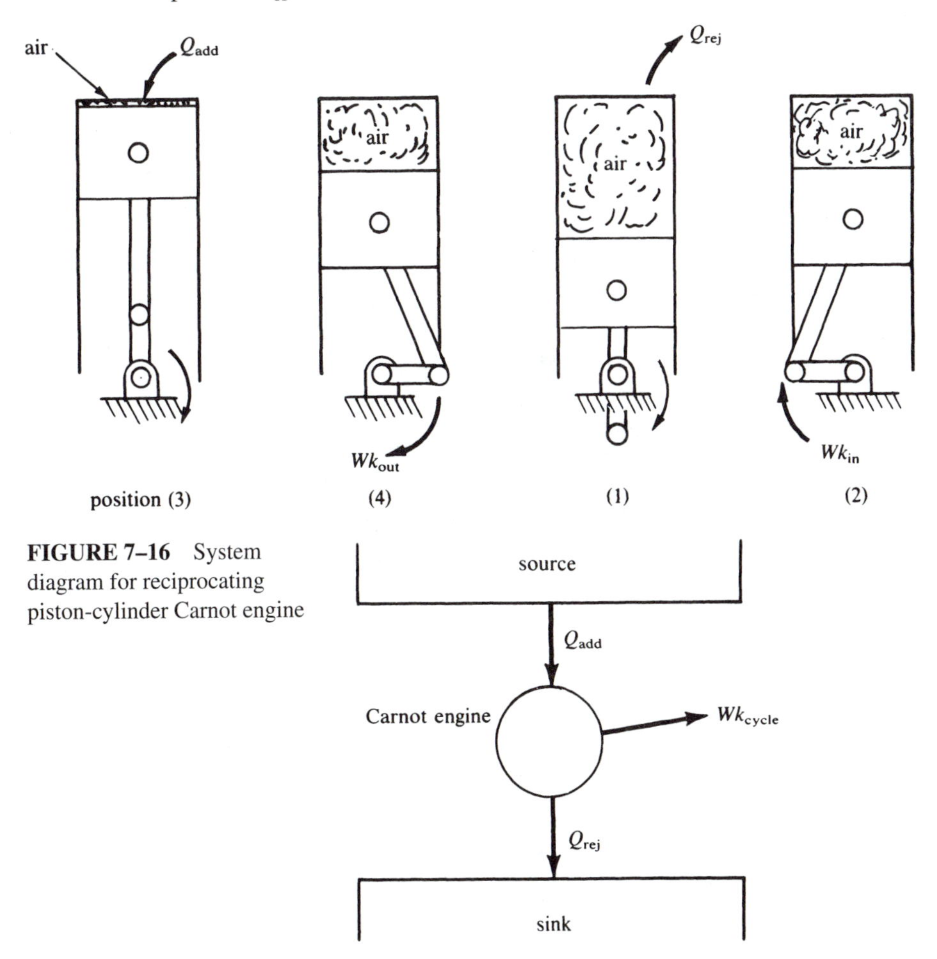

FIGURE 7–15 Carnot engine using the piston-cylinder mechanism

FIGURE 7–16 System diagram for reciprocating piston-cylinder Carnot engine

3–4 Reversible isothermal expansion at temperature T_H

4–1 Reversible adiabatic expansion from temperature T_H to T_L

The states correspond to those given in figure 7–15, and the processes are the same as those shown plotted on the p–V diagram in figure 7–3. The p–V diagram is repeated in figure 7–17 along with a T–S diagram of the Carnot cycle. Two particular problems are instructive in understanding how to predict a more detailed description of the Carnot engine or heat pump operating on the Carnot cycle: (1) given the pressure, temperature, and volume of the perfect gas at state 1 and the pressure (or temperature) and volume at state 3, determine the properties at all four states, the various work and heat terms, and the entropy changes; and (2) given the pressure, temperature, and volume at state 1, the amount of heat added, and the high temperature at which heat is added, determine the properties at all four states, the various work and heat terms, and the entropy changes. For either of these two problems, the sets of equations that follow may be used, since the perfect gas/constant specific heats assumptions are made. Given p_1, T_1, V_1, V_3, p_3 (or T_3), R, and c_v, it follows that $c_p = R + c_v$ and $k = c_p/c_v$, and we have the following states:

State 1: $$m = \frac{p_1 V_1}{RT_1}$$

State 2: $$T_2 = T_1 \qquad T_3 = \frac{p_3 V_3}{mR}$$

$$p_2 = p_3\left(\frac{T_2}{T_3}\right)^{k/(k-1)}$$

$$V_2 = \frac{mRT_2}{p_2} = V_3\left(\frac{p_3}{p_2}\right)^{1/k}$$

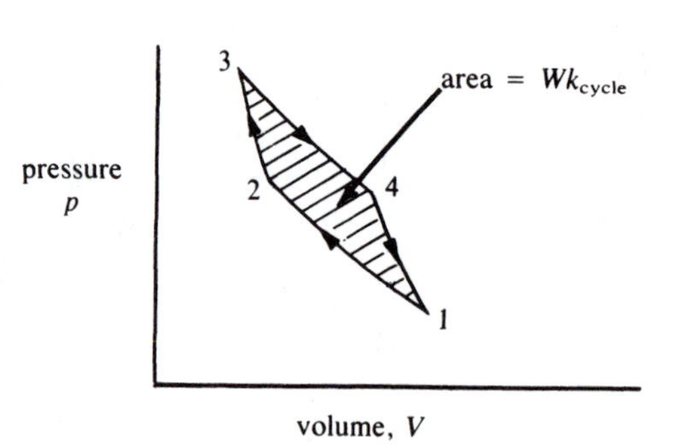

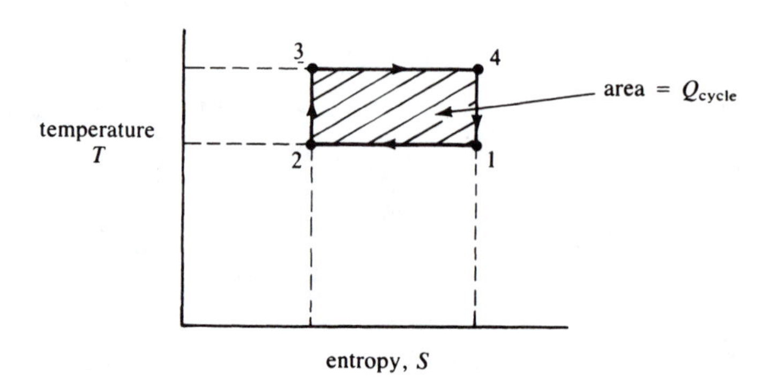

FIGURE 7–17
Pressure–volume and temperature–entropy diagrams for Carnot engine

State 4: $T_4 = T_3$

$$V_4 = V_1\left(\frac{T_1}{T_4}\right)^{1/(k-1)} = \frac{mRT_4}{p_4}$$

The work and heat terms are

$$Wk_{12} = Q_{12} = mRT_1 \ln\frac{V_2}{V_1}$$

$$Wk_{23} = \frac{mR}{1-k}(T_3 - T_2) = -mc_v(T_3 - T_2)$$

$$Q_{23} = 0$$

$$Wk_{34} = Q_{34} = mRT_3 \ln\frac{V_4}{V_3}$$

$$Wk_{41} = \frac{mR}{1-k}(T_1 - T_4) = mc_v(T_4 - T_1)$$

$$Q_{41} = 0$$

$$Q_{\text{add}} = Q_{34} = Q_H \qquad Q_{\text{rej}} = Q_{12} = Q_L$$

$$Q_{\text{net}} = Q_{34} + Q_{12}$$

$$Wk_{\text{net}} = Wk_{12} + Wk_{23} + Wk_{34} + Wk_{41} = Q_{\text{net}}$$

The energy and entropy changes are

$$\Delta U_{12} = mc_v(T_2 - T_1) = 0 = \Delta U_{34}$$

$$\Delta U_{23} = mc_v(T_3 - T_2) = -Wk_{23}$$

$$\Delta U_{41} = mc_v(T_1 - T_4) = -Wk_{41}$$

$$\Delta S_{12} = mR\ln\frac{V_2}{V_1} = mR\ln\frac{p_1}{p_2}$$

$$\Delta S_{23} = \Delta S_{41} = 0$$

$$\Delta S_{34} = mR\ln\frac{p_3}{p_4} = mR\ln\frac{V_4}{V_3}$$

The thermal efficiency $\eta_T = Wk_{\text{net}}/Q_{34}$. For Carnot heat pump or refrigeration devices,

$$Q_{\text{add}} = Q_{21} = -Q_{12} = -mRT_1\ln\frac{V_2}{V_1}$$

$$Q_{\text{rej}} = Q_{43} = -Q_{34} = -mRT_3\ln\frac{V_4}{V_3}$$

$$Wk_{\text{net}} = -Wk_{12} - Wk_{23} - Wk_{34} - Wk_{41} = Q_{21} + Q_{43}$$

$$\text{COP} = \frac{Q_{43}}{Wk_{\text{net}}}$$

$$\text{COR} = \frac{-Q_{21}}{Wk_{\text{net}}} = \text{COP} - 1$$

In appendix A–7 we describe a program, CARNOT, that is included on the soft diskette and that can conveniently solve for important properties, work and heat terms, energy and entropy changes, the thermal efficiency (for heat engines), and the coefficient of performance or refrigeration (for heat pump devices). These equations could also be used to solve by hand-calculator methods or by using *Engineering Equation Solver* (*EES*).

EXAMPLE 7–18 A Carnot engine is proposed that has the physical characteristics of a piston-cylinder while the gas contained within the cylinder is air. The piston has a diameter of 3 in and travels a distance of 4 in, reciprocating from within $^1/_4$ in of the cylinder and to $4^1/_4$ in. The arrangement is shown in figure 7–18. Heat proceeds to be added when the piston is at the extreme inward position 1 and proceeds to be rejected when the piston is in the extreme outward position 3. If the high-temperature region is at 1200°F and the low-temperature region is at 80°F, determine the engine's thermal efficiency; the pressures, temperatures, and volumes at each of the four states; the net work per cycle; heat added; and heat rejected. Assume that the pressure at state 1 or position 1 is 14.0 psia.

Solution The system is identified as the air in the cylinder, and the system diagram is sketched in figure 7–19. The *p–V* diagram is given in figure 7–20, and the physical positions denoted in figure 7–18 correspond to the numbers in this diagram. Figure 7–21 then shows the *T–S* diagram for the given Carnot engine.

We obtain the thermal efficiency from

$$\eta_T = \left(1 - \frac{T_L}{T_H}\right) \times 100$$

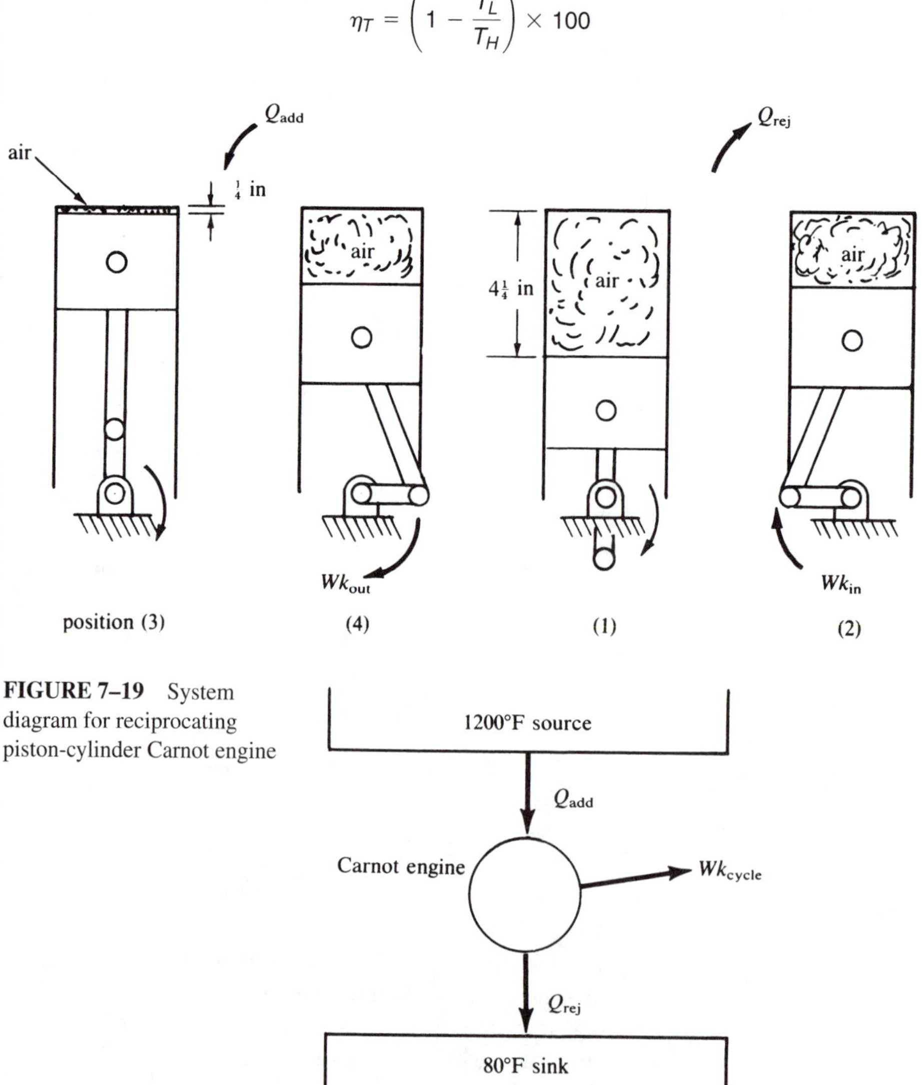

FIGURE 7–18 Carnot engine using the piston-cylinder mechanism

FIGURE 7–19 System diagram for reciprocating piston-cylinder Carnot engine

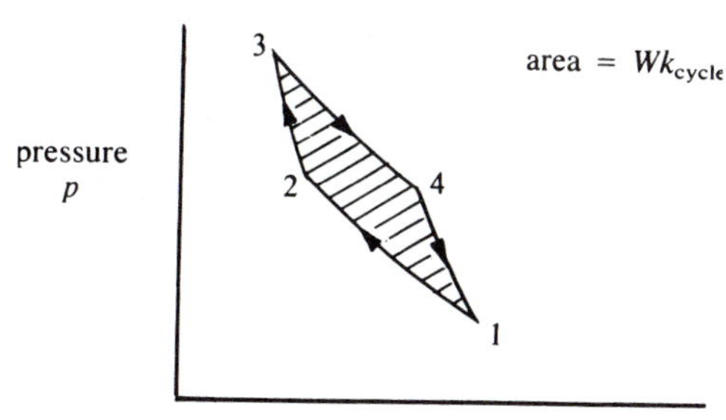

FIGURE 7–20
Pressure–volume diagram for engine of example 7–18

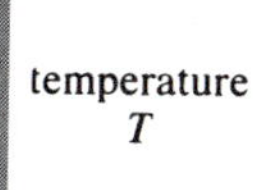

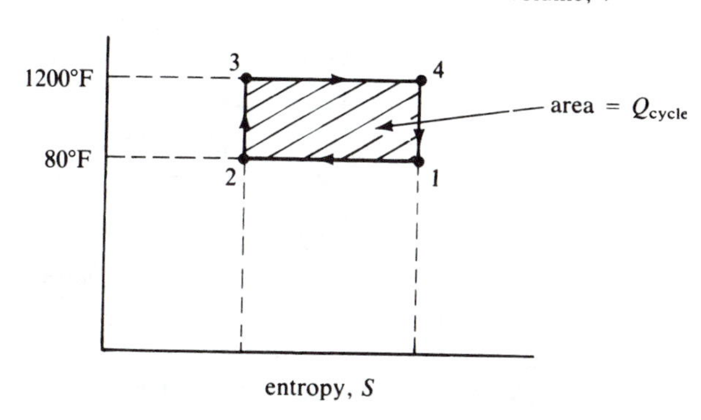

FIGURE 7–21
Temperature–entropy diagram for engine of example 7–18

which gives us

Answer

$$\eta_T = \left(1 - \frac{540}{1660}\right) \times 100 = 67.5\%$$

We now find the volumes, V_1 and V_3:

$$V_1 = (\text{piston face area})(4\tfrac{1}{4}\text{ in})$$
$$V_3 = (\text{piston face area})(\tfrac{1}{4}\text{ in})$$

Because the face area is circular, we then obtain

$$V_1 = (\pi)\left(\frac{3^2}{4}\text{ in}^2\right)(4.25\text{ in}) = 30.04\text{ in}^3$$
$$V_3 = (\pi)\left(\frac{3^2}{4}\text{ in}^2\right)(0.25\text{ in}) = 1.767\text{ in}^3$$

Converting to cubic feet units yields

$$V_1 = 30.04\text{ in}^3/1728\text{ in}^3/\text{ft}^3$$
$$= 0.01738\text{ ft}^3$$

and

$$V_3 = 0.001023\text{ ft}^3$$

The gas constant R is 53.36 ft · lbf/lbm · °R, and the specific heat c_v is 0.1718 Btu/lbm · °R from table B–4. Then, using the equations listed previously, we find, from either hand calculations or the program CARNOT,

$$c_p = 0.240382\text{ Btu/lbm}\cdot°\text{R} \quad K = 1.3992$$
$$m = 0.001216\text{ lbm}$$
$$T_2 = 540°\text{R} = 80°\text{F}$$

$$p_3 = 731 \text{ psia} \qquad p_2 = 14.25 \text{ psia}$$
$$V_2 = 0.0171 \text{ ft}^3$$
$$T_4 = 1660°\text{R} \qquad V_4 = 0.00104 \text{ ft}^3 \qquad p_4 = 719.2 \text{ psia}$$
$$Wk_{12} = -0.569 \text{ ft}\cdot\text{lbf} = -0.00073 \text{ Btu}$$
$$Wk_{23} = -182.03 \text{ ft}\cdot\text{lbf} = -0.23398 \text{ Btu}$$
$$Wk_{34} = 1.775 \text{ ft}\cdot\text{bf} = 0.00228 \text{ Btu}$$
$$Wk_{41} = 182.03 \text{ ft}\cdot\text{lbf} = 0.23398 \text{ Btu}$$
$$Q_{add} = 0.00228 \text{ Btu}$$
$$Q_{rej} = -0.00073 \text{ Btu}$$
$$Wk_{net} = 0.001462 \text{ Btu} = 1.136 \text{ ft}\cdot\text{lbf/cycle}$$
$$\Delta U_{23} = 0.23398 \text{ Btu} \qquad \Delta U_{12} = 0$$
$$\Delta U_{41} = -0.23398 \text{ Btu} \qquad \Delta U_{34} = 0$$
$$\Delta S_{12} = -0.001677 \text{ Btu/°R} \qquad \Delta S_{23} = 0$$
$$\Delta S_{34} = +0.001677 \text{ Btu/°R} \qquad \Delta S_{41} = 0$$

The thermal efficiency, calculated from the ratio of the net work to the heat added, is 67.5%, which is in agreement with the answer found from the temperature ratio equation. If the cycle is reversed and then used to describe a Carnot heat pump or refrigerator, the heat rejected (or heat added to a high-temperature region) is 0.00228 Btu, and the heat added (refrigeration effect) is 0.0073 Btu. The net work is 0.00155 Btu and is put into the device. The coefficient of performance, COP, is 1.48 and the COR is 0.48. Notice that the actual amount of net work is quite small for one cycle and the maximum pressure and temperature are high.

EXAMPLE 7–19 A reversible piston-cylinder device contains argon gas as it operates on a Carnot cycle. One thousand joules of heat is added at 700°C. The maximum volume is 3000 cm^3 where the pressure is 80 kPa and the temperature is −20°C. Determine the pressures, volumes, and temperatures at the four states; the net work; the heat transfers; and the thermal efficiency. Also, determine the refrigeration effect and the COR for the reversed system.

Solution This problem is one of the second type: the conditions at state 1 are given, the amount of heat added is given, and the high temperature is known. We have the following knowns:

$$p_1 = 80 \text{ kPa} \qquad T_1 = -20°\text{C} = 253 \text{ K} \qquad V_1 = 3000 \text{ cm}^3 = 0.003 \text{ m}^3$$
$$T_3 = T_4 = 700°\text{C} = 973 \text{ K} \qquad Q_{add} = 1000 \text{ J}$$

From table B–4, we find, for argon,

$$R = 0.208 \text{ kJ/kg}\cdot\text{K}$$
$$c_v = 0.312 \text{ kJ/kg}\cdot\text{K}$$

Using either hand calculations or the CARNOT program, we obtain the following results:

$$c_p = 0.520 \text{ kJ/kg}\cdot\text{K} \qquad k = 1.6666\ldots$$
$$m = 0.00456 \text{ kg} \qquad T_2 = T_1 = 253 \text{ K}$$
$$V_2 = 0.001015 \text{ m}^3 \left(= V_1\left[\frac{1}{\exp(Q_{add}/mRT_3)}\right]\right)$$
$$p_2 = 236.4 \text{ kPa} \qquad p_3 = 6856.4 \text{ kPa}$$
$$V_3 = 0.00013462 \text{ m}^3 \qquad T_4 = T_3 = 973 \text{ K}$$
$$p_4 = 2320 \text{ kPa} \qquad V_4 = 0.0003977 \text{ m}^3$$
$$Wk_{12} = -0.260021 \text{ kJ} = Q_{12} \qquad \Delta U_{12} = 0$$

$$Wk_{23} = -1.02451 \text{ kJ} = -\Delta U_{23} \qquad Q_{23} = 0$$
$$Wk_{34} = 1.000 \text{ kJ} = Q_{34} \qquad \Delta U_{34} = 0$$
$$Wk_{41} = 1.02451 \text{ kJ} = -\Delta U_{41} \qquad Q_{41} = 0$$
$$Wk_{net} = 0.739979 \text{ kJ}$$
$$\Delta S_{12} = -0.001028 \text{ kJ/K} \qquad \Delta S_{23} = 0$$
$$\Delta S_{34} = 0.001028 \text{ kJ/K} \qquad \Delta S_{41} = 0$$
$$\eta_T = 74.0\%$$

For the reversed cycle, the refrigeration effect is Q_{add}, or 0.260021 kJ. The COR is 0.35 and the COP is 1.35.

In examples 7–18 and 7–19, it would be a good review to check some hand-calculation answers against those found by using the program CARNOT. Also, by varying the type of gas or the amount of heat added per cycle and obtaining the results from the CARNOT program, you can see how the CARNOT cycle efficiency is independent of the gas or amount of heat. (See problems 7–58 and 7–62.)

No one has yet built a Carnot engine; there are at least two reasons:

1. Reversible cycles can be approached but never fully realized.
2. Although reversible adiabatic processes can be closely approximated in technology, the reversible isothermal process is difficult to put into practice and still have sufficient amounts of heat transfer in a reasonable time period.

The Carnot cycle/engine concepts are theoretical tools that have added immensely to the thermodynamic method; we will refer to them frequently as a standard of comparison.

7–11 SUMMARY

For cyclic devices, we have found that

$$Q_{net} = Wk_{net}$$
$$\Delta U = 0 \tag{7–1}$$

and the change in any property of the working medium of a cyclic device is zero over one complete cycle. The Carnot cycle was defined through the following four processes:

1–2 Reversible isothermal compression at temperature T_L

2–3 Reversible adiabatic compression from the low temperature T_L to a higher temperature T_H

3–4 Reversible isothermal expansion at temperature T_H

4–1 Reversible adiabatic expansion from temperature T_H to T_L

This cycle describes the operation of a Carnot heat engine. The reversed Carnot cycle describes the operation of a Carnot heat pump or refrigerator. Also, for the Carnot cycle, which operates between two temperature regions (T_L and T_H), we have

$$\frac{Q_H}{-Q_L} = \frac{Q_{add}}{-Q_{rej}} = \frac{T_H}{T_L} \tag{7–4}$$

and from this the entropy change was defined as

$$\Delta S = \sum \frac{\delta Q_{rev}}{T} \tag{7–5}$$

The entropy production, or rate of change of entropy, was defined as

$$\dot{S} = \frac{\dot{Q}_{rev}}{T} \tag{7–19}$$

For the Carnot cycle,

$$\Delta S_L = \frac{Q_L}{T_L}$$

$$\Delta S_H = \frac{Q_H}{T_H}$$

and

$$\Delta S_L + \Delta S_H = 0$$

Also,

$$Q_{\text{rev}} = \sum T\,\delta S \tag{7–9}$$

and the reversible heat can then be found from the area under a curve on the T–S (temperature–entropy) diagram, the curve being a description of a reversible process involving the heat Q_{rev}. The thermal efficiency of a heat engine was defined as

$$\eta_T = \frac{Wk_{\text{net}}}{Q_{\text{add}}} \tag{7–21}$$

and for the Carnot cycle,

$$\eta_T = 1 - \frac{T_L}{T_H} \tag{7–23}$$

For reversed cycles and the corresponding heat pumps or refrigerators, the coefficient of performance, COP, and coefficient of refrigeration, COR, are, respectively, defined as

$$\text{COP} = \frac{Q_{\text{rej}}}{Wk_{\text{net}}} \tag{7–25}$$

$$\text{COR} = \frac{Q_{\text{in}}}{-Wk_{\text{net}}} \tag{7–26}$$

and we also have

$$\text{COP} = 1 + \text{COR} \tag{7–27}$$

For the Carnot heat pumps and refrigerators,

$$\text{COP} = \frac{T_H}{T_H - T_L} \tag{7–28}$$

$$\text{COR} = \frac{T_L}{T_H - T_L} \tag{7–29}$$

The second law of thermodynamics requires that, for all heat engines and heat pumps,

1. There be at least two heat exchanges with two separate temperature regions.
2. The thermal efficiency be less than 100%.
3. $\Delta S_{\text{cycle}} = 0$, but $\Delta S_{\text{total}} \geq 0$ where $\Delta S_{\text{total}} = \Delta S_{\text{cycle}} + \Delta S_{\text{surrounding}}$.

Entropy is a measure of the disorder of a substance, or it is a measure of the availability of energy or potential to effect a change; that is, the greater the entropy, the less ordered or available is energy in a material. Entropy continues to increase as processes occur, whether intentionally through human design or unintentionally and spontaneously. Entropy changes are able to be found from some of the following:

1. For all reversible isothermal processes,

$$\Delta S = \frac{Q}{T} \tag{7–33}$$

2. For perfect gases with constant specific heats,

$$\Delta S = mc_v \ln \frac{T_2}{T_1} + \text{mR} \ln \frac{V_2}{V_1} \tag{7–38}$$

and

$$\Delta S = mc_p \ln \frac{T_2}{T_1} - mR \ln \frac{p_2}{p_1} \tag{7–44}$$

3. For perfect gases with variable specific heats,

$$\Delta S = m(\phi_1 - \phi_1) - mR \ln \frac{p_2}{p_1} \tag{7–46}$$

4. For incompressible liquids and solids,

$$\Delta S = mc \ln \frac{T_2}{T_1} \tag{7–52}$$

5. For pure substances,

$$\Delta S = m(s_2 - s_1) \tag{7–54}$$

The reversible adiabatic process is an isentropic process. For the isentropic process, $\Delta S = 0$, or entropy is constant. Then, for perfect gases with constant specific heats,

$$pv^k = \text{constant}$$

For nozzles, we have, for the outlet velocity,

$$\overline{V}_2 = \sqrt{2(h_2 - h_1)(1000) + \overline{V}_1^2} \qquad \text{SI units: } \overline{V} \text{ in m/s and } h \text{ in kJ/kg}$$

$$\overline{V}_2 = \sqrt{2(h_2 - h_1)(778)g_c + \overline{V}_1^2} \qquad \text{English units: } \overline{V} \text{ in ft/s and } h \text{ in Btu/lbm}$$

$$g_c = 32.17 \text{ lbm} \cdot \text{ft/lbf} \cdot \text{s}^2$$

and in many cases $\overline{V}_1$ may be neglected or set equal to zero. For perfect gases with constant specific heats, $h_2 - h_1 = c_p(T_2 - T_1)$.

The third law of thermodynamics states that it is impossible to reach absolute zero temperature with real, irreversible processes.

The CARNOT program is shown to be an assistance in analyzing the Carnot cycle as it applies to heat engines or heat pumps. Values for pressure, temperature, and volume can readily be found from standard process equations developed in earlier sections.

DISCUSSION QUESTIONS

Section 7–1

7–1 What is meant by a *heat engine*? Do you know of any heat engines?

7–2 Do all properties of a heat engine remain the same after the engine has gone through one complete cycle?

7–3 What is a *heat pump*?

Section 7–2

7–4 Is *entropy* a property? Why?

7–5 What is meant by the term *entropy generation*?

Section 7–3

7–6 Why can a heat engine never have a thermal efficiency of 100%? What if the engine is reversible?

Section 7–4

7–7 How are *refrigerators* and *heat pumps* different?

Section 7–5

7–8 Does the statement "heat cannot flow from cold to hot by itself" agree with the second law of thermodynamics?

7–9 What is a *Perpetual Motion Machine (PMM)*?

7–10 How are perpetual motion machines of the first kind (PMM1) different than perpetual motion machines of the second kind (PMM2)?

Section 7–6

7–11 What really is entropy?

7–12 How does equation 7–38 come about?

PRACTICE PROBLEMS

Section 7–1

Problems that use SI units are indicated by an (M) under the problem number; those that use English units are indicated by an (E).

7–1 (M) Determine the unknown quantities of the cycle indicated.

Process	Q, kJ	Wk, kJ	ΔU, kJ
1–2		0	−1.0
2–3	0	−0.4	
3–4	2.1		
4–1	0	1.3	

7–2 (E) Determine the unknown quantities of the cycle indicated.

Process	Q, Btu	Wk, Btu	ΔU, Btu	ΔU, °F	Δp, psia
1–2	−8	−8		185	100
2–3	−0.2		−0.3	390	
3–4	16	18		−200	−110
4–1	−0.3				−350

7–3 (M) Determine the unknown quantities of the cycle indicated.

Process	Q, kJ	Wk, kJ	ΔU, kJ
1–2	50		75
2–3		110	−20
3–1	−35		

7–4 (E) Determine the unknown quantities of the cycle indicated.

Process	Q, Btu	Wk, Btu	ΔE, Btu
1–2	0		1000
2–3	4000	0	
3–4	0	3200	
4–5		100	
5–1	1	−300	

Section 7–2

7–5 (E) Determine the work generated by a Carnot engine operation, between 2000°F and 100°F, if the entropy change is 10 Btu/°R.

7–6 (M) A Carnot engine operates at 200 rpm with 1 revolution/cycle and with an entropy change of 0.03 kJ/K. Determine the high temperature required to generate 100 kW of power. Assume T_L to be 27°C.

7–7 (M) Determine the heat added in the engine of problem 7–6.

7–8 (E) Determine the net heat and the heat rejected of the Carnot engine of problem 7–5.

7–9 (M) The entropy function of a certain gas is plotted in figure 7–22. Determine the energy required in the form of heat transfer to heat the material to 300 K from absolute zero temperature.

Section 7–3

7–10 (M) A Carnot engine operates between 1000°C and 200°C. If the entropy change is 0.01 kJ/kg · K, determine the cycle efficiency.

FIGURE 7–22

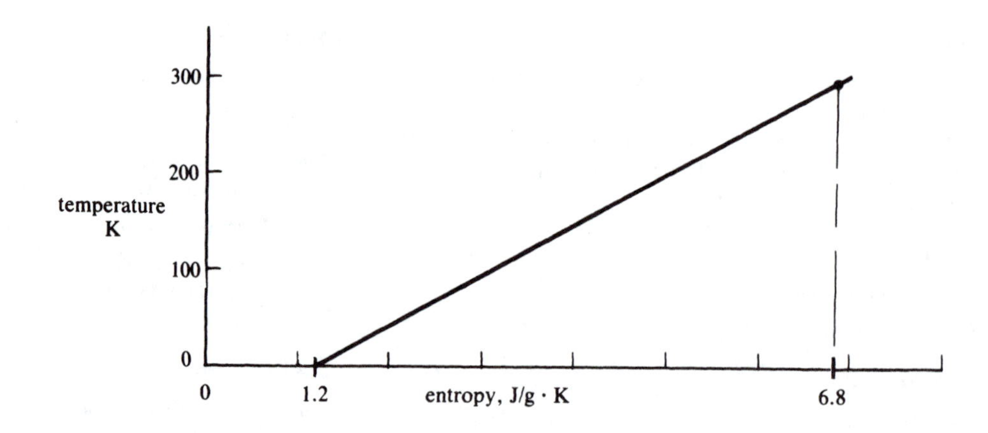

7–11 (M) At 1200°C, 400 kJ/s of heat is added to a Carnot engine. If the surroundings are at a temperature of 20°C, determine the cycle efficiency and the output power.

7–12 (M) What maximum efficiency may a heat engine have if it can exchange heat by radiation with the sun and outer space? Assume that the temperature of the sun is 10,000°C and that of outer space is −60°C.

7–13 (E) A Carnot engine receives heat from a boiler at 1800°R and rejects 20 BTU/s of heat at 600°R. Determine the following:
(a) Heat added from the boiler
(b) Power produced in hp

7–14 (E) A Carnot heat engine is designed to produce 70 hp at 3000 rpm. If the energy source is at 1500°F and the sink is at 90°F, determine the cycle efficiency and the ratio of heat transfer to the cycle, Q_{rej}/Q_{add}.

7–15 (E) Determine the rate of heat rejection of the engine in problem 7–14.

7–16 (E) An inventor claims an efficiency of 90% for an engine he has built. He claims that the highest temperature is 130°F. Is his claim possible? Discuss why or why not.

Section 7–4

7–17 (M) A Carnot refrigerator operates between −20°C and 20°C. Determine the refrigerating effect (Q_{add}) and the coefficient of refrigeration, COR, if the net work is 21,000 J.

7–18 (M) A Carnot heat pump, diagrammed in figure 7–7, operates between −15°C and 30°C with ammonia as a working fluid. Assume that the ammonia is a saturated vapor at point 4 and a saturated liquid at point 1. Determine the following:
(a) Heat added per kilogram of ammonia
(b) Heat rejected per kilogram
(c) Net work required per kilogram
(d) COP and COR
Compare the answers you obtained by using the appendix tables with the answers you obtained from *EES*.

7–19 (M) For the heat pump of problem 7–18, assume that 20 g/s of ammonia flows through the system. Determine the following:
(a) Power required
(b) Rate of heat added
(c) Rate of heat rejected

7–20 (E) A heat pump uses Refrigerant-22 as a working fluid. The cycle operates between −10°F and 70°F, and the flow rate of the refrigerant is 60 lbm/min. The device is similar to that shown in figure 7–7, but the compressor does not utilize the work developed by the expansion of the refrigerant from state 1 to state 2. If the refrigerant is saturated liquid at point 1 and saturated vapor at state 4, determine the following:
(a) Power required
(b) Rate of heat addition from the low-temperature source
(c) Rate of heat rejection to the high-temperature sink
(d) COP and COR
Compare the answers you obtained by using the appendix tables with the answers you obtained from *EES*.

Sections 7–5 and 7–6

7–21 An engine is proposed that extracts energy from seawater by heat transfer and then returns the cooled water to the ocean. No other heat transfer occurs with the engine. Does this violate any of our principles or laws of thermodynamics?

7–22 A device is proposed which is essentially a heat pump driven by a heat engine. A fluid is taken from a low-temperature area and, by means of the heat pump, is deposited in a high-temperature sink. This sink is subsequently used to furnish energy to drive the heat engine, which in turn rejects heat and fluid to the low-temperature region. Does this device violate the second law? Does this device violate any principle of thermodynamics?

7–23 (M) Seventy-two grams of air in a piston-cylinder is heated by 135 J while going through a reversible isothermal process at 70°C. Determine the entropy change and the specific entropy change of the air.

7–24 (E) During a reversible, isothermal process, 700 Btu of heat transfer is directed into a system at 80°F. Determine the entropy change of the system.

7–25 (E) On a summer day when the temperature is 80°F, a trash fire emits 30,000 Btu to the atmosphere. By how much has the entropy of the atmosphere increased?

7–26 (E) A heat engine operates at 60% thermal efficiency when it produces 35 hp and operates between 70°F and 2000°F. Determine the total entropy change of the engine and its surroundings. Does this engine violate the second law of thermodynamics?

7–27 (E) A heat pump is claimed to have a COP of 4.6. It operates between −40°F and 200°F, producing 300 Btu/min of heat. What is the entropy production of the pump and its surroundings? Does this device violate the second law of thermodynamics?

7–28 (M) An inventor claims that a new refrigerator can operate between −30°C and 200°C while removing 3000 J/s of heat and using 2 kW of power. Does this device violate the second law of thermodynamics?

7–29 (E) An adiabatic compressor is advertised as being able to take 50 psia, 500°F steam and compress it to 500 psia using 240 BTU/lbm of work. Is this possible?

Section 7–7

7–30 (M) Heat transfers from a system at 300°C to another system at 100°C at a rate of 10 kW. Determine the entropy generation, in kW/K, involved in this process.

7–31 (E) Helium gas is increased in pressure from 15 psia to 35 psia. Determine the specific entropy change of the helium if the compression is done reversibly under constant-temperature conditions.

7–32 (M) During an isometric process of 0.1 kg of nitric oxide (NO), the specific entropy decreases by 0.30 kJ/kg · K. If the NO was initially at −10°C, determine the final temperature.

7–33 (M) In a rigid stainless steel tank, 540 g of propane (C_3H_8) is heated from −20°C to 5°C. Determine the entropy change in kJ/K.

7–34 (M) During a constant-pressure process, 1.5 kg of sulfur dioxide (SO_2) is heated to 70°C from an initial temperature of 10°C. Determine the total and specific entropy changes.

7–35 (M) A drop in entropy of 100 J/kg · K is calculated for propylene gas as it is cooled to 20°C in an isobaric process. Determine the initial temperature.

7–36 (M) During an adiabatic process, argon gas exhibits an increase of specific entropy of 0.12 kJ/kg · K. What is the irreversible entropy increase of 20 g of argon?

7–37 (M) During a polytropic process involving air, the specific entropy decreases by 0.03 kJ/kg · K. The pressure ratio (final to initial pressure) is 12.1 : 1, and the initial temperature was 80°C. Obtain the polytropic exponent, *n*, and the final temperature.

7–38 (E) A polytropic process with an exponent *n* of 1.43 is found to describe an expansion through a gas turbine where the entrance pressure is 1200 kPa and the exhaust is 120 kPa. Assuming that the gas flowing through the turbine is air, determine the entropy change per kilogram.

7–39 (E) A perfect gas contained in a rigid container in 90°F surroundings exhibits a pressure drop from 28 psia to 20 psia. Assuming that the gas constant is 63 ft · lbf/lbm · °R and the specific heat, c_p, is 0.30 Btu/lbm · °R, calculate the temperature change and the specific entropy change for the gas.

7–40 (E) Acetylene gas is contained in a rigid steel tank while its pressure increases from 10 psig to 15 psig. Determine the change in specific entropy of the gas during this process.

7–41 (E) In a flexible container, 1.8 lbm of air is expanded at constant pressure from an initial volume of 20 ft^3 to 40 ft^3 finally. If the initial temperature is 195°F, determine the final temperature and the change in total and specific entropy.

7–42 (E) A gas having the same thermodynamic properties as those of air is contained in the combustion chamber of an internal combustion engine. The gas is at 1200°F and 200 psia, from which it expands polytropically to 15 psia. If the polytropic exponent is 1.51, determine the specific entropy change of the gas during this process.

7–43 (M) A 10-kg block of oak wood at 10°C is placed in a large oil bath at 110°C. Determine the entropy change of the wood and the oil. Use properties of engine oil for the oil, and neglect the fact that oil will soak into the wood.

7–44 (E) A 3-ton (6000-lbm) granite block is heated from 50°F to 80°F. Determine the entropy change of the granite.

7–45 (M) A 3000-kg cast iron housing at 400°C is quenched in 4000 kg of water at 10°C. What is the entropy change of the iron and the water?

7–46 (M) Steam is expanded from 2000 kPa to 500 kPa at 400°C. Determine the specific entropy change.

7–47 (M) Superheated steam is reversibly and adiabatically expanded from 3000 kPa and 480°C to 500 kPa. Determine the final temperature.

7–48 (M) Determine the final temperature of 70 lbm of steam at 75 psia and 1200°F contained in an insulated chest. The chest expands isentropically until the steam is at 10 psia.

7–49 (M) Mercury vapor is heated from a saturated liquid to a saturated vapor. Determine the specific entropy change if the pressure is

(a) 6 kPa
(b) 140 kPa
(c) 1000 kPa

7–50 (M) Refrigerant-22 is heated from saturated vapor at −20°C to superheated vapor at

(a) −5°C at 393 kPa
(b) 10°C at 218 kPa

Determine the specific entropy change for both processes.

7–51 (E) Determine the specific entropy change for Refrigerant-22 cooled from 50°F to 20°F at

(a) 5 psia
(b) 30 psia

7–52 (E) Steam is heated from saturated liquid at 15 psia to superheated steam at 200 psia and 600°F. Determine the specific entropy change.

7–53 (E) Steam is expanded from 300 psia and 800°F to a saturated vapor in a reversible adiabatic manner. Determine the final pressure of the steam.

Section 7–10

7–54 (M) A Carnot engine operating with a perfect gas between 800°C and 25°C has an operating pressure range between 20 kPa and 6000 kPa. Determine the work of the cycle, heat added, heat rejected, and efficiency if the gas has physical properties equivalent to those of air. Assume 1 kg of air and then determine the work and heat for 1 kg of air.

7–55 (M) A Carnot engine composed of a piston-cylinder uses 0.003 kg of air per cycle. The minimum volume of the enclosed cylinder is 50 cm^3 and the maximum is 1000 cm^3,

and the temperature range is 700°C to 50°C. If the engine is running at 400 cycles/min, determine the following:
(a) Power output
(b) Rate of heat addition
(c) Cycle efficiency
Compare the answers you obtained by using the appendix tables with the answers you obtained from *EES*.

7–56 (E) A Carnot heat engine operating at 3000 rpm with a perfect gas having the properties listed below has a heat source at 1500°F and a sink at 85°F. Determine the following:
(a) Cycle efficiency
(b) Work of the cycle
(c) Heat added and heat rejected

The gas properties known are

$$R = 48.3 \text{ ft} \cdot \text{lbf/lbm} \cdot {}^\circ\text{R}$$
$$c_p = 0.22 \text{ Btu/lbm} \cdot {}^\circ\text{R}$$
$$c_v = 0.157 \text{ Btu/lbm} \cdot {}^\circ\text{R}$$
$$k = 1.396$$
$$p_1 = 8 \text{ psia}$$
$$p_2 = 70 \text{ psia}$$

Assume 1 lbm.

7–57 (E) A reversible piston-cylinder device operates on a Carnot cycle and uses nitrogen as the working medium. The maximum volume is 460 in^3 when the pressure is 14.7 psia and the temperature is 85°F. The minimum volume is 4 in^3, and the maximum temperature is 2000°F. Determine the pressure, volume, and temperature at all four states; the work terms; the net work; the heat added; and the thermal efficiency.

7–58 Using known constant values for the lowest pressure, lowest volume, highest volume, and the two temperature regions, vary the type of perfect gas used in a Carnot heat engine (by using different R and c_v) to show that the thermal efficiency is independent of the type of gas. Use the program CARNOT or other method. What terms do vary with type of perfect gas?

7–59 (E) A Carnot heat pump uses nitrogen gas (which behaves like a perfect gas) and operates between 376°F and −15°F. The gas is at 15 psia when entering the compression and 270 psia when leaving. Determine the following for this cyclic device:
(a) Work of the compression per unit mass
(b) COP
(c) COR
Compare the answers you obtained by using the appendix tables with the answers you obtained from *EES*.

7–60 (M) A Carnot heat engine operates between 1000 K and 300 K, and uses helium as the working medium. The maximum volume is 3000 cm^3, and the minimum pressure is 100 kPa. There is 1.5 kJ of heat added per cycle. Determine all the pressures, volumes, and temperatures at the four states; the net work; the heat rejected; and the thermal efficiency.

7–61 (E) For a Carnot engine, the following conditions are known:

piston diameter = 3.5 in
piston travel = 4.25 in
piston clearance at top = 0.31 in
maximum temperature = 2800°F
minimum temperature = 100°F
minimum pressure = 5 psia

The working medium is argon. Determine the values for p, V, and T at all four states; the net work; the heat rejected; and the thermal efficiency.

7–62 Using known and constant values for p_1, V_1, T_1, and T_3 of a Carnot cycle, vary Q_{add} to show that the thermal efficiency of the cycle is independent of the heat added. Use the CARNOT program or other method, and determine which terms do depend on the heat added.

8 AVAILABILITY AND USEFUL WORK

Some concepts are presented here that have general value in determining the limitations imposed on thermodynamic systems as they execute processes. From the first law of thermodynamics, a general equation is developed for useful work. We see that useful work is precisely what the name implies—work that we can directly use. From this concept, we progress to a definition of availability and a discussion of its practical meaning. Emphasis is placed on evaluating the change in availability, which is equal to useful work, but theory is not ignored entirely. Example problems have been included that should aid in clarifying the energy–entropy–work relationships, including the ways in which energy progresses from an available to an unavailable condition. Available and unavailable energy are introduced as special cases of our availability function, and the free-energy functions are given as alternate statements resulting from availability.

New Terms

g'	Gibbs free energy	wk_{use}	Useful work
$\mathcal{G}$	Total Gibbs free energy	Λ	Function of availability
H'	Helmholtz free energy	Φ	Total availability
h'	Specific Helmholtz free energy	ϕ	Availability

8–1 USEFUL WORK

The simple homogeneous system can increase in volume and thereby provide work to be used external to the system. For a specific system, such as the piston-cylinder depicted in figure 8–1, we found that the work was described by the relationship

$$Wk_{cs} = \sum p\,\delta V \tag{8–1}$$

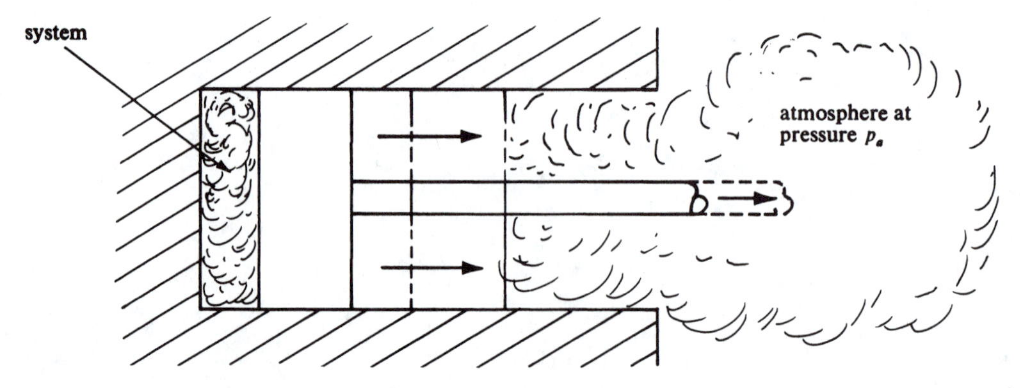

FIGURE 8–1 Piston-cylinder increasing the system volume

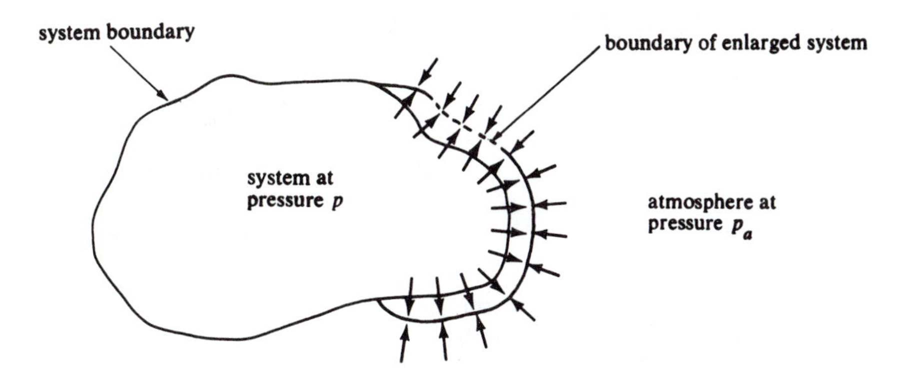

FIGURE 8–2 System increasing in volume

if the process was reversible. This equation then takes on various forms (see tables 6–4 and 6–5), depending on the type of process. The piston-cylinder is a detailed configuration of a general system, as shown in figure 8–2, which is capable of changing volume. However, almost all of the systems we can visualize will ultimately be used in an atmosphere (Earth, Mars, or somewhere else, rather than in a vacuum), and accordingly, the system must displace some of the atmosphere, namely, the amount by which the system itself changes volume. This requires work, which detracts from the $\sum p\,\delta V$ term; consequently, we say that **useful work**, Wk_{use}, is obtained from the equation

$$Wk_{use} = \sum p\,\delta V - p_a(V_2 - V_1) \tag{8–2}$$

for a reversible process. The term p_a is the atmospheric or surrounding pressure.

We can also recall the first law for the closed system:

$$E_2 - E_1 = Q - \sum p\,\delta V \tag{8–3}$$

we then substitute the result from equation (8–2) to obtain

$$E_2 - E_1 = Q - Wk_{use} - p_a(V_2 - V_1) \tag{8–4}$$

Since we are here considering ideal or reversible processes and assuming that heat transfers are conducted with the atmosphere at isothermal processes, we can then write

$$Q = T_a(S_2 - S_1) \tag{8–5}$$

where T_a is the atmospheric temperature. If we substitute this into equation (8–4) and rearrange slightly, we have

$$Wk_{use} = T_a(S_2 - S_1) - p_a(V_2 - V_1) - (E_2 - E_1) \tag{8–6}$$

We can easily include irreversible or real processes by using an inequality sign:

$$Wk_{use} \leq T_a(S_2 - S_1) - p_a(V_2 - V_1) - (E_2 - E_1) \tag{8–7}$$

Of course, we identify the "less than" with an irreversible process and the "equality" with an ideal or a reversible one.

CALCULUS FOR CLARITY 8–1

The useful work of a reversible process involving a closed system is

$$Wk_{use} = \int p\,dV - p_a(V_2 - V_1)$$

CALCULUS FOR CLARITY 8–1, continued

and then

$$\int p\,dV = Wk_{use} + p_a(V_2 - V_1)$$

The energy change of the system, given by equation (8–1) in differential form as

$$E_2 - E_1 = Q - \int p\,dV$$

becomes equation (8–4).

The reversible heat is

$$Q = \int T\,dS$$

and for a situation in which the heat transfer occurs reversibly and isothermally between a system and its surroundings, the temperature T is T_a, so that

$$Q = T_a \int dS = T_a(S_2 - S_1)$$

and equation (8–6) follows.

Various information can be drawn from either equation (8–6) or (8–7). For instance, from equation (8–6), we see that a decrease in a system's energy does not, of itself, assure a supply of useful work. The initial energy E_1 might very well be greater than the final energy E_2, but the terms $T_a(S_2 - S_1)$ may be negative or $p_a(V_2 - V_1)$ may be large; this could cancel any energy decreases.

Also, it is quite possible to get work out of a system during a single process and not have a decrease in energy, as the following example shows.

EXAMPLE 8–1 A piston-cylinder encloses a perfect vacuum, shown in figure 8–3. The pin holds the piston in the extended position. If the piston diameter is 15 cm, determine the useful work obtained from this device when the pin is removed.

Solution We identify the system as the enclosed vacuum and observe that the system has no entropy or energy values before, during, or after the process. We then obtain the useful work from

$$Wk_{use} = T_a(S_2 - S_1) - p_a(V_2 - V_1) - (U_2 - U_1) \quad \textbf{(8–6)}$$

and since

$$U_2 = 0 \qquad U_1 = 0 \qquad S_2 = 0 \qquad S_1 = 0$$

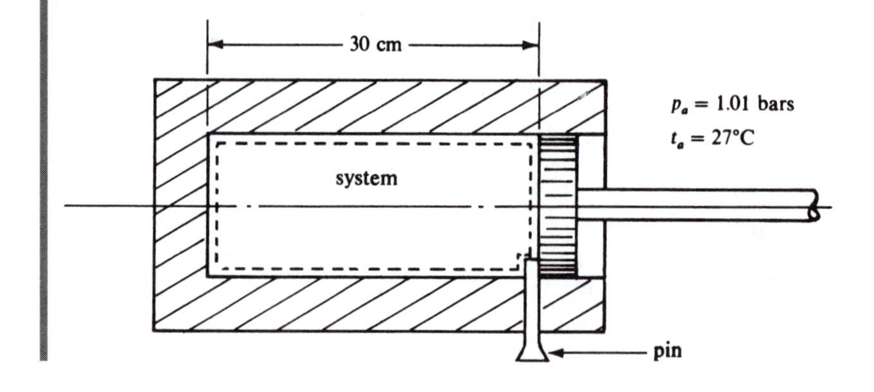

FIGURE 8–3 Vacuum system of example 8–1

we have

$$Wk_{use} = p_a(V_2 - V_1)$$

The initial volume is calculated as

$$V_1 = (\pi)(30\text{ cm})(7.5\text{ cm})^2 = 5300\text{ cm}^3$$
$$= 5.3 \times 10^{-3}\text{ m}^3$$

The final volume is zero because the atmosphere will push the piston into the cylinder completely. Then

$$Wk_{use} = (1.01 \times 10^5\text{ N/m}^2)(5.3 \times 10^{-3}\text{ m}^3)$$

Answer

$$= 535\text{ N}\cdot\text{m} = 535\text{ J}$$

and we see that a significant amount of work can be derived from an absolute vacuum.

8–2 AVAILABILITY

We have seen that the useful work obtained from a closed system is given by equation (8–6),

$$Wk_{use} = T_a(S_2 - S_1) - p_a(V_2 - V_1) - (E_2 - E_1)$$

or, per unit mass,

$$wk_{use} = T_a(s_2 - s_1) - p_a(v_2 - v_1) - (e_2 - e_1) \quad \textbf{(8–8)}$$

where wk_{use} is Wk_{use}/m. If we now group some properties together and define a new property, the **function of availability**, as

$$\Lambda = E + p_aV - T_aS \quad \textbf{(8–9)}$$

then, from equation (8–6), the useful work can also be given by

$$Wk_{use} = \Lambda_1 - \Lambda_2 = -\Delta\Lambda \quad \textbf{(8–10)}$$

For a given atmosphere, holding T_a and p_a constant, we can plot the property Λ. In figure 8–4 is plotted the surface of the Λ function along the entropy–volume axis. Notice that at one unique point, Λ is a minimum—when entropy has a value S_m, and volume a value V_m. This point, Λ_{min}, is a "lowest point," and intuitively, we see that a system will

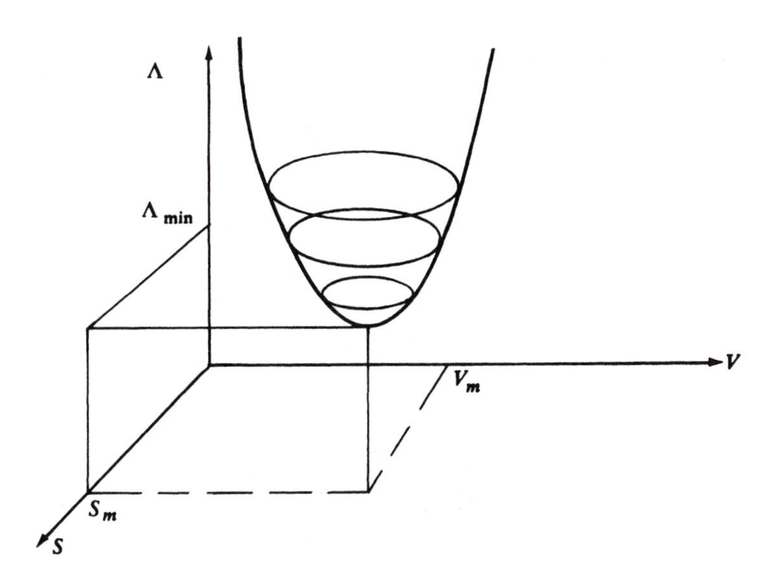

FIGURE 8–4 Graph of Λ function in Λ–V–S space

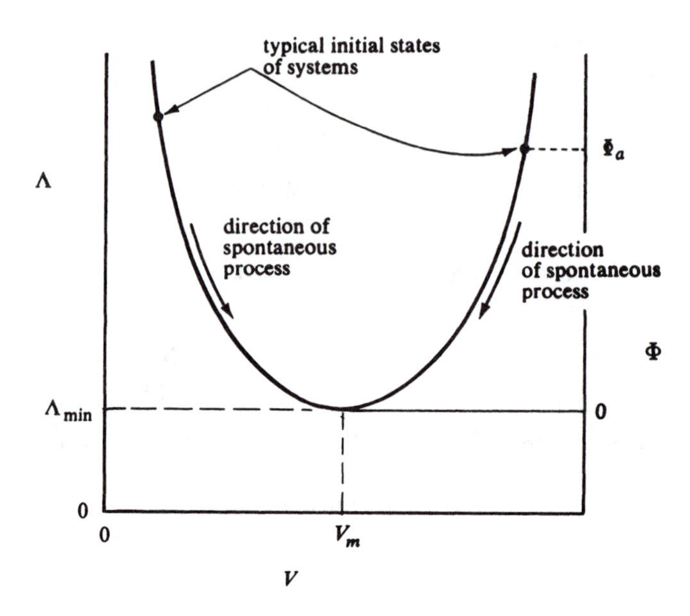

FIGURE 8–5 Graph of Λ or Φ and volume showing direction of a spontaneous or natural process of a system

gravitate toward the minimum point if it is at some state. In figure 8–5 is shown the Λ function for entropy having a constant value S_m. It is indicated that during a process where no work is put into the given system (natural process) and where the process is merely allowed to proceed in its course, the value of Λ invariably takes a lower value, or the system approaches the state given by Λ_{min}.

We will define the **total availability**, as

$$\Phi = \Lambda - \Lambda_{\text{min}} \tag{8–11}$$

and remark that Φ has a value of zero when it is at the state specified by the property values S_m, V_m, and E_m. Availability is plotted in the graph of figure 8–5 to show this relationship, and it can be seen that, because Λ_{min} represents a constant value,

$$\Phi_2 - \Phi_1 = \Lambda_2 - \Lambda_1$$

or, generally,

$$\Delta\Phi = \Delta\Lambda = (E_2 - E_1) + p_a(V_2 - V_1) - T_a(S_2 - S_1) \tag{8–12}$$

and using equation (8–10), we obtain

$$Wk_{\text{use}} = -\Delta\Phi \tag{8–13}$$

Again, we can include all processes, reversible and irreversible, by writing

$$Wk_{\text{use}} \leq -\Delta\Phi \tag{8–14}$$

The availability then gives us an upper limit or greatest expected output of a particular system. Let us look at a problem solved by the use of the availability concept.

EXAMPLE 8–2 Determine the vertical distance through which ice can raise itself in an atmosphere at 14.7 psia and 70°F. The energy required to change ice to water at 32°F is 144 Btu/lbm, and the specific heat of water can be taken as 1.0 Btu/lbm · °R.

Solution First, we must realize that the answer we will get is one that is a maximum value; for real irreversible processes, the vertical distance will be less than the calculated answer. The system is, obviously, the ice (or water) and may also be visualized as cubes of ice—1-lbm cubes as shown in figure 8–6. Also in this figure, we see how the ice is going to lift itself up; it acts like a low-temperature reservoir (but ultimately reaches equilibrium with the atmosphere) for a reversible heat engine, and the heat engine can then drive an appropriate device to hoist the ice, or water. This heat engine can, of course, run only until the ice has melted and attained a temperature of 70°F. We can easily identify the work to be used for each pound-mass of ice, assuming a gravitational acceleration of 32.2 ft/s^2 as 1 lbf $\times$ y, which is just the increase in potential energy of the water. The change in availability of the ice (or water) is obtained from the relationship

$$\Delta\Phi = \Delta\Lambda = \Lambda_a - \Lambda_1$$

Then

$$\Delta\Phi = E_a + p_a V_a - T_a S_a - E_1 - p_a V_1 + T_a S_1$$

The energy of the water is in the form of internal energy and potential energy (because it is being hoisted y feet up). We then write

$$\Delta\Phi = U_a - U_1 + W(y) + p_a(V_a - V_1) - T_a(S_a - S_1)$$

Since no appreciable volume change is exhibited by the water in going from ice to 70°F water, we can ignore the $p_a(V_a - V_1)$ term.

As we discussed in previous chapters, most materials require a proportionately large amount of energy to change phase, and this change occurs at constant temperature. This generally is true for the liquid–solid phase change as well as for the vapor–liquid phase change. The energy required to convert a solid to a liquid is frequently called the **latent heat of liquefaction** or just **latent heat**. We know it to be 144 Btu/lbm for water at 32°F, so our energy change can be found from the following:

$$\begin{aligned} U_a - U_1 &= (\text{latent heat}) + mc(T_a - T_1) \\ &= 144 \text{ Btu/lbm} + (1 \text{ lbm})(1 \text{ Btu/lbm} \cdot {}^\circ\text{R})(70^\circ\text{F} - 32^\circ\text{F}) \end{aligned}$$

FIGURE 8–6 Availability of an ice cube

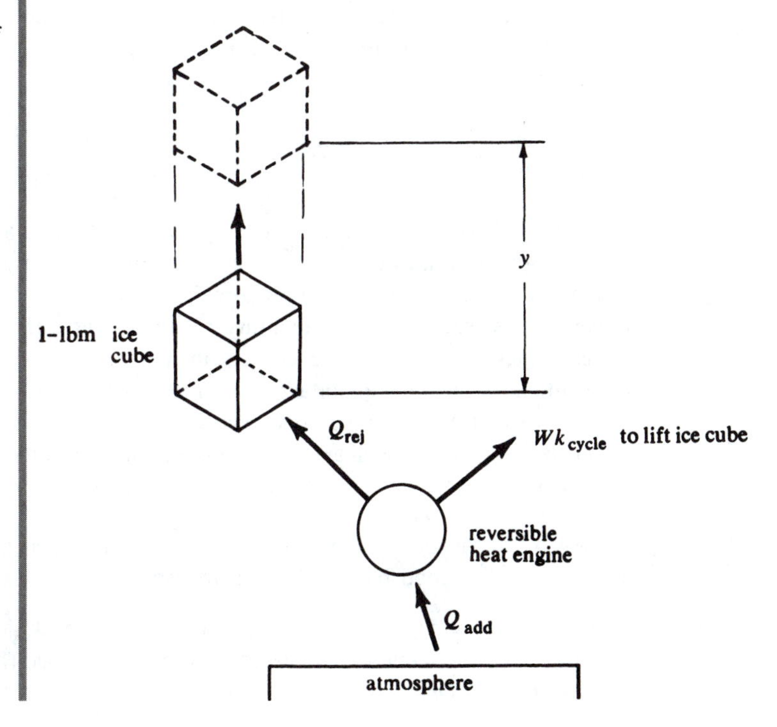

$$= 144 \text{ Btu/lbm} + 38 \text{ Btu}$$
$$= 182 \text{ Btu}$$

The energy change for each pound-mass of water is then 182 Btu. For the entropy change, we use the definition of entropy:

$$\Delta S = \frac{Q_{\text{rev}}}{T} = \frac{\text{latent heat}}{T_1} + \sum \frac{mc\,\delta T}{T}$$

From this, we obtain

$$\Delta S = 144 \text{ Btu}/492°\text{R} + (1 \text{ lbm})(1 \text{ Btu/lbm} \cdot °\text{R}) \sum \frac{\delta T}{T}$$
$$= 0.2927 \text{ Btu}/°\text{R} + (1 \text{ Btu}/°\text{R})\left(\ln \frac{530}{492}\right) = 0.2927 + 0.074$$
$$= 0.3667 \text{ Btu}/°\text{R}$$

For a change in availability of each pound-mass of the water, we then have

$$\Delta\Phi = 182 \text{ Btu} + (1 \text{ lbf})(y) - (530°\text{R})(0.3667 \text{ Btu}/°\text{R})$$
$$= -12.4 \text{ Btu} + 1 \text{ lbf}(y)$$

The useful work, which we derive directly from the ice, is zero (all our work is due to the reversible heat engine), so we find from equation (8–12) that

$$Wk_{\text{use}} = 0 = -\Delta\Phi$$

and

$$\Delta\Phi = -12.4 \text{ Btu} + (1 \text{ lbf})(y)$$

From this, we obtain

Answer

$$y = 12.4 \text{ Btu}/1 \text{ lbf} = 12.4 \times 778 = 9609 \text{ ft}$$

Notice in this problem that the actual lifting of the water is contingent on a number of mechanical apparatus: a reversible heat engine; a transmission to physically hoist the water; and some type of reversible heat transfer pipe from the heat engine to both atmosphere and water.

We concern ourselves next with some remarks of the degradation of energy to an unavailable state.

8–3 ENERGY DEGRADATION

Although we have seen that it is not energy but availability that is desirable, we will use the term **energy degradation** to imply a degrading of availability or a moving down the curve in figure 8–5. We lose availability in a real process, and in a reversible process, it merely remains the same. Loss of availability is subtle and difficult to detect, but predicting irreversibilites, degradation of energy, or increases in entropy in a general process is still more difficult. One of the important areas of research in thermodynamics is the development of more general and accurate equations or theorems that can allow these predictions of irreversibility—it is a crucial area since most of our present tools of thermodynamics can accurately be used only for reversible and static systems, which are indeed not the real world.

EXAMPLE 8–3

We seek a comparison of the maximum useful work obtainable from two systems that begin at identical states, proceed through different processes, and reach states that are not too different.

Solution

System 1, shown in figure 8–7, is an adiabatic chamber enclosing air in the left half. The right half is a vacuum separated from the left by a removable wall. The air pressure is 3.0 bar, the temperature is 40°C, and the volume is 0.5 m^3. The volume of the vacuum is also 0.5 m^3, so

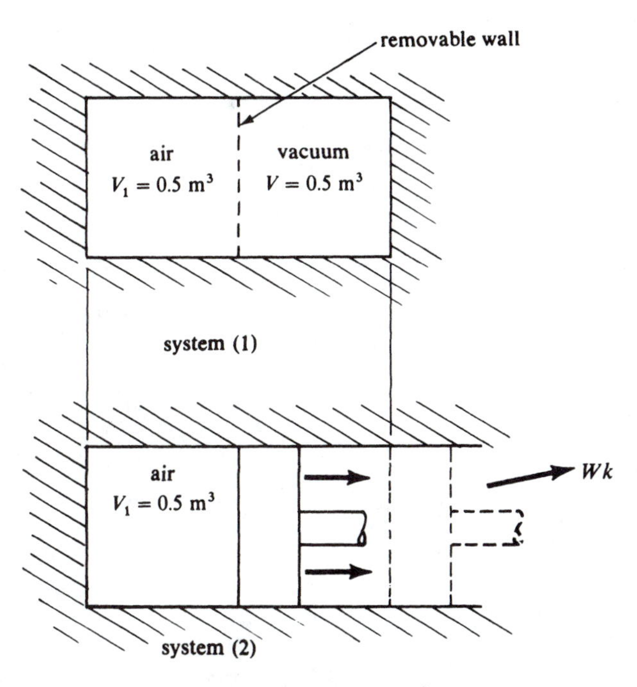

FIGURE 8–7 Two systems that are going through adiabatic processes with identical volumes but that will reach different states and have different amounts of work produced

when the removable wall is taken away, the air will occupy the full 1.0-m^3 volume. When this happens, the air pressure will drop to 1.5 bar, but the temperature will remain at 40°C. Why?

System 2 in figure 8–7 is a frictionless adiabatic piston-cylinder device that initially encloses 0.5 m^3 of air at 3.0-bar pressure and 40°C. The system then progresses through a reversible adiabatic process until the air fills a volume of 1.0 m^3.

Both systems are at final states that are slightly different; from system 2, we have extracted some useful work while system 1 has its initial energy value. Heat has not been transferred in either system, but we can ask, "Have we lost any availability in system 1 by not utilizing the expansion of the air from 0.5 m^3 to 1.0 m^3?" The answer, as we will see, is "Yes."

First let us look at system 1. We assume the air to be a perfect gas, so we have the following results:

$$\Delta s = s_2 - s_1 = R \ln \frac{V_2}{V_1}$$

$$= (287\ \text{J/kg}\cdot\text{K})\left(\ln \frac{1\ \text{m}^3}{0.5\ \text{m}^3}\right)$$

$$= 199\ \text{J/kg}\cdot\text{K} \tag{8–15}$$

We know that the work and heat are both zero, so

$$wk = 0 = q \tag{8–16}$$

From the first law, we must then have

$$\Delta u = u_2 - u_1 = 0 \tag{8–17}$$

Because the air acts as a perfect gas,

$$\Delta T = T_2 - T_1 = 0 \tag{8–18}$$

so, for the isothermal expansion,

$$p_2 = p_1 \frac{V_1}{V_2} = 1.5\ \text{bar}$$

For system 2, with air again acting in the system, we have these results:

$$\Delta S = 0$$
$$q = 0$$
$$wk = \frac{R}{1-k}(T_2 - T_1)$$
$$T_2 = T_1\left(\frac{V_1}{V_2}\right)^{k-1} = 237\text{ K} \qquad \textbf{(8–19)}$$

The work is computed, to give us

$$wk = \frac{287\text{ J/kg}\cdot\text{K}}{1-1.4}(237\text{ K} - 313\text{ K})$$
$$= 54{,}500\text{ J/kg} = 54.5\text{ kJ/kg}$$

Also, we get

$$\Delta u\text{–}wk = -54.5\text{ kJ/kg}$$

and

$$p_2 = p_1\left(\frac{V_1}{V_2}\right)^k = 1.14\text{ bar}$$

Obviously, the availability was the same for both systems at their initial states. To determine the availability of the systems at their final states, we need the change in availability. We find this change from equation (8–12):

$$\Delta\Phi = \Phi_2 - \Phi_1 = U_2 - U_1 + p_a(V_2 - V_1) - T_a(S_2 - S_1)$$

For system 1, the change is found to be

$$\Delta\Phi = 0 + p_a(V_2 - V_1) - T_a m(\Delta s)$$

We now assume that the surroundings are at a temperature of 27°C and a pressure of 1.01 bar. The system mass is computed from the perfect gas relation:

$$m = \frac{p_1V_1}{RT_1} = \frac{(3.9\times10^5\text{ N/m}^2)(0.5\text{ m}^3)}{(287\text{ J/kg}\cdot\text{K})(313\text{ K})}$$
$$= 1.67\text{ kg}$$

Thus, for system 1, the change in availability is

$$\Delta\Phi = (1.01\times10^5\text{ N/m}^2)(0.5\text{ m}^3)$$
$$- (300\text{ K})(1.67\text{ kg})(199\text{ J/kg}\cdot\text{K})$$
$$= -49.1\text{ kJ}$$

For system 2, the change in availability is

$$\Delta\Phi = m(u_2 - u_1) + p_a(V_2 - V_1) - Tm(\Delta s)$$
$$= (1.67\text{ kg})(-54.5\text{ kJ/kg}) + (1.01\times10^5\text{ N/m}^2)(0.5\text{ m}^3) - 0$$
$$= -91.0\text{ kJ} + 50.5\text{ kJ} = -40.5\text{ kJ}$$

Notice that because system 2 involved a reversible process, the change in availability is exactly equal in magnitude to the useful work obtained from the piston-cylinder. That is, from equation (8–6), the useful work is just +40.6 kJ. System 1, by going through an irreversible expansion, decreased its availability more than system 2, even though its energy remained constant. The apparent contradiction that gives system 2 a higher final availability than system 1 is due to the fact that the final temperature of system 2 is 237 K, or −36°C, well below the temperature of the surroundings, therefore providing a capability for heat transfer to those surroundings.

One of the practical conclusions to be drawn from example 8–3 is that a gas in a cylinder or any other source of availability must be used in a work process during the actual expansion or else the availability is lost forever. It is exactly like water flowing over a drop: if the potential energy of the water is not immediately converted into some other energy, such as electric power, then the water will flow down and dissipate (quite irreversibly) the kinetic energy and degrade the availability of the energy of that portion of water. Thus, engineers and technologists must be opportunistic in generating power for society by extracting energy from processes at the appropriate times and by better utilizing the naturally occurring processes of nature (such as tides, tornadoes, hurricanes, and winds) for energy sources.

One other task of those who furnish power or availability to society is the storage of such products. In mechanics, a flywheel rotates at high speed, storing kinetic energy for later use. Electric energy is stored in batteries or capacitors; thermal or internal energy is stored in material that is retained in insulated or adiabatic chambers. When the energy is put into storage, we lose some availability in the mere process of storing. The term **unavailable energy** has traditionally been associated with the amount of energy that is lost during a process and that can never again be put to useful work. **Available energy**, on the other hand, has been associated with that energy that can be reconverted to useful work; it is efficiently stored for future use. These two terms are tied up in the availability concept, but an example problem might provide some enlightenment.

EXAMPLE 8–4 A heat exchanger is a device that transfers thermal energy from one material to another. In figure 8–8 is shown such a heat exchanger which allows hot exhaust gases to transfer some of their energy to air flowing in the opposite direction. Outside the heat transfer, $\dot{Q}$, between the two streams, the system can be assumed to be adiabatic. Assume that 6.1 kg/s of hot gases enter at 500°C and leave at 200°C, all at a pressure of 1.01 bar. There are 6.0 kg/s of air flowing into the exchanger at 120°C and 1.01 bar. The specific heats at constant pressure are 0.7 kJ/kg · K for the gases and 1.007 kJ/kg · K for the air; both are assumed to be perfect gases having R values of 189 J/kg · K and 287 J/kg · K, respectively. If the atmospheric conditions are 27°C and 1.01 bar, determine the loss in availability per second of the system due to the heat exchange.

Solution We recognize that here we have an open system and that our concepts of useful work and availability were adapted to closed systems. It can be shown, however, that, for open, steady-flow systems, the rate of change in availability, $\delta\Phi/\delta t$, is obtained from

$$\dot{\Phi} = \dot{m}\left[(h_2 - h_1) + \frac{V_2^2 - V_1^2}{2g_c} + \frac{g}{g_c}(z_2 - z_1) - T_a(s_2 - s_1)\right] \tag{8–20}$$

The total rate of change in availability for this system is given by

$$\dot{\Phi} = \dot{\Phi}_{\text{air}} + \dot{\Phi}_{\text{gas}} \tag{8–21}$$

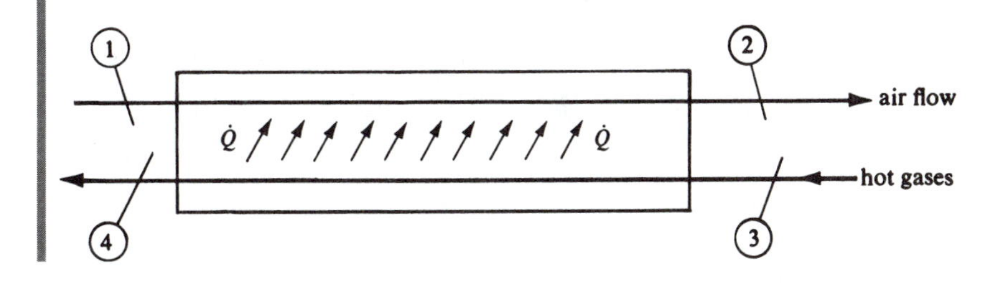

FIGURE 8–8 Heat exchanger

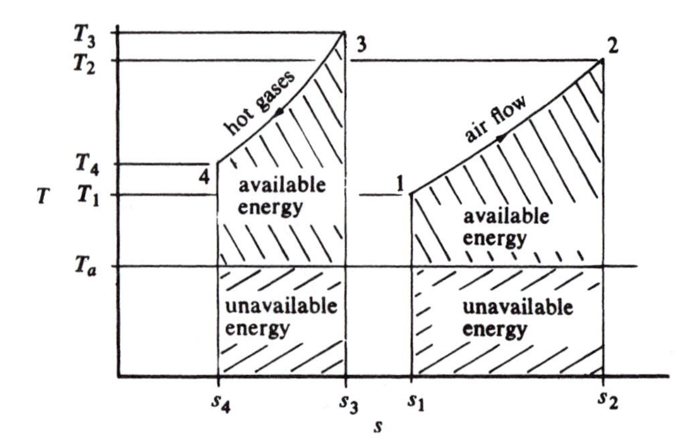

FIGURE 8–9 *T–s* diagram for heat exchanger

In figure 8–9 is shown the *T–s* diagram for the air and gas simultaneously. The total areas under each curve are equal and represent the heat transferred between the gas and air, $\dot{Q}$. That is, the total area under curve 3–4 is equal to the total area under 1–2. Notice also that the areas under the T_a abscissa line are labeled as "unavailable energy." This is a good way of visualizing equation (8–20) or (8–12), but in this graphical representation, the $p_a(V_2 - V_1)$ term is neglected. We see also that the "available energy" areas of figure 8–9 are equal to the availability per unit mass for this steady-flow problem. The generation of unavailable energy is then given by the expression

$$\dot{m}(T_a \, \Delta s) \tag{8–22}$$

For the gases, we determine the following properties:

$$\begin{aligned} h_4 - h_3 = c_p(T_4 - T_3) &= (0.7 \text{ kJ/kg} \cdot \text{K})(473 \text{ K} - 773 \text{ K}) \\ &= -210 \text{ kJ/kg} \\ s_4 - s_3 = c_p \ln \frac{T_4}{T_3} &= (0.7 \text{ kJ/kg} \cdot \text{K})\left(\ln \frac{473 \text{ K}}{773 \text{ K}}\right) \\ &= -0.344 \text{ kJ/kg} \cdot \text{K} \end{aligned}$$

From these results and using equation (8–20), we can calculate the rate of change of availability for the gas, assuming no kinetic or potential energy changes:

$$\begin{aligned} \dot{\Phi}_{\text{gas}} &= \dot{m}_{\text{gas}}[(h_4 - h_3) - T_a(s_4 - s_3)] \\ &= (6.1 \text{ kg/s})[-210 \text{ kJ/kg} + (300 \text{ K})(0.344 \text{ kJ/kg} \cdot \text{K})] \\ &= -651 \text{ kJ/s} \end{aligned}$$

The available energy per unit mass of the gas in the temperature–entropy diagram of figure 8–9 is equal to $h_4 - h_3 - T_a(s_4 - s_3)$, or -106.8 kJ/kg, and the unavailable energy is given by the term $T_a(s_4 - s_3)$ and equals -103.2 kJ/kg.

For the air, we have

$$h_2 - h_1 = c_p(T_2 - T_1) = (1.007 \text{ kJ/kg} \cdot \text{K})(T_2 - 393 \text{ K})$$

and the entropy change is given by

$$s_2 - s_1 = c_p \ln \frac{T_2}{T_1}$$

To determine the final air temperature, we notice from the first law of thermodynamics applied to the heat exchanger that the rate of increase in enthalpy of the air is exactly the same magnitude as the decrease of the gas. Therefore,

$$\dot{m}_{\text{air}}(h_2 - h_1) = \dot{m}_{\text{gas}}(h_3 - h_4)$$
$$(6.0 \text{ kg/s})(h_2 - h_1) = (6.1 \text{ kg/s})(210 \text{ kJ/kg})$$

and, using the relationship

$$h_2 - h_1 = c_p(T_2 - T_1)$$

we obtain

$$T_2 = T_1 + \frac{(6.1 \text{ kg/s})(210 \text{ kJ/kg})}{(6.0 \text{ kg/s})(1.007 \text{ kJ/kg}\cdot\text{K})}$$

Since T_1 is 393 K, we obtain

$$T_2 = 393 \text{ K} + 212 \text{ K} = 605 \text{ K}$$

Then the entropy change is computed:

$$s_2 - s_1 = (1.007 \text{ kJ/kg}\cdot\text{K})\left(\ln\frac{605 \text{ K}}{393 \text{ K}}\right) = 0.434 \text{ kJ/kg}\cdot\text{K}$$

Further, the enthalpy change is

$$h_2 - h_1 = (1.007 \text{ kJ/kg}\cdot\text{K})(605 \text{ K} - 393 \text{ K})$$
$$= 213 \text{ kJ/kg}$$

From equation (8–20), the rate of change of availability is

$$\dot{\Phi}_{\text{air}} = \dot{m}_{\text{air}}[(h_2 - h_1) - T_a(s_2 - s_1)]$$
$$= (6.0 \text{ kg/s})[213 \text{ kJ/kg} - (300 \text{ K})(0.434 \text{ kJ/kg}\cdot\text{K})]$$
$$= 497 \text{ kJ/s}$$

From equation (8–21), the total rate of change of availability for the heat exchanger is

$$497 \text{ kJ/s} - 651 \text{ kJ/s} = -154 \text{ kJ/s}$$

The available energy of the air per kilogram is given by

$$\frac{\dot{m}_{\text{air}}}{\dot{m}_{\text{gas}}}[(h_2 - h_1) - T_a(s_2 - s_1)]$$

Substituting values, we obtain

$$\frac{6.0 \text{ kg/s}}{6.1 \text{ kg/s}}[(213 \text{ kJ/kg}) - (300 \text{ K})(0.434 \text{ kJ/kg}\cdot\text{K})] = 81.4 \text{ kJ/kg}$$

The unavailable energy change per kilogram of gas is

$$\frac{\text{air}}{\dot{m}_{\text{gas}}}(T_a)(s_2 - s_1) = \frac{6.0 \text{ kg/s}}{6.1 \text{ kg/s}}(300 \text{ K})(0.434 \text{ kJ/kg}\cdot\text{K})$$
$$= 128.1 \text{ kJ/kg}$$

The total available energy change per kilogram of gas is obtained from the difference of the two designated areas under the curves in figure 8–9 or from the sum of the available energy changes of the gas and of the air. Thus, it is

$$-106.8 \text{ kJ/kg} + 81.4 \text{ kJ/kg} = -25.4 \text{ kJ/kg}$$

Similarly, the unavailable energy change per kilogram of gas is equal to the sum of the unavailable energy changes for the gas and air, or

$$-103.2 \text{ kJ/kg} + 128.1 \text{ kJ/kg} = 24.9 \text{ kJ/kg}$$

The decrease in available energy here corresponds to the decrease in availability or increase in unavailable energy. The total generation of unavailable energy is given by

$$\begin{aligned} \dot{m}_{\text{air}} T_a \, \Delta s_{\text{air}} + \dot{m}_{\text{gas}} T_a \, \Delta s_{\text{gas}} &= (6.0 \text{ kg/s})(300 \text{ K})(0.434 \text{ kJ/kg} \cdot \text{K}) \\ &\quad - (6.1 \text{ kg/s})(300 \text{ K})(0.344 \text{ kJ/kg} \cdot \text{K}) \\ &= 154 \text{ kJ/s} \end{aligned}$$

8–4 FREE ENERGY

For a system that is in thermal and pressure equilibrium with its surroundings, that is, $T = T_a$ and $p = p_a$, we can write the change in availability in the form

$$\Delta\Phi = (E_2 - E_1) + p(V_2 - V_1) - T(S_2 - S_1) \tag{8–23}$$

If there are no kinetic and potential energy changes, we have

$$E_2 - E_1 = U_2 - U_1$$

and then

$$\Delta\Phi = (U_2 - U_1) + p(V_2 - V_1) - T(S_2 - S_1) \tag{8–24}$$

The property identified with the quantity $U + pV - TS$ we call the **Gibbs free energy** $\mathcal{G}$ and write

$$\mathcal{G} = U + pV - TS \tag{8–25}$$

The Gibbs free energy has units of energy, and the **specific Gibbs free energy**, defined as

$$g' = \frac{\mathcal{G}}{m} = u + pv - Ts \tag{8–26}$$

has units of energy per unit mass. These properties are quite useful in analyzing chemical reactions, in predicting heats of combustion tabulated in table B–26, and, of course, in predicting the useful work obtainable from a system whose pressure and temperature are equal to constant atmospheric conditions. If this equilibrium holds for a system, then

$$\Delta\Phi = \Delta\mathcal{G} \tag{8–27}$$

and we can directly say that

$$Wk_{\text{use}} = -\Delta\mathcal{G} \tag{8–28}$$

For systems that are at constant volume, we predict the maximum useful work from

$$Wk_{\text{use}} = (E_2 - E_1) - T_a(S_2 - S_1) \tag{8–29}$$

and, again, if the system is in thermal equilibrium with the atmosphere at constant temperature $(T = T_a)$,

$$Wk_{\text{use}} = (E_2 - E_1) - T(S_2 - S_1) \tag{8–30}$$

We associate with the quantity $U - TS$ a property of the system called the **Helmholtz free energy** H', written

$$H' = U - TS \tag{8–31}$$

and obviously, for the foregoing conditions of the system with no kinetic or potential energy changes,

$$Wk_{\text{use}} = -\Delta H' \tag{8–32}$$

We can consider also the **specific Helmholtz free energy**, defined by

$$h' = \frac{H'}{m} = u - Ts \tag{8–33}$$

and thereby get an intensive property of the system.

The Gibbs and Helmholtz free energy properties are frequently used in thermodynamic literature, and this introduction to their derivation should be helpful to the engineer and technologist seeking the full use of thermodynamic tools.

EXAMPLE 8–5 Compute the specific Gibbs free energy and the specific Helmholtz free energy functions for saturated mercury vapor at 30 psia.

Solution

The specific Gibbs free energy function, g', can be computed from equation (8–26). First, we obtain the properties of saturated mercury vapor from table B–14. At 30 psia, we find that

$$\begin{aligned} h_g &= 147.2 \text{ Btu/lbm} \\ v_g &= 2.053 \text{ ft}^3\text{/lbm} \\ t &= 750.9°\text{F} = 1210.9°\text{R} \\ s_g &= 0.1331 \text{ Btu/lbm} \cdot °\text{R} \end{aligned}$$

so we then have

$$\begin{aligned} g' = h_g - Ts_g &= 147.2 \text{ Btu/lbm} - (1210.9°\text{R})(0.1331 \text{ Btu/lbm} \cdot °\text{R}) \\ &= -13.97 \text{ Btu/lbm} \end{aligned}$$

Answer

The specific Helmholtz free energy function can be computed from equation (8–33):

$$\begin{aligned} h' &= u_g - Ts_g = g' - pv_g \\ &= -13.97 \text{ Btu/lbm} - \frac{(30 \text{ psia})(2.053 \text{ ft}^3\text{/lbm})(144 \text{ in}^2\text{/ft}^2)}{778 \text{ ft} \cdot \text{lbf/Btu}} \\ &= -25.37 \text{ Btu/lbm} \end{aligned}$$

Answer

EXAMPLE 8–6 Determine the specific Gibbs and Helmholtz free energies for R32 at 100 psia and 160°F using *EES*.

Solution

The specific Gibbs and Helmholtz free energies are given by equations (8–26) and (8–33). Noting that the term $u + pv$ is the same as the enthalpy, we can then write in the *EES Equation* window, with English units,

gf = enthalpy (R32, T = 160, P = 100) − (160)*(entropy (R32, T = 160, P = 100))

hf = intEnergy (R32, T = 160, P = 100) − (160)*(entropy (R32, T = 160, P = 100))

Clicking the *Solve* option gives us

$$\begin{aligned} gf &= 161.6 \text{ Btu/lbm} \\ hf &= 139.2 \text{ Btu/lbm} \end{aligned}$$

Notice that *EES* asks for temperature in Fahrenheit, and even in the term *Ts*, the temperature is Fahrenheit. *EES* automatically converts this temperature to degrees Rankine before computing the free energies. When you use equations (8–26) or (8–33) with a hand calculation, however, be sure that the temperatures are degrees Rankine.

8–5 SUMMARY

Useful work is that work of a system which can be used in another system other than the surrounding atmosphere. Useful work from a system in an atmosphere at a pressure of p_a is given by

$$Wk_{use} = \sum p\,\delta V - p_a(V_2 - V_1) \tag{8–2}$$

Useful work is also given by the result

$$Wk_{use} \leq T_a(S_2 - S_1) - p_a(V_2 - V_1) - (E_2 - E_1) \tag{8–7}$$

Availability is defined by the equation

$$\Phi = E - E_a + p_a(V - V_a) - T_a(S - S_a) \tag{8–12}$$

where E_a, V_a, and S_a are the properties of the system when it is in equilibrium with the surroundings. The availability and useful work are related by the equality/inequality

$$Wk_{use} \leq -\Delta\Phi \tag{8–14}$$

In a process of a system, the available energy is the reversible work or heat, and the unavailable energy is the energy that cannot be used for work again. The change in the unavailable energy of a process is given by $mT_a(\Delta s_{total})$.

The Gibbs free energy is defined as

$$\mathcal{G} = U + pV - TS \tag{8–25}$$

and the Helmholtz free energy is defined as

$$H' = U - TS \tag{8–31}$$

DISCUSSION QUESTIONS

Section 8–1

8–1 What is meant by a *useful work*?

8–2 Is useful work increased or decreased by an increase in atmospheric pressure, or pressure which surrounds the system?

Section 8–2

8–3 What is the term *availability*?

8–4 Is *availability* a property of a system?

Section 8–3

8–5 Can energy be degraded without entropy increase?

8–6 Can you think of any process which does not degrade energy?

Section 8–4

8–7 What is meant by the term *free energy*?

PRACTICE PROBLEMS

Problems that use SI units are indicated by an (M) under the problem number; those that use English units are indicated by an (E). Assume that the atmospheric conditions are 27°C (80°F) and 101 kPa (14.7 psia) for the following problems.

Sections 8–1 and 8–2

8–1 (M) Determine the Λ function of 2 kg of H_2O at 240°C and
(a) 400-kPa pressure
(b) Saturated vapor
(c) Saturated liquid

8–2 (M) Determine the availability of a vacuum of 2 m^3.

8–3 (M) Determine the availability of the following gases at 500°C and 101 kPa, assuming each to be a perfect gas:
(a) Oxygen, O_2
(b) Methane, CH_4
(c) Carbon dioxide, CO_2
(d) Carbon monoxide, CO

8–4 (M) Determine the vertical distance through which 3 kg of ice at 0°C can lift itself at atmospheric conditions. The heat of fusion of water at 0°C is 152 kJ/kg.

8–5 (E) Determine the Λ function of 6 lbm of mercury vapor at 90 psia and
(a) Saturated vapor
(b) Saturated liquid
Assume that $v_f = 0.0013\ \text{ft}^3/\text{lbm}$.

8–6 (E) Determine the availability of 3 lbm of Refrigerant-22 at −20°F and
(a) Saturated vapor
(b) Saturated liquid

8–7 (E) Determine the maximum useful work obtainable from 1 lbm of the following perfect gases at 1000°F and 200 psia:
(a) Air
(b) Nitrogen, N_2
(c) Hydrogen, H_2
(d) Sulfur dioxide, SO_2

8–8 (E) Calculate the maximum useful work obtainable for
(a) 1 lbm H_2O at 700°F and 200 psia
(b) 1 lbm H_2O at 1000°F and 200 psia
(c) 1 lbm H_2O at 700°F and 300 psia

Section 8–3

8–9 (M) Calculate the generation or rate of change of availability for 200 kg/min of air flowing through a reversible adiabatic turbine from 650°C and 1500-kPa pressure, exhausting to 101 kPa. (*Hint:* Assume that there are no kinetic or potential energy changes and that the power generated in the turbine is all useful.)

8–10 (M) There is 0.2 kg of air contained in a frictionless piston-cylinder. The air is at a pressure of 2000 kPa and a temperature of 600°C. Through a reversible adiabatic process the air pressure is reduced to 100 kPa. Determine the change in availability of the air and the total change in availability of the universe.

8–11 (E) A piston-cylinder contains 0.5 lbm of oxygen gas at 30 psia and 100°F (figure 8–10). The device then compresses the gas in a reversible manner to a pressure of 300 psia. Assume that $n = 1.45$ and that the gas behaves as a perfect gas with constant specific heats. Determine the increase in availability of the oxygen and the total change in availability of the universe due to this compression process.

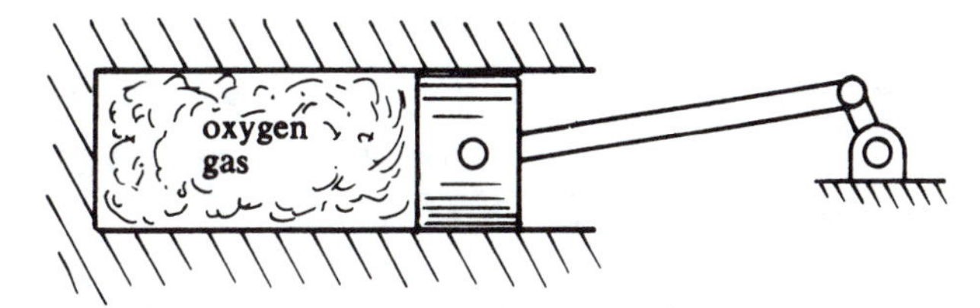

FIGURE 8–10

8–12 (E) Calculate the total rate of change of availability for 20 lb/s of air compressed in a reversible adiabatic manner from 100°F and 15 psia to 180 psia. (*Hint:* Assume no kinetic or potential energy changes. Obviously, the work put into the compressor is useful.)

8–13 (E) In the Φ–S–V space, determine expressions for the tangent or slope of the Φ surface in the plane of
(a) Constant entropy
(b) Constant volume
(c) Constant availability

Section 8–4

8–14 (M) Determine the specific Gibbs and the specific Helmholtz free energy functions for H_2O at 320°C and
(a) 1000-kPa pressure
(b) 10-kPa pressure
(c) Saturated liquid
(d) Saturated vapor

8–15 (E) Determine the Gibbs free energy for ammonia, NH_3, at 120°F and 20 psia. Use the appendix tables and compare with the answers you obtained by using *EES*.

8–16 (E) Determine the Helmholtz free energy for ammonia at 200°F and 60 psia. Use the appendix tables and compare with the answers you obtained by using *EES*.

8–17 (M) Using chart B–8, determine the Gibbs and Helmholtz free energies of R407c at 200 kPa and 70°C.

8–18 (M) Using chart B–9, determine the Gibbs and Helmholtz free energies of R502 at 100 kPa and 100°C. Compare your answers with what you obtained by using *EES*.

THE INTERNAL COMBUSTION ENGINE AND THE OTTO AND DIESEL CYCLES

9

The internal combustion (IC) engine, characterized by the reciprocating piston-cylinder gasoline engine, is probably the most common power-producing device in our society. It is utilized in all phases of transportation, for auxiliary electrical power generation, and in innumerable small, portable power tools. This engine is an example of what we have defined as a heat engine, and although the concept of an external heat source may not be clearly compatible with the practicality of an internal combustion (to the system) of fuel, we will see how this ambiguity can be eliminated.

We define the ideal Otto cycle and see how it fits into the analysis of the spark-ignited internal combustion engine. Initially, we lean heavily on the assumption that air is the working fluid, but use of the gas tables in solving Otto cycle problems will help to explain the departures of the air–fuel mixtures and exhaust gases from this assumption. A BASIC program, OTTO, is given to show the student a convenient way of analyzing the Otto cycle and its variations for applications to the Otto engines. The analogy between the actual engine and Otto cycle will be sharpened by using the polytropic process equations. The actual Otto cycle is then discussed with some of the more traditional modifications designed to improve the engine power, efficiency, fuel economy, or other engine performance parameters.

We introduce the diesel cycle and its adaption in the diesel engine. The air-standard analysis is treated thoroughly, and a comparison between the diesel and Otto cycles will better focus on the reasons for each and their traditional areas of application. The dual cycle is defined to better model the operation of actual diesel engines. A BASIC program, DIESEL, is presented to show the student a method of using the computer to assist in analyzing the diesel engines.

Finally, we discuss some of the design problems inherent in the reciprocating IC engine and the approaches for solving them. The engines described in this chapter have probably absorbed more engineering talent than any other comparable device. An overview of the shortcomings and the advantages of these engines should be useful to the student.

New Terms

bhp	Brake horsepower	p_r	Relative pressure
bmep	Brake mean effective pressure	r_c	Cutoff ratio
bsfc	Brake specific fuel consumption	r_v	Compression ratio
HHV	Higher heating value	sfc	Specific fuel consumption
ihp	Indicated horsepower	V_D	Displacement
imep	Indicated mean effective pressure	v_r	Relative volume
LHV	Lower heating value	η_{mech}	Mechanical efficiency
mep	Mean effective pressure	η_v	Volumetric efficiency
n	Number of cylinders in an engine		
N	Number of power cycles per unit time		

9–1 THE INTERNAL COMBUSTION ENGINE

An internal combustion (IC) engine is a heat engine that receives heat at a high temperature because of combustion or burning of a fuel inside the engine. The fuel is usually a hydrocarbon-type fuel such as gasoline, kerosene, or diesel fuel oil, but LP gas, methane (natural gas), and others are sometimes used. The fuel is normally mixed with air inside the engine; this mixture then burns rapidly, and the resulting exhaust gases are at a high temperature and pressure. This process is really not a heat transfer from a high-temperature region such as shown in figure 7–1, but is rather a release of chemical energy to provide a high-temperature, high-pressure gas inside the engine. The hot gases are then expanded in the engine, and this produces power or work. Finally, the gases are let out of the engine after reaching a low pressure and temperature, and this release can be compared to a heat transfer to a low-temperature region. Thus, the actual IC engine can be visualized as a form of the heat engine that we have considered in chapter 7.

The detailed manner in which the internal combustion engines operate to convert heat into work and the cycles through which these engines operate are discussed in the following sections. We will see that the piston-cylinder apparatus has been the most common mechanical device of the IC engines, although rotary mechanisms can also be adapted to perform as an IC engine. The Otto and diesel cycles are the two thermodynamic cycles that have been used to describe the operation of the IC engines. The dual cycle is sometimes used to better describe the actual diesel engine.

9–2 THE IDEAL OTTO CYCLE AND AIR-STANDARD ANALYSIS

The ideal Otto cycle is defined as the following set of reversible processes:

- 1–2 Adiabatic compression
- 2–3 Constant-volume heat addition
- 3–4 Adiabatic expansion
- 4–1 Constant-volume heat rejection

These processes are indicated in figure 9–1 where $p-V$ and $T-s$ diagrams are used to describe the Otto cycle. Notice that because these processes are all reversible, we can easily identify the enclosed area on the $p-V$ diagram as the net work of the cycle and the enclosed area on the $T-s$ diagram as the net heat added. To understand how these four processes manage to be descriptive of a real machine operation, let us see how the piston-cylinder can be used with an Otto cycle. Figure 9–2 shows the sequence of motions of the piston corresponding to those processes. If the piston is reciprocating in a continuous manner, processes 2–3 and 4–1 must be performed in a zero time period because there is no motion. This is a deviation of the ideal cycle from the actual case, but we will later see that this is not a large error.

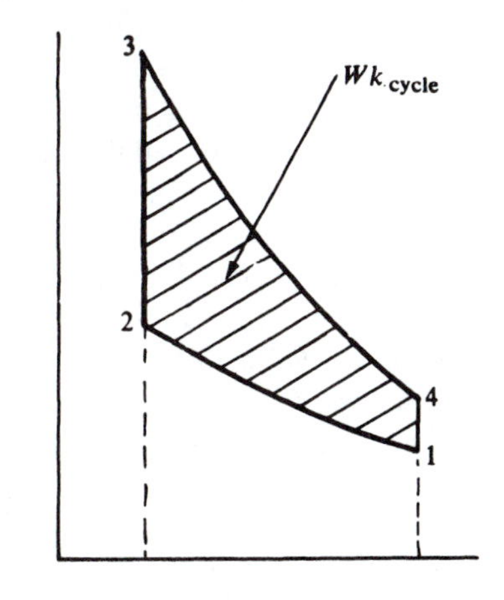

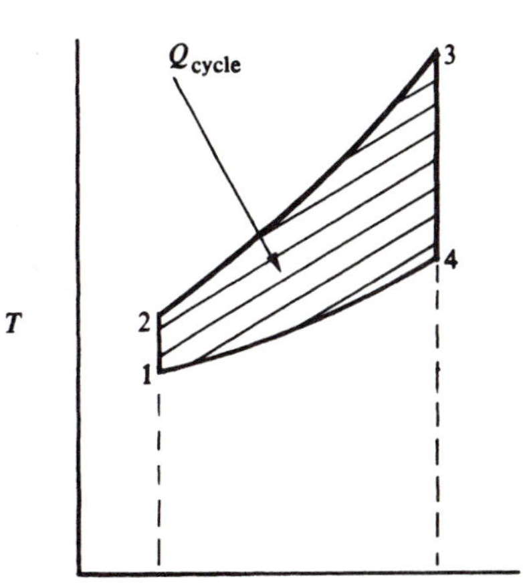

FIGURE 9–1 Property diagrams for ideal Otto cycle

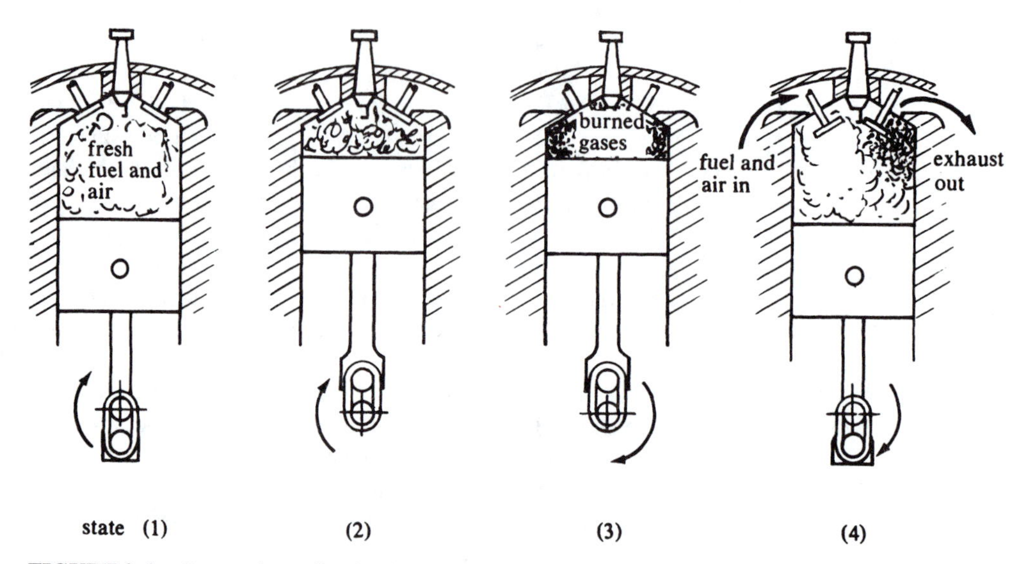

FIGURE 9–2 Otto cycle application in piston-cylinder device

Process 1–2 is a compression of a charge of air and unburned fuel. Frequently, the fuel is added to the air by means of a fuel injection near the end of the compression process 1–2. Fuel injection is usually accomplished by a pump that delivers the fuel at high pressure to a nozzle that sprays the fuel into the combustion chamber. Upon reaching state 2, the spark plug fires, initiating a chemical reaction between the fuel and air. This is the internal combustion that categorizes the cycle and releases energy from the fuel (or adds heat to the piston-cylinder), producing a high-temperature, high-pressure gas that drives the piston through an expansion process to state 4. Frequently, we will call this **spark ignition**, and the engine will be called an internal combustion spark-ignition (ICSI) engine. Heat is then removed by opening the exhaust and intake valves, discharging the burned exhaust gases, and just as quickly replacing this volume with a fresh charge of air and unburned fuel to proceed through the cycle again. This shuttling of gases allows us to have power produced on each cycle, called a **two-stroke cycle** or **two-stroke engine**. It is given this name because two strokes complete the cycle (up–down), but the quick gas shuttle requires imaginative design in valves. The configuration of figure 9–2 probably would not give a good exchange of fresh charge for exhaust, and much effort is expended in attempting to design a better two-stroke engine that does not waste fuel in discharging the exhaust gases. The prize is one power stroke per revolution of the engine. More often, though, the engine is changed from the Otto cycle by proceeding through one more revolution to rinse the exhaust gases and add fresh charges more effectively. This variation we call the **four-stroke cycle**, characterized by the opening of the exhaust valve only when the piston-cylinder achieves state 4. This valve remains open through process 1–5, shown in figures 9–3 and 9–4, during which all the exhaust is pushed out by the piston. At state 5, the exhaust valve closes and the intake valve opens. The fresh fuel–air mixture is then taken in by the retreating piston until state 1 is reached, the intake valve closes, and the normal Otto cycle can proceed. The four-stroke cycle is more commonly used because it allows for better control of the gases than the two-stroke cycle, but there is only one power stroke every other revolution, thus the term "four-stroke," since we have an in-out-in-out motion of the piston for each cycle. Theoretically, the four-stroke engine needs to rotate twice as fast as the two-stroke engine to achieve the same power. This does not generally hold true in practice, however, due to other complications which tend to degrade the attractiveness of the two-stroke cycle. In table 9–1,

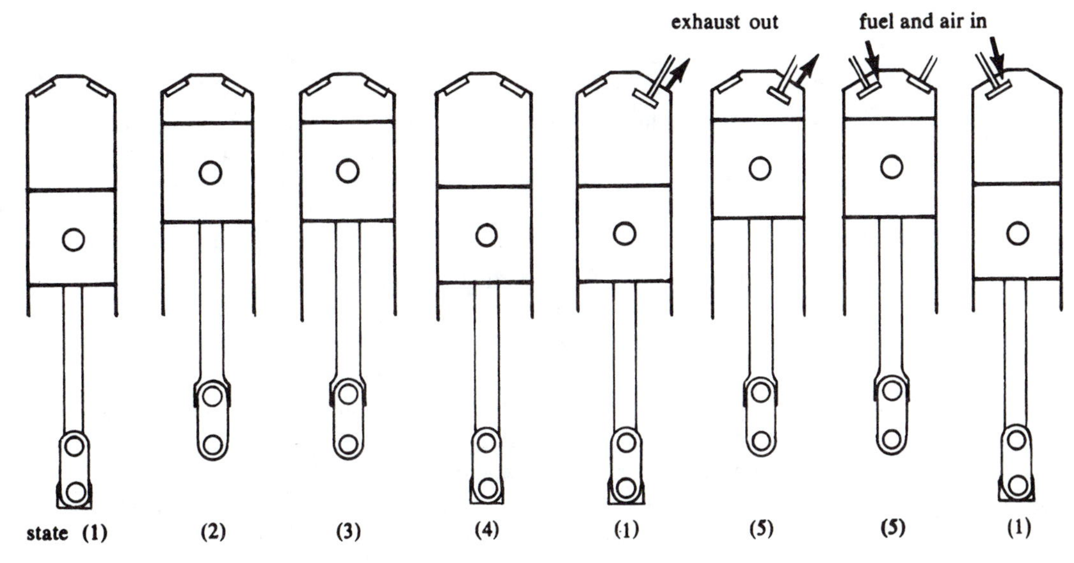

FIGURE 9–3 Four-stroke Otto cycle

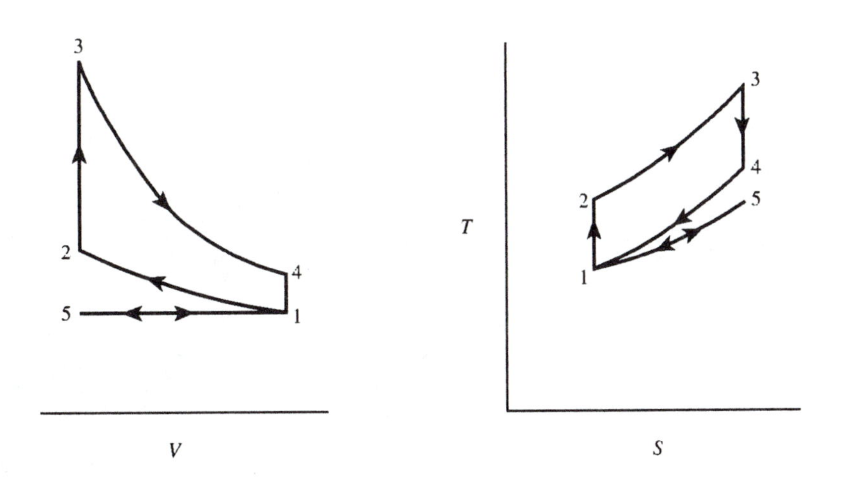

FIGURE 9–4 Property diagrams for the four-stroke Otto cycle

TABLE 9–1 Valve positions of Otto engine

	Four-Stroke Cycle		Two-Stroke Cycle	
Process	Intake Valve	Exhaust Valve	Intake Valve	Exhaust Valve
1–2	Closed	Closed	Closed	Closed
2–3	Closed	Closed	Closed	Closed
3–4	Closed	Closed	Closed	Closed
4–1	Closed	Open	Open	Open
1–5	Closed	Open		
5–1	Open	Closed		

somewhat of a recapitulation, it is indicated when exhaust and intake valves are open or closed. This is done for both the two- and four-stroke cycles and should aid in visualizing the motion of the spark-ignition, internal combustion, piston-cylinder engine.

Figure 9–5 shows a typical cutaway view of an internal combustion engine that operates on a cycle approximately like the Otto cycle. This figure shows the major components of the engine, and figure 9–6 shows a section of an actual IC engine with water cooling. This view shows the external characteristics of the typical water-cooled IC engine with a portion of it cut away to expose the major workings of the internal parts of the engine. Compare this figure with figure 9–5, and the component parts will be better visualized.

The Otto cycle is characterized by heat transfer during constant volume, but the real engine is continually experiencing heat transfers. Because of this, water is generally directed through cavities in the engine block to keep the cylinder and piston from reaching high temperatures, but the air-cooled engine in figure 9–7 is also effective. In this type of engine, water is not used, but air is forced around the cylinder block to provide connective heat transfer, thus keeping the engine cool.

Certain terms are used often and have precise meanings in IC engine analysis. Some of the terms and parameters most useful for us now are shown in figure 9–8. The **bore** refers to the

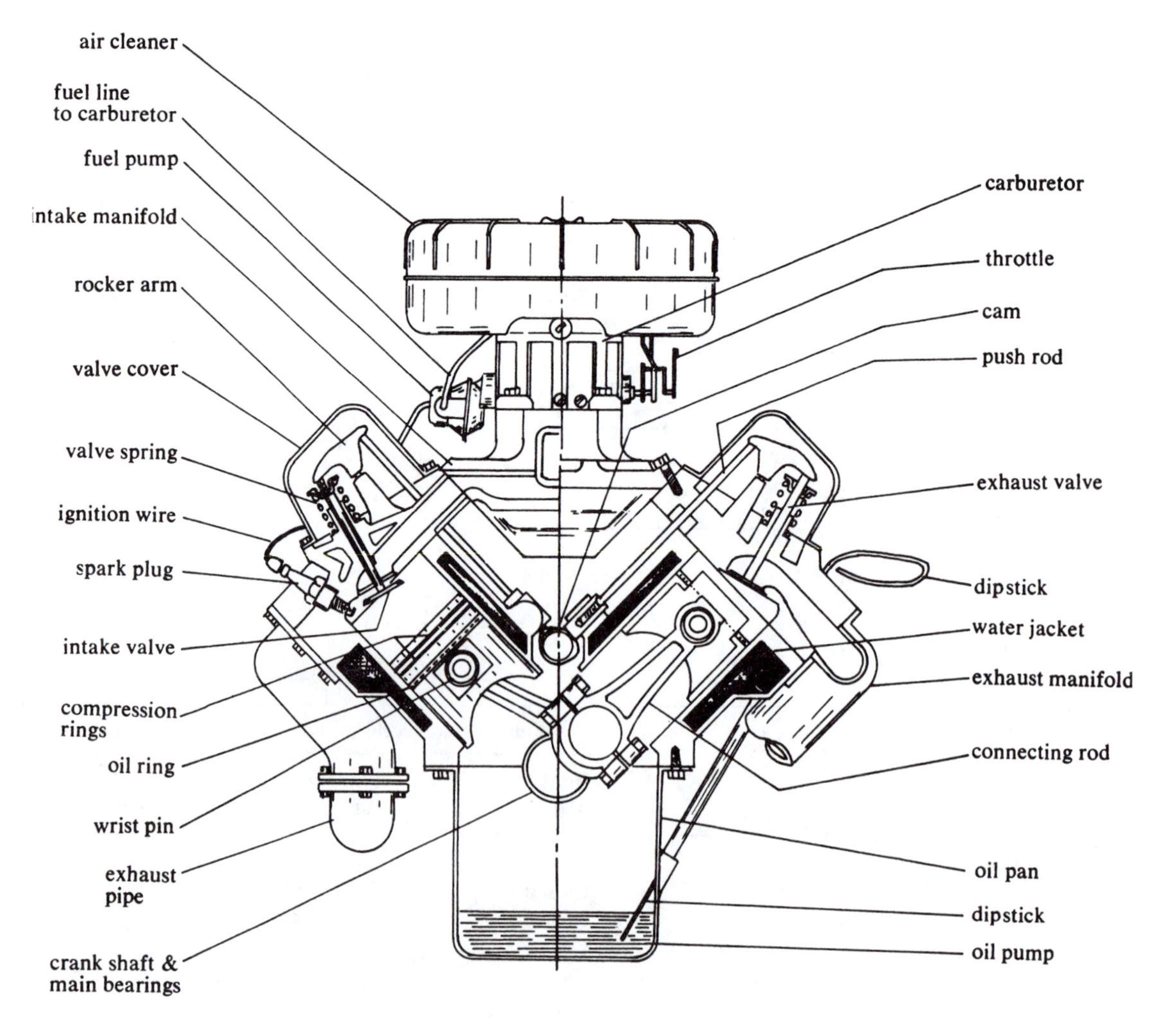

FIGURE 9–5 Typical internal combustion, spark-ignited engine, V configuration (cross-sectional view)

FIGURE 9–6 This is a cutaway view of the second-generation, 4.6-liter, four-cam V-8 Ford engine. (Courtesy of Ford Motor Company, Dearborn, MI)

FIGURE 9–7 The applications of small internal combustion engines often require more continuous power and torque. This engine includes a V-two arrangement and overhead valves to deliver more power at increased fuel economy. (Courtesy of the Kohler Company, Kohler, Wisconsin)

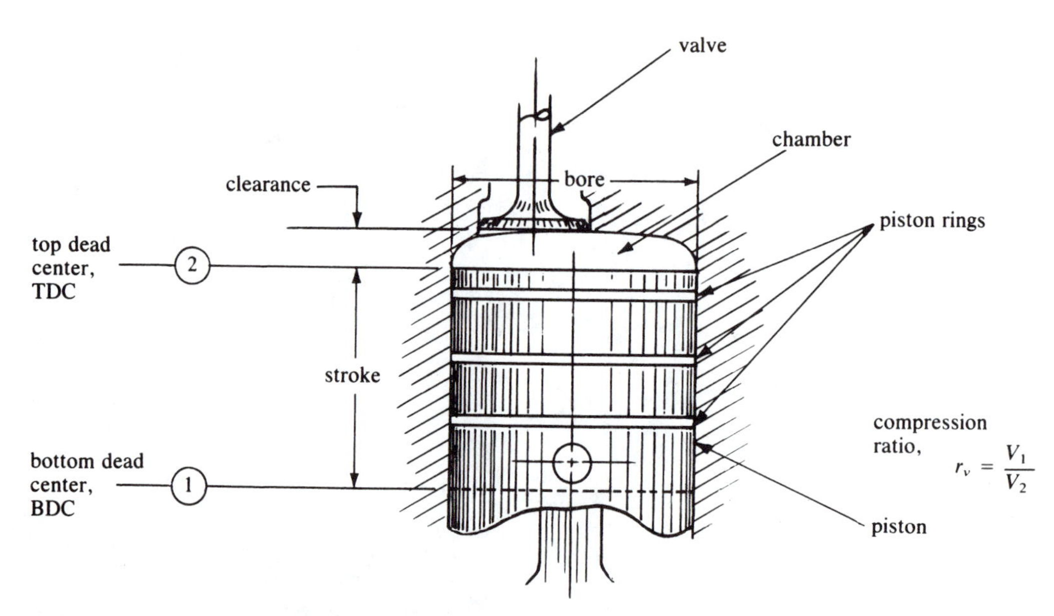

FIGURE 9–8 Common IC engine parameters

diameter of the cylinder, and the **piston** has a slightly smaller diameter than the bore so that it can slide freely in the cylinder. **Piston rings** provide for sealing between the piston and the cylinder so that gases remain inside the **chamber**. The **stroke** is the distance traveled by the piston from **top dead center** (TDC), when the piston is all the way in the cylinder, to **bottom dead center** (BDC), when the piston is retracted as far out as possible. The **clearance** is the minimum distance between the piston and the end of the cylinder, and the **clearance volume** is just that volume. We denote the clearance volume V_2 for most cases. The volume of the chamber when the piston is at *BDC* is denoted V_1, and the **compression ratio,** r_v, is given by

$$r_v = \frac{V_1}{V_2} \tag{9–1}$$

The **cylinder displacement**, V_D, is the volume swept by the piston from *TDC* to *BDC* and is

$$V_D = V_1 - V_2 \tag{9–2}$$

For multiple-piston engines, the engine displacement would be the cylinder displacement times the number of cylinders. Of course, if the engine displacement is given, the displacement for each cylinder would be the displacement divided by the number of cylinders.

EXAMPLE 9–1 An internal combustion, spark-ignited (ICSI) four-cylinder engine has a bore (piston diameter) of 3.0 in, a stroke or piston travel of 3.0 in, and a clearance of 0.4 in. Determine the compression ratio and the engine displacement.

Solution The compression ratio, computed from equation (9–1), can be determined after the two volumes V_1 and V_2 are computed. Referring to figure 9–8, we find the volume V_1 to be

$$\begin{aligned} V_1 &= (\pi)\left(\frac{\text{bore}}{2}\right)^2(\text{clearance} + \text{stroke}) \\ &= (\pi)(1.5\ \text{in})^2(0.4 + 3.0)\ \text{in} \\ &= 24.0\ \text{in}^3 \end{aligned}$$

Similarly, the volume V_2 is

$$V_2 = (\pi)\left(\frac{\text{bore}}{2}\right)^2(\text{clearance})$$
$$= (\pi)(1.5\text{ in})^2(0.4\text{ in}) = 2.83\text{ in}^3$$

and the compression ratio r_v is

Answer

$$r_v = \frac{24.0\text{ in}^3}{2.83\text{ in}^3} = 8.48$$

Frequently, the compression ratio is written 8.48:1, or 8.48 to 1.

The displacement V_D can be computed from equation (9–2) as

$$V_D = (V_1 - V_2)(n)$$

where n is the number of cylinders in the engine. For the engine of this problem, we have

Answer

$$V_D = (24.0 - 2.83)\text{ in}^3(4) = 84.68\text{ in}^3$$

Notice that the displacement is a measure of the volumetric size of the engine.

Frequently, the assumption is made that the air–fuel mixture in the combustion chamber is just air and that the exhaust gases after combustion behave as air would. We call this the **air-standard assumption**, and the study depending on this assumption is the **air-standard analysis**. As it turns out, the amount of air is usually about 20 times as much as the fuel, and the exhaust gases do have properties like those of air, so that the air-standard analysis is a reasonable approach to IC engines. Also, by assuming that air is a perfect gas with constant specific heats, we have more convenient equations without sacrificing accuracy to any great extent.

EXAMPLE 9–2 For the ICSI engine of example 9–1, assume that air is taken into the cylinders at 14.7 psia and 70°F. Determine the pressure and temperature of the air after compression to state 2 if the engine is operating on an Otto cycle.

Solution The compression from state 1 to state 2 is through a reversible adiabatic process. If we assume a perfect gas with constant specific heats, we may write

$$p_1 v_1^k = p_2 v_2^k$$

or

$$p_2 = p_1\left(\frac{V_1}{V_2}\right)^k = p_1 v_v^k$$
$$= (14.7\text{ psia})(8.48)^k$$

Since air is the working medium, we use $k = 1.4$ and obtain

Answer

$$p_2 = 293\text{ psia}$$

Also, we may now find the temperature by using the p–V–T relationships from chapter 6 or by using the perfect gas relation at state 2. If we use the p–V–T relationship, we get

$$T_2 = T_1\left(\frac{V_1}{V_2}\right)^{k-1} = T_1 r_v^{k-1}$$
$$= (70 + 460°\text{R})(8.48)^{0.4}$$

Answer

$$= 1246°\text{R}$$

Notice that the air is very warm after being compressed and would be warm enough to ignite some fuels without external sources such as spark plugs. If this temperature does become too high and ignites the fuel–air mixture, a condition called **preignition**, or "knock," results. Knock causes excessive wear and stresses in the engine and can be avoided only by reducing the compression ratio to reduce the temperature or by using a fuel that will not ignite at the operating temperature.

The Otto cycle defined by the four processes given at the beginning of this section can be analyzed by the methods described in chapter 6. If this air-standard analysis/perfect gas/constant specific heats assumptions are used, the relationships given in table 6–4 can be applied. In principle, the pressures, temperatures, and volumes and other properties at the four states (1, 2, 3, and 4) can be determined as well as the net work, heat added, heat rejected, and thermal efficiency. Let us see how this can be done.

EXAMPLE 9–3 An ICSI Otto cycle engine operates on a four-stroke cycle and has eight cylinders with a total displacement of 1200 cm^3 and a compression ratio of 6:1. The air entering the engine is at 27°C and 101-kPa pressure. The fuel–air mixture during combustion releases 3000 kJ/kg air of heat when the engine is under a load and running at 2200 rpm. Determine the p, V, and T properties at the four corners of the cycle and the power produced by the engine.

Solution The system diagram representing the engine and its heat and work terms is shown in figure 9–9. We assume air-standard analysis with perfect gas/constant specific heats. We have, from table B–4,

$$c_p = 1.007 \text{ kJ/kg}\cdot\text{K}$$
$$c_v = 0.719 \text{ kJ/kg}\cdot\text{K}$$
$$R = 0.287 \text{ kJ/kg}\cdot\text{K}$$

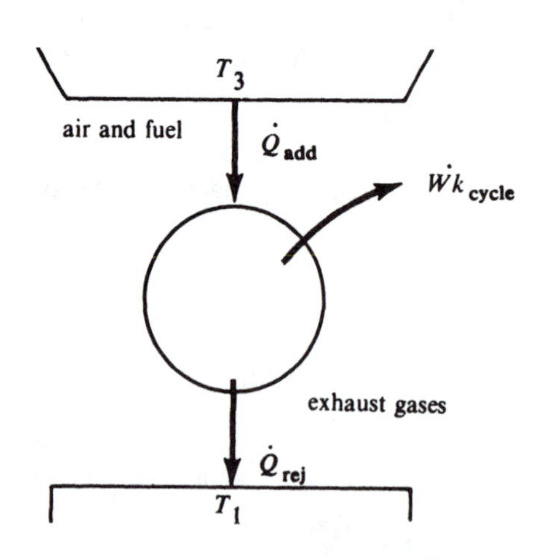

FIGURE 9–9 System diagram for Otto cycle

and

$$k = 1.399$$

To determine the properties at each of the four-cycle corners, let us first sketch the p–V and T–s diagrams. (See figure 9–10.) We see that $V_1 = V_4$ and $V_2 = V_3$. Let us then calculate the volumes from the dimensional data.

To obtain the volume V_1, we notice that

$$V_D = n(V_1 - V_2) = 1200 \text{ cm}^3$$

FIGURE 9–10 Typical Otto engine property diagrams

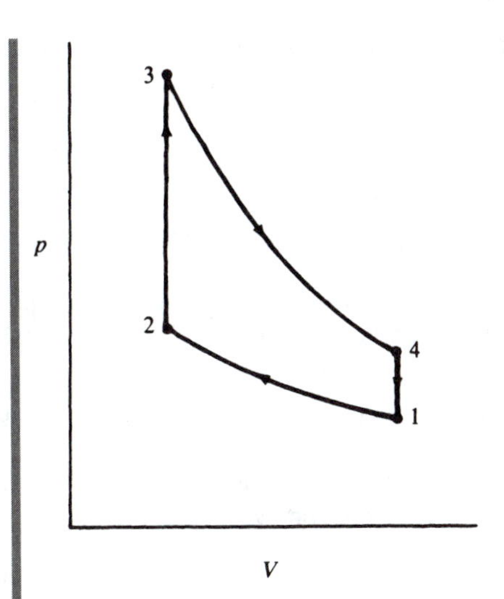

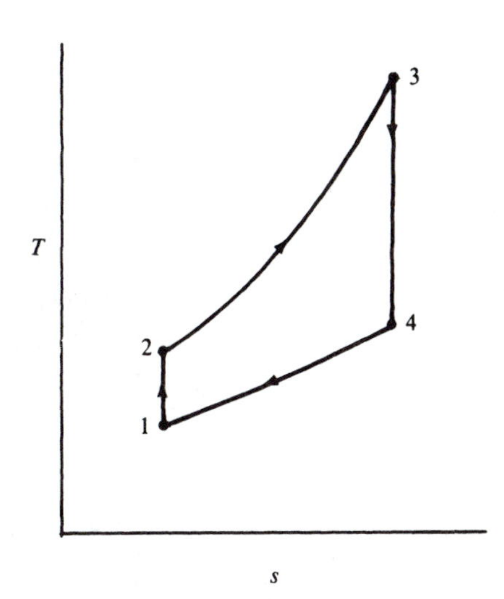

and since $n = 8$, it follows that

$$V_1 - V_2 = \frac{1200}{8} = 150 \text{ cm}^3$$

Further, the compression ratio is

$$r_v = \frac{V_1}{V_2} = 6$$

and if we substitute this into the preceding relation, we have

$$V_1 - \frac{V_1}{6} = 150 \text{ cm}^3$$

Solving for V_1, we get

Answer $$V_1 = 180 \text{ cm}^3$$

Thus, we may then easily obtain V_2:

Answer $$V_2 = \frac{V_1}{6} = 30 \text{ cm}^3$$

Since the cycle is an Otto cycle, we have

Answer $$V_3 = V_2 = 30 \text{ cm}^3$$

and

Answer $$V_4 = V_1 = 180 \text{ cm}^3$$

It is now convenient to calculate the mass of air entering the cylinders during one intake stroke of one cylinder. Using the perfect gas relationship at state 1, we find that

$$p_1V_1 = mRT_1$$

or

$$\begin{aligned} m &= \frac{p_1V_1}{RT_1} \\ &= \frac{(101 \text{ kN/m}^2)(180 \times 10^{-6} \text{ m}^3)}{(0.287 \text{ kN}\cdot\text{m/kg}\cdot\text{K})(300 \text{ K})} \\ &= 0.211 \times 10^{-3} \text{ kg} \end{aligned}$$

To obtain the pressure and temperature at state 2, we use

$$\frac{p_2}{p_1} = \left(\frac{V_1}{V_2}\right)^k = r_v^k$$

and by using $k = 1.399$, we have

Answer

$$p_2 = (101 \text{ kPa})(6)^{1.399} = 1239 \text{ kPa}$$

We may calculate the temperature at state 2 either from a process equation relating states 2 and 1 or from the perfect gas equation:

Answer

$$T_2 = \frac{p_2 V_2}{mR} = \frac{(1239 \text{ kN/m}^2)(30 \times 10^{-6} \text{ m}^3)}{(0.211 \times 10^{-3} \text{ kg})(0.287 \text{ kN} \cdot \text{m/kg} \cdot \text{K})} = 614 \text{ K}$$

Process 2–3 is a reversible, constant-volume process involving a heat transfer and no work, since the system is closed. Then we can write

$$u_3 - u_2 = q_{\text{add}} = 3000 \text{ kJ/kg air}$$

For perfect gases with constant specific heats,

$$u_3 - u_2 = c_v(T_3 - T_2) = 3000 \text{ kJ/kg air}$$

Since c_v is 0.719 kJ/kg · K, we have

$$T_3 - T_2 = \frac{3000 \text{ kJ/kg}}{0.719 \text{ kJ/kg} \cdot \text{K}}$$
$$= 4172 \text{ K} = T_3 - 614 \text{ K}$$

and

Answer

$$T_3 = 4172 \text{ K} + 614 \text{ K} = 4786 \text{ K}$$

Using the constant-volume process relations between states 2 and 3 yields

$$\frac{p_3}{p_2} = \frac{T_3}{T_2}$$

and then

$$p_3 = p_2 \frac{T_3}{T_2} = (1239 \text{ kPa})\left(\frac{4786 \text{ K}}{614 \text{ K}}\right)$$
$$= 9658 \text{ kPa}$$

The properties at state 4 are found with process equations for the reversible adiabatic process 3–4:

$$\frac{p_4}{p_3} = \left(\frac{V_3}{V_4}\right)^k = \left(\frac{1}{r_v}\right)^k = \frac{1}{6^{1.399}}$$

So

$$p_4 = 787 \text{ kPa}$$

For the constant-volume process 4–1, we have

$$\frac{p_4}{p_1} = \frac{T_4}{T_1}$$

$$T_4 = (300 \text{ K})\left(\frac{787 \text{ kPa}}{101 \text{ kPa}}\right) = 2338 \text{ K}$$

We can calculate the entropy change per unit mass from the equations

$$s = s_3 - s_2 = c_v \ln \frac{T_3}{T_2} = (0.719\ \text{kJ/kg}\cdot\text{K})\left(\ln \frac{4786\ \text{K}}{614\ \text{K}}\right)$$
$$= 1.48\ \text{kJ/kg}\cdot\text{K}$$

and

$$s_1 - s_4 = -1.48\ \text{kJ/kg}\cdot\text{K}$$

The net work per cylinder per cycle can be found by determining the enclosed area of the p–V diagram (figure 9–10 or 9–11):

$$\begin{aligned} Wk_{\text{cycle}} &= Wk_{34} + Wk_{12} \\ &= \frac{mR}{1-k}(T_4 - T_3 + T_2 - T_1) \\ &= \frac{(0.211 \times 10^{-3}\ \text{kg})(287\ \text{N}\cdot\text{m/kg}\cdot\text{K})}{-0.4} \\ &\quad \times (2338\ \text{K} - 4786\ \text{K} + 614\ \text{K} - 300\ \text{K}) \\ &= 323\ \text{N}\cdot\text{m/cycle} \end{aligned} \qquad \textbf{(9–3)}$$

The engine will then deliver, from eight cylinders,

Answer

$$Wk = (8)(323\ \text{N}\cdot\text{m/cycle}) = 2584\ \text{N}\cdot\text{m}$$

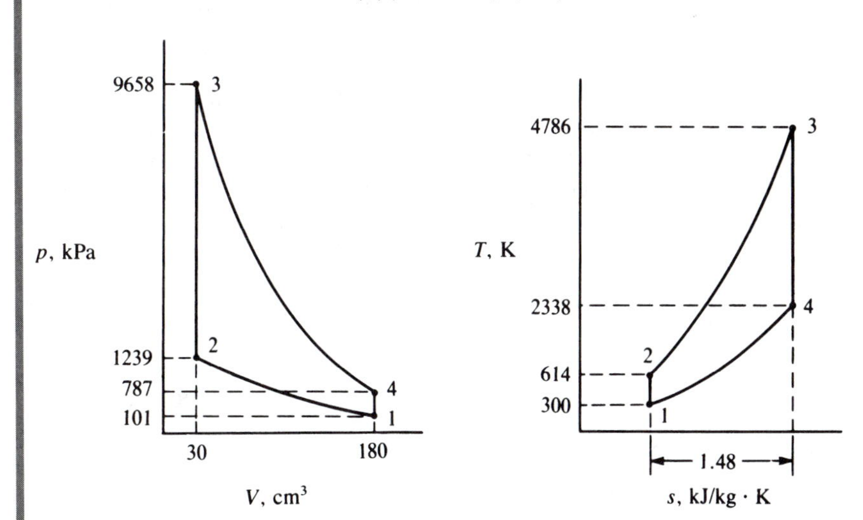

FIGURE 9–11 Otto engine diagrams with values from calculations

The power-produced $\dot{W}k$ will be the work produced per cycle times the number of cycles per time. This last term is 1100 cycles/min since the engine speed was 2200 rpm and the engine has a four-stroke cycle. Then

Answer

$$\begin{aligned} \dot{W}k &= (2584\ \text{N}\cdot\text{m/cycle})(1100\ \text{cycles/min}) \\ &= 2{,}842{,}400\ \text{N}\cdot\text{m/min} = 47.4\ \text{kJ/s} = 47.4\ \text{kW} \end{aligned}$$

EXAMPLE 9–4 A one-cylinder ICSI engine operating on an ideal two-stroke Otto cycle is to generate 5 hp at 4500 rpm with a compression ratio of 10. Assume that intake air is at 80°F and 14.7 psia. Determine the engine displacement required and the heat rejected if 1000 Btu/lbm of heat is added when the fuel–air mixture burns.

Solution

We use the air-standard analysis with constant specific heats and have

$$R = 53.36\ \text{ft} \cdot \text{lbf/lbm} \cdot {}^\circ\text{R}$$
$$c_v = 0.1718\ \text{Btu/lbm} \cdot {}^\circ\text{R}$$
$$c_p = 0.2404\ \text{Btu/lbm} \cdot {}^\circ\text{R}$$

and

$$k = 1.399$$

Since the engine is a two-stroke design, the work is extracted once for each revolution of the engine. Thus,

$$\dot{W}k = Wk_{\text{cycle}} \times 4500\ \text{rpm} = 5\ \text{hp}$$

and the work must be

$$Wk_{\text{cycle}} = \frac{5\ \text{hp}}{4500\ \text{cycles/min}} = \frac{(550\ \text{ft} \cdot \text{lbf/s} \cdot \text{hp})(5\ \text{hp})}{75\ \text{cycles/s}} = 36.7\ \text{ft} \cdot \text{lbf/cycle}$$

To compute the displacement of the engine, we must know the two volumes, V_1 and V_2. Since the compression ratio is to be 10, we have

$$V_1 = 10V_2$$

To find the volume V_1, we need to know the mass of air entering each cycle so that the perfect gas relation can be used at state 1. The mass is computed from equation (9–3), which requires each temperature to be determined. Proceeding, then, we have that $T_1 = 80 + 460 = 540°\text{R}$. Further, we may write

$$\left(\frac{T_2}{T_1}\right)^{k/k-1} = \left(\frac{V_1}{V_2}\right)^k$$

or

$$T_2 = T_1\left(\frac{V_1}{V_2}\right)^{k-1}$$

and then

$$T_2 = (540°\text{R})(10)^{0.399} = 1353°\text{R}$$

Now, using $c_v = 0.171\ \text{Btu/lbm} \cdot {}^\circ\text{R}$ and the relationship

$$q_{\text{add}} = c_v(T_3 - T_2)$$

we have

$$T_3 = \frac{q_{\text{add}}}{c_v} + T_2 = \frac{1000\ \text{Btu/lbm}}{0.1718\ \text{Btu/lbm} \cdot {}^\circ\text{R}} + 1353°\text{R} = 7174°\text{R}$$

Further, for process 3–4, we have

$$T_4 = T_3\left(\frac{V_3}{V_4}\right)^{k-1}$$

$$= 7174°\text{R}\left(\frac{1}{10}\right)^{0.399} = 2863°\text{R}$$

Then the work is

$$Wk_{\text{cycle}} = \frac{mR}{1-k}(T_4 - T_3 + T_2 - T_1)$$

$$36.7 \text{ ft}\cdot\text{lbf/cycle} = (m)\left(\frac{53.36 \text{ ft}\cdot\text{lbf/lbm}\cdot°\text{R}}{1 - 1.399}\right)$$
$$\times\ (2863°\text{R} - 7174°\text{R} + 1353°\text{R} - 540°\text{R})$$

and solving for the mass, we have

$$m = 7.85 \times 10^{-5} \text{ lbm/cycle}$$

The volume at state 1 can now be found from the perfect gas relation:

$$V_1 = mR\frac{T_1}{p_1}$$
$$= \frac{(7.85 \times 10^{-5} \text{ lbm})(53.36 \text{ ft}\cdot\text{lbf/lbm}\cdot°\text{R})(540°\text{R})}{(14.7 \text{ lbf/in}^2)(144 \text{ in}^2/\text{ft}^2)}$$
$$= 0.00106 \text{ ft}^3 = 1.84 \text{ in}^3$$

Then

$$V_2 = \frac{V_1}{10} = 0.184 \text{ in}^3$$

Since there is but one cylinder in the engine, its displacement is

Answer

$$V_D = V_1 - V_2 = 1.656 \text{ in}^3$$

Various dimensional arrangements of piston diameter or bore and the stroke could now be selected to provide this particular displacement.

The heat rejected per cycle is

$$Q_{\text{rej}} = Wk_{\text{cycle}} - Q_{\text{add}}$$

or

$$Q_{\text{rej}} = 36.7 \text{ ft}\cdot\text{lbf} - (1000 \text{ Btu/lbm})(778 \text{ ft}\cdot\text{lbf/Btu})(7.85 \times 10^{-5} \text{ lbm})$$

Answer

$$= -24.4 \text{ ft}\cdot\text{lbf/cycle}$$

Notice that the temperatures we have predicted at combustion are extremely high. At such high temperatures, many gases dissociate or ionize. This means that the gas molecules or atoms lose electrons, and thus the gas exhibits quite different characteristics from those of the normal air. We have not here accounted for such variations.

From example 9–4, you can see that many equations we have used in earlier chapters are applied to the Otto cycle. The analysis can be aided by using the computer. In chapter 7, we used the program CARNOT to quickly analyze the Carnot cycle. Here we can use the program OTTO (see appendix A–8 for the program description) to analyze the Otto cycle. The program is written to handle three types of problem statements:

1. Intake pressure and temperature, heat added per cycle, displacement (or bore and stroke), compression ratio, two- or four-stroke version, number of cylinders, and engine speed are known; to find the properties at the four states, the power, the rate of heat rejection, and the thermal efficiency (= net work/heat added).
2. Intake pressure and temperature, power produced, the displacement (or bore and stroke), compression ratio, two- or four-stroke version, number of cylinders, and engine speed are known; to find the properties at the four states, the rates of added and rejected heat, and the thermal efficiency.
3. Intake pressure and temperature, power produced, the maximum cycle temperature, compression ratio, two- or four-stroke version, and engine speed are known; to find the four states, the displacement, the added and rejected heats, and the thermal efficiency.

The program OTTO can be adapted or expanded to allow for other types of problems or to use computer graphics or other data handling. You may want to try example 9–3 or 9–4 with the use of OTTO and a personal computer.

The program *Engineering Equation Solver (EES)* can be used to solve many problems involving the Otto cycle. For instance, as demonstrated in earlier chapters, *EES* can solve a set of n equations for n unknowns, which is the situation in examples 9–3 and 9–4.

9–3 OTTO CYCLE EFFICIENCY

The efficiency of the Otto cycle, as indeed for any cycle, is defined as

$$\eta_T = \frac{Wk_{\text{cycle}}}{Q_{\text{add}}} \times 100 \qquad \textbf{(9–4)}$$

For the ideal Otto cycle and assuming a perfect gas working medium, we can reduce this to more specific equations. First, we have

$$Q_{\text{add}} = Q_{23} = U_3 - U_2 = mc_v(T_3 - T_2)$$

and in example 9–3, we saw that, from equation (9–3),

$$Wk_{\text{cycle}} = Wk_{12} + Wk_{34} = \frac{mR}{1-k}(T_2 - T_1 + T_4 - T_3)$$

Since $R = c_p - c_v$ and $k = c_p/c_v$, we can write

$$Wk_{\text{cycle}} = \frac{m(c_p - c_v)}{1 - (c_p/c_v)}(T_2 - T_1 + T_4 - T_3)$$

or

$$Wk_{\text{cycle}} = \frac{mc_v(c_p - c_v)}{c_p - c_v}(T_1 - T_2 + T_3 - T_4)$$
$$= mc_v(T_1 - T_4) + mc_v(T_3 - T_2)$$

We then obtain the efficiency by substituting these results into equation (9–4), yielding

$$\eta_T = \frac{mc_v(T_1 - T_4) + mc_v(T_3 - T_2)}{mc_v(T_3 - T_2)} \times 100$$

or

$$\eta_T = \left(1 - \frac{T_4 - T_1}{T_3 - T_2}\right) \times 100 \qquad \textbf{(9–5)}$$

Now, using some more algebraic manipulations, we obtain

$$\eta_T = \left[1 - \frac{T_1(T_4/T_1 - 1)}{T_2(T_3/T_2 - 1)}\right] \times 100$$

and

$$\frac{T_4}{T_3} = \left(\frac{V_3}{V_4}\right)^{k-1} = \left(\frac{V_2}{V_1}\right)^{k-1} = \frac{T_1}{T_2}$$

So

$$\frac{T_4}{T_1} = \frac{T_3}{T_2}$$

and we have

$$\eta_T = \left(1 - \frac{T_1}{T_2}\right) \times 100 \quad \textbf{(9–6)}$$

Equation (9–6) is frequently reduced further by noting that

$$\frac{T_1}{T_2} = \left(\frac{V_2}{V_1}\right)^{k-1} = \frac{1}{(r_v)^{k-1}}$$

which gives us

$$\eta_T = \left[1 - \frac{1}{(r_v)^{k-1}}\right] \times 100 \quad \textbf{(9–7)}$$

This result, as mentioned, is good only for ideal Otto cycles using perfect gases with constant specific heats. Notice that the efficiency of the Otto cycle is a function of the compression ratio. Thus, increasing the compression ratio will increase the efficiency of the engine an equivalent amount.

Notice from equation (9–7) that the efficiency is zero when the compression ratio is 1 and that the efficiency change is most rapid for compression ratios that are less than 10. Figure 9–12 shows a graph of thermal efficiency as the compression ratio of an ideal Otto engine changes. This graph gives a visual interpretation of the behavior of thermal efficiency as it is affected by the compression ratio.

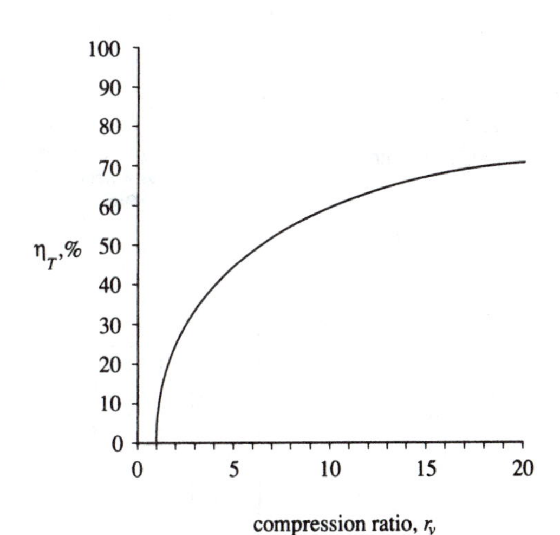

FIGURE 9–12 Otto cycle thermal efficiency as a function of compression ratio for a perfect gas with constant specific heats and $k = 1.399$

EXAMPLE 9–5 Determine the efficiency of the engine described in example 9–1.

Solution The engine in example 9–1 has a compression ratio of 8.48:1, so from equation (9–7), we have, assuming that $k = 1.399$,

Answer
$$\eta_T = \left[1 - \left(\frac{1}{8.48}\right)^{0.399}\right] \times 100 = 57.4\%$$

EXAMPLE 9–6 Determine the efficiencies of the engines in examples 9–3 and 9–4.

Solution Again, using equation (9–7), we readily obtain the cycle efficiency for the engine in example 9–3 having a compression ratio of 6:1. Thus,

Answer
$$\eta_T = \left[1 - \left(\frac{1}{6}\right)^{0.399}\right] \times 100 = 51.1\%$$

For the engine of example 9–4, the compression ratio was given as 10:1. We compute

Answer
$$\eta_T = \left[1 - \left(\frac{1}{10}\right)^{0.399}\right] \times 100 = 60.1\%$$

It is also interesting to compare the results of examples 9–5 and 9–6 to those of the ideal Carnot cycle. Operating between the high temperature of the engine, T_3, and the low temperature, T_1, the Carnot engine efficiency is calculated by using the values from example 9–3:

$$\eta_{\text{Carnot}} = \left(1 - \frac{T_1}{T_3}\right) \times 100 = \left(1 - \frac{300}{4786}\right) \times 100 = 93.7\%$$

For example 9–4,

$$\eta_{\text{Carnot}} = \left(1 - \frac{300}{4231}\right) \times 100 = 92.9\%$$

These comparisons are somewhat biased, however, for if we compare T–s diagrams of both Otto and Carnot cycles, we must conclude that for differential changes in entropy, both cycles have the same efficiency. We see this in figure 9–13. Using this criterion and example 9–3, we find the Carnot efficiency of the differential cycle to be

$$\eta_{\text{Carnot}} = \left(1 - \frac{T_1}{T_2}\right) \times 100 = \left(1 - \frac{300}{614}\right) \times 100$$
$$= 51.7\%$$

which is the same as for the Otto engine.

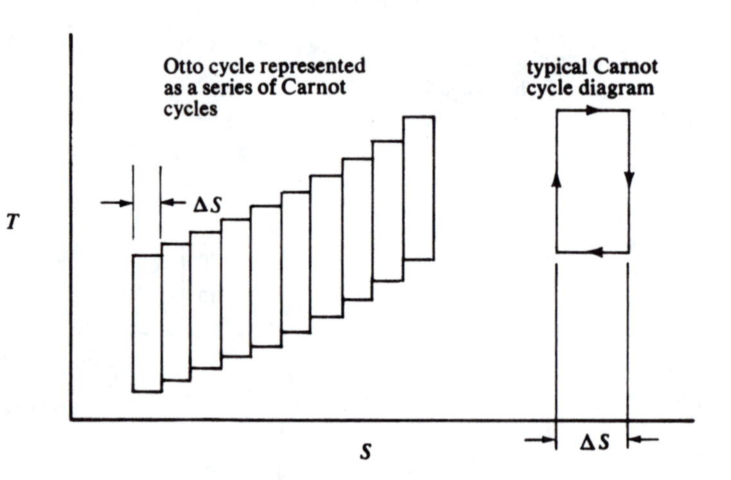

FIGURE 9–13 A T–s diagram comparing Carnot and Otto cycles

9–4 THE ACTUAL OTTO ENGINE

Using the tools developed previously, we will analyze the operation of an Otto engine that could conceivably be used to power an automobile, truck, tractor, or other mechanical device; that is, we will consider the actual or "practical" Otto engine. Before doing this, however, let us introduce additional terminology and cycle parameters, namely, *indicated horsepower* (ihp), *brake horsepower* (bhp), *fuel heating valve* (HV), *brake specific fuel consumption* (bsfc), and *mean effective pressure* (mep).

The indicated horsepower is determined from a p–V diagram of the test engine. This diagram can be obtained by using an engine indicator, which is a mechanism capable of measuring and recording the pressure in a cylinder while recording the piston position. The engine indicator is a common test device with a history of use dating to before 1900.

Recently, the cathode-ray oscilloscope (CRO) or oscilloscope has been used with an electromechanical pressure sensor to obtain the same p–V diagrams of operating IC engines. The diagrams are displayed on the oscilloscope screen and can be electronically recorded for future reference.

A detailed description of the indicator and the oscilloscope may be found in various reference literature. Figure 9–14 shows the typical pressure–volume diagram of an Otto engine, specifically of an engine with the throttle wide open. The curve can easily be seen to differ somewhat from the curve of the typical Otto cycle as shown in figure 9–1. The comparison is made easier by noting figure 9–15, where the ideal Otto cycle is superimposed on an actual Otto cycle. For further analysis of this example, we will use the gas tables to obtain air properties for the Otto cycle curves to calculate the area of the enclosed

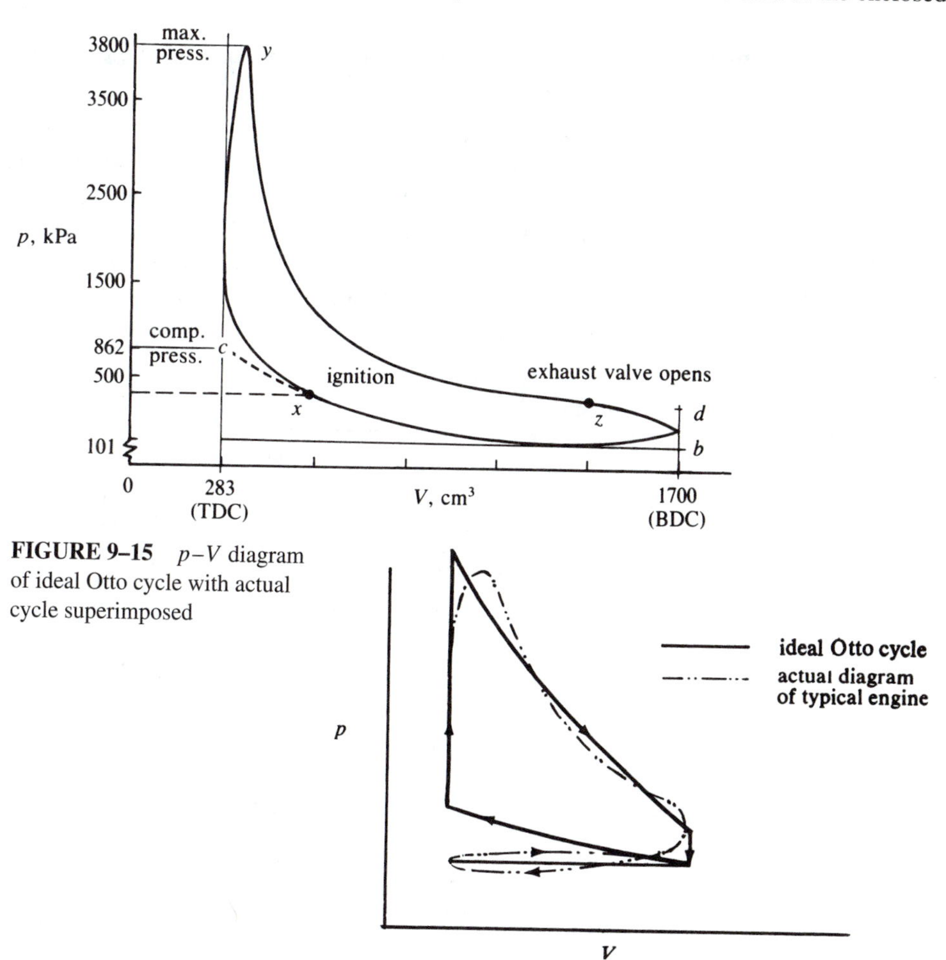

FIGURE 9–14 Typical p–V diagram for spark-ignition engines at wide-open throttle [from E. I. Obert, *Internal combustion engines*, 2d ed. (Scranton, Pa.: International Textbook Co., 1959); with permission of International Textbook Co.; revised to SI units by author]

FIGURE 9–15 p–V diagram of ideal Otto cycle with actual cycle superimposed

p–V diagram. This area, however, could be determined by using various geometric and mechanical devices (for example, the planimeter).

Also, we could determine the approximate area enclosed by the p–V diagram by using the computer and VISUAL BASIC program AREA for finding areas under curves. The enclosed area of the p–V diagram is work of the cycle, and the rate of this work, power, is referred to as **indicated horsepower**, ihp. It is not the power actually produced by the engine drive shaft but rather the power that could be produced at the piston of a frictionless engine. The ihp would be the net work (area enclosed by the p–V diagram) times the number of power strokes per unit time, N; that is,

$$\text{ihp} = (Wk_{\text{cycle}})(N) \qquad \textbf{(9–8)}$$

The net work can also be visualized as the displacement or volume change times an average pressure. We call this average pressure the **mean effective pressure** (mep), defined as

$$\text{mep} = \frac{Wk_{\text{cycle}}}{V_D} \qquad \textbf{(9–9)}$$

where V_D is the displacement given by equation (9–2). The mep has units of pressure and is sometimes called the **indicated mean effective pressure** (imep) to show that it is related to the indicated work. We can also combine equations (9–8) and (9–9) to obtain

$$\text{imep} = \frac{\text{ihp}}{V_D N} \qquad \textbf{(9–10)}$$

EXAMPLE 9–7 Determine the approximate area enclosed by the p–V diagram of figure 9–14. That is, determine the indicated work per cycle.

Solution We shall use the AREA program for calculating areas under curves (see appendix A–5) and apply it to the full cycle of figure 9–14. First, we need to select data points from the diagram and tabulate them. This has been done in table 9–2, where we have tabulated 15 points. Using a microcomputer and the AREA program, we find the net area to be 0.66496 kN · m or 0.66496 kJ. The same result can be obtained by hand calculations, as indicated in the remaining columns of table 9–2.

TABLE 9–2

State	p kPa	V cm^3	p_{ave}	δV	$p\,\delta V$ kJ
1	250	1700	177.5	−284	−0.05041
2	105	1416	112.5	−284	−0.03195
3	120	1132	185	−284	−0.05254
4	250	848	325	−284	−0.0923
5	400	564	500	−142	−0.0710
6	600	422	1050	−142	−0.1491
7	1500	280	2000	0	0
8	2500	280	3150	40	0.126
9	3800	320	2650	80	0.212
10	1500	400	1275	164	0.2901
11	1050	564	775	284	0.2201
12	500	848	470	284	0.13348
13	440	1132	420	284	0.11928
14	400	1416	325	284	0.0923
15	250	1700			

$\sum p\,\delta V = 0.66496$ kJ

Notice that the program for computing areas under curves works well for cycles as well as for its intended use of one process. Also, the listing in appendix A–5 shows that you can use only 20 data points or fewer as the program was written. If you need more than 20 data points, the two dimension statements, DIM X(20) and DIM Y(20), can be changed by replacing the "20" with whatever number of data points you need.

The actual power produced by an engine we call **brake horsepower**, bhp, and the mechanical efficiency, η_{mech}, is then obtained for the actual Otto engine from

$$\eta_{mech} = \frac{bhp}{ihp} \times 100 \tag{9–11}$$

The brake horsepower results are strictly from test data. We have no thermodynamic tools at this time to predict the value of bhp, although we can make some crude guesses from a knowledge of ihp. There are various test devices to measure bhp, all of which utilize some brake or resistance to measure engine output. Dynamometers, prony brakes, and water brakes are common devices for measuring bhp, but we will not here concern ourselves with the details of their workings. Using this measured value, bhp, however, we can determine a parameter called the **brake mean effective pressure**, bmep. We define this as

$$bmep = \frac{bhp}{V_D \times N} \tag{9–12}$$

The bmep is generally described in units of pressure, just like mep.

The fuel consumed by an engine is of major concern to both designer and user. The **specific fuel consumption**, sfc, is the amount of fuel used to produce an amount of work, or the rate of fuel consumption while developing a given power. In general,

$$sfc = \frac{\dot{m}_f}{\dot{W}k_{cycle}} \tag{9–13}$$

and, in particular, the amount of fuel used per bhp per hour is called the **brake specific fuel consumption**, bsfc, and is calculated from

$$bsfc = \frac{m_f}{(bhp)t} = \frac{\dot{m}_f}{bhp} \tag{9–14}$$

where the rate of fuel consumption $\dot{m}_f$ is equal to the mass of fuel m consumed during a time period t while producing power bhp. The amount of fuel used in relation to the amount of *clean air* required is, of course, meager (about 15 parts of air per part fuel). We could discuss a brake-specific air consumption. However, **volumetric efficiency**, η_v, defined as the ratio of the mass of air taken into the engine cylinder, $\dot{m}_a$, to the mass that theoretically could have been taken in at atmospheric conditions, $\dot{m}_t$, or

$$\eta_v = \frac{\dot{m}_a}{\dot{m}_t} \tag{9–15}$$

is a common parameter used to describe engine performance. Notice that volumetric efficiency is not a volume efficiency, but rather a mass efficiency.

The **theoretical air flow**, $\dot{m}_t$, is found from the relationship

$$\dot{m}_t = \frac{NV_D p_a}{RT_a} \tag{9–16}$$

where p_a and T_a are the atmospheric pressure and temperature and R is the gas constant. The **actual mass flow**, $\dot{m}_a$, must be measured while the engine is operating.

The chemical reaction between the air and fuel produces heat, which in turn increases the temperature and pressure of the compressed gases and ultimately results in work *out*. The heat released during the air–fuel reaction is called the *heating value* of the fuel. When this reaction occurs, water is formed (either in a vapor or in a liquid), and if we allow the water to condense to a liquid, the heating value is called the **higher heating value**, HHV, and the **lower heating value**, LHV, results if we retain the water in a vapor state. We can write

$$\text{HHV} - \text{LHV} = \text{heat of vaporization of water formed during reaction} \quad \textbf{(9–17)}$$

and note that the heating values are tabulated in table B–26. The heating values are based on the mass of fuel; that is, the HHV of propane (for example) is 50,349 kJ/kg of fuel or 21,646 Btu/lbm of fuel. If there were 20 kg of air for each kilogram of fuel (or 20 lbm of air per pound-mass of fuel), the HHV would be 50,349/20 kJ/kg air = 2517.45 kJ/kg air. The English units would be 1082.3 Btu/lbm air. The combustion of fuels and a description of methods to determine the proper amounts of air required to burn fuels are subjects of a later chapter. A discussion of fuel selection is beyond the scope of this book, but it should be noted that the octane number, one of the most common scales for rating fuels, is based on an arbitrary scale where *n*-octane has an octane number of 100 and heptane has an octane number of zero. Other fuels usually fall between these two in octane number. The octane number is a measure of the antiknock or preignition characteristics of fuel, the higher number indicating a high resistance to preignition and therefore a good fuel to use in an ICSI engine.

EXAMPLE 9–8

An ICSI engine is found to have an air flow of 2.4 kg/min when running at 3000 rpm under a constant loaded condition. The engine is a four-stroke version having 2000 cm^3 displacement, and the atmospheric conditions are 100 kPa and 20°C. Determine the volumetric efficiency.

Solution

The volumetric efficiency is determined from equation (9–15), and the theoretical mass flow is given by equation (9–16):

$$\dot{m}_t = \frac{NV_D p_a}{RT_a} = \frac{(3000/2 \text{ cycle/min})(2000 \text{ cm}^3)(100 \text{ kN/m}^2)}{(0.287 \text{ kN}\cdot\text{m/kg})(293 \text{ K})(10^6 \text{ cm}^3/\text{m}^3)} = 3.57 \text{ kg/min}$$

The volumetric efficiency is then

$$\eta_v = \frac{2.4 \text{ kg/min}}{3.57 \text{ kg/min}}$$

Answer

$$= 67.2\%$$

EXAMPLE 9–9

Determine the mean effective pressure for the engines in examples 9–3 and 9–4.

Solution

For the engine of example 9–3, we found that the cyclic work is 2584 N · m. Thus,

$$\text{mep} = \frac{2584 \text{ N}\cdot\text{m}}{1200 \text{ cm}^3 \times 10^{-6} \text{ m}^3/\text{cm}^3}$$

Answer

$$= 21.5 \times 10^5 \text{ N/m}^2 = 2150 \text{ kPa}$$

For the engine of example 9–4, we have the cyclic work indicated as 36.7 ft · lbf and the displacement as 1.656 in^3. We then find that

$$\text{mep} = \left(\frac{36.7 \text{ ft}\cdot\text{lbf}}{1.656 \text{ in}^3}\right)(12 \text{ in/ft})$$

Answer

$$= 266 \text{ psi}$$

Frequently, there are not good engine indicator diagrams or other information available for a particular engine. This is especially true if the engine is a new design, and under these circumstances, the ideal Otto cycle must be used as a first approximation. The ideal Otto cycle, however, is not a completely accurate description of the operation of an ICSI engine. To better describe the operation of these engines, we may make the following modifications to the ideal Otto cycle analysis:

1. Assume variable specific heats for the air-standard analysis.
2. Assume that the compression and expansion processes are reversible and polytropic with the process equation
$$pv^n = C$$
instead of $pv^k = C$ for a reversible adiabatic process.
3. Assume irreversible adiabatic processes for the compression and expansion, and use the adiabatic efficiency concept.

Let's consider each of these modifications in turn.

Variable Specific Heat Analysis

The assumption of constant specific heats for a perfect gas allowed for straightforward equations relating the various state properties in the Otto cycle. If variable specific heats are accepted (still using the air-standard analysis), we need to use table B–6 for the air properties, and these properties are all dependent on the temperature. If the temperature of the air is known, the enthalpy, internal energy, relative pressure (p_r), relative volume (v_r), and ϕ can be determined. The ϕ function is defined from chapter 7 as

$$\phi = \sum \frac{c_p\,\delta T}{T} \tag{9–18}$$

and p_r and v_r by the equations

$$p_r = e^{\phi/R} \quad \text{or} \quad \phi = R \ln p_r$$
$$v_r = \frac{RT}{p_r} \tag{9–19}$$

It can be shown that, for a reversible adiabatic process, the following relationships hold between the pressures and volumes:

$$\frac{p_2}{p_1} = \frac{p_{r_2}}{p_{r_1}}$$

and

$$\frac{v_2}{v_1} = \frac{v_{r_2}}{v_{r_1}} \tag{9–20}$$

EXAMPLE 9–10

An Otto engine operates with a compression ratio of 7:1, where the intake conditions are 40°F and 14.5 psia. Determine the temperature and pressure after compression and the work of compression per pound-mass of air. Use air-standard analysis and variable specific heats.

Solution

Using the air table B–6, we find that, at state 1 (intake), where $T_1 = 500°\text{R}$ and $p_1 = 14.5$ psia,

$$u_1 = 85.20 \text{ Btu/lbm}$$
$$p_{r_1} = 1.0590$$

and

$$v_{r_1} = 174.90$$

Then we use the reversible adiabatic process relationships of equation (9–20) to determine the conditions at state 2 after compression:

$$v_{r_2} = v_{r_1}\frac{v_2}{v_1}$$
$$= (174.90)\left(\frac{1}{7}\right) = 24.99$$

The temperature at state 2 is 1074°R or 614°F from table B–6. The pressure at state 2 can be found from the perfect gas equation, $p_2 = RT_2/v_2$, but $v_2 = v_1/7 = RT_1/7p_1$, so that

$$p_2 = RT_2\frac{7p_1}{RT_1} = 7p_1\frac{T_2}{T_1}$$

Answer

$$= (7)(14.5\text{ psia})\left(\frac{1074°\text{R}}{500°\text{R}}\right) = 218.022\text{ psia}$$

The work of compression for the reversible adiabatic compression is the difference in internal energies between states 1 and 2:

$$wk_{12} = u_1 - u_2$$

From table B–6, we find that $u_2 = 185.9$ Btu/lbm at 1074°R. Then

Answer

$$wk_{12} = 85.20\text{ Btu/lbm} - 185.9\text{ Btu/lbm} = -100.7\text{ Btu/lbm}$$

We can do a complete analysis of the Otto engine by using variable specific heats, the air-standard analysis, and the relationships

$$q_{\text{add}} = q_{23} = u_3 - u_2$$
$$wk_{\text{expansion}} = wk_{34} = u_3 - u_4$$

and

$$q_{\text{rej}} = q_{41} = u_1 - u_4$$

where the internal energies are related to the temperatures as shown by table B–6. Process 3–4 would be analyzed similarly to example 9–10 except that the process is a reversible adiabatic expansion.

Polytropic Compression and Expansion Analysis

A more accurate analysis of the Otto engine than that using variable specific heats with the simple Otto cycle is the assumption of reversible polytropic processes for the compression and expansion strokes. The analysis follows the simple Otto cycle except that the equation $pv^n = C$ is used for the two polytropic processes and the exponent n must be determined or known. Frequently, the exponent is different for each of the two processes, so we denote these as n_{12} for the compression (from state 1 to 2) and n_{34} for the expansion from 3 to 4. If the p–V diagram is available, the exponents can be determined from the following relationships:

$$n_{12} = \frac{\ln(p_1/p_2)}{\ln(V_2/V_1)}$$
$$n_{34} = \frac{\ln(p_3/p_4)}{\ln(V_4/V_3)} \quad \textbf{(9–21)}$$

If the p–V diagram is not known, the exponents must be approximated. For reversible polytropic processes of the IC engines, the polytropic exponent will nearly always have a value between 1.0 and k $(=c_p/c_v)$ because heat will be transferring out of the system or engine during these processes (so that the processes are not adiabatic with $n = k$), yet the temperatures will be increasing (so that the processes are not isothermal with $k = 1.0$). This means that the exponents must be greater than 1.0 but less than k. If the exponents were greater than k, the heat would be transferring into the engine, and this would be a violation of the second law of thermodynamics; that is, heat transfer is spontaneous from a cold region to a hot region.

EXAMPLE 9–11 For the engine of example 9–3, assume that the compression process is polytropic with $n_{12} = 1.3$, and assume that the expansion 3–4 is also polytropic with $n_{34} = 1.25$. If all else is the same as in the previous engine, determine the work of the cycle, the power produced at 2200 rpm, and the heat rejected per cycle.

Solution The system under analysis is the same as that in example 9–3, so the volumes are the same. The engine can be described by the system diagram as shown in figure 9–9, and the air in the chamber at the beginning of compression, state 1, is at 101 kPa and 27°C. The compression ratio is 6:1 and the volumes are

$$V_1 = V_4 = 180 \text{ cm}^3$$
$$V_2 = V_3 = 30 \text{ cm}^3$$

The pressure at state 2 is computed from the equation

$$p_2 = p_1\left(\frac{V_1}{V_2}\right)^{n_{12}}$$
$$= (101 \text{ kPa})(6)^{1.3} = 1037 \text{ kPa}$$

and the temperature at state 2 is

$$T_2 = \frac{p_2V_2}{mR} = \frac{(1037 \text{ kPa})(30 \times 10^{-6} \text{ m}^3)}{(0.211 \times 10^{-3} \text{ kg})(287 \text{ N}\cdot\text{mg/kg}\cdot\text{K})} = 514 \text{ K}$$

The temperature at state 3 can now be computed from the relationship

$$q_{add} = c_v(T_3 - T_2) = 3000 \text{ kJ/kg}$$

and then

$$T_3 = \frac{3000 \text{ kJ/kg}}{c_v} + T_2$$
$$= \frac{3000 \text{ kJ/kg}}{0.719 \text{ kJ/kg}\cdot\text{K}} + 514 \text{ K}$$
$$= 4687\ K$$

From the process equation

$$\frac{T_4}{T_3} = \left(\frac{V_3}{V_4}\right)^{n_{34-1}}$$

we can now calculate the temperature at state 4:

$$T_4 = (4687 \text{ K})\left(\frac{1}{6}\right)^{1.25-1} = 2995 \text{ K}$$

The work is then easily computed from a variation of equation (9–3):

$$Wk_{cycle} = \frac{mR}{1 - n_{43}}(T_4 - T_3) + \frac{mR}{1 - n_{21}}(T_2 - T_1)$$
$$= (0.211 \times 10^{-3} \text{ kg})(287 \text{ J/kg}\cdot\text{K})$$
$$\times \left(\frac{2995 \text{ K} - 4687 \text{ K}}{1 - 1.25} + \frac{514 \text{ K} - 300 \text{ K}}{1 - 1.3}\right)$$
$$= 366.7 \text{ J/cycle}$$

The power produced by the engine will be the work for eight cylinders running at 2200 rpm but at 1100 cycles/min. Thus,

$$\dot{W}k = (8)(366.7 \text{ J/cycle} \times 1100 \text{ cycles/min})$$
$$= 323 \times 10^4 \text{ J/min} = 3230 \text{ kJ/min}$$

Answer
$$= 53.8 \text{ kW}$$

We can compute the heat rejected per cycle from the balance

$$Q_{\text{rej}} = Wk_{\text{cycle}} - Q_{\text{add}}$$

which gives us

$$Q_{\text{rej}} = (366.5 \text{ J})(8) - (3000 \text{ kJ/kg})(0.211 \times 10^{-3} \text{ kg})(8)$$

Answer
$$= -2.13 \text{ kJ}$$

This value for the rejected heat is the amount of heat lost by the engine during three processes: 1–2, 3–4, and 4–1. The simple Otto cycle has only one heat rejection, during process 4–1, but the polytropic processes also involve heat.

A comparison of these results with the answers from example 9–3 indicates a decrease in work or power produced by the engine. This is a typical result of using polytropic processes to represent better the actual engine—the ideal Otto cycle is just not quite as accurate as the polytropic cycle, provided that the exponents n are accurate.

Figure 9–16 shows the property diagrams resulting from example 9–11. Notice that entropy changes occur in all four of the cycle processes, and, if we desired, the magnitudes of these changes could be calculated from equations listed in tables 6–4 and 6–5. For process 1–2, we have

$$\begin{aligned}\Delta S_{12} &= mc_v \ln \frac{T_2}{T_1} + mR \ln \frac{V_2}{V_1} \\ &= (0.211 \times 10^{-3} \text{ kg})(0.719 \text{ kJ/kg} \cdot \text{K})\left(\ln \frac{514 \text{ K}}{300 \text{ K}} \right) \\ &\quad + (0.211 \times 10^{-3} \text{ kg})(0.287 \text{ kJ/kg} \cdot \text{K}) \ln \frac{1}{6} \\ &= -0.0268 \text{ J/K}\end{aligned}$$

FIGURE 9–16

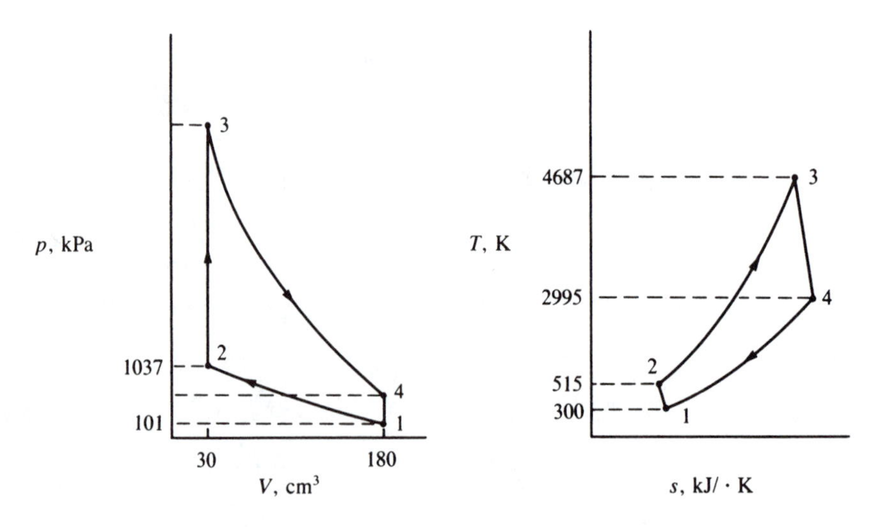

and for process 3–4,

$$\Delta S_{34} = mc_v \ln \frac{T_4}{T_3} + mR\, ln \frac{V_4}{V_3}$$
$$= (0.211 \times 10^{-3}\ \text{kg})(0.719\ \text{kJ/kg}\cdot\text{K})\left(\ln \frac{2995\ \text{K}}{4687\ \text{K}}\right)$$
$$+ (0.211 \times 10^{-3}\ \text{kg})(0.287\ \text{kJ/kg}\cdot\text{K}) \ln 6$$
$$= 0.04056\ \text{J/K}$$

These entropy changes are indicated in the $T-S$ diagram of Figure 9–16.

The computer program OTTO described in appendix A–8 will compute various parameters of the Otto cycle based on polytropic compression and expansion processes.

Irreversible Adiabatic Compression and Expansion Analysis

Without resorting to actual measurements of an operating engine or using involved modeling methods, the use of irreversible adiabatic processes to describe the compression and expansions in an IC engine is the most nearly accurate method for analyzing these devices. We use adiabatic efficiencies defined previously as

$$\eta_{\text{compression}} = \frac{wk_{cs}}{wk_{act}} \qquad \textbf{(9–22)}$$

and

$$\eta_{\text{expansion}} = \frac{wk_{act}}{wk_{cs}} \qquad \textbf{(9–23)}$$

where wk_{cs} is the boundary work of a reversible adiabatic process and wk_{act} is the actual work involved in the process. Since the cycles are accepted as irreversible for this analysis, we should determine by how much they are irreversible. As we saw in chapter 7, the entropy change of all cyclic devices is zero, and for reversible ones, the total entropy change is also zero. On the other hand, for irreversible cyclic devices, the total entropy change will be positive (nonzero) because the entropy change of the surroundings will be positive. The surroundings of a cyclic device or heat engine consist of at least one high-temperature region and one low-temperature region. Now, consider the simple Otto cycle and its $T-S$ diagram, shown in figure 9–1 and redrawn in figure 9–17. For the irreversible cycle, we have the superimposed diagram, *assuming the same heat input.* The entropy increases for processes 1–2 and 3–4 reflect their irreversible nature and are given by

$$\Delta S_{12} = mc_v \ln \frac{T_2}{T_{2'}}$$
$$\Delta S_{34} = mc_v \ln \frac{T_4}{T_{4'}} \qquad \textbf{(9–24)}$$

where $T_{2'}$ and $T_{4'}$ are the temperatures at states 2 and 4 if the two processes 1–2 and 3–4 would be reversible and adiabatic, beginning at the states 1 and 3. For an actual engine operating on such a cycle as that indicated by figure 9–17 (the irreversible one), the engine would develop less net work than the simple Otto engine, even though the enclosed area of the irreversible cycle is greater than for the simple Otto cycle. The shaded area of the $T-S$ diagram represents the net heat of the simple Otto engine, and it is equal to the net work as well. The area enclosed by the irreversible cycle includes an area outside on both sides of the shaded area which can be interpreted as the minimum irreversible heat transfer to the surroundings. If this heat is transferred to a region in a reversible manner, the total entropy change of the lower-temperature region would be $S_4 - S_1$, and that of the high-temperature region would be $S_2 - S_3$. Obviously, from figure 9–17, the minimum total entropy change

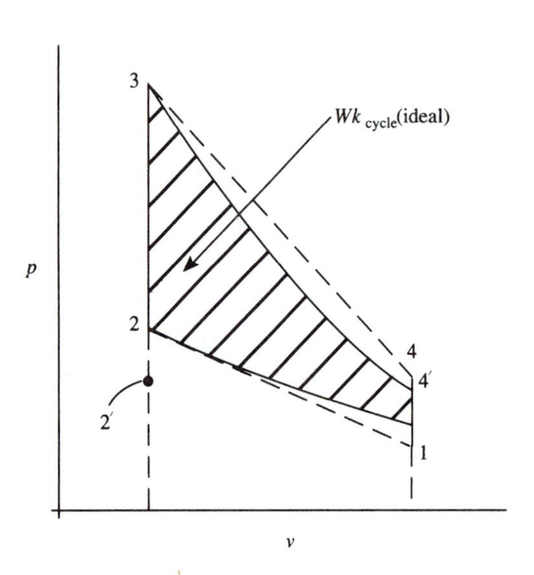

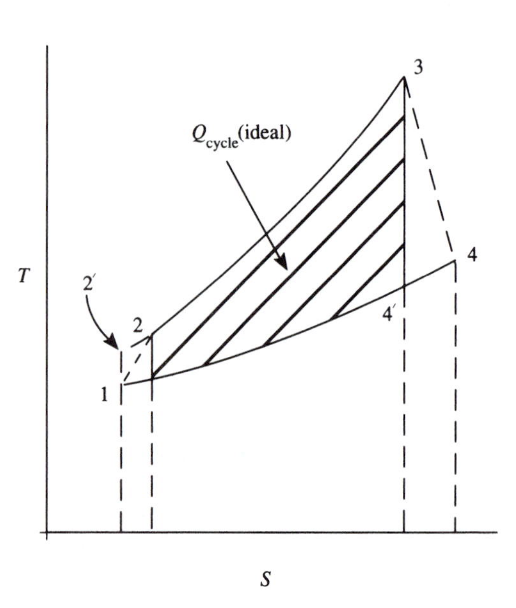

FIGURE 9–17 Irreversible Otto cycle diagrams

is given by the sum $S_2 - S_1 + S_4 - S_3$. The entropy generation of the irreversible engine is then

$$\dot{S}_{\text{gen}} \geq (\Delta S_{12} + \Delta S_{34})N \tag{9–25}$$

We will demonstrate the methods of the irreversible compression and expansion analysis with an example.

EXAMPLE 9–12 An irreversible four-stroke cycle engine having a compression ratio of 8.5:1, a displacement of 225 in^3, and intake conditions of 14.3 psia and 70°F is to deliver 200 bhp at 3000 rpm. The adiabatic efficiencies are found to be 88% for both expansion and compression, and the mechanical efficiency is 85%. Determine the mep, the rate of heat added and rejected, the thermal efficiency, and the rate of entropy generation.

Solution Since the engine is to deliver 200 bhp (brake horsepower) with a mechanical efficiency of 85%, the indicated horsepower or power produced by the working substance (air) is found from equation (9–11):

$$\text{ihp} = \frac{\text{bhp}}{\eta_{\text{mech}}} = \frac{200\ \text{hp}}{0.85} = 235.3\ \text{hp}$$

The mep is then found from equation (9–10):

$$\text{mep} = \frac{\text{ihp}}{V_D N} = \frac{(235.3\ \text{hp})(550\ \text{ft}\cdot\text{lbf/s}\cdot\text{hp})(60\ \text{s/min})(12\ \text{in/ft})}{(225\ \text{in}^3)(1500\ \text{cycles/min})}$$

Answer

$$= 276\ \text{psi}$$

The heat terms can be found after we determine the temperatures at the four states. We use air-standard analysis with constant specific heats, although a more elaborate analysis could be made by using variable specific heats. At state 1, we have $p_1 = 14.3$ psia and $T_1 = 530°\text{R}$. For a reversible adiabatic process between 1 and 2, we can find $T_{2'}$, with $k = 1.399$:

$$T_{2'} = T_1\left(\frac{V_1}{V_2}\right)^{k-1} = (530°\text{R})(8.5)^{0.399} = 1244.8°\text{R}$$

The ideal work, wk_{cs}, is then found from

$$wk_{cs} = u_1 - u_2 = c_v(T_1 - T_2)$$

Using 0.1718 Btu/lbm·°R for c_v, we obtain −122.8 Btu/lbm for the work. The actual work, wk_{12}, is then calculated from equation (9–22),

$$\begin{aligned} wk_{12} &= \frac{wk_{cs}}{\eta_{\text{compression}}} \\ &= \frac{-122.8 \text{ Btu/lbm}}{0.88} \\ &= -139.5 \text{ Btu/lbm} \end{aligned}$$

and the actual temperature at state 2 can be found from the equation

$$wk_{12} = c_v(T_1 - T_2) = (0.1718 \text{ Btu/lbm}\cdot°\text{R})(530°\text{R} - T_2)$$

and

$$T_2 = 530°\text{R} + \frac{139.5}{0.1718°\text{R}}$$

Then

$$T_2 = 1342.3°\text{R}$$

Using a similar analysis for the expansion process, 3–4, we find that

$$wk_{34} = (\eta_{\text{expansion}})(c_v)(T_3 - T_{4'})$$

where $T_{4'}$ is the temperature of a state resulting from a reversible adiabatic process between V_3 and V_4, and $T_{4'}$ is related to T_3 by the equation

$$\begin{aligned} T_{4'} &= T_3\left(\frac{V_3}{V_4}\right)^{k-1} = T_3\left(\frac{V_2}{V_1}\right)^{k-1} \\ &= T_3\left(\frac{1}{8.5}\right)^{0.399} = 0.4258T_3 \end{aligned}$$

The actual work of process 3–4 is then

$$\begin{aligned} wk_{34} &= (0.88)(0.1718 \text{ Btu/lbm}\cdot°\text{R})(T_3 - 0.4258T_3) \\ &= (0.88)(0.1718)(0.5742T_3) = 0.0868T_3 \end{aligned}$$

The net work per pound-mass of air is $wk_{\text{net}} = wk_{12} + wk_{34}$ and is related to the power by the equation

$$wk_{\text{net}} = \frac{\text{ihp}}{mN}$$

where m is the mass of air per cycle. We can find the mass from the perfect gas equation, $m = p_1V_1/RT_1$, after finding the intake volume, V_1. To find V_1, we use the compression ratio $V_1/V_2 = 8.5$ and the displacement $V_D = V_1 - V_2$ to solve for V_1. We have

$$225 \text{ in}^3 = 0.1302 \text{ ft}^3 = V_1 - V_2 = V_1 - \frac{V_1}{8.5}$$

or

$$V_1 = 0.1476 \text{ ft}^3$$

Then the mass is

$$\begin{aligned} m &= \frac{(14.3 \text{ lbf/in}^2)(144 \text{ in}^2/\text{ft}^2)(0.1476 \text{ ft}^3)}{(53.36 \text{ ft}\cdot\text{lbf/lbm}\cdot°\text{R})(530°\text{R})} \\ &= 0.0107 \text{ lbm/cycle} \end{aligned}$$

The net work is then

$$wk_{\text{net}} = \frac{(235.3\ \text{hp})(33{,}000\ \text{ft}\cdot\text{lbf/min}\cdot\text{hp})}{(0.0107\ \text{lbm/cycle})(1500\ \text{cycles/min})}$$
$$= 483{,}794\ \text{ft}\cdot\text{lbf/lbm} = 621.8\ \text{Btu/lbm}$$

If we set this equal to the two work terms, we obtain

$$621.8\ \text{Btu/lbm} = -139.5\ \text{Btu/lbm} - 0.0868T_3$$

and, solving for T_3, we find that $T_3 = 8771°\text{R}$. The actual work of process 3–4 is given by

$$wk_{34} = wk_{\text{net}} - wk_{12} = 621.8\ \text{Btu/lbm} - (-139.5\ \text{Btu/lbm})$$
$$= 761.3\ \text{Btu/lbm}$$

and this is equal to $c_v(T_3 - T_4)$. We then have, for the temperature at state 4,

$$T_4 = T_3 - \frac{wk_{34}}{c_v}$$
$$= 8771°\text{R} - \frac{761.3\ \text{Btu/lbm}}{0.1718\ \text{Btu/lbm}\cdot°\text{R}}$$
$$= 4339.7°\text{R}$$

The rates of heat transfers, $\dot{Q}_{\text{add}}$ and $\dot{Q}_{\text{rej}}$, can now be found. We have

$$\dot{Q}_{\text{add}} = \dot{m}c_v(T_3 - T_2)$$

and

$$\dot{Q}_{\text{rej}} = \dot{m}c_v(T_1 - T_4)$$

where the mass flow rate $\dot{m}$ is the mass per cycle times the number of cycles per unit time, N. Then

$$\dot{m} = mN = (0.0107\ \text{lbm/cycle})(1500\ \text{cycles/min})$$
$$= 16.05\ \text{lbm/min}$$
$$= 0.2675\ \text{lbm/s}$$

and the rates of heat transfers are

$$Q_{\text{add}} = (0.2675\ \text{lbm/s})(0.1718\ \text{Btu/lbm}\cdot°\text{R})(8771°\text{R} - 1342.3°\text{R})$$

Answer
$$= 341.4\ \text{Btu/s} = 482.9\ \text{hp}$$

and

$$Q_{\text{rej}} = (0.2675\ \text{lbm/s})(0.1718\ \text{Btu/lbm}\cdot°\text{R})(530°\text{R} - 4339.7°\text{R})$$

Answer
$$= -175.1\ \text{Btu/s} = -247.6\ \text{hp}$$

The sum of these two heat transfer rates is 235.3 hp, in agreement with our previous results.

The thermal efficiency is

$$\eta_T = \frac{\dot{W}k_{\text{net}}}{\dot{Q}_{\text{add}}} = \frac{235.3\ \text{hp}}{482.9\ \text{hp}}$$

Answer
$$= 48.7\%$$

The entropy generation is at least as great as the following amount, from equation (9–25):

$$\dot{S}_{gen}(\Delta s_{12} + \Delta s_{34})\dot{m} = \left(c_v \ln\frac{T_2}{T_{2'}} + c_v \ln\frac{T_4}{T_{4'}}\right)\dot{m}$$

$$= \left(\ln\frac{T_2}{T_{2'}} + \ln\frac{T_4}{T_{4'}}\right)\dot{m}c_v$$

The temperature, $T_{4'}$, is $T_3(V_3/V_4)^{k-1}$, or 3734.3°R. Then

$$\dot{S}_{gen} = \left[\ln\frac{1342.3°\text{R}}{1244.8°\text{R}} + \ln\frac{4339.7°\text{R}}{3734.3°\text{R}}\right](0.1718\ \text{Btu/lbm}\cdot°\text{R})(0.2675\ \text{lbm/s})$$

Answer
$$= 0.01037\ \text{Btu/s}\cdot°\text{R}$$

The computer may be used to advantage on problems of this length. The program OTTO would allow you to complete these types of problems more easily and quickly.

The designs of Otto engines have been altered to increase thermodynamic efficiency by increasing the piston stroke or decreasing the cylinder clearance volume. Equation (9–7) is invariably cited in justifying the increase of compression ratios in Otto engines. However, although this relation is approximately correct for the actual cycles, the combustion of fuels is made less effective in some cases, and frequently a phenomenon known as *preignition* or *knock* is observed in high-compression engines. Knock occurs because all fuel–air mixtures have a temperature at which they spontaneously combust or "explode." This temperature is reached in some cylinder chambers before the spark plug can ignite the mixture, and consequently a sharp explosion ensues, producing a noise (knock) and a sharp reduction in power. This may be noted in figure 9–14 by visualizing the effects of the burning of fuel before the ignition point x (to the right on curve xb).

To prevent preignition, the shape of the combustion chamber is frequently revised to provide more continuous, expanding combustion of the fuel–air mixture with an elimination of potential corners or "hot spots" which could cause knock. The most common method of preventing preignition has been to add "ignition depressants," such as tetrathyllead, to fuel. It is also worth mentioning that the fuel required in high-compression engines ($r_v \geq 6$, approximately) is a more refined extraction from crude oil than that required for low-compression engines or steady-state combustion processes. Water injected into the combustion chamber during compression has an effect similar to that of the additives (tetraethyllead) in reducing preignition. With the recent concern with lead emissions, in exhaust of IC engines and the elimination of lead from fuels, lower compression ratios are being used in engine designs. The efficiencies of these lower-compression-ratio engines has been improved over the older, higher-compression-ratio engines by utilizing solid-state ignition for better spark timing control, by using microprocessors and sensors to better control the fuel–air mixture entering the combustion chamber, and by using fuel injection. Fuel injection is a method of supplying fuel to the combustion chamber at a high pressure immediately before the fuel–air mixture is to be ignited by the spark plug. Fuel injectors are quite accurate in controlling the amount of fuel, and they are often monitored and controlled by computers. The fuel injector replaces the carburetor, a device that mixes air with fuel by convection and supplies this mixture to the combustion chamber.

A method for increasing the power of an internal combustion engine is through supercharging. This is an arrangement in which an air–fuel mixture or air is supplied to the combustion chamber under a pressure that is higher than atmospheric. Supercharging is achieved by using a pump or compressor, and it can be used with carburetors or fuel injectors. If the pressure at which the air–fuel mixture is supplied is less than atmospheric, the process is called **scavenging**.

There have been other modifications to achieve higher engine efficiency or greater output which are beyond the purposes of this text.

9–5 THE DIESEL ENGINE AND AIR-STANDARD ANALYSIS

The diesel engine is an internal combustion engine, similar to the Otto engine but operating at higher compression ratios, which ignites the fuel–air mixture by the compression process; the heat addition is accomplished during a longer time period. The ideal diesel engine operates on the ideal diesel cycle, defined by the following four reversible processes:

1–2 Adiabatic compression
2–3 Constant-pressure heat addition
3–4 Adiabatic expansion
4–1 Constant-volume heat rejection

Figure 9–18 shows the diagrams illustrating the paths of these processes on $p-V$ and $T-S$ coordinates, analogous to the Otto cycle. The enclosed area on the $p-V$ plane is the work of the cycle, and the enclosed area on the $T-S$ plane is the net heat added; the areas must be equal. The mechanism normally used to carry out these processes is the piston-cylinder device in figure 9–19. It has the same basic configuration and operation as that depicted in figure 9–2

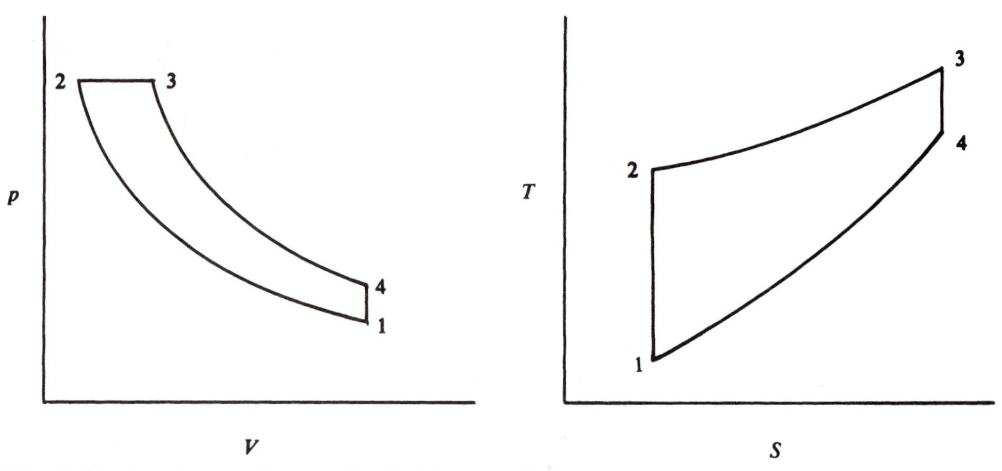

FIGURE 9–18 Property diagrams for ideal diesel cycle

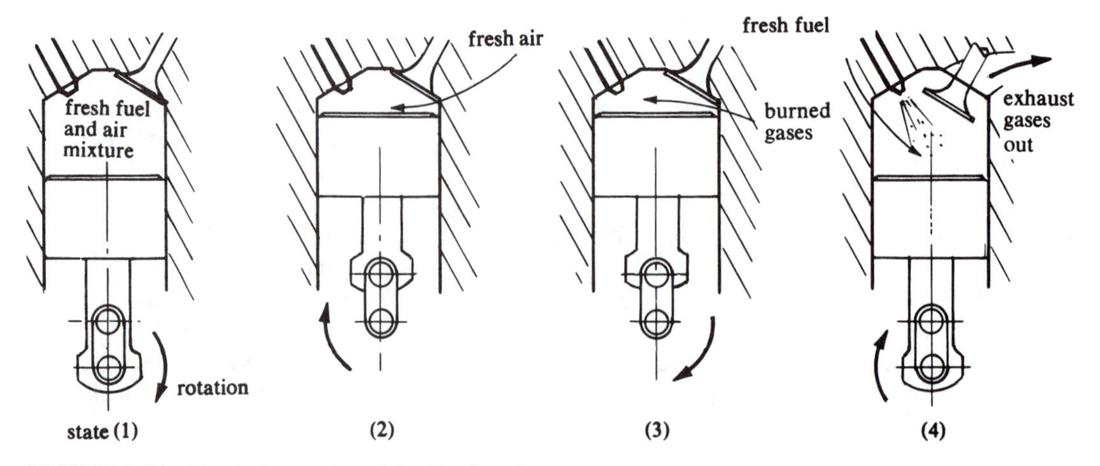

FIGURE 9–19 Physical operation of the diesel engine

except that no spark plug is present in the diesel cycle and the fuel is introduced by a fuel injector, which is a nozzle that sprays fuel into the combustion chamber, as shown in figure 9–19. Process 1–2 is a compression of a fresh charge of air and fuel (or just air); reaching state 2, the gases combust spontaneously due to the high pressure, and the process continues at constant pressure to state 3. Most commonly, only air is introduced and compressed in process 1–2, and fuel is injected at a pressure near that of state 2. Process 2–3 classifies the diesel engine as an internal combustion, compression-ignited engine (IC-CI engine).

Process 3–4 is an expansion of the exhaust gases until the piston reaches BDC (bottom dead center) at state 4. Then the exhaust is rejected into the atmosphere in process 4–1. Simultaneously, air and fuel (or just fresh air) is injected into the chamber to be ready for a new cycle. This constitutes the two-stroke diesel cycle with the exhaust and intake strokes of the piston together taking one revolution of the engine.

An actual diesel engine is shown in figure 9–20. This illustration serves to give the reader a better picture of the major components of a diesel engine. Notice the counterweighted crankshaft located at the bottom of the engine. This is connected to the piston by means of a connecting rod. Four valves situated at the top of the cylinder are used for exhausting the spent gases. Midway down the cylinder, a row of openings around the periphery of the cylinder can be seen. These are the intake openings where fresh air is introduced into the cylinder when the piston is at the bottom of its stroke. Fuel is injected from above after the compression stroke. The fuel injector used for this can be seen directly between the two exposed valves.

In figure 9–21 are shown the details of the critical parts that convert thermal energy to work through the piston-cylinder device. This view illustrates the manner of operation of the intake of air and fuel and the exhausting of the spent gases. At the top center of the figure are shown two three-lobed rotors which pump air into the center cavity where it is introduced into the cylinder through the peripheral openings in the cylinder walls when the

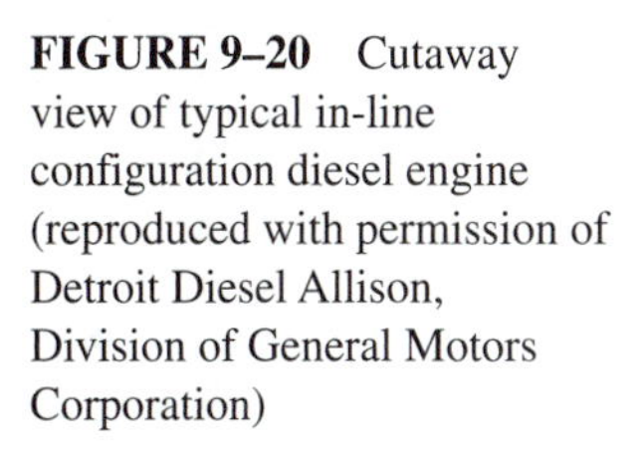

FIGURE 9–20 Cutaway view of typical in-line configuration diesel engine (reproduced with permission of Detroit Diesel Allison, Division of General Motors Corporation)

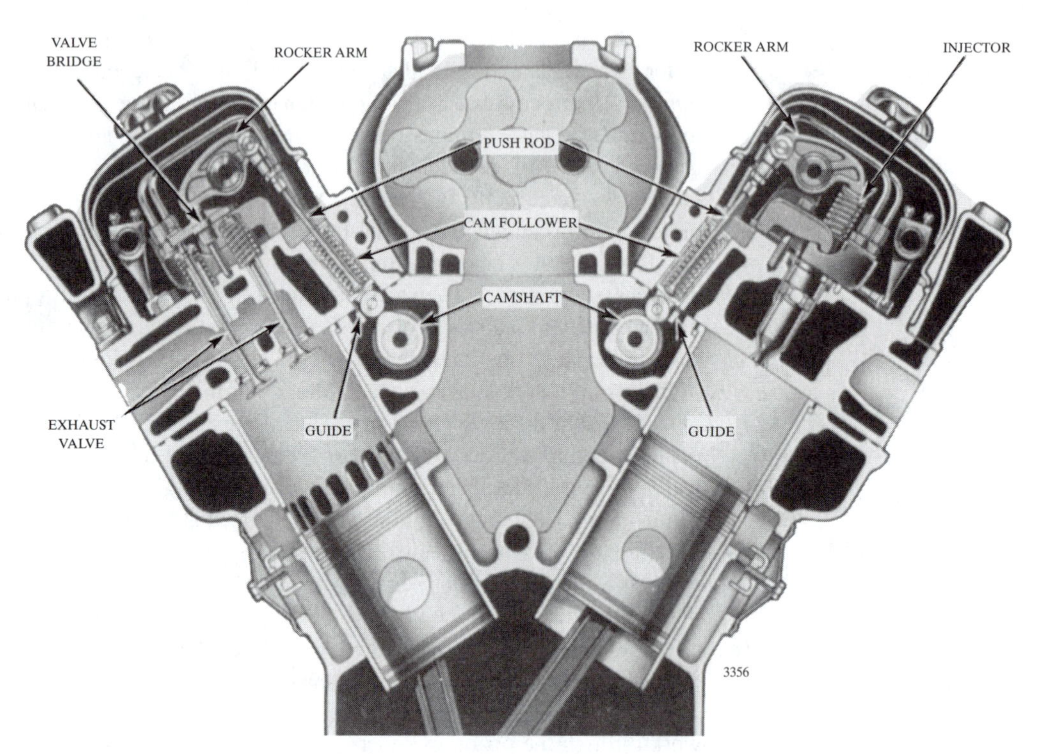

FIGURE 9–21 Cross-sectional view of typical V-configuration two-stroke-cycle diesel engine (reproduced with permission of Detroit Diesel Allison, Division of General Motors Corporation)

piston is at the bottom of its travel. The exhaust valves are actuated through a linkage system by the camshaft to allow the exhaust to escape from the cylinder at an appropriate time during the cycle. Notice that fuel is introduced into the cylinder through the injector while fresh air is introduced separately through openings in the cylinder walls. Finally, an external view of a typical diesel engine is shown in figure 9–22. This view shows the external components of a typical diesel engine. The large six-bladed fan is needed to draw air past a radiator to cool the engine when running. The radiator is not shown in this figure.

Although the diesel engine is commonly used in heavy applications such as trucks, tractors, and stationary power plants, a lighter engine has been used for some time in automotive applications. The diesel, however, seems to be gaining in popularity here due to its economy and durability.

To simplify calculations, the diesel engine can be analyzed with the assumption that air is the only working medium and is a perfect gas. This is the air-standard analysis, and it ignores the fuel, except as a heat source. This was also done with the Otto engine. The work of a cycle, as was mentioned, is the enclosed area of the p–V plane, which gives us the relation

$$Wk_{cycle} = Wk_{12} + Wk_{23} - Wk_{34} \tag{9–26}$$

For an air-standard analysis we have, assuming constant specific heats,

$$Wk_{12} = \frac{1}{1-k}(p_2V_2 - p_1V_1) \tag{9–27}$$

or

$$Wk_{12} = \frac{mR}{1-k}(T_2 - T_1) \tag{9–28}$$

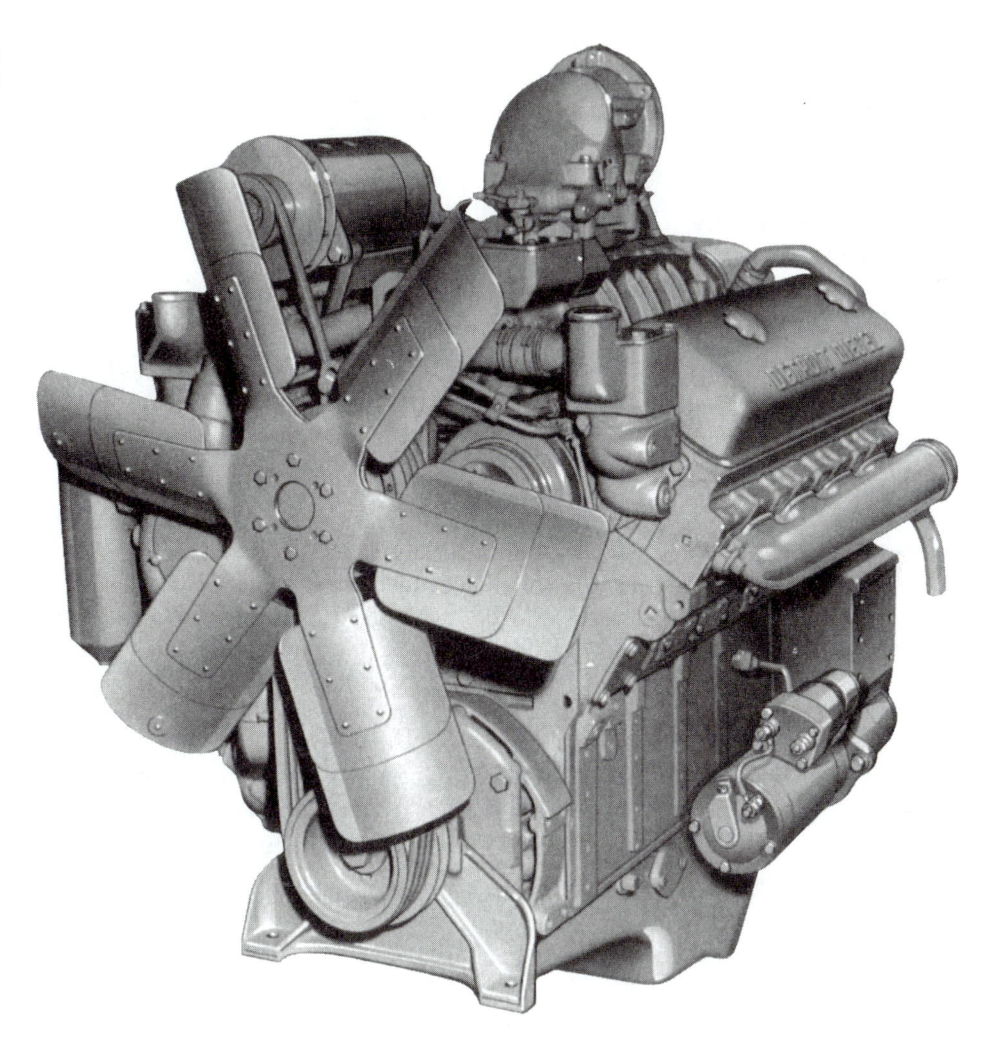

FIGURE 9–22 Diesel engine (reproduced with permission of Detroit Diesel Allison, Division of General Motors Corporation)

Also,

$$Wk_{23} = p_2(V_3 - V_2) \tag{9–29}$$

and

$$Wk_{34} = \frac{mR}{1 - k}(T_4 - T_3) \tag{9–30}$$

The compression ratio, r_v, of a diesel engine is the same as for an Otto engine:

$$r_v = \frac{V_1}{V_2} \tag{9–31}$$

The cutoff ratio is defined as

$$r_c = \frac{V_3}{V_2} \tag{9–32}$$

These results can give us the net work of an ideal cycle, provided we know the properties at the four cycle corners. If we wish to fit the diesel cycle to actual engines and find that processes 1–2 and 3–4 are not adiabatic, polytropic diesel processes can easily be substituted. Then the work can be obtained from the foregoing equations, but k will need to be

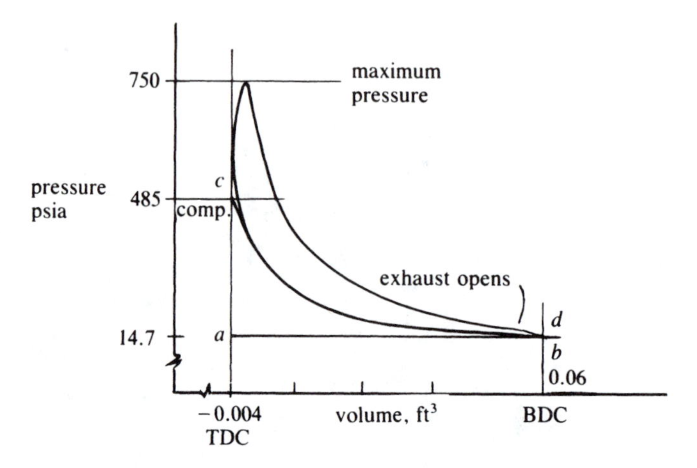

FIGURE 9–23 Typical $p–V$ diagram for compression-ignition engines operating at full load, slow speed, and mechanical injection [from E. F. Obert, *Internal combustion engines*, 2d ed. (Scranton, Pa.: International Textbook Co., 1959); with permission of International Textbook Co.]

replaced by the polytropic exponent n. An analysis of the actual diesel cycle closely parallels the Otto cycle analysis. The terms bhp, ihp, mep, bmep, bsfc, stroke, bore, and $p–V$ diagrams are all determined in a similar manner for the diesel as for the Otto engine. A typical $p–V$ diagram from an actual engine is shown in figure 9–23, and it should be compared to the ideal diesel cycle diagram of figure 9–18, as well as the diagram for the actual ICSI engine in figure 9–14.

EXAMPLE 9–13 An ideal diesel cycle engine operates with a compression rate of 20:1, taking in air at 101 kPa and 27°C. There is 0.05 kg of fuel added to each kilogram of air, and the mixture behaves like air. During combustion, the air–fuel mixture releases 2100 kJ/kg. Determine the optimum cutoff ratio r_c by using gas table B–6.

Solution The cutoff ratio is defined by equation (9–32):

$$r_c = \frac{V_3}{V_2}$$

Since process 2–3 is isobaric (i.e., takes place under constant pressure), we may write

$$\frac{V_3}{V_2} = \frac{T_3}{T_2}$$

We now determine these two temperatures with the aid of the data from table B–6. First, noting that the compression process 1–2 is reversible and adiabatic, we have, from equation (9–20),

$$\frac{V_2}{V_1} = \frac{v_{r_2}}{v_{r_1}} = \frac{1}{20}$$

From table B–6, we find that $v_{r_1} = 144.3$ at $T_1 = 300$ K. Then

$$v_{r_2} = \frac{144.3}{20} = 7.215$$

and from table B–6, we find that T_2 must be 932.5 K. Also, the enthalpy at state 2 from this table is 969.5 kJ/kg. The heat added during process 2–3 is derived from the energy equation and is

$$\begin{aligned} q_{\text{add}} &= \Delta u + wk_{cs} \\ &= \Delta u + p\,\Delta v \\ &= \Delta h = h_3 - h_2 \end{aligned}$$

from which we can find the enthalpy at state 3:

$$h_3 = q_{add} + h_2$$
$$= 2100 \text{ kJ/kg} + 969.5 \text{ kJ/kg} = 3069.5 \text{ kJ/kg}$$

From table B–6, we find the temperature at state 3 to be

$$T_3 = 2645 \text{ K}$$

The cutoff ratio may then be found from the equation

Answer

$$r_c = \frac{T_3}{T_2} = \frac{2645 \text{ K}}{932.5 \text{ K}} = 2.84$$

EXAMPLE 9–14 The p–V diagram shown in figure 9–23 is developed from a two-stroke, six-cylinder diesel engine operating at 1200 rpm. We use an air-standard analysis and know that

$$\text{bhp} = 115 \text{ hp} \qquad \text{at 1200 rpm}$$

and that the fuel usage at 1200 rpm is 75.0 lbm/h. Determine the Wk_{cycle}, ihp, mechanical efficiency, mep, and bsfc.

Solution

We extract the following property values from the curves of the p–V diagram in figure 9–23:

$$p_1 = 14.7 \text{ psia} \qquad V_1 = 0.06 \text{ ft}^3 = V_4$$
$$p_2 = 750 \text{ psia} \qquad V_2 = 0.004 \text{ ft}^3$$
$$p_3 = 750 \text{ psia} \qquad V_3 = 0.008 \text{ ft}^3 \quad \text{(estimated)}$$
$$p_4 = 20 \text{ psia} \quad \text{(estimated)} \qquad T_1 = 70°\text{F} \quad \text{(assumed)}$$

We may analyze this engine by using the BASIC program to compute the enclosed area of the p–V diagram or by using polytropic processes for the compression and expansion strokes. Let us use the polytropic process method. We obtain

$$\frac{P_1}{P_2} = \frac{V_2^{n_{12}}}{V_1}$$

$$n_{12} = \frac{\ln(p_1/p_2)}{\ln(V_2/V_1)}$$

$$= \frac{\ln(14.7/750)}{\ln(0.004/0.06)} = 1.45$$

$$\frac{p_3}{p_4} = \left(\frac{V_4}{V_3}\right)^{n_{34}}$$

and

$$n_{34} = \frac{\ln(p_3/p_4)}{\ln(V_4/V_3)} = \frac{\ln(750/20)}{\ln(0.06/0.008)} = \frac{3.62}{2.01} = 1.80$$

The compression ratio for this engine is 15:1, and the cutoff ratio, defined as V_3/V_2, can be computed as

$$r_c = \frac{V_3}{V_2} = \frac{0.008}{0.004} = 2$$

The amount of air introduced into the cylinder for each power stroke is obtained from the perfect gas equation,

$$p_1 V_1 = mRT_1$$

or

$$m = \frac{p_1 V_1}{RT_1} = \frac{(14.7 \text{ lbf/in}^2)(144 \text{ in}^2/\text{ft}^2)(0.06 \text{ ft}^3)}{(53.3 \text{ ft}\cdot\text{lbf/lbm}\cdot{}^\circ\text{R})(460 + 70){}^\circ\text{R}}$$
$$= 0.0045 \text{ lbm/cycle-cylinder}$$

Since the engine is a two-stroke cycle arrangement, the number of cycles equals the number of revolutions.

The work per cycle or revolution is obtained after we determine the temperature at states 2, 3, and 4. We have

$$T_2 = T_1\left(\frac{V_1}{V_2}\right)^{n_{12}-1} = 530^\circ\text{R}\left(\frac{0.06}{0.004}\right)^{0.45} = 1793^\circ\text{R}$$

$$T_4 = T_1\frac{p_4}{p_1} = 530^\circ\text{R}\left(\frac{20.0}{14.7}\right) = 721^\circ\text{R}$$

and

$$T_3 = T_2\frac{V_3}{V_2} = 1793^\circ\text{R}\left(\frac{0.008}{0.004}\right) = 3586^\circ\text{R}$$

Check each process to make sure that you understand how these equations are applicable.

Now the work is calculated:

$$Wk_{\text{cycle}} = \frac{mR}{1 - n_{12}}(T_2 - T_1) + p_2(V_3 - V_2) + \frac{mR}{1 - n_{34}}(T_4 - T_3)$$
$$= \left(\frac{0.0045 \text{ lbm} \times 53.3 \text{ ft}\cdot\text{lbf/lbm}\cdot{}^\circ\text{R}}{1.00 - 1.45}\right)(1793 - 530){}^\circ\text{R}$$
$$+ (750 \text{ lbf/in}^2 \times 144 \text{ in}^2/\text{ft}^2)(0.008 - 0.004) \text{ ft}^3$$
$$+ \left(\frac{0.0045 \text{ lbm} \times 53.3 \text{ ft}\cdot\text{lbf/lbm}\cdot{}^\circ\text{R}}{1.0 - 1.8}\right)(721 - 3586){}^\circ\text{R}$$
$$= (-673 + 432 + 859) \text{ ft}\cdot\text{lbf/cycle}$$

Answer

$$= 618 \text{ ft}\cdot\text{lbf/cycle} = 0.794 \text{ Btu/cycle}$$

The total work for all six cylinders is

Answer

$$Wk_{\text{cycle}} = 6 \times 0.794 = 4.764 \text{ Btu/cycle}$$

and the indicated horsepower at 1200 rpm is obtained from

$$\text{ihp} = Wk_{\text{cycle}}N = 4.764 \times 1200 \text{ Btu/min} = 5717 \text{ Btu/min}$$

Answer

$$= (5717 \text{ Btu/min})\left(\frac{60 \text{ min/h}}{2545 \text{ Btu/hp}\cdot\text{h}}\right) = 134.8 \text{ hp}$$

Since the brake horsepower is 115 hp at 1200 rpm, the mechanical efficiency, from equation (9–4), is

Answer

$$\eta_T = \frac{\text{bhp}}{\text{ihp}} \times 100 = \frac{115}{134.8} \times 100 = 85.3\%$$

The mean effective pressure is obtained from equation (9–9):

$$\text{mep} = \frac{Wk_{\text{cycle/cylinder}}}{\text{cylinder displacement}} = \frac{618 \text{ ft} \cdot \text{lbf/cycle}}{(0.06 - 0.004) \text{ ft}^3}$$

Answer $$= 11{,}036 \text{ lbf/ft}^2 = 76.6 \text{ psi}$$

We calculate the bmep from equation (9–12):

$$\text{bmep} = \frac{\text{bhp}}{(\text{displacement})(N)(n)} = \frac{115 \text{ hp} \times 33{,}000 \text{ ft} \cdot \text{lbf/min} \cdot \text{hp}}{(0.054 \text{ ft}^3)(6 \text{ cylinders})(1200 \text{ rpm})}$$

Answer $$= 9760 \text{ lbf/ft}^2 = 67.8 \text{ psi}$$

The fuel consumption bsfc is calculated from equation (9–14):

$$\text{bsfc} = \frac{\dot{m}}{\text{bhp}} = \frac{75 \text{ lbm/h}}{115 \text{ bhp}}$$

Answer $$= 0.652 \text{ lbm/bhp} \cdot \text{h}$$

9–6 THE DIESEL–OTTO COMPARISON

The most common parameter for comparing various engines is thermodynamic efficiency. Let us derive relationships for the efficiency of the diesel engine and see how they compare with the Otto cycle. We first use the standard definition of thermodynamic efficiency:

$$\eta_T = \frac{Wk_{\text{cycle}}}{Q_{\text{add}}} \times 100 \tag{9–4}$$

Notice that, for the diesel cycle with a perfect gas having constant specific heats,

$$Q_{\text{add}} = mc_p(T_3 - T_2) \tag{9–33}$$

and

$$\begin{aligned} Wk_{\text{cycle}} &= Q_{\text{add}} + Q_{\text{rej}} \\ &= mc_p(T_3 - T_2) + mc_v(T_1 - T_4) \end{aligned}$$

The efficiency can then be written

$$\eta_T = \frac{mc_p(T_3 - T_2) + mc_v(T_1 - T_4)}{mc_p(T_3 - T_2)} \times 100$$

$$= \left(1 + \frac{c_v}{c_p}\frac{T_1 - T_4}{T_3 - T_2}\right) \times 100 = \left(1 - \frac{1}{k}\frac{T_4 - T_1}{T_3 - T_2}\right) \times 100 \tag{9–34}$$

This can be put in other terms by a little more manipulation, yielding

$$\eta_T = \left[1 - \frac{1}{(r_v)^{k-1}}\frac{r_c^k - 1}{k(r_c - 1)}\right] \times 100 \tag{9–35}$$

For the polytropic processes in the diesel cycle, equation (9–35) does not hold, but equation (9–4) does apply. A comparison now between the diesel and Otto cycles shows that, at a given compression ratio, the Otto engine is theoretically more efficient [compare equation

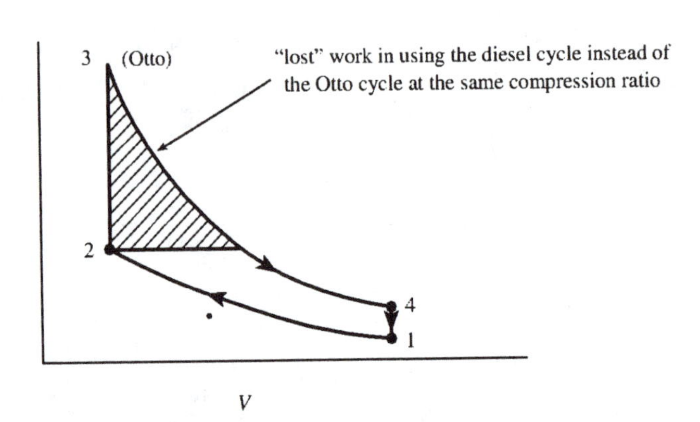

FIGURE 9–24 *p–V* diagram comparison of Otto and diesel cycles at the same compression ratio

(9–35) to equation (9–7)] than the diesel. This is further exemplified by the superimposed Otto cycle on the diesel cycle in figure 9–24. It is shown that the shaded area represents work realized by the Otto engine and not by the diesel. The primary reason for developing the diesel, however, has been that higher compression ratios could be achieved. The fuel in an Otto engine exhibits preignition at elevated compression ratios, but in the diesel, it is just this preignition that causes the fuel to burn. Compression ratios can then be increased to values limited only by material strength and dynamics of the piston-cylinder. Most diesel engines have thermodynamic efficiencies that are higher than those of Otto engines, because of the increased compression. As a side benefit, the diesel can generally utilize a less refined fuel than the Otto. In fact, the inventor and first developer of the diesel, Rudolph Diesel, visualized coal dust as the primary fuel, but this has since been replaced by liquid fuels. At any rate, the diesel can generally use a less expensive fuel and use it more efficiently than the Otto engine.

As a review, we list the equations of the Otto and diesel cycles in table 9–3, which can be used for elementary thermodynamic analyses. The computer program DIESEL described in appendix A–9 and available on a soft diskette allows for quicker analysis of the diesel cycle problems.

9–7 THE DUAL CYCLE

The simple diesel cycle frequently does not represent well the operation of the ICCI engine or diesel engine. A method of better modeling the actual diesel engine is with the dual cycle, which has a heat addition at constant volume (like the Otto cycle) plus a heat addition at constant pressure. The percentage of heat added during each of these two processes is the unique feature of the dual cycle that allows it to better accommodate the actual diesel engine. We define the dual cycle as the following five processes:

1–2 Reversible adiabatic compression
2–3 Reversible isometric heat addition
3–4 Reversible isobaric heat addition
4–5 Reversible adiabatic expansion
5–1 Reversible isometric heat rejection

The *p–V* diagram of a dual cycle is shown in figure 9–25, and you can see that there are five distinct processes. If the heat is all added at constant volume (isometric), the dual cycle is just the same as the ideal Otto cycle, and if the heat is all added at constant pressure (isobaric), the dual cycle is equivalent to the diesel cycle. Frequently, the diesel engine operates on a dual cycle, where the fraction of heat added at constant volume is from 40% to 60% of the total heat added. The computer program DIESEL includes the option of analyzing the diesel engine with a dual cycle, and this is considered in the next section.

TABLE 9–3 Equations for IC engines (assuming perfect gases with constant specific heats)

Term	Otto	Diesel
Q_{add}	$mc_v(T_3 - T_2)$	$mc_p(T_3 - T_2)$
Q_{rej}	$mc_v(T_1 - T_4)$	$mc_v(T_1 - T_4)$
Wk_{cycle} (ideal)	$\frac{mR}{1-k}(T_2 - T_1 + T_4 - T_3)$	$\frac{mR}{1-k}(T_2 - T_1 + T_4 - T_3) + p_2(V_3 - V_2)$
Wk_{cycle} (polytropic)	$\frac{mR}{1-n_{21}}(T_2 - T_1) + \frac{mR}{1-n_{43}}(T_4 - T_3)$	$\frac{mR}{1-n_{21}}(T_2 - T_1) + \frac{mR}{1-n_{43}}(T_4 - T_3) + p_2(V_3 - V_2)$
r_v	V_1/V_2	V_1/V_2
r_c	NA	V_3/V_2
η_T (ideal)	$1 - \frac{1}{(r_v)^{k-1}}$	$1 - \frac{1}{(r_v)^{k-1}}\frac{r_c^k - 1}{k(r_c - 1)}$
mep or imep	$\frac{Wk_{cycle}}{\text{displacement}} = \frac{\text{ihp}}{\text{displacement} \times N}$	
bmep	$\frac{\text{bhp}}{\text{displacement} \times N}$	
η_{mech}	$\frac{\text{bhp}}{\text{ihp}} \times 100$	
ihp	$Wk_{cycle}N$	
N	Engine speed in rpm or cycles/min for two-stroke versions and engine speed divided by 2 for the four-stroke versions	

NA, not applicable.

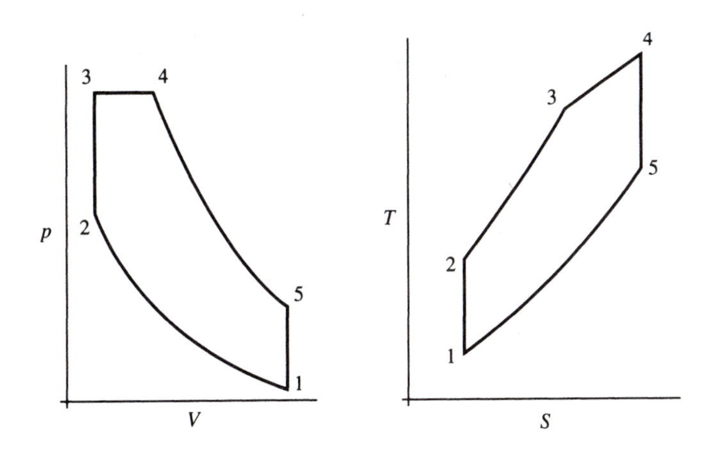

FIGURE 9–25 Property diagrams for an ideal dual-cycle IC engine

9–8 COMPUTER-AIDED ANALYSIS

In the previous sections of this chapter, we have considered various problems relating to the Otto and diesel engines. Many of the problems were very long and time consuming to solve, and much of the effort involved rote use of straightforward process relationships and repetitive calculations. These types of activities are well suited to the computer. Appendices A–8 and A–9 describe two programs in VISUAL BASIC to give you examples of how you may use the computer to assist in better analyzing the internal combustion engines. Both of the programs are based on an air-standard analysis with constant specific heats, although properties for perfect gases other than air can be used. Provisions are made so that you can use the modifications of polytropic processes or the irreversible adiabatic processes.

The programs require certain inputs, and then the computer can proceed to determine the various parameters of either the Otto or the diesel engines. The programs can be used to best advantage when you vary one particular parameter, say, the compression ratio, and keep all the other input parameters constant. The repetitive calculations can be handled quickly by a microcomputer, and you can obtain the variation of the output parameters as they are affected by the varied input parameter. The existing programs, OTTO and DIESEL, could be expanded to allow for graph plotting of the parameters (p–V, T–S, or Wk–r_v, or others as well) or to make the program accommodate variable specific heats. Many other elaborations could also be done to make the programs better allow for analysis and designing of IC engines.

EXAMPLE 9–15

Consider a diesel engine operating on a dual cycle with intake conditions of 100 kPa and 10°C. The engine has a compression ratio of 19.5:1 and a displacement of 7000 cm^3 (mL) or 7 L, and it operates from 800 to 2000 rpm. It is a four-stroke version, and at 800 rpm, 12 kJ/cycle of heat is added at constant volume. No heat is added at constant pressure at 800 rpm. At increased engine speeds, the amount of heat added at constant volume remains at 12 kJ/cycle, but additional heat is now added at constant pressure so that, at 2000 rpm, the cutoff ratio is 3.4:1. Assuming that the cutoff ratio varies directly with the engine speed, determine the power as a function of the engine speed.

Solution

We use a dual cycle for the engine and have the following data:

$$r_v = \frac{V_1}{V_2} = 19.5 \qquad V_1 - V_2 = 7000 \text{ cm}^3$$

$$p_1 = 100 \text{ kPa} \qquad T_1 = 10°\text{C} = 283 \text{ K}$$

$$r_c = 3.4 \text{ at } 2000 \text{ rpm} \qquad (\text{or } 1000 \text{ cycles/min})$$

Table 9–4 illustrates the way in which the cutoff ratio (and therefore the amount of heat added at constant pressure) varies with the engine speed. As an example, the following computations are shown for an engine speed of 900 rpm:

$$N = 450 \text{ cycles/min} \qquad V_1 = 7378.38 \text{ cm}^3$$

$$V_2 = \frac{V_1}{19.5} = 378.38 \text{ cm}^3$$

$$\dot{m} = \frac{p_1 V_1}{R T_1} N = 4.088 \text{ kg/min}$$

$$T_2 = T_1 \left(\frac{V_1}{V_2} \right)^{k-1} = 928.5 \text{ K}$$

TABLE 9–4 Results of example 9–15

Speed, rpm	r_c	$\dot{Q}_v$, kJ/s	$\dot{Q}_p$, kJ/s	$\dot{Q}_{add}$, kJ/s
800	1.0	80	0	80.0
900	1.2	90	37.9	127.9
1000	1.4	100	84.3	184.3
1100	1.6	110	139.0	249.2
1200	1.8	120	202.2	322.2
1300	2.0	130	273.8	403.8
1400	2.2	140	353.9	493.9
1500	2.4	150	442.3	592.3
1600	2.6	160	539.2	699.2
1700	2.8	170	644.5	814.5
1800	3.0	180	758.3	938.3
1900	3.2	190	880.4	1070.4
2000	3.4	200	1011.0	1211.0

The rate of heat addition at constant volume, $\dot{Q}_v$ is 90 kJ/s or 5400 kJ/min. The heat added at constant volume is

$$q_{add} \text{ (at constant volume)} = c_v(T_3 \, 2 \, T_2) = \frac{\dot{Q}_v}{N}$$
$$= \frac{5400 \text{ kJ/min}}{450 \text{ cycles/min}}$$
$$= 12 \text{ kJ/cycle}$$

and also

$$q_{add} = c_v(T_3 - 928.5 \text{ K}) = \frac{(12 \text{ kJ/cycle})(450 \text{ cycles/min})}{4.088 \text{ kg/min}}$$

Using $c_v = 0.719$ kJ/kg · K for air from table B–4, we solve for T_3:

$$T_3 = 2766 \text{ K}$$

At 900 rpm, $r_c = 1.2$, so $T_4 = T_3(r_c) = 3319$ K and

$$q_{add} \text{ (at constant pressure)} = c_p(T_4 \, 2 \, T_3)$$
$$= (1.007 \text{ kJ/kg} \cdot \text{K})(3319 \; 2 \; 2766 \text{ K}) = 556.9 \text{ kJ/kg}$$

The rate of heat addition at constant pressure is

$$\dot{Q}_p = (556.9 \text{ kJ/kg})\left(\frac{4.099 \text{ kg/min}}{60 \text{ s/min}}\right)$$
$$= 37.9 \text{ kJ/s} = 37.9 \text{ kW}$$

Then

$$\dot{Q}_{add} = 90 \text{ kW} + 37.9 \text{ kW} = 127.9 \text{ kW}$$

Also,

$$V_3 = V_2 = 287.38 \text{ cm}^3 \qquad V_4 = (V_3)(r_c) = 454.1 \text{ cm}^3$$

and

$$T_5 = T_4\left(\frac{V_4}{V_5}\right)^{k-1} = 1090 \text{ K}$$

TABLE 9–5 Results of example 9–15

Speed, rpm	Power, kW
800	55.5
900	88.4
1000	133.6
1100	168.3
1200	215.0
1300	266.9
1400	321.3
1500	380.6
1600	443.6
1700	510.5
1800	582.1
1900	654.8
2000	732.1

The rate of heat rejection is

$$Q_{rej} = \dot{m}c_v(T_1 - T_5) = 39.5 \text{ kW}$$

so the net power of the cycle is

$$\dot{W}k_{net} = 127.9 \text{ kW} - 39.5 \text{ kW} = 88.4 \text{ kW} \qquad \text{at 900 rpm}$$

The remaining results can be obtained from the use of the program DIESEL with a microcomputer. The program is best utilized for this problem if a type 1 problem is selected and the dual-cycle option is also selected. The data in table 9–5 can be obtained on a personal computer if air properties are used, 100% mechanical efficiency is indicated, and ideal processes are assumed for the cycle.

Many of the Otto, diesel, and dual cycle analyses can be conveniently completed with *EES*.

Further analysis could be made of this engine or other IC engines; for instance, the thermal efficiency could be found as the speed changes, the maximum temperature experienced in the engines (at state 4) could be found for different speeds, or the heat rejection as speed changes could be determined. The method shown in the last example would be equally applicable for the Otto engine as well.

9–9 ENGINE DESIGN CONSIDERATIONS

In previous sections, we have discussed the detailed calculations for predicting the energy transfers of the internal combustion engines, both Otto and diesel. In addition, the Otto cycle has been discussed from the standpoint of modifications to make its performance more desirable. Here let us consider very briefly some general aspects of IC engine design, presented only to give the reader an overview of the inherent problems and advantages of these engines. There are a number of books, some of which are listed in appendix C, which can furnish finer detail. Here we treat (1) engine balancing, (2) speed limitation, (3) gas flows and manifolding, (4) engine timing, (5) cooling systems, (6) advantages of the IC engine, and (7) rotary engine designs.

Engine Balancing

The piston-cylinder device which reciprocates invariably transfers its mechanical work to a rotating shaft by means of a crank. This mechanism is shown in figure 9–26 with a counterweight superimposed to show its position on the crank. The counterweight is used to prevent extreme imbalance in the crank and rotating shaft when the power stroke of the

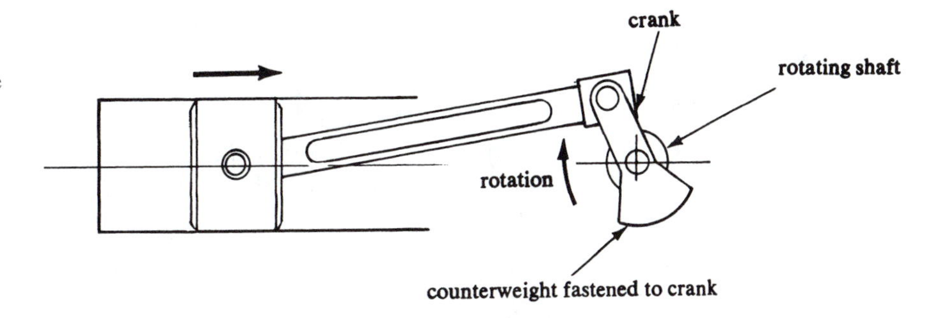

FIGURE 9–26 Piston-cylinder crank mechanism of the reciprocating engine of the Otto or diesel cycles

piston produces a force on the crank. This method of balancing is complemented by reducing the mass of pistons, valves, connecting rods, and other moving parts. However, a perfectly balanced reciprocating engine is impossible to achieve.

Speed Limitations

A device that has an imbalance (even a small amount) will be subject to dynamic effects that become extremely intolerable at high speeds. Therefore, reciprocating engines have definite inherent speed limitations due to their characteristics of the piston-cylinder crank mechanism. In addition, the Otto and diesel engines utilize intake and exhaust valves which have response limitations, in either mechanical or hydraulic actuation. That is, valves can physically be opened in a finite time, which will produce an upper limit to engine speeds. Of course, we have not mentioned irreversible effects, which increase with speed and can become overwhelming in the expansion process where the gases are expected to produce the work of the engine.

The manner in which engine speed affects engine performance is seen in figure 9–27. Here is a typical performance chart that is developed from engine tests. Thermodynamics can give insights into why the parameters behave as they do, but it cannot now accurately predict the full shapes of the curves. Notice that the individual curves do not have maximum or minimum values at the same speeds; that is, maximum power is developed at a higher speed than efficiency in the engine of figure 9–27. The optimization process, or selection of the speed that gives the all-around best performance, is a problem that must be considered in all engine design or selection methods. We will not pursue this further except to say that the example problems we have considered in this chapter are essentially single-speed conditions, and the full analysis of engines must include the consideration of a performance chart as illustrated in figure 9–27.

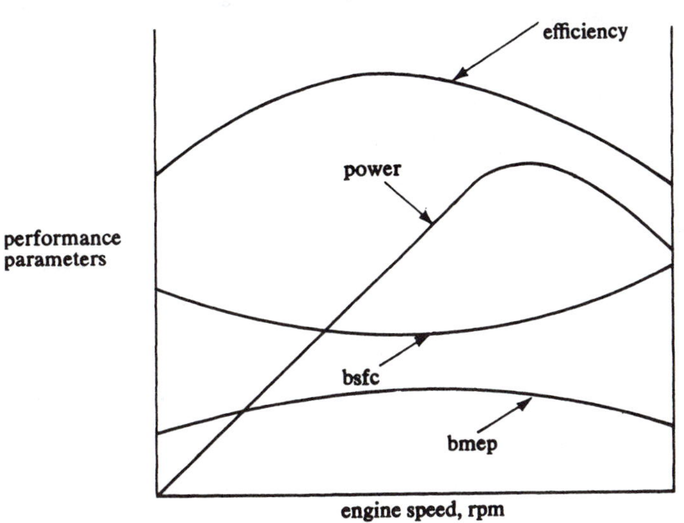

FIGURE 9–27 Typical performance chart of a reciprocating internal combustion engine

Gas Flows and Manifolding

Fuel injection is designed to replace carburetors and valves in supplying fuel to cylinders; however, the latter are more common. With the use of carburetors, however, there exists a problem of equal fuel–air mixture to the cylinders, as illustrated in figure 9–28. Due to fluid or gas friction and the different distances, cylinders 2 and 3 will obviously get better service than cylinder 1 or 4. The design challenge here is to provide equal flow paths to each cylinder. This applies not only to the carburetor but also to the exhaust pipe, which tends to retard the flow of exhaust gases from the cylinder.

Engine Timing

With the spark-ignition engine there exists the problem of synchronizing the spark to the piston position. The spark from the spark plug comes slightly before the piston has fully compressed the gases (at TDC) in general. As speeds are varied, the amount of time between the spark and TDC must also vary, which requires a proper mechanical–electrical coordination of the peripheral engine components such as the distributor, the ignition points, and the carburetor.

Cooling Systems

The heat engine must reject heat to the surroundings, and the Otto and diesel engines are no exceptions. They commonly reject more than three-fourths of the heat added, and if this heat is not taken away from the engine block quickly, the engine will reach extremely high temperatures and permanently damage the materials. Air forced around the engine is the most obvious method of cooling, but this requires a large surface area to be effective. It is generally achieved by putting cooling fins or vanes in the engine block (see figure 9–29)—an effective, but expensive, process.

Water cooling is a common method of removing the rejected heat from the engine block. This is an effective method for large engines as well as small, but it requires additional equipment and thus reduces the reliability of the engine. Radiators, water pumps, and hoses are some of the items readily identified with a water cooling system.

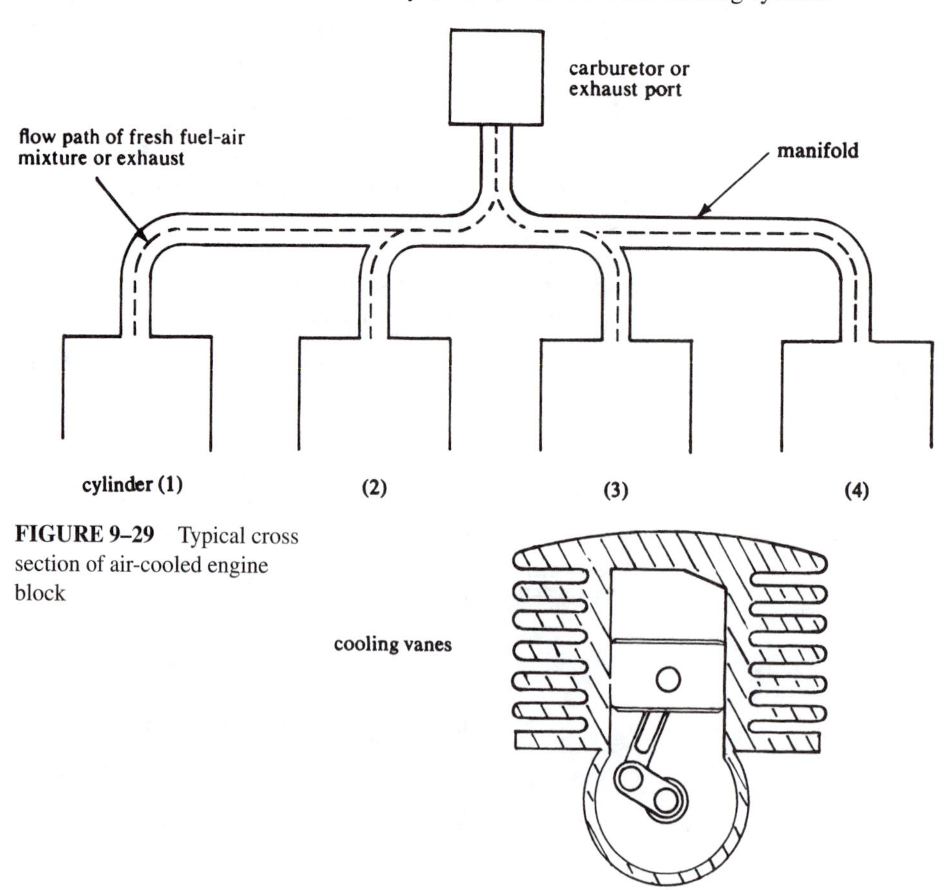

FIGURE 9–28 Typical configuration of a manifold and the flow pattern

FIGURE 9–29 Typical cross section of air-cooled engine block

Advantages of the IC Reciprocating Engine

We have pointed out a few of the problems of the IC engine, but let us reiterate its advantages. The reciprocating IC engine is an engine that is easily started, can produce its nominal power almost immediately upon starting, can be stopped easily and quickly, and can provide fast changes in speed (acceleration and deceleration).

The diesel engine is well suited for providing large amounts of power at slow and medium speeds. It is frequently designed with large components to better withstand high compression ratios and thus provide an engine with extended lifetime.

Rotary Engine Designs

The rotary engine has always been an appealing alternative to the reciprocating piston-cylinder engine. The most recent designs of a rotary engine, operating on the Otto or diesel cycles, is the Wankel engine. A cross section of this engine is shown in figure 9–30, and it can be seen that a three-lobed element rotates within a slightly "dog-boned" cavity. The rotating element traps a charge of air–fuel mixture and compresses that charge. The charge then burns or combusts to release energy, and the exhaust gases then expand

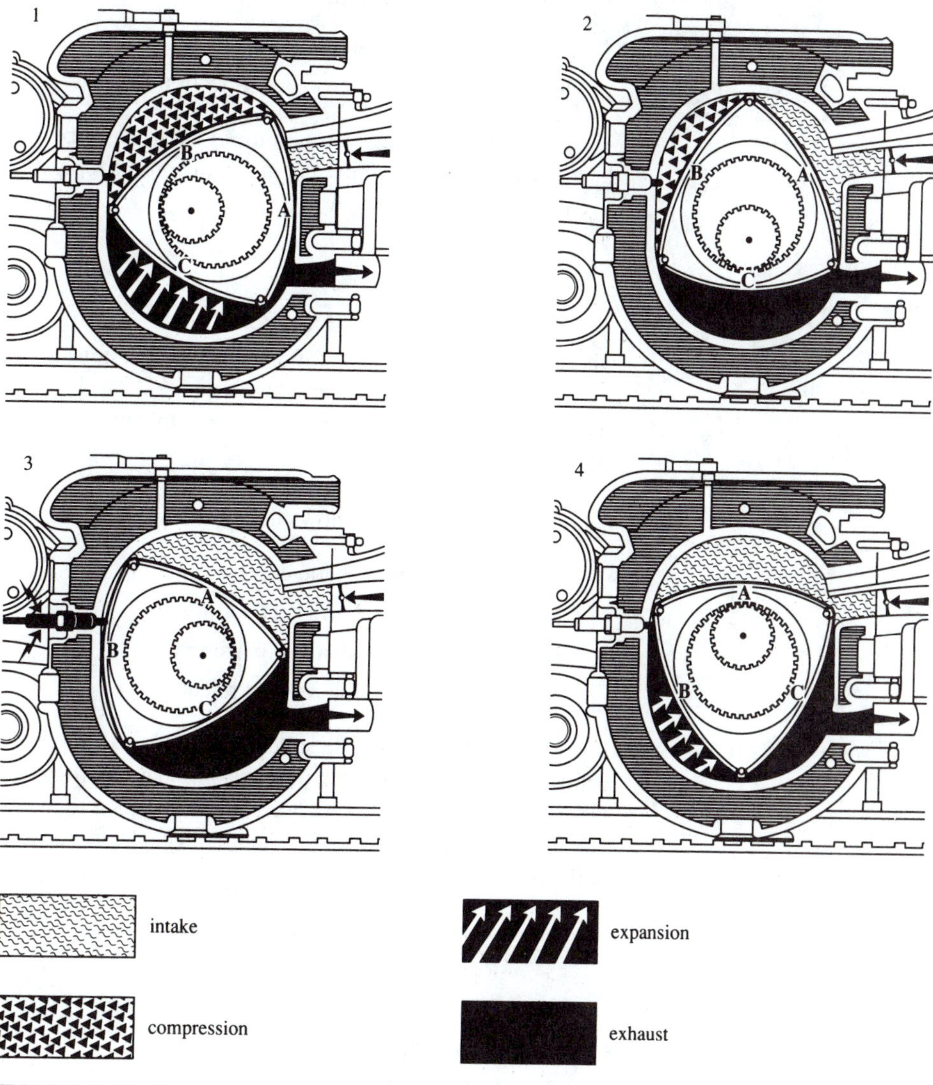

FIGURE 9–30 Cross section of the four-stroke Wankel engine [from Jan P. Norbye, *The Wankel engine* (Radnor, Pa.: Chilton Book Company), copyright 1971 by the author; reprinted with the permission of the publisher, Chilton Book Company, Radnor, Pa.]

to provide work to rotate the element. The analysis of the Wankel engine closely follows that for the normal Otto cycle except that there are always three separate charges of air–fuel proceeding through cycles simultaneously. Thus, for each rotor, there is the equivalent of three cylinders of a piston-cylinder engine. The engine can run at high speeds because there are no reciprocating parts, and it has a very high power-to-weight ratio. Its primary fault is a chronic sealing problem of the three-lobed element which has never been completely solved.

9–10 SUMMARY

In this chapter, the internal combustion (IC) engines were considered. The IC engine has two classifications: spark-ignition (ICSI) or compression-ignition (ICCI). The ICSI engine was analyzed with the Otto cycle, defined by the following processes:

1–2 Reversible adiabatic compression
2–3 Reversible isometric heat addition
3–4 Reversible adiabatic expansion
4–1 Reversible isometric heat rejection

The air-standard analysis assumes that all gases passing through the engine are air. If the air is assumed to be a perfect gas with constant specific heats, the thermal efficiency is given by the equation

$$\eta_T = 1 - \frac{1}{(r_v)^{k-1}} \times 100 \qquad \textbf{(9–7)}$$

where r_v is the compression ratio. With modifications to the Otto cycle, one can often better model the operation of an actual engine. The three modifications considered in this chapter were variable specific heats, reversible polytropic processes, and irreversible adiabatic processes.

The compression-ignition engine is usually referred to as the diesel engine. The ideal operation of this engine follows the diesel cycle, defined by the following processes:

1–2 Reversible adiabatic compression
2–3 Reversible isobaric heat addition
3–4 Reversible adiabatic expansion
4–1 Reversible isometric heat rejection

Modifications to the diesel cycle can be made to better describe the operation of the actual diesel engine. In particular, the variable specific heats, the polytropic processes, and the irreversible adiabatic processes can be adapted to the diesel cycle. Also, the dual cycle was introduced to provide a better approximation to the actual diesel engine, and this cycle was defined by five processes:

1–2 Reversible adiabatic compression
2–3 Reversible isometric heat addition
3–4 Reversible isobaric heat addition
4–5 Reversible adiabatic expansion
5–1 Reversible isometric heat rejection

DISCUSSION QUESTIONS

Section 9–1

9–1 What is meant by an *internal combustion engine*?

Section 9–2

9–2 What is the difference between *two-stroke engines* and *four-stroke engines*?

9–3 What is meant by the term *clearance*?

9–4 What is meant by *compression ratio*?

9–5 What is *air standard analysis*?

Section 9–3

9–6 How does the specific heats ratio, k, affect thermal efficiency of an Otto engine?

Section 9–4

9–7 What is the difference between *indicated horsepower* and *brake horsepower*?

9–8 What is meant by mean effective pressure?

9–9 What is *volumetric efficiency*?

Section 9–5

9–10 What is the *cutoff ratio* in a diesel cycle?

Section 9–6

9–11 Why does the Otto cycle have greater thermal efficiency than the diesel cycle for the same compression ratio?

9–12 Why is the diesel usually more efficient than the Otto cycle engine?

Section 9–7

9–13 What is the *dual cycle*?

Section 9–9

9–14 What is a *manifold* used for?

9–15 Why has the *Wankel* or rotary engine not been more successful?

PRACTICE PROBLEMS

Section 9–2

Problems that use SI units are indicated by an (M) under the problem number; those that use English units are indicated by an (E).

9–1 (M) Determine the cylinder displacement and engine displacement of an engine having six cylinders with a bore of 102 mm and a stroke of 120 mm.

9–2 (M) Determine the compression ratio of an engine with a bore of 98 mm, a stroke of 100 mm, and a clearance volume of 75 cm^3.

9–3 (M) A two-stroke four-cylinder ideal Otto engine operates at 4800 rpm. Air is taken in at 101 kPa and 20°C. The volume in each cylinder is 500 cm^3 before compression and 90 cm^3 after compression. If 2500 J/g air of heat is added, determine

(a) Compression ratio
(b) p, V, and T at the four corners
(c) Entropy change during heat addition process
(d) Wk_{cycle}
(e) Q_{rej}
(f) Power of engine
(g) Sketch the p–V and T–s diagrams.

9–4 (M) A four-stroke eight-cylinder Otto engine operates at 2500 rpm, and the compression ratio is 6:1. The cylinder volume is 2000 cm^3 before compression, and the intake air is at 100 kPa and 30°C. If the heat addition is 900 kJ/kg, determine

(a) Wk_{cycle}, or power
(b) Heat rejected per cycle

9–5 (E) Determine the displacement of an eight-cylinder engine having a bore of 3.25 in and a stroke of 3 in.

9–6 (E) Determine the stroke of an engine with a compression ratio of 8:1, a bore of 2.5 in, and a clearance volume of 2 in^3.

9–7 (E) A six-cylinder automobile engine has cylinders with a bore of 2.5 inches, a stroke of 2.25 inches, and a clearance volume (per cylinder) of 1.6 cubic inches. Determine the engine displacement and the compression ratio for this engine.

9–8 (E) An ideal Otto engine with six cylinders operates on a four-stroke cycle with a compression ratio of 7.5:1. Given that $p_1 = 14.7$ psia, $T_1 = 100°F$, $V_1 = 104\ ft^3$, and the heat added is 1000 Btu/lbm, determine for an air-standard analysis

(a) p and V at the four corners of the cycle
(b) T at the four corners
(c) Wk_{cycle}
(d) Q_{rej}

Plot the p–V and T–s diagrams, labeling the coordinate points at 1, 2, 3, and 4.

9–9 (E) A four-stroke two-cylinder Otto engine has a bore of 24 in, a stroke of 20 in, and a compression ratio of 7.2:1. The air taken in is at 14.7 psia and 40°F, and the compression and expansion processes are reversible and adiabatic ($k = 1.4$). If 1400 Btu/lbm air is added and the engine speed is 320 rpm, determine
(a) Wk_{cycle}
(b) $\dot{Q}_{rej}$

Section 9–3

9–10 (M) For the engine of problem 9–3, determine the thermodynamic efficiency.

9–11 (M) For the engine of problem 9–4, determine the thermodynamic efficiency.

9–12 (M) Determine the thermal efficiency of a six-cylinder engine that has a displacement of 630 cm^3 and a clearance volume of 70 cm^3.

9–13 (E) For an eight-cylinder engine that has a stroke of 4 in, a bore of 3.7 in, and a clearance volume of 4.6 in^3, determine the thermal efficiency.

9–14 (E) For the engine of problem 9–8, determine the mep.

9–15 An Otto engine has a compression of 6.8:1. What is its thermodynamic efficiency?

Section 9–4

9–16 (M) If the fuel consumption is 5 kg/h and the bhp is 190 hp for the engine of problem 9–4, determine the mechanical efficiency, the bmep, and the bsfc.

9–17 (M) A 3-liter (L) six-cylinder Otto engine develops a mean effective pressure of 1500 kPa when operating at 3000 rpm. The engine is a four-stroke engine having a compression ratio of 8:1. Determine the power produced by the engine when operating at 3000 rpm and the rate of heat addition.

9–18 (M) A five-cylinder four-stroke cylinder engine having a cylinder bore of 5 cm and stroke of 4.5 cm operates on an ideal air standard Otto cycle. If the engine is found to have an intake air flow of 340 g/min when operating at 2000 rpm, what is its volumetric efficiency?

9–19 (M) A particular four-cylinder four-stroke internal combustion engine has a volumetric efficiency that varies with engine speed as shown in figure 9–31. The engine displacement is 2.4 liters. Determine the mass flow of intake air at engine speeds of 2000 rpm, 4000 rpm, and 6000 rpm if the air is assumed to be at 100 kPa and 20°C.

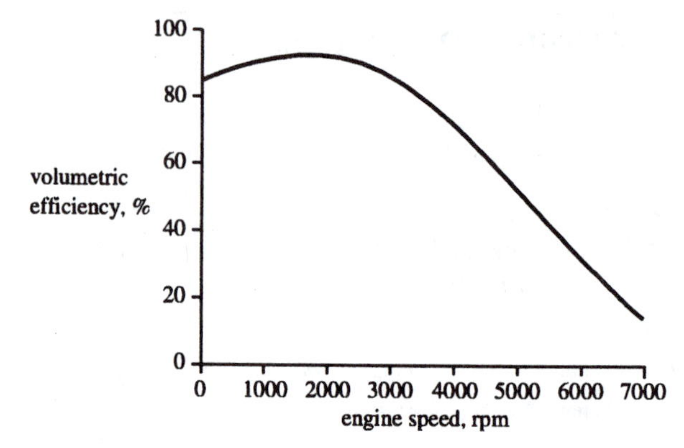

FIGURE 9–31

9–20 (M) A four-stroke four-cylinder engine operating at 6000 rpm produces the p–V diagram shown in figure 9–32. Determine the following quantities as closely as possible, assuming a lowest temperature of 15°C:
(a) Power
(b) $\dot{Q}_{add}$
(c) Thermodynamic efficiency
(d) $\dot{Q}_{rej}$

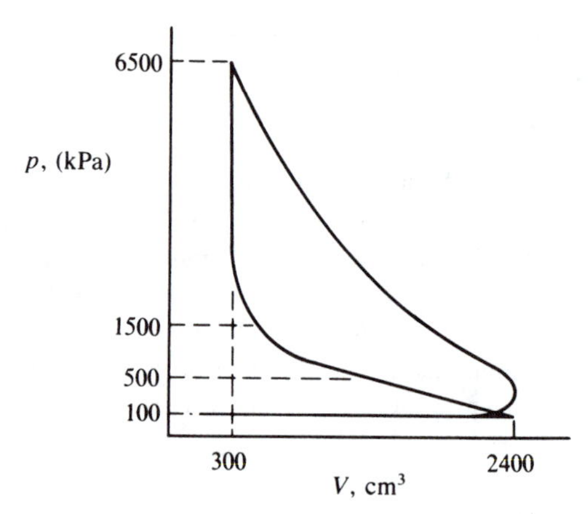

FIGURE 9–32

9–21 (M) For the engine of problem 9–20, assume that the bhp is 480 hp and the fuel consumption is 600 g/min. Determine
(a) Mechanical efficiency
(b) bmep
(c) bsfc

9–22 (E) An Otto engine operating on an ideal four-stroke air-standard cycle is required to develop 100 hp at 5000 rpm.

It must take air in at 14.7 psia and 90°F, and it is limited to four cylinders and a compression ratio of 6.5:1 and a maximum bore of 2.5 in. If the proposed fuel to be used can deliver 850 Btu/lbm air, calculate

(a) p, v, and T at the four corners (Use gas table B–6.)
(b) Wk_{cycle}
(c) Stroke (Use gas table B–6.)
(d) $\dot{Q}_{rej}$ (Use gas table B–6.)

9–23 (E) An eight-cylinder, four-stroke engine is operating at 6000 rpm when the p–V diagram is as shown in figure 9–33. Determine

(a) ihp
(b) Thermodynamic efficiency
(c) mep (or imep)

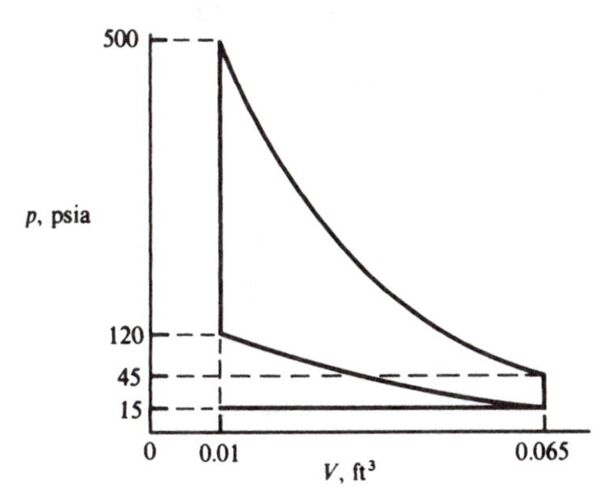

FIGURE 9–33

9–24 (E) If the bhp is found to be 450 hp for the engine of problem 9–23 and the fuel consumption is 400 lbm/h, determine

(a) bmep
(b) bsfc
(c) Mechanical efficiency

9–25 (E) A two-stroke, one-cylinder ICSI engine has a bore of $2\frac{1}{2}$ in and a stroke of $2\frac{3}{4}$ in. It is operating at 3000 rpm and indicates 3.5 hp. The intake conditions are 14.7 psia and 40°F. The compression ratio is 6.5:1. Using polytropic exponents $n_{12} = 1.25$ and $m_{34} = 1.30$ and air-standard analysis, determine

(a) mep
(b) Rate of heat addition
(c) Thermal efficiency
(d) Entropy changes for compression and expansion

This problem may be solved by using the program OTTO.

9–26 (M) A two-stroke, two-cylinder engine has a 5-cm bore and a 6-cm stroke. The compression ratio is 7:1. It delivers a 3 kW at 2800 rpm with intake conditions of 100 kPa and 17°C. Use air-standard analysis and polytropic compression and expansion. Use $n_{12} = 1.3$ and $n_{34} = 1.35$. Determine

(a) mep
(b) Rate of heat addition
(c) Thermal efficiency
(d) Entropy changes for compression and expansion

This problem may be solved by using the program OTTO.

9–27 (M) A six-cylinder, four-stroke ICSI engine operates at 1900 rpm while delivering 265 hp. The engine has a displacement of 5000 mL, its compression ratio is 8:1, and its compression and expansion strokes are adiabatic with adiabatic efficiency of 80% for both processes. Intake conditions are 100 kPa and 27°C. Determine

(a) mep
(b) Rate of heat addition
(c) Thermal efficiency
(d) Minimum entropy generation

This problem may be solved by using the program OTTO.

9–28 (E) An eight-cylinder, four-stroke ICSI engine is to deliver 300 hp at 3600 rpm. The intake conditions are 14.7 psia and 60°F. If its mep is not to exceed 350 psi, and if its compression and expansion strokes are adiabatic with 85% efficiency for both of these processes, determine

(a) Minimum engine displacement
(b) Rate of heat addition
(c) Thermal efficiency
(d) Minimum entropy generation

This problem may be solved, by using the program OTTO.

Sections 9–5, 9–6, and 9–7

9–29 (M) A five-cylinder 3.6-L four-stroke diesel engine has a supercharger that provides intake air under pressure and thereby results in situations when the volumetric efficiency of the engine is greater than 100%. If the volumetric efficiency is found to be 110% when the engine is running at 3400 rpm and at standard air conditions, determine the mass flow of intake air in kg/min.

9–30 (M) A three-cylinder two-stroke cycle diesel engine having a displacement of 2.7 L and operating at 2000 rpm is found to have intake air mass flow rate of 5 kg/min at 20°C and 100 kPa. Determine the volumetric efficiency for the engine at this speed.

9–31 (M) A two-stroke diesel engine has six cylinders, a compression ratio of 15:1, a cutoff ratio of 3:1, and a bore and stroke of 14 cm and 12 cm, respectively. The intake air is at 27°C and 101 kPa. Determine

(a) p, V, and T at the cycle corners
(b) Entropy change during process 1–2
(c) Entropy change during process 2–3
(d) Wk_{cycle}
(e) Thermodynamic efficiency

9–32 (M) If the engine of problem 9–31 consumes 0.9 kg/min of fuel when running at 3000 rpm, and delivers 800 bhp, determine
(a) bsfc
(b) bmep
(c) Mechanical efficiency

9–33 (M) There is 80 kJ/cycle of heat supplied to an ideal, four-stroke, two-cylinder diesel engine when running at 300 rpm. The following data are given for each cylinder:

$$p_1 = 101 \text{ kPa}$$
$$T_1 = 27°\text{C}$$
$$V_1 = 0.04 \text{ cm}^3$$
$$p_2 = 6500 \text{ kPa}$$

Using the air table B–6, determine
(a) Wk_{cycle}
(b) mep
(c) $\dot{Wk}$
(d) Thermodynamic efficiency

9–34 (M) An ideal, four-stroke, three-cylinder diesel engine uses 0.4 kg of air at 100 kPa and 20°C on each cycle. The compression ratio is 15:1 and the cutoff ratio is 2.6:1. Using the air table B–6, determine
(a) Q_{add}
(b) Q_{rej}
(c) Wk_{cycle}
(d) Thermodynamic efficiency
(e) mep

9–35 (E) A diesel engine produces 1500 bhp at 280 rpm. It has eight cylinders, an 18-in bore, an 18-in stroke, and a two-stroke cycle. If 12 lbm/min of fuel with an LHV of 18,400 Btu/lbm is used and the imep is 82 psi, determine
(a) Mechanical efficiency
(b) Thermodynamic efficiency
(c) bmep
(d) bsfc

9–36 (E) There are 0.05 lbm of fuel with an LVH of 18,000 Btu/lbm and 0.9 lbm of air supplied to a diesel engine. Given that

$$p_1 = 14.7 \text{ psi}$$
$$T_1 = 135°\text{F}$$
$$r_v = 14$$

determine r_c by using the air tables.

9–37 (E) There is 91 Btu/cycle supplied to an ideal diesel engine, and 0.5 lbm of air at 14.7 psia and 130°F is supplied every cycle. At the end of compression, the pressure is 560 psia. If the engine is running at 900 rpm, determine (using the air table B–6), assuming a two-stroke cycle:
(a) r_v
(b) r_c
(c) Wk_{cycle}
(d) $\dot{Wk}_{cyc}$
(e) Thermodynamic efficiency
(f) mep
Use of the program DIESEL is recommended.

9–38 (E) A four-stroke diesel engine operates on a dual cycle with 600-in^3 engine displacement. It develops 500 hp at 2500 rpm and has a compression ratio of 19:1 and a cutoff ratio of 3.0 : 1. Atmospheric conditions are 14.7 psia and 70°F. Determine
(a) Rate of heat addition
(b) Thermal efficiency
(c) Maximum temperature and pressure in the cycle
Use of the program DIESEL is recommended.

9–39 (M) A dual-cycle, four-stroke diesel engine has a cutoff ratio of 2:1 and a compression ratio of 20:1. It develops 300 kW at 2000 rpm under inlet conditions of 100 kPa and 17°C and maximum cycle temperature of 3500°C. Determine
(a) Maximum cycle temperature and pressure
(b) Rate of heat addition
(c) Thermal efficiency
Use of the program DIESEL is recommended.

9–40 From the efficiency equation

$$\eta_T = \frac{Wk_{cycle}}{Q_{add}} \times 100$$

derive the relation

$$\eta_T = \left[1 - \frac{1}{(r_v)^{k-1}}\left(\frac{r_c^k - 1}{k(r_c - 1)}\right)\right] \times 100$$

for an ideal diesel cycle.

9–41 Derive a relation for the thermodynamic efficiency of a diesel engine having polytropic processes for compression and expansion. Assume that the polytropic exponents are equal (i.e., $n_{21} = n_{43}$).

9–42 Compare the thermodynamic efficiencies of an Otto engine having a compression ratio of 10.5:1 and a diesel engine having a compression ratio of 15:1 and a cutoff ratio of 2.5 : 1. Which is higher?

Section 9–8

Use of the computer programs OTTO and DIESEL with a microcomputer is recommended for assistance in solving the following problems:

9–43 Determine how the work, thermal efficiency, and heat rejection vary with increasing heat addition to an Otto engine.

9–44 Determine how the work and thermal efficiency vary as adiabatic efficiency decreases from 90% to 40% for an irreversible Otto cycle engine.

9–45 Determine how the thermal efficiency varies with specific heats for the Otto and diesel cycles.

9–46 Determine how the work varies with increasing percent of heat added at constant volume for a dual-cycle diesel engine. (*Hint:* Use a constant cutoff ratio, as well as all other parameters of the engine.)

10 GAS TURBINES, JET PROPULSION, AND THE BRAYTON CYCLE

The jet engine, which extracts energy from a gas turbine, has become a popular power-producing device, replacing the reciprocating IC engine wherever higher power per unit engine mass, smoother operation, or increased maintainability is demanded. We study the thermodynamics of the jet engine and other adaptations of the gas turbine in this chapter.

The ideal Brayton cycle, a thermodynamic heat engine cycle, is defined, and its operation is compared to that of the jet engine. To give the reader a better appreciation of the actual engines, the mechanics of the gas turbine, the compressor, the combustor, and nozzles and diffusers are presented. These devices are the essential components of the common jet engine, and a knowledge of their individual functions helps to make the Brayton cycle more believable.

The gas turbine analyses are done using the air-standard analysis, with air being considered as a perfect gas. We use many of the perfect gas processes introduced in chapter 6 and determine how power is mechanically extracted in a gas turbine.

We consider three modifications to the Brayton cycle air-standard analysis to better describe the actual gas turbine operation: (1) use of air tables to account for specific heat values, which may vary with temperature; (2) reversible polytropic compression and expansion processes; and (3) irreversible adiabatic compression and expansion.

Regenerative heating, the most common attempt to increase the efficiency of the Brayton cycle, is considered in the context of a Brayton cycle application.

Finally, the rocket engine is discussed, even though it is neither a gas turbine nor a complete heat engine. An elementary thermodynamic analysis of the solid and liquid rockets is then made.

New Terms

F_I	Impulse (thrust)	r_p	Pressure ratio
I_{sp}	Specific impulse	η_D	Diffuser efficiency
$\dot{m}_F$	Mass flow rate of fuel	η_N	Nozzle efficiency
$\dot{m}_a/\dot{m}_F$	Air/fuel ratio		

10–1 THE IDEAL BRAYTON CYCLE AND THE GAS TURBINE ENGINE

The ideal Brayton cycle is defined by the following four reversible processes:

1–2 Adiabatic compression from state 1 to state 2

2–3 Constant pressure heat addition

3–4 Adiabatic expansion

4–1 Constant-pressure heat rejection to state 1

Figure 10–1 shows the p–V and T–s diagrams of a typical Brayton cycle with the numbers denoting equilibrium states that correspond to those in the definition of the cycle. A comparison can be made between these diagrams and those for the Otto cycle in figure 9–1. The similarities

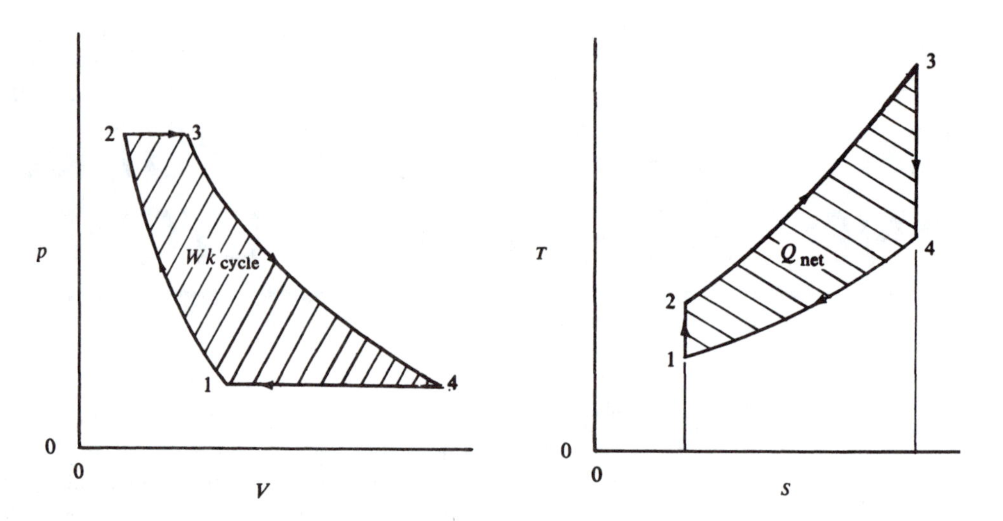

FIGURE 10–1 Property diagrams of ideal Brayton cycle

and the differences should be readily apparent. The Brayton cycle is most often used to describe the operation of the gas turbine engine, an engine that has been used to power vehicles such as aircraft, where it is commonly referred to as a jet or turbojet engine. The gas turbine is also used in stationary applications for electrical power generation, either as a standby unit for providing additional power occasionally or as a continuous electrical power generator.

Figure 10–2 shows a typical arrangement of the critical components of the aircraft gas turbine, or jet engine. The major components of the typical gas turbine are listed in a cutaway view in figure 10–2: the fans, the compressor, the combustion chamber or combustor, the turbine and turbine nozzles, and the exhaust section. In the upcoming two sections,

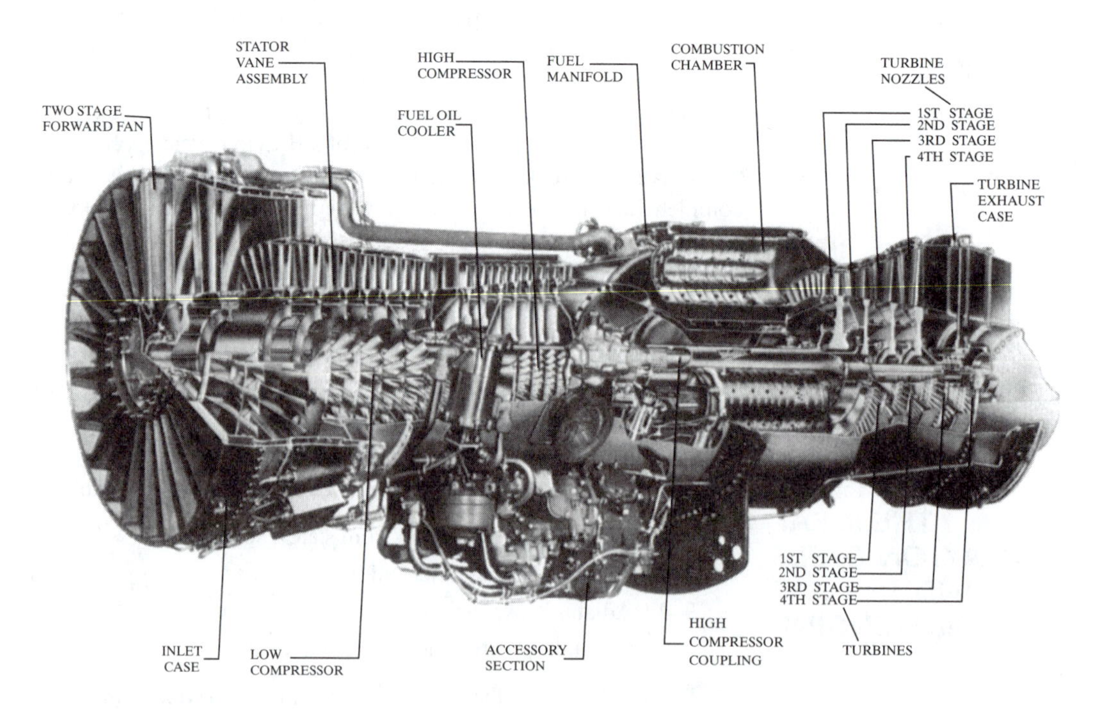

FIGURE 10–2 Typical aircraft gas turbine engine (from "The Aircraft Gas Turbine Engine and Its Operation," August 1970 ed.; with permission of Pratt & Whitney Aircraft Division)

we consider in detail the major components: the turbine, the compressor, and the combustor. In section 10–5, we consider the complete engine and its cycle. Keep in mind that this is only an example and that there are many other engines which have physical characteristics different from those of the gas turbine, but which could be approximated by the Brayton cycle.

A schematic diagram of the gas turbine operating on a Brayton cycle is shown in figure 10–3. If we apply the steady-flow energy equation to the individual components and assume reversible processes and no significant kinetic or potential energy changes, we obtain the following:

1. For the compression 1–2,

$$h_2 - h_1 = -wk_{\text{comp}} \tag{10–1}$$

2. For the combustion 2–3,

$$h_3 - h_2 = q_{\text{add}} \tag{10–2}$$

3. For the turbine expansion 3–4,

$$h_4 - h_3 = -wk_{\text{turb}} \tag{10–3}$$

For the enclosed area in the p–V diagram of figure 10–1, which is the net work of the cycle, we get

$$wk_{\text{cycle}} = wk_{\text{turb}} + wk_{\text{comp}} \tag{10–4}$$

or

$$wk_{\text{cycle}} = h_3 - h_4 + h_1 - h_2 \tag{10–5}$$

The enclosed area in the T–s diagram must be the net heat added to the cycle, and

$$q_{\text{net}} = q_{\text{add}} + q_{\text{rej}} \tag{10–6}$$

This also can be identified as

$$q_{\text{net}} = wk_{\text{cycle}} \tag{10–7}$$

If we assume a perfect gas as the working medium, then equations (10–1), (10–2), (10–3), and (10–5) can be further altered. In particular, the thermodynamic efficiency

$$\eta_T = \frac{Wk_{\text{cycle}}}{Q_{\text{add}}} \times 100 \tag{10–8}$$

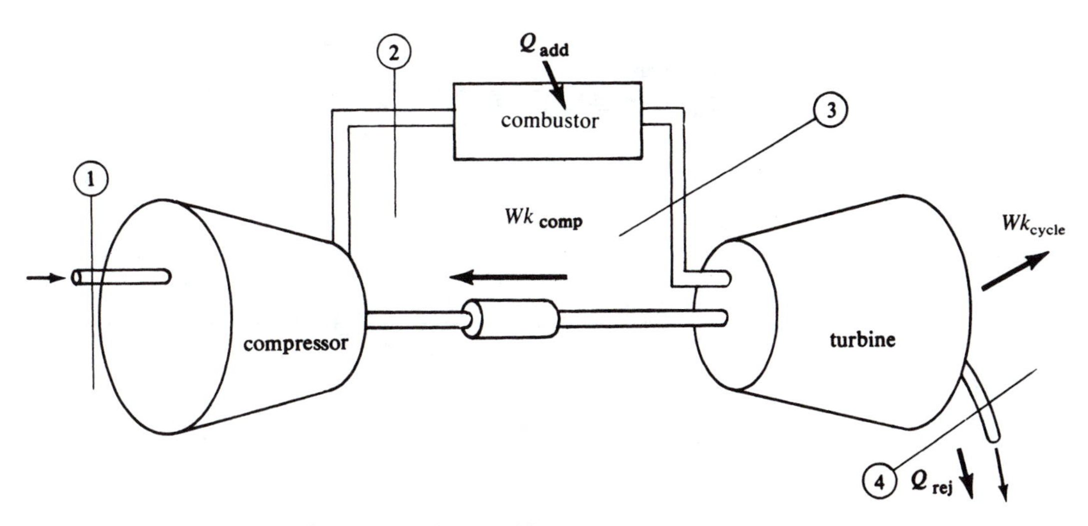

FIGURE 10–3 Schematic of Brayton cycle gas turbine

reduces to

$$\eta_T = \left[1 - \frac{1}{(r_p)^{(k-1)/k}}\right] \times 100 \tag{10–9}$$

for perfect gases and reversible conditions (see problem 10–6), where r_p is the pressure ratio given by p_2/p_1 or p_3/p_4.

EXAMPLE 10–1

A 1000-hp gas turbine operates with a perfect gas on the Brayton cycle and has an inlet pressure of 14.7 psia. The combustor pressure is 100 psia, and the ratio of the specific heats of the perfect gas is 1.4. Determine the thermal efficiency of the engine and the rate of heat addition that is required.

Solution

We assume that the perfect gas has constant specific heats, and the efficiency can then be found from equation (10–9), where r_p = 100 psia/14.7 psia = 6.803. Then

Answer

$$\eta_T = 1 - \frac{1}{(6.803)^{0.4/1.4}} = 42.2\%$$

The rate of heat addition may be found from the definition of thermal efficiency, equation (10–8). We have

$$\dot{Q}_{add} = \frac{\dot{W}k_{cycle}}{\eta_T} = \frac{1000\text{ hp}}{0.422} = 2370\text{ hp}$$

or, expressed in more customary heat transfer units,

$$Q_{add} = \frac{2370\text{ hp}\cdot\text{Btu}}{1.41\text{ hp}\cdot\text{s}}$$

Answer

$$= 1680\text{ Btu/s}$$

10–2 THE GAS TURBINE

The manner by which thermal energy is converted into mechanical energy, thereby providing work in the gas turbine, is shown in the schematic diagram of figure 10–4. Here the high velocity stream of gas (or liquid) is directed against the paddle wheel, thus inducing a rotation of the wheel. This arrangement is called an **impulse turbine** and is one of the two

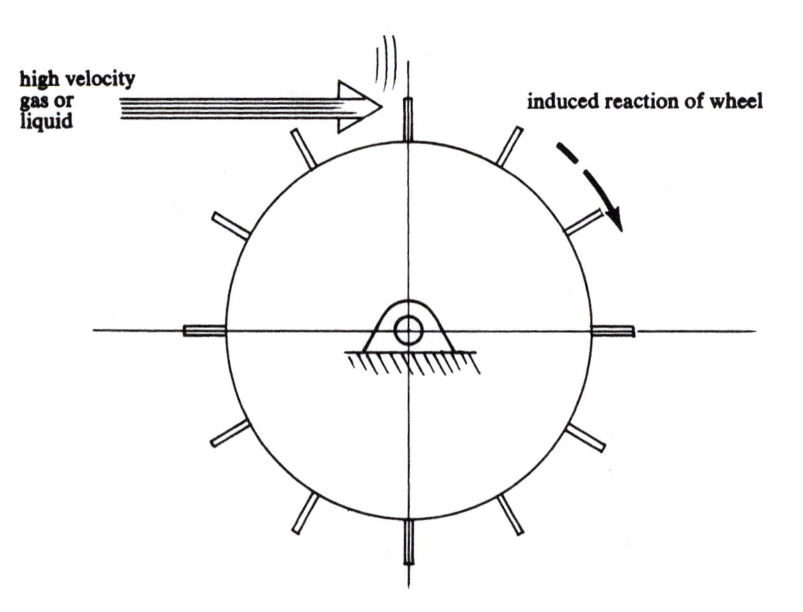

FIGURE 10–4 Basic principle of impulse turbine

FIGURE 10–5 Typical impulse turbine

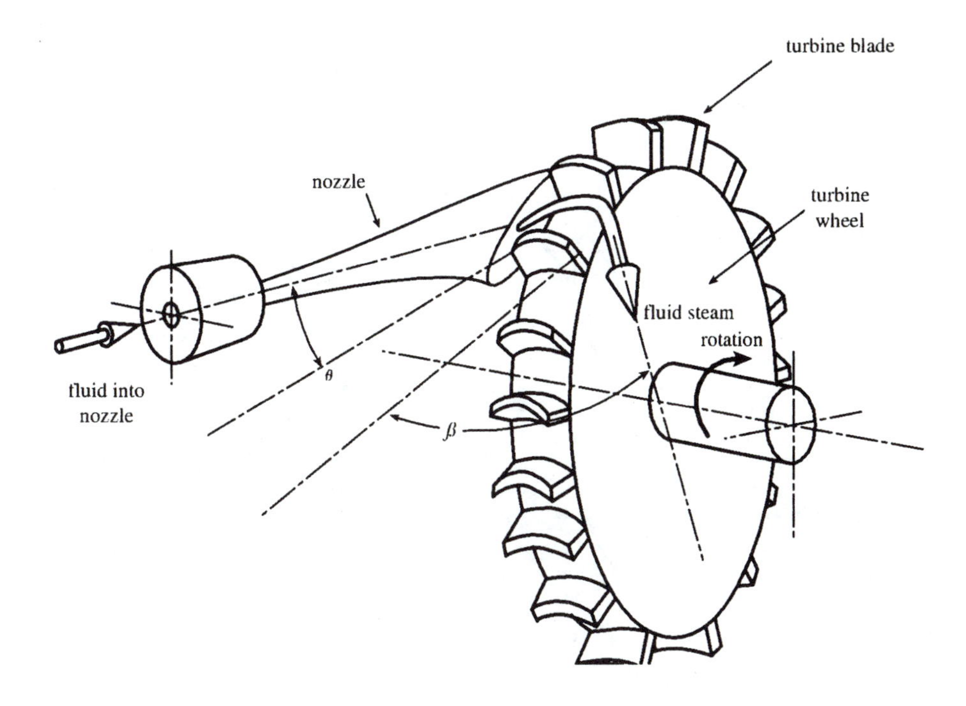

types of turbines. Figure 10–5 shows an isometric diagram of how an actual impulse turbine may be configured. The other type of turbine, the **reaction turbine**, is depicted in figure 10–6; here the fluid is ejected at a high velocity from nozzles attached to the turbine wheel. This causes a reaction that propels the turbine wheel in the opposite direction of the fluid stream. Both types of turbines are utilized in practice, and if more than one wheel is used in a turbine (then called a **multistage turbine**), the reaction principle has many inherent advantages. Figure 10–7 shows a typical turbine stage in which entrance stationary nozzle vanes act to increase the kinetic energy of hot gases and direct them toward the rotating turbine blades. The hot gases pass through this section and are redirected toward another downstream turbine stage or to the atmosphere.

FIGURE 10–6 Reaction turbine

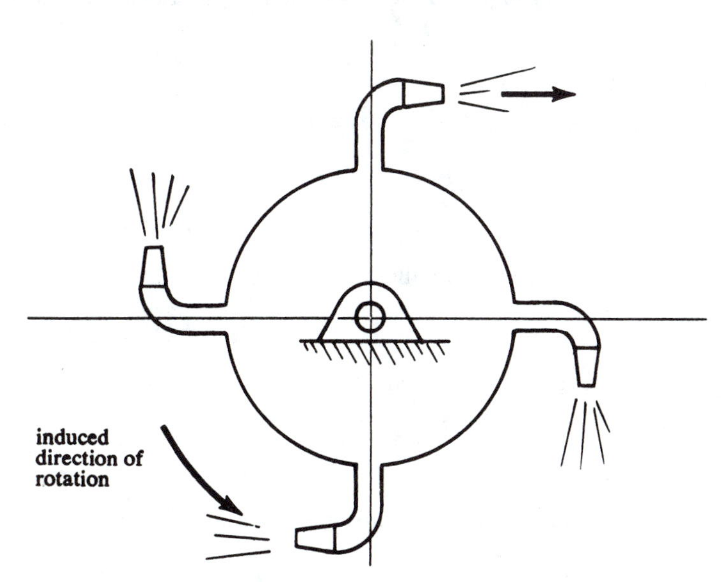

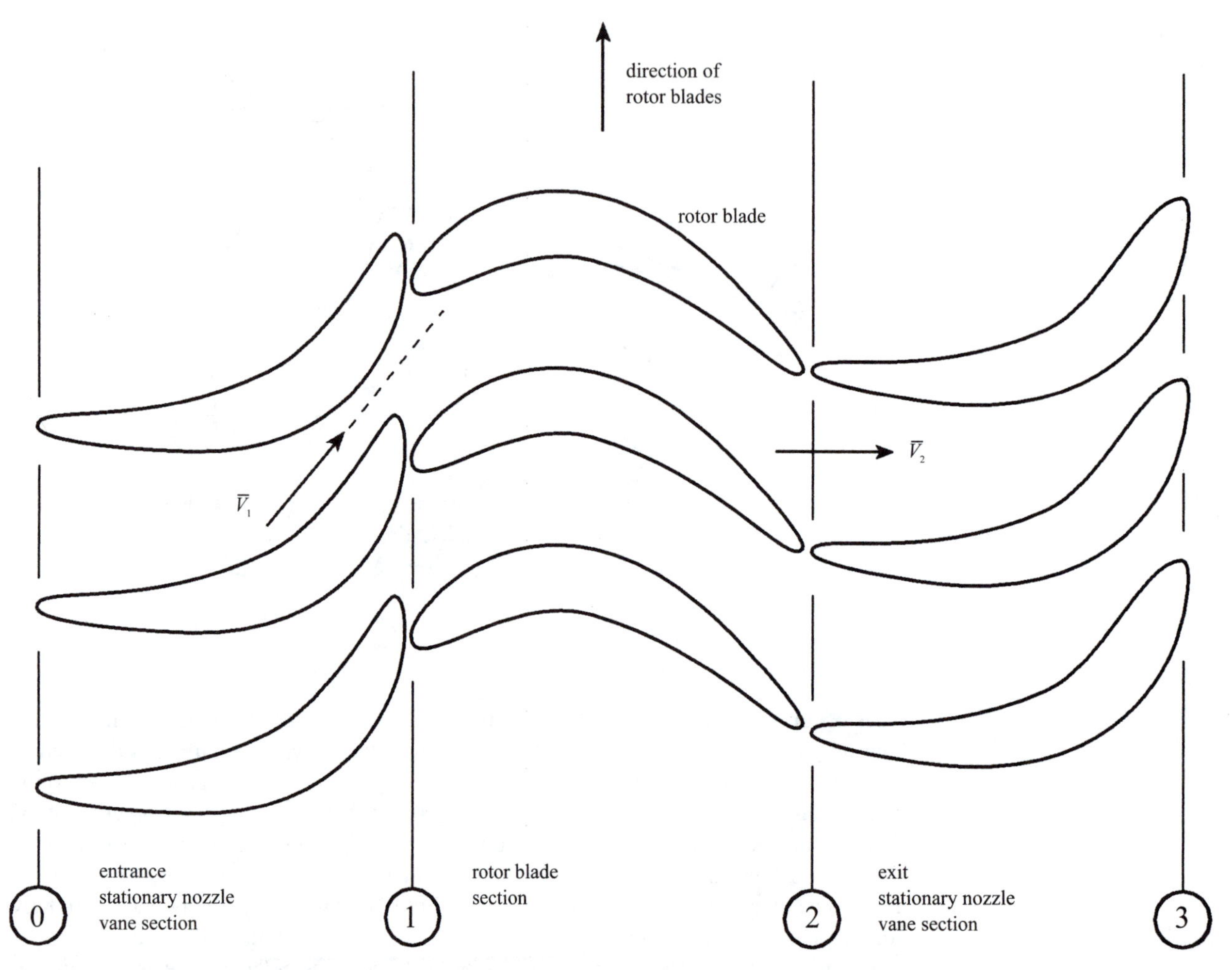

FIGURE 10–7 Typical gas or steam turbine stage. Hot gases enter an entrance nozzle section from a supply or upstream stage, are directed against rotor blades, and leave at a lower pressure and temperature through exit nozzle vanes to another downstream stage or the atmosphere

For an adiabatic turbine, neglecting potential energy changes, the energy equation is

$$\frac{\overline{V}_3^2 - \overline{V}_0^2}{2} + h_3 - h_0 = -wk_{os} \tag{10–10}$$

and we can simplify this by noting that $\overline{V}_0$ and $\overline{V}_3$ are negligible. Then

$$h_2 - h_1 = wk_{os} \tag{10–11}$$

which is the work produced in a turbine if the process is adiabatic and if kinetic energy changes are neglected. The power is then found from

$$\dot{m}(h_2 - h_1) = -\dot{w}k_{os} \tag{10–12}$$

This equation is good with either SI or English units.

10–3 COMBUSTORS AND COMPRESSORS

The high-temperature gases furnished to the gas turbine nozzle are produced in the combustor. A combustor is merely a chamber at constant pressure which allows for the burning or combustion of fuel and air. It can be visualized as a pipe with fuel injectors situated as shown in figure 10–8. The flame wall is a continuous, steady-state chemical reaction of fuel and air, producing the hot gases for the turbine. We can consider then the flame wall as a heat addition process, and the entire combustor can be represented by a system diagram as shown in figure 10–9.

Notice in figure 10–9 that, to conserve mass, we must have

$$\dot{m}_2 = \dot{m}_1 + \dot{m}_F \tag{10–13}$$

The fuel/air ratio, m_F/m_1, is generally about 1:30, but it can vary greatly from this because of the fuel, air conditions, or operating requirements. The first law, written for the combustor, is

$$\dot{m}_2 h_2 + \dot{m}_F h_F - \dot{m}_1 h_1 = \dot{Q}_{\text{add}} \tag{10–14}$$

assuming that no kinetic or potential energy changes occur. The work or power is obviously zero in the combustor, and if we neglect the fuel mass, equation (10–14) can be written

$$\dot{m}_2(h_2 - h_1) = \dot{Q}_{\text{add}} \tag{10–15}$$

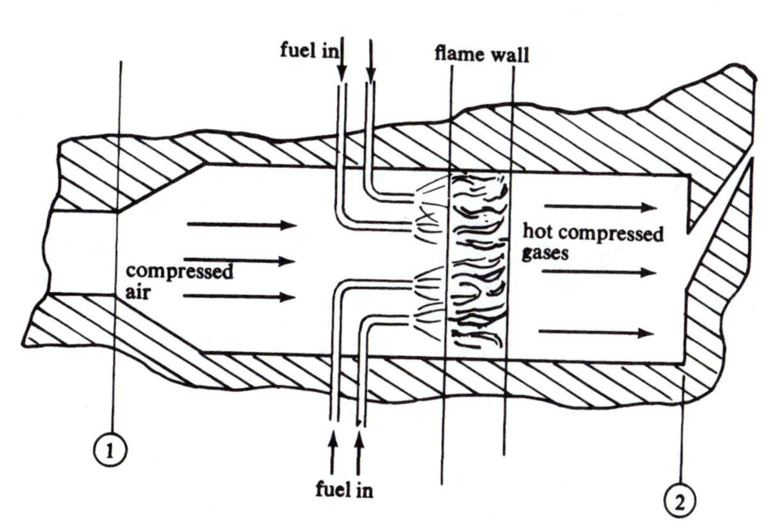

FIGURE 10–8 Typical combustor

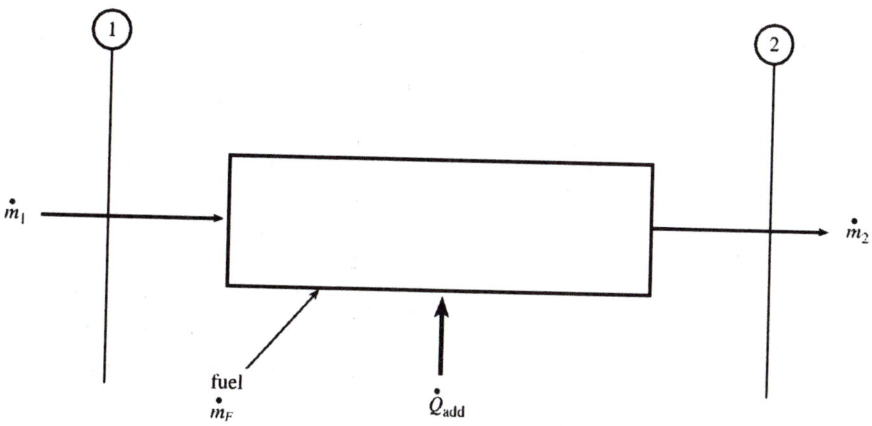

FIGURE 10–9 System diagram of combustor

or

$$h_2 - h_1 = q_{add} \tag{10–16}$$

where q is based on unit mass of air flowing through the combustor.

EXAMPLE 10–2 A combustor of a gas turbine receives 30 kg/s of air at 2000 kPa and 700 K. If the heat addition due to the fuel combustion is 20,000 kJ/s, determine the temperature of the air leaving the combustor.

Solution The energy balance on the combustor is given by equation (10–15), and if we assume that air is a perfect gas with constant specific heats, we have

$$\dot{m}(c_p)(T_2 - T_1) = \dot{Q}_{add}$$

where T_2 is the exit temperature of the combustor. Solving for T_2 with $c_p = 1.007$ kJ/kg·K yields

$$T_2 = T_1 + \frac{\dot{Q}_{add}}{\dot{m}c_p} = 700\text{ K} + \frac{20{,}000\text{ kJ/s}}{(30\text{ kg/s})(1.007\text{ kJ/kg}\cdot\text{K})}$$

Answer

$$= 1362\text{ K}$$

An alternative solution to this problem could be obtained by using table B–6 with variable specific heats. Then $h_1 = 713.3$ kJ/kg at 700 K (from table B–6), and we have

$$h_2 = h_1 + \frac{\dot{Q}_{add}}{\dot{m}}$$

$$= 713.3\text{ kJ/kg} + \frac{20{,}000\text{ kJ/s}}{30\text{ kg/s}}$$

$$= 1380.0\text{ kJ/kg}$$

By interpolation on table B–6, the T_2 is 1287 K. The difference between these two answers is about 5.8%.

The gases flowing through the combustor are at an essentially constant high pressure. To get the air or gases up to a high pressure as they enter the combustor, a pump or compressor must be used. We will see in the discussion of the ram jet that other methods exist to compress the gas, but characteristically, in the pump or compressor, gases are received at a low velocity, are accelerated through blades or fans to a high velocity, and finally are restricted to increase the gas density and pressure while reducing the velocity to a low value again. There are many varied configurations to produce compressed gases, but here we will consider only the typical one used with a gas turbine. It is essentially a reversed turbine, which has been considered in some detail already. Gases are taken in at low pressure, as indicated in the system diagram of figure 10–10, and are compressed through a series of turbine-type blades and nozzles to get a high-pressure gas. The work to drive the compressor is normally supplied by a shaft from the gas turbine itself. The first law of thermodynamics can take the form

$$\frac{\overline{V}_2^2 - \overline{V}_1^2}{2} + h_2 - h = q - wk_{comp} \tag{10–17}$$

for the compressor, and if the walls are assumed to be adiabatic, we can eliminate q. The velocity terms are also neglected in some cases, although this can represent a significant error for high-speed jet engine aircraft.

At any rate, if we do make these simplifying assumptions, then

$$h_2 - h_1 = -wk_{comp} \tag{10–18}$$

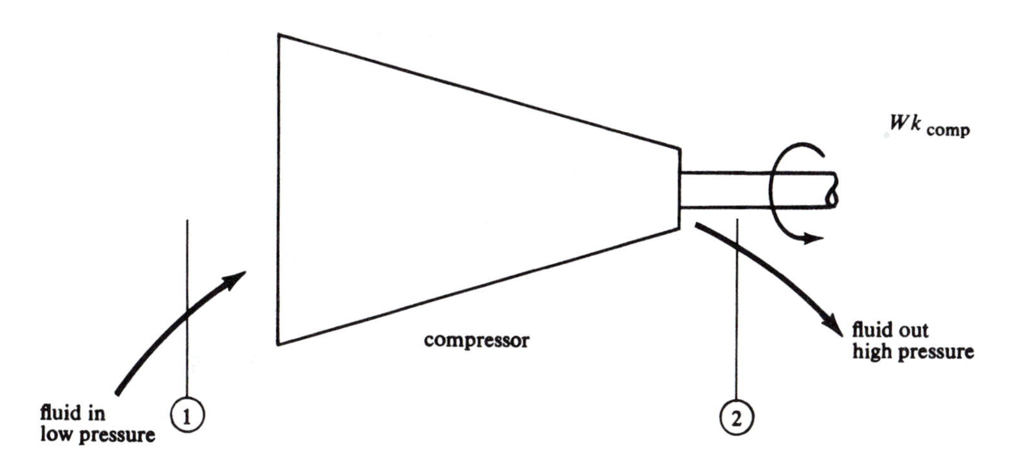

FIGURE 10–10 System diagram for compressor

and we can then also determine work, if we have a reversible compressor, from the area to the left of a curve in a *p–v* diagram. This is seen in figure 10–11, where the area is found from equation (6–29):

$$wk_{comp} = \frac{k}{1-k}(p_2v_2 - p_1v_1)$$

By using the polytropic relation for work, equation (6–16), or

$$wk_{comp} = \frac{n}{1-n}(p_2v_2 - p_1v_1)$$

we can consider *nonadiabatic* compressors as well, but then we must use equation (10–17) instead of (10–18) for any energy balances.

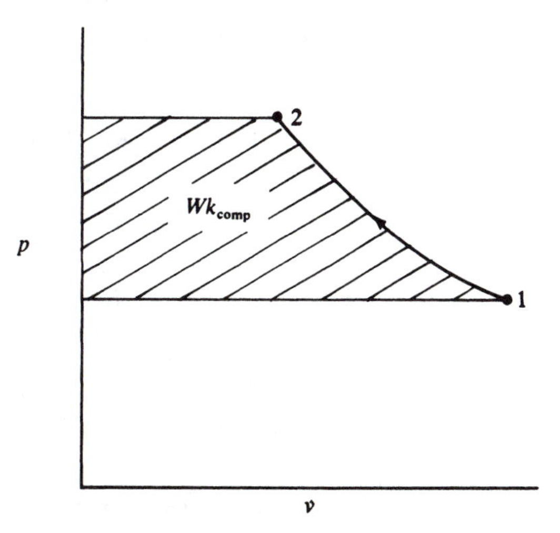

FIGURE 10–11 *p–v* diagram of compressor

EXAMPLE 10–3 A reversible polytropic compressor (with $n = 1.2$) receives air at 1.2 kg/m^3 and 100 kPa and compresses it to 2100 kPa. Determine the work done per kilogram of air.

Solution If we neglect kinetic and potential energies of the air as it progresses through the compressor, we may use the equation for compressor work:

$$wk = \frac{n}{1-n}(p_2v_2 - p_1v_1)$$

The specific volume v_1 is the reciprocal of the density, so

$$v_1 = \frac{1}{\rho_1} = \frac{1}{1.2 \text{ kg/m}^3} = 0.833 \text{ m}^3/\text{kg}$$

From the polytropic equation $p_1 v_1^n = p_2 v_2^n$, we find v_2:

$$v_2 = v_1\left(\frac{p_1}{p_2}\right)^{1/n}$$
$$= (0.833 \text{ m}^3/\text{kg})\left(\frac{100 \text{ kPa}}{2100 \text{ kPa}}\right)^{1/1.2}$$
$$= 0.0659 \text{ m}^3/\text{kg}$$

The compressor work is then

$$wk = \left(\frac{1.2}{-0.2}\right)(2100 \text{ kN/m}^2 \times 0.0659 \text{ m}^3/\text{kg} - 100 \text{ kN/m}^2 \times 0.833 \text{ m}^3/\text{kg})$$
$$= -330.54 \text{ kN} \cdot \text{m/kg}$$

Answer
$$= -330.54 \text{ kJ/kg}$$

10–4 NOZZLES AND DIFFUSERS

Fluids such as gases and liquids are carriers of energy, and during their flow through pipes, conduits, channels, or other conveyors, it is frequently desired to convert the energy to other forms. For instance, high-temperature gases (containing internal energy) may be accelerated to increase the kinetic energy by forcing the gases through a restriction called a **nozzle**. The kinetic energy increase would be done at the expense of internal energy or enthalpy. Similarly, a high-pressure liquid may be accelerated by passing it through a nozzle, as illustrated in figure 10–12. Here, the kinetic energy is increased at the expense of a pressure (and enthalpy) drop in the fluid. These two examples are typical of the use of **converging nozzles**, nozzles that have a decreasing cross-sectional area downstream in the flow. There are, however, other types of nozzles: **diverging nozzles** or **diffusers**, and **converging–diverging nozzles** or **deLaval nozzles**. Diffusers are commonly used to decelerate high-velocity gases or liquids from regions of low pressure to those at higher pressure.

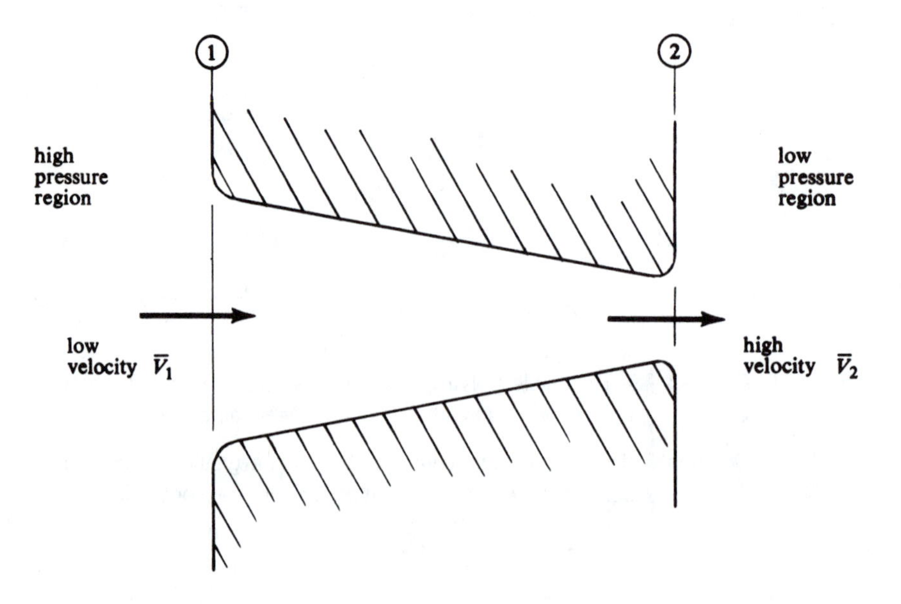

FIGURE 10–12 Typical converging nozzle

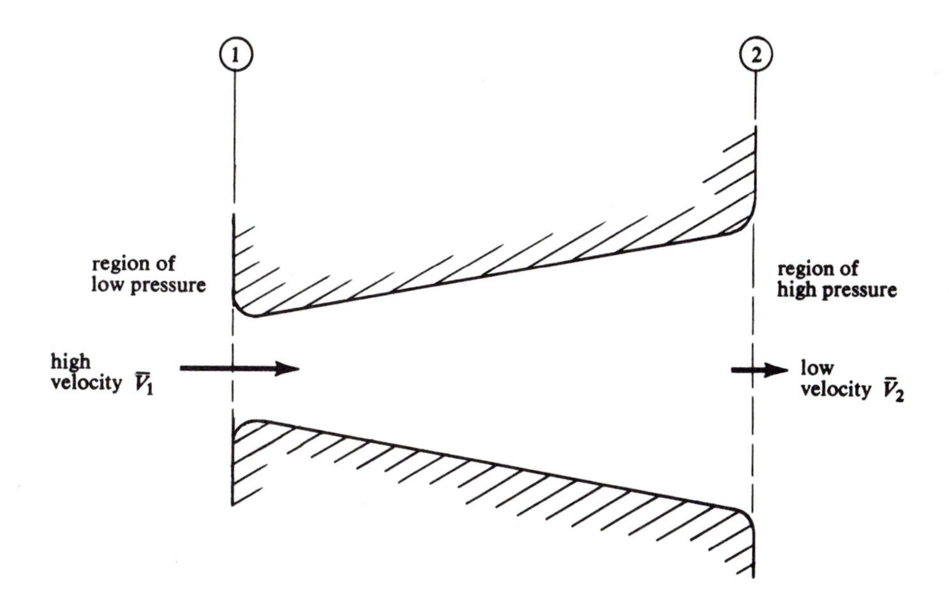

FIGURE 10–13 Typical diffuser

That is, diffusers compress fluids at the expense of kinetic energy. A typical one is shown in figure 10–13.

Fluids that travel through nozzles at velocities less than sonic (less than the speed of sound) behave rather predictably. No uncommon situations arise, and their use for converging nozzles and diffusers can intuitively be seen. However, gases traveling at sonic or supersonic velocities are subject to complicating events such as shock waves and accelerations in diffuser sections. As a consequence of supersonic flow of compressible gases, deLaval nozzles are used to accelerate these fluids. Figure 10–14 shows the normal configuration of a deLaval nozzle and typical velocity and pressure curves in the passage. Notice that, for subsonic flow (figure 10–14a), the fluid is accelerated very slightly. In fact, if the section between the throat (smallest cross-sectional area) and station 2 were removed, we would have a converging nozzle with much better accelerating characteristics.

For supersonic flow (figure 10–14b), we see that this velocity increases throughout the fluid flow in the nozzle. Normally, the velocity of the fluid at the throat of a nozzle will be sonic if the velocities in the divergent section are supersonic. The solid curve (I) for velocity represents the supersonic flow condition, where the gases leave the nozzle at a high supersonic velocity. At some place external to the nozzle, a shock wave will be present, which quickly decelerates the gases to subsonic speed. The discontinuous curve (II) represents a condition in which a shock wave occurs in the diverging section of the nozzle. The shock wave can be seen to be a jump discontinuity in pressure and velocity and is normally described as an irreversible adiabatic process. The velocity downstream of the shock wave is subsonic and decreases to the exit in the normal fashion of subsonic diffusers. A more complete description and analysis of nozzle flows of compressible or incompressible fluids can be found in many good fluid mechanics or gas dynamics texts.

Let us now review the meaning of nozzles. In table 10–1 are listed the energy conversion for the three types of nozzles we have considered: converging, diverging, and deLaval.

We have seen before that gas turbines are visualized as devices converting the internal energy of a high-temperature, high-pressure gas into rotational kinetic energy by means of turbine blades. The nozzle, however, is required to convert gas energy into kinetic energy, and the turbine merely converts the kinetic energy to rotational kinetic energy. In steam turbine operations, the nozzle serves the same useful purpose as it does for gas turbine.

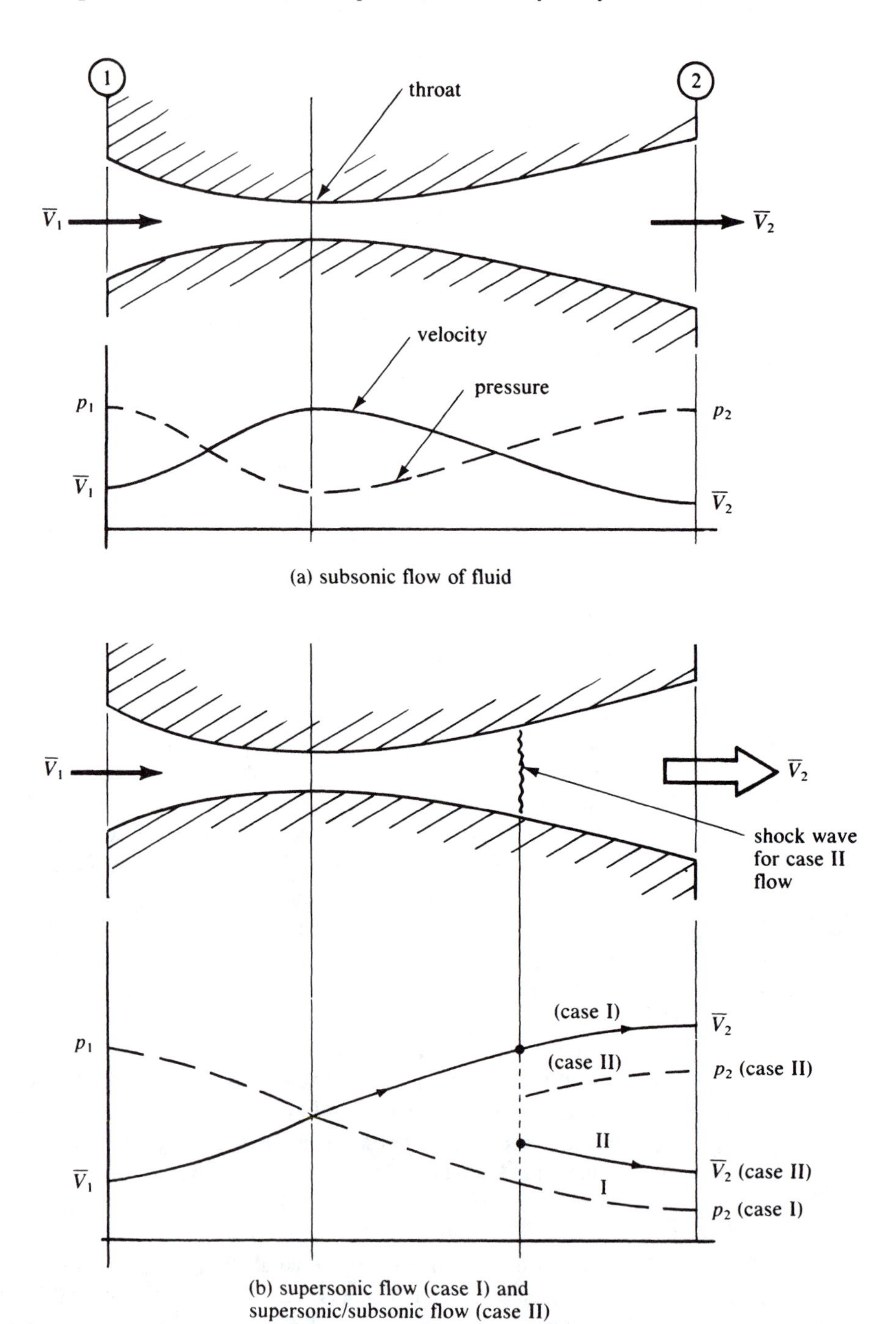

FIGURE 10–14 Flow through a deLaval nozzle

TABLE 10–1 Energy conversions in nozzles

	Energy Converted		
Nozzle Type	**From**	**To**	**Type of Flow**
Converging	Internal or enthalpy	Kinetic	Subsonic or supersonic
Diverging, diffuser	Kinetic	Internal or enthalpy	Subsonic or supersonic
DeLaval	Internal or enthalpy	Kinetic	Supersonic

In propelling high-speed aircraft, the nozzle accelerates the exhaust gases, thus causing a forward thrust in the aircraft.

Let us now analyze the nozzle from a thermodynamic viewpoint. The nozzles we have considered are normally operated under steady-flow conditions so that the equation

$$\rho_1 A_1 \bar{V}_1 = \rho_2 A_2 \bar{V}_2 \tag{10–19}$$

will then be descriptive of the nozzle.

The work done on or by any of the nozzles is zero, and commonly, the flow is assumed to be adiabatic. Then, when potential energy changes are negligible, the first law energy equation for steady flow, per unit mass, becomes

$$\frac{1}{2}(\bar{V}_2^2 - \bar{V}_1^2) = h_1 - h_2 \tag{10–20}$$

Frequently, the term **stagnation state** is used to describe the states in the nozzle. *Stagnation* means a state in which the fluid has been stopped, or is stagnant, through a reversible adiabatic process. The velocity is zero at a stagnation state. At the inlet state 1, the stagnation enthalpy is

$$h_1^* = h_1 + \frac{1}{2}\bar{V}_1^2 \tag{10–21}$$

and, for English units,

$$h_1^* = h_1 - \frac{1}{2g_c}\bar{V}_1^2 \tag{10–22}$$

For perfect gases with constant specific heats, these equations can be written

$$T_1^* = \begin{cases} T_1 + \dfrac{1}{2c_p}\bar{V}_1^2 & \text{(SI units)} \quad (10–23) \\ T_1 + \dfrac{1}{2c_p g_c}\bar{V}_1^2 & \text{(English units)} \quad (10–24) \end{cases}$$

so

$$h_1^* = c_p T_1^* \tag{10–25}$$

where we call T^* the stagnation temperature. A similar equation could be written for state 2:

$$T_2^* = T_2 + \frac{\bar{V}_2^2}{2c_p} \qquad \text{(SI units)} \tag{10–26}$$

Equation (10–20) can be written as

$$\frac{1}{2}\bar{V}_2^2 = h_1 + \frac{1}{2}\bar{V}_1^2 - h_2 = h_1^* - h_2 \tag{10–27}$$

or

$$\bar{V}_2 = \sqrt{2(h_1^* - h_2)} \tag{10–28}$$

This can be reduced to

$$\bar{V}_2 = \sqrt{2c_p(T_1^* - T_2)} \tag{10–29}$$

or, in English units,

$$\bar{V}_2 = \sqrt{2g_c c_p(T_1^* - T_2)} \tag{10–30}$$

for gases having constant specific heats.

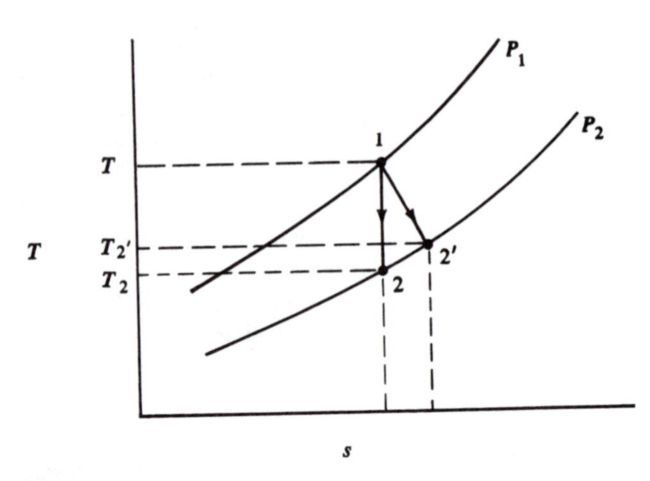

FIGURE 10–15 *T–s* diagram of flow through a converging nozzle

Notice that these equations are nearly the same as those in example 7–15 for an isentropic nozzle. In that example, the inlet velocity was found to be small compared to the exit velocity, so that it could be ignored, or we would say that $T_1^* = T_1$ because $\overline{V}_1 = 0$.

Nozzles are assumed to be adiabatic, and if they are also reversible, the flow of fluid would be conducted at constant entropy. On a *T–s* diagram, shown in figure 10–15, the reversible adiabatic flow through a converging nozzle would be represented by path 1–2. We say that this is 100% efficient and note that an expansion to the same pressure p_2 conducted in an irreversible manner, as in path 1–2′, would reduce the nozzle efficiency η_N. We define nozzle efficiency by the equation

$$\eta_N = \frac{h_1^* - h_{2'}}{h_1^* - h_2} \times 100 \tag{10–31}$$

where $h_{2'}$ is the actual enthalpy of the exiting fluid. For perfect gases having constant specific heats, equation (10–31) becomes

$$\eta_N = \frac{T_1^* - T_{2'}}{T_1^* - T_2} \times 100 \tag{10–32}$$

EXAMPLE 10–4 Air at a stagnation state of 277°C and 1000 kPa enters a converging nozzle. Determine the exit velocity if the nozzle is reversible and the pressure on the exit side is 101 kPa.

Solution We may use equation (10–29) because the system, a converging nozzle, is conducting air which we assume has constant specific heats. The exit temperature can be computed from the perfect gas relationship and the reversible adiabatic process equation $pv^k = C$. From the equation

$$\frac{p_1}{p_2} = \left(\frac{T_1}{T_2}\right)^{k/(k-1)}$$

we solve for T_2:

$$T_2 = T_1\left(\frac{p_2}{p_1}\right)^{(k-1)/k} = T_1^*\left(\frac{p_2}{p_1}\right)^{(k-1)/k}$$

Setting $k = 1.4$, we find that

$$T_2 = (277°\text{C} + 273\text{ K})\left(\frac{101\text{ kPa}}{1000\text{ kPa}}\right)^{0.4/1.4}$$

$$= 286\text{ K}$$

Then, from equation (10–29), we obtain

$$\bar{V}_2 = \sqrt{(2)(1.007\ \text{kJ/kg}\cdot\text{K})(550\ \text{K} - 286\ \text{K})(10^3\ \text{N}\cdot\text{m/kJ})}$$

Answer

$$= 729\ \text{m/s}$$

EXAMPLE 10–5 Air enters a diffuser at 1500 ft/s, 600°R, and 15 psia. If 3 lbm/s of air flows through the diffuser, determine the entrance area of the diffuser, the pressure at the exit, and the Mach number of the air at the entrance. Assume that the velocity at the exit is negligible and that the diffuser efficiency is 90%.

Solution The entrance area can be calculated from the continuity equation (10–19),

$$\rho_1 A_1 \bar{V}_1 = \dot{m}_1$$

or

$$A_1 = \frac{\dot{m}_1}{\bar{V}_1 \rho_1}$$

We have

$$v_1 = \frac{RT_1}{p_1} = \frac{1}{\rho_1}$$

or

$$\rho_1 = \frac{p_1}{RT_1} = \frac{15\ \text{lbf/in}^2}{(53.3\ \text{ft}\cdot\text{lbf/lbm}\cdot{}^\circ\text{R})(600^\circ\text{R})}$$

$$= \frac{(15)(144)}{(53.3)(600)}\ \text{lbm/ft}^3 = 0.0675\ \text{lbm/ft}^3$$

Then

$$A_1 = \frac{3\ \text{lbm/s}}{(1500\ \text{ft/s})(0.0675\ \text{lbm/ft}^3)}$$

Answer

$$= 0.0296\ \text{ft}^2 = 4.26\ \text{in}^2$$

If we assume that the diffuser, shown in figure 10–16, is adiabatic, we may apply equation (10–20) in the form

$$\frac{1}{2g_c}\bar{V}_1^2 = h_2 - h_1 = c_p(T_{2'} - T_1)$$

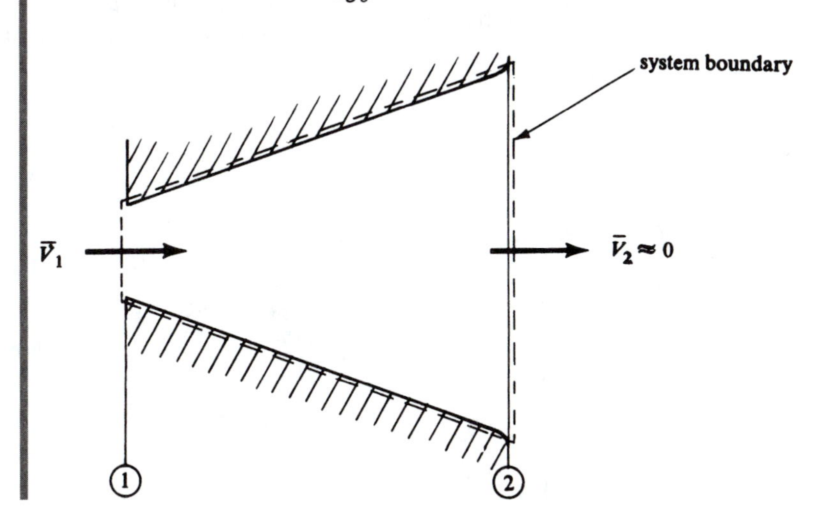

FIGURE 10–16 Adiabatic diffuser

Then

$$T_{2'} = T_1 + \frac{1}{2c_p g_c}\bar{V}_1^2$$

$$= 600°\text{R} + \frac{1}{(2)(0.24 \text{ Btu/lbm}\cdot°\text{R})(32.2 \text{ ft}\cdot\text{lbm/s}^2\cdot\text{lbf})}\left[\frac{1500^2 \text{ ft}^2/\text{s}^2}{778 \text{ ft}\cdot\text{lbf/Btu}}\right]$$

$$= 787°\text{R}$$

Notice now that because the diffuser is not reversible, but rather is 90% efficient, we can see on the T–s diagram of the process (figure 10–17) that the actual temperature $T_{2'}$ is higher than the reversible adiabatic T_2. That is, we define the diffuser efficiency in a manner similar to the nozzle efficiency, or

$$\eta_D = \frac{T_2 - T_1}{T_{2'} - T_1} \times 100 \tag{10–33}$$

where the temperatures at state 2 are effectively the stagnation temperatures and where the gas has constant specific heat values. The primed subscript 2′ represents the actual temperature at state 2′, and the unprimed, the ideal temperature. Then

$$90\% = \frac{T_2 - 600}{787 - 600} \times 100$$

or

$$T_2 = 768°\text{R}$$

and, using the reversible isentropic relation,

$$\frac{p_1}{p_2} = \left(\frac{T_1}{T_2}\right)^{k/(k-1)}$$

we may obtain the exit pressure. With $k = 1.4$, we have

$$p_2 = p_1\left(\frac{T_2}{T_1}\right)^{k/(k-1)} = (15 \text{ psia})\left(\frac{768}{600}\right)^{1.4/0.4}$$

Answer

$$= 35.6 \text{ psia}$$

The **Mach number** is used frequently in gas dynamics or fluid mechanics analysis. It is defined by the relationship

$$\text{Mach number} = \frac{\text{velocity in fluid with respect to the system boundary}}{\text{sonic velocity in the fluid at that point}}$$

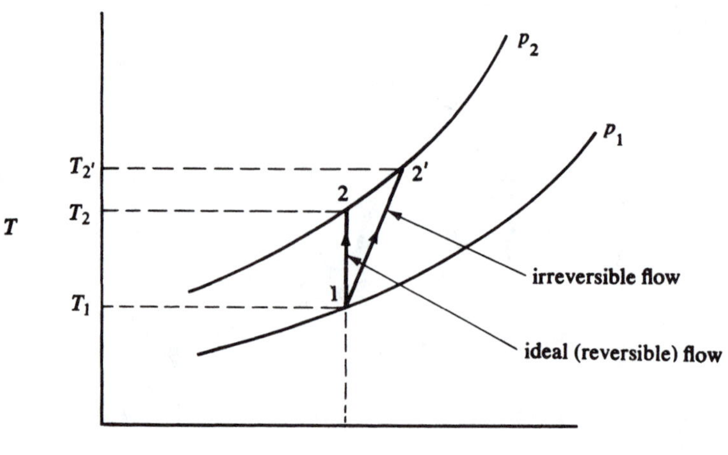

FIGURE 10–17 T–s diagram for diffuser flow in example 10–5

The sonic velocity is given by

$$\bar{V}_s = \sqrt{g_c kRT}$$

for perfect gases, so

$$\text{Mach number} = \frac{\bar{V}}{\sqrt{g_c kRT}} \tag{10–34}$$

At the entrance to our diffuser, we have

$$\bar{V}_1 = 1500 \text{ ft/s}$$
$$T_1 = 600°\text{R}$$

so

$$\text{Mach number} = \frac{1500 \text{ ft/s}}{\sqrt{(32.2 \text{ ft} \cdot \text{lbm/s}^2 \cdot \text{lbf})(1.4)(53.3 \text{ ft} \cdot \text{lbf/lbm} \cdot °\text{R})(600°\text{R})}}$$

Answer $= 1.25$

Notice that a Mach number equal to unity (1) would describe sonic velocity conditions. In our problem, the velocity is greater than sonic, so that is called *supersonic*. We have, in general, that

	(implies)	
$M = 1$	———	sonic velocity
$M > 1$	———	supersonic
$M < 1$	———	subsonic

Also, observe that the Mach number is dependent on the air or fluid temperature as well as upon the velocity, and the condition of flow (whether subsonic, sonic, or supersonic) is determined not only by velocity but also by the fluid properties of temperature, pressure, or volume as well.

For the deLaval nozzle, which can operate with supersonic conditions in the diverging section, the Mach number is 1.0 (the fluid velocity is the same as the speed of sound) at the throat. In fact, the Mach number cannot be greater than 1.0 at the throat for any situation of the nozzle. Thus, referring to figure 10–14, if the gases passing through the deLaval nozzle are supersonic in the diverging section (Mach number 1.0), the Mach number is exactly 1.0 at the throat.

10–5 THE GAS TURBINE ENGINE AND AIR-STANDARD ANALYSIS

The gas turbine engine discussed in section 10–1 operates in a manner that can be described by the Brayton cycle. If the fuel that is added in the combustor is neglected and air is assumed to be the working fluid throughout the engine, we call the analysis an air-standard analysis. If the air is assumed to be a perfect gas with constant specific heats, the thermal efficiency of the gas turbine is given by equation (10–9). Other parameters of the engine, such as temperatures and pressures, can be obtained from the process equations of perfect gases with constant specific heats given in chapter 6. We will give an example to show how the analysis of a gas turbine can be made.

EXAMPLE 10–6 A gas turbine operates through a pressure ratio of 15:1, and the inlet conditions are 14.7 psia and 60°F. The engine is required to deliver 1000 hp when operating in accordance with an air-standard Brayton cycle with perfect gas and constant specific heats. If the maximum allowable temperature in the engine is 2500°R, determine the pressures and temperatures at the four states, the mass flow rate of air required, and the rates of heat addition and rejection.

Solution

For the ideal Brayton cycle with a pressure ratio of 15:1, we can determine the thermal efficiency from equation (10–9). Assuming that k is 1.399, we find that

$$\eta_T = 1 - \frac{1}{(15)^{0.399/1.399}} = 53.8\%$$

From the general definition of thermal efficiency, we find the rate of heat addition:

$$\dot{Q}_{add} = \frac{\dot{W}k}{\eta_T} = \frac{1000 \text{ hp}}{0.538}$$

Answer

$$= 1858.7 \text{ hp} = 1314 \text{ Btu/s}$$

The rate of heat rejection is just

$$\dot{Q}_{rej} = \dot{W}k - \dot{Q}_{add} = 1000 \text{ hp} - 1858.7 \text{ hp}$$

Answer

$$= -858.7 \text{ hp} = -607.05 \text{ Btu/s}$$

The pressures and temperatures can be found by using the various process equations. At state 1,

$$p_1 = 14.7 \text{ psia}$$
$$T_1 = 60°\text{F} = 520°\text{R}$$

Using figure 10–1 as a reference and the definition of the Brayton cycle, we obtain, at state 2,

$$p_2 = (r_p)(p_1) = (15)(14.7 \text{ psia}) = 220.5 \text{ psia}$$
$$T_2 = T_1\left(\frac{p_2}{p_1}\right)^{(k-1)/k}$$

Using k as 1.399 gives us

$$T_2 = (520°\text{R})(15)^{0.399/1.399} = 1125.7°\text{R}$$

The maximum temperature must occur at state 3, so

$$T_3 = 2500°\text{R}$$

and

$$p_3 = p_2 = 220.5 \text{ psia}$$

At state 4, we have

$$p_4 = p_1 = 14.7 \text{ psia}$$

and

$$T_4 = T_3\left(\frac{p_4}{p_3}\right)^{(k-1)/k}$$
$$= (2500°\text{R})\left(\frac{1}{15}\right)^{0.399/1.399}$$
$$= 1154.8°\text{R}$$

The mass flow rate of air can be found from

$$\dot{Q}_{add} = \dot{m}c_p(T_3 - T_2)$$

or

$$\dot{Q}_{rej} = \dot{m}c_p(T_1 - T_4)$$

Using the first of these two equations, we have

$$= \frac{\dot{Q}_{add}}{c_p(T_3 - T_2)} = \frac{1318 \text{ Btu/s}}{(0.2404 \text{ Btu/lbm} \cdot \text{R})(2500°\text{R} - 1125.7° \text{R})}$$

Answer

$$= 3.989 \text{ lbm/s}$$

Example 10–6 shows how an ideal Brayton cycle may be used to help in analyzing the gas turbine engine. Modifications to the Brayton cycle may be made to better describe the actual gas turbine. Three methods of modifying the air-standard Brayton cycle are as follows:

1. Use of air tables, such as table B–6, to account for variable specific heats of the perfect gas.
2. Use of reversible polytropic compression and expansion processes in the compressor and gas turbine.
3. Use of irreversible adiabatic compression and expansion. The adiabatic efficiency must be used with this modification.

These alterations to an ideal cycle are the same as those suggested for modifying the Otto cycle to better describe the operation of the internal combustion piston-cylinder engines treated in section 9–4. Let us now consider examples that use these modifications. We will look at one example that uses the air table approach and one that uses the irreversible process approach.

EXAMPLE 10–7 An ideal air-standard gas turbine operates with a pressure ratio of 18 : 1 and inlet conditions of 101 kPa and 7°C. Maximum combustor temperature is 2800 K, and air flow is 10 kg/s. Accounting for the variations in specific heat of air, determine the power produced, the rates of heat addition and rejection, the pressures and temperatures at the four states, and the thermal efficiency.

Solution

At state 1, we have

$$p_1 = 101 \text{ kPa}$$
$$T_1 = 7°\text{C} = 280 \text{ K}$$
$$h_1 = 280.1 \text{ kJ/kg}$$

and

$$p_{r_1} = 1.089$$

For the reversible adiabatic process 1–2, we can write

$$\frac{p_{r_2}}{p_{r_1}} = \frac{p_2}{p_1}$$

so

$$p_{r_2} = p_{r_1}\left(\frac{p_2}{p_1}\right) = (1.089)(18) = 19.602$$

From table B–6, using linear interpolation between temperatures, we obtain

$$T_2 = 629.56 \text{ K}$$
$$h_2 = 638.2 \text{ kJ/kg}$$

The pressure at state 2 is $p_2 = p_1(r_p) = 1818$ kPa. The temperature at state 3 is the maximum temperature, 2800 K, and from table B–6, we read

$$h_3 = 3268.4 \text{ kJ/kg}$$
$$p_{r_3} = 9159$$

The pressure at state 3 is the same as at 2, 1818 kPa. For the reversible adiabatic process 3–4, we have

$$p_{r_4} = p_{r_3}\left(\frac{p_4}{p_3}\right) = (9159)\left(\frac{1}{18}\right) = 508.833$$

and from table B–6, again using linear interpolation, we find that

$$T_4 = 1438.3 \text{ K}$$

and

$$h_4 = 1561.59 \text{ kJ/kg}$$

The net power produced is

$$\begin{aligned}\dot{W}k &= \dot{m}(h_1 - h_2 + h_3 - h_4)\\ &= (10 \text{ kg/s})(280.1 \text{ kJ/kg} - 638.2 + 3268.4 - 1561.59)\\ &= 13{,}487.1 \text{ kW}\end{aligned}$$

Answer

The heat rates are

Answer

$$\dot{Q}_{\text{add}} = \dot{m}(h_3 - h_2) = 26{,}302 \text{ kW}$$

and

Answer

$$\dot{Q}_{\text{rej}} = \dot{m}(h_1 - h_4) = 12{,}814.9 \text{ kW}$$

The thermal efficiency is

Answer

$$\eta_T = \frac{\dot{W}k}{\dot{Q}_{\text{add}}} = 51.3\%$$

EXAMPLE 10–8 Consider the gas turbine of example 10–6 with the modification that the adiabatic efficiency of the turbine is 88% and that of the compressor is 86%. Determine the pressures, volumes, and temperatures at the four states; the mass flow of air; the rates of heat addition and rejection; the thermal efficiency; and the entropy generation.

Solution

Since the gas turbine is not ideal, we cannot use the thermal efficiency equation (10–9) for this problem. We proceed by first determining the various properties at the four states. At state 1,

$$p_1 = 14.7 \text{ psia}$$
$$T_1 = 520°\text{R}$$

The properties at state 2 are affected by the adiabatic efficiency of the compressor, so that we write

$$\eta_{\text{comp}} = 0.86 = \frac{wk_{os}}{wk_{\text{comp}}}$$

and, for perfect gases with constant specific heats,

$$\begin{aligned}\eta_{\text{comp}} &= \frac{c_p(T_1 - T_{2s})}{c_p(T_1 - T_2)}\\ &= \frac{T_1 - T_{2s}}{T_1 - T_2}\end{aligned}$$

For example 10–6, we found that $T_{2s} = 1125.7°\text{R}$ (for $\eta_{\text{comp}} = 100\%$), so

$$T_2 = -\frac{T_1 - T_{2s}}{\eta_{\text{comp}}} + T_1$$

or

$$T_2 = \frac{T_{2s} - T_1}{\eta_{comp}} - T_1 = 1224.3°R$$

The pressure at state 2 is still 220.5 psia, because the pressure ratio is 15:1. At state 3, we have

$$p_3 = p_2 = 220.5 \text{ psia}$$
$$T_3 = 2500°R$$

For state 4, we must account for the adiabatic efficiency of the turbine with the calculations:

$$\eta_{turb} = 0.88 = \frac{wk_{turb}}{wk_{os}} = \frac{c_p(T_3 - T_4)}{c_p(T_3 - T_{4s})} = \frac{T_3 - T_4}{T_3 - T_{4s}}$$

where T_{4s} is the temperature at state 4 if process 4 − 1 had been reversible and adiabatic. We found T_{4s} to be 1154.8°R in example 10–6, so we can then find T_4:

$$T_4 = (\eta_{turb})(T_{4s} - T_3) + T_3 = (0.88)(1154.8°R - 2500°R) + 2500°R = 1316.2°R$$

The net cycle work per pound-mass of air is the sum of the compressor and turbine work terms:

$$wk_{cycle} = wk_{comp} + wk_{turb} = c_p(T_1 - T_2) + c_p(T_3 - T_4)$$
$$= (0.2404 \text{ Btu/lbm·°R})(520°R - 1224.3°R) + (0.2404 \text{ Btu/lbm·°R}) \times (2500°R - 1316.2°R)$$
$$= 115.3 \text{ Btu/lbm}$$

The power is then the net work times the mass flow of air, or

$$\dot{W}k = \dot{m}wk_{cycle} = 1000 \text{ hp} = 706.9 \text{ Btu/s}$$

and the mass flow is

$$\dot{m} = \frac{\dot{W}k}{wk_{cycle}} = \frac{706.9 \text{ Btu/s}}{115.3 \text{ Btu/lbm}} = 6.13 \text{ lbm/s}$$

Answer

The added heat is

$$\dot{Q}_{add} = \dot{m}c_p(T_3 - T_2) = (6.13 \text{ lbm/s})(0.2404 \text{ Btu/lbm·°R})(2500°R - 1224.3°R)$$

Answer

$$= 1879.9 \text{ Btu/s}$$

and the thermal efficiency is

Answer

$$\eta_T = \frac{\dot{W}k}{\dot{Q}_{add}} = \frac{706.9 \text{ Btu/s}}{1879.9 \text{ Btu/s}} = 37.6\%$$

The rate of heat rejection can be found from the difference between the net power and heat added or from the equation

Answer

$$\dot{Q}_{rej} = \dot{m}c_p(T_1 - T_4)$$

From this equation, we find that $\dot{Q}_{rej}$ is −1173.3 Btu/s.

For a check, the difference between the power and added heat is equal to -1173 Btu/s, which is essentially the same answer. The entropy generation will be at least as great as the quantity $\dot{m}(\Delta s_{12} + \Delta s_{34})$, where

$$\Delta s_{12} = s_2 - s_{2s} = c_p \ln\left(\frac{T_2}{T_{2s}}\right)$$

and

$$\Delta s_{34} = s_4 - s_{4s} = c_p \ln\left(\frac{T_4}{T_{4s}}\right)$$

Substituting values into these equations, we obtain

$$\Delta s_{12} = (0.2404\ \text{Btu/lbm}\cdot{}^\circ\text{R})\left(\ln\frac{1224.3^\circ\text{R}}{1125.7^\circ\text{R}}\right)$$
$$= 0.020185\ \text{Btu/lbm}\cdot{}^\circ\text{R}$$

and

$$\Delta s_{34} = (0.2404\ \text{Btu/lbm}\cdot{}^\circ R)\left(\ln\frac{1316.2^\circ\text{R}}{1154.8^\circ\text{R}}\right)$$
$$= 0.031450\ \text{Btu/lbm}\cdot{}^\circ\text{R}$$

The entropy generation is then

Answer

$$\dot{S}_{\text{gen}} \geq (6.13\ \text{lbm/s})(0.020185\ \text{Btu/lbm}\cdot{}^\circ\text{R} + 0.031450)$$
$$\geq 0.31652\ \text{Btu/s}\cdot{}^\circ\text{R}$$

Figure 10–18 shows these results on p–v and T–s diagrams.

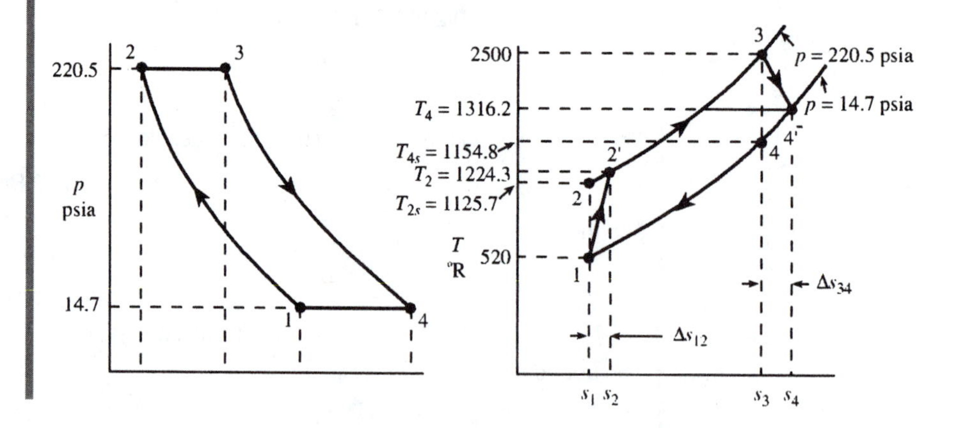

FIGURE 10–18 Property diagrams for irreversible gas turbine cycle (not to scale)

Notice in example 10–8 how the mass flow has increased from that in example 10–6 due to the inefficiencies of the turbine and compressor.

If we wish to consider the fuel and modify the air-standard analysis to account for this term, we may use a mass and energy balance for the combustor. This complicates the analysis somewhat but gives better descriptions of the actual operation of the gas turbine. If you look at figure 10–9, you can see that the mass balance for the combustor is given by the equation

$$\dot{m}_2 = \dot{m}_1 - \dot{m}_F$$

where $\dot{m}_2$ is the mass flow of exhaust gases leaving the combustor and $\dot{m}_F$ is the fuel mass flow. The term $\dot{m}_1$ is the compressed air coming from the compressor (state 2 in the Brayton cycle notation). The air/fuel ratio is equal to the ratio $\dot{m}_1/\dot{m}_F$ and is a term often used in combustion processes such as those occurring in the combustor. The air/fuel ratio is usually of the order of 20:1 up to 40:1. In chapter 14, the air/fuel ratio will be considered in some detail.

The energy balance of the combustor is given by the equation

$$\dot{Q}_{add} = \dot{m}_2 h_2 - \dot{m}_1 h_1 - \dot{m}_F h_F \tag{10–35}$$

where h_F is the enthalpy of the liquid fuel. Frequently, the gas turbine uses kerosene as a fuel. Jet fuel is a derivative of kerosene and is also used in gas turbines. For kerosene, one may use as a first approximation, from *Gas Tables* by Keenan and Kaye,* for h_F of octane,

$$h_F = (0.5T - 287)\ \text{Btu/lbm fuel} \tag{10–36}$$

where T is in degrees Rankine. For SI units, with T measured on the Kelvin scale, equation (10–36) is

$$h_F = (2.1T - 668)\ \text{kJ/kg fuel} \tag{10–37}$$

The $\dot{Q}_{\text{add}}$ term can be interpreted as the heat released by the fuel combustion, and if it were divided by the mass flow of the fuel, it would be equal to the heating value of the fuel: either the lower heating value (LHV) or the higher heating value (HHV). In section 9–4, we discussed the difference between the LHV and the HHV. The value for the LHV and HHV for kerosene is often taken to be equal to that for *n*-octane; thus LHV = 19,256 Btu/lbm kerosene. If the mass and energy balance equations for the combustor are combined, we have

$$\frac{\dot{m}_1}{\dot{m}_F}(h_2 - h_1) - h_F = \text{LHV} \tag{10–38}$$

When using this equation, notice that the term $(h_2 - h_1)$ would be written as $(h_3 - h_2)$ for the Brayton cycle notations given in figure 10–1.

EXAMPLE 10–9 A combustor operates by burning kerosene with an air/fuel ratio of 32:1. If the inlet air temperature is 700°F, the fuel temperature is 40°F, and the exhaust gases have properties like those of air, determine the exhaust temperature.

Solution We assume perfect gases with constant specific heats (except for the fuel) and modify equation (10–38) to read

$$\frac{\dot{m}_1}{\dot{m}_F}(C_p)(T_2 - T_1) - H_F = \text{LHV}$$

For a fuel temperature of 40°F or 500°R, we find the enthalpy of the fuel:

$$h_F = (0.5)(500) - 287 = -37\ \text{Btu/lbm fuel}$$

Then the exhaust temperature T_2 can be computed:

$$T_2 = \frac{\text{LHV} + h_F}{(\dot{m}_1/\dot{m}_F)(c_p)} + T_1$$

$$= \frac{19{,}256\ \text{Btu/lbm kerosene} - 37\ \text{Btu/lbm kerosene}}{(32\ \text{lbm air/lbm kerosene})(0.2404\ \text{Btu/lbm air})} + 700°\text{F} + 460°\text{R}$$

Answer $= 3658.3°\text{R}$

---

* From J. H. Keenan and J. Kaye, *Gas Tables* (New York: John Wiley & Sons, Inc., 1948), with permission of the authors and publisher.

10–6 REGENERATIVE CYCLES

The gas turbine operates in a manner such that the exhaust temperatures are often very much higher than the inlet air temperatures. It is possible to transfer some of this thermal energy in the exhaust to the inlet air and thus reduce the amount of heat that would be required to be added. This could result in a substantial savings of fuel, and the technique for extracting thermal energy from the exhaust and recycling it in the engine is called **regenerative heating**. The gas turbine is then said to be operating on a **regenerative cycle**. Figure 10–19 is a schematic of one possible arrangement to accomplish regenerative heating. Also, in figure 10–20 are shown the property diagrams corresponding to the engine of figure 10–19.

In each property diagram of figure 10–20, the shaded area represents the heat added by virtue of the regeneration process. From an application of the steady-flow energy equations to the components, we obtain the following:

1. For the compressor,

$$h_2 - h_1 = -wk_{\text{comp}} \tag{10–39}$$

2. For the regenerator,

$$h_5 - h_6 = h_3 - h_2 \tag{10–40}$$

3. For the combustor,

$$h_4 - h_3 = q_{\text{add}} \tag{10–41}$$

4. For the turbine,

$$h_4 - h_5 = -wk_{\text{turb}} \tag{10–42}$$

The thermodynamic efficiency can then be written

$$\eta_T = \frac{wk_{\text{net}}}{q_{\text{add}}} = \frac{h_4 - h_5 + h_1 - h_2}{h_4 - h_3} \times 100 \tag{10–43}$$

Now, if we assume a perfect gas medium with constant specific heats, we can make the substitution $h = c_pT$ and obtain

$$\eta_T = \frac{T_4 - T_5 + T_1 - T_2}{T_4 - T_3} \times 100 \tag{10–44}$$

or if we have $T_3 = T_5$,

$$\eta_T = \left(1 - \frac{T_2 - T_1}{T_4 - T_5}\right) \times 100 \tag{10–45}$$

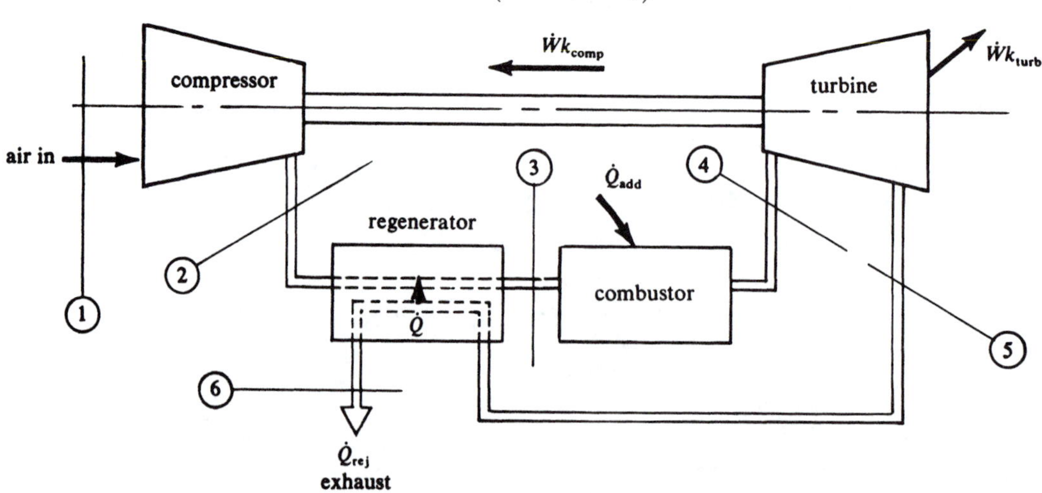

FIGURE 10–19 Regenerative heating Brayton cycle gas turbine

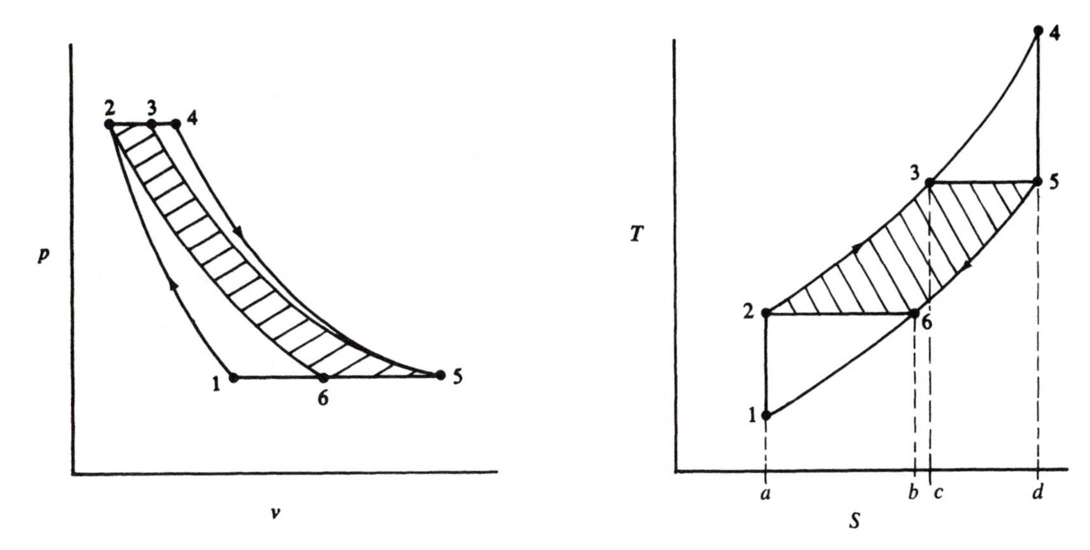

FIGURE 10–20 Regenerative heat Brayton cycle property diagrams

This can be revised further by using some algebra to obtain

$$\eta_T = \left(1 - \frac{T_1 T_2/T_1 - 1}{T_4 1 - T_5/T_4}\right) \times 100 = \left[1 - \left(\frac{T_1}{T_4}\right)(r_p)^{(k-1)/k}\right] \times 100 \qquad \textbf{(10–46)}$$

where

$$r_p = \frac{p_2}{p_1} = \frac{p_4}{p_5} \quad \text{and} \quad \frac{T_2}{T_1} = \frac{T_4}{T_5}$$

This interesting result shows that an increase in compression or pressure ratio will decrease the efficiency of a regenerative Brayton cycle engine. This is directly opposite to the result of the simple Brayton cycle, as given by equation (10–9), where increases in pressure ratios increase the cycle efficiency. This should provide enough indication that any given engine needs to be analyzed from a fresh outlook. Keep the basic thermodynamic tools as foundations, and, from this base, make assumptions and restrictive developments in context with the actual situation.

EXAMPLE 10–10 A reversible, regenerative gas turbine uses 1.0 kg/s of air. Its operating pressure ratio is 15:1, and the compressor inlet conditions are 101-kPa pressure and 17°C. If the turbine exhaust is at 700 K, determine the following from an air-standard analysis, assuming perfect gas behavior:

(a) Thermodynamic efficiency.
(b) Power developed.
(c) Head added and rejected.

Solution (a) We calculate the efficiency from equation (10–46); first, however, we need to determine T_4:

$$\frac{T_4}{T_5} = \left(\frac{p_4}{p_5}\right)^{(k-1)/k} = (r_p)^{(k-1)/k}$$

We can then substitute this into the efficiency equations to obtain

$$\eta_T = \left(1 - \frac{T_1}{T_5}\right) \times 100$$

which yields

Answer
$$\eta_T = \left(1 - \frac{290\ \text{K}}{700\ \text{K}}\right) \times 100 = 58.5\%$$

Notice that the efficiency of a simple Brayton cycle gas turbine operating with the same pressure ratio is

$$\eta_T = \left[1 - \frac{1}{(r_p)^{(k-1)/k}}\right] \times 100 = \left(1 - \frac{1}{15^{0.286}}\right) \times 100 = 53.9\%$$

We therefore gain 4.6% efficiency by using the reheater on this engine.

(b) The power developed can be obtained from the relation $\dot{W}k = \dot{m}wk_{\text{cycle}}$. For air-standard analysis,

$$wk_{\text{cycle}} = c_p(T_4 - T_5 + T_1 - T_2)$$

and

$$T_4 = T_5\left(\frac{p_4}{p_5}\right)^{(k-1)/k} = (700\ \text{k})(15)^{0.286} = 1519\ \text{K}$$

Similarly, for the temperature, T_2,

$$T_2 = T_1\left(\frac{p_2}{p_1}\right)^{(k-1)/k} = (290\ \text{K})(15)^{0.286} = 629\ \text{K}$$

and then

$$\begin{aligned} wk_{\text{cycle}} &= (1.007\ \text{kJ/kg}\cdot\text{K})(1519\ \text{K} - 700\ \text{K} + 290\ \text{K} - 629\ \text{K}) \\ &= 483\ \text{kJ/kg} \end{aligned}$$

The delivered power is then computed from

Answer
$$\begin{aligned} \dot{W}k_{\text{cycle}} &= (1.0\ \text{kg/s})(483\ \text{kJ/kg}) \\ &= 483\ \text{kW} \end{aligned}$$

(c) We can calculate the heat added in at least two ways:

$$q_{\text{add}} = \frac{wk_{\text{cycle}}}{\eta_T} \times 100$$

or

$$q_{\text{add}} = c_p(T_4 - T_5)$$

From the first relationship,

Answer
$$q_{\text{add}} = \frac{483\ \text{kJ/kg}}{58.5\%} \times 100 = 826\ \text{kJ/kg}$$

From the second method, we have

Answer
$$\begin{aligned} q_{\text{add}} &= (1.007\ \text{kJ/kg}\cdot\text{K})(1519\ \text{K} - 700\ \text{K}) \\ &= 825\ \text{kJ/kg} \end{aligned}$$

Thus, there is close agreement between the two methods.

The heat rejected can be computed from the balance:

Answer
$$\begin{aligned} q_{\text{rej}} &= wk_{\text{cycle}} - q_{\text{add}} \\ &= 483\ \text{kJ/kg} - 825\ \text{kJ/kg} \\ &= -342\ \text{kJ/kg} \end{aligned}$$

The gas turbine, operating in accordance with the Brayton cycle, is a device that is capable of high performance. Operating efficiencies of 40% to 50% have been achieved, and it is capable of high bursts of power (with an accompanied decrease in efficiency). Mechanically, it is quite simple to operate and maintain. It is inherently stable, and by rotating symmetrically about a single axis, it can easily be balanced. Its main detriment is its limitation on a high-temperature and consequent efficiency limitation; that is, T_3 in the simple cycle or T_4 in the regenerative cycle cannot exceed maximum temperatures that materials can constantly withstand. The cycle is in steady state, and the upper temperature is retained for long periods of time—unlike the Otto cycle. (See chapter 9.) The developing of materials that can withstand ever-higher temperature, of course, helps the push for higher gas turbine efficiency, but a limit is still there. Additionally, a gas turbine can represent a relatively high initial investment in construction, which can be considered a drawback, but more constant criticism is directed at its slow acceleration characteristics for ground vehicles. The gas turbine is a device that functions best at a constant, optimum speed (as is true of most all engines), but the lack of acceptable acceleration in propelling surface vehicles has not been proven, and, in fact, the exact opposite has frequently been demonstrated.

10–7 JET PROPULSION

The gas turbine has been used to produce electrical power and to provide power for vehicles. The power from the engine can often be extracted through a rotating shaft—through shaft work or power. This method of power transmission is not always used, however, and particularly in aircraft applications, the power is often applied through jet propulsion. If an aircraft is driven by a turbo-prop engine, then the power is extracted through a rotating shaft that drives a propeller, which in turn drives the aircraft. If, on the other hand, the aircraft is driven by a turbojet or jet engine, the power is extracted by jet propulsion, and for this application the gas turbine provides only enough power to drive the compressor. This situation means that the gases leaving the turbine are much hotter than they would be if all the power were extracted through a rotating shaft. These exhaust gases are then directed through an exhaust nozzle (frequently a deLaval nozzle, discussed in section 10–4), and the gases leaving the nozzle are then at a high velocity. The nozzle acts to increase the speed or velocity of the exhaust gases, and this causes a reaction on the vehicle or aircraft. Figure 10–21 shows the mechanical principle involved in this phenomenon, called jet propulsion. Inlet air to the gas turbine compressor has a velocity $\overline{V}_1$ (or the aircraft may have a velocity $\overline{V}_1$ opposite in direction to the velocity shown) and, after passing through the engine, leaves at a velocity $\overline{V}_4$, which is considerably higher than $\overline{V}_1$. The acceleration of the gases, $(\overline{V}_4 - \overline{V}_1)/t$, times the mass of the gases is the force required for that acceleration, F, and in English units,

$$F = m\frac{\overline{V}_4 - \overline{V}_1}{tg_c} = \dot{m}\frac{(\overline{V}_4 - \overline{V}_1)}{g_c} \tag{10–47}$$

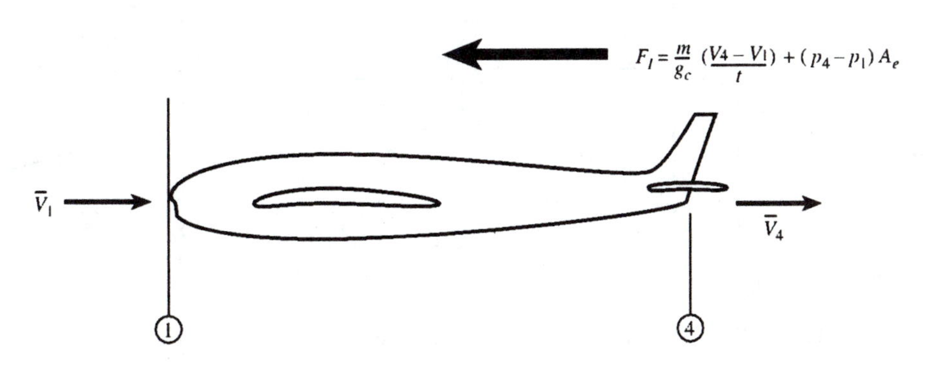

FIGURE 10–21 Reaction principle of jet aircraft propulsion

Every force has an opposing force, and we call this force F_I, the impulse that actually drives the aircraft. In addition, the difference in pressure between the entrance at atmospheric pressure and the exit can cause a force given by $(p_4 - p_1)A_e$, where A_e is the area of the exhaust nozzle. Adding this to the impulse force of the preceding equation, we obtain

$$F_I = m\frac{\overline{V}_4 - \overline{V}_1}{tg_c} + (p_4 - p_1)A_e \tag{10–48}$$

Frequently, the specific impulse, defined by

$$I_{sp} = \frac{F_I}{\dot{m}} = \frac{\overline{V}_4 - \overline{V}_1}{g_c} + \frac{(p_4 - p_1)A_e}{\dot{m}} \tag{10–49}$$

is referred to in the literature. This quantity provides a measure of the propulsive capability of a particular design. In SI units, the term g_c is 1 m · kg/N · s^2 in equations (10–48) and (10–49).

If fuel mass flow is accounted for, equations (10–48) and (10–49) must be modified to the increased mass of the exhaust gases. Frequently, it turns out that the increased mass at the exhaust can be neglected, and only at the combustor, where the temperature will change due to the fuel accounting, must any change in the analysis occur. Let us consider a gas turbine engine used as a turbojet engine. We will use the air-standard analysis with fuel accounting in the combustor.

EXAMPLE 10–11 An aircraft is propelled by a reversible gas turbine, or turbojet engine, shown in figure 10–22. The aircraft is flying at 40,000-ft altitude under standard day atmospheric conditions at 500 mph. Assume that the compressor has a pressure ratio of 10:1 and a fuel/air of 1:35, and that the fuel is kerosene. Assume also that the kerosene has the same LHV as octane (C_8H_{18}) and burns completely, releasing the full LHV in the combustor. The fuel is at 50°F. If the engine operates on an ideal Brayton cycle and the gases are ideal, determine the values of *p*, *v*, *T*, and s for the Brayton cycle corners. Then determine the thermodynamic efficiency and the mass flow rate of air required to power the aircraft if the wind resistance (drag) is 600 lbf and the flight is level.

Also, determine the properties of the exhaust gases leaving the turbine and entering the exhaust nozzle at state 3′.

Solution Figure 10–23 shows the *p–v* and *T–s* diagrams for the jet engine. The standard day atmospheric condition fixes the properties at state 1. From table B–3, they are p_1 = 5.5584 in Hg,

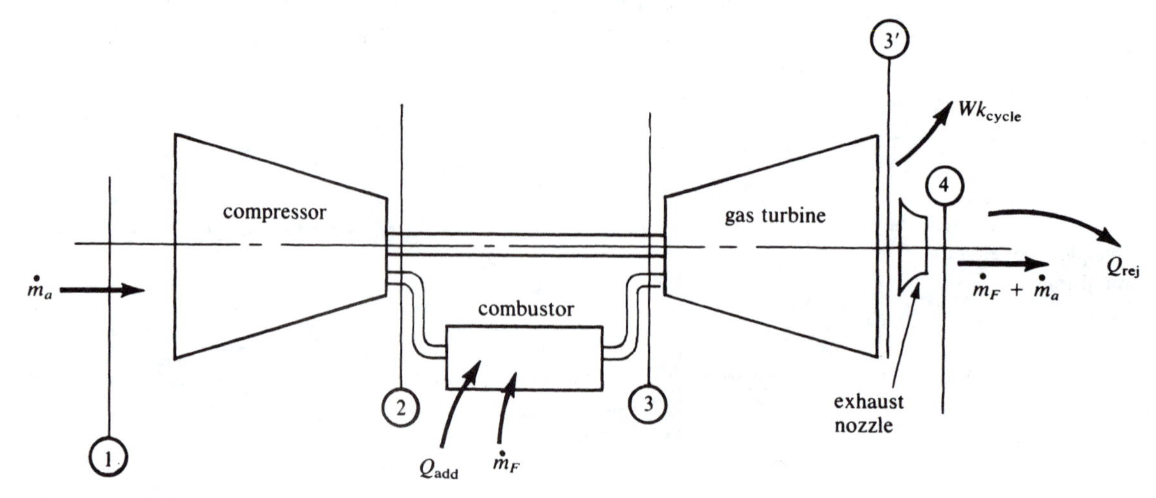

FIGURE 10–22 Turbojet engine

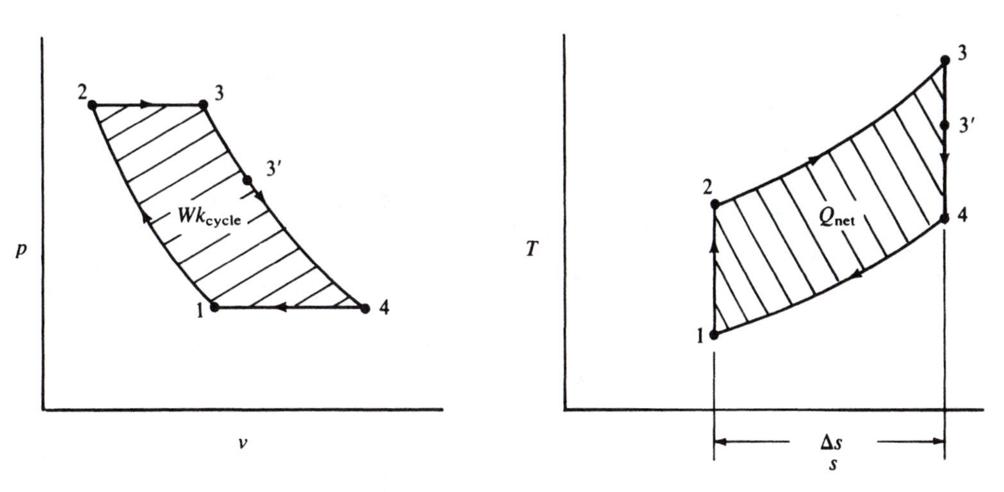

FIGURE 10–23 Property diagrams for reversible jet engine

$\rho_1 = 0.018895$ lbm/ft^3, and $T_1 = 389.97$°R. The pressure at state 2 is obtained from the compressor pressure ratio $p_2/p_1 = 10$, which gives us

$$p_2 = 10p_1 = 55.584 \text{ in Hg}$$

Also,

$$p_3 = p_2 = 55.584 \text{ in Hg} \times 0.491 \text{ psi/in Hg} = 27.3 \text{ psia}$$

and

$$p_4 = p_1 = 2.73 \text{ psia}$$

Since the engine is operating in accordance with an ideal Brayton cycle, we can write

$$p_1 v_1^k = p_2 v_2^k$$

and

$$\frac{p_1}{p_2} = \left(\frac{v_2}{v_1}\right)^k$$

or

$$v_2 = \left(\frac{p_1}{p_2}\right)^{1/k} v_1$$

But

$$v_1 = \frac{1}{\rho_1} = \frac{1 \text{ ft}^3}{0.018895 \text{ lbm}} = 52.9 \text{ ft}^3/\text{lbm}$$

so

$$v_2 = 52.9 \text{ ft}^3/\text{lbm} \left(\frac{2.73}{27.3}\right)^{1/1.4}$$

Answer

$$= 10.2 \text{ ft}^3/\text{lbm}$$

Using the perfect gas relation,

$$T_2 = \frac{p_2 v_2}{R}$$

we have

$$T_2 = \frac{(27.3 \text{ lbf/in}^2)(10.2 \text{ ft}^3/\text{lbm})(144 \text{ in}^2/\text{ft}^2)}{53.3 \text{ ft}\cdot\text{lbf/lbm}\cdot{}^\circ\text{R}}$$

Answer

$$= 752.3^\circ\text{R}$$

To determine the properties at state 3, we use equation (10–38) for perfect gases with constant specific heats:

$$\frac{\dot{m}_2}{\dot{m}_F}(c_p)(T_3 - T_2) - h_F = \text{LHV}$$

For air-standard analysis $c_p = 0.2404$ Btu/lbm·°R. Also, from equation (10–36), we find h_F, the fuel enthalpy, to be −32 Btu/lbm kerosene at 50°F. Using a value of 19,256 Btu/lbm kerosene for the LHV, we then solve for T_3:

$$T_3 = \frac{\text{LHV} + h_F}{c_p(\dot{m}_2/\dot{m}_F)} + T_2$$

$$= \frac{19{,}256 \text{ Btu/lbm} - 32 \text{ Btu/lbm}}{(0.2404 \text{ Btu/lbm}\cdot{}^\circ\text{R})(35 \text{ lbm/lbm})} + 752.3^\circ = 3037.06^\circ\text{R}$$

From the perfect gas relation,

$$v_3 = \frac{RT_3}{p_3} = \frac{(53.3 \text{ ft}\cdot\text{lbf/lbm}\cdot{}^\circ\text{R})(3037.06^\circ\text{R})}{(144 \text{ in}^2/\text{ft}^2)(27.3 \text{ lbf/in}^2)}$$

Answer

$$= 41.177 \text{ ft}^3/\text{lbm}$$

Using isentropic expansion equations for process 3–4, we have

$$\frac{p_4}{p_3} = \left(\frac{v_3}{v_4}\right)^k$$

$$v_4 = \left(\frac{p_3}{p_4}\right)^{1/k} v_3 = (10)^{1/1.4}(41.177)$$

Answer

$$= 213.275 \text{ ft}^3/\text{lbm}$$

We can determine the temperature at state 4 from the perfect gas relation:

Answer

$$T_4 = \frac{p_4 v_4}{R} = \frac{(2.73)(144)(213.275)}{53.3} = 1573.03^\circ\text{R}$$

The entropy change is obtained from process 2–3 or process 4–1. Both of these processes are isobaric, and we can write

$$\Delta s = s_4 - s_1 = s_3 - s_2 = c_p \ln\frac{T_4}{T_1} = c_p \ln\frac{T_3}{T_2}$$

$$= (0.2404 \text{ Btu/lbm}\cdot{}^\circ\text{R})\left(\text{in}\ \frac{1573.03}{389.97}\right)$$

Answer

$$= 0.33528 \text{ Btu/lbm}\cdot{}^\circ\text{R}$$

We can calculate the thermodynamic efficiency from either equation (10–8) or (10–9). From the latter,

Answer

$$\eta_T = \left(1 - \frac{1}{10^{0.4/1.4}}\right) \times 100 = 48.2\%$$

For a constant drag force, the power requirement is given by the equation

$$Wk = F\overline{V}$$

where F is the force (in this case, the drag) and $\bar{V}$ is the velocity. We then have

$$\dot{W}k = (600 \text{ lbf})\left(500 \times \frac{5280}{3600} \text{ ft/s}\right)$$
$$= 440{,}000 \text{ ft}\cdot\text{lbf/s}$$
$$= 440{,}000 \times \frac{1}{550}\text{hp} = 800 \text{ hp}$$

The mass flow rate of air required can be obtained from the relationship

$$\dot{W}k = \dot{m}wk_{\text{cycle}} = 800 \text{ hp}$$

The work of the Brayton cycle can be obtained from equations (10–6) and (10–7):

$$wk_{\text{cycle}} = q_{\text{add}} + q_{\text{rej}}$$

We have

$$q_{\text{add}} = 19{,}256 \text{ Btu/lbm fuel}$$
$$= \frac{1}{35} \times 19{,}256 \text{ Btu/lbm air} = 550.17 \text{ Btu/lbm air}$$

The heat rejected is just process 4–1, occurring in the surroundings:

$$q_{\text{rej}} = h_1 - h_4 = c_p(T_1 - T_4)$$
$$= (0.2404 \text{ Btu/lbm}\cdot{}^\circ\text{R})(389.97^\circ\text{R} - 1573.03)$$
$$= -284.41 \text{ Btu/lbm air}$$

So

$$wk_{\text{cycle}} = 265.76 \text{ Btu/lbm}$$

and the mass flow of air is

$$\dot{m}_{\text{air}} = \frac{800 \text{ hp}}{(1.41 \text{ hp/Btu/s})(265.76 \text{ Btu/lbm})}$$

Answer
$$= 2.135 \text{ lbm/s}$$

The properties of the gases leaving the turbine and entering the exhaust nozzle can be found by observing that the gas turbine produces only enough power for the compressor. We set the turbine work equal to the compressor work (except for the sign):

$$wk_{\text{turbine}} = -wk_{\text{comp}}$$

For perfect gases with constant specific heats,

$$c_p(T_3 - T_{39}) = c_p(T_2 - T_1)$$

Then $T_3 - T_{3'} = T_2 - T_1$, and solving for $T_{3'}$ yields

$$T_{3'} = T_3 + T_1 - T_2 = 2674.73^\circ\text{R}$$

Since the turbine is reversible and adiabatic, the pressure $p_{3'}$ can be found from the relationship

$$p_{3'} = \text{p}_3\left(\frac{T_{3'}}{T_3}\right)^{k/(k-1)}$$

Answer
$$= 17.50 \text{ psia}$$

The turbojet does not conveniently adapt to regenerative heating cycles for increased efficiencies; however, there are two techniques that provide the turbojet engine with quick bursts of power and increased performance for short durations: **after-burning** and **water injection**. The former involves an additional combustion of fuel immediately behind the turbine, which increases the temperature of the already hot exhaust gases and induces an increased exit velocity. As we have seen from equation (10–48), the force on the aircraft is increased and, thus, drives it faster.

Water injection is a scheme whereby water is sprayed into the exhaust from the turbine. Although this process does not increase the gas temperature, it increases the mass of the expelled gases. This requires equation (10–48), namely,

$$F_I = m\frac{\overline{V}_4 - \overline{V}_1}{tg_c} + (p_4 - p_1)A_e$$

to be read as

$$F_I = \frac{m_4\overline{V}_4 - m_1\overline{V}_1}{tg_c} + (p_4 - p_1)A_e \tag{10–50}$$

which can then be seen as an increase in the thrust force F_I due to m_4 being greater than m_1.

10–8 ROCKETS

The rocket engine is included here because it is frequently compared to the jet engine as a means of producing power. It has some distinct advantages over the gas turbine engine:

1. It can develop maximum power at zero velocity.
2. It is not "air breathing" as are all other gas turbine or jet engines.
3. It can develop much higher thrust-to-weight ratios.

These characteristics have made the rocket engine the primary means of power for travel beyond the Earth's atmosphere. The rocket engine has a distinct advantage when compared to other heat engines, however, because the rocket is not a true cyclic device; that is, it is not a heat engine but a device executing processes that *never* return the engine to its initial state. This means the rocket need not be limited to the second law of thermodynamics—it may produce work with only one heat exchange. This is precisely what is done, and in figure 10–24 the cross section of a typical engine shown indicates the physical processes being conducted. There are only two: combustion of a fuel (either solid or liquid), normally done at constant pressure; and expansion of the hot exhaust gases through a nozzle. Ideally, this expansion is reversible and adiabatic. In figure 10–25 are drawn the process curves of these two ideal processes. As can be seen, the work available to propel the rocket

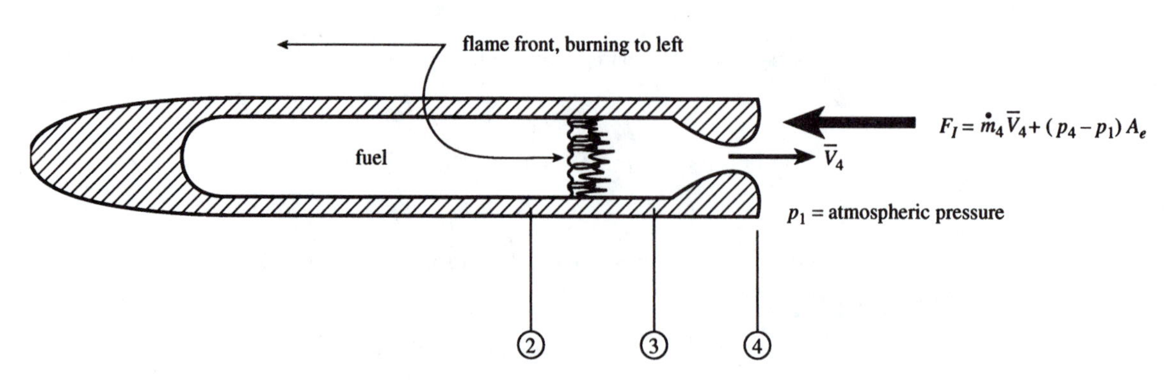

FIGURE 10–24 Cross section of typical rocket

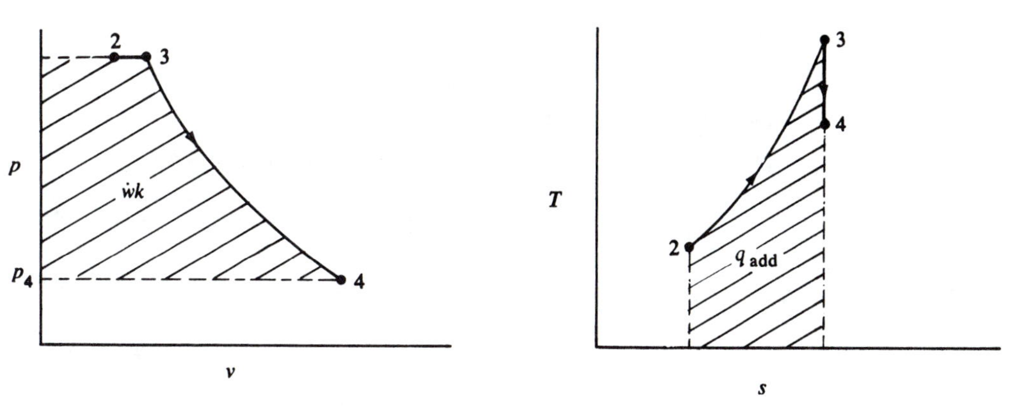

FIGURE 10–25 Property diagrams for rocket engine

is the area to the left of the p–v curve and the heat added is the area under the T–s curve. The major part of the rocket is the nozzle, which accelerates the hot gases, and we will look at an example problem concerned with just this. First, however, let us write the general equation for thrust induced, F_I, from equation (10–48):

$$F_I = m\frac{\overline{V}_4 - \overline{V}_1}{tg_c} + (p_4 - p_1)A_e$$

In the rocket, however, no mass is entering the system, so $\overline{V}_1 = 0$, yielding

$$F_I = \frac{m(\overline{V}_4)}{tg_c} + (p_4 - p_1)A_e \tag{10–51}$$

The specific impulse of the rocket is then written

$$I_{sp} = \frac{F_I}{\dot{m}} = \frac{\overline{V}_4}{g_c} + \frac{p_4 - p_3}{\dot{m}}A_e \tag{10–52}$$

EXAMPLE 10–12 A rocket burns 200 lbm/s of fuel and oxidizer at 2700°F and 300 psia. The nozzle exhaust area is 2 ft^2, and the pressure is assumed to be atmospheric (14.7 psia) at the exit. Determine the exhaust velocity, the specific impulse, the impulse, and the power produced by the engine if the nozzle is reversible and adiabatic, and if we assume air-standard conditions.

Solution For a reversible adiabatic nozzle, we write the steady-flow energy equation

$$h_4 - h_3 + \frac{\overline{V}_4^2 - \overline{V}_3^2}{2g_c} = 0$$

Since $\overline{V}_3 \cong 0$, we have

$$\overline{V}_4 = \sqrt{2g_c(h_3 - h_4)}$$

Invoking air-standard conditions gives us

$$h_3 - h_4 = c_p(T_3 - T_4)$$

Then, for reversible adiabatic processes,

$$T_4 = T_3\left(\frac{p_4}{p_3}\right)^{(k-1)/k} = (3160°\text{R})\left(\frac{14.7}{300}\right)^{0.286}$$
$$= 1334°\text{R}$$

and the velocity can be calculated:

$$\bar{V}_4 = \sqrt{2 \times 32.2 \text{ ft} \cdot \text{lbm/s}^2 \cdot \text{lbf} \times 0.24 \text{ Btu/lbm} \cdot {}^\circ\text{R} \times 778 \text{ ft} \cdot \text{lbf/lbm } (3160 - 1334)^\circ\text{R}}$$

Answer $= 4685 \text{ ft/s}$

The specific impulse is obtained from equation (10–52), assuming that $p_4 = p_3$:

Answer
$$I_{sp} = \frac{4685 \text{ ft/s}}{32.2 \text{ ft} \cdot \text{lbm/lbf} \cdot \text{s}^2} + 0 = 145 \text{ lbf} \cdot \text{s/lbm}$$

The impulse F_I can be quickly calculated from equation (10–51):

$$F_I = \frac{\dot{m}}{g_c}(V_4) = \dot{m}I_{sp} = 200 \text{ lbm/s} \times 145 \text{ lbf} \cdot \text{s/lbm}$$

Answer $= 29{,}000 \text{ lbf}$

Using our common equations, we obtain the power from the equation

$$\dot{W}k = \dot{m}wk$$

where

$$wk = c_p(T_3 - T_4)$$

Then

$$\dot{W}k = (200 \text{ lbm/s})(0.24\text{Btu/lbm} \cdot {}^\circ\text{R})(3160 - 1334)^\circ\text{R}$$

or

Answer
$$\dot{W}k = 87{,}648 \text{ Btu/s} = 123{,}584 \text{ hp}$$

10–9 COMPUTER-AIDED ANALYSIS OF THE GAS TURBINE

An analysis of the gas turbine done with the Brayton cycle can be aided by the use of the digital computer or microcomputer. If you are involved in only one particular problem of the gas turbine, the hand calculation methods presented in earlier examples in this chapter will probably be most efficient, but if you want to do an analysis of how parameters of the gas turbine may vary as one particular parameter (or more than one) varies, the computer can be a helpful tool. Even if you are doing only one problem, using a program such as *EES* can be helpful and often less time consuming. In appendix A–10, the program BRAYTON is described, which, when used with a microcomputer, can assist in analyzing the gas turbine. The program uses the Brayton cycle air-standard analysis with perfect gases and constant specific heats. It also provides for the modifications of assuming polytropic processes or for the irreversible adiabatic processes in the compressor and turbine. Provisions are also made for the regenerative cycle discussed in section 10–6. The program is contained on a soft diskette that also contains other programs referred to in this book. It should be emphasized that the program is intended only to provide the reader with exposure to those conditions where the computer will be a significant assistance to the solution of simple gas turbines; it is not meant to be a detailed program for final designs of a gas turbine. The program BRAYTON can be used to solve three types of problems:

1. Given the pressure ratio, inlet conditions (p_1, T_1), maximum temperature allowed, and power produced, to find the properties at the four states, the rates of heat addition and rejection, the mass flow rate of gas, and thermal efficiency.
2. Given the pressure ratio, inlet conditions, mass flow of gas, and maximum temperature allowed, to find the properties at the four states, the power generated, the rates of heat added and rejected, and the thermal efficiency.
3. Given the inlet conditions, maximum temperature, power produced, and rate of heat addition, to find the thermal efficiency, the properties at the four states, and the rate of heat rejection.

For irreversible adiabatic modifications, the adiabatic efficiency must be known; the program then allows for determination of entropy generation. The student can expand the program BRAYTON to incorporate other elaborations of the gas turbine operation; for instance, an accounting of the variation of specific heats could be included in the program, an accounting of different fuels could be made, or an accounting for the turbojet application could be included.

10–10 SUMMARY

The operation of the gas turbine engine can be described by the Brayton cycle, which is defined by the following four processes:

1–2 Reversible adiabatic compression through a compressor
2–3 Reversible isobaric heat addition in a combustor
3–4 Reversible adiabatic expansion through a turbine
4–1 Reversible isobaric heat rejection

The air-standard analysis can be applied to the Brayton cycle, and for perfect gases with constant specific heats, the thermal efficiency of the gas turbine operating on a Brayton cycle is

$$\eta_T = 1 - \frac{1}{(r_p)^{(k-1)/k}} \times 100 \qquad \textbf{(10–9)}$$

where r_p is the pressure ratio over the compressor or the turbine. The combustor is a constant-pressure chamber in which heat is added by a continuous combustion of a fuel mixed with air. The maximum temperatures in the gas turbine engine occur between the combustor and the gas turbine.

Nozzles are used to convert the thermal energy in the exhaust gases into kinetic energy. Nozzles are used to direct gases against the turbine blades where the kinetic energy is converted into angular kinetic energy. Nozzles can also be used in the exhaust to accelerate the exhaust gases and provide for jet propulsion. The deLaval nozzle can accelerate gases to speeds greater than that of sound, or to supersonic speeds. If the gases passing through the deLaval nozzle reach supersonic speeds, the speed of the gases at the narrowest section in the nozzle (called the throat) are at sonic velocity. The gases are then at Mach 1, where the Mach number is defined as the ratio of the speed or velocity of a gas to the speed of sound in that gas.

The analysis of gas turbines is done with the Brayton cycle: If a better description of the operation of the engine is required, the Brayton cycle can be modified. Three modifications discussed in this chapter are variable specific heats, reversible polytropic compressions and expansions, and irreversible adiabatic compressions and expansions. For an accounting of fuel addition to the combustor, we defined the air/fuel ratio, $\dot{m}_a/\dot{m}_F$, and with the fuel lower heating value (LHV), we wrote the mass/energy balance for the combustor:

$$\frac{\dot{m}_a}{\dot{m}_F}(\Delta h) - h_F = \text{LHV} \qquad \textbf{(10–38)}$$

Here, h_F is the enthalpy of the liquid fuel.

The gas turbine may have improved efficiency by using the regenerative heating cycle. Regenerative heating uses the hot exhaust gases to heat the inlet air or the air entering the combustor. The efficiency of a regenerative cycle using a regeneration at the combustor is given by

$$\eta_T = 1 - \left(\frac{T_1}{T_4}\right)(r_p)^{(k-1)/k} \times 100 \qquad \textbf{(10–46)}$$

For applications of jet propulsion, the gas turbine produces power only for the compressor, and the hot exhaust gases are expanded through an exhaust nozzle that accelerates the gases. This produces a thrust on the aircraft or vehicle due to the reaction of acceleration of the gases. The thrust developed in jet propulsion is given by the equation

$$F_I = m\frac{\overline{V}_4 - \overline{V}_1}{tg_c} + (p_4 - p_1)A_e \quad (10\text{–}48)$$

and the specific impulse is defined as

$$I_{sp} = \frac{F_I}{\dot{m}} = \frac{\overline{V}_4 - \overline{V}_1}{g_c} + \frac{(p_4 - p_1)A_e}{\dot{m}} \quad (10\text{–}49)$$

The rocket is a special form of jet propulsion that is not a steady-state heat engine.

DISCUSSION QUESTIONS

Section 10–1

10–1 What is meant by the *Brayton cycle*?

10–2 What is the significance of the *pressure ratio* in a Brayton cycle?

Section 10–2

10–3 How are *reaction* and *impulse* turbines different?

10–4 What is *multistaging* in a turbine?

Section 10–3

10–5 What do you think is a problem in designing a *combustor*?

Section 10–4

10–6 Why do you think a *converging nozzle* accelerates incompressible fluids or fluids that are traveling at a speed less than that of the speed of sound, but decelerates compressible fluids that are traveling at speeds greater than sonic?

10–7 What is a *deLaval nozzle*?

10–8 What is the *Mach number*?

Section 10–6

10–9 What purpose does *regeneration* serve in the Brayton cycle?

Section 10–7

10–10 What is the difference between *gas turbines* and *jet propulsion*?

10–11 What is meant by the term *specific impulse*?

Section 10–8

10–12 How is a *rocket* different than a *jet engine*?

PRACTICE PROBLEMS

Sections 10–1

Problems that use SI units are indicated by an (M) under the problem number; those that use English units are indicated by an (E).

10–1 Determine the thermodynamic efficiency of an ideal gas turbine operating with a pressure ratio of 20.

10–2 Determine the thermodynamic efficiency of an ideal Brayton cycle engine if the pressure increase across the compressor is 18 times the inlet pressure.

10–3 If the pressure ratio across the compressor of an ideal Brayton cycle engine is 22:1 and the working medium is air, determine the temperature ratio across the compressor.

10–4 (E) A 4000-hp ideal gas turbine engine operates between 15- and 160-psia pressures. Determine the rate of heat addition to the cycle.

10–5 Starting from the steady-flow energy equation, state the necessary assumptions for arriving at equations (10–1), (10–2), and (10–3).

10–6 (E) Prove that the thermodynamic efficiency of an ideal Brayton cycle is given by

$$\eta_T = \left[1 - \frac{1}{(r_p)^{(k-1)/k}}\right] \times 100$$

for a perfect gas working medium.

Sections 10–2 and 10–3

10–7 (M) Air at 1000 K and 500 kPa expands through a turbine to 100 kPa. What is the temperature of the air leaving the turbine?

10–8 (M) A turbine receives 60 kg/min of air at 20 m/s at 1200 K. The exhaust air is at 600 K and 10 m/s. Determine the work produced by the turbine per unit mass of air and the power produced.

10–9 (M) A gas turbine is designed to produce 3000 hp when using air at a maximum inlet temperature of 650 K and assuming a heat loss of 70 kJ/s. If the exhaust temperature is desired to be no greater than 200°C, determine the mass flow rate of air required.

10–10 (M) A turbine expands air from a region where the pressure and temperature are 2000 kPa and 1000°C to a region at 101 kPa. Neglecting kinetic and potential energy changes of the air and assuming that the process through the turbine is polytropic with an exponent $n = 1.5$, determine
(a) wk_{turb}
(b) q
(c) Wk_{turb} if the mass flow of air is 6 kg/s

10–11 (M) If 60,000 kg/h of air is supplied to a combustor at 500 kPa and 160°C and the leaving gases are to be at 800°C, determine the rate of heat addition required by the combustion of fuel and air. Assume the kinetic energy changes and the mass of fuel to be negligible.

10–12 (M) A compressor receives air at 101 *kPa*, 40°C, and 6 m/s. If the gases are compressed reversibly and adiabatically to 900 kPa and they leave the compressor at 90-m/s velocity, determine the amount of work per kilogram of air that is required by the compressor. If the process is reversible and polytropic with $n = 1.34$ and kinetic energy changes are neglected, determine the pressure of the air leaving the compressor and the work required by the compressor per kilogram of air. Assume that the temperature T_2 is 540 K.

10–13 (E) An ideal turbine receives 30 lbm/min of air at 2100°R and exhausts it at 700°F. Determine the power produced by the turbine if kinetic energy changes are neglected.

10–14 (E) A gas turbine has an adiabatic efficiency of 85%. If air at 1800°F and 300 psia expands through the turbine to 15 psia, determine the outlet air temperature and the work done per pound-mass by the air.

10–15 (E) A turbine reversibly and adiabatically expands air from 280 psia to 15 psia and 800°F. Determine the inlet air temperature and the work gained from the turbine per pound-mass of air if kinetic and potential energy changes are neglected in the turbine.

10–16 (E) Air enters a combustor at 100 psia, 400°F, and 200ft/s. Fuel having a heating value of 17,000 Btu/lbm is burned, and the gases leaving the combustor (assuming they have properties similar to those of air) are at 1700°F and 400 ft/s. Determine the amount of fuel required per pound-mass of air, $\dot{m}_F/\dot{m}_a$.

10–17 (E) Air having a density of 0.06 lbm/ft^3 and a pressure of 14.7 psia is compressed to 0.10 lbm/ft^3. Determine the pressure of the air leaving the compressor and the work per pound-mass of air required if the process is reversible and polytropic with $n = 1.34$ and the velocity changes of the air flow through the compressor are neglected.

Section 10–4

10–18 (M) Air enters a reversible adiabatic converging nozzle at 5 m/s, 1000 K, and 1200 kPa. Determine the velocity of air leaving the nozzle if the pressure there is 110 kPa.

10–19 (M) Determine the entrance and exit areas of the nozzle in problem 10–18 if the mass flow rate is to be 120 kg/min.

10–20 (M) Air at 500-m/s velocity enters a diffuser at 80 kPa. If the diffuser is reversible and adiabatic and the exit pressure is 240 kPa, determine the inlet air temperature. The air is quiescent when it leaves the diffuser.

10–21 (M) An adiabatic diffuser is used to direct oxygen gas into a compressor. Oxygen is furnished at 650 K and 760 mm Hg through an area of 10 cm^2. The diffuser is 88% efficient when the oxygen leaves at 1000 K and 3000 mm Hg. Determine the entrance velocity and the mass flow of oxygen.

10–22 (E) A nozzle has 85% efficiency and operates between 180 psia and 12 in Hg pressure. Air flows in at 1600°R and leaves at 3200 ft/s. Determine the mass flow of air through the nozzle if the entrance area is 0.5 in^2.

10–23 (E) For the nozzle of problem 10–22, determine the exit area of the nozzle.

10–24 (E) A reversible adiabatic diffuser receives helium gas at 800°R and 1950-ft/s velocity. If the exhaust pressure must be 50 psia, determine the entrance pressure of the helium. Assume that the exhaust velocity is zero.

10–25 (E) A diffuser having 82% efficiency is operating between 15- and 35-psia pressures. Air enters at 1650°R and leaves at 2200°R. Determine the entering air velocity and the mass flow rate if the entrance area is 0.02 ft^3.

10–26 Determine the Mach number of the flow at the entrance in problems 10–20, 10–21, 10–22, and 10–25.

10–27 For a supersonic deLaval nozzle having a throat area and temperature given as follows, determine the volume flow through the nozzle:
(a) Area = 20 cm^2 and T = 420 K
(b) Area = 3 in^2 and T = 850°R

Section 10–5

10–28 (M) If the inlet air temperature to the compressor of problem 10–3 is 20°C, determine the work required of the compressor.

10–29 (M) An ideal Brayton cycle engine operates with a pressure ratio of 8, the working medium is air, and the compressor inlet temperature is 30°C. If the entropy change across the combustor is 0.7 kJ/kg·K, determine the pressure, specific volume, and temperature at the four corners of the cycle. Assume that the inlet pressure to the compressor is 101 kPa.

10–30 (E) If the inlet air temperature to the compressor of problem 10–2 (assuming air to be the working medium) is 100°F, determine the compressor work.

10–31 Sketch the p–V and T–s diagrams for an ideal Brayton cycle engine. Then, if the heat added is 300 Btu/lbm air, $r_p = 10$, $p_1 = 14.7$ psia, $T_1 = 100$°F, and the working medium is air, determine the properties p, r, and T at the four corners of the cycle.

10–32 (M) Air enters an ideal stationary gas turbine engine at 20°C and 10^6 N/m^2. The pressure ratio is 8, and the air leaves the combustor at 1100 K. Assume that the mass of fuel is negligible and the mass flow of air is 100 kg/s. Determine
(a) Cycle thermodynamic efficiency
(b) Heat added per unit time
(c) Heat rejected per unit time
(d) Power developed

10–33 (E) An ideal Brayton cycle gas turbine using air operates with a pressure ratio of 7.5 to 1. The entrance conditions are 14.7 psia and 60°F. If 500 Btu/lbm of heat are added in the combustor, using the gas tables, determine
(a) p, v, and T at the four cycle corners
(b) Δs for compression and expansion
(c) wk_{cycle}
(d) Thermodynamic efficiency

10–34 (E) An ideal closed gas turbine engine operates with air at 14.7 psia and 50°F. After compression the air is at 860°R. Then, the gases entering the turbine are at 1400°F. If the combustor is burning octane fuel, determine
(a) Air/fuel ratio
(b) wk_{turb}
(c) wk_{comp}
(d) Thermodynamic efficiency

10–35 (E) For the gas turbine cycle shown in figure 10–26, the following data apply:

$\eta_{comp} = 90\%$ $\qquad p_2 = 100$ psia
$\eta_{turb} = 90\%$ $\qquad T_3 = 2500$°R
$\eta_{combust} = 65\%$ $\qquad$ fuel LHV = 15,000 Btu/lbm
$p_1 = 14$ psia $\qquad \dot{W}k_{turb} = 7000$ hp
$T_1 = 65$°F

Determine
(a) $\dot{m}_1$ and $\dot{m}_F$
(b) Thermodynamic efficiency

10–36 (E) Using the gas tables, solve problem 10–34.

10–37 (M) An ideal Brayton cycle gas turbine operates with a pressure ratio of 7.6:1. The entering air is at a pressure of 101 kPa, and the temperature is 27°C. The air/fuel ratio is 35:1, and the fuel has an LHV of 45,000 kJ/kg. Determine
(a) p–v and T–s diagrams
(b) q_{add}
(c) wk_{turb}
(d) wk_{comp}
(e) wk_{cycle}
(f) Thermodynamic efficiency

10–38 (E) Using the gas tables, solve problem 10–35.

10–39 (M) A gas turbine operates with a pressure ratio of 19:1 and inlet conditions of 100 kPa and 17°C. The maximum combustor temperature is 3300 K, the mass flow of air is 8.6 kg/s, and the compressor and turbine have adiabatic efficiencies of 90%. Determine the power; the rate of heat addition; the thermal efficiency; the p's, v's, and T's at the four states; and the entropy generation. Use of the program BRAYTON may be helpful.

10–40 (E) A gas turbine has a pressure ratio of 20:1, inlet conditions of 40°F and 14.7 psia, and a maximum temperature of 3500°R. The mass flow of air is 6.0 lbm/s, and the adiabatic efficiency for the turbine is 85% and for the compressor is 90%. Determine the power, the rate of

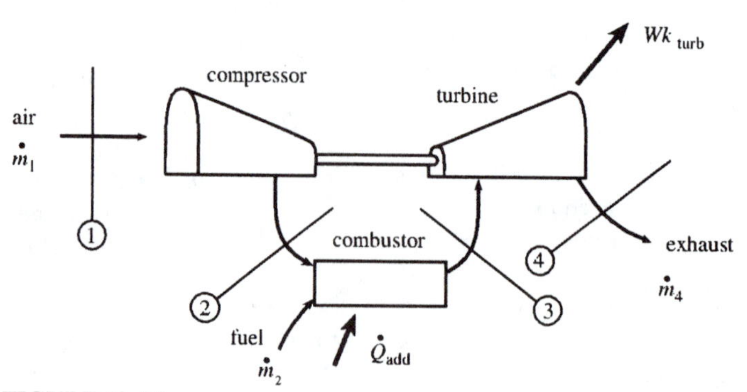

FIGURE 10–26

heat addition, the thermal efficiency, the properties at the four states, and the entropy generation. Use of the program BRAYTON may be helpful.

10–41 (M) A reversible gas turbine has polytropic compression and expansion processes. For the compression, the exponent n is 1.32, and for the expansion, n is 1.28. The pressure ratio is 20:1, inlet conditions are 98 kPa and 27°C, and the maximum temperature is 2000 K. If 5000 kW of power is produced, determine the mass flow of air, the rate of heat addition, the thermal efficiency, the properties at the four stages, and the rate of heat rejection. Use of the program BRAYTON may be helpful.

10–42 (E) A gas turbine may be approximated by a reversible polytropic Brayton cycle. The polytropic exponents are 1.2 for the compression and 1.3 for the expansion. The inlet conditions are 14.6 psia and 50°F, the maximum temperature allowed is 2000°F, 4500 hp of power is produced, and 6300 Btu/s of heat are added. Determine the thermal efficiency, the mass flow of air required, the properties at the four states, and the heat rejection rate. Use of the program BRAYTON may be helpful.

Section 10–6

10–43 (E) An ideal regenerative cycle gas turbine takes air at 14.0 psia and 20°F and produces 8000 hp. If the pressure ratio is 15:1 and the maximum temperature is 1800°F, determine the properties at the four states, the mass flow of air, the rates of heat addition and rejection, and the thermal efficiency.

10–44 (M) An ideal regenerative cycle gas turbine operates with air at 101 kPa and 7°C, a pressure ratio of 20:1, and a maximum allowable temperature of 2400 K. If the engine produces 10,000 kW, determine the pressure and temperatures at the four states, the mass flow of air, the rates of heat addition and rejection, and the thermal efficiency.

10–45 (M) The effectiveness of a regenerator is defined by the ratio r_g given by

$$r_g = \frac{(T_3 - T_2)\dot{m}_2}{(T_5 - T_6)(\dot{m}_2 + \dot{m}_{\text{fuel}})} \times 100\%$$

where the subscripts refer to figure 10–19 and the specific heats are constant and equal for air and the exhaust. Determine the effectiveness of a generator that receives air at 700 kPa and 260°C. The air going to the combustor is at 620°C, and the gases leaving the turbine are at 800°C. Assume that the air/fuel ratio is 30:1. Plot the T–s diagram of a cycle that could be using this regenerator, indicating the significant temperature data.

10–46 (E) A regenerative gas turbine uses 200,000 lbm/h of air and 4000 lbm/h of fuel having an LHV of 17,500 Btu/lbm. The air is taken in at 14.7 psia and 65°F and is compressed reversibly and adiabatically to 120 psia. A regenerator with an effectiveness of 60% (see problem 10–45) heats the air to 1100°R. Assume that the turbine is reversible and adiabatic. Then determine

(a) $\dot{Q}_{\text{add}}$
(b) $\dot{Wk}_{\text{cycle}}$
(c) Exhaust temperature
(d) $\dot{Q}_{\text{rej}}$
(e) Cycle thermodynamic efficiency

Section 10–7

10–47 (M) An aircraft flying at 8000 m in a standard day atmosphere is traveling with an airspeed of 200 m/s. The craft is propelled by an ideal gas turbine which operates on a pressure ratio of 9:1 and which burns kerosene. If the air/fuel ratio is 40:1, calculate

(a) wk_{turb}
(b) wk_{comp}
(c) wk_{cycle}
(d) q_{rej}

10–48 (M) Using the gas table, solve problem 10–47.

10–49 (M) An ideal turbojet engine operates with pressure ratio of 16:1, a maximum temperature of 2200 K, and 10 kg/s of air flow. When the engine is operating at an altitude of 10,000 m and with an intake air velocity of 300 m/s and exhaust at 550 m/s, determine

(a) Thrust, F_I
(b) Specific impulse, I_{sp}
(c) Required power
(d) Rate of heat addition

Neglect the pressure difference, $p_4 - p_1$, between intake and exhaust.

10–50 (E) An ideal turbojet engine operates to fly an aircraft at 60,000-ft altitude with a pressure ratio of 21:1, an intake air velocity of 890 ft/s, and a maximum operating temperature of 2000°R. If 8000 hp is required to move the aircraft, determine the exhaust gas velocity, the thrust, and the specific impulse.

Section 10–8

10–51 (M) A rocket engine burns 700 kg/s of fuel and oxidizer at 1300°C and 20 atm pressure. The nozzle is frictionless and adiabatic and has a throat area equal to 0.6 m^3. Assume that the exhaust gases have a density of 0.004 g/cm^3 and $c_p = 0.24$ cal/g · K. If the rocket is operating in an atmosphere at 760 mm Hg, determine

(a) Exhaust temperature
(b) Impulse
(c) Power produced by the rocket

10–52 (M) Determine the specific impulse of a rocket that burns 3600 lbm of fuel and air per second. The rocket nozzle has a cross-sectional area of 10 ft^2, allowing the exhaust gases to expand from the combustion chamber at 600 psia to a 10-psia atmospheric pressure surrounding. Assume that the density of gases leaving the nozzle is 0.05 lbm/ft^3.

Section 10–9

Use of the BRAYTON program is recommended to assist in solving the following problems.

10–53 Consider a simple gas turbine operating on the Brayton cycle. Determine how the rate of heat addition, thermal efficiency, and mass flow vary if the pressure ratio alone is changed.

10–54 Consider an irreversible gas turbine and determine how the mass flow of working fluid (air), rate of heat addition, and thermal efficiency vary if the adiabatic efficiencies of the compressor and turbine vary. You may assume the efficiency to be the same for both turbine and compressor.

10–55 Consider an ideal regenerative gas turbine. Determine how the thermal efficiency, power, and rate of heat rejection vary if the pressure ratio is changed.

10–56 A turbojet engine operates at constant altitude, pressure ratio, and mass flow rate. How does the exhaust velocity vary if the thrust changes?

11 STEAM POWER GENERATION AND THE RANKINE CYCLE

Electrical power represents the form in which the greatest amount of energy is used directly by society. However, the overwhelming majority of this power is generated by steam and the steam turbine. Fossil fuels (coal, oil, and gas) or nuclear fuels provide the chemical availability to produce the steam, which in turn drives the steam turbine and electric generator. This cycle has demonstrated the highest thermodynamic efficiency for vast power production in a myriad of technical devices, and it appears destined to be the major source of mechanical power for quite some time. A typical steam turbine power plant is shown in figure 11–1, where the raw material is coal, transported by river barge or railroads. When burned, the coal produces thermal energy, which boils water to drive steam turbines and, in turn, electrical generators. The workings of this system constitute the typical Rankine or steam turbine cycle, which we examine in this chapter.

The Rankine cycle, an idealization, is defined and compared to the operation of the actual steam turbine cycle. The major components of the closed system cycle—boiler or steam generator, condenser, feedwater pumps, and steam turbine—are discussed to show the physical processes corresponding to the Rankine cycle and to assist in the cycle analysis.

The phase changes, so important in steam power applications, are discussed along with the concept of superheated steam. Property diagrams (T–s and p–v) are used to help expand the details of the phase change beyond the introduction of section 5–1.

FIGURE 11–1 Modern steam turbine, electric power generating station (reproduced with permission of Dayton Power and Light Company, Dayton, Ohio)

The steam tables and Mollier diagram are used to identify the properties of steam at given states (the perfect gas assumption is never used with steam except at very high temperatures) to give the reader a storehouse of ready and useful data for problem solving. Some traditional steam turbines operating on the Rankine cycle are then analyzed with the thermodynamic tools.

Modifications of the Rankine steam turbine cycle to increase power output or efficiency are introduced with treatments of the regenerative and reheat cycles. Finally, some general considerations are made regarding the design of future steam turbine cycles.

New Terms

AF Air/fuel ratio

$\dot{m}_{sr}$ Steam rate

η_{boiler} Boiler efficiency

11–1 THE RANKINE CYCLE

The thermodynamic cycle that most properly describes the workings of the ideal steam turbine is the Rankine cycle, defined by four reversible processes:

1–2 Adiabatic compression of liquid (water)
2–3 Isobaric heat addition to convert liquid to a vapor (steam)
3–4 Adiabatic expansion of vapor to low pressure
4–1 Isobaric heat rejection to condense vapor to liquid

These four processes are the same combination that describes the gas turbine or Brayton cycle (see chapter 10), but here we are involved with phase changes in the working media, which lend unique characteristics to the Rankine cycle not present in the Brayton cycle. In figure 11–2 are depicted the typical property diagrams of the Rankine cycle, and, in figure 11–3, the schematic of the major components of the steam turbine cycle is described by these property diagrams. Notice that a saturation line is drawn in the property diagrams to give the reader an idea of the approximate relation of the process to the phase change of H_2O which occurs “under” the saturation line.

For the ideal steam turbine cycle of figures 11–2 and 11–3, we can apply the steady-flow energy equation to the individual components to arrive at the following pertinent equations (neglecting kinetic and potential energy changes):

$$q_{add} = h_3 - h_2 \quad \textbf{(11–1)}$$

$$wk_{turb} = h_3 - h_4 \quad \textbf{(11–2)}$$

$$q_{rej} = h_1 - h_4 \quad \textbf{(11–3)}$$

$$wk_{pump} = h_1 - h_2 \quad \textbf{(11–4)}$$

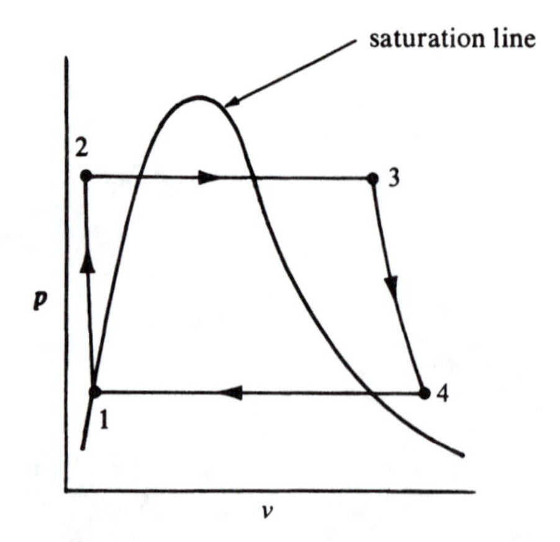

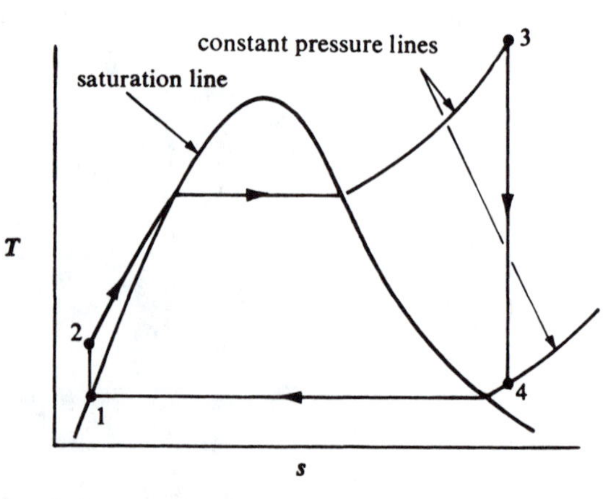

FIGURE 11–2 Property diagrams of the Rankine cycle

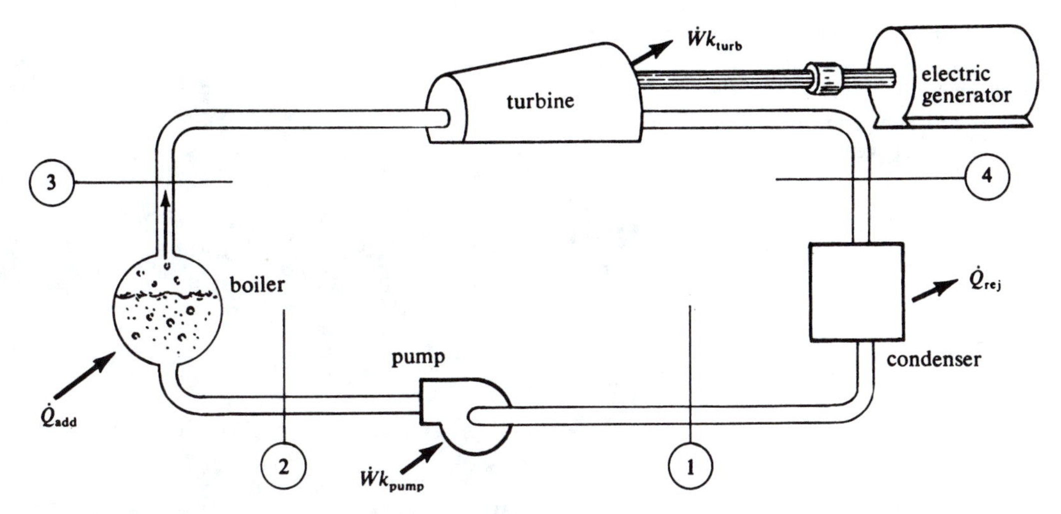

FIGURE 11–3 Typical closed steam turbine cycle

For the full cycle,

$$wk_{\text{cycle}} = wk_{\text{turb}} + wk_{\text{pump}} \tag{11–5}$$

and by using the results, we obtain

$$wk_{\text{cycle}} = h_3 - h_4 + h_1 - h_2 \tag{11–6}$$

Also, we have, for thermodynamic efficiency,

$$\eta_T = \frac{wk_{\text{cycle}}}{q_{\text{add}}} \times 100 = \frac{h_3 - h_4 + h_1 - h_2}{h_3 - h_2} \times 100$$
$$= \left(1 - \frac{h_4 - h_1}{h_3 - h_2}\right) \times 100 \tag{11–7}$$

We will now look at the individual components: the boiler, the condenser, the pump, and the steam turbine, to visualize equations (11–1) through (11–7). In this chapter, no perfect gas assumption is made, so c_v or c_p and temperatures generally are not used. Rather, we use steam tables to seek enthalpy values.

11–2 BOILERS AND STEAM GENERATORS

Steam is generated in a boiler or steam generator. Liquid H_2O is supplied to the boiler, and because of a heat addition, the water becomes vapor (or steam). A typical boiler unit is shown in figure 11–4, which indicates that our initial concept of a "teapot" is not quite correct. Tubes carry water from a lower tank past a combustion furnace and into an upper tank. From here (at which point the water is steam), superheater pipes pass the saturated steam through the furnace again and elevate the steam temperature well beyond a saturated vapor state. In the construction of a boiler (a steam generating unit), the combustion chamber is normally an integral part of the system. Whether coal, oil, gas, or nuclear fuel is used, the intimate relation between combustion and the water system prevents undue waste of heat. That is, the heat transfer to the water is maximized by design of the total unit. In addition, the furnace-type steady combustion provides the most nearly complete fuel combustion and consequently gives the best release of chemical energy from the fuel to be used in generating steam.

FIGURE 11–4 Typical steam generating boiler: field-erected industrial boiler (reproduced with permission of Combustion Engineering, Inc., Windsor, Conn.)

In the best combustion, we can approximate the heat added to the water in the boiler by the lower fuel heating value (LHV). That is, in figure 11–3, which depicts a boiler,

$$q_{\text{add}} \leq \text{LHV}\frac{\dot{m}_{\text{fuel}}}{\dot{m}_{\text{steam}}} \tag{11–8}$$

where LHV is based on a unit mass of fuel. We can also calculate heat added to the water from equation (11–1):

$$q_{\text{add}} = h_3 - h_2$$

With these two results, we can define the boiler efficiency as

$$\eta_{\text{boiler}} = \frac{h_3 - h_2}{\text{LHV}}\frac{\dot{m}_{\text{steam}}}{\dot{m}_{\text{fuel}}} \tag{11–9}$$

The boiler is an important component in the steam cycle, and with the search for new energy sources, it has been the center of attention.

Another term used to describe the steam generator and its operation is the steam rate, $\dot{m}_{sr}$. The steam rate is defined by the equation

$$\dot{m}_{sr} = \frac{\dot{m}_{\text{steam}}}{\dot{W}k_{\text{cycle}}} \tag{11–10}$$

and it provides a measure of the capacity of the steam generator, even though it depends on the power extracted by the turbine or the net power, $\dot{W}k_{\text{cycle}}$.

11–3 STEAM TURBINES

Steam at a high pressure and temperature has a large amount of internal energy, but a device is needed to convert this energy into mechanical work or power. The steam turbine is the machine that does this conversion, and it operates on exactly the same principles as the historic waterwheel or the gas turbine. Readers are advised to read section 10–2 at this time if they have not already done so, as the description of the gas turbine and its operation is analogous to that of the steam turbine. Although the steam turbine is occasionally constructed as a reaction turbine, the majority of the machines are impulse-reaction turbines. In this device, steam is supplied from a boiler, and its internal energy is then converted into kinetic energy in a nozzle, which directs the steam against buckets attached to a turbine wheel. This steam is diverted by the buckets into an exhaust passage or another nozzle, and a turbine wheel. The diversion of steam causes the turbine wheel to rotate and thus produce shaft work, which can easily be utilized by mechanically connecting an electric generator or other device to the turbine shaft. For a frictionless turbine allowing reversible adiabatic expansion of the steam, we have found that, by writing the steady-flow energy equation, neglecting kinetic and potential energy changes, equation (11–2) resulted:

$$wk_{\text{turb}} = h_3 - h_4$$

The work can also be obtained by using steam velocity values:

$$wk = (\overline{V}_{1x} - \overline{V}_{2x})\overline{V}_b$$

The power is found by multiplying the work by the mass flow of steam through the turbine:

$$\dot{W}k_{\text{turb}} = \dot{m}(h_3 - h_4)$$

The expansion of steam through a turbine is most commonly conducted in a manner to achieve approximately saturated vapor at the turbine exhaust. That is, at state 4, we wish to have extracted as much energy as possible from the steam in producing turbine work, but if any moisture or liquid particles are in the steam while it is anywhere in the turbine, these particles can act as abrasives and thus seriously damage parts of the turbine. Additionally, moisture in a turbine can induce corrosion of vital parts. For this reason, we desire a steam that is a vapor ready to condense (saturated vapor) at the turbine exit, or with small amounts of moisture.

For turbines that have irreversible effects (this includes *all* turbines), the work produced will be less than in the ideal case. Perhaps the best way to see what happens here is through a T–s diagram. In figure 11–5 is shown the ideal expansion from pressure p_3 to p_4. As indicated, the steam has just achieved a saturated vapor state at the exhaust, at state $4s$. In the irreversible process, fluid friction during expansion of steam between the same pressures p_3 and p_4 causes an entropy *increase*, and thus the exhaust steam is still superheated at a pressure p_4, but at a temperature T_4. The work for the irreversible process is then obtained from

$$wk_{\text{turb}} = h_3 - h_4 \tag{11–11}$$

with an entropy increase of $s_4 - s_3$ in the steam.

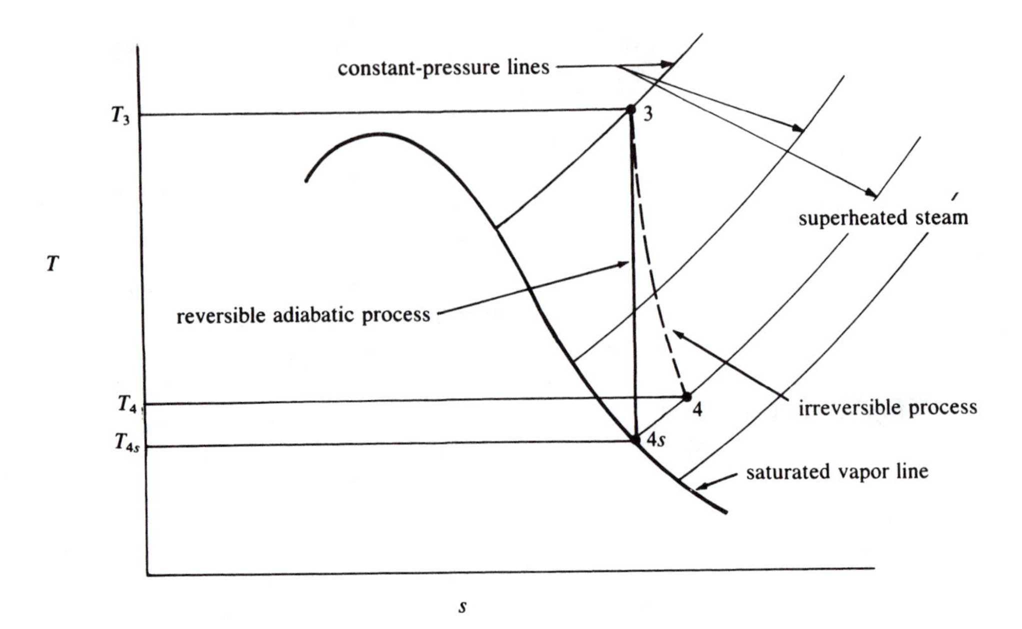

FIGURE 11–5 Expansion processes through a turbine

We use the reversible adiabatic equation as the ideal process and use the adiabatic efficiency, defined earlier by the equation

$$\eta_{\text{turb}} = \frac{wk_{\text{act}}}{wk_{\text{ideal}}} \times 100 = \frac{h_3 - h_4}{h_3 - h_{4s}} \times 100 \tag{11–12}$$

where h_{4s} represents the enthalpy after a reversible adiabatic expansion through the turbine.

EXAMPLE 11–1 A steam turbine receives steam at 8 MPa and 720°C. If the exhaust pressure of the turbine is 100 kPa and the adiabatic efficiency is 92%, determine the actual turbine work per kilogram of steam. Compare your answer with one you obtained by using *EES*.

Solution We may use equation (11–12) and solve for the actual work:

$$wk_{\text{act}} = (\eta_{\text{turb}})(h_3 - h_{4s})$$

From table B–12, we find h_3 at 8 MPa and 720°C to be 3929 kJ/kg. Also, the entropy at that state is 7.329 kJ/kg·K, and the entropy at state 4*s* will also be 7.329 kJ/kg·K. The pressure at state 4*s* is the same as at the exhaust, state 4, which is 100 kPa. We find, from table B–11, (and table B–12), that $s_{4s} < s_g$ at 100 kPa, s_g being 7.360 kJ/kg·K at 100 kPa. The quality at state 4*s* is then found to be

$$\begin{aligned} x_{4s} &= \frac{s_{4s} - s_f}{s_{fg}} \\ &= \frac{7.329 - 1.303 \text{ kJ/kg}\cdot\text{K}}{6.057 \text{ kJ/kg}\cdot\text{K}} \\ &= 99.488\% \end{aligned}$$

Next, the enthalpy h_{4s} is determined:

$$\begin{aligned} h_{4s} &= x_{4s}h_{fg} + h_f \\ &= (0.99488)(2258 \text{ kJ/kg}) + 417 \text{ kJ/kg} \\ &= 2663.44 \text{ kJ/kg} \end{aligned}$$

The actual turbine work can then be obtained:

$$wk_{act} = (0.92)(3929 - 2663.44 \text{ kJ/kg})$$

Answer

$$= 1164.3 \text{ kJ/kg}$$

Using *EES Equation* windows and checking that SI units with temperature in degrees Celsius and pressure in kPa, we enter the following equations:

$$h_1 = \text{enthalpy (steam}, T = 720, p = 8000)$$
$$s_1 = \text{entropy (steam}, T = 720, p = 8000)$$
$$s_{2s} = s_1$$
$$h_{2s} = \text{enthalpy (steam}, p = 100, s = s_{2s})$$
$$w = 0.92*(h1 - h2s)$$

Then, clicking the *Calculate* option and *Solve* gives an answer that is in agreement with the answer obtained by using the appendix tables,

$$w = 1164 \text{ (kJ/kg)}$$

We have seen some examples like this earlier, and the methods of solving them should be familiar to you.

11–4 PUMPS

In the Rankine steam power cycle, the steam is exhausted from the turbine to a low pressure (perhaps at atmospheric pressure or lower), and the boiler is furnished liquid water at a high pressure. To raise the steam or water from a low to a high pressure and then force it into the boiler is the task of the pump. In some cycles, we may call the device a **compressor**, but here we are handling liquid water, and **pump** is a more common term. Pumps are continuous-flow mechanisms that are essentially reversed turbines, or they are piston-cylinder devices that provide intermittent flow. In either case, we may consider the pump to be a steady-flow open system when observed over a sufficient time span. Applying the steady-flow energy equation, neglecting kinetic and potential energy changes, and assuming an adiabatic compression in the pump, we obtain equation (11–4):

$$wk_{pump} = h_1 - h_2$$

The power required to convey $\dot{m}$ flow rate of water through the pump is determined from the equation

$$\dot{W}k_{pump} = \dot{m}(h_1 - h_2) \qquad \textbf{(11–13)}$$

If the pump is not an ideal one, its adiabatic efficiency is defined as

$$\eta_{pump} = \frac{\dot{m}(h_1 - h_{2s})}{\dot{W}k_{act}}$$

and h_{2s} is the enthalpy if the pump were reversible and adiabatic, or ideal.

In many cases the water passing through the pump is saturated liquid. In this state, water is nearly incompressible (the density or specific volume remains constant with changes in pressure), so we can write the general equation for work in an open system as

$$wk = -\sum v\,\delta p$$

and since $v =$ constant, we have

$$wk_{pump} = -v \sum \delta p = -v(p_2 - p_1) \qquad \textbf{(11–14)}$$

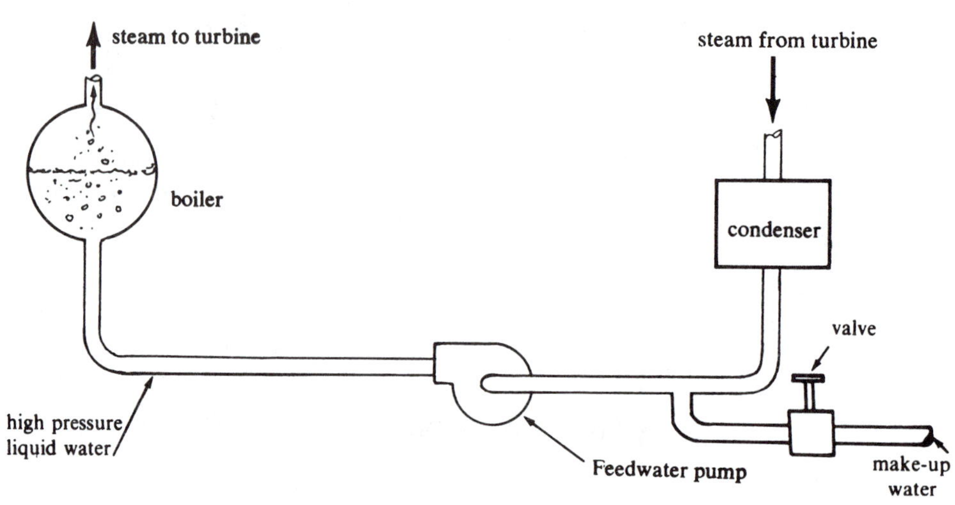

FIGURE 11–6 Pumps in a steam turbine cycle

A discrepancy may occur between the answers of equations (11–4) and (11–14) for pump work, in which case the answer obtained from (11–4) should be used. The reason for this discrepancy is that in equation (11–14), the specific volume is assumed constant, which is incorrect for all real fluids or gases. If we identified v the average (v_{av}), we would be correct to use equation (11–14) and obtain agreement with equation (11–4), but the average specific volume is difficult to predict in some problems.

Pumps are normally driven by electric motors or small steam turbines. In either case, the pump work must be accountable in determining the net cycle work. Also, if some of the H_2O is lost due to leakage (usually through turbine seals) in the cycle, fresh feed water can be added to the cycle as shown in figure 11–6.

11–5 CONDENSERS

Steam exhausted from a turbine in a Rankine cycle is pumped back into the boiler to recycle the water. If the steam exhausted by the turbine were furnished directly to the feedwater pump, however, the ideal work required by the pump to deliver high-pressure steam to the boiler would be exactly the work obtained from the reversible adiabatic expansion in the turbine operating between the same pressures. In effect, then, one steam turbine would be using all its output to drive the pump, and the cycle work would be zero. If there were irreversibilities in the cycle, the steam turbine work would not be enough to drive the pump, and we would then need to add work from the surroundings. To prevent these undesirable results, a condenser receives the exhaust steam, condenses this steam to a liquid, and then feeds this liquid water to the pump. This device is shown in the cycle of figure 11–3, where it is indicated that heat is rejected. This is the precise role of the condenser—it transfers heat to the surroundings so that the heat engine can operate. We have seen that the second law of thermodynamics demands a transfer of heat between two regions for any cyclic heat engine. Of course, a mechanical device does not obey a law. Rather, a law obeys the operation of a mechanical device, but without the condenser, the steam turbine cycle is involved (ideally) in only one heat transfer, that at the boiler. In any case, if the condenser is used as an open steady-flow system and if kinetic and potential energy changes in the system are neglected, the first law of thermodynamics gives us

$$q_{rej} = h_1 - h_4 \tag{11–3}$$

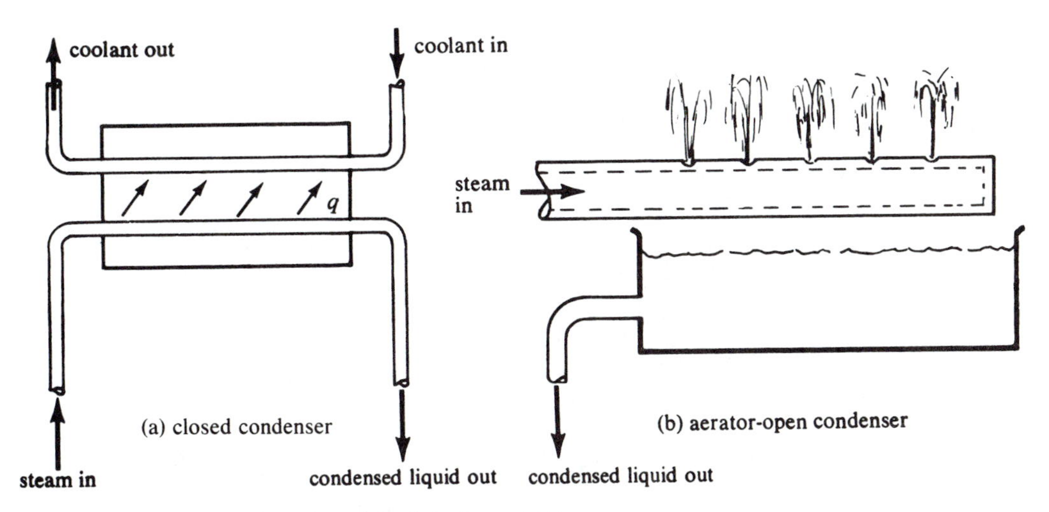

FIGURE 11–7 Types of condensers in their elemental form

The rate of heat transfer $\dot{Q}_{rej}$ can be obtained from

$$\dot{Q}_{rej} = \dot{m}(h_1 - h_4) \tag{11–15}$$

where $\dot{m}$ is the mass flow through the condenser. The mechanical reason that the condenser reduces the pump work, and thus makes the steam turbine practical, is that the steam vapor is converted (condensed) into a liquid. Consequently, the state of the medium leaving the condenser should ideally be a saturated liquid, and then enthalpy h_1 so determined.

The condenser can be built in a number of different physical configurations. It may be a closed condenser, as shown in figure 11–7a, which transfers heat from the steam to a coolant across an intermediate boundary. The coolant may be air, river water, or some other fluids, but it is not in direct contact with the working medium (H_2O). Most condensers are constructed in this general manner. A variation of this device is the cooling tower, which relies on heat transfer to the surrounding air.

A second method of manifesting the phase change of steam vapor to liquid is the aerator or open condenser, illustrated in figure 11–7b. This device literally exhausts the steam into the surroundings; due to this intimate mixing, the heat transfer is rapid and the condensing of the steam to liquid is equally rapid. The drawback to this type of condenser, although it is less expensive than the closed condenser, is that the exhaust pressure of the turbine cannot be less than the atmospheric pressure. The closed condenser can operate at pressures less than atmospheric and thus allow for exhaust vacuum pressures in the turbine and increase cycle efficiencies.

11–6 STEAM AS A WORKING FLUID

If you have not already read and studied the material of chapter 5, it would be good to go back and review sections 5–1, 5–2, and 5–5. We will be making use of much of the material presented in these three sections.

The working medium used in the steam turbine cycle is water. Chemically, we symbolize it as H_2O and then call it a pure substance if no impurities are present. Water, of course, can take the three common phases with which we are familiar—solid (or ice), liquid, and vapor (or steam). The phase diagram of water is shown in figure 11–8. A simplified version of this diagram is shown in figure 5–2. Notice that at the triple point all phases converge and can exist in equilibrium, and beyond the critical point (to the right of it) the liquid and vapor phases cannot be differentiated. For our purposes, the liquid–vapor phase change and the vapor or steam phase

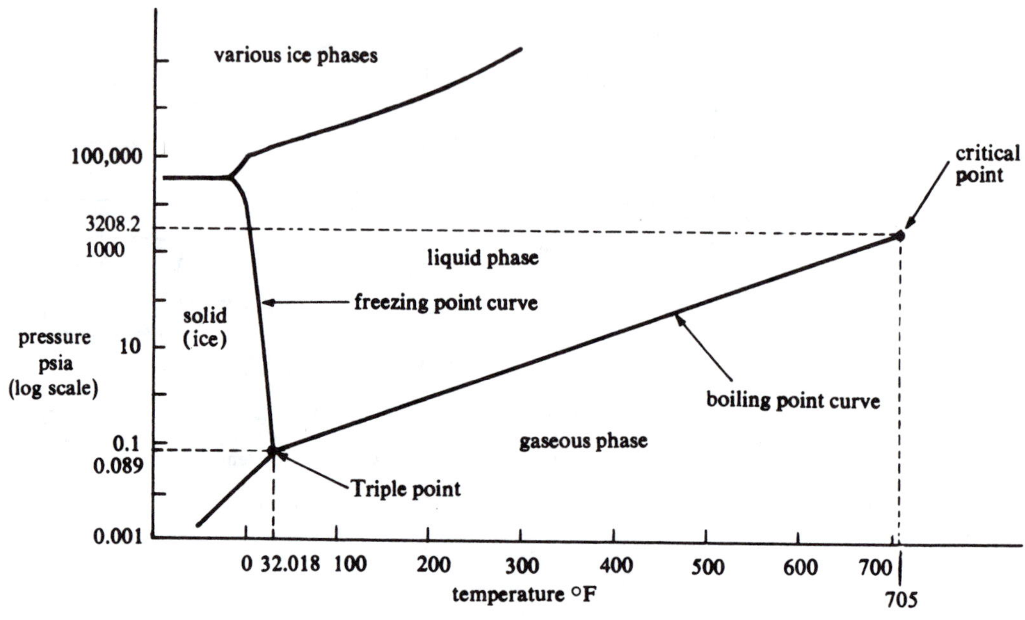

FIGURE 11–8 Phase diagram for water (H_2O) (not to scale)

itself will be sufficient. Suppose, for instance, that we have a constant-pressure container of water. If we add energy to the contents, monitoring the temperatures as we add energy, we notice that the water begins to have an increased temperature. In figure 11–9, we see the result of this plotted on a *T–E* diagram, indicated by the curve *A–B*. At point *B*, the water begins to boil; that is, it changes from a liquid to a vapor. If the boiler pressure is 101 kPa, the temperature at point *B* will be 373 K; but if the pressure is higher (or lower) than 101 kPa, the boiling temperature will be greater than (or less than) the 373 K. Various isobaric lines are shown in the graph of figure 11–9 to show this effect.

Once the water begins to boil, we may keep adding energy without changing the temperature until we reach a point when all the water is now converted to vapor or steam. This is the state *C* on the isobaric line, and if we continue adding energy, the temperature increases again

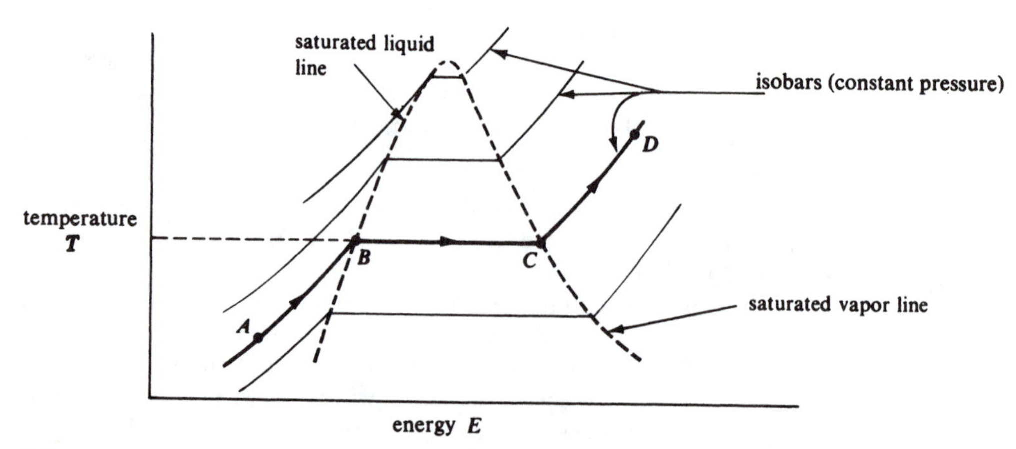

FIGURE 11–9 Temperature–energy relation of material going through the liquid–gas phase change at various pressures

for the steam. Any particular state beyond point C is called a **superheated steam state** (such as point D), and the unique points B and C are called the **saturated liquid** and **saturated vapor state**, respectively. We could connect the various saturated liquid and vapor points for differing pressures, as indicated by the dashed line in figure 11–9, labeled the *saturation line*. In figures 11–10 and 11–11 are shown property diagrams in which this saturation line is clearly indicated.

A more detailed view of the p–v diagram of water, including the solid phase of water, is shown in figure 5–5. At the top of the dome-shaped saturation line is shown the **critical point**, which is the point where the phase change becomes undefined. The T–s diagram of figure 11–10 depicts isobaric lines, as in figure 11–9. Also, the area under the T–s diagram curves should be visualized as "heat transfers," as we have always done.

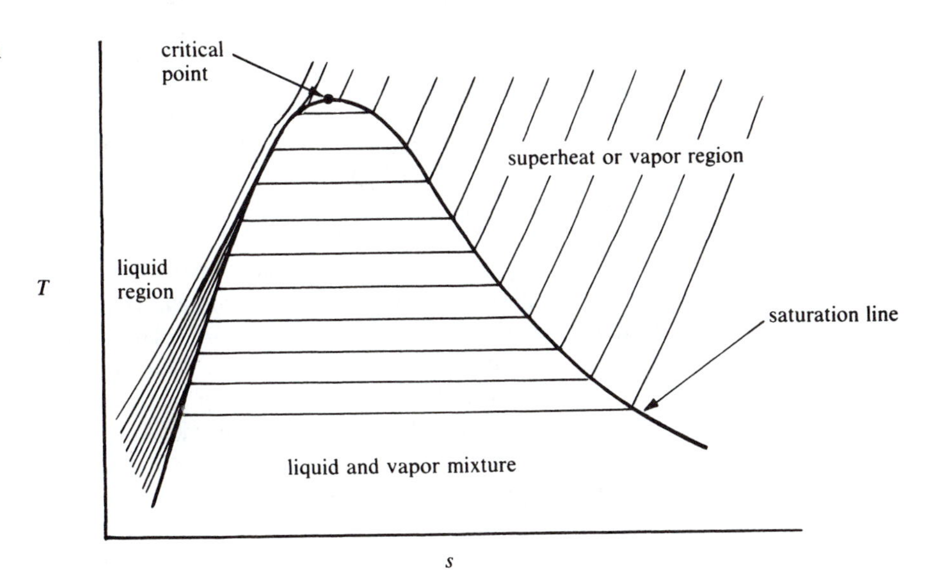

FIGURE 11–10 T–s diagram of steam–water phase change

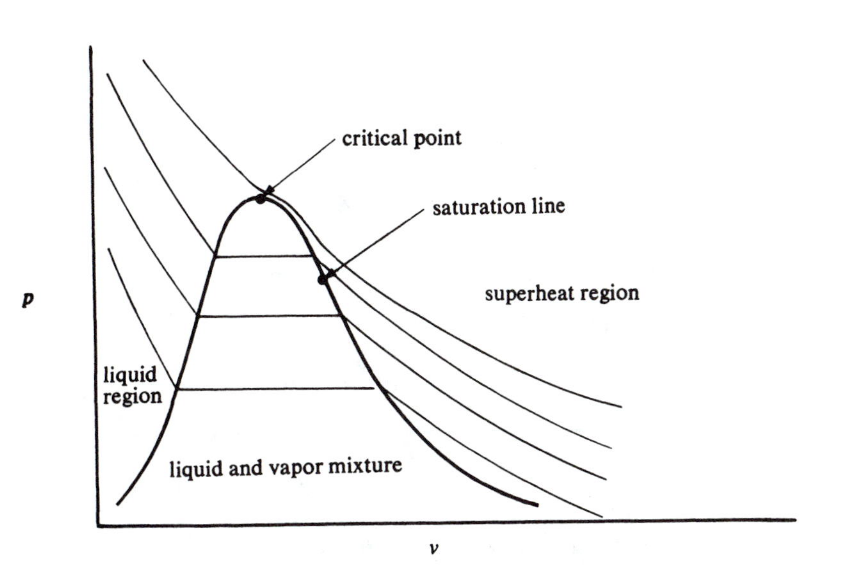

FIGURE 11–11 p–v diagram of steam–water phase change

In section 5–2, we discussed the following terms, which we will now use often:

Superheated steam or vapor region of water

Saturated vapor or steam that is all vapor but is not superheated

Saturated liquid or water that is liquid but is just ready to boil or vaporize

Quality, a ratio of the mass of vapor to the total mass of steam that is boiling or in the phase transition between liquid and vapor

Compressed liquid or water that is below the boiling point; also sometimes called subcooled liquid

Appendix tables B–11, B–12, and B–13 will furnish the data for the properties of water at saturated conditions, superheated steam conditions, and high-temperature liquid conditions. In table B–11 are listed the saturation properties as functions of temperature as well as of pressure. The properties specific volume, enthalpy, and entropy are tabulated. Notice that the subscript f denotes saturated liquid and g denotes saturated vapor. For enthalpy and entropy, the third and fourth terms are tabulated, h_{fg} and s_{fg}. These are the differences of the saturated states and are defined as

$$h_{fg} = h_g - h_f \tag{11–16}$$

and

$$s_{fg} = s_g - s_f \tag{11–17}$$

The term h_{fg} is obviously the measure of the energy added or lost during the vapor–liquid phase change. It is referred to as the **heat of vaporization**, **heat of condensation**, **latent heat**, **latent heat of vaporization**, or **latent heat of condensation**.

Through the property data above, we may determine the precise values of the properties of a state intermediate between saturated liquid and saturated vapor, provided that we know the quality χ of the state. We defined quality previously (see section 5–5) as the ratio of the vapor mass to the mixture mass. The state we wish to define, state a in this case, is depicted in figure 11–12 by use of the entropy property. Here we can write

$$100 \times \frac{s_a - s_f}{s_g - s_f} = \chi\% = \text{quality at state } a \text{ in percent}$$

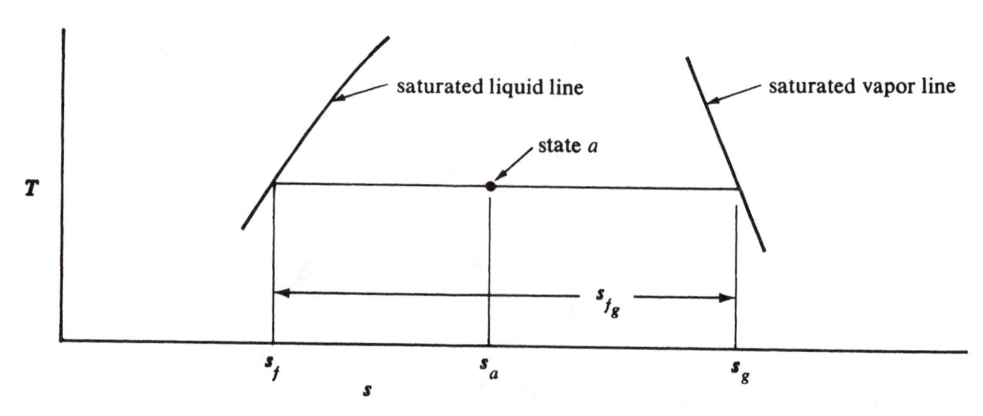

FIGURE 11–12 Relation between the entropy of a state intermediate between a saturated liquid and vapor and the entropy of the saturation points

and then

$$s_a = \frac{\chi}{100}(s_g - s_f) + s_f$$

or

$$s_a = \frac{\chi}{100}s_{fg} + s_f \tag{11–18}$$

Similarly, we can write

$$h_a = \frac{\chi}{100}h_{fg} + h_f \tag{11–19}$$

$$v_a = \frac{\chi}{100}(v_g - v_f) + v_f \tag{11–20}$$

and

$$u_a = \frac{\chi}{100}(u_g - u_f) + u_f \tag{11–21}$$

Internal energy can easily be calculated from the data in tables B–11 and B–12 by recalling that $u = h - pv$. The superheated steam properties are given in table B–12, where the pressure and temperature together determine the particular state. Specific volume, enthalpy, and entropy are tabulated at each of these points.

Subcooled liquid properties are listed in an abbreviated table B–13, that furnishes the data of differences between saturated liquid and the specified subcooled or compressed state properties. In section 5–5, we looked at various examples of using property tables for pure substances. Here we will review those methods as they apply to steam power generation.

EXAMPLE 11–2 A steam-generating unit supplies 1000 lbm/min of superheated steam at 400 psia and 800°F. Determine the entropy, enthalpy, and specific heat of the steam.

Solution From table table B–12 we obtain the answers:

$$s = 1.6850 \text{ Btu/lbm} \cdot {}^\circ\text{R}$$

$$h = 1417.0 \text{ Btu/lbm}$$

Answer

$$v = 1.8151 \text{ ft}^3/\text{lbm}$$

EXAMPLE 11–3 Steam is condensing at a pressure of 0.08 MPa. If the quality of the steam is 30% at a particular instant, determine the temperature, enthalpy, entropy, and internal energy of the steam. Compare your answer with the one you obtained by using *EES*.

Solution The steam is in the phase change, so we obtain the required data from the saturated steam table B–11. At a pressure of 0.08 MPa, the temperature is 93.5°C. We then use equations (11–19), (11–18), and (11–21):

$$h_a = \frac{\chi}{100}h_{fg} + h_f$$

$$s_a = \frac{\chi}{100}s_{fg} + s_f$$

$$u_a = \frac{\chi}{100}(u_g - u_f) + u_f$$

Next, from table B–11, we find that

$$h_{fg} = 2273 \text{ kJ/kg} \qquad h_f = 392 \text{ kJ/kg} \qquad h_g = 2665 \text{ kJ/kg}$$
$$s_{fg} = 6.201 \text{ kJ/kg}\cdot\text{K} \qquad s_f = 1.233 \text{ kJ/kg}\cdot\text{K} \qquad v_g = 2.09 \text{ m}^3\text{/kg}$$
$$u_f = h_f - v_f p \qquad v_f = 0.001039 \text{ m}^3\text{/kg}$$
$$= 392 \text{ kJ/kg} - (0.001039 \text{ m}^3\text{/kg})(80 \text{ kN/m}^2)$$
$$\simeq 392 \text{ kJ/kg}$$
$$u_g = h_g - v_g p$$
$$= 2665 \text{ kJ/kg} - (2.09 \text{ m}^3\text{/kg})(80 \text{ kN/m}^2)$$
$$= 2497.8 \text{ kJ/kg}$$

From these results, we compute

Answer
$$h_a = (0.30)(2273 \text{ kJ/kg}) + 392 \text{ kJ/kg} = 1074 \text{ kJ/kg}$$

Answer
$$s_a = (0.30)(6.201 \text{ kJ/kg}\cdot\text{K}) + 1.233 \text{ kJ/kg}\cdot\text{K} = 3.093 \text{ kJ/kg}\cdot\text{K}$$

Answer
$$u_a = (0.30)(2497.8 - 392) \text{ kJ/kg} + 392 \text{ kJ/kg} = 1023.74 \text{ kJ/kg}$$

Using *EES Equation* windows and checking that SI units with temperature in degrees Celsius and pressure in kPa, we enter the following equations:

$$T = \text{temperature (steam, } p = 80, x = 0.3)$$
$$h = \text{enthalpy (steam, } p = 80, x = 0.3)$$
$$s = \text{entropy (steam, } p = 80, x = 0.3)$$
$$U = \text{intenergy (steam, } p = 80, x = 0.3)$$

Then, clicking the *Calculate* option and *Solve* gives answer that are in good agreement with the answer obtained by using the appendix tables:

$$T = 93.48 \quad (°\text{C})$$
$$h = 1074 \quad (\text{KJ/kg})$$
$$s = 3.093 \quad (\text{KJ/kg}\cdot\text{K})$$
$$U = 1023 \quad (\text{kJ/kg})$$

EXAMPLE 11–4 A feedwater pump compresses saturated liquid water at 200°F to 400 psia and 200°F. Determine the enthalpy of the water leaving the pump. Compare your answer with the one you obtained by using *EES*.

Solution The enthalpy of the entering fluid is easily found from table B–11:

$$h_f = 168.1 \text{ Btu/lbm}$$

Note that the pressure is 11.53 psia at this point. The difference between the enthalpy of the exit water and that at the inlet is obtained from table B–13:

$$h - h_f = +0.88 \text{ Btu/lbm}$$

Then

$$h = +0.88 + h_f = 0.88 + 168.1$$

Answer
$$= 168.98 \text{ Btu/lbm}$$

Using *EES Equation* windows and checking that English units with temperature in degrees Fahrenheit and pressure in psia, we enter the following equations:

$$h = \text{enthalpy}\,(\text{steam}, T = 200, p = 400)$$

Then, clicking the *Calculate* option and *Solve* gives an answer, $h = 169$ (Btu/lbm), that is in good agreement with the answer obtained by using the appendix tables.

EXAMPLE 11–5 Determine the work done on each kilogram of steam for a pump isentropically compressing water from a saturated liquid at 90°C to 10 MPa. Compare your answer with the one you obtained by using *EES*.

Solution From equation (11–4),

$$wk_{\text{pump}} = h_1 - h_2$$

we may use the values for enthalpy listed in tables B–11 and B–13. From table B–11, we find that at 90°C,

$$h_f = h_1 = 377 \text{ kJ/kg}$$

and then, from table B–13,

$$h - h_f = 10.2 \text{ kJ/kg} \quad \text{at 90°C and 10.0 bar } (= 10 \text{ Mpa})$$

or

$$h = h_2 = 377 \text{ kJ/kg} + 10.2 \text{ kJ/kg} = 387.2 \text{ kJ/kg}$$

Thus, the pump work is simply

$$wk_{\text{pump}} = 377 \text{ kJ/kg} - 387.2 \text{ kJ/kg}$$

Answer

$$= -10.2 \text{ kJ/kg}$$

This is the value listed in table B–13 as the pump work. An alternative solution to obtaining the pump work is to assume a constant-volume compression and write

$$wk_{\text{pump}} = (v_{\text{av}})(p_1 - p_2)$$

where v_{av} is the average value of the specific volume of the steam. For this case, $p_1 = 70$ kPa, $p_2 = 10$ MPa, and v_{av} is obtained approximately as follows:

$$v_{\text{av}} = \frac{v_1 + v_2}{2} = \frac{0.00104 + 0.00145 \text{ m}^3/\text{kg}}{2} = 0.00125 \text{ m}^3/\text{kg}$$

Then the pump work is

$$wk_{\text{pump}} = (0.00125 \text{ m}^3/\text{kg})(99.3 \times 10^5 \text{ N/m}^2)$$

Answer

$$= -12.4 \text{ kJ/kg}$$

Notice that there is a discrepancy between the two methods. The first answer should be accepted as the more accurate, but either method will give results that are acceptable for most engineering work.

Using *EES Equation* windows and checking that SI units with temperature in degrees Celsius and pressure in kPa, we enter the following equations:

$$h_f = \text{enthalpy (steam, } T = 90, p = x = 0.0)$$
$$s_f = \text{entropy (steam, } T = 90, x = 0.0)$$
$$s = s_f$$
$$h = \text{enthalpy (steam, } p = 10000, s = s)$$
$$w = h_f - h$$

Then, clicking the *Calculate* option and *Solve* gives an answer $w = -10.26$ (kJ/kg), that is in good agreement with the answer obtained by using the appendix tables.

EXAMPLE 11–6 A boiler supplies 1000 kg/m steam at 3 MPa and 720°C. It burns coal having a lower heating value of 25,000 kJ/kg coal. If the efficiency of the boiler is 90% and the water supplied to the boiler has an enthalpy of 600 kJ/kg, determine the rate of coal consumption.

Solution Let us first define the generating unit and the fluxes involved in it. (See figure 11–13.) We find the heat added from

$$\dot{Q}_{add} = \dot{m}_{steam}(h_3 - h_2)$$

and from table B–12 we find that

$$h_3 = 3957 \text{ kJ/kg}$$

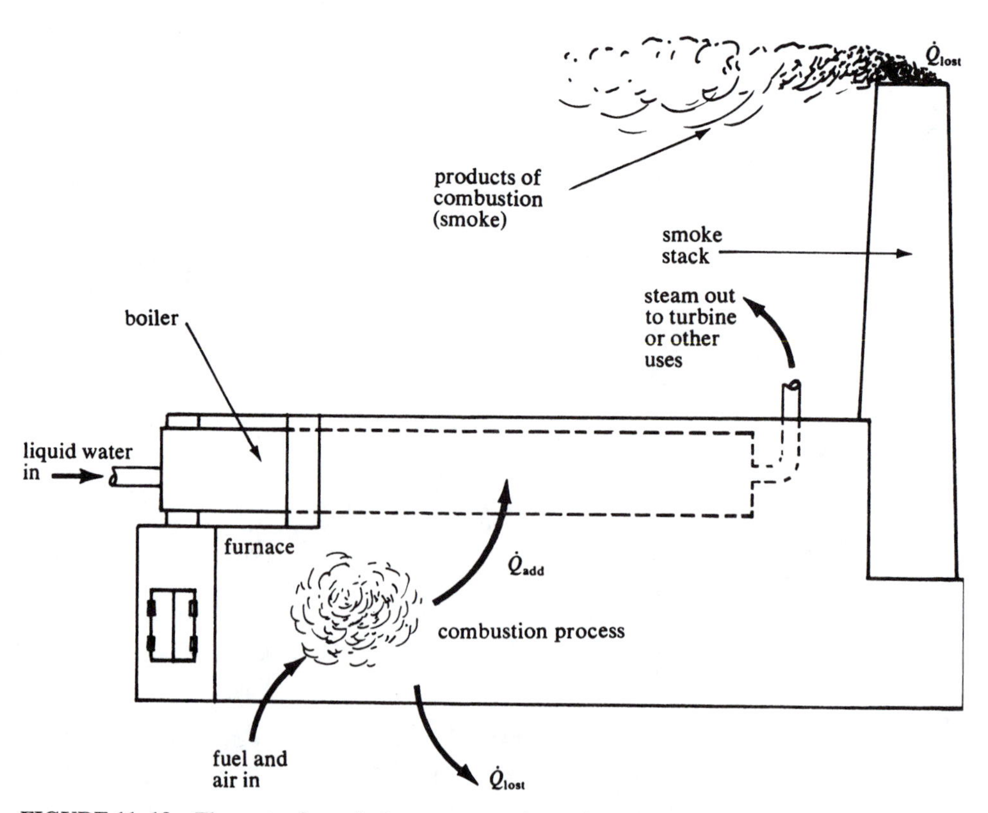

FIGURE 11–13 Elements of a typical steam generating unit

Then

$$\dot{Q}_{add} = (1000\ \text{kg/m})(3957\ \text{kJ/kg} - 600\ \text{kJ/kg})$$
$$= 3{,}357{,}000\ \text{kJ/m}$$

The boiler efficiency was defined by equation (11–9), which can be written

$$\eta_{boiler} = \frac{\dot{Q}_{add}}{\dot{m}_{fuel}(\text{LHV})} \times 100$$

For this problem, we have

$$90\% = \frac{3{,}375{,}000\ \text{kJ/m}}{\dot{m}_{fuel}(25{,}000\ \text{kJ/kg})} \times 100$$

and the mass of fuel used is then

Answer

$$\dot{m}_{fuel} = 149.2\ \text{kg/m}$$

Many people prefer to extract data from graphs or charts rather than from tables. The steam tables contain the required data to analyze properly the behavior of steam in the vapor–liquid, superheated, and subcooled regions, and the Mollier diagram duplicates this information in chart form. This diagram is included in appendix B in a reduced form on chart B–3. Mollier diagrams are quite popular in sizes fourfold to tenfold larger than given here, but the basic method for use of this chart can be grasped from this small version. Notice that the Mollier diagram is essentially an *h–s* diagram with various isolines, as illustrated in figure 11–14. Notice that the saturation line and the critical

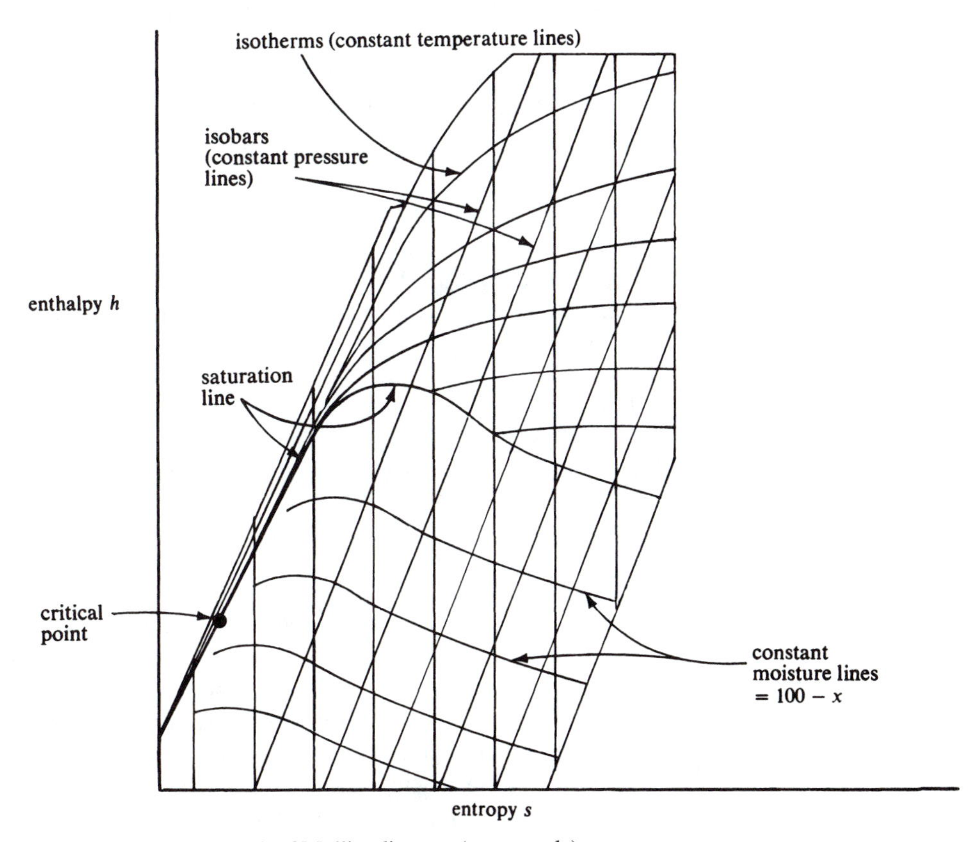

FIGURE 11–14 Sketch of Mollier diagram (not to scale)

point are identified on the sketch as well as on the detailed Mollier diagram. The region below the saturation line, the phase change, is crisscrossed by isobars and constant-moisture lines. Using these lines, we may easily find enthalpy and entropy from a known pressure and quality, since the moisture is related to quality by the equation

$$\text{moisture percentage} = 100 - x \tag{11-22}$$

The isobars extend into the superheated steam region above the saturation line. This region is also mapped by isotherms, and between a given temperature and pressure, the enthalpy and entropy may be found from the Mollier diagram.

One of the most useful characteristics of the Mollier diagram is the tracing of reversible adiabatic steam turbine processes. These follow vertical lines (constant entropy), so from the given initial and final pressures, we may easily extract enthalpy from the chart. There are other conveniences to be discovered from this useful graph, and those who have trouble relating tabulated data with actual physical processes may find the Mollier diagram much more convenient and more descriptive of the steam processes. We use both the steam table data and the Mollier diagram data in the following example problems.

EXAMPLE 11–7 Determine from the Mollier diagram the percent moisture in steam expanded isentropically from 800°F and 1000 psia to 2 psia.

Solution Referring to chart B–3, we find the entropy of steam at 800°F and 1000 psia to be 1.56 Btu/lbm · °R. Following this entropy line vertically down the chart until it intersects the 2.0-psia line, we then read

Answer 79.8% vapor or a moisture percentage of 100% − 79.8% = 20.2%

EXAMPLE 11–8 A steam turbine expands steam at 5000 kPa and 640°C isentropically to 0.2 bar. Determine the work done by the turbine through this expansion.

Solution From figure 11–3, the work of the turbine is

$$wk_{\text{turb}} = h_3 - h_4$$

We will obtain the enthalpy values from the Mollier diagram, although we could find them from table B–11 or B–12. From chart B–3 (the Mollier diagram), we find that $h_3 = 3750$ kJ/kg. Notice that the SI pressure units on chart B–3 are the "bar." The bar is defined as

$$1 \text{ bar} = 100 \text{ kPa}$$

The state at (4) can be found by following the vertical entropy line down the chart, beginning at state 3 and ending at the intersection with the 0.2-bar line. The enthalpy reads $h_4 = 2390$ kJ/kg. The work of the turbine is then

$$wk_{\text{turb}} = 3750 \text{ kJ/kg} - 2390 \text{ kJ/kg}$$

Answer

$$= 1360 \text{ kJ/kg}$$

Finally, the properties of steam can be found from various commercially available computer software packages for microcomputers or mainframe computers. These programs normally compute the steam properties from algebraic expressions or equations rather than store and retrieve tabulated values of the properties at fixed pressures and temperatures. The programs usually require of the user the values for two properties, so using the computer packages for steam involves the same thinking as is required for using the steam tables. The primary advantage of computer routines results from using them as subroutines or larger programs that address the

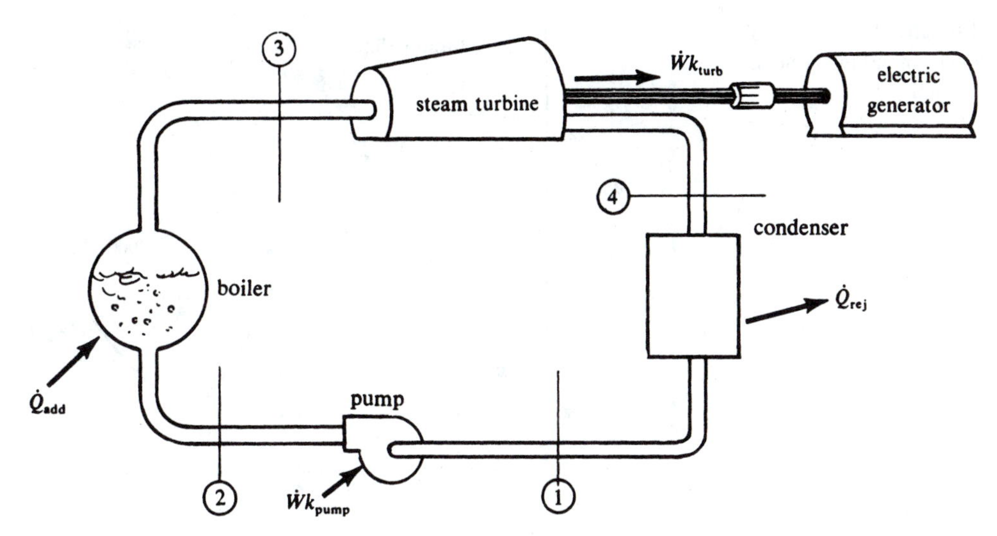

FIGURE 11–15 Simple steam turbine power generator

Rankine cycle or other thermodynamic cycles. The program RANKINE uses a version of these steam property programs as a subroutine. RANKINE, described in appendix A–11, treats the steam turbine power cycles and is available on soft diskette from the publisher.

11–7 ANALYSIS OF STEAM POWER GENERATION CYCLES

The simple steam turbine power generator, shown in figure 11–15, is identical to the typical closed steam turbine cycle shown in figure 11–3. The simple steam turbine operates on the Rankine cycle, which was defined earlier. A study of figure 11–15 shows that the device satisfies the requirements for calling it a heat engine: heat is added in the boiler, heat is rejected in the condenser, and work is produced by a steam turbine. Also, work is required of a pump so that the net work would be the sum of the turbine and pump work (where turbine work is positive and pump work is negative).

The following example problems should tie together the various components in the simple steam turbine cycle and serve to show the use of the steam data in analyzing the complete cycle.

EXAMPLE 11–9 An ideal steam turbine uses 5000 kg/h of steam. The steam is superheated to 560°C and 2000 kPa when supplied to the turbine. The exhaust temperature is 60°C, and the fluid leaving the condenser is assumed to be a saturated liquid. Determine the power generated by the cycle, the rate of heat rejected, the amount of coal consumed if the boiler unit is assumed to be 100% effective while the LHV of coal is 30,000 kJ/kg, and the steam rate.

Solution This heat engine is sketched in figure 11–15, and the property diagrams are depicted in figures 11–16 and 11–17. The property data labeled in the diagrams are either given in the problem statement or easily obtained from the steam tables. The power generated by the cycle is obtained from

$$\dot{W}k_{\text{cycle}} = \dot{m}wk_{\text{cycle}}$$

where

$$wk_{\text{cycle}} = wk_{\text{turb}} + wk_{\text{pump}}$$

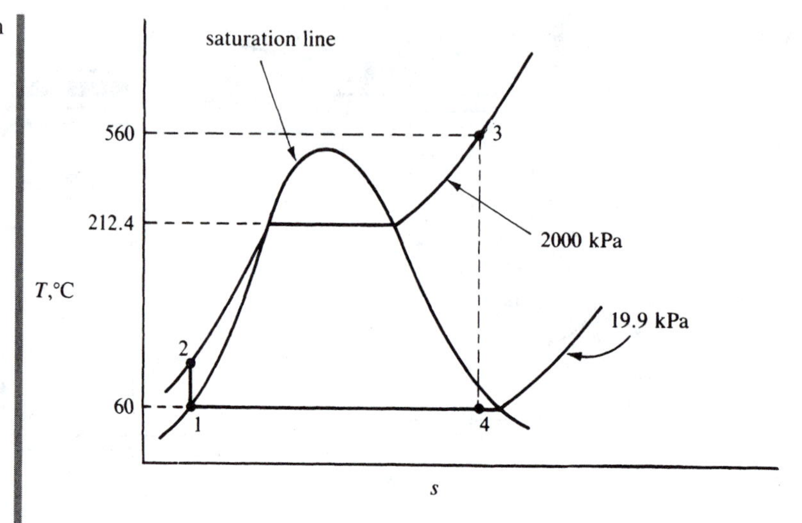

FIGURE 11–16 *T–s* diagram of steam turbine cycle of example 11–9

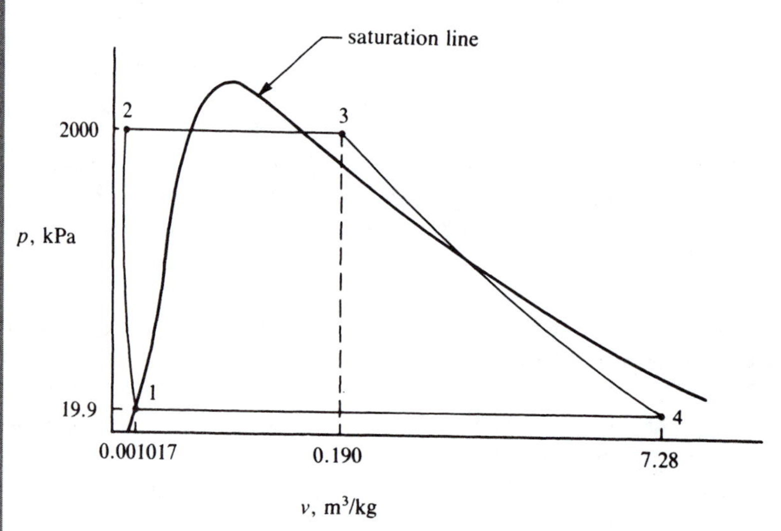

FIGURE 11–17 *p–V* diagram of steam turbine cycle of example 11–9

The ideal cycle involves adiabatic processes in the pump and turbine, so we use equations (11–2) and (11–4). From table B–11, at 60°C, we obtain the value

$$h_1 = h_{f1} = 251 \text{ kJ/kg}$$

From table B–12, at 560°C and 2000-kPa (2.0-MPa) pressure,

$$h_3 = 3600 \text{ kJ/kg}$$

and from table B–13, we interpolate for a value of $h - h_f$. Thus, at a temperature of 60°C, for a pressure of 2.0 MPa,

$$h_2 = h_1 + 2.0 \text{ kJ/kg} = 253 \text{ kJ/kg}$$

From the Mollier diagram, we find h_4 by first locating the position of state 3 in the superheated region. Then we follow the entropy line vertically down to the pressure of 19.9 kPa. This technique is shown in figure 11–18, where we also see that the moisture can be determined. We read

$$h_4 = 2460 \text{ kJ/kg}$$

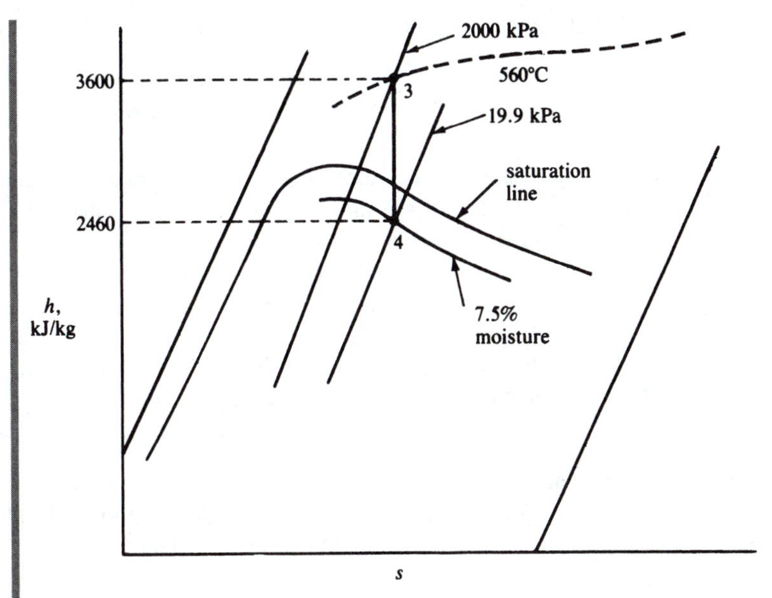

FIGURE 11–18 Mollier diagram of reversible adiabatic expansion in example 11–9 (not to scale)

and the moisture is 7.5%, so $\chi = 92.5\%$. Then

$$\begin{aligned} wk_{\text{cycle}} &= h_3 - h_4 + h_1 - h_2 \\ &= 3600 \text{ kJ/kg} - 2460 \text{ kJ/kg} + 251 \text{ kJ/kg} - 253 \text{ kJ/kg} \\ &= 1138 \text{ kJ/kg} \end{aligned}$$

The power generated by the steam turbine is then

$$\begin{aligned} \dot{W}k_{\text{cycle}} &= \dot{m}wk_{\text{cycle}} = (500 \text{ kg/h})(1138 \text{ kJ/kg}) \\ &= 5.69 \times 10^6 \text{ kJ/h} \end{aligned}$$

or

Answer
$$\dot{W}k_{\text{cycle}} = 1580 \text{ kW}$$

From equation (11–3),we obtain

$$\begin{aligned} q_{\text{rej}} &= h_1 - h_4 = 251 \text{ kJ/kg} - 2460 \text{ kJ/kg} \\ &= -2209 \text{ kJ/kg} \end{aligned}$$

and then

Answer
$$\dot{Q}_{\text{rej}} = \dot{m}q_{\text{rej}} = 11.045 \times 10^6 \text{ kJ/h}$$

We get the rate of heat added from

$$\begin{aligned} \dot{Q}_{\text{add}} &= \dot{m}q_{\text{add}} = \dot{m}(h_3 - h_2) \\ &= (5000 \text{ kg/h})(3600 \text{ kJ/kg} - 253 \text{ kJ/kg}) \\ &= 16.735 \times 10^6 \text{ kJ/h} \end{aligned}$$

Since the boiler has a 100% effectiveness, we may write

$$\dot{m}_{\text{fuel}}(\text{LHV}) = \dot{Q}_{\text{add}} = 165.735 \times 10^6 \text{ kJ/h}$$

Hence, for an LHV of 30,000 kJ/kg,

Answer
$$\dot{m}_{\text{fuel}} = 557.83 \text{ kg/h}$$

The thermodynamic efficiency can be obtained from

$$\eta_T = \frac{Wk_{\text{cycle}}}{\dot{Q}_{\text{add}}} \times 100 = 34\%$$

The steam rate may be computed from equation (11–10):

Answer

$$\dot{m}_{sr} = \frac{\dot{m}_{steam}}{\dot{W}k_{cycle}} = \frac{500 \text{ kg/h}}{1580 \text{ kW}} = 3.16 \text{ kg/kW} \cdot \text{h}$$

The Rankine cycle is a description of the operation of a simple steam turbine power generator. To make the analysis more descriptive of the actual operation, however, the Rankine cycle may be modified by considering the inefficiencies or irreversibilities of the boiler, turbine, and pump. The following example considers the same device as that of example 11–9 but with modifications to the simple Rankine cycle analysis.

EXAMPLE 11–10 Consider the steam turbine power generator of example 11–9 with exhaust steam from the turbine at 99% quality. Determine the power produced, the turbine efficiency, the steam rate, and the minimum entropy generation. Compare your answers with the one you obtained by using *EES*.

Solution The system is defined in figure 11–15; however, our property diagrams given in figures 11–19 and 11–20 deviate slightly at state 4 from those of example 11–9. The enthalpy values correspond to the data of example 11–9, except at state 4:

$$h_1 = 251 \text{ kJ/kg}$$
$$h_2 = 253 \text{ kJ/kg}$$
$$h_3 = 3600 \text{ kJ/kg}$$

FIGURE 11–19

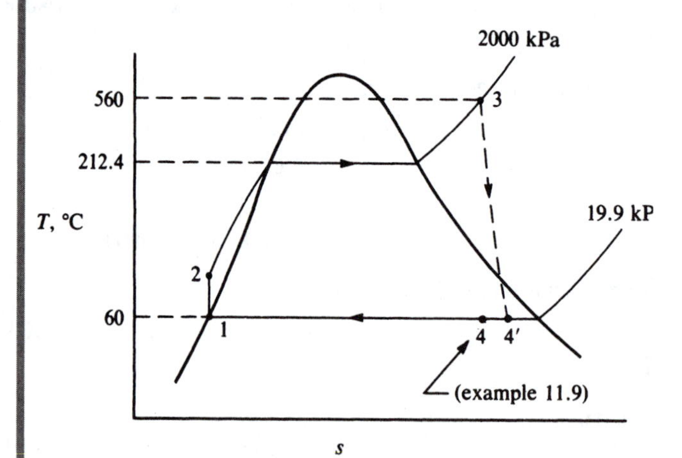

FIGURE 11–20

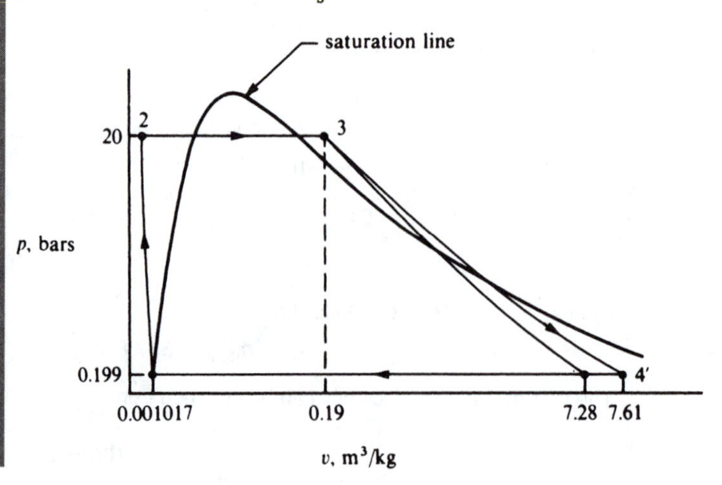

At state 4, we find the enthalpy from the Mollier diagram: $p_4 = 0.199$ bar and $\chi_4 = 99\%$, so

$$h_4 = 2600 \text{ kJ/kg}$$

Then the power is computed from the equation

$$\begin{aligned}\dot{W}k_{cycle} &= \dot{m}wk_{cycle} = \dot{m}(h_3 - h_4 + h_1 - h_2)\\ &= (5000 \text{ kg/h})(3600 \text{ kJ/kg} - 2600 \text{ kJ/kg} + 251 \text{ kJ/kg} - 253 \text{ kJ/kg})\\ &= 4.99 \times 10^6 \text{ kJ/h}\end{aligned}$$

Answer

or

Answer

$$\dot{W}k_{cycle} = 1386 \text{ kW}$$

During the irreversible expansion 3–4, there is an increase in the entropy equal to

$$s_{4'} - s_3$$

The value for s_3 is found to be 7.596 kJ/kg·K from table B–12, and $s_{4'}$ is 7.78 kJ/kg·K from the Mollier diagram. Thus, the entropy increases by 0.184 kJ/kg·K.

The efficiency of the turbine is the same as the adiabatic efficiency defined in chapter 6, which is

$$\eta_{turb} = \frac{h_3 - h_{4'}}{h_3 - h_4} \times 100$$

where Wk_{os} is the work of a reversible adiabatic turbine and Wk_{act} is the actual work. For our example, the reversible adiabatic work is the same as that of the turbine of example 11–9:

$$Wk_{os} = h_3 - h_4 = 3600 \text{ kJ/kg} - 2460 \text{ kJ/kg} = 1140 \text{ kJ/kg}$$

The actual work, Wk_{act}, is

$$Wk_{act} = h_3 - h_4 = 3600 \text{ kJ/kg} - 2600 \text{ kJ/kg} = 1000 \text{ kJ/kg}$$

Then

Answer

$$\eta_{turb} = \frac{1000}{1140} \times 100 = 87.7\%$$

The steam rate is

$$\dot{m}_{sr} = \frac{\dot{m}_{steam}}{\dot{W}k_{cycle}} = \frac{5000 \text{ kg/h}}{1386 \text{ kW}} = 3.6 \text{ kg/kW·h}$$

The minimum entropy generation will be the entropy generation of the turbine:

$$\begin{aligned}\dot{S}_{gen} &\geq (\dot{m}_{steam})(s_{4'} - s_3)\\ &\geq (5000 \text{ kg/h})(0.184 \text{ kJ/kg·K})\\ &\geq 920 \text{ kJ/K·h}\end{aligned}$$

Answer

Using *EES Equation* windows and checking that SI units with temperature in degrees Celsius and pressure in kPa, we enter the following equations:

$h1$ = enthalpy (steam, T = 60, x = 0.0)
$s1$ = entropy (steam, T = 60, x = 0.0)
$s2$ = $s1$
$h2$ = enthalpy (steam, p = 2000, s = $s2$)
$h3$ = enthalpy (steam, T = 560, p = 2000)
$s3$ = entropy (steam, T = 560, p = 2000)

$$h4 = \text{enthalpy}\,(\text{steam}, T = 60, x = 0.99)$$
$$s4 = \text{entropy}\,(\text{steam}, T = 60, p = 0.99)$$
$$m = 5000/3600$$
$$w = m{*}(h3 - h4 + h1 - h2)$$
$$s4s = s3$$
$$h4s = \text{enthalpy}\,(\text{steam}, T = 60, s = s4s)$$
$$\text{eff} = (h3 - h4)/(h3 - hs)$$
$$\text{SR} = m{*}3600/w$$
$$\text{sm} = m{*}3600{*}(s4 - s3)$$

Then, clicking the *Calculate* option and *Solve* gives answers that are in reasonable agreement with the answer obtained by using the appendix tables and the Mollier diagram:

$w = 1408$	(kW = power produced)
$eff = 0.9271$	(= 92.71% turbine efficiency)
$\text{SR} = 3.552$	(kg/kW-h = steam rate)
$sm = 1199$	(kJ/K-h = minimum entropy generation)

Clearly, the efficiency of the steam cycle decreases with inefficiencies in the turbine. Methods of increasing the efficiencies are considered in the sections that follow.

The program RANKINE can aid in the analysis of the simple steam turbine and the steam turbine with irreversibilities. RANKINE is available on soft diskette through the publisher, and the student is encouraged to use the computer to bet ter understand the operations of the steam power generators. The program RANKINE will give results that may be different from the hand computation answers shown in examples 11–9 and 11–10. These differences are nearly always the result of the fact that the values of the properties computed through the subroutines are not in very close agreement with the values obtained from the steam tables. Results from the program RANKINE, however, should be within 5% of the results obtained from hand calculations.

The principal use of the computer with programs such as RANKINE is to aid in better analyzing the operation of steam turbines if one or more of the parameters or properties is varied; for instance, RANKINE can aid in determining how the thermal efficiency of the turbine cycle changes with decreased condenser pressure while keeping the boiler pressure and temperature constant. The variation in power generation caused by an increase or decrease in the turbine adiabatic efficiency could be another study that could be computer aided. Similar studies are suggested in the practice problem section of this chapter.

11–8 THE REHEAT CYCLE

The most obvious method to increase the efficiency of the steam turbine cycle, apparent from the T–s diagram of figure 11–19, is to increase the steam pressure and temperature—thereby increasing the turbine work by the same amount as the additional heat added (ideally). There are, unfortunately, upper limits to operating pressures and temperatures, generally determined by the boiler, turbine, pump, pipe, and condenser materials. Consequently, to achieve higher thermodynamic efficiencies than those of the simple Rankine cycle, the reheat device is used. This involves the addition of energy to steam after it has expanded through a high-pressure turbine and is subsequently expanding through a low-pressure turbine. The typical cycle is sketched in figure 11–21; the property diagrams, in figure 11–22. The cycle is still closed, and the system is all the water contained in the various components. Notice in the property diagrams that the shaded areas represent the additional heat and work of the reheat cycle over the simple Rankine cycle. Of course, these diagrams are descriptive of the reversible

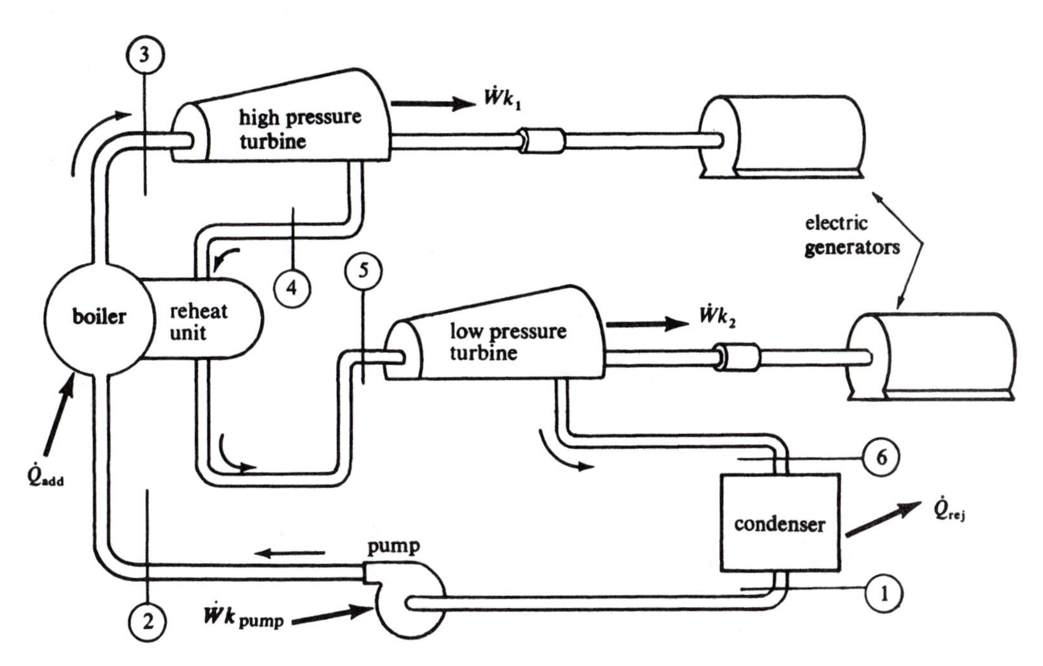

FIGURE 11–21 Reheat cycle

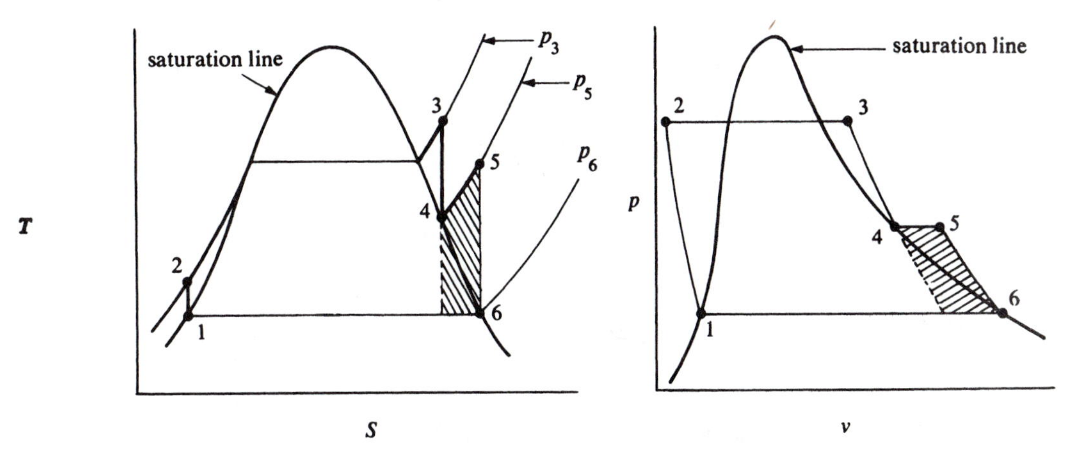

FIGURE 11–22 Reheat cycle property diagram

processes only, and with this restriction, we can determine the following relationships for the ideal reheat cycle:

$$q_{add} = h_3 - h_2 + h_5 - h_4 \quad \text{(11–23)}$$

$$q_{rej} = h_1 - h_6 \quad \text{(11–24)}$$

$$wk_{pump} = h_1 - h_2 \quad \text{(11–25)}$$

$$wk_{turb} = h_3 - h_4 + h_5 - h_6 \quad \text{(11–26)}$$

The reheat cycle is best adapted to systems in which superheated steam at very high pressure and temperature is used so that the initial expansion through the high-pressure turbine will leave a sufficiently high steam pressure for reheating. Let us now look at an example of this engine.

EXAMPLE 11–11 Steam at 2000 psia and 1000°F enters the high-pressure turbine stages of a reheat cycle and expands reversibly and adiabatically to 200 psia. This steam is then directed through the reheater, from which it emerges at a temperature of 950°F. The low-pressure turbine expands the steam reversibly and adiabatically to 1 psia. If 7 lbm/s of steam is used, determine the power produced, the rate of heat added, the cycle efficiency, and the steam rate. Assume that the circulating pump operates ideally and that the fluid leaving the condenser is saturated liquid.

Solution We will obtain the power from the relation $\dot{W}k_{\text{cycle}} = \dot{m}wk_{\text{cycle}}$, where we have, for the ideal cycle,

$$wk_{\text{cycle}} = h_3 - h_4 + h_5 - h_6 + h_1 - h_2$$

and the subscripts refer to the physical points of the reheat cycle of figure 11–21. We shall assume here that the engine fits the same description. Figure 11–23 shows the T–s diagram of the cycle with certain states defined by given temperatures. From table B–12, we obtain the following:

$$h_3 = 1474.1 \text{ Btu/lbm}$$
$$h_5 = 1503.2 \text{ Btu/lbm} \qquad \text{(by interpolation)}$$

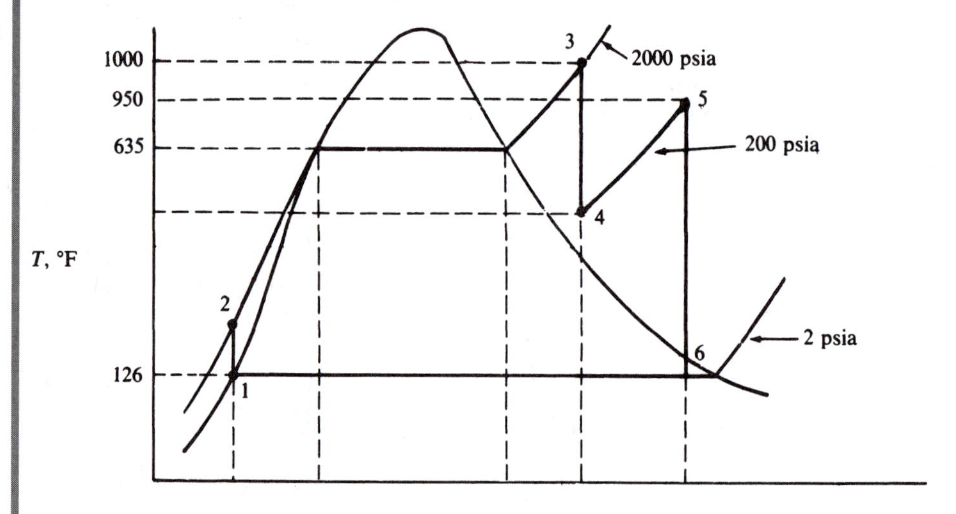

FIGURE 11–23 T–s diagram of reheat cycle (not to scale)

Also, because $s_4 = s_3$, we can see that, at state 4, the steam is still superheated (from table B–12 or the Mollier diagram), and we obtain

$$h_4 \approx 1211 \text{ at } 200 \text{ psia}$$

We have $s_5 = s_6$, and because $s_5 = 1.8235$ Btu/lbm · °R,

$$1.8235 = \frac{x}{100} s_{fg6} + s_{f_6}$$

where

$$s_{f_6} = 0.1326 \text{ Btu/lbm} \cdot \text{°R at 1 psia} \qquad \text{table B–11}$$

and

$$s_{fg6} = 1.9781 \text{ Btu/lbm} \cdot \text{°R} \qquad \text{table B–11}$$

Then

$$\frac{\chi}{100} = \frac{1.8235 \text{ Btu/lbm}\cdot{}^\circ\text{R} - 0.1326 \text{ Btu/lbm}\cdot{}^\circ\text{R}}{1.9781 \text{ Btu/lbm}\cdot{}^\circ\text{R}} = 0.8548$$

and the enthalpy at state 6 is

$$h_6 = \frac{\chi}{100} h_{fg6} + h_{f6} = (0.8548)(1036.1 \text{ Btu/lbm}) + 69.73 \text{ Btu/lbm} = 955.388 \text{ Btu/lbm}$$

From table B–11, we find that

$$h_1 = h_{f1} = 69.73 \text{ Btu/lbm} = h_{f6}$$

The pump work can be computed from equation (11–4),

$$wk_{pump} = h_1 - h_2$$

or from equation (11–14),

$$wk_{pump} = -\sum v\delta p \approx -v_{av}(p_2 - p_1)$$

From equation (11–14), we first approximate the average specific volume by v_1. Then

$$v_1 \approx v_{f1} \approx 0.016136 \text{ ft}^3\text{/lbm}$$

and the pump work is

$$wk_{pump} = -(0.016136 \text{ ft}^3\text{/lbm})(2000 - 1.0 \text{ lbf/in}^2)\left(\frac{144 \text{ in}^2\text{/ft}^2}{778 \text{ ft}\cdot\text{lbf/Btu}}\right) = 5.97 \text{ Btu/lbm}$$

From this, we have

$$h_2 = h_1 - wk_{pump} = 69.73 - (-5.97) = 75.70 \text{ Btu/lbm}$$

and we can then calculate the net work of the cycle:

$$\begin{aligned} wk_{cycle} &= h_3 - h_4 + h_5 - h_6 + h_1 - h_2 \\ &= 1474.1 \text{ Btu/lbm} - 1211 \text{ Btu/lbm} + 1503.2 \text{ Btu/lbm} \\ &\quad - 955.388 \text{ Btu/lbm} + 69.73 \text{ Btu/lbm} - 75.70 \text{ Btu/lbm} \\ &= 804.94 \text{ Btu/lbm} \end{aligned}$$

The power is computed from

Answer

$$\dot{W}k_{cycle} = \dot{m}wk_{cycle} = (7 \text{ lbm/s})(804.94 \text{ Btu/lbm}) = 5634.58 \text{ Btu/s} = 7944.76 \text{ hp}$$

The rate of heat addition to the cycle is obtained from

$$\dot{Q}_{add} = \dot{m}q_{add}$$

where we compute q_{add} from the equation

$$q_{add} = h_3 - h_2 + h_5 - h_4 = 1474.1 \text{ Btu/lbm} - 75.70 \text{ Btu/lbm}$$

$$+ 1503.2 \text{ Btu/lbm} - 1211 \text{ Btu/lbm}$$
$$= 1690.6 \text{ Btu/lbm}$$

Then

Answer
$$\dot{Q}_{add} = (7 \text{ lbm/s})(1690.6 \text{ Btu/lbm}) = 11{,}834.2 \text{ Btu/s}$$

We compute the cycle efficiency from the equation

Answer
$$\eta_T = \frac{\dot{W}k_{cycle}}{\dot{Q}_{add}} \times 100 = \frac{5634.58 \text{ Btu/s}}{11{,}834.2 \text{ Btu/s}} \times 100 = 47.6\%$$

The steam rate is obtained by the defining equation (11–10):

$$\dot{m}_{sr} = \frac{\dot{m}_{steam}}{\dot{W}k_{cycle}} = \frac{7 \text{ lbm/s}}{7944.76 \text{ hp}} \times 3600 \text{ s/h}$$

Answer
$$= 3.17 \text{ lbm/hp} \cdot \text{h}$$

The program RANKINE includes the reheat cycle as an option in analyzing the steam turbine generator.

11–9 THE REGENERATIVE CYCLE

A method that can increase steam cycle efficiency without increasing the superheated steam pressure and temperature is the regenerative heating process, which is essentially a method of adding heat at a higher temperature. Regenerative heating, in practice, is the process of expansion of steam in the turbine well into the phase-change region. Moisture is withdrawn mechanically from the turbine to reduce the effects of wear and corrosion. In figure 11–24 are shown the physical features of the regenerative steam turbine cycle, and the property diagrams corresponding to this cycle are depicted in figure 11–25. The saturated liquid from the condenser is fed to a mixing chamber by a pump (c). The chamber allows the liquid to be heated by mixing with steam bled from the turbine; then this mixture is fed to the next chamber and ultimately back to the boiler for recirculation. In the cycle of figure 11–24, two mixing chambers are utilized, but more could be added. In general, the first regenerative withdraw, corresponding to state 4 in figures 11–24 and

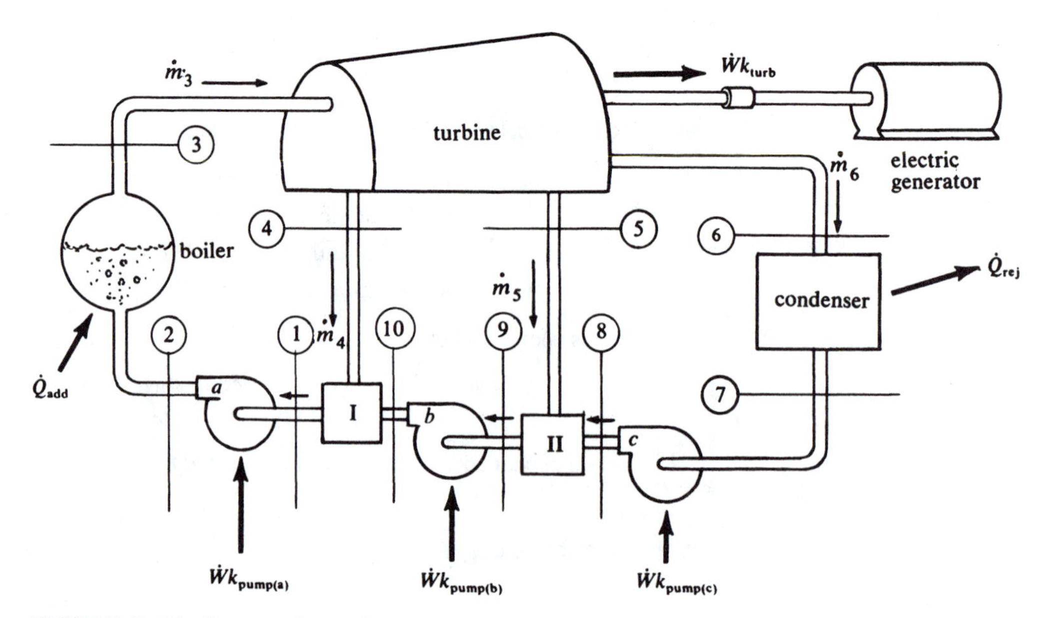

FIGURE 11–24 Regenerative cycle

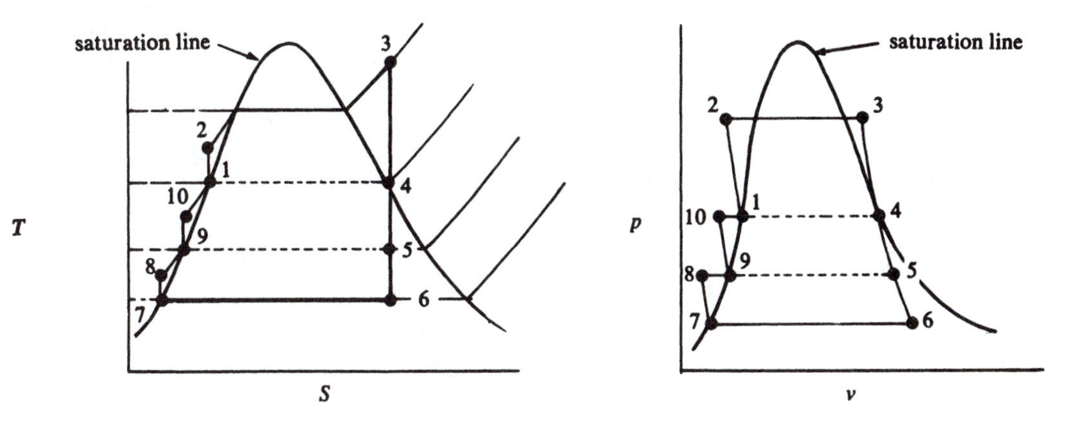

FIGURE 11–25 Property diagrams of regenerative cycle

11–25, is at a superheated state. The diagram in figure 11–25 indicates that this withdrawal is done at a saturated vapor condition.

The computational techniques for the regenerative cycle are identical to those of the simple Rankine cycle with the following major deviation: mass must be accounted for throughout the cycle in a more detailed manner than in the simple Rankine cycle analysis. If we look at the conservation of mass through the turbine, we find that

$$\dot{m}_3 = \dot{m}_4 + \dot{m}_5 + \dot{m}_6 \tag{11–27}$$

and using this relationship, we can obtain the total power produced by the turbine from the equation

$$\dot{W}k_{\text{turb}} = \dot{m}_3(wk_{\text{turb}(3-4)}) + (\dot{m}_5 + \dot{m}_6)(wk_{\text{turb}(4-5)}) + (m_6)(wk_{\text{turb}(5-6)}) \tag{11–28}$$

where the $wk_{\text{turb}(3-4)}$ term is the work produced per unit mass of steam expanding from state 3 to state 4. Disregarding kinetic and potential energy changes, we have, for an ideal turbine with regeneration,

$$\dot{W}k_{\text{turb}} = \dot{m}_3(h_3 - h_4) + (\dot{m}_5 + \dot{m}_6)(h_4 - h_5) + \dot{m}_6(h_5 - h_6) \tag{11–29}$$

If we look at the mixing chambers, we can write the steady-flow energy equation for them. Neglecting kinetic or potential energy changes, we have, for chamber I,

$$\dot{m}_1 h_1 = \dot{m}_4 h_4 + \dot{m}_{10} h_{10}$$

but

$$\dot{m}_{10} = \dot{m}_1 - \dot{m}_4$$

so

$$\dot{m}_1 h_1 = \dot{m}_4 h_4 + (\dot{m}_1 - \dot{m}_4) h_{10} \tag{11–30}$$

and if we solve for the fraction $\dot{m}_4/\dot{m}_1$ [by dividing every term in equation (11–30) by $\dot{m}_1$], we obtain

$$\frac{\dot{m}_4}{\dot{m}_1} = \frac{h_1 - h_{10}}{h_4 - h_{10}} \tag{11–31}$$

For mixing chamber II, using conservation of energy and mass, we find that

$$\frac{\dot{m}_5}{\dot{m}_9} = \frac{h_9 - h_8}{h_5 - h_8} \tag{11–32}$$

The fraction leaving the mixing chamber should be saturated liquid for proper operation of regenerative heating. As a result of this fact, h_9 should be equal to h_f at the pressure of the water at state 9, and h_1 should be h_f at pressure p_1.

Using these relations, we can analyze quite precisely the expected flow rates from a knowledge of the enthalpy values. The following example shows some of the numerical computation involved in a regenerative heating cycle.

EXAMPLE 11–12 An ideal regenerative steam cycle operates with 200,000 kg/h of steam. The steam is at 8.0-MPa pressure and 560°C as it enters the turbine. The exhaust steam is at 20 kPa. Assume that steam is extracted at two stages for regeneration. The first stage corresponds to saturated vapor, and the second is at 150 kPa. Determine the amounts of steam extracted for regeneration in the two states, the power produced, and the thermodynamic efficiency. Compare your answers with the one you obtained by using *EES*.

Solution The system is shown in figure 11–24 because the cycle includes two stages of regeneration. The *T–s* diagram is sketched in figure 11–26 to show the various states and the properties. From table B–12, we find that $h_3 = 3544$ kJ/kg. The enthalpies at states 4, 5, and 6 can then be obtained from table B–11 or the Mollier diagram. Using the Mollier diagram, we find that $h_4 = 2740$ kJ/kg, $h_5 = 2574$ kJ/kg, and $h_6 = 2275$ kJ/kg. From table B–11, we find that $h_7 = 251$ kJ/kg, $h_9 = 467$ kJ/kg, and $h_1 = 601$ kJ/kg, which are enthalpies of the saturated liquids at the corresponding pressures. Using a form of equation (11–4) gives us

$$h_2 = h_1 - wk_{\text{pump } a}$$
$$h_{10} = h_9 - wk_{\text{pump } b}$$
$$h_8 = h_7 - wk_{\text{pump } c}$$

and from equation (11–14) we get

$$\begin{aligned} wk_{\text{pump } a} &\simeq -v_{f1}(p_2 - p_1) \\ &= -(0.00138\ \text{m}^3/\text{kg})(8000\ \text{kPa} - 390\ \text{kPa}) \\ &= -10.5\ \text{kJ/g} \end{aligned}$$

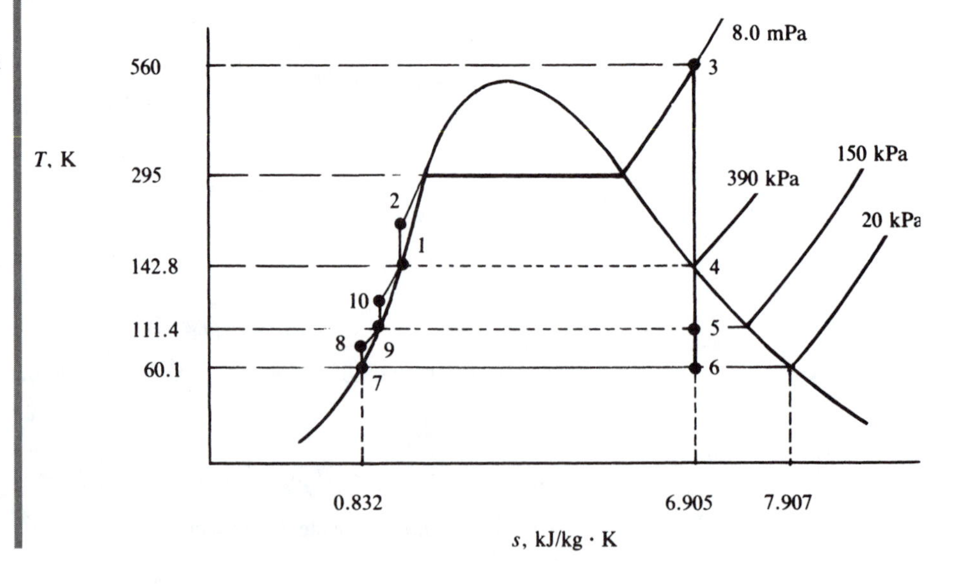

FIGURE 11–26 *T–s* diagram for two-stage regenerative cycle (not to scale)

The enthalpy at state 2 is then

$$h_2 = 601 \text{ kJ/kg} - (-10.5 \text{ kJ/kg}) = 611.5 \text{ kJ/kg}$$

Similarly, we find that

$$wk_{\text{pump } b} = -0.259 \text{ kJ/kg}$$

or

$$h_{10} = 467 \text{ kJ/kg} - (-0.259 \text{ kJ/kg}) \simeq 467 \text{ kJ/kg}$$

and

$$wk_{\text{pump } c} = -0.1365 \text{ kJ/kg}$$

which yields

$$h_8 = 251 \text{ kJ/kg} - (-0.1365 \text{ kJ/kg}) \simeq 251 \text{ kJ/kg}$$

From these data, we can then proceed with our analysis. Using equation (11–31), we can obtain the flow rate of steam at state 4, in the form

$$\dot{m}_4 = \frac{\dot{m}_1(h_1 - h_{10})}{h_4 - h_{10}} = \frac{(2 \times 10^5 \text{ kg/h})(601 \text{ kJ/kg} - 467 \text{ kJ/kg})}{2740 \text{ kJ/kg} - 467 \text{ kJ/kg}}$$

Answer

$$= 11{,}790 \text{ kg/h}$$

The mass flow of steam at state 9 is the same as $\dot{m}_1 - \dot{m}_4$, or

$$\dot{m}_9 = 188{,}210 \text{ kg/h}$$

Then, using equation (11–32), we solve for $\dot{m}_5$:

$$\dot{m}_5 = \frac{\dot{m}_9(h_9 - h_8)}{h_5 - h_8}$$

$$= \frac{(200{,}000 \text{ kg/h} - 11{,}790 \text{ kg/h})(467 \text{ kJ/kg} - 251 \text{ kJ/kg})}{2574 \text{ kJ/kg} - 251 \text{ kJ/kg}}$$

Answer

$$= 17{,}500 \text{ kg/h}$$

We have $\dot{m}_3 = \dot{m}_1 = 200{,}000$ kg/h and

$$\dot{m}_6 = \dot{m}_1 - \dot{m}_4 - \dot{m}_5 = 170{,}710 \text{ kg/h}$$

The power, computed from equation (11–29), is

$$\dot{W}k_{\text{turb}} = \dot{m}_3(h_3 - h_4) + (\dot{m}_5 + \dot{m}_6)(h_4 - h_5) + \dot{m}_6(h_5 - h_6)$$

which gives us

$$\begin{aligned}\dot{W}k_{\text{turb}} &= (200{,}000 \text{ kg/h})(3544 - 2740) \\ &\quad + (188{,}210 \text{ kg/h})(2740 - 2574) \\ &\quad + (170{,}710 \text{ kg/h})(2574 - 2275) = 243{,}085.15 \text{ MJ/h}\end{aligned}$$

Answer

$$= 67.524 \text{ MW}$$

We obtain the net power produced by subtracting the power of the pumps; that is,

$$\dot{W}k_{\text{cycle}} = \dot{W}k_{\text{turb}} + \dot{W}k_{\text{pump } a} + \dot{W}k_{\text{pump } b} + \dot{W}k_{\text{pump } c}$$

We have

$$\dot{W}k_{\text{pump }a} = \dot{m}\dot{W}k_{\text{pump }a} = (200{,}000 \text{ kg/h})(-10.5 \text{ kJ/kg})$$
$$= -2.1 \times 10^6 \text{ kJ/h}$$
$$= -583.3 \text{ kW}$$

Similarly, we find that

$$\dot{W}k_{\text{pump }b} = -13.54 \text{ kW}$$

and

$$\dot{W}k_{\text{pump }c} = -6.45 \text{ kW}$$

Then

$$\dot{W}k_{\text{cycle}} = 67{,}524 \text{ kW} - 583.3 \text{ kW} - 13.54 \text{ kW} - 6.45 \text{ kW}$$

Answer
$$= 66{,}921 \text{ kW}$$

The rate of heat addition is found from the equation

$$\dot{Q}_{\text{add}} = \dot{m}_1(h_3 - h_2)$$

so

$$\dot{Q}_{\text{add}} = (200{,}000 \text{ kg/h})(3544 \text{ kJ/kg} - 611.5 \text{ kJ/kg})$$
$$= 586.5 \times 10^6 \text{ kJ/h} = 162{,}916.7 \text{ kW}$$

The thermodynamic efficiency is then calculated from the equation

$$\eta_T = \left(\frac{\dot{W}k_{\text{cycle}}}{\dot{Q}_{\text{add}}}\right) \times 100$$

yielding

Answer
$$\eta_T = 41.1\%$$

Using *EES Equation* windows and checking that SI units with temperature in degrees Celsius and pressure in kPa, we enter the following equations:

$h1$ = enthalpy (steam, p = 390, x = 0.0)
$s1$ = entropy (steam, p = 390, x = 0.0)
$s2$ = $s1$
$h2$ = enthalpy (steam, p = 8000, s = $s2$)
$h3$ = enthalpy (steam, T = 560, p = 8000)
$s3$ = entropy (steam, T = 560, p = 8000)
$h4$ = enthalpy (steam, s = $s3$, x = 1.0)
$h5$ = enthalpy (steam, p = 150, s = $s3$)
$h6$ = enthalpy (steam, p = 20, s = $s3$)
$h7$ = enthalpy (steam, p = 20, x = 0.0)
$s7$ = entropy (steam, p = 20, x = 0.0)
$s8$ = $s7$
$h8$ = enthalpy (steam, p = 150, s = $s8$)
$h9$ = enthalpy (steam, p = 150, x = 0.0)
$s9$ = entropy (steam, p = 150, x = 0.0)
$s10$ = $s9$

$$h10 = \text{enthalpy}(\text{steam}, p = 390, s = s10)$$
$$m = 200000/3600$$
$$m4 = m*(h1 - h10)/(h4 - h10)$$
$$m5 = (m - m4)*(h9 - h8)/(h5 - h8)$$
$$m6 = m - m4 - m5$$
$$wt = m*(h3 - h4) + (m5 + m6)*(h4 - h5) + m6*(h5 - h3)$$
$$wp = m*(h1 - h2) + (m5 + m6)*(h9 - h10) + m6*(h7 - h8)$$
$$w = wt + wp$$
$$qa = m*(h3 - h2)$$
$$eff = w/qa$$

Then, clicking the *Calculate* option and *Solve* gives answers that are in reasonable agreement with those obtained by using the appendix tables and the Mollier diagram:

$m4 = 3.269$ (kg/s = mass flow extracted at high pressure)
$m5 = 4.858$ (kg/s = mass flow extracted at low pressure)
$w = 67109$ (kw power produced)
$eff = 0.4114$ (= 41.14% thermal efficiency)

The computer program RANKINE includes the regenerative cycle as an option for analysis of the steam turbine. Provisions are included for up to five extractions of steam and five mixing chambers.

11–10 THE REHEAT–REGENERATIVE CYCLE

The reheat–regenerative steam cycle operates with a higher efficiency than any of the steam power cycles operating between similar temperature and pressure regimes. For this reason, most modern electric power generating systems using steam power utilize this cycle. The cycle components are generally a high-pressure turbine, a reheater, a low-pressure turbine, a condenser, regenerative heaters, pumps, and a steam generator. The regenerative heaters are used with either or both of the high- and low-pressure turbines. A typical arrangement for the reheat–regenerative cycle is shown in figure 11–27. The particular cycle shown has one reheat, one regeneration for the high-pressure turbine, and two regenerations for the low-pressure turbine. The temperature–entropy diagram in figure 11–28 indicates some of the typical states corresponding to the cycle shown in figure 11–27. Remember that regeneration can occur in the superheat region for states 13 and 14 and that the exhaust from the turbine at state 4 could be superheated steam, although this would not be most effective in general. The analysis of a reheat–regenerative cycle follows that of the individual reheat and regenerative cycles. The enthalpy values at the individual states can be found by the techniques discussed and demonstrated previously. The mass flow rates through the regenerative stages are determined in the same manner as in the regenerative cycle.

EXAMPLE 11–13

A reheat–regenerative steam power cycle operates with three regenerations, two on the high-pressure turbine and one on the low-pressure turbine. Steam at a rate of 10 lbm/s is supplied at 850°F and 1200 psia, with regeneration stages at 600 psia and 400 psia. Reheat occurs at 200 psia with steam supplied to the low-pressure turbine at 800°F. Regeneration steam is then extracted at 30 psia, and exhaust is at 1 psia. Determine the power produced in the cycle and the cycle efficiency. Compare your answers with those you obtained by using *EES*.

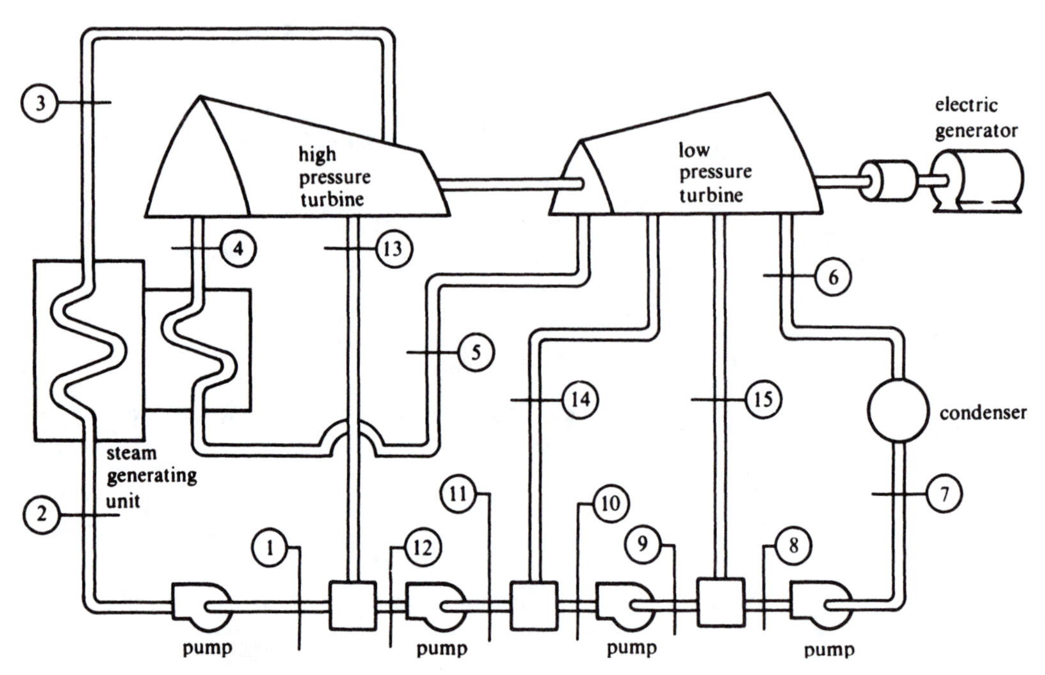

FIGURE 11–27 Typical elements of a reheat–regenerative cycle

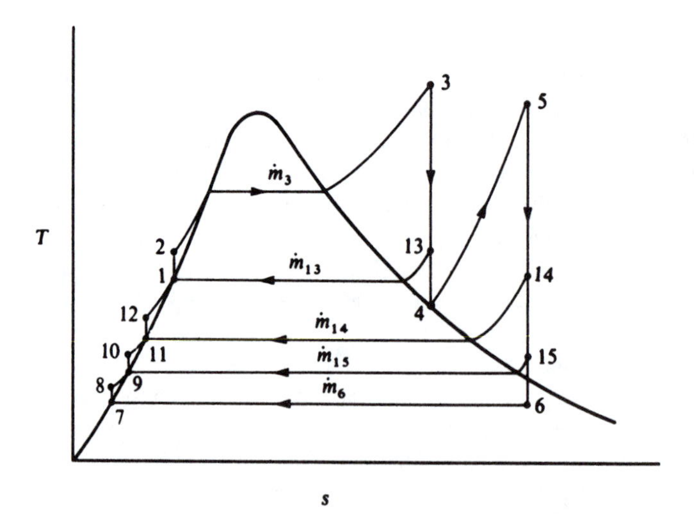

FIGURE 11–28
Temperature–entropy diagram for typical reheat–regenerative cycle

Solution Using the cycle shown in figure 11–29 as a representation of the cycle under consideration, we proceed to determine the enthalpy values at the various station points. Using the Mollier diagram (chart B–3), we find that

$$h_3 = 1410 \text{ Btu/lbm} \qquad h_{13} = 1326 \text{ Btu/lbm}$$
$$h_4 = 1216 \text{ Btu/lbm} \qquad h_{14} = 1272 \text{ Btu/lbm}$$
$$h_5 = 1425 \text{ Btu/lbm} \qquad h_{15} = 1215 \text{ Btu/lbm}$$
$$h_6 = 987 \text{ Btu/lbm}$$

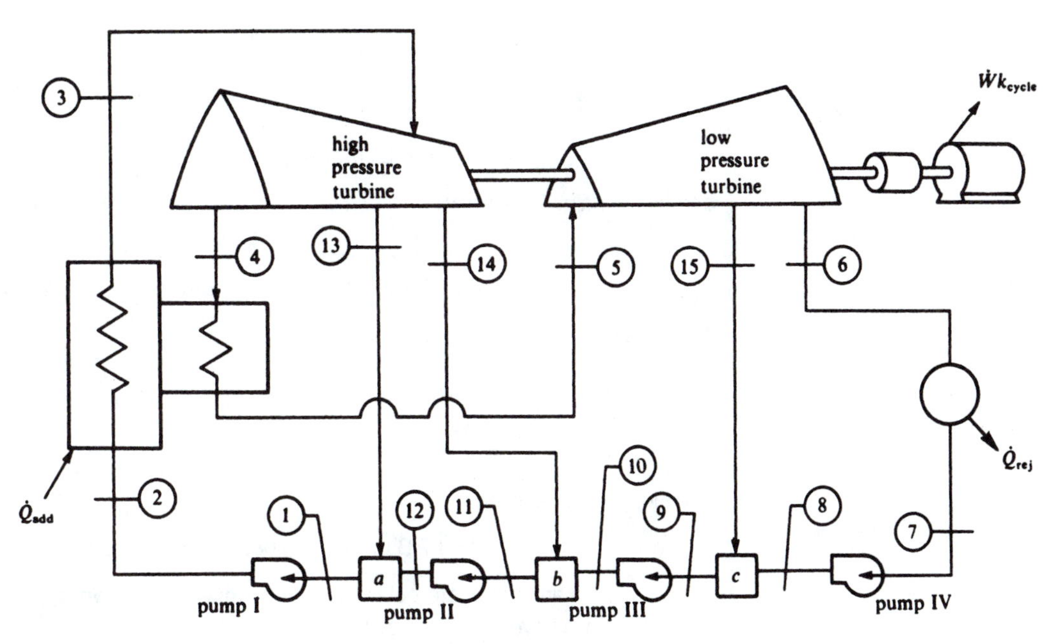

FIGURE 11–29 Reheat–regenerative cycle of example 11–13

From table B–11,

$$h_7 = h_f = 69.7 \text{ Btu/lbm}$$
$$h_9 = h_f \text{ (at 30 psia)} = 218.9 \text{ Btu/lbm}$$
$$h_{11} = h_f \text{ (at 400 psia)} = 424.2 \text{ Btu/lbm}$$
$$h_1 = h_f \text{ (at 600 psia)} = 471.7 \text{ Btu/lbm}$$

The work required by the pumps may be approximated by the equation

$$wk_{\text{pump}} = -v_{\text{av}}(\Delta p)$$

Then, for pump I, the average specific volume is near 0.02 ft^3/lbm between 600 and 1200 psia. The work is then simply

$$wk_{\text{pump I}} = -(0.02 \text{ ft}^3/\text{lbm})(1200 \text{ psia} - 600 \text{ psia})\left(\frac{144 \text{ in}^2/\text{ft}^2}{778 \text{ ft}\cdot\text{lbf/Btu}}\right)$$
$$= -2.22 \text{ Btu/lbm}$$

The enthalpy at state 2 is

$$h_2 = h_1 - wk_{\text{pump I}} = 471.7 \text{ Btu/lbm} + 2.22 \text{ Btu/lbm}$$
$$= 473.92 \text{ Btu/lbm}$$

Similarly,

$$wk_{\text{pump II}} = -(0.02 \text{ ft}^3/\text{lbm})(600 \text{ psia} - 400 \text{ psia})\left(\frac{144 \text{ in}^2/\text{ft}^2}{778 \text{ ft}\cdot\text{lbf/Btu}}\right)$$
$$= -0.714 \text{ Btu/lbm}$$

and the enthalpy at state 12 is

$$h_{12} = h_{11} - wk_{\text{pump II}} = 424.2 \text{ Btu/lbm} + 0.714 \text{ Btu/lbm}$$
$$= 424.91 \text{ Btu/lbm}$$

The work for pump III is

$$wk_{\text{pump III}} = -(0.0170\ \text{ft}^3/\text{lbm})(400\ \text{psia} - 30\ \text{psia})\left(\frac{144\ \text{in}^2/\text{ft}^2}{778\ \text{ft}\cdot\text{lbf/Btu}}\right)$$
$$= -1.164\ \text{Btu/lbm}$$

and the enthalpy for state 10 is

$$h_{10} = h_9 - wk_{\text{pump III}}$$
$$= 218.9\ \text{Btu/lbm} + 1.164\ \text{Btu/lbm} = 220\ \text{Btu/lbm}$$

The work of pump IV is

$$wk_{\text{pump IV}} = -(0.0166\ \text{ft}^3/\text{lbm})(30\ \text{psia} - 1.0\ \text{psia})\left(\frac{144\ \text{in}^2/\text{ft}^2}{778\ \text{ft}\cdot\text{lbf/Btu}}\right)$$
$$= -0.089\ \text{Btu/lbm}$$

and the enthalpy at state 8 is

$$h_8 = 69.7\ \text{Btu/lbm} + 0.089\ \text{Btu/lbm} = 69.8\ \text{Btu/lbm}$$

The mass flow rates are obtained from the methods shown in section 11–9. Thus, for regenerator *a*, we have as an energy balance

$$\dot{m}_1 h_1 = \dot{m}_{12} h_{12} + \dot{m}_{13} h_{13}$$

and as a mass balance

$$\dot{m}_{12} = \dot{m}_{14} + \dot{m}_{10} = \dot{m}_{14} + \dot{m}_1 - \dot{m}_{13} - \dot{m}_{14}$$

Combining these relationships and rearranging the equations, we have an equation like (11–31) or (11–32):

$$\dot{m}_{13} = \frac{\dot{m}_1(h_1 - h_{12})}{h_{13} - h_{12}}$$

Since $\dot{m}_1 = 10$ lbm/s, we can compute the mass flow at state 13, namely,

$$\dot{m}_{13} = (10\ \text{lbm/s})\left(\frac{471.7\ \text{Btu/lbm} - 424.94\ \text{Btu/lbm}}{1326\ \text{Btu/lbm} - 424.94\ \text{Btu/lbm}}\right) = 0.52\ \text{lbm/s}$$

and the mass flow rate at state 12 is

$$\dot{m}_{12} = 10\ \text{lbm/s} - 0.52\ \text{lbm/s} = 9.48\ \text{lbm/s}$$

From a similar analysis of regenerator *b*, we find that

$$\dot{m}_{14} = 1.93\ \text{lbm/s}$$

Thus,

$$\dot{m}_4 = 7.55\ \text{lbm/s}$$

and from regenerator *c* we obtain

$$\dot{m}_{15} = \dot{m}_9 \frac{h_9 - h_8}{h_{15} - h_8}$$
$$= 1.30\ \text{lbm/s}$$

where $\dot{m}_9 = \dot{m}_4$. Also,

$$\dot{m}_6 = \dot{m}_4 - \dot{m}_{15} = 6.25\ \text{lbm/s}$$

The power produced from the cycle is computed from the equation

$$\begin{aligned}\dot{W}k_{cycle} &= \dot{W}k_{turb} + \dot{W}k_{pumps}\\ &= \dot{m}_{13}(h_3 - h_{13}) + \dot{m}_{14}(h_3 - h_{14}) + \dot{m}_4(h_3 - h_4) + \dot{m}_{15}(h_5 - h_{15})\\ &\quad + \dot{m}_6(h_5 - h_6) + \dot{m}_7 wk_{pump\ IV}\\ &\quad + \dot{m}_9 wk_{pump\ III} + \dot{m}_{11} wk_{pump\ II} + \dot{m}_1 wk_{pump\ I}\\ &= (0.52\ \text{lbm/s})(1410 - 1326) + (1.93)(1410 - 1272)\\ &\quad + (7.55)(1410 - 1216) + (1.30)(1425 - 1215) + (6.25)(1425 - 987)\\ &\quad + (6.25)(-0.089) + (7.55)(-1.164) + (9.48)(-0.714) + (10)(-2.22)\end{aligned}$$

Answer $$= 4746.9\ \text{Btu/s} = 6693.1\ \text{hp}$$

We may convert this answer to electrical power units to obtain

Answer $$\dot{W}k_{cycle} = 4995\ \text{kW}$$

The heat rejected by the condenser is

$$\begin{aligned}\dot{Q}_{rej} &= \dot{m}_6(h_7 - h_6)\\ &= (6.25\ \text{lbm/s})(69.7\ \text{Btu/lbm} - 987\ \text{Btu/lbm})\\ &= -5733.1\ \text{Btu/s}\end{aligned}$$

and the heat added is obtained from the balance

$$\dot{Q}_{add} = \dot{W}k_{cycle} - \dot{Q}_{rej} = 10{,}480\ \text{Btu/s}$$

The efficiency is computed in the usual manner, yielding

Answer $$\eta_T = \frac{\dot{W}k_{cycle}}{\dot{Q}_{add}} = 45.3\%$$

Using *EES Equation* windows and checking that English Units with temperature in degrees Fahrenheit and pressure in psia, we enter the following equations:

$h1$ = enthalpy (steam, p = 600, x = 0.0)
$s1$ = entropy (steam, p = 600, x = 0.0)
$s2$ = $s1$
$h2$ = enthalpy (steam, p = 1200, s = $s2$)
$h3$ = enthalpy (steam, T = 850, p = 1200)
$s3$ = entropy (steam, T = 850, p = 1200)
$s4$ = $s3$
$h4$ = enthalpy (steam, p = 200, s = $s4$)
$h5$ = enthalpy (steam, T = 800, p = 200)
$s5$ = entropy (steam, T = 800, p = 200)
$s6$ = $s5$
$h6$ = enthalpy (steam, p = 1, s = $s6$)
$h7$ = enthalpy (steam, p = 1, x = 0.0)
$s7$ = entropy (steam, p = 1, x = 0.0)
$s8$ = $s7$
$h8$ = enthalpy (steam, p = 30, s = $s8$)
$h9$ = enthalpy (steam, p = 30, x = 0.0)
$s9$ = entropy (steam, p = 30, x = 0.0)

```
s10 = s9
h10 = enthalpy (steam, p = 400, s = s10)
h11 = enthalpy (steam, p = 400, x = 0.0)
s11 = entropy (steam, p = 400, x = 0.0)
s12 = s11
h12 = enthalpy (steam, p = 600, s = s12)
h13 = enthalpy (steam, p = 600, s = s3)
h14 = enthalpy (steam, p = 400, s = s3)
h15 = enthalpy (steam, p = 30, s = s3)
  m = 10
m13 = m*(h1 - h12)/(h13 - h12)
m14 = (m - m13)*(h11 - h10)/(h14 - h10)
m12 = m - m13
m10 = m - m13 - m14
  m = m10
 m4 = m10
 m5 = m4
m15 = m5*(h9 - h8)/(h15 - h8)
 m6 = m5 - m15
 m8 = m6
wth = m*(h3 - h13) + (m - m13)*(h13 - h14) + (m - m13 - m14)*(h14 - h4)
wtl = m5*(h5 - h15) + (m5 - m15)*(h15 - h6)
 wp = m*(h1 - h2) + m12*(h11 - h12) + m9*(h10 - h9) + m8*(h7 - h8)
  w = 1.054*(wth + wtl + wp)
 qa = m*(h3 - h2) + m4*(h5 - h4)
eff = w/(qa*1.054)
```

Then, clicking the *Calculate* option and *Solve* gives answers that are in reasonable agreement with those obtained using the appendix tables and the Mollier diagram:

$w = 5289$ (kw power produced)

$eff = 0.4573$ (= 45.73% thermal efficiency)

The reheat–regenerative cycle has been used in various forms and combinations of regeneration. In figure 11–30 is shown a typical moderate-sized steam turbine performance diagram. Although some of the notation on the diagram may be unclear to you, if you study the figure, you can see that the cycle is a reheat–regenerative one with one regeneration on the high-pressure turbine coinciding with the exhaust of that turbine, two regenerations in an intermediate-pressure turbine stage, and two regenerations in a low-pressure stage. The low-pressure turbine stage is diagrammed as two turbines operating together, and the steam flow is split between these two stages. Mass flows, pressures, temperatures, and enthalpies are indicated on the diagram, and even an accounting of various leakages of steam around the turbine is made. Leakages are usually internal to the system and so represent recirculating steam that does not do work but that returns to the boiler.

The computer program RANKINE includes the reheat–regenerative cycles as an option in analyzing the steam turbine engine. The program RANKINE is limited in its scope and should be considered as an educational tool only and not a program for final designs of modern steam turbine power generation.

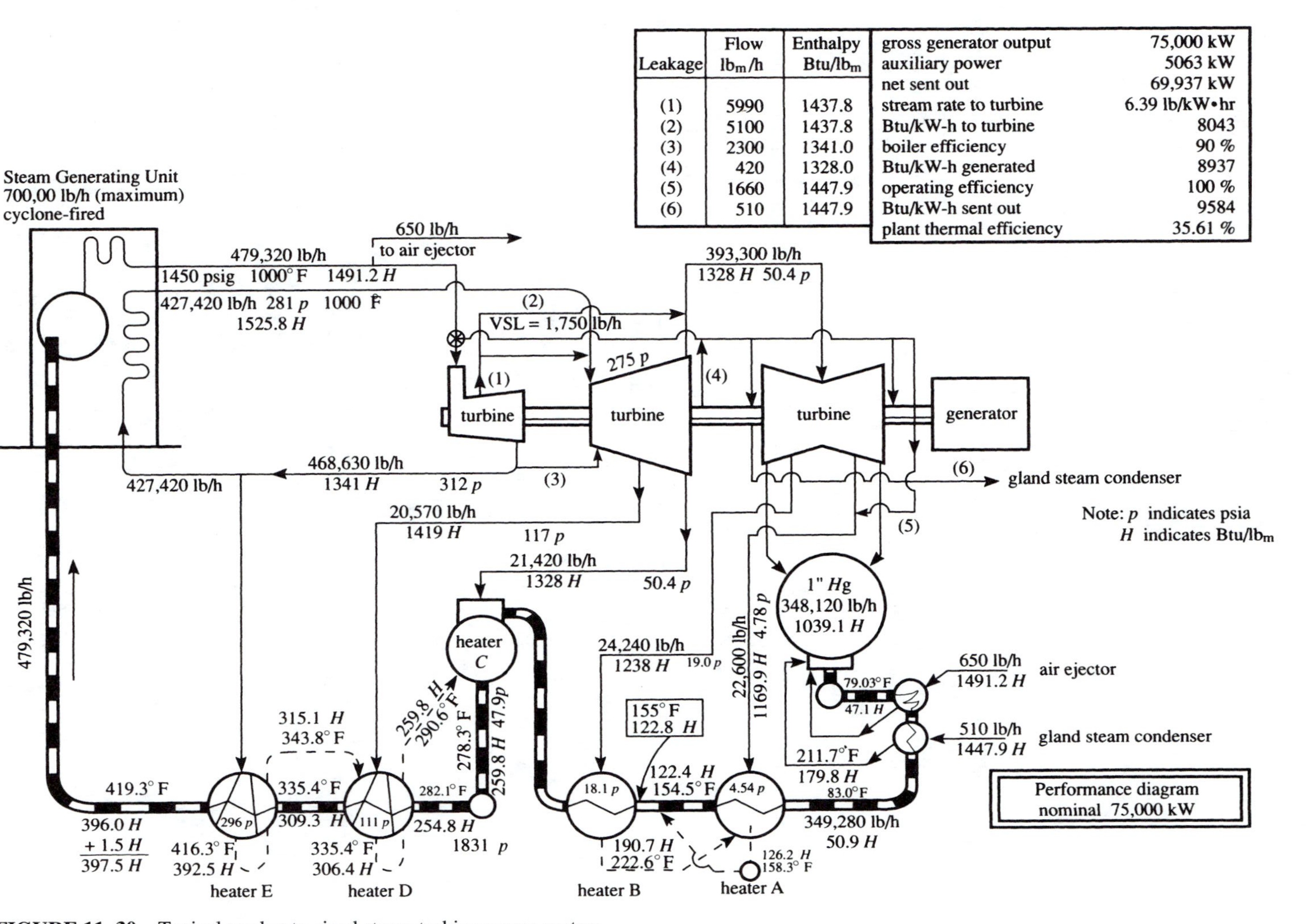

Leakage	Flow lb_m/h	Enthalpy Btu/lb_m
(1)	5990	1437.8
(2)	5100	1437.8
(3)	2300	1341.0
(4)	420	1328.0
(5)	1660	1447.9
(6)	510	1447.9

gross generator output	75,000 kW
auxiliary power	5063 kW
net sent out	69,937 kW
stream rate to turbine	6.39 lb/kW•hr
Btu/kW-h to turbine	8043
boiler efficiency	90 %
Btu/kW-h generated	8937
operating efficiency	100 %
Btu/kW-h sent out	9584
plant thermal efficiency	35.61 %

FIGURE 11–30 Typical moderate-sized steam turbine power system

11–11 OTHER CONSIDERATIONS OF THE RANKINE CYCLE

We have seen that the Rankine cycle in operation is a closed steam turbine cycle and that we can improve its efficiency by using reheat or regenerative heating devices. Steam turbines operating with both reheaters and regenerators are the most efficient power-producing devices available for large power demands. Increased improvements in thermodynamic efficiencies can also be realized with higher superheated steam pressures and temperatures, but, of course, an upper limit is set here by material properties. The upper limits for continuous operations seem to be around 8000-kPa (1200-psia) pressures with 650°C (1200°F) temperatures. Higher temperatures and pressures are always considered in newer designs, but generally there are significant compromises of safety.

One of the most attractive aspects of the steam turbine cycle is its independence of the source of thermal energy. Whereas internal combustion engines are critically dependent on their working media (requiring a precise refined fuel), the steam turbine merely requires energy to boil water. This energy can come from coal, petroleum, or natural gas combustion; from high-temperature exhaust gases of some other engine; from solar energy; or from nuclear reactors. Beyond the steam generating unit there are few, if any, design alterations needed to accommodate these different types of heat sources. We can see then that the steam turbine cycle is well adapted to convert thermal energy to mechanical and electrical energy, and it allows us to hope that future power requirements may be met.

The combustion of fuel (if this type of heat source is utilized for generating steam) can be accomplished under well-controlled conditions and thus effect more efficient combustion and less obnoxious exhaust or products of combustion.

The present technologies of steam turbine coal burning power cycles are achieving thermal efficiencies of the order of 30% to 35%, which is below that predicted by the ideal cycle analyses. With the use of more efficient pulverized coal boilers, scrubbers, and other devices to reduce the nitric oxides and sulphur dioxides, efficiencies of around 45% have been achieved, and this seems to be about the most advanced state-of-the-art technology. Higher efficiencies seem to be very possible and practical, and methods for achieving them are continuously being studied. This is not only economically promising, but promising in the reduction of negative impacts on the natural environment. In other words, reducing thermal, chemical, and biological pollution is good for the environment and for corporations.

An attractive use of the steam turbine cycle is "behind" or in tandem with another engine operating at a much higher temperature. This sort of arrangement is called **cogeneration**, and the use of gas turbines in combination with the steam turbine is being developed. Figure 11–31 shows how such a cogeneration cycle would be configured. Note that power is derived from both the gas turbine and the steam turbine, with the steam turbine using steam produced from hot exhaust gases of the gas turbine. The overall thermal efficiency of the cogeneration cycle is usually higher than that of either the gas turbine cycle or the steam turbine cycle alone. A commercial version of this type of cogeneration cycle is shown in figure 11–32.

It has been proposed to use magnetohydrodynamics as the leading engine of a cogeneration cycle with a steam turbine cycle. The magnetohydrodynamic engine operates at very high temperatures, producing high amounts of power, and resulting in hot exhaust gases. The steam generator could then produce superheated steam from this exhaust and get additional power that might be lost. The MHD converter produces electrical power, and the two separate units can then operate as an efficient generating station.

The steam turbine has been characterized as a constant-speed device, probably because its greatest development has come in large power units. In small units, however, the steam turbine may be applicable to variable-speed applications, although this utilization has not been fully attempted.

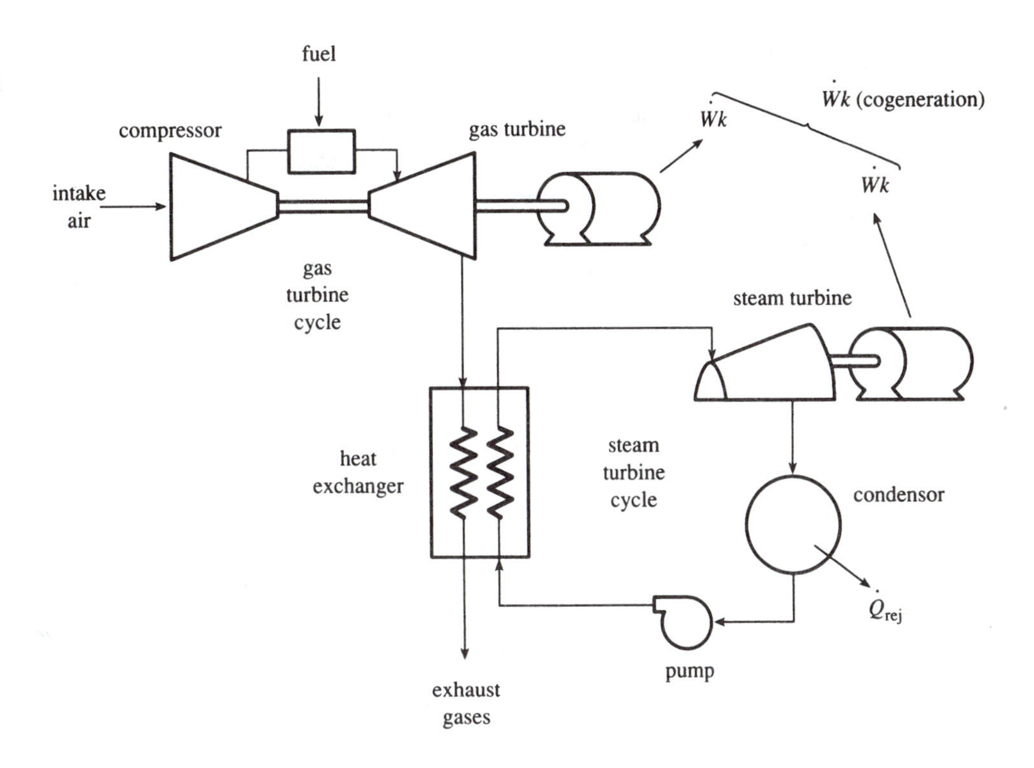

FIGURE 11–31 Typical cogeneration cycle composed of a gas turbine and a steam turbine cycle

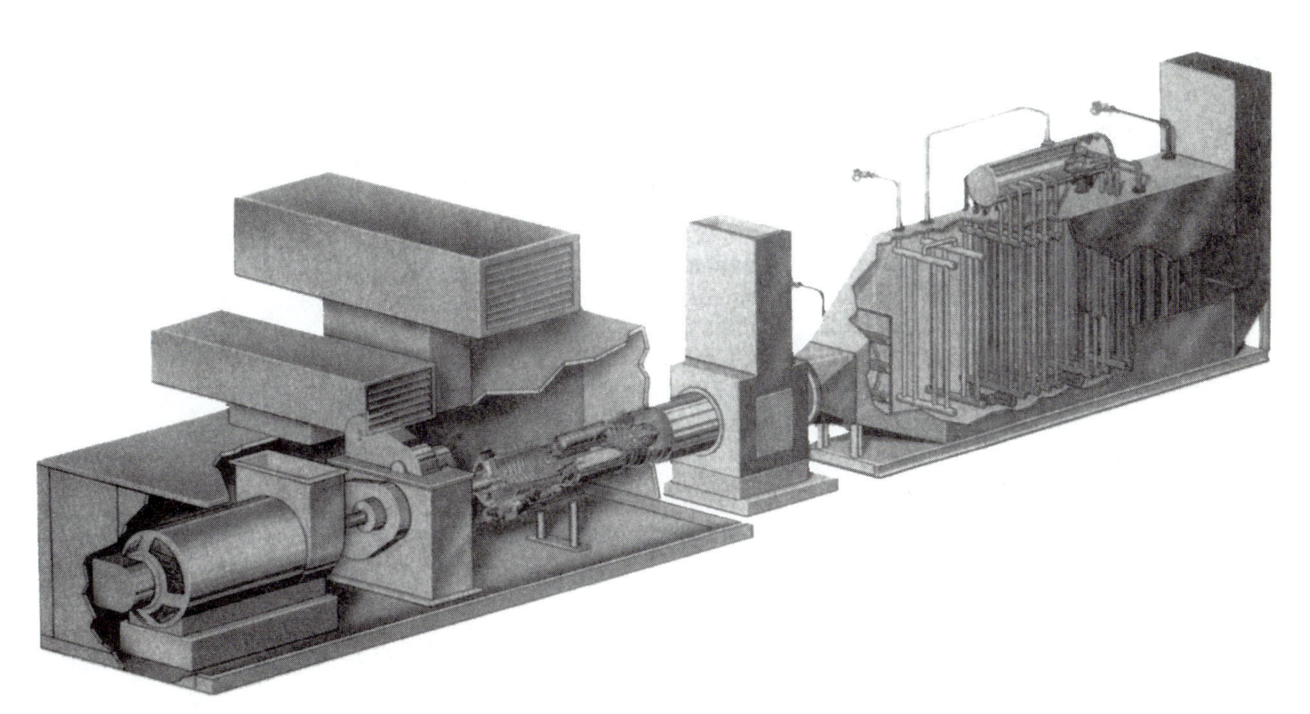

FIGURE 11–32 Model 501-KB5 gas turbine cogeneration system (courtesy of General Motors Corporation, Allison Gas Turbine Division, Indianapolis, IN)

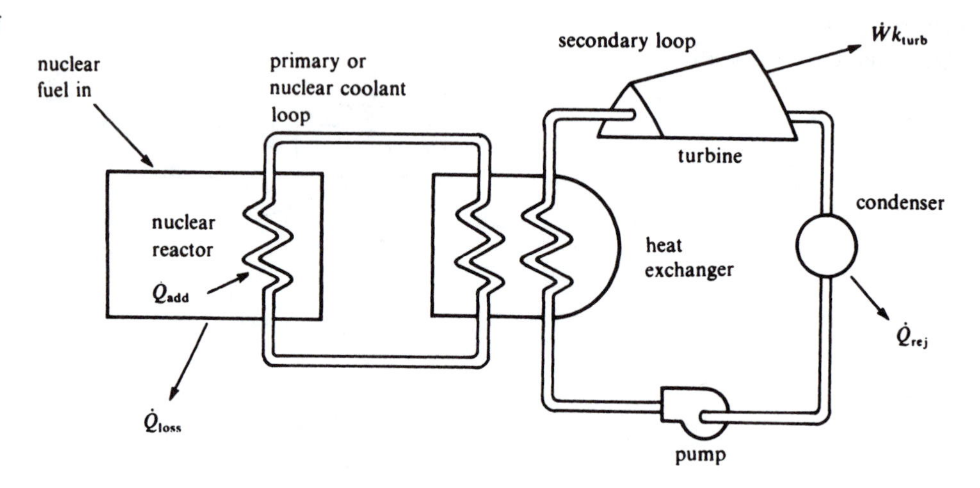

FIGURE 11–33 Elements of nuclear power cycle

Finally, the promise of nuclear power as a practical source of power can, at this time, be realized only with the steam power cycles that we have considered. The basic elements of a typical cycle for this application are shown in figure 11–33. The nuclear reactor, which generates the thermal energy during the reaction, has heat carried away from it by a nuclear coolant. This coolant generally is separate and distinct from the steam power cycle. The coolant passes through a heat exchanger, in the typical system, and transfers heat to the steam. This coolant then returns to the reactor to receive more thermal energy. Some of the materials that have been used as coolants for this operation are water, liquid sodium, helium, and carbon dioxide.

Although the nuclear power plant is becoming a feasible and common mode of electrical power generation, its total practicality and safety are still in doubt. It is not clear that the total energy required to produce the processed nuclear fuel and the special handling of the fuel and waste products are offset by the power actually produced. In many ways, the customary fuels—coal, gas, and oil—are still subsidizing the operations of the nuclear power plants.

11–12 SUMMARY

The steam turbine power generators were discussed, and the Rankine cycle was given as a description of the operation of these devices. The Rankine cycle for steam as the working fluid was defined as

- 1–2 Adiabatic compression of liquid water
- 2–3 Isobaric heat addition to convert liquid to a vapor
- 3–4 Adiabatic expansion of vapor to low pressure
- 4–1 Isobaric heat rejection to condense vapor to liquid

Pumps or feedwater pumps compress water and increase the water pressure so that the water can enter the boiler. Pump work is approximated by the product $v_{av}\,\Delta p$ and is equal to the enthalpy change across the pump. Boilers or steam generators transfer heat from the combustion of fuel into the thermal energy in steam. The boiler efficiency is defined as

$$\eta_{boiler} = \frac{h_3 - h_2}{\text{LHV}} \frac{\dot{m}_{steam}}{\dot{m}_{fuel}} \tag{11–9}$$

and a measure of the capacity of a boiler is its steam rate:

$$\dot{m}_{sr} = \frac{\dot{m}_{steam}}{\dot{W}k_{cycle}} \tag{11–10}$$

Steam turbines convert thermal energy in the steam into shaft work. This conversion is accomplished through an expansion from high to low pressure, either reversibly or irreversibly. Steam turbines are normally considered to be adiabatic devices.

Condensers receive steam as it exhausts from the turbine, and this steam is then condensed to a saturated liquid by removal of heat. The Mollier diagram is an enthalpy-entropy diagram that can be used to determine the properties of steam. It is sometimes used instead of the steam tables.

The simple steam turbine cycle can be described more accurately by using a modified Rankine cycle that includes inefficiencies in the pumps, turbines, and boilers. The reheat cycle is a variation of the simple Rankine cycle, which allows more heat to be added at a high temperature. The reheat cycle uses a high-pressure turbine, a reheat section in a steam generator, and a low-pressure turbine. The thermal efficiencies of reheat cycles are usually higher than those of comparable simple Rankine cycle devices.

Regenerative heating is another manner of increasing the thermal efficiency of the steam turbine. Regenerative heating recirculates some of the thermal energy in the steam instead of rejecting that energy in the condenser. There may be more than one regenerative stage in a steam turbine.

The reheat–regenerative steam turbine is the state of the art in steam turbine power generation. This system uses both reheat and regeneration, and the thermal efficiencies are the highest obtained. The computer can be of assistance in analyzing the various steam turbine power cycles.

DISCUSSION QUESTIONS

Section 11–1

11–1 What is meant by the *Rankine cycle*?

Section 11–2

11–2 What is meant by *steam rate*?

11–3 How would you suggest ways to improve boiler efficiency?

Section 11–3

11–4 What is a *steam turbine*?

11–5 What is *adiabatic efficiency* as it applies to steam turbines?

Section 11–4

11–6 What purpose do *pumps* serve in the Rankine cycle?

Section 11–5

11–7 What purpose do *condensers* serve in the Rankine cycle?

Section 11–6

11–8 What is the relationship between *moisture* and quality?

Section 11–7

11–9 What is the *Mollier diagram*?

Section 11–8

11–10 What is *reheat* in a Rankine cycle?

11–11 How many *reheats* would normally be used in a Rankine cycle?

Section 11–9

11–12 What is *regeneration* in the Rankine cycle?

11–13 What components are needed for a regeneration?

Section 11–10

11–14 What purpose is served by using both reheat and regeneration?

Section 11–11

11–15 Is dealing with environmental concerns such as thermal and chemical pollution good economics?

11–16 What defines the technology of *cogeneration*?

11–17 What are the components of a *nuclear power cycle*?

PRACTICE PROBLEMS

Problems that use SI units are indicated by an (M) under the problem number; those that use English units are indicated by an (E). Mixed unit problems are indicated by a (C).

Section 11–1

11–1 A closed-cycle steam power plant delivers 10 MW of electrical power. If 10,000 W is required to drive water-recirculating pumps and if 1.2×10^6 Btu/min of heat is added to the cycle, determine the amount of heat rejected and the cycle efficiency.

11–2 Sketch the p–v and T–s diagrams for a Rankine cycle, and indicate which processes have work and heat present.

11–3 What are the source of power and the working medium in the common steam turbine cycle?

Sections 11–2 and 11–6

11–4 (M) There is 20 kg/s of water furnished to a boiler at 1500 kPa and as a saturated liquid. If the steam leaving the boiler is at 1500 kPa and 1400 kJ of heat is added per kilogram of water, determine
(a) Rate of heat addition
(b) Steam temperature

11–5 (M) A steam generator operates with an effectiveness of 85% when using coal as a fuel. If the incoming water is a saturated liquid at 1000 kPa and if there is 2000 kg of steam produced per hour at 400°C, determine the amount of coal needed per hour. Assume that the LHV of coal is 28,000 kJ/kg.

11–6 (E) A boiler must supply steam at 250 psia and 650°F. If the water supplied to the boiler is saturated liquid at 3 psia, determine the amount of heat added per unit mass of steam.

11–7 (E) A gas-fired steam generator uses a fuel with an LHV of 16,000 Btu/lbm. A burning rate of 6 lbm/min of fuel is expected, and the boiler is assumed to have an efficiency of 90%. If water is supplied to the boiler at conditions comparable to a saturated liquid at 130°F and if the steam produced is at 400 psia and 600°F, determine the mass flow of water through this unit.

Sections 11–3 and 11–6

11–8 (M) A turbine reversibly and adiabatically expands steam from 4000 kPa and 640°C to a saturated vapor. Determine the final steam temperature and the work produced per kilogram of steam.

11–9 (M) A turbine receives steam at 4000 kPa and 800°C. It expands the steam to 100 kPa and 240°C. Determine the work done per kilogram of steam and the turbine efficiency.

11–10 (E) In a turbine, 700 lbm/min of steam is isentropically expanded from 350 psia and 700°F to 10 psia. Estimate the power produced by the turbine.

11–11 (E) A turbine has an efficiency of 90% in expanding steam from 300 psia and 650°F to 3 psia. Determine the work done and the final steam temperature.

Sections 11–4 and 11–6

11–12 (C) Determine the size of motor you would recommend to drive the following pumps:
(a) A feedwater pump rated at 30,000 kg/min of water at 90°C from 70-kPa pressure to 1000-kPa pressure
(b) A feedwater pump rated at 50,000 lbm/min of water at 180°F from 2 to 380 psia

11–13 (C) Saturated liquid water is pumped into a steam generator. Determine the enthalpy of the steam entering the generator, if the water has a constant density through the pump, for the following conditions:
(a) Density of water is 917 kg/m^3, inlet pressure to pump is 500 kPa, and outlet pressure is 4000 kPa.
(b) Density of water is 60 lbm/ft^3, inlet pressure to pump is 5 psia, and outlet pressure is 600 psia

11–14 (C) For the two pumps given, determine the work done by the pumps per unit of mass of water in delivering saturated liquid water at a low pressure to a steam generator at a high pressure. Use the steam tables in appendix B, and assume that the pumps are reversible and adiabatic.
(a) Pump inlet pressure = 476 kPa; pump discharge pressure = 6000 kPa
(b) Pump inlet pressure = 10 psia; pump discharge pressure = 600 psia

11–15 What is the general equation to calculate pump work, starting from the steady-flow energy and assuming an adiabatic process as the only restriction?

Sections 11–5 and 11–6

11–16 (M) A condenser receives steam at 80 kPa and 95% quality. Determine the heat rejected per unit mass of steam if the exit condition is a saturated liquid.

11–17 (M) If a condenser is to reject 2000 kJ/s of heat and the mass (M)–flow of water is 80 kg/min, determine the steam quality of that steam entering the condenser if the pressure is 20 kPa and the water leaving the condenser is a saturated liquid.

11–18 (E) A sealed condenser is used to reject heat from a steam turbine cycle in which 10 lbm/s of steam is converted from a saturated vapor to a saturated liquid at 10 psia (figure 11–34). The coolant is water at 15 psia and has a

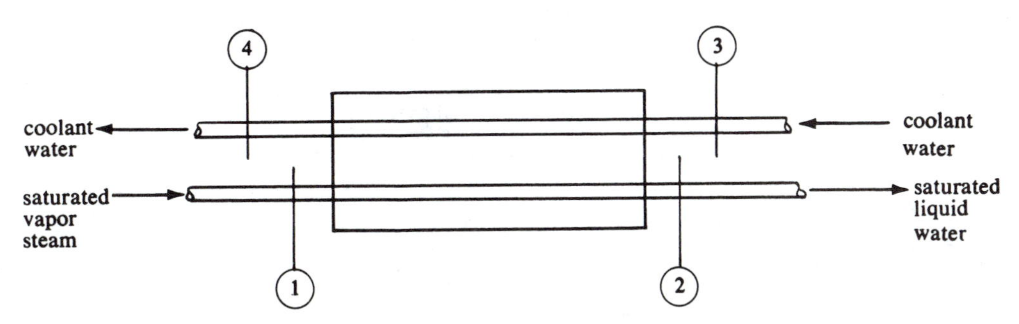

FIGURE 11–34

mass flow rate of 200 lbm/s. If the specific heat of the coolant is assumed to be 1 Btu/lbm · °R, determine the temperature increase of the coolant through the condenser.

11–19 (E) A condenser rejects heat from a steam turbine cycle and thereby converts the working medium (steam) from a saturated vapor to a saturated liquid at 120°F. Determine the entropy decrease per pound-mass of steam in the condenser and the entropy increase of the atmosphere. What is the total increase in entropy of the universe for the condenser process?

11–20 (M) A power plant is located on a river, and river water is used to cool the steam flowing through the condenser of the steam turbine power system. The plant produces 100 MW of electrical power and operates at 36% efficiency. If the river water flowing through the condenser is to not have an increase of temperature of more than 2°C, determine the flow rate of river water required.

Use the Mollier diagram for problems 11–21 through 11–24.

11–21 (C) Enthalpy of steam at
(a) 400°C and 1200 kPa
(b) 1000°F and 150 psia

11–22 (C) Entropy and enthalpy of steam at
(a) 200 kPa and 6% moisture
(b) 30 psia and 5% moisture

11–23 (C) Temperature and enthalpy of steam at
(a) 10 kPa and $s = 7.80$ kJ/kg · K
(b) 1.5 psia and $s = 1.9$ Btu/lbm · °R

11–24 (C) Enthalpy and pressure of steam with
(a) Temperature of 260°C and entropy of 5.80 kJ/kg · K
(b) Temperature of 500°R and entropy of 1.65 Btu/lbm · °R

Section 11–7

11–25 (M) A steam turbine produces 700 hp through a reversible adiabatic expansion from 2000 kPa and 480°C to 80°C. Neglecting pump work, determine the steam rate of the cycle.

11–26 (M) A 200 MW steam turbine power generation system operates on an ideal Rankine cycle. The steam enters the turbine at 640°C and 14 MPa, and it leaves at 86°C to the condenser. Determine
(a) The mass flow rate of steam required for the generation of 200 MW of power
(b) Cycle efficiency

11–27 (M) An ideal closed-cycle steam turbine uses 5500 kg/h of steam. The steam leaves the boiler at 4000-kPa pressure and is expanded reversibly and adiabatically to 60 kPa and 95% quality in the turbine. The water leaving the condenser is a saturated liquid. Using the Mollier diagram, compute
(a) Turbine power produced
(b) Heat rejected
(c) Pump work
(d) Heat added
(e) Cycle efficiency and steam rate

11–28 (M) A steam generating unit supplies 6000 kg/h of steam at 7 MPa and 800°C. Assume that the generator burns pulverized coal having an LHV of 30,000 kJ/kg and that the heat transfer effectiveness is 88%. A turbine expands this steam at 95% efficiency to 30°C, and a condenser allows the steam to revert to a saturated liquid. If the pumps are assumed to be 100% efficient, determine
(a) Heat added to cycle
(b) Heat added to water in boiler
(c) Work of turbine
(d) Heat rejected
(e) Net cycle work
(f) Overall cycle efficiency
(g) Steam rate
(h) Minimum entropy generation

11–29 (E) A 200-MW power station operates on a closed-cycle steam turbine, as shown in figure 11–35. The mechanical efficiency of the turbine–electric generator unit is 95%, and the turbine efficiency in extracting energy from steam is 90%. The steam expands from 300 psia

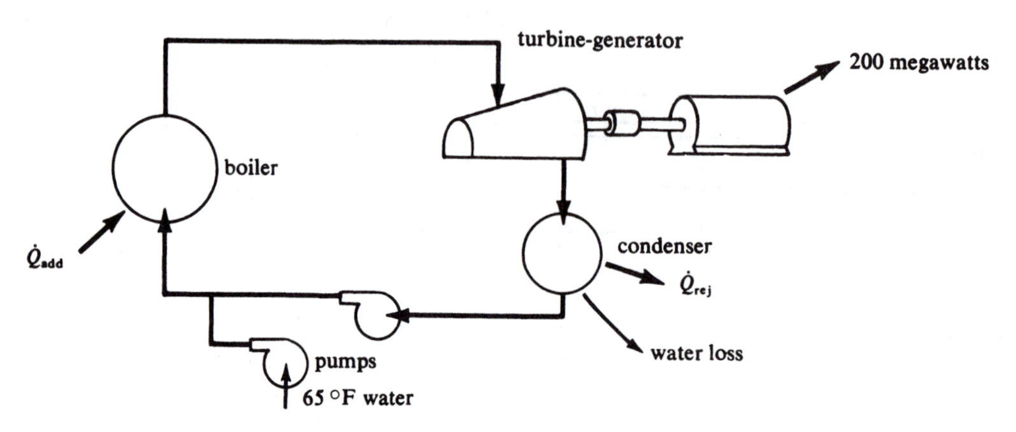

FIGURE 11–35

and 1050°F to 12 in Hg in the turbine and is condensed to a saturated liquid in the condenser. The condenser is assumed to lose 5% of the steam to the atmosphere. A feedwater pump inserts 65°F water into the system after the circulating pump. Determine

(a) Mass flow rate of steam through turbine
(b) Heat added to water in boiler
(c) Heat rejected in condenser
(d) Work of both pumps (assume water to be incompressible at all points)
(e) Cycle efficiency
(f) Minimum entropy generation

11–30 (E) A steam turbine cycle produces a net power of 10 MW. The turbine generator unit has a mechanical efficiency of 100% while steam is expanded reversibly and adiabatically through the turbine from 190 psia and 900°F to 5 psia. The water leaving the condenser is saturated liquid and is incompressible. If the boiler unit has a heat transfer efficiency of 90% and burns coal having an LHV of 11,000 Btu/lbm, determine

(a) Pump work per unit mass of steam
(b) Power produced by turbine
(c) Rate of coal consumption (lbm/h)
(d) Rate of heat rejection

11–31 (C) Determine how the condenser pressure affects the power generated in a steam turbine by varying the turbine exhaust pressure while holding turbine inlet conditions the same. Use of the program RANKINE or *EES* is recommended for this problem.

11–32 Determine how turbine adiabatic efficiency affects the amount of steam required, or the steam rate, to produce the same output power. Do this study by varying the adiabatic efficiency of the turbine, using the same inlet and exhaust pressures and the same inlet steam temperature. Use of the program RANKINE or *EES* is recommended for this problem.

Section 11–8

11–33 (M) In an ideal turbine reheat cycle, the turbine reversibly and adiabatically expands 60 kg/s of steam from 3000 kPa and 880°C to 400 kPa. The steam is then reheated to 880°C and expanded further to 10 kPa. Determine

(a) Steam rate
(b) Cycle efficiency

11–34 (E) For the reheat cycle shown in figure 11–36, determine

(a) Turbine output (horsepower)
(b) Rate of heat added
(c) Steam rate
(d) Cycle efficiency

11–35 (C) Determine how the thermal efficiency of a steam turbine reheat cycle is affected by the pressure of the reheat stage. Do this study by varying the exhaust pressure of the high-pressure turbine. Use the same pressure and temperature at the inlet to the high-pressure turbine and the same exhaust pressure of the low-pressure turbine. Use of the program RANKINE or *EES* is recommended for this problem.

Section 11–9

11–36 (M) Steam at 7-MPa pressure and 560°C is supplied to a regenerative steam turbine with a single extraction. If the cycle is ideal, the steam is expanded in the turbine to 10-kPa pressure, and the steam is extracted at 500 kPa, determine, per unit mass of steam,

(a) Amount of steam extracted per unit time
(b) wk_{turb}
(c) Cycle efficiency

11–37 (C) For the ideal regenerative cycle shown in figure 11–37, determine

(a) $\dot{m}_4$ and $\dot{m}_5$
(b) $\dot{W}k_c$, $\dot{W}k_b$, and $\dot{W}k_a$,
(c) $\dot{W}k_{turb}$
(d) Cycle efficiency
(e) Q_{rej}

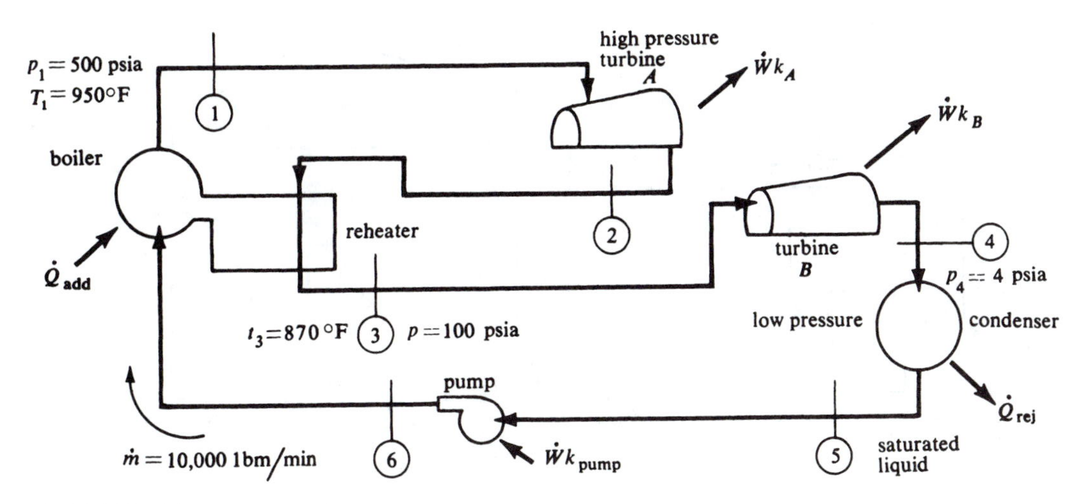

FIGURE 11–36

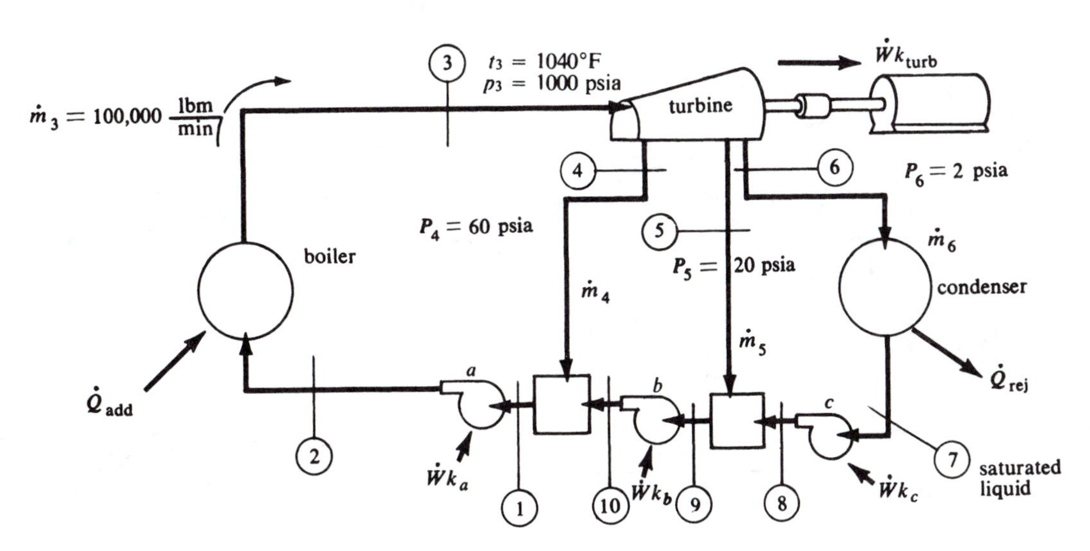

FIGURE 11–37

11–38 (C) Determine how the number of regenerative heating stages affects the thermal efficiency of a steam turbine. Do this study by using the same boiler pressure and temperature, the same exhaust pressure to the condenser, and by varying the number of regenerative heating stages from one to five stages. For each situation, use regenerative pressures equally spaced between the boiler pressure and the condenser pressure. Use of the program or *EES* is recommended for this problem.

Section 11–10

11–39 For the reheat–regenerative cycle shown in figure 11–38, determine

(a) Cycle efficiency
(b) Power produced
(c) Steam rate

11–40 For the reheat–regenerative cycle shown in figure 11–39, determine

(a) $\dot{W}k_{turb}$
(b) $\dot{W}k_{cycle}$
(c) $\dot{Q}_{rej}$
(d) Cycle efficiency
(e) Steam rate

11–41 (E) Determine the adiabatic efficiency of the high-pressure turbine of the steam turbine cycle in figure 11–30.

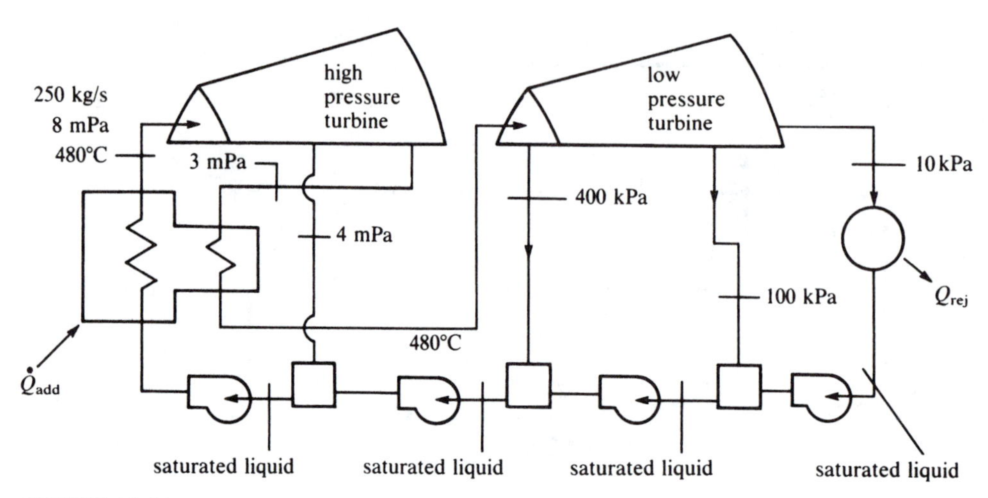

FIGURE 11–38

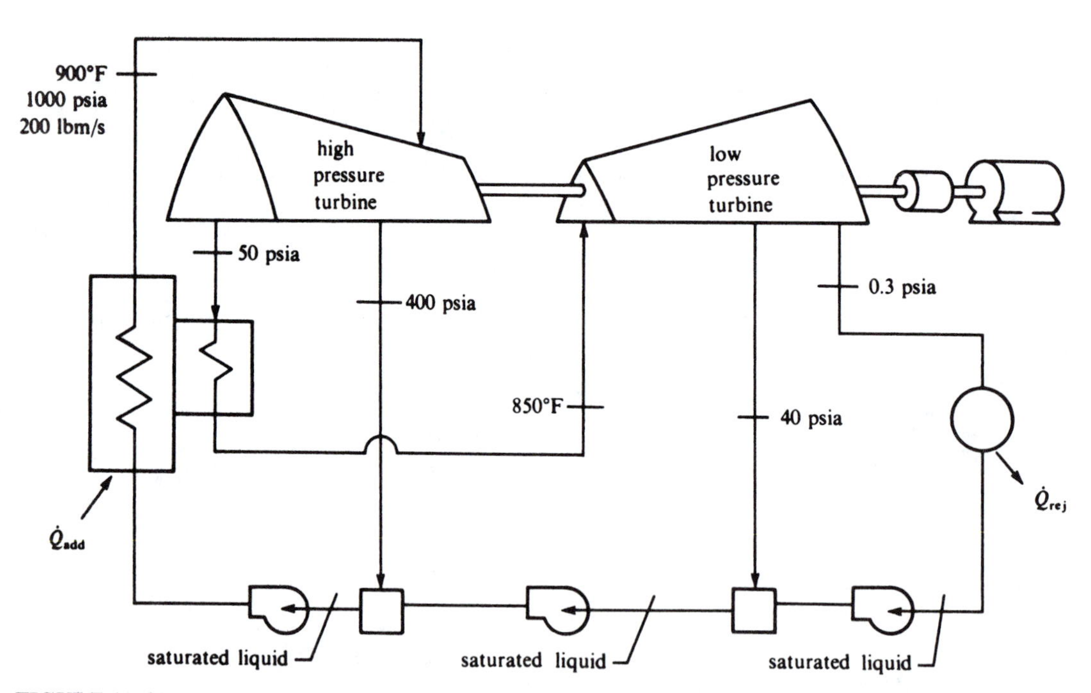

FIGURE 11–39

12 REFRIGERATION AND HEAT PUMPS

In this chapter, we consider the reversed heat engines and their operating cycles. The various devices that utilize the reversed heat engine cycles provide the mechanisms for today's air conditioners, refrigerators, freezers, and some heaters. Although all of these devices can be classified as heat pumps, it is the heating devices that are generally referred to as heat pumps.

We consider first the reversed Carnot cycle in order to introduce the basic concepts of refrigeration and some of the terminology associated with them and the heat pumps. Next we introduce the vapor compression refrigeration cycles. Vapor compression is the most common method used by today's actual refrigeration and heat pump devices and is therefore an important system for students to know. The various working media or refrigerants are discussed, and examples of the analyses of some vapor compression systems are shown.

The reversed Brayton cycle is used to analyze systems that provide air conditioning for many modern aircraft. Absorption refrigeration systems are introduced to give the student some idea of the operation of systems that do not require electricity or other external power sources.

Cryogenics is defined, and gas liquefaction is discussed. A simple outline of the purposes of cryogenics and the methods of achieving very low temperatures are given. Finally, the heat pump as applied to the problem of heating is considered, and some of the operating characteristics are presented.

12–1 THE REVERSED CARNOT CYCLE

We have discussed the Carnot cycle as a heat engine operating between two temperature regions and producing work. In reverse, the Carnot cycle transfers energy from a lower-temperature region to a higher-temperature one; we have already been introduced to this device, called the *heat pump*. With this arrangement, the Carnot cycle is defined by the following four reversible processes:

1–2 Isothermal heat addition
2–3 Adiabatic compression
3–4 Isothermal heat rejection
4–1 Adiabatic expansion to initial state 1

The property diagrams are given in figure 12–1 and the system diagram in figure 12–2. Notice that the enclosed areas in the property diagrams must be equal for the reversible cycle, so that

$$Wk_{\text{cycle}} = Q_{\text{net}} = \sum Q = Q_{\text{add}} + Q_{\text{rej}} \tag{12–1}$$

For the Carnot heat pump, we have

$$\begin{aligned} Q_{\text{add}} &= T_L\,\Delta S \\ Q_{\text{rej}} &= -T_H\,\Delta S \end{aligned} \tag{12–2}$$

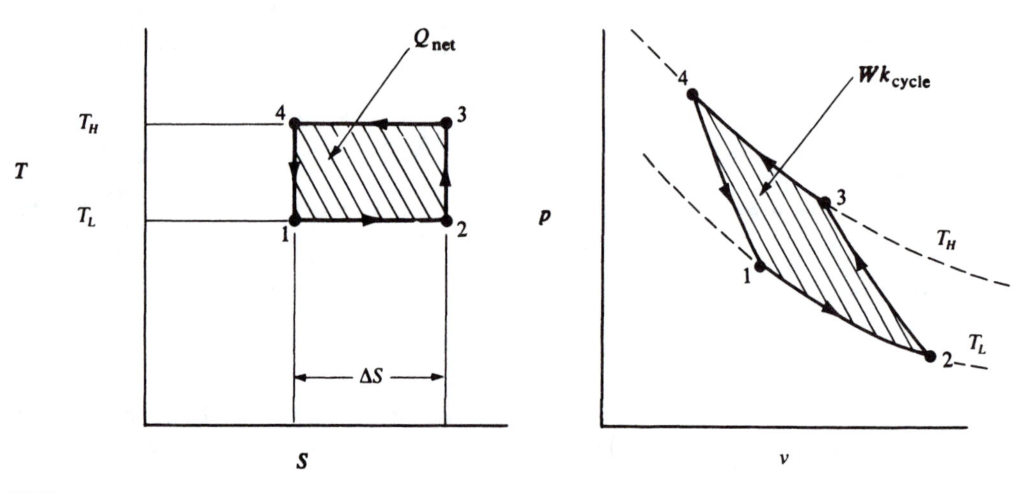

FIGURE 12–1 Property diagrams for Carnot heat pump

where ΔS is the absolute value of the entropy change during the isothermal heat rejection and isothermal heat addition. Then

$$Q_{net} = (T_L - T_H)\,\Delta S = Wk_{cycle} \quad \textbf{(12–3)}$$

and the coefficient of performance COP, defined by equation (7–25), namely,

$$\text{COP} = \frac{\text{heat rejected}}{\text{net cycle work}} = \frac{Q_{rej}}{Wk_{cycle}}$$

reduces to

$$\text{COP} = \frac{T_H}{T_H - T_L} \quad \textbf{(12–4)}$$

FIGURE 12–2 System diagram of Carnot heat pump

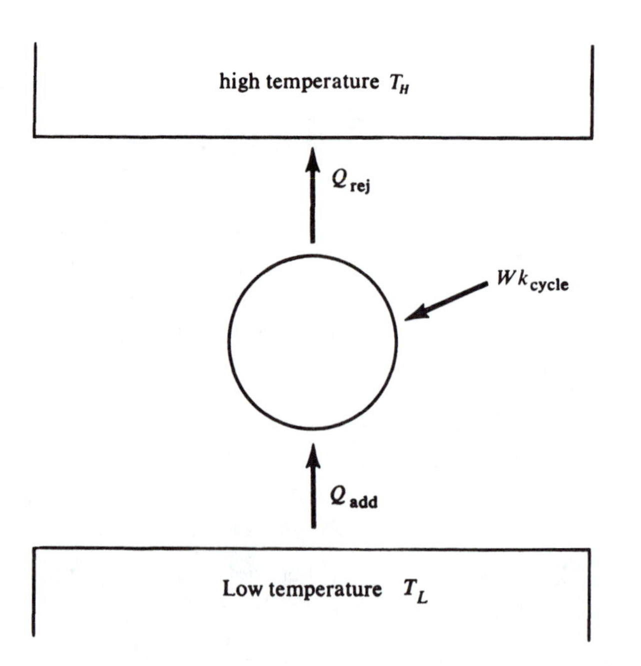

Earlier, we defined the coefficient of refrigeration, COR, by equation (7–26):

$$\text{COR} = \frac{Q_{\text{add}}}{-Wk_{\text{cycle}}}$$

For the Carnot heat pump,

$$\text{COR} = \frac{T_L}{T_H - T_L} \tag{12–5}$$

One of the common units in the English system for describing the refrigerating capacity of a heat pump device (such as an air conditioner or freezer) is the *ton*.

> **1 ton of refrigeration:** ***Amount of energy removed from 1 ton of water at 32°F and 14.7 psia, in converting from a pure liquid to a solid (ice) over a period of 24 hours.***

Thus, the unit ton is a description of the rate of heat transfer, and we can give it further meaning by noting that 144 Btu is required to convert 1 lbm of saturated liquid H_2O at 32°F and 14.7 psia into ice at 32°F. Then

$$\begin{aligned} 1 \text{ ton refrigeration} &= 144 \text{ Btu/lbm} \times 2000 \text{ lbm/ton} \times 1 \text{ day/24 h} \\ &= 12{,}000 \text{ Btu/h} \\ &= 200 \text{ Btu/min} \\ &= 3.33 \text{ Btu/s} \\ &= 3.5098 \text{ kW} \end{aligned}$$

Often, the term BTUH is used to describe the capacity of an air conditioner or refrigerator. This unit should be read as Btu per hour or Btu/h. Figure 12–3 shows a modern room air conditioner whose cooling capacity is described in BTUH units, to be read as Btu per hour.

FIGURE 12–3 Carrier's International™ Series Straight Cool CM and GM models, with cooling capacities ranging from 8800 to 27,500 BTUH, are designed for medium- and larger-sized rooms (courtesy of United Technologies, Carrier, Syracuse, NY)

EXAMPLE 12–1 A room needs to be kept at 25°C when the atmospheric temperature is 50°C. If the room requires a 2-ton air conditioner, determine the minimum amount of power required and the minimum amounts of heat put in the surroundings due to this operation.

Solution We note that the best device conceivable for air conditioning would be a reversible Carnot heat engine. For this device, from equation (12–5), the coefficient of refrigeration is

$$\text{COR} = \frac{25°\text{C} + 273\ \text{K}}{(50°\text{C} + 273\ \text{K}) - (25°\text{C} + 273\ \text{K})} = 11.9$$

Then, since the added heat $\dot{Q}_{\text{add}}$ is given as 2 tons (= 7.0196 kW), we can write, from equation (7–26),

$$11.9 = \frac{\dot{Q}_{\text{add}}}{-\dot{W}k_{\text{cycle}}}$$

and we then get

$$\dot{W}k_{\text{cycle}} = \frac{-7.0196\ \text{kW}}{11.9} = -0.5899\ \text{kW}$$

Recall that 0.746 kW = 1.0 hp, so we can convert this answer to

Answer
$$\dot{W}k_{\text{cycle}} = -0.79\ \text{hp}$$

The amount of heat put into the surroundings is simply the heat rejected or, from equation (12–1),

$$\dot{Q}_{\text{rej}} = \dot{W}k_{\text{cycle}} - \dot{Q}_{\text{add}}$$

and substituting values into this equation gives

$$\dot{Q}_{\text{rej}} = -0.5899\ \text{kW} - 7.0196\ \text{kW}$$

Answer
$$= -7.6095\ \text{kW}$$

Interestingly, although the room is cooled for a time, the net effect of an air conditioner or any other heat pump device is to increase the temperature of the surroundings or the universe, which includes the room itself. Further, there is an increasing interest in the application of the heat pump as a heating device. The ability of this mechanism to transfer heat from a cold to a hot or warmer environment represents a significant characteristic for heating.

EXAMPLE 12–2 A Carnot heat pump is proposed to heat a dwelling in cold climates. If the expected minimum climatic temperature is −30°C and the dwelling is to be kept at 27°C, determine the minimum power required. It is anticipated that 10 kW of heat is required.

Solution The heat pump is used in heating when the rejected heat $\dot{Q}_{\text{rej}}$ is the desired output and is utilized to heat a dwelling or other facility. Here we use 10 kW of heat for heating, and from the COP, we have

$$\frac{\dot{Q}_{\text{rej}}}{\dot{W}k_{\text{cycle}}} = \frac{T_H}{T_H - T_L}$$

Then we obtain, for the work,

$$\dot{W}k_{\text{cycle}} = \frac{\dot{Q}_{\text{rej}}(T_H - T_L)}{T_H} = \frac{(10\ \text{kW})(300\ \text{K} - 243\ \text{K})}{27°\text{C} + 273\ \text{K}}$$

Answer
$$= 1.9\ \text{kW}$$

Notice that the heat pump provides more energy as heat than is supplied as work. Thus, compared with, say, electric resistance heating or an oil or gas furnace, the heat pump is more efficient. However, there are some design problems associated with the heat pump which detract significantly from its value as a heating device. We discuss this further in section 12–7.

12–2 THE VAPOR COMPRESSION CYCLES

Today there are a large number of devices operating on the vapor compression cycle. Nearly all refrigerators, freezers, air conditioners, and heat pumps use systems that can be described as vapor compression units, so the study of vapor compression cycles is an important part of applied thermodynamics. The vapor compression cycle is defined by the following processes:

1–2 Reversible constant-pressure heat addition during a phase change of the working medium or refrigerant

2–3 Reversible adiabatic compression

3–4 Reversible constant-pressure heat rejection in which the working medium condenses to a saturated liquid

4–1 Throttling expansion at constant enthalpy to a low pressure

This cycle has several distinguishing features. First, the working medium that is progressing through the cycle has an evaporation process at low temperature and pressure and condensation to a saturated liquid at high temperature and pressure. Also, while three of the four processes are reversible, an irreversible throttling process makes the cycle an irreversible one.

The typical pressure–volume diagram for the ideal reversible vapor compression cycle is depicted in figure 12–4, and the corresponding temperature–entropy diagram is shown in figure 12–5. Notice that a vapor compression can be performed with dry vapor (superheated) or a mixture of saturated vapor and liquid. As the descriptions of the cycles imply, the dry compression cycle involves a compression process 2–3 with a dry vapor, while the wet vapor compression cycle involves a mixture of vapor and liquid through the compression.

The dry compression cycle seems to be more popular in application to actual refrigeration devices even though the wet compression more closely approximates the reversed Carnot cycle; that is, the COP of the wet compression cycle would be expected to exceed the COP of the dry compression, both operating between the same pressures. The reason for the success of the dry compression cycle is that compressors typically perform better with a pure vapor than with a mixture of vapor and liquid.

Another tool often used to evaluate and analyze vapor compression cycles is the *pressure–enthalpy diagram*. Figure 12–6 shows these diagrams for normal dry and wet

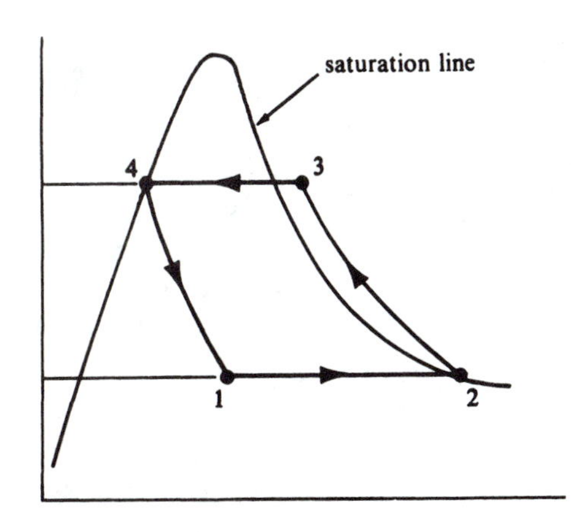

FIGURE 12–4 p–v diagram for ideal vapor compression refrigeration cycle

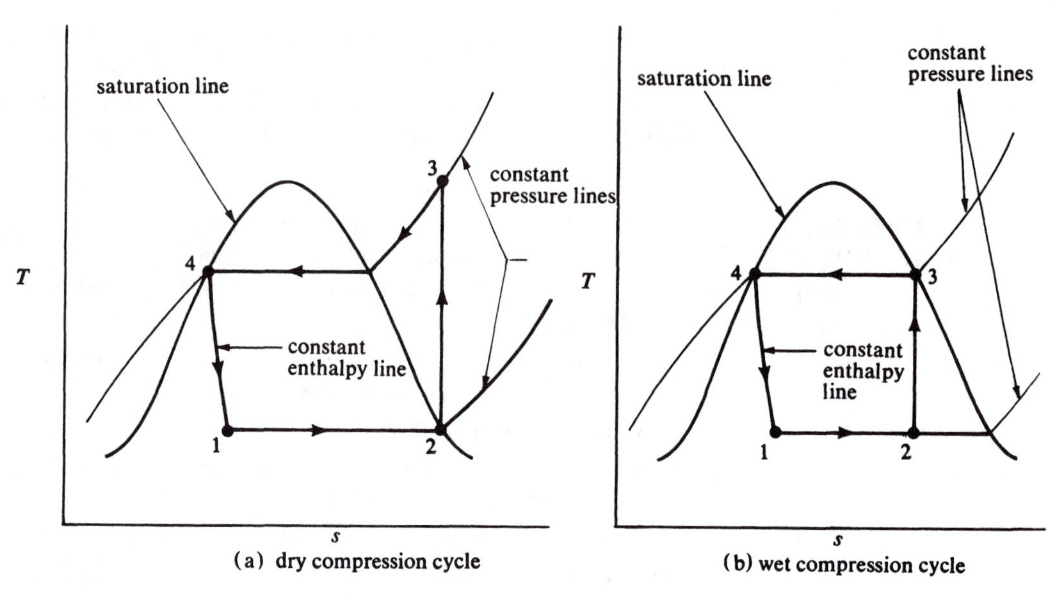

FIGURE 12–5 $T-s$ diagram for ideal refrigeration cycles

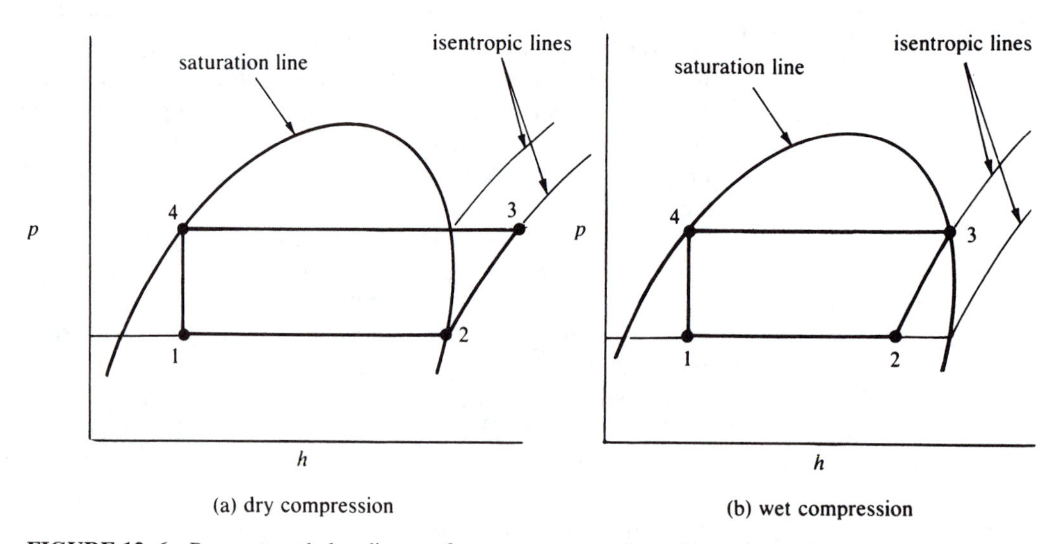

FIGURE 12–6 Pressure–enthalpy diagram for vapor compression refrigeration cycle

vapor compression refrigeration cycles, with the saturation line again determining the limits of the cycle as it does in the T–s diagram. From this observation, we may expect that a critical decision in the design of a refrigeration unit is in the selection of the working medium or *refrigerant*.

Refrigerants are normally selected for vapor compression cycles by the following criteria:

1. Economical
2. Nontoxic or harmless to the surroundings
3. Nonflammable
4. High latent heat of condensation (h_{fg}) at refrigerating temperature
5. Low saturation pressure at operating temperature

Thermodynamically, criterion 4 represents the most significant aspect of working media. A high latent heat of condensation reflects the capability for a high amount of heat addition to the cycle refrigerant per unit of refrigerant mass. In table 12–1 are listed the heats of condensation h_{fg} for some common refrigerants at typical saturation pressures, in order that a quick comparison can be made. Remember, however, that the heat of vaporization alone should not determine the most desirable refrigerants. Even some properties other than those listed here, such as viscosity, solubility, or thermal conductivity, could enter into the requirements for selecting a working medium for refrigeration cycles.

Some of the substances used for refrigeration processes are

Ammonia
Sulfur dioxide
Carbon dioxide
Methyl chloride
Dichlorodifluoromethane (Refrigerant-12, R-12)
Chlorodifluoromethane (Refrigerant-22, R-22)
Propane
Butane
Dichloromethane (carrene)

Refrigerants R-12 and R-22 have been identified as contributors to the degradation of the Earth's environment, particularly as causes of the depletion of the ozone (O_3) at upper elevations and specifically over the north and south poles. Ozone at upper elevations of the atmosphere seems to act to reduce the sun's radiation, and if this ozone is eliminated, there could be very serious worldwide damage to the Earth's environment and to human beings. As a consequence of this new information, R-12 is being discontinued for use by responsible manufacturers and by governments and their agencies. Even though the use of R-12 will eventually be eliminated, the engineer and the engineering technologist need to understand its behavior as a refrigerant. Tables B–15 through B–20 present the thermodynamic properties of ammonia, R-12, and R-22 for quantitative analysis of vapor compression cycles. If the reader prefers to obtain information from charts rather than tables, chart B–4 presents the properties of R-12 on a pressure–enthalpy, p–h, diagram, and chart B–5 presents the properties of R-22.

TABLE 12–1 Heat of condensation h_{fg} of some common refrigerants

		Saturation Pressure		Saturation Temperature		h_{fg}	
Refrigerant	**Formula**	**kPa**	**psia**	**°C**	**°F**	**kJ/kg**	**Btu/lbm**
Ammonia	NH_3	332	48.21	−6	20	1240	533.1
Sulfur dioxide	SO_2	118	17.18	−6	20	384	165.3
Carbon dioxide	CO_2	1005	145.8	−40	−40	317	136.5
Methyl chloride	CH_3Cl	202	29.3	−6	20	412	178.4
R-12	CCl_2F_2	246	35.7	−6	20	158	67.9
R-22	$CHClF_2$	240	34.7	−20	−5	221	94.9
R-22	$CHClF_2$	398	57.7	−6	20	211	90.5
R-134a	CF_3CFH_2	125	18.17	−21	−6	215	92.5
R-123	CCl_2HCF_3	20	2.90	−13.3	−10	182	78.2

Tetrafluoroethane (R-134a) and trifluoroethane (R-123) have been used as replacements for R-12 and R-22. The saturation and superheat properties of R-134a are given in table B–22, and the saturation and superheat properties of R-123 are given in table B–21. In table 12–1, it can be seen that the heats of vaporization of R-134a and R-123 compare favorably with those of the other common refrigerants. Chart B–6 presents properties of R-123 on a pressure–enthalpy diagram, and chart B–7 presents properties of R-134a for readers who prefer to obtain data from charts instead of tables. Although the precision of the data is not as good with charts as it is with tabulated data, the accuracy is sufficient for most engineering and technological applications, and often the information can be gotten more conveniently than with tables.

In addition, there have been mixtures or *blends* of two or three refrigerants that have been found to be attractive for certain applications. Two of these blends that we will consider are R-407*c* and R-502. R-407c is a precise mixture of 23% R-32, 25% R-125, and 52% R-134a, and the saturation properties for this refrigerant are given in table B–23. The $p-h$ diagram for R-407c is shown in chart B–8. R-502 is a mixture of 48.8% R-22 and 51.2% R-115, and its saturation properties are listed in table B–24. The $p-h$ diagram for R-502 is given in chart B–9.

The typical physical components that constitute a vapor compression refrigeration cycle are shown in figure 12–7. Here we see that an evaporator or cooling coil absorbs energy or accepts heat addition from an external region by utilizing a working medium during its phase change. (See figure 12–5 or 12–6.) The refrigerant collects energy and is a saturated vapor (or near to that state) as it enters the compressor. The compressor increases the pressure and, more important, the temperature of the refrigerant as it leaves. Here the working medium is probably a superheated vapor or, in the case of a wet vapor compression, a saturated vapor. The condenser allows the refrigerant to release much of its energy in a heat rejection. Then the refrigerant returns through an expander or restriction to the evaporator.

By applying the steady-flow energy equation to the various components, we obtain the following (making appropriate assumptions on the individual items):

1. For the evaporator,

$$h_2 - h_1 = q_{\text{add}} \tag{12–6}$$

2. For the compressor,

$$h_3 - h_2 = -wk_{\text{comp}} \tag{12–7}$$

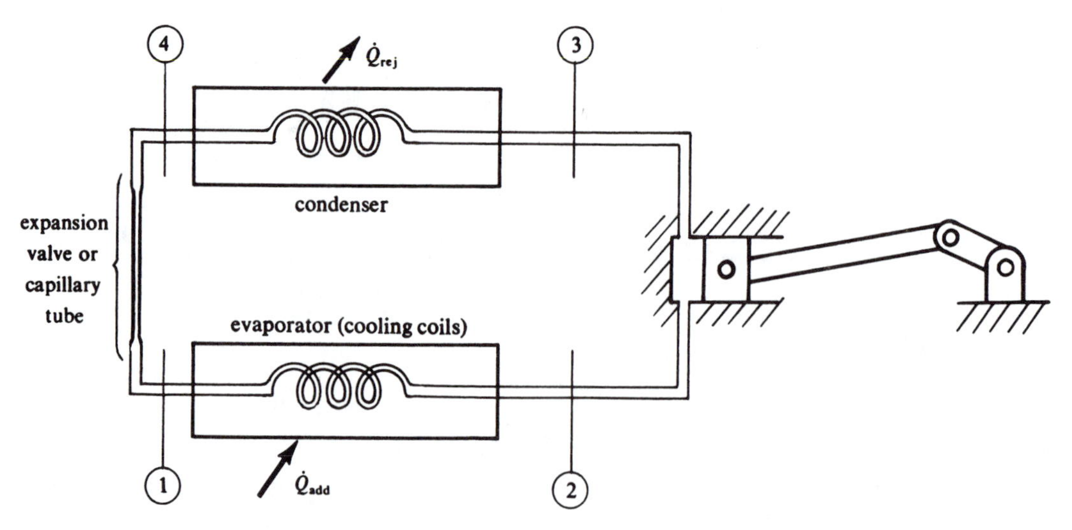

FIGURE 12–7 Sketch of typical vapor compression refrigerator

3. For the condenser,

$$h_4 - h_3 = q_{rej} \tag{12–8}$$

4. For the expander,

$$h_1 - h_4 = 0 \tag{12–9}$$

The coefficient of performance will be

$$\text{COP} = \frac{\dot{Q}_{rej}}{\dot{W}k_{cycle}} = \frac{q_{rej}}{wk_{cycle}}$$

which can be written as

$$\text{COP} = \frac{h_4 - h_3}{h_2 - h_3} \tag{12–10}$$

for the vapor compression cycle. Also, the coefficient of refrigeration is

$$\text{COR} = \frac{h_2 - h_1}{h_2 - h_3} \tag{12–11}$$

Notice the difference between these last two equations and those applicable to the Carnot heat pump [equations (12–4) and (12–5)].

12–3 ANALYSIS OF VAPOR COMPRESSION REFRIGERATION SYSTEMS

The analysis of vapor compression refrigeration systems can begin only after the specific application of the system is defined. These devices can be used in varied applications requiring different operating temperatures and heat transfer rates. We begin with the wet or dry compression cycles and demonstrate how they may be used to help analyze the vapor compression systems.

EXAMPLE 12–3

A refrigerator operates with ammonia as the working medium, and the system removes 1000 Btu/h from a freezer compartment at 26°F. If the heat is rejected to a room that is at 85°F and the refrigerator can be described by an ideal wet vapor compression cycle with saturated vapor ammonia leaving the compressor, determine

(a) COP and COR
(b) The flow rate of the ammonia through the cycle
(c) The power required

Solution

This refrigerator can be described by the sketch given in figure 12–7 and the property diagrams for the cycle as those given in figures 12–4, 12–5b, and 12–6b.

(a) From the ammonia table (table B–19,) we obtain directly, or through interpolation, the values

$$h_3 = h_g \text{ at } 85°\text{F} = 631.4 \text{ Btu/lbm}$$
$$h_4 = h_f \text{ at } 85°\text{F} = 137.8 \text{ Btu/lbm}$$
$$h_1 = h_4 \text{ and } s_3 = 1.1919 \text{ Btu/lbm} \cdot °\text{R}$$

Since $s_2 = s_3$, we can write

$$s_2 = \chi(s_{g2} - s_{f2}) + s_{f2} = 1.1919$$

Also from table B–19, because $t_2 = 26°\text{F}$, we have

$$s_{g2} = 1.2762 \quad h_{fg2} = 548.1 \text{ Btu/lbm}$$
$$s_{f2} = 0.1573 \quad h_{f2} = 71.3 \text{ Btu/lbm}$$

and then we can calculate the quality χ:

$$\chi = \frac{1.1919 - 0.1573}{1.2762 - 0.1573} = 0.925$$

Hence,

$$\begin{aligned} h_2 &= \chi(h_{fg2}) + h_{f2} \\ &= (0.925)(548.1) + 71.3 = 578.2 \text{ Btu/lbm} \end{aligned}$$

and from equation (12–6),

$$q_{add} = h_2 - h_1 = 578.2 - 137.8 = 440.4 \text{ Btu/lbm}$$

From equation (12–8),

$$q_{rej} = h_4 - h_3 = 137.8 - 631.4 = -493.6 \text{ Btu/lbm}$$

From equations (12–10) and (12–11), we get

Answer

$$\text{COP} = \frac{-493.6}{578.2 - 631.4} = 9.278$$

and

Answer

$$\text{COR} = \frac{440.4}{578.2 - 631.4} = 8.278$$

(b) From the given condition that

$$\dot{Q}_{add} = 1000 \text{ Btu/h}$$

we can write

$$\dot{Q}_{add} = \dot{m}q_{add} = \dot{m}(440.4 \text{ Btu/lbm})$$

Then

Answer

$$\dot{m} = \frac{1000 \text{ Btu/h}}{440.4 \text{ Btu/lbm}} = 2.27 \text{ lbm/h}$$

(c) The power required can be determined from the relationship

$$\begin{aligned} \dot{W}k_{cycle} &= \dot{m}wk_{cycle} \\ &= (2.27 \text{ lbm/h})(578.2 - 631.4 \text{ Btu/lbm}) \\ &= 120.8 \text{ Btu/h} \end{aligned}$$

or

Answer

$$\begin{aligned} \dot{W}k_{cycle} &= (120.8 \text{ Btu/lbm})(1/2545 \text{ Btu/hp} \cdot \text{h}) \\ &= 0.047 \text{ hp} \end{aligned}$$

EXAMPLE 12–4 An air conditioner operates in 40°C weather to keep a room at 20°C by withdrawing 12,000 kJ/h of heat. Compare the answers you obtained by using the appendix table with those you obtained by using *EES*.

(a) Assuming that the evaporator and condenser have perfect heat conduction, that the cycle is an ideal dry compression cycle, and that the working medium is Refrigerant-22, determine the rate of heat rejected to the atmosphere, the power required, the COP, and the COR.

(b) If the air conditioner is redesigned so that it now uses R-134a, determine the rate of heat rejected, the power required, the COP, and the COR.

Solution The air conditioner is physically describable by the sketch in figure 12–7, and the operating cycle is depicted by the property diagrams in figures 12–6a, 12–5a, and 12–4. To determine the heat rejection, the power, and the coefficient of performance, we must determine the enthalpy values. From table B–17, we obtain the following values:

$$h_2 = h_g \text{ at } 20^\circ\text{C} = 411.9 \text{ kJ/kg}$$
$$s_2 = 1.72462 \text{ kJ/kg} \cdot \text{K}$$

The pressure at state 3 must be the saturation pressure at 40°C or 1.5335 MPa. Then we may obtain h_3 from a double interpolation.

We have at state 3 that $s_3 = s_2 = 1.72462 \text{ kJ/kg} \cdot \text{K}$ and $p_3 = 1.5335$ MPa. Using table B–18, we interpolate once at 1.288 MPa between saturation properties and at 55°C to determine enthalpy at an entropy of 1.72462 kJ/kg · K. This value is found to be 420.49 kJ/kg. Then we interpolate once more at 1.571 MPa (to have values for enthalpy both above and below 1.5335 MPa). The enthalpy here is 425.14 kJ/kg. The final interpolation between 1.571 and 1.288 MPa, with $s = 1.72462 \text{ kJ/kg} \cdot \text{K}$, and solving for enthalpy, gives

$$\frac{h_3 - 425.14}{420.49 - 425.14} = \frac{1.5335 - 1.571}{1.2880 - 1.571}$$

so that

$$h_3 = 424.52 \text{ kJ/kg}$$

Also,

$$h_4 = h_f \text{ at } 40^\circ\text{C} = 249.686 \text{ kJ/kg}$$

and

$$h_1 = h_4$$

We can now proceed to compute the required terms. The heat rejected can be obtained from equation (12–8), or

$$q_{\text{rej}} = h_4 - h_3 = 249.686 \text{ kJ/kg} - 424.52 \text{ kJ/kg}$$
$$= -174.8 \text{ kJ/kg}$$

and the rate of heat rejection is then

$$\dot{Q}_{\text{rej}} = \dot{m} q_{\text{rej}}$$

The mass flow rate is calculated from

$$\dot{Q}_{\text{add}} = \dot{m} q_{\text{add}} = \dot{m}(h_2 - h_1) = 12.000 \text{ kJ/h}$$

or

$$\dot{m} = \frac{12{,}000 \text{ kJ/h}}{411.9 \text{ kJ/kg} - 249.686 \text{ kJ/kg}}$$
$$= 73.98 \text{ kg/h}$$

so

$$\dot{Q}_{\text{rej}} = (73.98 \text{ kg/h})(-174.8 \text{ kJ/kg})$$

Answer

$$= -12{,}931.7 \text{ kJ/h}$$

The power is computed from

$$\dot{W}k_{\text{cycle}} = \dot{m} wk_{\text{cycle}} = \dot{m}(h_2 - h_3)$$
$$= (73.98 \text{ kg/h})(411.9 \text{ kJ/kg} - 424.52 \text{ kJ/kg})$$

Answer

$$= -933.63 \text{ kJ/h}$$

We may write this answer in more familiar units:

Answer
$$\dot{W}k_{cycle} = -0.259 \text{ kW}$$

From equations (12–10) and (12–11), we calculate

$$\text{COP} = \frac{-12{,}933.63 \text{ kJ/h}}{-933.63 \text{ kJ/h}}$$

Answer
$$= 13.85$$

and

$$\text{COR} = \frac{-12{,}000 \text{ kJ/h}}{-933.63 \text{ kJ/h}}$$

Answer
$$= 12.85$$

Using the *EES Equation* window, checking that SI units with temperature in degrees Celsius and pressure in kPa, we write the following equations:

$$h2 = \text{enthalpy (R22, } T = 20, x = 1.0)$$
$$s2 = \text{entropy (R22, } T = 20, x = 1.0)$$
$$s3 = s2$$
$$h3 = \text{enthalpy (R22, } P = 1533.5, s = s3)$$
$$h4 = \text{enthalpy (R22, } T = 40, x = 0.0)$$
$$h1 = h4$$
$$qa = 12{,}000$$
$$m = qa/(h2 - h1)$$
$$qr = m*(h4 - h3)$$
$$w = m*(h2 - h3)$$
$$\text{COP} = qr/w$$
$$\text{COR} = qa/w$$

Then, clicking the *Calculation* option and then *Solve* gives answers that are in good agreement with those obtained from the appendix tables,

$$q_r = -12{,}934 \text{ (kJ/hr)}$$
$$w = -933.9 \text{ (kJ/hr)}$$
$$\text{COP} = 13.85$$
$$\text{COR} = 12.85$$

(b) From the saturation properties section of table B–22, we read

$$h_2 = h_g \quad \text{at } 20°\text{C} = 259.0 \text{ kJ/kg}$$
$$s_2 = s_g = 0.912 \text{ kJ/kg} \cdot \text{K}$$

From table B–22, the saturation pressure at 40°C can be read as the pressure at state 3:

$$p_3 = 1017 \text{ kPa}$$

From the superheat section of table B–22, we may determine the enthalpy at state 3. First, we notice that the pressure and entropy are known at state 3:

$$s_3 = s_2 = 0.912 \text{ kJ/kg} \cdot \text{K}$$

By a double interpolation, we find that

$$h_3 \approx 271 \text{ kJ/kg}$$

From the saturation section of table B–22, we read

$$h_4 = h_f \quad \text{at } 40°\text{C} = 105.7 \text{ kJ/kg}$$

and

$$h_1 = h_4$$

The heat added is

$$q_{add} = h_2 - h_1 \approx 153.3 \text{ kJ/kg}$$

and the mass flow is

$$\dot{m} = \frac{\dot{Q}_{add}}{q_{add}} = \frac{12{,}000 \text{ kJ/h}}{153.3 \text{ kJ/kg}} = 78.3 \text{ kg/h}$$

The rate of heat rejection is

$$\dot{Q}_{rej} = \dot{m}(h_4 - h_3) = (78.3 \text{ kg/h})(105.7 - 271 \text{ kJ/kg})$$

Answer
$$= -12{,}943 \text{ kJ/h}$$

The required power is

$$\dot{W}k_{cycle} = \dot{m}(h_2 - h_3) = (78.3 \text{ kg/h})(259.0 - 271 \text{ kJ/kg})$$

Answer
$$= -939.6 \text{ kJ/h} = -0.261 \text{ kW}$$

The coefficients are

Answer
$$\text{COP} = \frac{-12{,}943}{-939.6} = 13.78$$

and

Answer
$$\text{COR} = 12.78$$

These results compare favorably with the performance expected with R-22.

Using the *EES Equation* window, checking that SI units with temperature in degrees Celsius and pressure in kPa, we write the following equations:

$$h2 = \text{enthalpy}\,(\text{R134a}, T = 20, x = 1.0)$$
$$s2 = \text{entropy}\,(\text{R134a}, T = 20, x = 1.0)$$
$$s3 = s2$$
$$h3 = \text{enthalpy}\,(\text{R134a}, P = 1017, s = s3)$$
$$h4 = \text{enthalpy}\,(\text{R134a}, T = 40, x = 0.0)$$
$$h1 = h4$$
$$qa = 12{,}000$$
$$m = qa/(h2 - h1)$$
$$qr = m*(h4 - h3)$$
$$w = m*(h2 - h3)$$
$$\text{COP} = qr/w$$
$$\text{COR} = qa/w$$

Then, clicking the *Calculation* option and then *Solve* gives answers that are in good agreement with those obtained from the appendix tables,

$$q_r = -12{,}929 \text{ (kJ/hr)}$$
$$w = -928.8 \text{ (kJ/hr)}$$
$$\text{COP} = 13.92$$
$$\text{COR} = 12.92$$

Notice in examples 12–3 and 12–4 that the temperature of the refrigerant (ammonia in example 12–3 and R-22 and R-134a in example 12–4) drops to a lower temperature in the throttling process 1–4. Also, these analyses could be elaborated more by using the irreversibilities in the compressor and by recognizing that temperature differences must occur in the evaporator and the condenser if heat is to be transferred in those two devices.

12–4 THE REVERSED BRAYTON CYCLE OR AIR CYCLE

Any heat engine can, in principle, be reversed and act as a heat pump or refrigeration unit. The gas turbine discussed in chapter 10 has been used in just this way to provide air conditioning to aircraft passengers. It is then sometimes called an air cycle or air turbine cycle. A schematic of a simple air cycle used in modern turbojet aircraft for air-conditioning purposes is shown in figure 12–8.

If you study the schematic closely and compare figure 12–8 with figure 10–3, you can see that the turbine looks like a reversed Brayton cycle device. We see that the compressor is a reversed turbine (requiring input power), an air turbine is a reversed compressor (supplying some power to the compressor), and an air cooler is a reversed combustor that rejects heat. In actual practice, the air cycle frequently is modified so that the air at state 2 is removed from the outlet of a turbojet engine compressor at high pressure and temperature, and the air turbine supplies power to a fan or small compressor that is used to move air past the air cooler, thereby increasing the effectiveness of the heat removal in the air cooler. A schematic of an actual system might look like figure 12–9. Air at state 2 is then cooled by ambient air (taken in from outside the aircraft, for instance) and then expanded through a small air turbine. The air turbine exhaust air is now simply cool air that is furnished to the aircraft compartments after it is mixed in a controlled proportion with air that passed through a throttle to state 6. An accurate analysis of the system, however, requires that an accounting be made of the work or power required to supply air to state 2 from the engine. For these purposes, figure 12–8 is more useful. The operation of the air cycle of figure 12–8 can be described by a reversed Brayton cycle, defined as

1–2 Reversible adiabatic compression of air
2–3 Reversible isobaric heat removal
3–4 Reversible adiabatic expansion of air
4–1 Reversible isobaric heat addition

The property diagrams for this cycle are shown in figure 12–10. It should be noted that this is a gas or vapor cycle and that process 4–1 is an idealized process where cool air is then heated by persons or some other heat source. The air is then presumed to flow back outside into the atmosphere before returning to the cycle. Also, the schematic of the actual

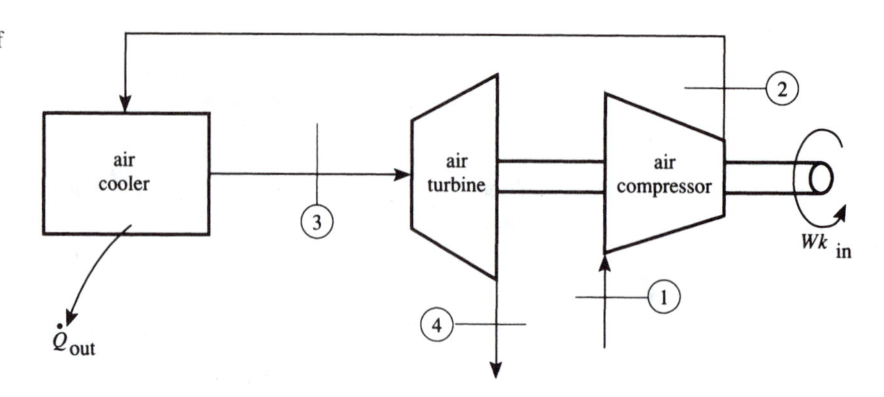

FIGURE 12–8 Schematic of simple air-cycle refrigerator

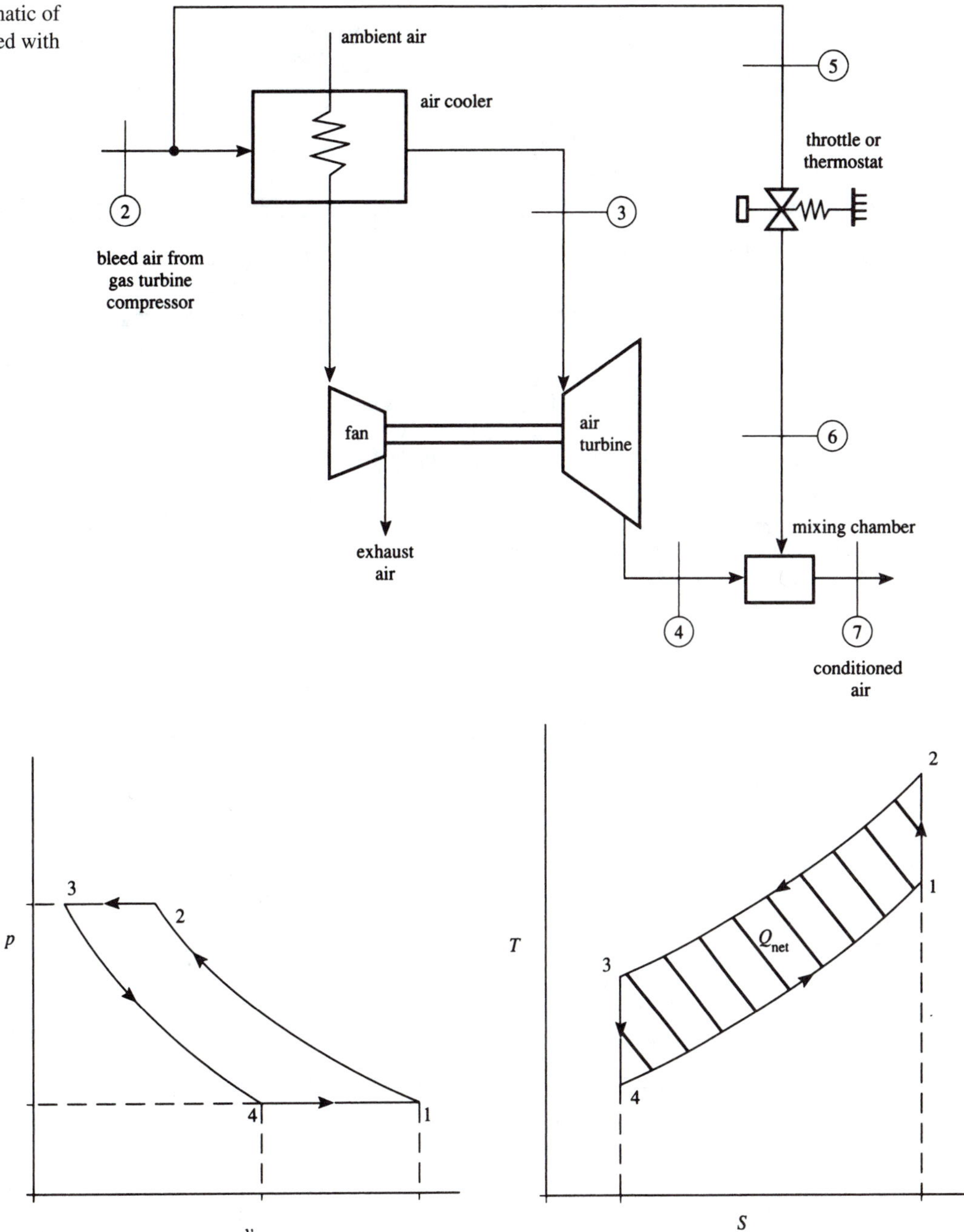

FIGURE 12–9 Schematic of air-cycle refrigerator used with gas turbine

FIGURE 12–10 Property diagrams for an air-cycle refrigerator

cycle of figure 12–9 shows that the turbine work is used entirely to overcome friction and induce the flow of cooling air through the air cooler. The air turbine does not, in the situation of figure 12–9, provide a positive work term to the cycle as would be indicated by process 3–4 of the $p-v$ diagram in figure 12–10.

In principle, we can adapt the closed cycle to the air turbine cycle and call it an air refrigerator. Such a device is shown in the schematic of figure 12–11, and it can be seen how

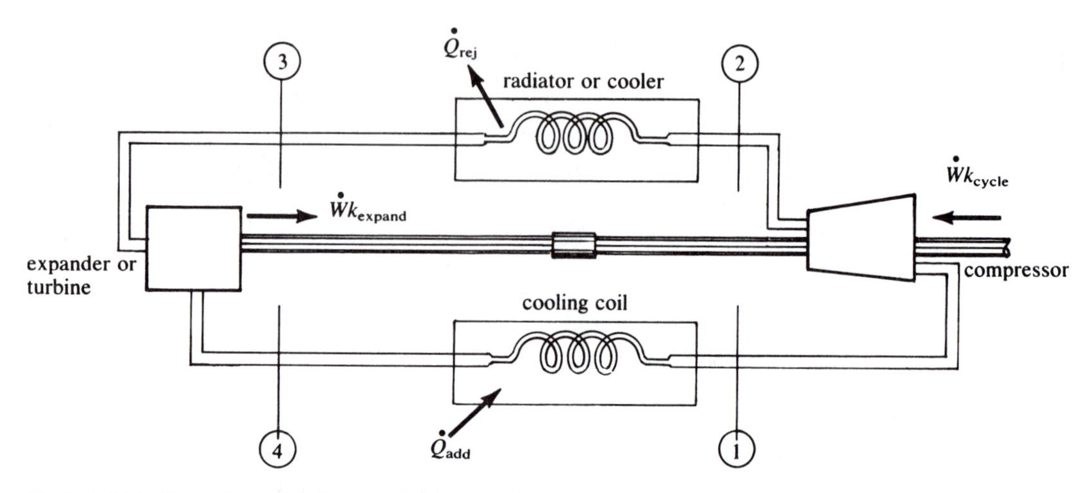

FIGURE 12–11 Schematic of gas refrigeration cycle

heat is added to the cycle in a cooling coil. The analysis of a closed-cycle device would be the same as for an open one.

We should briefly review the method of analysis of the various components making up the air cycle or air refrigerator. We assume that air acts as a perfect gas and that its specific heats are constant. Then, for the ideal cycle, with reversible adiabatic processes through the compressor and turbine, we get, for any perfect gas,

$$Wk_{\text{comp}} = \frac{-k}{k-1}(p_2V_2 - p_1V_1) \tag{12–12}$$

and

$$Wk_{\text{turb}} = \frac{-k}{k-1}(p_4V_4 - p_3V_3) \tag{12–13}$$

which, respectively, reduce to

$$Wk_{\text{comp}} = mc_p(T_1 - T_2) \tag{12–14}$$

and

$$Wk_{\text{turb}} = mc(T_3 - T_4) \tag{12–15}$$

for perfect gases with constant specific heats.

The power of the turbine and compressor would be

$$\dot{W}k_{\text{turb}} = \dot{m}c_p(T_3 - T_4)$$

and

$$\dot{W}k_{\text{comp}} = \dot{m}c_p(T_1 - T_2)$$

The heat rejected in the air cooler of figure 12–8 or the cooler of figure 12–11 can be written

$$\dot{Q}_{\text{rej}} = \dot{m}c_p(T_3 - T_2)$$

The heat added, or the cooling load of the refrigeration system, is given by

$$\dot{Q}_{\text{add}} = \dot{m}c_p(T_1 - T_4)$$

The cooling load would be the capacity of the refrigerator or air conditioner to provide that amount of heat removal, and the heat pump units could provide heating in an amount given by the value of Q_{rej}. The heat and work terms can be set equal to each other for the cycle, so that

$$Q_{\text{add}} + Q_{\text{rej}} = Wk_{\text{turb}} + Wk_{\text{comp}}$$

EXAMPLE 12–5 A 10-kW reversible air-cycle air conditioner is used to cool an aircraft compartment to 17°C when the ambient outside air temperature is 37°C. The pressure ratio across the compressor is 15:1. Determine the mass flow of air required, the power required to operate the air conditioner, the coefficient of performance (COP), and the coefficient of refrigeration (COR).

Solution Referring to figure 12–8 for the air-conditioner system and figure 12–10 for the property diagrams, we have

$$\dot{Q}_{\text{add}} = 10 \text{ kW} = \dot{m}c_p(h_1 - h_4)$$

Assuming perfect gas and constant specific heats gives

$$\dot{Q}_{\text{add}} = \dot{m}c_p(T_1 - T_4)$$

Also, $T_1 = 37°\text{C}$, $T_4 = 17°\text{C}$, and $c_p = 1.007 \text{ kJ/kg·K}$. Then the mass flow of air is found from the calculation

$$\dot{m} = \frac{\dot{Q}_{\text{add}}}{c_p(T_1 - T_4)} = \frac{10 \text{ kW}}{(1.007 \text{ kJ/kg·K})(37 - 17°\text{C})}$$

Answer

$$= 0.4965 \text{ kg/s}$$

The work of compression is done in a reversible and adiabatic manner so that we can write, for a perfect gas with constant specific heats,

$$T_2 = T_1\left(\frac{p_2}{p_1}\right)^{(k-1)/k}$$

and for $k = 1.399$,

$$T_2 = (37 + 273)\left(\frac{15}{1}\right)^{0.399/1.399} = 671.1 \text{ K}$$

The work of compression is then

$$wk_{\text{comp}} = c_p(T_1 - T_2) = (1.007 \text{ kJ/kg·K})(310 - 671.1 \text{ K})$$
$$= -363.63 \text{ kJ/kg}$$

The compressor power is

$$\dot{W}k_{\text{comp}} = \dot{m}wk_{\text{comp}} = -180.54 \text{ kW}$$

The turbine is also reversible and adiabatic, so we can write

$$T_3 = T_4\left(\frac{p_3}{p_4}\right)^{(k-1)/k}$$
$$= (290 \text{ K})\left(\frac{15}{1}\right)^{0.399/1.399} = 627.8 \text{ K}$$

and the turbine power is

$$\dot{W}k_{\text{turb}} = \dot{m}c_p(T_3 - T_4)$$
$$= (0.4965 \text{ kg/s})(1.007 \text{ kJ/kg·K})(627.8 - 290 \text{ K})$$
$$= 168.89 \text{ kW}$$

The net power, or power required to operate the air conditioner, is

Answer

$$\dot{W}k_{net} = 168.89 \text{ kW} - 180.54 \text{ kW} = -11.65 \text{ kW}$$

The rejected heat rate is

$$\dot{Q}_{rej} = \dot{W}k_{net} - \dot{Q}_{add} = -11.65 \text{ kW} - 10 \text{ kW}$$
$$= -21.65 \text{ kW}$$

and, finally,

$$\text{COP} = \frac{\dot{Q}_{rej}}{\dot{W}k_{net}} = \frac{-21.65 \text{ kW}}{-11.65 \text{ kW}}$$

Answer

$$= 1.858$$

and

$$\text{COR} = \frac{-\dot{Q}_{add}}{\dot{W}k_{net}} = \frac{-10 \text{ kW}}{-11.65 \text{ kW}}$$

Answer

$$= 0.858$$

Notice in example 12–5 that the COP and COR values are lower than those in the vapor compression systems. This indicates that the air-cycle refrigeration devices are not as efficient or effective as the vapor compression systems. The air-cycle devices can be justified if there are convenient and readily available supplies of compressed air. Such situations occur around the operation of turbojet or turboprop engines. In fact, let us now consider the system of example 12–5 but with the variations given in figure 12–9.

EXAMPLE 12–6 An air-cycle air conditioner is used by extracting air from a turbojet engine at 671.1 K and 1500 kPa. This air is to be furnished to an aircraft compartment at 100 kPa and 17°C. The air leaving the turbine must be no colder than 0°C to prevent icing of the system. Assuming that there is a 10-kW cooling load on the system and that a 20-kW compressor is required to move adequate amounts of ambient air for cooling the compressed air, as shown in figure 12–9, determine the mass flow of extracted or "bleed" air and the mass flow of ambient air required if the temperature increase of the cooling air is not to exceed 20°C.

Solution We refer to figure 12–9 and notice that the cooling load or heat added to the cycle must be written

$$\dot{Q}_{add} = \dot{m}_7 c_p (T_1 - T_7)$$

where $\dot{m}_7$ is the total mass flow of air and T_1 is an external ambient air temperature. If we assume that $T_1 = 37°C = 310$ K, as in example 12–5, then the calculations are identical to those in example 12–5, and

Answer

$$\dot{m}_7 = 0.4965 \text{ kg/s}$$

For the mixing chamber, we write

$$\dot{m}_7 = \dot{m}_6 + \dot{m}_4 \quad \text{(mass balance)}$$

and

$$h_7 \dot{m}_7 = h_6 \dot{m}_6 + h_4 \dot{m}_4 \quad \text{(energy balance)}$$

For a perfect gas with constant specific heats, this last equation becomes

$$\dot{m}_7 c_p T_7 = \dot{m}_6 c_p T_6 + \dot{m}_4 c_p T_4$$

or

$$\dot{m}_7 T_7 = \dot{m}_6 T_6 + \dot{m}_4 T_4 \quad \text{(reduced form of energy balance)}$$

The temperatures are $T_7 = 17°C = 290$ K, $T_4 = 0°C = 273$ K, and $T_6 = T_5 = T_2 = 671.1$ K. Therefore, combining the mass balance equation with the reduced form of the energy balance, we find that

$$\dot{m}_4(T_4 - T_6) = \dot{m}_7(T_7 - T_6)$$

or

$$\dot{m}_4 = \frac{\dot{m}_7(T_7 - T_6)}{T_4 - T_6} = (0.4965 \text{ kg/s})\left(\frac{290 \text{ K} - 671.1 \text{ K}}{273 \text{ K} - 671.1 \text{ K}}\right)$$
$$= 0.4753 \text{ kg/s}$$

The temperature entering the turbine, T_3, can now be found from the equation

$$\dot{W}k_{\text{turb}} = \dot{m}_4 c_p(T_3 - T_4)$$

and then

$$T_3 = \frac{\dot{W}k_{\text{turb}}}{\dot{m}_4 c_p} + T_4$$
$$= \frac{20 \text{ kJ/s}}{(0.4753 \text{ kg/s})(1.007 \text{ kJ/kg}\cdot\text{K})} + 273 \text{ K}$$
$$= 314.8 \text{ K}$$

The throttle between the cooler and the turbine allows for a reduction in pressure of the bleed air. The heat rejected in the bleed air is given by the result

$$\dot{Q}_{\text{rej}} = \dot{m}_4 c_p(T_3 - T_2)$$
$$= (0.4753 \text{ kg/s})(1.007 \text{ kJ/kg}\cdot\text{K})(314.8 \text{ K} 2\, 671.1 \text{ K})$$

Answer

$$= -170.53 \text{ kW}$$

The air cooler transfers the rejected heat to the ambient air so that, for an energy balance of the air cooler,

$$\dot{m}_4 c_p(T_2 - T_3) = 170.53 \text{ kJ/s} = \dot{m}_{\text{air}} c_p(\Delta T)_{\text{air}}$$

or, since $(\Delta T)_{\text{air}} = 20$ K,

$$\dot{m}_{\text{air}} = \frac{170.53 \text{ kJ/s}}{(1.007 \text{ kJ/kg}\cdot\text{K})(20 \text{ K})}$$
$$= 8.467 \text{ kg/s}$$

The analysis of the reversed gas turbine or air turbine cycle can be modified to provide a more accurate description of the operation of the systems if irreversibilities in the turbine and compressor are considered. The following example of a closed cycle demonstrates this method.

EXAMPLE 12–7 An air refrigeration cycle uses a radiator and a cooling coil to operate between 10°F and 400°F as shown in figure 12–11. Assume that the pressure ratio is 5:1 and that the adiabatic efficiencies of the turbine and compressor are 90%. Determine the COP and COR of the device if it acts as a refrigerator, cooling a space at 10°F, and has a cooling load of 2 tons. Also determine the minimum entropy generation of the system.

Solution We refer to figure 12–10 and note that the modifications of irreversible turbines and compressors change the *T–s* diagram to that shown in figure 12–12. We have that $T_2 = 400°F = 860°R$ and $T_4 = 10°F = 470°R$. The temperature at a state 1*s* where a

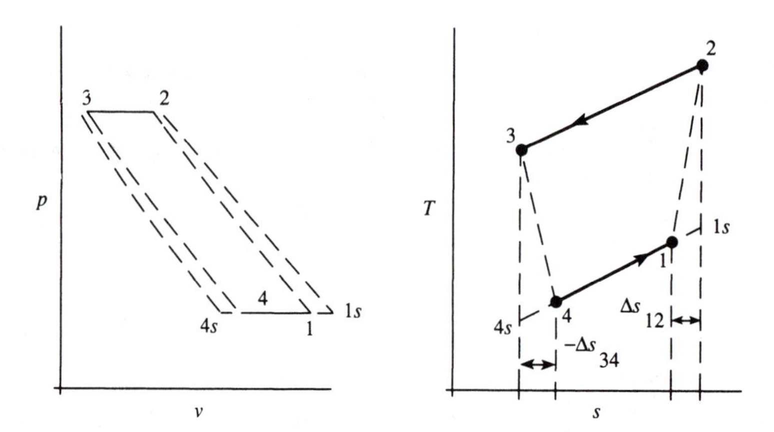

FIGURE 12–12 Property diagrams of irreversible air-cycle refrigerator

reversible adiabatic compression would occur in the compressor is determined from the equation

$$T_{1s} = T_2\left(\frac{p_1}{p_2}\right)^{(k-1)/k}$$

For $k = 1.399$ and $p_1/p_2 = {}^1\!/_5$, we get

$$T_{1s} = (860\text{ K})\left(\frac{1}{5}\right)^{0.399/1.399} = 543.44°\text{R}$$

Using the definition of adiabatic efficiency for a compressor, we find the actual temperature at state 1. We have

$$\eta_{\text{comp}} = \frac{wk_{\text{ideal}}}{wk_{\text{act}}} = \frac{c_p(T_{1s} - T_2)}{c_p(T_1 - T_2)}$$

and

$$T_1 = \frac{T_{1s} - T_2}{\eta_{\text{comp}}} + T_2$$

$$= \frac{543.44°\text{R} - 860°\text{R}}{0.90} + 860°\text{R}$$

$$= 508.3°\text{R}$$

The actual work of the compressor is

$$wk_{\text{act}} = c_p(T_1 - T_2) = -84.55\text{ Btu/lbm}$$

and the entropy increase in the compressor is

$$\Delta s_{12} = c_p \ln\frac{T_{1s}}{T_1}$$

$$= (0.2404\text{ Btu/lbm}\cdot°\text{R})\left(\ln\frac{543.44°\text{R}}{508.3°\text{R}}\right)$$

$$= 0.016070\text{ Btu/lbm}\cdot°\text{R}$$

The temperature T_{4s} at a state corresponding to a reversible adiabatic expansion through the turbine is found from

$$T_{4s} = T_3\left(\frac{p_4}{p_3}\right)^{(k-1)/k}$$

$$= T_3\left(\frac{1}{5}\right)^{0.399/1.399} = 0.6319T_3$$

Using the definition of adiabatic efficiency of a turbine with a perfect gas and constant specific heats yields

$$\eta_{\text{turb}} = \frac{T_3 - T_4}{T_3 - T_{4s}} = \frac{T_3 - T_4}{T_3 - 0.6319T_3}$$

We can then solve for T_3 to obtain

$$0.90 = \frac{T_3 - 470°\text{R}}{0.3681T_3}$$

and

$$T_3 = 702.8°\text{R}$$

Then

$$T_{4s} = 444.1°\text{R}$$

and the entropy change in the turbine is

$$\begin{aligned}\Delta S_{34} &= c_p \ln \frac{T_4}{T_{4s}} \\ &= (0.2404\ \text{Btu/lbm} \cdot °\text{R})\left(\ln \frac{470°\text{R}}{444.1°\text{R}}\right) \\ &= 0.0136\ \text{Btu/lbm} \cdot °\text{R}\end{aligned}$$

The actual turbine work is

$$\begin{aligned}wk_{\text{turb}} &= c_p(T_3 - T_4) \\ &= (0.2404\ \text{Btu/lbm} \cdot °\text{R})(628.1°\text{R} - 470°\text{R}) \\ &= 38.01\ \text{Btu/lbm}\end{aligned}$$

so the net work is

$$\begin{aligned}wk_{\text{net}} &= wk_{\text{turb}} + wk_{\text{comp}} \\ &= 38.01\ \text{Btu/lbm} + (-84.55\ \text{Btu/lbm}) \\ &= -46.54\ \text{Btu/lbm}\end{aligned}$$

The heat rejected is

$$\begin{aligned}q_{\text{rej}} &= c_p(T_3 - T_2) \\ &= (0.2404\ \text{Btu/lbm} \cdot °\text{R})(628.1°\text{R} - 860°\text{R}) \\ &= -55.75\ \text{Btu/lbm}\end{aligned}$$

and the heat added is

$$\begin{aligned}q_{\text{add}} &= c_p(T_1 - T_4) = (0.2404\ \text{Btu/lbm} \cdot °\text{R})(508.3°\text{R} - 470°\text{R}) \\ &= 9.21\ \text{Btu/lbm}\end{aligned}$$

The COP is then determined from its defining equation,

$$\begin{aligned}\text{COP} &= \frac{q_{\text{rej}}}{wk_{\text{net}}} \\ &= \frac{-55.75\ \text{Btu/lbm}}{-46.54\ \text{Btu/lbm}} = 1.198\end{aligned}$$

Answer

and the COR from

$$\text{COR} = \frac{q_{\text{add}}}{wk_{\text{net}}} = \frac{-9.21}{-46.54} = 0.198$$

Answer

The minimum entropy generation will be

$$\dot{S}_{\text{min}} = \dot{m}(\Delta s_{12} + \Delta s_{34})$$

The mass flow is determined from the equation

$$\dot{Q}_{add} = \dot{m}q_{add} = 2 \text{ tons} = 6.66 \text{ Btu/s} = \dot{m}(21.23 \text{ Btu/lbm})$$

or

$$\dot{m} = \frac{6.66}{21.23} = 0.3137 \text{ lbm/s} = 1129.3 \text{ lbm/h}$$

Then

Answer

$$\dot{S}_{min} = (1129.3 \text{ lbm/h})(0.016085 \text{ Btu/lbm} \cdot {}^\circ\text{R}) + 0.0136 \text{ Btu/lbm} \cdot {}^\circ\text{R}$$
$$= 33.52 \text{ Btu/}^\circ\text{R} \cdot \text{h}$$

The program BRAYTON, which is used with a microcomputer to aid in solving gas turbine problems, can also be used by the student to aid in solving the reversed gas turbine cycle problems.

12–5 AMMONIA ABSORPTION REFRIGERATION

The ammonia absorption cycle represents an attempt to reduce the mechanical work required to compress refrigerants and replace this energy demand by a heat transfer process. The typical cycle, shown in figure 12–13, indicates that the condenser, the expansion valve, and the evaporator are components having identical functions in the ammonia absorption and vapor compression cycles. That is, the condenser serves to remove heat from the system; the expansion valve allows isenthalpic (constant-enthalpy) expansion of the ammonia to a low pressure; and the evaporator allows for heat addition to the low-pressure ammonia from a region being refrigerated or cooled. In the ammonia absorption cycle, the compressor is replaced by an aux-

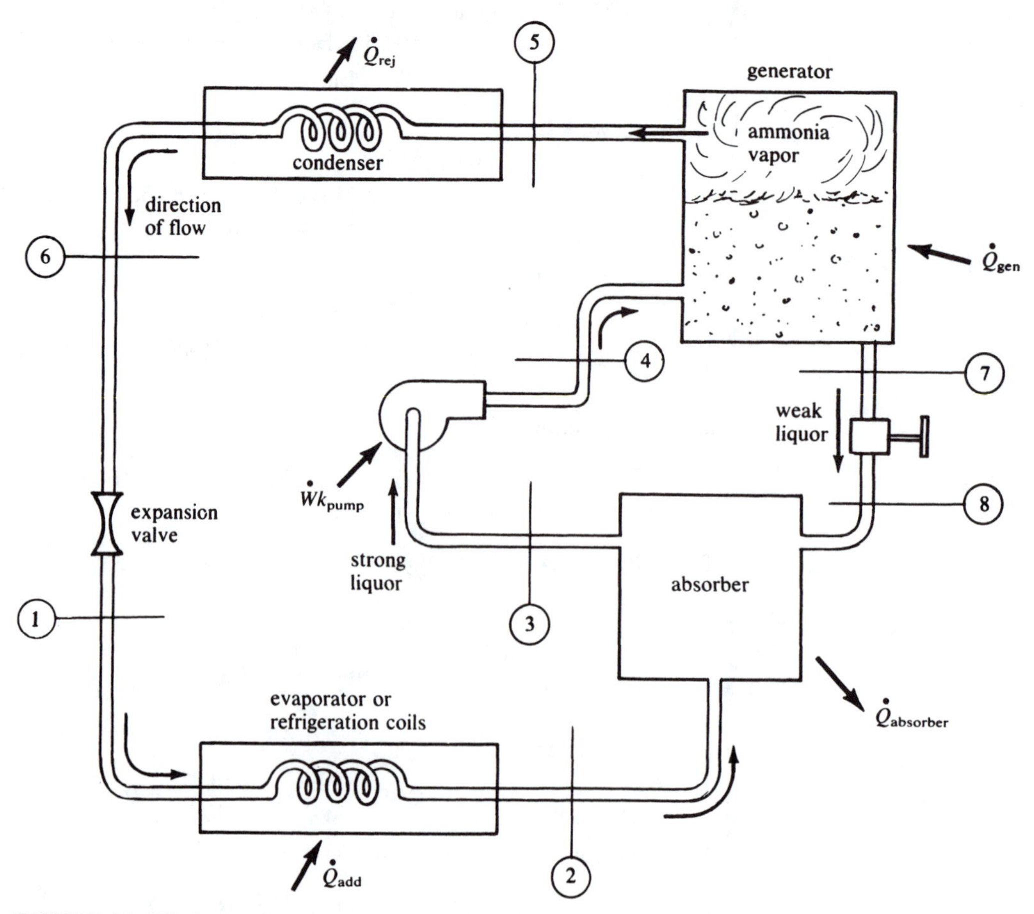

FIGURE 12–13 Ammonia absorption refrigeration cycle

iliary water system. The absorber receives a low-pressure mixture predominantly of water with some ammonia (weak liquor) and low-pressure ammonia vapor. These two fluids are mixed together in the absorber so that the ammonia is "absorbed" or dissolved in the weak liquor. The fluid flowing out of the absorber is then a strong solution of water and ammonia (strong liquor) and is pumped to a generator under high pressure. Heat is supplied to the generator by steam, electric coils, or other appropriate means, which increases the temperature of the water–ammonia liquor. A consequence of this is that ammonia vapor boils off and recirculates to the condenser, while the remaining liquor returns to the absorber. All of this operates because ammonia is more soluble in cold water than in hot water, and the absorber and the generator provide the means of taking advantage of this phenomenon.

Notice in figure 12–13 that the system requires one more heat exchanger than does a vapor compression system. A special ammonia absorption refrigeration system design, called the Electrolux-Servel type, requires no pumps or other moving parts and only one heat input. Such a system, shown in figure 12–14, uses ammonia and water as in the basic

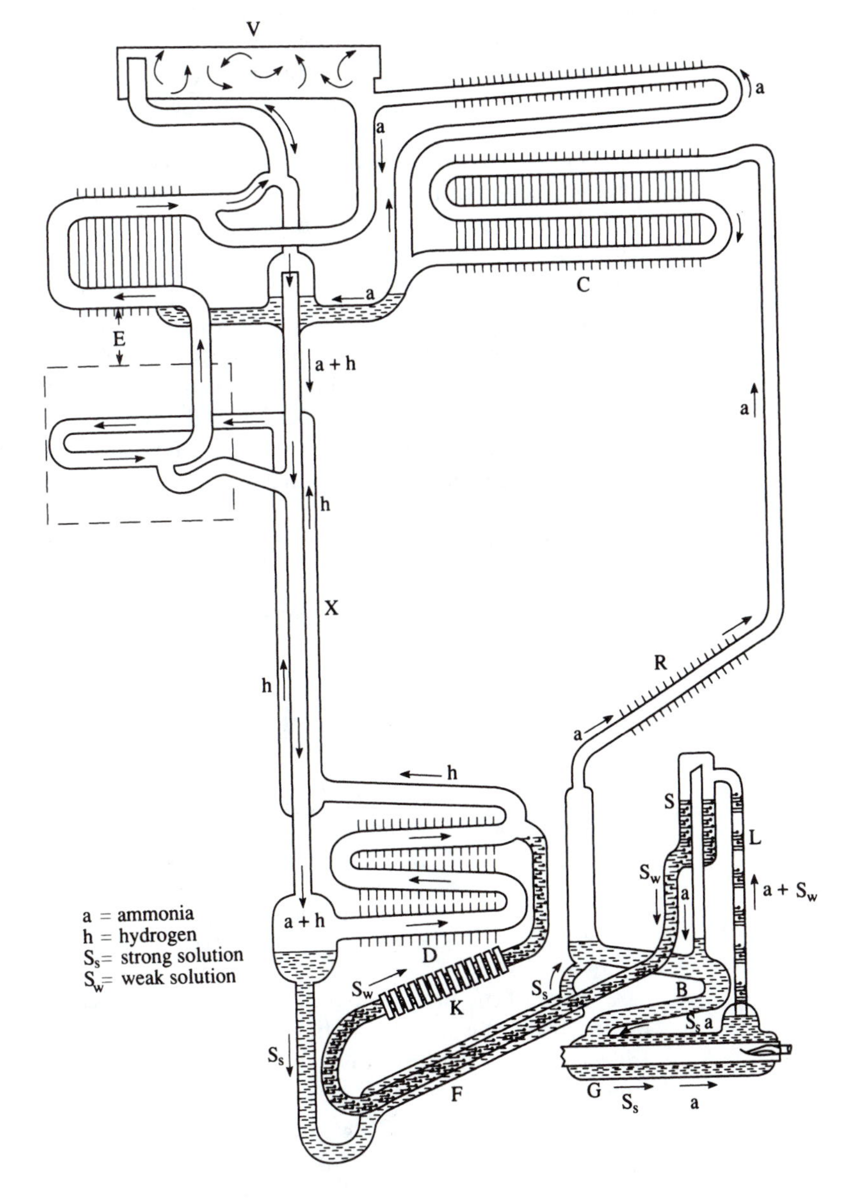

FIGURE 12–14 Diagrammatic layout of Electrolux-Servel absorption machine: B, analyzer; C, condenser; D, absorber; E, evaporator; F, exchanger; G, generator; K, forecooler; R, rectifier; S, liquid–vapor separator; V, hydrogen reserve vessel; X, heat exchanger [from T. Baumeiston et al., eds., *Mark's standard handbook for mechanical engineers*, 8th ed. (New York: McGraw-Hill Book Company, 1978), reprinted with permission of McGraw-Hill Book Company]

ammonia absorption systems but also contains hydrogen. Hydrogen in a mixture with water and ammonia compensates for the pump, to create a recirculating flow through the system. The Electrolux-Servel process may be an option for solar-powered refrigeration units since the only input is heat, and this could be done with a solar collector. The Electrolux-Servel units have used propane or other fuels to provide a heat source.

For a complete analysis of the ammonia absorption refrigeration cycles, the reader is referred to publications specifically concerned with refrigeration.

12–6 CRYOGENICS AND GAS LIQUEFACTION

Cryogenics is a branch of science and technology concerned with substances at temperatures below 120 K (−153°C) or 216°R (−244°F). Because such low temperatures are generally not encountered in or near the Earth, cryogenics is involved with man-made or artificially refrigerated states. There are at least five specific applications of cryogenics that we can list:

1. Liquefaction of gases
2. Superconductivity and super fluid behavior
3. Cryogenic freezing of biological materials
4. Production of ultra-high-vacuum chambers
5. Reduction of noise in electronic communication through conductors

For the purpose of this book, we will concern ourselves only with gas liquefaction. This application demonstrates the principles of thermodynamics and is an important technology of which the engineer or technologist should have some knowledge. Liquefied gases are used as propellants or fuels for rockets (hydrogen and oxygen), as refrigerants for cold storage of various commercial products (nitrogen and carbon dioxide), and as fluids that have nearly no resistance to flow (helium). Natural gas is pumped through pipelines over long distances and is sometimes cryogenically cooled to a liquid to reduce the power needed for pumping. Air, helium, hydrogen, nitrogen, and oxygen are some of the gases that have been liquefied. Table 12–2 lists the boiling temperature T_{sat} or temperature at which the listed gas exhibits the liquid–gaseous phase change at 1 atm of pressure (101 kPa), and it can easily be seen that very low temperatures are required.

TABLE 12–2 Saturation temperature of gases at 101 kPa (14.7 psia)

Substance	Saturation Temperature °C	Saturation Temperature °F
Air	−194.2	−317.6
Helium	−213.3	−352.0
Hydrogen	−252.7	−422.9
Nitrogen	−195.8	−320.4
Oxygen	−183.0	−297.4
Chlorine	−34.6	−30.3
H_2O (water)	100.0	212.0

Source: Condensed from the International Critical Tables with permission of the National Academy of Sciences. Revised to SI units by author.

The most common and typical method of producing liquid gases (such as oxygen, hydrogen, and air) is depicted in figure 12–15. Here the gas itself is a working medium (refrigerant) and is converted to a saturated liquid in a heat exchanger. After passing through a throttle or expansion valve, the liquid gas (state $1f$) is removed, and vapor is recirculated to the compressor. Makeup gas is required to keep the system in steady state.

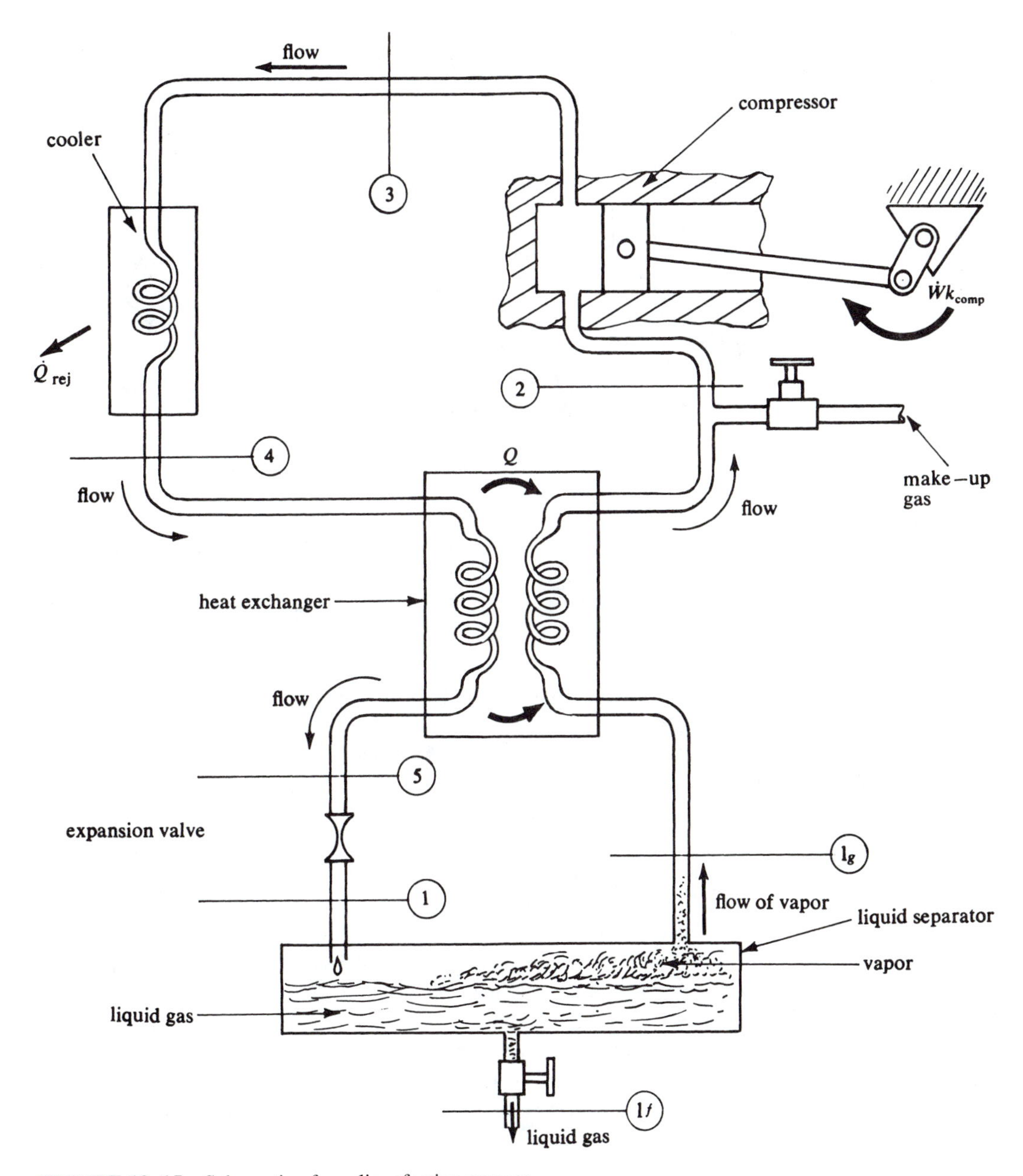

FIGURE 12–15 Schematic of gas-liquefaction process

In figure 12–16 is presented the temperature–entropy diagram of the gas-liquefaction process, assuming reversible and ideal situations for all the components. In this diagram, the temperature at state 4 (entering the high-pressure side of the heat exchanger) is indicated as being equal to that at state 2, but in actual cases, T_4 must be greater than T_2 if heat transfer is to be effected over a finite period of time. Also, at state 5, the gas may not be a saturated liquid but may be a mixture of liquid and vapor, a saturated vapor, or even somewhat superheated. However, if the cycle is to have meaning, the expansion 5–4 through the expansion valve must result in a mixture of liquid and vapor. This allows for the separation of the mixture and appearance of liquid gas at low pressure.

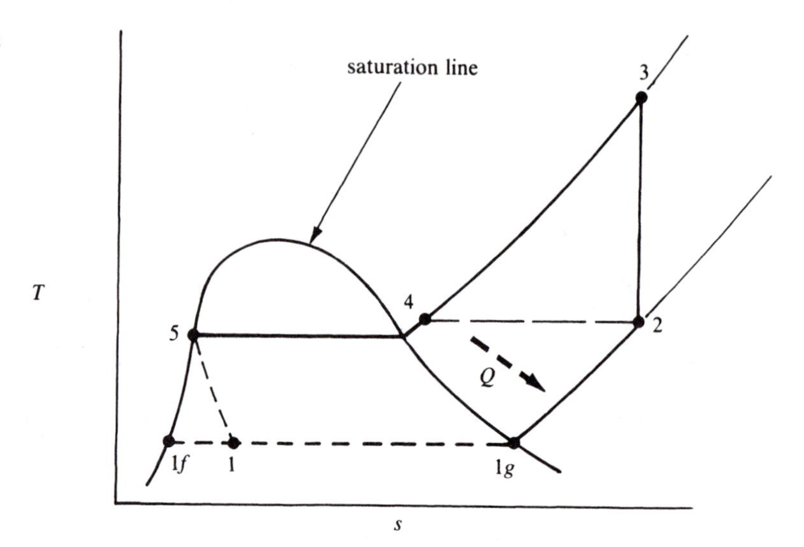

FIGURE 12–16 *T–s* diagram of gas liquefaction

In some situations, it is desirable to use more than one cycle for liquefying gases. A cascade cycle, shown in figure 12–17, can be used to achieve very low temperatures, and in fact this system is often used to produce dry ice (solid carbon dioxide). Dry ice is at −78°C (−110°F) at atmospheric pressure and so is not strictly in a cryogenic state, but it has many commercial uses. Other cascade cycles involve two or more compressors, and flash tanks have been developed.

12–7 HEAT PUMPS

The use of thermodynamic devices to transfer heat from one environment to another has been a significant contribution to our modern life-style. The previous sections in this chapter have been concerned mainly with the application of transferring heat out of an area so that the temperature there may be lowered. In this section, we discuss the application of heating or transferring heat into an area, such as a dwelling or chamber, from a cooler external environment. Today these devices are referred to as *heat pumps*, although we have used this term for all reversed heat engines.

The heat pump can be designed to operate as an air conditioner in warm seasons and a heating unit during the colder seasons. The device has the capability of providing more energy in the form of heat than is put into the system as work. Therefore, it represents a very attractive alternative to the common electric resistance heating. The primary difficulty associated with the heat pump is its inability to function well over a wide range of temperatures. Thus, if it is designed to operate in moderate weather, it is generally inadequate for very cold climates, and, conversely, a design for cold weather use would be uneconomical and even unacceptable in warmer weather. As a total heating system, the heat pump requires some major technological improvements which, as of now, are unavailable.

A device that has been developed into a useful heat pump in temperate climates is the vapor compression air conditioner (see section 12–2), which can be conveniently adjusted to act as a heat pump. Such an arrangement requires that the heat rejection from the device be directed into the area requiring heat and that the added heat (to the heat pump) be taken from the outside or surroundings. This situation reverses the operation of the normal air-conditioner unit.

The heat pump can operate on any of the refrigeration cycles and use most of the common refrigerants. The devices are commonly driven by an electric motor or similar power mechanism. The rejected heat $\dot{Q}_{rej}$ from the cycle represents the heating effect produced by the heat pump, and, in a typical application, heat is transferred from a low-temperature atmosphere to a high-temperature dwelling.

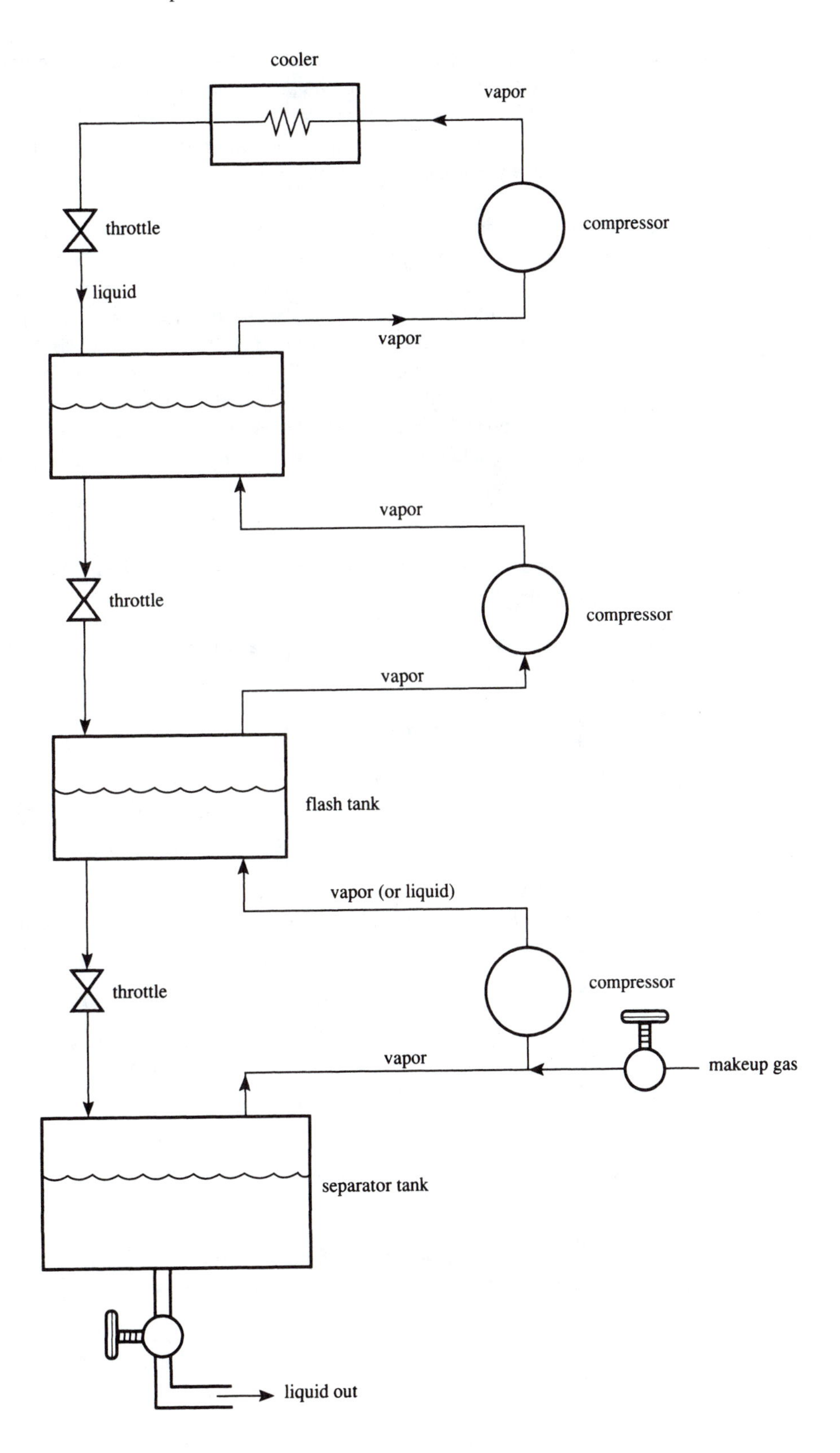

FIGURE 12–17 Cascade cycle for gas liquefaction

EXAMPLE 12–8 A heat pump using a dry vapor compression cycle with Refrigerant-22 as the working medium is rated at 6 kW and has an optimum condition of 20°C at 0°C outside air temperature. Determine the power required to drive the heat pump and the flow rate of the refrigerant through the system. Compare the answers you obtained by using the appendix table with those you obtained by using *EES*.

Solution

This device has a dry vapor compression cycle, so we may obtain the enthalpy values at the four corners of the cycle as indicated in figure 12–6. From table B–17, we find $h_4 = h_1 =$ 224.084 kJ/kg, which is the enthalpy of saturated liquid at 20°C. The pressure here, at state 4, and at state 3 is 0.90993 MPa. We assume ideal heat transfer in the heat pump and, further, find that $h_2 = h_g$ at 0°C = 405.4 kJ/kg. At 0°C, the pressure is 0.49757 MPa bars, and the entropy at state 2 is s_g (0°C) or 1.75179 kJ/kg · K. At state 3, the entropy must also have this value so, from table B–18, and using $p_3 = 0.90993$ MPa pressure and $s_3 = 1.75179$ kJ/kg · K, we interpolate to find $h_3 = 420$ kJ/kg. We may then use equation (12–7) to determine the required work. To determine the power, we use a variation of equation (12–8) to first obtain the mass flow rate. We have

$$\dot{m}(h_4 - h_3) = \dot{Q}_{\text{rej}} = -6.0 \text{ kW}$$

because the rejected heat represents the heat rating. Then

$$\dot{m} = \frac{-6.0 \text{ kW}}{224.084 \text{ kJ/kg} - 420 \text{ kJ/kg}}$$

Answer

$$= 0.0306 \text{ kg/s}$$

The power is therefore

$$\dot{W}k_{\text{cycle}} = \dot{m}(h_2 - h_3)$$
$$= (0.0306 \text{ kg/s})(405.4 \text{ kJ/kg} - 420 \text{ kJ/kg})$$

Answer

$$= -0.45 \text{ kJ/s} = -0.45 \text{ kW}$$

Using the *EES Equation* window, and making sure that we have SI Units with temperature in degrees Celsius and pressure in kPa, we write the following equations:

$$h3 = \text{enthalpy (R22, } T = 20, x = 1.0)$$
$$s3 = \text{entropy (R22, } T = 20, x = 1.0)$$
$$s2 = s3$$
$$h2 = \text{enthalpy (R22, } T = 0, s = s2)$$
$$h4 = \text{enthalpy (R22, } T = 20, x = 0.0)$$
$$qr = -60$$
$$m = qr/(h4 - h3)$$
$$w = m*(h2 - h3)$$

Then, clicking the *Calculation* option and then *Solve* gives answers that are in good agreement with those obtained from the appendix table:

$$\dot{m} = 0.032 \text{ (kh/s)}$$
$$w = -0.4469 \text{ (kW)}$$

EXAMPLE 12–9 A heat pump is designed for 10°F climate and to deliver 50,000 Btu/h into a 70°F dwelling. If this device operates on a wet compression cycle with Refrigerant-12 as the working medium, determine the flow rate of the refrigerant through the system and the flow rate required if the outside temperature drops to −20°F and if 150,000 Btu/h is required. Assume that the temperature in the dwelling is 60°F at the lower outside temperature. Compare the answers you obtained by using the appendix table with those you obtained by using *EES*.

Solution For the wet compression heat pump cycle, we refer to figure 12–6b, and the enthalpies, from table B–15, are

$$h_1 = h_4 = 24.050 \text{ Btu/lbm} = h_f \quad \text{at } 70°\text{F}$$
$$h_3 = 84.359 \text{ Btu/lbm} = h_g \quad \text{at } 70°\text{F}$$

The flow rate may then be computed from the relationship

$$\dot{m}(h_4 - h_3) = \dot{Q}_{\text{rej}} = 50{,}000 \text{ Btu/h}$$

so

$$\dot{m} = \frac{-50{,}000 \text{ Btu/h}}{24.050 \text{ Btu/lbm} - 84.359 \text{ Btu/lbm}}$$

Answer

$$= 829.06 \text{ lbm/h}$$

At the colder condition, we find that

$$h_1 = h_4 = 21.766 \text{ Btu/lbm} = h_f \quad \text{at } 60°\text{F}$$
$$h_3 = 83.409 \text{ Btu/lbm} = h_g \quad \text{at } 60°\text{F}$$

and

$$\dot{m} = \frac{-150{,}000 \text{ Btu/h}}{21.766 \text{ Btu/lbm} - 83.409 \text{ Btu/lbm}}$$

Answer

$$= 2433.4 \text{ lbm/h}$$

Using the *EES Equation* window, and making sure that we have English units with temperature in degrees Fahrenheit and pressure in psia, we write the following equations:

$$h3 = \text{enthalpy}\,(\text{R12}, T = 70, x = 1.0)$$
$$h4 = \text{enthalpy}\,(\text{R12}, T = 70, x = 0.0)$$
$$qrl = -50000$$
$$ml = qrl/(h4 - h3)$$
$$qrl = -150000$$
$$h3h = \text{enthalpy}\,(\text{R12}, T = 60, x = 1.0)$$
$$h4h = \text{enthalpy}\,(\text{R12}, T = 60, x = 0.0)$$
$$mh = qrh/(h4h - h3h)$$

Then, clicking the *Calculation* option and then *Solve* gives answers that are in good agreement with those obtained by using the appendix table:

$$ml = 829 \text{ (lbm/h)}$$
$$mh = 2433 \text{ (lbm/h)}$$

12–8 SUMMARY

The reversed Carnot cycle is a method of transferring heat from a cold region at T_L to a warm region at T_H. The coefficient of performance (COP) is defined as

$$\text{COP} = \frac{Q_{\text{rej}}}{Wk_{\text{cycle}}}$$

and, for reversed Carnot cycles, this is

$$\text{COP} = \frac{T_H}{T_H - T_L} \tag{12–4}$$

The coefficient of refrigeration (COR) is defined as

$$\text{COR} = \frac{-Q_{\text{add}}}{Wk_{\text{cycle}}}$$

and, for the Carnot cycle, this is

$$\text{COR} = \frac{T_L}{T_H - T_L} \tag{12–1}$$

The term *ton of refrigeration* is defined as

$$1 \text{ ton} = 200 \text{ Btu/min} = 3.33 \text{ Btu/s}$$

The vapor compression refrigeration systems use a refrigerant as a working fluid and utilize a cycle made up of a compression, constant-pressure condensation, a throttle, and a constant-pressure evaporation. The vapor compression refrigeration cycle is an irreversible cycle because the throttling process is irreversible, but systems that operate on the vapor compression cycle are used in a variety of applications. Some common refrigerants are ammonia, sulfur dioxide, Refrigerant-12 (R-12), and Refrigerant-22 (R-22). Refrigerant-134a (R-134a) and Refrigerant-123 (R-123) have been used to replace R-12 and other environmentally dangerous refrigerants.

The reversed gas turbine cycle or air turbine cycle is used to describe the operation of air conditioners in those situations where a supply of high-pressure, high-temperature air is readily available. Such situations occur around the operation of turbojet engines for aircraft. Practical air turbine cycles are usually open cycles, although the closed-cycle operations are conceivable.

Ammonia absorption refrigeration uses the characteristic that ammonia is more soluble in cold water than hot water to transfer heat from a cold region to a warm or hot region. The ammonia absorption systems use heat as the principal external input, but some systems use a small recirculating pump as well. A variation of the ammonia absorption systems is the Electrolux-Servel system, which uses hydrogen in the ammonia– water mixture and thereby can eliminate the small pump. It then uses only heat as an external input.

Cryogenics is a branch of science concerned with materials at temperatures below 120 K or 216°R. Gas liquefaction is a particularly important application of cryogenics where naturally occurring gases are cooling to their liquid states. A simple gas liquefaction process requires an isothermal or adiabatic compression, a cooler, a throttle, and a liquid separator. For steady-state operation of such systems, makeup gas must be continually introduced into the device to replace the liquid gas removed. For very low temperature conditions, the cascade cycle can be used.

The heat pumps are generally an application of vapor compression refrigeration systems where the rejected heat from the cycle is used as a heat supply and the added heat is taken from a cold surrounding.

DISCUSSION QUESTIONS

Section 12–1

12–1 What is meant by 1 ton of refrigeration?

12–2 What is meant by coefficient of performance (COP)?

12–3 What is meant by coefficient of refrigeration (COR)?

Section 12–2

12–4 What is meant by the term *vapor compression refrigeration*?

12–5 What is a blend?

Section 12–3

12–6 What is a *throttle*?

12–7 What is an *expansion valve*?

12–8 What is a *capillary tube*?

Section 12–4

12–9 What is gas refrigeration?

Section 12–5

12–10 What is "absorbed" in an absorption refrigeration system?

Section 12–6

12–11 What defines the technology *cryogenics*?

12–12 What is lowest temperature that can be reached in a cryogenic system?

Section 12–7

12–13 What is a heat pump?

PRACTICE PROBLEMS

Problems that use SI units are indicated by an (M) under the problem number; those that use English units are indicated by an (E). Mixed unit problems are indicated by a (C).

Section 12–1

12–1 (M) A Carnot heat pump operates between a high-temperature region of 80°C and a low-temperature region of −10°C. Determine the COP and COR.

12–2 (M) A reversible Carnot heat engine receives heat at −5°C and requires 0.1 kW for each kilowatt of heat received. Determine the COP.

12–3 (M) For the heat pump of problem 12–2 determine
(a) Temperature of sink for rejected heat
(b) COR

12–4 (M) A 200-kW Carnot freezer operates between −10°C and 30°C. Determine the rate of heat rejected to the atmosphere and the power required.

12–5 (E) A reversed Carnot cycle operates between 100°F and 10°F. Determine the COR and COP.

12–6 (E) A Carnot air conditioner has a COP of 4 and a capacity of 5 tons. Determine the power requirement of this device.

12–7 (E) A Carnot freeze is used to freeze water at 14.7 psia. What is the maximum COR achieved by this device if the minimum atmospheric temperature is 28°C?

12–8 Sketch the *p–V*, *T–s*, and *h–s* diagrams of the ideal Carnot cycle.

Sections 12–2 and 12–3

12–9 (M) Saturated liquid ammonia is throttled through an expansion valve from 45°C to 190 kPa. Determine the enthalpy and the quality of the ammonia leaving the expansion valve.

12–10 (M) At a rate of 6.0 kg/s, Refrigerant-22 enters an evaporator as 12% vapor and at 131.7 kPa. If the refrigerant leaves with a quality of 88%, determine the rate of heat addition.

12–11 (M) An ideal dry compression refrigerator operates on the cycle diagrammed in figure 12–5a. If the working medium is Refrigerant-22 and the pressures are 1190 kPa and 296 kPa, determine
(a) wk_{comp}
(b) q_{add}
(c) q_{rej}

12–12 (M) If the refrigerator of problem 12–11 is redesigned to use R-134a, determine the compression work, the heat added, and the heat rejected.

12–13 (E) Determine the heat added per pound-mass of ammonia if the ammonia has 20% quality and is at −20°F entering the evaporator and is a saturated vapor upon leaving.

12–14 (E) An ideal wet compression air conditioner using Refrigerant-22 uses a cycle as diagrammed in figure 12–5b. If the refrigerant is at 30°F in the evaporator and 110°F in the condenser, determine
(a) wk_{comp}
(b) q_{add}
(c) q_{rej}

12–15 (C) Calculate the COP for
(a) Cycle of problem 12–11
(b) Cycle of problem 12–14

12–16 (M) A typical R-134a dry vapor compression refrigerator has a COR of about 1.3 when operating between −15°C and 30°C. For a refrigerator rated at 0.1 tons of refrigeration, determine the power required and the mass flow rate of R-134a. Assume that the refrigerant is saturated liquid leaving the condenser.

12–17 (M) Saturated liquid Refrigerant-22 at 25°C is expanded to 105 kPa through an expansion valve. It leaves the evaporator at 100% quality and 105 kPa pressure. If 12 kg/min of the refrigerant flow through the device, what is its rating? Is this a wet or a dry compression cycle?

12–18 (M) A freezer using Refrigerant-22 removes 1500 kJ/h from a compartment at −2°C. The refrigerant is at −5°C in the evaporator, and the quality is 85% leaving the evaporator. If the freezer operates on an ideal wet compression cycle, determine
(a) $\dot{m}$ of refrigerant
(b) $\dot{Q}_{rej}$
(c) $\dot{Wk}_{cycle}$
(d) COP and COR

12–19 (M) If the freezer of problem 12–18 is revised so that it can use R-134a, determine the mass flow of the R-134a required, the rate of heat rejection, the power required, the COP, and the COR. Assume that R-134a is a saturated vapor at 80°C after a reversible adiabatic compression from −5°C.

12–20 (E) A wet compression cycle refrigerator using ammonia as a working medium operates such that the ammonia is 15°F in the evaporator and 90°F in the condenser. Determine the power required per ton of refrigeration.

12–21 (E) A freezer using Refrigerant-22 removes 1500 Btu/h from a compartment at 28°F. The refrigerant is at 23°F in the evaporator, and the quality is 85% leaving the evaporator. If the freezer operates on an ideal wet compression cycle, determine
(a) $\dot{m}$ of refrigerant
(b) $\dot{Q}_{rej}$
(c) $\dot{Wk}_{cycle}$
(d) COP and COR

12–22 (E) If the freezer of problem 12–21 is revised so that it can use R-123, determine the mass flow of R-123, the rate of heat rejected, the required power, the COP, and the COR. Assume that R-123 is a saturated vapor at 23°F before a reversible adiabatic compression to 180°F.

12–23 A chiller is a large air conditioner. A certain chiller system uses R-407c at 1.4 MPa pressure at the condenser and 600 kPa at the evaporator. If it operates on a reversible dry compression cycle, determine the COP. Note: The refrigerant has a temperature change in the evaporator, called a temperature glide, which is discussed in chapter 13. This phenomena will not affect your answer.

12–24 A chiller operates on 30 lbm/s of R-502 between 90°F at the condenser and 20°F at the evaporator. If the cycle is reversible ideal wet, determine the COP and the COR. Then determine the power required to operate the system.

12–25 A refrigeration unit operates on an ideal wet compression cycle and uses R-502. Operating between −6°C at the evaporator and 40°C at the condenser, what is the heat rejection per kg R-502? What is the compressor work?

Section 12–4

12–26 A reversed ideal Brayton cycle refrigerator operates with a pressure ratio of 18:1. Determine the COP and COR.

12–27 (M) An ideal gas refrigerator keeps a freezer at −10°C when operating and rejects heat to a surrounding at 35°C. The freezer unit requires a ½-hp electric motor to run a compressor, and the pressure ratio of the compressor is 16:1, assuming air to be the working medium. Determine
(a) Capacity rating of the freezer in kilowatts
(b) COP

12–28 (M) For the unit of problem 12–27, determine
(a) COR
(b) Rate of heat rejected to surrounding
(c) Mass flow rate of air

12–29 (E) A reversed ideal Brayton cycle refrigerator uses 5000 lbm/h of air with a pressure ratio of 10:1. If the refrigerator is rated at 2 tons, determine the power required by the unit.

12–30 (M) An ideal air-cycle air conditioner is used to cool an aircraft cabin to 15°C. The cooling load is 15 kW when outside air is 35°C, and the pressure ratio is 18:1 across the turbine. Determine the mass flow rate of air required, the required power to operate the system, and the coefficient of refrigeration, COR.

12–31 (E) An ideal air-cycle air conditioner cools an aircraft compartment to 60°F. The cooling load is 6 tons when the outside air is 98°F. For a turbine pressure ratio of 16:1, determine the mass flow rate of required air, the necessary power to operate the system, and the COP and COR.

12–32 (M) Consider the actual air-cycle air conditioner of problem 12–30, with the exception that the turbine is used to drive a fan to circulate ambient air as in figure 12–9. The turbine outlet air temperature is not to be less than −5°C, and the turbine generates 5 kW. The turbine is reversible and adiabatic. Determine the mass flow of bleed air from a gas turbine compressor and the mass flow of ambient air required if that air is not to increase by more than 30 K in the air cooler.

12–33 (E) An air refrigerating machine uses 100 lbm/min of air to cool a storage room. The air is 75°F as it leaves the radiator and 30°F as it enters the compressor. Assume that the air is reversibly and adiabatically compressed from 15 psia to 55 psia in the compressor and expanded reversibly and adiabatically through the turbine. Determine

(a) Temperature of air entering the radiator
(b) Temperature of air entering cooling coils
(c) $\dot{Q}_{net}$
(d) COP and COR
(e) Net work rate to cycle

12–34 (E) Use the same cycle as in problem 12–33, with the exceptions that the compression of the air is reversible and polytropically done with $n = 1.34$ and the expansion is reversible and polytropic with $n = 1.45$. Determine

(a) Temperature of air entering the radiator
(b) Temperature of air entering the cooling coil
(c) $\dot{Q}_{net}$
(d) COP
(e) Refrigeration rating of device

Section 12–7

12–35 (M) A heat pump operates on an ideal wet compression refrigeration cycle between 25°C and 10°C. The machine is designed to use R-123. If the compressor is driven by an electric motor capable of delivering $^1/_3$ hp to the compressor, determine the maximum rate of heating in kW and the COP.

12–36 (E) One particular problem associated with heat pumps is their dropping off of heating rate as the low temperature decreases. A particular heat pump, operating with R-134a and on an ideal dry vapor compression cycle and with condenser outlet temperature of 70°F, is found to have refrigerant flow rates as follows:

(a) 0.2 lbm/s at evaporator temperature of 30°F
(b) 0.17 lbm/s at evaporator temperature of 10°F
(c) 0.08 lbm/s at evaporator temperature of −10°F

Determine the heating rate in Btu/h and the COP for these three evaporator temperatures.

12–37 (M) A heat pump rated at 15 kW operates with a maximum flow rate of 4.0 kg/min of Refrigerant-22. The cycle is of the wet vapor compression type and can operate in climatic temperatures to −10°C. Determine the maximum inside temperature expected from this cycle.

12–38 (E) A heat pump provides 65,000 Btu/h of heat to a dwelling at 80°F when the outside temperature is 0°F. The device operates on a dry compression cycle and uses Refrigerant-22 as the working medium. Determine the minimum flow rate of the refrigerant, the COP, and the power required.

12–39 (E) If the heat pump of problem 12–38 is adapted to use R-134a, determine the flow rate of R-134a, the required power, and the COP.

12–40 (M) A 10-kW heat pump is to operate on a wet vapor compression cycle and use R-123. The refrigerant high pressure is 130.49 kPa, and the low pressure is 32.73 kPa. Determine the minimum required mass flow of R-123 and the coefficient of performance. Assume the R-123 is saturated vapor at the beginning of the compression.

12–41 (M) A heat pump operates on an ideal dry vapor compression refrigeration cycle, supplying 10 kW of heat to an indoor environment at 20°C. The heat comes from outside air which can be as low as −10°C. Determine the amount of power required to drive the heat pump and the maximum mass flow rate of ammonia required.

12–42 A heat pump is designed to operate on R-502 between −10°C and 40°C. What will be the rate of heat supplied per kg of R-502 if the cycle is ideal dry compression?

12–43 A heat pump is designed to operate on R-407c between 280 psia and 36 psia. The cycle is reversible and wet. Determine the COR.

MIXTURES

In this chapter, the concept of mixtures and the elementary methods of analyzing mixtures are presented. Mass analysis and molar analysis are explained. We consider a mixture of perfect gases and define partial pressure, mass fractions, and other terms necessary for an accurate, useful description of the system. Then the common mixture of air and water vapor is considered in terms of the thermodynamics of meteorology.

Applications to heating and air conditioning that require temperature or humidity controls are developed.

Chemical potential is defined in an intuitive manner and related to the water–air interaction. To give the reader the barest concept of the mechanism of mixing, *diffusion* is introduced through *Fick's law* and *mass balance*.

New Terms

C	Concentration ratio	Ω	Mole fraction
β	Relative humidity	ω	Specific humidity or humidity ratio
μ	Chemical potential	$\mathcal{D}$	Diffusivity
ψ	Mass fraction		

13–1 MIXTURE ANALYSIS

Earlier we studied pure substances, which are materials that can be described by one chemical species or one type of molecule or atom. Two or more pure substances mixed together are termed a **mixture**. Earlier we considered mixtures of saturated liquids and saturated vapors in the saturation region or phase-change region of pure substances. Here we consider mixtures of two or more pure substances. Also, we limit ourselves, in this chapter, to mixtures of pure substances that do not react with each other; that is, the substances remain the same pure substances even after mixing. In chapter 14, we consider mixtures that do react with each other and change into different pure substances upon mixing.

Let us now consider an important mixture that we use every day—air. Air is a mixture of oxygen, nitrogen, argon, and some other substances (some of which we would like to eliminate). We may determine the amounts of each substance, called constituents or components of the mixture, by measuring their amounts. If we measure the mass of each constituent and divide that mass by the total mass of the mixture, we call that fraction the mass fraction, ψ. If we denote the mass fraction of constituent i by ψ_2, then

$$\psi_i = \frac{m_i}{m_T} \tag{13–1}$$

where m_i is the mass of the i constituent and m_T is the total mass of the mixture. Also, the total mass of the mixture is the sum of all the m_i's, or

$$m_T = \sum m_i \tag{13–2}$$

We call the listing of the m_i's and the ψ_i's of all the constituents in a given mixture the **mass analysis** of the mixture.

If we were to measure the number of moles (kg · mol or lbm · mol) of the constituents of a mixture, which a chemist would probably do, we would find that the mole fraction of the ith constituent is

$$\Omega_i = \frac{N_i}{N_T} \qquad \textbf{(13–3)}$$

where N_i is the number of moles of the ith constituent and N_T is the total number of moles in the mixture. The total number of moles is also the sum of the number of moles of all the constituents:

$$N_T = \sum N_i \qquad \textbf{(13–4)}$$

The listing of the moles and mole fractions of a mixture is called the **molar analysis** of the mixture. The mass analysis and molar analysis can be converted between each other by the molecular mass concept. Using air as an example, we demonstrate the mass and molar analyses.

EXAMPLE 13–1 A sample of air is found to have the molar analysis given in table 13–1. Determine the mass analysis of the sample.

TABLE 13–1 Molar analysis of an air sample

Component	g · mol
Nitrogen, N_2	9.30
Oxygen, O_2	2.52
Carbon dioxide, CO_2	0.04
Argon, Ar	0.18
Impurities	0.02

Solution The molar analysis of table 13–1 shows that the major component of the air sample is nitrogen, with 9.30 g · mol. The total number of moles is just the sum of the components:

$$N_T = 9.30 + 2.52 + 0.04 + 0.18 + 0.02 = 12.06 \text{ g}\cdot\text{mol}$$

The mole fraction of each component is then determined from equation (13–3). For instance, the mole fraction of nitrogen is

$$\Omega_{N_2} = \frac{9.30}{12.06} = 0.771 = 77.1\%$$

Table 13–2 lists the mole fractions of all the components in the air sample. Also, the mass of each component is the number of moles times the molecular mass (MW) of that substance. Nitrogen occurs as a diatomic atom of air (N_2), having two atoms in one molecule, so that the molecular mass or weight is MW $= 2 \times 14.0 = 28.0$ g/g · mol, where the atomic weight or mass of nitrogen was found from table B–1 to be 14.0. Then the mass of nitrogen in the sample is

$$m_{N_2} = N_{N_2} \times MW_{N_2} = (9.30 \text{ g}\cdot\text{mol})(28.0 \text{ g/g}\cdot\text{mol})$$
$$= 260.4 \text{ g} = 0.2604 \text{ kg}$$

Similar results can be found for the masses of the oxygen (O_2), carbon dioxide (CO_2), and argon (Ar) by using atomic masses from table B–1, determining the molecular weights MW by using the chemical formulas, and computing the masses.

TABLE 13–2 Molar/mass analysis of air sample of example 13–1

Component	MW_i g/g · mol	Moles N_i g · mol	Mole Fraction Ω_i	Mass m_i g	Mass Fraction ψ_i
N_2	28.0	9.30	0.771	260.4	0.744
O_2	32.0	2.52	0.209	80.64	0.2304
CO_2	44.0	0.04	0.003	1.76	0.0050
Ar	40.0	0.18	0.015	7.20	0.0206
Impurities	—	0.02	0.002	—	—
Total		12.06	1.000	350.0	1.0000

The masses of all the components are listed in table 13–2. The total mass of the mixture is the sum of the masses of all the components:

$$m_T = 350.00 \text{ g}$$

The mass fraction, ψ_i, for each component is computed from equation (13–1), and these results are listed in table 13–2. Notice that the mass of the impurities was neglected, so the mass fraction was eliminated for those impurities. If a molecular mass had been given for the impurities, that term could have been included in the mass analysis.

The engineer or technologist is often involved with liquids or solids that are mixtures of identifiable substances. For instance, coal, a widely used fuel, is identified by its amount of carbon content through a mass analysis (sometimes called an ultimate analysis). Table 13–3 lists the ultimate analyses of some typical coals. Bituminous coal is the most common type, although subbituminous and lignite are also used in large amounts. Anthracite is a hard coal that burns with a hot, smokeless flame. It is rather scarce and is therefore a relatively expensive coal. The entry "ash" in the table represents solids that would not normally burn. It is the residue left after coal has burned completely.

Sometimes the ultimate analysis (or mass analysis) needs to be converted to a molar analysis for purposes of combustion studies. In chapter 14, we consider combustion processes, but here we demonstrate the conversion of an ultimate analysis to a molar analysis.

TABLE 13–3 Ultimate analysis of some typical coal samples from data published by the U.S. Bureau of Mines

			Ultimate Analysis						
Coal Variety	State	County	Sulfur S	Carbon C	Hydrogen H_2	Nitrogen N_2	Oxygen O_2	Ash	Heating Value Btu/lbm
I. Anthracite	Pennsylvania	Schuykill	0.5	82.5	2.5	1.2	4.9	8.4	12,970
II. Bituminous	West Virginia	Fayette	0.8	85.1	5.0	1.6	4.9	2.6	14,930
	Illinois	Williamson	0.9	68.5	5.1	1.1	16.3	8.1	12,015
III. Subbituminous	Wyoming	Sweetwater	0.7	48.1	5.1	1.3	38.1	6.7	7,830
IV. Lignite	South Dakota	Perkins	2.2	38.0	6.6	0.5	44.4	8.3	6,307

EXAMPLE 13–2 Determine the number of moles of carbon in 1 kg of a typical coal from Williamson County, Illinois, and determine the molar analysis. Assume that ash has the same molecular weight as silicon.

Solution From table 13–3, we find that Williamson County coal has 68.5% carbon by mass. Therefore, we can say that the carbon in 1 kg of coal is 0.685 kg. The number of moles of carbon in 1 kg of coal is

$$\frac{0.685\ \text{kg}}{\text{MW}} = \frac{0.685\ \text{kg}}{12.0\ \text{kg/kg}\cdot\text{mol}}$$
$$= 0.057083\ \text{kg}\cdot\text{mol/kg} = 57.083\ \text{g}\cdot\text{mol/kg coal}$$

The molar analysis requires a conversion of all the components in the coal to moles. Table 13–4 lists the results of that conversion in the column headed "N_i." The total number of moles per kilogram of coal is 91.031 g · mol. Finally, the mole fractions, Ω_i, are computed from equation (13–3), and these results are presented in table 13–4.

TABLE 13–4 Results of example 13–2: molar analysis of coal

Component	Mass g/kg coal	MW g/g · mol	N_i g · mol/kg coal	Ω_i g · mol/g · mol
Sulfur, S	9.0	32.1	0.280	0.003
Carbon, C	685.0	12.0	57.083	0.627
Hydrogen, H_2	51.0	2.016	25.298	0.278
Nitrogen, N_2	11.0	28.0	0.393	0.004
Oxygen, O_2	163.0	32.0	5.094	0.056
Ash	81.0	28.1	2.883	0.032
Totals	1000.0		91.031	1.000

It is sometimes necessary to know the enthalpy, internal energy, entropy, or specific heat of mixtures based on the total mass of the mixture. The enthalpy of a mixture per unit mass of the mixture, for instance, is given by the equation

$$h = \sum \psi_i h_i \tag{13–5}$$

where h_i is the enthalpy of the ith component of the mixture per unit mass of that component, and ψ_i is its mass fraction. The value for enthalpy is then just the sum of all the products, $\sum \psi_i h_i$, as indicated by equation (13–5). Similar equations can be written for internal energy, entropy, and specific heat:

$$u = \sum \psi_i u_i$$

$$s = \sum \psi_i s_i \tag{13–6}$$

$$c_p = \sum \psi_i c_{pi}$$

If the properties of a mixture need to be expressed per mole of the mixture, the equations become, respectively,

$$\bar{h} = \sum \Omega_i \bar{h}_i$$

$$\bar{u} = \sum \Omega_i \bar{u}_i$$

$$\bar{s} = \sum \Omega_i \bar{s}_i \tag{13–7}$$

$$\bar{c}_p = \sum \Omega_i \bar{c}_{p_i}$$

where the overbar "¯" is used to indicate that the property is per mole instead of per unit mass. Notice that the component properties $\bar{h}_i$, $\bar{u}_i$, $\bar{s}_i$, and $\bar{c}_{pi}$ are also determined per mole of that particular component. Let us now consider an example using these concepts.

EXAMPLE 13–3 Dry silica sand is usually considered to be a mixture of silicon and air. The silicon is in solid particles (grains of sand), and the air is a perfect gas surrounding the sand. Determine the specific heat at constant pressure, c_p, of dry silica sand if the mass analysis is 95% silicon and 5% air.

Solution Using equation (13–6), we have

$$c_p = \psi_{\text{air}} c_{p_{\text{air}}} + \psi_{\text{Si}} c_{p_{\text{Si}}}$$

From the mass analysis, $\psi_{\text{air}} = 0.05$ and $\psi_{\text{Si}} = 0.95$. Also, from table B–8, $c_{p_{\text{Si}}} = 0.691$ kJ/kg·K and $c_{p_{\text{air}}} = 1.007$ kJ/kg·K. Then

$$c_p = (0.95)(0.691 \text{ kJ/kg·K}) + (0.05)(1.007 \text{ kJ/kg·K})$$

Answer

$$= 0.7068 \text{ kJ/kg·K}$$

The specific heat value for sand listed in table B–8 is 0.80 kJ/kg·K, which is the average value of various types of sand. Our answer agrees well with this value.

13–2 PERFECT GAS MIXTURES

In this section, we assume that all the mixture components are perfect gases, so we can use the perfect gas equation. Suppose that we take 1 m^3 of oxygen at 101 kPa and 27°C and mix it with 1 m^3 of nitrogen at 101 kPa and 27°C in a 2-m^3 container. After these two gases have been sufficiently mixed, we notice that each gas is occupying the full volume of 2 m^3 and that the temperature is 27°C. For this final condition, then, if we assume that both gases are perfect, we can write

$$p_O V = N_O R_u T$$

for the oxygen and

$$p_N V = N_N R_u T$$

for the nitrogen, where the number of moles of oxygen is N_O, the number of moles of nitrogen is N_N, and R_u is the universal gas constant. If we add these two equations together, we have

$$(p_N + p_O)(V) = (N_N + N_O)R_u T$$

If we call $N_T = N_N + N_O$ the total number of moles and $p_T = p_N + p_O$ the total pressure, then

$$(p_T)(V) = N_T R_u T$$

Now, if we consider the ratio p_O/p_T, we can see from the preceding equations that

$$\frac{p_O}{p_T} = \frac{N_O R_u T/V}{N_T R_u T/V} = \frac{N_O}{N_T} = \Omega_O = \text{mole fraction of oxygen}$$

Also,

$$\frac{p_N}{p_T} = \frac{N_N}{N_T} = \Omega_N = \text{mole fraction of nitrogen}$$

so that we can write

$$p_O = \Omega_O p_T \quad \text{and} \quad p_N = \Omega_N p_T$$

We call the pressures p_O and p_N the partial pressures of the oxygen and nitrogen, respectively.

In general, for a mixture of perfect gases, the partial pressure of the ith component is

$$p_i = \Omega_i p_T \tag{13–8}$$

where p_T is the pressure of the mixture, or the total pressure. Also, the sum of the partial pressures is equal to the total pressure:

$$p_T = \sum p_i \tag{13–9}$$

Equation (13–9) is sometimes called *Dalton's law of partial pressures.*

EXAMPLE 13–4 A sample of air is found to have a molar analysis of 74.9% nitrogen, 23.9% oxygen, and 1.2% argon. The sample was taken at 100-kPa pressure and 28°C. Calculate the partial pressures of the components.

Solution Using equation (13–8), we obtain, for nitrogen,

Answer $$p_N = (0.749)(100 \text{ kPa}) = 74.9 \text{ kPa}$$

Similarly, for oxygen, we find that

Answer $$p_O = (0.239)(100 \text{ kPa}) = 23.9 \text{ kPa}$$

and for argon,

Answer $$p_{Ar} = (0.012)(100 \text{ kPa}) = 1.2 \text{ kPa}$$

EXAMPLE 13–5 Two pounds-mass of air at 80°F and 15 psia is found to be composed of 75.5% (by mass) nitrogen, 23.2% oxygen, and 1.3% argon. Calculate the partial pressures of each component.

Solution To calculate the partial pressures of the mixture components, we must first determine the molar analysis of the mixture. These results are given in table 13–5.

TABLE 13–5 Mass/molar analysis of air sample of example 13–5

Component	ψ_i lbm/lbm	m_i lbm	MW lbm/lbm · mol	N_i lbm · mol	Ω_i lbm · lbm/lbm · mol
Nitrogen, N_2	0.755	1.51	28.0	0.0539	0.781
Oxygen, O_2	0.232	0.464	32.0	0.0145	0.210
Argon, Ar	0.013	0.026	40.0	0.00065	0.009
Total	1.000	2.000		0.06905	1.000

Using equation (13–8) gives us

$$p_i = \Omega_i p$$

We know that the air pressure p is 15 psia. We then find the partial pressure of nitrogen, p_N, from

Answer $$p_N = (0.781)(15 \text{ psia}) = 11.715 \text{ psia}$$

Similarly, for oxygen,

Answer $$p_O = (0.210)(15) = 3.15 \text{ psia}$$

and for argon,

Answer $$p_{Ar} = (0.009)(15) = 0.135 \text{ psia}$$

Mixing two or more substances together is an irreversible process. To understand why this is so, just consider whether it is possible to "un-mix" or separate the components of a mixture. Such processes can be done, but they require work or heat to accomplish the separation—it cannot occur by itself or spontaneously. Mixing can occur without any effort provided that the components are in the same container and not separated by an actual boundary or wall. We can obtain some idea of the amount by which mixing processes are irreversible by calculating the entropy change involved in the mixing. We call this the entropy of mixing, and it can be computed from the equation

$$\Delta S = \sum m_i \, \Delta s_i \tag{13–10}$$

For perfect gases with constant specific heats, the Δs_i's can be determined from equations developed in chapter 6 and 7:

$$\Delta s_i = c_{pi} \ln \frac{T}{T_i} - R_i \ln \frac{p_T}{p_i}$$

Since $p_i/p_T = \Omega_i$, the mole fraction of the ith component is

$$\Delta s_i = c_{pi} \ln \frac{T}{T_i} - R_i \ln \Omega_i \tag{13–11}$$

EXAMPLE 13–6 At 27°C and 101-kPa pressure, 1.0 m^3 of oxygen is mixed with 1.0 m^3 of nitrogen. The mixing process occurs in a 2.0-m^3 container also at 27°C. Determine the entropy change due to this process.

Solution From equation (13–10), we have

$$\Delta S = m_O \, \Delta s_O + m_N \, \Delta s_N$$

where the subscripts O and N denote the properties of oxygen and nitrogen, respectively. From the perfect gas laws, we have

$$m_O = \frac{p_O V_O}{R_O T_O} = \frac{(1.01 \times 10^5 \text{ N/m}^2)(1.0 \text{ m}^3)}{(260 \text{ N} \cdot \text{m/kg} \cdot \text{K})(300 \text{ K})} = 1.295 \text{ kg}$$

and

$$m_N = \frac{p_N V_N}{R_N T_N} = \frac{(1.01 \times 10^5 \text{ N/m}^2)(1.0 \text{ m}^3)}{(297 \text{ N} \cdot \text{m/kg} \cdot \text{K})(300 \text{ K})} = 1.134 \text{ kg}$$

Using equation (13–11), we have

$$\Delta s_O = c_{p_o} \ln \frac{T}{T_O} - R_O \ln \Omega_O$$

and

$$\Delta s_N = c_{p_N} \ln \frac{T}{T_N} - R_N \ln \Omega_N$$

In this process of mixing oxygen with nitrogen at 27°C, we have $T = T_O = T_N = 300$ K. Then

$$\Delta s_O = -R_O \ln \Omega_O \quad \text{and} \quad \Delta s_N = -R_N \ln \Omega_N$$

The mole fractions can be found from

$$N_N = \frac{p_N V_N}{R_u T} = \frac{m_N}{\text{MW}} = \frac{1.134 \text{ kg}}{28.0 \text{ kg/kg} \cdot \text{mol}}$$
$$= 0.04050 \text{ kg} \cdot \text{mol}$$

and

$$N_O = \frac{m_O}{MW} = \frac{1.295 \text{ kg}}{32.0 \text{ kg/kg} \cdot \text{mol}}$$
$$= 0.040469 \text{ kg} \cdot \text{mol}$$

The total moles are 0.080969 kg · mol, so that

$$\Omega_N = \frac{N_N}{N_T} = \frac{0.04050}{0.080969} = 0.5002$$
$$\Omega_O = \frac{N_O}{N_T} = 0.4998$$

Then the entropy changes are

$$\Delta s_N = -R_N \ln \Omega_N = (0.297 \text{ kJ/kg} \cdot \text{K}) \ln(0.5002)$$
$$= 0.2057 \text{ kJ/kg} \cdot \text{K}$$

and

$$\Delta s_O = -R_O \ln \Omega_O = -(0.260 \text{ kJ/kg} \cdot \text{K}) \ln(0.4998)$$
$$= 0.1803 \text{ kJ/kg} \cdot \text{K}$$

The total entropy change, or entropy of mixing, is then

$$\Delta S = m_N \Delta s_N + m_O \Delta s_O = (1.134 \text{ kg})(0.2057 \text{ kJ/kg} \cdot \text{K})$$
$$+ (1.295 \text{ kg})(0.1803 \text{ kJ/kg} \cdot \text{K})$$
$$= 0.46675 \text{ kJ/K}$$

13–3 WATER AND AIR MIXTURES AND THE PSYCHOMETRIC CHART

Although air is a mixture of nearly perfect gases, it is commonly considered to be a homogeneous, single component having distinct properties associated with it. These properties, such as the gas constant R, would be the "average" property, or the "total" property in the case of pressure. In the cases where we consider air a uniform, homogeneous substance, we are assuming it to be dry; that is, no water vapor is mixed with the air. Dry air is generally assumed to be composed of 78% nitrogen, 21% oxygen, and 1% argon by volume. In reality, however, air seeks an equilibrium with its surroundings which are commonly liquid water. (The sea and sky are common neighbors.) When equilibrium is achieved, the rate of evaporation of the liquid water equals the rate of condensation of water vapor (see figure 13–1)—the water

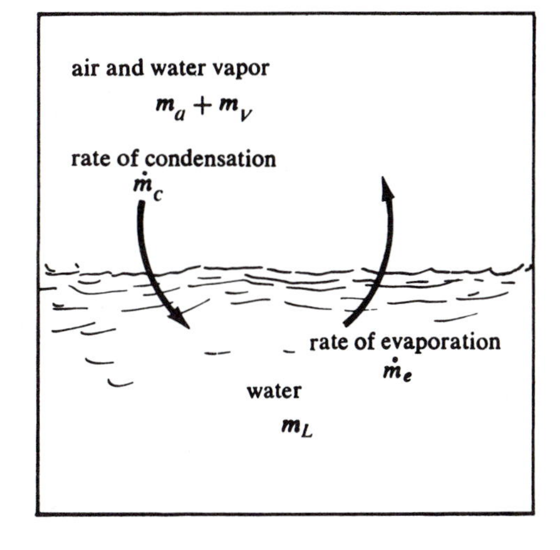

FIGURE 13–1 Equilibrium of liquid water and air in a closed container

vapor being a gas mixed with the air. To describe properly the air–water mixture, we may define, for some precise volume of this mixture, the mass ratio of water vapor to air, or

$$\omega = \frac{m_v}{m_a} \tag{13–12}$$

This ratio is called **specific humidity** or **humidity ratio**, where m_v is the mass of water vapor and m_a is the mass of air. If we assume that the water vapor acts as a perfect gas, the specific humidity can be written

$$\omega = \frac{p_v V/R_v T}{p_a V/R_a T} = \frac{R_a p_v}{R_v p_a} = 0.622\frac{p_v}{p_a} \tag{13–13}$$

Now, when water and its vapor are in equilibrium, we say that the water is in a phase change (the liquid–vapor phase change, in fact). This liquid and vapor mixture is called **saturated**, and, depending on the quality (see chapter 5 for a definition of *quality*), it is somewhere between a saturated liquid and a saturated vapor. If the water vapor is in a mixture with air and is in equilibrium with liquid water, the water vapor is a saturated vapor, and we call the mixture **saturated air**.

If the air–water vapor mixture is saturated air, the partial pressure of the water vapor p_v is equal to the saturation pressure p_g, corresponding to the temperature of the air. The saturated steam tables (table B–11) can be used to obtain the corresponding saturation pressure at a given temperature. Air that is not saturated is either dry air or some intermediate condition between dry air and saturated air. To describe the relative condition, we define **relative humidity** as

$$\beta = \frac{p_v}{p_g} \times 100\% \tag{13–14}$$

which gives us the conditions

$$\beta = 100\% \text{ for saturated air } (p_v = p_g)$$

and

$$\beta = 0\% \text{ for dry air } (p_v = 0)$$

Air that has a relative humidity between 100% and 0% is called **unsaturated air** and is a mixture of dry air and superheated water vapor. By using equations (13–13) and (13–14), we see that relative and specific humidities are related by

$$\omega = 0.622\beta\left(\frac{p_g}{p_a}\right)\frac{1}{100} \tag{13–15}$$

The units of specific humidity are mass of water vapor per mass of dry air. In SI units, this is usually expressed in grams of water vapor per kilogram of dry air. The amount of water vapor in an air–water vapor mixture is normally much less than the amount of dry air, so using the units kg/kg would give an inconveniently small number. In English units, the customary units of specific humidity are grains of water vapor per pound-mass of dry air (grains/lbm da). The grain is a mass unit defined as

$$1 \text{ lbm} = 7000 \text{ grains}$$

so that

$$1 \text{ grain} = 1/7000 \text{ lbm}$$

Unsaturated air could become saturated through a temperature change. The temperature at which an air–water vapor mixture becomes saturated air is called the **dew point temperature**. Figure 13–2 shows, on a *T–s* diagram, an example state *a* that is unsaturated

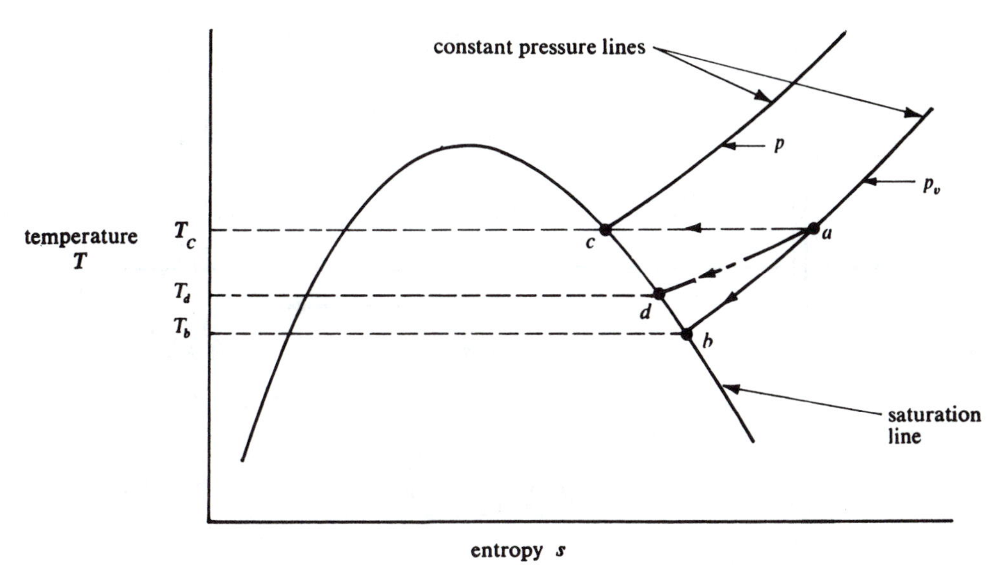

FIGURE 13–2 Temperature–entropy diagram for water vapor

air. By reducing the temperature at constant partial vapor pressure p_v, we reach the dew point temperature at state b.

Another manner of reaching a saturation condition from an unsaturated state is to increase the partial vapor pressure p_v at constant temperature until the pressure p_g is reached. This process is indicated by the curve a–c in figure 13–2. Notice that an increase in vapor pressure p_v is normally associated with a corresponding reduction in partial air pressure p_a, because, by Dalton's law, we must have $p = p_a + p_v$, where p is the atmospheric pressure.

For further descriptions of the air–water mixture, the dry-bulb and wet-bulb temperatures are commonly given. The dry-bulb temperature is that which is recorded in normal atmospheric conditions. The wet-bulb temperature, however, is measured in an atmosphere of saturated air and ideally is measured in a saturated condition. This process, an adiabatic saturation process, is depicted by curve a–d in figure 13–2, in which T_d would represent the wet-bulb temperature of air having a dry-bulb temperature T_a. In this process, no heat is transferred out of the air–water vapor mixture (thus the term *adiabatic*), but saturated vapor is added to the mixture to increase the vapor pressure.

The wet- and dry-bulb temperatures are recorded by a psychrometer such as that shown in figure 13–3. The air flow is induced either by moving the psychrometer in the air (which is then called a **sling psychrometer**, shown in figure 13–4, because it is normally rotated by hand) or by using a fan to force air through a stationary psychrometer. When measured in either of these two manners, however, the wet-bulb temperature is not precisely that which would be achieved through an adiabatic saturation process, but is rather achieved through a saturation process with minor heat transfer.

The enthalpy of an air–water vapor mixture is usually given as a value based on unit mass of dry air. This convention allows for more convenient analysis, as we shall see in the next section. Consider state 1 of figure 13–3. We can write

$$h_1 = c_pT_1 + \omega_1 h_{g_1}$$

and

$$h_3 = c_pT_3 + \omega_3 h_{g_3} \tag{13–16}$$

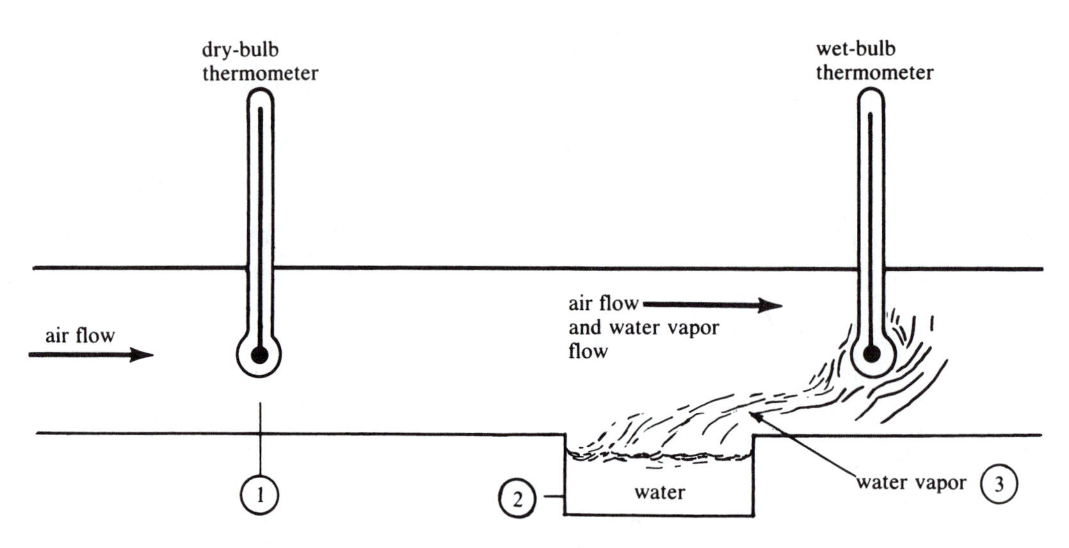

FIGURE 13–3 General features of wet- and dry-bulb psychrometer

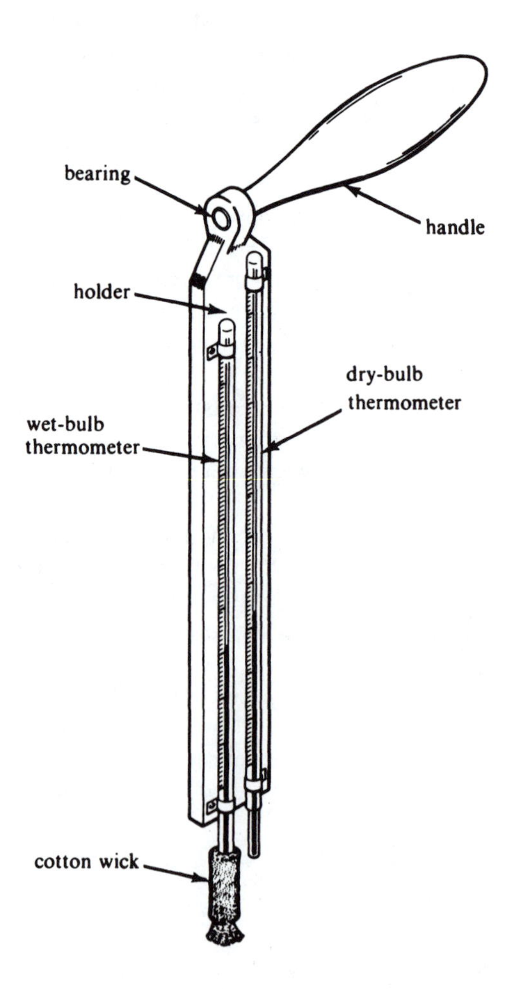

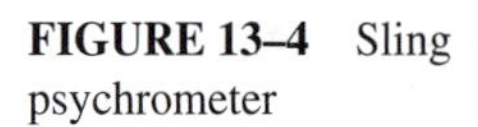
FIGURE 13–4 Sling psychrometer

Also, for a mass and energy balance of the flow in figure 13–3 and for the sling psychrometer of figure 13–4, we derive the relationship

$$\omega_1 = \frac{c_p(T_{wb} - T_{db}) + \omega_3 h_{fg_3}}{h_{g_1} - h_{f_3}} \tag{13–17}$$

where c_p = specific heat of air, 1.007 kJ/kg · K or 0.2404 Btu/lbm · R
T_{wb} = wet-bulb temperature, equal to T_3
T_{db} = dry-bulb temperature, equal to T_1
ω_3 = specific humidity at state 3, where the relative humidity is 100% = $0.622\, p_g/(p - p_g)$ and p_g is the saturation pressure at the wet-bulb temperature
h_{fg_3} = heat of vaporization found from table B–11 at the wet-bulb temperature
h_{g_1} = enthalpy of saturated vapor at the dry-bulb temperature
h_{f_3} = enthalpy of saturated liquid at the wet-bulb temperature

Equation (13–17) can be used to determine the specific humidity at the conditions of some air sample after using a sling psychrometer to measure the wet- and dry-bulb temperatures. All other terms in the equation can be found from steam tables. Also, all of the various properties and terms of the air–water vapor mixture, given in the following list, can be correlated and determined once two of the properties are known:

Specific humidity
Relative humidity
Vapor pressure
Dry- and wet-bulb temperature
Dew point temperature
Enthalpy

We can use equations (13–13) through (13–17) and steam tables to make this correlation. Another more convenient and rapid method of determining the same set of properties is with the psychrometric chart sketched in figure 13–5. In the sketch, you can see the basic coordinates and the graphical relationship between the properties. In chart B–10, the SI and English versions of the psychrometric chart are presented for the customary range used by engineers and technologists.

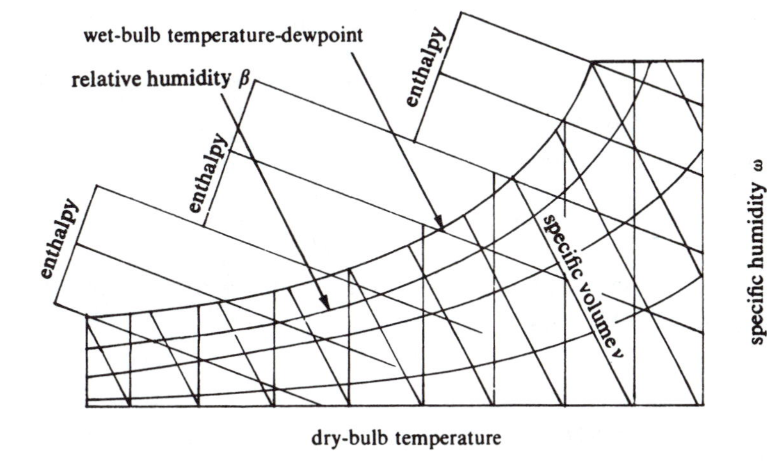

FIGURE 13–5 Sketch of the basic parameters constituting a psychrometric chart

The charts given in figure 13–5 and in appendix B are based on an atmospheric pressure of 14.7 psia or 101 kPa. If the atmospheric pressure deviates greatly from this value, the chart must be corrected. (See the following references in appendix C–1: Lee and Sears, *Thermodynamics*, and Mooney, *Mechanical Engineering Thermodynamics*.)

The enthalpy values from the psychrometric chart are based on unit mass of dry air, as given in equation (13–16). The specific volume, or inverse of the density, is also based on unit mass of dry air, and the mixture is assumed to be a perfect gas. Next we look at three examples of how to use the ideas and equations that we have been discussing.

EXAMPLE 13–7 Unsaturated air is at 20°C. Find the partial pressure required of the water vapor to convert this mixture to the saturated state at 20°C.

Solution: Using the steam table B–11 or the psychrometric chart B–10, we obtain $p_g = 2.3$ kPa at 20°C, so the pressure of the vapor, p_v, must be 2.3 kPa for the saturation state of air at 20°C.

EXAMPLE 13–8 Atmospheric air at 14.7 psia has a temperature of 85°F and a wet-bulb temperature of 70°F. Determine the vapor pressure, the relative humidity, the specific humidity, and the enthalpy of the atmospheric air.

Solution: Because the atmospheric pressure is standard, we may directly use the data of the psychrometric chart. The dry-bulb temperature can be assumed to be 85°F. Then, using chart B–10, we obtain the vapor pressure:

Answer
$$p_v = 0.286 \text{ psia}$$

Also from the chart, the relative humidity is

Answer
$$\beta = 48\%$$

and the specific humidity is

$$\omega = 86 \text{ grains/lbm}$$

There are 7000 grains in 1 lbm, so we get

Answer
$$\omega = \frac{86}{7000} \text{ lbm vapor/lbm air} = 0.0123$$

Also, we should be able to get the preceding answers to correlate in equation (13–15), or

$$\omega = 0.622\beta \frac{p_g}{p_a}\left(\frac{1}{100}\right)$$

where, from equation (13–14),

$$p_g = \frac{p_v}{\beta} \times 100$$
$$= \text{saturation pressure at dry-bulb temperature}$$

From table B–11, we have $p_g = 0.5959$ psia at a temperature of 85°F. Then, using the value of p_v above from the psychrometric chart, we have

$$p_g = \frac{0.286 \text{ psia}}{48} \times 100 = 0.596 \text{ psia}$$

which agrees with our result from the saturated steam table. The enthalpy is read directly from the psychrometric chart as

Answer
$$h = 34.1 \text{ Btu/lbm air}$$

The enthalpy is based on the unit of mass of dry air, so the total enthalpy must be determined by using only the mass of dry air, not the mixture mass.

EXAMPLE 13–9 An air–water vapor mixture is at 70% relative humidity and 17.5°C. Determine the dew point temperature and the specific volume.

Solution: From chart B–10 we read, at a dry-bulb temperature of 17.5°C and relative humidity of 70%,

Answer

$$\text{dew point temperature, } T_{dp} = 12.5°\text{C}$$

and

Answer

$$\text{specific volume, } v = 0.836 \text{ m}^3/\text{kg da}$$

Although the humidity, dew point, and temperature can all be determined from the sling psychrometer, it has become more convenient to use psychrometers that have digital electronic readouts or that can provide an electronic signal for automatic humidity controls. One type of such a device, shown in figure 13–6, can sense the humidity and dew point of an air–water vapor mixture within two minutes, without requiring the operator to "sling" the device. Other devices that incorporate an electrical or electronic circuit to provide automatic controls for regulating the humidity in a region are sometimes referred to as **humidistats**.

13–4 PROCESSES OF AIR–WATER MIXTURES

Engineers and technologists encounter air–water vapor mixtures in many of their professional experiences. In this section, we consider five processes that involve air–water vapor mixtures which are important and often encountered: (1) air drying processes, (2) heating and humidification processes, (3) evaporative cooling processes, (4) dehumidifying processes, and (5) adiabatic mixing of two air streams. We consider these processes in order.

Air Drying Processes

Drying processes have many important applications in technology and in design. In an air drying process, water or some other liquid is removed from a system (such as concrete, grain, or clothes) and transferred to air. The process, sometimes called **evaporation**, can be

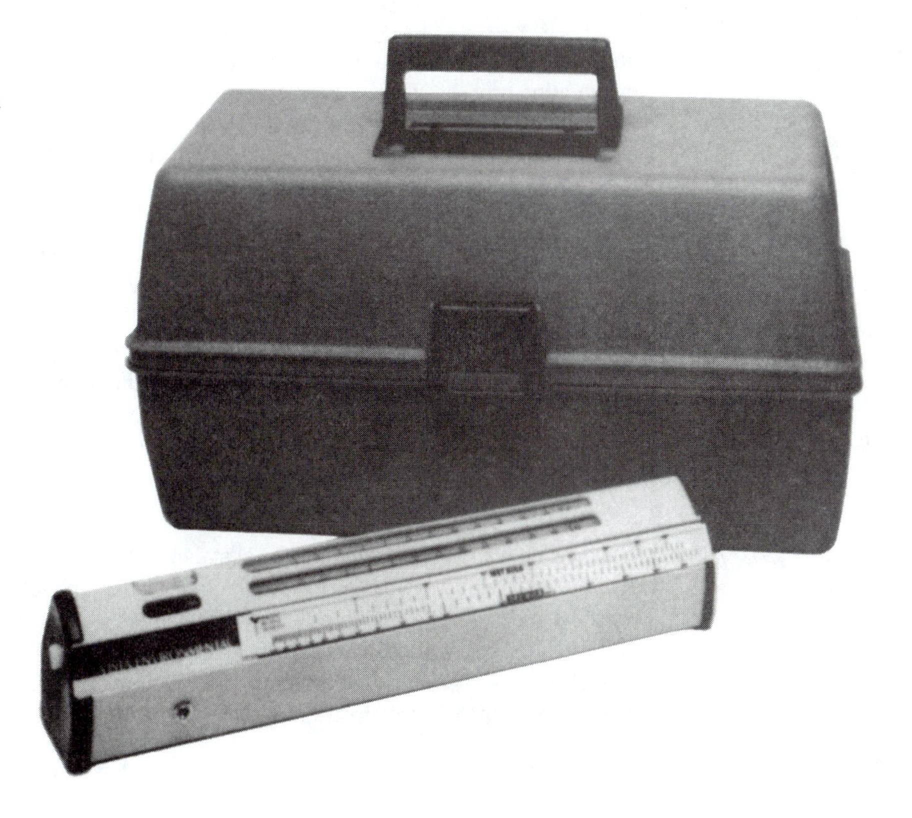

FIGURE 13–6 Portable automatic psychrometer with carrying case (courtesy of Vista Environmental)

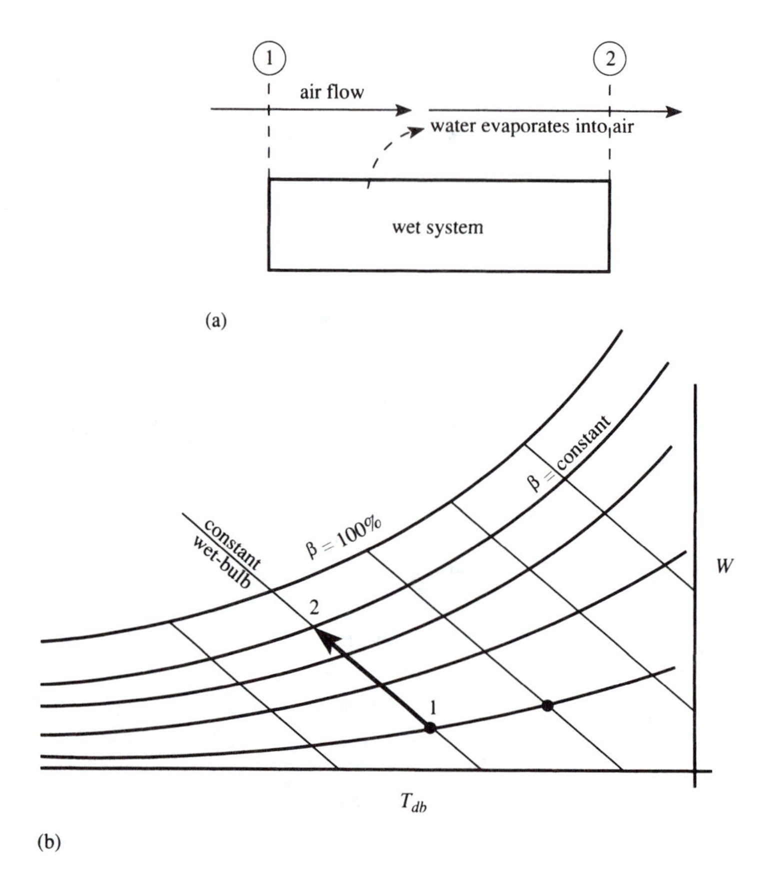

FIGURE 13–7 Air drying process: (a) system drying by air flow; (b) psychrometric chart to show process in (a)

analyzed with the aid of the psychrometric chart. Figure 13–7 shows a typical air drying or evaporation process. The air interacts with a system beginning at state 1 and removes water from the system during its change to state 2. The process follows approximately along a constant-enthalpy (called **isenthalpic**) line. If the water that is evaporating has a temperature equal to the wet-bulb temperature, the process will follow a constant wet-bulb temperature line. The reason the process follows this path is that it is assumed to be an adiabatic process, and from the steady-flow energy equation applied to the air, the enthalpy will remain nearly constant. The amount of water that evaporates into the air from some system is given by the difference in the specific humidities between states 2 and 1. Notice also that the reverse process, from state 2 to state 1, is impossible; that is, one cannot increase air temperature and dehumidify at the same time without heat transfer.

EXAMPLE 13–10 Air at 100°F and 40% relative humidity is used in a clothes dryer. Determine the amount of water that can be removed per minute if the air flow is 2000 ft^3/min at the dryer exhaust. Figure 13–8 is a schematic of the device.

Solution The maximum amount of water that the airstream can absorb will be that when the exhaust air is at 100% relative humidity. Using the psychrometric chart and referring to figures 13–7 and 13–8, we find at the inlet state 1, where $T_1 = 100°\text{F}$ and $\beta_1 = 40\%$, that $\omega_1 = 115.4$ grains/lbm da. Also, at the exhaust state 2, where $\beta_2 = 100\%$, and $T_{wb_1} = T_{wb_2}$, $\omega_2 = 150$ grains/lbm da. The water that can be absorbed by the airstream per lbm dry air is then $\omega_2 - \omega_1 = 34.6$ grains/lbm da. The rate at which water can be removed would be

$$\dot{m}_w = \dot{m}_{da}(\omega_2 - \omega_1)$$

FIGURE 13–8 Clothes dryer

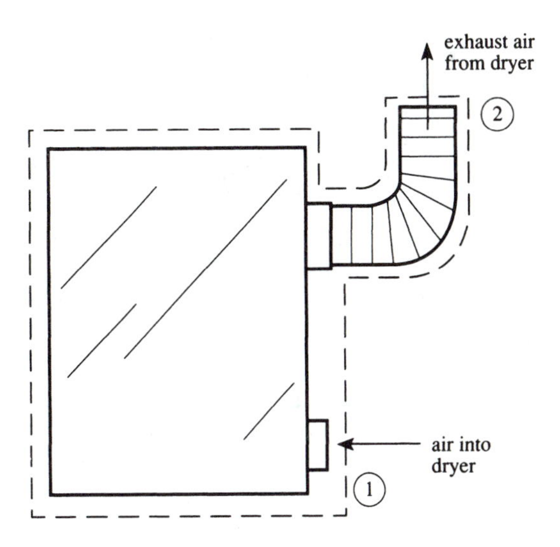

where $\dot{m}_{da}$ is the flow rate of dry air. The volume flow rate for air is 2000 ft^3/min, so the mass flow rate, $\dot{m}_{da}$, will be the volume flow rate divided by the specific volume of the air:

$$\dot{m}_{da} = \frac{\dot{V}}{v} = \frac{\dot{V}_2}{v_2}$$

From the psychrometric chart, $v_2 = 14.04$ ft^3/lbm da, so that

$$\dot{m}_{da} = (2000\text{ ft}^3/\text{min})/(14.04\text{ ft}^3/\text{lbm da}) = 142.45\text{ lbm/min}$$

The drying rate, or amount of water that can be absorbed by the air, is

$$\dot{m}_w = \frac{(142.45\text{ lbm/min})(34.6\text{ grains/lbm da})}{7000\text{ grains/lbm}}$$
$$= 0.704\text{ lbm water/min}$$

Heating and Humidification Processes

In many climates, it is desirable to increase the air temperature. In these situations (such as in winter seasons where air is cold and dry), the air, after being heated, is so dry as to be uncomfortable. As a result of such processes, it is often advisable to add moisture to the air, or humidify the air. The heating and humidification process is a common one in many technological applications. A schematic of a system that is capable of performing these processes is shown in figure 13–9, and figure 13–10 illustrates the processes on a psychrometric chart. Typically, the process involves a heating stage and an evaporation stage. Sometimes, if the water is introduced as high-temperature steam, there can be a heat transfer occurring in the evaporation of the steam, and then a state 3′ at a higher temperature could be reached. This is indicated in figure 13–10. We now consider examples of the heating and humidification process.

EXAMPLE 13–11 Assume that air enters a humidifier at 101 kPa, 20°C, and 40% relative humidity and that it should be 25°C and 50% relative humidity upon leaving the unit. Determine the amount of heat required in the heating unit and the amount of water required per kilogram of dry air. Assume that the entering liquid water is at 15°C.

Solution From figure 13–9, the steady-flow energy equation for the heater/humidifier can be written as

$$m_a h_3 - (m_a h_1 + m_w h_w) = Q$$

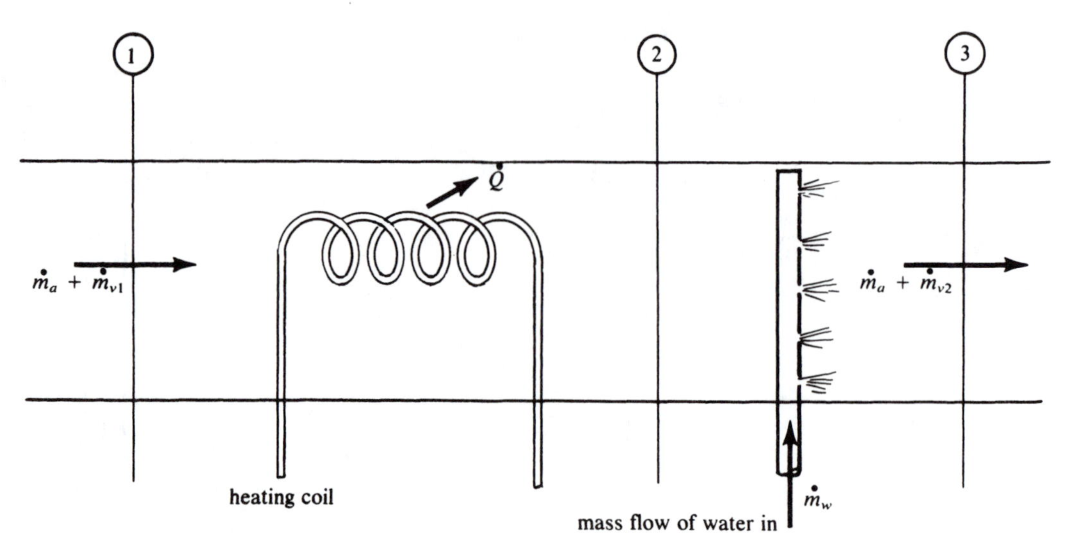

FIGURE 13–9 Heating and humidifying process

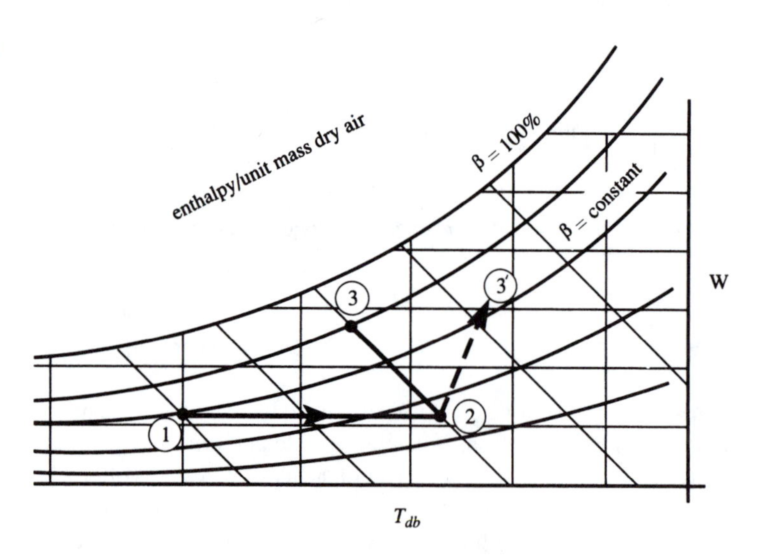

FIGURE 13–10 Heating and humidifying process on a psychrometric chart

and the mass flow equation as

$$\dot{m}_a + \dot{m}_{v1} + \dot{m}_w = \dot{m}_a + \dot{m}_{v3}$$

From the psychrometric chart, we read

$$h_3 = 52 \text{ kJ/kg}$$
$$h_1 = 36 \text{ kJ/kg}$$
$$\omega_3 = 10 \text{ g/kg dry air}$$
$$= 0.010 \text{ kg/kg dry air}$$

and

$$\omega_1 = 6 \text{ g/kg} = 0.006 \text{ kg/kg dry air}$$

Using the steady-flow energy equation, we find the heat required per kilogram of dry air:

$$\frac{Q}{m_a} = q = h_3 - h_1 - \frac{m_w}{m_a} h_w$$

$$= 52 \text{ kJ/kg} - 36 \text{ kJ/kg} - \frac{m_w}{m_a} h_w$$

From the mass flow equation, we get

$$\dot{m}_w = \dot{m}_{v3} - \dot{m}_{v1} = \dot{m}_a \omega_3 - \dot{m}_a \omega_1$$

Then

$$\frac{m_w}{m_a} = \omega_3 - \omega_1 = 0.010 \text{ kg/kg dry air} - 0.006 \text{ kg/kg dry air}$$

$$= 0.004 \text{ kg/kg dry air}$$

and from table B–11, we obtain the enthalpy of the 15°C water that is evaporated:

$$h_w = 63 \text{ kJ/kg}$$

Then

Answer

$$q = 69 \text{ kJ/kg} - 53 \text{ kJ/kg} - (0.004 \text{ kg/kg})(63 \text{ kJ/kg})$$

$$= 15.75 \text{ kJ/kg}$$

The amount of water has been determined and is the 0.004 kg/kg dry air computed previously.

EXAMPLE 13–12 Consider a heater/humidifier with a heater rated at 30 kW. The unit operates with an air flow of 1.054 m^3/s at 0°C and 60% relative humidity. The air leaving the unit is to be at 27°C and 60% relative humidity. Determine the condition of the water that must be added to achieve the desired outlet conditions, whether steam or liquid, and at what temperature.

Solution Referring to figures 13–9 and 13–10, we have

$$\dot{Q} = \dot{m}_{da}(h_2 - h_1) = 30 \text{ kW} = 30 \text{ kJ/s}$$

and $\dot{m}_{da} = \dot{V}/v$. The specific volume of the air is approximately 0.775 m^3/kg da from chart B–10. Then

$$\dot{m}_{da} = \frac{1.054 \text{ m}^3/\text{s}}{0.775 \text{ m}^3/\text{kg da}} = 1.36 \text{ kg da/s}$$

Also from chart B–10, we read $h_1 = 7$ kJ/kg da, so that

$$h_2 = \frac{\dot{Q}}{\dot{m}_{da}} + h_1$$

$$= \frac{30 \text{ kJ/s}}{1.36 \text{ kg da/s}} + 7 \text{ kJ/kg da}$$

$$= 29 \text{ kJ/kg da}$$

Then, at the outlet, we find that $h_3 = 63$ kJ/kg da and $\omega_3 = 13.4$ g/kg da. The wet-bulb temperature will be 21.5°C. Referring now to figure 13–10, we can see that water added at room temperature and evaporated into the air will not provide the necessary temperature. Therefore, we write the energy balance (see example 13–11):

$$q = h_3 - h_2 - (\omega_3 - \omega_2)h_w = 0$$

Then

$$h_w = \frac{h_3 - h_2}{\omega_3 - \omega_2} = \frac{63\ \text{kJ/kg} - 29\ \text{kJ/kg}}{0.0134\ \text{kg/kg} - 0.0024\ \text{kg/kg}} = 3090.9\ \text{kJ/kg}$$

For water with an enthalpy of 3090.9 kJ/kg, we assume that it is steam, and from table B–11, we may assume superheated steam at 150 kPa. By interpolating from table B–12, we find that the steam would be at 309°C if its enthalpy were 3090.9 kJ/kg at 150 kPa. Therefore, we recommend

Answer

$$T_w = 309°\text{C} \qquad \text{and} \qquad p = 150\ \text{kPa}$$

Evaporative Cooling Processes

Some climatic conditions make humidification desirable as a cooling process. For hot and very dry (relative humidities of 10% or less) climates, the evaporative cooling process takes advantage of this phenomenon. The process is the same as the drying process or humidification process except that the air is now conditioned so that its dry-bulb temperature is reduced and the relative and specific humidities are increased to make more comfortable air.

EXAMPLE 13–13 Air at 105°F and 0% relative humidity is to be furnished at 75°F. Determine the amount of water required per pound-mass of dry air for evaporative cooling, and find the relative humidity of that air at 75°F. A schematic of the system is shown in figure 13–11.

Solution Referring to figure 13–11, we see that the amount of water required per pound-mass of dry air is

$$m_w = \omega_2 - \omega_1 = \omega_2$$

since $\omega_1 = 0$. From the psychrometric chart, we read

Answer

$$\omega_2 = 50\ \text{grains/lbm da} = m_w$$

and

Answer

$$\beta_2 = 39\%$$

The air supplied in example 13–13 is cool due to evaporation of the water, and it is also more comfortable for humans because of the moderate relative humidity. In chapter 16, we consider the concept of comfort in more detail.

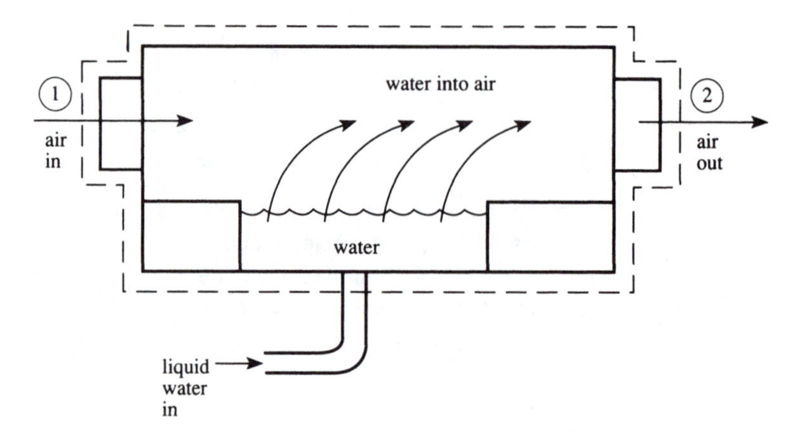

FIGURE 13–11 Evaporative cooler

Dehumidifying Processes Dehumidification involves a reduction in the specific humidity or water in the air, usually to provide a more comfortable condition. The most common physical method of removing water from air is to cool the air to the dew point and then cool that air further until the appropriate specific humidity is reached. In figure 13–12, this process is shown on a psychrometric chart. The process begins at state 1, progresses to stage 4 or the dew point, and continues to state 3, where the final specific humidity has been reached. If the air is too cold at state 3, a heating process can be included to bring the temperature up to the desired value. Process 3–2 would be such a heating process. In figure 13–13 is shown a system used to accomplish a dehumidification/heating process. Finally, as a reference from previous discussions of temperature–entropy diagrams, a temperature–entropy diagram of the water vapor that is involved with the process is shown in figure 13–14, corresponding to the states of figures 13–12 and 13–13.

FIGURE 13–12 Psychrometric chart for dehumidifying process

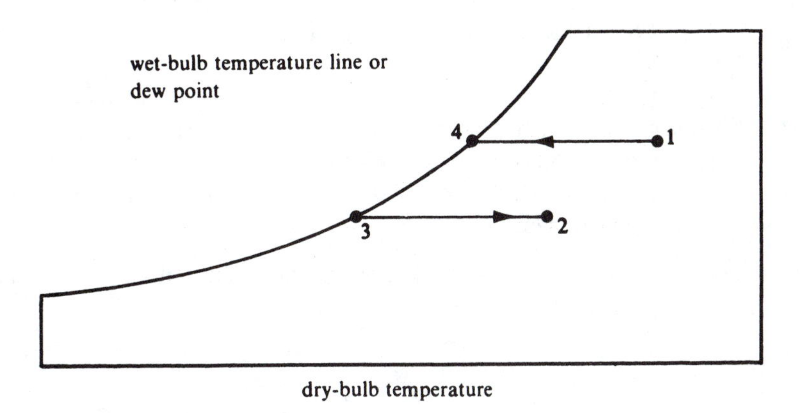

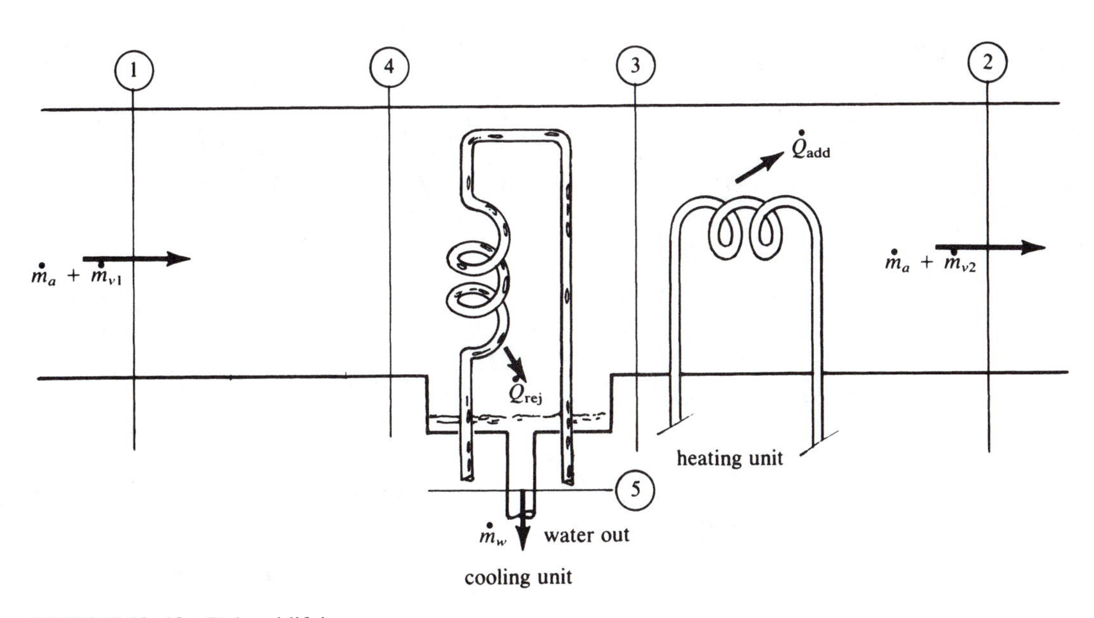

FIGURE 13–13 Dehumidifying process

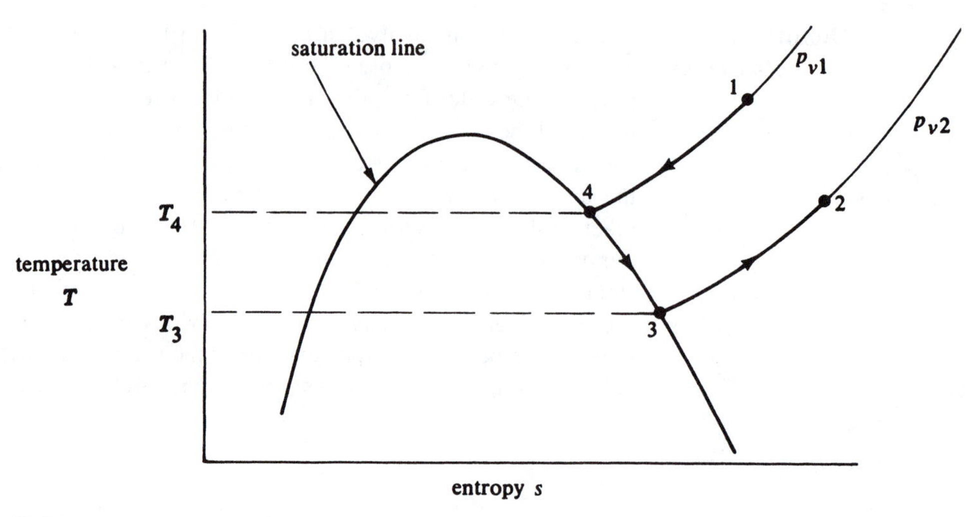

FIGURE 13–14 Dehumidifying process on a temperature–entropy plane

EXAMPLE 13–14 Assume that air is taken in at 14.7 psia, 80°F, and 80% relative humidity. If the air is to be cooled to 72°F and reduced to 60% relative humidity, determine the heat rejected in the cooling unit, the heat added in the heating unit, and the amount of water condensed per pound-mass of dry air.

Solution Applying the steady-flow energy equation to the cooling and heating units separately, we obtain

$$m_a h_3 + m_w h_5 - m_a h_1 = Q_{\text{rej}}$$

and

$$m_a h_2 - m_a h_3 = Q_{\text{add}}$$

Per unit mass of dry air, we have

$$h_3 + \frac{m_w}{m_a} h_5 - h_1 = q_{\text{rej}}$$

and

$$h_2 - h_3 = q_{\text{add}}$$

From the psychrometric chart, we read

$$h_1 = 38.8 \text{ Btu/lbm da}$$
$$h_2 = 28.4 \text{ Btu/lbm da}$$

and

$$\omega_1 = \frac{125}{7000} = 0.0178 \text{ lbm/lbm da}$$
$$\omega_2 = \frac{70}{7000} = 0.010 \text{ lbm/lbm da}$$

The amount of water condensed per pound-mass of dry air is the decrease in specific humidity:

$$\frac{m_w}{m_a} = \omega_1 - \omega_2 = 0.0178 - 0.010$$

Answer

$$= 0.0078 \text{ lbm condensate/lbm da}$$

The enthalpy at state 5, the condensed water, is given by $h_5 = h_f$ at the dew point of the incoming air mixture. The dew point is found from the psychrometric chart to be 73.4°F, and then $h_f = 41.45 = h_5$ from table B–11.

The method of reading the dew point corresponding to state 1 is indicated in figure 13–12, and in the same manner, we obtain the dew point of the dehumidified air at state 2; $t_3 = 57.2$°F and the enthalpy $h_3 = 24.6$. We can now calculate the heat transfer as

Answer
$$q_{rej} = 24.6 + (0.0078)(41.45) - 38.8 = -13.9 \text{ Btu/lbm da}$$

and

Answer
$$q_{add} = 28.4 - 24.6 = 3.8 \text{ Btu/lbm da}$$

Adiabatic Mixing of Two Air Streams

The mixing or combining of two air streams is a common occurrence in many engineering and technological applications, such as in the heating, cooling, or ventilating of buildings. Consider the schematic of two air streams, (1) and (2), combined to provide a mixed air stream (3) as shown in the schematic of figure 13–15. Under the assumption that the mixing occurs at constant pressure and is adiabatic, the energy balance can be written as

$$\dot{m}_1 h_1 + \dot{m}_2 h_2 = \dot{m}_3 h_3 \tag{13–18}$$

where the mass flow rates are mass of dry air per unit time and the enthalpies are those of the dry air plus the water vapor per unit mass of dry air. The mass balance can be written as

$$\dot{m}_1 + \dot{m}_2 = \dot{m}_3 \tag{13–19}$$

for the dry air and as

$$\dot{m}_1 \omega_1 + \dot{m}_2 \omega_2 = \dot{m}_3 \omega_3 \tag{13–20}$$

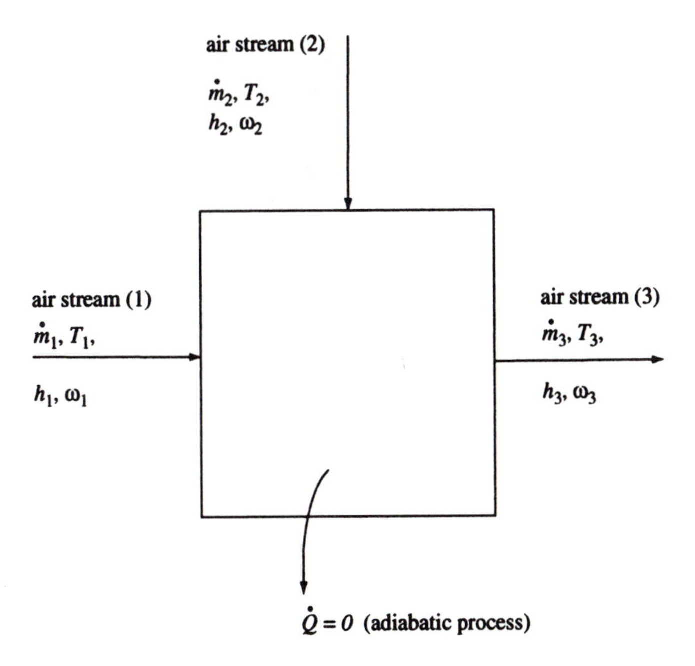

FIGURE 13–15 Adiabatic mixing of two air streams

for the moisture in the air, substituting equation 13–19 into equation 13–18 and eliminating the mass flow of state (2) gives

$$\frac{\dot{m}_1}{\dot{m}_3} = \frac{h_3 - h_2}{h_1 - h_2} \tag{13–21}$$

Similarly, combining equations 13–19 and 13–20 gives

$$\frac{\dot{m}_1}{\dot{m}_3} = \frac{\omega_3 - \omega_2}{\omega_1 - \omega_2} \tag{13–22}$$

It is clear that equations 13–21 and 13–22 relate the specific humidities and the enthalpies, so that

$$\frac{h_3 - h_2}{h_1 - h_2} = \frac{\omega_3 - \omega_2}{\omega_1 - \omega_2} \tag{13–23}$$

and this relationship is linear. On a psychrometric chart, as shown in figure 13–16, the mixed air stream (3) can be determined by constructing a straight line between the two inlet air stream states, (1) and (2). State (3) is then located on this line at a point between (1) and (2), the distance between (3) and (2) being to the mass flow of (1) as the distance between (1) and (2) is to the mass flow of (3). This is given by equation 13–21 or

$$\frac{h_1 - h_2}{\dot{m}_3} = \frac{h_3 - h_2}{\dot{m}_1} \tag{13–24}$$

This means that state (3) will be nearer that inlet state having the greater mass flow rate. If the two inlet mass flows are the same amount, then state (3) will be exactly in the middle between the two inlet states.

EXAMPLE 13–15 1000 cubic feet per minute (cfm) of air at 90°F and 60% relative humidity are mixed with 3000 cfm of air at 70°F and 40% relative humidity. Determine the mixed air stream temperature, relative humidity, and volume flow rate in cfm.

Solution The problem statement gives volume flow rates of the incoming air streams. Using equations 4–6 and 4–11, we have

$$\dot{m} = \rho A\overline{V} = \rho\dot{V} = \frac{\dot{V}}{v}$$

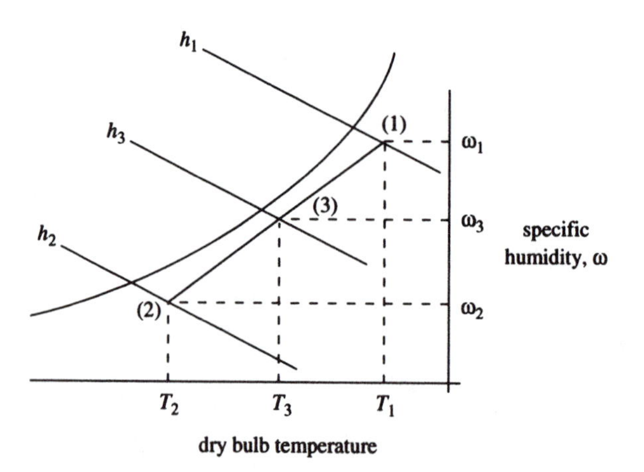

FIGURE 13–16 Adiabatic mixing of two air streams on a psychometric chart

From the psychrometric chart, we find, at the inlet states, that
For state (1),

$T_1 = 90°F$, $\nu_1 = 14.26$ ft^3/lbm dry air, $h_1 = 42$ BTU/lbm dry air, $\beta_1 = 60\%$, and $\omega_1 = 0.0184$ lbm water vapor/lbm dry air.

For state (2),

$T_2 = 70°F$, $\nu_2 = 13.48$ ft^3/lbm dry air, $h_2 = 23.7$ BTU/lbm dry air, $\beta_2 = 40\%$, and $\omega_2 = 0.0062$ lbm water vapor/lbm dry air.

Then the mass flow rates are

$$\dot{m}_1 = \frac{1000\ \text{ft}^3/\text{min}}{14.26\ \text{ft}^3/\text{lbm}} = 70.126\ \text{lbm/min} \approx 70\ \text{lbm/min}$$

and

$$\dot{m}_2 = \frac{3000\ \text{ft}^3/\text{min}}{13.48\ \text{ft}^3/\text{lbm}} = 222.55\ \text{lbm/min} \approx 223\ \text{lbm/min}$$

Next, using the conservation of mass, or equation 13–19, we get

$$\dot{m}_3 = \dot{m}_1 + \dot{m}_2 \approx 70\ \text{lbm/min} + 223\ \text{lbm/min} = 293\ \text{lbm/min}$$

Using equation 13–24, we can now solve for h_3:

$$h_3 = (h_1 - h_2)\left(\frac{\dot{m}_1}{\dot{m}_2}\right) + h_2 = (42\ \text{Btu/lbm} - 23.7\ \text{Btu/lbm})\left(\frac{70\ \text{lbm/min}}{293\ \text{lbm/min}}\right) + 23.7\ \text{Btu/lbm}$$

$$h_3 = 28.07\ \text{Btu/lbm}$$

If a straight line is now constructed on a psychrometric chart between states (1) and (2), then the exiting flow state (3) can be determined by locating the enthalpy h_3 and intersecting this enthalpy line with the straight line. The intersection is state (3), having properties

Answer $T_3 = 75°F$ (dry-bulb) and relative humidity, $\beta_3 \approx 50\%$.

Also, the specific volume is $\nu_3 = 13.66$ ft^3/lbm dry air, so that the volumetric flow rate is

Answer $$\dot{V}_3 = \dot{m}_3\nu_3 = (293\ \text{lbm/min})(13.66\ \text{ft}^3/\text{lbm}) = 4002.38\ \text{ft}^3/\text{min}$$

Notice that the volumetric flow rate is nearly the same as the sum of the two inlet volumetric flow rates, 3000 cfm + 1000 cfm = 4000 cfm. In many problems involving heating, ventilating, or air conditioning, where two air streams are mixed, the air is assumed to be incompressible so that the volumetric flow rates can be summed; that is, $V_1 + V_2 = V_3$.

Finally, in this last example, there are other methods that could have been used to obtain the same answers. For instance, instead of solving for the enthalpy, h_3, to determine the state (3) on the psychrometric chart, one could solve for the specific humidity at state 3, ω_3, by using equation 13–22 and then proceeding to locate state (3) on the straight line between states (1) and (2) with the specific humidity. Also, one could determine state (3) without using the psychrometric chart by solving for both the enthalpy and specific humidity at state (3) and, by using the saturation values for steam from the steam tables and an iteration process, arrive at the answers for the dry-bulb temperature, relative humidity, and the specific volume of the mixed air stream, state (3).

The five processes that we considered here—drying, heating and humidification, evaporative cooling, dehumidifying, and adiabatic mixing of two air streams—are important for engineers and technologists to recognize and be able to understand. The last three of these processes are often classified as heating and air-conditioning processes, and we consider them further in chapter 16 when we discuss the subject of heating and air-conditioning.

13–5 CHEMICAL POTENTIAL

In section 13–4, we considered evaporation and mixing of air and water. One method of better understanding the mechanism of mixing and evaporation is through the concept of equilibrium. **Equilibrium** is a relative term that requires precise definition to become meaningful. We have earlier seen that two bodies or systems are in **thermal equilibrium** if their temperatures are equal. **Mechanical equilibrium** is achieved between two systems when their pressures are equal at the interface. (Mechanical equilibrium is static or quasi-static). In figure 13–17, we see two systems, A and B, which we can consider as liquids, solids, or gases. They are thermal equilibrium when $T_A = T_B$ and in static mechanical equilibrium when $p_A = p_B$. There is also one other form of equilibrium, called **chemical equilibrium**, which we must consider for a complete definition of equilibrium. These two systems, which we assume are each mixtures of the same components, 1 and 2, are said to be in chemical equilibrium when the chemical potential of component 1 in system A (μ_{A1}) is equal to the chemical potential of component 1 in system B (μ_{B1}) and, in a similar manner, when $\mu_{B2} = \mu_{B1}$. That is, two systems that interface one another attain chemical equilibrium when both have the same components and these components have the same chemical potential in both systems. It can be shown that the chemical potential of a component 1 in a state A is

$$\mu_{A1} = h_{A1} - T_A s_A$$

and for any general component i,

$$\mu_{Ai} = h_{Ai} - T_A s_{Ai} \tag{13–25}$$

Chemical equilibrium can be visualized as a balance of the mass flow of a component, say, i, from a system A to another system B with the mass flow of the same component from system B to A. Chemical equilibrium between A and B is then achieved when all the components of the two systems balance the mass flow rates. It is this balance that is reflected in the criterion for chemical equilibrium:

$$\mu_{Ai} = \mu_{Bi} \tag{13–26}$$

$$\mu_{A1} = h_{A1} - T_A s_{A1}$$

$$\mu_{A2} = h_{A2} - T_A s_{A2}$$

$$\mu_{B1} = h_{B1} - T_B s_{B1}$$

$$\mu_{B2} = h_{B2} - T_B s_{B2}$$

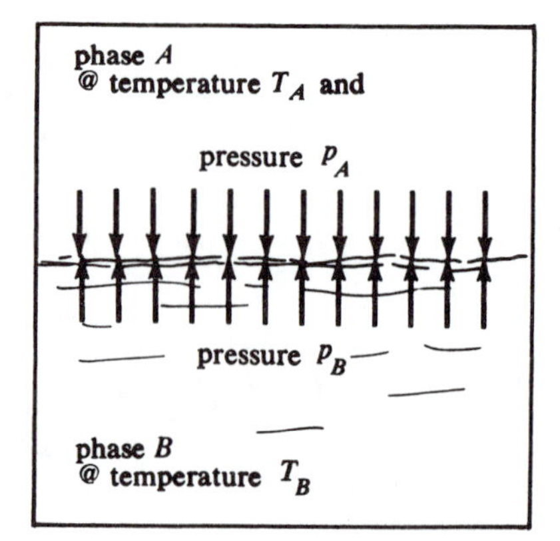

FIGURE 13–17 Equilibrium between two systems

EXAMPLE 13–16 Determine the chemical potential of saturated liquid water at 160°C.

Solution The system (water) is a one-component mixture, and its chemical potential can be computed from equation (13–25). From table B–11, we have

$$h_{A1} = 676 \text{ kJ/kg}$$
$$s_{A1} = 1.942 \text{ kJ/kg}\cdot\text{K}$$

so

$$\mu_{A1} = 676 \text{ kJ/kg} - (433 \text{ K})(1.942 \text{ kJ/kg}\cdot\text{K})$$

Answer
$$= -165 \text{ kJ/kg}$$

EXAMPLE 13–17 Air at 50°F and 50% relative humidity is in contact with lake water at 50°F. Determine the situation when the air (air–water vapor mixture) and the lake water are in chemical equilibrium, based on the chemical potential of the water.

Solution For the situation given, the water in the air is a superheated vapor at 50°F and with a partial pressure given by

$$p_V = (\beta)p_g = (0.50)(0.178 \text{ psia}) = 0.089 \text{ psia}$$

where p_g was read from table B–11 at 50°F. The superheat tables for water in this text do not show values for the vapor at 50°F and 0.089 psia; however, the values are

$$h = 1083 \text{ Btu/lbm}$$
$$s = 2.980 \text{ Btu/lbm}\cdot{}^\circ\text{R}$$

so that

$$\mu = (1083 \text{ Btu/lbm}) - (460 + 50^\circ\text{R})(2.980 \text{ Btu/lbm}\cdot{}^\circ\text{R})$$
$$= -436.8 \text{ Btu/lbm}$$

The lake water that interfaces with the air can be assumed to be a saturated liquid, and the chemical potential for this is

$$\mu = h_f - (510^\circ\text{R})s_f$$
$$= (18.05 \text{ Btu/lbm}) - (510^\circ\text{R})(0.361 \text{ Btu/lbm}\cdot{}^\circ\text{R})$$
$$= -0.361 \text{ Btu/lbm}$$

where h_f and s_f were read from table B–11 at 50°F. It can be seen that equilibrium is not present for the condition given by the air at 50% relative humidity. Equilibrium will be reached when the water in the air becomes saturated vapor (and so that the air is at 100% relative humidity). This will be the situation when the criteria of equation (13–26) will first be satisfied. To see this, we calculate the chemical potential for saturated vapor steam at 50°F:

$$\mu = h_g - Ts_g$$
$$= (1083.4 \text{ Btu/lbm}) - (510^\circ\text{R})(2.1262 \text{ Btu/lbm}\cdot{}^\circ\text{R})$$
$$= -0.962 \text{ Btu/lbm}$$

With better precision in h_g, s_g, and T, this value would be equal to the chemical potential for saturated liquid, −0.361 Btu/lbm.

The concepts of chemical equilibrium and chemical potential have applications in understanding a variety of processes. For instance, how substances can be combined to obtain uniform mixtures or how substances are soluble are better understood with the concept of chemical potential. The development of absorption refrigeration, for instance, introduced in chapter 12, requires the understanding of how substances can be mixed together in certain processes and separated in other processes, depending on temperature and chemical potentials. There are numerous other situations in which the concepts of chemical equilibrium and chemical potential are helpful.

13–6 DIFFUSION

In section 13–4, we considered evaporation, drying, and mixing of air and water. In section 13–5, we saw that if the chemical potential of a substance is not constant, the substance will tend to move to a region of low chemical potential until chemical equilibrium is reached. Chemical equilibrium is defined as a condition when the chemical potential is the same throughout. These studies do not, however, tell us how long it takes to reach equilibrium or how fast a substance moves to other regions, such as in mixing. For instance, it is important to know the rate of evaporation of water into air under prescribed conditions if proper sizing of dryers, air conditioners, heaters, or coolers is to be made. A method of obtaining some idea of the rate of movement of substances from one place to another is with the concept of diffusion. We consider diffusion in the context of air–water vapor mixtures, and extensions to other systems can then be made.

Suppose that two systems are brought into contact with each other and are in thermal and mechanical equilibrium. Then, if we consider the systems A and B as shown in figure 13–18a, it follows that

$$T_A = T_B \qquad \text{and} \qquad p_A \,(\text{at } x = 0) = p_B \,(\text{at } x = 0)$$

Let us assume that system A is dry air and system B is pure water so that the mixture A can be described as 100% air and B as 100% water, as depicted in the graph of figure 13–18a.

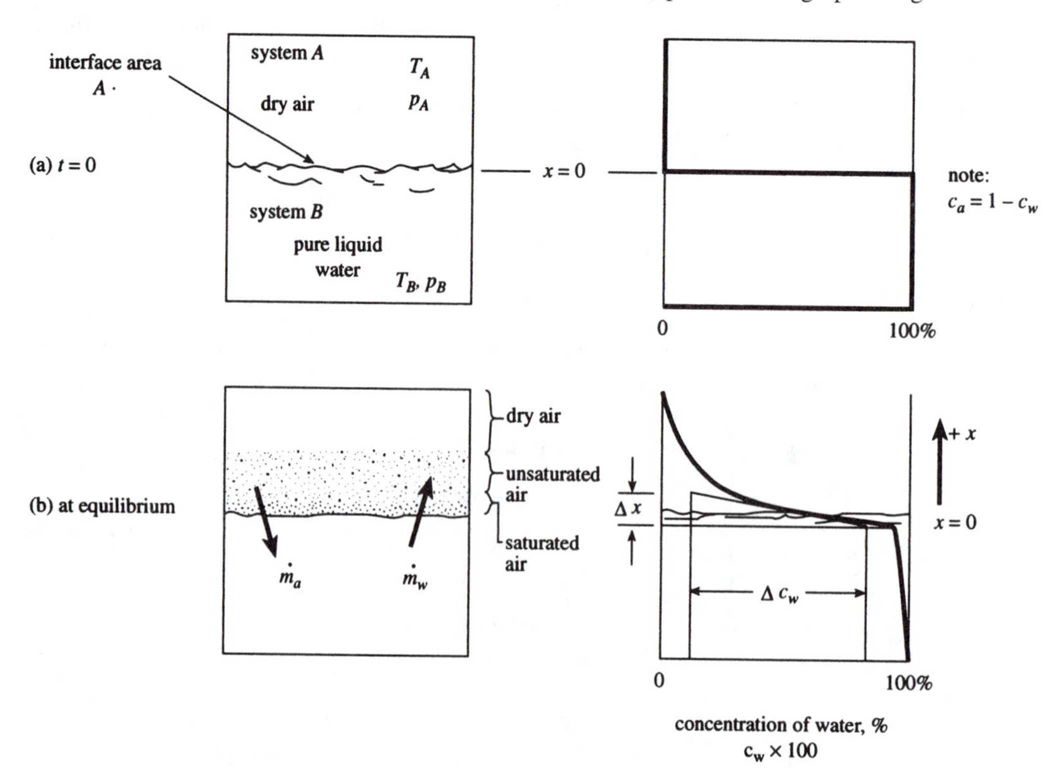

FIGURE 13–18 (a) Closed system of pure water and dry air, initial condition (time $t = 0$); (b) closed system of pure water and dry air after a finite period of time has elapsed

As we have seen in section 13–5, however, for complete equilibrium to be achieved, the chemical potential of the air in A must equal the chemical potential of the air in the water ($\mu_{A(\text{air})} = \mu_{B(\text{air})}$). Similarly, $\mu_{A(\text{water})} = \mu_{B(\text{water})}$ must hold, and these two statements imply that water must be present in A and air present in B. As a consequence of this natural action, water will migrate to A and air will migrate to B. The description of the mass flow rates of this migration are given by *Fick's law of diffusion*, namely,

$$\text{mass flow rate of water to } A = \dot{m}_w = -\rho_w A_i \mathcal{D}_{wA} \frac{\delta c_w}{\delta x} \tag{13–27}$$

and

$$\text{mass flow rate of air to } B = \dot{m}_a = -\rho_a A_i \mathcal{D}_{aB} \frac{\delta c_a}{\delta x} \tag{13–28}$$

where A_i is the cross-sectional area of the interface between the two systems; $\mathcal{D}_{wA}$ is the diffusivity or coefficient of diffusion of water in system A; $\mathcal{D}_{aB}$ is the diffusivity of air in system B; and c is the **concentration ratio**, defined as the ratio of the substance density in a mixture to its density in the pure, unmixed state. Thus, for the example of air interfacing with water, the concentration ratio, or just concentration, of water in the air would be

$$c_w = \frac{\rho_{wa}}{\rho_{wv}}$$

where ρ_{wv} is the density of saturated vapor water at the air temperature. For air mixed in liquid water,

$$c_a = \frac{\rho_a \dot{w}}{\rho_a}$$

We can approximate equation (13–27) by taking finite increments of x, so that

$$\dot{m}_w \approx -\rho_w A_i \mathcal{D}_{wA} \frac{\Delta c_w}{\Delta x} \tag{13–29}$$

and

$$\dot{m}_a \approx -\rho_a A_i \mathcal{D}_{aB} \frac{\Delta c_a}{\Delta x} \tag{13–30}$$

where Δc is the change in concentration in a distance Δx. Initially, we will see an infinite mass flow rate of water to the system A and air to the system B, since

$$\Delta C_a = c_{aB} = 0 - 1 = -1$$
$$\Delta C_{Bw} = c_{wA} = 0 - 1 = -1$$

But $\Delta x = 0$ for both, so from equations (13–29) and (13–30), we obtain

$$\dot{m}_w \approx -\rho_w A_i \mathcal{D}_{wA} \left(\frac{1}{0}\right) = \infty$$

and

$$\dot{m}_a \approx \rho_a A_i \mathcal{D}_{aB} \left(\frac{1}{0}\right) = \infty$$

This mass flow rate will occur only for an infinitesimal time, as mixing of air and water will occur in both systems A and B. After some period of time, system A will be a nonuniform mixture of dry air and water. Near the interface, the mixture will be saturated air, but further up, the concentration of water will progressively decrease so that the mixture could be described as almost dry air. Similarly, air will diffuse into the water, system B, but the concentration will decrease with depth in the system. This state is depicted by figure 13–18b, where the concentration is plotted as a function of the distance from the interface. Notice that $\Delta c/\Delta x$ represents the slope of the curve of concentration at a particular point. Steady-state conditions will be reached when the curve of concentration remains unchanged with time, but until this condition is achieved, the curve can be expected to change so that the magnitude of $\Delta c/\Delta x$ decreases. That is, diffusion will continue to seek an equality of concentration between systems.

Diffusion can occur between a liquid and a gas (as we have just indicated), between a gas and a gas, a liquid and a liquid, a liquid and a solid, a gas and a solid, or a solid and a solid. Of course, the rates of mass flow vary extremely between these combinations, but the mass flow can be predicted through the coefficient of diffusion. Table 13–6 shows the order of magnitudes of $\mathscr{D}$ for the various combinations of systems. Remember that the numbers are merely approximate values to describe the variation in diffusion. It can easily be seen from the table that diffusion of a solid to a solid is much less than that of a gas to a gas or any other combination.

Table 13–7 gives some nominal values of diffusivity $\mathscr{D}$ for common materials.

TABLE 13–6 Orders of magnitude for diffusivity at 27°C and 101 kPa

Diffusing Phase (Transported Material) i	Diffused Phase, A (System Invaded by Material)	$\mathscr{D}_{ia}$ cm^2/s
Gas	Gas	0.1
Gas	Liquid	1×10^{-5}
Gas	Solid	1×10^{-10}
Liquid	Gas	1×10^{-5}
Liquid	Liquid	1×10^{-5}
Liquid	Solid	1×10^{-15}
Solid	Gas	1×10^{-10}
Solid	Liquid	1×10^{-15}
Solid	Solid	1×10^{-20}

TABLE 13–7 Diffusivity of some common materials at 20°C and 1 atm pressure

A	B	$\mathscr{D}_{ab}$	
Diffusing Material	Diffused Material	cm^2/s	ft^2/h
Ammonia	Air	0.236	0.914
Carbon dioxide	Air	0.164	0.636
Water vapor	Air	0.256	0.992
Ethyl ether	Air	0.093	0.360
Helium	SiO_2	0.3×10^{-10}	$1.16 \times 10-9$
Helium	Pyrex	4.5×10^{-11}	1.74×10^{-10}
Bismuth	Pb	1.1×10^{-16}	4.26×10^{-16}
Mercury	Pb	2.5×10^{-15}	9.67×10^{-15}
Ethanol	Water	1.13×10^{-5}	4.37×10^{-5}
Water (liquid)	*n*-Butanol	1.25×10^{-5}	4.80×10^{-5}

Source: Data abstracted from M. Jakob and G. Hawkins, *Elements of heat transfer*, 3d ed. (New York: John Wiley & Sons, Inc., 1957), p. 276; and from R. B. Bird, W. Stewart, and E. Lightfoot, *Transport phenomena* (New York: John Wiley & Sons, Inc., 1960), pp. 504–5; with permission of John Wiley & Sons, Inc.

EXAMPLE 13–18 Consider air at 50°F and 50% relative humidity interfacing with water at 50°F and 14.7 psia. Estimate the evaporation rate of water into the air if it is found that the air has 80% relative humidity at 2 in above the water surface and has 100% at the water surface.

Solution We will estimate the evaporation rate by using Fick's law for water vapor diffusion into air and assume that rate is equal to the evaporation of water into the air. From table 13–7, diffusivity is 0.992 ft²/h for water vapor into air. For the density of water vapor, we use the reciprocal of the specific volume of saturated vapor at 50°F:

$$\rho = \frac{1}{1704.8\ \text{ft}^3/\text{lbm}} = 0.000587\ \text{lbm/ft}^3$$

The concentration slope, $\Delta c/\Delta x$, can be approximated by

$$\frac{\Delta c}{\Delta x} = \frac{0.8 - 1.0}{2\ \text{in}} = -0.1/\text{in} = -1.2/\text{ft}$$

Using equation (13–27) for a unit area, we have

$$\begin{aligned} \dot{m}_w &- (0.000587\ \text{lbm/ft}^3)(0.992\ \text{ft}^2/\text{h})(-1.2/\text{ft}) \\ &= 0.000699\ \text{lbm/ft}^2\cdot\text{h} \end{aligned}$$

Therefore, the evaporation rate of water is 0.000699 lbm/h as a first approximation. We could assume steady-state conditions and then predict the time required for complete evaporation of the water into the air. Of course, as the water level was lowered, the conditions would probably change, and the evaporation rate would then also probably change.

13–7 MIXTURE PHASE CHANGE BEHAVIOR

So far in this chapter we have considered mixtures that do not change phase. Here we will discuss briefly the behavior of some mixtures as they go through the liquid–vapor phase change. Similar behavior may be expected for the solid–liquid and the solid–vapor phase changes, but we will not consider those in this text. We will also limit our discussion to mixtures where the components do not chemically react.

Consider a binary mixture of two substances, say A and B, with boiling temperatures T_A for substance A and T_B for substance B. If the mixture temperature is determined for various mass fractions of A—call it ψ_A—then the mixture will behave similarly to that shown in figure 13–19, figure 13–20, or figure 13–21 If the mixture behaves like that shown in either figure 13–19 or figure 13–20, it is called an *azeotropic mixture* and if it behaves like that shown figure 13–21, then it is called a *zeotropic mixture*. Notice in all three of these figures that there is identified a *boiling* (or *bubble) curve* and a *dew point curve*. The *boiling curve* represents the temperature at which a particular mixture will begin to boil or form bubbles if the mixture is all liquid. The *dew point curve* represents the temperature at which all of the mixture has become a vapor if the mixture is being heated. For the situation where a vapor mixture is being cooled, the *dew point curve* represents the temperature at which the mixture begins to condense or become a liquid. Also, for the case of a cooling mixture, the boiling curve then represents the temperature at which all of the mixture has condensed from the vapor state to the liquid. For a particular *azeotropic mixture* having a mass fraction of A of ψ_{A1} as indicated in figure 13–19 and figure 13–20, the boiling and dew point curves coincide at the *azeotropic point*, meaning that the *azeotropic mixture* will go through the liquid–vapor phase change at the temperature T_{gaz}. It is clear from figure 13–19 and figure 13–20 that the phase change temperature is lower (figure 13–19) or higher (figure 13–20) than the phase change temperature of either of the components. A precise mixture of water and ethanol is an *azeotropic mixture*. Also, the

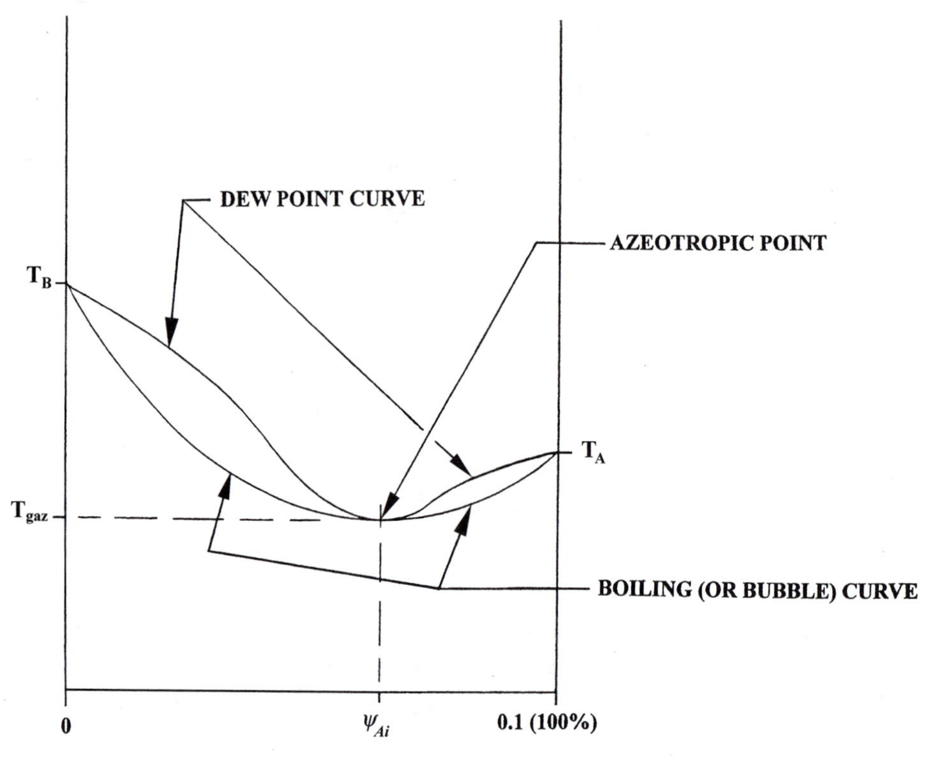

FIGURE 13–19
Temperature-composition diagram for an azeotropic mixture having a minimum phase change temperature

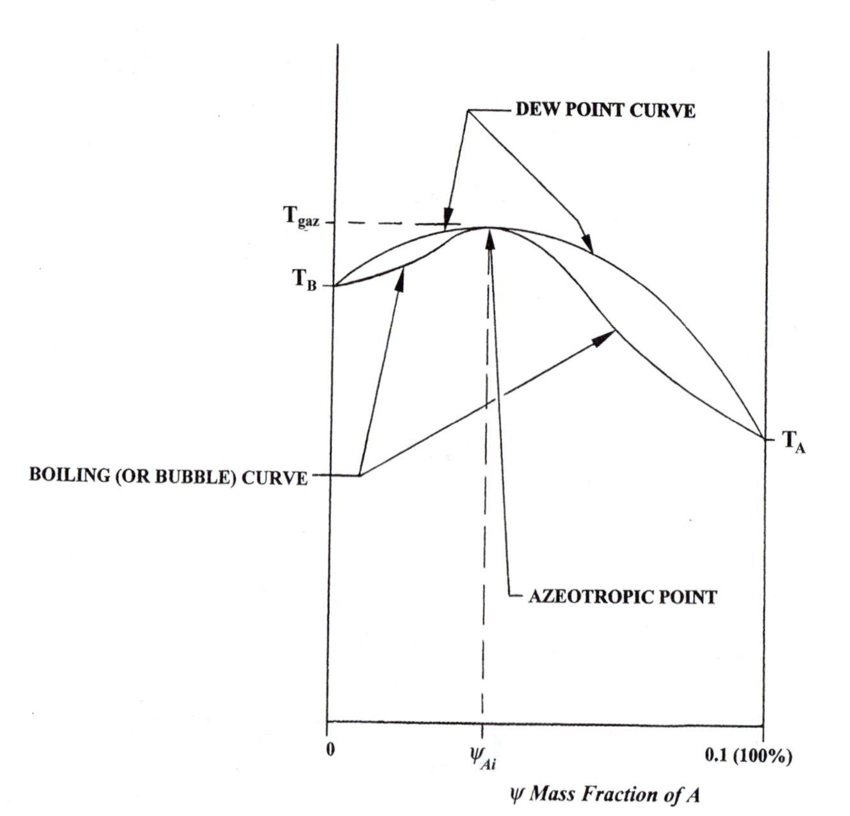

FIGURE 13–20
Temperature-composition diagram for an azeotropic mixture having a maximum phase change temperature

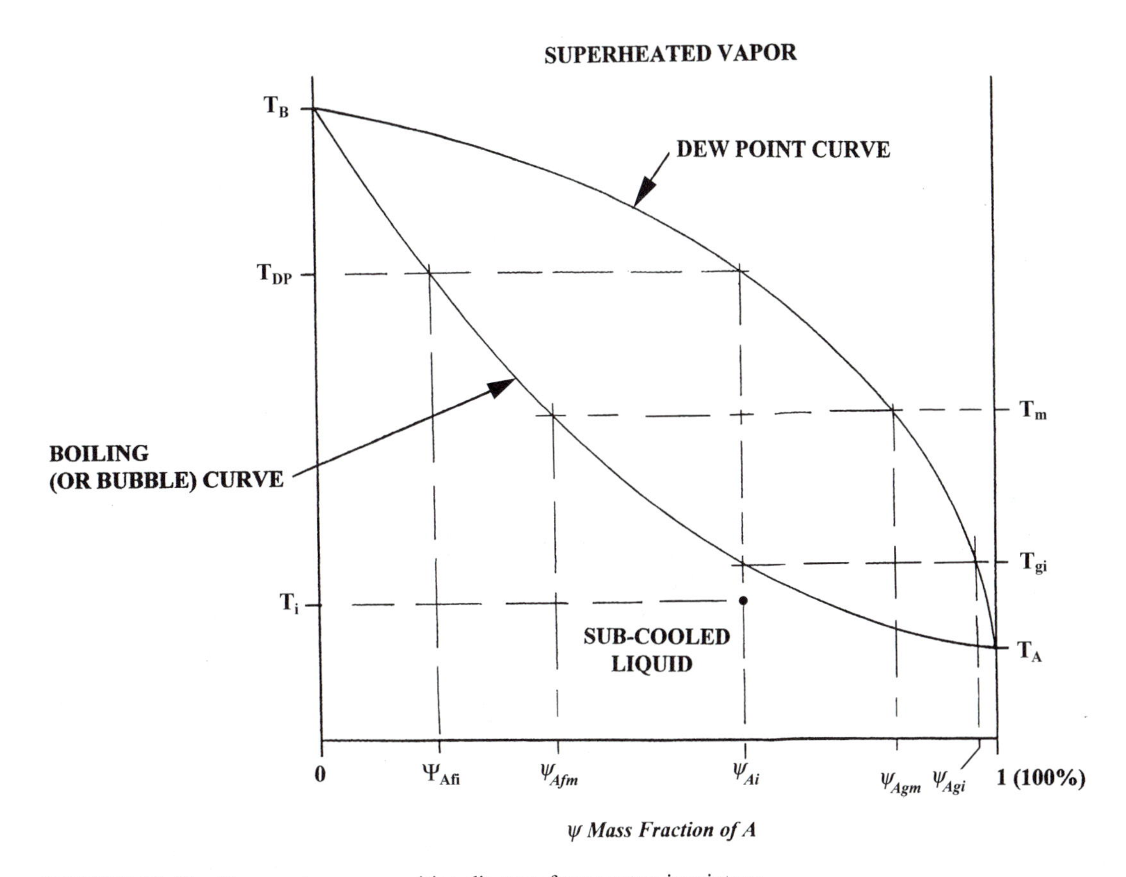

FIGURE 13–21 Temperature-composition diagram for a zeotropic mixture

refrigerant R-502, which was mentioned in chapter 12, is an *azeotropic mixture*. There are other mixtures or blends of substances which behave as *azeotropic mixtures*.

The behavior of *zeotropic mixtures* follow that shown in figure 13–21. Some examples of *zeotropic mixtures* are ammonia/water and R-407c. To demonstrate this behavior more clearly, consider a binary mixture of A and B having a mass fraction ψ_{Ai} and which is a liquid at temperature T_i. Place an amount of the liquid mixture in a frictionless piston-cylinder container, as shown in figure 13–22a, so that the pressure can remain constant at all times. Now the mixture is heated to T_{gi} as indicated in figure 13–21 figure 13–22b at which time bubbles begin to form. These bubbles will be mostly component A, the mass fraction being determined by constructing a horizontal line over to the right until it intersects with the *dew point curve*. The mass fraction of these bubbles will then be ψ_{Agi}. Now, if the mixture continues to be heated, the temperature rises, and for example, at T_m the mixture will have some saturated liquid and some saturated vapor. The saturated vapor will have a mass fraction of A of ψ_{Agm}, and for the saturated liquid, the mass fraction of A will be ψ_{Afm}, as indicated in figure 13–21 and figure 13–22c. Notice that the vapor will have a higher concentration of A and the liquid a higher concentration of B.

Recall from chapter 5 that the term *quality* was used to describe the ratio of the mass of saturated vapor to the total mixture mass in a substance going through a phase change, such as boiling or condensing. Here, with the binary mixture, the quality at the temperature T_m can be determined by first writing the mass of the mixture as

$$m_f + m_g = m_T \tag{13–31}$$

FIGURE 13–22 Constant pressure container

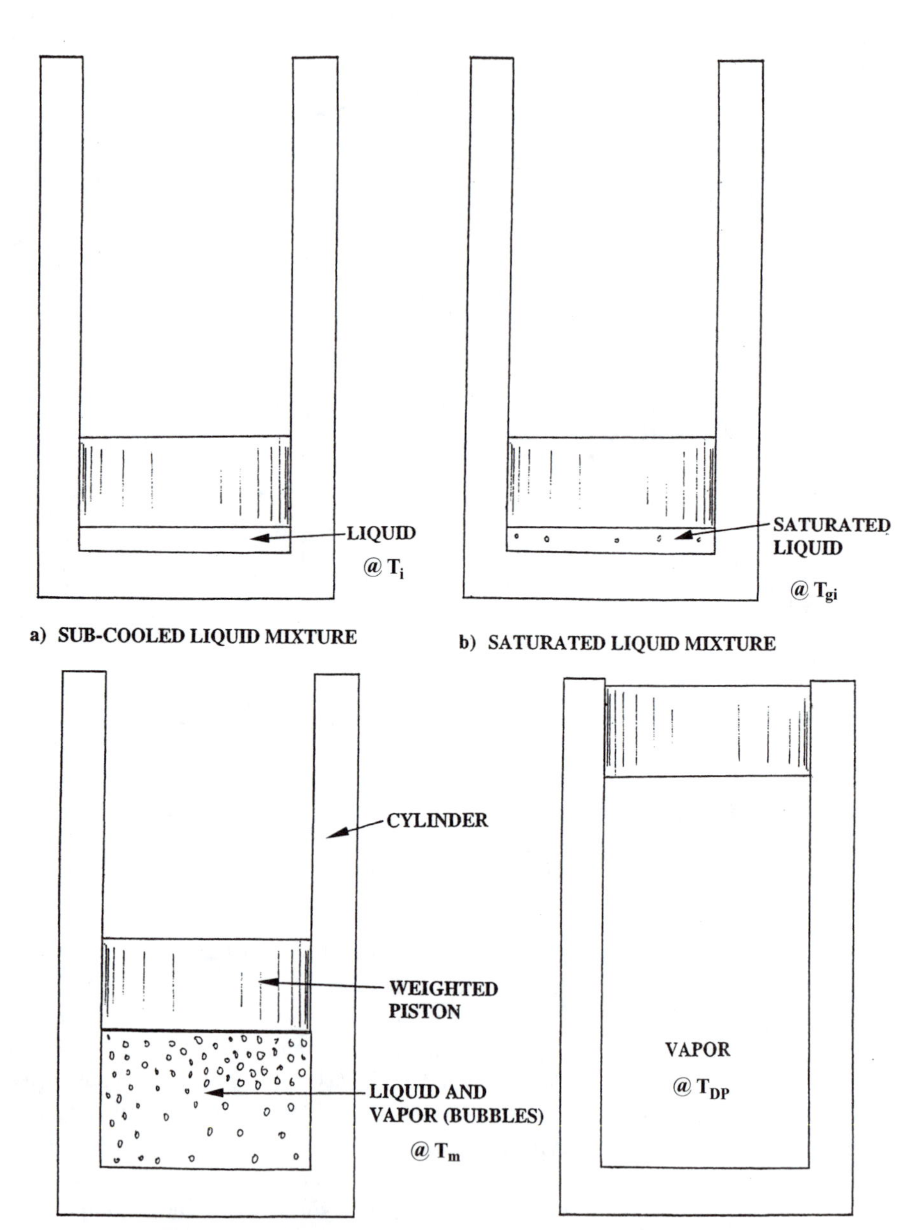

where m_f is the mass of saturated liquid and m_g is the mass of saturated vapor. The mass of component A will be conserved, so that

$$m_T \psi_{Ai} = m_f \psi_{Afm} + m_g \psi_{Agm} \tag{13–32}$$

By eliminating m_f in equation 13–32 by replacing it with the result from equation 13–31 in the form

$$m_f = m_T - m_g$$

and manipulating the equation, we get the mixture quality at temperature T_m:

$$\chi_m = \frac{m_g}{m_T} = \frac{\psi_{Ai} - \psi_{Afm}}{\psi_{Agm} - \psi_{Afm}} \tag{13–33}$$

If the mixture continues to be heated at constant pressure, it eventually will reach the temperature T_{DP} at which point all of the mixture will be a saturated vapor except for perhaps a drop or so of liquid. These liquid drops will have a high concentration of B with a mass fraction of A given by ψ_{ADP}, as indicated in figure 13–21 and figure 13–22d. Any further heating of the mixture results in a superheated mixture of the beginning liquid concentrations of A and B.

During the boiling of the mixture, the pressure remained constant, however, the temperature changed from T_{gi} on the bubble curve to T_{DP} on the dew point curve. This temperature change is called a *temperature glide*.

Notice that an *azeotropic mixture* will behave as a *zeotropic mixture* when the mass fractions are not precisely those corresponding to the *azeotropic point*.

EXAMPLE 13–19 R-407c is a zeotropic mixture. Determine the temperature glide as it condenses from a saturated vapor to a saturated liquid at 300 kPa. Then determine the temperature glide if it boils at 300 kPa.

Solution: Referring to the saturated R-407c table B–23, we find that the saturated vapor is at −11.69°C (the dew point) and is at −18.22°C at the bubble temperature. Thus, the temperature glide is

Answer

$$\text{temperature glide} = -18.22 - (-11.69) = -6.53°\text{C}$$

The temperature glide for the case of boiling would be

Answer

$$+6.53°\text{C}$$

EXAMPLE 13–20 The temperature–composition diagram for water/ammonia at standard atmospheric pressure, 14.7 psia, is shown in Figure 13–23. Determine the quality of the water/ammonia mixture at 100°F when the sub-cooled liquid state has 40% ammonia.

Solution: From figure 13–23, we see that the mixture is in the saturation region, between a saturated liquid and a saturated vapor. The initial mass fraction of ammonia is

$$\psi_{Ai} = 0.4$$

The mass fraction of ammonia in the saturated liquid is

$$\psi_{Afm} = 0.26$$

and, for the mass fraction of ammonia in the saturated vapor,

$$\psi_{Agm} = 0.956$$

The quality of the mixture at 100°F is determined from equation 13–33 as

$$\chi_m = \frac{\psi_{Ai} - \psi_{Afm}}{\psi_{Agm} - \psi_{Afm}} = \frac{0.4 - 0.26}{0.956 - 0.26}$$

or

Answer

$$\chi_m = 0.201(20.1\%)$$

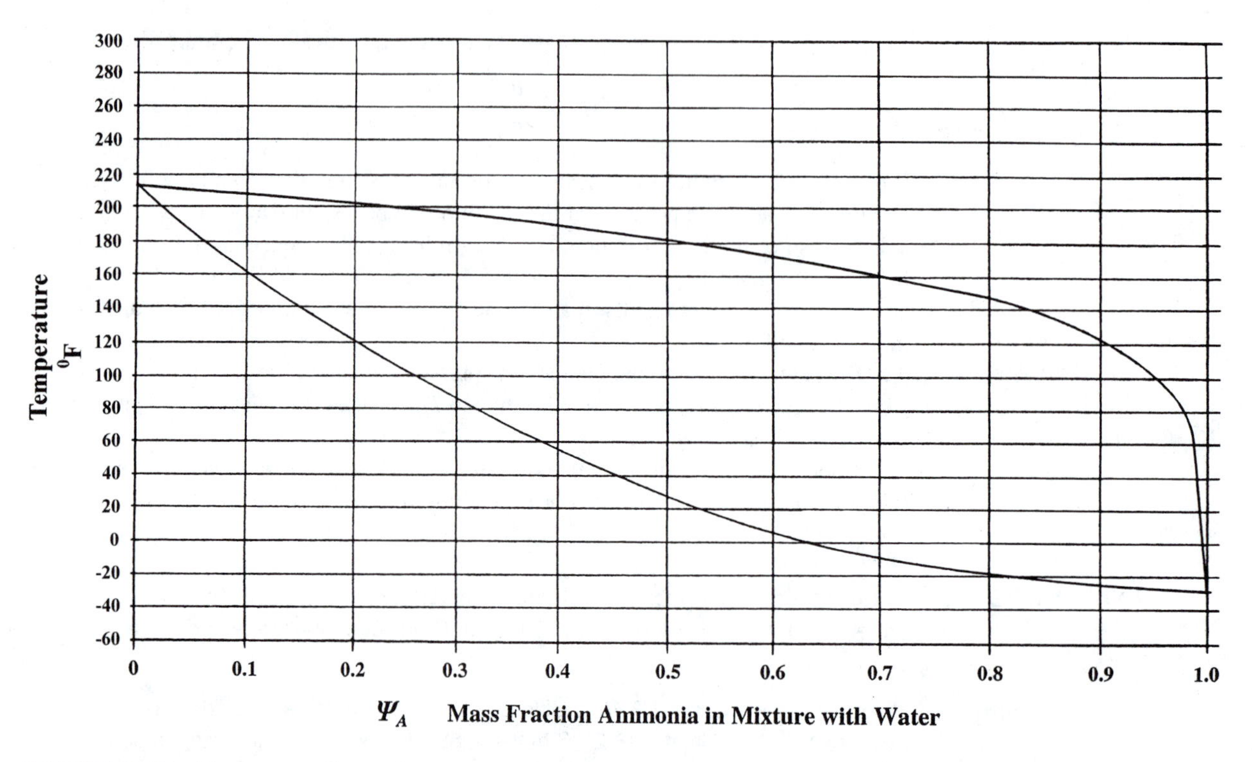

FIGURE 13–23 Temperature-Composition Diagram for Water/Ammonia at Standard Atmosphere of 14.7 psia

13–8 SUMMARY

Mixtures are two or more substances occupying the same region. The mass fraction of a substance i in a mixture is

$$\psi_i = \frac{m_i}{m_T} \tag{13–1}$$

and the total mass is

$$m_T = \sum m_i \tag{13–2}$$

The listing of the mass fractions of a mixture is called the mass analysis. The mole fraction of a substance i is given by

$$\Omega_i = \frac{N_i}{N_T} \tag{13–3}$$

and the total number of moles in a mixture is

$$N_T = \sum N_i \tag{13–4}$$

A molar analysis is the listing of all the molar fractions of all the mixture components. Molar and mass analysis can be converted to the other by the molecular weight or molecular mass.

Properties of mixtures based on a unit mass of the mixture are

$$h = \sum \psi_i h_i \tag{13–5}$$

$$\left.\begin{aligned} u &= \sum \psi_i u_i \\ s &= \sum \psi_i s_i \\ c_p &= \sum \psi_i c_{p_i} \end{aligned}\right\} \tag{13–6}$$

and, per unit mole of the mixture,

$$\begin{aligned} \bar{h} &= \sum \Omega_i \bar{h}_i \\ \bar{u} &= \sum \Omega_i \bar{u}_i \\ \bar{s} &= \sum \Omega_i \bar{s}_i \\ \bar{c}_p &= \sum \Omega_i \bar{c}_{p_i} \end{aligned} \tag{13–7}$$

For perfect gas mixtures, we may define the partial pressure as

$$p_i = \Omega_i p_T \tag{13–8}$$

where the total mixture pressure is

$$p_T = \sum p_i \tag{13–9}$$

The entropy of mixing of two or more pure substances that are nonreacting is computed from the equation

$$\Delta S = \sum m_i \Delta s_i \tag{13–10}$$

Air–water vapor mixtures have important applications to engineering and technology. They are often approximated as perfect gas mixtures. The specific humidity ω is the ratio of the mass of water in the air to the mass of dry air:

$$\omega = \frac{m_w}{m_{da}} = 0.622 \frac{p_v}{p_a} \tag{13–13}$$

The vapor pressure of the water in the air, p_v, is the partial pressure of the water. The relative humidity is given by

$$\beta = \frac{p_v}{p_g} \tag{13–14}$$

where p_g is the saturation pressure of water corresponding to the air temperature. If air is cooled until the relative humidity is 100%, that temperature is called the **dew point temperature**.

Psychrometers are used to measure humidity in air. The dry-bulb temperature T_{db} is the air temperature, and the wet-bulb temperature is that for the same air after being fully saturated with water to 100% relative humidity. The specific humidity may be computed from the equation

$$\omega = \frac{c_p(T_{wb} - T_{db}) + \omega_3 h_{fg_3}}{h_{g_1} - h_{f_3}} \tag{13–17}$$

The psychrometric chart is useful for determining the various properties of air–water vapor mixtures.

Four processes involving air–water mixtures are particularly important: drying processes, heating and humidification processes, evaporative cooling processes, and dehumidifying processes.

Chemical equilibrium is a concept used to explain the motion of a pure substance from one region to another. The chemical potential of a substance i is given by

$$\mu_i = h_i - T_{s_i} \tag{13–25}$$

and is used to define chemical equilibrium.

Fick's law of diffusion,

$$\dot{m} = -\rho A \mathfrak{D} \frac{\delta c}{\delta x} \tag{13–27}$$

can be used to predict the rate at which a substance moves from one region to another. $\mathfrak{D}$, called the **diffusivity**, is determined experimentally.

Mixtures that go through phase changes can behave either as *zeotropic mixtures* or as *azeotropic mixtures. Zeotropic mixtures* exhibit a *temperature glide* or temperature change at constant pressure when progressing through a phase change. *Azeotropic mixtures* exhibit temperature glides through a phase change as well, except at one unique mixture composition where the mixture does not change temperature during the constant pressure phase change. The temperature–composition at which this occurs is called the azeotropic point.

DISCUSSION QUESTIONS

Section 13–1

13–1 What is a mixture?

13–2 Is there a limit to how many components or substances are in a mixture?

13–3 What is meant by the term *mass fraction*?

13–4 What is a *mole fraction*?

13–5 How do mixture properties differ from a substance's properties?

Section 13–2

13–6 What is meant by the term *partial pressure*?

13–7 Why does mixing create entropy?

Section 13–3

13–8 Is the common air–water mixture a perfect gas mixture? Why or why not?

13–9 What is humidity?

13–10 What is meant by *dew point*?

13–11 What is meant by *wet-bulb temperature*?

13–12 What is meant by *dry-bulb temperature*?

13–13 How is relative humidity different than humidity ratio?

Section 13–4

13–14 What happens during evaporation?

13–15 What is the difference between drying and humidifying?

13–16 Why does evaporative cooling work?

13–17 What is meant by *adiabatic mixing*?

Section 13–5

13–18 How is chemical potential used?

Section 13–6

13–19 What is diffusion?

13–20 What is Fick's law of diffusion?

Section 13–7

13–21 Why do you think there are temperature glides?

13–22 Why do you think some mixtures behave as zeotropic mixtures?

13–23 Do azeotropic mixtures experience temperature glides during phase changes?

PRACTICE PROBLEMS

Sections 13–1

Problems that use SI units are indicated by an (M) under the problem number; those that use English units are indicated by an (E). Mixed unit problems are indicated by a (C).

13–1 (M) A mixture is made of 7 kg of helium, 3 kg of neon, and 6 kg of argon. Determine the mass fractions and the mole fractions of each of the components of the mixture.

13–2 (E) There are 2 lbm · mol of xenon, 4 lbm · mol of argon, and 6 lbm · mol of helium mixed together. Determine the mass and molar analyses.

13–3 (E) Using the data of table 13–3, determine the molar analysis of Perkins County, South Dakota, lignite. Assume that ash has a molecular mass equal to that of silicon.

13–4 (M) Determine the partial pressures of the constituents in a mixture of 0.05 kg of oxygen, 0.10 kg of CO_2, 0.10 kg of argon, and 0.062 kg of helium at 200-kPa pressure.

13–5 (M) Calculate the entropy increase as 8 g of water vapor mix with 70 g of air at 1 atm pressure and 20°C.

13–6 (E) Determine the partial pressure of water vapor in a mixture of 8% (by mass) H_2O, 80% air, and 12% CO_2 if the mixture is at 14.7 psia and 80°F.

13–7 (E) Determine the entropy increase due to a mixing of 2 ft^3 of helium at 10 psia and 60°F and 3 ft^3 of argon at 20 psia and 60°F in a 3-ft^3 container at 60°F.

Sections 13–2

13–8 (E) Determine the specific heat of a mixture of 100 lbm of sand and 50 lbm of water.

13–9 (M) Determine the specific heat at constant volume of a mixture of argon and neon. The mole fractions are 45% for argon and 55% for neon.

13–10 (E) At 60°F and 14.7 psia, 2 lbm · mol of carbon dioxide (CO_2) is mixed with 3 lbm · mol of argon (Ar). Determine the entropy increase if the final mixture pressure is 14.7 psia.

Section 13–3

13–11 (C) Obtain the saturation temperature of steam (water vapor) at
(a) 20 psia
(b) 200 kPa

13–12 (C) Obtain the saturation pressure of water vapor at
(a) 60°F
(b) 80°C

13–13 (C) Determine the enthalpy and specific humidity for air having a relative humidity of 60% and a dry-bulb temperature of
(a) 90°F
(b) 12°C

13–14 (M) Air at 101-kPa pressure and 20°C has a wet-bulb temperature of 15°C. Determine
(a) Vapor pressure
(b) ω
(c) β

13–15 (E) Atmospheric air is at a pressure of 29.4 mm Hg and 80°F. If the wet-bulb temperature is 68°F, determine
(a) ω
(b) β
(c) Dew point temperature

Section 13–4

13–16 (M) How much moisture per hour can a 10-m^3/s stream of air at 35°C and 20% relative humidity absorb?

13–17 (E) A stream of air has a flow rate of 300 ft^3/min, and the air is at 100°F and 30% relative humidity. How much water can it absorb in a drying process?

13–18 (M) A drying process requires that 10 kg of water be evaporated from a surface in 1 h. If air at 40°C and 30% relative humidity is available, determine the mass flow and volume flow rates required for the process.

13–19 (M) A humidifier operates with air entering at 101 kPa and 25°C. If the wet-bulb temperature is 15°C and the leaving air must be at 20°C and 50% relative humidity, determine the amount of heat transfer required and the water required per unit mass of air. Assume that water is at 20°C.

13–20 (E) A heater-humidifier has a heater rated at 5800 Btu/min. A total of 5 lbm/s of air at 35°F and 80% relative humidity needs to be heated to 75°F and 65% relative humidity. Determine the conditions of water necessary to accomplish this process.

13–21 (M) A heater-humidifier with a 75-kW heater handles 2 kg/s of air at 5°C and 60% relative humidity. The supply air is to be at 25°C and 60% relative humidity. Determine the conditions of water at 150 kPa needed to achieve this process.

13–22 (M) Air at 35°C and 10% relative humidity is to be cooled by evaporative cooling to 25°C. Determine the amount of water needed per kilogram of dry air and the relative humidity of the cooled air.

13–23 (E) A wet 4-foot by 6-foot beach towel is hung on a clothesline to dry. The air temperature is reported to be 85°F with 40% relative humidity. The air is moving at 5 miles per hour. If, at a certain time during the drying of the beach towel, the air temperature is measured as 70°F, 100% relative humidity, immediately downstream from the towel and the air is assumed to be all passing through the towel, what is the rate at which water is evaporating from the towel? That is, what is the towel's drying rate in mass of water vapor per unit time?

13–24 (E) There is 1500 3 ft^3/min of air at 92°F, 14.7 psia, and 10% relative humidity to be conditioned to 70°F and 60% relative humidity. Determine the amount of heat transfer required and the rate of water flow necessary to humidify this air.

13–25 (M) A refrigerator of 40 liter (L) capacity has its door opened 50 times per day, on the average. Each time the door is opened, half of the air in the refrigerator flows out and is replaced with room air at 30°C and 60% relative humidity. The air in the refrigerator is set at 2°C and is assumed to be 100% relative humidity. Estimate the amount of water that condenses in the refrigerator each day due to the air exchanges from opening and closing the door.

13–26 (M) Hot air at 40°C has a measured wet-bulb temperature of 15°C. It is desired that the air be cooled to 20°C. How much water per kilogram of dry air is required for an evaporative cooler, and what is the relative humidity of the cooled air?

13–27 (E) Air at 110°F and 5% relative humidity is to be cooled to 80°F. Determine the amount of water needed for an evaporative cooler and the relative humidity of the cooled air. There is 1000 ft^3/min of air flow through the system to be cooled.

13–28 (M) A dehumidifier having a maximum cooler capacity of 360 kJ/min provides a 6 kg of air per minute at 25°C and 55% relative humidity. If the incoming air is at 35°C, determine the maximum relative humidity it may have and still allow the humidifier to furnish the 6 kg/min of air at 25°C and 55% relative humidity.

13–29 (E) Air at 14.7 psia, 90°F, and 95% relative humidity is to be cooled and dehumidified to 75°F and 60% relative humidity. Determine the amount of water removed per pound-mass of dry air, the amount of cooling required (in Btu/lbm da), and the amount of heating required (in Btu/lbm da).

13–30 (M) Air at 101 kPa, 40°C, and 60% relative humidity is to be cooled to 17°C and 60% relative humidity. If the flow rate of air is 300 m^3/min, determine the amount of water removed per minute, the rate of cooling required in a cooling coil, and the rate of heat needed in a heating coil.

13–31 (E) 5000 ft^3/min (cfm) of 40°F air at 80% relative humidity is mixed adiabatically with 10,000 cfm of 80°F air at 60% relative humidity. Determine the dry-bulb temperature, the relative humidity, and the volumetric flow rate of the mixed air stream.

13–32 (M) A warm air stream at 30°C and 35% relative humidity is to be mixed with 1 m^3/s of a cold air stream so that 3 m^3/s of air at 25°C and 40% relative humidity is furnished. Assuming that the air is incompressible, estimate the amount of warm air needed to be mixed with the cold air stream and then determine the dry-bulb temperature and relative humidity of the cold air stream.

Section 13–5

13–33 (C) Determine the chemical potential of saturated water vapor at
(a) 500°F (b) 240°C

13–34 (C) Determine the chemical potential of saturated ammonia vapor at
(a) 600 kPa (b) 23.74 psia

13–35 (C) Calculate the chemical potential of saturated Refrigerant 22 liquid at
(a) 60°C (b) 140°F

13–36 (E) Determine the chemical potential of water at 70°F which is in chemical equilibrium with an air–water mixture at 70°F and 60% relative humidity.

Section 13–6

13–37 (M) Predict the concentration of ammonia in air 1 cm from an interface of vapor ammonia and air if the interface area is 15 cm^2 and the evaporation rate of ammonia is 0.002 g/min. Assume that the air and ammonia system is 20°C.

13–38 (M) Liquid mercury (Hg) is contained in a lead beaker as shown in figure 13–24. Calculate the amount of mercury lost through diffusion to the beaker after 48 h if the concentration of mercury is 2% at a distance 0.01 cm into the lead beaker wall from the inside surface. Assume that the system is at a temperature of 20°C.

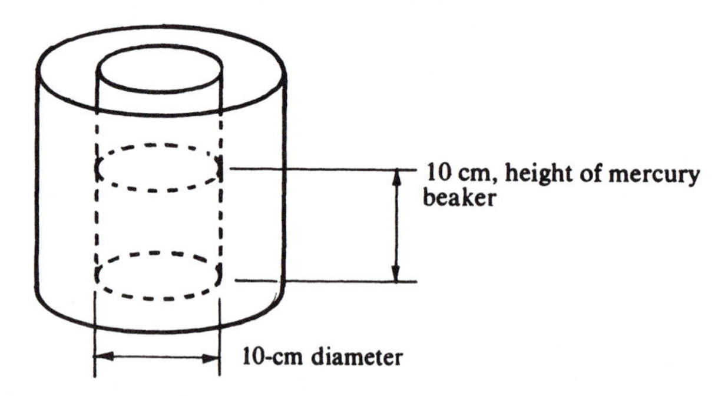

FIGURE 13–24

13–39 (E) A 3-ft-radius sphere contains helium gas at 20 psia and 95°F. Determine the amount of helium lost through diffusion in 24 h if the sphere is made of Pyrex and if it is assumed that at $^1/_8$ in from the inside wall, helium is totally absent.

13–40 (E) Water is evaporating at 3 lbm/h in a 30-in^3 container. Estimate the specific humidity 1 in from the water–air surface in the air if the dry-bulb temperature is 70°F. Assume that the water–air surface interface area is 9 in^2.

Section 13–7

13–41 What is the temperature glide of a mixture of 80% ammonia in water when it boils at atmospheric pressure?

13–42 What is the enthalpy of R-502 at 10°C and 60% quality?

13–43 What is the quality of a mixture of 20% ammonia in water at 150°F?

REACTING MIXTURES AND COMBUSTION

14

In this chapter, the combustion process is discussed as an important example of a mixture of chemically reacting substances. Some important fuels involved in combustion processes are identified, and then combustion is analyzed with chemical reaction equations.

Topics relating to the air–fuel ratio in combustion, heats of formation, heats of combustion and heating values, and combustion temperatures are included in this chapter.

Entropy generation in combustion processes is presented to show the degree of irreversibility of these processes.

New Term

Q_c Heat of combustion

14–1 THE COMBUSTION PROCESS

When pure substances are combined, they make up what we call a mixture. In chapter 13, we considered nonreacting mixtures in which the pure substances did not change their chemical composition after being mixed together. Here we consider mixtures of two or more substances that do react and create new or different substances from the original components. The science of chemistry involves the study of the reactions involved when two or more pure substances (or chemicals) are mixed together. We limit ourselves in this book to one class of those reactions: combustion of hydrocarbon fuels. For engineers and technologists, the combustion processes are important and have a wide range of applications. Combustion occurs in steam generation, in the combustion chamber of gas engines, in the combustor of turbojets or gas turbine engines, in the gas or fuel oil furnace, and in the burning or incineration of waste. There are other applications of the combustion process.

Combustion occurs, for instance, when hydrogen and oxygen combine to form water. We may write the chemical reaction equation for this as

$$H_2 + 0.5O_2 \longrightarrow H_2O \tag{14–1}$$

so that combustion is an oxidation process occurring at a rapid rate. Another example of a combustion process is the combination of carbon with oxygen to produce carbon dioxide:

$$C + O_2 \longrightarrow CO_2 \tag{14–2}$$

As a third example, consider the combustion of octane fuel (C_8H_{18}) with oxygen to produce carbon dioxide and water:

$$C_8H_{18} + 12.5O_2 \longrightarrow 8CO_2 + 9H_2O \tag{14–3}$$

These three examples of combustion demonstrate some features that can be identified with combustion processes:

1. Combustion is associated with a pure substance reacting with oxygen gas, O_2. Combustion is a special form of the oxidation processes of chemistry.
2. The final product resulting from the combustion is carbon dioxide if carbon is the fuel or part of the original mixture.
3. The final product resulting from the combustion is water if hydrogen is part of the original mixture.

Notice in equation (14–1) that the reaction involves 1 mol of hydrogen gas (two atoms of hydrogen) and $\frac{1}{2}$ mol of oxygen gas to create 1 mol of water. The equation is chemically balanced because the number of atoms are the same on both sides of the equation, even though the number of moles has decreased $(1 + \frac{1}{2} = 1\frac{1}{2})$. For the combustion of carbon in equation (14–2), we can see again that the number of carbon atoms is conserved and so are the oxygen atoms. A more careful study of equation (14–3) can show that here again the hydrogen, oxygen, and carbon atoms are conserved. We call equations (14–1), (14–2), and (14–3) **stoichiometric equations**; such equations can be written for all reacting mixtures, provided that the chemical formulas are known. The left side of the equation is sometimes referred to as a **stoichiometric mixture**. The components on the right side of the equation are called the products or products of combustion. It is important that the stoichiometric equations be balanced if the equations are to be useful for our studies of combustion processes. We will now demonstrate an example of balancing a reaction equation to obtain a stoichiometric equation.

EXAMPLE 14–1 Methane gas, CH_4, is burned with pure oxygen. Write the stoichiometric equation for the combustion of methane.

Solution Methane has carbon and hydrogen, so we expect that the products will be carbon dioxide and water. We write

$$CH_4 + O_2 \longrightarrow CO_2 + H_2O$$

We see that the carbon atoms are balanced, so we have 1 mol of carbon dioxide from 1 mol of methane. There are four atoms of hydrogen in a methane molecule, so we need 2 mol of water to have four atoms on the right side as well. Finally, we see that there are four atoms of oxygen on the right side after identifying 2 mol of water (two oxygens in the carbon dioxide and two in the 2 mol of water). There must therefore be 2 mol of oxygen gas to balance the oxygen on both sides of the equation. The stoichiometric equation is

Answer

$$CH_4 + 2O_2 \longrightarrow CO_2 + 2H_2O$$

Notice in example 14–1 and in the stoichiometric equation that the analysis involves moles instead of mass. We could convert the equation to mass balances by using the molecular masses. We demonstrate that now.

EXAMPLE 14–2 Determine the mass fraction of octane, C_8H_{18}, and oxygen, O_2, in a stoichiometric mixture.

Solution We refer to equation (14–3) for the stoichiometric equation for octane. We see that the mole fractions of octane and oxygen are, respectively,

$$\Omega_{\text{octane}} = \frac{1}{13.5} = 0.074$$

and

$$\Omega_{\text{oxygen}} = \frac{12.5}{13.5} = 0.926$$

We then determine the masses of octane and oxygen in the stoichiometric mixture, sum the two values to find the total mass, and determine the mass fractions by using equation (13–1):

$$\psi_i = \frac{m_i}{m_T}$$

The results are tabulated in table 14–1, and it can be seen that 22.16 g of octane is combined with 77.84 g of oxygen to obtain a stoichiometric mixture. Of course, any mass ratio of 22.16 parts octane to 77.84 parts oxygen will produce the same sort of mixture.

TABLE 14–1 Stoichiometric mixture analysis

Component	Octane Ω_i mol/mol	MW g/g · mol	m_i g/g · mol	ψ_i g/g%
Octane	0.074	114	8.436	22.16
Oxygen	0.926	32	29.632	77.84
Total	1.000		38.068	100.00

14–2 FUELS

In combustion processes, a mixture of oxygen and another substance reacts to form carbon dioxide or water. The substances that react with oxygen are often called **fuels**, and the fuel and oxygen are called the **reactants**. As we said before, the components that are formed in a combustion process are called the **products of combustion** or just the **products**. The engineer and technologist should have some understanding of the various fuels to predict their heating values. We have referred to various fuels in earlier chapters, and here we give a brief discussion of some fuels.

Coal has been an important energy source for many years, and it still is. As we saw in chapter 13, coal is a mixture of substances, the most important one of which is carbon. Coal is mined from the Earth as a solid mixed with other materials. It is washed, separated from rocks and other impurities, and is then available as a fuel. As table 13–3 indicates, coal contains other substances even after the washing and separating processes. When the stoichiometric equation for coal combustion is written, however, coal is usually treated as pure carbon. Table B–26 lists the heating value for carbon (graphite), and this value may be used for coal, unless more detailed information on the coal is available, such as that of table 13–3.

Coal combustion can be accomplished in a number of ways. Coal can be burned as large chunks or pieces to provide a slow-burning fuel. It can also be crushed into small particles and burned to provide relatively hot fires. The small particles mix better with oxygen than do large pieces and so give more rapid combustion. Coal can also be pulverized into a dust to provide even better mixing with oxygen and give a somewhat more controllable fire than with the larger particles. Finally, coal may be burned in fluidized beds. Here the coal is crushed into small particles about 1 to 2 cm in diameter and mixed with limestone. Then air is forced from below through the bed of coal and limestone. The resulting combustion can be slower, with lower temperatures, than in other methods of burning, but with more complete combustion. Of the various coal combustion processes, it has been suggested that fluidized-bed combustion provides combustion products that are least detrimental to human beings and the environment.

Coal is used primarily as a fuel in the production of electrical energy, although some coal is used to provide heating for other applications.

Crude oil, pumped from wells in petroleum-producing areas of the world, is a very complex mixture of substances from which various petroleum products can be separated or distilled. Nearly all of these products can be identified as hydrocarbons, that is, chemicals that contain hydrogen and carbon in their structure. They represent a large source of energy and are therefore important fuels for combustion studies. Fuels derived from crude oil can be classified into the following petroleum fuel groups:

1. Gasoline
2. Distillates and residuals
3. Liquefied petroleum (LP) gas
4. Kerosene and jet fuel

Gasoline is a mixture of a number of hydrocarbons, and the mixture generally vaporizes or boils at a low temperature relative to other hydrocarbon fuels. It is used as a fuel in spark-ignition internal combustion (SIIC) engines for transportation and other applications. Gasoline is required not to detonate or ignite before being triggered by a controlled electrical spark, so that the ignition characteristics of gasoline are important. The octane rating is an arbitrary scale that has been invented to provide some idea of how well a particular grade of gasoline can be expected to perform during combustion. An octane number of 100 means that the gasoline has ignition characteristics equivalent to 2-2-4 trimethylpentane (or *n*-octane, C_8H_{18}, as listed in tables B–25 and B–26). This would be the most desirable fuel for spark-ignition engines, and the least desirable fuel would be one with an octane rating of 0, or an octane number of zero. A hydrocarbon that arbitrarily has an octane rating of zero is *n*-heptane, C_7H_{16}. Gasolines are not one particular hydrocarbon, however, but are complex mixtures of many of those listed in table B–26 as well as some unidentified compounds. For the purpose of this book, we will be concerned with the combustion of one identifiable hydrocarbon fuel.

Distillates are hydrocarbon mixtures having generally higher vapor pressures and boiling temperatures than those of gasolines. They are often used for residential heating as fuel oils and for light diesel engines in which compression ignition occurs. Residual fuels are used in larger diesel engines and for some heating applications. Distillates and residuals are sometimes called fuel oil or diesel fuel. Diesel fuels are required to self-ignite, so their characteristics should be different from those for gasolines, which are required not to self-ignite. The cetane number gives some idea of the self-igniting characteristics of diesel fuel and is defined with the two substances hexadecane (or cetane) and heptamethylnonane. Hexadecane, $C_{16}H_{34}$, has been identified as a substance with the most desirable characteristics as a diesel fuel and has been assigned cetane number 100. The cetane number of heptamethylnonane, $C_{16}H_{16}$, is 15. Most diesel fuels have cetane ratings around 70 to 95.

LP gases, including natural gas, are naturally occurring vapors as well as commercially refined vapors and liquids from petroleum. Natural gas has a high percentage of methane, CH_4, and for our purposes, we assume it to be entirely methane. LP gas is often composed of propane gas, C_3H_{18}. Butane gas is usually made up of about 95% the chemical butane, C_4H_{10}.

Biomass has become a significant fuel in certain areas of technology. The term **biomass** usually means wood and wood products, such as paper. The major identifiable chemical in wood is cellulose, $C_6H_{10}O_5$, and the analysis of waste from society (refuse or garbage) often shows high percentages of cellulose. Cellulose is therefore involved in a wide variety of technological situations: as a fuel in biomass energy systems, as a fuel in wood-burning applications, and as a disposable substance in waste incineration.

Finally, alcohols are a class of chemical substances that have been suggested for replacement of the depleting petroleum-based fuels. Alcohols can be derived from organic or biological materials such as wood, corn, and grains. The two most commonly occurring alcohols are methyl alcohol, CH_3OH, and ethyl alcohol, C_2H_5OH, although alcohol as a fuel would probably be a complex mixture of various alcohols. Also, alcohols tend to attract water, so that an alcohol fuel could have significant amounts of water in an analysis.

14–3 AIR/FUEL RATIOS

Fuels that are involved in combustion processes react with oxygen to produce products of combustion. It is common to mix a fuel with air rather than oxygen because pure oxygen is not a readily available material, as is air. Furthermore, pure oxygen can be a dangerous substance that could be a fire hazard. Therefore, if air is mixed with a fuel to promote a combustion process, the stoichiometric equation discussed in section 14–1 needs to be revised to account for the air instead of oxygen. The accepted standard molar analysis for air, or standard air, is 1 mol of oxygen to 3.76 mol of nitrogen. Therefore, if we burn carbon with standard air instead of pure oxygen, the combustion equation is

$$C + O_2 + 3.76N_2 \longrightarrow CO_2 + 3.76N_2 \quad \textbf{(14–4)}$$

and, for the combustion of hydrogen in air,

$$H_2 + 0.5O_2 + 1.88N_2 \longrightarrow H_2O + 1.88N_2 \quad \textbf{(14–5)}$$

Notice that the nitrogen gas is assumed to be an inert substance. In combustion analysis, we will see that the nitrogen absorbs much of the thermal energy or heating value of the fuel in the combustion process. Also, in actual combustion processes, the nitrogen can reach high enough temperatures that it may react with the oxygen to create nitric oxides: NO, NO_2, NO_3, and so on. These nitric oxides can then combine with free hydrogen atoms to create nitric acids, which are suspected as a source of atmospheric pollution.

It is convenient in combustion analysis to use the term **air/fuel ratio**, defined as

$$AF = \frac{\text{mass of air}}{\text{mass of fuel}} \quad \textbf{(14–6)}$$

The air/fuel ratio may be determined for a combustion process involving stoichiometric amounts of oxygen and fuel and revised to account for the nitrogen in the standard air, as shown in equations (14–4) and (14–5). In these situations, we say that the process has 100% theoretical air.

EXAMPLE 14–3 Determine the AF ratio for methane with 100% theoretical air.

Solution

The stoichiometric equation for methane is

$$CH_4 + O_2 \longrightarrow CO_2 + 2H_2O$$

and, for 100% theoretical air,

$$CH_4 + O_2 + 3.76N_2 \longrightarrow CO_2 + 2H_2O + 3.76N_2$$

The mass of air reacting with 1 kg·mol of methane is

$$32 \text{ kg } O_2 + 3.76 \times 28.0 \text{ kg } N_2 = 137.28 \text{ kg air}$$

The mass of the fuel (methane) is 16 kg of methane. This is from the chemical formula for methane, CH_4, where 12 kg of carbon + 4 kg of hydrogen (approximately) = 16 kg. The air/fuel ratio is then

Answer

$$AF = \frac{137.28 \text{ kg air}}{16 \text{ kg methane}} = 8.58 \text{ kg/kg}$$

This problem could be solved in English units to give the same AF ratio.

The air/fuel ratio in actual combustion processes is frequently adjusted or changed from that predicted with 100% theoretical air. If less air is supplied than 100%, the air/fuel ratio is said to be “rich,” and the combustion is likely to be incomplete, although the combustion can begin at a lower temperature with a rich mixture. If the AF ratio is increased to greater than 100%, it is said to be “lean,” and there is excess air. The best fuel economies in IC engines are often achieved with lean mixtures. Excess air is often given in percentage units, and we can write

$$\text{excess air\%} = \text{theoretical air\%} - 100\% \quad \textbf{(14–7)}$$

Therefore, for 100% theoretical air, the excess air is zero. For 250% theoretical air, the excess air would be 150%, and so on. Clearly, greater excess air means greater AF ratios as well.

EXAMPLE 14–4 Propane fuel, C_3H_8, is burned in 150% theoretical air. Determine the percent excess air and the AF ratio.

Solution From equation (14–7), we have, for the excess air,

Answer

$$\text{excess air\%} = 150\% - 100\% = 50\%$$

The reaction equation for propane is

$$C_3H_8 + 5O_2 + 18.8N_2 \longrightarrow 3CO_2 + 4H_2O + 18.8N_2$$

where $5 \times 3.76 = 18.8$ mol of N_2, and for 150% theoretical air, the reaction equation becomes

$$C_3H_8 + 7.5O_2 + 28.8N_2 \longrightarrow 3CO_2 + 4H_2O + 28.2N_2 + 2.5O_2$$

The mass of air is the mass of oxygen plus the mass of nitrogen:

$$\text{mass of air} = 7.5 \text{ mol } O_2 \times 32.0 \text{ lbm/mol} + 28.2 \text{ mol } N_2 \times 28.0 \text{ lbm/mol}$$
$$= 1029.6 \text{ lbm air/lbm} \cdot \text{mol } C_3H_8$$

The mass of fuel is

$$\text{mass of fuel} = 12.0 \times 3 + 8 \times 1.007 = 44.064 \text{ lbm propane}$$

Then the AF ratio is

$$\text{AF} = \frac{1029.6 \text{ lbm}}{44.064 \text{ lbm}} = 23.366 \text{ lbm/lbm}$$

or

Answer

$$23.366 \text{ kg/kg}$$

Notice that the excess air allows for oxygen gas to be present in the products of combustion.

14–4 HEAT OF FORMATION

Chemical reactions frequently have energy transfers associated with them, and these energy transfers are nearly always in the form of heat. Chemical molecules can be visualized as substances constructed from naturally occurring atoms through an energy transfer, or heat of formation. For instance, we know that acetylene gas, C_2H_2, is composed of two carbon atoms and two hydrogen atoms. The heat of formation is the amount of energy required to form acetylene gas from carbon and hydrogen gas (H_2) is 226,731 kJ/kg · mol, or 97,542 Btu/lbm · mol. Water is composed of one hydrogen molecule and half an oxygen molecule (O_2), or two hydrogen atoms and one oxygen atom. The energy associated with the formation of water from hydrogen and oxygen is −285,830 kJ/kg · mol or −122,967 Btu/lbm · mol. The minus sign indicates that energy or heat is given off during the formation of water from hydrogen and oxygen. In table B–25, the values of the heat of formation, $\Delta\bar{h}_f^\circ$, are given for some substances. The heat of formation is an enthalpy change, so the notation $\Delta\bar{h}$ is consistent with our notation. The subscript f refers to the formation, and the superscript ° refers to a standard state of 298 K, 101 kPa, or 77°F, 14.696 psia.

Notice in table B–25 that some atoms have zero heats of formation. For instance, carbon, hydrogen gas (H_2), and nitrogen gas (N_2) all have zero heats of formation. This indicates that they are naturally occurring and do not have any energy associated with their formation at the standard state. Notice also that the single atoms of nitrogen, hydrogen, and oxygen have large positive heats of formation, which means that these atoms are not naturally occurring and much energy is required to produce them.

The heats of formation listed in table B–25 are all given for the substances at 298 K, 101 kPa, or 77°F, 14.696 psia, and the values can be treated as the values of the enthalpy of that substance at the standard temperature of 298 K or 77°F. In the following section, we will see that this treatment of the heat of formation allows us to calculate the heat associated with many combustion processes.

14–5 COMBUSTION ANALYSIS

Combustion processes can be visualized as the mixing of a fuel and air to produce products of combustion. These processes are often carried out at constant pressure in either a closed system or an open one. The general problem of combustion is sketched in the schematic of figure 14–1. From this schematic and the conservation of energy, we have

$$\dot{Q}_c = \sum \dot{m}_3 h_3 - \dot{m}_1 h_1 - \dot{m}_2 h_2 \qquad \textbf{(14–8)}$$

where $\dot{Q}_c$ is the heat given off by the combustion process, called the **heat of combustion**. Because there is often more than one product of combustion, the enthalpy at state 3 (the products) is denoted as a sum of all the products of combustion, $\sum \dot{m}_3 h_3$.

Frequently, the heat of combustion is determined for 1 mol of a fuel. Equation (14–8) should then be written

$$q_c = \sum N_3 \bar{h}_3 - \bar{h}_1 - N_2 \bar{h}_2 \qquad \textbf{(14–9)}$$

where $\sum N_3 \bar{h}_3$ is the sum of the products of combustion for 1 mol of fuel. The enthalpies $\bar{h}_3$, $\bar{h}_1$, and $\bar{h}_2$ are based on 1 mol of the substance.

The heat of combustion is usually determined from the process when the fuel and air are at standard pressure and temperature, 298 K and 101 kPa. If the products of combustion are also assumed to be at the standard pressure and temperature, the heat of combustion is called the heating value of the fuel. In actual combustion processes, the products are usually at a high temperature, so the heat of combustion, $\dot{Q}_c$ or q_c, could represent a potential heat transfer if the exhaust products are cooled back to standard temperature. The enthalpy values in equations (14–8) and (14–9) can be set equal to the heats of formation, such as those listed in table B–25.

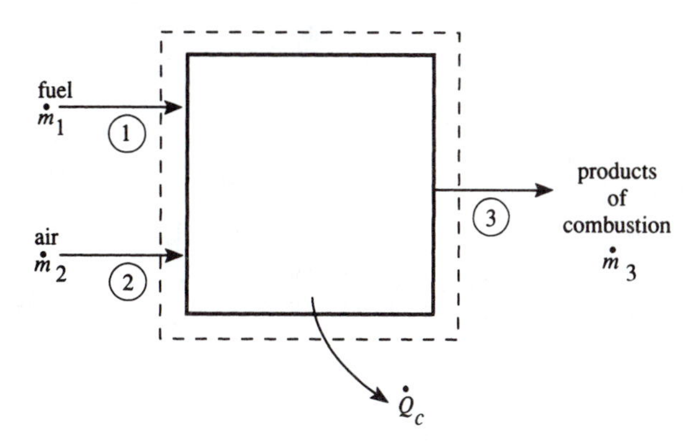

FIGURE 14–1 General schematic of a combustion process

Also, the heating value can be interpreted in two ways: (1) the products contain water as a vapor or (2) the products contain water as a liquid. If the products of combustion are assumed to contain liquid water, the heating value is called the **higher heating value**, HHV, and if the water is assumed to be a vapor or saturated vapor, the heating value is the **lower heating value**, LHV. The HHV and the LHV are listed for some fuels in table B–26, where the HHV and LHV are the absolute values of the heat combustion. They are related through the heat of vaporization of water at 298 K (77°F) by the equation

$$\text{HHV} = \text{LHV} + 2442\left(\frac{m_{u'}}{m_F}\right) \tag{14–10}$$

where HHV and LHV are in kJ/kg fuel, m_w is the mass of water in the products, and m_F is the mass of fuel. In English units, the equation would be

$$\text{HHV} = \text{LHV} + 1050\left(\frac{m_{u'}}{m_F}\right) \tag{14–11}$$

EXAMPLE 14–5 Determine the higher and lower heating values, HHV and LHV, for *n*-butane. Use the table B–25 values for enthalpies of the terms for a combustion analysis, and compare the results to the heating values listed in table B–26.

Solution

Using table B–26, we obtain the heating values:

Answer

$$\text{HHV} = 49{,}504 \text{ kJ/kg fuel}$$
$$= 21{,}283 \text{ Btu/lbm fuel}$$

Answer

$$\text{LHV} = 45{,}718 \text{ kJ/kg}$$
$$= 19{,}655 \text{ Btu/lbm}$$

Now, using a combustion analysis, we first write the combustion equation for *n*-butane and 100% theoretical air:

$$C_4H_{10} + 6.5O_2 + 24.44N_2 \longrightarrow 4CO_2 + 5H_2O + 24.44N_2$$

The heat of combustion q_c can be calculated from equation (14–9). This calculation is demonstrated in table 14–2.

The total heat of combustion is found to be −2,657,033 kJ/kg · mol of *n*-butane fuel. Because the fuel has a mass of 58.08 kg/kg · mol, then the heat of combustion −2,657,033/58.08 = −45,747.8 kJ/kg of fuel. As you can see in table 14–2, the fuel and air are listed first; then the products are listed; and their molecular masses or weights, MW_i, are listed in column 1. Column 2 lists the number of moles of each substance for 1 mol of fuel, and column 3 gives the heats of formation from table B–25. The total enthalpy of each substance for

TABLE 14–2 Heat of combustion calculation for *n*-butane and 100% theoretical air

Substance	(1) MW_i kg/kg · mol	(2) N_i kg · mol	(3) $\Delta\bar{h}_f^\circ$ kJ/kg · mol	(4) $N_i\bar{h}_i$ kJ/kg · mol fuel
n-Butane, C_4H_{10}	58.08	1	−126,147	−126,147
Oxygen, O_2	32.0	6.5	0	0
Nitrogen, N_2	28.0	24.44	0	0
CO_2	44.0	4	−393,520	−1,574,080
H_2O	18.016	5	−241,820	−1,209,100
N_2	28.0	24.44	0	0
Total				−2,657,033

1 mol of fuel is listed in column 4. The sum of column 4 is the total heat of the combustion, found by using the summation of equation (14–9), that is, subtracting the fuel and air terms from the sum of the products. The heating value, as an absolute value of 45,747.8 kJ/kg · mol, is in agreement with the value given in table B–26 for the lower heating value.

Notice that the heat of formation of vapor water, or gas, was used in table 14–2, which accounts for the fact that the lower heating value was found. The higher heating value is obtained if the heat of formation of liquid water, −285,830 kJ/kg · mol, is used. The sum of the terms in column 4 then is −2,875,083 kJ/kg · mol fuel or, per mass units, −49,502.1 kJ/kg *n*-butane. The absolute value, 49,502.1 kJ/kg, is also in agreement with the HHV listed in table B–26.

Now let us consider a fuel whose heating value is not listed in table B–26, but which we can determine from the data in table B–25 and from the ideas just considered.

EXAMPLE 14–6 Determine the HHV and the LHV of methyl alcohol.

Solution We write the reaction equation for methyl alcohol and 100% theoretical air:

$$CH_3OH + 1.5O_2 + 5.64N_2 \longrightarrow CO_2 + 2H_2O + 5.64N_2$$

Then we obtain values for the heats of formation from table B–25. Table 14–3 lists the substance, its molecular weight, the number of moles per mole of alcohol, the heat of formation, and the total enthalpy per mol of alcohol. The heat of combustion is the sum of the total enthalpy terms, where the fuel and air terms are subtracted from the products terms. The result is −274,726 Btu/lbm · mol alcohol, with the absolute value being 274,726 Btu/lbm · mol. The mass of methyl alcohol is 32.032 lbm/lbm · mol, so that the heating value is 274,726/32.032 = 8576.6 Btu/lbm alcohol. This is the lower heating value because the heat of formation of water in its gas phase was used in the calculations. If the heat of formation of liquid water is used, −122,967 Btu/lbm · mol, then the result of the sum of the total enthalpy terms is −312,594 Btu/lbm · mol alcohol, and the higher heating value would be 312,594 Btu/lbm · mol or 9758.8 Btu/lbm methyl alcohol.

In examples 14–5 and 14–6, the nitrogen was included in the analysis but did not contribute to the calculations of the heating values. The purpose of including the nitrogen at this time was to show you a general method of solving combustion problems. In the next section, we consider methods used for predicting combustion temperatures, and the nitrogen will contribute to that analysis. Also, the nitrogen in the products of combustion reduces the partial pressure of the water in those products, so the temperature at which the water will condense or become a liquid will be lowered. As we have just seen, the phase of the water in the products affects the heat of combustion to some extent, so, in the following example, we consider in some detail how you can decide whether the water is a liquid or a gas in the

TABLE 14–3 Heat of combustion for methyl alcohol and 100% theoretical air

Substance	MW_i lbm/lbm · mol	N_i lbm · mol	$\Delta\bar{h}_f^\circ$ Btu/lbm · mol	$N_i\bar{h}_i^\circ$ Btu/lbm · mol fuel
CH_3OH	32.032	1	−102,636	−102,636
O_2	32.0	1.5	0	0
N_2	28.0	5.64	0	0
CO_2	44.0	1	−169,296	−169,296
H_2O	18.016	2	−104,033	−208,066
N_2	28.0	5.64	0	0
Total				−274,726

products, or whether some fraction of the water is condensed. We will use the mixture analyses of chapter 13 for some of this analysis here.

EXAMPLE 14–7 Methyl alcohol is burned with 100% theoretical air at 14.696 psia and 77°F. Determine the partial pressure of the water in the products of combustion if the CO_2, H_2O, and N_2 are all acting as perfect gases. Then determine the dew point of the products, and discuss this result.

Solution

From example 14–6, we found that the products of combustion included 5.64 mol of N_2, 2 mol of H_2O, and 1 mol of CO_2. If these components are all perfect gases, the mole fractions can be used to determine the partial pressure of the water in that mixture. The mole fraction of water is $\Omega_{H_2O} = 2/8.64 = 0.23148$ mol of H_2O per mol of mixture. The partial pressure of the water is then

$$p_{H_2O} = \Omega_{H_2O} p_T$$
$$= (0.23148)(14.696 \text{ psia})$$
$$= 3.40185 \text{ psia}$$

For water at 3.40185 psia, the dew point temperature is the same as the saturation temperature at that pressure, so that, interpolating from table B–11, we obtain

Answer

$$T_{dp} = 138.1°\text{F}$$

In our analysis of example 14–6, we assumed that the products were at 77°F so that the mixture has gone below the dew point temperature. Therefore, water has condensed out of the mixture, and, in fact, the water continues to condense out of the products until the partial pressure of the remaining water vapor is equal to that corresponding to the saturation pressure of water at 77°F.

From tables B–11, that pressure is found to be 0.4776 psia by an interpolation. The mole fraction of the water vapor must then be

$$\Omega_g = \frac{0.4776 \text{ psia}}{14.696 \text{ psia}} = 0.0325$$

and the mole fraction is also given by

$$\Omega_g = \frac{N_g}{N_g + 6.64 \text{ mol}}$$

where N_g is the number of moles of water that remain as a gas in the products. Solving for N_g, we obtain $N_g = 0.222$ lbm · mol, and thus, of the 2 lbm · mol of water in the products, 1.778 lbm · mol condenses out as liquid and 0.222 lbm · mol remains as a gas. The analysis of example 14–6 could be revised to include this information and thereby yield a more precise heat of combustion of methyl alcohol at 77°F, 14.696 psia. These results are given in table 14–4, revised from table 14–3. The heat of combustion of the methyl alcohol is therefore −278,929.348 Btu/lbm · mol CH_3OH or −8707.8 Btu/lbm CH_3OH. Notice that somewhat more heat is predicted from this type of analysis compared to the lower heating value, but there is less than the higher heating value.

TABLE 14–4 Calculation of heat of combustion of methyl alcohol at 77°F, 14.696 psia, with an accounting of water condensing in the products of combustion

Substance	N_i lbm · mol	$\Delta\bar{h}_f^\circ$ Btu/lbm · mol	$N_i\bar{h}_i^\circ$ Btu/lbm · mol fuel
CH_3OH	1	−102,636	−102,636
O_2	1.5	0	0
N_2	5.64	0	0
CO_2	1	−169,296	−169,296
$H_2O(l)$	1.778	−104,033	−184,970.674
$H_2O(g)$	0.222	−122,967	−27,298.674
N_2	5.64	0	0
Total			−278,929.348

The combustion process can be analyzed with assistance from the computer and computer programs. As an example of how this may be done, the program COMBUST, described in appendix A–12 and included on a soft diskette available from the publisher, can be used with a microcomputer to determine the LHV, the HHV, the partial pressure of the water vapor, the dew point temperature, the amount of water condensed out of the products upon cooling to standard temperature, and the actual heat of combustion for that process. The program requires the following input in order that it run to completion:

1. Fuel
2. Percent theoretical air
3. Temperature of the fuel and air
4. Temperature of the products of combustion

The program allows you to use fuel, air, and the products of combustion at states other than the standard temperature of 298 K or 77°F. In the next section, we see how the temperatures other than standard state are accounted for in the combustion analysis.

14–6 ADIABATIC COMBUSTION TEMPERATURE

In the application of combustion processes to technology, the engineer and technologist frequently need to know the highest temperature associated with that combustion. For instance, in the study of the Otto and diesel engines, we saw that fuel was burned internally to the engine, and the products of combustion were then expanded in a piston-cylinder device to produce power. The gas turbine used a combustor to provide hot gases that were then expanded in a steady-flow turbine to produce power. In the steam turbine power cycle, the steam was produced in a boiler or steam generator, which allowed transfer of heat from combustion products from coal or other fuels to water. In all these examples, one must have a knowledge of the combustion temperature to be able to analyze the heat engine operations. Here we discuss how the maximum temperatures may be predicted.

If we consider a combustion process involving a fuel and air, as diagrammed in figure 14–1, we expect a heat transfer $\dot{Q}_c$, the heat of combustion, during the process. The calculation of $\dot{Q}_c$ then followed after assuming the temperature of the products. If we assume the heat of combustion is zero, the process will be an adiabatic combustion process. The final temperature of the products can then be determined from equation (14–8) or equation (14–9), where the heat is zero. The temperature found from this calculation is called the adiabatic flame temperature or the adiabatic combustion temperature. In combustion processes, heat or thermal energy is given off, so the products of combustion are cooler than predicted by the adiabatic process, but the products are always warmer than the surroundings. Any heat transfer that does occur will tend to cool the products of combustion, so the maximum amount of heat involved would be that when the products cool to the surrounding temperature. We now demonstrate how to find the adiabatic combustion temperature.

EXAMPLE 14–8

Acetylene gas is burned in 100% theoretical air at 101 kPa. Determine the maximum temperature expected of the products of combustion, assuming that the acetylene gas and air are supplied at 298 K.

Solution

We write the combustion equation as

$$C_2H_2 + 2.5O_2 + 9.4N_2 \longrightarrow 2CO_2 + H_2O + 9.4N_2$$

and then find the heat of formation for each substance. We assume that the water will all be in the vapor state because the temperature of the products is expected to be above the dew point. In table 14–5, these results are listed, together with the enthalpies of the products per mole of acetylene. Another column (column 5) lists the difference in enthalpy values between the standard state (column 4) and the actual state at a temperature T_3. Notice that the fuel and air were supplied at the standard-state temperature, so the value in column 5 is

TABLE 14–5 Calculation for combustion of acetylene gas with 100% theoretical air and with adiabatic conditions

Substance	(1) MW$_i$	(2) N_i kg · mol	(3) $\Delta\bar{h}_f^\circ$ kJ/kg · mol	(4) $N_i\bar{h}_i^\circ$ kJ/kg · mol fuel	(5) $(\bar{h}_i - \bar{h}_i^\circ)N_i$ kJ/kg · mol fuel
C_2H_2	26.016	1	226,731	226,731	0
O_2	32.0	2.5	0	0	0
N_2	28.0	9.4	0	0	0
CO_2	44.0	2	−393,520	−787,040	$mc_p(T_3 - 298\text{ K})$
H_2O	18.016	1	−241,820	−241,820	$mc_p(T_3 - 298\text{ K})$
N_2	28.0	9.4	0	0	$mc_p(T_3 - 298\text{ K})$

zero for each of those terms. Also, the products are all assumed to be perfect gases having constant specific heat values. The values for the specific heats are given in table B–4.

From the results of table 14–5 and the combustion equation (14–8), or from conservation of energy applied to the combustion process, set $\dot{Q}_c = 0$. Then the term $\sum m_3h_3$ is equal to the sum of the fuel and air terms, $m_1h_1 + m_2h_2$. This equality will be the sum of columns 4 and 5 of the products, set equal to the sum of the fuel and air terms of columns 4 and 5:

$$\begin{aligned}(44.0\text{ kg/kg}\cdot\text{mol})&(2\text{ kg}\cdot\text{mol})(c_p)_{CO_2}(T_3 - 298\text{ K})\\ &+ (18.016)(1)(c_p)_{H_2O}(T_3 - 298\text{ K})\\ &+ (28.0)(9.4)(c_p)_{N_2}(T_3 - 298\text{ K})\\ &= 787{,}040\text{ kJ/mol} + 241{,}820\text{ kJ/mol} + 226{,}731\text{ kJ/mol}\end{aligned}$$

From table B–4, the specific heats are

$$\begin{aligned}(c_p)_{CO_2} &= 0.844\text{ kJ/kg}\cdot\text{K}\\ (c_p)_{H_2O} &= 1.864\text{ kJ/kg}\cdot\text{K}\\ (c_p)_{N_2} &= 1.040\text{ kJ/kg}\cdot\text{K}\end{aligned}$$

and solving the equation for T_3 gives

Answer

$$T_3 = 3588.5\text{ K} = 3315.5°\text{C}$$

This result tells us that the water in the products is in the vapor state, so the original assumption of a perfect gas for all the products is reasonable. Also, notice that the nitrogen affected the calculation for the combustion temperature. If excess air were supplied to the process, even more nitrogen gas would have been involved, and the combustion temperature would have been less than that predicted in the calculation with 100% theoretical air.

The program COMBUST, described in appendix A–12 and available on soft diskette, can be used conveniently and rapidly to find the adiabatic combustion temperature for hydrocarbon fuels.

14–7 ENTROPY GENERATION IN COMBUSTION

All combustion processes are irreversible, so any combustion analysis should include a prediction of the amount of the entropy generation. If we refer to figure 14–1 again, we can see that the entropy generation can be written

$$\dot{S}_{gen} = \sum \dot{m}_3s_3 - \dot{m}_1s_1 - \dot{m}_2s_2 + \dot{S}_{surr} \qquad \textbf{(14–12)}$$

where $\sum \dot{m}_3 s_3$ is the sum of the entropy rates of each of the products of combustion, $\dot{m}_1 s_1$ is the entropy rate of the fuel, $\dot{m}_2 s_2$ is the entropy rate of the air supplied to the combustion, and $\dot{S}_{\text{surr}} \geq (-\dot{Q}_c)/T_{\text{surr}}$. We assign a negative sign to $\dot{Q}_c$ because $\dot{Q}_c$ is calculated as the heat of the system, and the heat to or from the surroundings is the negative of that value. We could then write, for the entropy generation,

$$\dot{S}_{\text{gen}} \geq \sum \dot{m}_3 s_3 - \dot{m}_1 s_1 - \dot{m}_2 s_2 - \frac{\dot{Q}_c}{T_{\text{surr}}} \tag{14–13}$$

For 1 mol of fuel, the entropy increase is

$$\Delta S_{\text{gen}} \geq \sum N_3 \bar{s}_3 - \bar{s}_1 - N_2 \bar{s}_2 - \frac{q_c}{T_{\text{surr}}} \tag{14–14}$$

where q_c is determined from equation (14–9) and as demonstrated in previous examples. For perfect gases with constant specific heats, the entropy values are given by the equation

$$s_i = s_i^{\circ} + c_{p_i} \ln \frac{T_i}{T_o} - R_i \ln \Omega_i \tag{14–15}$$

where s_i° is the entropy of the component i at standard-state temperature and pressure (listed in table B–25 as absolute entropy), T_i is the temperature of the component, T_o is 298 K or 537°R, R_i is the gas constant, and Ω_i is the mole fraction of the component in the products or in the air if O_2 and N_2 in the supplied air are being evaluated. Equation (14–15) can be revised to read, in moles,

$$\bar{s}_i = \bar{s}_i^{\circ} + \overline{c_{p_i}} \ln \frac{T_i}{T_o} - R_u \ln \Omega_i \tag{14–16}$$

where R_u is the universal gas constant, 8.31 kJ/kg · mol · °R or 1.985 Btu/lbm mol · °R.

We now consider an example of how to predict entropy generation for combustion processes.

EXAMPLE 14–9 Coal is burned at a rate of 1600 kg/h, and 62.5% of the coal is carbon; 110% theoretical air is supplied to the combustion process. Determine the entropy generation associated with the process if the products of combustion, the fuel, and the supplied air are at 298 K, 101 kPa. The surrounding temperature is also 298 K.

Solution The amount of carbon in the coal is 62.5% of 1600 kg/h or 1000 kg/h. The combustion equation for the coal with 110% theoretical air is

$$C + 1.1O_2 + 4.136N_2 \longrightarrow CO_2 + 4.136N_2 + 0.1O_2$$

The analysis of the combustion and the determination of the heat of combustion, q_c, are given in columns 1, 2, 3, and 4 of table 14–6. The heat of combustion is found to be −393,520 kJ/kg · mol coal. The heat transfer rate is then

$$\begin{aligned}\dot{Q}_c &= \frac{\dot{m} q_c}{MW_{\text{coal}}} \\ &= \frac{(1000\ \text{kg/h})(-393{,}520\ \text{kJ/kg} \cdot \text{mol})}{12\ \text{kg/kg} \cdot \text{mol}} \\ &= -9109.3\ \text{kJ/s}\end{aligned}$$

The absolute entropy values of the various components are listed in table 14–6, column 5, obtained from table B–25. The mole fractions are then given in column 6. The coal is assumed

TABLE 14–6 Combustion analysis of coal with 110% theoretical air and including entropy generation analysis

Substance	(1) MW_i	(2) N_i kg · mol	(3) $\Delta\bar{h}_f^\circ$ kJ/kg · mol	(4) $N_i\bar{h}_i^\circ$ kJ/kg · mol	(5) $\bar{s}_i^\circ$ kg · mol · K	(6) Ω_i	(7) $N_i\bar{s}_i$ kJ/kg · mol · K
Carbon, C	12.0	1	0	0	5.74	1	5.74
O_2	32.0	1.1	0	0	205.04	0.21	239.81
N_2	28.0	4.136	0	0	191.50	0.79	800.15
CO_2	44.0	1	−393,520	−393,520	213.67	0.191	227.43
O_2	32.0	0.1	0	0	205.04	0.019	23.80
N_2	28.0	4.136	0	0	191.50	0.790	800.15
Total				−393,520			5.68

to be a solid, so that its mole fraction is 1.0 and is not contributing to the perfect gas mixture in the supplied air. The entropy values per mole of coal or carbon are given in column 7, calculated from equation (14–16). For instance, for CO_2, the entropy is

$$S_{CO_2} = N_{CO_2}\bar{s}_{CO_2} = N_{CO_2}[\bar{s}^\circ_{CO_2} - R_u \ln(\Omega_{CO_2})]$$
$$= (1)[213.67 \text{ kJ/kg mol} \cdot \text{K} - (8.315 \text{ kJ/kg mol} \cdot \text{K})(\ln(0.191))]$$
$$= 227.43 \text{ kJ/kg mol coal} \cdot \text{K}$$

and the other entropy results follow.

Then, adding column 7 according to equation (14–14), we find that

$$\Delta s_{gen} \geq \sum N_3\bar{s}_3 - \bar{s}_1 - N_2\bar{s}_2 - \frac{q_c}{T_{surr}}$$
$$\geq (5.68 \text{ kJ/kg} \cdot \text{mol} \cdot \text{K}) + \frac{393{,}520 \text{ kJ/kg} \cdot \text{mol}}{298 \text{ K}}$$
$$\geq 1326.22 \text{ kJ/kg} \cdot \text{mol} \cdot \text{K}$$

The entropy generation is then given by equation (14–12) or (14–13) and is the same as

$$\dot{S}_{gen} = \frac{\dot{m}_{coal}(\Delta s_{gen})}{MW_{coal}}$$
$$\geq \frac{(1000 \text{ kg/h})(1326.22 \text{ kJ/kg} \cdot \text{mol} \cdot \text{K})}{120 \text{ kg/kg} \cdot \text{mol}}$$
$$\geq 110{,}518.3 \text{ kJ/h} \cdot \text{K}$$

Answer
$$\geq 30.70 \text{ kJ/K} \cdot \text{s}$$

Let us now look at the adiabatic combustion process and compute the entropy generation for that kind of process. If the process were reversible (and adiabatic), the entropy generation would be zero, but we will see that it is positive for combustion processes, indicating that it is an irreversible process.

EXAMPLE 14–10 Methane is burned with 110% theoretical air. If the methane and air are at 77°F and 14.696 psia, and if the combustion occurs in an adiabatic manner, determine the entropy generation per mole of methane. Assume that the surroundings are at 537°R (77°F).

Solution This combustion process involves no heat transfer, so the entropy generation is just that of the system. The combustion equation for methane and 110% theoretical air is

$$CH_4 + 2.2O_2 + 8.272N_2 \longrightarrow CO_2 + 2H_2O + 0.2O_2 + 8.272N_2$$

TABLE 14–7 Combustion analysis of methane burned with 110% theoretical air, including analysis of entropy generation

Component	(1) MW_i	(2) N_i $\frac{\text{lbm}\cdot\text{mol}}{\text{mol}}$	(3) $\Delta\bar{h}^\circ_{fi}$ Btu/lbm mol	(4) $N_i\bar{h}^\circ_i$ $\frac{\text{Btu}}{\text{lbm}\cdot\text{mol}}$	(5) $(\bar{h}_i - \bar{h}^\circ_i)N_i$ Btu/lbm · mol	(6) $\bar{s}^\circ_i$ $\frac{\text{Btu}}{\text{lbm}\cdot\text{mol}\cdot{}^\circ\text{R}}$	(7) Ω_i	(8) $N_i\bar{s}_i$ $\frac{\text{Btu}}{\text{lbm}\cdot\text{mol}\cdot{}^\circ\text{R}}$
CH_4	16.032	1	−32,202	−32,202	0	44.52	1	44.52
O_2	32.0	2.2	0	0	0	49.01	0.21	114.64
N_2	28.0	8.272	0	0	0	45.75	0.79	382.31
CO_2	44.0	1	−169,296	−169,296	$(1)(44)(0.2015)(T_3 - 537)$	51.07	0.0872	75.06
H_2O	18.016	2	−104.033	−208.666	$(2)(18.016)(0.4452)(T_3 - 537)$	45.11	0.1743	131.80
O_2	32.0	0.2	0	0	$(0.2)(32)(0.219)(T_3 - 537)$	49.01	0.0174	14.44
N_2	28.0	8.272	0	0	$(8.272)(28.0)(0.2483)(T_3 - 537)$	45.75	0.7211	508.02
								$\sum N_i\bar{s}_i = 187.85$

We continue the combustion analysis by listing the MW_i, N_i, $\Delta\bar{h}^\circ_i$, $N_i k^\circ_i$ and $(h_i \cdot h^\circ_i)N_i$ for each term, so that we can then find the adiabatic combustion temperature, T_3. The results are listed in table 14–7, columns 1 through 5. The products of combustion are all assumed to be perfect gases with constant specific heats, and the specific heats were found from table B–4. The adiabatic combustion temperature, T_3, was found by adding the terms in columns 4 and 5 for the products and setting them equal to the terms in columns 4 and 5 for the methane and air. The result for T_3 is then solved from that equality and is found to be 4654.87°R, or about 4655°R.

The standard-state entropies, s°_i, are read from table B–25 and are listed in column 6 of table 14–7. The mole fractions are given in column 7, and the entropy of each component per mole of methane in column 8. The values of s_i were determined from equation (14–16) with T_i set to T_3 = 4655°R. For instance, for water in the products,

$$(N_i\bar{s}_i)_{H_2O} = N_{H_2O}\left[\bar{s}^\circ_{H_2O} + (18.016)(c_p)_{H_2O}\left(\ln\frac{4655°\text{R}}{537°\text{R}}\right) - (1.987\ \text{Btu/lbm}\cdot\text{mol}\cdot°\text{R})\ln(0.1743)\right]$$

$$= 131.80\ \text{Btu/°R}\cdot\text{lbm}\cdot\text{mol methane}$$

Other results follow and are listed in column 8 of table 14–7.

The total entropy generation of the combustion per mole of CH_4 is the sum of the products of combustion terms in column 8 minus the terms in column 8 for the methane and air. The result is

Answer

$$\Delta S_{gen} = 187.85\ \text{Btu/°R}\cdot\text{lbm}\cdot\text{mol methane}$$

14–8 SUMMARY

Stoichiometric combustion equations are chemical reaction equations for oxygen and a fuel. The stoichiometric equation for hydrogen is

$$H_2 + 0.5O_2 \longrightarrow H_2O \tag{14–1}$$

and for *n*-octane is

$$C_8H_{18} + 12.5O_2 \longrightarrow 8CO_2 + 9H_2O \tag{14–3}$$

Combustion involves a hydrocarbon and oxygen to produce water or carbon dioxide. Hydrocarbon fuels are complex mixtures of chemical compounds; however, some of the most important fuels can be approximated by certain chemical formulas. Coal can be considered as carbon, gasoline can be *n*-octane, diesel fuel can be cetane, LP gas can be propane or butane, natural gas can be methane, and wood can be identified as cellulose. Alcohol is usually methyl or ethyl alcohol.

The air/fuel ratio, AF ratio, is defined as the ratio of the mass of air supplied to a combustion process to the mass of fuel. Air is assumed, for combustion purposes, to be composed of 1 mol of oxygen gas to 3.76 mol of nitrogen gas.

Theoretical air is the percentage of supplied air based on stoichiometric air. One hundred percent theoretical air means that the amount of air supplied is given by the stoichiometric equation. Excess air is the percentage of air greater than stoichiometric amounts actually supplied in a combustion process. Excess air is the theoretical air supplied minus 100%.

The heat of formation of a compound or molecule is the energy required to form that substance from the naturally occurring substances. A negative heat of formation implies that heat is given off during the combustion or forming of that material, and a positive value means that energy or heat is required to create those substances. The heat of formation can be interpreted as the enthalpy of the substance at the standard-state temperature and pressure, 298 K (77°F) and 101 kPa (14.696 psia).

The heat of combustion $\dot{Q}_c$ is determined from the energy balance

$$\dot{Q}_c = \sum \dot{m}_3 h_3 - \dot{m}_1 h_1 - \dot{m}_2 h_2 \qquad \textbf{(14–8)}$$

The products of combustion are usually CO_2, H_2O, and N_2. If excess air is supplied during a combustion, there will be oxygen and additional nitrogen in the products, and if there is less than 100% theoretical air, there will be some of the fuel in the products.

The higher heating value HHV of a fuel is the absolute value of the heat of combustion when the fuel, air, and products of combustion are all at standard-state conditions and the water in the products is in the liquid phase. The lower heating value LHV is the absolute value of the heat of combustion if the water in the products is in the gaseous or vapor phase. The LHV and HHV are related through the heat of vaporization of water at 298 K:

$$\text{HHV} = \text{LHV} + 2442\left(\frac{m_w}{m_F}\right) \quad \text{(SI units)} \qquad \textbf{(14–10)}$$

$$\text{HHV} = \text{LHV} + 1050\left(\frac{m_w}{m_F}\right) \quad \text{(English units)} \qquad \textbf{(14–11)}$$

If the products of combustion are perfect gases, the partial pressure of water vapor in the products is

$$p_w = \Omega_w p_T$$

where Ω_w is the mole fraction of water in the products. If the dew point of the products of combustion is below the temperature of those products, the water is all vapor. If the dew point is above the temperature of the products of combustion, the water will condense out of the products until the partial pressure of the water corresponds to that of the saturation pressure at the temperature of the products of combustion.

The adiabatic combustion process is an ideal process where no heat is transferred in the combustion. The adiabatic combustion temperature is the temperature of the products of combustion resulting from this process and represents the maximum possible temperature in the combustion.

Entropy generation in combustion processes can be determined from the equations

$$\dot{S}_{gen} \geq \sum \dot{m}_3 s_3 - \dot{m}_1 s_1 - \dot{m}_2 s_2 - \frac{\dot{Q}_c}{T_{surr}} \quad \textbf{(14–13)}$$

and, per mole of fuel,

$$\Delta S_{gen} \geq \sum N_3 \bar{s}_3 - \bar{s}_1 - N_2 \bar{s}_2 - \frac{q_c}{T_{surr}} \quad \textbf{(14–14)}$$

DISCUSSION QUESTIONS

Section 14–1

14–1 What is meant by the term *stoichiometric*?

14–2 What is *combustion*?

14–3 What is *oxidation*? Is it the same as *combustion*?

Section 14–2

14–4 What is meant by the term *fuel*?

14–5 What is meant by the term *reactant*?

14–6 What is a *hydrocarbon*?

Section 14–3

14–7 Why would one want to know the *air/fuel ratio?*

14–8 What does the term *excess air* designate?

Section 14–4

14–9 What is meant by the term *heat of formation*?

14–10 Can elements be formed in any other way than by heat?

Section 14–5

14–11 What is meant by the term *heat of combustion*?

14–12 Why is the phase of water in the products of combustion so important in the heating value?

Section 14–6

14–13 What does the *adiabatic flame temperature* represent?

14–14 What is the actual combustion temperature?

PRACTICE PROBLEMS

Problems that use SI units are indicated by an (M) under the problem number; those that use English units are indicated by an (E). Mixed unit problems are indicated by a (C).

Section 14–1

14–1 (C) Write the stoichiometric equation for acetylene gas, C_2H_2, and oxygen, O_2.

14–2 (C) Write the stoichiometric equation for propane, C_3H_8, and oxygen.

14–3 (C) Write the stoichiometric equation for ethyl alcohol, C_2H_5OH, and oxygen.

14–4 (C) Determine the mass analysis for the stoichiometric mixture of carbon and oxygen.

14–5 (C) Determine the mass analysis for cellulose, $C_6H_{10}O_5$, and oxygen in a stoichiometric mixture.

14–6 (C) Determine the mass analysis for the stoichiometric mixture of methyl alcohol, CH_3OH, and oxygen.

Section 14–3

14–7 (C) Determine the air/fuel ratio for the burning of *n*-butane in 100% theoretical air.

14–8 (C) If 100% theoretical air is mixed with benzene, determine the air/fuel ratio.

14–9 (C) Cetane is burned in 100% excess air. Determine the percent theoretical air and the air/fuel ratio.

14–10 (C) Carbon monoxide is burned in 50% excess air. Determine the percent theoretical air and the air/fuel ratio.

Section 14–4

14–11 (C) Determine the heat of formation of the following:
(a) Carbon dioxide
(b) Ethyl alcohol
(c) Cellulose
(d) Coal (carbon)
(e) Free nitrogen (N)

Section 14–5

14–12 (C) Determine the HHV and LHV for *n*-octane by using the values from table B–19 and checking with values from table B–18.

14–13 (C) Determine the HHV and LHV for acetylene gas by using the combustion analysis with values from table B–18 and comparing to the values from table B–19.

14–14 (C) Determine the HHV and LHV for cellulose, $C_6H_{10}O_5$.

14–15 (C) Determine the partial pressure of water vapor in the products of combustion of propane with 100% theoretical air. Then determine the dew point of the products. Assume that the water, CO_2, and N_2 are all perfect gases and that the pressure is 101 kPa or 14.696 psia.

14–16 (C) Ethylene is burned in 125% theoretical air at 14.69 psia or 101 kPa. Determine the partial pressure of the water vapor in the products of combustion and the dew point of the products. Assume that the water, CO_2, and N_2 are perfect gases.

14–17 (C) If the products of combustion in problem 14–15 are cooled to 25°C (77°F), determine the amount of water that has condensed and determine the actual heat of combustion.

14–18 (C) If the products of combustion in problem 14–16 are cooled to 77°F (25°C), determine the amount of water that condenses out and the actual heat of combustion.

14–19 (C) Determine how the heats of combustion (HHV, LHV, and actual) are affected by the amount of theoretical air, varying the percent theoretical air from 50% to 200%. For this analysis, you may wish to use the computer program COMBUST or a similar program with a microcomputer. Determine the effect for the following fuels:
(a) Methane
(b) Benzene
(c) Acetylene

Section 14–6

14–20 (M) Determine the adiabatic combustion temperature for *n*-octane with 100% theoretical air. Assume that the fuel and air are at 298 K and 101 kPa and that the combustion occurs at constant pressure.

14–21 (E) Determine the adiabatic combustion temperature for ethyl alcohol with 100% theoretical air. Assume that the fuel and air are at 77°F and 14.696 psia.

14–22 (M) Determine the adiabatic combustion temperature for cellulose at 101 kPa. Assume that the cellulose and air are at 298 K and 101 kPa and that there is 100% theoretical air.

14–23 (E) Determine the adiabatic combustion temperature for methane and 100% theoretical air at 14.696 psia and 77°F.

14–24 (C) Determine how the adiabatic combustion temperature is affected by the amount of air, from 50% theoretical air to 200%. Use of the computer with a program such as COMBUST is recommended for this problem. Consider this effect with the following hydrocarbon fuels:
(a) Propane
(b) *n*-Octane
(c) Carbon (coal)

Section 14–7

14–25 (E) Determine the rate of entropy generation in the burning of 2000 lbm/h of carbon in 125% theoretical air where the carbon, air, and products of combustion are all at 77°F and 14.696 psia.

14–26 (M) Determine the entropy generation per kg · mol of propane burning in 150% theoretical air. Assume that the air, fuel, and products of combustion are at 298 K and 101 kPa. Also assume that the water is all condensed out of the products. (*Hint:* If the water condenses out, its mole fraction is 1 but is not part of the gaseous products of combustion.)

14–27 (M) Determine the rate of entropy generation for the burning of 30 kg/min of *n*-octane with 95% theoretical air and for adiabatic combustion at 101 kPa. The air and *n*-octane are at 298 K.

14–28 (E) Determine the entropy generation per lbm · mol of cellulose in 90% theoretical air and for adiabatic combustion at 14.696 psia. Assume that the cellulose and air are at 77°F and $\bar{s}_i^\circ$ = 20 Btu/lbm · mol · °R for cellulose.

HEAT TRANSFER

In this chapter, the three modes of heat transfer—conduction, convection, and radiation—are introduced and discussed. Conduction is presented through Fourier's law of heat conduction, with a discussion of thermal conductivity. Conductance and thermal resistance, including *R*-values, are then presented with applications to one-dimensional heat conduction through walls and radial heat transfer through pipes and tubes.

Convection heat transfer is presented with Newton's law of heating/cooling, and the general types of problems that we can solve by convection are outlined.

Some combined conduction and convection heat transfer problems are presented: the heat flow through a wall or pipe with an accounting for the convection heat transfer on both sides of the conducting region, the heat flow through fins, and the heat flow from a small object where conduction is much greater than the surrounding convection heat transfer, called lumped heat capacity.

Forced convection heat transfer is discussed briefly for flow past flat plates, around objects, and through pipes and tubes, with emphasis on the evaluation of the convection coefficient.

New Terms

Bi	Biot number	t_c	Time constant
Gr	Grashof number	α (alpha)	Absorptivity
K_h	Convective heat transfer coefficient	β_p	Volumetric expansion
P	Perimeter	ϵ (epsilon)	Emissivity
Pr	Prandtl number	μ_k (mu)	Viscosity
$\dot{q}_A$	Heat transfer per unit area	κ (kappa)	Thermal conductivity
$\dot{q}_L$	Heat transfer per unit length	ρ_r (rho)	Reflectivity
Re_L	Reynolds number based on length	σ (sigma)	Stefan–Boltzmann constant
Re_D	Reynolds number based on diameter	τ (tau)	Transmissivity
R_T	Thermal resistance	$\mathcal{J}$	Radiosity
R_V	Thermal resistance for unit area, *R*-value	$\mathcal{U}$	Overall heat transfer coefficient

Natural convection heat transfer is presented through the chimney effect, and some important applications are given: convection heating or cooling of a pool of liquid, and heating of a fluid or gas along a vertical wall.

Radiation heat transfer is also discussed briefly. Blackbody and graybody radiation is discussed, and Kirchhoff's law of radiation is derived. Absorptivity, emissivity, reflectivity, and transmissivity are all defined. Solar energy as a form of radiation is discussed, and the shape factors for various surface configurations are given. Radiation between two and three bodies is analyzed, and computer aids are demonstrated.

Heat exchangers are discussed and analyzed with the LMTD method.

15–1 CONDUCTION HEAT TRANSFER

Heat flows across the boundary of a system, or through systems, because of a temperature difference and is always from hot to colder regions. Figure 15–1 shows a schematic of how you can visualize heat flow from a hot surface A through a solid material to a colder surface. Some solids seem to be able to conduct heat faster than others, and a measure of this ability to conduct heat is called the **thermal conductivity**, κ. For the heat flow $\dot{Q}$ of figure 15–1, we can use an equation called **Fourier's law of conduction**:

$$\dot{Q} = -\frac{\kappa A \, \Delta T}{\Delta x} \tag{15–1}$$

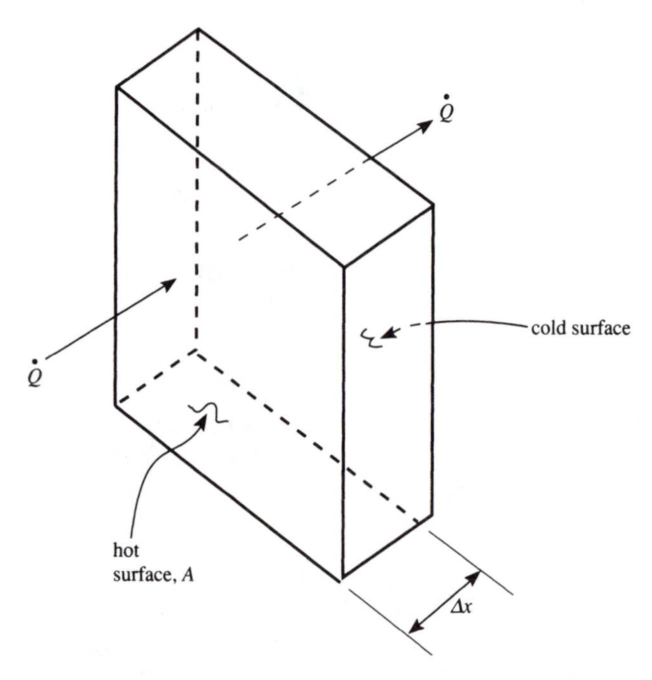

FIGURE 15–1 Conduction heat transfer through a solid

Where, the minus sign indicates that heat flow is in the direction of lower temperatures. Thus, if we say that heat flow from hot to cold is positive, the ΔT term in equation (15–1) will be negative, and the minus sign in front of the right side of the equation gives a positive $\dot{Q}$. Notice that $\dot{Q}$ is a rate of heat transfer with kW, kJ/s, Btu/s, or similar units. Also, notice that A is the area crossed by the heat flow and Δx is the thickness through which the heat flows. It is often more convenient to determine the heat flow per unit area, $\dot{q}_A = \dot{Q}/A$, given by

$$\dot{q}_A = \frac{-\kappa \, \Delta T}{\Delta x} \tag{15–2}$$

CALCULUS FOR CLARITY 15–1

Fourier's law of conduction is rigorously defined by the equation

$$\dot{Q}_x = -\kappa_x A \frac{\partial T}{\partial x} \tag{15–3}$$

where $\dot{Q}_x$ is the conduction heat transfer in the x-direction, κ_x is the thermal conductivity in the x-direction of the conducting material, and $\partial T/\partial x$ is the rate of change in temperature, or the **temperature gradient**, in the x-direction. From equation (15–3), it can be seen that the conduction heat transfer through a material may

CALCULUS FOR CLARITY 15–1, continued

also occur in the y- and z-directions, denoted $\dot{Q}_y$ and $\dot{Q}_z$, respectively. Therefore, conduction heat transfer is a vector having three components, $\dot{Q}_x$, $\dot{Q}_y$, and $\dot{Q}_z$, in the three dimensions. If the conduction heat transfer occurs in one direction only, say, the x-direction, then Fourier's law of conduction is

$$\dot{Q} = -\kappa A \frac{dT}{dx} \tag{15–4}$$

where κ is the thermal conductivity in the x-direction. The heat transfer indicated in figure 15–1 is most accurately one-dimensional heat transfer.

The thermal conductivity of a conducting material may be different in the three directions, x, y, and z, so it must be distinguished from the x-direction thermal conductivity as κ_y or κ_z for the y- and z-directions. If thermal conductivity is the same in all directions for the material, that material is said to be **isotropic**. Studies and analyses involving conduction heat transfer nearly always assume isotropic material so that thermal conductivity is identified as one value, κ, as given in equation (15–4).

For the heat transfer of the solid shown in figure 15–1, if the heat transfer into the hot surface is the same as that from the cold surface, and if the temperature does not change with time in the solid, then the transfer occurs in steady state, and equation (15–4) may be written

$$\dot{Q}\,dx = -\kappa A\,dT$$

If the thermal conductivity is constant, then, after integrating both sides of this equation, we have

$$\dot{Q}\,\Delta x = -\kappa A\,\Delta T$$

which can be arranged to become equation (15–1).

From equation (15–1) or (15–2) we can see that thermal conductivity has units of energy–length per unit time–area–temperature difference; that is,

$$\kappa = \frac{-\dot{Q}\,\Delta x}{A\,\Delta T} \tag{15–5}$$

Some commonly used units for thermal conductivity are W/m·K (or W/m·°C) and Btu·in/h·ft^2·°R (or Btu·in/h·ft^2·°F). Because thermal conductivity and heat transfer involve temperature difference (instead of temperature values), the units of thermal conductivity can interchange Kelvin (K) and Celsius (°C) units or Rankine (°R) and Fahrenheit (°F). Table B–8 lists thermal conductivity values for selected materials in SI units of W/m·K. Conversion of these values to English units or other SI units, such as W/cm·K, can be made with conversions listed in the table on the front inside cover. It is informative to study the range of values of thermal conductivity. Those materials that conduct heat well, such as copper and aluminum, are called heat **conductors** and have comparatively large values of thermal conductivity. Substances or materials that do not conduct heat well at all are called **insulators** or insulation and have small values for thermal conductivity. Mineral wool, for instance, is used as a building insulation and has a κ-value of 0.039 W/m·K (an approximate value given in table B–8), and copper has κ-value of 400 W/m·K. As you can see, then, there is a difference of approximately four orders of magnitude in the values between insulators and conductors.

It has been observed that denser materials tend to have greater thermal conductivity values than do less dense materials. There are exceptions to this rule, however; many dense materials, such as stone and concrete, have κ-values in between those of conductors and insulators. Thermal conductivity can be visualized as a transfer of energy between molecules or atoms, much as electricity is conducted in an electric wire. The **Wiedemann–Franz law** states that the thermal conductivity of a material is related to the electrical conductivity of that material at the same temperature, the relationship being that the ratio of the electrical conductivity to the thermal conductivity is proportional to the absolute value of the temperature. There are, however, exceptions to the Wiedemann–Franz law, but the analogy of electrical to thermal conduction is a useful one for analysis. This analogy is best made by defining the thermal resistance of a material as

$$R_T = \frac{\Delta T}{\dot{Q}} \tag{15–6}$$

By comparing this equation to (15–1), we can see that the thermal resistance is related to the thermal conductivity by

$$R_T = \frac{\Delta x}{\kappa A} \tag{15–7}$$

where the direction of heat flow is expected to go from hot to cold and the temperature difference in equation (15–6) is presumed to be an absolute value. By writing equation (15–6) as $\dot{Q}R_T = \Delta T$, we can see an analogy to the important electrical equation of Ohm's law (see chapter 17), which says that $\mathscr{E} = \mathscr{I}\mathscr{R}$ Here, $\mathscr{I}$ is the electrical current and is comparable to heat flow, Q; $\mathscr{R}$ is electrical resistance and is comparable to thermal resistance; and $\mathscr{E}$ is electrical potential or voltage and is comparable to a temperature difference. As we will see, using thermal resistance often gives a more convenient form of analyzing heat transfer by conduction, convection, and radiation.

A term used in building construction and in heating and air-conditioning applications is the ***R*-value**, defined

$$1\ R\text{-value} = 1\ \text{h} \cdot \text{ft}^2 \cdot {}^\circ\text{R/Btu} = 1\ \text{h} \cdot \text{ft}^2 \cdot {}^\circ\text{F/Btu}$$

Using conversion factors listed in the table on the inside front cover, we can see that an R-value of 1 is equal to 0.1761 $\text{m}^2 \cdot {}^\circ\text{C/W}$ and an R-value of 10 is 1.761 $\text{m}^2 \cdot {}^\circ\text{C/W}$. The R-value can be used in equation (15–2), and then the heat flow, $\dot{q}_A$, can be found from

$$\dot{q}_A = \frac{\Delta T}{R_v} \tag{15–8}$$

where the temperature difference is an absolute value.

EXAMPLE 15–1

A brass plate 3 in thick by 3 ft high and 2 ft wide conducts heat. One side of the plate is at 800°F and the other is at 200°F. Determine the total rate of heat transfer through the plate, the thermal resistance of the brass plate, and its R-value.

Solution

The heat flows across an area of $3 \times 2 = 6\ \text{ft}^2$. The thermal conductivity of brass is given as 114 W/m · K in table B–8, and in the table on the inside front cover, we find the conversion factor of 6.9348 Btu · in/h · ft² · °F = 1 W/m · K. The thermal conductivity of brass, in English units, is then

$$\begin{aligned}\kappa_{\text{brass}} &= 114\ \text{W/m} \cdot \text{K} \times 6.9348\ \text{Btu} \cdot \text{in} \cdot \text{m} \cdot \text{K/h} \cdot \text{ft}^2 \cdot {}^\circ\text{F} \cdot \text{W} \\ &= 790.57\ \text{Btu} \cdot \text{in/h} \cdot \text{ft}^2 \cdot {}^\circ\text{F}\end{aligned}$$

The rate of heat transfer is then found from equation (15–1):

$$\dot{Q} = -(790.57 \text{ Btu} \cdot \text{in/h} \cdot \text{ft}^2 \cdot {}^\circ\text{F})(6 \text{ ft}^2)\left(\frac{200 - 800^\circ\text{F}}{3 \text{ in}}\right)$$

Answer

$$= 948{,}684 \text{ Btu/h} = 263.5 \text{ Btu/s}$$

The thermal resistance of the brass plate is given by equation (15–7):

$$R_T = \frac{\Delta}{\kappa A} = \frac{3 \text{ in}}{(790.57 \text{ Btu} \cdot \text{in/h} \cdot \text{ft}^2 \cdot {}^\circ\text{F})(6 \text{ ft}^2)}$$

Answer

$$= 0.000632 \text{ h} \cdot {}^\circ\text{F/Btu}$$

The R-value is then

$$R_v = \frac{\Delta x}{\kappa} = R_T A = (0.000632 \text{ h} \cdot {}^\circ\text{F/Btu})(6 \text{ ft}^2)$$

Answer

$$= 0.00379 \text{ h} \cdot \text{ft}^2 \cdot {}^\circ\text{F/Btu} = R\text{-value of } 0.00379$$

EXAMPLE 15–2 Compare the R-values of single-strength window glass (3 mm thick or $\frac{1}{8}$ in thick), pine boards 20 mm ($\frac{3}{4}$ in thick), concrete 300 mm (12 in thick), and Styrofoam 25 mm (1 in thick).

Solution The R-values are determined from equation (15–8) after obtaining thermal conductivity from table B–8. In table 15–1, column 1, are listed the κ-values from table B–8. In column 2 are listed the English unit values for thermal conductivity, and in column 4 are given the results found by using equation (15–8).

TABLE 15–1 Comparison of R-values for four materials given in example 15–2

Material	κ W/m · K	κ Btu · in/h · ft^2 · °F	Δx in	R-value h · ft^2 · °F/Btu
Window glass	1.40	9.70875	0.125	0.0129
Pine board	0.15	1.04022	0.750	0.72
Concrete	1.40	9.70875	12.0	1.236
Styrofoam	0.029	0.2011	1.0	4.972

Notice in example 15–2 that the R-value depends not only on the thermal conductivity values but also on the thickness.

In many situations where heat transfer occurs, there are two or more different materials through which the heat must pass or be conducted. In these cases, the heat transfer is given by the equation

$$\dot{Q} = \frac{\Delta T_{\text{overall}}}{\sum R_T} \tag{15–9}$$

where $\Delta T_{\text{overall}}$ is the temperature difference over all the materials involved and $\sum R_T$ is the sum of the thermal resistances of all the materials. The following example will serve to demonstrate the use of equation (15–9).

EXAMPLE 15–3 A thermopane, shown in figure 15–2, is constructed of two single-strength sheets of window glass, each 3 mm thick, separated by a 6-mm air gap. The air is sealed in between the glass so that it cannot move but acts as a heat-conducting material. Determine the thermal resistance of a thermopane window of area 1.5 m × 1 m. If the temperature difference across the window is 20°C, determine the rate of heat transfer through the window. Also, compare the R-value of thermopanes to that of single-strength glass 3 mm thick.

FIGURE 15–2 Cross section of thermopane window

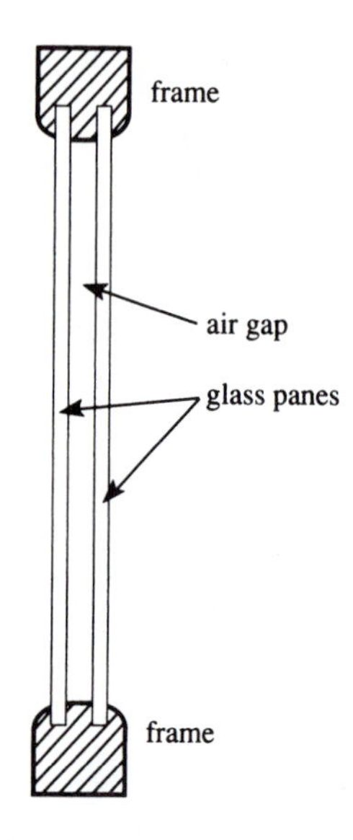

Solution

In figure 15–2, we can see that the heat flow must cross both panes of glass and the air gap. The total thermal resistance, $\sum R_T$, is the sum of the thermal resistances of the two glass panes and the air gap:

$$\sum R_T = 2 \times R_{T_{\text{glass}}} + R_{T_{\text{air}}}$$

Using equation (15–7), we find that

$$\sum R_T = 2 \times \frac{0.003 \text{ m}}{(1.4 \text{ W/m} \cdot \text{K})(1.5 \text{ m}^2)} + \frac{0.006 \text{ m}}{(0.026 \text{ W/m} \cdot \text{K})(1.5 \text{ m}^2)}$$

Answer

$$= 2 \times 0.00143 \text{ K/W} + 0.15385 \text{ K/W} = 0.15671 \text{ K/W}$$

We find the heat transfer by using equation (15–9):

$$\dot{Q} = \frac{20°\text{C}}{0.15671 \text{ K/W}} = 127.6 \text{ W}$$

The R-value of the thermopane is the total thermal resistance, $\sum R_T$, multiplied by the area, 1.5 m^2. Then

Answer

$$R_v = \left(\sum R_T\right) \times A = 0.235065 \text{ m}^2 \cdot \text{K/W}$$

Converting this answer to English units gives us

$$R_v = \frac{0.235065 \text{ m}^2 \cdot \text{K/W}}{0.176 \text{ Btu} \cdot \text{m}^2 \cdot \text{K/h} \cdot \text{ft}^2 \cdot °\text{F} \cdot \text{W}}$$

Answer

$$= 1.336 \quad \text{or an } R = \text{value of about 1.3}$$

The R-value of a single-strength glass was found in example 15–2 to be about 0.0129, so a comparison of this glass with a thermopane shows that thermopanes may be expected to have about 100 times the thermal resistance of single-strength glass. For thermopanes to operate properly, it is crucial that the air in the gap be completely sealed. If the seal is broken, heat will flow out by air flow through convection. If this occurs, the thermopane loses its great advantage over regular window glass.

Heat transfer through a window, a wall, or any other solid is affected not only by conduction but also by **convection heat transfer**, which we will consider later. The thermal resistance to convection heat transfer can be significant and is sometimes referred to as the *film resistance*. This film resistance acting on the outer surfaces of a window, when added to the thermal resistance of the window pane (or panes, for thermopanes) and the potential air gap, results in a ratio of thermal resistances that is less than 100 but still often greater than 30. The method of including film resistance is demonstrated in example 15–8, and the comparison of thermal resistances with and without film resistance for a window is sought in practice problems 15–19 and 15–20. In practice, the thermopanes, triple panes (three panes of glass with two air gaps), and quad panes (four panes of glass and three air gaps) provide significantly higher thermal resistances than does a single-pane window but not as great as that indicated in example 15–3.

Conduction does not always occur over a material whose area is constant or uniform. A more general definition of Fourier's law of conduction is

$$\dot{Q} = -\frac{\kappa A\, \delta T}{\delta x} \tag{15–10}$$

where T is a temperature difference over a very small distance δx through which heat flows. For a cylinder, a tube, or a pipe, where heat flows radially outward or inward, as diagrammed in figure 15–3, it can be shown that, from equation (15–10), the heat conduction over a length L is

$$\dot{Q} = \frac{2\pi\kappa L(T_i - T_o)}{\ln(r_o/r_i)} = \frac{2\pi\kappa L(T_i - T_o)}{\ln(D_o/D_i)} \tag{15–11}$$

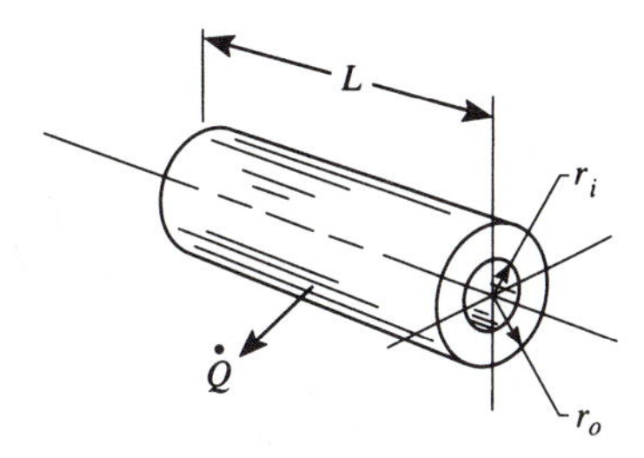

FIGURE 15–3 Heat flow in a radial direction through a tube or pipe

where T_i is the temperature of the inside radius, r_i; T_o is the outside temperature at r_o; and D_i and D_o are the inside and outside diameters of the tube or pipe. In equation (15–11), the heat is assumed to be positive if it flows outward and negative if it flows inward. The thermal resistance of the system shown in figure 15–3 is

$$R_T = \frac{\ln(r_o/r_i)}{2\pi\kappa L} \tag{15–12}$$

and for the thermal resistance per foot or meter of length, the thermal resistance would be $\ln(r_o/r_i)/2\pi\kappa$. The heat transfer through a pipe or tube per foot or meter is

$$\dot{q}_L = \frac{\dot{Q}}{L} = \frac{2\pi\kappa(T_i - T_o)}{\ln(r_o/r_i)} \tag{15–13}$$

CALCULUS FOR CLARITY 15–2

Equation (15–11) is derived from the exact definition of Fourier's law of conduction, equation (15–4), where the heat transfer occurs radially instead of in the *x*-direction. Thus, referring to the schematic of radial heat transfer in figure 15–4, we have

$$\dot{Q} = -\kappa A \frac{dT}{dr} \tag{15–14}$$

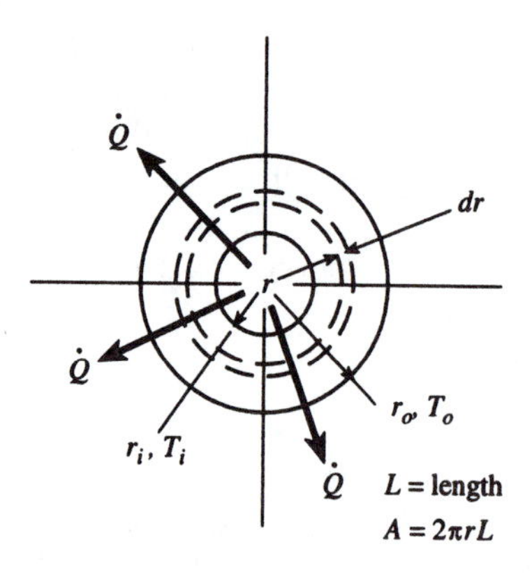

FIGURE 15–4 Radial conduction heat transfer for T_i greater than T_o

The surface area is $A = 2\pi\kappa rL$, so that we have

$$\dot{Q} = -2\pi\kappa rL \frac{dT}{dr}$$

Assuming steady-state conditions, we have Q as a constant, or the same at all radii, so we may write

$$\frac{\dot{Q}}{r} dr = -[2\kappa\pi L]\, dT$$

Integrating both sides of this equation from r_i, where $T = T_i$, to r_o, where $T = T_o$, gives

$$\dot{Q} \ln \frac{r_o}{r_i} = -2\kappa\pi L(T_o - T_i) = 2\kappa\pi L(T_i - T_o)$$

This equation is the same as (15–11) after transposing $\ln(r_o/r_i)$ to the right side.

EXAMPLE 15–4 A 2-in–outside diameter (OD) by 1.5-in–inside diameter (ID) wrought iron water line is buried in the ground. If water flow through the pipe is such that the inside temperature of the pipe is 62°F and the ground is 50°F surrounding the outside of the pipe, determine the heat flow through the pipe per unit of length and the direction of the heat transfers.

Solution The heat flow is from the inside to the outside, so that the water is losing heat. The heat loss per unit length of pipe is found from equation (15–13):

$$\dot{q}_L = \frac{2\pi\kappa(T_i - T_o)}{\ln(D_o/D_i)}$$

$$= \frac{2\pi(51\ \text{W/m}\cdot\text{K})(0.5779\ \text{Btu}\cdot\text{m}\cdot\text{K/W}\cdot\text{h}\cdot\text{ft}\cdot{}^\circ\text{F})(62^\circ\text{F} - 50^\circ\text{F})}{\ln(2\ \text{in}/1.5\ \text{in})}$$

Answer $= 7724.5\ \text{Btu/h}\cdot\text{ft}$

Heat transfer through pipes and tubes often involves concentric materials. For instance, insulated steam lines have an insulating material wrapped around the pipe to retard heat losses. Then the heat must cross through more material that has a large thermal resistance. We may use equation (15–9) with the resistances of the individual materials found from equation (15–12). We now demonstrate how problems of concentric cylinders may be analyzed.

EXAMPLE 15–5 A 10-cm-OD by 6-cm-ID steam line delivers superheated steam at 1000°C. The line is made of steel and is wrapped with 10 cm of asbestos insulation and 1 cm of plaster covering the asbestos. The cross section of the line is shown in figure 15–5. If the inside steam line temperature is 980°C and the heat losses are expected to be 770 W/m, determine the outside surface temperature of the plaster.

FIGURE 15–5 Steam line

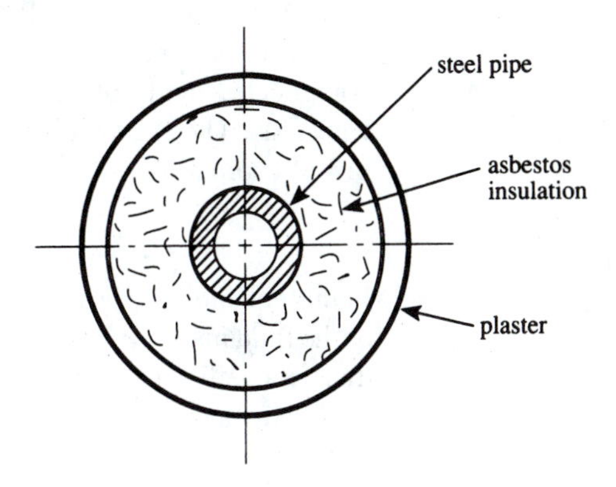

Solution The heat loss per unit length of pipe is given by the equation

$$\dot{q}_L = \frac{\Delta T_{\text{overall}}}{\sum R_{T_L}}$$

where R_{T_L} is the resistance per unit length. Using equation (15–12), we find that

$$\sum R_{T_L} = \frac{\ln(5\text{ cm}/3\text{ cm})}{2\pi\kappa_{\text{steel}}} + \frac{\ln(15\text{ cm}/5\text{ cm})}{2\pi\kappa_{\text{asbestos}}} + \frac{\ln(16\text{ cm}/15\text{ cm})}{2\pi\kappa_{\text{plaster}}}$$

From table B–20, the thermal conductivity values for the three materials—steel, asbestos, and plaster—are, respectively, $\kappa_{\text{steel}} = 43\text{ W/m}\cdot\text{K}$, $\kappa_{\text{asbestos}} = 0.156\text{ W/m}\cdot\text{K}$, and $\kappa_{\text{plaster}} = 0.107\text{ W/m}\cdot\text{K}$. Substituting these into the preceding equation, we obtain

$$\sum R_{T_L} = 1.2187\text{ m}\cdot\text{K/W}$$

and the heat loss is said to be 120 W/m. The overall temperature difference is then

$$\Delta T_{\text{overall}} = (\dot{q}_L)\left(\sum R_{T_L}\right)$$
$$= (770\text{ W/m})(1.2187\text{ m}\cdot\text{K/W}) = 938.4\text{ K}$$

The outside temperature is then

$$T_o = T_i - \Delta T_{\text{overall}} = 980°\text{C} - 938.4°\text{C}$$

Answer

$$= 41.6°\text{C}$$

15–2 CONVECTION HEAT TRANSFER

In section 15–1, we discussed one of the modes of heat transfer, conduction, which involves the transfer of heat through molecular motion. A second mode of heat transfer is convection, which depends on mass motion of a fluid or gas to convey the heat. Convection heat transfer usually occurs between a solid surface and a fluid or gas flowing past that surface. Figure 15–6 shows three examples of how heat may be transferred by convection. In figure 15–6a, a gas or liquid flows across a surface of a plate. If the plate is hot, heat flows into the fluid from the plate; otherwise, heat will flow from a hot fluid into a cold plate. In figure 15–6b, convection is indicated between a cylindrical surface and a fluid that surrounds the surface, and in figure 15–6c, convection occurs between a vertical plate and air or another gas. Here (in figure 15–6c) the flow of air is upward because as the air becomes hotter, it rises, and cold air replaces the air at the bottom. This is the principle of natural or free convection, which we discuss further in section 15–5. Convection heat transfer involves the flow of a fluid, the conduction of heat into or out of the fluid, and the flow of heat across a solid boundary. As a result of these actions, convection is a complicated mode of heat transfer that has been difficult to describe in mathematical equations. The accepted equation used to describe convection heat transfer is **Newton's law of cooling/heating**:

$$\dot{Q} = K_h A(T_w - T_\infty) \tag{15–15}$$

In this equation, T_w is a wall or surface temperature, T_∞ is the average or bulk temperature of the fluid involved in the heat transfer, A is the surface area, and K_h is the convection heat transfer coefficient. In many books on heat transfer, K_h is denoted as h, but we will write it as K_h, to avoid confusion with enthalpy. The units of K_h are usually $\text{W/m}^2 \cdot {}^\circ\text{C}$ or $\text{Btu/h} \cdot \text{ft}^2 \cdot {}^\circ\text{F}$.

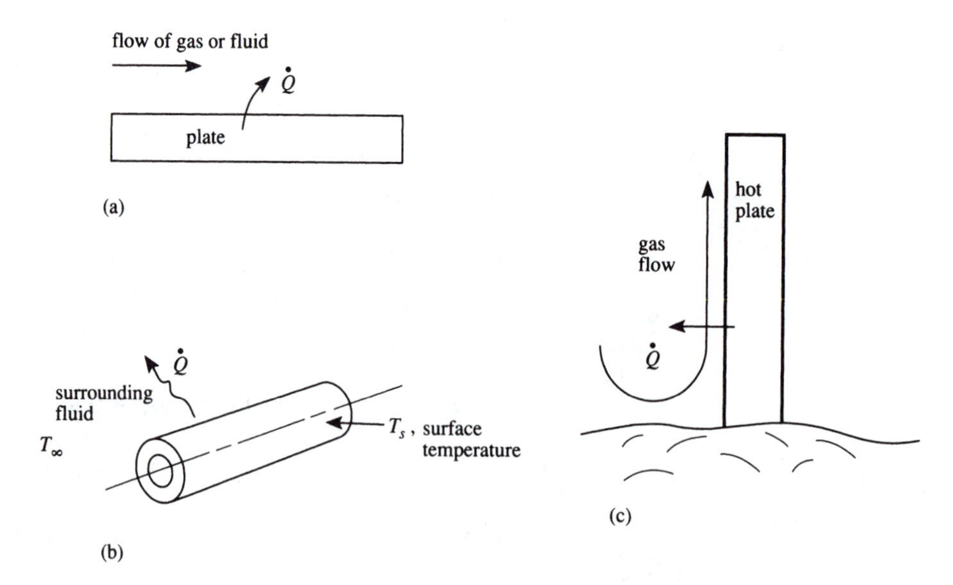

FIGURE 15–6 Examples of convection heat transfer: (a) convection heat transfer between flat plate and flowing fluid or gas; (b) convection heat transfer between a cylinder and a surrounding fluid; (c) natural convection heat transfer

Equation (15–15) is written as if the wall temperature, T_w, is greater than the fluid temperature. If the opposite situation occurs, heat will be transferring from the fluid, and the temperature difference in equation (15–15) would then be written $(T_\infty - T_w)$. In table 15–2 are given some typical values of K_h, depending on the type of heat transfer involved. In sections 15–4 and 15–5, we will see some ways of predicting K_h more precisely.

TABLE 15–2 Approximate values for the convective heat transfer coefficient, K_h

Type of Convection	K_h W/m²·°C	K_h Btu/h·ft²·°F
Air at 6 ft/s over a flat plate	12	2.1
Air at 95 ft/s over a flat plate	75	13.2
Air flow past a 2-in-diameter cylinder at 160 ft/s	180	32
Water flow at 1 ft/s in a 1-in-diameter tube	3500	616
Natural convection		
a. Vertical plate 1 ft high into air	4.5	0.79
b. Horizontal cylinder 2 inches in diameter into air	6.5	1.14

EXAMPLE 15–6 Determine the heat transfer from a 2-m-high by 4-m-long vertical wall at 60°C into still air at 20°C.

Solution An approximate solution can be obtained by using equation (15–15) and a value for K_h from table 15–2. Using $K_h = 4.5\ \text{W/m}^2\cdot°\text{C}$ and $A = 2\ \text{m} \times 4\ \text{m} = 8\ \text{m}^2$, we find that

$$\dot{Q} = K_h A(T_w - T_\infty) = (4.5\ \text{W/m}^2\cdot°\text{C})(8\ \text{m}^2)(60 - 20°\text{C})$$

Answer

$$= 1440\ \text{W}$$

In section 15–5, we consider in detail how a more precise value for K_h can be found and therefore a more reliable answer obtained for the heat transfer.

EXAMPLE 15–7 A mercury-in-glass thermometer at 70°F is placed in water at 180°F. Determine the rate of heat transfer to the thermometer per unit area if K_h is 157 Btu/h·ft²·°F.

Solution Using Newton's law of cooling, we can find the heat transfer per unit area. We have

$$\dot{q}_A = \frac{\dot{Q}}{A} = K_h(T_\infty - T_w) = (157\ \text{Btu/h}\cdot\text{ft}^2\cdot°\text{F})(180 - 70°\text{F})$$

Answer

$$= 17{,}270\ \text{Btu/h}\cdot\text{ft}^2$$

15–3 COMBINED CONDUCTION–CONVECTION APPLICATIONS

Nearly all heat transfer occurs with more than just one mode of heat transfer involved. Many important engineering and technological problems involve conduction and convection at the same time. Here we discuss some of those situations.

Heat transfer across a building wall involves convection heat transfer between inside air and the inside wall surface, conduction through the wall, and then convection at the outside surface and outside air. In this situation, it is convenient to use thermal resistances for the conduction and the convection. Thermal resistance for convection is defined by the relationship

$$R_T = \frac{1}{K_h A} \tag{15–16}$$

Then the heat transfer can be determined from equation (15–9), $\dot{Q} = \Delta T_{\text{overall}}/\sum R_T$, where the $\sum R_T$ now includes the thermal resistance of the convection as well as thermal resistance of the conduction. The following two examples illustrate this idea.

EXAMPLE 15–8 An outside building wall has the cross section shown in figure 15–7. On the inside, the convective heat transfer coefficient is 2.0 Btu/h·ft²·°F, and on the outside it is 3.1 Btu/h·ft²·°F. The inside air temperature is 65°F, and the outside air temperature is 20°F. Determine the heat loss through the wall per unit area and also determine the inside and outside wall temperatures and the temperature drop across the Styrofoam.

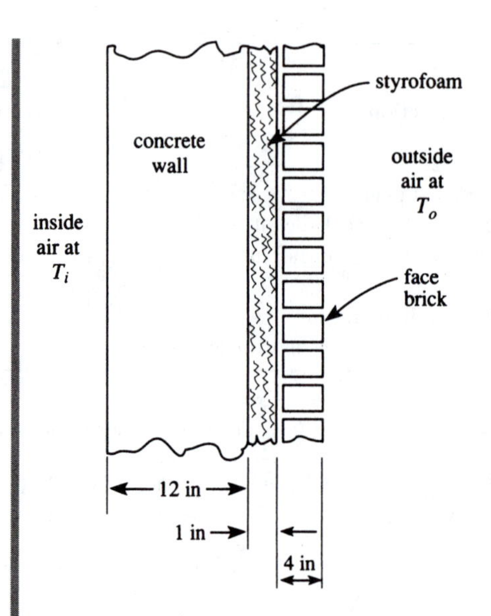

FIGURE 15–7 Outside building wall with conduction and convection heat transfer

Solution The heat transfer across the wall per unit area is

$$\dot{q}_A = \frac{\dot{Q}}{A} = \frac{\Delta T_{\text{overall}}}{\left(\sum R_T\right)A}$$

where $\Delta T_{\text{overall}} = 65°F - 20°F = 45°F$. The total thermal R-value is

$$\sum R_T A = \frac{1}{K_{h_i}} + \left(\frac{\Delta x}{\kappa}\right)_{\text{concrete}} + \left(\frac{\Delta x}{\kappa}\right)_{\text{Styrofoam}} + \left(\frac{\Delta x}{\kappa}\right)_{\text{brick}} + \frac{1}{K_{h_o}}$$

$$= \frac{1}{2.0 \text{ Btu/h} \cdot \text{ft}^2 \cdot °\text{F}} + \frac{12 \text{ in}}{1.4 \times 6.9348 \text{ Btu} \cdot \text{in/h} \cdot \text{ft}^2 \cdot °\text{F}}$$

$$+ \frac{1 \text{ in}}{0.029 \times 6.9348 \text{ Btu} \cdot \text{in/h} \cdot \text{ft}^2 \cdot °\text{F}}$$

$$+ \frac{4 \text{ in}}{0.70 \times 6.9348 \text{ Btu} \cdot \text{in/h} \cdot \text{ft}^2 \cdot °\text{F}} + \frac{1}{3.1 \text{ Btu/h} \cdot \text{ft}^2 \cdot °\text{F}}$$

$$= 0.5 + 1.236 + 4.972 + 0.824 + 0.323$$

$$= 7.855 \text{ h} \cdot \text{ft}^2 \cdot °\text{F/Btu}$$

and the heat transfer is

Answer

$$\dot{q}_A = \frac{45°\text{F}}{7.855 \text{ h} \cdot \text{ft}^2 \cdot °\text{F/Btu}} = 5.728 \text{ Btu/h} \cdot \text{ft}^2$$

The wall temperatures may be found by noticing that the heat transfer crosses all the materials. We can then write

$$\dot{q}_A = \frac{T_i - T_{\text{inside wall}}}{1/K_{h_i}} = 5.728 \text{ Btu/h} \cdot \text{ft}^2$$

Therefore,

$$T_{\text{inside wall}} = (T_i - 5.728 \text{ Btu/h} \cdot \text{ft}^2 \cdot °\text{F})\left(\frac{1}{K_{h_i}}\right)$$

$$= 65°\text{F} - (5.728)(0.5)$$

Answer

$$= 62.1°\text{F}$$

Also, at the outside,

$$\dot{q}_A = \frac{T_{\text{outside wall}} - T_o}{1/K_{h_o}} = 5.728 \text{ Btu/h} \cdot \text{ft}^2 \cdot {}^\circ\text{F}$$

We find that

Answer

$$T_{\text{outside wall}} = 21.9^\circ\text{F}$$

The temperature drop across the Styrofoam is found by using the equation

$$\dot{q}_A = \frac{(\Delta T)_{\text{Styrofoam}}}{(R_{T_{\text{Styrofoam}}})A}$$

Then

$$\begin{aligned}(\Delta T)_{\text{Styrofoam}} &= \dot{q}_A (R_{T_{\text{Styrofoam}}})A \\ &= (5.728 \text{ Btu/h} \cdot \text{ft}^2)\left(\frac{1 \text{ in}}{0.029} \times 6.9348 \text{ Btu} \cdot \text{in/h} \cdot \text{ft}^2 \cdot {}^\circ\text{F}\right) \\ &= 28.5^\circ\text{F}\end{aligned}$$

Answer

The total temperature drop across the wall, including the air, is 45°F, and the Styrofoam accounts for 28.5°F of that drop. Also, the temperature of the inside surface of the Styrofoam is found from the equation

$$\begin{aligned}T_{\text{inside Styrofoam}} &= T_{\text{inside wall}} - (\dot{q}_A)(R_T A)_{\text{concrete}} \\ &= 62.1^\circ\text{F} - (5.728 \text{ Btu/h} \cdot \text{ft}^2)(1.236 \text{ h} \cdot \text{ft}^2 \cdot {}^\circ\text{F/Btu}) \\ &= 55^\circ\text{F}\end{aligned}$$

and the outside of the Styrofoam is then 55°F = 28.5°F = 26.5°F.

Let us now consider conduction and convection in a tube.

EXAMPLE 15–9 A water-to-water tube heat exchanger uses 6-cm-OD stainless steel tubes with a 5-cm ID, as shown in figure 15–8. Cool water at 10°C flows through the tube with $K_{h_i} = 4000 \text{ W/m}^2 \cdot {}^\circ\text{C}$, and hot water at 90°C surrounds $K_{h_o} = 2000 \text{ W/m}^2 \cdot {}^\circ\text{C}$. Determine the total thermal resistance per unit length of tube and the heat transfer per unit length.

FIGURE 15–8 Tube heat exchanger

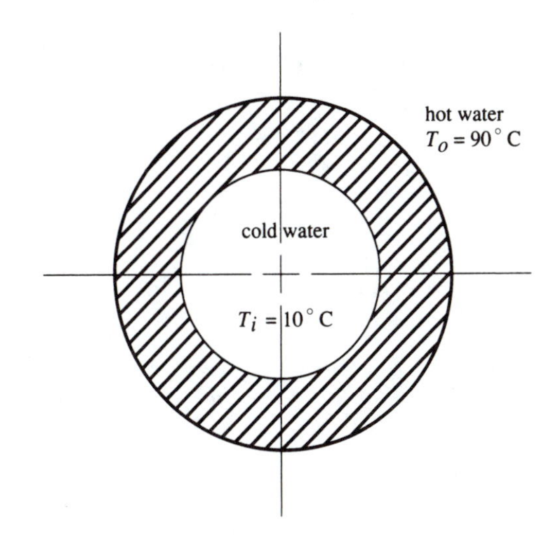

Solution

The total resistance of the tube heat exchanger per tube length is the sum of the thermal resistances of the inside convection resistance, the conduction through the tube, and the convection on the outside of the tube. We write

$$\sum R_{T_L} = \frac{1}{A_i K_{h_i}} - \frac{\ln(r_o/r_i)}{2\pi\kappa_{\text{stainless steel}}} + \frac{1}{A_o K_{h_o}}$$

where $A_i = 2\pi(r_i) = 2\pi(2.5\text{ cm}) = 15.7\text{ cm} = 0.157\text{ m}$
$A_o = 2\pi(r_o) = 2\pi(3.0\text{ cm}) = 18.85\text{ cm} = 0.1885\text{ m}$

and from table B–20,

$$\kappa_{\text{stainless steel}} = 14.0\text{ W/m}\cdot\text{K}$$

Then

$$\sum R_{T_L} = \frac{1}{(0.157\text{ m})(4000\text{ W/m}^2\cdot\text{K})} + \frac{\ln(3/2.5)}{2\pi \times 14.0\text{ W/m}\cdot\text{K}} + \frac{1}{(0.1885\text{ m})(2000\text{ W/m}^2\cdot\text{K})}$$
$$= 0.00632\text{ m}\cdot\text{K/W}$$

Answer

$$= 6.32\text{ m}\cdot\text{K/kW}$$

The heat transfer unit length of tube is

$$\dot{q}_L = \frac{\Delta T_{\text{overall}}}{\sum R_{T_L}} = \frac{(90 - 10°\text{C})}{6.32\text{ m}\cdot\text{K/W}}$$

Answer

$$= 12.66\text{ kW/m}$$

Engineers and technologists frequently need to provide for high rates of heat transfer for cooling, for drying, or for other purposes. Convection heat transfer at a surface is a limiting factor in many of these conditions, so the designer must try to increase heat transfer by one of three methods: larger temperature differences, large K_h values, or greater surface areas for convection. A successful method of increasing surface areas is through the use of fins. A fin is a thin section of material extending out from a base surface, as shown in figure 15–9.

FIGURE 15–9 Square fin

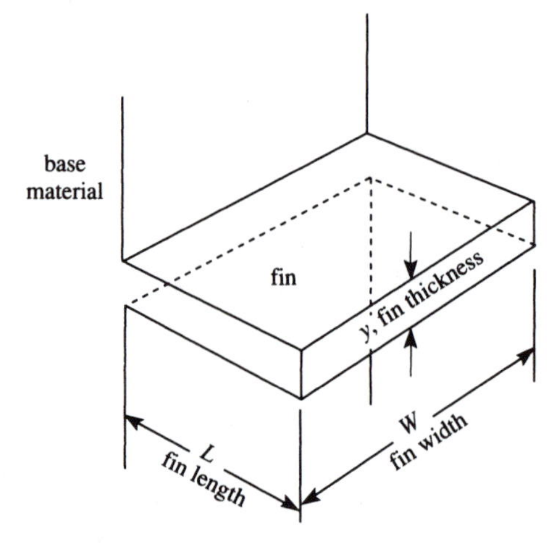

In this figure, you can see that the cross section is rectangular, and it is the same size over its extended length. This is called a square fin and is the simplest geometry involved with fins. You have probably seen fins on compressor cases, on air-cooled engine blocks, on radiators, or on other devices. Here we discuss square fins and indicate how you may analyze fins with more complicated shapes.

If you look at the cross section or cutaway of a square fin shown in figure 15–10, you can see that heat is conducted down the fin and is then convected out along the fin surface. The fin temperature is denoted as T_o at the base, and its temperature steadily decreases along its length. Some heat will be convected out the end of the fin unless the fin is so long that the temperature of the end of the fin is the same as the surrounding temperature, T_∞. If the fin is long enough that its end temperature is T_∞, it can be shown that the fin temperature at a point on the fin a distance x from the base is

$$T = T_\infty + (T_o - T_\infty)e^{-\sqrt{K_h P/\kappa A}x} \quad \textbf{(15–17)}$$

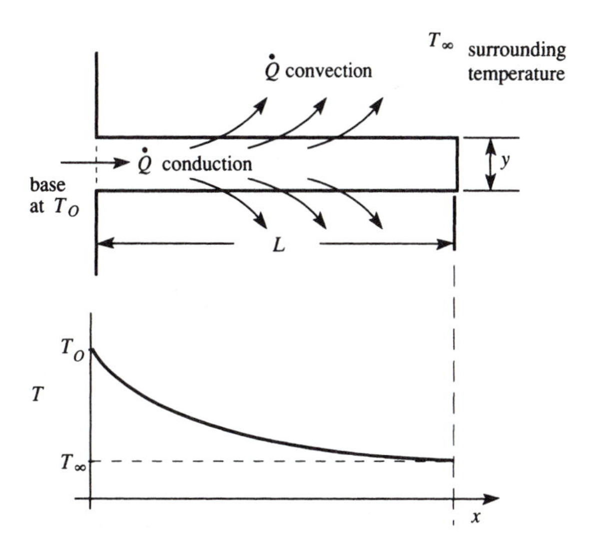

FIGURE 15–10 Heat transfer through a fin

where P is the fin perimeter, $2w + 2y$, and A is the area $w \times y$. In figure 15–10 is sketched the approximate shape of the curve for fin temperature along its length as predicted by equation (15–17). Using equation (15–17) and Fourier's law of heat conduction, equation (15–4), we can show that the heat flow through the fin is given by the equation

$$\dot{Q}_{\text{fin}} = \sqrt{K_h \kappa A P}\,(T_o - T_\infty) \quad \textbf{(15–18)}$$

CALCULUS FOR CLARITY 15–3

The temperature distribution and heat transfer of the square fin shown in figures 15–9 and 15–10 can be determined by considering an energy balance of a small element of the fin, as shown in figure 15–11. Assuming that the fin is at a higher temperature than its surroundings, there is conduction heat transfer into the element at a surface located at position x and conduction out at $x + \Delta x$. Convection heat transfer leaves the fin on the fin outer surface between the locations x and $x + \Delta x$, as shown in figure 15–11. The energy balance may be written

$$\dot{Q}_x = \dot{Q}_{x+\Delta x} + \dot{Q}_{\text{convection}}$$

CALCULUS FOR CLARITY 15–3, continued

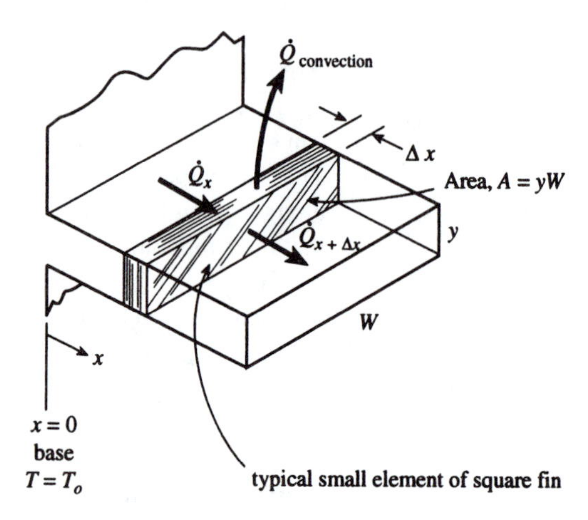

FIGURE 15–11 Energy balance of a small element of a square fin

Using Fourier's law of conduction and Newton's law of cooling, this equation becomes

$$-\kappa A\left[\frac{dT}{dx}\right]_x = -\kappa A\left[\frac{dT}{dx}\right]_{x+\Delta x} + K_h P\,\Delta x\,(T - T_\infty) \quad \textbf{(15–19)}$$

where A is the cross-sectional area of the fin, yW, as indicated in figures 15–9 and 15–11; P is the perimeter of the fin, $2y + 2W$; and the subscript notation x and $x + \Delta x$ on the two terms dT/dx indicates that the derivative of T with respect to x may change between x and $x + \Delta x$. This equation may be rearranged to read

$$\left(\frac{1A}{\Delta x}\right)\left(\left[\frac{dT}{dx}\right]_{x+\Delta x} - \left[\frac{dT}{dx}\right]_x\right) + \left\{\frac{K_h P}{\kappa A}\right\}(T - T_\infty) \quad \textbf{(15–20)}$$

If Δx is made very small, or if equation (15–20) is taken in the limit as Δx approaches zero, then equation (15–20) becomes

$$\frac{d^2T}{dx^2} = \left\{\frac{K_h P}{\kappa A}\right\}(T - T_\infty) \quad \textbf{(15–21)}$$

This is a second-order linear differential equation, provided that K_h and κ are constant. In order to obtain a solution to the temperature T from this equation, it is convenient to introduce the term T^* as

$$T^* = T - T_\infty \quad \textbf{(15–22)}$$

which gives

$$\frac{d^2T^*}{dx^2} = \frac{d^2T}{dx^2}$$

Equation (15–21) can be revised to read

$$\frac{d^2T^*}{dx^2} = \left\{\frac{K_h P}{\kappa A}\right\}T^* \quad \textbf{(15–23)}$$

CALCULUS FOR CLARITY 15–3, continued

and a solution of this equation is

$$T^* = C_1 e^{nx} + C_2 e^{-nx} \quad \textbf{(15–24)}$$

where C_1 and C_2 are constants to be determined and

$$n = \sqrt{(K_h P/\kappa A)}$$

We can verify this temperature by substituting equation (15–24) and its second derivative into equation (15–23). The solution to the temperature as given by equation (15–24) is realistic only if C_1 is equal to zero. This condition is true because the fin length x may become very large, making the temperature very large, and therefore $C_1 = 0$. The temperature at the base is T_o and $x = 0$ at the base. Substituting this into equation (15–24) gives

$$T_o - T_\infty = C_2 e^{-n(0)} = C_2$$

The temperature distribution becomes

$$T - T_\infty = (T_o - T_\infty)e^{-nx}$$

which is equation (15–17). We determine the heat transfer of the fin, $\dot{Q}_{\text{fin}}$, by noticing that it is the conduction heat transfer at the base:

$$\dot{Q}_{\text{fin}} = -\kappa A\left[\frac{dT}{dx}\right]_{x=0} = -[\kappa A](T_o - T_\infty)(-n)e^{-n(0)}$$
$$= -[\kappa A](T_o - T_\infty)\left\{-\sqrt{(K_h P/\kappa A)}\right\}$$

This equation is the same as (15–18), after the constants are combined.

Equation (15–18) is correct only for very long fins where the end temperature is equal to the surrounding temperature. For short fins or more complicated geometry of fins, more involved equations are necessary to predict the heat flow. The fin efficiency is an important parameter used in fin analysis, and it is defined as

$$\eta_{\text{fin}} = \frac{\dot{Q}_{\text{fin}}}{\dot{Q}_o} \quad \textbf{(15–25)}$$

where $\dot{Q}_o$ is the heat convected from the fin if it were at its base temperature, T_o, over all its surface. Then

$$\dot{Q}_o = K_h PL(T_o - T_\infty) \quad \textbf{(15–26)}$$

For a long fin where the heat transfer can be described by equation (15–18), the fin efficiency is

$$\eta_{\text{fin}} = \frac{1}{L}\sqrt{\frac{\kappa A}{K_h P}} \quad \textbf{(15–27)}$$

Equation (15–27) predicts the fin efficiency for very long fins, but it is instructive to plot the fin efficiency as the length of the fin is changed. Figure 15–12 shows the plot of this relationship, and it can be seen that the efficiency increases with shorter fins. In fact, according to equation (15–27), the efficiency increases to more than 100%, an impossible situation. As the fin length is decreased to a length where the fin efficiency approaches 100%, equation (15–27) is not applicable, and other, more involved, expressions need to be used to predict the efficiency.

FIGURE 15–12 Fin efficiency of a long fin for different fin lengths

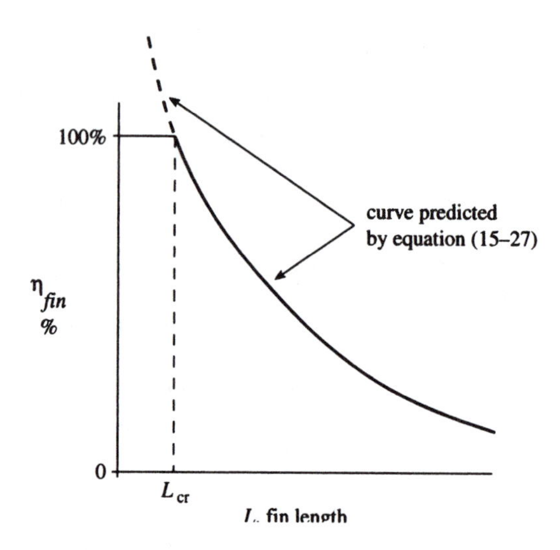

Using the relationship of equation (15–27), we can say the critical length, L_{cr}, is that length when the fin efficiency is 100%. From equation (15–27), this means that $L_{cr}\sqrt{K_h P/\kappa A}$, and for shorter fins, it is assumed that the efficiency is 100%, as sketched in figure 15–12. In this discussion, of course, we are always assuming that the fin length L or L_{cr} is still long enough that the temperature of the end of the fin is that of the surroundings, so the results are only approximations to actual fins. The way in which the fin efficiency depends on the fin length, however, seems to be a reasonable approximation of actual fin behavior. Figure 15–13 shows the relationship of fin efficiency of a square fin to the parameter $L_c^{3/2}(K_h/\kappa A_m)^{1/2}$. Notice that the fin width, w, is not a factor in this relationship, but the fin is assumed to be such that the width is much larger than the length, L. Also, L_c is defined as $L + y/2$ and is not the critical fin length previously discussed.

Fins circumferentially fitted around a cylinder as shown in figure 15–14 are often used. The efficiency of these types of fins is given by the graph in figure 15–15.

FIGURE 15–13 Efficiency of square fin

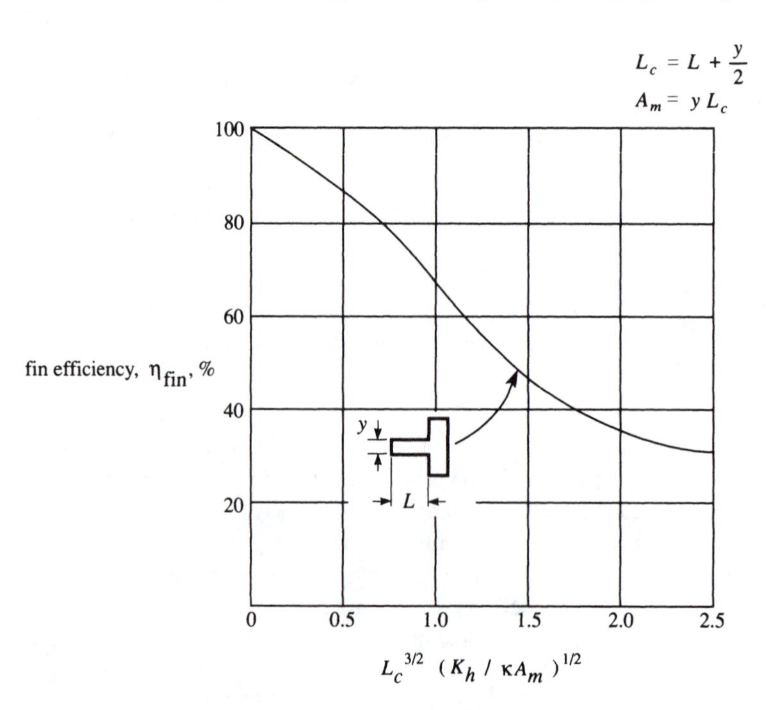

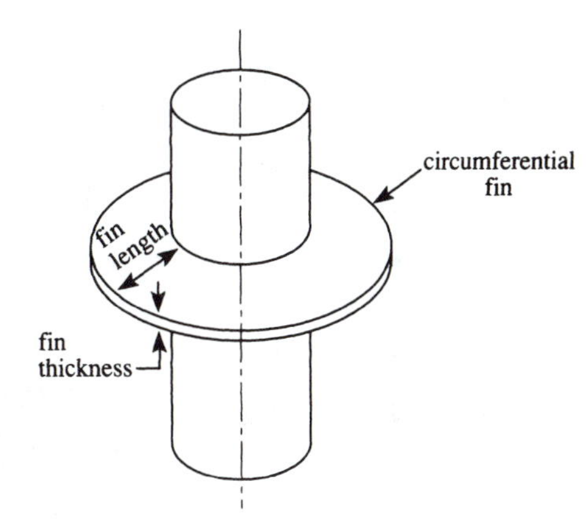

FIGURE 15–14 Circumferential fin

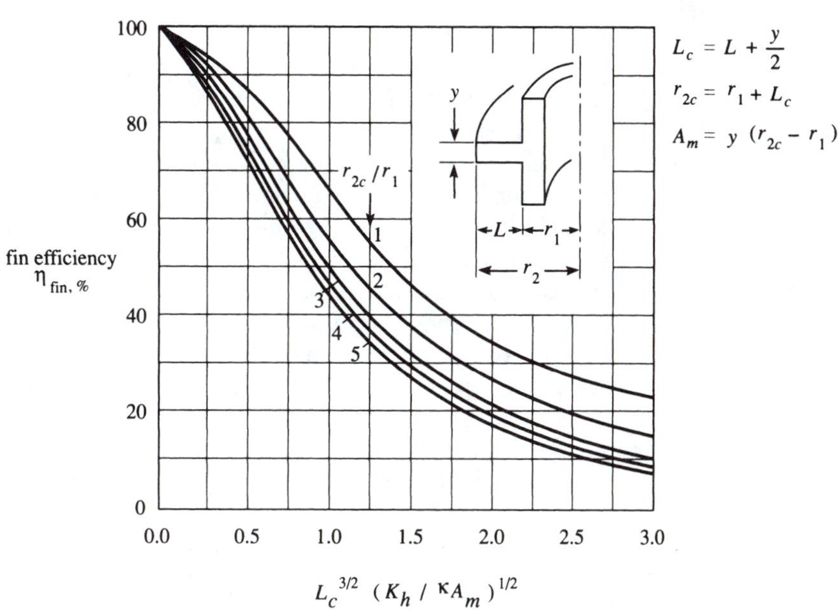

FIGURE 15–15 Efficiency of a circumferential fin

EXAMPLE 15–10 A circumferential steel fin is 8 cm long and is joined to a 20-cm-diameter pipe. The fin is 3 mm thick, and the convective heat transfer coefficient is 35 W/m$^2\cdot$K. Determine the fin efficiency and the rate of heat transfer for the fin if the pipe wall temperature is 300°C and the surroundings are at 20°C.

Solution We use figure 15–15 to find the fin efficiency. First we find

$$L_c = L + \frac{y}{2} = 8\text{ cm} + \frac{0.3\text{ cm}}{2} = 8.15\text{ cm} = 0.0815\text{m}$$

$$r_{2c} = r_1 + L_c = 10\text{ cm} + 8.15\text{ cm} = 18.15\text{ cm} = 0.1815\text{ m}$$

$$A_m = y(r_{2c} - r_1) = (0.003\text{ m})(0.1815\text{ m} - 0.10\text{ m})$$

$$= 0.0002445\text{ m}^2$$

$$\kappa = 43\text{ W/m}\cdot\text{K} \qquad \text{(from table B–20 for steel)}$$

Then we find

$$L_c^{3/2}\left(\frac{K_h}{\kappa A_m}\right)^{1/2} = (0.0815\text{ m})^{3/2}\left(\frac{35\text{ W/m}^2\cdot\text{K}}{43\text{ W/m}\cdot\text{K} \times 0.0002445\text{ m}}\right)^{1/2}$$

$$= 1.342$$

The ratio $r_{2c}/r_1 = 1.815$, so that, using figure 15–15, we find that $\eta_{fin} = 42\%$ (approximately). Notice that some visual interpolation is required to determine the efficiency. The heat transfer from a fin at temperature $T_o = 300°C$, $\dot{Q}_o$, is $\dot{Q}_o = K_h A(T_o - T_\infty)$. The fin area is, neglecting the fin thickness,

$$A = 2\pi(r_2^2 - r_1^2) = 2\pi[(18\text{ cm})^2 - (10\text{ cm})^2]$$
$$= 1407.4\text{ cm}^2 = 0.14074\text{ m}^2$$

Then

$$\dot{Q}_o = (35\text{ W/m}^2\cdot\text{K})(0.14074\text{ m}^2)(300°\text{C} - 20°\text{C})$$
$$= 1379.3\text{ W}$$

and the actual heat transfer from the fin is

Answer
$$\dot{Q}_{actual} = (\eta_{fin})(\dot{Q}_o) = (0.42)(1379.3\text{ W})$$
$$= 579.3\text{ W}$$

Many times more than one fin is used to cool a surface or otherwise increase heat transfer. In such situations, it is important to provide enough space between the fins so that proper convective heat transfer can occur. Thus, if 5.793 kW of heat is to be transferred, you can use 10 fins like those in example 15–10, but there should be sufficient space between each fin, say, 3 cm or 10 times the fin thickness, to provide enough air for convective heat transfer over the fin area. Fins are often used in environments where a buildup of soot, grease, or other material on the fin could retard heat transfer. Without adequate fin spacing, this buildup could reduce heat transfer to significantly less than that predicted by the method demonstrated in example 15–10.

We consider one more important application of conduction/convection heat transfer—the heating or cooling of a small solid object in which the rate of conduction in the solid is much more than the convection around the object. For this situation, the object is small enough and the heat conducts through it fast enough that the temperature of the object is the same as its surface temperature, T_s. In figure 15–16 is sketched the physical situation involved in this type of problem. In any actual case, the solid object will have a temperature at its surface that is higher than the inside temperature if it is being heated, and the surface will be cooler if the object is being cooled. Considering the object in figure 15–16

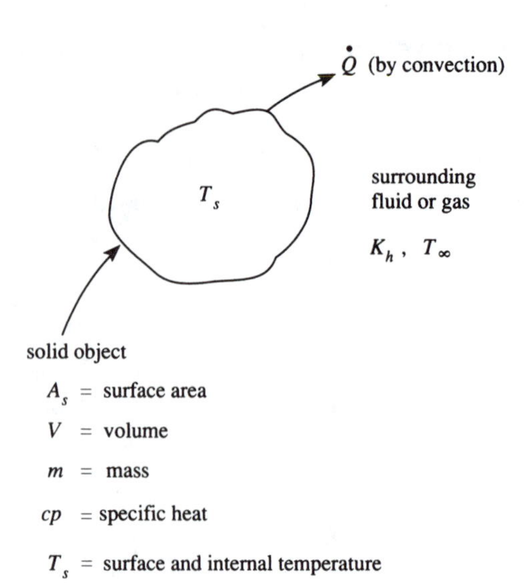

FIGURE 15–16 Lumped heat capacity system

having surface area A_s, volume V, mass m, convective heat transfer coefficient K_h, and thermal conductivity κ, it can be shown that a convenient parameter for studying the heat transfer is with the Biot number defined as

$$\mathrm{Bi} = \frac{K_h V}{A_s \kappa} \tag{15–28}$$

If the Biot number is less than about $^1/_{10}(0.1)$, the temperature of the object can be considered the same throughout, and the analysis for that condition is called the **lumped heat capacity** problem. From the first law of thermodynamics, we have, for the object,

$$\dot{Q} = mc_p \frac{\delta T_s}{\delta t}$$

where δT_s is the temperature change of the object during a time δt. The heat transfer rate, $\dot{Q}$, is convective heat transfer, so we can write

$$\dot{Q} = K_h A_s (T_\infty - T_s) = mc_p \frac{\delta T_s}{\delta t}$$

It can then be shown that the object temperature T_s is given by the equation

$$T_s = T_\infty + (T_o - T_\infty)e^{-(K_h A_s / mc_p)t} \tag{15–29}$$

where T_o is the temperature of the object at the beginning of the heat transfer when $t = 0$. A plot of T_s against time t is shown in figure 15–17, and it can be seen that T_s becomes closer and closer to T_∞ as the time increases. From equation (15–29), you can see that T_s never exactly equals T_∞ until t is infinitely large, but we know that, in real cases, the temperature of objects becomes the same as that of their surroundings after some time. The **time constant** is used to describe how long an object requires to reach a temperature nearly the same as its surroundings. Thermometers, thermocouples, and thermistors, in particular, are often described partly by their time constant, because these devices are expected to sense their surroundings' temperature. The time constant is defined as

$$t_c = \frac{mc_p}{K_h A_s} \tag{15–30}$$

and if we substitute this into equation (15–29), we obtain

$$T_s = T_\infty + (T_o - T_\infty)e^{-(1.0)} = T_\infty + (0.3679)(T_o - T_\infty)$$

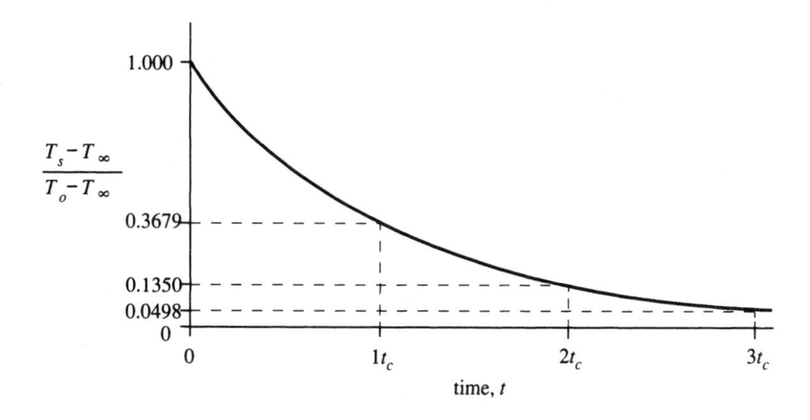

FIGURE 15–17
Temperature (T_s) of an object based on lumped heat capacity

The time constant t_c is a time period at which the original temperature difference between the object and its surroundings has been reduced by 63.21% (100% − 36.79%). The units of t_c are usually seconds or milliseconds. For large objects, the minute may be a more appropriate unit for the time constant. Notice that, after three time constants, the object is within 95% (100% − 4.98%) of the surrounding temperature based on the original temperature difference. For many engineering and technological problems, three time constants would be a good approximation of when thermal equilibrium is reached between an object and its surroundings.

CALCULUS FOR CLARITY 15–4

An energy balance of the lumped heat capacity system shown in figure 15–16 can be written

$$\dot{Q} - \dot{W}k = \frac{dU}{dt}$$

The heat transfer is by convection, and the power is done at constant pressure, $\dot{W}k = p\,dV/dt$. Thus,

$$\dot{Q} = K_h A_s (T_\infty - T_s) = \frac{dH}{dt} = mc_p \frac{dT}{dt}$$

It is convenient to define the temperature difference $T' = T_\infty = T_s$ so that

$$\frac{dT'}{dt} = -\frac{dT}{dt}$$

The energy balance then becomes

$$K_h A_s T' = -mc_p \frac{dT'}{dt}$$

which can be rearranged to give

$$-\left[\frac{K_h A_s}{mc_p}\right] dt = \left[\frac{1}{T'}\right] dT'$$

Integrating both sides of this equation gives

$$-\left[\frac{K_h A_s}{mc_p}\right] t + C = \ln T'$$

At time zero, or $t = 0$, the temperature $T = T_o$, and the temperature difference can be denoted $T'_o = T_\infty - T_o$. The constant C must be

$$C = \ln T'_o$$

which gives

$$-\left[\frac{K_h A_s}{mc_p}\right] t = \ln T' - \ln T'_o = \ln\left[\frac{T'}{T'_o}\right]$$

for the solution. Taking the antilog of both sides then results in equation (15–29),

$$e^{-t/tc} = \frac{T'}{T'_o} = \frac{(T_\infty - T_s)}{(T_\infty - T_o)}$$

where t_c is the time constant given by equation (15–30).

EXAMPLE 15–11 A 4-mm-diameter by 20-cm-long mercury-in-glass thermometer at 15°C is placed in a glass of water. The water is known to be at 75°C. Within 5% (3.75°C), determine the time constant of the thermometer and the time before the thermometer indicates 75°C. Assume that $K_h = 160\ \text{W/m}^2\cdot\text{K}$ for the thermometer in the water and that the thermometer has the same properties as those of glass.

Solution The time constant of the thermometer is given by equation (15–30). From table B–8, we find that $c_p = 750\ \text{W}\cdot\text{s/kg}\cdot\text{K}$ and $\rho = 2500\ \text{kg/m}^3$ for glass. The mass is equal to the density times the volume, and the volume is

$$V = \frac{\pi(\text{diameter})^2(\text{length})}{4}$$
$$= \frac{(\pi)(0.004\ \text{m})^2(0.2\ \text{m})}{4} = 0.0000025\ \text{m}^3$$

so the mass

$$m = (2500\ \text{kg/m}^3)(0.0000025\ \text{m}^3) = 0.00625\ \text{kg}$$

The surface area of the thermometer is

$$A_s = \pi(\text{diameter})(\text{length}) = (\pi)(0.004\ \text{m})(0.20\ \text{m}) = 0.0025\ \text{m}^2$$

The time constant is then

$$t_c = \frac{mc_p}{K_h A_s}$$
$$= \frac{(0.00625\ \text{kg})(750\ \text{W}\cdot\text{s/kg}\cdot\text{K})}{(160\ \text{W/m}^2\cdot\text{K})(0.0025\ \text{m}^2)}$$

Answer
$$= 11.7\ \text{s}$$

When the thermometer reaches within 5% of the water temperature, we have $T_s = 75°\text{C} - 3.75°\text{C} = 71.25°\text{C}$. From equation (15–29), we can then solve for the time, obtaining

$$T_s - T_\infty = (T_o - T_\infty)e^{[-(K_h A_s/mc_p)t]}$$
$$-3.75°\text{C} = (15°\text{C} - 75°\text{C})e^{(-1/11.7\ \text{s})(t)}$$

and

$$\frac{-3.75°\text{C}}{-60°\text{C}} = 0.0625 = e^{[-(0.0855)(t)]}$$

so that

$$\ln(0.0625) = -(0.0855)(t) = -2.772589$$

Therefore,

Answer
$$t = \frac{-2.772589}{0.0855} = 32.4\ \text{s}$$

The thermometer requires more than $^1\!/_2$ min to have a temperature within 5% of the actual water temperature.

EXAMPLE 15–12 Common bricks are heated in an oven (or kiln) to dry and to harden them. Consider a brick that is 4 in by 8 in by 2 in with a convective heat transfer coefficient of 0.8 $\text{Btu/h}\cdot\text{ft}^2\cdot°\text{F}$ at 60°F when the brick is in an oven at 600°F. Determine the time constant of the brick and the time required for the brick to reach 500°F.

Solution

The brick has a surface area of $A_s = (4 + 4 + 2 + 2 \text{ in}) \times (8 \text{ in}) + 4 \text{ in} \times 2 \text{ in} \times 2 = 112 \text{ in}^2 = 0.78 \text{ ft}^2$. The volume of the brick is $4 \text{ in} \times 2 \text{ in} \times 8 \text{ in} = 64 \text{ in}^3 = 0.037 \text{ ft}^3$. From table B–8, the thermal conductivity of brick is $0.70 \text{ W/m}\cdot\text{K} = 0.405 \text{ Btu/h}\cdot\text{ft}\cdot{}^\circ\text{F}$. The density is $1600 \text{ kg/m}^3 = 99.75 \text{ lbm/ft}^3$. We now calculate the Biot number,

$$\text{Bi} = \frac{K_h V}{A_s \kappa} = \frac{(0.8 \text{ Btu/h}\cdot\text{ft}^2\cdot{}^\circ\text{F})(0.037 \text{ ft}^3)}{(0.78 \text{ ft}^2)(0.405 \text{ Btu/h}\cdot\text{ft}\cdot{}^\circ\text{F})}$$
$$= 0.093$$

so that we can use the lumped heat capacity analysis. The time constant is

$$t_c = \frac{mc_p}{K_h A_s} = \frac{\rho V c_p}{K_h A_s}$$

The specific heat is found to be $0.84 \text{ kJ/kg}\cdot\text{K} = 0.20 \text{ Btu/lbm}\cdot{}^\circ\text{R}$ for the brick, so

$$t_c = \frac{(99.75 \text{ lbm/ft}^3)(0.037 \text{ ft}^3)(0.20 \text{ Btu/lbm}\cdot{}^\circ\text{R})}{(0.80 \text{ Btu/h}\cdot\text{ft}^2\cdot{}^\circ\text{F})(0.78 \text{ ft}^2)}$$

Answer

$$= 1.18 \text{ h} = 4258.6 \text{ s}$$

Using equation (15–29), we find the time for the brick to reach 500°F:

$$500^\circ\text{F} = 600^\circ\text{F} + (60^\circ\text{F} - 600^\circ\text{F})e^{(-t/4258.6)}$$

A little algebraic manipulation then yields

$$e^{(-t/4258.6)} = \frac{-100^\circ\text{F}}{-540^\circ\text{F}} = 0.185$$

$$\frac{-t}{4258.6 \text{ s}} = \ln(0.185) = -1.686$$

so that

Answer

$$t = 7181.7 \text{ s} = 1.99 \text{ h}$$

There are many problems involving heat transfer of a solid object with its surroundings where the Biot number is greater than 0.1, and for those types of problems, the reader should refer to heat transfer textbooks.

15–4 FORCED CONVECTION

In this section and the next, we discuss how the convective heat transfer coefficient, K_h, may be predicted for some different conditions. Table 15–2 lists approximate values for K_h, but these values may be far different from those of an actual situation being studied; and here we will see how a more precise value can be found. Convective heat transfer involves a fluid flowing past a solid surface across which heat moves. Forced convection means that the fluid flow is caused by some external source of power other than gravity. Some examples of sources for the flow of fluids are fans, pumps, and wind. Forced convection is also present on solid surfaces or objects moving in still air, such as an aircraft flying or an automobile moving along a road. Here the source of the power to create forced convection situations is from an aircraft engine or an auto engine. The general situation of forced convection is illustrated in figure 15–18, where a fluid interacts with a flat plate. A boundary layer of fluid that is disturbed because of the motion involved is indicated as a volume of fluid near the plate. The fluid outside the boundary layer is assumed to have a speed or velocity of $\overline{V}_\infty$, and the fluid within the boundary layer has a range of velocities from $\overline{V}_\infty$ to zero where the fluid is touching the plate. Here it is assumed that the plate is not moving

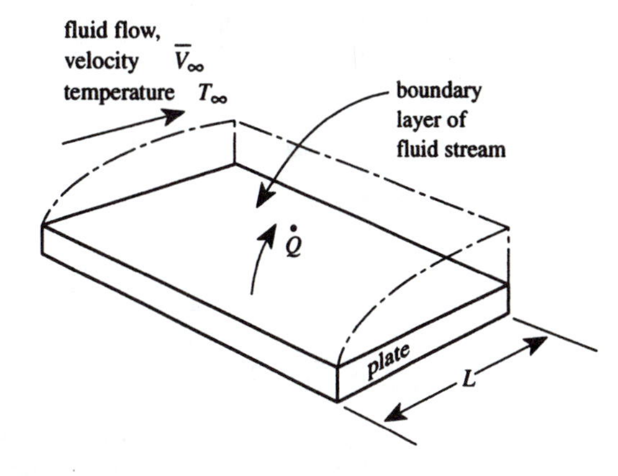

FIGURE 15–18 Forced convection heat transfer with a flat plate

so that its velocity is zero. The case where the plate is moving and the air or fluid is not moving can be visualized as exactly the same problem in reverse. Within the boundary layer, heat is transferred by forced convection to or from the fluid because of conduction in the fluid and because of mixing of the fluid itself in the boundary layer. The fluid motion in the boundary layer may be described in one of two ways: laminar flow, where the fluid moves in parallel paths to the plate and does not mix with itself; or turbulent flow, where it is swirling about the boundary layer. Laminar and turbulent flows are important qualitative concepts used in the analysis of fluid flow and heat transfer problems. The convective heat transfer coefficient, K_h, can be seen to be identified with the boundary layer because that is where so much of the heat transfer occurs. Many experimental and analytic studies have been devoted to the evaluation of K_h and the study of heat transfer in the boundary layer. A common method used to predict whether flow is laminar or turbulent is through the idea of the Reynolds number, Re_L. Referring to figure 15–18, we see that

$$\text{Re}_L = \frac{\rho \bar{V}_\infty L}{\mu_k} \tag{15–31}$$

where μ_k is the fluid viscosity. Values for viscosity of selected fluids are listed in table B–20. A high Reynolds number indicates more nearly turbulent fluid flow; a low number, more nearly laminar flow. For fluid flow past a plate as shown in figure 15–18, the flow can be expected to be turbulent if Re_L is about 50,000 (unitless number) or more and is laminar if Re_L is less than 50,000. If the plate is sufficiently long that turbulent flow occurs, the following equation can be shown to give a reasonable value for K_h:

$$K_h = (\kappa)^{2/3}(c_p \mu_k)^{1/3} \frac{0.037\,\text{Re}_L^{0.8} - 850}{L} \tag{15–32}$$

The following restrictions to equation (15–32) should be observed:

1. Constant plate temperature
2. Reynolds number (Re_L) below 10,000,000 but greater than 50,000

EXAMPLE 15–13 A 30-km/h wind at −10°C is blowing across a flat horizontal building roof. The roof is 80 m by 80 m and has a surface temperature of 20°C. Estimate the heat loss from the roof to the air.

Solution The Reynolds number should first be calculated from equation (15–31). The various terms in that equation are

$$\bar{V}_\infty = 30 \text{ km/h} = 8.33 \text{ m/s}$$

and, from table B–8

$$\rho = 1.178 \text{ kg/m}^3$$
$$\kappa = 0.026 \text{ W/m}\cdot\text{K}$$
$$\mu_k = 0.0181 \text{ g/m}\cdot\text{s} = 0.0000181 \text{ kg/m}\cdot\text{s}$$
$$c_p = 1.007 \text{ kJ/kg}\cdot\text{K} = 1007 \text{ J/kg}\cdot\text{K}$$

Then

$$\text{Re}_L = \frac{(1.178 \text{ kg/m}^3)(8.33 \text{ m/s})(80 \text{ m})}{0.0000181 \text{ kg/m}\cdot\text{s}} = 43{,}371{,}227 \approx 0.434 \times 10^8$$

Since Re_L is greater than 50,000, the flow is turbulent, and we may use equation (15–32) to approximate K_h, even though Re_L is greater than 10,000,000:

$$K_h = (0.026 \text{ W/m}\cdot\text{K})^{2/3}(1007 \text{ J/kg}\cdot\text{K} \times 0.0000181 \text{ kg/m}\cdot\text{s})^{1/3} \times \left[\frac{(0.037)(0.434 \times 10^8)^{0.8} - 850}{80 \text{ m}}\right] = 13.5 \text{ W/m}^2\cdot\text{K}$$

The heat loss may then be estimated from equation (15–15):

$$\dot{Q} = K_h A(T_w - T_\infty) = (13.5 \text{ W/m}^2\cdot\text{K})(80 \times 80 \text{ m}^2)[20°\text{C} - (-10°\text{C})] = 2{,}592{,}000 \text{ W} = 2.59 \text{ MW}$$

Answer

Other situations that would prevent us from using equation (15–32) for a prediction of K_h are treated in heat transfer textbooks and reference books.

The problem of flow of a fluid around an object through forced convection has important engineering and technical applications. Consider forced convection of a fluid around a cylinder, for instance, as shown in figure 15–19. It can be shown that a reasonable value for K_h, as found from previous studies and experiments, can be determined from the equation

$$K_h = \frac{\kappa}{D}(C)(\text{Re}_D)^n\left[\frac{c_p\mu_k}{\kappa}\right]^{1/3} \tag{15–33}$$

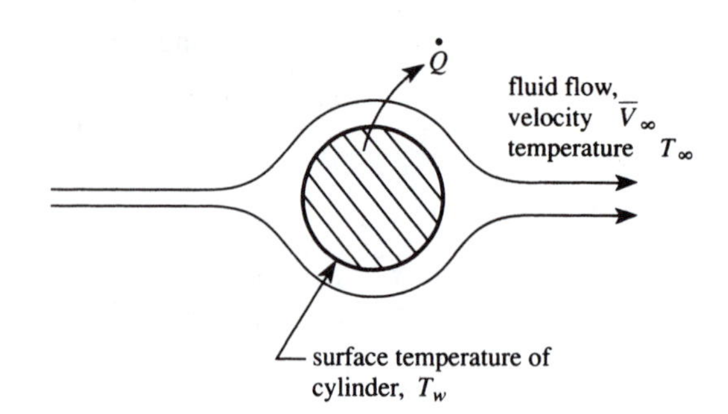

FIGURE 15–19 Flow of a fluid around a cylinder

where C and n are constants that can be found in table 15–3 and

$$\text{Re}_D = \frac{\rho \overline{V}_\infty D}{\mu_k}$$

is a Reynolds number based on the cylinder diameter, D, instead of the Reynolds number as given by equation (15–31). Notice that table 15–3 also gives values for C and n for objects other than cylinders. The values for D for those objects that are not cylinders are defined in the table. The quantity $c_p\mu_\kappa/\kappa$ is called the Prandtl number, Pr, and is a dimensionless quantity like the Reynold's number. The reader can verify from properties given in table B–8 of the appendix that the Prandtl number for air at 300 K (27°C) is about 0.701 and for water at 300 K is about 5.89.

EXAMPLE 15–14 Superheater tubes of 6-cm diameter are arranged across the flow of exhaust gases in a superheater section of a steam generator. If the exhaust gases are moving at 15 m/s and have properties the same as those of carbon dioxide, determine K_h and the heat transfer between the gases and the superheater tubes if the tube surface temperature is 1100 K. The exhaust gases are at 1200 K.

Solution The convective heat transfer coefficient can be found by using equation (15–33) and table 15–3. The Reynolds number is

$$\mathrm{Re}_D = \frac{\rho \bar{V}_\infty D}{\mu_k}$$

and the properties of the gas are found from table B–8:

$$c_p = 0.844\ \mathrm{kJ/kg \cdot K}$$
$$\kappa = 0.017\ \mathrm{W/m \cdot K}$$
$$\mu_k = 0.0148\ \mathrm{g/m \cdot s}$$
$$\rho = 1.81\ \mathrm{kg/m^3} \quad \text{at 300 K}$$

TABLE 15–3 Constants C and n for use in equation (15–33)

Cross Section	Re_D	C	n
cylinder	0.4 to 4	0.989	0.33
	4 to 40	0.911	0.385
	40 to 4,000	0.683	0.466
	4,000 to 40,000	0.193	0.618
	40,000 to 400,000	0.0266	0.805
diamond (D)	5×10^3 to 1×10^5	0.2460	0.588
square (D)	5×10^3 to 1×10^5	0.102	0.675
plate (D)	4×10^3 to 1.5×10^4	0.228	0.731

We assume that the gas acts as a perfect gas at constant pressure so that the density at 1200 K is $\rho = 1.81\ \text{kg/m}^3\ (300\ \text{K}/1200\ \text{K}) = 0.4525\ \text{kg/m}^3$. Then

$$\text{Re}_D = \frac{(0.4525\ \text{kg/m}^3)(15\ \text{m/s})(0.06\ \text{m})}{0.0000148\ \text{kg/m}\cdot\text{s}}$$
$$= 27{,}516$$

From table 15–3, we read $C = 0.193$ and $n = 0.618$, and from equation (15–33), we find that

$$K_h = \left(\frac{0.017\ \text{W/m}\cdot\text{K}}{0.06\ \text{m}}\right)(0.193)(27{,}516)^{0.618}\left(\frac{0.844\ \text{kJ/kg}\cdot\text{K} \times 0.0000148\ \text{kg/m}\cdot\text{s}}{0.017\ \text{W/m}\cdot\text{K}}\right)^{1/3}$$

Answer

$$= 27.35\ \text{W/m}^2\cdot\text{K}$$

The heat transfer between the gases and the superheater tube per unit length of tube is

$$\dot{q}_L = K_h\left(\frac{A_s}{L}\right)(T_w - T_\infty)$$
$$= (27.35\ \text{W/m}^2\cdot\text{K})(\pi \times 0.06\ \text{m})(100°\text{C})$$

Answer

$$= 515\ \text{W/m}$$

Convective heat transfer of fluid past more than one tube is often more complicated because of the effect of adjoining tubes. Situations such as these are treated in heat transfer textbooks.

The heat transfer between fluid flowing inside a tube and the walls of that tube can be analyzed in a manner much like the flow past a flat plate. The Reynolds number for internal flow such as this is defined as $\text{Re}_D = \rho\overline{V}_\infty D/\mu_k$, where D is the inside diameter of the tube or pipe and $\overline{V}_\infty$ is the average velocity of the fluid. A common equation used to predict K_h for convection heat transfer between a hot or a cold fluid and the tube walls is

$$K_h = \frac{\kappa}{D}(0.023)(\text{Re}_D)^{0.8}\left(\frac{c_p\mu_k}{\kappa}\right)^n \qquad \textbf{(15–34)}$$

where n is 0.4 if the fluid is being heated and 0.3 if the fluid is being cooled. Notice that the quantity $c_p\mu_\kappa/\kappa$ is the Prandtl number, Pr, discussed earlier. The limitations on equation (15–34) are as follows:

1. The tube or pipe must have smooth walls.
2. The temperature of the tube must be constant.
3. Re_D must be greater than 2500.

Equations for predicting K_h values for other flow conditions of forced convection can be found in heat transfer books.

EXAMPLE 15–15 Water flows through a smooth pipe of 2-in diameter at 3 ft/s. The pipe is 25 ft long, and the water enters at 50°F. If the pipe wall temperature is 190°F, determine the temperature of the water leaving the pipe.

Solution Figure 15–20 shows a sketch of the problem, which involves the use of the steady-flow energy equation in the form

$$\dot{Q} = \dot{m}(h_2 - h_1) = \dot{m}c_p(T_2 - T_1)$$

The heat transfer is accomplished by forced convection, so we can write

$$K_hA_s(T_w - T_\infty) = \dot{m}c_p(T_2 - T_1)$$

FIGURE 15–20 Convection heat transfer in a tube

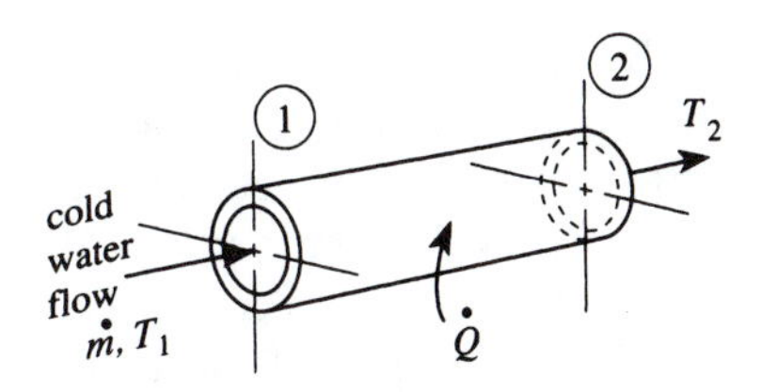

In this equation, we must recognize that T_∞ is changing value along the pipe, so we can approximate it by the average of inlet and outlet water temperatures:

$$T_\infty = \frac{T_1 + T_2}{2}$$

Then the energy equation becomes

$$K_h A_2 \left(T_w - \frac{T_1 + T_2}{2} \right) = \dot{m} c_p (T_2 - T_1)$$

We can then solve for the outlet temperature, T_2, after determining the values for the other terms in this equation:

$$A_s = \pi \left(\frac{2}{12}\ \text{ft} \right)(25\ \text{ft}) = 13.09\ \text{ft}^2$$

From table B–8,

$$c_p = 4.18\ \text{kg/kg} \cdot \text{K} = 0.998\ \text{Btu/lbm} \cdot {}^\circ\text{R}$$
$$\kappa = 0.608\ \text{W/m} \cdot \text{K} = 0.3521\ \text{Btu/h} \cdot \text{ft} \cdot {}^\circ\text{F}$$
$$\mu_k = 0.857\ \text{g/m} \cdot \text{s} = 0.000575\ \text{lbm/ft} \cdot \text{s}$$
$$\rho = 62.4\ \text{lbm/ft}^3$$

The Reynolds number is

$$\text{Re}_D = \frac{(62.4\ \text{lbm/ft}^3)(3\ \text{ft/s})(2/12\ \text{ft})}{0.000575\ \text{lbm/ft} \cdot \text{s}} = 54{,}261$$

so that, using equation (15–34) and $n = 0.4$ for heated water, we obtain

$$K_h = \left(\frac{0.35\ \text{Btu/h} \cdot \text{ft} \cdot {}^\circ\text{F}}{\frac{1}{6}\ \text{ft}} \right)(0.023)(54.261)^{0.8} \left(\frac{0.998 \times 0.000575}{0.351} \right)^{0.4}$$
$$= 22.8\ \text{Btu/h} \cdot \text{ft}^2 \cdot {}^\circ\text{F} = 0.00634\ \text{Btu/s} \cdot \text{ft}^2 \cdot {}^\circ\text{F}$$

The mass flow rate is

$$\dot{m} = \rho \bar{V}_\infty A = (62.4\ \text{lbm/ft}^3)(3\ \text{ft/s})(\pi)\left(\frac{1}{12}\ \text{ft} \right)^2 = 4.084\ \text{lbm/s}$$

Substituting these values into the energy balance equation yields

$$(0.00634\ \text{Btu/s} \cdot \text{ft}^2 \cdot {}^\circ\text{F})(13.09\ \text{ft}^2)\left(190^\circ\text{F} - \frac{50^\circ\text{F} + T_2}{2} \right) = (4.084\ \text{lbm/s})(0.998\ \text{Btu/lbm} \cdot {}^\circ\text{R})(T_2 - 50^\circ\text{F})$$

This equation reduces to

$$15.768 - 2.075 - 0.0415 T_2 = 4.0758 T_2 - 203.79$$

so

Answer

$$T_2 = 52.82^\circ\text{F}$$

15–5 NATURAL CONVECTION

Natural convection is the upward motion of a liquid or gas as its density decreases while being heated. Examples of natural convection are water in a pot that is heating on a stove burner as the hot water rises and the cooler water drops down to the bottom, still air as it rises from a hot road surface or hot roof of a building, and the air in a room that is heated by a space heater or radiator.

All natural convection situations have some common effects:

1. Heat is added to a portion of a fluid or gas.
2. The fluid or gas decreases in density as it is heated and its temperature increases.
3. A buoyant force or upward force acts to make the less dense, warm fluid rise in a region where gravity is present.

The problem of determining the natural convection heat transfer coefficient, K_h, between a warm vertical plate or wall and a fluid or gas, as shown in figure 15–21, has been studied, and the heat transfer is usually described by equation (15–15),

$$\dot{Q} = K_h A_s (T_w - T_\infty) \quad \textbf{(15–15)}$$

FIGURE 15–21 Natural convection heat transfer at a hot vertical wall or surface

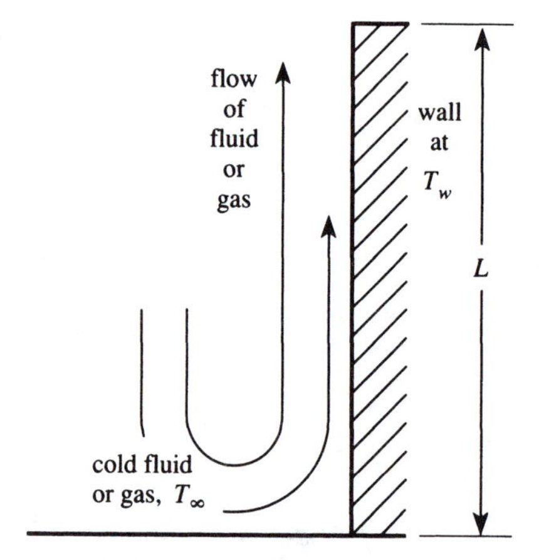

where T_∞ is the average temperature of the fluid or gas. For a vertical surface such as a room wall or building wall, the value for the convective heat transfer coefficient, K_h, can be found from the equation

$$K_h = \frac{4\kappa}{3L}[0.508(\mathrm{Pr})^{1/2}(0.952 + \mathrm{Pr})^{-1/4}\mathrm{Gr}^{1/2}] \quad \textbf{(15–35)}$$

In this equation, Pr is the Prandtl number $c_p\mu_\kappa/\kappa$,

$$\mathrm{Gr} = \frac{g(T_w - T_\infty)L^3\rho^2}{\mu_k^2}\beta_p$$

is the **Grashof number**, and the term β_p is called the volumetric expansion of the fluid. It is the change in density of the fluid as the temperature changes per unit density, the density determined at the average temperature over which the fluid is changing. The volumetric expansion is equal to the inverse of the fluid's temperature, $1/T_\infty$, for a perfect gas. Thus, for perfect gases, the Grashof number is

$$\mathrm{Gr} = \frac{g(T_w - T_\infty)L^3\rho^2}{\mu_k^2 T_\infty}$$

Table B–10 of the appendix lists volumetric expansions for some fluids at atmospheric pressure, which can be used to approximate the Grashof number. This table lists values at only one temperature for most of the fluids, so the reader needs to assume these values for other reasonable temperatures as well. Because air is often involved in natural convective heat transfer, equation (15–35) can be simplified further by substituting the various properties of air into that equation. The result is that $K_h = 1.42((T_w - T_\infty)/L)^{1/4}$ if the value for the term $(\text{Gr})(c_p\mu_k/\kappa)$ is less than about 1×10^9, and $K_h = 1.31(T_w - T_\infty)^{1/3}$ if $(\text{Gr})(c_p\mu_k/\kappa)$ is greater than 1×10^9. Studies of natural convection for situations other than for vertical wall heat transfer have been conducted, and the values for K_h under those situations are usually determined from equations that are similar to equation (15–35). In table 15–4 are listed some simplified equations for determining K_h for air. Equations for other gases can be found in heat transfer textbooks.

An important application of natural convections is that of a chimney or smokestack. Figure 15–22 shows the side view of a typical chimney that allows hot gases to rise through

TABLE 15–4 Simplified equations for natural convection heat transfer coefficient, K_h, for air and some surfaces ($K_h = \text{W/m}^2 \cdot {}^\circ\text{C}$)

	$\mathbf{Gr}(c_p\mu_k/\kappa)$	
Surface	$<1 \times 10^9$	$>1 \times 10^9$
Vertical wall or cylinder (height L)	$1.42\left(\frac{T_w - T_\infty}{L}\right)^{1/4}$	$1.31(T_w - T_\infty)^{1/3}$
Horizontal cylinder (diameter D)	$1.32\left(\frac{T_w - T_\infty}{D}\right)^{1/4}$	$1.24(T_w - T_\infty)^{1/3}$
Horizontal plate (length L) heated plate facing up or cooling plate facing down	$1.32\left(\frac{T_w - T_\infty}{L}\right)^{1/4}$	$1.52(T_w - T_\infty)^{1/3}$
Horizontal plate (length L) cooling plate facing up or heating plate facing down	$0.59\left(\frac{T_w - T_\infty}{L}\right)^{1/4}$	$0.59\left(\frac{T_w - T_\infty}{L}\right)^{1/4}$

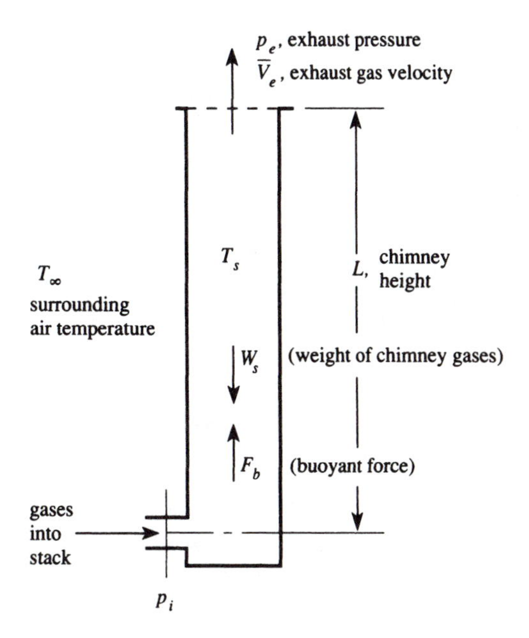

FIGURE 15–22 Chimney and the chimney effect

the vertical opening or passage. Chimneys are usually circular or square in cross section, and through the buoyancy or chimney effect, they produce a slight negative pressure in the chimney, which in turn creates a flow of the gases upward through the chimney. In most small furnace systems or stoves, the chimney is an important element that draws air through the combustion chamber. In large systems such as steam power plants, a chimney serves to place the exhaust gases at a high enough elevation to be dispersed into the upper atmosphere.

Referring now to figure 15–22, if we consider a sum of the forces acting on the gas in the chimney, we have

$$\sum F_v = F_b - W_s - (p_e - p_i)A_e = 0 \tag{15–36}$$

If the chimney is operating properly, the buoyant force, F_b, is greater than the weight of the gases in the chimney, W_s, so the gases will rise. The difference in pressure between the top and bottom of the chimney, $p_e - p_i = \Delta p$, is the same as the chimney draft pressure, and it can be measured at the base of the chimney. The complete sum of forces acting vertically on the chimney gases shows that the buoyant force overcomes the weight of the gases and creates the draft pressure. There is also friction on the sides of the chimney, which usually contributes to a small resistance to the flow. The weight of the chimney gases is equal to $(\rho_s g)(A_e L) = (p_s/RT_s)(g)(A_e L)$. The buoyant force is equal to the weight of surrounding air that could occupy the chimney, $(\rho_\infty g)(A_e L) = (p_\infty/RT_\infty)(g)(A_e L)$. From equation (15–36), the draft pressure is then

$$\Delta p = \frac{F_b - W_s}{A_e} = \left(\frac{p_\infty}{RT_\infty} g A_e L - \frac{p_s}{RT_s} g A_e L\right)\frac{1}{A_e}$$

But $p_\infty \approx p_s$, so this equation reduces to

$$\Delta p = \frac{p_\infty}{R}(gL)\left(\frac{1}{T_\infty} - \frac{1}{T_s}\right) \quad \text{(SI units, kPa)} \tag{15–37}$$

In English units, equation (15–37) is usually written as

$$\Delta p = \frac{p_\infty}{Rg_c}(gL)\left(\frac{1}{T_\infty} - \frac{1}{T_s}\right) \quad (\text{in psi or lbf/ft}^2)$$

This relationship tells us that the chimney gases must be at a higher temperature than the surroundings if an upward draft is to be obtained. Using the conservation of energy equation or first law of thermodynamics for the chimney gases, we can find the exit velocity of the exhaust gases:

$$\bar{V}_e = \begin{cases} \sqrt{\dfrac{2\,\Delta p}{\rho}} & \text{(SI units, m/s)} \\ \sqrt{\dfrac{2\,\Delta p g_c}{\rho}} & \text{(English units, ft/s)} \end{cases} \tag{15–38}$$

The mass flow of exhaust gases is then

$$\dot{m}_e = \rho_e \bar{V}_e A_e \tag{15–39}$$

Let us now consider natural convection heat transfer in a chimney.

EXAMPLE 15–16 A 20-ft-high chimney with a 1-ft-diameter cross section removes 300°F exhaust gases having the same properties as those of air. The surrounding air temperature is 60°F, the air pressure is 14.7 psia, and g is 32.17 ft/s^2. Assuming that the chimney temperature is 90°F throughout, estimate the heat loss of the chimney gases and the volume flow rate of those gases.

Solution

We use equation (15–15) to predict the heat loss and determine K_h from the relationships listed in table 15–4. The analysis requires that the gas temperature, T_s, be known, so we assume that it is the average of the inlet and the outlet values, $(T_i + T_e)/2$. It is known that $T_i = 300°\text{F}$, and if T_e is estimated, the solution will follow. Then, by using an energy balance, we can directly calculate the value for T_e, and if that answer is the same as the assumed T_e value, the solution is complete. If the T_e from the energy balance is not equal to that assumed, another assumed value for T_e (some value between the two) should be used to find T_s and then the calculations repeated until T_e matches the assumed value for T_e. This process is called *iteration* and is used often in complicated problems. Usually, one or two iterations will give an acceptable solution to a problem. We will assume that $T_e = 200°\text{F}$ and then $T_s = (300°\text{F} + 200°\text{F})/2 = 250°\text{F}$. The Grashof number needs to be determined so that a relationship for K_h can be selected from table 15–4. From table B–8,

$$c_p = 1.007\ \text{kJ/kg}\cdot\text{K} = 0.2404\ \text{Btu/lbm}\cdot°\text{R}$$
$$\kappa = 0.026\ \text{W/m}\cdot\text{K} = 0.00000417\ \text{Btu/s}\cdot\text{ft}\cdot°\text{F}$$
$$\mu_k = 0.0181\ \text{g/m}\cdot\text{s} = 0.00001215\ \text{lbm/ft}\cdot\text{s}$$

From the perfect gas law,

$$\rho_s = \frac{p_s}{RT_s} = \frac{14.7 \times 144\ \text{lbf/ft}^2}{(53.36\ \text{ft}\cdot\text{lbf/lbm}\cdot°\text{R})(250 + 460°\text{R})}$$
$$= 0.05587\ \text{lbm/ft}^3$$

The Grashof number is

$$\text{Gr} = (32.17\ \text{ft}/^2)(250 - 60°\text{F})(20\ \text{ft})^3(0.05587\ \text{lbm/ft}^3)^2$$
$$\div\ (0.00001215\ \text{lbm/ft}\cdot\text{s})^2(710°\text{R})$$
$$= 14.56 \times 10^{11}$$

The product $(\text{Gr})(c_p\mu_k/\kappa) = 1.019 \times 10^{12}$, so, from table 15–4, we select

$$K_h = 1.31(T_s - T_\infty)^{1/3}$$

Because K_h will be in SI units when using this equation, we convert $T_s - T_\infty$ to SI units: $250°\text{F} - 90°\text{F} = 160°\text{F} = 88.9°\text{C}$. Then

$$K_h = 1.31(88.9)^{1/3} = 5.85\ \text{W/m}^2\cdot\text{K} = 0.000286\ \text{Btu/s}\cdot\text{ft}^2\cdot°\text{F}$$

From equation (15–15),

$$\dot{Q}_{\text{loss}} = (0.000286\ \text{Btu/s}\cdot\text{ft}^2\cdot°\text{F})(\pi \times 1\ \text{ft} \times 20\ \text{ft})(160°\text{F})$$
$$= 2.88\ \text{Btu/s}$$

We now check to see if the gases will be 200°F when leaving the chimney. The velocity of the gases is predicted by equation (15–38), and Δp is given by equation (15–37):

$$\Delta p = \frac{(14.7 \times 144\ \text{lbf/ft}^2)(32.17\ \text{ft}/^2)(20\ \text{ft})}{(53.36\ \text{ft}\cdot\text{lbf/lbm}\cdot\text{R})(32.17\ \text{ft}\cdot\text{lbm/lbf}\cdot\text{s}^2)}\left(\frac{1}{520°\text{R}} - \frac{1}{710°\text{R}}\right)$$
$$= 0.408\ \text{lbf/ft}^2$$

The exit velocity is then

$$\bar{V}_e = \sqrt{\frac{2\,\Delta p\, g_c}{\rho}} = \sqrt{\frac{(2)(0.408\ \text{lbf/ft}^2)(32.17\ \text{lbm}\cdot\text{ft/lbf}\cdot\text{s}^2)}{0.05587\ \text{lbm/ft}^3}}$$
$$= 21.7\ \text{ft/s}$$

The volume flow rate is

$$\dot{V} = \bar{V}_e A_e = (21.7\ \text{ft/s})(\pi)(0.5\ \text{ft})^2 = 17.04\ \text{ft}^3/\text{s}$$

and the mass flow is

$$\dot{m}_e = \rho_e \bar{V}_e A_e = \rho \dot{V} = (0.05587\ \text{lbm/ft}^3)(17.04\ \text{ft}^3/\text{s}) = 0.952\ \text{lbm/s}$$

Then, from the first law of thermodynamics for the chimney gases, we get

$$\dot{Q}_{\text{loss}} = \dot{m}c_p(T_i - T_e) = 2.88\ \text{Btu/s}$$

Solving for T_e yields

$$T_e = T_i - \frac{2.88\ \text{Btu/s}}{\dot{m}c_p} = 300°\text{F} - \frac{2.88\ \text{Btu/s}}{(0.952\ \text{lbm/s})(0.2404\ \text{Btu/lbm}\cdot°\text{F})} = 287.4°\text{F}$$

From this result, we see that the assumption that the temperature of the exhaust gases leaving the chimney was 200°F was low. We should at this point assume a new value for T_e, say, 286°F, and repeat the calculations. This second iteration would then give the following results:

$$T_s = 293°\text{F}$$
$$\rho = 0.0543\ \text{lbm/ft}^3$$
$$\text{Gr} = 13.3 \times 10^{11}$$
$$K_h = 0.00031\ \text{Btu/s}\cdot\text{ft}^2\cdot°\text{F}$$

Answer

$$\dot{Q} = 3.12\ \text{Btu/s}$$
$$\bar{V}_e = 18.69\ \text{ft/s}$$
$$\Delta p = 0.468\ \text{lbf/ft}^2$$

Answer

$$\dot{m} = 0.99\ \text{lbm/s}$$

Consequently, $T_e = 286.8°\text{F}$, which is in reasonable agreement with the second assumed value for T_e of 286°F.

In chimney operations, the heat loss should not be so great as to reduce the exhaust temperatures and thereby reduce the flow rate of those gases.

Another application of natural convection heat transfer is the **Trombé wall**, shown in figure 15–23. This arrangement has been used to heat air by solar energy as it enters or recirculates in large buildings. Trombé walls, or glass walls, have been fitted to older buildings and to new structures, where they have been found to operate well on tall buildings of

FIGURE 15–23 Glass or Trombé wall

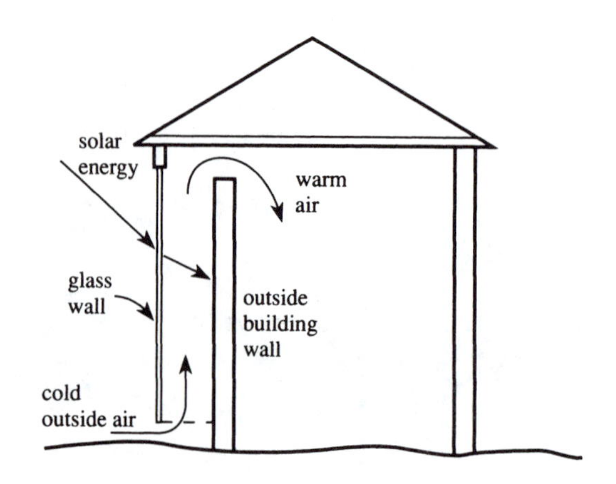

four stories or more. The operation of the glass wall relies on solar energy to heat the air moving between the glass and the outside building wall. Some of the solar energy is absorbed by the building wall, and some is transferred to the air. As a result of being heated, the air rises through the passageway by natural convection. One mode of operation of the Trombé wall uses outside air entering the passageway at the bottom, heated in the wall section, and entering the building at the top, as indicated in figure 15–23.

Natural convection is used in many industrial heating processes where liquids are heated or cooled in tanks. The interested student should refer to heat transfer and fluid mechanics textbooks for more complete treatment of this subject.

15–6 RADIATION HEAT TRANSFER

Up to now, we have considered only two of the modes of heat transfer, conduction and convection, which rely on energy transfer through materials that are in contact with surrounding media. The third mode of heat transfer, radiation, relies on electromagnetic energy radiation and functions best when the heat exchange occurs over a vacuum. Also, whereas conduction and convection do not conduct heat well over long distances, radiation can be transmitted with no theoretical limit to the distance. For instance, solar energy is a form of radiation heat transfer, and it travels around 93 million miles, or about 148 million kilometers, to reach the Earth. Now we will consider some properties of radiation and the materials involved in its transfer.

Radiation heat transfer, as we said, is a form of electromagnetic (EM) radiation. Light is a form of EM radiation and is described as being the visible part of the spectrum of EM radiation. The spectrum of EM radiation in terms of the wavelength is shown in figure 15–24. Notice that the spectrum is the logarithm of the wavelength, log λ, and the thermal radiation band can be seen as extending from 100 μm (1×10^{-4} m or $\log 10^{-4} = -4$) to 0.15 μm [0.15×10^{-6} m or $\log (0.15 \times 10^{-6}) = -6.82$]. It can be shown that a perfect radiator of thermal energy, called a blackbody, emits thermal energy proportional to the surface area and the surface temperature to the fourth power:

$$\dot{E}_b = A\sigma T^4 \tag{15–40}$$

Here, A is the blackbody surface area, T is the surface temperature, and s is the **Stefan–Boltzmann constant**, 5.67×10^{-8} W/m$^2 \cdot$K^4.

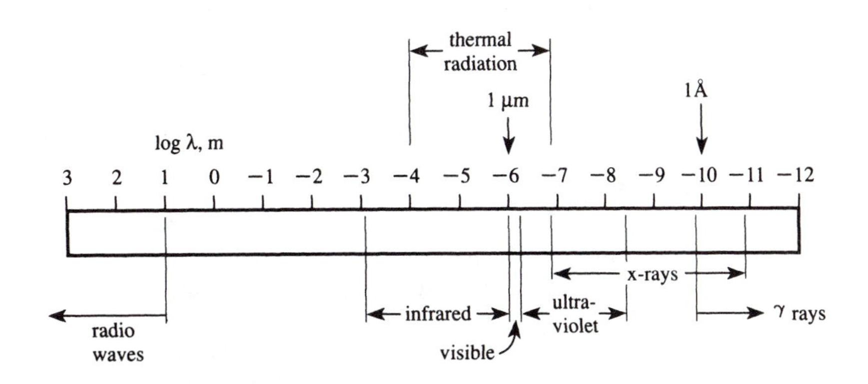

FIGURE 15–24
Electromagnetic spectrum

Consider now a spherical cavity as shown in figure 15–25; A_1 and A_2 are the surface areas of a blackbody inside the cavity and the surface area of the cavity; and T_1 and T_2 are their surface temperatures. Blackbody 1 emits energy $\dot{E}_{b_1} = A_1\sigma T_1^4$ and receives from surface 2, an amount $A_1\sigma T_2^4$. The net heat transfer of blackbody 1 is

$$\dot{Q}_{net} = A_1\sigma(T_2^4 - T_1^4)$$

If T_1 is more than T_2, the blackbody cools down because of the net loss of heat until T_1 and T_2 are equal. If T_1 is less than T_2, net heat will be added to the blackbody until T_1 and T_2 are again equal.

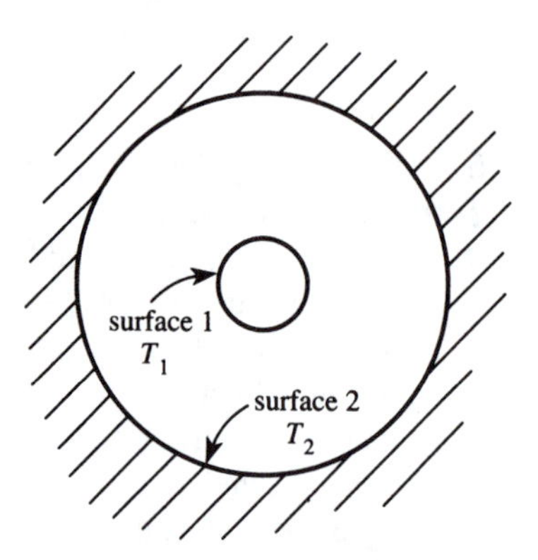

FIGURE 15–25 Blackbody inside another blackbody

A blackbody is a perfect (and therefore unrealistic) material surface, but some surfaces behave in a manner close to a blackbody. A convenient concept for determining how a real surface compares to a blackbody is the gray body. A gray body is defined as a body whose emitted thermal energy can be described by the equation

$$\dot{E}_g = A\epsilon\sigma T^4 \quad \textbf{(15–41)}$$

where ϵ is called the **emissivity**. Emissivity has a unitless value between 1.0 and 0 and is the fraction of energy emitted by a gray body compared to a blackbody. Emissivities for various surfaces are listed in table B–9. The rate of energy absorbed by a gray body can be defined as

$$\dot{E}_{abs} = A\alpha\sigma T^4 \quad \textbf{(15–42)}$$

where α is called **absorptivity**. Absorptivity has a value between 1.0 and 0. If we now consider body 1 in figure 15–25 as a gray body instead of a blackbody, we obtain, for the net heat transfer to or from surface 1,

$$\dot{Q}_{net} = A_1\epsilon_1\sigma T_1^4 - A_1\alpha_1\sigma T_2^4$$

If surfaces 1 and 2 are allowed to reach thermal equilibrium, $T_1 = T_2$ and $\dot{Q}_{net} = 0$. From this result, we must have that $\epsilon_1 = \alpha_1$ or else the temperatures would be different. The statement

$$\alpha = \epsilon \quad \textbf{(15–43)}$$

is called **Kirchhoff's law of radiation** or **Kirchhoff's identity** and is an important relationship for thermal radiation analysis.

Consider now a gray body receiving thermal radiation on its surface as shown in figure 15–26. Since the body is gray, the thermal radiation directed against it per unit area can follow one of three paths: (1) be absorbed as $\dot{E}_{abs}$, (2) be reflected as $\dot{E}_r$, or (3) be transmitted through the material as $\dot{E}_t$. If the gray body is an opaque material, no energy is transmitted, and we have the relationship

$$\alpha + \rho_r = 1.0 \quad \textbf{(15–44)}$$

where ρ_r is the reflectivity, defined as $\rho_r = \dot{E}_r/\dot{E}_a$, from figure 15–26.

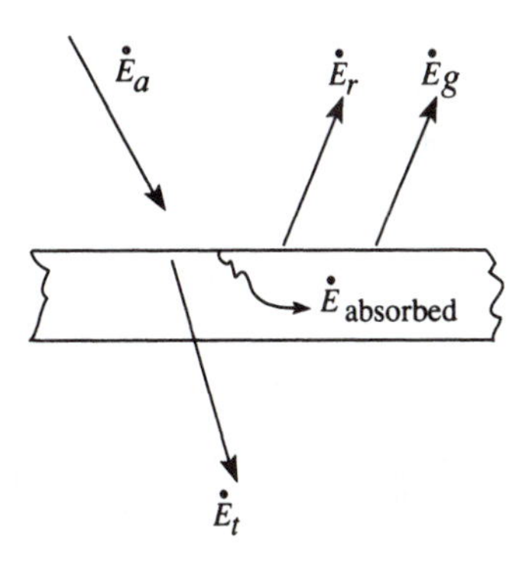

FIGURE 15–26 Paths of thermal radiation directed onto a gray surface

For transparent materials, the thermal radiation can be transmitted and reflected, but not absorbed:

$$\rho_r + \tau = 1.0 \qquad \textbf{(15–45)}$$

Here, τ is the **transmissivity**, defined as $\tau = \dot{E}_t/\dot{E}_a$ from figure 15–26. All real material surfaces involve absorption, reflection, and transmission of thermal energy, so the equation

$$\alpha + \rho_r + \tau = 1.0 \qquad \textbf{(15–46)}$$

is a more nearly correct description of reality.

EXAMPLE 15–17 Determine the emissivity and transmissivity of window glass if it is found that the reflectivity is 0.02.

Solution: Using table B–9, we find the emissivity to be 0.92 for glass. From Kirchhoff's, identity, we have

Answer
$$\epsilon = \alpha = 0.92$$

Then, using equation (15–46), we find the transmissivity:

Answer
$$\tau = 1.0 - 0.92 - 0.02 = 0.06$$

Window glass is an important and interesting material. This example does not really describe the actual behavior of glass, which acts as a transparent material to solar energy at high frequencies or short wavelengths. At low frequencies or long wavelengths, it acts as an opaque material. As a result of this behavior, glass allows the sun's radiation to pass through, but if that energy is then absorbed by low-temperature walls and objects and reradiated at low frequencies, the thermal radiation will not pass back through the glass. Glass, in fact, is an appropriate material for windows. Consider again figure 15–26. The net thermal radiation leaving the surface is denoted as $\dot{Q}_{\text{net}}$, and for an opaque gray body, this is equal to $\dot{E}_r + \dot{E}_g - \dot{E}_a$. Using Kirchhoff's identity [equation (15–43)] and the definition of reflectivity, we obtain

$$\dot{Q}_{\text{net}} = \frac{\dot{E}_b/A - \mathcal{J}}{(1 - \epsilon)/\epsilon A} \qquad \textbf{(15–47)}$$

where $\mathcal{J}$ is called the **radiosity** and is the total radiation that leaves the surface per unit area. Equation (15–47) gives a convenient form for calculating the radiation of an opaque gray surface. For blackbodies, we found that $\mathcal{J} = \dot{E}_b/A = \sigma T^4$, so equation (15–47) should not be used for blackbodies. Equation (15–47) can also be interpreted by the thermal

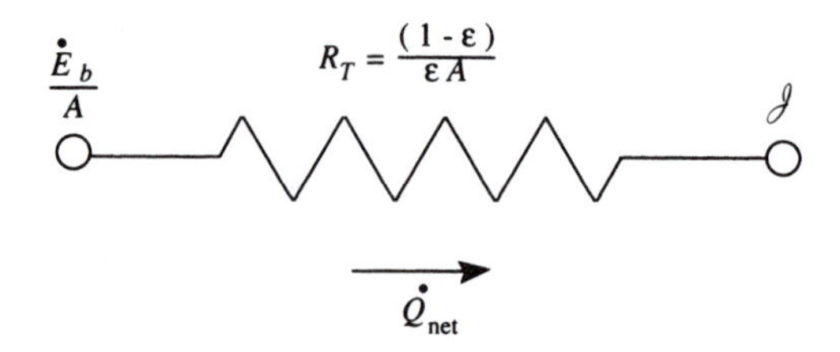

FIGURE 15–27 Electrical analogy of thermal radiating opaque gray surface

resistance concept if the term $(1 - \epsilon)/\epsilon A$ is the thermal resistance. Then we can use the electrical analogy as sketched in figure 15–27 to see that the net heat is like a current flow between points $\dot{E}_b/A$ and $\mathcal{J}$.

EXAMPLE 15–18 A 10-m-by-20-m furnace wall at 1200 K acts as an opaque gray body with $\epsilon = \alpha = 0.80$. If there is a net heat flow of 1000 kW, determine the radiosity.

Solution The furnace wall thermal resistance is

$$R_T = \frac{1-\epsilon}{\epsilon A} = \frac{1-0.8}{(0.8)(200\ \text{m}^2)}$$
$$= 0.00125\ \text{m}^{-2}$$

The net heat $\dot{Q}_{net}$ is -1000 kW, so, using equation (15–47), we find that

$$\mathcal{J} = -R_T\dot{Q}_{net} + \frac{\dot{E}_b}{A}$$
$$= -(0.00125\ \text{m}^{-2})(-1000\ \text{kW}) + \alpha T^4$$
$$= 1.25\ \text{kW/m}^2 + (5.67 \times 10^{-11}\ \text{kW/m}^2\cdot\text{K}^4)(1200\ \text{K})^4$$

Answer
$$= 118.8\ \text{kW/m}^2$$

Before we analyze the heat exchange between surfaces, we need to recognize another complicating feature of radiation heat transfer. Two surfaces that exchange heat by radiation must "see" each other, as diagrammed in figure 15–28. The shape factor, F_{12}, is the fraction of surface 1 that sees surface 2, and F_{21} is the fraction of surface 2 that sees surface 1. The value for a shape factor will be between 1.0 and 0. An important relationship between any two surfaces is the **reciprocity law**:

$$A_1F_{12} = A_2F_{21} \tag{15–48}$$

The sum of all shape factors of one surface must be 1.0; that is,

$$F_{11} + F_{12} + F_{13} + \cdots = 1.0 \tag{15–49}$$

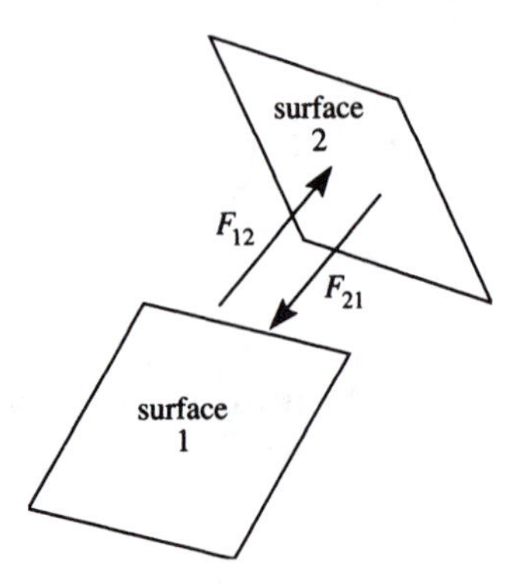

FIGURE 15–28 Shape factor concept

If surface 1 is convex, F_{11} will be zero; otherwise, it will be a value between 1.0 and 0. Shape factors have been determined for many geometric shapes, and four of the most common ones are shown in figures 15–29, 15–30, 15–31, and 15–32. Other shape factor information can be found in the literature. Using the results of these three figures, we can derive shape factors for other configurations, as the following example demonstrates.

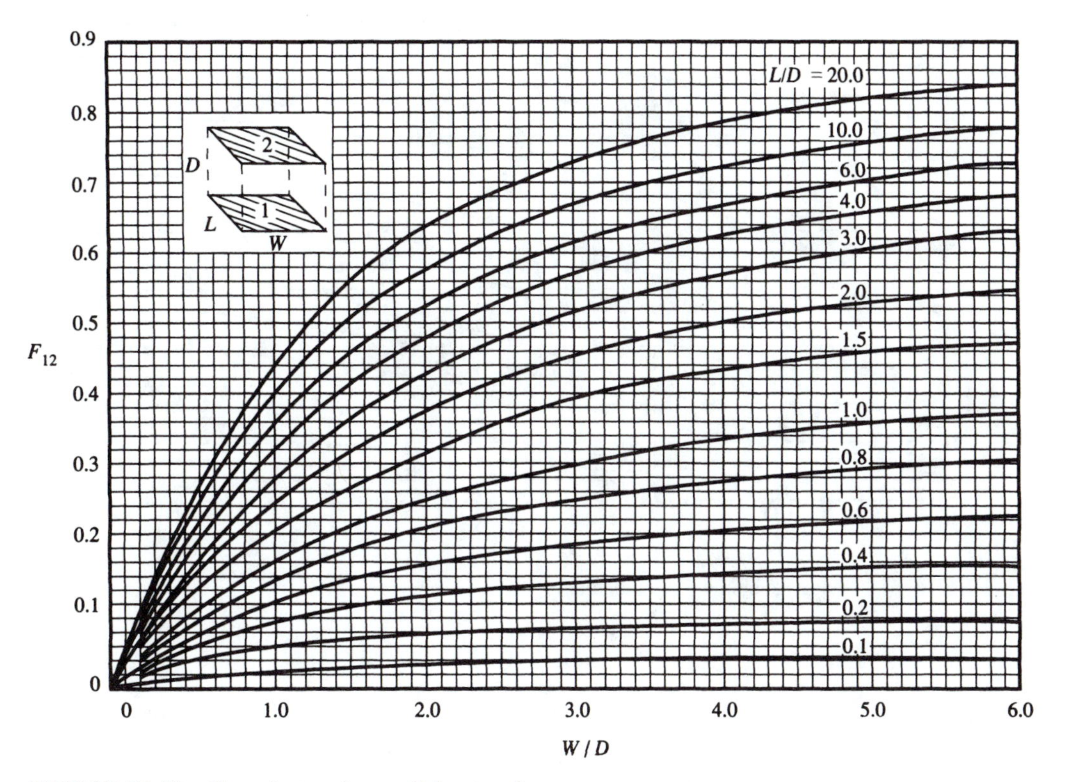

FIGURE 15–29 Shape factors for parallel rectangles

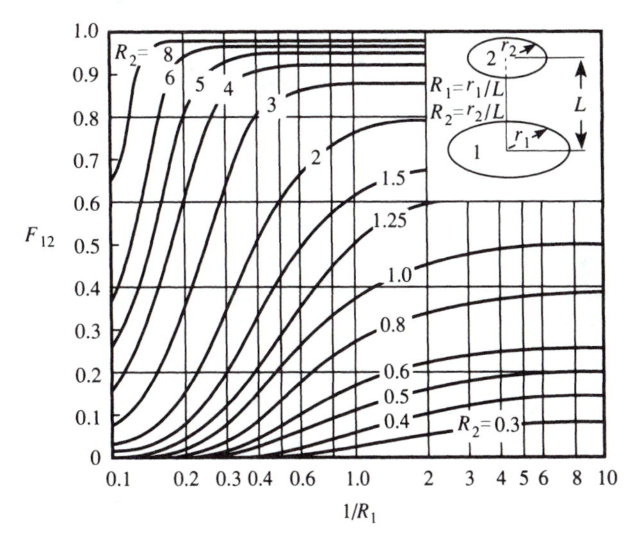

FIGURE 15–30 Shape factors for parallel concentric disks

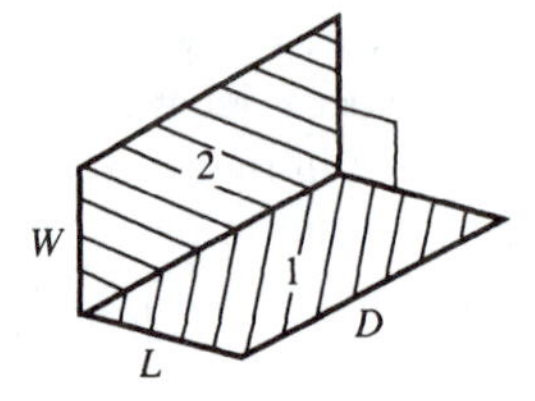

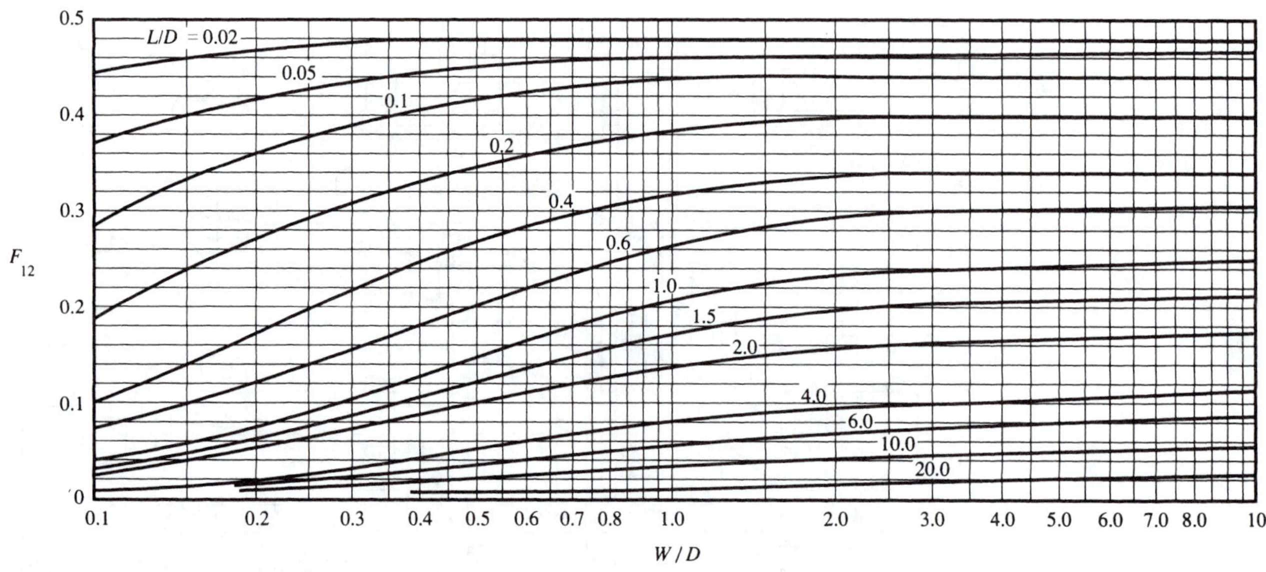

FIGURE 15–31 Shape factors for perpendicular rectangles

FIGURE 15–32 Shape factors for two concentric cylinders of the same length, one inside the other

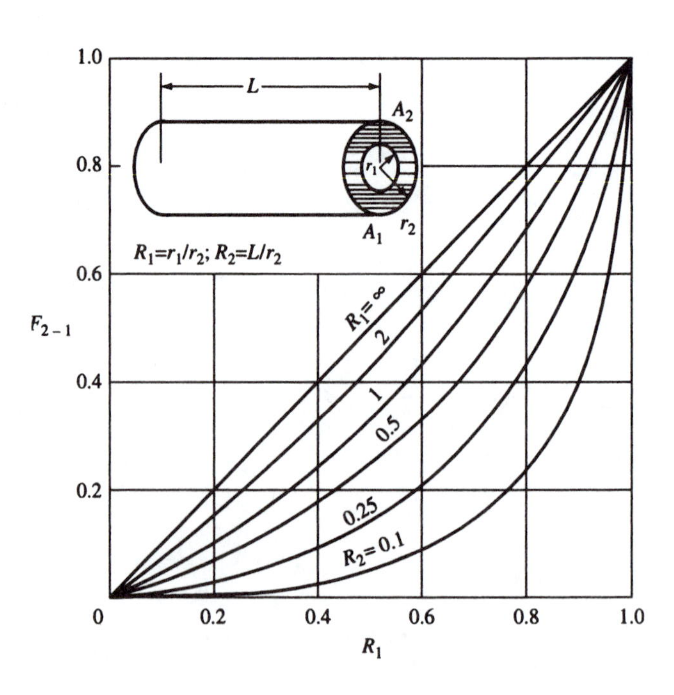

EXAMPLE 15–19 Determine the shape factor F_{21} between the window and floor as shown in figure 15–33. Also determine the shape factors F_{24} and F_{12}, where surface 4 is the surrounding surface. The area designated as 2 on figure 15–31 is the area $A_2 + A_3$ on figure 15–33. To find F_{21}, we may use figure 15–31 to find F_{31} and $F_{(3+2)(1)}$. Then

$$(A_3 + A_2)F_{(3+2)(1)} = A_3F_{31} + A_2F_{21}$$

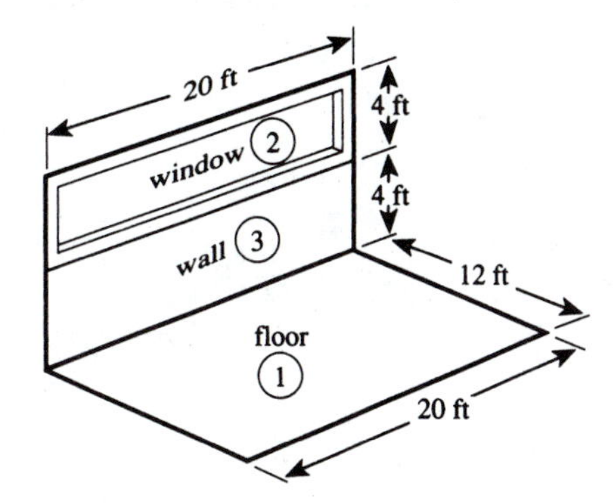

FIGURE 15–33 Window in a room

and

$$F_{21} = \frac{(A_3 + A_2)F_{(3+2)(1)} - A_3F_{31}}{A_2}$$

From figure 15–33, we have $A_3 = A_2 = 80\ \text{ft}^2$ and $A_1 = 240\ \text{ft}^2$, and for figure 15–31, $L = 12$ ft, $W_3 = 4$ ft, $W_{3+2} = 8$ ft, $D = 20$ ft, $W_3/D = 4/20 = 0.2$, $L/D = 0.6$, $W_{3+2}/D = 0.4$. Then

$$F_{13} = 0.128 \text{ and } F_{(3+2)(1)} = 0.192$$

The shape factor F_{21} is then

$$F_{21} = \frac{(80\ \text{ft}^2 + 80\ \text{ft}^2)(0.192) - (80\ \text{ft}^2)(0.128)}{80\ \text{ft}^2}$$

Answer
$$= 0.256$$

For surface 2, we may write

$$F_{22} + F_{21} + F_{23} + F_{24} = 1.0$$

but $F_{22} = F_{23} = 0$, so that

Answer
$$F_{24} = 1.0 - 0.256 = 0.744$$

Using the reciprocity relation of equation (15–48), we have

$$F_{12} = \frac{A_2}{A_1}F_{21}$$

$$= \left(\frac{80\ \text{ft}^2}{240\ \text{ft}^2}\right)(0.256)$$

Answer
$$= 0.0853$$

The amount of radiation that leaves one surface (1) and reaches another surface (2) is given by the relationship $\mathcal{J}_1A_1F_{12}$, and the amount of radiation that leaves 2 and reaches 1 is $\mathcal{J}_2A_2F_{21}$. The net radiation between 1 and 2 is then (using $\dot{Q}_{net}$ as positive if it is leaving surface 1)

$$\dot{Q}_{net} = \mathcal{J}_1A_1F_{12} - \mathcal{J}_2A_2F_{21} = (\mathcal{J}_1 - \mathcal{J}_2)A_2F_{21}$$
$$= (\mathcal{J}_1 - \mathcal{J}_2)A_1F_{12}$$

Using the resistance analogy, we have

$$\dot{Q}_{\text{net}} = \frac{\mathcal{J}_1 - \mathcal{J}_2}{1/A_2F_{21}} = \frac{\mathcal{J}_1 - \mathcal{J}_2}{1/A_1F_{12}} \tag{15–50}$$

For the case of two large parallel plates that can see only each other, we can use the circuit or network shown in figure 15–34. The net heat flow is

$$\dot{Q}_{\text{net}_{1-2}} = \frac{\dot{E}_{b_1}/A_1 - \dot{E}_{b_2}/A_2}{\dfrac{1-\epsilon_1}{\epsilon_1 A_1} + \dfrac{1}{A_1F_{12}} + \dfrac{1-\epsilon_2}{\epsilon_2 A_2}}$$

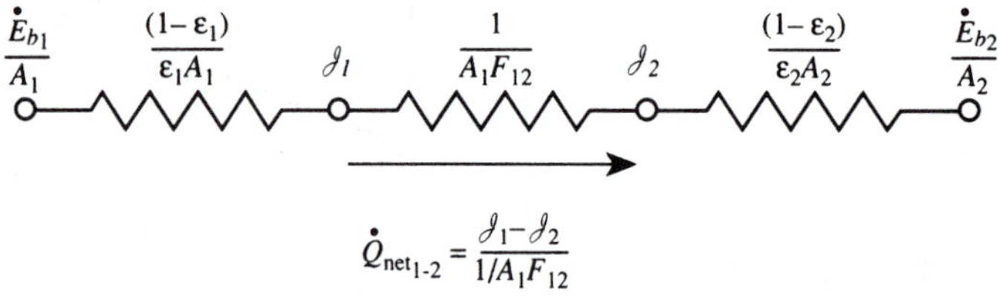

$\dfrac{\dot{E}_{b1}}{A_1}$ — $\dfrac{(1-\varepsilon_1)}{\varepsilon_1 A_1}$ — $\mathcal{J}_1$ — $\dfrac{1}{A_1F_{12}}$ — $\mathcal{J}_2$ — $\dfrac{(1-\varepsilon_2)}{\varepsilon_2 A_2}$ — $\dfrac{\dot{E}_{b2}}{A_2}$

$$\dot{Q}_{\text{net}_{1\text{-}2}} = \frac{\mathcal{J}_1 - \mathcal{J}_2}{1/A_1F_{12}}$$

FIGURE 15–34 Circuit for two surface radiation heat transfer

but $F_{12} = F_{21} = 1.0$ and $A_2 = A_1$, so that this equation reduces to

$$\dot{Q}_{\text{net}_{1-2}} = \frac{\sigma A(T_1^4 - T_2^4)}{1/\epsilon_1 + 1/\epsilon_2 - 1} \tag{15–51}$$

and if the plates are blackbodies,

$$\dot{Q}_{\text{net}_{1-2}} = \sigma A(T_1^4 - T_2^4) \tag{15–52}$$

For the situation where a pipe is inside another tube or pipe, such as shown in figure 15–35 for the radiation heat transfer between the two surfaces, A_1 and A_2, the circuit of figure 15–34 can be used. Because A_1 and A_2 are not equal, the net heat transfer can be written

$$\dot{Q}_{\text{net}_{1-2}} = \frac{\sigma A_1(T_1^4 - T_2^4)}{1/\epsilon_1 + (A_1/A_2)(1/\epsilon_2 - 1)} \tag{15–53}$$

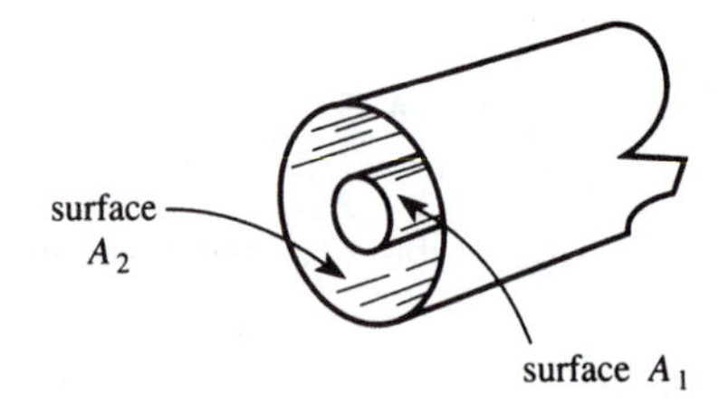

FIGURE 15–35 Radiation heat transfer between two cylinders, one inside the other

For those cases where A_1 is much smaller than A_2, equation (15–53) becomes

$$\dot{Q}_{\text{net}_{1-2}} = \sigma\epsilon_1 A_1(T_1^4 - T_2^4) \tag{15–54}$$

EXAMPLE 15–20 A 12-cm-diameter cast iron steam pipe having a surface temperature of 120°C is concentrically located inside a concrete conduit of 24-cm diameter. Determine the net heat transfer due to radiation per unit length between the steam pipe and the conduit if the conduit is at 25°C.

Solution This is a problem of two surfaces that can see only each other, so the net heat is given by equation (15–53). The areas are

$$A_1 = \pi D_1 L = \pi(0.12\text{ m})L \quad \text{and} \quad A_2 = \pi(0.24\text{ m})L$$

Answer From table B–9, we find that, for cast iron, $\epsilon_1 = 0.44$, and for concrete, $\epsilon_2 = 0.90$. Then

$$\frac{\dot{Q}_{net_{1-2}}}{L} = \frac{(5.67 \times 10^{-8}\ \text{W/m}^2 \cdot \text{K}^4)(\pi)(0.12\ \text{m})[(393\ \text{K})^4 - (298\ \text{K})^4]}{(1/0.44) + (1/2)(1/0.9 - 1)}$$
$$= 146.6\ \text{W/m}$$

If three surfaces are exchanging heat through radiation, the problems become more complicated than two-surface problems. One important problem involving three surfaces is when radiation exchange occurs between two of the surfaces and the third is a blackbody at some known temperature, such as room temperature or atmospheric temperature. This situation often occurs in large rooms or chambers when two objects are exchanging radiant heat. There are other problems that can be approximated by this type of problem. The circuit diagram for three-surface radiation with surface 3 a blackbody is sketched in figure 15–36. The general approach to solving these types of problems is first to determine the radiosities and then solve for the net heat transfers. Referring to figure 15–36, we use the idea that the radiosity nodes are in thermal equilibrium so that the sum of the net heat flows at each of the radiosity nodes is zero. This gives the two equations

$$\frac{\sigma T_1^4 - \mathcal{J}_1}{(1 - \epsilon_1)/\epsilon_1 A_1} + \frac{\mathcal{J}_2 - \mathcal{J}_1}{1/A_1 F_{12}} + \frac{\mathcal{J}_3 - \mathcal{J}_1}{1/A_1 F_{13}} = 0$$
$$\frac{\sigma T_2^4 - \mathcal{J}_2}{(1 - \epsilon_2)/\epsilon_2 A_2} + \frac{\mathcal{J}_1 - \mathcal{J}_2}{1/A_1 F_{12}} + \frac{\mathcal{J}_3 - \mathcal{J}_2}{1/A_2 F_{23}} = 0$$

Then, with the use of some algebra and the relationships $\mathcal{J}_3 = \sigma T_3^4$, $F_{12} + F_{13} = 1.0$, and $F_{21} + F_{23} = 1$, these two equations become

$$A_{11}\mathcal{J}_1 + A_{12}\mathcal{J}_2 = B_1$$
$$A_{21}\mathcal{J}_1 + A_{22}\mathcal{J}_2 = B_2 \qquad \textbf{(15–55)}$$

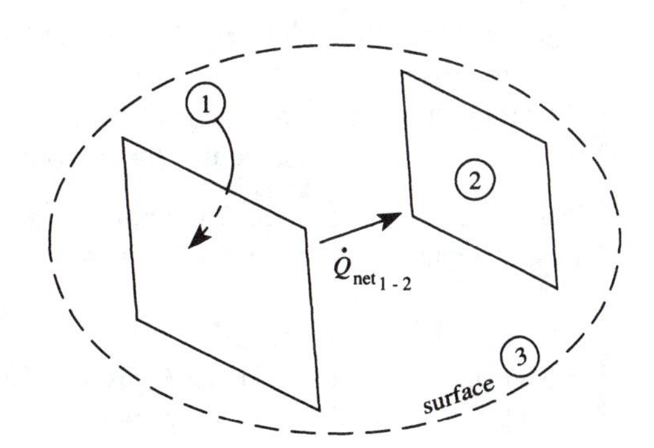

FIGURE 15–36 Three-surface radiation with surface 3 a blackbody

where

$$A_{11} = \frac{1}{1-\epsilon_1} \qquad B_1 = \frac{\epsilon_1\sigma}{1-\epsilon_1}T_1^4 + F_{13}\sigma T_3^4$$

$$A_{12} = -F_{12}$$

$$A_{21} = -F_{21} \qquad B_2 = \frac{\epsilon_2\sigma}{1-\epsilon_2}T_2^4 + F_{23}\sigma T_3^4$$

$$A_{22} = \frac{1}{1-\epsilon_2}$$

Using algebra or rules of matrices, we can show that

$$\mathcal{J}_1 = \frac{A_{22}B_1 - A_{12}B_2}{A_{11}A_{22} - A_{12}A_{21}}$$

$$\mathcal{J}_2 = \frac{-A_{21}B_1 + A_{11}B_2}{A_{11}A_{22} - A_{12}A_{21}} \tag{15–56}$$

Then the net radiation from surfaces 1 and 2 is

$$\dot{Q}_{net_1} = \frac{\sigma T_1^4 - \mathcal{J}_1}{(1-\epsilon_1)/\epsilon_1 A_1}$$

$$\dot{Q}_{net_2} = \frac{\sigma T_2^4 - \mathcal{J}_2}{(1-\epsilon_2)/\epsilon_2 A_2} \tag{15–57}$$

$$\dot{Q}_{net_3} = \dot{Q}_{net_1} + \dot{Q}_{net_2} \tag{15–58}$$

Equation (15–55) through (15–58) can be solved by hand calculators. However, a computer program is included on soft diskette which allows for microcomputer assistance if the following terms are given: ϵ_1, ϵ_2, A_1, A_2, T_1, T_2, T_3, and F_{12}.

EXAMPLE 15–21 A space heater has a 2-ft-by-3-ft heating surface at 1200°F, and its emissivity is 0.6. Determine the net heat transfer to a 2-ft-by-3-ft sheet of plywood at 80°F placed 2 ft away from the heater. Then determine the net heat transfer to the plywood sheet if it is moved back so that it is 10 ft away. Assume that the surroundings are a blackbody at 70°F.

Solution We can use equation (15–55) to determine the radiosities and then the net heat transfers. First, we must determine the various terms in those two equations:

$$A_1 = A_2 = 2 \times 3 = 6 \text{ ft}^2$$

From figure 15–29, the value for F_{12} is found to be approximately 0.248, so that $F_{13} = 0.752$. From table B–9, the emissivity of plywood is taken to be 0.92, and the temperatures are

$$T_1 = 1660°\text{R}$$
$$T_2 = 540°\text{R}$$
$$T_3 = 530°\text{R}$$

Then equations (15–55) become

$$2.5\mathcal{J}_1 - 0.248\mathcal{J}_2 = 19{,}921.84 \text{ Btu/h} \cdot \text{ft}^2$$
$$-0.248\mathcal{J}_1 + 12.5\mathcal{J}_2 = 1804.71 \text{ Btu/h} \cdot \text{ft}^2$$

Solving for the radiosities as given by equations (15–56), we have

$$\mathcal{J}_1 = 7998.8 \text{ Btu/h} \cdot \text{ft}^2$$
$$\mathcal{J}_2 = 303.1 \text{ Btu/h} \cdot \text{ft}^2$$

Then the net heat transfers are

$$\dot{Q}_{net_1} = \frac{\sigma T_1^4 - \mathcal{J}_1}{(1 - \epsilon_1)/\epsilon_1 A_1} = \frac{(0.174 \times 10^{-8})(1660)^4 - 7998.8}{(0.40)/(0.6 \times 6)} = 46{,}922.4 \text{ Btu/h}$$

The net heat flow from the plywood is

$$\dot{Q}_{net_2} = \left[\frac{\sigma T_2^4 - \mathcal{J}_2}{(1 - \epsilon_2)/\epsilon_2 A_2}\right] = \left[\frac{(0.174 \times 10^{-8})(540)^4 - 303.1}{0.08/(0.92)(6)}\right]$$

Answer

$$= -10{,}705.1 \text{ Btu/h}$$

The negative sign means that the heat is coming into the plywood from the heater and the surroundings.

Also, the net heat to the surroundings would be found by using equation (15–58):

$$\dot{Q}_{net_3} = \dot{Q}_{net_1} + \dot{Q}_{net_2} = 46{,}922.4 \text{ Btu/h} - 10{,}705.1 \text{ Btu/h} = 36{,}217.3 \text{ Btu/h}$$

If the plywood is now moved back so that it is 10 ft away from the heater, the shape factor F_{12} is then found from figure 15–29 to be about 0.0175. The shape factor F_{13} is then 0.9825, and by using equations (15–55) and (15–56), we find that the radiosities are $\mathcal{J}_1 = 7983$ Btu/h and $\mathcal{J}_2 = 158.1$ Btu/h. The net heat transfers are found from equations (15–57) to be

$$\dot{Q}_{net_1} = 47{,}064.6 \text{ Btu/h}$$

and

Answer

$$\dot{Q}_{net_2} = -700.13 \text{ Btu/h}$$

The net heat transfer to the surroundings, $\dot{Q}_{net_3}$, is 46,364.47 Btu/h. Notice how much less heat is radiated to the plywood when it is moved 10 ft away. At this location, the radiation is 700.13 Btu/h, compared to 10,705.1 Btu/h at the closer location.

By using the program RADIATE (described in appendix A–13) with a microcomputer, you can solve example 15–21 and obtain nearly the same answers.

15–7 HEAT EXCHANGERS

In this book, heat exchangers have been considered often. The steam generator or boiler (see chapter 11) used in electric power generation is a heat exchanger. Condensers and evaporators are also heat exchangers, as are radiators of automobile engines. Hot water or steam radiators used to heat rooms or buildings are examples of heat exchangers, and many other devices used in society and in technology are heat exchangers.

In this section, we consider heat exchangers and in particular determine the methods used to predict proper sizes of these devices for given applications. Figure 15–37 shows some general arrangements for heat exchangers, particularly the shell-and-tube type and its variations. The shell and tube is a common form of heat exchanger that is used in a wide variety of processes. The concepts and methods that we will now develop can be used for other types of heat exchangers, but the specific details may be different. The heat exchangers shown in figure 15–37 usually operate with cold and hot fluids and exchange heat either by **parallel flow**, where the fluids enter the heat exchanger at the same end and flow in the same direction, or by **counterflow**, where the fluids flow in opposite directions in the heat exchanger. A one-pass counterflow shell-and-tube heat exchanger is shown in figure 15–38. The steady-flow energy equation for the heat exchanger is

$$\dot{m}_H(h_1 - h_2) = \dot{m}_C(h_4 - h_3) \quad \textbf{(15–59)}$$

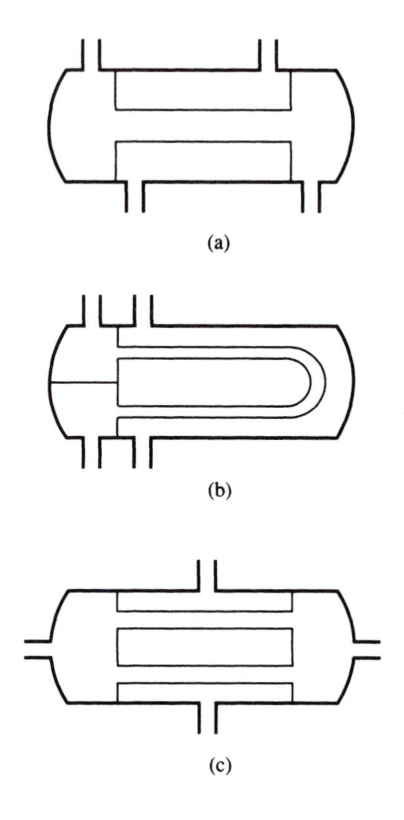

FIGURE 15–37 Some heat exchangers: (a) one-pass shell-and-tube; (b) two-pass shell-and-tube; (c) cross-flow

For fluids that are incompressible, this equation reduces to

$$\dot{m}_H c_p(T_1 - T_2) = \dot{m}_C c_p(T_4 - T_3) \tag{15–60}$$

The thermal resistance concepts for conduction/convection heat transfer over a tube or pipe, as developed in example 15–9, apply to the shell-and-tube heat exchanger where the heat transfer, $\dot{Q}$, is between the two fluids. From example 15–9, we have

$$\dot{Q} = \frac{\Delta T_{\text{overall}}}{\sum R_T}$$

where $\sum R_T = \dfrac{1}{A_i K_{h_i}} + \dfrac{\ln(r_o/r_i)}{2\pi L \kappa_{\text{tube}}} + \dfrac{1}{A_o K_{h_o}}$

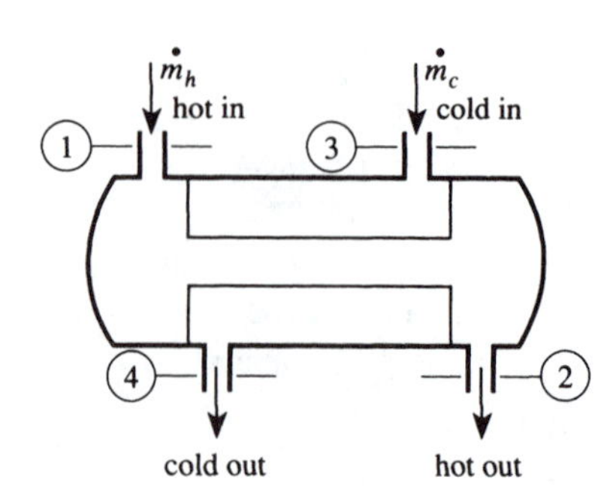

FIGURE 15–38 One-pass shell-and-tube counterflow heat exchanger

In heat exchanger design, it is customary to use the term **conductance** or **overall conductance**, which is defined as

$$\mathcal{U} = \frac{1}{A_0 \sum R_T}$$

so that

$$\dot{Q} = \mathcal{U} A_o \, \Delta T_{\text{overall}} \tag{15–61}$$

where A_o is the outside area of the tube. The heat transfer, $\dot{Q}$, can also be set equal to each side of equation (15–59) or (15–60). Then

$$\mathcal{U} A_o \, \Delta T_{\text{overall}} = \dot{m}_H c_p (T_1 - T_2) = \dot{m}_C c_p (T_4 - T_3) \tag{15–62}$$

and this equation is used to predict the size of heat exchangers. The overall temperature difference, $\Delta T_{\text{overall}}$, is the temperature difference between the hot and the cold fluids. In actual heat exchangers, this temperature difference is changing throughout the system, and a method used to determine the value of the difference is with the **log mean temperature difference** (LMTD), defined as

$$\text{LMTD} = \frac{(T_2 - T_3) - (T_1 - T_4)}{\ln[(T_2 - T_3)/(T_1 - T_4)]} = \Delta T_{\text{overall}} \tag{15–63}$$

This equation can be used for both counterflow and parallel flow. If the LMTD is undefined, such as when $T_1 = T_4$, $T_2 = T_3$ or $T_2 - T_3 = T_1 - T_4$, the $\Delta T_{\text{overall}}$ can be found from the approximation

$$\Delta T_{\text{overall}} \approx \frac{T_1 + T_2}{2} - \frac{T_3 + T_4}{2} \tag{15–64}$$

For heat exchanger arrangements other than that of figure 15–38, a correction factor, F_c, can be defined for use in equation (15–62), so that

$$\mathcal{U} A_o F_c (\Delta T_{\text{overall}}) = \dot{m}_H c_p (T_1 - T_2) = \dot{m}_C c_p (T_4 - T_3) \tag{15–65}$$

The correction factor, F_c, can be determined from figures 15–39 and figure 15–40 for some other heat exchanger arrangements. For heat exchangers involving phase changes of one or both of the fluids, such as condensers, evaporators, and boilers, the correction factor, F_c, has a value of 1.0. Further charts and information on correction factors treating more involved heat exchangers are given in heat transfer books and in the standards of the Tubular Exchange Manufacturers Association (TEMA).

EXAMPLE 15–22 Determine the overall conductance, $\mathcal{U}$, for an air-to-air heat exchanger made of a Teflon® tube, 12-cm OD by 10-cm ID, concentrically located inside a 30-cm-diameter steel shell. Use K_{h_o} of 100 W/m$^2 \cdot$K and K_{h_i} of 2000 W/m$^2 \cdot$K

Solution The overall conductance is given by

$$\mathcal{U} = \frac{1}{\dfrac{A_o}{A_i K_{h_i}} + \dfrac{A_o \ln(r_o/r_i)}{2\pi L \kappa_{\text{tube}}} + \dfrac{1}{K_{h_o}}}$$

$$= \frac{1}{\dfrac{D_o}{D_i K_{h_i}} + \dfrac{D_o \ln(r_o/r_i)}{2\kappa_{\text{tube}}} + \dfrac{1}{K_{h_o}}}$$

FIGURE 15–39 Correction factor for two-pass shell-and-tube heat exchanger

F_c correction factor for equation (15-65)

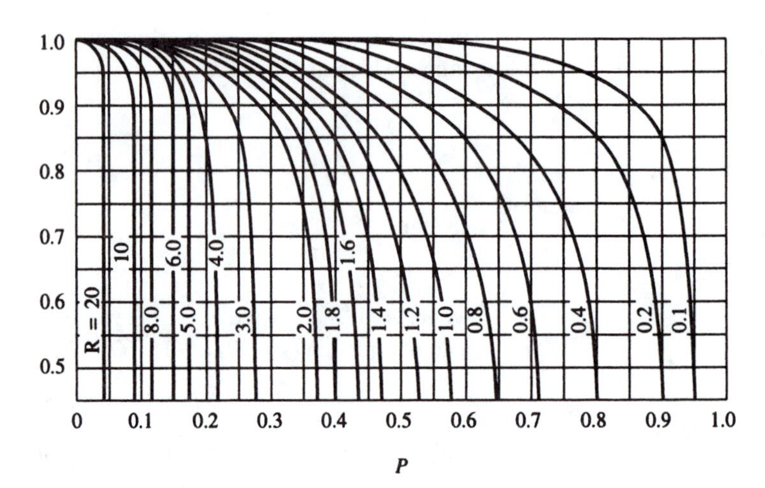

$$P = \frac{T_{co} - T_{ci}}{T_{hi} - T_{ci}}$$

$$R = \frac{T_{hi} - T_{ho}}{T_{co} - T_{ci}}$$

T_{co} = outlet cold fluid

T_{ci} = inlet cold fluid

T_{ho} = outlet hot fluid

T_{hi} = inlet hot fluid

FIGURE 15–40 Correction factor for cross-flow shell-and-tube heat exchanger

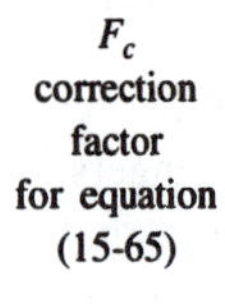

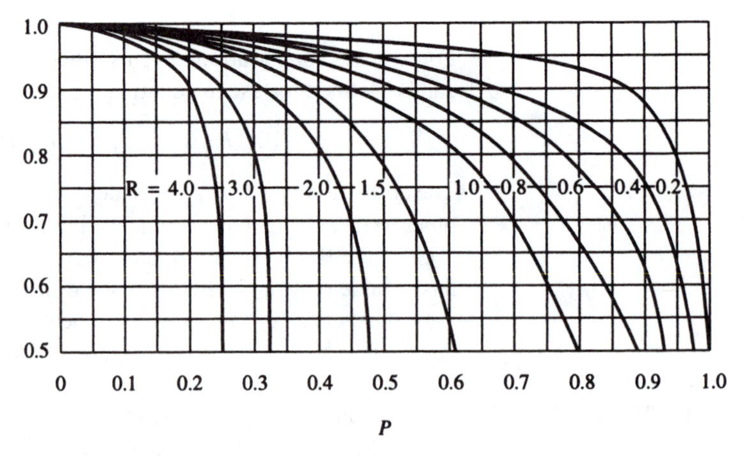

$$P = \frac{T_{co} - T_{ci}}{T_{hi} - T_{ci}}$$

$$R = \frac{T_{hi} - T_{ho}}{T_{co} - T_{ci}}$$

T_{co} = outlet cold fluid

T_{ci} = inlet cold fluid

T_{ho} = outlet hot fluid

T_{hi} = inlet hot fluid

Using a value of 0.35 W/m·K for the thermal conductivity of Teflon® from table B–8, and substituting the values for the terms into this equation, we have

$$\mathcal{U} = \frac{1}{\dfrac{0.12\text{ m}}{(0.10\text{ m})(2000\text{ W/m}^2\cdot{}^\circ\text{C})} + \dfrac{(0.12\text{ m})\ln(6/5)}{2(0.35\text{ W/m}\cdot\text{K})} + \dfrac{1}{100\text{ W/m}^2\cdot\text{K}}}$$

Answer

$$= 23.89\text{ W/m}^2\cdot\text{K} = 23.89\text{ W/m}^2\cdot{}^\circ\text{C}$$

EXAMPLE 15–23 For the air-to-air heat exchanger of example 15–22, assume counterflow of cold air at −10°C and warm air at 20°C. Also assume that the cold air leaves at 10°C through the inside of the tube at 2 m/s. Determine the length of the tube required for the heat exchanger.

Solution The length of the tube for the heat exchanger can be found if the heat exchanger area is determined. The area can be found from equation (15–62), after computing the heat transfer from the mass flow/enthalpy change term, $\dot{m}_C c_p(T_4 - T_3)$. The air density

$$\rho_4 = \frac{p_4}{RT_4} = \frac{100\text{ kN/m}^2}{(0.287\text{ kN}\cdot\text{m/kg}\cdot\text{K})(280\text{ K})} = 1.2444\text{ kg/m}^3$$

The area, A_4, is $\pi r_i^2 = \pi(0.05\text{ m})^2 = 0.00785\text{ m}^2$, and the velocity, $\overline{V}_4$, is 2 m/s. Then the mass flow rate is

$$\dot{m}_4 = \rho_4 \overline{V}_4 A_4 = (1.2444\text{ kg/m}^3)(2\text{ m/s})(0.00785\text{ m}^2) = 0.0195\text{ kg/s} = \dot{m}_C$$

From table B–4, we find $c_p = 1.007$ kJ/kg·K, so, for the heat transfer,

$$\dot{Q} = \dot{m}_C c_p(T_4 - T_3) = (0.0195\text{ kg/s})(1.007\text{ kJ/kg}\cdot\text{K})[10^\circ\text{C} - (-10^\circ\text{C})] = 0.394\text{ kJ/s} = 0.394\text{ kW}$$

Also, if we assume that the mass flow of the hot air is about the same as that of the cold air, we find that $T_1 - T_2 = T_4 - T_3$, so after noting that $T_1 = 20^\circ$C, we find that $T_2 = 0^\circ$C. The overall temperature difference is approximately

$$\Delta T_{\text{overall}} = 10^\circ\text{C}$$

and the surface area of the heat exchanger can then be found from equation (15–62):

$$A_o = \frac{\dot{Q}}{\mathcal{U}\,\Delta T_{\text{overall}}} = \frac{0.394\text{ kW}}{(0.02389\text{ kW/m}^2\cdot\text{K})(10\text{ K})} = 1.649\text{ m}^2$$

The tube length is then

Answer

$$L = \frac{A_o}{\pi D_o} = \frac{1.649\text{ m}^2}{(\pi)(0.12\text{ m})} = 4.37\text{ m}$$

EXAMPLE 15–24 Determine the necessary area required for a cross-flow heat exchanger for hot and cold water. Assume that $\mathcal{U}$ is 240 Btu/h·ft²·°F, the hot water enters at 190°F and leaves at 130°F, and 300 lbm/min of cold water enters at 68°F and leaves at 90°F. Determine the amount of hot water flowing through the heat exchanger per minute, and find the area of the heat exchanger.

Solution

Using the energy balance of equation (15–60), we have

$$\dot{m}_H = \frac{\dot{m}_C c_p(T_4 - T_3)}{c_p(T_1 - T_2)} = \dot{m}_C \frac{T_4 - T_3}{T_1 - T_2}$$

$$= (300 \text{ lbm/min})\left(\frac{90°\text{F} - 68°\text{F}}{190°\text{F} - 130°\text{F}}\right) = 110 \text{ lbm/min}$$

The heat exchanger surface area A_o can be found from equation (15–65):

$$\dot{Q} = \dot{m}_C c_p(T_4 - T_3) = \mathcal{U} A_o F_c(\text{LMTD})$$

From figure 15–39, we determine

$$P = \frac{90°\text{F} - 68°\text{F}}{190°\text{F} - 68°\text{F}} = 0.18$$

$$R = \frac{190°\text{F} - 130°\text{F}}{90°\text{F} - 68°\text{F}} = 2.73$$

and the correction factor, F_c, is found from figure 15–39 to be about 0.96.

The LMTD is found from equation (15–63):

$$\text{LMTD} = \Delta T_{\text{overall}} = \frac{(130°\text{F} - 68°\text{F}) - (190°\text{F} - 90°\text{F})}{\ln[(130°\text{F} - 68°\text{F})/(190°\text{F} - 90°\text{F})]}$$

$$= 79.49°\text{F}$$

The heat transfer is

$$\dot{Q} = \dot{m}_C c_p(T_4 - T_3)$$

$$= (300 \text{ lbm/min})(60 \text{ min/h})(1.00 \text{ Btu/lbm}\cdot°\text{F})(90°\text{F} - 68°\text{F})$$

$$= 396{,}000 \text{ Btu/h}$$

and the surface area of the heat exchanger is

$$A_o = \frac{\dot{Q}}{\mathcal{U} F_c(\text{LMTD})}$$

$$= \frac{396{,}000 \text{ Btu/h}}{(240 \text{ Btu/h}\cdot\text{ft}^2\cdot°\text{F})(0.96)(79.49°\text{F})}$$

Answer

$$= 21.62 \text{ ft}^2$$

Heat exchangers can be analyzed by using other methods, and the interested student can find some of these methods described in heat transfer books.

15–8 SUMMARY

There are three modes of heat transfer: conduction, convection, and radiation. Conduction heat transfer in one direction through a slab is given by

$$\dot{Q} = -\frac{\kappa A\,\Delta T}{\Delta x} \tag{15–1}$$

or

$$\dot{Q} = \frac{\Delta T}{R_T}$$

where κ is the thermal conductivity and R_T is the thermal resistance. Conduction heat transfer in a radial direction through a tube or pipe is given by

$$\dot{Q} = \frac{2\pi\kappa L(T_i - T_o)}{\ln(r_o/r_i)} \tag{15–11}$$

Convection heat transfer is described by the equation

$$\dot{Q} = K_h A(T_w - T_\infty) \tag{15–15}$$

Combined conduction/convection heat transfer situations occur over building walls and through tubes and pipes. Fins are also devices that conduct heat and by convection can increase the amount of heat transfer on a given surface area. For square fins that are long enough that the fin end temperature is the same as the surrounding temperature, the fin temperature along its length is given by

$$T = T_\infty + (T_o - T_\infty)e^{\sqrt{K_h P/\kappa A}x} \tag{15–17}$$

and the heat transfer through this type of fin is

$$\dot{Q} = \sqrt{K_h \kappa AP}(T_o - T_\infty) \tag{15–18}$$

where T_o is the fin base temperature. Fin efficiency is defined as

$$\eta_{\text{fin}} = \frac{\dot{Q}_{\text{fin}}}{\dot{Q}_o} \tag{15–25}$$

where $\dot{Q}_o = K_h A_{\text{fin}}(T_o - T_\infty)$

For very long fins, the fin efficiency is

$$\eta_{\text{fin}} = \frac{1}{L}\sqrt{\frac{\kappa A}{K_h P}} \tag{15–27}$$

Circumferential fins are used frequently and are usually analyzed through the fin efficiency relationship.

For solid objects that have conduction and convection and where the parameter $K_h V/A_s \kappa \leq 0.1$, lumped heat capacity analysis can be used. For the lumped heat capacity, the temperature of the object is the same as that at the surface and inside, and is given by the equation

$$T_s = T_\infty + (T_o - T_\infty)e^{-(K_k A_s/mc_p)t} \tag{15–29}$$

where T_o is the temperature of the object as it is first surrounded by a fluid at another temperature T_∞. The time constant is defined as

$$t_c = \frac{mc_p}{K_h A_s} \tag{15–30}$$

and is the time when the difference in temperature between the object and the surroundings has been reduced by 63.31%.

Forced convection heat transfer is due to fluid motion caused by some external force other than gravity. The convective heat transfer coefficient over a flat plate is approximated by

$$K_h = \frac{\kappa^{2/3}}{L}(c_p\mu_k)^{1/3}(0.037Re_L^{0.8} - 850) \tag{15–32}$$

For flow around objects,

$$K_h = \frac{\kappa}{D}(C)\left(\frac{\rho\bar{V}_\infty D}{\mu_k}\right)^n\left(\frac{c_p\mu_k}{\kappa}\right)^{1/3} \tag{15–33}$$

where C and n are empirical constants. For flow through tubes and smooth pipes where the Reynolds number, Re_D, is greater than about 2500,

$$K_h = \left(\frac{\kappa}{D}\right)(0.023)(\mathrm{Re}_D)^{0.8}\left(\frac{c_p\mu_k}{\kappa}\right)^n \tag{15–34}$$

where n is 0.4 if the fluid is heated and 0.3 if the fluid is cooled.

Natural convection occurs if a gas or a fluid rises due to decreased density through heating. For convection heat transfer with air and a surface, the simplified equation for predicting the convective heat transfer coefficient K_h is

$$K_h = C\left(\frac{T_w - T_\infty}{L}\right)^n$$

or

$$K_h = C(T_w - T_\infty)^n$$

depending on the value of the product, $(Gr)(c_p\mu_k/\kappa)$, where the Grashof number is

$$Gr = \frac{g(T_w - T_\infty)L^3\rho^2}{\mu_k^2 T_\infty}$$

for perfect gases. The chimney effect, or induced air or gas flow upward due to density decrease, is described by the draft pressure

$$\Delta p = \frac{p_\infty g L}{R}\left(\frac{1}{T_\infty} - \frac{1}{T_s}\right) \tag{15–37}$$

and the velocity of the leaving gases is

$$\overline{V}_e = \begin{cases} \sqrt{\dfrac{2\,\Delta p}{\rho}} & \text{(SI units, m/s)} \\ \sqrt{\dfrac{2\,\Delta p g_c}{\rho}} & \text{(English units, ft/s)} \end{cases} \tag{15–38}$$

The chimney or smokestack and the glass or Trombé wall are two applications of natural convection heat transfer.

Radiation heat transfer is a form of electromagnetic radiation. For blackbodies, the rate of energy emitted is

$$\dot{E}_b = A\sigma T^4 \tag{15–40}$$

where σ is the Stefan–Boltzmann constant. Gray-body radiation is defined as

$$\dot{E}_g = A\epsilon\sigma T^4 \tag{15–41}$$

where ϵ is the emissivity. Kirchhoff's identity says that $\alpha = \epsilon$, where α is the absorptivity. For gray opaque bodies,

$$\alpha + \rho_r = 1.0 \tag{15–44}$$

where ρ_r is the reflectivity. For transparent bodies,

$$\tau + \rho_r = 1.0 \tag{15–45}$$

where τ is the transmissivity. For real materials and surfaces,

$$\alpha + \tau + \rho_r = 1.0 \tag{15–46}$$

The net heat transfer leaving a surface or system is

$$\dot{Q}_{\text{net}} = \frac{\dot{E}_b/A - \mathcal{J}}{(1 - \epsilon)/\epsilon A} \tag{15–47}$$

where $\mathcal{J}$ is the radiosity or total heat leaving a surface per unit area.

The shape factor F_{12} is the fraction of surface 1 that sees surface 2. For any surface seeing a number of other surfaces,

$$F_{11} + F_{12} + F_{13} + F_{14} + \cdots = 1.0 \tag{15–49}$$

For flat or convex surfaces, which cannot see themselves, $F_{11} = 0$. The reciprocity law says that

$$A_1 F_{12} = A_2 F_{21}$$

For two surfaces that can see only each other,

$$\dot{Q}_{\text{net}_{1-2}} = \frac{\sigma A_1 (T_1^4 - T_2^4)}{1/\epsilon_1 + (A_1/A_2)(1/\epsilon_2 - 1)} \tag{15–53}$$

For cases where A_2 is much larger than A_1, but surfaces 1 and 2 can see only each other,

$$\dot{Q}_{\text{net}_{1-2}} = \sigma \epsilon_1 A_1 (T_1^4 - T_2^4) \tag{15–54}$$

and if $A_1 = A_2$, then

$$\dot{Q}_{\text{net}_{1-2}} = \frac{\sigma A_1 (T_1^4 - T_2^4)}{1/\epsilon_1 + 1/\epsilon_2 - 1} \tag{15–51}$$

Problems involving radiation heat transfer between three surfaces or more require the simultaneous solution of sets of equation or matrix operations.

Heat exchangers can be analyzed with the LMTD method, where

$$\dot{Q} = \mathcal{U} A_o F_c(\text{LMTD}) = \dot{m}_C c_p (\Delta T)_{\text{cold}} = \dot{m}_H c_p (\Delta T)_{\text{hot}}$$

in which $\mathcal{U}$ is called the **overall conductance** of the heat exchanger, A_o is the surface area of the heat exchanger, and LMTD is the overall temperature difference, defined as

$$\text{LMTD} = \frac{(T_{h_0} - T_{c_i}) - (T_{h_i} - T_{c_o})}{\ln[(T_{h_o} - T_{c_i})/(T_{h_i} - T_{c_o})]} \tag{15–63}$$

where T_{h_o} = outlet temperature of hot fluid
T_{h_i} = inlet temperature of hot fluid
T_{c_o} = outlet temperature of cold fluid
T_{c_i} = inlet temperature of cold fluid

If the LMTD is undefined for a particular situation, the overall temperature difference in the heat exchanger can be approximated by the equation

$$\Delta T_{\text{overall}} = \frac{T_{h_i} + T_{h_o}}{2} - \frac{T_{c_o} + T_{c_i}}{2} \tag{15–64}$$

The correction factor F_c is 1.0 for single-pass counterflow heat exchangers and is less than 1.0 for others.

DISCUSSION QUESTIONS

Section 15–1

15–1 How are *thermal conductivity* and *thermal resistance* different?

15–2 What does *Fourier's law* describe?

15–3 What is meant by an *R-value*?

Section 15–2

15–4 How is *convection heat transfer* different than *conduction heat transfer*?

15–5 What affects the *convection heat transfer coefficient*? Is it ever a constant?

Section 15–3

15–6 What is a *fin*?

15–7 What is the ideal heat transfer at a fin, based on the definition of the fin efficiency?

15–8 What does the term *time constant* represent for a system that is being cooled or heated?

Section 15–4

15–9 How are *forced* and *free or natural* convection different?

15–10 What is the Reynolds number?

15–11 What is the Prandtl number?

Section 15–5

15–12 When does the *chimney effect* occur?

15–13 What is the *Grashof number*?

Section 15–6

15–14 What is a *blackbody*?

15–15 Define *emissivity*.

15–16 Define *absorptivity*.

15–17 What is *Kirchhoff's law of radiation*?

15–18 Define *reflectivity*.

15–19 Define *transmissivity*.

15–20 What is a *gray body*?

15–21 Define *shape factor*.

15–22 What is *reciprocity*?

15–23 What is meant by *radiosity*?

Section 15–7

15–24 What is a *parallel flow* heat exchanger?

15–25 What is a *counter flow* heat exchanger?

15–26 What does *conductance* have to do with heat exchanger design?

15–27 What does the term *LMTD* mean?

PRACTICE PROBLEMS

Section 15–1

Problems that use SI units are indicated by an (M) under the problem number; those that use English units are indicated by an (E).

15–1 (M) The rate of heat loss from a 0.1-m^2 area of a furnace wall 40 cm thick is 150 W. If the inside of the furnace is at 1000°C and the wall is composed of fire brick, estimate the outside wall temperature.

15–2 (E) Sawdust is used to insulate an ice storage chamber. If a 2-ft-thick wall of sawdust surrounds the chamber, the inside surface is 25°F, and the outside surface is 80°F, determine the rate of heat loss through sawdust for 1 ft^2 of area. Also determine the *R*-value for 2-ft-thick sawdust.

15–3 (M) The combustion chamber of an internal combustion engine is at a temperature of 1100°C when fuel is burned in the chamber. Assuming that the engine is made of iron and has an average thickness of 6.5 cm between the combustion chamber and the outside surface, determine the heat transfer rate per unit of wall area when combustion of fuel occurs. The temperature around the engine is 26°C, but the outside surface of the engine is 38°C.

15–4 (E) For figure 15–41, the outside brick wall is at a temperature of 15°F at a certain time in severe winter weather. The inside wallboard, which is separated from the brick wall by a 6-in air gap, is at 78°F. Determine, approximately, the heat rate per square foot area of wall for this condition, and indicate the direction, whether from brick to wallboard, or vice versa.

15–5 (E) For problem 15–4, assume that during summer weather the brick wall temperature is 80°F. Find the heat transfer rate through the air gap per unit area and the direction of it under this condition.

15–6 (M) Determine the *R*-value and the rate of heat flow per unit area of the wall shown in figure 15–42 if the inside temperature T_i is 20°C and the outside temperature T_o is 38°C.

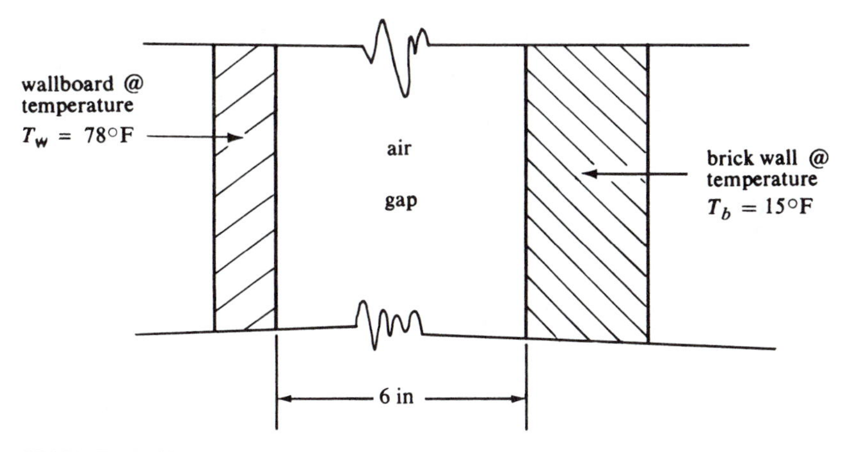

FIGURE 15–41

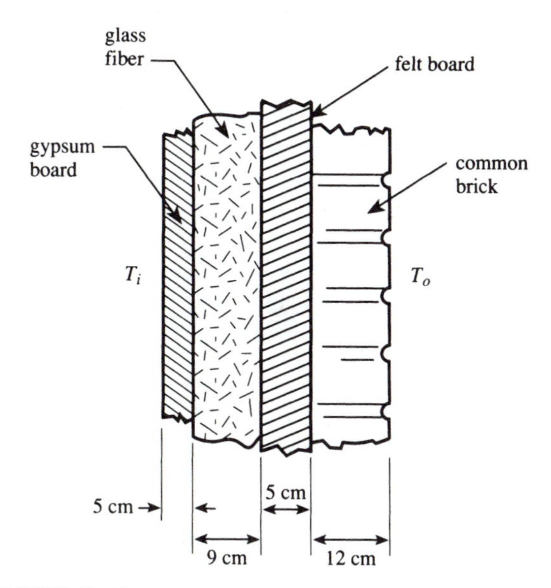

FIGURE 15–42

15–7 (E) Determine the heat transfer rate per foot length through a copper pipe that has an outside diameter of 2 in and an inside diameter of $1\,{}^{1}\!/_{2}$ in. The pipe contains 180°F ammonia and is surrounded by 80°F air.

15–8 (M) An insulated steam line has 2 cm of rock wool insulation surrounding a 5-cm-OD iron pipe with a 0.5-cm wall. Determine the heat loss through the line if steam at 110°C is in the line and the surrounding temperature is 10°C.

15–9 (E) Boiler tubes in the superheater section of a steam generating unit are surrounded by combustion gases at 1100°F, and the steam flowing through the tubes is at 900°F. If the tubes are made of iron, have an inside diameter of 8 in, and have ${}^{1}\!/_{2}$-in-thick walls, determine the heat transfer rate to the steam.

15–10 (M) Teflon® tubing of 4-cm OD and 2.6-cm ID conducts 2 W/m of heat through its walls from the outside to the inside when the outside surface is 80°C. Determine the thermal resistance per unit length and the inside surface temperature.

15–11 (E) Triple-pane window glass has been used in some building construction. Triple panes are the same as thermopanes (see example 15–3), except that one more pane of glass is used and an additional air gap is included, as shown in figure 15–43. Determine the R-value for triple-pane glass.

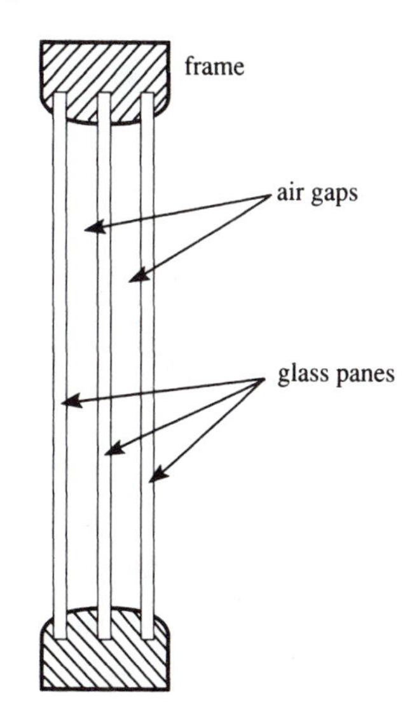

FIGURE 15–43 Triple-pane window

Section 15–2

15–12 (M) The cross section of a supersonic aircraft wing is shown in figure 15–44, with the flow of air around the wing indicated by lines called **streamlines**. The ambient atmospheric

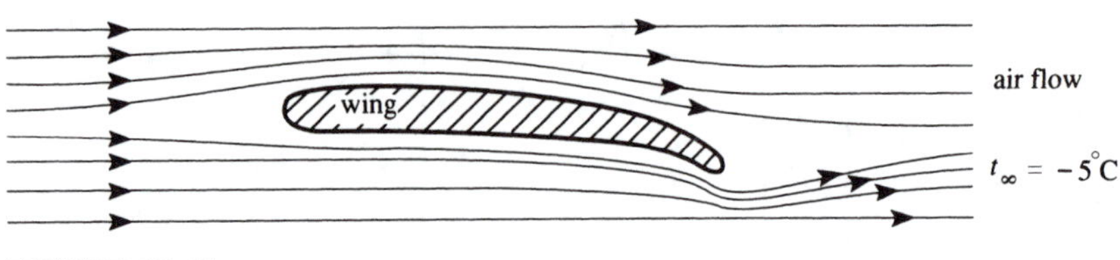

FIGURE 15–44

temperature of the undisturbed air t_∞ is $-5°C$, and the temperature of the surface of the wing t_s is found to be 480°C. Assuming that the coefficient of heat transfer is $4.2\ kW/m^2 \cdot K$, find the rate of convection heat transfer per unit of wing area. (*Note:* The answer should be in kW/m^2 units.)

15–13 (E) A 2-in-OD boiler pipe containing water is heated by forcing hot air, at 800°F, around the pipe. If the coefficient of heat transfer is found to be $10\ Btu/h \cdot ft^2 \cdot °F$, and if the pipe's outer surface temperature is 285°F, find the rate of heat transfer to the pipe per foot of length.

15–14 (M) Air flows past a 5-cm-diameter electric power line at approximately 50 m/s. If the power line has a surface temperature of 10°C and the air is at $-10°C$, determine the heat loss of the power line per unit length.

15–15 (E) Water at 55°F flows through a copper tube of 1-in ID at approximately 1 ft/s. Determine the heat transfer per foot of the tube length if the tube ID is 190°F.

Section 15–3

15–16 (M) Determine the total thermal resistance per unit area for the wall shown in figure 15–45 with the indicated convective heat transfer coefficients at the right and left wall surfaces.

15–17 (E) Determine the total thermal resistance per unit length for the tube shown in figure 15–46 with the indicated convective heat transfer coefficients.

15–18 (M) For the wall shown in figure 15–45, determine the overall temperature difference if the heat transfer is $10\ W/m^2$ across the wall.

15–19 (M) Determine the thermal resistance of the single-pane glass and the thermopane of example 15–3 if the convective heat transfer coefficient K_h is $100\ W/m^2 \cdot K$ on each side of the outer window surfaces. Then determine the ratio of the thermal resistance of the thermopane to that of the single-pane glass. In example 15–3, this ratio

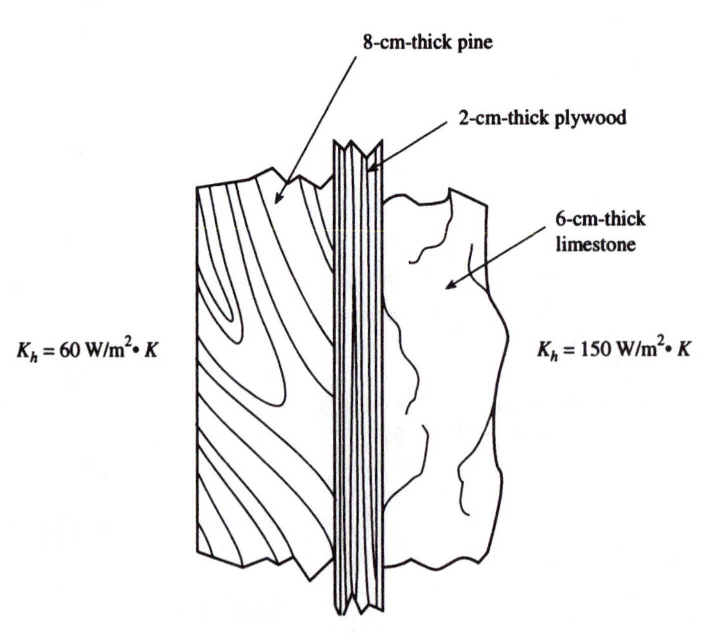

FIGURE 15–45

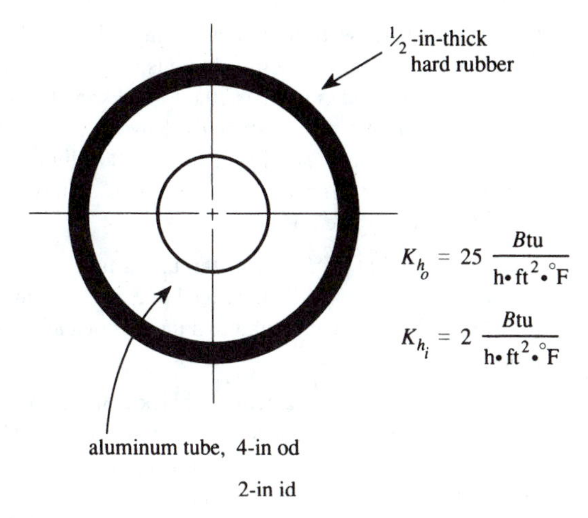

FIGURE 15–46

was determined to be 103.6 (= 1.336/0.0129), so what can you say about the effect of accounting for convection heat transfer to heat transfer through solids?

15–20 (M) For the triple-pane window of problem 15–11, determine the thermal *R*-value if convection heat transfer is accounted for. Use $K_h = 85 \text{ W/m}^2 \cdot \text{K}$ on one side and $120 \text{ W/m}^2 \cdot \text{K}$ on the other side. Compare this total *R*-value to that for a single pane of glass having the same convective heat transfer coefficients.

15–21 (E) Estimate the heat transfer per unit length of a Teflon® pipe, 6-in OD and 5.9-in ID, if air at 0°F flows through the pipe and air at 55°F surrounds the pipe. Use an estimated value for the convective heat transfer coefficients on the inside and outside pipe surfaces.

15–22 (M) Plot the temperature over a brass fin from its base to the tip assuming that the fin is very long, 10 mm thick, and 1 m wide; has a K_h of $30 \text{ W/m}^2 \cdot \text{K}$ and a base temperature of 200°C; and is surrounded by a gas at 80°C.

15–23 (E) A square aluminum fin is $^1/_8$ in thick, 3 ft wide, and 4 in long; has a K_h of $4.5 \text{ Btu/h} \cdot \text{ft}^2 \cdot °\text{F}$ and a base temperature of 500°F; and is surrounded by a gas at 100°F. Determine the fin efficiency and the heat transfer through the fin.

15–24 (M) A square stainless steel fin is 2 cm thick and 8 cm long, and it has a K_h of $100 \text{ W/m}^2 \cdot \text{K}$ and a base temperature of 400°C. Determine the fin efficiency and the heat transferred by the fin per unit width if the surrounding temperature is 20°C.

15–25 (E) A circumferential cast iron fin is 1 in thick, 3 in long, and 3 in diameter, and it has a K_h of $1.3 \text{ Btu/h} \cdot \text{ft}^2 \cdot °\text{F}$. The base temperature of the fin is 250°F, and the surrounding temperature is 100°F. Determine the fin efficiency and the heat transferred by the fin.

15–26 (M) An electrical power transformer housing has six flat, square fins wrapped around the housing as shown in figure 15–47. Each fin is 1 cm thick and 2 cm long, extending out from the housing surface. Neglecting heat transfer at the four corners, determine the total heat transfer from the six fins if the housing outer surface is 40°C and the air surrounding the transformer is 20°C. The average convective heat transfer coefficient, K_h, is $20 \text{ W/m}^2 \cdot \text{K}$, and the housing and fins are cast iron.

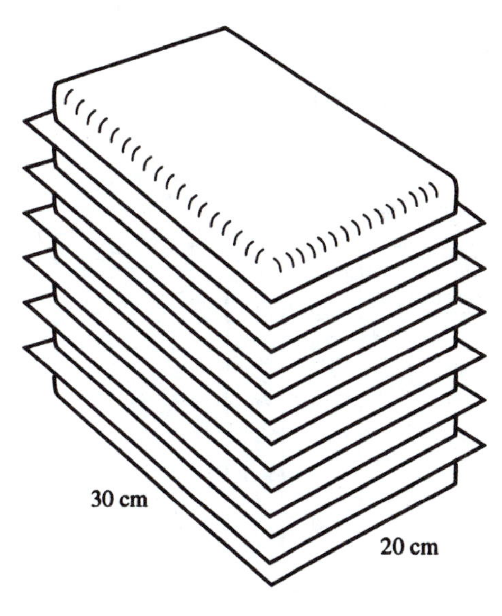

FIGURE 15–47

15–27 (M) A thermistor is used to measure temperatures. Assume that the thermistor is a cylinder of 1-mm diameter and 5-mm length and that it has the same properties as those of stainless steel. Determine the thermistor time constant when it is in water where K_h is $50 \text{ W/m}^2 \cdot \text{K}$. If the water is at 85°C and the thermistor is at 5°C when it is placed in the water, determine the time before it reaches 80°C.

15–28 (E) Granite stone is used to store solar energy. Assume that the stone is spherical and of 3-in diameter and that K_h is $4.5 \text{ Btu/h} \cdot \text{ft}^2 \cdot °\text{F}$. Determine the time constant for granite stones. If the stones are surrounded by 50°F air and suddenly the temperature of the air drops to 45°F, how long before the stone is at 46°F?

15–29 (M) A thermocouple is used to sense temperatures inside a furnace. If the thermocouple is a sphere of 3-mm diameter and has the same properties as those of carbon steel, and if K_h is $35 \text{ W/m}^2 \cdot \text{K}$, determine the time constant for the thermocouple.

Section 15–4

15–30 (E) Air at 120°F is blown across a 40-ft-by-40-ft concrete slab at 50°F with a velocity of 20 ft/s. Determine the convective heat transfer coefficient, K_h, and the heat transferred to the slab.

15–31 (M) Wind at 20 km/h blows across a 20-m-wide paved highway when the air temperature is 30°C and the pavement is 50°C. Determine the rate of heat transfer from the pavement per unit length due to forced convection.

15–32 (E) A wind of 10 mph and 40°F blows across a chimney as shown in figure 15–48. The chimney is 30 in × 30 in and 20 ft tall and has an average surface temperature of 70°F. Determine the heat loss due to convection.

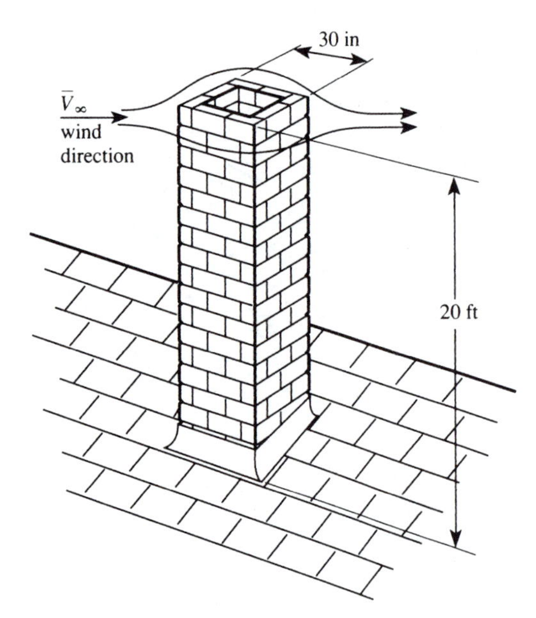

FIGURE 15–48

15–33 (M) Cold water at 5°C and moving at 2 m/s passes around a 3-cm-diameter pipe at 80°C shown in figure 15–49. Determine the heat transfer due to convection per length of the pipe.

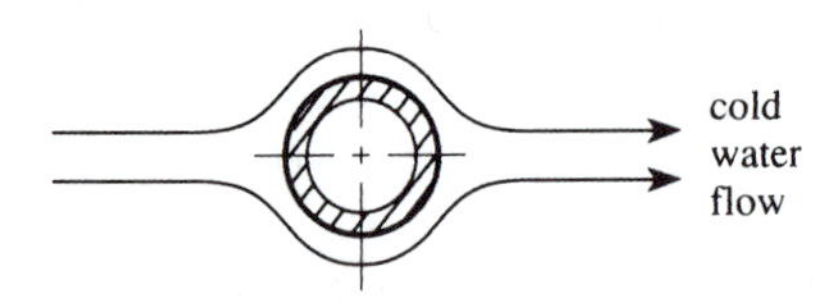

FIGURE 15–49

15–34 (E) Oil at 150°F is passed through a cooling tube at 4 ft/s. The cooling tube surface temperature is 60°F, the inside diameter is 1 in, and the tube is 20 ft long. Estimate what the oil temperature will be when it leaves the cooling tube, assuming that viscosity of oil is 0.007 lbm/ft·s and that its other properties are the same as those in table B–20 for engine oil.

15–35 (M) Ethylene glycol at 30°C passes through a tube of 3-cm Diameter and 30-m length at a speed of 3 m/s. Estimate the average tube wall temperature if the glycol leaves the tube at 60°C.

15–36 (M) A fan is used to cool an electronic black control box by blowing air past the box as shown in figure 15–50. The box is 2 cm by 2 cm by 12 cm and has outside surface temperature of 50°C. The air is at 20°C and moves at 5 m/s. Determine the convective heat transfer coefficient K_h and the rate of cooling in kW. Neglect the heat transfer at the two ends of the box.

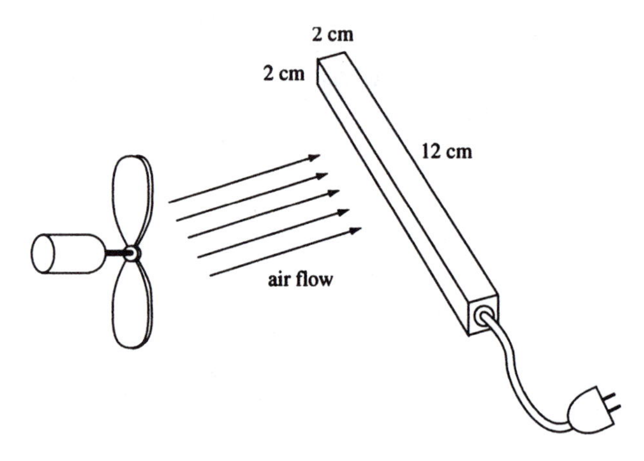

FIGURE 15–50

Section 15–5

15–37 (E) A large furnace has 20-ft-high vertical walls, is 50 ft long, and is at 600°F. The air temperature outside the furnace is 80°F. Estimate the heat loss through the wall due to natural convection.

15–38 (M) A Trombé wall 60 m wide has been installed on a 30-story building (3 m/story). On a certain day, the outside building wall is heated to 50°C due to solar energy passing through the glass Trombé wall, and the outside air temperature is 0°C. Estimate K_h for this application if the air enters the building at 20°C but enters the Trombé wall passageway at 0°C.

15–39 (E) A smokestack is 5 ft in diameter and 60 ft high. When the stack gas temperature is 350°F and the outside air temperature 50°F, estimate the draft pressure and the volume flow rate of stack gases.

15–40 (M) For the Trombé wall of problem 15–38, assume that the vertical passageway is 1.5 m wide. Estimate the amount of air that could be introduced into the building.

15–41 (E) If your skin temperature is 80°F and you extend your arm out horizontally in still air that is at 40°F, estimate the heat loss from your arm.

15–42 (M) A 2-cm-diameter steel bar at 800°C is air cooled by laying the bar on a horizontal rack. Estimate the heat transfer from the bar if the air temperature is 25°C.

15–43 (M) 15-cm by 30-cm flat sheets of steel are heated to 250°C and then quenched in an oil bath at 50°C as shown in figure 15–51. Using thermal properties of unused engine oil and machine oil for volumetric expansion, determine K_h and the heat transfer from one plate (both sides) in kW at the instant that the sheet is hung in the oil.

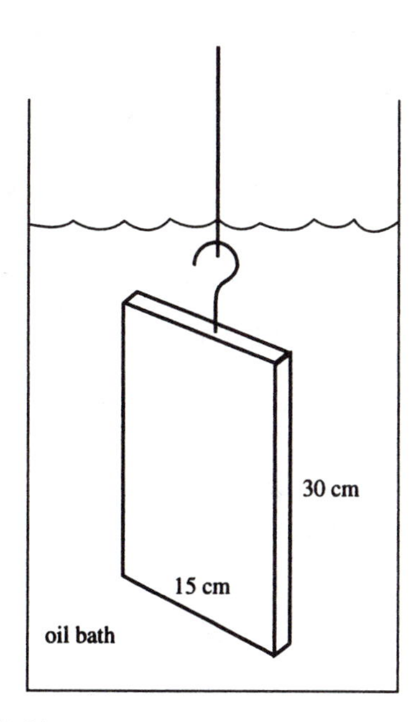

FIGURE 15–51

15–44 (E) For the stack of problem 15–39, estimate the heat loss of the gases and the maximum inlet temperature if the gases leaving the stack can be no more than 300°F. Assume that the stack wall temperature is 90°F.

Section 15–6

15–45 Determine the reflectivity of polished steel if it is an opaque gray material.

15–46 (M) A building roof at 40°C acts as an opaque gray body with $\epsilon = 0.80$. If the net heat flow to the roof is $1\ \text{kW/m}^2$, determine the total radiation leaving the roof, or the radiosity.

15–47 (E) The inside wall of the cylinder of an IC engine is 850°F. The engine is made of cast iron, and the net heat flow to the cylinder wall is $1000\ \text{Btu/h} \cdot \text{ft}^2$. Determine the radiosity of the cylinder wall.

15–48 (M) Determine the shape factor between two 10-cm-diameter disks facing each other and separated by 15 cm. Then find the shape factor between each disk and the surroundings.

15–49 (E) Determine the shape factor for an 8-ft-by-20-ft wall to a floor abutting the wall at 90°. The floor is 20 ft by 20 ft. What is the shape factor between the floor and the surroundings other than to the vertical wall?

15–50 (M) Determine the shape factor for surfaces 1 and 2 shown in figure 15–52.

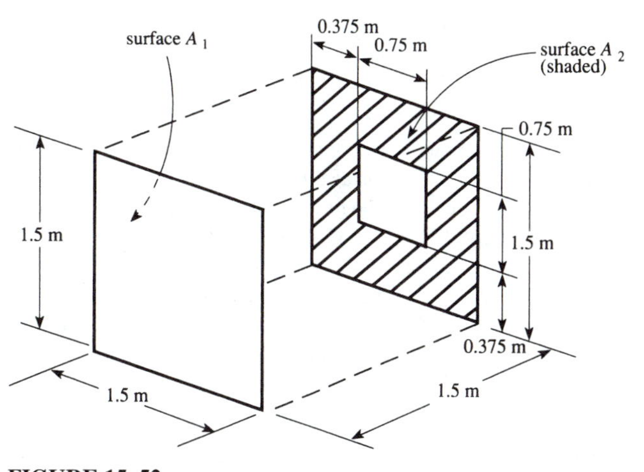

FIGURE 15–52

15–51 (E) A short rod 4 inches in diameter and 12 inches long is placed concentrically inside of a cylinder having inside diameter of 8 inches as shown in figure 15–53. Determine the shape factors F_{12}, F_{21}, and F_{13}. As indicated in the figure, surface (1) is that of the short rod, surface (2) is that of the inside of the cylinder, and surface (3) is the annular area at the ends between the rod and cylinder.

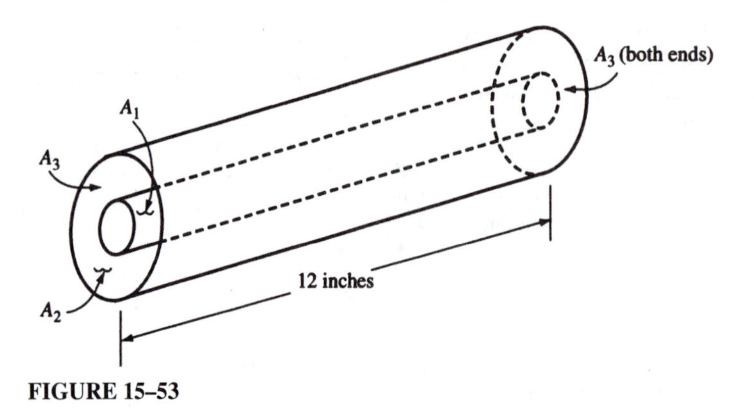

FIGURE 15–53

15–52 (E) The surface of the sun seems to have a temperature of around 10,000°F. What would you guess the rate of heat emission from the sun per square foot area to be?

15–53 (M) A radiation pyrometer is a device utilizing radiant heat to measure the temperature of a surface. Assume that a pyrometer has an area of 5 cm^2 and a temperature of 20°C when it is directed toward a fire having a temperature of 1100°C. Determine the net rate of heat transfer toward the pyrometer if blackbody radiation is assumed.

15–54 (E) A 2-ft-by-1-ft pan of water is placed outside at night when the sky has an effective blackbody temperature of −60°F. If the water is 65°F, determine the radiation heat loss of the water.

15–55 (M) A rectangular heat treat furnace is 1 m high, 2 m wide, and 2 m long (inside dimensions). The top surface is the heat source for the furnace and has a surface temperature of 2000°C with an emissivity of 0.86. The sides and floor are blackbody radiators with surface temperatures of 1500°C, and the front is open with an effective temperature of 100°C and an emissivity of 0.98. Determine the rate of heat required for steady-state conditions when the front is left open.

15–56 (M) A 12-in-diameter pot of water at 110°F is positioned 3 in above a 6-in-diameter electric burner that has a surface temperature of 1000°F. The burner emissivity is 0.70. The pot is polished stainless steel with a lower surface temperature of 95°F. Determine the heat transfer to the pot, assuming that the surroundings are blackbodies at 80°F.

Section 15–7

15–57 (E) Determine the overall conductance of a counterflow one-pass shell-and-tube heat exchanger where copper tubing of 6-in OD and $5\,^1\!/_2$-in ID is used. Water with a K_h of 200 Btu/h · ft^2 · °F flows around the tube, and ethylene glycol with a K_h of 550 Btu/h · ft^2 · °F is in the tube.

15–58 (M) Approximating an automobile radiator as a water-to-air, cross-flow shell-and-tube heat exchanger, determine the surface area of tubing required of a radiator if 1 kg/s water enters the radiator at 90°C and leaves at 78°C. Air passes around the radiator tubes with an increased temperature of 30°C, entering at 15°C. Assume a $\mathcal{U}$ of 450 W/m^2 · °C, and then determine the minimum amount of air required.

15–59 (E) A two-pass shell-and-tube heat exchanger has an overall conductance of 280 Btu/h · ft^2 · °F when hot water at 150°F enters at 15 lbm/s and cold water at 65°F enters at 30 lbm/s. The hot water leaves at 100°F. Determine the surface area required of the heat exchanger.

15–60 (M) An oil cooler for a stationary diesel engine is a shell-and-tube two-pass heat exchanger having an overall heat transfer coefficient of 500 W/m^2 · °C. The oil mass flow rate is 5 kg/s, entering at 90°C and cooling to 70°C as it leaves the cooler. Water is used as the coolant, entering at 20°C at a rate of 3 kg/s. Determine the water outlet temperature and the effective surface area of the cooler.

HEATING AND AIR CONDITIONING

16

In this chapter, we introduce some of the important parameters and concepts used in the analysis of heating and air-conditioning systems. Human comfort is discussed, the degree-day term is defined, and heating and cooling loads are explained. Concepts and methods developed in earlier parts of the book are used. It would be advantageous for the reader to review chapters 12, 13, and 15 before beginning this chapter. General problems of heating closed living areas are analyzed, and similar analyses of air-conditioned spaces are conducted.

New Terms

DD Degree-day

16–1 PARAMETERS IN HEATING AND AIR CONDITIONING

The engineer and technologist usually encounter problems that are defined in precise terms and quantities. This situation allows for quantitative analyses and results that can be interpreted clearly. In the study of heating and air-conditioning systems, however, human behavior and comfort must be accommodated in any analysis, and this can be frustrating to designers of equipment and architects of the confines. Heating and air conditioning are concerned principally with providing a comfortable area in which people can conduct their activities.

Human comfort involves the mental and physical state of persons, and so is a subject that extends well beyond the purposes of this book. However, in terms of the technical concepts of this book, the humidity and temperature of air can be used to define a comfort region. Figure 16–1 is a graph of temperature (dry-bulb temperature) and relative humidity

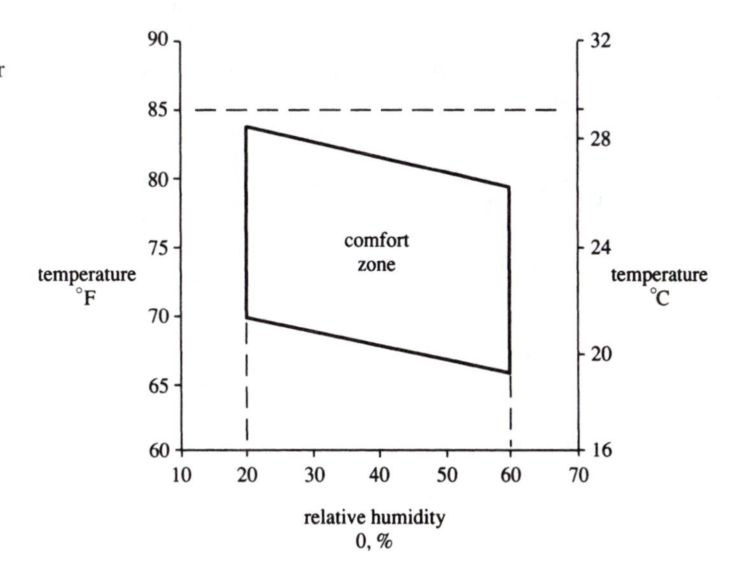

FIGURE 16–1 Suggested comfort zone in heating and air conditioning

TABLE 16–1 Approximate amounts of body heat loss for various human activities

Type of Activity	Body Heat Loss W	Body Heat Loss Btu/h
Sleeping	88	300
Sitting, reading	117	400
Typing, active sitting	146	500
Walking, light exercise	205	700
Walking down stairs	439	1500
Active labor	527	1800
Carpenter work, climbing activity	878	3000
Jogging up stairs	1318	4500

that indicates the variation of these two terms that we can expect without being uncomfortable. The results shown in the figure assume that persons are lightly clothed and engaged in sitting activities, such as in a classroom or sitting at a desk. The comfort zone would be different for persons engaged in strenuous work or sleeping. The draft or amount of air motion and the surrounding surface temperatures (which would result in radiation heat transfers) are important parameters that can affect the condition of human comfort and which heating and air-conditioning designers should consider. Also, the health of persons could change the comfort zone. A person's basal metabolism and age, and the type of clothing worn, are further considerations that could make the comfort zone change from that given in figure 16–1. For our purposes, we will use only the dry-bulb temperature and relative humidity to define comfort zones.

Human beings require energy to live and conduct their activities. Nearly all of this energy is finally converted into irreversible work, and this becomes heat from a person's body. From past experiments it has been shown that the rate of heat given off by people depends on the type of activity being engaged in. In table 16–1 are listed some approximate amounts of heat generated by a person performing different activities. The values listed are intended to provide only an estimate of the average heat given off; however, the values give some important information for the design of heating and air-conditioning systems.

Ventilation must be considered in the design of heating and air-conditioning systems. If people are to occupy an area, they must have adequate supplies of oxygen and air free of contaminates, such as carbon monoxide gas. Fresh outside air is usually the source of this air, and the air inside an occupied area must be exchanged over some time period. The term **air change per hour** is often used to indicate that all the air in an occupied area is to be removed and replaced with fresh air a fixed number of times each hour. If air is to be changed every 10 minutes, for example, this is the same as 6 air changes per hour (1 air change per 10 min $\times$ 60 min/h $=$ 6 air changes per hour). In table 16–2 are listed the recommended air changes per hour for some human activities and situations. These numbers are only es-

TABLE 16–2 Range of air changes per hour for ventilating an area occupied by people

Type of Occupancy	Air Changes per Hour
Libraries, churches, retail stores, bowling lanes, sports arenas	6 to 30
Auditoriums, theaters, classrooms, kitchens	8 to 35
Cafeterias, restaurants, offices, hospitals, garages	6 to 20
Machine shops, school and industrial shops	5 to 15

timates and are continually being revised. For instance, economic considerations indicate that fewer air changes would be desirable; health concerns would strive for more air changes per hour. The numbers from table 16–2 do not indicate the number of persons occupying a given region. Occupancies that can have the higher densities of persons (churches, stores, and sports arenas) have greater ventilation requirements than those where low human densities may occur. Many designers use other methods to arrive at parameters for deciding how much ventilation is necessary or desirable. One method that can indicate ventilation requirements is the amount of air required per person engaged in a certain activity. This would tend to eliminate human density factors and provide better sensitivity to the individual. The interested reader may wish to refer to heating and ventilating books for further discussion of this problem.

For the purposes of this book, the air change per hour values taken from table 16–2 should be interpreted as the ventilation required and may include recirculated inside air. Fresh outside air could be expected to be one half to one third of the total air change given in the table.

The concept of the **degree-day** (DD) has been used to help analyze heating requirements for living areas. The degree-day is defined as the average daily temperature in degrees Fahrenheit from 65°F for one day. That is,

$$DD = (65°F - T_{ave})(1 day) \qquad \textbf{(16–1)}$$

Thus, if the average daily temperature is 45°F for a certain day, the DD value for that day is 20 degree-days. The DD is often given in monthly periods or even in yearly periods. For designing purposes, it is convenient to know the number of degree-days expected for each month at a given geographic location. Thus, if a location has a daily average temperature of 45°F for one continuous month, the DD value for that month would be $(65 - 45)(30 \text{ days/month}) = 600$ degree-days for that month. If the 45°F persisted for one year, the DD value would be 7200 degree-days for the year. Table 16–3 lists some monthly DD and yearly DD data for various geographic locations in the continental United States. Similar data can be obtained for Canada and other parts of the world. Also, figure 16–2 shows the normal yearly DD lines for the continental United States and parts of Canada. The lines may be interpolated for detailed DD data; however, local conditions may create unexpected variations from the results given in figure 16–2. Another limitation of the DD concept is its lack of recognition of extreme conditions that are likely to occur. While the month of December has higher DD values than September in all locations of the United States, the minimum temperatures that one should design for are not recognized in the DD values. Also, chill factors caused by high winds and gusts when the temperature is low can create major variations in the DD interpretation. Furthermore, some studies have been made that indicate that the 65°F reference used in equation (16–1) should be replaced by a lower value (for energy conservation purposes) such as 59°F or a base temperature that varies from month to month. In section 16–2, we analyze heating systems that an engineer or technologist may encounter, and we limit ourselves to the basic concepts and quantities given here.

The DD concept can be used to predict cooling loads in climates where the average temperatures are above some agreed-upon comfortable value. Usually, for cooling loads, the DD is given for one day by

$$DD = (T_{ave} - 85°F)(1 day) \qquad \textbf{(16–2)}$$

although 85°F may be a point for debate. In section 16–3, we analyze air-conditioning and cooling systems by using the DD concept to help determine the operation of those systems.

TABLE 16–3 Average degree-days by month and year for various cities in the continental United States

City	State	July	Aug.	Sept.	Oct.	Nov.	Dec.	Jan.	Feb.	Mar.	Apr.	May	June	Year
Birmingham	Alabama	0	0	17	118	438	614	614	485	381	128	25	0	2820
Montgomery	Alabama	0	0	0	55	267	458	483	360	265	66	0	0	1954
Flagstaff	Arizona	49	78	243	586	876	1135	1231	1014	949	687	465	212	7525
Phoenix	Arizona	0	0	0	13	182	360	425	275	175	62	0	0	1492
Little Rock	Arkansas	0	0	10	110	405	654	719	543	401	122	18	0	2982
Los Angeles	California	0	0	17	41	140	253	328	244	212	129	68	19	1451
Sacramento	California	0	0	17	75	321	567	614	403	317	196	85	5	2600
San Francisco	California	189	177	110	128	237	406	462	336	317	279	248	180	3069
Denver	Colorado	0	5	103	385	711	958	1042	854	797	492	260	60	5673
Hartford	Connecticut	0	14	101	384	699	1082	1178	1050	871	528	201	31	6139
Wilmington	Delaware	0	0	47	282	585	927	983	876	698	396	110	6	4910
Washington	District of Columbia	0	0	32	231	510	831	884	770	606	314	80	0	4258
Miami	Florida	0	0	0	0	5	48	57	48	15	0	0	0	173
Tallahassee	Florida	0	0	0	31	209	366	385	287	203	38	0	0	1519
Tampa	Florida	0	0	0	0	60	163	201	148	102	0	0	0	674
Atlanta	Georgia	0	0	8	107	387	611	632	515	392	135	24	0	2811
Savannah	Georgia	0	0	0	38	225	412	424	330	238	43	0	0	1710
Boise	Idaho	0	0	135	389	762	1054	1169	868	719	453	249	92	5890
Chicago	Illinois	0	0	28	161	492	784	856	683	523	182	47	0	3756
Springfield	Illinois	0	0	56	259	666	1017	1116	907	713	350	127	14	5225
Indianapolis	Indiana	0	0	59	247	642	986	1051	893	725	375	140	16	5134
Des Moines	Iowa	0	6	89	346	777	1178	1308	1072	849	425	183	41	6274
Topeka	Kansas	0	0	42	242	630	977	1088	851	669	295	112	13	4919
Louisville	Kentucky	0	0	41	206	549	849	911	762	605	270	86	0	4279
New Orleans	Louisiana	0	0	0	5	141	283	341	223	163	19	0	0	1175
Portland	Maine	15	56	199	515	825	1237	1373	1218	1039	693	394	117	7681
Baltimore	Maryland	0	0	29	207	489	812	880	776	611	326	73	0	4203
Boston	Massachusetts	0	7	77	315	618	998	1113	1002	849	534	236	42	5791
Detroit	Michigan	0	8	96	381	747	1101	1203	1072	927	558	251	60	6404
Grand Rapids	Michigan	0	20	105	394	756	1107	1215	1086	939	546	248	58	6474
Sault Ste. Marie	Michigan	109	126	298	639	1005	1398	1587	1442	1302	846	499	224	9475
Duluth	Minnesota	66	91	277	614	1092	1550	1696	1448	1252	801	487	200	9574
International Falls	Minnesota	70	118	356	716	1230	1733	1922	1618	1395	834	437	171	10,600
Minneapolis	Minnesota	8	17	157	459	960	1414	1562	1310	1057	570	259	80	7853
Jackson	Mississippi	0	0	0	69	310	503	535	405	299	81	0	0	2202
Saint Louis	Missouri	0	0	38	202	570	893	983	792	620	270	94	7	4469
Butte	Montana	115	174	450	744	1104	1442	1575	1294	1172	804	561	325	9760
Great Falls	Montana	24	50	273	524	894	1194	1311	1131	1008	621	359	166	7555
Lincoln	Nebraska	0	7	79	310	741	1113	1240	1000	794	377	172	32	5865
Reno	Nevada	27	61	165	443	744	986	1048	804	756	519	318	165	6036

EXAMPLE 16–1 Determine the total DD value for the three-month period December through February for Boston, Massachusetts; Minneapolis, Minnesota; San Francisco, California; and Miami, Florida. Then determine the average daily temperature for the three-month period at these four locations.

Solution Using table 16–3, we can determine the DD value for each of the cities over the three-month period December through February. We read the monthly DD values and add them together. The results are given in table 16–4. Also, the average daily temperature may be found by using equation (16–1) with 90 days for the time period. Then

$$T_{ave} = 65°F - \frac{DD}{90 \text{ days}}$$

City	State	July	Aug.	Sept.	Oct.	Nov.	Dec.	Jan.	Feb.	Mar.	Apr.	May	June	Year
Concord	New Hampshire	11	57	192	527	849	1271	1392	1226	1029	660	316	82	7612
Newark	New Jersey	0	0	47	301	603	961	1039	932	760	450	148	11	5252
Albuquerque	New Mexico	0	0	10	218	630	899	970	714	589	289	70	0	4389
Albany	New York	0	6	98	388	708	1113	1234	1103	905	531	202	31	6319
Buffalo	New York	16	30	122	433	753	1116	1225	1128	992	636	315	72	6838
New York	New York	0	0	39	263	561	908	995	904	753	456	153	18	5050
Asheville	North Carolina	0	0	50	262	552	769	794	678	572	285	105	5	4072
Raleigh	North Carolina	0	0	10	118	387	651	691	577	440	172	29	0	3075
Bismarck	North Dakota	29	37	227	598	1098	1535	1730	1464	1187	657	355	116	9033
Cincinnati	Ohio	0	0	42	222	567	880	942	812	645	314	108	0	4532
Cleveland	Ohio	0	9	60	311	636	995	1101	977	846	510	223	49	5717
Columbus	Ohio	0	0	59	299	654	983	1051	907	741	408	153	22	5277
Oklahoma City	Oklahoma	0	0	12	149	459	747	843	630	472	169	38	0	3519
Portland	Oregon	13	14	85	280	534	701	791	594	515	347	199	70	4143
Philadelphia	Pennsylvania	0	0	33	219	516	856	933	837	667	369	93	0	4523
Pittsburgh	Pennsylvania	0	0	56	298	612	924	992	879	735	402	137	13	5048
Providence	Rhode Island	0	7	68	330	624	986	1076	972	809	507	197	31	5607
Charleston	South Carolina	0	0	0	34	214	410	445	363	260	43	0	0	1769
Columbia	South Carolina	0	0	0	76	308	524	538	443	318	77	0	0	2284
Rapid City	South Dakota	32	24	193	500	891	1218	1361	1151	1045	615	357	148	7535
Sioux Falls	South Dakota	16	21	155	472	984	1414	1575	1274	1023	558	276	80	7848
Chattanooga	Tennessee	0	0	24	169	477	710	725	588	467	179	45	0	3384
Memphis	Tennessee	0	0	13	98	392	639	716	574	423	131	20	0	3006
Nashville	Tennessee	0	0	22	154	471	725	778	636	498	186	43	0	3513
Amarillo	Texas	0	0	37	240	594	859	921	711	586	298	99	0	4345
Brownsville	Texas	0	0	0	0	59	159	219	106	74	0	0	0	617
Dallas	Texas	0	0	0	53	299	518	607	432	288	75	0	0	2272
Houston	Texas	0	0	0	7	181	321	394	265	184	36	0	0	1388
Salt Lake City	Utah	0	0	61	330	714	995	1119	857	701	414	208	64	5463
Burlington	Vermont	19	47	172	521	858	1308	1460	1313	1107	681	307	72	7865
Norfolk	Virginia	0	0	5	118	354	636	679	602	464	220	41	0	3119
Richmond	Virginia	0	0	31	181	456	750	787	675	529	254	57	0	3720
Seattle	Washington	49	45	134	329	540	679	753	602	558	396	246	107	4438
Spokane	Washington	17	28	205	508	879	1113	1243	988	834	561	330	146	6852
Charleston	West Virginia	0	0	60	250	576	834	887	750	632	310	110	8	4417
Green Bay	Wisconsin	32	58	183	515	945	1392	1516	1336	1132	696	347	107	8259
Madison	Wisconsin	10	30	137	419	864	1287	1417	1207	1011	573	266	79	7300
Milwaukee	Wisconsin	11	24	112	397	795	1184	1302	1117	961	606	335	100	6944
Cheyenne	Wyoming	33	39	241	577	897	1125	1225	1044	1029	717	462	173	7562
Sheridan	Wyoming	27	41	239	578	957	1271	1392	1170	1035	645	387	161	7903

For Boston, Massachusetts, in December, the DD value is 998°F-day, in January it is 1113, and in February it is 1002. The total DD for the three months is 3113°F-day, and then

$$
\begin{aligned}
T_{ave} &= 65°F - \frac{3113°F\text{-day}}{90 \text{ days}} \\
&= 30.4°F
\end{aligned}
$$

Similar calculations can be made for Minneapolis, San Francisco, and Miami. The results of these calculations are given in table 16–4.

One parameter in heating and air conditioning that is often neglected is the solar radiation from the sun. The sun has been recognized as a source of heat for centuries; however, it

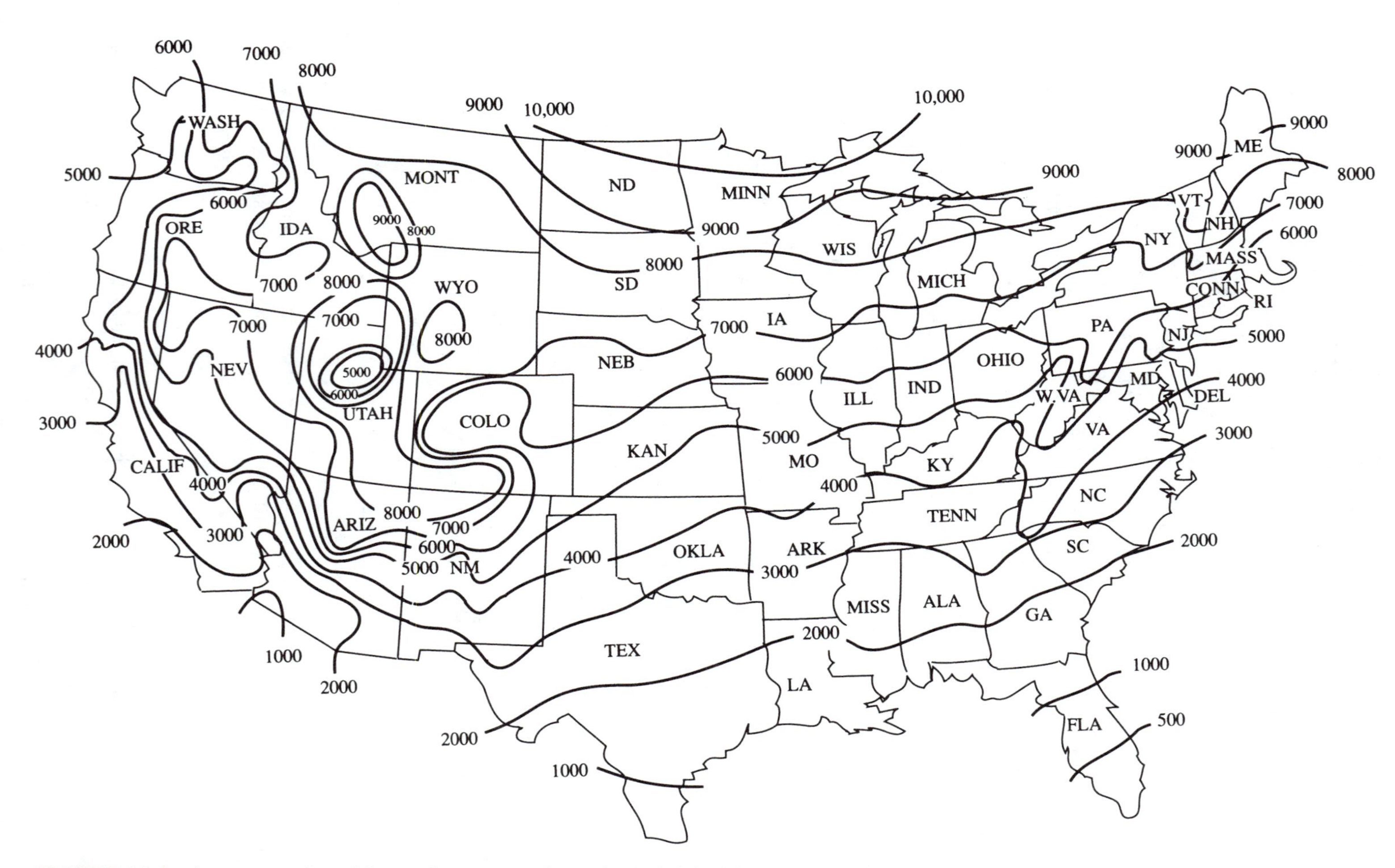

FIGURE 16–2 Average number of degree-days per year in continental United States [*Handbook of air conditioning, heating, and ventilation*, 3d ed. (New York: The Industrial Press, 1979); reprinted with permission of publisher]

TABLE 16–4 Total degree-days for some locations over the three-month period December through February

City	DD for December Through February degree-days	Average Daily Temperature °F
Boston	3113	30.4
Minneapolis	4286	17.4
San Francisco	1204	51.6
Miami	153	63.3

is often ignored or neglected in heating or cooling studies. In heating systems, the designer can use the sun's radiation or solar energy to augment or replace many of the conventional heating systems or even the cooling systems. Also, by eliminating north-facing windows in the Northern Hemisphere (south-facing windows in the Southern Hemisphere), one can reduce the heat losses due to windows. Unfortunately, many buildings are designed without regard to position with respect to the sun. Only after a structure is completed is the impact of solar energy appreciated. A building and its environment that incorporates many passive solar energy concepts is diagrammed in figure 16–3. In the winter months, the sun's energy

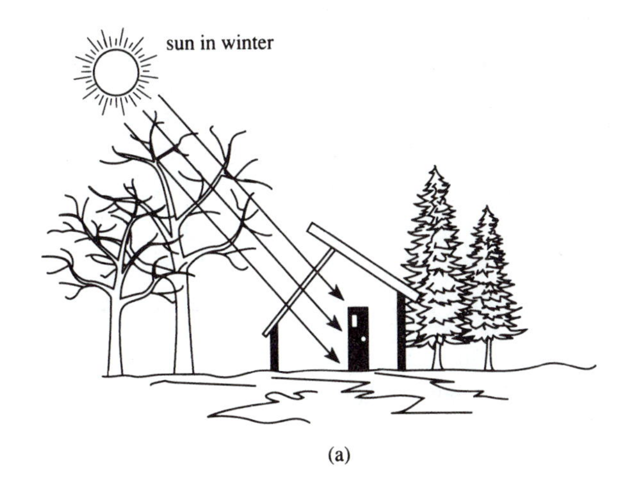

(a)

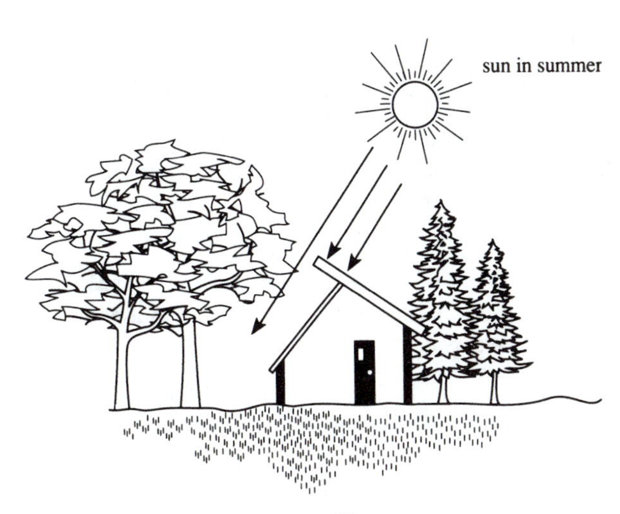

(b)

FIGURE 16–3 Passive solar energy incorporated into a small building: (a) winter season; (b) summer season

is free to enter the building through windows and is absorbed in a massive collector wall. The wall could be constructed of bricks or stone or be a water or liquid tank. During the summer months when heat is not required, the sun's energy is deflected by an overhanging roof, and the collector wall is still at a moderate temperature. Trees outside the structure should be selected to provide shelter and shade in the summer; in the winter, some of the trees will lose their foliage, which permits the sun the reach the building. Some of these ideas can be incorporated into large structures. In chapter 15, we discussed how Trombé walls may be incorporated into new or existing tall or multistoried buildings.

16–2 ANALYSIS OF SPACE HEATING

The term **space heating** refers to the area or space that needs to be heated or kept at some prescribed temperature. A large office building, an office or room in a building, a home, an automobile, or any volume may require space heating.

The general problem in the design and analysis of space heating systems is diagrammed in figure 16–4, where it is usually expected that steady-state conditions occur. Then

$$Q_{\text{load}} = Q_H + Q_{\text{gain}} \tag{16–3}$$

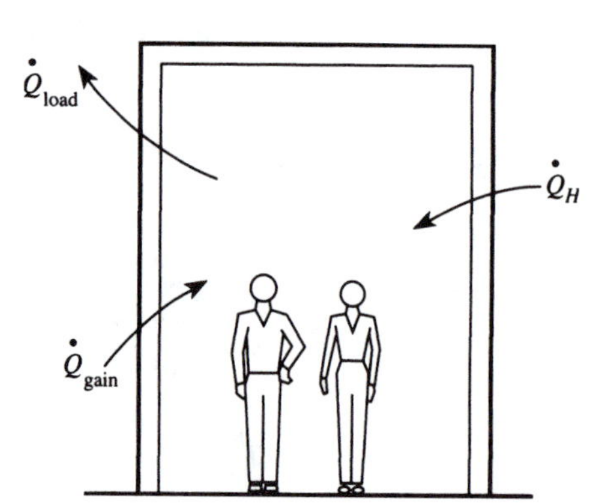

FIGURE 16–4 General problem in space heating

where Q_H is the heat required of a space heating system and Q_{load} is called the **heating load** of the system or space. Determining the heating load involves at least the following items:

1. Heat losses through the walls, ceiling (or roof), and floor
2. Heat loss due to ventilation or air changes
3. Heat losses through windows or doors

The term Q_{gain} is the **heat gain** of the space, which is caused by

1. Body heat of people or other occupants
2. Heat from lights, appliances, or other power equipment
3. Solar energy inputs from windows
4. Stored heat from collector walls and other reservoirs of thermal energy

The amount of **space heating**, or heat required of some external heating system, is Q_H, the difference between Q_{load} and Q_{gain}. A method of approach to the analysis of space heating problems is the following:

1. Define the room, building, or space to be heated. Identify its geographic location, calendar time (whether winter, summer, or combination of seasons), outside walls, partitions, and vertical dimensions.
2. Identify the expected number of human occupants and their activities while in the space. Identify any animals or items that are to occupy the space.

3. Identify electric lighting systems, appliances that use energy, locations of windows and solar energy inputs, and any thermal storage systems.
4. Determine the heating load.
5. Determine the heat gain and predict the necessary space heating, Q_H.

An example will show how to proceed with the analysis.

EXAMPLE 16–2 A small, one-story office building is 30 ft by 60 ft, located in Denver, Colorado. The walls have an *R*-value of 25; the ceiling, 35. There are expected to be 3 to 10 persons occupying the office at any time during a working day of 10 hours. Estimate the necessary space heating for the month of January, neglecting any solar energy inputs or losses of radiant heat through windows and doors.

Solution The solution to this problem will require numerous assumptions and approximations. We begin by defining the space area and volume:

1. We assume that the walls are 8 ft high, so that the volume of the office is 30 ft × 60 ft × 8 ft = 14,400 ft^3. The wall area is $[(30\text{ ft} \times 8\text{ ft}) + (60\text{ ft} \times 8\text{ ft})] \times 2 = 1440\text{ ft}^2$, and the roof area is 30 ft × 60 ft = 1800 ft^2. The DD value in Denver, Colorado, in January is 1042°F-day.

2. We will assume the minimum number of persons, three, and that they are engaged in seated activities. The heat gain of the persons can be 500 Btu/h.

3. Assume that there is 500 W of lighting and 500 W of power equipment.

4. The heating load is now determined from

$$Q_{\text{load}} = Q_{\text{walls}} + Q_{\text{ventilation}}$$

where $Q_{\text{walls}} = \dfrac{\text{DD}}{\sum R_T}$

The thermal resistance $\sum R_T$ is the *R*-value divided by the surface areas:

$$\sum R_T = \frac{25\text{ h}\cdot\text{ft}^2\cdot{}^\circ\text{F/Btu}}{1440\text{ ft}^2} + \frac{35\text{ h}\cdot\text{ft}^2\cdot{}^\circ\text{F/Btu}}{1800\text{ ft}^2}$$
$$= 0.0368\text{ h}\cdot{}^\circ\text{F/Btu}$$

Then

$$Q_{\text{walls}} = \frac{1042^\circ\text{F-day} \times 24\text{ h/day}}{0.0368\text{ h}\cdot{}^\circ\text{F/Btu}}$$
$$= 679{,}565\text{ Btu}$$

The air changes per hour are estimated at 9 from table 16–2. Also, we will assume that one third of the ventilation requirement will be fresh outside air, or 3 air changes per hour. The volume flow rate of air is then

$$= 3\text{ changes/h} \times 14{,}400\text{ ft}^3\text{/change}$$
$$= 43{,}200\text{ ft}^3\text{/h}$$

Using air density of 0.08 lbm/ft^3 and specific heat of 0.2404 Btu/lbm · °R, we approximate the ventilation heat by the equation

$$Q_{\text{vent}} = \dot{V}\rho c_p(\text{DD}_{\text{adj}})$$

where DD_{adj} is 10/24 of the DD value for the month of January in Denver. The 10/24 fraction accounts for the fact that the working day is only 10 hours and heating is required only during those hours. Then

$$\begin{aligned} Q_{vent} &= (\dot{V})(\text{density})(\text{specific heat})(DD_{adj}) \\ &= (43{,}200\ \text{ft}^3/\text{h})(0.08\ \text{lbm/ft}^3)(0.2404\ \text{Btu/lbm}\cdot{}^\circ\text{R}) \\ &\quad \times (1042 \times 24\ {}^\circ\text{F}\cdot\text{h})(10/24) \\ &= 8{,}657{,}169\ \text{Btu} \end{aligned}$$

The heating load is then the sum of the wall and ventilation heat terms:

$$Q_{load} = 9{,}336{,}734\ \text{Btu}$$

5. The heat gain is estimated from the equation

$$Q_{gain} = Q_{persons} + Q_{other}$$

The heat gain of the human beings, $Q_{persons}$, is found by assuming active sitting of the persons, and from table 16–1, the expected heat loss is 500 Btu/h per person. Then, for three persons over 30 days of 10 hours per day, we estimate the heat gain to the building:

$$\begin{aligned} Q_{persons} &= 3 \times 500\ \text{Btu/h} \times 10\ \text{h} \times 30\ \text{days} \\ &= 450{,}000\ \text{Btu} \end{aligned}$$

The other heat gains are due to lighting and the power equipment:

$$\begin{aligned} Q_{other} &= 1000\ \text{W} \times 3.414\ \text{Btu/W}\cdot\text{h} \times 10\ \text{h/day} \times 30\ \text{days} \\ &= 1{,}024{,}200\ \text{Btu} \end{aligned}$$

The heat gain is then

$$Q_{gain} = 1{,}024{,}200 + 450{,}000 = 1{,}474{,}200\ \text{Btu}$$

The required space heating is then the difference between the heat load and the heat gain:

Answer

$$Q_H = 9{,}336{,}734\ \text{Btu} - 1{,}474{,}200\ \text{Btu} = 7{,}862{,}534\ \text{Btu}$$

Further elaborations could be made, but they are beyond the purposes of this book.

Many analysts of heating systems determine the heating load by using only the ventilating heat loss and assuming that the wall heat losses are balanced by the solar gains, the heat from human beings, and the appliance or other equipment heat gains. From example 16–2, the ventilating heat loss can be the largest component in the heating load. If the number of air changes per hour were reduced, the heating requirement would be reduced proportionately. The health and safety problems associated with fewer air changes, however, may eliminate any energy and economic benefits that may be obtained by reducing the ventilation. One method used to reduce the ventilation heat losses is an air-to-air heat exchanger. This device provides a means to transfer the heat (thermal energy) in the exhaust air into fresh intake air. Air-to-air heat exchangers typically have overall conductances (see section 15–7) of around 10 to 40 $\text{W/m}^2\cdot{}^\circ\text{C}$ (2 to 8 $\text{Btu/h}\cdot\text{ft}^2\cdot{}^\circ\text{F}$) and are capable of transferring nearly all of the potential thermal energy to the fresh air.

EXAMPLE 16–3 A private condominium in Baltimore, Maryland, is 10 m by 20 m by 10 m and has two stories, as sketched in Figure 16–5. Assume that the front and rear walls have an overall thermal resistance of 100 $\text{m}^2\cdot{}^\circ\text{C/W}$ and that the roof has a thermal resistance of 150 $\text{m}^2\cdot{}^\circ\text{C/W}$. Estimate the required heat during February. Then an air-to-air heat exchanger is installed which

FIGURE 16–5
Condominium

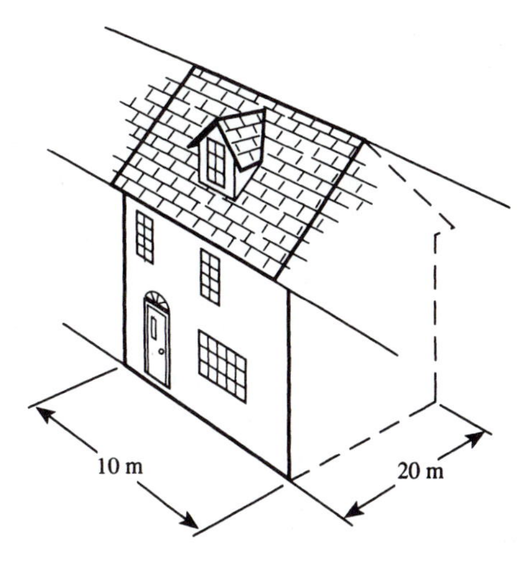

provides intake air at 13°C. Estimate the required heat for February and the amount of heat saved by installing the air-to-air heat exchanger.

Solution

The condominium has a total volume of approximately 10 m × 20 m × 10 m = 2000 m^3, an outside wall area of 10 m × 10 m × 2 = 200 m^2, and a projected roof area of 200 m^2 in a horizontal plane. We assume that the condominium is attached to other buildings on the remaining sides and that these walls are at the temperature of the occupied area, so that there will be no heat loss through these walls. Also, we assume there will be three persons occupying the condominium and that each person generates 130 W of heat. The appliances, lights, and other power equipment are assumed to generate 750 W of heat. The heat gain is (with 30 days per month)

$$Q_{\text{gain}} = (3\text{ persons} \times 130\text{ W/person} + 750\text{ W})(30\text{ days} \times 24\text{ h/day} \times 3600\text{ s/h})$$
$$= 2{,}954{,}880\text{ kJ/month}$$

The heating load Q_{load} is computed from the wall heat loss, the roof heat loss, and the ventilation heat loss. The DD is found from table 16–3 to be 776°F-day or 431°C-day. The total thermal resistance of the walls is $(100\text{ m}^2 \cdot {}^\circ\text{C/W})/(200\text{ m}^2) = 0.5^\circ\text{C/W}$, and that of the roof, $(150\text{ m}^2 \cdot {}^\circ\text{C})/(200\text{ m}^2) = 0.75^\circ\text{C/W}$. The heat losses through the walls and roof are then

$$Q = \frac{\text{DD}}{\sum R_T}$$
$$= \frac{431^\circ\text{C-day}}{0.5 + 0.75^\circ\text{C/W}}$$
$$= 344.8\text{ W}$$

and

$$Q = (344.8\text{ W})(24\text{ h/day} \times 3600\text{ s/h}) = 29{,}791\text{ kJ}$$

We calculate the ventilating load by first assuming that there will be five air changes per hour and that fresh air will be two air changes of the five. Air density is 1.17 kg/m^3, and the specific heat of air is assumed to be 1.007 kJ/kg · K. Then the volume flow rate, is

$$\dot{V} = 2\text{ air changes/h} \times 2000\text{ m}^3 = 4000\text{ m}^3\text{/h}$$

and the ventilating heat loss is

$$\begin{aligned} Q_{\text{vent}} &= (\dot{V})(\rho)(c_p)(\text{DD}) \\ &= (4000\ \text{m}^3/\text{h})(1.17\ \text{kg/m}^3)(1.007\ \text{kJ/kg}\cdot\text{K})(431^\circ\text{C}\cdot\text{day})(24\ \text{h/day}) \\ &= 48{,}748{,}789\ \text{kJ/month} \end{aligned}$$

The required heat for the condominium will then be

Answer

$$\begin{aligned} Q_H &= 48{,}748{,}789\ \text{kJ} + 29{,}791\ \text{kJ} - 2{,}954{,}880\ \text{kJ} \\ &= 45{,}823{,}700\ \text{kJ/month} \end{aligned}$$

If an air-to-air heat exchanger is now installed in the condominium, the ventilation heat loss will be the only term in the calculation to be affected. Instead of using the DD value, we will use a temperature difference of 18.7°C − 13°C = 5.7°C, where the 18.7°C is the inside temperature of the condominium and 13°C is the intake air temperature after passing through the air-to-air heat exchanger. Then

$$\begin{aligned} Q_{\text{vent}} &= (4000\ \text{m}^3/\text{h})(1.17\ \text{kg/m}^3)(1.007\ \text{kJ/kg}\cdot\text{K})(5.7^\circ\text{C})(24\ \text{h/day})(30\ \text{days/month}) \\ &= 19{,}341{,}167\ \text{kJ/month} \end{aligned}$$

The required heat for the condominium with an air-to-air heat exchanger is then

Answer

$$\begin{aligned} Q_H &= 19{,}341{,}167\ \text{kJ} + 29{,}791\ \text{kJ} - 2{,}954{,}880\ \text{kJ} \\ &= 16{,}416{,}078\ \text{kJ/month} \end{aligned}$$

The heat that was saved is

Answer

$$Q_{\text{saved}} = 29{,}407{,}622\ \text{kJ/month}$$

The reader should notice here that further analysis can readily be done. We could determine the amount of fuel required based on some heating value of the fuel, and we could make an economic analysis to justify the expense of installing and maintaining an air-to-air heat exchanger. The computer can be an advantage in detailed analysis of heating systems.

16–3 ANALYSIS OF AIR CONDITIONING AND REFRIGERATION

Air-conditioning and refrigeration systems are intended to remove heat from a region or space. The general problem is diagrammed in figure 16–6, which shows that the heat removed, Q_H, must at least balance the heat gain, Q_{gain}, and the cooling load, Q_{load}. Notice how this problem is different from the space heating problem diagrammed in figure 16–4. In refrigeration and air conditioning, the heat removal is given by

$$Q_H = Q_{\text{load}} + Q_{\text{gain}} \tag{16–4}$$

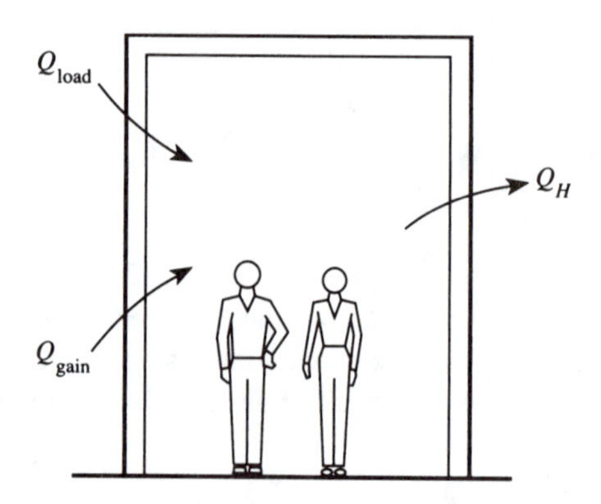

FIGURE 16–6 General problem in space cooling

The methods used to determine the heating loads and heat gains are the same as those used for space heating problems, the only difference being that the heating load is into the space or region. The degree-day concept can be used for determining the heating loads; however, the DD is more conveniently found from equation (16–2) because tables are not readily available for high-temperature climatic conditions. The average daily temperature must be assumed or previously determined before equation (16–2) can be used to determine the DD value.

EXAMPLE 16–4 Assume that the average daily temperature is 90°F for Springfield, Illinois, during August and that the average nighttime temperature is 75°F. Determine the DD value for cooling load calculations for Springfield in August.

Solution Using $T_{ave} = 90°F$ for 12 h and $T_{ave} = 75°F$ for the remaining 12 h in a day, we find the DD value for one day from equation (16–2) to be

$$\text{DD} = (90°\text{F} - 85°\text{F})(12/24) + (0)(12/24) = 2.5 \text{ degree-days}$$

We ignore the negative DD value for the night. Then for a 30-day month, the DD value is $2.5 \times 30 = 75$ degree-days for the month of August in Springfield, Illinois.

The suggested method of approaching the problems of space heating can be used equally well for refrigeration and air-conditioning system analysis. We list the outline for that method:

1. Define the region to be cooled or refrigerated.
2. Identify all occupants and their activities in the region.
3. Identify appliances, power equipment, windows, and thermal storage units.
4. Determine the cooling loads.
5. Determine the heat gains and the required cooling.

Often the humidity needs to be controlled in air-conditioning problems, so that an analysis may include cooling/dehumidifying processes or evaporative cooling processes. These processes were discussed in chapter 12.

EXAMPLE 16–5 Estimate the cooling load, Q_{load}, for a 20-ft-by-20-ft-by-8-ft-high retail sales store in Des Moines, Iowa, during July. Assume an average daily temperature of 92°F and a nighttime temperature of 68°F. Assume that the walls and ceiling have an R-value of 10. Then estimate the heat gains from the occupants and power equipment.

Solution We can determine the cooling load from the equation

$$Q_{load} = Q_{walls} + Q_{vent}$$

where $Q_{load} = \dfrac{\text{DD}}{\sum R_T}$

$$Q_{vent} = \dot{m}c_p(\text{DD})$$

The DD value is estimated by using equation (16–2):

$$\text{DD} = (92 - 85)(12/24) + (0)(12/24) = 3.5 \text{ degree-days/day}$$

For one month, DD $= 3.5 \times 30 = 105$ degree-days. The total thermal resistance is

$$\sum R_T = \frac{10 \text{ h} \cdot \text{ft}^2 \cdot °\text{F/Btu}}{20 \text{ ft} \times 20 \text{ ft}} + \frac{10 \text{ h} \cdot \text{ft}^2 \cdot °\text{F/Btu}}{20 \text{ ft} \times 8 \text{ ft} \times 4}$$
$$= 0.0406 \text{ h} \cdot °\text{F/Btu}$$

and the heat gain through the walls and ceiling is

$$Q_{walls} = \frac{(105°\text{F}\cdot\text{days})(24\ \text{h/day})}{0.0406\ \text{h}\cdot°\text{F/Btu}}$$
$$= 62{,}069\ \text{Btu}$$

We assume that there will be 10 air changes per hour and that the volume of one air change is $20 \times 20 \times 8 = 3200\ \text{ft}^3$. Using an air density of 0.075 lbm/ft^3 and specific heat of 0.2404 Btu/lbm · °R, we can find the ventilation heat gain:

$$Q_{vent} = \dot{V}\rho c_p(\text{DD})$$
$$= (3200\ \text{ft}^3/\text{change} \times 10\ \text{changes/h})(0.075\ \text{lbm/ft}^3)(0.2404\ \text{Btu/lbm}\cdot°\text{R})$$
$$\times\ (105°\text{F}\cdot\text{day} \times 12\ \text{h/day})$$
$$= 726{,}970\ \text{Btu}$$

The cooling load is then the sum of the wall heat gain and the ventilation heat gain:

Answer
$$Q_{load} = 789{,}039\ \text{Btu}$$

The heat gain must be estimated from the heat given off by those persons occupying the store and the heat given off by any power equipment. We can assume that there will be 15 persons in the store and that each person generates 700 Btu/h. The power equipment and lighting may generate another 1000 W of heat, or 3414 Btu/h. The heat gain from these two sources is then

$$Q_{gain} = (700\ \text{Btu/h}\cdot\text{person} \times 15\ \text{persons} + 3414\ \text{Btu/h})(12\ \text{h/day})(30\ \text{days/month})$$

Answer
$$= 5{,}009{,}040\ \text{Btu/month}$$

EXAMPLE 16–6 For the air-conditioning requirement of example 16–5, assume that the relative humidity is 86% during July and that the relative humidity is to be no more than 60% after cooling. Determine the amount of water condensed out of the air per day. Neglect the moisture from the persons occupying the store or any moisture that may be involved with the power equipment.

Solution The amount of water removed per pound-mass of dry air will be the difference in the specific humidities of the intake air and the conditioned air, $\omega_1 - \omega_2$. From chart B–10, we find at $T_1 = 92°\text{F}$ and $\phi_1 = 86\%$ that $\omega_1 = 194$ grains/lbm = 0.0277 lbm/lbm. At the conditioned state, $T_2 = 85°\text{F}$, $\phi_2 = 60\%$ and $\omega_2 = 110$ grains/lbm = 0.0157 lbm/lbm. The total amount of water condensed out of the air in one day will be

$$(\omega_1 - \omega_2)(\dot{m}_{da}) = (0.0277\ \text{lbm/lbm} - 0.0157\ \text{lbm/lbm}) \times (3200\ \text{ft}^3/\text{change}$$
$$\times\ 10\ \text{changes/day} \times 12\ \text{h/day} \times 0.075\ \text{lbm/ft}^3)$$

Answer
$$= 345.6\ \text{lbm/day}$$

This is a significant amount of water, so drainage should be provided.

EXAMPLE 16–7 A three-bedroom house in southern California is to be cooled with an evaporative cooler. The house is a single-level structure 20 m by 15 m by 4 m high. Estimate the required cooling, assuming that three persons live in the house. Also determine the amount of water required per day for the evaporator during August. Assume an average daily temperature of 40°C, an average relative humidity of 5%, and an average nighttime temperature of 25°C, and assume that 27°C is the thermostat setting in the house. The house has thermal resistance of 60 $\text{m}^2\cdot°\text{C/W}$ in the walls and ceiling.

Solution The house volume is approximately equal to 20 m × 15 m × 4 m = 1200 m^3, and the outside surface area in the walls and roof is (20 m × 4 m × 2 + 15 m × 4 m × 2 + 20 m × 15 m) = 580 m^2. Assume that there will be five air changes per hour and that the three persons will each generate 130 W of heat. Appliances and lighting are assumed to generate another 1000 W of heat. The heat gain from the occupants and the appliances is then

$$Q_{gain} = (130\text{ W/person} \times 3\text{ persons} + 1000\text{ W})(24\text{ h/day} \times 30\text{ days} \times 3600\text{ s/h})$$
$$= 3{,}602{,}880\text{ kJ}$$

The cooling DD can be estimated from equation (16–2), with T_{ave} = 40°C and 27°C for the reference temperature:

$$DD = (40 - 27)(12/24) = 6.5°\text{C}\cdot\text{day/day}$$

For August,

$$DD = 6.5 \times 30 = 195°\text{C}\cdot\text{day}$$

The wall and ceiling heat gains are determined from the equation

$$Q_{wall} = \frac{DD}{\sum R_T}$$

where the thermal resistance is (60 m$^2\cdot$°C/W)/580 m^2 = 0.103°C/W. Then

$$Q_{wall} = \frac{195°\text{C}\cdot\text{day}}{0.103°\text{C/W}}(3600\text{ s/h})(24\text{ h/day}) = 163{,}573\text{ kJ}$$

The required cooling is

Answer

$$Q_H = Q_{gain} + Q_{wall} = 3{,}766{,}453\text{ kJ}$$

The water supplied to the evaporator must be equal to the mass of dry air times the difference in specific humidity between the conditioned air and the intake air:

$$\dot{m}_{water} = \dot{m}_{da}(\omega_2 - \omega_1)$$

From chart B–10 at 40°C, 5% relative humidity, ω_1 = 2.5 g/kg = 0.0025 kg/kg. At 27°C and an isenthalpic (constant-enthalpy) evaporation from state 1, ω_2 = 0.007 kg/kg. The mass of dry air per day is

$$\dot{m}_{da} = \dot{V}\rho = (1200\text{ m}^3\text{/change} \times 5\text{ changes/h} \times 12\text{ h/day})(1.178\text{ kg/m}^3)$$
$$= 84{,}816\text{ kg/day}$$

The required water is then

$$\dot{m}_{water} = (84{,}816\text{ kg/day})(0.007\text{ kg/kg} - 0.0025\text{ kg/kg})$$

Answer

$$= 381.7\text{ kg/day}$$

16–4 SUMMARY

Heating and air conditioning require some information about human comfort and climatic conditions. Human comfort can be defined in a region of temperature and relative humidity. People generate heat in an amount that depends on their type of activity, and they also need adequate amounts of air, or ventilation. One way of designing for ventilation is by changing the air in a space a prescribed number of times per hour, or air changes per hour. Climatic conditions can be accounted for by using the degree-day concept, defined as

$$DD = (65°\text{F} - T_{ave})(n\text{ days}) \quad \textbf{(16–1)}$$

where n is the number of days, usually 1, 30 for one month, or 360 for one year. This equation provides the design parameter DD for predicting heat losses through walls and roofs of structures and for heat losses due to ventilation. For heating, or space heating, the required heat to a space is given by

$$Q_{load} = Q_H + Q_{gain} \tag{16–3}$$

where Q_{load} = heat losses in walls and ceilings + heat losses due to ventilation + heat losses due to radiation

Q_{gain} = heat gain from people or other occupants of a space + heat gain in lighting, appliances, and power equipment + solar or other radiation heat gains + stored energy released from thermal storage masses

For air-conditioning problems, we can use cooling/heating processes to condense water from the air in humid climates. In arid climates, we can use evaporative cooling processes to achieve comfortable air conditioning. The required cooling in a refrigeration or air-conditioning system can be written

$$Q_H + Q_{load} + Q_{gain} \tag{16–4}$$

where Q_{load} = heat gain to space through walls and roofs or ceilings

Q_{gain} = heat gain from people or other occupants + heat gain from appliances and power equipment + solar and other radiation heating

The degree-day concept may be used for cooling problems, but the term must be defined as

$$\mathrm{DD} = (T_{ave} - 85°\mathrm{F})(n \text{ days})$$

where T_{ave} is the average daily temperature for n days. Sometimes the reference temperature is other than 85°F.

DISCUSSION QUESTIONS

16–1 Why is the amount of body heat loss associated with the type of activity?

16–2 How many air changes per hour would you like to have in your bedroom when you are asleep?

16–3 Based on the data of figure 16–2, what state of the United States has the greatest variation in annual degree-days?

16–4 How is space heating different from space cooling?

16–5 It has been pointed out that using the degree-day concept for space cooling is not a good way to accurately predict the instantaneous cooling load. Why is this so?

PRACTICE PROBLEMS

Section 16–1

16–1 Determine the DD for New York, New York, during the period January through March. Then determine the average daily temperature for that period in New York.

16–2 Determine the DD for one year in Seattle, Washington, and convert it to SI units. Then determine the average daily temperature in Seattle in degrees Celsius.

16–3 Estimate the expected heat given off by 300 persons in an auditorium while watching a play.

16–4 Recommend the number of air changes per hour for a railroad passenger car based on the values given in table 16–2.

Section 16–2

16–5 Estimate the heat loss in your apartment, room, or home for the month of December based on the DD values from table 16–3 or figure 16–2.

16–6 Estimate the required heat for one year in a local library.

16–7 Estimate the size of an air-to-air heat exchanger for a three-bedroom house in your home town.

16–8 Estimate the required yearly heat for a 40-story building 30 m by 60 m by 140 m high. Assume that the building is fitted with a south-facing Trombé wall, that it is located in Toronto, Ontario, Canada, and that the Trombé wall has a net solar input of 1260 kJ/m^2-day. The 30-m-long wall faces south.

Section 16–3

16–9 Estimate the cooling load of a health clinic in Phoenix, Arizona, during July if the average daily temperature is 100°F (38°C) for 12 h and 65°F for the remaining 12 h.

16–10 Estimate the amount of water required for an evaporative cooler for a motel room in San Antonio, Texas, during June. Assume an average daily temperature of 95°F and a nighttime temperature of 65°F. Assume that the average relative humidity is 5% during the day. Assume two air changes per hour, a room size of 4800 ft^3, room temperature of 76°F and 30% relative humidity.

16–11 Estimate the cooling load for a movie theater in Columbus, Ohio, during July and August. Assume average daily temperatures of 38°C and 80% relative humidity. Assume that the average nighttime temperature is 20°C and 85% relative humidity. Then determine the amount of water condensed from the air per day. Neglect moisture from persons, appliances, or equipment.

16–12 Estimate the cooling load and size of an air conditioner required for a three-bedroom house in Omaha, Nebraska, during July. Assume an average daily temperature of 92°F and 85% relative humidity. Then determine the expected amount of water condensed from the air per day.

16–13 Estimate the cooling load for a Styrofoam picnic cooler having 2-cm-thick walls and outside dimensions of 60 cm × 15 cm × 20 cm. The cooler is filled with picnic supplies at 4°C and is on a New Jersey beach for 3 h when the average temperature is 40°C.

16–14 Estimate the cooling load (in kJ or Btu) required to freeze 140 kg (308 lbm) of food from 20°C (67°F) and keep it frozen at 0°C (32°F) for one month.

16–15 (E) A classroom in a university is 30 feet by 40 feet by 8 feet high, and all of the walls, the floor, and the ceiling are inside, so no heat gains or losses occur with the surroundings. When there are 25 students and one instructor in the classroom, how much cooling needs to be done to prevent the room from having an increase in temperature?

16–16 (E) For the classroom of problem 16–15, if air at 55°F is supplied for ventilation and cooling, how much air, in lbm/h, is required to match the cooling load? Assume that the room temperature is 75°F.

OTHER POWER DEVICES

17

In this chapter, we are concerned with some diversified examples of systems to which the concepts of thermodynamics can be applied. Of course, any volume in space can be identified as a system and thus be subject to the laws of thermodynamics, but the examples presented here are typical devices (or systems) that might be encountered by engineers and technologists and thus should be considered from a thermodynamic approach. *Electric generators, motors*, and *batteries* are described, and a brief discussion is given to present the electrical analogies of work, heat, and energy. We then consider the operation of the *hydrogen–oxygen fuel cell* and some simple *thermoelectric devices*. Finally, *magnetohydrodynamic* and *electrohydrodynamic* systems are described in very brief terms.

Engineers and technologists must apply their knowledge and expertise to all phases of social needs, and to emphasize this trend, some *biological systems* are identified for thermodynamic analysis. Finally, the Stirling cycle is treated.

New Terms

$\mathscr{E}$	Electric potential	$\mathscr{R}$	Resistance
$\mathscr{F}$	Faraday's constant	$\mathscr{T}$	Thomson coefficient
$\mathscr{I}$	Electrical current	$\mathscr{V}$	Voltage
$\mathscr{Q}$	Electrical charge	α	Seebeck coefficient

17–1 ELECTRIC GENERATORS, MOTORS, AND BATTERIES

Elementary to a consideration of electrical devices is an understanding of the concept of **current**. We visualize atoms as comprising a nucleus surrounded by electrons, *e*. Each electron, when stripped from its nuclear influence, possesses a charge, called a *coulomb*; the amount of charge possessed by each electron has been found to be 1.6×10^{-19} coulomb (C). If a metal or other material contains an abundance of free electrons, it is said to have a negative charge, and if there is a shortage of electrons, it is positive. Free electrons will migrate from an area in which they are dense to an area in which they are scarce; that is, electrons flow from negative to positive, as shown in figure 17–1, where we also see that current $\mathscr{I}$ is depicted as flowing from positive to negative. Current is defined by the equation

$$\mathscr{I} = \frac{\delta \mathscr{Q}}{\delta t} \qquad \textbf{(17–1)}$$

where $\mathscr{Q}$ is the amount of charge and where the unit of current, coulomb/s, is called the *ampere*. Since electrons possess negative charges, we then say that the positive current of electricity must travel from positive to negative.

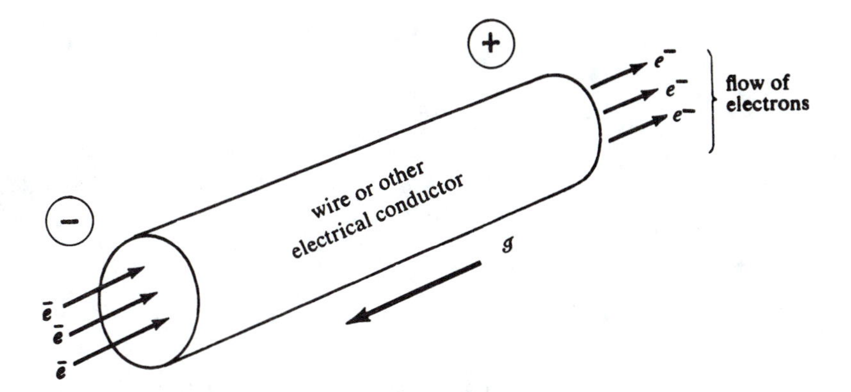

FIGURE 17–1 Electric current flow

Another concept we need to consider briefly in electrical phenomena is that of the electrical potential $\mathscr{E}$. We write

$$\mathscr{E} = \frac{\delta E_e}{\delta \mathscr{Q}} \tag{17–2}$$

where E_e is the electrical energy. The unit for the electrical potential can easily be seen to be Btu/coulomb or, more commonly, joule/coulomb. The volt is defined as the joule/coulomb.

The essential components of a common electric generator are a **rotor** and a **stator**. The stator is a stationary housing consisting of magnets or electromagnets, as shown in figure 17–2. The rotor is situated between the stator magnets so that as it rotates, it interrupts the magnetic lines of force and thus induces an electric current through a conducting wire. Frequently, the stator is composed of the conducting wires, and the rotor is then the source of magnetic fields, but here we will consider the arrangement as shown in figure 17–2. We first write the first law for the generator:

$$\Delta E_e = Q_f - Wk_{\text{gen}} \tag{17–3}$$

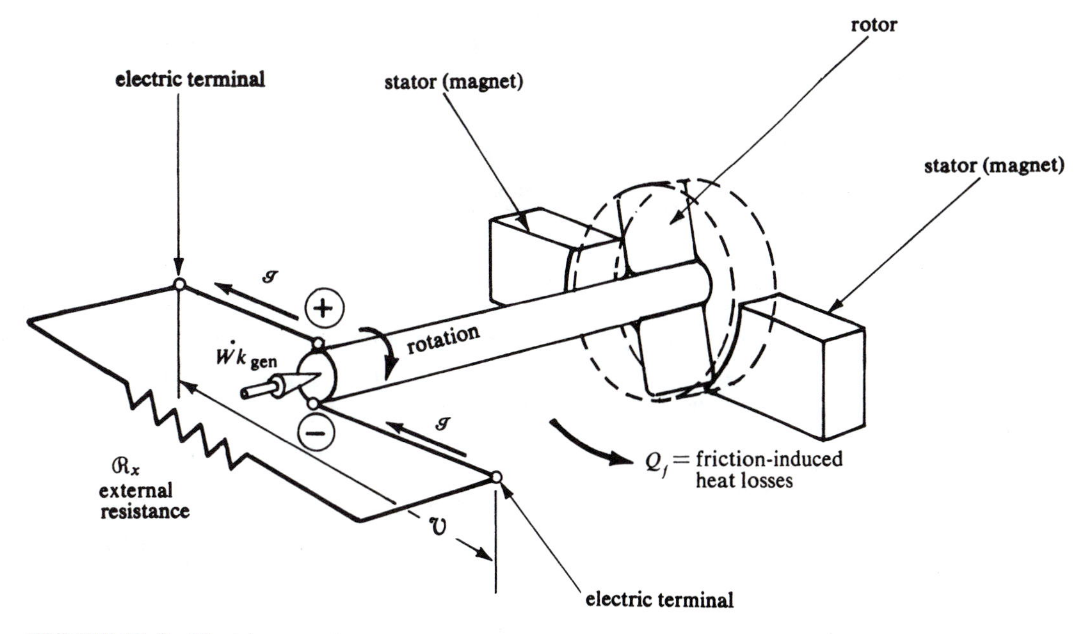

FIGURE 17–2 Electric generator

Then, if we assume the process to be reversible, $Q_f = 0$. This gives us

$$\Delta E_e = -Wk_{\text{gen}}$$

and from equation (17–2), we obtain

$$\Delta E_e = \sum \delta E_e = \sum \mathscr{E}\,\delta \mathscr{Q} = -Wk_{\text{gen}} \qquad \textbf{(17–4)}$$

Since

$$\dot{E}_e = \frac{\delta E_e}{\delta t} = \mathscr{E}\frac{\delta \mathscr{Q}}{\delta t}$$

we can use the definition of current, equation (17–1), to find

$$\mathscr{E}\mathscr{I} = -\dot{W}k_{\text{gen}} \qquad \textbf{(17–5)}$$

If we measure the potential or voltage across the terminals, its value will be $\mathscr{E}$, provided that there exists no external connection between the poles. If, on the other hand, we insert a load or resistance $\mathscr{R}_x$ between the poles, as depicted in figure 17–2, the voltage will be less than $\mathscr{E}$. **Ohm's law** states that

$$\text{potential drop} = \text{resistance} \times \text{current} \qquad \textbf{(17–6)}$$

so, calling the voltage across the resistance $\mathscr{V}$, we have

$$\mathscr{V} = \mathscr{R}_x \mathscr{I} \qquad \textbf{(17–7)}$$

where $\mathscr{R}_x$ is the resistance measured in ohms, $\mathscr{I}$ is the current in amperes, and $\mathscr{V}$ is the voltage. The current is also flowing through the generator rotor, so that a potential drop, due to the conducting wire resistance $\mathscr{R}_I$, will be present. Then

$$\mathscr{V} = \mathscr{E} - \mathscr{I}\mathscr{R}_I \qquad \textbf{(17–8)}$$

for the closed-circuit generator.

The efficiency of the generator is given by

$$\eta = \frac{\mathscr{V}\mathscr{I}}{\dot{W}k_{\text{gen}}} \times 100 \qquad \textbf{(17–9)}$$

and is normally in the range 85% to 95%. Irreversibilities such as friction have been neglected here, so that a more realistic correction may be made to actual generators by writing

$$\dot{W}k_{\text{gen}} = \mathscr{E}\mathscr{I} + \dot{W}k_{\text{friction}} \qquad \textbf{(17–10)}$$

Electric motors can be visualized as reversed generators. A voltage is applied across the terminals so that a current exists in the rotor, which in turn induces an electric field, $\mathscr{E}_{\text{ind}}$. The interaction between the electric and magnetic fields then produces a rotation and torque through the rotor. For the motor, we have

$$\mathscr{E}_{\text{ind}} = \mathscr{V} - \mathscr{I}\mathscr{R}_I \qquad \textbf{(17–11)}$$

and the mechanical efficiency is

$$\eta_{\text{mech}} = \frac{\dot{W}k_{\text{motor}}}{\mathscr{V}\mathscr{I}} \times 100 \qquad \textbf{(17–12)}$$

where the electric power used by the motor is $\mathscr{V}\mathscr{I}$ The motor output power is given by

$$\dot{W}k_{\text{motor}} = \mathscr{E}_{\text{ind}}\mathscr{I} - \dot{W}k_{\text{friction}} \qquad \textbf{(17–13)}$$

Whereas the electric generator and the motor involve energy transfers between electrical and mechanical systems, the common electrical battery utilizes a transfer of energy between chemical and electrical systems.

The **Daniel cell** is a device that produces electrical energy from chemical reactions. It is not exactly like the common batteries, but its workings involve the essential concepts of the typical battery. The Daniel cell, shown in figure 17–3, is composed of zinc sulfate ($ZnSO_4$) and copper sulfate ($CuSO_4$) solutions separated by a barrier that prevents mixing of these two solutions, yet which allows ions (such as Cu^{2+}, Zn^{2+}, and $SO_4{}^{2-}$) to pass. A solid bar of zinc (anode) is placed in the $ZnSO_4$ solution, and a bar of copper (cathode) in the $CuSO_4$ solution. The chemical reaction at the anode is

$$Zn \rightarrow Zn^{2+} + 2e^- \qquad \textbf{(17–14)}$$

and at the cathode is

$$Cu^{2+} + 2e^- \rightarrow Cu \qquad \textbf{(17–15)}$$

The total chemical reaction for the Daniel cell is found by adding equations (17–14) and (17–15):

$$\underbrace{Zn + Cu^{2+}}_{\text{reactants}} \rightarrow \underbrace{Cu + Zn^{2+}}_{\text{products}} \qquad \textbf{(17–16)}$$

FIGURE 17–3 Daniel cell battery

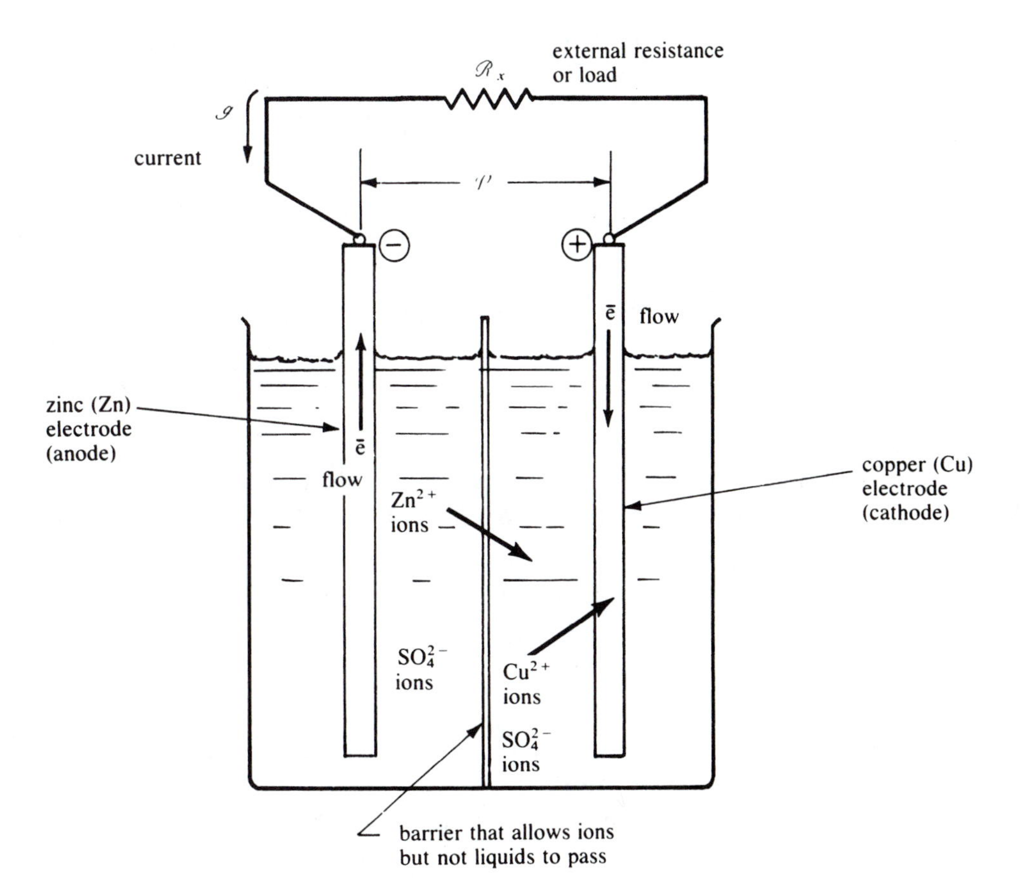

The result is that the cathode becomes plated with copper, the anode loses zinc, and the electrons are allowed to flow through an external load. The maximum work or electrical energy obtainable from chemical reaction (17–16) is given by equation (8–28):

$$Wk_{\text{use}} = -\Delta\mathcal{G}$$

Since the useful work in also the maximum work obtainable for useful purposes, we write

$$Wk_{\text{max}} = -\Delta\mathcal{G}$$

where the Gibbs free energy can be shown to be calculated from the summation

$$\Delta\mathcal{E} = \underbrace{\sum \Delta\mathcal{E}^{\circ}{}_f}_{\text{products}} - \underbrace{\sum \Delta\mathcal{E}^{\circ}{}_f}_{\text{reactans}} \tag{17–17}$$

where $\Delta\mathcal{G}_f^{\circ}$ represents the amount of the Gibbs free energy required to form the particular molecule from the elements, for example, the energy required to form sodium chloride (NaCl) from sodium (Na) and chlorine (Cl) atoms. Table B–25 lists values of the Gibbs free energies of formation $\Delta\mathcal{G}_f^{\circ}$ for various materials. Of course, the value of the Gibbs free energy of formation is zero for the natural elements nonionized.

The work done to convey the electrons from the cathode to the anode is also equal to $n\mathcal{F}^{\circ}\mathcal{V}^{\circ}$, where n represents the gram-moles of electrons; $\mathcal{F}$ is the **Faraday constant**, given by

$$\mathcal{F} = 96{,}500 \text{ C/g} \cdot \text{mol}$$

and $\mathcal{V}^{\circ}$ is the potential across the anode and cathode when the circuit is open, that is, when the resistance $\mathcal{R}_x$ is disconnected. Then we have

$$n\mathcal{F}\mathcal{V}^{\circ} = -\Delta\mathcal{G} \tag{17–18}$$

and the actual output voltage $\mathcal{V}$ is

$$\mathcal{V} = \mathcal{V}^{\circ} - \mathcal{I}\mathcal{R}_I \tag{17–19}$$

where $\mathcal{R}_I$ is the Daniel cell resistance. The power obtainable from the cell is

$$\mathcal{V}\mathcal{I} = \mathcal{I}^2\mathcal{R}_x \tag{17–20}$$

and, for the efficiency, we write

$$\eta_{\text{battery}} = \frac{\mathcal{V}\mathcal{I}}{\Delta\mathcal{G}} \tag{17–21}$$

EXAMPLE 17–1 Determine the maximum work obtainable from a Daniel cell, the open-circuit voltage, and the operating voltage if 0.1 A of current is drawn and the internal resistance is 0.05Ω.

Solution: We calculate the maximum work from the sum of the Gibbs free energy values. Using equation (17–17) and the values from table B–25, we have

$$\Delta\mathcal{G} = \text{Cu} + \text{Zn}^{2+} - \text{Zn} + \text{Cu}^{2+}$$

or

$$\Delta\mathcal{G} = \underbrace{0 + (-147{,}290)}_{\text{products}} - \underbrace{0 - 65{,}020}_{\text{reactants}} = -212{,}310 \frac{\text{kJ}}{\text{kg} \cdot \text{mol}}$$

Then

Answer

$$Wk_{\text{max}} = -\Delta\mathcal{G} = 212{,}310 \text{ kJ/kg} \cdot \text{mol}$$

The open-circuit voltage can be calculated from equation (17–18), namely,

$$\mathcal{V}^{\circ} = \frac{-\Delta\mathcal{G}}{n\mathcal{F}} = \frac{212{,}310 \text{ kJ/kg} \cdot \text{mol}}{96{,}500n}$$

and $n = 2$ g · mol of electrons per gram-mole of reaction, so

$$\mathcal{V}^{\circ} = \frac{212{,}310 \text{ kJ/kg} \cdot \text{mol}}{2 \text{ g} \cdot \text{mol/g mol} \times 96{,}500 \text{ C/g} \cdot \text{mol}}$$

Answer

$$= 1.10 \text{ J/C} = 1.10 \text{ V}$$

If we now use equation (17–19), we can calculate the operating voltage when 0.1 A of current is drawn:

$$\mathcal{V} = 1.10 \text{ V} - (0.1 \text{ A})(0.05 \ \Omega)$$

Answer

$$= 1.095 \text{ V}$$

Example 17–1 and the preceding discussion are items and areas that engineers and engineering technologists in particular may encounter in practice—they were presented to illustrate the application of the thermodynamic concepts to electrical devices. In no way, however, should readers infer that this brief presentation is a sufficient treatment of electrical concepts; nor should they assume that other thermodynamic ideas such as entropy cannot be applied to electrical phenomena.

17–2 FUEL CELLS

The fuel cell has been receiving increased attention due to its attractiveness as a "clean' source of electrical power as well as the promise for replacing the internal combustion engine used in automobiles, buses, and trucks. The fuel cell is an electrochemical devise that can continuously convert chemical energy into electrical energy of power as long as a fuel and a reactant are supplied. The reason this is possible is that, during a chemical reaction, there is commonly an electron and proton exchange or rearrangement. Thus if a fuel and a reactant, such as oxygen, chemically react without combustion occurring and if the intermediate free electrons can be conducted through a circuit, the result is the fuel cell. There is usually some heat that accompanies this reaction, however the temperature will remain essentially constant so that the fuel cell efficiency is not limited by the second law of thermodynamics or the Carnot efficiency, as given by equation (7–23). Many fuels have been proposed for use in a fuel cell, including biological or organic fuels, but the most successful combination has been hydrogen and oxygen.The fuel cell using these two fuels is called the *hydrogen–oxygen fuel cell.*

The fuel cell was invented by William R. Grove, who first reported it in 1839. Thus, it has been known but not "in existance" for a long time, but its technology is still not fully developed, due to the problems associated with the demands on the materials for anode, cathode, and electrolyte, as well as the physical configurations. A brief discussion of types of fuel cells may explain this situation. At least six types of hydrogen–oxygen fuell cells have been developed:

1. Proton exchange membrane fuel cell (PEMFC)
2. Direct methanol fuel cell (DMFC)
3. Alkaline fuel cell (AFC)
4. Phosphoric acid fuel cell (PAFC)
5. Molten carbonate fuel cell (MCFC)
6. Solid oxide fuel cell (SOFC)

The configuration of the basic element of the PEMFC is shown in figure 17–4, where it can be seen that the anode and cathode are separated by a solid membrane that acts as an electrolyte.

The electrolyte allows hydrogen ions (H^+) to pass from the anode to the cathode. The chemical reactions can be written as

$$H_2 \rightarrow 2H^+ + 2e^- \qquad \text{at the anode} \qquad \textbf{(17–22)}$$

and

$$1/2\ O_2 + 2H^+ + 2e^- \rightarrow H_2O \qquad \text{at the cathode} \qquad \textbf{(17–23)}$$

The net chemical reaction for both the anode and the cathode is

$$H_2 + 1/2\ O_2 - H_2O \qquad \textbf{(17–24)}$$

From equation (8–28), we find the maximum work to be

$$Wk_{\text{max}} = -\Delta\mathcal{G}$$

and the open-circuit voltage from equation (17–18) is

$$\mathcal{V}^{\circ} = \frac{-\Delta\mathcal{G}}{n\mathcal{F}}$$

EXAMPLE 17–2 Calculate the maximum work obtainable from the hydrogen–oxygen fuel cell, and calculate the open-circuit voltage.

Solution From table B–25, we have

$$\Delta\mathcal{G}_f^{\circ} = \begin{cases} -237{,}180 & \text{for } H_2O \text{ (the products)} \\ 0 & \text{for } H_2 \text{ and } O_2 \text{ (the reactants)} \end{cases}$$

so, using equation (17–17), we obtain

$$\Delta\mathcal{G} = -237{,}180 \text{ kJ/kg} \cdot \text{mol}$$

Since the molecular weight (MW) of water is 18 g/g · mol, we have

$$\Delta\mathcal{G} = -\frac{237{,}180}{18} = -13{,}177 \text{ kJ/kg } H_2O$$

Then

Answer

$$Wk_{\text{fuel cell}} = 13{,}177 \text{ kJ/kg}$$

The open-circuit voltage is

$$\mathcal{V}^{\circ} \frac{-\Delta\mathcal{G}}{n\mathcal{F}} = \frac{237{,}180}{(2)(96{,}500)}$$
$$= 1.23 \text{ J/C}$$

Answer

$$= 1.23 \text{ V}$$

It has been found that platinum (Pt) seems to be an ideal material for making the two reactions indicated by equations (17–22) and (17–23) proceed; it is often the material used for the anode and the cathode. Since it is expensive, it is now coated in a thin layer onto a carbon or other material. PEMFC requires pure hydrogen and oxygen gases to operate; however, air can often be used, and if methanol or other fuels are used for the hydrogen supply, a process needs to occur to separate the hydrogen prior to entering the fuel cell. Devices called *reformers* are often used for this. The operating temperatures of the PEMFC are approximately 50°C to 90°C. The arrangement shown in figure 17–4 is sometimes referred to as the membrane electrode assembly (MEA) and, by stacking many of these units in series and parallel, higher voltages and higher power can be obtained. The PEMFC was used in the first NASA manned spacecraft and shows promise for wider usage in stationary power plants as well as in automobiles an other vehicles.

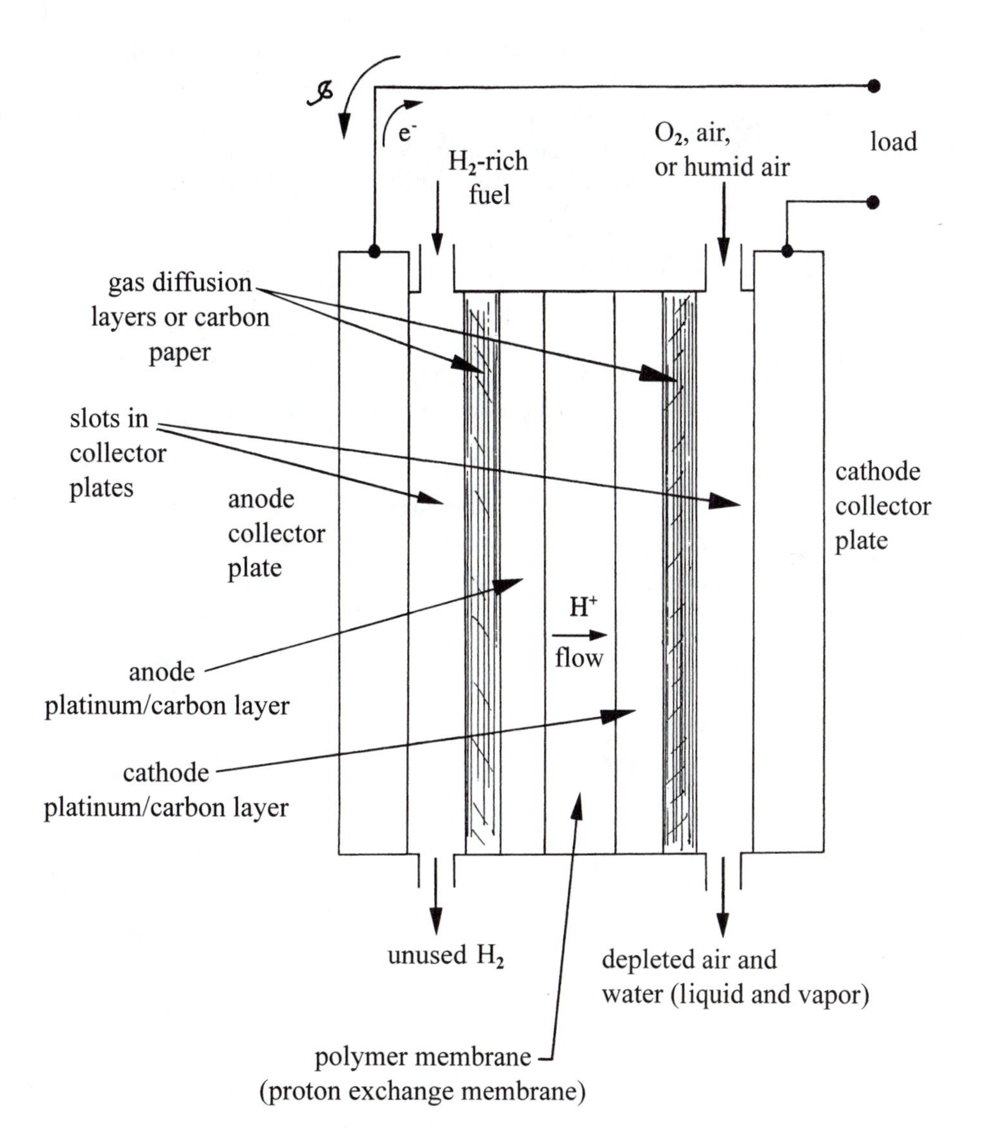

FIGURE 17–4 The typical proton exchange membrane fuel cell (PEMFC)

The DMFC is similar to the PEMFC, but is uses methanol as the fuel and air as the reactant. This is an advantage, in that hydrogen is not required as a fuel, but a major problem associated with the DMFC is the slow reaction of the methanol at the anode, which is

$$CH_3OH + H_2O \rightarrow CO_2 + 6H^+ + 6e^- \qquad \textbf{(17–25)}$$

The chemical reaction at the cathode is

$$\frac{3}{2}O_2 + 6H^+ + 6e^- \rightarrow 3H_2O \qquad \textbf{(17–26)}$$

and the net chemical reaction for the direct methanol fuel cell is

$$CH_3OH + \frac{3}{2}O_2 + H_2O \rightarrow CO_2 + 3H_2O \qquad \textbf{(17–27)}$$

This fuel cell needs to have water supplied to it and, since it uses a hydrocarbon (methanol alcohol), produces carbon dioxide.

The alkaline fuel cell (AFC) was the fuel cell used in the manned Apollo mission to the moon and one of the arrangements for the basic element of the AFC in shown in figure 17–5. Here the electrolyte is a liquid such as potassium hydroxide (KOH), which is pumped around the anode and cathode in steady flow. The anode and cathode can be constructed of nickel or silver with suitable coatings for allowing water and ions to pass and to act as catalysts for the reactions. The chemical reaction at the anode is

$$2H_2 + 4OH^- \rightarrow 4H_2O + 4e^- \quad \textbf{(17–28)}$$

and at the cathode

$$O_2 + 4e^- + 2H_2O \rightarrow 4OH^- \quad \textbf{(17–29)}$$

The overall fuel cell reaction can then be written as

$$O_2 + 2H_2 \rightarrow 2H_2O \quad \textbf{(17–30)}$$

If air is used instead of oxygen, any carbon dioxide in the air will degrade the KOH electrolyte and reduce the performance of the fuel cell. The operating temperatures of the

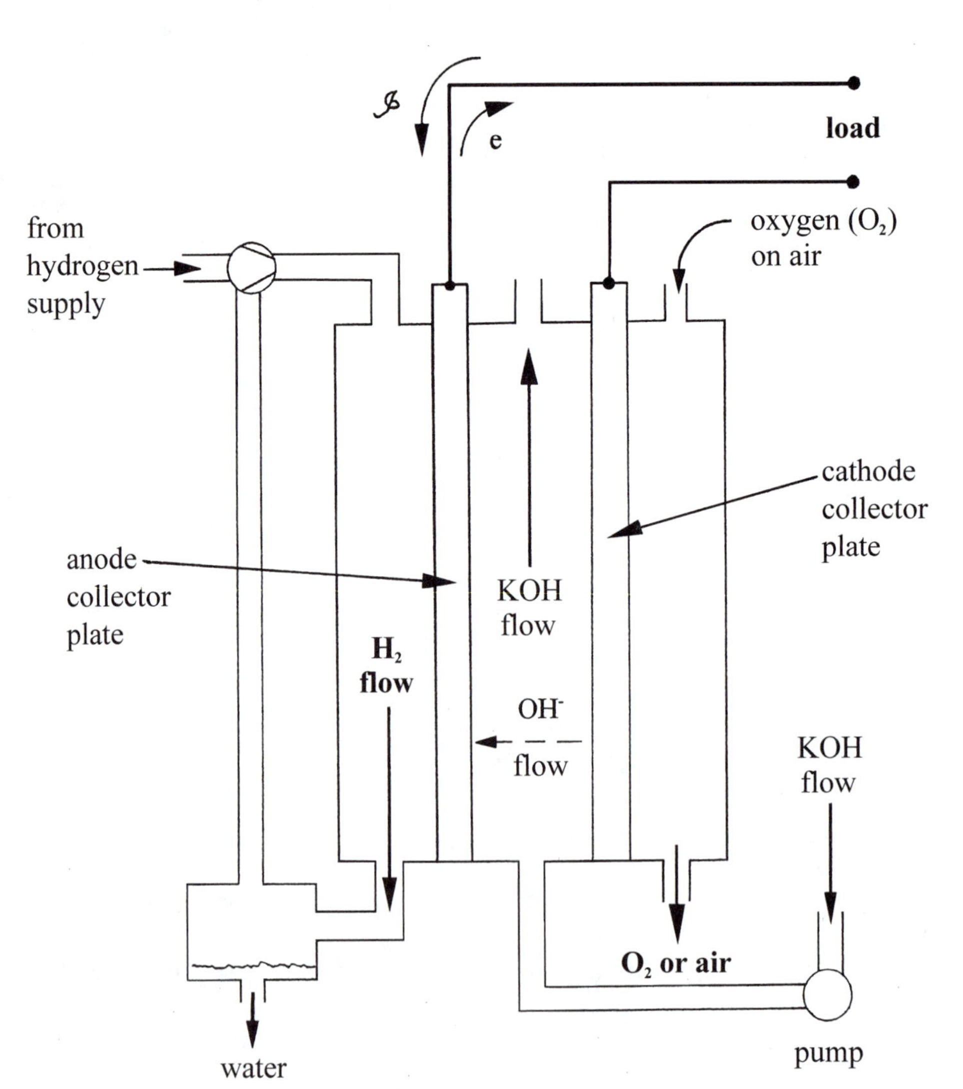

FIGURE 17–5 Typical alkaline fuel cell

AFC are around 60°C to 120°C. As stated, the AFC has demonstrated a niche of usefulness but is not being as intensively studied or developed as others.

The phosphoric acid fuel cell (PAFC) has been shown to have potential for large power generators, with the largest unit generating 11 MW of power and many units generating from 1 to 5 MW. It utilizes phosphoric acid, H_3PO_4, as an electrolyte. The electrodes, anode, and cathode are typically carbon with platinum coating for the catalyst of the chemical reactions. The reactions are the same as for the PEMFC, equations (17–22), (17–23), and (17–24). The operating temperatures of the PAFC are typically around 200°C. The PAFC is being seriously developed for stationary power generation.

The molten carbonate fuel cell (MCFC) utilizes as an electrolyte a molten (liquid) mixture of lithium (Li) and potassium (K) or other alkali metal carbonates. This mixture is usually retained in a ceramic matrix of $LiAlO_2$ and the operating temperatures are around 650°C. The typical arrangement of the MEA is shown in figure 17–6. Because of the high

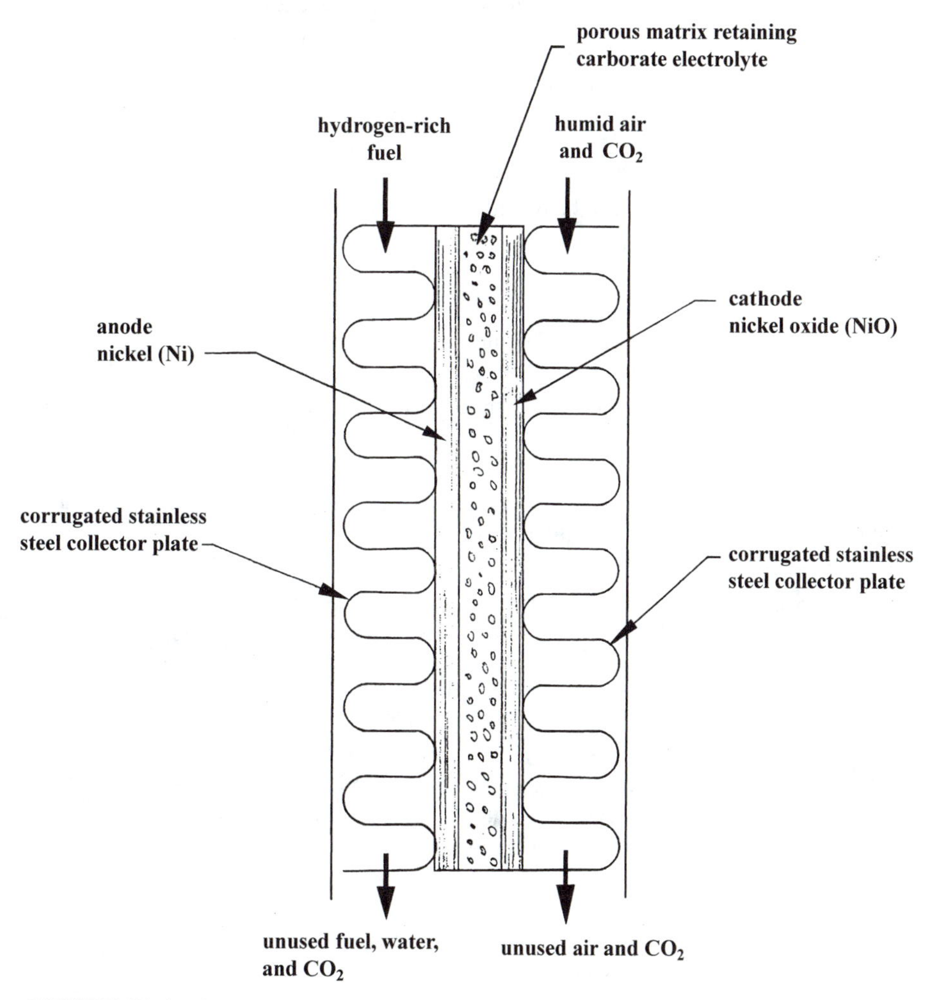

FIGURE 17–6 Cross section of a molten carbonate fuel cell (MCFC) membrane electrode assembly (MEA). A fuel cell would contain many MEAS. The stainless steel collector plates will be electrically connected to adjacent collector plates and together form a means of obtaining significant electrical power from the fuel cell.

temperatures, the electrodes are usually constructed of stainless steel with a catalyst coating of nickel (Ni). The reaction at the anode is

$$2H_2 + 2CO_3^- \rightarrow 2H_2O + 2CO_2 + 4e^- \quad \textbf{(17–31)}$$

and at the cathode

$$O_2 + 2CO_2 + 4e^- \rightarrow 2CO_3^- \quad \textbf{(17–32)}$$

with the overall fuel cell reaction being

$$H_2 + \frac{1}{2}O_2 + CO_2 \rightarrow H_2O + CO_2 \quad \textbf{(17–33)}$$

The MCFC, as with all fuel cells, operates by requiring the electrons to pass through an external load, but the electrolyte conducts carbonate ions (CO_3^-).

The solid oxide fuel cell (SOFC) operates at the highest fuel cell temperature, around 800°C to 1100°C, with eliminates the need for catalysts at the anode and cathode. The electrodes have typically been constructed of porous ceramics with suitable coatings. The electrolyte is usually yttria (Y_2O_3) stabilized zirconia (ZrO_2). The reaction at the anode is

$$H_2 + O^- \rightarrow H_2O + 2e^- \quad \textbf{(17–34)}$$

at the cathode

$$\frac{1}{2}O_2 + 2e^- \rightarrow O^- \quad \textbf{(17–35)}$$

with the overall fuel reaction being

$$H_2 + \frac{1}{2}O_2 \rightarrow H_2O \quad \textbf{(17–36)}$$

Thus, fuel cells have been a tantalizing means for providing power based on the simple concept of hydrogen and oxygen forming water. The major technological problems that have been associated with them since 1839 are materials for the anode, the cathode, and the electrolyte, as well as having a reliable and practical source of hydrogen and oxygen.

17–3 THERMOELECTRIC DEVICES

We have seen that resistance in a conducting wire produces a voltage drop given by Ohm's law, equation (17–7):

$$\mathcal{V} = \mathcal{I}\mathcal{R}$$

The power presented by this is given by

$$\mathcal{V}\mathcal{I} = \mathcal{I}^2\mathcal{R} \quad \textbf{(17–37)}$$

and this power is dissipated in a resistor $\mathcal{R}$ in the form of an increase in the temperature of the resistor. If the resistor is in equilibrium and steady-state conditions with the surroundings, the power $\mathcal{I}^2\mathcal{R}$ is reflected in heat transfer to the surroundings, as shown in figure 17–7. We call this effect **Joule heating**, $\dot{Q}_J$, which represents the most common means of electrical energy dissipation. Joule heating is used to advantage in any electric heater and is present in any conductor of electricity having resistance.

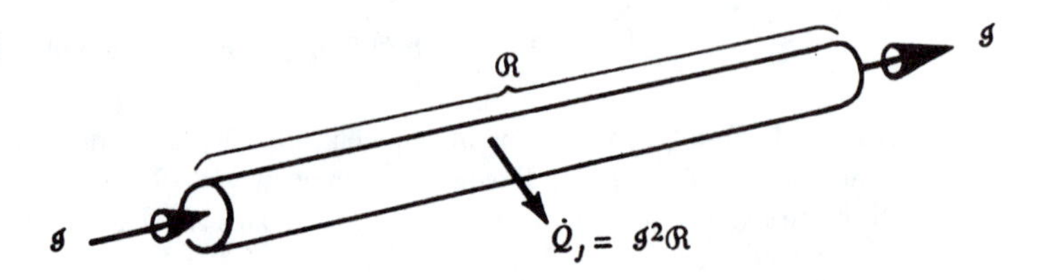

FIGURE 17–7 Joule heating

If two wires or conductors, A and B, composed of different materials, are joined as shown in figure 17–8 and a current is impressed through them, another heat transfer can be induced. This heat, called **Peltier heating**, is due to thermoelectric phenomena called the *Peltier* and *Seebeck effects* and is the basis for the thermocouple. The heat emanating from the joint, $\dot{Q}$, is given by the sum of the Joule and Peltier heats:

$$\dot{Q} = \mathcal{I}^2\mathcal{R} + \mathcal{I}(\alpha_A - \alpha_B) \tag{17–38}$$

FIGURE 17–8 Peltier effect

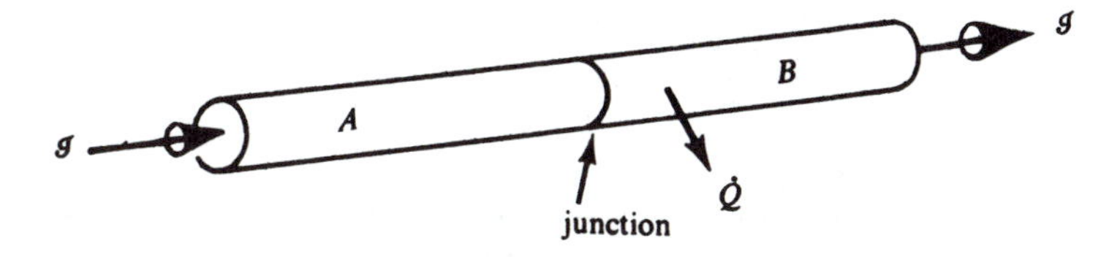

Here, α_A and α_B are the Seebeck coefficients of wires A and B. The Seebeck coefficients are the mathematical values allowing for the Peltier heat and are generally found to be functions of the material temperature. A further discussion of the Seebeck coefficient can be found in publications devoted to direct energy conversion devices. It may be noted that the Peltier effect can be reversed by impressing heat transfer into the joint and thus inducing a current in the wires. This is the essential operation of the temperature-measuring device called the **thermocouple**. In the thermocouple, the current is measured, allowing for a prediction of the joint temperature.

Another thermoelectric phenomenon found in all actual conductors is the **Thomson effect**. A wire or other electric conductor subject to a temperature change in its volume, as shown in figure 17–9, will have a current induced in it. The heat transfer due to this effect is called the **Thomson heat** and is generally written as $\mathcal{I}_{\text{ind}}\mathcal{T}\ \Delta T$, where $\mathcal{T}$ is the Thomson coefficient (determined by experimental means) and ΔT is $T_1 - T_2$. The heat transfer on the sides of the conductor is then

$$\dot{Q} = \mathcal{I}_{\text{ind}}^2\mathcal{R} + \mathcal{I}_{\text{ind}}\mathcal{T}\ \Delta T \tag{17–39}$$

FIGURE 17–9 Thomson effect

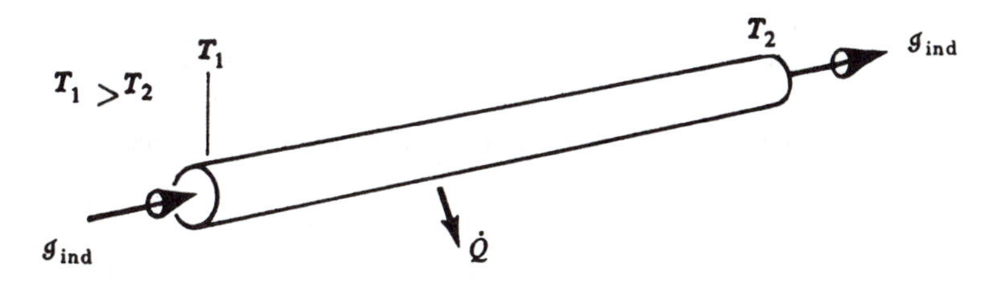

Semiconductor materials are utilized in some useful thermoelectric devices, called **thermoelectric generators**. These materials (semiconductors) exhibit a selectivity in allowing current flow, generally by prohibiting current to pass in one of the two directions, and they also have wide varieties of values for Seebeck coefficients. Figure 17–10 shows a thermoelectric device composed of two distinct types of semiconductor materials, N-type and P-type. The bar labeled A is a conductor of electricity and a receiver of external heat transfer. From the concept of Peltier heating discussed previously, we see that in this case the Peltier heat is added to the junctions between A and the N-type material and P-type material. At these two joints, we then have, from equation (17–38),

$$\dot{Q}_{\text{add}} = \mathcal{I}_{\text{ind}}^2\mathcal{R} + \mathcal{I}_{\text{ind}}(\alpha_n - \alpha_A) + \mathcal{I}_{\text{ind}}(\alpha_A - \alpha_p) \tag{17–40}$$

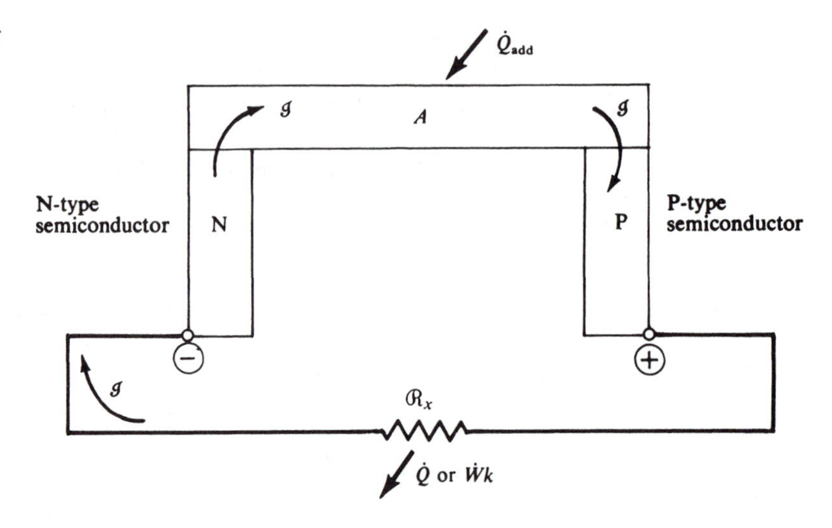

FIGURE 17–10 Elements of a simple thermoelectric generator

where $\mathcal{I}_{\text{ind}}$ is the current induced in the circuit due to the *addition* of Peltier heat. The semiconductor materials, N- and P-types, have properties such that α_n is negative and α_p is positive. This means that the Seebeck effect at the two junctions, instead of canceling, will add and give us, from equation (17–40),

$$\dot{Q}_{\text{add}} = \mathcal{I}_{\text{ind}}^2 \mathcal{R} - \mathcal{I}_{\text{ind}}(|\alpha_n| + |\alpha_p|) \tag{17–41}$$

Thermoelectric generators are attractive as sources of electric power since they are not inherently dependent on the source of thermal energy. Heat may be transferred from a nuclear reactor, steam, hot exhaust gas, the sun, or numerous other sources.

17–4 MAGNETO-HYDRODYNAMICS

The electric generator produces electric power by passing a conducting wire through a magnetic field. A magnetohydrodynamic (MHD) generator, such as shown in figure 17–11, produces electric power in much the same way except that the conductor passing through the magnetic field is a hot gas or fluid. The basic operation of this device is indicated in figure 17–12, where hot gases (ionized gases if possible) flow through a chamber having two sides that conduct electricity (sides *a-b-c-d* and *f-e-g-h*) and two sides that are insulated (sides *b-e-g-c* and *a-f-h-d*). An electric field $\mathcal{E}_e$ is applied across the two conducting sides, and this field induces a magnetic field having lines of force perpendicular to the gas flow. The flow of hot gases through these fields induces

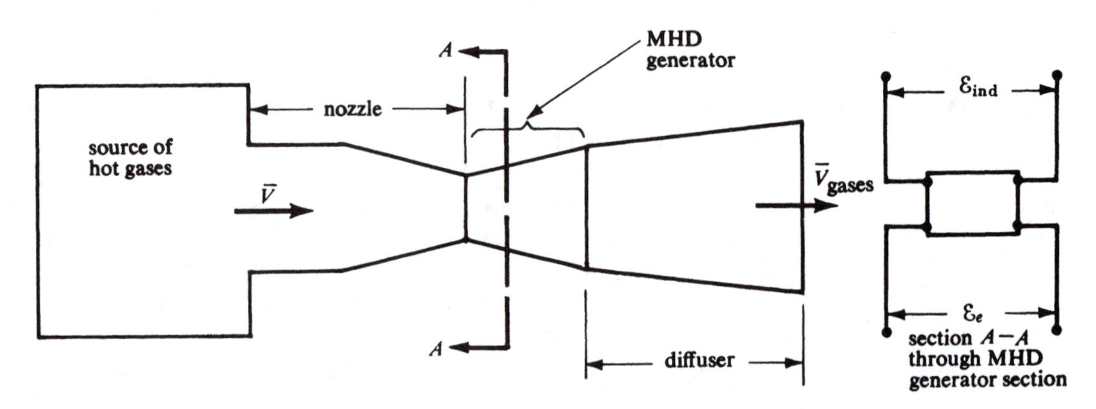

FIGURE 17–11 Sketch of magnetohydrodynamic power generator

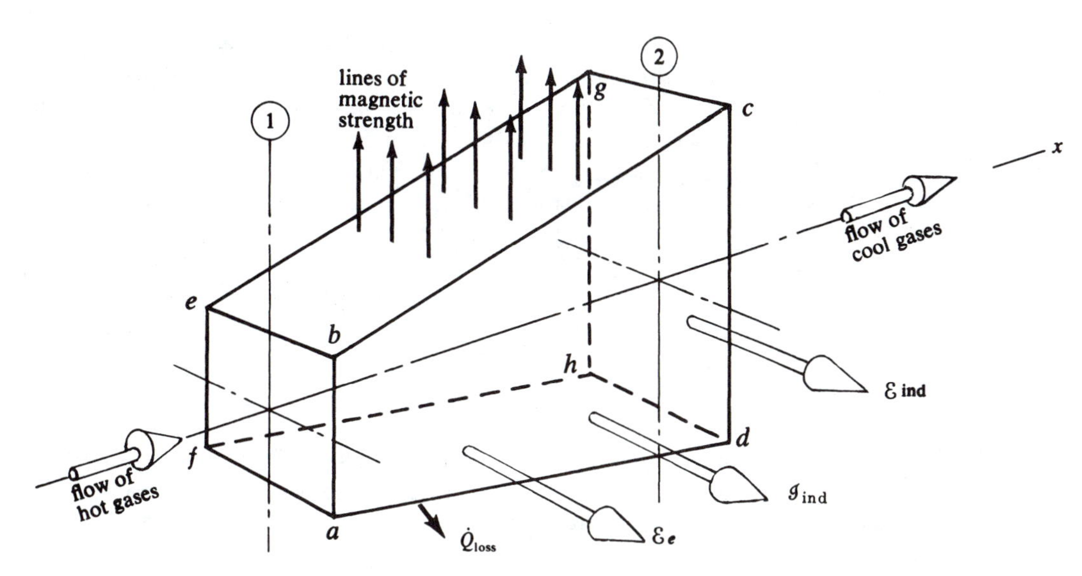

FIGURE 17–12 Basic operation of MHD generator

an electric potential $\mathscr{E}_{\text{ind}}$ of significant strength. By applying the first law of thermodynamics to this system, we have

$$\dot{m}\frac{\overline{V}_2^2 - V_1^2}{2} + \dot{m}(h_2 - h_1) = \dot{Q}_{\text{loss}} - \mathcal{I}_{\text{ind}}\mathscr{E}_{\text{ind}} + \mathcal{I}_{\text{ind}}\mathscr{E}_e \qquad \textbf{(17–42)}$$

where $\mathcal{I}_{\text{ind}}$ is the induced current due to the MHD effect and $\dot{m}$ is the mass flow rate of the gases.

The MHD generator represents a device that can utilize very high temperature gases with present materials and produce significant amounts of power. In addition, the exhaust gases from the device can be used to provide a heat source for a conventional heat engine, such as a gas turbine or a steam turbine. When used in this manner in tandem with other devices, the MHD generator can represent a significant improvement in thermodynamic efficiencies.

Another interesting aspect of the phenomena being considered here is the reversed procedure; that is, if magnetic and electric fields are applied across a conduit containing a fluid that can conduct electricity, a velocity or flow is induced in the fluid. This technique is responsible for the concept of pumps having no moving parts.

The primary disadvantages of the MHD device are twofold:

1. The gas or fluid must have a sufficiently low resistance and must be a good electrical conductor. Most gases do not have these properties.
2. The source of hot gases must be exceedingly large if significant amounts of power are to be generated.

Significant treatises exist to which the reader may refer for a more complete treatment of magnetohydrodynamic devices.

17–5 BIOLOGICAL SYSTEMS

Engineers and technologists have been applying engineering and scientific concepts to inorganic, inanimate objects or systems without too much hesitation. On the other hand, organic, biological systems have been avoided with few exceptions. The following two systems, the muscle and the heart, are biological devices that have been investigated from engineering viewpoints and are presented here to set examples for application of thermodynamic concepts to other biological systems heretofore avoided.

The muscle represents an organic system that converts chemical energy to mechanical work, and this is achieved at essentially constant temperature. Many have attempted to explain scientifically the details of this conversion with no great success, and indeed the muscle represents the only known device that directly converts chemical energy to work. It appears that the source of chemical energy resides in a macromolecular substance called adenosine triphosphate (ATP) and other simpler phosphates. This reaction releases energy to be used in work and heat. The muscle system is shown in figure 17–13, along with a weight W representing an external force applied to the muscle. It is viewed in a process of contracting or shortening while exerting a force; the actual process has been simplified by replacing the muscle force by weight W. We may apply the first law to the muscle as follows:

$$H_2 - H_1 = Q - Wk \tag{17–43}$$

In this equation, H_2 is the enthalpy of the ATP and phosphates and H_1 is the enthalpy of the resulting chemical, adenosine diphosphate (ADP). The work obtained from a muscle is Wk, and the maximum work is predicted from equation (8–28):

$$Wk_{\max} = -\Delta\mathcal{G}$$

We may consider the mechanical efficiency of the muscle system to be

$$\eta_{\text{mech}} = \frac{Wk}{\Delta\mathcal{G}} \times 100 \tag{17–44}$$

and this seems to be around 30% to 40% for most healthy muscles. The actual work obtained from a muscle is, of course, dependent on the amount of concentration or distance through which force is applied. If the force applied to the muscle is more than the muscle can move, there will be no contraction of the muscle and no work done. (The muscle will use up $\Delta\mathcal{G}$ however.) We can also say that the weight W will have no velocity in this case. Now, if the force applied to the muscle is reduced by using a smaller weight, W, the muscle will then pull the weight through a distance and with some velocity. Continuing, if the

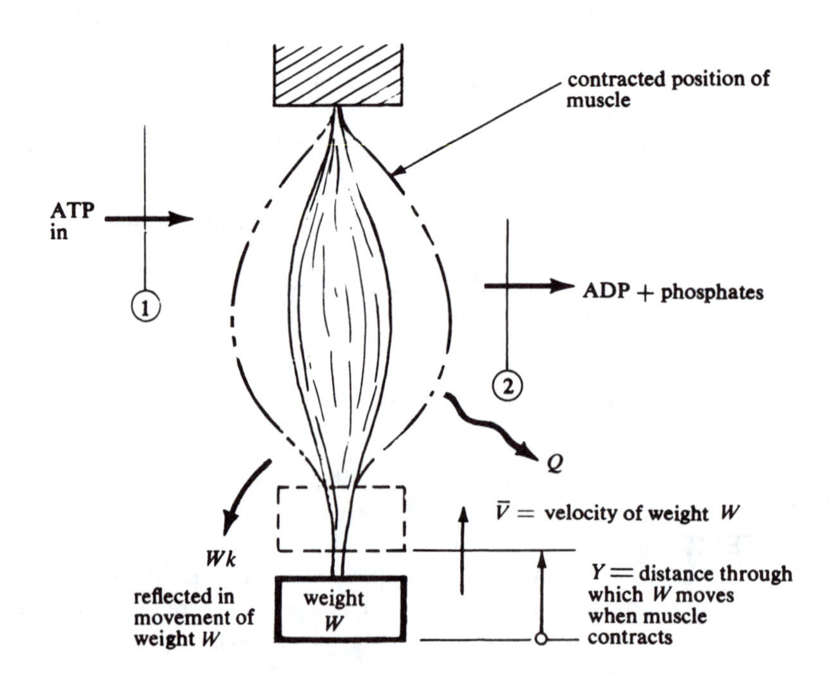

FIGURE 17–13
Muscle as a system

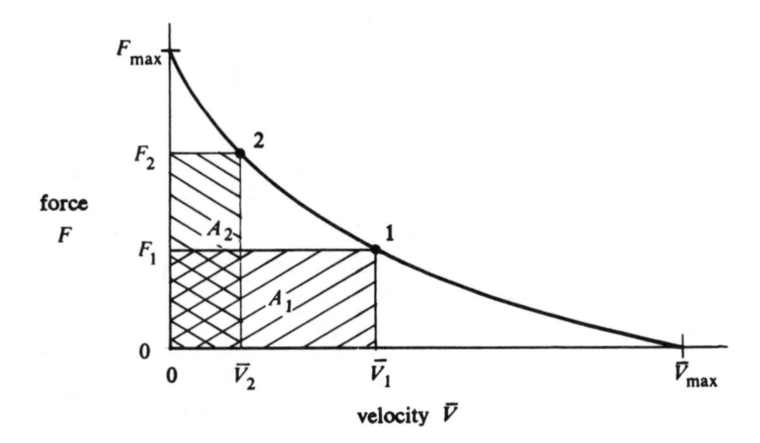

FIGURE 17–14 Force–velocity for typical muscle contraction

force is reduced further, the weight will be pulled faster by the muscle until it happens that when no weight or force is applied to the muscle, contraction will proceed at a rapid rate. This relationship, the force applied to the muscle versus the velocity at which the weight moves, is shown in figure 17–14. The velocity $\bar{V}_{max}$ represents the velocity of muscle contraction when no force $(F = 0)$ is applied, and the force F_{max} represents the greatest amount of force a muscle can pull. The power derived from a muscle pulling a constant force or weight is

$$\dot{W}k = F\bar{V} \tag{17–45}$$

and is represented by the rectangular area under the curve in figure 17–14. As an example, the power derived from a muscle working under a pull of F_1 and with velocity $\bar{V}_1$ is the area A_1. Similarly, for a force F_2 and velocity $\bar{V}_2$, the power is the area A_2.

The heart is an organ that is composed of various muscles acting together. The actual configuration of the heart is quite complex, and blood flowing through it is equally complicated, as indicated in figure 17–15, but for our purposes we can consider it to be a form of pump. The heart functions as essentially two separate blood pumps. The heavy arrows in the diagram indicate directions of blood flow. Oxygen-poor blood comes from the body to the *right atrium*. When the heat expands, this blood is drawn into the *right ventricle*. Then the blood is expelled into the pulmonary artery when the heart contracts. The pulmonary artery sends the blood to the lungs, where oxygen is supplied to the blood. The oxygen-rich blood returns to the heart through the pulmonary veins and into the *left atrium*. Upon expansion of the heart, the blood flows into the *left ventricle*, and then, when the heart contracts, the *left ventricle* reduces in volume and sends the oxygen-rich blood into the body via the *aorta*.

We can simplify the heart to the system as sketched in figure 17–16, and the energy equation gives us, approximately,

$$\dot{m}\frac{\bar{V}_2^2 - \bar{V}_1^2}{2} + (h_1 - h_2)\dot{m} = \dot{Q} - \dot{W}k \tag{17–46}$$

where $\dot{m}$ is the mass flow of the blood. It should be mentioned that blood flow through the heart is not steady, although over a long period of time it may be approximated as such; nor is the flow of blood as simple an analogy as water flowing through a rigid pipe. The mechanism of the heart is so complex that we may never fully understand its workings, but our simplistic approach is better than none in attempting to analyze the heart.

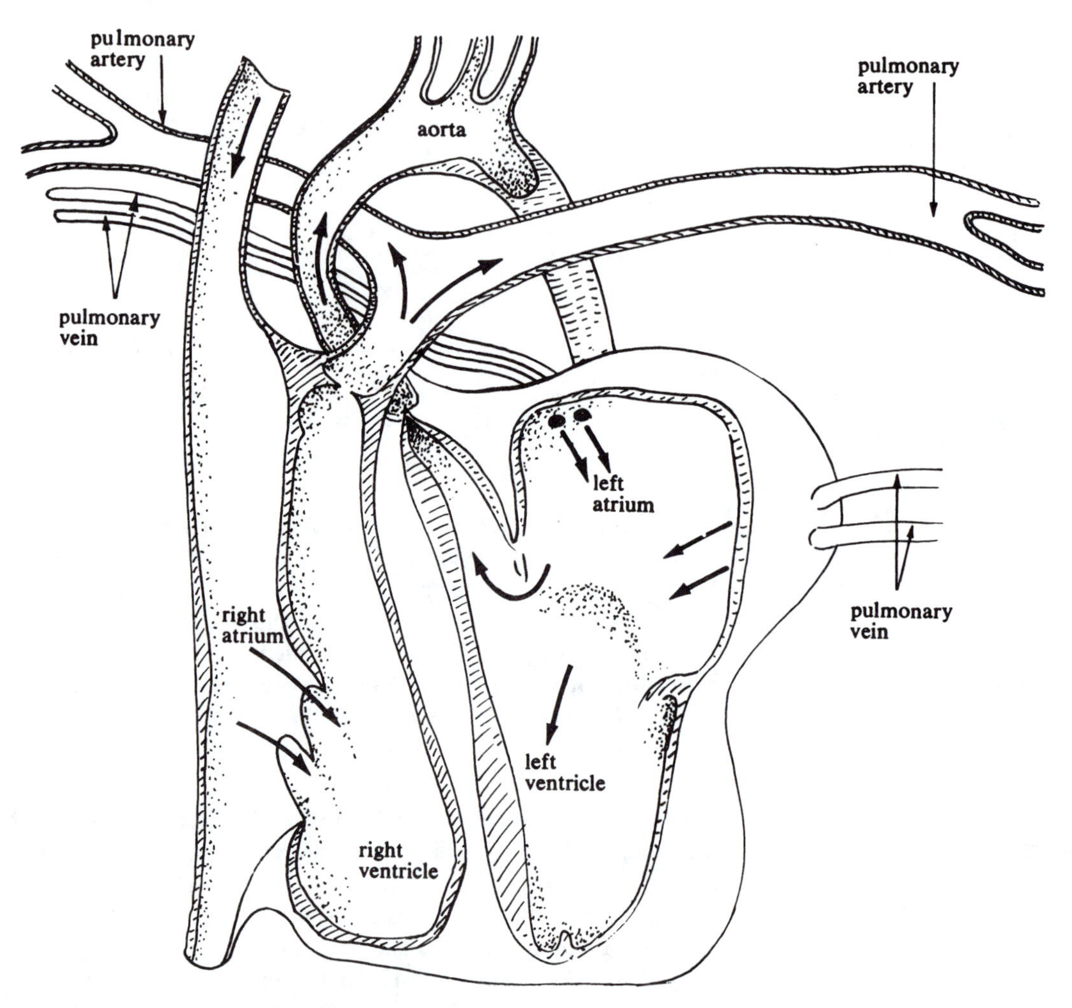

FIGURE 17–15 Cross section of heart [revised from W. I. Keeton, *Biological science* (New York: W. W. Norton Company, Inc., 1967), p. 246; originally from B. G. King and M. J. Showers, *Human anatomy and physiology*, 6th ed. (Philadelphia: W. B. Saunders Company, 1969); with permission of W. W. Norton & Company, Inc., W. B. Saunders Company, and B. G. King]

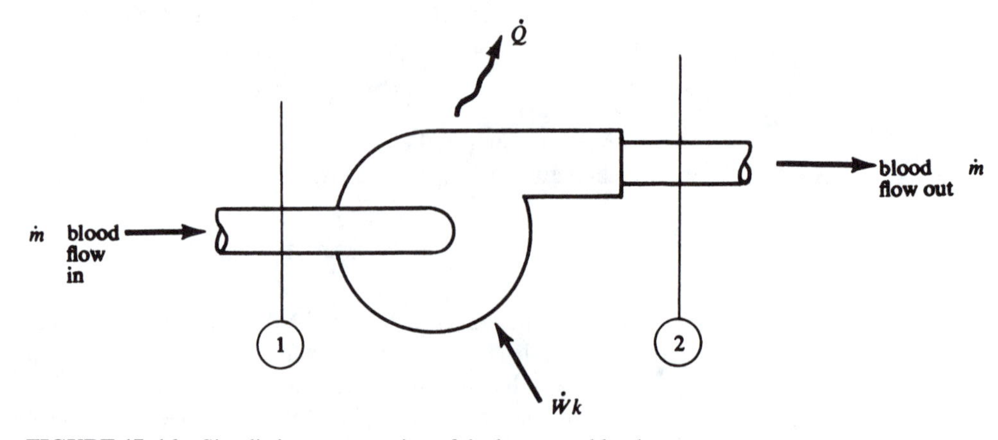

FIGURE 17–16 Simplistic representation of the heart as a blood pump

FIGURE 17–17
Pressure–volume diagram for the heart as it operates for one complete cycle

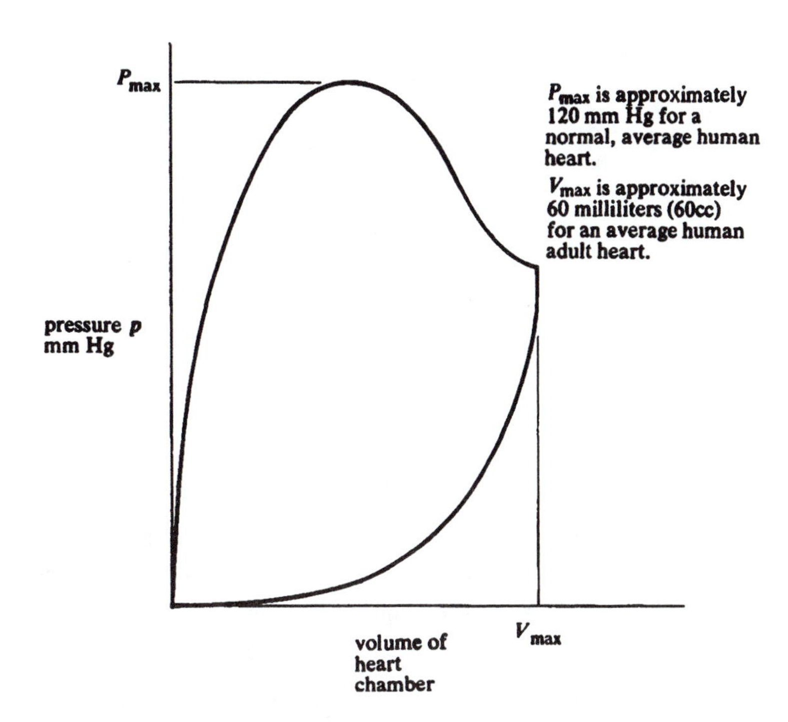

The pressure–volume diagram of the blood passing through the heart is shown in figure 17–17. The enclosed area is frequently referred to as the *heart output*. Of course, the diagram shown is only representative, and the maximum pressure shown varies from one heart to another. An average maximum blood pressure for a healthy, medium-aged human heart seems to be 120 mm Hg. Recall that the maximum work, given by equation (8–28), is

$$Wk_{\max} = -\Delta\mathcal{G}$$

where $-\Delta\mathcal{G}$ is the change in the Gibbs free energy of the ATP/ADP reaction mentioned previously as the source of energy for muscles. Also, the work can be determined by computing the area enclosed by the *p–V* diagram.

EXAMPLE 17–3 Using figure 17–17 as a description of a heart under certain conditions, and assuming that $p_{\max}$ is 120 mm Hg and $V_{\max}$ is 60 cm^3, determine the approximate work done by the heart for each pulse. Then, assuming a pulse rate of the heart of 80 beats per minute, determine the power requirement of the heart.

Solution We shall determine various data points or states on the *p–V* diagram and use the AREA program for computing areas under curves to calculate the work. From figure 17–17, we determine, approximately, the values shown in table 17–1.

The required work is then 4365 mm Hg · cm^3/pulse. This can be converted to energy units by

$$Wk = (4365 \text{ mm Hg}\cdot\text{cm}^3/\text{pulse})(0.133 \text{ kN/m}^2\cdot\text{mm Hg})(10^{-6} \text{ m}^3/\text{cm}^3)$$

Answer
$$= 0.58 \times 10^{-3} \text{ kJ/pulse}$$

The power is just the pulse rate times the work:

$$Wk = (0.58 \times 10^{-3} \text{ kJ/pulse})(80/60 \text{ pulses/s})$$

Answer
$$= 0.773 \text{ W}$$

TABLE 17–1

State	*p*, mm Hg	*V*, cm^3
1	0	0
2	6	30
3	25	45
4	80	60
5	90	45
6	110	40
7	120	30
8	115	20
9	80	0
1	0	0

17–6 STIRLING CYCLE DEVICES

The **Stirling cycle** is a thermodynamic cycle defined by the four reversible processes:

1–2 Reversible isothermal compression at a low temperature T_L

2–3 Reversible isometric (constant-volume) heat addition

3–4 Reversible isothermal expansion at a high temperature T_H

4–1 Reversible isometric heat rejection

The Stirling cycle processes are shown in figure 17–18 on *p–v* and *T–S* diagrams for perfect gas working medium. By using the process equations developed in this book, we can show the thermal efficiency of the Stirling cycle engines operating with perfect gases to be the same as that of Carnot heat engines:

$$\eta_T = 1 - \frac{T_L}{T_H} \qquad \textbf{(17–47)}$$

The heat transfer terms are

$$Q_{\text{add}} = Q_{34} = T_H(S_4 - S_3) \qquad \textbf{(17–48)}$$

and

$$Q_{\text{rej}} = Q_{12} = T_L(S_2 - S_1) \qquad \textbf{(17–49)}$$

where

$$(S_4 - S_3) = (S_1 - S_2) = mR \ln \frac{v_1}{v_2} = mR \ln \frac{P_2}{P_1}$$

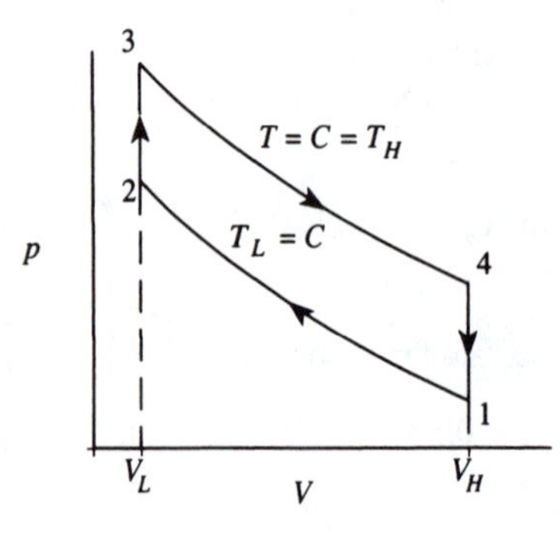

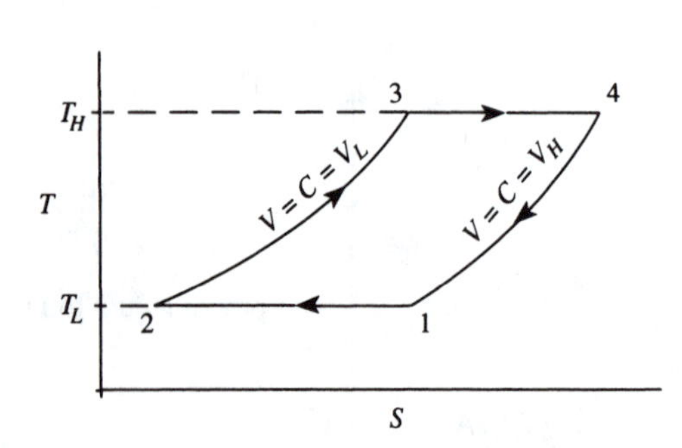

FIGURE 17–18 Stirling cycle property diagrams for a vapor working medium

The heat transfers occurring during isometric processes 2–3 and 4–1 are of the same magnitude but opposite direction or sign. That is, process 2–3 has heat addition or positive heat, and process 4–1 has heat rejection or negative heat. In the Stirling cycle devices, these two heat terms (Q_{23} and Q_{41}) are usually accounted for by a regenerative process inside the engine or refrigerator. The net work is the same as the net heat:

$$Wk_{\text{net}} = Q_{\text{add}} + Q_{\text{rej}} \tag{17–50}$$

The Stirling cycle has been used to describe some mechanical devices that have been used to generate power or to provide refrigerating effects. In the late nineteenth century, a device called an **air engine** was used to produce moderate amounts of power, on the order of 10 to 30 hp. The operation of the air engine can be visualized by referring to the schematic shown in figure 17–19. In the position shown, the air acting on the power piston is at a state equivalent to that of state 2 of the Stirling cycle. The displacement piston then moves to the left and displaces some of the cold air into the hot chamber. This motion closes the cold air path to the piston and opens the path for hot air to act on the power piston. This state would be approximated by state 3 of the Stirling cycle and in figure 17–19, would be represented by 90° rotation of the flywheel in a counterclockwise direction. The hot air then pushes the power piston out and rotates the flywheel through a power stroke, reaching a position where the flywheel has rotated another 90° counterclockwise and the air is approximated by state 4 of the Stirling cycle. Another 90° flywheel rotation will then bring the displacer piston back to the original position shown in figure 17–19 and the air to a state like state 1 of the Stirling cycle. The power piston is still extended out of the cylinder, and it moves into the cylinder through yet another 90° flywheel rotation, returning the engine to the original state of figure 17–19. In the operation of the air engine shown in the figure, regeneration during processes 2–3 and 4–1 are mixing processes of hot and cold air created by the displacer piston motion. The air engine has had low thermal efficiencies and was replaced by other engines in the early twentieth century. Recently, efforts have been made to adopt the Stirling engine principle to innovative configurations, mostly using piston-cylinder devices and regenerative heating chambers. The thermal efficiencies have still been found to be low, but its attractiveness as an external combustion device, capable of using low-grade fuels or external heat sources, has caused interest in it to continue. It could be adapted to solar energy applications, nuclear power applications, geothermal power applications (using geysers or hot springs from naturally occurring sources), and others. The Stirling

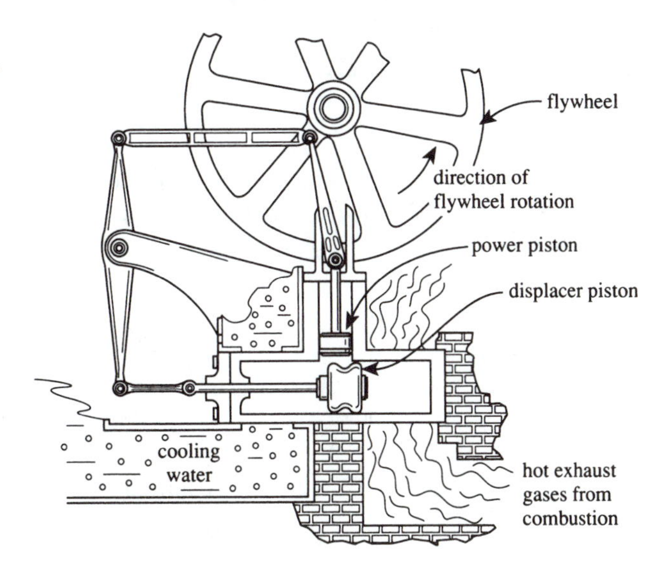

FIGURE 17–19 Partial schematic of a version of an air engine using the Stirling cycle

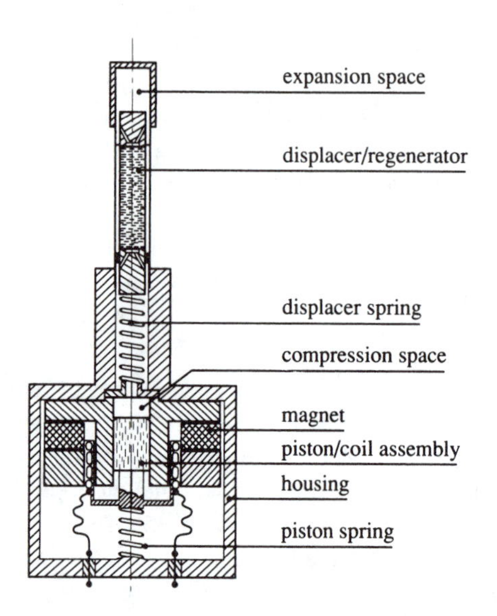

FIGURE 17–20 Schematic of a small Stirling refrigerator (from A. K. de Jonge, "A small free-piston Stirling refrigerator," *Proceedings of 14th Intersociety Energy Conversion Engineering Conference*, Boston, Massachusetts, 1979; reprinted with permission of the American Chemical Society)

cycle engine has been suggested as a power source for artificial hearts, where it could conceivably derive power over the temperature differentials experienced in the human body.

The reversed Stirling engine, or Stirling refrigerator, has been used in a range of applications. In situations where cryogenic temperatures are expected (see chapter 12), small Stirling refrigerators have been successfully used. figure 17–20 shows a schematic cross section of such a device that uses helium as the working medium (instead of air as in the air engine) and is driven by the electromagnet and two springs. The expansion space is capable of cooling a small space to approximately 80 K (−193°C) with a cooling load of around 1 W. Lower temperatures and higher cooling loads have been achieved. Also, laboratory-scale Stirling refrigerators have been built and used to provide cooling temperatures in the range 80 to 160 K with cooling loads of around 1 to 2 kW.

EXAMPLE 17–4

The reversible Stirling air engine shown in figure 17–19 operates between 300°C and 30°C. The power piston has a bore of 6 cm and a stroke of 4 cm. The minimum volume of air is 150 cm^3, and the pressure at the beginning of the compression is 100 kPa. Determine the thermal efficiency, the work produced per cycle, and the maximum pressure in the cycle.

Solution

Since the engine is reversible, we may calculate the thermal efficiency from equation (17–47):

Answer

$$\eta_T = 1 - \frac{T_L}{T_H} = 1 - \frac{303}{573} = 47.1\%$$

The work is determined from equation (17–50), and the heat transfer terms are from equations (17–48) and (17–49). We should also notice that the heat terms are related by the equation $Q_{\text{add}}/Q_{\text{rej}} = -T_H/T_L$, of which we will make use. We find the volumes to be

$$\begin{aligned} V_2 &= 150 \text{ cm}^3 \\ V_1 - V_2 &= \pi(\text{bore})^2(\text{stroke}) \\ &= (\pi)(3 \text{ cm})^2(4 \text{ cm}) = 113 \text{ cm}^3 \end{aligned}$$

so that $V_1 = 263$ cm^3. Then

$$\begin{aligned} s_2 - s_1 &= R \ln \frac{V_2}{V_1} = (0.287 \text{ kJ/kg}\cdot\text{K})\left(\ln \frac{150 \text{ cm}^3}{263 \text{ cm}^3}\right) \\ &= -0.161 \text{ kJ/kg}\cdot\text{K} \end{aligned}$$

The mass of the air is

$$m = \frac{p_1 V_1}{R T_1} = \frac{(100 \text{ kN/m}^2)(0.000263 \text{ m}^3)}{(0.287 \text{ kN}\cdot\text{m/kg}\cdot\text{K})(303 \text{ K})} = 0.0003024 \text{ kg}$$

The rejected heat is

$$Q_{rej} = Q_{12} = m T_L (s_2 - s_1) = (0.0003024 \text{ kg})(303 \text{ K})(-0.161 \text{ kJ/kg}\cdot\text{K}) = -0.01475 \text{ kJ/cycle}$$

and the added heat is

$$Q_{add} = -\frac{T_H}{T_L}(Q_{rej}) = -(573 \text{ K}/303 \text{ K})(-0.01475 \text{ kJ/cycle}) = 0.0279 \text{ kJ/cycle}$$

The net work is then

$$Wk_{net} = Q_{add} + Q_{rej} = 0.0279 - 0.01475 \text{ kJ/cycle}$$

Answer

$$= 0.01315 \text{ kJ/cycle}$$

The maximum pressure (at state 3) can be found for the reversible Stirling engine by noticing that the pressure will be related as follows:

$$\frac{p_2}{p_1} = \frac{V_1}{V_2} \quad \text{(isothermal process)}$$

and

$$\frac{p_3}{p_2} = \frac{T_3}{T_2} \quad \text{(isometric process)}$$

Then

$$p_3 = p_2 \frac{T_3}{T_2} = p_1 \frac{V_1}{V_2}\frac{T_3}{T_2} = (100 \text{ kPa})\left(\frac{263}{150}\right)\left(\frac{573}{303}\right)$$

Answer

$$= 331.6 \text{ kPa}$$

17–7 SUMMARY

This chapter examines some examples of thermodynamics applied to nontraditional areas. It is shown that thermodynamics can be used to help analyze basic electric power systems, batteries, fuel cells, thermoelectric devices, magnetohydrodynamic systems, biological systems, and Stirling cycle devices.

The concepts of thermodynamics and the methods of analysis can be applied to any system where energy, work, or heat is involved. The limitations of thermodynamic applications are only those imposed by the hesitation of the readers.

DISCUSSION QUESTIONS

17–1 Why is the positive direction for electric current usually defined as opposite to actual electron flow in an electrical conductor?

17–2 What is a fuel cell?

17–2 Is a fuel cell an energy storage device, like a battery?

17–3 What gives the *thermocouple* the ability to sense temperature?

17–4 Does a magnetohydrodynamic (MHD) power generator seem to be a good candidate for a cogeneration system?

17–5 Can you think of any physical system that does not obey the laws of thermodynamics?

17–6 How is the term *external combustion*, as applied to a heat engine, different from the internal combustion engines of chapter 9?

PRACTICE PROBLEMS

Problems that use SI units are indicated by an (M) under the problem number; those that use English units are indicated by an (E).

Section 17–1

17–1 An electric generator is found to have an efficiency of 93% when delivering 60 A of electric current. If 15 hp is required to drive the generator, determine the generator's voltage output.

17–2 An electric generator has an internal resistance of 0.03 Ω and delivers 100 A at 120 V. Determine the induced electric potential and the generator efficiency.

17–3 An electric motor has an internal resistance of 0.06 Ω and operates at 120 V and 60 A. Determine the motor output power if no friction exists.

17–4 A 3-hp electric motor dissipates 2 W of frictional resistance and operates at 220 V. If internal resistance is 0.09 Ω, determine the motor current and efficiency.

17–5 A Daniel cell is built and found to have an internal resistance of 0.06 Ω. If 1 A of current is drawn, calculate the maximum work, the open-circuit voltage, and the operating voltage.

17–6 The electric cell is proposed which uses a zinc anode, a silver (Ag) cathode, and zinc sulfate and silver sulfate (Ag_2SO_4) solutions. In all other respects, it resembles the Daniel cell. The cathode reaction is

$$2Ag^+ + 2e^- \longrightarrow Ag$$

Determine the maximum work obtainable from this cell and its open-circuit voltage.

Section 17–2

17–7 If the internal resistance of a hydrogen–oxygen fuel cell is 0.05 Ω, determine the operating voltage if 2 A is drawn.

17–8 A fuel cell is proposed which uses sodium (Na) and chlorine (Cl) and produces sodium chloride (NaCl). Determine the maximum work obtainable and the open-circuit voltage.

17–9 Octane (C_8H_{18}) in the gaseous phase is utilized as a fuel in a fuel cell, which also uses oxygen. The chemical reaction is $C_8H_{18} - 12.5O_2 \longrightarrow 8CO_2 + 9H_2O$. Determine the maximum work obtainable, the open-circuit voltage, and the operating voltage when 16 A is drawn with no internal resistance.

17–10 The chemical reactions of the direct methanol fuel cell (DMFC) can be described by equation (17–27). Determine the maximum work per mole of methanol that could be expected from the DMFC. Also determine the open-circuit voltage.

Section 17–3

17–11 A 10-Ω resistor carries 8 A. Determine the steady-state Joule heating.

17–12 A resistor radiates 50 Btu/h when a current of 0.7 A passes through it. Determine the resistance in ohms.

17–13 An electric heater is required to conduct 200 Btu/min. Determine the resistance and current if 120 V is applied across the heater coil.

17–14 A PN-type thermoelectric generator receives 10 cal/s of heat from an external supply. If the resistance of the generator is 0.05 Ω and the Seebeck coefficients are $\alpha_n = -0.13$ V and $\alpha_p = 0.20$ V, determine the induced current and the voltage of the device.

Section 17–4

17–15
(E) Hot gases having properties of air enter an MHD generator at 2000°F and 700 ft/s. If they leave the generator at 1050°F and 400 ft/s, estimate the work generated per pound-mass of gases if heat losses are neglected.

17–16 (M) An MHD power generator receives 1200 g/s of 1600°C gases having a velocity of 35 m/s and discharges them at 10 m/s and 650°C. A potential of 300 V is applied across the generator section where the gases have a resistance of 1.5 Ω. If heat losses are 500 cal/s and the gases have properties of air, determine the net power produced by the MHD generator.

Section 17–5

17–17 (M) A muscle lifts 50 g through 2 cm. If the efficiency is found to be 32%, what is the change in Gibbs free energy of the ATP–ADP reaction?

17–18 (M) A heart is pumping 5000 g/min of blood, which has a density of 1 g/cm^3. If the thermal effects are neglected (no temperature and $Q = 0$), and if the velocity change is negligible between entering and leaving blood, estimate the power produced by the heart to pump this blood if the pressure is 110 mm Hg leaving and 0 mm Hg entering.

17–19 (M) A muscle is found to have the force–velocity relationship:

$$F \text{ (grams)} = 60 - 6\overline{V}$$

where $\overline{V}$ is in cm/s. Write a computer program to calculate the power as the velocity changes. Then plot the results on a power–velocity diagram, and determine the velocity at which maximum power is produced. Note that velocity cannot be negative.

17–20 (M) A heart that has a volume of 65 cm^3 is using 7.5 kJ/min of power to pump blood to a pressure of 170 mm Hg. Predict the average or mean effective blood pressure in the heart if the heart is pumping at 120 beats per minute.

17–21 (M) For the ventricular pressure–volume diagram of a heart shown in figure 17–21, determine the work and power of the heart if the heart beat is 95 beats/minute.

Section 17–6

17–22 (E) A reversible Stirling air engine operates between 80°F and 740°F. Its power piston has a 6-in bore and 5-in stroke, and the maximum volume is 300 in^3. The minimum pressure in the cycle is 12 psia. Determine the thermal efficiency, the work per cycle, and the maximum pressure in the cycle.

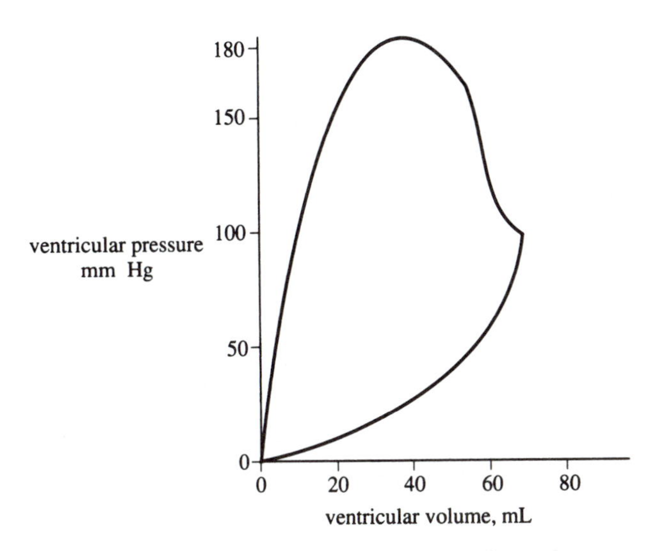

FIGURE 17–21 Ventricular pressure–volume diagram for a heart

17–23 (M) A reversible Stirling air engine operates between 40°C and 200°C, and its volumes are 2000 cm^3 and 4000 cm^3. The maximum pressure in the cycle is 300 kPa. Determine the thermal efficiency, the minimum pressure in the cycle, and the work per cycle.

17–24 (M) A reversible Stirling refrigerator uses helium as the working medium. Assume that the helium acts as a perfect gas with constant specific heats. Determine the amount of helium required in the system to provide 2 W of cooling at 80 K if the high temperature is 180 K. The refrigerator has a minimum volume of 110 mL and a maximum of 190 mL. Also determine the power required to drive the device.

MATHEMATICAL RELATIONSHIPS

A

A–1 GEOMETRIC FORMULAS

The following is a compilation of some of the equations used for determining areas and volumes of certain geometric shapes encountered in this book. The material is not intended to be a complete handbook of formulas, and the student is directed to mathematical textbooks and handbooks for such listings or for information that is not given here.

a. Rectangle

$$\text{Area, } A = ab$$

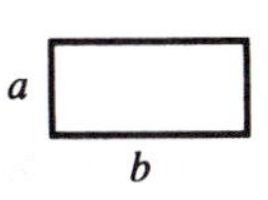

where a and b are the lengths of two adjacent sides.

b. Parallelogram

$$\text{Area, } A = bh$$

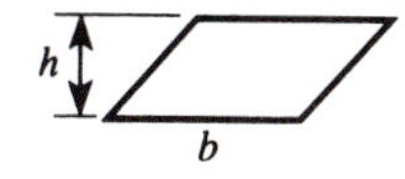

c. Trapezoid

$$\text{Area, } A = \frac{1}{2}h(a + b)$$

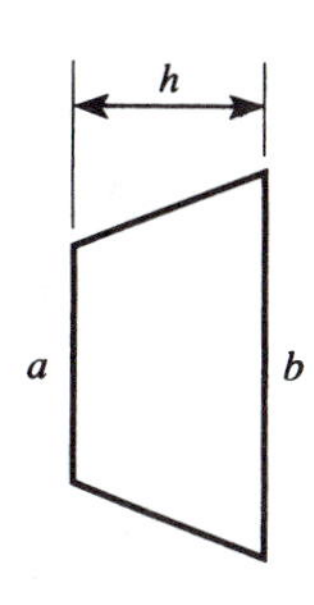

d. Triangle

$$\text{Area, } A = \frac{1}{2}bh$$

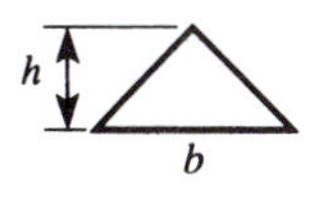

e. Circle

$$\text{Area, } A = \pi r^2 = \frac{\pi d^2}{4}$$

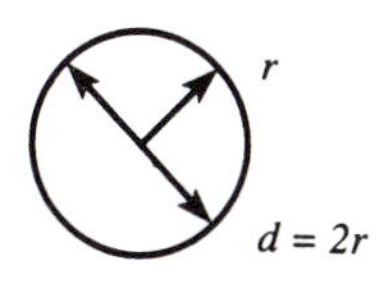

f. Sphere

$$\text{Surface area, } A = 4\pi r^2 = \pi d^2$$

$$\text{Volume, } V = \frac{4}{3}\pi r^3 = \frac{1}{6}\pi d^3$$

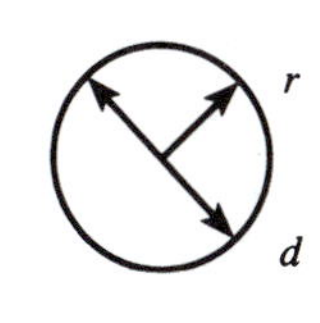

g. Cylinder

$$\text{Surface area, } A = 2\pi r(r + l)$$

$$\text{Volume, } V = \pi r^2 l = \frac{\pi d^2 l}{4}$$

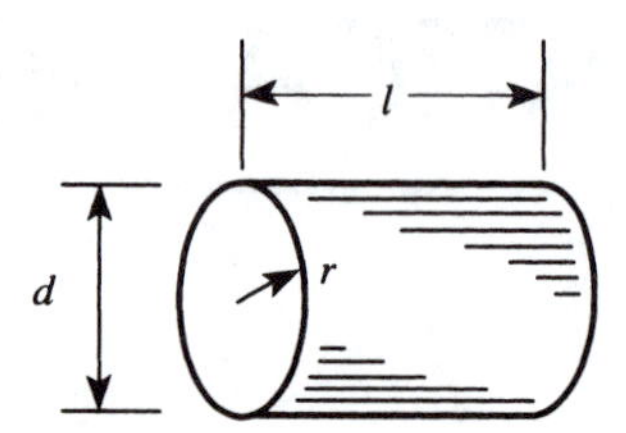

Note: If the radius is small compared with the length, we may use the following formula for the surface area:

$$\text{Surface area, } A = 2\pi rl = \pi dl$$

A–2 LOGARITHMIC RELATIONSHIPS

The following definitions and algebraic relationships are useful for obtaining a better understanding of certain derivations in this book and for manipulating some of the mathematical relations found in thermodynamics:

Definition

$$\text{logarithm }(X) = \log X = N$$

where $10^N = X$.
Also,

$$\log Y = M \qquad 10^M = Y$$

It follows, then, that

$$\log X + \log Y = \log XY$$

$$(X)(Y) = 10^{N+M}$$

$$\log X - \log Y = \log \frac{X}{Y}$$

$$\frac{X}{Y} = 10^{N-M}$$

Using x and y as variables, independent of the exponents n and m, we have

$$\frac{x^n}{y^m} = x^n y^{-m}$$

$$\left(\frac{x}{y}\right)^n \frac{x}{y} = \left(\frac{x}{y}\right)^{n+1}$$

$$\left(\frac{x}{y}\right)^n \frac{y}{x} = \left(\frac{x}{y}\right)^n \left(\frac{x}{y}\right)^{-1} = \left(\frac{x}{y}\right)^{n-1}$$

Definition

$$e = 1 + \frac{1}{1!} + \frac{1}{2!} + \frac{1}{3!} + \frac{1}{4!} + \cdots + \frac{1}{n!} + \cdots$$

where $1! = 1$

$2! = 1 \times 2 = 2$

$3! = 1 \times 2 \times 3 = 6$

$4! = 1 \times 2 \times 3 \times 4 = 24$

and so on. The value of n is set equal to each of the whole numbers up to a value approaching infinity. The numerical value of e has been computed and found to be

$$e = 2.71828\ldots$$

Definition

$$e^L = X$$

and

$$\text{natural logarithm}\,(X) = \ln X = L$$

It can be shown that the following relationship exists between the common and natural logarithms:

$$\ln X = (2.302585\ldots)\log X$$

A–3 RAISING NUMBERS TO POWERS

The following problem frequently arises in thermodynamic calculations. Given a value for x and n, determine the result of x^n. Using logarithms, we have

$$\log x^n = n(\log x) = y$$
$$\text{antilog}\, y = x^n$$

By using a calculator or log table, we can obtain the log of x. Multiplying this value by n yields a new value, y. Using log tables or calculator, we can obtain the antilog of y. This value corresponds to the desired quantity, x^n.

A–4 MATHEMATICAL FUNCTIONS AND AREAS UNDER CURVES

Many times in thermodynamics a property or variable depends for its value on another property or on time or a place. Let us call y the property or variable that depends on another variable, x. The mathematical shorthand of saying that y depends on x or that y is a function of x is to write

$$y = f(x)$$

Now the exact way in which y depends on x must be given or be determined in some way. Suppose that we consider a spring; here the force required to shorten the spring must continually be increased if the spring is to continue to shorten. We say that the spring force, F, is a function of the spring shortening, say x. Normally, if the spring is a linear spring, the exact relationship between F and x is

$$F = bx$$

where b is a spring constant. Thus, the variable F is directly related to x, or $F = f(x) = bx$. The simplest function $f(x)$ is one in which the dependent variable or property is a constant or has a constant value. This is a redundant case but one that occurs frequently in thermodynamics, for instance, the constant-pressure process, the constant-temperature process, and so on. We may plot y (or F or any other dependent variable) as a function of, say, x on graph paper or simply on an x–y coordinate system. Then the line (called a curve whether it is straight or not) that results from this graphic or geometric description of $y = f(x)$ or plots of y versus x has many applications and interpretations in thermodynamics. The area directly under the curve can represent a thermodynamic property change, work, or heat. For instance, if y is a force and x is the displacement or distance through which that force moves, the area under the curve represents work, or if y is an absolute temperature and x is entropy, the area might be the reversible heat transfer. Thus, being able to determine the area under a curve is a useful skill to possess in thermodynamics.

Next is a compilation of some of the more frequently encountered situations of $y = f(x)$ and the particular values for the resulting areas under those curves. The student can verify that sometimes the areas, A, have negative or minus values. Such situations occur if the x-value decreases or moves from right to left as shown in the figures, and the fact that the area could be negative is in agreement with the ideas and sign conventions of thermodynamics.

a. y constant:

$$y = C$$
$$A = C(x_2 - x_1)$$

Using integral calculus, we have

$$A = \int y\,dx = \int C\,dx = C\int dx = Cx + \text{constant}$$

Evaluating between x_1 and x_2 gives

$$A = C(x_2 - x_1)$$

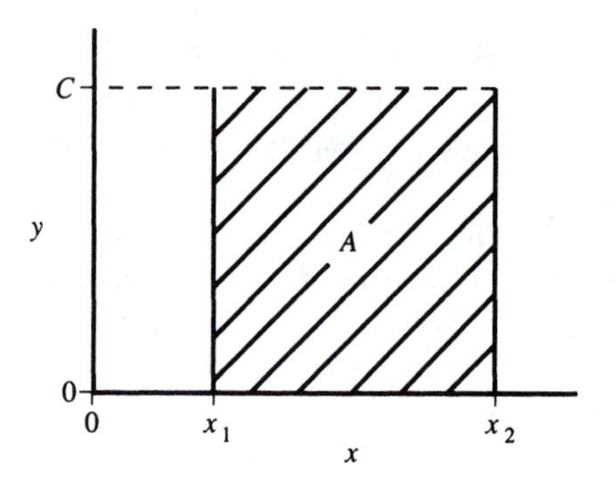

b. y varies with x:

$$y = Bx$$
$$A = \frac{1}{2}(y_1 + y_2)(x_2 - x_1)$$
$$= \frac{1}{2}B(x_2^2 - x_1^2)$$

Using integral calculus, we have

$$A = \int y\,dx = \int Bx\,dx = B\int x\,dx = \left[\frac{1}{2}\right]Bx^2 + \text{constant}$$

Evaluating between x_1 and x_2 gives

$$A = \left[\frac{1}{2}\right]B(x_2^2 - x_1^2)$$

In general, for $y = Bx^n$ where n is an exponent, the area is

$$A = \int Bx^n\,dx = B\int x^n\,dx = \left[\frac{B}{(n+1)}\right]x^{n+1} + \text{constant}$$

Evaluating between x_1 and x_2 gives

$$A = \left[\frac{B}{(n+1)}\right](x_2^{n+1} - x_1^{n+1})$$

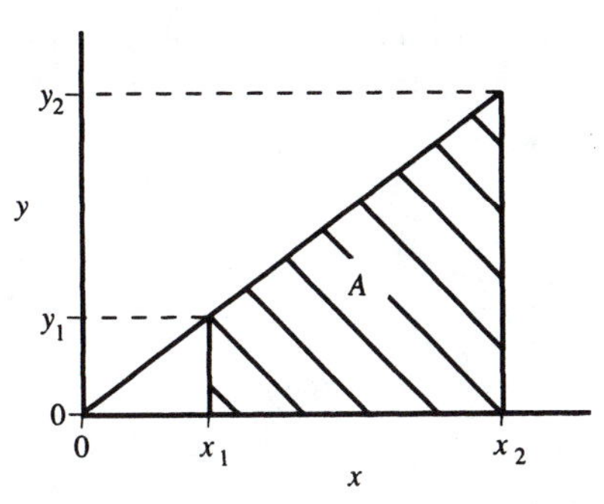

c. y varies with x, but is not zero at $x = 0$:

$$y = C + Bx$$

$$A = \frac{1}{2}(y_1 + y_2)(x_2 - x_1)$$

$$= \left[\frac{1}{2}\right] B(x_2^2 - x_1^2) + C(x_2 - x_1)$$

Using integral calculus, we have

$$A = \int y\,dx = \int (C + Bx)\,dx = \int C\,dx + \int Bx\,dx$$

$$= \left[\frac{1}{2}\right] Bx^2 + Cx + \text{constant}$$

Evaluating between x_1 and x_2 gives

$$A = \left[\frac{1}{2}\right] B(x_2^2 - x_1^2) + C(x_2 - x_1)$$

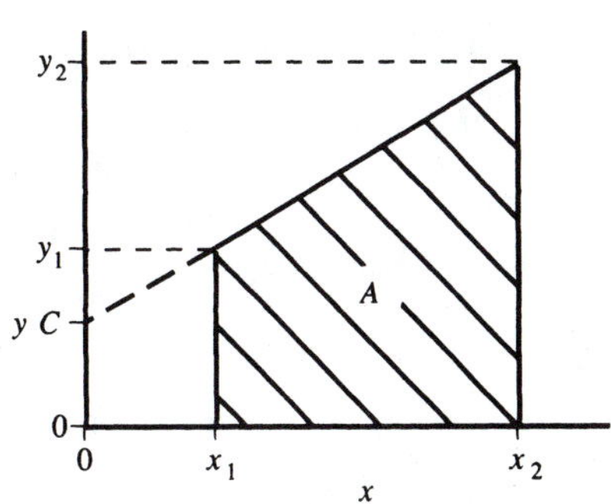

d. y varies inversely with x:

$$y = \frac{B}{x}$$

$$A = B \ln \frac{x_2}{x_1}$$

$$= B \ln \frac{y_1}{y_2}$$

Using integral calculus, we have

$$A = \int y\,dx = \int B\frac{dx}{x} = B\int\frac{dx}{x} = B \text{ In } x + \text{constant}$$

Evaluating between x_1 and x_2 gives

$$A = B \ln\frac{x_2}{x_1}$$

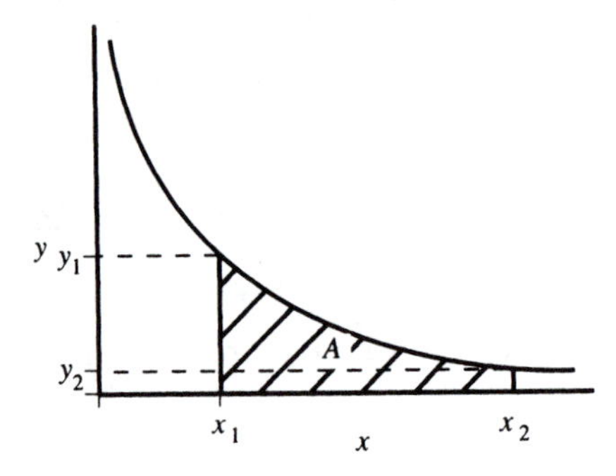

e. y varies inversely with x^n:

$$y = \frac{B}{x^n}$$

$$A = \frac{B}{1-n}(x_2^{1-n} - x_1^{1-n})$$

$$= \frac{1}{1-n}(y_2x_2 - y_1x_1)$$

Using integral calculus, we have

$$A = \int y\,dx = \int B\frac{dx}{x^n} = B\int x^{-n}\,dx$$

$$= \left[\frac{B}{(1-n)}\right]x^{1-n} + \text{constant}$$

Evaluating between x_1 and x_2 gives

$$A = \left[\frac{B}{(1-n)}\right](x_2^{1-n} - x_1^{1-n})$$

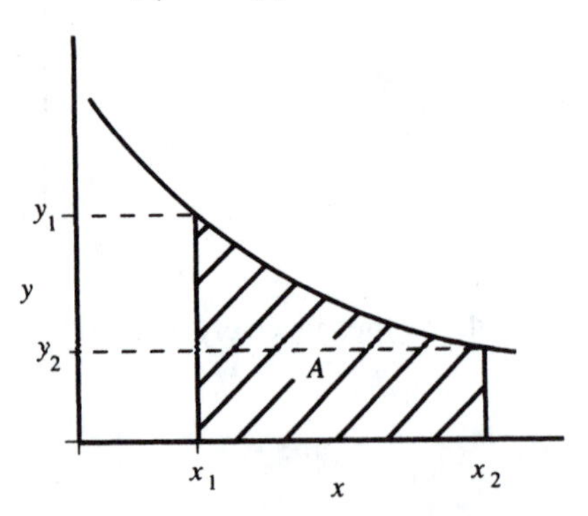

If there does not exist an equation $y = f(x)$, but the values of y at various values of x are known, such as in the property tables of appendix B in this book, we may plot those data on an

x–y coordinate and compute the approximate area under the curve connecting those points by means of the trapezoidal rule area routine presented in chapter 1. For instance, we may plot pressure as the y-coordinate and specific volume as the x-coordinate. Then plotting values for specific volumes from, say, the steam tables at particular pressures, we can get a pressure-specific volume plot from which an area under a curve may be calculated from a specific volume v_1 to another value v_2. Figure A–1 shows an example of such a situation, and the total area under the curve, A, is approximately equal to the sum of a number of very small areas, δA's. In this example, there happen to be nine small areas, or nine steps of δv. Each of the δA's is computed from the trapezoid area equation applied to the particular δA for which the pressure values correspond to the specific volumes at that point. The total area is, for this particular example, equal to the sum of nine δA's or $p\ \delta v$'s:

$$A = \sum_{i=l}^{9} \delta A_i$$
$$= \delta A_1 + \delta A_2 + \delta A_3 + \delta A_4 + \delta A_5 + \delta A_6 + \delta A_7 + \delta A_8 + \delta A_9$$
$$\delta A = p\delta v$$

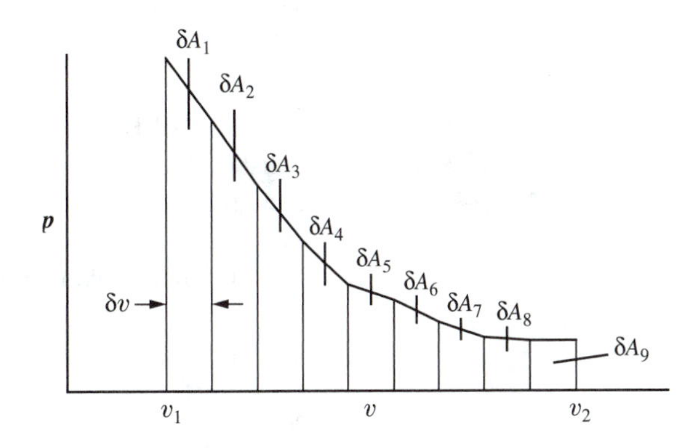

FIGURE A–1 Example of computing the area under a curve by the approximate method of summing a number of small areas

If one wants to make the approximate solution for the area under a curve more nearly exact or precise, the step δx can be made smaller; then more of these steps are required to find the total sum or total area. If δx is made nearly zero, so that an infinite number of these steps are needed, one would find the exact area under a curve. For engineering purposes, a reasonable number of steps of δx is from, say, 5 to perhaps 50. For the purposes of this book, using 10 steps of δx is probably enough to give an acceptable answer. Readers interested in a more complete discussion of how to handle the concept of making δx very small while requiring that the number of steps approach an infinite number, thus finding the exact solution to the area under the curve, are directed to any good calculus textbook.

A–5 PROGRAM AREA

The program AREA is a VISUAL BASIC program used to calculate the area under a curve. The next listing gives the program on a compact disc (CD) available from the publisher. The program can be used on microcomputers and requires you to give the number of points on the curve and the values of x and y for each of those points. The area under the curve is then computed and printed out. The following program can be used for 20 or fewer points, (more points can be used if the number 20 in the first two lines is changed to whatever number of points you wish to use):

```
100 DIM X(20)
110 DIM Y(20)
120 PRINT"THIS PROGRAM COMPUTES THE AREA UNDER A CURVE."
130 PRINT"IT USES A TRAPEZOID AREA OF SEGMENTS OF THE AREA"
140 PRINT"AND SUMS THOSE AREA-SEGMENTS TO OBTAIN THE TOTAL AREA."
150 PRINT"INPUT THE NUMBER OF POINTS ON THE CURVE, N. N MUST BE"
160 PRINT"20 OR LESS"
170 INPUT N
180 NI=N-1
190 FOR I=1 TO N
200   PRINT"INPUT Y(";I;"),X(";I;")"
210   INPUT Y(I),X(I)
220 NEXT I
230 A=0.0
240 FOR I=1 TO NI
250   A=A+(Y(I)+Y(I+1))*(X(I+1)-X(I))/2
260 NEXT I
270 PRINT"AREA UNDER THE CURVE IS:"
280 PRINT A
290 END
```

A–6 PROGRAM DELHCO

The program DELHCO is a VISUAL BASIC program contained on a CD available from the publisher. It can be used to compute the change in enthalpy of carbon monoxide if the initial and final temperatures are known in degrees Rankine. The program uses a trapezoid rule area summation and asks for the number of intervals wanted. The answer is given in Btu/lbm carbon monoxide. Following is a listing of the program DELHCO:

```
100 DIM T(100)
110 DIM CP(100)
120 PRINT"THIS PROGRAM COMPUTES THE CHANGE IN ENTHALPY OF"
130 PRINT"CARBON MONOXIDE BEHAVING AS A PERFECT GAS AND WHEN"
140 PRINT"ITS SPECIFIC HEAT, CP, IS GIVEN BY THE EQUATION"
150 PRINT"CP = 9.46 - 3290/T + 1070000/T2 BTU/LBM-MOLE-R"
160 PRINT"THE FIRST TEMPERATURE T1 IS, IN RANKINE"
170 INPUT T1
180 PRINT"THE SECOND TEMPERATURE T2 IS, IN RANKINE"
190 INPUT T2
200 PRINT"HOW MANY INTERVALS DO YOU WANT?"
210 INPUT NI
220 N=NI+1
230 D=(T2-T1)/NI
240 DH=0.0
250 T(1)=T1
260 FOR I=2 TO N
270 T(I)=T(I-1)+D
280 NEXT I
290 FOR I=1 TO N
300 CP(I)=9.46-(3290.)/T(I)+(1.07ES)/((T(I))*(T(I))
310 NEXT I
320 FOR I=1 TO NI
330 DH=DH+(CP(I)+CP(I+1))*(D)/2.
340 NEXT I
350 DH=DH/28.011
360 PRINT"THE ENTHALPY CHANGE OF CARBON MONOXIDE IN BTU/LBM IS:"
370 PRINT DH
380 END
```

A–7 PROGRAM CARNOT

The program CARNOT is a VISUAL BASIC program contained on a CD available from the publisher. The program allows for computer assistance in analyzing problems involving the Carnot heat cycle. The program requires that you know certain parameters of the Carnot cycle; the computer can then follow through the program to determine pressures, volumes, and temperatures at all four states of the cycle; the various work terms and net work; the heat added in the cycle; the heat rejected in the cycle; the internal energy changes; the entrophy changes and the thermal efficiency. It also gives the COP and COR for the reversed Carnot cycle, the refrigerating effect, and the amount of work required to drive an equivalent heat pump or refrigerator. Two specific types of problems can be treated with this program: (1) when you know the lowest temperature and pressure, greatest and least volumes, and highest temperature; and (2) when you know the lowest pressure, highest and lowest temperature, greatest volume, and heat added in one cycle.

A–8 PROGRAM OTTO

The program OTTO is a VISUAL BASIC program contained on a CD available from the publisher. The program allows for computer assistance in analyzing some problems involving the Otto power cycle. The program requires that you know certain parameters of the Otto cycle; the computer can then follow through the program to determine the pressures, volumes, and temperatures at the four corners of the cycle; the various work terms; the heat added; the heat rejected; the internal energy changes; the various entropy changes; and the thermal efficiency. There are provisions for determining the specific fuel consumption, power produced, and mean effective pressure. Three specific types of problems can be considered with the program OTTO: (1) when the lowest pressure and temperature, compression ratio, displacement, number of cylinders, rate of heat addition, and engine speed are known; (2) when the lowest temperature and pressure, compression ratio, displacement, number of cylinders, engine speed, and net power output are known; and (3) when the lowest pressure and temperature, compression ratio, number of cylinders, engine speed, net power, and maximum cycle temperature are known. Provisions for analyzing polytropic processes and irreversible processes, where the adiabatic efficiencies must be known, are included in the program.

A–9 PROGRAM DIESEL

The program DIESEL is a VISUAL BASIC program contained on a CD available from the publisher. The program allows for computer assistance in analyzing some problems involving the Diesel and the dual power cycles. The working medium is treated as a perfect gas with constant specific heat values. The program requires that you know certain parameters of the Diesel or dual cycle; the computer can then follow through the program routine to determine the pressures, volumes, and temperatures at the four (or five in the case of the dual cycle) corners; the various work and heat terms; the internal energy changes; the entropy changes; and the thermal efficiency. The specific fuel consumption, mean effective pressure, and output power can also be determined. Three specific types of problems can be considered with the program DIESEL: (1) when the lowest pressure and temperature, compression ratio, displacement, number of cylinders, engine speed, and net power are known; (2) when the lowest pressure and temperature, compression ratio, displacement, number of cylinders, engine speed, and rate of heat addition are known; and (3) when the lowest pressure and temperature, compression ratio, number of cyclinders, engine spped, net power, and heat addition are known. Provisions are also made in the program to handle polytropic or irreversible adiabatic processes.

A–10 PROGRAM BRAYTON

The program BRAYTON is a VISUAL BASIC program included on a CD available from the publisher. The program allows for computer assistance in analyzing some problems involving the Brayton or gas turbine power cycle. The program uses a perfect gas with constant specific heats as the working medium and can be used to solve three specific types of problems:

1. Given the pressure ratio, inlet pressure and temperature, maximum allowable temperature, and power produced, the program provides routines to determine the various properties at

the four states of the Brayton cycle, rates of heat addition and rejection, mass flow rate of air or other gas, and thermal efficiency.

2. Given the pressure ratio, inlet pressure and temperature, mass flow rate of the gases through the cycle, and maximum allowable temperature, the computer will determine the properties at the four states of the Brayton cycle, rates of heat addition and rejection, power generated, and thermal efficiency.

3. Given the inlet pressure and temperature, maximum allowable temperature, the power produced, and rate of heat addition, the computer will determine the properties at the four states of the Brayton cycle, thermal efficiency, and rate of heat rejection.

Provisions are included in the program for considering the polytropic or irreversible adiabatic compressions and expansions.

A–11 PROGRAM RANKINE

The program RANKINE is a VISUAL BASIC program included on a CD available from the publisher. The program allows for computer assistance in analyzing problems involving the Rankine or steam turbine power cycle. The program incorporates subroutines for evaluating steam properties in the superheated, saturated vapor, saturated liquid, or the saturation region states. It can be used to consider the following types of specific problems:

1, 2. Given the simple Rankine cycle, where the boiler pressure and temperature, condenser pressure, fuel heating value, adiabatic efficiencies of the pump and turbine, boiler efficiency, and either the net output power or mass flow of steam is known, the computer can follow through the program to determine the pump power, turbine power, heat of rejection, heat added, steam rate, thermal efficiency, and entropy generation.

3, 4. Given the simple reheat cycle, where the boiler pressure and temperature, condenser pressure, adiabatic efficiencies of the pump and of the high- and low-pressure turbine, boiler efficiency, fuel heating value, reheat pressure, temperature of the steam entering the low-pressure turbine, and either the net output power or mass flow of steam is known, the computer can determine the same sorts of parameters for the simple reheat cycle as for the simple Rankine cycle.

5, 6. Given the simple regenerative cycle, where the boiler pressure and temperature, condenser pressure, adiabatic efficiencies of the pump and turbine, fuel heating value, number of regenerative stages (as many as five stages can be used) and their pressures, and either the net output power or mass flow of steam is known, the computer will determine the same sorts of parameters for the simple regenerative cycle as for the simple Rankine cycle.

7, 8. Given the reheat–regenerative cycle, in which the boiler pressure and temperature, condenser pressure, adiabatic efficiencies of the pump and turbines, fuel heating value, number of regenerative stages (as many as five in both high- and low-pressure turbines) and their pressures, and either the net power output or mass flow rate of the steam is known, the computer will determine the same sort of parameters as for the simple Rankine cycle.

A–12 PROGRAM COMBUST

The program COMBUST is a VISUAL BASIC program contained on a CD available from the publisher. The program can be used with a microcomputer to assist in analyzing some combustion processes of certain fuels. The program is written to handle the following fuels: carbon (C), hydrogen (H_2), methane (CH_4), acetylene gas (C_2H_2), ethylene (C_2H_4), propane gas (C_3H_8), *n*-butane (C_4H_{10}), benzene (C_6H_6), *n*-octane (C_8H_{18}), methyl alcohol (CH_3OH), ethyl alcohol (C_2H_5OH), and cellulose ($C_6H_{10}O_5$). The program requires that you know the percentage of theoretical air supplied, fuel temperature, air temperature, and temperature of the products of combustion. The computer can then follow through the program to determine the higher and lower heating values, partial pressure of the water in the products, dew point temperature of the products, amount of water condensed out of the products if they are cooled to 25°C, actual heat of the combustion, and adiabatic combustion temperature.

A–13 PROGRAM RADIATE

The program RADIATE is a VISUAL BASIC program contained on a CD available from the publisher. The program can be used with a microcomputer to assist in analyzing the radiation between two gray-body surfaces that are convex and can see a surrounding that behaves as a blackbody at a given known temperature. The program requires that the emissivities of the two gray-body surfaces and the surface temperatures of the two surfaces and the surroundings be known. The shape factor between the two gray surfaces must also be known, and then the computer can follow through the program to determine the net radiation to each of the two gray surfaces and to the surroundings.

B

TABLES AND CHARTS

TABLE B–1 Table of the elements

IA	H											IIIA	IVA	VA	VIA	VIIA	O
1																1	2
H																H	He
1.008																1.008	4.003
3	4											5	6	7	8	9	10
Li	Be											B	C	N	O	F	Ne
6.94	9.01											10.8	12.0	14.0	16.0	19.0	20.2
11	12											13	14	15	16	17	18
Na	Mg											Al	Si	P	S	Cl	Ar
23.0	24.3	IIIB	IVB	VB	VIB	VIIB		VII(B)		IB	IIB	27.0	28.1	31.0	32.1	35.45	40.0
19	20	21	22	23	24	25	26	27	28	29	30	31	32	33	34	35	36
K	Ca	Sc	Ti	V	Cr	Mn	Fe	Co	Ni	Cu	Zn	Ga	Ge	As	Se	Br	Kr
39.1	40.1	45.0	47.9	50.9	52.0	54.9	55.8	58.9	58.7	63.5	65.4	69.7	72.6	74.9	79.0	79.9	83.8
37	38	39	40	41	42	43	44	45	46	47	48	49	50	51	52	53	54
Rb	Sr	Y	Zr	Nb	Mo	Tc	Ru	Rh	Pd	Ag	Cd	In	Sn	Sb	Te	I	Xe
85.5	87.6	88.9	91.2	92.9	95.9	(97)	101.1	102.9	106.4	107.9	112.4	114.8	118.7	121.8	127.6	126.9	131.3
55	56	(57-	72	73	74	75	76	77	78	79	80	81	82	83	84	85	86
Cs	Ba	71)	Hf	Ta	W	Re	Os	Ir	Pt	Au	Hg	Tl	Pb	Bi	Po	At	Rn
132.9	137.3		178.5	180.9	183.9	186.2	190.2	192.2	195.1	197.0	200.6	204.4	207.2	209.0	(210)	(210)	(222)
87	88	(89-															
Fr	Ra	103)															
(223)	(226)																
			57	58	59	60	61	62	63	64	65	66	67	68	69	70	71
			La	Ce	Pr	Nd	Pm	Sm	Eu	Gd	Tb	Dy	Ho	Er	Tm	Yb	Lu
			139	140	141	144	(145)	150	152	157	159	163	165	167	169	173	175
			89	90	91	92	93	94	95	96	97	98	99	100	101	102	103
			Ac	Th	Pa	U	Np	Pu	Am	Cm	Bk	Cf	Es	Fm	Md	No	Lw
			(227)	232	(231)	238	(237)	(244)	(243)	(247)	(247)	(251)	(254)	(253)	(256)	(254)	(257)

TABLE B–2 Gravitational acceleration near the Earth
SI units

Altitude Above Sea Level, m	Location on Earth's surface, degrees latitutde					
	Equator, 0°	20°	40°	60°	80°	90°
	meters per second per second					
0	9.7804	9.7865	9.8018	9.8191	9.8307	9.8322
300	9.7795	9.7856	9.8009	9.8182	9.8298	9.8313
600	9.7786	9.7847	9.8000	9.8173	9.8289	9.8304
1,000	9.7774	9.7835	9.7988	9.8161	9.8277	9.8292
1,500	9.7759	9.7820	9.7976	9.8161	9.8277	9.8292
2,000	9.7749	9.7809	9.7961	9.8144	9.8259	9.8276
3,000	9.7714	9.7775	9.7928	9.8101	9.8226	9.8232
30,000	9.6879	9.6964	9.7116	9.7290	9.7415	9.7421

English units

Altitude Above Sea Level, ft	Location on Earth's surface, degrees latitude					
	Equator, 0°	20°	40°	60°	80°	90°
	feet per second per second					
0	32.088	32.108	32.158	32.215	32.253	32.258
1,000	32.085	32.105	32.155	32.212	32.250	32.255
2,000	32.082	32.102	32.152	32.209	32.247	32.252
4,000	32.076	32.096	32.146	32.203	32.241	32.246
6,000	32.070	32.090	32.140	32.203	32.241	32.246
8,000	32.064	32.084	32.134	32.191	32.229	32.234
10,000	32.058	32.078	32.128	32.185	32.226	32.228
100,000	31.780	31.808	31.858	31.915	31.953	31.958

Source: U.S. Coast and Geodetic Survey, 1912.

TABLE B–3 U.S. standard atmospheric conditions*

SI Units

Altitude, m	Temperature, K	Pressure, kPa	Density, kg/m^3
−500	291.400	107.478	1.2849
0	288.150	101.325	1.2250
500	284.900	95.4612	1.1673
1,000	281.651	89.8762	1.1117
2,000	275.154	79.5014	1.0066
4,000	262.166	61.6604	0.81935
6,000	249.187	47.2176	0.66011
8,000	236.215	35.6516	0.52579
10,000	223.252	26.4999	0.41351
20,000	216.650	5.5293	0.088910
40,000	250.350	0.287143	0.039957
60,000	255.772	0.0224606	0.0030592
80,000	180.65	0.0010366	0.00001999
90,000	180.65	0.00016438	0.00000317

English units

Altitude, ft	Temperature, °R	Pressure, in Hg	Density, lbm/ft^3
−1000	522.236	31.0185	0.078737
0	518.670	29.9213	0.076474
500	516.887	29.3846	0.075362
1,000	515.104	28.8557	0.074261
2,000	511.538	27.8212	0.072098
4,000	504.408	25.8426	0.067917
6,000	497.279	23.9798	0.063925
8,000	490.152	22.2276	0.060116
10,000	483.025	20.5808	0.056483
20,000	447.415	13.7612	0.040773
40,000	389.970	5.5584	0.018895
60,000	389.970	2.1354	0.007259
80,000	397.693	0.8273	0.002758
100,000	408.572	0.3290	0.001068
120,000	433.578	0.1358	0.000415
140,000	463.923	0.0595	0.000170
200,000	456.999	0.0058	0.000017

Source: Abstracted from *U.S. standard atmosphere*, prepared under sponsorship of NASA, USAF, and USWB. Printed by U.S. Government Printing Office, 1962.

* U.S. Standard Atmosphere: ideal air devoid of moisture or dust rotating with the Earth. The data represent approximately 45° latitude annual average conditions in the United States.

TABLE B–4 Gas constants (SI units)

Substance	Symbol	M	R, $\frac{J}{kg \cdot K}$	c_p, kJ/kg·K at 25°C	c_v, kJ/kg·K at 25°C	k, $\frac{c_p}{c_v}$
Acetylene	C_2H_2	26.038	320	1.687	1.368	1.234
Air		28.967	287	1.007	0.719	1.399
Ammonia	NH_3	17.032	488	2.096	1.607	1.304
Argon	Ar	39.944	208	0.521	0.312	1.668
Benzene	C_6H_6	78.114	106	1.045	0.939	1.113
n-Butane	C_4H_{10}	58.124	143	1.676	1.533	1.093
Isobutane	C_4H_{10}	58.124	143	1.666	1.523	1.094
1-Butene	C_4H_8	56.108	148	1.527	1.374	1.111
Carbon dioxide	CO_2	44.011	189	0.844	0.655	1.288
Carbon monoxide	CO	28.011	297	1.040	0.744	1.399
Carbon tetrachloride	CCl_4	153.839				
n-Deuterium	D_2	4.029				
Dodecane	$C_{12}H_{26}$	170.340	49	1.646	1.597	1.031
Ethane	C_2H_6	30.070	277	1.751	1.475	1.188
Ethyl ether	$C_4H_{10}O$	74.124				
Ethylene	C_2H_4	28.054	297	1.552	1.256	1.236
Freon, F-12	CCl_2F_2	120.925	69	0.573	0.504	1.136
Helium	He	4.003	2079	5.196	3.117	1.667
n-Heptane	C_7H_{16}	100.205	83	1.656	1.573	1.053
n-Hexane	C_6H_{14}	86.178	96	1.660	1.564	1.062
Hydrogen	H_2	2.016	4124	14.302	10.178	1.405
Hydrogen sulfide	H_2S	34.082				
Mercury	Hg	200.610				
Methane	CH_4	16.043	519	2.227	1.708	1.304
Methyl fluoride	CH_3F	34.035				
Neon	Ne	20.183	412	1.030	0.618	1.667
Nitric Oxide	NO	30.008	277	0.995	0.718	1.386
Nitrogen	N_2	28.016	297	1.040	0.743	1.400
Octane	C_8H_{18}	114.232	73	1.653	1.581	1.046
Oxygen	O_2	32.000	260	0.917	0.657	1.396
n-Pentane	C_5H_{12}	72.151	115	1.666	1.551	1.074
Isopentane	C_5H_{12}	72.151	115	1.663	1.548	1.074
Propane	C_3H_8	44.097	189	1.667	1.478	1.128
Propylene	C_3H_6	42.081	198	1.519	1.279	1.187
Sulfur dioxide	SO_2	64.066	130	0.621	0.491	1.264
Water vapor	H_2O	18.016	462	1.864	1.402	1.329
Xenon	Xe	131.300	63	0.158	0.095	1.667

Source: Data selected from J. F. Masi, *Trans. ASME*, 76:1067 (October, 1954); National Bureau of Standards (U.S.). Circ. 500, February 1952; "Selected Values of Properties of Hydrocarbons and Related Compounds," American Petroleum Institute Research Project 44, Thermodynamics Resarch Center, Texas A & M University, College Station, Texas (Loose Leaf Data Sheets, extant 1972).

TABLE B–4 (continued) Gas constants (English units)*

Substance	Symbol	M	R, $\frac{ft \cdot lbf}{lbm \cdot °R}$	c_p, $\frac{Btu}{lbm \cdot °R}$ at 77°F	c_v, $\frac{Btu}{lbm \cdot °R}$ at 77°F	k, $\frac{c_p}{c_v}$
Acetylene	C_2H_2	26.038	59.39	0.4030	0.3267	1.234
Air		28.967	53.36	0.2404	0.1718	1.399
Ammonia	NH_3	17.032	90.7	0.5006	0.3840	1.304
Argon	Ar	39.944	38.73	0.1244	0.0746	1.668
Benzene	C_6H_6	78.114	19.78	0.2497	0.2243	1.113
n-Butane	C_4H_{10}	58.124	26.61	0.4004	0.3662	1.093
Isobutane	C_4H_{10}	58.124	26.59	0.3979	0.3637	1.094
1-Butene	C_4H_8	56.108	27.545	0.3646	0.3282	1.111
Carbon dioxide	CO_2	44.011	35.12	0.2015	0.1564	1.288
Carbon monoxide	CO	28.011	55.19	0.2485	0.1776	1.399
Carbon tetrachloride	CCl_4	153.839				
n-Deuterium	D_2	4.029				
Dodecane	$C_{12}H_{26}$	170.340	9.074	0.3931	0.3814	1.031
Ethane	C_2H_6	30.070	51.43	0.4183	0.3522	1.188
Ethyl ether	$C_4H_{10}O$	74.124				
Ethylene	C_2H_4	28.054	55.13	0.3708	0.3000	1.236
Freon, F-12	CCl_2F_2	120.925	12.78	0.1369	0.1204	1.136
Helium	He	4.003	386.33	1.241	0.7446	1.667
n-Heptane	C_7H_{16}	100.205	15.42	0.3956	0.3758	1.053
n-Hexane	C_6H_{14}	86.178	17.93	0.3966	0.3736	1.062
Hydrogen	H_2	2.016	766.53	3.416	2.431	1.405
Hydrogen sulfide	H_2S	34.082				
Mercury	Hg	200.610				
Methane	CH_4	16.043	96.40	0.5318	0.4079	1.304
Methyl fluoride	CH_3F	34.035				
Neon	Ne	20.183	76.58	0.2460	0.1476	1.667
Nitric Oxide	NO	30.008	51.49	0.2377	0.1715	1.386
Nitrogen	N_2	28.016	55.15	0.2483	0.1774	1.400
Octane	C_8H_{18}	114.232	13.54	0.3949	0.3775	1.046
Oxygen	O_2	32.000	48.29	0.2191	0.1570	1.396
n-Pentane	C_5H_{12}	72.151	21.42	0.3980	0.3705	1.074
Isopentane	C_5H_{12}	72.151	21.42	0.3972	0.3697	1.074
Propane	C_3H_8	44.097	35.07	0.3982	0.3531	1.128
Propylene	C_3H_6	42.081	36.72	0.3627	0.3055	1.187
Sulfur dioxide	SO_2	64.066	24.12	0.1483	0.1173	1.264
Water vapor	H_2O	18.016	85.80	0.4452	0.3349	1.329
Xenon	Xe	131.300	11.78	0.03781	0.02269	1.667

Source: E. F. Obert, *Concepts of thermodynamics* (New York: McGraw-Hill Book Company), copyright 1960. Used with permission of McGraw-Hill Book Company.

* 1 Btu/lbm · °R = 1 calorie/g · K.

TABLE B–5 Critical properties

Substance	Ref. date	Symbol	M	T_c, K	p_c, atm	v_c, cm^3/g mol	v_c, ft^3/ mol	z_c
Acetylene	1928	C_2H_2	26.038	309.5	61.6	113		0.274
Air	1917		28.967	132.41	37.25	93.25		
Ammonia	1920	NH_3	17.032	405.4	111.3	72.5	1.16	0.243
Argon	1910	Ar	39.944	150.72	47.996	75	1.20	0.291
Benzene	1948	C_6H_6	78.114	562.6	48.6	260	4.17	0.274
n-Butane	1939	C_4H_{10}	58.124	425.17	37.47	255	4.08	0.274
Isobutane	1910	C_4H_{10}	58.124	408.14	36.00	263	4.21	0.283
1-Butene	1950	C_4H_8	56.108	419.6	39.7	240	3.84	0.277
Carbon dioxide	1950	CO_2	44.011	304.20	72.90	94	1.51	0.275
Carbon monoxide	1936	CO	28.011	132.91	34.529	93	1.49	0.294
Carbon tetrachloride	1931	CCl_4	153.839	556.4	45.0	276		0.272
n-Deuterium	1951	D_2	4.029	38.43	16.421			
Dodecane	1953	$C_{12}H_{26}$	170.340	659	17.9		11.5	0.237
Ethane	1939	C_2H_6	30.070	305.43	48.20	148	2.37	0.285
Ethyl ether	1929	$C_4H_{10}O$	74.124	467.8	35.6	282.9		
Ethylene	1939	C_2H_4	28.054	283.06	50.50	124	1.99	0.270
Freon, F-12	1957	CCl_2F_2	120.925	385.16	40.6	217	3.47	0.279
Helium	1936	He	4.003	5.19	2.26	58	0.929	0.308
n-Heptane	1937	C_7H_{16}	100.205	540.17	27.00	426	6.82	0.260
n-Hexane	1946	C_6H_{14}	86.178	507.9	29.94	368	5.89	0.264
Hydrogen	1951	H_2	2.016	33.24	12.797	65	1.04	0.304
Hydrogen sulfide	1948	H_2S	34.082	373.7	88.8	98	1.57	0.284
Mercury	1953	Hg	200.610					
Methane	1953	CH_4	16.043	190.7	45.8	99	1.59	0.290
Methyl fluoride	1932	CH_3F	34.035	317.71	58.0			
Neon	1936	Ne	20.183	44.39	26.86	41.7	0.668	0.308
Nitric oxide	1951	NO	30.008	179.2	65.0	58	0.929	0.256
Nitrogen	1951	N_2	28.016	126.2	33.54	90	0.144	0.291
Octane	1931	C_8H_{18}	114.232	569.4	24.64	486	7.77	0.256
Oxygen	1948	O_2	32.000	154.78	50.14	74	1.19	0.292
n-Pentane	1899	C_5H_{12}	72.151	469.78	33.31	311	4.98	0.269
Isopentane	1910	C_5H_{12}	72.151	461.0	32.92	308	4.93	0.268
Propane	1940	C_3H_8	44.097	370.01	42.1	200	3.20	0.277
Propylene	1953	C_3H_6	42.081	365.1	45.40	181	2.90	0.274
Sulfur dioxide	1945	SO_2	64.066	430.7	77.8	122		0.269
Water vapor	1934	H_2O	18.016	647.27	218.167	56	0.897	0.230
Xenon	1951	Xe	131.300	289.81	58.0	118.8	1.90	0.290

Source: E. F. Obert, *Comcepts of thermodynamics* (New York: McGraw-Hill Book Company), copyright 1960. Used with permission of McGraw-Hill Book Company.

TABLE B–6 Properties of air at low pressure (SI units)

T, K	h, kJ/kg	p_r	u, kJ/kg	v_r	ϕ, kJ/kg·K
60	59.7	0.00503	42.4	7961	0.903
80	79.7	0.01372	56.8	3887	1.191
100	99.8	0.02990	71.1	2230	1.414
120	119.8	0.0565	85.3	1416	1.597
140	139.8	0.0968	99.7	964	1.752
160	159.9	0.1543	114.0	691	1.885
180	179.9	0.2328	128.3	516	2.003
200	200.0	0.336	142.6	397	2.109
220	220.0	0.469	156.8	313	2.204
240	240.0	0.636	171.1	252	2.292
260	260.1	0.841	185.5	206.3	2.372
280	280.1	1.089	199.8	171.5	2.446
300	300.2	1.386	214.1	144.3	2.512
320	320.3	1.738	228.1	122.8	2.578
340	340.4	2.15	242.9	105.5	2.641
360	360.6	2.63	257.2	91.4	2.699
380	380.8	3.18	271.7	79.8	2.753
400	401.0	3.81	286.2	70.1	2.805
420	421.3	4.52	300.7	61.9	2.855
440	444.6	5.33	315.3	55.0	2.902
460	462.0	6.25	330.0	49.1	2.947
480	482.5	7.27	344.8	44.0	2.991
500	503.0	8.41	359.5	39.6	3.033
550	554.8	11.86	396.9	30.9	3.131
600	607.0	16.28	434.8	24.6	3.222
650	659.8	21.9	473.3	19.83	3.307
700	713.3	28.8	512.4	16.21	3.386
750	767.3	37.4	552.1	13.39	3.461
800	821.9	47.8	592.3	11.17	3.531
850	877.2	60.3	633.2	9.40	3.598
900	932.9	75.3	674.6	7.97	3.662
950	989.2	93.1	716.6	6.81	3.723
1000	1046.0	114.0	759.0	5.85	3.781
1100	1161.1	167.1	845.3	4.39	3.891
1200	1277.8	238	933.4	3.36	3.992
1300	1395.0	331	1022.9	2.62	4.087
1400	1515.4	451	1113.6	2.07	4.175
1500	1636.0	602	1205.5	1.662	4.258
1600	1757.5	791	1298.3	1.349	4.337
1700	1880.1	1025	1392.2	1.106	4.411
1800	2003.3	1310	1486.8	0.916	4.482
1900	2127.4	1655	1582.1	0.766	4.549
2000	2252.1	2068	1678.1	0.645	4.613
2200	2503.2	3138	1871.8	0.468	4.732
2400	2756.6	4607	2067.8	0.347	4.843
2600	3011.7	6575	2265.5	0.264	4.947
2800	3268.4	9159	2464.8	0.2039	5.040
3000	3526.5	12490	2665.5	0.1602	5.129
3200	3786.0	16720	2867.5	0.1276	5.212
3600	4308.2	28569	3275.0	0.0840	5.366

TABLE B–6 (continued) Properties of air at low pressure (English units)*

T, °R	h, Btu/lbm	p_r	u, Btu/lbm	v_r	ϕ, Btu/lbm · °R
200	47.67	.04320	33.96	1714.9	0.36303
250	59.64	.09415	42.50	983.6	0.41643
300	71.61	.17795	51.04	624.5	0.46007
350	83.57	.3048	59.58	425.4	0.49695
400	95.53	.4858	68.11	305.0	0.52890
450	107.50	.7329	76.65	227.45	0.55710
500	119.48	1.0590	85.20	174.90	0.58233
525	125.47	1.2560	89.48	154.84	0.59403
550	131.46	1.4779	93.76	137.85	0.60518
575	137.47	1.7269	98.05	123.34	0.61586
600	143.47	2.005	102.34	110.88	0.62607
625	149.49	2.313	106.64	100.08	0.63589
650	155.50	2.655	110.94	90.69	0.64533
675	161.54	3.032	115.26	82.47	0.65443
700	167.56	3.446	119.58	72.25	0.66321
725	173.60	3.900	123.91	68.86	0.67169
750	179.66	4.396	128.25	63.20	0.67991
800	191.81	5.526	136.97	53.63	0.69558
850	204.01	6.856	145.74	45.92	0.71037
900	216.26	8.411	154.57	39.64	0.72438
950	228.58	10.216	163.46	34.45	0.73771
1000	240.98	12.298	172.43	30.12	0.75042
1050	253.45	14.686	181.47	26.48	0.76259
1100	265.99	17.413	190.58	23.40	0.77426
1150	278.61	20.51	199.78	20.771	0.78548
1200	291.30	24.01	209.05	18.514	0.79628
1250	304.08	27.96	218.40	16.563	0.80672
1300	316.94	32.39	227.83	14.868	0.81680
1350	329.88	37.35	237.34	13.391	0.82658
1400	342.90	42.88	246.93	12.095	0.83604
1450	356.00	49.03	256.60	10.954	0.84523
1500	369.17	55.86	266.34	9.948	0.85416
1550	382.42	63.40	276.17	9.056	0.86285
1600	395.74	71.73	286.06	8.263	0.87130
1650	409.13	80.89	296.03	7.556	0.87954
1700	422.59	90.95	306.06	6.924	0.88758
1750	436.12	101.98	316.16	6.357	0.89542
1800	449.71	114.03	326.32	5.847	0.90308
1900	477.09	141.51	346.85	4.974	0.91788
2000	504.71	174.00	367.61	4.258	0.93205
2200	560.59	256.6	409.78	3.176	0.95868
2400	617.22	367.6	452.70	2.419	0.98331
2600	674.49	513.5	496.26	1.8756	1.00623
3000	790.68	941.4	585.04	1.1803	1.04779
3500	938.40	1829.3	698.48	0.7087	1.09332
4000	1088.26	3280	814.06	0.4518	1.13334
4500	1239.86	5521	931.39	0.3019	1.16905
5000	1392.87	8837	1050.12	0.20959	1.20129
5500	1547.07	13568	1170.04	0.15016	1.23068
6000	1702.29	20120	1291.00	0.11047	1.25769
6500	1858.44	28974	1412.87	0.08310	1.28268

Source: J. H. Keenan and J. Kaye, *Gas tables* (New York: John Wiley & Sons, Inc., 1948) with permission of authors and publisher. Conversion to SI by author.

* Per pound-mass.

TABLE B–7 Specific heat equations for some common gases*

Gas or Vapor	Equation, c_p, Btu/mol · °R	Range, °R	Maximum % Error
O_2	$c_p = 11.515 - \frac{172}{\sqrt{T}} + \frac{1530}{T}$	540–5000	1.1
N_2	$c_p = 9.47 - \frac{3.47(10^3)}{T} + \frac{1.16(10^6)}{T^2}$	540–9000	1.7
CO	$c_p = 9.46 - \frac{3.29(10^3)}{T} + \frac{1.07(10^6)}{T^2}$	540–9000	1.1
H_2O	$c_p = 19.86 - \frac{597}{\sqrt{T}} + \frac{7500}{T}$	540–5400	1.8
CO_2	$c_p = 16.2 - \frac{6.53(10^3)}{T} + \frac{1.41(10^6)}{T^2}$	540–6300	0.8

Gas or Vapor	*a*	*b*	*c*	*d*	Range, K	Maximum % Error
Air	6.713	0.4697	1.147	−0.4696	273–1800	0.72
	6.557	1.477	−0.2148	0	273–3800	1.64
CO	6.726	0.4001	1.283	−0.5307	273–1800	0.89
	6.480	1.566	−0.2387	0	273–3800	1.86
CO_2	5.316	14.285	−8.362	1.784	273–1800	0.67
	$c_p = 18.036$	$-0.00004474T$	$-158.08(T)^{-1/2}$		273–3800	2.65
H_2	6.952	−0.4576	0.9563	−0.2079	273–1800	1.01
	6.424	1.039	−0.07804	0	273–3800	2.14
H_2O	7.700	0.4594	2.521	−0.8587	273–1800	0.53
	6.970	3.464	−0.4833	0	273–3800	2.03
O_2	6.085	3.631	−1.709	0.3133	273–1800	1.19
	6.732	1.505	−0.1791	0	273–3800	3.24
N_2	6.903	−0.3753	1.930	−6.861	273–1800	0.59
	6.529	1.488	−0.2271	0	273–3800	2.05
NH_3	6.5846	6.1251	2.3663	−1.5981	273–1500	0.91
CH_4	4.750	12.00	3.030	−2.630	273–1500	1.33
C_3H_8	−0.966	72.79	−37.55	7.580	273–1500	0.40
C_4H_{10}	0.945	88.73	−43.80	8.360	273–1500	0.54
C_6H_6	−8.650	115.78	−75.40	18.54	273–1500	0.34
C_2H_2	5.21	22.008	−15.59	4.349	273–1500	1.46
CH_3OH	4.55	21.86	−2.91	−1.92	273–1000	0.18

Source: E. F. Obert, *Concepts of thermodynamics* (New York: McGraw-Hill Book Company), copyright 1960: with permission of McGraw-Hill Book Company.

* $c_p = a + b(10^{-3})T + c(10^{-6})T^2 + d(10^{-9})T^3 (T, \mathrm{K})$ = cal/g mol · K or closely Btu/lbm mol · °R

TABLE B–8 Thermal properties of selected materials (SI units)

Material	Temperature, T, K	Density, ρ, kg/m^3	Specific, Heat, c_p, $kJ/kg \cdot K$	Thermal Conductivity, κ, $W/m \cdot K$	Viscosity, μ_k, $g/m \cdot s$
		Nonmetallic solids			
Asbestos	300	2,100	1.047	0.156	
Asphalt	300	2,115	0.920	0.062	
Brick					
Common	300	1,600	0.840	0.70	
Fire clay	1,000	2,640	0.960	1.50	
Masonry	300	1,700	0.837	0.66	
Clay	300	1,460	0.880	1.30	
Concrete	300	2,300	0.880	1.40	
Cork board	300	160	1.90	0.045	
Ebonite	300		1.382		
Felt board	300			0.035	
Glass	300	2,500	0.750	1.40	
Glass fiber	300	220		0.035	
Glass wool	300	40	0.70	0.038	
Gypsum board (plaster)	300	817	1.084	0.107	
Ice	273	920	2.00	2.20	
Mineral wool (batts)	300			0.039	
Polystyrene (Styrofoam)	300	37		0.029	
Polyurethane Foam, rigid	300	40		0.023	
Rock wool	300	160		0.040	
Rubber					
Foam	300	70		0.032	
Hard	300	1,190	2.01	0.16	
Sand	300	1,515	0.80	0.30	
Saw dust	300	215		0.059	
Skin (human)	300		3.60	0.37	
Snow					
Loose	273	110		0.05	
Packed	273	500		0.19	
Soil	300	2,050	1.840	0.52	
Stone					
Granite	300	2,630	0.775	2.79	
Limestone	300	2,320	0.810	2.15	
Sandstone	300	2,150	0.745	2.90	
Teflon®	300	2,200	1.00	0.35	
Wood (cross-grain)					
Balsa	300	140		0.055	
Pine (yellow)	300	640	2.805	0.15	
Oak	300	545	2.385	0.17	
Plywood	300	550	1.200	0.12	
Wool	300	90		0.036	
Vermiculite					
Loose	300	80	0.835	0.068	

TABLE B–8 (continued) Thermal properties of selected materials (SI units)

Material	Temperature, T, K	Density, ρ, kg/m^3	Specific, Heat, c_p, $kJ/kg \cdot K$	Thermal Conductivity, κ, $W/m \cdot K$	Viscosity, μ_k, $g/m \cdot s$
		Gases			
Air	300	1.178	1.007	0.026	0.0181
Ammonia	300	0.689	2.096	0.025	0.0102
Carbon dioxide	300	1.81	0.844	0.017	0.0148
Carbon monoxide	300	1.16	1.040	0.025	0.0182
Hydrogen	300	0.0817	14.302	0.190	0.0090
Nitrogen	300	1.15	1.040	0.026	0.0176
Oxygen	300	1.31	0.911	0.027	0.0200
Steam	800	0.272	1.186	0.067	0.914
Methane	300	0.655	2.227	0.020	0.0134
		Metallic solids			
Aluminum	300	2,700	0.902	236	
Brass	300	8,520	0.382	114	
Copper	300	8,900	0.385	400	
Iron					
Wrought	300	7,850	0.460	51	
Cast	300	7,270	0.420	39	
Lead	300	11,340	0.129	36	
Silicon	300	2,330	0.691	159	
Steel					
Carbon	300	7,800	0.473	43	
Stainless	300	7,900	0.477	14	
		Liquids			
Ammonia	300	605	4.80	0.520	0.22
Benzene	300	895	1.81	0.159	0.65
Engine oil (unused)	300	886	1.89	0.145	712.0
Ethyl alcohol	300	788	2.395	0.168	1.2
Methyl alcohol	300	793	2.55	0.202	0.53
Ethylene glycol	300	1,116	2.385	0.257	21.4
Glycerine	300	1,260	2.39	0.287	1490
Water	300	997	4.18	0.608	0.857
		Liquid metals			
Bismuth	600	9,997	0.145	16.4	1.61
Lead	700	10,476	0.157	17.4	2.15
Lithium	600	498	4.190	48.0	0.57
Mercury	600	12,816	0.134	14.2	0.84
Potassium	600	766	0.783	44.0	0.23
Sodium	600	871	1.309	76.0	0.32

TABLE B–8 (continued) Thermal properties of selected materials (English units)

Material	**Temperature,** T, °R	**Density,** ρ, lbm/ft^3	**Specific Heat,** c_p, Btu/lbm°F	**Thermal Conductivity,** κ, Btu/h-ft-°F	**Viscosity,** μ_k, lbm/ft-h
		Nonmetallic Solids			
Asbestos	540	131.1	0.250	0.090	
Asphalt	540	132.0	0.220	0.036	
Brick					
Common	540	99.89	0.201	0.405	
Fire Clay	1800	164.8	0.229	0.867	
Masonry	540	106.1	0.200	0.381	
Clay	540	91.15	0.210	0.751	
Concrete	540	143.6	0.210	0.809	
Cork board	540	3.746	0.454	0.026	
Ebonite	540		0.330		
Felt board	540			0.020	
Glass	540	156.1	0.179	0.809	
Glass fiber	540	13.73		0.020	
Glass wool	540	2.497	0.167	0.022	
Gypsum board (plaster)	540	51.00	0.259	0.062	
Ice	492	57.44	0.478	1.271	
Mineral wool (batts)	540			0.023	
Polystyrene (Styrofoam)	540	2.310		0.017	
Polyurethane foam, rigid	540	2.497		0.013	
Rock wool	540	9.989		0.023	
Rubber					
Foam	540	4.370		0.018	
Hard	540	74.29	0.480	0.092	
Sand	540	94.58	0.191	0.173	
Saw dust	540	13.42		0.034	
Skin (human)	540		0.860	0.214	
Snow					
Loose	492	6.867		0.029	
Packed	492	31.21		0.110	
Soil	540	128.0	0.439	0.301	
Stone					
Granite	540	164.2	0.185	1.612	
Limestone	540	144.8	0.193	1.242	
Sandstone	540	134.2	0.178	1.676	
Teflon®	540	137.3	0.239	0.202	
Wood (cross-grain)					
Balsa	540	8.740		0.032	
Pine (yellow)	540	39.95	0.670	0.087	
Oak	540	34.02	0.570	0.098	
Plywood	540	34.34	0.287	0.069	
Wool	540	5.619		0.021	
Vermiculite					
Loose	540	4.994	0.199	0.483	

TABLE B–8 (continued) Thermal properties of selected materials (English units)

Material	Temperature, T, °R	Density, ρ, lbm/ft^3	Specific Heat, c_p, Btu/lbm°F	Thermal Conductivity, κ, Btu/h-ft-°F	Viscosity, μ_k, lbm/ft-h
		Gases			
Air	540	0.0735	0.241	0.015	0.0438
Ammonia	540	0.0430	0.501	0.014	0.0247
Carbon dioxide	540	0.1130	0.202	0.010	0.0358
Carbon monoxide	540	0.0724	0.248	0.014	0.0440
Hydrogen	540	0.0051	0.816	0.110	0.0218
Nitrogen	540	0.0717	0.248	0.015	0.0426
Oxygen	540	0.0818	0.218	0.016	0.0484
Steam	1440	0.0170	0.283	0.039	2.2110
Methane	540	0.0409	0.532	0.012	0.0324
		Metallic Solids			
Aluminum	540	168.6	0.215	136.38	
Brass	540	531.9	0.091	65.88	
Copper	540	555.6	0.092	231.16	
Iron					
Wrought	540	490.1	0.110	29.47	
Cast	540	453.9	0.100	22.54	
Lead	540	708.0	0.0308	20.80	
Silicon	540	145.5	0.165	91.89	
Steel					
Carbon	540	487.0	0.113	24.85	
Stainless	540	493.2	0.114	8.09	
		Liquids			
Ammonia	540	37.77	1.146	0.301	0.532
Benzene	540	55.87	0.432	0.092	1.572
Engine oil (unused)	540	55.31	0.451	0.084	1722.4
Ethyl alcohol	540	49.19	0.572	0.097	2.903
Methyl alcohol	540	49.51	0.609	0.032	1.282
Ethyl glycol	540	69.67	0.570	0.149	51.768
Glycerine	540	78.66	0.571	0.166	3604.4
Water	540	62.24	1.000	0.351	2.073
		Liquid metals			
Bismuth	1080	624.1	0.0346	9.478	3.895
Lead	1260	654.0	0.0375	10.06	5.201
Lithium	1080	31.09	1.0008	27.74	1.379
Mercury	1080	800.1	0.032	8.206	2.032
Potassium	1080	47.82	0.187	25.43	0.556
Sodium	1080	54.38	0.313	43.92	0.774

TABLE B–9 Approximate emissivities of some material surfaces

Material	Temperature, K	Emissivity, ϵ
Aluminum		
Polished	300–900	0.04–0.06
Oxidized	400–800	0.20–0.33
Brass		
Polished	350	0.09
Dull	300–600	0.22
Copper		
Polished	300–500	0.04–0.05
Oxidized	600–1000	0.5–0.8
Iron		
Cast	300	0.44
Wrought	300–500	0.28
Rusted	300	0.61
Lead		
Polished	300– 500	0.06–0.08
Rough	300	0.43
Stainless steel		
Polished	300–1000	0.17–0.3
Oxidized	600–1000	0.5–0.6
Steel		
Polished	300–500	0.08–0.14
Oxidized	300	0.05
Asphalt pavement	300	0.90
Brick (common)	300	0.94
Cloth	300	0.75–0.9
Concrete	300	0.90
Glass (window)	300	0.92
Ice	273	0.95–0.99
Paint		
Aluminum	300	0.45
Black	300	0.88
Oil (all colors)	300	0.94
White	300	0.90
Red primer	300	0.93
Paper (white)	300	0.93
Sand	300	0.90
Human skin	300	0.95
Snow	273	0.80–0.96
Soil (earth)	300	0.93–0.9
Water	273–373	0.95
Wood		
Beech	300	0.94
Oak	300	0.90

TABLE B–10 Volumetric expansion of some liquids at atmospheric pressure

Substance	**Temperature Kelvin (K)**	$\beta_p \times 10^3\ K^{-1}$
Ammonia	223	1.93
Benzene	293	1.237
Bromine	293	1.132
Ethyl Alcohol	293	1.120
Ethyl Glycol	293	0.6375
Gasoline	313	1.496
Gasoline	423	1.752
Kerosene (T-1)	293	0.955
Mercury	325	0.1819
Methyl Alcohol	293	1.199
Oil, Machine	303	0.638
Oil, Transformer	273	0.680
Propane	243	1.9
Sodium	373	0.275
Water	283	0.070
	293	0.182
	323	0.449
	373 (liquid)	0.752

Source: I. Grigoriev and E. Melilikhov, *Handbook of Physical Quantities*, (Boca Raton, Florida: CRC Press), copyright 1997; by permission of CRC Press, Inc.

TABLE B–11 Saturated steam table: temperature (SI units)

Temp., °C,	Pres-sure, kpa,	Specific volume, m^3/kg		Enthalpy, kJ/kg			Entropy, kJ/kg · K		
T	p	v_f	v_g	h_f	h_{fg}	h_g	s_f	s_{fg}	s_g
0.01[a]	0.6	0.0010002	206.3	0.00	2501	2501	0.000	9.154	9.154
5	0.9	0.0010001	147.2	21.1	2489	2510	0.076	8.948	9.024
10	1.2	0.0010004	106.4	42.0	2477	2519	0.151	8.748	8.899
20	2.3	0.001002	57.8	83.9	2454	2537	0.296	8.371	8.667
30	4.2	0.001004	32.9	125.7	2430	2556	0.437	8.015	8.452
40	7.4	0.001008	19.6	167.5	2406	2574	0.572	7.684	8.256
50	12.3	0.001012	12.0	209	2383	2592	0.704	7.371	8.075
60	19.9	0.001017	7.68	251	2358	2609	0.831	7.077	7.908
70	31.2	0.00102	5.05	293	2333	2626	0.955	6.799	7.754
80	47.4	0.00103	3.41	335	2308	2643	1.075	6.537	7.612
90	70.1	0.00104	2.36	377	2282	2659	1.193	6.286	7.479
100	101.3	0.00104	1.67	419	2257	2676	1.307	6.048	7.355
110	143.3	0.00105	1.210	461	2230	2691	1.418	5.821	7.239
120	198.5	0.00106	0.892	504	2202	2706	1.528	5.602	7.130
130	270.1	0.00107	0.668	546	2175	2721	1.635	5.397	7.027
140	361.4	0.00108	0.509	589	2145	2734	1.739	5.191	6.930
150	476	0.00109	0.393	632	2114	2746	1.842	4.996	6.838
160	618	0.00110	0.307	676	2082	2758	1.943	4.808	6.751
170	792	0.00111	0.243	719	2050	2769	2.042	4.625	6.667

TABLE B–11 (continued) Saturated steam table: temperature (SI units)

Temp., °C,	Pres-sure, MPa,	Specific volume, m³/kg		Enthalpy, kJ/kg			Entropy, kJ/kg · K		
T	p	v_f	v_g	h_f	h_{fg}	h_g	s_f	s_{fg}	s_g
180	1.003	0.00113	0.194	763	2015	2778	2.140	4.446	6.586
190	1.255	0.00114	0.156	808	1978	2786	2.236	4.271	6.507
200	1.555	0.00115	0.127	852	1941	2793	2.331	4.101	6.432
210	1.908	0.00117	0.104	898	1900	2798	2.425	3.933	6.358
220	2.320	0.00119	0.0860	944	1858	2802	2.518	3.767	6.285
230	2.798	0.00121	0.0715	990	1813	2803	2.610	3.603	6.213
240	3.348	0.00123	0.0597	1038	1765	2803	2.702	3.441	6.143
250	3.978	0.00125	0.0501	1086	1715	2801	2.793	3.279	6.072
260	4.694	0.00128	0.0422	1135	1661	2796	2.885	3.116	6.001
270	5.51	0.00130	0.0356	1185	1605	2790	2.976	2.954	5.930
280	6.42	0.00133	0.0301	1237	1543	2780	3.068	2.789	5.857
290	7.45	0.00137	0.0255	1290	1476	2766	3.161	2.622	5.783
300	8.59	0.00140	0.0216	1345	1404	2749	3.255	2.450	5.705
310	9.87	0.00145	0.0183	1402	1325	2727	3.351	2.272	5.623
320	11.29	0.0015	0.0154	1462	1238	2700	3.450	2.085	5.535
330	12.87	0.0016	0.0130	1526	1140	2666	3.552	1.889	5.441
340	14.61	0.0016	0.0108	1595	1027	2622	3.661	1.675	5.336
350	16.54	0.0017	0.0088	1671	0894	2565	3.779	1.433	5.212
360	18.67	0.0019	0.0069	1762	0719	2481	3.916	1.137	5.053
370	21.05	0.0022	0.0049	1893	0438	2331	4.114	0.681	4.795
374.15[b]	22.13	0.00326	0.00326	2100	0	2100	4.430	0.000	4.430

Source: Data abstracted from *ASME steam tables*, copyright 1967, and reproduced with permission of the American Society of Mechanical Engineers.

[a]Triple point.

[b]Critical point.

TABLE B–11 (continued)

Temp., °C,	Pres-sure, MPa,	Specific volume, m^3/kg		Enthalpy, kJ/kg			Entropy, kJ/kg · K		
T	p	v_f	v_g	h_f	h_{fg}	h_g	s_f	s_{fg}	s_g
6.9	1.0	0.0010001	129.9	29	2484	2513	0.105	8.875	8.98
17.5	2.0	0.0010007	87.9	73	2460	2533	0.261	8.459	8.72
28.9	4.0	0.001004	39.5	121	2433	2554	0.423	8.047	8.49
36.2	6.0	0.001006	23.7	152	2415	2567	0.521	7.809	8.33
41.5	8.0	0.001009	18.1	174	2402	2576	0.593	7.637	8.23
45.8	10	0.001010	13.4	192	2392	2584	0.649	7.500	8.149
54.0	15	0.001014	10.0	226	2373	2599	0.755	7.252	8.007
60.1	15	0.001017	7.65	251	2358	2609	0.832	7.075	7.907
75.9	40	0.001026	3.99	318	2318	2636	1.026	6.644	7.670
86.0	60	0.001033	2.73	360	2293	2653	1.145	6.386	7.531
93.5	80	0.001039	2.09	392	2273	2665	1.233	6.201	7.434
99.6	100	0.00104	1.69	417	2258	2675	1.303	6.057	7.360
100.0	101.3	0.00104	1.67	419	2257	2676	1.307	6.048	7.355
111.4	150	0.00105	1.16	467	2226	2693	1.434	5.780	7.223
120.2	200	0.00106	0.89	505	2202	2707	1.530	5.597	7.127
143.6	400	0.00108	0.462	605	2133	2738	1.777	5.120	6.897
158.8	600	0.00110	0.316	657	2086	2757	1.931	4.830	6.761
170.4	800	0.00111	0.240	721	2048	2769	2.046	4.617	6.663

TABLE B–11 (continued) Saturated steam table: pressure (SI units)

Pressure, MpA, p	Temp., °C, T	Specific volume, m^3/kg		Enthalpy, kJ/kg			Entropy, kJ/kg · K		
		v_f	v_g	h_f	h_{fg}	h_g	s_f	s_{fg}	s_g
1.0	179.9	0.00113	0.195	763	2015	2778	2.138	4.449	6.587
2.0	212.4	0.00118	0.0996	909	1890	2799	2.447	3.893	6.340
3.0	233.8	0.00122	0.0667	1008	1796	2804	2.646	3.540	6.186
4.0	250.3	0.00125	0.0498	1088	1713	2801	2.796	3.274	6.070
5.0	263.9	0.00129	0.0394	1154	1640	2794	2.921	3.052	5.973
6.0	275.6	0.00132	0.0324	1214	1571	2785	3.027	2.863	5.890
7.0	285.8	0.00135	0.0274	1267	1505	2772	3.122	2.692	5.814
8.0	295.0	0.00138	0.0235	1317	1441	2758	3.208	2.537	5.745
9.0	304.1	0.00142	0.0205	1364	1379	2743	3.287	2.391	5.678
10	311.0	0.00145	0.0180	1408	1317	2725	3.360	2.155	5.515
11	318.0	0.00149	0.0160	1450	1255	2705	3.430	2.123	5.553
12	324.6	0.00153	0.0143	1491	1194	2685	3.496	1.996	5.492
13	330.8	0.00157	0.0128	1532	1130	2662	3.560	1.871	5.431
14	336.6	0.00161	0.0115	1571	1067	2638	3.624	1.748	5.372
15	342.1	0.00166	0.0104	1610	1001	2611	3.685	1.625	5.310
16	347.3	0.00171	0.0093	1650	932	2582	3.746	1.501	5.247
17	352.3	0.00177	0.0084	1690	858	2548	3.807	1.370	5.177
18	357.0	0.00184	0.0075	1732	778	2510	3.871	1.236	5.107
19	361.4	0.0019	0.0067	1776	690	2466	3.938	1.089	5.027
20	365.7	0.0020	0.0059	1827	583	2410	4.015	0.913	4.928
21	369.8	0.0022	0.0050	1888	449	2336	4.108	0.695	4.803
22	373.7	0.0027	0.0037	2016	152	2168	4.303	0.288	4.591
22.13[b]	374.15	0.00326	0.00326	2100	0	2100	4.430	0.000	4.430

TABLE B–11 (continued) Saturated steam table: temperature (English units)

Temp., °F, T	Pres-sure, psia, p	Specific volume, ft³/lbm		Enthalpy, Btu/lbm			Entropy, Btu/lbm · R		
		v_f	v_g	h_f	h_{fg}	h_g	s_f	s_{fg}	s_g
32.018[a]	0.08865	0.0160	3302.4	0.003	1075.5	1075.5	0.0000	2.1873	2.1873
36.00	0.104	0.0160	2839.0	4.008	1073.2	1077.2	0.0081	2.1651	2.1732
40.00	0.122	0.0160	2445.8	8.027	1071.0	1079.0	0.0162	2.1432	2.1594
50.00	0.178	0.0160	1704.8	18.05	1065.3	1083.4	0.0361	2.0901	2.1262
60.00	0.256	0.0160	1207.6	28.06	1059.7	1087.7	0.0555	2.0391	2.0946
80.00	0.506	0.0161	633.3	48.04	1048.4	1096.4	0.0932	1.9426	2.0359
100.00	0.949	0.0161	350.4	68.0	1037.1	1105.1	0.1295	1.8530	1.9825
101.74	1.000	0.0161	333.60	69.7	1036.1	1105.8	0.1326	1.8455	1.9781
120.00	1.693	0.0162	203.26	89.0	1025.6	1113.6	0.1646	1.7693	1.9339
140.00	2.889	0.0163	123.00	108.0	1014.0	1122.0	0.1985	1.6910	1.8895
160.00	4.741	0.0164	77.29	128.0	1002.2	1130.2	0.2313	1.6174	1.8487
162.24	5.000	0.0164	73.532	130.2	1000.9	1131.1	0.2349	1.6094	1.8443
180.00	7.51	0.0165	50.225	148.0	990.2	1138.2	0.2631	1.5480	1.8111
193.21	10.00	0.0166	38.420	161.2	982.1	1143.3	0.2836	1.5043	1.7879
200.00	11.53	0.0166	33.639	168.1	977.9	1146.0	0.2940	1.4824	1.7764
212.00	14.696	0.0167	26.799	180.2	970.3	1150.5	0.3121	1.4447	1.7568
213.03	15.000	0.0167	26.290	181.2	969.7	1150.9	0.3137	1.4415	1.7552
220.00	17.19	0.0168	23.148	188.2	965.2	1153.4	0.3241	1.4201	1.7442
227.96	20.00	0.0168	20.087	196.2	960.1	1156.3	0.3358	1.3962	1.7320
240.00	24.97	0.0169	16.321	208.5	952.1	1160.6	0.3533	1.3609	1.7142
250.34	30.00	0.0170	13.744	218.9	945.2	1164.1	0.3682	1.3313	1.6995

TABLE B–11 (continued) Saturated steam table: temperature (English units)

Temp., °F, T	Pres-sure, psia, p	Specific volume, ft^3/lbm		Enthalpy, Btu/lbm			Entropy, Btu/lbm · R		
		v_f	v_g	h_f	h_{fg}	h_g	s_f	s_{fg}	s_g
260.00	35.43	0.0171	11.762	228.8	938.6	1167.4	0.3819	1.3043	1.6862
267.25	40.00	0.0172	10.497	236.1	933.6	1169.8	0.3921	1.2844	1.6765
280.00	49.20	0.0173	8.644	249.2	924.6	1173.8	0.4098	1.2501	1.6599
292.71	60.00	0.0174	7.174	262.2	915.4	1177.6	0.4273	1.2167	1.6440
300.00	67.01	0.0175	6.466	269.7	910.0	1179.7	0.4372	1.1979	1.6351
312.04	80.00	0.0176	5.471	282.1	901.0	1183.1	0.4534	1.1675	1.6208
327.82	100.00	0.0177	4.431	298.5	888.6	1187.2	0.4743	1.1284	1.6027
340.00	118.0	0.0179	3.788	311.3	878.8	1190.1	0.4902	1.0990	1.5892
358.43	150.0	0.0181	3.014	330.6	863.4	1194.1	0.5141	1.0554	1.5695
380.00	195.7	0.0184	2.335	353.6	844.5	1198.0	0.5416	1.0057	1.5473
400.97	250.0	0.0187	1.843	376.1	825.0	1201.1	0.5679	0.9585	1.5264
420.00	308.8	0.0189	1.500	396.9	806.2	1203.1	0.5915	0.9165	1.5080
444.60	400.0	0.0193	1.161	424.2	780.4	1204.6	0.6217	0.8630	1.4847
500.00	680.9	0.0204	0.675	487.9	714.3	1202.2	0.6890	0.7443	1.4333
518.21	800.0	0.0209	0.569	509.8	689.6	1199.4	0.7111	0.7051	1.4163
550.00	1045.4	0.0218	0.424	549.4	641.8	1191.2	0.7501	0.6356	1.3856
567.19	1200.0	0.0223	0.362	571.9	613.0	1184.8	0.7714	0.5969	1.3683
600.00	1543.2	0.0236	0.267	617.1	550.6	1167.7	0.8134	0.5196	1.3330
650.00	2208.4	0.0268	0.162	696.4	425.0	1121.4	0.8837	0.3830	1.2667
700.00	3094.3	0.0366	0.0752	822.5	172.7	995.2	0.9901	0.1490	1.1390
*705.47	3208.2	0.0508	0.0508	906.0	000.0	906.0	1.0612	0.0000	1.0612

TABLE B–11 (continued)

Pressure psia, p	Temp., °F, T	Specific volume ft^3/lbm		Enthalpy, Btu/lbm			Entropy, Btu/lbm • °R		
		v_f	v_g	h_f	h_{fg}	h_g	s_f	s_{fg}	s_g
†.08865	32.018	0.016022	3302.4	0.003	1075.5	1075.5	0.0000	2.1873	2.1873
0.50	79.59	0.016071	641.5	47.62	1048.6	1096.3	0.0925	1.9446	2.0370
1.0	101.74	0.016136	333.6	69.73	1036.1	1105.8	0.1326	1.8455	1.9781
5.0	162.24	0.0164	73.532	130.2	1000.9	1131.1	0.2349	1.6094	1.8443
10.0	193.21	0.0166	38.420	161.2	982.1	1143.3	0.2836	1.5043	1.7879
14.696	212.00	0.0167	26.799	180.2	970.3	1150.5	0.3121	1.4447	1.7568
15.0	213.03	0.0167	26.290	181.2	969.7	1150.9	0.3137	1.4415	1.7552
20.0	227.96	0.0168	20.087	196.3	960.1	1156.3	0.3358	1.3962	1.7320
30.0	250.34	0.0170	13.744	218.9	945.2	1164.1	0.3682	1.3313	1.6995
40.0	267.25	0.0172	10.497	236.1	933.6	1169.8	0.3921	1.2844	1.6765
50.0	281.02	0.0173	8.515	250.2	923.9	1174.1	0.4112	1.2473	1.6585
60.0	292.71	0.0174	7.174	262.2	915.4	1177.6	0.4273	1.2167	1.6440
70.0	302.93	0.0175	6.205	272.7	907.8	1180.6	0.4411	1.1905	1.6316
80.0	312.04	0.0176	5.471	282.1	900.9	1183.1	0.4534	1.1675	1.6208
90.0	320.28	0.0177	4.895	290.7	894.6	1185.3	0.4643	1.1470	1.6113
100.0	327.82	0.0177	4.431	298.5	888.6	1187.2	0.4743	1.1284	1.6027
120.0	341.27	0.0179	3.728	312.6	877.8	1190.4	0.4919	1.0960	1.5879
140.0	353.04	0.0180	3.219	325.0	868.0	1193.0	0.5071	1.0681	1.5752
160.0	363.55	0.0182	2.834	336.1	859.0	1195.1	0.5206	1.0435	1.5641
180.0	373.08	0.0183	2.531	346.2	850.7	1196.9	0.5328	1.0215	1.5543
200.0	381.80	0.0184	2.287	355.5	842.8	1198.3	0.5438	1.0016	1.5454

TABLE B–11 (continued) Saturated steam table: pressure (English units)

Pressure psia, p	Temp., °F, T	Specific volume ft^3/lbm		Enthalpy, Btu/lbm			Entropy, Btu/lbm · °R		
		v_f	v_g	h_f	h_{fg}	h_g	s_f	s_{fg}	s_g
220.0	389.88	0.0185	2.086	364.2	835.4	1199.6	0.5540	0.9834	1.5374
240.0	397.39	0.0186	1.918	372.3	828.4	1200.6	0.5634	0.9665	1.5299
260.0	404.44	0.0187	1.774	379.9	821.6	1201.5	0.5722	0.9508	1.5230
280.0	411.07	0.0188	1.650	387.1	815.1	1202.3	0.5805	0.9361	1.5166
300.0	417.35	0.0189	1.543	394.0	808.9	1202.9	0.5882	0.9223	1.5105
350.0	431.73	0.0191	1.326	409.8	794.2	1204.0	0.6059	0.8909	1.4968
400.0	444.60	0.0193	1.161	424.2	780.2	1204.6	0.6217	0.8630	1.4847
450.0	456.28	0.0195	1.032	437.3	767.5	1204.8	0.6360	0.8378	1.4738
500.0	467.01	0.0198	0.928	449.5	755.1	1204.7	0.6490	0.8148	1.4639
600.0	486.20	0.0201	0.770	471.7	732.0	1203.7	0.6723	0.7738	1.4461
700.0	503.08	0.0205	0.656	491.6	710.2	1201.8	0.6928	0.7377	1.4304
800.0	518.21	0.0209	0.569	509.8	689.6	1199.4	0.7111	0.7051	1.4163
900.0	531.95	0.0212	0.5009	526.7	669.7	1196.4	0.7279	0.6753	1.4032
1000.0	544.58	0.0216	0.4460	542.6	650.4	1192.9	0.7434	0.6476	1.3910
1200.0	567.19	0.0223	0.3625	571.9	613.0	1184.8	0.7714	0.5969	1.3683
1400.0	587.07	0.0231	0.3018	598.9	576.5	1175.3	0.7966	0.5507	1.3474
1600.0	604.87	0.0239	0.2555	624.2	540.3	1164.5	0.8199	0.5076	1.3274
1800.0	621.02	0.0247	0.2186	648.5	503.8	1152.3	0.8417	0.4662	1.3079
2000.0	635.80	0.0257	0.1883	672.1	466.2	1138.3	0.8625	0.4256	1.2881
2500.0	668.11	0.0286	0.1307	731.7	361.6	1093.3	0.9139	0.3206	1.2345
*3208.2	705.47	0.05078	0.05078	906.0	000.0	906.0	1.0612	0.0000	1.0612

TABLE B–12 Steam properties of the superheated states in SI units: v (m^3/kg); h (kJ/kg); s (kJ/kg · K)

Pressure, kPa, p (Sat. temp., °C)		Sat. vapor	Temperature, °C, T 80	160	240	320	400	480	560	640	720	800	880	960
1.0 (6.9)	v	129.9	164.0	201.1	238.3	275.4	312.6	349.8	386.9	424.1				
	h	2513	2651	2803	2958	3117	3280	3448	3619	3796				
	s	8.98	9.406	9.793	10.121	10.408	10.665	10.902	11.122	11.325				
10 (45.8)	v	13.4	16.3	20.0	23.7	27.4	30.2	34.8	38.5	42.2				
	h	2584	2649	2802	2957	3117	3280	3448	3619	3796				
	s	8.149	8.337	8.727	9.056	9.343	9.601	9.838	10.058	10.262				
40 (75.9)	v	3.99	4.09	4.98	5.91	6.84	7.77	8.69	9.61	10.54				
	h	2636	2645	2800	2956	3116	3279	3447	3619	3795				
	s	7.670	7.690	8.086	8.415	8.703	8.962	9.199	9.419	9.622				
100 (99.6)	v	1.69		1.98	2.36	2.73	3.10	3.472	3.84	4.21	4.58	4.95	5.32	5.69
	h	2675		2796	2954	3114	3278	3446	3618	3795	3974	4175	4348	4539
	s	7.360		7.654	7.988	8.281	8.541	8.777	8.995	9.195	9.384	9.563	9.733	9.895
150 (111.4)	v	1.16		1.32	1.57	1.82	2.07	2.31	2.56	2.81	3.06	3.30	3.55	3.79
	h	2693		2793	2952	3112	3277	3445	3617	3794	3973	4157	4347	4539
	s	7.223		7.462	7.799	8.092	8.353	8.589	8.807	9.007	9.197	9.376	9.546	9.708
200 (120.2)	v	0.89		0.984	1.18	1.36	1.55	1.73	1.91	2.10	2.30	2.48	2.66	2.85
	h	2707		2790	2950	3111	3276	3445	3617	3794	3973	4157	4347	4539
	s	7.127		7.324	7.663	7.957	8.219	8.456	8.673	8.873	9.064	9.242	9.414	9.575
300 (133.5)	v	0.61		0.651	0.780	0.906	1.03	1.16	1.28	1.40	1.53	1.65	1.77	1.89
	h	2725		2783	2946	3109	3274	3444	3616	3793	3972	4157	4246	4538
	s	6.992		7.126	7.470	7.766	8.030	8.268	8.486	8.686	8.876	9.056	9.226	9.388
400 (143.6)	v	0.46		0.484	0.583	0.678	0.772	0.866	0.959	1.052	1.145	1.237	1.330	1.422
	h	2738		2776	2941	3106	3273	3443	3615	3792	3971	4156	4346	4538
	s	6.897		6.980	7.332	7.631	7.895	8.134	8.352	8.553	8.774	8.923	9.093	9.255
500 (151.8)	v	0.37		0.384	0.464	0.541	0.617	0.692	0.767	0.841	0.916	0.990	1.062	1.138
	h	2749		2767	2937	3104	3272	3441	3614	3791	3971	4156	4346	4537
	s	6.822		6.864	7.224	7.525	7.791	8.030	8.249	8.450	8.640	8.819	8.990	9.152

TABLE B–12 (continued) Steam properties of the superheated states in SI units: v (m^3/kg); h (kJ/kg); s (kJ/kg · K)

Pressure, mPa, p (Sat. temp., °C)		Sat. vapor	Temperature, °C, T											
			80	**160**	**240**	**320**	**400**	**480**	**560**	**640**	**720**	**800**	**880**	**960**
1.0 (179.9)	v	0.195			0.227	0.268	0.307	0.345	0.382	0.420	0.457	0.494	0.531	0.569
	h	2778			2918	3091	3263	3435	3609	3787	3968	4154	4344	4536
	s	6.587			6.877	7.189	7.461	7.703	7.924	8.127	8.318	8.498	8.669	8.831
2.0 (212.4)	v	0.0996			0.108	0.131	0.151	0.171	0.190	0.209	0.228	0.247	0.265	0.284
	h	2799			2875	3065	3246	3423	3600	3780	3963	4150	4340	4533
	s	6.340			6.491	6.837	7.122	7.371	7.596	7.802	7.994	8.174	8.346	8.509
3.0 (233.8)	v	0.0667			0.0683	0.085	0.099	0.113	0.126	0.139	0.152	0.164	0.177	0.188
	h	2804			2823	3038	3229	3411	3592	3773	3957	4145	4337	4531
	s	6.186			6.225	6.615	6.916	7.172	7.400	7.608	7.803	7.984	8.157	8.320
4.0 (250.3)	v	0.498				0.062	0.073	0.084	0.094	0.104	0.113	0.123	0.132	0.142
	h	2801				3010	3210	3399	3583	3766	3952	4141	4334	4528
	s	6.070				6.446	6.762	7.028	7.259	7.470	7.666	7.848	8.023	8.187
5.0 (263.9)	v	0.0394				0.048	0.058	0.067	0.075	0.083	0.090	0.098	0.106	0.113
	h	2794				2980	3193	3386	3574	3759	3946	4136	4329	4524
	s	5.973				6.304	6.640	6.912	7.148	7.362	7.558	7.742	7.916	8.080
6.0 (275.6)	v	0.0324				0.039	0.047	0.055	0.062	0.069	0.075	0.082	0.088	0.094
	h	2785				2948	3174	3373	3564	3751	3940	4132	4326	4521
	s	5.890				6.177	6.535	6.815	7.056	7.271	7.471	7.655	7.830	7.994
7.0 (285.8)	v	0.0274				0.032	0.040	0.047	0.053	0.059	0.064	0.070	0.075	0.081
	h	2772				2913	3155	3360	3554	3744	3935	4127	4322	4518
	s	5.814				6.058	6.442	6.731	6.976	7.194	7.395	7.580	7.755	7.919
8.0 (295.0)	v	0.0235				0.027	0.034	0.040	0.046	0.051	0.056	0.061	0.066	0.071
	h	2758				2874	3135	3347	3544	3736	3929	4122	4318	4515
	s	5.745				5.943	6.358	6.657	6.905	7.126	7.329	7.515	7.690	7.856
9.0 (304.1)	v	0.0205				0.023	0.030	0.036	0.040	0.046	0.050	0.054	0.058	0.063
	h	2743				2829	3114	3334	3534	3728	3922	4117	4314	4512
	s	5.678				5.827	6.280	6.589	6.843	7.092	7.270	7.459	7.635	7.801

Source: Data abstracted from *ASME steam tables*, copyright 1967, and reproduced with permission of the American Society of Mechanical Engineers.

TABLE B–12 (continued)

Pressure, MPa, p (Sat. temp., °C)		Sat. vapor	Temperature, °C, T											
			80	160	240	320	400	480	560	640	720	800	880	960
10	v	0.0180				0.0193	0.0265	0.0316	0.0362	0.0404	0.0446	0.0486	0.0525	0.0564
(311.0)	h	2725				2778	3093	3320	3524	3719	3915	4111	4309	4507
	s	5.515				5.705	6.207	6.527	6.786	7.011	7.217	7.406	7.585	7.752
11	v	0.0160				0.0163	0.0236	0.0284	0.0327	0.0366	0.0404	0.0441	0.0477	0.0512
(318.0)	h	2705				2719	3071	3305	3513	3711	3909	4103	4305	4504
	s	5.553				5.579	6.138	6.469	6.733	6.961	7.168	7.357	7.536	7.704
12	v	0.0143					0.0211	0.0258	0.0298	0.0334	0.0369	0.0403	0.0436	0.0469
(324.6)	h	2685					3049	3291	3503	3703	3903	4102	4301	4500
	s	5.492					6.071	6.415	6.684	6.915	7.123	7.314	7.493	7.661
13	v	0.0128					0.0191	0.0235	0.0273	0.0307	0.0340	0.0371	0.0402	0.0432
(330.8)	h	2662					3026	3277	3493	3696	3897	4097	4297	4497
	s	5.431					6.006	6.364	6.638	6.872	7.082	7.274	7.454	7.623
14	v	0.0115					0.0173	0.0216	0.0252	0.0284	0.0307	0.0344	0.0373	0.0401
(336.6)	h	2638					3000	3262	3482	3688	3841	4092	4293	4493
	s	5.372					5.942	6.314	6.594	6.832	6.992	7.238	7.418	7.587
15	v	0.0104					0.0157	0.0199	0.0233	0.0264	0.0286	0.0320	0.0348	0.0374
(342.1)	h	2611					2973	3248	3472	3680	3855	4087	4289	4490
	s	5.310					5.878	6.268	6.554	6.794	6.956	7.204	7.385	7.554
16	v	0.0093					0.0143	0.0184	0.0217	0.0247	0.0267	0.0300	0.0326	0.0351
(347.3)	h	2582					2945	3233	3461	3672	3829	4082	4284	4486
	s	5.247					5.816	6.223	6.515	6.758	6.922	7.171	7.353	7.523
17	v	0.0084					0.0131	0.0171	0.0203	0.0231	0.0251	0.0282	0.0306	0.0330
(352.3)	h	2548					2915	3218	3450	3664	3873	4077	4280	4482
	s	5.177					5.753	6.179	6.477	6.725	6.942	7.140	7.323	7.493
18	v	0.0075					0.0119	0.0160	0.0190	0.0217	0.0242	0.0266	0.0289	0.0311
(357.0)	h	2510					2884	3206	3440	3656	3867	4072	4276	4479
	s	5.107					5.688	6.137	6.441	6.691	6.911	7.110	7.294	7.465

TABLE B–12 (continued) Steam properties of the superheated states in SI units: v (m^3/kg); h (kJ/kg); s (kJ/kg · K)

Pressure, MPa, p (Sat. temp., °C)		Sat. vapor	Temperature, °C, T											
			80	160	240	320	400	480	560	640	720	800	880	960
19	v	0.0067					0.0109	0.0149	0.0179	0.0205	0.0229	0.0251	0.0273	0.0295
(361.4)	h	2466					2851	3188	3429	3648	3861	4068	4272	4476
	s	5.027					5.622	6.095	6.407	6.660	6.881	7.082	7.266	7.437
20	v	0.0059					0.0100	0.0140	0.0169	0.0194	0.0217	0.0238	0.0259	0.0280
(365.7)	h	2410					2816	3170	3418	3640	3855	4063	4268	4473
	s	4.928					5.553	6.055	6.374	6.631	6.853	7.056	7.240	7.412
21	v	0.0050					0.0091	0.0132	0.0160	0.0184	0.0206	0.0216	0.0246	0.0266
(369.8)	h	2336					2778	3152	3407	3632	3849	4058	4264	4470
	s	4.803					5.481	6.015	6.342	6.603	6.826	7.031	7.215	7.388
22	v	0.0037					0.0083	0.0124	0.0151	0.0175	0.0196	0.0216	0.0235	0.0254
(373.7)	h	2168					2736	3135	3396	3624	3843	4053	4260	4466
	s	4.591					5.406	5.975	6.312	6.576	6.801	7.007	7.192	7.365
22.13	v	0.00326					0.0082	0.0123	0.0150	0.0174	0.0195	0.0215	0.0234	0.0253
(374.15)	h	2100					2730	3133	3395	3623	3842	4052	4259	4466
	s	4.430					5.395	5.970	6.308	6.572	6.798	7.004	7.189	7.362
23	v						0.0075	0.0117	0.0143	0.0166	0.0187	0.0206	0.0225	0.0243
	h						2690	3117	3385	3616	3837	4048	4256	4463
	s						5.324	5.936	6.283	6.549	6.776	6.983	7.170	7.343
25	v						0.0060	0.0104	0.0130	0.0152	0.0171	0.0189	0.0206	0.0223
	h						2579	3080	3362	3600	3824	4038	4248	4457
	s						5.137	5.860	6.225	6.498	6.729	6.938	7.126	7.300

TABLE B–12 (continued)

Pressure, psia, p (Sat. temp., °F)		Sat. vapor	Temperature, °F, T										
			200	300	400	500	600	700	800	1000	1200	1400	1500
1	v	333.60	392.50	452.3	511.90	571.5	631.1	690.7					
(101.74)	h	1105.8	1150.2	1195.7	1241.8	1288.6	1336.1	1384.5					
	s	1.9781	2.0209	2.1152	2.1722	2.2237	2.2708	2.3144					
4	v	90.63	97.79	112.86	127.85	142.79	157.71	172.62	187.53	217.33	247.13	276.92	291.82
(152.97)	h	1127.3	1149.0	1195.0	1241.4	1288.3	1335.9	1384.3	1433.6	1534.8	1639.6	1748.0	1803.5
	s	1.8625	1.8967	1.9617	2.0190	2.0707	2.1178	2.1615	2.2022	2.2767	2.3440	2.4057	2.4347
10	v	38.42	38.84	44.98	51.03	57.04	63.03	69.00	74.98	86.91	98.84	110.76	116.72
(193.21)	h	1143.3	1146.6	1193.7	1240.6	1287.8	1335.5	1384.0	1433.4	1534.6	1639.5	1747.9	1803.4
	s	1.7879	1.7928	1.8593	1.9173	1.9692	2.0166	2.0603	2.1011	2.1757	2.2430	2.3046	2.3337
15	v	26.290		29.899	33.963	37.985	41.986	45.978	49.964	57.926	65.882	73.833	77.807
(213.03)	h	1150.9		1192.5	1239.9	1287.3	1335.2	1383.8	1433.2	1534.5	1639.4	1747.8	1803.4
	s	1.7552		1.8134	1.8720	1.9242	1.9717	2.0155	2.0563	2.1309	2.1982	2.2599	2.2890
20	v	20.087		22.356	25.428	28.457	31.466	34.465	37.458	43.435	49.405	55.370	58.352
(227.96)	h	1156.3		1191.4	1239.2	1286.9	1334.9	1383.5	1432.9	1534.3	1639.3	1747.8	1803.3
	s	1.7320		1.7805	1.8397	1.8921	1.9397	1.9836	2.0244	2.0991	2.1665	2.2282	2.2572
30	v	13.744		14.810	16.892	18.929	20.945	22.951	24.952	28.943	32.927	36.907	38.896
(250.34)	h	1164.1		1189.0	1237.8	1286.0	1334.2	1383.0	1432.5	1534.0	1639.0	1747.6	1803.2
	s	1.6995		1.7334	1.7937	1.8467	1.8946	1.9386	1.9795	2.0543	2.1217	2.1834	2.2125
50	v	8.515		8.769	10.062	11.306	12.529	13.741	14.947	17.350	19.746	22.137	23.332
(281.02)	h	1174.1		1184.1	1234.9	1284.1	1332.9	1382.0	1431.7	1533.4	1638.6	1747.3	1802.9
	s	1.6585		1.6720	1.7349	1.7890	1.8374	1.8816	1.9227	1.9977	2.0652	2.1270	2.1561
75	v	5.816			6.645	7.494	8.320	9.135	9.945	11.553	13.155	14.752	15.550
(307.60)	h	1181.9			1231.2	1281.7	1331.3	1380.7	1430.7	1532.7	1638.1	1746.9	1802.6
	s	1.6259			1.6868	1.7424	1.7915	1.8361	1.8774	1.9526	2.0202	2.0821	2.1113

TABLE B–12 (continued) Steam properties of the superheated states in English units: v (ft^3/lbm); h (Btu/lbm); s (Btu/lbm · °R)

Pressure, psia, p (Sat. temp., °F)		Sat. vapor	Temperature, °F, T										
			200	300	400	500	600	700	800	1000	1200	1400	1500
100	v	4.431			4.935	5.588	6.216	6.833	7.443	8.655	9.860	11.060	11.659
(327.82)	h	1187.2			1227.4	1279.3	1329.6	1379.5	1429.7	1532.0	1637.6	1746.5	1802.2
	s	1.6027			1.6516	1.7088	1.7586	1.8036	1.8451	1.9205	1.9883	2.0502	2.0794
150	v	3.014			3.2208	3.6799	4.1112	4.5298	4.9421	5.7568	6.5642	7.3671	7.7674
(358.43)	h	1194.1			1219.1	1274.3	1326.1	1376.9	1427.6	1530.5	1636.5	1745.7	1801.7
	s	1.5695			1.5993	1.6602	1.7115	1.7573	1.7992	1.8751	1.9431	2.0052	2.0344
200	v	2.287			2.3598	2.7247	3.0583	3.3783	3.6915	4.3077	4.9165	5.5209	5.8219
(381.80)	h	1198.3			1210.1	1269.0	1322.6	1374.3	1425.5	1529.1	1635.4	1745.0	1800.9
	s	1.5454			1.5593	1.6242	1.6773	1.7239	1.7663	1.8426	1.9109	1.9732	2.0025
250	v	1.843				2.1504	2.4262	2.6872	2.9410	3.4382	3.9278	4.4131	4.6546
(400.97)	h	1201.1				1263.5	1319.0	1371.6	1423.4	1527.6	1634.4	1744.2	1800.2
	s	1.5264				1.5951	1.6502	1.6976	1.7405	1.8173	1.8858	1.9482	1.9776
300	v	1.543				1.7665	2.0044	2.2263	2.4407	2.8585	3.2688	3.6746	3.8764
(417.35)	h	1202.9				1257.7	1315.2	1368.9	1421.3	1526.2	1633.3	1743.4	1799.6
	s	1.5105				1.5703	1.6274	1.6758	1.7192	1.7964	1.8652	1.9278	1.9572
350	v	1.326				1.4913	1.7028	1.8970	2.0832	2.4445	2.7980	3.1471	3.3205
(431.73)	h	1204.0				1251.5	1311.4	1366.2	1419.2	1524.7	1632.3	1742.6	1798.9
	s	1.4968				1.5483	1.6077	1.6571	1.7009	1.7787	1.8477	1.9105	1.9400
400	v	1.161				1.2841	1.4763	1.6499	1.8151	2.1339	2.4450	2.7515	2.9037
(444.60)	h	1204.6				1245.1	1307.4	1363.4	1417.0	1523.3	1631.2	1741.9	1798.2
	s	1.4847				1.5282	1.5901	1.6406	1.6580	1.7632	1.8325	1.8955	1.9250
500	v	0.928				0.9919	1.1584	1.3037	1.4397	1.6992	1.9507	2.1977	2.3200
(467.01)	h	1204.7				1231.2	1299.1	1357.7	1412.7	1520.3	1629.1	1740.3	1796.9
	s	1.4639				1.4921	1.5652	1.6176	1.6578	1.7371	1.8069	1.8702	1.8998
600	v	0.770				0.7944	0.9456	1.0726	1.1892	1.4093	1.6211	1.8284	1.9309
(486.20)	h	1203.7				1215.9	1290.3	1351.8	1408.3	1517.4	1627.0	1738.8	1795.6
	s	1.4461				1.4590	1.5329	1.5884	1.6351	1.7155	1.7859	1.8494	1.8792

TABLE B–12 (continued)

Pressure, psia, p (Sat. temp., °F)		Sat. vapor	Temperature, °F, T										
			200	300	400	500	600	700	800	1000	1200	1400	1500
700	v	0.656					0.7928	0.9072	1.0102	1.2023	1.3858	1.5647	1.6530
(503.08)	h	1201.8					1281.0	1345.6	1403.7	1514.4	1624.8	1737.2	1794.3
	s	1.4304					1.5090	1.5673	1.6154	1.6970	1.7679	1.8318	1.8617
800	v	0.569					0.6774	0.7828	0.8759	1.0470	1.2093	1.3669	1.4446
(518.21)	h	1199.4					1271.1	1339.3	1399.1	1511.4	1622.7	1735.7	1792.9
	s	1.4163					1.4869	1.5484	1.5980	1.6807	1.7522	1.8164	1.8464
900	v	0.501					0.5869	0.6858	0.7713	0.9262	1.0720	1.2131	1.2825
(531.95)	h	1196.4					1260.6	1332.7	1394.4	1508.5	1620.6	1734.1	1791.6
	s	1.4032					1.4659	1.5311	1.5822	1.6662	1.7382	1.8028	1.8329
1000	v	0.446					0.5137	0.6080	0.6875	0.8295	0.9622	1.0901	1.1529
(544.58)	h	1192.9					1249.3	1325.9	1389.6	1505.4	1618.4	1732.5	1790.3
	s	1.3910					1.4457	1.5149	1.5677	1.6530	1.7256	1.7905	1.8207
1200	v	0.362					0.4016	0.4905	0.5615	0.6845	0.7974	0.9055	0.9584
(567.19)	h	1184.8					1224.2	1311.5	1379.7	1499.4	1614.2	1729.4	1787.6
	s	1.3683					1.4061	1.4851	1.5415	1.6298	1.7035	1.7691	1.7996
1400	v	0.302					0.3176	0.4059	0.4712	0.5809	0.6798	0.7737	0.8195
(587.07)	h	1175.3					1194.1	1296.1	1369.3	1493.2	1609.9	1726.3	1785.0
	s	1.3474					1.3652	1.4575	1.5182	1.6096	1.6845	1.7508	1.7815
1600	v	0.255						0.3415	0.4032	0.5031	0.5915	0.6748	0.7153
(604.87)	h	1164.5						1279.4	1358.5	1486.9	1605.6	1723.2	1782.3
	s	1.3274						1.4312	1.4968	1.5916	1.6678	1.7347	1.7657
1800	v	0.219						0.2906	0.3500	0.4426	0.5229	0.5980	0.6343
(621.02)	h	1152.3						1261.1	1347.2	1480.6	1601.2	1720.1	1779.7
	s	1.3079						1.4054	1.4768	1.5753	1.6528	1.7204	1.7516

TABLE B–12 (continued) Steam properties of the superheated states in English units: v (ft^3/lbm); h (Btu/lbm); s (Btu/lbm · °R)

Pressure, psia, p (Sat. temp., °F)		Sat. vapor	Temperature, °F, T										
			200	300	400	500	600	700	800	1000	1200	1400	1500
2000	v	0.188						0.2488	0.3072	0.3942	0.4680	0.5365	0.5695
(635.80)	h	1138.3						1240.9	1335.4	1474.1	1596.9	1717.0	1771.1
	s	1.2881						1.3794	1.4578	1.5603	1.6391	1.7075	1.7389
2500	v	0.131						0.1681	0.2293	0.3068	0.3692	0.4259	0.4529
(668.11)	h	1093.3						1176.7	1303.4	1457.5	1585.9	1709.2	1770.4
	s	1.2345						1.3076	1.4129	1.5269	1.6094	1.6796	1.7116
3000	v	0.085						0.0982	0.1759	0.2484	0.3033	0.3522	0.3753
(695.33)	h	1020.3						1060.5	1267.0	1440.2	1574.8	1701.4	1763.8
	s	1.1619						1.1966	1.3692	1.4976	1.5841	1.6561	1.6888
3200	v	0.0566							0.1588	0.2301	0.2827	0.3291	0.3510
(705.08)	h	931.6							1250.9	1433.1	1570.3	1698.3	1761.2
	s	1.0832							1.3515	1.4866	1.5749	1.6477	1.6806

TABLE B–13 Isentropic work of compression (ideal pump work) for water at saturated liquid, $(h - h_f\ \text{kJ/kg})_f$ (SI units)

Pressure, *p*		Saturation temperature, °C						
bars	**MPa**	**0**	**30**	**90**	**150**	**210**	**270**	**330**
20	2.0	1.9	1.9	2.1	1.9	0.2		
30	3.0	2.6	2.8	2.8	2.8	1.9		
40	4.0	3.9	4.2	4.4	4.2	2.3		
80	8.0	7.4	7.7	7.7	7.9	6.0	2.8	
100	10.0	9.8	10.2	10.2	10.5	9.3	6.5	
120	12.0	11.9	12.1	12.3	12.1	11.6	9.1	
140	14.0	14.2	14.0	14.2	14.4	14.2	11.6	3.0
160	16.0	15.8	15.8	16.1	16.8	16.5	13.7	5.4
180	18.0	17.7	17.7	18.1	18.8	18.6	16.3	8.4
200	20.0	20.0	20.0	20.5	21.2	20.9	19.5	11.9
220	22.0	22.1	22.1	22.6	24.2	24.2	22.1	14.2
250	25.0	25.1	25.1	25.4	26.3	27.0	25.6	18.6

Source: Data abstracted from *ASME steam tables*, copyright 1967, and reproduced with permission of the American Society of Mechanical Engineers.

TABLE B–13 (continued) Steam table: compressed liquid (English units)

Pressure, psia, p		Temperature, °F, T						
		32	100	200	300	400	500	600
200	$(v - v_f) \times 10^5$ ft^3/lbm	−1.2	−1.0	−0.7	−1.0			
	$(h - h_f)$ Btu/lbm	+0.61	+0.52	+0.42	+0.26			
	$(s - s_f) \times 10^3$ Btu/lbm · °F	+0.0	−0.1	−0.2	−0.3			
400	$(v - v_f) \times 10^5$ ft^3/lbm	−2.2	−1.9	−1.7	−2.0	−2.0		
	$(h - h_f)$ Btu/lbm	+1.21	+1.05	+0.88	+0.63	+0.17		
	$(s - s_f) \times 10^3$ Btu/lbm · °F	+0.0	−0.2	−0.5	−0.6	−0.4		
600	$(v - v_f) \times 10^5$ ft^3/lbm	−3.2	−3.0	−3.7	−4.0	−4.0		
	$(h - h_f)$ Btu/lbm	+1.82	+0.58	+1.33	+1.00	+0.39		
	$(s - s_f) \times 10^3$ Btu/lbm · °F	+0.0	−0.3	−0.7	−1.0	−1.0		
1000	$(v - v_f) \times 10^5$ ft^3/lbm	−5.2	−5.0	−5.7	−7.0	−9.0	−7.0	
	$(h - h_f)$ Btu/lbm	+3.02	+1.37	+2.24	+1.74	+0.86	−0.11	
	$(s - s_f) \times 10^3$ Btu/lbm · °F	+0.1	−0.6	−1.2	−1.7	−2.0	−1.4	
2000	$(v - v_f) \times 10^5$ ft^3/lbm	−11.2	−10.0	−10.7	−14.0	−20.0	−29.0	−32.0
	$(h - h_f)$ Btu/lbm	+6.01	+5.26	+4.51	+3.62	+2.09	−0.37	−2.62
	$(s - s_f) \times 10^3$ Btu/lbm · °F	+0.2	−1.2	−2.4	−3.5	−4.6	−5.6	−4.3
3000	$(v - v_f) \times 10^5$ ft^3/lbm	−16.2	−14.0	−15.7	−21.0	−31.0	−48.0	−88.0
	$(h - h_f)$ Btu/lbm	+8.97	+7.88	+6.79	+5.52	+3.37	−0.38	−7.02
	$(s - s_f) \times 10^3$ Btu/lbm · °F	+0.2	−1.8	−3.6	−5.2	−7.0	−9.4	−12.5

Source: Data abstracted from *ASME steam tables*, copyright 1967, and reproduced with permission of the American Society of Mechanical Engineers.

TABLE B–14 Mercury vapor properties (SI units)*

Pressure, kPa, p	Temp., °C, T	Specific volume, m^3/kg, v_g	Enthalpy, kJ/kg		Entropy, kJ/kg·K	
			h_f	h_g	s_f	s_g
3	208.0	6.75	32.49	330.3	0.0884	0.708
6	231.6	3.48	36.17	331.9	0.0958	0.682
9	245.7	2.45	38.36	332.6	1.1002	0.667
12	257.1	1.83	40.15	333.3	0.1035	0.656
15	265.8	1.49	41.52	334.0	0.1061	0.649
20	278.0	1.13	43.43	334.7	0.1096	0.638
30	295.8	0.77	46.19	335.9	0.1146	0.624
40	309.1	0.588	48.29	336.8	0.1182	0.613
50	320.0	0.478	48.96	337.3	0.1211	0.606
60	329.1	0.403	49.75	337.5	0.1235	0.600
80	343.2	0.318	53.61	338.9	0.1271	0.590
100	355.9	0.254	55.57	339.6	0.1303	0.582
120	366.0	0.216	57.15	340.3	0.1327	0.576
140	375.3	0.184	58.62	341.0	0.1350	0.570
100	383.2	0.164	59.85	341.5	0.1369	0.566
200	397.1	0.133	62.01	342.2	0.1401	0.559
250	411.7	0.108	64.29	343.3	0.1435	0.551
300	423.9	0.091	66.20	344.0	0.1463	0.545
350	434.6	0.079	67.87	344.7	0.1487	0.540
400	444.1	0.070	69.34	345.4	0.1507	0.535
500	460.7	0.057	71.92	346.3	0.1543	0.530
600	474.8	0.048	74.13	347.3	0.1573	0.523
700	487.2	0.042	76.06	348.0	0.1599	0.517
800	498.3	0.037	77.71	348.9	0.1621	0.514
900	508.3	0.0336	79.20	349.6	0.1642	0.510
1000	517.4	0.0304	80.83	350.1	0.1660	0.506
1100	526.3	0.0278	82.18	350.8	0.1677	0.503
1200	534.3	0.0257	83.43	351.2	0.1693	0.500

* h_f and s_f are measured from 32°F or 0°C.

TABLE B–14 (continued) Mercury vapor properties (English units)

Pressure, psia, p	Temp., °F, T	Specific volume, ft^3/lbm, v_g	Enthalpy, Btu/lbm			Entropy, Btu/lbm·°F		
			h_f	h_{fg}	h_g	s_f	s_{fg}	s_g
0.4	402.3	114.5	13.81	128.1	141.9	0.02094	0.1486	0.1696
0.6	426.1	78.23	14.70	127.6	142.3	0.02195	0.1441	0.1660
0.8	443.8	59.71	15.36	127.2	142.6	0.02269	0.1408	0.1635
1.0	458.1	48.45	15.89	126.9	142.8	0.02328	0.1382	0.1615
1.5	485.1	33.14	16.90	126.3	143.2	0.02436	0.1337	0.1580
2	505.2	25.31	17.65	125.8	143.5	0.02514	0.1304	0.1556
3	535.4	17.34	18.78	125.2	144.0	0.02629	0.1258	0.1521
4	558.0	13.26	19.62	124.7	144.3	0.02714	0.1225	0.1497
5	576.2	10.77	20.30	124.3	144.6	0.02780	0.1200	0.1478
6	591.4	9.096	20.87	123.9	144.8	0.02834	0.1179	0.1462
7	605.0	7.882	21.37	123.6	145.0	0.02882	0.1161	0.1450
8	616.8	6.963	21.81	123.4	145.2	0.02923	1.1146	0.1439
9	627.5	6.244	22.21	132.2	145.4	0.02960	0.1133	0.1429
10	637.3	5.661	22.58	122.9	145.5	0.02993	0.1121	0.1420
15	676.5	3.892	24.04	122.1	146.1	0.03124	0.1074	0.1387
20	706.2	2.983	25.15	121.4	146.6	0.03220	0.1041	0.1363
25	730.4	2.429	26.05	120.9	146.9	0.03297	0.1016	0.1345
30	750.9	2.053	26.81	120.4	147.2	0.03360	0.09953	0.1331
35	769.0	1.781	27.49	120.0	147.5	0.03416	0.09774	0.1319
40	784.8	1.576	28.08	119.7	147.8	0.03464	0.09621	0.1308
45	799.3	1.414	28.62	119.4	148.0	0.03507	0.09486	0.1299
50	812.5	1.284	29.11	119.1	148.2	0.03546	0.09364	0.1291
60	836.1	1.086	29.99	118.6	148.6	0.03614	0.09154	0.1276
70	856.6	0.9436	30.75	118.1	148.9	0.03672	0.08976	0.1264
80	874.8	0.8349	31.43	117.7	149.1	0.03725	0.08824	0.1254
90	891.6	0.7497	32.06	117.3	149.4	0.03771	0.8687	0.1245
100	906.9	0.6811	32.63	117.0	149.6	0.03813	0.08565	0.1237
120	934.4	0.5767	33.60	116.4	150.1	0.03887	0.08353	0.1224
140	958.3	0.5012	34.55	115.9	150.4	0.03951	0.08175	0.1212
160	979.9	0.4438	35.35	115.4	150.8	0.04007	0.08019	0.1202
180	999.6	0.3990	36.09	115.0	151.1	0.04058	0.07881	0.1193

Source: T. Baumeister and L. S. Marks, eds., *Standard handbook for mechanical engineers*, 7th ed. (New York: McGraw-Hill Book Company), copyright 1967; with permission of McGraw-Hill Book Company, Conversion to SI by author.

TABLE B–15 Dichlorodifluoromethane, CCl_2F_2 (R-12), properties of the saturation states (SI units)

Temperature, K	Pressure, MPa	Specific volume, m^3/kg		Enthalpy, kJ/kg		Entropy, kJ/kg · K	
		v_f	v_g	h_f	h_g	s_f	s_g
170	0.000867	0.000593	13.460	328.51	523.56	3.7732	4.9205
180	0.00218	0.000602	5.6757	336.96	527.97	3.8215	4.8826
190	0.00488	0.000612	2.6673	345.40	532.48	3.8671	4.8517
200	0.00995	0.000622	1.3713	353.87	537.07	3.9105	4.8265
210	0.01877	0.000633	0.75991	362.37	541.72	3.9520	4.8060
220	0.03311	0.000644	0.44844	370.94	546.40	3.9918	4.7893
230	0.05519	0.000656	0.27905	379.59	551.09	4.0302	4.7758
240	0.08761	0.000668	0.18161	388.36	555.77	4.0674	4.7649
243.36	0.10133	0.000673	0.15861	391.33	557.33	4.0797	4.7618
250	0.13334	0.000682	0.12278	397.25	560.42	4.1036	4.7562
260	0.19566	0.000696	0.08574	406.31	565.01	4.1389	4.7493
270	0.27811	0.000711	0.06154	415.53	569.52	4.1735	4.7438
280	0.38448	0.000728	0.04520	424.95	573.91	4.2075	4.7395
290	0.51870	0.000746	0.03386	434.58	578.16	4.2409	4.7360
300	0.68491	0.000767	0.02578	444.43	582.21	4.2739	4.7332
310	0.88742	0.000790	0.01988	454.53	586.01	4.3065	4.7306
320	1.1308	0.000815	0.01549	464.91	589.49	4.3388	4.7281
330	1.4198	0.000846	0.01214	475.61	592.54	4.3710	4.7253
340	1.7600	0.000883	0.00955	486.71	595.02	4.4033	4.7218
360	2.6188	0.000989	0.00582	510.84	597.20	4.4699	4.7098
380	3.7764	0.001232	0.00305	542.34	588.78	4.5515	4.6737
384.95	4.125	0.001792	0.001792	566.9	566.9	4.614	4.614

TABLE B–15 (continued) Dichlorodifluoromethane, CCl_2F_2 (R-12), properties of the saturation states (English units)

Temperature, F	Pressure, psia	Specific volume, ft³/lbm		Enthalpy, Btu/lbm		Entropy, Btu/lbm · R	
		v_f	v_g	h_f	h_g	s_f	s_g
−140	0.25623	0.009658	110.46	−20.652	61.896	−0.056123	0.20208
−120	0.64190	0.009816	46.741	−16.565	64.052	−0.043723	0.19359
−100	1.4280	0.009985	22.164	−12.466	66.248	−0.032005	0.18683
−80	2.8807	0.010164	11.533	−8.3451	68.467	−0.020862	0.18143
−60	5.3575	0.010357	6.4774	−4.1919	70.693	−0.010214	0.17714
−40	9.3076	0.010564	3.8750	0.0000	72.913	0.000000	0.17373
−20	15.267	0.010788	2.4429	4.2357	75.110	0.009831	0.17102
−10	19.189	0.010906	1.9727	6.3716	76.196	0.014617	0.16989
0	23.849	0.011030	1.6089	8.5207	77.271	0.019323	0.16888
10	29.335	0.001116	1.3241	10.684	78.335	0.023954	0.16798
20	35.736	0.011296	1.0988	12.863	79.385	0.028515	0.16719
30	43.148	0.011438	0.91880	15.058	80.419	0.033013	0.16648
40	51.667	0.011588	0.77357	17.273	81.436	0.037453	0.16586
50	61.394	0.011746	0.65537	19.507	82.433	0.041839	0.16530
60	72.433	0.011913	0.55839	21.766	83.409	0.046180	0.16479
70	84.888	0.012089	0.47818	24.050	84.359	0.050482	0.16434
80	98.870	0.012277	0.41135	26.365	85.282	0.054751	0.16392
100	131.86	0.012693	0.30794	31.100	87.029	0.063227	0.16315
120	172.35	0.013174	0.23326	36.013	88.610	0.071680	0.16241
140	221.32	0.013746	0.17799	41.162	89.967	0.080205	0.16159
160	279.82	0.014449	0.13604	46.633	91.006	0.088927	0.16053
180	349.00	0.015361	0.10330	52.562	91.561	0.098039	0.15900
200	430.09	0.016659	0.076728	59.203	91.278	0.10789	0.15651
220	524.43	0.018986	0.053140	67.246	89.036	0.11943	0.15149
223.6	596.9	0.02870	0.02870	78.86	78.86	0.1359	0.1359

Source: Data abstracted and adapted from "Thermodynamic Properties of Freon-22," copyright 1955 and 1956, with permission of E. I. du Pont de Nemours & Co.

TABLE B–16 Dichlorodifluoromethane, CCl_2F_2 (R-12), properties of **TABLE B–16** Dichlorodifluoromethane, CCl_2F_2 (R-12), properties of the superheated states (SI units): v (m^3/kg); h (kJ/kg · K); s (kJ/kg · K)

Pressure, MPa, p (Sat. temp., K)		Saturated vapor properties	Temperature, °K, T										
			240	260	280	300	320	340	360	380	400	420	440
0.01	v	1.3664	1.6438	1.7825	1.9209	2.0591	2.1971	2.3351	2.4729	2.6108	2.7485	2.8863	3.0240
(200.08)	h	537.10	557.71	568.73	580.17	592.02	604.25	616.83	629.73	642.94	656.42	670.18	684.18
	s	4.8264	4.9201	4.9642	5.0066	5.0475	5.0869	5.1250	5.1619	5.1976	5.2322	5.2657	5.2983
0.04	v	0.39813	0.40596	0.44162	0.47689	0.51195	0.54687	0.58163	0.61633	0.65100	0.68564	0.72020	0.75477
(223.58)	h	548.07	556.98	568.16	579.72	591.65	603.93	616.55	629.49	642.72	656.23	670.00	684.01
	s	4.7842	4.8226	4.8673	4.9101	4.9513	4.9909	5.0292	5.0661	5.1019	5.1365	5.1701	5.2027
0.1013	v	0.15861		0.17099	0.18551	0.19978	0.21388	0.22787	0.24178	0.25564	0.26945	0.28324	0.29700
(243.36)	h	557.33		566.96	578.77	590.87	603.28	615.99	629.00	642.28	655.83	669.63	683.67
	s	4.7618		4.8001	4.8438	4.8856	4.9256	4.9641	5.0013	5.0372	5.0719	5.1056	5.1383
0.2	v	0.08407			0.09164	0.09926	0.10669	0.11400	0.12123	0.12839	0.13552	0.14261	0.14968
(260.60)	h	565.28			577.17	589.58	602.20	615.07	628.19	641.57	655.19	669.04	683.13
	s	4.7489			4.7930	4.8358	4.8769	4.9155	4.9530	4.9891	5.0241	5.0579	5.0906
0.4	v	0.04355				0.04752	0.05158	0.05548	0.05929	0.06304	0.06673	0.07039	0.07402
(281.28)	h	574.46				586.78	599.90	613.13	626.51	640.08	653.85	667.83	682.01
	s	4.7391				4.7814	4.8237	4.8638	4.9021	4.9387	4.9741	5.0082	5.0411
0.6	v	0.02939				0.03014	0.03312	0.03593	0.03862	0.04123	0.04379	0.04631	0.04880
(295.15)	h	580.27				583.66	597.42	611.06	624.74	638.53	652.47	666.58	680.87
	s	4.7345				4.7459	4.7903	4.8316	4.8707	4.9080	4.9438	4.9782	5.0114
0.8	v	0.02208					0.02382	0.02610	0.02825	0.03031	0.03231	0.03426	0.03618
(305.91)	h	584.50					594.70	608.86	622.89	636.94	651.06	665.31	679.71
	s	4.7316					4.7643	4.8072	4.8473	4.8853	4.9215	4.9563	4.9897
1.0	v	0.01760					0.01816	0.02016	0.02201	0.02374	0.02538	0.02703	0.02861
(314.85)	h	587.73					591.68	606.50	620.95	635.27	649.61	664.01	678.53
	s	4.7294					4.7418	4.7863	4.8280	4.8663	4.9035	4.9387	4.9725
1.5	v	0.01147						0.01208	0.01359	0.01493	0.01618	0.01736	0.01850
(332.50)	h	593.19						599.56	615.52	630.79	645.76	660.63	675.50
	s	4.7245						4.7434	4.7890	4.8303	4.8687	4.9050	4.9396

TABLE B–16 (continued) Dichlorodifluoromethane, CCl_2F_2 (R-12), properties of the superheated states (SI units): v (m^3/kg); h (kJ/kg · K); s (kJ/kg · K)

Pressure, MPa, p (Sat. temp., K)		Saturated vapor properties	Temperature, °K, T										
			240	**260**	**280**	**300**	**320**	**340**	**360**	**380**	**400**	**420**	**440**
2.0	v	0.00823							0.00924	0.01046	0.01152	0.01250	0.01343
	h	596.15							608.93	625.69	641.55	657.02	672.32
(346.22)	s	4.7190							4.7551	4.8005	4.8412	4.8789	4.9145
2.5	v	0.00621							0.00640	0.00769	0.00869	0.00957	0.01038
	h	597.17							600.02	619.69	636.89	653.15	668.98
(357.56)	s	4.7118							4.7196	4.7729	4.8170	4.8567	4.8935
3.0	v	0.00509								0.00573	0.00676	0.00759	0.00833
	h	596.26								612.17	631.60	648.95	665.45
(367.27)	s	4.7017								4.7433	4.7942	4.8365	4.8749
4.0	v	0.00224									0.00418	0.00505	0.00575
	h	574.75									617.77	639.26	657.74
(383.23)	s	4.6354									4.7462	4.7987	4.8416
5.0	v										0.00221	0.00346	0.00417
	h										590.94	626.91	648.92
	s										4.6711	4.7592	4.8105
6.0	v											0.00233	0.00309
	h											610.66	638.80
	s											4.7137	4.7793

TABLE B–16 (continued) Dichlorodifluoromethane, CCl_2F_2 (R-12), properties of the superheated states (English units): v (ft^3/lbm); h (Btu/lbm); s (Btu/lbm · °R)

Pressure, psia, p (Sat. temp., °F)		Saturated vapor properties	Temperature, °F, T										
			−30	0	30	60	90	120	150	180	210	240	270
3.0	v	11.207	12.594	13.500	14.403	15.302	16.198	17.093	17.987	18.879	19.771		
	h	68.589	74.754	78.714	82.803	87.015	91.345	95.785	100.33	104.98	109.72		
(−78.76)	s	0.18118	0.19632	0.20523	0.21384	0.22219	0.23029	0.23815	0.24580	0.25324	0.26048		
6.0	v	5.9284	6.2379	6.7012	7.1600	7.6155	8.0685	8.5198	8.9696	9.4184	9.8644		
	h	71.099	74.512	78.515	82.638	86.876	91.226	95.683	100.24	104.90	109.65		
(−56.08)	s	0.17648	0.18453	0.19353	0.20222	0.21062	0.21875	0.22665	0.23431	0.24177	0.24902		
12.0	v	3.0583		3.3006	3.5381	3.7720	4.0034	4.2329	4.4610	4.6880	4.9142	5.1397	
	h	74.015		78.110	82.302	86.594	90.986	95.477	100.06	104.74	109.51	114.36	
(−30.00)	s	0.17230		0.18150	0.19034	0.19884	0.20706	0.21501	0.22272	0.23021	0.23749	0.24458	
25.0	v	1.5430			1.6531	1.7723	1.8888	2.0032	2.1161	2.2279	2.3388	2.4491	
	h	77.504			81.547	85.965	90.455	95.021	99.667	104.39	109.20	114.058	
(2.35)	s	0.16868			0.17715	0.18591	0.19431	0.20240	0.21021	0.21778	0.22512	0.23225	
30.0	v	1.2969			1.3625	1.4644	1.5633	1.6600	1.7553	1.8494	1.9426	2.0351	2.1271
	h	78.452			81.245	85.716	90.246	94.843	99.513	104.26	109.08	113.97	118.94
(11.11)	s	0.16789			0.17371	0.18257	0.19104	0.19918	0.20704	0.21463	0.22200	0.22915	0.23609
35.0	v	1.2111			1.1548	1.2442	1.3306	1.4148	1.4975	1.5789	1.6595	1.7394	1.8187
	h	79.272			80.937	85.463	90.034	94.663	99.357	104.12	108.96	113.87	118.84
(18.92)	s	0.16727			0.17071	0.17968	0.18823	0.19643	0.20432	0.21195	0.21934	0.22651	0.23347
40.0	v	0.98748			0.99865	1.0789	1.1560	1.2309	1.3041	1.3761	1.4472	1.5176	1.5874
	h	79.999			80.622	85.206	89.819	94.480	99.200	103.99	108.84	113.76	118.74
(25.93)	s	0.16677			0.16804	0.17712	0.18575	0.19401	0.20195	0.20961	0.21702	0.22420	0.23118
50.0	v	0.79830				0.84713	0.91134	0.97313	1.0332	1.0920	1.1499	1.2070	1.2636
	h	81.249				84.676	89.380	94.110	98.882	103.71	108.59	113.54	118.55
(38.15)	s	0.16597				0.17271	0.18151	0.18988	0.19791	0.20563	0.21310	0.22032	0.22733
60.0	v	0.67020				0.69210	0.74970	0.80110	0.85247	0.90252	0.95157	0.99988	1.0476
	h	82.299				84.126	88.929	93.731	98.558	103.43	108.35	113.32	118.35
(48.64)	s	0.16537				0.16892	0.17791	0.18641	0.19453	0.20233	0.20984	0.21710	0.22414

TABLE B–16 (continued) Dichlorodifluoromethane, CCl_2F_2 (R-12), properties of the superheated states (English units): v (ft^3/lbm); h (Btu/lbm); s (Btu/lbm · °R)

Pressure, psia, p (Sat. temp., °F)		Saturated vapor properties	Temperature, °F, T										
			−30	0	30	60	90	120	150	180	210	240	270
80.0	v	0.50684					0.54281	0.58556	0.62623	0.66543	0.70356	0.74090	0.77762
	h	84.002					87.981	92.945	97.891	102.85	107.84	112.87	117.95
(66.21)	s	0.16450					0.17190	0.18070	0.18902	0.19696	0.20458	0.21193	0.21903
100.0	v	0.40681					0.41876	0.45562	0.49009	0.52291	0.55457	0.58538	0.61553
	h	83.350					86.964	92.116	97.197	102.26	107.32	112.42	117.54
(80.76)	s	0.16389					0.16685	0.17597	0.18452	0.19262	0.20036	0.20780	0.21497
120.0	v	0.33891						0.36841	0.39896	0.42766	0.45508	0.48158	0.50739
	h	86.459						91.237	96.471	101.64	106.79	111.95	117.13
(93.29)	s	0.16341						0.17184	0.18065	0.18892	0.19679	0.20432	0.21157
140.0	v	0.28966						0.30549	0.33350	0.35939	0.38387	0.40734	0.43008
	h	87.388						90.297	95.709	101.00	106.25	111.47	116.70
(104.35)	s	0.16300						0.16808	0.17718	0.18566	0.19367	0.20130	0.20862
160.0	v	0.25224						0.25764	0.28404	0.30797	0.33032	0.35157	0.37203
	h	88.179						89.283	94.906	100.34	105.68	110.98	116.27
(114.30)	s	0.16264						0.16454	0.17400	0.18270	0.19086	0.19860	0.20600
200.0	v	0.19934							0.21370	0.23535	0.25496	0.27323	0.29060
	h	89.429							93.141	98.921	104.49	109.96	115.38
(131.74)	s	0.16195							0.16801	0.17737	0.18588	0.19387	0.20145
300.0	v	0.12649								0.13482	0.15249	0.16761	0.18129
	h	91.168								94.556	101.08	107.14	112.97
(166.18)	s	0.16008								0.16537	0.17534	0.18419	0.19235

Source: Data abstracted and adapted from "Thermodynamic Properties of Freon-12," copyright 1955 and 1956, with permission of E. I. du Pont de Nemours & Co.

TABLE B–17 Monochlordifluoromethane, $CHClF_2$ (R-22), properties of saturation states (SI units)

Temp, °C	Pressure, MPa	Specific volume, 10^3 m³/kg		Enthalpy, kJ/kg		Entropy, kJ/kg · K	
		v_f	v_g	h_f	h_g	s_f	s_g
−50	0.06439	0.69526	324.557	144.959	383.921	0.77919	1.85000
−45	0.08271	0.70219	256.990	150.153	386.282	0.80216	1.83708
−40	0.10495	0.70936	205.745	155.414	388.609	0.82490	1.82504
−35	0.13168	0.71680	166.400	160.747	390.896	0.84743	1.81380
−30	0.16348	0.72452	135.844	166.140	393.138	0.86976	1.80329
−25	0.20098	0.73255	111.859	171.606	395.330	0.89190	1.79342
−20	0.24483	0.74091	92.8432	177.142	397.467	0.91386	1.78415
−15	0.29570	0.74964	77.6254	182.749	399.544	0.93564	1.77540
−10	0.35430	0.75876	65.3399	188.426	401.555	0.95725	1.76713
−5	0.42135	0.76831	55.3394	194.176	403.496	0.97870	1.75928
0	0.49757	0.77834	47.1354	200.000	405.361	1.00000	1.75179
5	0.58378	0.78889	40.3556	205.899	407.143	1.02116	1.74463
10	0.68070	0.80002	34.7136	211.877	408.835	1.04218	1.73775
15	0.78915	0.81180	29.9874	217.937	410.430	1.06309	1.73109
20	0.90993	0.82431	26.0032	224.084	411.918	1.08390	1.72462
25	1.0439	0.83765	22.6242	230.324	413.289	1.10462	1.71827
30	1.1919	0.85193	19.7417	236.664	414.530	1.12530	1.71200
35	1.3548	0.86729	17.2686	243.114	415.627	1.14594	1.70576
40	1.5335	0.88392	15.1351	249.686	416.561	1.16659	1.69946
45	1.7290	0.90203	13.2841	256.396	417.308	1.18730	1.69305
50	1.9423	0.92193	11.6693	263.264	417.839	1.20811	1.68643
55	2.1744	0.94400	10.2521	270.318	418.116	1.22910	1.67208
60	2.4266	0.96878	9.00062	277.594	418.089	1.25038	1.67208
65	2.6999	0.99702	7.88749	285.142	417.687	1.27206	1.66402
70	2.9959	1.02987	6.88899	293.038	416.809	1.29436	1.65504
75	3.3161	1.06916	5.98334	301.399	415.299		

Source: Data abstracted from "Thermodynamic Properties of Freon-22," copyright 1964, with permission of E. I. du Pont de Nemours & Co.

TABLE B–17 (continued) Monochlordifluoromethane, $CHClF_2$ (R-22), properties of saturation states (English units)

Temp, °F	Pressure, psia	Volume, ft^3/lbm Liquid, v_f	Volume, ft^3/lbm Vapor, v_g	Enthalpy, Btu/lbm Liquid, h_f	Enthalpy, Btu/lbm Latent, h_{fg}	Enthalpy, Btu/lbm Vapor, h_g	Entropy, Btu/(lbm)·°R Liquid, s_f	Entropy, Btu/(lbm)·°R Vapor, s_g
−100	2.3983	0.010664	18.433	−14.564	107.935	93.371	−0.03734	0.26274
−90	3.4229	0.010771	13.235	−12.216	106.759	94.544	−0.03091	0.25787
−80	4.7822	0.010881	9.6949	−9.838	105.548	95.710	−0.02457	0.25342
−70	6.5522	0.010995	7.2318	−7.429	104.297	96.868	−0.01832	0.24932
−60	8.818	0.011113	5.4844	−4.987	103.001	98.014	−0.01214	0.24556
−50	11.674	0.011235	4.2224	−2.511	101.656	99.144	−0.00604	0.24209
−40	15.222	0.011363	3.2957	0.000	100.257	100.257	0.00000	0.23888
−30	19.573	0.011495	2.6049	2.547	98.801	101.348	0.00598	0.23591
−25	22.086	0.011564	2.3260	3.834	98.051	101.885	0.00894	0.23451
−20	24.845	0.011634	2.0926	5.131	97.285	102.415	0.01189	0.23315
−15	27.865	0.011705	1.6895	6.436	96.502	102.939	0.01483	0.23184
−10	31.162	0.011778	1.6825	7.751	95.704	103.455	0.01776	0.23058
−5	34.754	0.011853	1.5177	9.075	94.889	103.964	0.02067	0.22936
0	38.657	0.011930	1.3723	10.409	94.056	104.465	0.02357	0.22817
5	42.888	0.012008	1.2434	11.752	93.206	104.958	0.02645	0.22703
10	47.464	0.012088	1.1290	13.104	92.338	105.442	0.02932	0.25592
15	52.405	0.012171	1.0272	14.466	91.451	105.917	0.03218	0.22484
20	57.727	0.012255	0.93631	15.837	90.545	106.383	0.03503	0.22379
25	63.450	0.012342	0.85500	17.219	89.620	106.839	0.03787	0.22277
30	69.591	0.012431	0.78208	18.609	88.674	107.284	0.04070	0.22178
40	83.206	0.012618	0.65753	21.422	86.720	108.142	0.04632	0.21986
50	98.727	0.012815	0.55606	24.275	84.678	108.935	0.05190	0.21803
60	116.31	0.013025	0.47272	27.172	82.540	109.712	0.05745	0.21627
70	136.12	0.013251	0.40373	30.116	80.298	110.414	0.06296	0.21456
80	158.33	0.013492	0.34621	33.109	77.943	111.052	0.06846	0.21288
90	183.09	0.013754	0.20789	36.158	75.461	111.619	0.07394	0.21122
100	210.60	0.014038	0.25702	39.267	72.838	112.105	0.07942	0.20956
120	274.60	0.014694	0.19238	45.705	67.077	112.782	0.09042	0.20613
140	351.94	0.015518	0.14418	52.528	60.403	112.931	0.10163	0.20235
160	444.53	0.016627	0.10701	59.948	52.316	112.263	0.11334	0.19776
180	554.78	0.018332	0.07679	68.498	41.570	110.068	0.12635	0.19133
200	686.35	0.022436	0.047438	80.862	21.990	102.853	0.14460	0.17794
204.81	721.91	0.030525	0.030525	91.329	0.000	91.329	0.16016	0.16016

TABLE B–18 Monochlorodifluoromethane, $CHClF_2$ (R-22), properties of superheated states (SI units): v (10^3 m^3/kg); h (kJ/kg); s (kJ/kg · K)

Pressure, MPa, p (Sat. temp. °C)		Sat. properties	Temperature, °C, T										
			−80	−65	−50	−35	−20	−5	10	25	40	55	70
0.0049	v	3580	3778	4075	4371	4667	4692	5257	5552				
	h	364.4	369.5	377.5	385.7	394.2	403.0	412.0	421.3				
(−90)	s	1.998	2.026	2.065	2.103	2.140	2.176	2.211	2.244				
0.0091	v	2018	2039	2201	2361	2522	2682	2842	3002	3162	3322		
	h	368.3	369.4	377.3	385.6	394.1	402.9	411.9	421.3	430.8	440.7		
(−82)	s	1.961	1.966	2.006	2.044	2.081	2.117	2.151	2.185	2.218	2.250		
0.0159	v	1199		1255	1348	1440	1532	1624	1716	1807	1900		
	h	372.3		377.1	385.4	393.9	402.7	411.8	421.2	430.7	440.6		
(−74)	s	1.927		1.951	1.990	2.027	2.063	2.097	2.131	2.164	2.197		
0.0205	v	940.9		965.0	1036.9	1108.4	1179.7	1250.8	1321.8	1392.6	1463.3		
	h	374.2		376.9	385.2	393.8	402.6	411.7	421.1	430.7	440.5		
(−20)	s	1.912		1.926	1.964	2.001	2.037	2.072	2.106	2.139	2.171		
0.0263	v	746.3		750.1	806.5	862.6	918.4	974.0	1029.5	1084.9	1140.1	1195.3	
	h	376.2		376.7	385.1	393.7	402.5	411.6	421.0	430.6	440.5	450.6	
(−66)	s	1.898		1.901	1.940	1.977	2.013	2.048	2.082	2.115	2.147	2.179	
0.0334	v	598.0			633.8	678.3	722.5	766.5	810.3	854.1	897.8	941.3	
	h	378.1			384.9	393.5	402.4	411.5	420.9	430.5	440.4	450.5	
(−62)	s	1.885			1.916	1.954	1.990	2.025	2.059	2.092	2.124	2.156	
0.0420	v	483.6			502.8	538.4	573.9	609.1	644.1	679.1	713.9	748.7	
	h	380.1			384.6	393.3	402.2	411.3	420.7	430.4	440.3	450.4	
(−58)	s	1.873			1.893	1.931	1.967	2.002	2.036	2.070	2.102	2.134	
0.0522	v	394.6			402.3	431.3	460.0	488.4	516.7	544.9	573.0	601.1	
	h	382.0			384.3	393.0	401.9	411.3	420.6	430.2	440.1	450.3	
(−53)	s	1.861			1.871	1.909	1.950	1.981	2.015	2.048	2.081	2.112	
0.0644	v	324.6			324.6	348.3	371.7	395.0	418.1	441.1	463.9	486.7	509.5
	h	383.9			383.9	392.7	401.7	410.9	420.4	430.0	440.0	450.1	460.5
(−50)	s	1.850			1.850	1.888	1.925	1.960	1.994	2.028	2.060	2.092	2.123

TABLE B–18 (continued) Monochlorodifluoromethane, $CHClF_2$ (R-22), properties of superheated states (SI units): v (10^3 m^3/kg); h (kJ/kg); s (kJ/kg · K)

Pressure, MPa, p (Sat. temp. °C)		Sat. properties	Temperature, °C, T										
			−50	−35	−20	−5	10	25	40	55	70	85	100
0.0788	v	269.0		283.4	302.8	322.0	341.0	359.9	378.7	397.4	416.0		
	h	385.8		392.3	401.4	410.6	420.1	429.8	439.8	450.0	460.4		
(−46)	s	1.840		1.868	1.904	1.940	1.974	2.008	2.040	2.072	2.103		
0.096	v	224.6		232.2	248.4	264.4	280.2	295.9	311.4	326.9	342.4		
	h	387.7		391.9	401.0	410.3	419.8	429.6	439.6	449.8	460.2		
(−39)	s	1.830		1.848	1.885	1.920	1.955	1.989	2.021	2.053	2.084		
0.110	v	197.0		200.9	215.1	229.1	242.9	256.6	270.2	283.8	297.2		
	h	389.0		391.5	400.7	410.0	419.6	429.4	439.4	449.6	460.1		
(−39)	s	1.823		1.833	1.870	1.906	1.841	1.975	2.007	2.039	2.070		
0.132	v	166.4		166.4	178.5	190.3	202.0	213.5	225.0	236.3	247.6	258.9	
	h	390.9		390.9	400.2	409.6	419.2	429.1	439.1	449.4	459.8	470.5	
(−35)	s	1.814		1.814	1.852	1.888	1.923	1.957	1.989	2.021	2.053	2.083	
0.157	v	141.4			148.9	159.0	169.0	178.8	188.4	198.1	207.6	217.1	
	h	392.7			399.6	409.1	418.8	428.7	438.8	449.1	459.6	470.3	
(−31)	s	1.805			1.833	1.870	1.905	1.939	1.972	2.004	2.035	2.066	
0.185	v	120.8			124.9	133.6	142.1	150.5	158.8	167.0	175.1	183.2	
	h	394.5			398.9	408.5	418.3	428.4	438.4	448.7	459.3	470.1	
(−27)	s	1.797			1.815	1.852	1.888	1.922	1.955	1.987	2.019	2.049	
0.218	v	103.7			105.3	112.8	120.2	127.4	134.6	141.6	148.6	155.5	
	h	396.2			398.1	407.9	417.7	427.8	438.0	448.4	459.0	469.8	
(−23)	s	1.790			1.797	1.835	1.871	1.905	1.938	1.971	2.002	2.033	
0.254	v	89.5				95.70	102.13	108.41	114.59	120.68	126.70	132.67	138.60
	h	397.9				407.1	417.1	427.2	437.5	447.9	458.6	469.4	480.4
(−19)	s	1.782				1.818	1.854	1.889	1.922	1.955	1.987	2.018	2.048
0.296	v	77.6				81.50	87.16	92.66	98.05	103.35	108.59	113.8	118.9
	h	399.5				406.2	416.4	426.6	436.9	447.5	458.1	469.0	480.1
(−15)	s	1.775				1.801	1.838	1.873	1.907	1.939	1.971	2.002	2.033

Source: Data abstracted from "Thermodynamic Properties of Freon-22," copyright 1964; used with permission of E. I. du Pont de Nemours & Co.

TABLE B–18 (continued)

Pressure, MPa, p (Sat. temp. °C)		Sat. properties	Temperature, °C, T										
			−5	10	25	40	55	70	85	100	115	130	145
0.393 (−7)	v	59.1	59.71	64.20	68.51	71.71	76.81	80.84	84.82	88.74	92.64		
	h	402.7	404.1	414.6	425.0	435.6	446.3	457.1	468.1	479.3	490.6		
	s	1.762	1.786	1.805	1.842	1.876	1.909	1.942	1.973	2.004	2.033		
0.451 (−3)	v	51.9		55.36	59.23	62.97	66.61	70.18	73.69	77.16	80.59		
	h	404.3		413.5	424.1	434.8	445.6	456.5	467.6	478.8	490.2		
	s	1.756		1.790	1.826	1.861	1.895	1.927	1.959	1.990	2.019		
0.514 (1)	v	45.7		47.86	51.35	54.71	57.97	61.15	64.27	67.35	70.38		
	h	405.7		412.2	423.1	433.9	444.8	455.8	466.9	478.2	489.7		
	s	1.750		1.774	1.811	1.846	1.880	1.913	1.945	1.976	2.006		
0.584 (5)	v	40.4		41.46	44.64	47.68	50.61	53.46	56.25	59.00	61.70		
	h	407.1		410.9	421.9	432.9	443.9	455.0	466.3	477.6	489.9		
	s	1.745		1.758	1.796	1.832	1.866	1.899	1.931	1.963	1.993		
0.660 (9)	v	35.8		35.96	38.89	41.66	44.31	46.89	49.40	51.85	54.27	56.66	
	h	408.5		409.3	420.6	431.8	443.0	454.2	465.5	477.0	488.5	500.2	
	s	1.739		1.742	1.781	1.817	1.852	1.886	1.918	1.949	1.980	2.009	
0.744 (13)	v	31.8			33.93	36.48	38.90	41.23	43.50	45.72	47.90	50.04	
	h	409.8			419.1	430.5	441.9	453.2	464.7	476.2	487.9	499.6	
	s	1.734			1.766	1.803	1.838	1.872	1.905	1.936	1.967	1.997	
0.836 (17)	v	28.3			29.64	32.00	34.22	36.35	38.42	40.43	42.00	44.33	
	h	411.0			417.4	429.1	440.7	452.2	463.8	475.4	487.1	499.0	
	s	1.728			1.750	1.788	1.825	1.859	1.892	1.924	1.954	1.984	
0.936 (21)	v	25.3			25.90	28.10	30.16	32.12	34.01	35.84	37.63	39.38	
	h	412.2			415.5	427.5	439.3	51.0	462.7	474.5	486.3	498.2	
	s	1.723			1.734	1.774	1.811	1.846	1.879	1.911	1.942	1.972	
1.044 (25)	v	22.6			22.6	24.7	26.6	28.4	30.2	31.8	33.48	35.08	36.64
	h	413.3			413.3	425.7	437.8	449.7	461.6	473.5	485.4	497.4	509.6
	s	1.718			1.718	1.759	1.797	1.832	1.866	1.899	1.930	1.960	1.990

TABLE B–18 (continued) Monochlorodifluoromethane, $CHClF_2$ (R-22), properties of superheated states (SI units): v (10^3 m^3/kg); h (kJ/kg); s (kJ/kg · K)

Pressure, MPa, p (Sat. temp. °C)		Sat. properties	Temperature, °C, T										
			55	70	85	100	115	130	145	160	175	190	205
1.288 (33)	v	18.2	20.80	22.38	23.88	25.31	26.69	27.8	29.34				
	h	415.2	434.3	446.7	459.0	471.2	483.4	495.6	507.9				
	s	1.708	1.768	1.806	1.841	1.847	1.906	1.937	1.967				
1.571 (41)	v	14.7	16.24	17.67	18.99	20.24	21.43	22.58	23.69	24.78			
	h	416.7	429.7	443.0	455.8	468.4	480.9	493.4	506.0	518.5			
	s	1.698	1.739	1.778	1.815	1.849	1.882	1.914	1.944	1.974			
1.898 (49)	v	11.97	12.58	13.93	15.14	16.26	17.30	18.31	19.27	20.21			
	h	417.8	423.9	438.3	451.9	465.1	478.0	490.8	503.6	516.4			
	s	1.688	1.707	1.750	1.788	1.824	1.858	1.891	1.922	1.952			
2.273 (57)	v	9.733		10.902	12.050	13.075	14.021	14.911	15.761	16.679	17.372		
	h	418.1		432.2	447.0	461.0	474.5	487.8	500.9	514.0	527.1		
	s	1.677		1.719	1.761	1.799	1.834	1.868	1.900	1.931	1.960		
2.700 (65)	v	7.887		8.351	9.515	10.494	11.370	12.179	12.940	13.667	14.366		
	h	417.7		424.0	440.8	455.9	470.2	484.1	497.7	511.1	524.5		
	s	1.664		1.683	1.731	1.772	1.810	1.845	1.878	1.909	1.939		
3.185 (73)	v	6.336			7.367	8.357	9.197	9.950	10.646	11.302	11.927	12.529	
	h	416.0			432.5	449.5	465.0	479.6	493.8	507.7	521.4	535.0	
	s	1.649			1.696	1.743	1.783	1.820	1.855	1.887	1.919	1.948	
3.735 (81)	v	4.988			5.413	6.538	7.383	8.106	8.756	9.358	9.926	10.47	
	h	412.3			419.8	441.1	458.5	474.3	489.2	503.7	517.8	531.8	
	s	1.630			1.651	1.709	1.755	1.795	1.831	1.865	1.897	1.928	
4.037 (85)	v	4.358			4.358	5.707	6.580	7.298	7.933	8.514	9.058	9.573	10.068
	h	409.1			409.1	435.7	454.5	471.1	486.6	501.4	515.8	530.0	544.1
	s	1.617			1.617	1.690	1.739	1.781	1.819	1.854	1.886	1.918	1.947

TABLE B–18 (continued) Monochlorodifluoromethane, $CHClF_2$ (R-22), properties of superheated states (English units): v (ft^3/lbm); h (Btu/lbm); s (Btu/lbm · °R) (saturation properties in parentheses)

Absolute pressure, psi

Temp., °F	5 — 19.7411* (−78.62°F) v	h	s
	(9.3011)	(95.871)	(0.25283)
−80			
−70	9.5237	97.018	0.25581
−60	9.7810	98.362	0.25921
−50	10.038	99.721	0.26257
−40	10.293	101.094	0.26588
−30	10.549	102.482	0.26915
−20	10.803	103.885	0.27238
−10	11.058	105.302	0.27557
0	11.311	106.735	0.27872
10	11.565	108.182	0.28183
20	11.818	109.643	0.28491
30	12.070	111.120	0.28796
40	12.323	112.611	0.29097
50	12.575	114.117	0.29396
60	12.826	115.638	0.29691
70	13.078	117.174	0.29984
80	13.329	118.724	0.30274
90	13.580	120.288	0.30561
100	13.831	121.867	0.30845
110	14.082	123.461	0.31128
120	14.333	125.069	0.31407
130	14.583	126.691	0.31685
140	14.833	128.327	0.31960
150	15.084	129.977	0.32233
160	15.334	131.642	0.32504
170	15.584	133.320	0.32772
180	15.834	135.012	0.33039
190	16.083	136.718	0.33304
200	16.333	138.437	0.33566
210	16.583	140.170	0.33827
220	16.832	141.916	0.34086
230	17.082	143.676	0.34343
240			

Temp., °F	10 — 9.561* (−55.59°F) v	h	s
	(4.8778)	(98.515)	(0.24339)
−50	4.9518	99.291	0.24590
−40	5.0838	100.690	0.24927
−30	5.2152	102.101	0.25260
−20	5.3460	103.526	0.25588
−10	5.4762	104.963	0.25911
0	5.6060	106.414	0.26230
10	5.7353	107.878	0.26545
20	5.8643	109.356	0.26856
30	5.9929	110.847	0.27164
40	6.1212	112.353	0.27468
50	6.2492	113.872	0.27769
60	6.3769	115.404	0.28067
70	6.5044	116.951	0.28362
80	6.6316	118.512	0.28654
90	6.7586	120.086	0.28943
100	6.8855	121.674	0.29229
110	7.0122	123.276	0.29513
120	7.1387	124.892	0.29794
130	7.2651	126.552	0.30073
140	7.3913	128.165	0.30349
150	7.5174	129.822	0.30623
160	7.6434	131.493	0.30895
170	7.7693	133.177	0.31165
180	7.8951	134.875	0.31432
190	8.0208	136.586	0.31697
200	8.1464	138.310	0.31961
210	8.2719	140.048	0.32222
220	8.3974	141.799	0.32482
230	8.5228	143.562	0.32739
240	8.6481	145.339	0.32995
250	8.7734	147.129	0.33249

Temp., °F	15 — 0.304 (−40.57°F) v	h	s
	(3.3412)	(100.194)	(0.23906)
−40	3.3463	100.276	0.23925
−30	3.4365	101.712	0.24263
−20	3.5261	103.159	0.24596
−10	3.6152	104.618	0.24924
0	3.7037	106.088	0.25248
10	3.7918	107.570	0.25567
20	3.8794	109.065	0.25882
30	3.9667	110.571	0.26192
40	4.0537	112.0911	0.26500
50	4.1404	113.623	0.26803
60	4.2268	115.168	0.27103
70	4.3129	116.727	0.27400
80	4.3989	118.298	0.27694
90	4.4846	119.882	0.27985
100	4.5701	121.480	0.28273
110	4.6554	123.091	0.28559
120	4.7406	124.715	0.28841
130	4.8256	126.352	0.29121
140	4.9105	128.003	0.29399
150	4.9952	129.667	0.29674
160	5.0799	131.344	0.29947
170	5.1644	133.034	0.30217
180	5.2488	134.737	0.30486
190	5.3332	136.454	0.30752
200	5.4174	138.183	0.31016
210	5.5016	139.925	0.31278
220	5.5857	141.680	0.31538
230	5.6697	143.448	0.31797
240	5.7537	145.229	0.32053
250	5.8376	147.022	0.32307
260	5.9125	148.828	0.32560
270			

*in Hg

TABLE B–18 (continued) Monochlorodifluoromethane, $CHClF_2$ (R-22), properties of superheated states (English units): v (ft^3/lbm); h (Btu/lbm); s (Btu/lbm · °R) (saturation properties in parentheses)

Temp.,	20		
	5.304		
	(−29.12°F)		
°F	*v*	*h*	*s*
	(2.5527)	(101.444)	(0.23566)
−30			
−20	2.6156	102.785	0.23874
−10	2.6841	104.266	0.24207
0	2.7521	105.756	0.24535
10	2.8196	107.257	0.24858
20	2.8867	108.769	0.25177
30	2.9534	110.292	0.25491
40	3.0198	111.826	0.25801
50	3.0858	113.372	0.26107
60	3.1516	114.930	0.26410
70	3.2171	116.500	0.26709
80	3.2823	118.082	0.27005
90	3.3474	119.667	0.27298
100	3.4122	121.284	0.27588
110	3.4769	122.904	0.27874
120	3.5414	124.536	0.28159
130	3.6058	126.182	0.28440
140	3.6700	127.840	0.28719
150	3.7341	129.510	0.28995
160	3.7981	131.194	0.29269
170	3.8619	132.890	0.29540
180	3.9257	134.599	0.29810
190	3.9894	136.321	0.30077
200	4.0529	138.055	0.30342
210	4.1164	139.802	0.30865
220	4.1799	141.562	0.30865
230	4.2432	143.334	0.31124
240	4.3065	145.119	0.31381
250	4.3697	146.916	0.31636
260	4.4329	148.725	0.31889
270	4.4960	150.547	0.32141
280	4.5591	152.381	0.32390

Temp.,	25		
	10.304		
	(−19.73°F)		
°F	*v*	*h*	*s*
	(2.0704)	(102.444)	(0.23308)
−20			
−10	2.1251	103.907	0.23637
0	2.1808	105.419	0.23970
10	2.2360	106.939	0.24297
20	2.2908	108.469	0.24619
30	2.3452	110.008	0.24937
40	2.3992	111.558	0.25250
50	2.4529	113.118	0.25559
60	2.5063	114.689	0.25864
70	2.5594	116.271	0.26166
80	2.6123	117.865	0.26464
90	2.6650	119.470	0.26758
100	2.7175	121.087	0.27050
110	2.7698	122.716	0.27338
120	2.8219	124.357	0.27624
130	2.8738	126.010	0.27907
140	2.9257	127.675	0.28187
150	2.9774	129.533	0.28464
160	3.0289	131.043	0.28739
170	3.0804	132.745	0.29012
180	3.1318	134.460	0.29282
190	3.1830	136.187	0.29550
200	3.2342	137.926	0.29815
210	3.2853	139.678	0.30079
220	3.3363	141.443	0.30340
230	3.3873	143.219	0.30600
240	3.4382	145.008	0.30857
250	3.4890	146.809	0.31113
260	3.5397	148.622	0.31367
270	3.5905	150.447	0.31619
280	3.6411	152.284	0.31869
290	3.6917	154.134	0.32117

Temp.,	30		
	15.304		
	(−11.71°F)		
°F	*v*	*h*	*s*
	(1.7439)	(103.279)	(0.23101)
−10	1.7521	103.541	0.23159
0	1.7997	105.076	0.23497
10	1.8467	106.616	0.23828
20	1.8933	108.165	0.24154
30	1.9395	109.721	0.24475
40	1.9853	111.286	0.24792
50	2.0308	112.861	0.25104
60	2.0760	114.445	0.25412
70	2.1209	116.040	0.25176
80	2.1655	117.645	0.26016
90	2.2100	119.261	0.26312
100	2.2542	120.888	0.26606
110	2.2982	122.526	0.26896
120	2.3421	124.176	0.27183
130	2.3858	125.837	0.27467
140	2.4294	127.510	0.27748
150	2.4728	129.194	0.28027
160	2.5162	130.891	0.28303
170	2.5594	132.600	0.28576
180	2.6024	134.320	0.28848
190	2.6455	136.053	0.29116
200	2.6884	137.797	0.29383
210	2.7312	139.554	0.29647
220	2.7739	141.323	0.29909
230	2.8166	143.104	0.30169
240	2.8592	144.897	0.30427
250	2.9018	146.701	0.30684
260	2.9443	148.518	0.30938
270	2.9867	150.347	0.31190
280	3.0291	152.187	0.31441
290	3.0715	154.040	0.31689

TABLE B–18 (continued)

Absolute pressure, psi

Temp., °F	40 25.304 (1.63°F) v	h	s
	(1.3285)	(104.627)	(0.22780)
10	1.3594	105.953	0.23064
20	1.3959	107.541	0.23399
30	1.4319	109.134	0.23728
40	1.4675	110.732	0.24051
50	1.5028	112.337	0.24369
60	1.5378	113.950	0.24682
70	1.5724	115.570	0.24991
80	1.6068	117.199	0.25296
90	1.6410	118.837	0.25596
100	1.6749	120.485	0.25893
110	1.7087	122.142	0.26187
120	1.7423	123.810	0.26477
130	1.7757	125.487	0.26764
140	1.8090	127.176	0.27048
150	1.8421	128.875	0.27329
160	1.8751	130.585	0.27607
170	1.9080	132.306	0.27883
180	1.9407	134.039	0.28156
190	1.9734	135.783	0.28426
200	2.0060	137.538	0.28694
210	2.0385	139.304	0.28960
220	2.0709	141.082	0.29223
230	2.1033	142.872	0.29485
240	2.1356	144.673	0.29744
250	2.1678	146.486	0.30001
260	2.2000	148.310	0.30257
270	2.2321	150.145	0.30510
280	2.2641	151.992	0.30761
290	2.2961	153.851	0.31011
300	2.3281	155.721	0.31259
310	2.3600	157.602	0.31505

Temp., °F	50 35.304 (12.61°F) v	h	s
	(1.0744)	(105.692)	(0.22535)
10	—	—	—
20	1.0968	106.897	0.22788
30	1.1269	108.529	0.23125
40	1.1564	110.163	0.23455
50	1.1857	111.800	0.23780
60	1.2145	113.443	0.24099
70	1.2431	115.091	0.24413
80	1.2714	116.745	0.24722
90	1.2994	118.406	0.25027
100	1.3272	120.075	0.25328
110	1.3548	121.752	0.25625
120	1.3822	123.438	0.25918
130	1.4095	125.133	0.26208
140	1.4366	126.838	0.26495
150	1.4635	128.552	0.26778
160	1.4903	130.276	0.27059
170	1.5170	132.010	0.27337
180	1.5436	133.755	0.27611
190	1.5701	135.510	0.27884
200	1.5965	137.276	0.28153
210	1.6228	139.053	0.28421
220	1.6491	140.840	0.28686
230	1.6752	142.638	0.28948
240	1.7013	144.448	0.29209
250	1.7274	146.268	0.29467
260	1.7533	148.100	0.29723
270	1.7792	149.943	0.29978
280	1.8051	151.797	0.30230
290	1.8309	153.661	0.30480
300	1.8567	155.537	0.30729
310	1.8824	157.424	0.30976
320	1.9081	159.322	0.31221

Temp., °F	60 45.304 (22.03°F) v	h	s
	(0.90222)	(106.569)	(0.22337)
30	0.92300	170.904	0.22612
40	0.94863	109.577	0.22950
50	0.97385	111.249	0.23282
60	0.99871	112.923	0.23607
70	1.0232	114.600	0.23927
80	1.0475	116.281	0.24241
90	1.0715	117.967	0.24551
100	1.0952	119.658	0.24855
110	1.1187	121.356	0.25156
120	1.1420	123.061	0.25453
130	1.1652	124.774	0.25746
140	1.1882	126.495	0.26035
150	1.2111	128.225	0.26321
160	1.2338	129.963	0.26604
170	1.2564	131.711	0.26884
180	1.2788	133.468	0.27161
190	1.3012	138.235	0.27435
200	1.3235	137.012	0.27706
210	1.3457	138.799	0.27975
220	1.3678	140.596	0.28241
230	1.3898	142.403	0.28505
240	1.4118	144.221	0.28767
250	1.4337	146.050	0.29027
260	1.4556	147.889	0.29284
270	1.4773	149.739	0.29539
280	1.4991	151.600	0.29792
290	1.5208	153.471	0.30044
300	1.5424	155.353	0.30293
310	1.5640	157.246	0.30541
320	1.5856	159.149	0.30786
330	1.6071	161.063	0.31030

TABLE B–18 (continued) Monochlorodifluoromethane, $CHClF_2$ (R-22), properties of superheated states (English units): v (ft^3/lbm); h (Btu/lbm); s (Btu/lbm · °R) (saturation properties in parentheses)

Temp.,	70		
	55.304		
	(30.32°F)		
°F	v	h	s
	(0.77766)	(107.312)	(0.22172)
30	—	—	—
40	0.79981	108.972	0.22507
50	0.82224	110.682	0.22846
60	0.84428	112.391	0.23178
70	0.86598	114.098	0.23503
80	0.88736	115.807	0.23823
90	0.90846	117.519	0.24137
100	0.92932	119.234	0.24446
110	0.94995	120.953	0.24751
120	0.97038	122.679	0.25051
130	0.99063	124.410	0.25347
140	1.0107	126.148	0.25639
150	1.0306	127.893	0.25928
160	1.0504	129.647	0.26213
170	1.0701	131.408	0.26495
180	1.0896	133.178	0.26774
190	1.1091	134.957	0.27050
200	1.1284	136.745	0.27323
210	1.1477	138.543	0.27594
220	1.1669	140.350	0.27862
230	1.1860	142.166	0.28127
240	1.2050	143.993	0.28390
250	1.2239	145.830	0.28650
260	1.2428	147.677	0.28909
270	1.2617	149.534	0.29165
280	1.2805	151.401	0.29419
290	1.2992	153.279	0.29672
300	1.3179	155.167	0.29922
310	1.3365	157.066	0.30170
320	1.3551	158.975	0.30416
330	1.3737	160.894	0.30661
340	1.3923	162.824	0.30904

Temp.,	80		
	65.304		
	(37.76°F)		
°F	v	h	s
	(0.68318)	(107.954)	(0.22029)
40	0.68782	108.347	0.22107
50	0.70822	110.098	0.22454
60	0.72820	111.843	0.22793
70	0.74780	113.584	0.23125
80	0.76708	115.323	0.23450
90	0.78605	117.061	0.23770
100	0.80477	118.801	0.24083
110	0.82325	120.544	0.24392
120	0.84152	122.290	0.24696
130	0.85960	124.040	0.24995
140	0.87751	125.796	0.25290
150	0.89256	127.558	0.25582
160	0.91286	129.326	0.25869
170	0.93034	131.102	0.26154
180	0.94770	132.885	0.26435
190	0.96495	134.677	0.26712
200	0.98209	134.476	0.26987
210	0.99915	132.284	0.27259
220	1.0161	140.101	0.27529
230	1.0330	141.928	0.27795
240	1.0498	143.763	0.28060
250	1.0666	145.608	0.28322
260	1.0833	147.463	0.28581
270	1.0999	149.328	0.28838
280	1.1165	151.202	0.29094
290	1.1330	153.087	0.29347
300	1.1495	154.981	0.29598
310	1.1659	156.885	0.29847
320	1.1823	158.800	0.30094
330	1.1987	160.725	0.30339
340	1.2150	162.660	0.30583

Temp.,	90		
	73.304		
	(44.53°F)		
°F	v	h	s
	(0.60897)	(108.516)	(0.21903)
50	0.61924	109.496	0.22096
60	0.63766	111.280	0.22443
70	0.65568	113.056	0.22781
80	0.67334	114.827	0.23112
90	0.69069	116.594	0.23437
100	0.70777	118.360	0.23755
110	0.72495	120.127	0.24068
120	0.74120	121.894	0.24376
130	0.75760	123.665	0.24678
140	0.77383	125.439	0.24977
150	0.78989	127.218	0.25271
160	0.80581	129.002	0.25561
170	0.82159	130.793	0.25848
180	0.83725	132.589	0.26131
190	0.85279	134.393	0.26411
200	0.86824	136.205	0.26687
210	0.88359	138.024	0.26961
220	0.89885	139.851	0.27232
230	0.91403	141.687	0.27500
240	0.92914	143.532	0.27766
250	0.94418	145.385	0.28029
260	0.95916	147.248	0.28289
270	0.97408	149.120	0.28548
280	0.98894	151.002	0.28804
290	1.0038	152.893	0.29058
300	1.0185	154.794	0.29309
310	1.0332	156.704	0.29559
320	1.0479	158.624	0.29807
330	1.0626	160.554	0.30053
340	1.0772	162.494	0.30297
350	1.0917	164.444	0.30540

TABLE B–18 (continued)

Absolute pressure, psi

Temp., °F	100		
	85.304		
	(50.77°F)		
	v	h	s
	(0.54908)	(109.013)	(0.21790)
60	0.56498	110.700	0.22117
70	0.58177	112.514	0.22463
80	0.59818	114.319	0.22801
90	0.61425	116.117	0.23131
100	0.63003	117.911	0.23454
110	0.64555	119.702	0.23771
120	0.66084	121.492	0.24083
130	0.67592	123.284	0.24389
140	0.69081	125.077	0.24691
150	0.70554	126.874	0.24988
160	0.72011	128.674	0.25281
170	0.73454	130.480	0.25570
180	0.74885	132.290	0.25855
190	0.76304	134.107	0.26137
200	0.77712	135.931	0.26415
210	0.79111	137.761	0.26691
220	0.80510	139.599	0.26963
230	0.81883	141.445	0.27233
240	0.83257	143.299	0.27500
250	0.84624	145.161	0.27764
260	0.85985	147.032	0.28026
270	0.87340	148.912	0.28285
280	0.88689	150.800	0.28542
290	0.90033	152.698	0.28797
300	0.91372	154.605	0.29050
310	0.92707	156.522	0.29300
320	0.94038	158.448	0.29549
330	0.95365	160.383	0.29796
340	0.96688	162.328	0.30040
350	0.98008	164.238	0.30283
360	0.99324	166.248	0.30525

Temp., °F	110		
	95.304		
	(56.55°F)		
	v	h	s
	(0.49969)	(109.456)	(0.21687)
60	0.50526	110.101	0.21812
70	0.52109	111.956	0.22165
80	0.53651	113.798	0.22510
90	0.55156	115.628	0.22846
100	0.56631	117.451	0.23174
110	0.58078	119.269	0.23496
120	0.59501	121.083	0.23812
130	0.60901	122.897	0.24122
140	0.62282	124.710	0.24427
150	0.63646	126.525	0.24727
160	0.64994	128.342	0.25023
170	0.66327	130.163	0.25314
180	0.67648	131.988	0.25602
190	0.68956	133.818	0.25886
200	0.70254	135.654	0.26166
210	0.71542	137.496	0.26443
220	0.72821	139.345	0.26717
230	0.74091	141.201	0.26989
240	0.75354	143.064	0.27257
250	0.76609	144.935	0.27522
260	0.77858	146.814	0.27785
270	0.79101	148.702	0.28046
280	0.80338	150.598	0.28304
290	0.81570	152.502	0.28560
300	0.82798	154.416	0.28813
310	0.84020	156.339	0.29065
320	0.85239	158.271	0.29314
330	0.86453	160.212	0.29561
340	0.87664	162.162	0.29807
350	0.88871	164.121	0.30050
360	0.90075	166.091	0.30292

Temp., °F	120		
	105.304		
	(61.95°F)		
	v	h	s
	(0.45822)	(109.853)	(0.21593)
70	0.47032	111.381	0.21884
80	0.48494	113.262	0.22236
90	0.49918	115.128	0.22578
100	0.51309	116.982	0.22912
110	0.52670	118.827	0.23239
120	0.54005	120.667	0.23559
130	0.55318	122.503	0.23873
140	0.56610	124.337	0.24182
150	0.57884	126.171	0.24485
160	0.59142	128.005	0.24784
170	0.60384	129.842	0.25078
180	0.61613	131.682	0.25368
190	0.62830	133.526	0.25654
200	0.64036	135.375	0.25936
210	0.65232	137.229	0.26215
220	0.66418	139.088	0.26490
230	0.67596	140.954	0.26763
240	0.68766	142.827	0.27033
250	0.69929	144.707	0.27299
260	0.71085	146.595	0.27564
270	0.72235	148.490	0.27825
280	0.73379	150.394	0.28084
290	0.74517	152.306	0.28341
300	0.75651	154.226	0.28595
310	0.76780	156.155	0.28848
320	0.77905	158.092	0.29098
330	0.79026	160.039	0.29346
340	0.80144	161.995	0.29592
350	0.81257	163.959	0.29836
360	0.82368	165.933	0.30078
370	0.83475	167.916	0.30319

TABLE B–18 (continued) Monochlorodifluoromethane, $CHClF_2$ (R-22), properties of superheated states (English units): v (ft^3/lbm); h (Btu/lbm); s (Btu/lbm · °R) (saturation properties in parentheses)

Temp., °F	140		
	125.304		
	(71.83°F)		
	v	h	s
	(0.39243)	(110.535)	(0.21425)
70	—	—	—
80	0.40342	112.143	0.21725
90	0.41646	114.086	0.22082
100	0.42911	116.009	0.22428
110	0.44143	117.915	0.22766
120	0.45346	119.809	0.23096
130	0.46524	121.694	0.23418
140	0.47679	123.573	0.23734
150	0.48814	125.447	0.24044
160	0.49932	127.319	0.24348
170	0.51033	129.189	0.24648
180	0.52121	131.060	0.24943
190	0.53195	132.933	0.25233
200	0.54258	134.808	0.25520
210	0.55309	136.686	0.25802
220	0.56351	138.569	0.26081
230	0.57385	140.456	0.26357
240	0.58409	142.349	0.26629
250	0.59427	144.247	0.26899
260	0.60437	146.152	0.27165
270	0.61441	148.064	0.27429
280	0.62439	149.983	0.27690
290	0.63432	151.909	0.27949
300	0.64419	153.843	0.28205
310	0.65402	153.784	0.28459
320	0.66380	157.734	0.28711
330	0.67354	159.692	0.28960
340	0.68325	161.658	0.29208
350	0.69292	163.633	0.29453
360	0.70255	165.616	0.29696
370	0.71216	167.608	0.29938
380	0.72173	169.609	0.30178

Temp., °F	160		
	145.304		
	(80.71°F)		
	v	h	s
	(0.34249)	(111.095)	(0.21276)
80	—	—	—
90	0.35387	112.984	0.21623
100	0.36568	114.986	0.21984
110	0.37710	116.961	0.22334
120	0.38820	118.917	0.22674
130	0.39901	120.856	0.23006
140	0.40958	122.783	0.23330
150	0.41992	124.701	0.23647
160	0.43008	126.612	0.23958
170	0.44006	128.519	0.24263
180	0.44989	130.423	0.24563
190	0.45958	132.326	0.24858
200	0.46914	134.229	0.25149
210	0.47859	136.133	0.25435
220	0.48794	138.040	0.25718
230	0.49720	139.949	0.25997
240	0.50637	141.863	0.26272
250	0.51546	143.781	0.26544
260	0.52447	145.704	0.26814
270	0.53342	147.632	0.27080
280	0.54231	149.567	0.27343
290	0.55115	151.507	0.27604
300	0.55993	153.455	0.27862
310	0.56866	155.410	0.28117
320	0.57735	157.372	0.28371
330	0.58599	159.341	0.28622
340	0.59460	161.319	0.28870
350	0.60317	163.304	0.29117
360	0.61170	165.297	0.29362
370	0.62020	167.298	0.29604
380	0.62868	169.308	0.29845
390	0.63712	171.326	0.30084

Temp., °F	180		
	165.304		
	(88.81°F)		
	v	h	s
	(0.30323)	(111.555)	(0.21142)
90	0.30461	111.808	0.21188
100	0.31587	113.904	0.21566
110	0.32669	115.959	0.21930
120	0.33713	117.983	0.22282
130	0.34724	119.983	0.22624
140	0.35708	121.964	0.22957
150	0.36668	123.930	0.23282
160	0.37607	125.885	0.23600
170	0.38527	127.831	0.23912
180	0.39430	129.770	0.24217
190	0.40319	131.706	0.24518
200	0.41194	133.638	0.24813
210	0.42057	135.570	0.25103
220	0.42910	137.501	0.25390
230	0.43752	139.434	0.25672
240	0.44586	141.369	0.25951
250	0.45411	143.307	0.26226
260	0.46229	145.249	0.26497
270	0.47040	147.195	0.26766
280	0.47845	149.145	0.27031
290	0.48644	151.102	0.27294
300	0.49437	153.063	0.27554
310	0.50225	155.032	0.27811
320	0.51009	157.006	0.28066
330	0.51788	158.988	0.28319
340	0.52564	160.976	0.28569
350	0.53335	162.972	0.28817
360	0.54103	164.976	0.29063
370	0.54868	166.986	0.29307
380	0.55630	169.005	0.29549
390	0.56389	171.032	0.29789

TABLE B–18 (continued)

Absolute pressure, psi

Temp., °F	200 185.304 (96.27°F) v	h	s
	(0.27150)	(111.934)	(0.21018)
100	0.27553	112.750	0.21165
110	0.28596	114.900	0.21545
120	0.29595	117.004	0.21911
130	0.30556	119.073	0.22265
140	0.31487	121.114	0.22608
150	0.32390	123.133	0.22942
160	0.33270	125.134	0.23268
170	0.34130	127.122	0.23586
180	0.34972	129.100	0.23898
190	0.35798	131.070	0.24203
200	0.36609	133.034	0.24503
210	0.37408	134.995	0.24798
220	0.38196	136.953	0.25089
230	0.38973	138.910	0.25374
240	0.39741	140.867	0.25656
250	0.40500	142.826	0.25934
260	0.41251	144.787	0.26209
270	0.41995	146.751	0.26480
280	0.42733	148.719	0.26747
290	0.43465	150.691	0.27012
300	0.44190	152.668	0.27274
310	0.44911	154.650	0.27533
320	0.45627	156.637	0.27790
330	0.46339	158.631	0.28044
340	0.47046	160.631	0.28296
350	0.47749	162.638	0.28545
360	0.48449	164.652	0.28792
370	0.49146	166.672	0.29037
380	0.49839	168.700	0.29280
390	0.50530	170.736	0.29521
400	0.51218	172.779	0.29760

Temp., °F	240 225.304 (109.67°F) v	h	s
	(0.22327)	(112.487)	(0.20793)
110	0.22360	112.564	0.20806
120	0.23318	114.873	0.21208
130	0.24225	117.113	0.21591
140	0.25089	119.299	0.21959
150	0.25920	121.443	0.22314
160	0.26721	123.554	0.22657
170	0.27498	125.639	0.22991
180	0.28253	127.702	0.23316
190	0.28990	129.749	0.23633
200	0.29710	131.783	0.23944
210	0.30416	133.807	0.24248
220	0.31109	135.822	0.24547
230	0.31790	137.832	0.24841
240	0.32461	139.838	0.25130
250	0.33122	141.842	0.25414
260	0.33775	143.844	0.25694
270	0.34420	145.847	0.25970
280	0.35059	147.850	0.26243
290	0.35690	149.855	0.26512
300	0.36316	151.863	0.26778
310	0.36936	153.874	0.27041
320	0.37551	155.889	0.27301
330	0.38161	157.908	0.27559
340	0.38767	159.932	0.27814
350	0.39369	161.962	0.28066
360	0.39967	163.997	0.28315
370	0.40562	166.037	0.28563
380	0.41153	168.084	0.28808
390	0.41741	170.138	0.29051
400	0.42327	172.198	0.29292
410	0.42910	174.265	0.29531
420			

Temp., °F	280 265.304 (121.52°F) v	h	s
	(0.18821)	(112.814)	(0.20586)
130	0.19583	114.912	0.20944
140	0.20427	117.294	0.21345
150	0.21224	119.600	0.21726
160	0.21983	121.848	0.22092
170	0.22711	124.051	0.22445
180	0.23413	126.217	0.22786
190	0.24092	128.354	0.23118
200	0.24753	130.468	0.23441
210	0.25396	132.564	0.23756
220	0.26025	134.644	0.24064
230	0.26641	136.713	0.24366
240	0.27246	138.773	0.24663
250	0.27840	140.825	0.24954
260	0.28424	142.873	0.25241
270	0.29000	144.917	0.25523
280	0.29569	146.958	0.25801
290	0.30130	148.999	0.26075
300	0.30685	151.040	0.26345
310	0.31234	153.082	0.26612
320	0.31778	155.125	0.26876
330	0.32317	157.172	0.27137
340	0.32851	159.221	0.27395
350	0.33381	161.274	0.27650
360	0.33907	163.332	0.27902
370	0.34429	165.394	0.28152
380	0.34948	167.460	0.28400
390	0.35463	169.533	0.28645
400	0.35976	171.611	0.28888
410	0.36486	173.695	0.29129
420	0.36994	175.784	0.29368
430	0.37499	177.881	0.29605

TABLE B–18 (continued) Monochlorodifluoromethane, $CHClF_2$ (R-22), properties of superheated states (English units): v (ft^3/lbm); h (Btu/lbm); s (Btu/lbm · °R) (saturation properties in parentheses)

Temp., °F	290			Temp., °F	300		
	275.304				285.304		
	(124.29°F)				(126.98°F)		
	v	h	s		v	h	s
	(0.18088)	(112.865)	(0.20536)		(0.17400)	(112.904)	(0.20487)
130	0.18600	114.312	0.20783	130	0.17670	113.688	0.20620
140	0.19445	116.755	0.21194	140	0.18520	116.199	0.21042
150	0.20239	119.110	0.21583	150	0.19313	118.607	0.21441
160	0.20992	121.398	0.21955	160	0.20062	120.938	0.21820
170	0.21712	123.635	0.22313	170	0.20776	123.210	0.22184
180	0.22404	125.830	0.22659	180	0.21460	125.436	0.22534
190	0.23073	127.992	0.22995	190	0.22120	127.625	0.22874
200	0.23722	130.128	0.23321	200	0.22759	129.784	0.23204
210	0.24354	132.244	0.23639	210	0.23379	131.919	0.23525
220	0.24970	134.342	0.23950	220	0.23984	134.035	0.23839
230	0.25573	136.426	0.24255	230	0.24575	136.136	0.24145
240	0.26164	138.500	0.24553	240	0.25154	138.225	0.24446
250	0.26745	140.566	0.24846	250	0.25722	140.304	0.24741
260	0.27315	142.625	0.25135	260	0.26280	142.375	0.25031
270	0.27877	144.680	0.25418	270	0.26829	144.441	0.25316
280	0.28432	146.732	0.25697	280	0.27370	146.503	0.25597
290	0.28979	148.782	0.25973	290	0.27904	148.563	0.25873
300	0.29519	150.831	0.26244	300	0.28431	150.621	0.26146
310	0.30054	152.881	0.26512	310	0.28952	152.679	0.26415
320	0.30583	154.932	0.26777	320	0.29467	154.738	0.26681
330	0.31107	156.986	0.27039	330	0.29978	156.798	0.26944
340	0.31626	159.041	0.27297	340	0.30483	158.861	0.27203
350	0.32141	161.101	0.27553	350	0.30985	160.926	0.27460
360	0.32652	163.164	0.27807	360	0.31482	162.995	0.27714
370	0.33160	165.231	0.28057	370	0.31975	165.068	0.27965
380	0.33664	167.303	0.28306	380	0.32465	167.145	0.28214
390	0.34165	169.380	0.28551	390	0.32952	169.227	0.28460
400	0.34662	171.463	0.28795	400	0.33436	171.314	0.28705
410	0.35157	173.551	0.29037	410	0.33917	173.407	0.28947
420	0.35650	175.645	0.29276	420	0.34395	175.505	0.29186
430	0.36139	177.745	0.29513	430	0.34871	117.609	0.29424

TABLE B–19 Properties of ammonia (NH_3) at saturation (SI units)

Temp., °C, T	Pressure, kPa, p	Specific volume, m^3/kg v_f	v_g	Enthalpy, kJ/kg h_f	h_g	Entropy, kJ/kg · K s_f	s_g
−70	10.9	0.0014	9.21	−129.1	1335.4	−0.595	6.619
−65	16.0	0.0014	6.47	−107.9	1344.4	−0.490	6.490
−60	22.0	0.0014	4.70	−86.5	1353.3	−0.389	6.368
−55	30.0	0.0014	3.48	−65.1	1362.6	−0.289	6.255
−50	41.0	0.0014	2.63	−43.7	1371.6	−0.193	6.150
−45	55.0	0.0014	2.01	−21.9	1380.9	−0.096	6.054
−40	72.0	0.0014	1.55	0.0	1390.0	0.000	5.962
−35	93.0	0.0015	1.22	22.3	1397.9	0.096	5.874
−30	120.0	0.0015	0.963	44.7	1405.6	0.205	5.792
−25	152.0	0.0015	0.772	67.2	1413.0	0.281	5.702
−20	190.0	0.0015	0.624	89.8	1420.0	0.368	5.623
−15	236.0	0.0015	0.509	112.3	1426.5	0.456	5.548
−10	291.0	0.0015	0.418	135.4	1433.0	0.544	5.476
−5	355.0	0.0015	0.347	158.2	1438.9	0.632	5.422
0	429.0	0.0016	0.290	181.2	1444.4	0.716	5.338
5	516.0	0.0016	0.243	204.4	1449.6	0.799	5.275
10	615.0	0.0016	0.205	227.7	1454.2	0.883	5.213
15	728.0	0.0016	0.175	251.4	1458.6	0.963	5.154
20	857.0	0.0016	0.149	275.2	1462.6	1.043	5.095
25	1000.0	0.0017	0.129	298.9	1465.8	1.126	5.041
30	1170.0	0.0017	0.110	323.1	1468.9	1.206	4.982
35	1350.0	0.0017	0.0955	347.5	1471.4	1.281	4.932
40	1550.0	0.0017	0.0832	371.9	1473.3	1.361	4.878
45	1780.0	0.0018	0.0726	396.8	1474.5	1.436	4.823
50	2030.0	0.0018	0.0635	421.9	1474.7	1.516	4.773

TABLE B–19 (continued) Properties of ammonia (NH_3) at saturation (English units)

Temp., °F, T	Pressure, psia, p	Specific Volume, ft^3/lbm v_f	v_g	Enthalpy, Btu/lbm h_f	h_g	Entropy, Btu/lbm·°R s_f	s_g
−100	1.24	0.022	182.90	−61.5	571.4	−0.1579	1.6025
−90	1.86	0.022	124.28	−51.4	575.9	−0.1309	1.5667
−80	2.74	0.022	86.54	−41.3	580.1	−0.1036	1.5336
−70	3.94	0.023	61.65	−31.1	584.4	−0.0771	1.5026
−60	5.55	0.023	44.73	−20.9	588.8	−0.0514	1.4747
−50	7.67	0.023	33.08	−10.5	593.2	−0.0254	1.4487
−40	10.41	0.023	24.86	0.0	597.6	0.0000	1.4242
−30	13.90	0.023	18.97	10.7	601.4	0.0250	1.4001
−20	18.30	0.024	14.68	21.4	605.0	0.0497	1.3774
−10	23.74	0.024	11.50	32.1	608.5	0.0738	1.3558
0	30.42	0.024	9.116	42.1	611.8	0.0975	1.3352
10	38.51	0.024	7.309	53.8	614.9	0.1208	1.3157
20	48.21	0.025	5.910	64.7	617.8	0.1437	1.2969
30	59.74	0.025	4.825	75.7	620.5	0.1663	1.2790
40	73.32	0.025	3.971	86.8	623.0	0.1885	1.2618
50	89.19	0.026	3.294	97.9	625.2	0.2105	1.2453
60	107.60	0.026	2.751	109.2	627.3	0.2322	1.2294
70	128.80	0.026	2.312	120.5	629.1	0.2537	1.2140
80	153.00	0.027	1.955	132.0	630.7	0.2749	1.1991
90	180.60	0.027	1.661	143.5	632.0	0.2958	1.1846
100	211.90	0.027	1.419	155.2	633.0	0.3166	1.1705
110	247.00	0.028	1.217	167.0	633.7	0.3372	1.1566
120	286.40	0.028	1.047	179.0	634.0	0.3576	1.1427
125	307.80	0.029	0.973	185.1	634.0	0.3679	1.1358

Source: The American Society of Heating, Refrigerating, and Air Conditioning Engineers, "Bulletin on Thermodynamic Properties of Refrigerants" (New York: ASHRAE, 1969); with permission of ASHRAE.

TABLE B–20 Ammonia properties at superheat (datum of −40°C) (SI units):
v (m^3/kg); h (kJ/kg); s (kJ/kg · K)

Pressure, MPa, p (Sat. temp., °C)		Sat. properties	Temperature, °C, T									
			−40	−20	0	20	40	60	80	100	120	140
0.04	v	2.66	2.82	3.07	3.33	3.58	3.83	4.08	4.33			
(50.3)	h	1371.2	1398.8	1437.9	1479.8	1521.9	1564.5	1607.5	1651.2			
	s	6.155	6.305	6.439	6.608	6.753	6.896	7.025	7.151			
0.07	v	1.58		1.73	1.88	2.02	2.17	2.30	2.45			
(40.4)	h	1389.3		1434.4	1477.5	1520.3	1563.1	1606.3	1650.3			
	s	5.970		6.146	6.318	6.469	6.615	6.745	6.875			
0.10	v	1.14		1.20	1.32	1.42	1.52	1.62	1.72			
(−33.6)	h	1400.0		1431.2	1474.9	1518.4	1561.7	1604.9	1649.4			
	s	5.849		5.966	6.142	6.293	6.439	6.573	6.699			
0.20	v	0.595			0.649	0.699	0.755	0.805	0.855	0.905	0.986	
(−18.9)	h	1421.4			1466.5	1511.9	1556.6	1600.9	1645.6	1690.8	1736.6	
	s	5.606			5.778	5.937	6.083	6.222	6.351	6.477	6.582	
0.30	v	0.397			0.425	0.461	0.497	0.532	0.566	0.600	0.634	0.667
(−9.2)	h	1434.0			1457.5	1505.2	1551.4	1596.8	1641.9	1687.7	1733.8	1780.1
	s	5.468			5.572	5.723	5.874	6.016	6.146	6.276	6.393	6.523
0.40	v	0.310			0.313	0.341	0.370	0.396	0.422	0.448	0.478	0.499
(−1.9)	h	1442.4			1447.5	1498.2	1545.9	1592.6	1638.7	1684.7	1731.5	1778.0
	s	5.367			5.384	5.564	5.719	5.866	6.000	6.125	6.247	6.364

Source: The American Society of Heating, Refrigerating, and Air Conditioning Engineers, "Bulletin on Thermodynamic Properties of Refrigerants" (New York: ASHRAE, 1969); with permission of ASHRAE.

TABLE B–20 (continued) Ammonia properties at superheat (datum of −40°C) (SI units): v (m^3/kg); h (kJ/kg); s (kJ/kg · K)

Pressure, MPa, p (Sat. temp., °C)		Sat. properties	Temperature, °C, T									
			−40	−20	0	20	40	60	80	100	120	140
0.50	v	0.251				0.270	0.293	0.315	0.334	0.351	0.378	0.398
(3.8)	h	1448.6				1490.7	1540.7	1588.0	1635.2	1681.9	1728.2	1776.1
	s	5.288				5.439	5.598	5.707	5.882	6.012	6.134	6.251
0.60	v	0.210				0.222	0.242	0.260	0.278	0.291	0.313	0.330
(8.1)	h	1453.5				1483.3	1534.9	1581.9	1631.5	1678.7	1726.1	1773.6
	s	5.234				5.326	5.497	5.707	5.786	5.941	6.037	6.159
0.70	v	0.181				0.187	0.204	0.220	0.236	0.247	0.266	0.281
(13.9)	h	1457.7				1475.1	1532.8	1579.1	1627.7	1675.7	1723.3	1771.2
	s	5.167				5.225	5.401	5.602	5.698	5.836	5.958	6.075
0.80	v	0.160				0.162	0.177	0.192	0.206	0.216	0.233	0.246
(17.8)	h	1461.0				1467.2	1523.1	1574.9	1624.2	1672.6	1718.9	1769.2
	s	5.120				5.141	5.326	5.485	5.631	5.765	5.891	6.008
0.90	v	0.142					0.155	0.169	0.181	0.194	0.205	0.217
(21.7)	h	1463.8					1516.8	1570.1	1620.3	1669.6	1717.9	1766.6
	s	5.079					5.250	5.418	5.564	5.698	5.823	5.945
1.0	v	0.129					0.139	0.151	0.163	0.174	0.185	0.195
(24.9)	h	1465.8					1511.0	1565.4	1616.8	1652.9	1715.7	1764.7
	s	5.041					5.187	5.355	5.506	5.644	5.769	5.891

TABLE B–20 (continued)

Pressure, MPa, p (Sat. temp., °C)		Sat. properties	Temperature, °C, T									
			−40	−20	0	20	40	60	80	100	120	140
1.2	v	0.108					0.113	0.124	0.134	0.144	0.155	0.163
(30.9)	h	1469.3					1497.9	1555.6	1609.1	1660.3	1710.3	1760.1
	s	4.970					5.066	5.246	5.401	5.543	5.673	5.795
1.4	v	0.0924					0.0943	0.1043	0.1136	0.1217	0.1299	0.1380
(36.3)	h	1471.9					1484.2	1545.9	1601.2	1653.8	1705.2	1755.4
	s	4.919					4.957	5.146	5.309	5.455	5.585	5.711
1.6	v	0.0805						0.0899	0.0980	0.1055	0.1123	0.1199
(41.1)	h	1473.3						1535.4	1593.3	1647.5	1699.6	1751.0
	s	4.865						5.058	5.225	5.376	5.510	5.640
1.8	v	0.0718						0.0780	0.0855	0.0924	0.0993	0.1055
(45.4)	h	1474.5						1524.5	1585.2	1641.0	1694.0	1746.1
	s	4.823						4.974	5.150	5.305	5.443	5.573
2.0	v	0.0643						0.0687	0.0762	0.0824	0.0886	0.0943
(49.4)	h	1474.7						1513.1	1562.6	1634.2	1688.7	1741.7
	s	4.781						4.899	5.079	5.242	5.384	5.514

TABLE B–20 (continued) Ammonia properties at superheat (datum of −40°F) (English units): v (ft^3/lbm); h (Btu/lbm); s (Btu/lbm · °R)

Pressure, psia, p		Temperature, °F, T											
		−40	−20	0	20	40	60	80	100	120	160	200	300
5.0	v	52.36	54.97	57.55	60.12	62.69	65.24	67.79	70.33	72.87	77.95		
	h	600.3	610.4	620.4	630.4	640.4	650.5	660.6	670.7	680.9	701.6		
	s	1.5149	1.5385	1.5608	1.5821	1.6026	1.6223	1.6413	1.6598	1.6778	1.7122		
10.0	v	25.90	27.26	28.58	29.90	31.20	32.49	33.78	35.07	36.35	38.90	41.45	
	h	597.8	608.5	618.9	629.1	639.3	649.5	659.7	670.0	680.3	701.1	722.2	
	s	1.4293	1.4542	1.4773	1.4992	1.5200	1.5400	1.5593	1.5779	1.5960	1.6307	1.6637	
20.0	v			14.09	14.78	15.45	16.12	16.78	17.43	18.08	19.37	20.66	
	h			615.5	626.4	637.0	647.5	658.0	668.5	678.9	700.0	721.2	
	s			1.3907	1.4138	1.4356	1.4562	1.4760	1.4950	1.5133	1.5485	1.5817	
30.0	v			9.25	9.731	10.20	10.65	11.10	11.55	11.99	12.87	13.73	
	h			611.9	623.5	634.6	645.5	656.2	666.9	677.5	698.8	720.3	
	s			1.3371	1.3618	1.3845	1.4059	1.4261	1.4456	1.4642	1.4998	1.5334	
40.0	v				7.203	7.568	7.922	8.268	8.609	8.945	9.609	10.27	11.88
	h				620.4	632.1	643.4	654.4	665.3	676.1	697.7	719.4	774.6
	s				1.3231	1.3470	1.3692	1.3900	1.4098	1.4288	1.4648	1.4987	1.5766
50.0	v					5.988	6.280	6.564	6.843	7.117	7.655	8.185	9.489
	h					629.5	641.2	652.6	663.7	674.7	696.6	718.5	774.0
	s					1.3169	1.3399	1.3613	1.3816	1.4009	1.4374	1.4716	1.5500
60.0	v					4.933	5.184	5.428	5.665	5.897	6.352	6.787	7.892
	h					626.8	639.0	650.7	662.1	673.3	695.5	717.5	773.7
	s					1.2913	1.3152	1.3373	1.3581	1.3778	1.4148	1.4493	1.5281

TABLE B–20 (continued)

Pressure, psia, p		Temperature, °F, T											
		−40	−20	0	20	40	60	80	100	120	160	200	300
70.0	v					4.177	4.401	4.615	4.822	5.025	5.420	5.807	6.750
	h					623.9	636.6	648.7	660.6	671.8	694.3	716.6	772.7
	s					1.2688	1.2937	1.3166	1.3378	1.3579	1.3954	1.4302	1.5095
80.0	v						3.812	4.005	4.190	4.371	4.722	5.063	5.894
	h						634.3	646.7	658.7	670.4	693.2	715.6	772.1
	s						1.2745	1.2981	1.3199	1.3404	1.3784	1.4136	1.4933
90.0	v						3.353	3.529	3.698	3.862	4.178	4.484	5.228
	h						631.8	644.7	657.0	668.9	692.0	714.7	771.5
	s						1.2571	1.2814	1.3038	1.3247	1.3633	1.3988	1.4789
100.0	v						2.985	3.149	3.304	3.454	3.743	4.021	4.695
	h						629.3	642.6	655.2	667.3	690.8	713.7	770.8
	s						1.2409	1.2661	1.2891	1.3104	1.3495	1.3854	1.4660
120.0	v							2.576	2.712	2.842	3.089	3.326	3.895
	h							638.3	651.6	664.2	688.4	711.8	769.6
	s							1.2386	1.2628	1.2850	1.3254	1.3620	1.4435
160.0	v								1.969	2.075	2.272	2.457	2.895
	h								643.9	657.8	683.5	707.9	767.1000
	s								1.2186	1.2429	1.2859	1.3240	1.4076
200.0	v								1.520	1.612	1.780	1.935	2.295
	h								635.6	650.9	678.4	703.9	764.5
	s								1.1809	1.2077	1.2537	1.2935	1.3791
250.0	v									1.240	1.386	1.518	1.815
	h									641.5	671.8	698.8	761.3
	s									1.1690	1.2195	1.2617	1.3501
300.0	v										1.123	1.239	1.496
	h										664.7	693.5	758.1
	s										1.1894	1.2344	1.3257

TABLE B–21 Properties of R-123 (CCl_2HCF_3) at saturation (SI units)

Temperature, °C,	Pressure, kPa,	Specific volume, m^3/kg		Enthalpy, kJ/kg		Entropy, kJ/kg·K	
		v_f	v_g	h_f	h_g	s_f	s_g
−40	3.78	0.0006	3.333	167.2	356.3	0.8705	1.6814
−35	5.17	0.0006	2.488	171.4	359.2	0.8883	1.6765
−30	6.98	0.0006	1.880	175.5	362.0	0.9050	1.6723
−25	9.29	0.0006	1.439	179.4	364.9	0.9211	1.6687
−20	12.22	0.0006	1.115	183.4	367.9	0.9369	1.6656
−15	15.88	0.0006	0.8726	187.4	370.8	0.9526	1.6631
−10	20.41	0.0006	0.6906	191.5	373.8	0.9683	1.6610
−5	25.97	0.0007	0.5516	195.7	376.7	0.9841	1.6593
0	32.73	0.0007	0.4444	200.0	379.7	1.0000	1.6580
5	40.86	0.0007	0.3614	204.4	382.8	1.0161	1.6572
10	50.57	0.0007	0.2962	209.0	385.8	1.0323	1.6567
15	62.06	0.0007	0.2447	213.7	388.8	1.0487	1.6565
20	75.55	0.0007	0.2035	218.5	391.9	1.0652	1.6566
25	91.29	0.0007	0.1705	223.4	394.9	1.0817	1.6570
30	109.52	0.0007	0.1437	228.4	398.0	1.0983	1.6577
35	130.49	0.0007	0.1219	233.5	401.0	1.1149	1.6587
40	154.48	0.0007	0.1039	238.7	404.1	1.1316	1.6598
45	181.76	0.0007	0.0891	243.9	407.1	1.1481	1.6612
50	212.61	0.0007	0.0768	249.2	410.2	1.1646	1.6627
55	247.33	0.0007	0.0665	254.6	413.2	1.1811	1.6644
60	286.23	0.0007	0.0578	260.0	416.2	1.1974	1.6662
65	329.62	0.0007	0.0504	265.5	419.2	1.2163	1.6681
70	377.81	0.0007	0.0442	271.0	422.1	1.2297	1.6701
75	431.13	0.0008	0.0388	276.6	425.1	1.2456	1.6721
80	489.93	0.0008	0.0342	282.2	427.9	1.2614	1.6742
85	554.55	0.0008	0.0303	287.8	430.8	1.2771	1.6763
90	625.35	0.0008	0.0268	293.4	433.5	1.2926	1.6784
95	702.68	0.0008	0.0239	299.1	436.2	1.3080	1.6805
100	786.93	0.0008	0.0213	304.8	438.9	1.3232	1.6825
110	977.71	0.0008	0.0170	316.4	443.9	1.3533	1.6862
120	1200.91	0.0009	0.0136	328.1	448.6	1.3830	1.6895
130	1459.98	0.0009	0.0109	340.0	452.7	1.4125	1.6920
140	1758.60	0.0009	0.0088	352.3	456.3	1.4419	1.6935
150	2100.93	0.0010	0.0070	365.1	458.9	1.4718	1.6935
160	2492.01	0.0010	0.0055	378.7	460.3	1.5026	1.6910
170	2938.83	0.0011	0.0042	393.7	459.3	1.5357	1.6837
180	3454.38	0.0013	0.0028	412.6	451.0	1.5764	1.6613

Source: Data abstracted from "Thermodynamic Properties of HCFC-123, 1993" with permission of E. I. du Pont de Nemours & Co.

TABLE B–21 (continued)

Temperature, °F,	Pressure, psia,	Specific volume, ft³/lbm		Enthalpy, Btu/lbm		Entropy, Btu/lbm · °R	
		v_f	v_g	h_f	h_g	s_f	s_g
−40	0.548	0.0099	53.476	0.0	81.3	0.0000	0.1938
−30	0.776	0.0100	38.610	2.0	82.8	0.0047	0.1925
−20	1.080	0.0101	32.329	3.9	80.2	0.0091	0.1915
−10	1.478	0.0102	21.142	5.8	79.7	0.0134	0.1905
0	1.993	0.0102	16.000	7.7	79.2	0.0176	0.1898
10	2.651	0.0103	12.255	9.6	78.6	0.0217	0.1892
20	3.480	0.0104	9.515	11.6	78.1	0.0259	0.1887
30	4.513	0.0105	7.663	13.7	91.1	0.0301	0.1883
40	5.785	0.0106	5.921	15.8	92.6	0.0344	0.1881
50	7.334	0.0107	4.746	18.0	94.0	0.0387	0.1879
60	9.203	0.0108	3.839	20.2	95.5	0.0430	0.1879
70	11.436	0.0109	3.133	22.5	96.9	0.0474	0.1879
80	14.081	0.0110	2.578	24.9	98.4	0.0518	0.1880
90	17.186	0.0111	2.138	27.3	99.9	0.0562	0.1883
100	20.803	0.0112	1.786	29.7	101.3	0.0606	0.1885
110	24.988	0.0113	1.502	32.2	102.8	0.0650	0.1889
120	29.796	0.0114	1.271	34.8	104.2	0.0694	0.1893
130	35.285	0.0116	1.082	37.3	105.7	0.0738	0.1897
140	41.515	0.0117	0.9255	39.9	107.1	0.0781	0.1902
150	48.549	0.0118	0.7959	42.5	108.5	0.0824	0.1907
160	56.450	0.0120	0.6875	45.2	109.9	0.0867	0.1912
170	65.284	0.0121	0.5963	47.8	111.3	0.0909	0.1918
180	75.118	0.0123	0.5192	50.5	112.7	0.0951	0.1923
190	86.023	0.0124	0.4536	53.2	114.0	0.0993	0.1929
200	98.069	0.0126	0.3975	55.9	115.3	0.1034	0.1934
220	125.885	0.0130	0.3076	61.4	117.8	0.1114	0.1945
240	159.180	0.0134	0.2400	66.9	120.2	0.1194	0.1954
260	198.621	0.0140	0.1883	72.6	122.2	0.1272	0.1962
280	244.917	0.0146	0.1478	78.4	124.0	0.1350	0.1967
300	298.868	0.0154	0.1155	84.5	125.4	0.1429	0.1967
320	361.442	0.0164	0.0887	91.0	126.1	0.1511	0.1961
340	434.013	0.0182	0.0652	98.2	125.5	0.1599	0.1940
360	519.422	0.0231	0.0389	108.3	119.6	0.1720	0.1858

Source: Data abstracted from "Thermodynamic Properties of HCFC-123, 1993" with permission of E. I. du Pont de Nemours & Co.

TABLE B–21 (continued) Properties of R-123 (CCl_2HCF_3) in superheat (SI units):
v (m^3/kg); h (kJ/kg); s (kJ/kg · K)

Pressure, kPa, p (Sat. temp., °C)		Saturation properties	Temperature, °C, T									
			0	**20**	**40**	**60**	**80**	**100**	**120**	**140**	**160**	**180**
20	v	0.7036	0.7334	0.7897	0.8454	0.9005	0.9555	1.011	1.065	1.120		
(−10.41)	h	373.5	380.3	393.6	407.3	421.4	435.9	450.8	466.1	481.8		
	s	1.661	1.686	1.733	1.779	1.822	1.864	1.906	1.946	1.984		
60	v	0.2525		0.2583	0.2778	0.2968	0.3155	0.3342	0.3528	0.3713	0.3898	
(14.16)	h	388.3		392.4	406.4	420.7	435.3	450.3	465.6	481.4	497.5	
	s	1.657		1.671	1.717	1.761	1.804	1.845	1.885	1.924	1.962	
100	v	0.1565			0.1641	0.1760	0.1875	0.1989	0.2103	0.2215	0.2328	0.2441
(27.48)	h	396.4			405.4	419.9	434.7	449.7	465.1	480.9	497.0	513.5
	s	1.657			1.687	1.732	1.775	1.816	1.856	1.895	1.934	1.971
150	v	0.1069			0.1072	0.1155	0.1235	0.1313	0.1390	0.1467	0.1543	0.1619
(39.11)	h	403.6			404.2	419.0	433.9	449.1	464.5	480.3	496.5	513.0
	s	1.660			1.662	1.708	1.751	1.793	1.833	1.872	1.911	1.948
200	v	0.0814				0.0852	0.0914	0.0974	0.1034	0.1092	0.1150	0.1209
(48.03)	h	409.0				418.0	433.1	448.4	463.9	479.8	496.0	512.5
	s	1.662				1.690	1.734	1.776	1.816	1.856	1.894	1.931
250	v	0.0658				0.0670	0.0722	0.0771	0.0820	0.0867	0.0915	0.0962
(55.36)	h	413.4				417.0	432.3	447.7	463.3	479.2	495.4	511.9
	s	1.665				1.675	1.720	1.762	1.803	1.843	1.881	1.918
300	v	0.0552					0.0593	0.0636	0.0677	0.0718	0.0758	0.0798
(61.65)	h	417.2					431.5	447.0	462.7	478.6	494.9	511.4
	s	1.667					1.708	1.751	1.792	1.832	1.870	1.907
400	v	0.0418					0.0432	0.0466	0.0498	0.0530	0.0561	0.0592
(72.14)	h	423.4					429.7	445.5	461.4	477.4	493.7	510.4
	s	1.671					1.689	1.733	1.774	1.814	1.852	1.890

TABLE B–21 (continued)

Pressure, kPa, p (Sat. temp., °C)		Saturation properties	Temperature, °C, T										
			100	120	140	160	180	200	220	240	260	280	300
500	v	0.0336	0.0363	0.0391	0.0417	0.0443	0.0469	0.0494	0.0519				
(80.81)	h	428.4	444.0	460.0	476.2	492.6	509.3	526.3	543.6				
	s	1.765	1.717	1.759	1.800	1.838	1.876	1.913	1.948				
600	v	0.0280	0.0295	0.0319	0.0342	0.0364	0.0386	0.0408	0.0430	0.0451			
(88.26)	h	432.6	442.3	458.6	475.0	491.5	508.2	525.3	542.6	560.3			
	s	1.678	1.704	1.747	1.787	1.826	1.864	1.901	1.937	1.972			
800	v	0.0209	0.0226	0.0247	0.0248	0.0266	0.0283	0.0301	0.0317	0.0334			
(100.74)	h	439.3	439.6	456.4	472.4	489.1	506.0	523.2	540.7	558.5			
	s	1.683	1.687	1.731	1.767	1.807	1.845	1.882	1.918	1.953			
1000	v	0.0166		0.0174	0.0191	0.0206	0.0221	0.0236	0.0250	0.0264	0.0278		
(111.07)	h	444.4		452.4	469.7	486.7	503.8	521.1	538.7	556.6	574.8		
	s	1.687		1.707	1.750	1.790	1.829	1.866	1.903	1.938	1.973		
1500	v	0.0106			0.0122	0.0126	0.0138	0.0149	0.0160	0.0170	0.0180	0.0190	
(131.42)	h	453.3			461.6	480.1	497.9	515.8	533.7	551.9	570.4	589.2	
	s	1.692			1.713	1.756	1.797	1.835	1.872	1.909	1.944	1.978	
2000	v	0.0075				0.0084	0.0095	0.0105	0.0114	0.0106	0.0114	0.0121	0.0128
(147.19)	h	458.3				472.0	491.4	510.0	528.5	555.6	579.6	603.7	628.0
	s	1.694				1.726	1.770	1.810	1.848	1.995	2.047	2.098	2.145
3000	v	0.0041					0.0048	0.0059	0.0067	0.0074	0.0081	0.0088	0.0094
(171.27)	h	458.8					473.5	496.8	517.3	537.1	556.8	576.6	596.7
	s	1.682					1.715	1.765	1.808	1.847	1.885	1.921	1.957

Source: Data abstracted from "Thermodynamic Properties of HCFC-123, 1993" with permission of E. I. du Pont de Nemours & Co.

TABLE B–21 (continued) Properties of R-123 (CCl_2HCF_3) in superheat (English units): v (ft^3/lbm); h (Btu/lbm); s (Btu/lbm · °R)

Pressure, psia, p (Sat. temp., °F)		Saturation properties	Temperature, °F, T									
			40	80	120	160	200	240	280	320	360	400
5	v	6.784	6.873	7.466	8.047	8.621	9.193	9.762	10.332	10.900		
(34.07)	h	91.7	92.7	99.2	105.8	112.7	119.7	127.0	134.5	142.2		
	s	0.1882	0.1901	0.2026	0.2145	0.2259	0.2370	0.2477	0.2581	0.2682		
10	v	3.552		3.677	3.978	4.273	4.563	4.852	5.141	5.429	5.716	
(63.77)	h	96.0		98.7	105.5	112.4	119.5	126.8	134.3	142.0	149.9	
	s	0.1879		0.1930	0.2051	0.2166	0.2278	0.2385	0.2489	0.2590	0.2689	
20	v	1.853			1.942	2.097	2.248	2.397	2.546	2.619	2.840	2.986
(97.91)	h	101.0			104.9	112.0	119.1	126.5	134.0	137.8	149.6	157.7
	s	0.1885			0.1953	0.2071	0.2183	0.2291	0.2396	0.2447	0.2596	0.2692
40	v	0.959				1.007	1.089	1.169	1.247	1.324	1.401	1.477
(137.68)	h	106.8				110.9	118.3	125.7	133.3	141.1	149.0	157.1
	s	0.1901				0.1969	0.2084	0.2193	0.2299	0.2401	0.2500	0.2596
60	v	0.648					0.7017	0.7586	0.8137	0.8677	0.9212	0.9742
(164.15)	h	110.5					117.4	124.9	132.6	140.4	148.4	156.5
	s	0.1915					0.2021	0.2133	0.2239	0.2342	0.2441	0.2538
80	v	0.488					0.5065	0.5527	0.5966	0.6392	0.6811	0.7224
(184.6)	h	113.3					116.3	124.1	131.9	139.7	147.7	155.9
	s	0.1926					0.1972	0.2087	0.2194	0.2298	0.2398	0.2495
100	v	0.390						0.4285	0.4659	0.5018	0.5367	0.5711
(201.52)	h	115.5						123.2	131.1	139.0	147.1	155.3
	s	0.2037						0.2048	0.2158	0.2262	0.2363	0.2461
120	v	0.323						0.3450	0.3784	0.4099	0.4404	0.4701
(216.07)	h	117.4						122.3	130.3	138.3	146.4	154.7
	s	0.1943						0.2014	0.2126	0.2232	0.2333	0.2431

TABLE B–21 (continued)

Pressure, psia, p (Sat. temp., °F)		Saturation properties	Temperature, °F, T									
			240	280	320	360	400	440	480	520	560	600
140	v	0.275	0.2846	0.3155	0.3440	0.3713	0.3978	0.4238	0.4493	0.4743		
(228.91)	h	118.9	121.3	129.5	137.6	145.8	154.1	162.5	171.1	179.9		
	s	0.1949	0.1983	0.2098	0.2205	0.2307	0.2406	0.2502	0.2595	0.2686		
160	v	0.239		0.2680	0.2944	0.3194	0.3435	0.3670	0.3900	0.4125		
(240.45)	h	120.2		128.6	136.9	145.1	153.5	161.9	170.6	179.4		
	s	0.1954		0.2071	0.2180	0.2283	0.2383	0.2479	0.2573	0.2664		
180	v	0.210		0.2306	0.2556	0.2789	0.3012	0.3227	0.3438	0.3644	0.3845	
(250.97)	h	121.3		127.7	136.1	144.4	152.8	161.4	170.0	178.9	187.8	
	s	0.1959		0.2046	0.2157	0.2261	0.2361	0.2458	0.2553	0.2645	0.2734	
200	v	0.187		0.2001	0.2244	0.2463	0.2671	0.2872	0.3068	0.3258	0.3444	
(260.64)	h	122.3		126.7	135.3	143.7	152.2	160.8	169.5	178.4	187.4	
	s	0.1962		0.2022	0.2136	0.2241	0.2342	0.2439	0.2534	0.2627	0.2717	
240	v	0.151		0.1528	0.1768	0.1971	0.2159	0.2337	0.2510	0.2677	0.2839	
(278.02)	h	123.9		124.4	133.6	142.3	150.9	159.6	168.4	177.4	186.5	
	s	0.1966		0.1973	0.2095	0.2204	0.2306	0.2405	0.2501	0.2594	0.2685	
280	v	0.126			0.1417	0.1614	0.1789	0.1953	0.2110	0.2260	0.2406	0.2548
(293.34)	h	125.0			131.7	140.8	149.6	158.4	167.3	176.4	185.5	194.8
	s	0.1968			0.2055	0.2169	0.2274	0.2374	0.2471	0.2565	0.2657	0.2746
320	v	0.106			0.1139	0.1342	0.1509	0.1662	0.1807	0.1946	0.2080	0.2209
(307.09)	h	125.7			129.3	139.1	148.2	157.2	166.2	175.3	184.6	194.0
	s	0.1966			0.2013	0.2135	0.2244	0.2346	0.2444	0.2539	0.2631	0.2722
360	v	0.089			0.0896	0.1123	0.1288	0.1434	0.1571	0.1700	0.1824	0.1944
(319.57)	h	126.1			126.2	137.3	146.8	155.9	165.1	174.3	183.7	193.1
	s	0.1961			0.1963	0.2102	0.2215	0.2319	0.2418	0.2515	0.2608	0.2699
400	v	0.076				0.0940	0.1108	0.1250	0.1380	0.1502	0.1619	0.1730
(331.0)	h	126.0				135.2	145.2	154.6	163.9	173.3	182.7	192.3
	s	0.1953				0.2067	0.2187	0.2293	0.2394	0.2492	0.2586	0.2678
500	v	0.046				0.0536	0.0772	0.0912	0.1032	0.1142	0.1245	0.1343
(355.77)	h	122.2				126.4	140.8	151.2	160.9	170.6	180.3	190.1
	s	0.1891				0.1943	0.2116	0.2233	0.2339	0.2440	0.2537	0.2631

Source: Data abstracted from "Thermodynamic Properties of HCFC-123, 1993" with permission of E. I. du Pont de Nemours & Co.

TABLE B–22 Properties of R-134a (CH_2FCF_3) at saturation (SI units)

Temperature, °C,	Pressure, kPa,	Specific volume, $m^3/kg \times 10^3$		Enthalpy, kJ/kg		Entropy, kJ/kg · K	
		v_f	v_g	h_f	h_g	s_f	s_g
−40	51.14	0.7069	361.4	148.4	374.3	0.7967	1.7655
−35	66.07	0.7142	284.3	154.6	377.4	0.8231	1.7586
−30	84.29	0.7217	226.0	160.9	380.6	0.8492	1.7525
−25	106.32	0.7294	181.7	167.3	383.7	0.8750	1.7470
−20	132.67	0.7375	147.4	173.7	386.8	0.9005	1.7422
−15	163.90	0.7458	120.7	180.2	389.8	0.9257	1.7379
−10	200.60	0.7545	99.6	186.7	392.9	0.9507	1.7341
−5	243.39	0.7637	82.8	193.3	395.9	0.9755	1.7308
0	292.93	0.7732	69.3	200.0	398.8	1.0000	1.7278
5	349.87	0.7833	58.3	206.8	401.7	1.0244	1.7252
10	414.92	0.7938	49.4	213.6	404.5	1.0485	1.7229
15	488.78	0.8050	42.1	220.5	407.3	1.0726	1.7208
20	572.25	0.8167	36.0	227.5	410.0	1.0964	1.7189
25	666.06	0.8293	30.9	234.6	412.6	1.1202	1.7171
30	771.02	0.8427	26.6	241.8	415.1	1.1439	1.7155
35	887.91	0.8570	23.0	249.2	417.5	1.1676	1.7138
40	1017.61	0.8725	20.0	256.6	419.8	1.1912	1.7122
45	1161.01	0.8893	17.4	264.2	421.9	1.2148	1.7105
50	1319.00	0.9076	15.1	217.9	423.8	1.2384	1.7086
55	1492.59	0.9277	13.2	279.8	425.6	1.2622	1.7064
60	1682.76	0.9501	11.5	287.9	427.1	1.2861	1.7039
65	1890.54	0.9753	10.0	296.2	428.3	1.3102	1.7009
70	2117.34	1.004	8.7	304.8	429.1	1.3347	1.6971
75	2364.31	1.038	7.5	313.7	429.5	1.3597	1.6924
80	2632.97	1.078	6.5	322.9	429.2	1.3854	1.6863
85	2925.11	1.128	5.5	332.8	428.1	1.4121	1.6782
90	3242.87	1.194	4.6	343.4	425.5	1.4406	1.6668
95	3589.44	1.295	3.7	355.6	420.5	1.4727	1.6489
100	3969.94	1.535	2.7	373.2	407.0	1.5187	1.6092
101	4051.35	1.766	2.2	383.0	396.0	1.5447	1.5794

Source: Data abstracted from "Thermodynamic Properties of HFC-134a, 1993" with permission of E. I. du Pont de Nemours & Co.

TABLE B–22 (continued) Properties of R-134a (CH_2FCF_3) at saturation (English units)

Temperature, °F,	Pressure, psia,	Specific volume, ft^3/lbm		Enthalpy, Btu/lbm		Entropy, Btu/lbm · °R	
		v_f	v_g	h_f	h_g	s_f	s_g
−40	7.417	0.0113	5.790	0.0	97.2	0.0000	0.2316
−30	9.851	0.0115	4.437	3.0	98.7	0.0070	0.2297
−20	12.885	0.0116	3.447	6.0	100.2	0.0139	0.2281
−10	16.620	0.0117	2.712	9.0	101.7	0.0208	0.2267
0	21.163	0.0119	2.158	12.1	103.1	0.0275	0.2255
10	26.625	0.0120	1.736	15.2	104.6	0.0342	0.2244
20	33.129	0.0122	1.409	18.4	106.0	0.0408	0.2235
30	40.800	0.0124	1.154	21.6	107.4	0.0473	0.2227
40	49.771	0.0125	0.9523	24.8	108.8	0.0538	0.2220
50	60.180	0.0127	0.7916	28.0	110.2	0.0602	0.2214
60	72.167	0.0129	0.6622	31.4	111.5	0.0666	0.2208
70	85.890	0.0131	0.5570	34.7	112.8	0.0729	0.2203
80	101.494	0.0134	0.4709	38.1	114.0	0.0792	0.2199
90	119.138	0.0136	0.3999	41.6	115.2	0.0855	0.2194
100	138.996	0.0139	0.3408	45.1	116.3	0.0918	0.2190
110	161.227	0.0142	0.2912	48.7	117.4	0.0981	0.2185
120	186.023	0.0145	0.2494	52.4	118.3	0.1043	0.2181
130	213.572	0.0148	0.2139	56.2	119.2	0.1106	0.2175
140	244.068	0.0152	0.1835	60.0	119.9	0.1170	0.2168
150	277.721	0.0157	0.1573	64.0	120.5	0.1234	0.2160
160	314.800	0.0162	0.1344	68.1	120.8	0.1299	0.2150
170	355.547	0.0168	0.1143	72.4	120.9	0.1366	0.2137
180	400.280	0.0176	0.0964	76.9	120.7	0.1435	0.2119
190	449.384	0.0186	0.0800	81.8	119.8	0.1508	0.2093
200	503.361	0.0201	0.0645	87.3	118.0	0.1589	0.2054
210	563.037	0.0232	0.0474	94.4	113.4	0.1693	0.1976
213	582.316	0.0259	0.0394	98.4	109.5	0.1750	0.1915

Source: Data abstracted from "Thermodynamic Properties of HFC-134a, 1993" with permission of E. I. du Pont de Nemours & Co.

TABLE B–22 (continued) Properties of R-134a (CH_2FCF_3) in superheat (SI units):
v (m^3/kg); h (kJ/kg); s (kJ/kg · K)

Pressure, kPa, p (Sat. temp., °C)		Saturation properties	Temperature, °C, T									
			−20	0	20	40	60	80	100	120	140	160
80	v	0.2375	0.2499	0.2720	0.2934	0.3147	0.3357	0.3566	0.3774	0.3981		
(−31.09)	h	379.9	388.5	404.5	421.0	438.1	455.8	474.2	493.1	512.7		
	s	1.754	1.789	1.849	1.908	1.964	2.019	2.072	2.125	2.176		
100	v	0.1925	0.1983	0.2163	0.2337	0.2509	0.2679	0.2847	0.3014	0.3180		
(−26.34)	h	382.8	387.9	404.0	420.6	437.7	455.5	473.9	492.9	512.5		
	s	1.748	1.769	1.830	1.888	1.945	2.000	2.054	2.106	2.157		
150	v	0.1312		0.1420	0.1540	0.1658	0.1773	0.1888	0.200	0.2112		
(−17.12)	h	388.5		402.7	419.6	436.9	454.9	473.4	492.4	512.1		
	s	1.740		1.793	1.853	1.910	1.966	2.020	2.072	2.123		
200	v	0.0999		0.1048	0.1142	0.1232	0.1321	0.1408	0.1493	0.1578	0.1663	
(−10.08)	h	392.8		401.4	418.5	436.1	454.2	472.8	491.9	511.7	532.0	
	s	1.734		1.766	1.827	1.885	1.941	1.995	2.048	2.099	2.150	
250	v	0.0807		0.0824	0.0902	0.0977	0.1049	0.1120	0.1189	0.1258	0.1326	
(−4.29)	h	396.3		400.0	417.5	435.3	453.5	472.2	491.4	511.2	531.6	
	s	1.730		1.744	1.806	1.865	1.921	1.976	2.029	2.080	2.131	
300	v	0.0677			0.0742	0.0806	0.0868	0.0928	0.0986	0.1044	0.1102	
(0.66)	h	399.2			411.4	434.4	452.8	471.6	490.9	510.8	531.2	
	s	1.728			1.788	1.848	1.905	1.959	2.013	2.065	2.115	
400	v	0.0512			0.0542	0.0593	0.0641	0.0688	0.0733	0.0777	0.0821	0.0864
(8.91)	h	403.9			414.2	432.7	451.4	470.4	489.9	509.9	530.4	551.4
	s	1.723			1.759	1.820	1.878	1.934	1.987	2.039	2.090	2.140
500	v	0.0411			0.0421	0.0465	0.0505	0.0543	0.0581	0.0617	0.0653	0.0688
(15.71)	h	407.7			411.8	430.8	449.9	469.2	488.9	509.0	529.6	550.7
	s	1.721			1.735	1.798	1.857	1.913	1.967	2.020	2.071	2.121

TABLE B–22 (continued)

Pressure, kPa, p (Sat. temp., °C)		Saturation properties	Temperature, °C, T										
			40	**60**	**80**	**100**	**120**	**140**	**160**	**180**	**200**	**220**	**240**
600	v	0.0343	0.0379	0.0414	0.0447	0.0479	0.0510	0.0540	0.0570				
(21.54)	h	410.8	428.9	448.4	468.0	487.8	508.1	528.8	550.0				
	s	1.718	1.778	1.838	1.895	1.950	2.003	2.054	2.105				
800	v	0.0257	0.0271	0.0300	0.0327	0.0352	0.0376	0.0400	0.0423	0.0446			
(31.29)	h	415.7	424.8	445.2	465.4	485.7	506.3	527.2	548.6	570.5			
	s	1.715	1.745	1.808	1.866	1.922	1.976	2.025	2.079	2.128			
1000	v	0.0203	0.0204	0.0231	0.0254	0.0276	0.0296	0.0316	0.0335	0.0353			
(39.35)	h	419.5	420.2	441.8	462.7	483.5	504.4	525.6	547.2	569.2			
	s	1.712	1.715	1.782	1.843	1.900	1.954	2.007	2.058	2.108			
1500	v	0.0131		0.0136	0.0156	0.0173	0.0189	0.0203	0.0217	0.0230	0.0243		
(55.2)	h	425.7		431.7	455.2	477.5	499.4	521.4	543.5	565.9	588.7		
	s	1.706		1.725	1.793	1.855	1.912	1.966	2.019	2.069	2.118		
2000	v	0.0093			0.0106	0.0121	0.0134	0.0146	0.0158	0.0168	0.0178	0.0188	
(67.47)	h	428.8			446.1	470.8	494.1	516.9	539.7	562.6	585.7	609.1	
	s	1.700			1.749	1.817	1.878	1.935	1.988	2.040	2.090	2.139	
3000	v	0.0053				0.0067	0.0079	0.0089	0.0098	0.0106	0.0114	0.0121	0.0128
(86.22)	h	427.6				453.5	481.8	507.2	531.6	555.6	579.6	603.7	628.0
	s	1.676				1.747	1.820	1.884	1.941	1.995	2.047	2.098	2.145
4000	v	0.0025					0.0050	0.0060	0.0068	0.0075	0.0082	0.0088	0.0093
(100.37)	h	404.4					465.7	495.9	522.7	548.2	573.2	598.1	623.0
	s	1.602					1.763	1.839	1.902	1.959	2.013	2.065	2.114

Source: Data abstracted from “Thermodynamic Properties of HFC-134a, 1993” with permission of E. I. du Pont de Nemours & Co.

TABLE B–22 (continued) Properties of R-134a (CH_2FCF_3) in superheat (English units): v (ft^3/lbm); h (Btu/lbm); s (Btu/lbm · °R)

Pressure, psia, p (Sat. temp., °F)		Saturation properties	Temperature, °F, T									
			0	20	40	60	80	100	120	140	160	180
10	v	4.37	4.713	4.936	5.160	5.379	5.599	5.817	6.035	6.250	6.464	6.680
(−29.5)	h	98.8	104.2	108.0	111.8	115.8	119.8	123.9	128.0	132.3	136.7	141.1
	s	0.2297	0.2419	0.2499	0.2578	0.2655	0.2731	0.2805	0.2879	0.2951	0.3022	0.3093
20	v	2.28	2.291	2.412	2.530	2.646	2.760	2.873	2.984	3.095	3.205	3.315
(−22.4)	h	102.8	103.3	107.2	111.1	115.1	119.2	123.4	127.6	131.9	136.3	140.8
	s	0.2258	0.2268	0.2351	0.2432	0.2511	0.2588	0.2664	0.2738	0.2811	0.2883	0.2954
40	v	1.18			1.212	1.277	1.339	1.400	1.459	1.517	1.575	1.632
(29.0)	h	107.3			109.6	113.8	118.1	122.4	126.7	131.1	135.6	140.1
	s	0.2228			0.2275	0.2357	0.2437	0.2515	0.2592	0.2666	0.2739	0.2811
60	v	0.79				0.8179	0.8636	0.9073	0.9497	0.9910	1.031	1.071
(49.8)	h	110.2				112.4	116.9	121.3	125.8	130.3	134.8	139.4
	s	0.2214				0.2258	0.2342	0.2422	0.2500	0.2577	0.2651	0.2724
80	v	0.60					0.6245	0.6602	0.6943	0.7271	0.7590	0.7902
(65.9)	h	112.3					115.6	120.2	124.8	129.4	134.0	138.7
	s	0.2205					0.2267	0.2351	0.2432	0.2510	0.2586	0.2660
100	v	0.48					0.4795	0.5109	0.5403	0.5684	0.5953	0.6214
(79.1)	h	113.9					114.1	119.0	123.7	128.5	133.2	137.9
	s	0.2199					0.2203	0.2291	0.2375	0.2455	0.2533	0.2608
120	v	0.40						0.4104	0.4371	0.4621	0.4858	0.5086
(90.5)	h	115.3						117.7	122.6	127.5	132.3	137.2
	s	0.2194						0.2238	0.2324	0.2407	0.2487	0.2564
140	v	0.34							0.3627	0.3857	0.4073	0.4278
(100.5)	h	116.4							121.4	126.5	131.5	136.4
	s	0.2190							0.2279	0.2364	0.2446	0.2524

TABLE B–22 (continued)

Pressure, psia, p (Sat. temp., °F)		Saturation properties	Temperature, °F, T									
			120	140	160	180	200	220	240	260	280	300
160	v	0.29	0.3061	0.3280	0.3481	0.3670	0.3850	0.4024	0.4192	0.4355	0.4517	0.4673
(109.5)	h	117.3	120.2	125.4	130.5	135.6	140.6	145.7	150.7	155.8	160.9	166.0
	s	0.2186	0.2235	0.2324	0.2409	0.2489	0.2566	0.2641	0.2715	0.2786	0.2856	0.2925
180	v	0.26	0.2613	0.2825	0.3017	0.3195	0.3363	0.3523	0.3678	0.3828	0.3974	0.4117
(117.7)	h	118.1	118.8	124.3	129.6	134.8	139.9	145.0	150.1	155.2	160.4	165.6
	s	0.2182	0.2193	0.2287	0.2374	0.2456	0.2535	0.2611	0.2685	0.2757	0.2828	0.2897
200	v	0.23		0.2457	0.2643	0.2813	0.2971	0.3122	0.3266	0.3405	0.3540	0.3672
(125.2)	h	118.8		123.0	128.5	133.9	139.1	144.3	149.5	154.7	159.9	165.1
	s	0.2178		0.2250	0.2340	0.2425	0.2505	0.2583	0.2658	0.2731	0.2802	0.2871
240	v	0.19		0.1885	0.2071	0.2233	0.2380	0.2517	0.2646	0.2769	0.2888	0.3003
(138.7)	h	119.8		120.2	126.3	132.0	137.5	142.9	148.2	153.5	158.8	164.1
	s	0.2169		0.2176	0.2276	0.2366	0.2451	0.2531	0.2609	0.2683	0.2756	0.2827
280	v	0.16			0.1646	0.1810	0.1952	0.2081	0.2200	0.2313	0.2421	0.2525
(150.6)	h	120.5			123.7	129.9	135.7	141.4	146.9	152.3	157.7	163.1
	s	0.2160			0.2211	0.2310	0.2400	0.2484	0.2564	0.2641	0.2715	0.2787
320	v	0.13				0.1482	0.1625	0.1750	0.1864	0.1969	0.2069	0.2165
(161.3)	h	120.9				127.5	133.8	139.7	145.5	151.1	156.6	162.1
	s	0.2148				0.2254	0.2351	0.2439	0.2523	0.2602	0.2678	0.2751
360	v	0.11				0.1210	0.1363	0.1489	0.1600	0.1701	0.1795	0.1884
(171)	h	120.9				124.6	131.6	138.0	144.0	149.8	155.5	161.1
	s	0.2135				0.2193	0.2301	0.2396	0.2483	0.2565	0.2643	0.2718
400	v	0.10				0.0966	0.1145	0.1276	0.1386	0.1484	0.1575	0.1659
(179.9)	h	120.7				120.7	129.1	136.0	142.4	148.4	154.3	160.0
	s	0.2119				0.2119	0.2249	0.2352	0.2444	0.2529	0.2609	0.2686
500	v	0.07					0.0665	0.0867	0.0990	0.1088	0.1174	0.1252
(199.4)	h	118.1					118.7	130.0	137.8	144.6	151.0	157.2
	s	0.2057					0.2065	0.2234	0.2348	0.2444	0.2532	0.2614

Source: Data abstracted from "Thermodynamic Properties of HFC-134a, 1993" with permission of E. I. du Pont de Nemours & Co.

TABLE B–23 Saturated R-407c (R-32, 23%/R-125, 25%/R134a, 52%) (SI units)

Pressure, kPa	Temperature °C Bubble	Temperature °C Dew	Specific volume, cm³/g v_f	v_g	Enthalpy, kJ/kg h_f	h_g	Entropy, kJ/kg·K s_f	s_g
20	−72.79	−65.14	0.668	990.7	103.81	372.02	0.5934	1.9078
60	−54.16	−46.88	0.707	353.9	127.48	383.20	0.7061	1.8553
100	−44.04	−36.97	0.723	219.0	140.53	389.13	0.7643	1.8333
140	−36.78	−29.85	0.735	159.4	150.03	393.28	0.8050	1.8196
180	−31.00	−24.20	0.745	125.6	157.65	396.51	0.8367	1.8098
220	−26.15	−19.45	0.754	103.8	164.11	399.16	0.8630	1.8022
260	−21.95	−15.34	0.762	88.47	169.75	401.41	0.8855	1.7960
300	−18.22	−11.69	0.769	77.12	174.80	403.36	0.9053	1.7908
360	−13.28	−6.86	0.779	64.67	181.53	405.88	0.9313	1.7843
420	−8.94	−2.62	0.788	55.67	187.50	408.02	0.9539	1.7789
500	−3.84	2.36	0.800	46.93	194.61	410.45	0.9803	1.7728
600	1.74	7.80	0.813	39.19	202.46	412.97	1.0089	1.7665
800	11.05	16.87	0.837	29.34	215.84	416.85	1.0562	1.7564
1000	18.75	24.35	0.859	23.33	227.18	419.69	1.0949	1.7483
1200	25.37	30.77	0.880	19.25	237.16	421.81	1.1281	1.7412
1400	31.21	36.41	0.901	16.30	246.17	423.38	1.1575	1.7348
1600	36.46	41.47	0.921	14.06	254.46	424.53	1.1839	1.7288
1800	41.24	46.07	0.942	12.30	262.19	425.30	1.2081	1.7229
2000	45.65	50.29	0.963	10.87	269.48	425.75	1.2305	1.7172
2200	49.73	54.19	0.985	9.69	276.42	425.91	1.2515	1.7113
2400	53.55	57.82	1.007	8.69	283.08	425.79	1.2714	1.7054
2600	57.14	61.22	1.030	7.83	289.51	425.39	1.2903	1.6992
2800	60.53	64.41	1.055	7.09	295.77	424.72	1.3086	1.6928
3000	63.74	67.41	1.081	6.43	301.91	423.76	1.3262	1.6860
3200	66.80	70.25	1.110	5.85	307.97	422.48	1.3435	1.6787
3400	69.71	72.94	1.142	5.31	314.01	420.85	1.3605	1.6707
3600	72.50	75.49	1.177	4.83	320.10	418.79	1.3775	1.6618
3800	75.18	77.91	1.218	4.37	326.32	416.20	1.3948	1.6518
4000	77.75	80.19	1.267	3.93	332.81	412.89	1.4126	1.6400
4200	80.24	82.33	1.331	3.50	339.86	408.43	1.4319	1.6254
4400	82.67	84.30	1.424	3.03	348.17	401.74	1.4546	1.6048
4652.8	86.08	86.08	1.976	1.976	375.52	375.52	1.5298	1.5298

TABLE B–23 (continued) Saturated R-407c (R-32, 23%/R-125, 25%/R134a, 52%) (English units)

Pressure, psia	Temperature °F		Specific volume, ft^3/lbm		Enthalpy, Btu/lbm		Entropy, Btu/lbm · °R	
	Bubble	Dew	v_f	v_g	h_f	h_g	s_f	s_g
3.0	−98.09	−84.34	0.0109	15.38	−17.78	97.46	−0.04551	0.26746
6.0	−77.61	−64.27	0.0112	8.033	−11.59	100.40	−0.02887	0.25939
10.0	−60.71	−47.71	0.0114	4.975	−6.42	102.81	−0.01565	0.25386
14.0	−48.60	−35.84	0.0116	3.626	−2.68	104.50	−0.00643	0.25042
18.0	−38.99	−26.43	0.0117	2.861	0.32	105.82	0.00075	0.24795
22.0	−30.94	−18.55	0.0118	2.367	2.84	106.91	0.00667	0.24603
26.0	−23.97	−11.72	0.0119	2.020	5.04	107.83	0.01175	0.24448
30.0	−17.80	−5.67	0.0120	1.763	7.01	108.64	0.01620	0.24318
36.0	−9.64	2.31	0.0122	1.482	9.62	109.69	0.02203	0.24156
42.0	−2.47	9.32	0.0123	1.278	11.93	110.60	0.02709	0.24021
50.0	5.95	17.55	0.0124	1.080	14.67	111.63	0.03299	0.23872
60.0	15.13	26.52	0.0126	0.9050	17.68	112.72	0.03935	0.23718
80.0	30.44	41.47	0.0129	0.6824	22.79	114.44	0.04984	0.23479
100.0	43.08	53.79	0.0132	0.5463	27.09	115.75	0.05840	0.23294
120.0	53.93	64.35	0.0135	0.4542	30.84	116.80	0.06569	0.23140
140.0	63.50	73.65	0.0137	0.3876	34.21	117.64	0.07209	0.23006
160.0	72.10	81.99	0.0139	0.3371	37.28	118.33	0.07782	0.22885
180.0	79.92	89.57	0.0142	0.2975	40.13	118.90	0.08303	0.22774
200.0	87.13	96.53	0.0144	0.2654	42.79	119.36	0.08783	0.22669
220.0	93.81	102.98	0.0146	0.2390	45.30	119.73	0.09230	0.22568
240.0	100.06	108.99	0.0148	0.2168	47.68	120.01	0.09648	0.22470
260.0	105.93	114.64	0.0151	0.1979	49.95	120.23	0.10042	0.22374
280.0	111.48	119.95	0.0153	0.1815	52.14	120.38	0.10417	0.22278
300.0	116.74	124.98	0.0155	0.1672	54.25	120.47	0.10774	0.22183
320.0	121.74	129.75	0.0158	0.1546	56.29	120.50	0.11116	0.22087
340.0	126.52	134.30	0.0160	0.1434	58.27	120.48	0.11446	0.21989
360.0	131.09	138.63	0.0163	0.1333	60.20	120.40	0.11765	0.21890
380.0	135.48	142.78	0.0165	0.1242	62.10	120.26	0.12074	0.21787
400.0	139.70	146.75	0.0168	0.1160	63.96	120.06	0.12375	0.21681
500.0	158.69	164.41	0.0184	0.0832	72.98	118.13	0.13803	0.21072
600.0	175.04	179.00	0.0210	0.0583	82.49	113.66	0.15262	0.20158
674.84	186.94	186.94	0.0317	0.0317	98.82	98.82	0.17754	0.17754

TABLE B–24 Saturated R-502 (R-22, 48.8%/R-115, 51.2%) (SI units)

Temperature, °C	Pressure kPa	Specific volume, cm^3/g		Enthalpy, kJ/kg		Entropy, kJ/kg · K	
		v_f	v_g	h_f	h_g	s_f	s_g
−60.00	48.72	0.655	318.29	139.94	318.11	0.7546	1.5905
−50.00	81.42	0.668	197.26	148.77	323.16	0.7950	1.5765
−40.00	129.64	0.683	127.69	158.09	328.15	0.8357	1.5651
−30.00	197.86	0.699	85.77	167.89	333.03	0.8767	1.5558
−20.00	291.01	0.716	59.46	178.15	337.76	0.9178	1.5483
−10.00	414.30	0.735	42.34	188.87	342.31	0.9589	1.5420
−6.00	473.26	0.743	37.21	193.27	344.07	0.9754	1.5399
0.00	573.13	0.756	30.84	200.00	346.63	1.0000	1.5368
2.00	609.65	0.761	29.01	202.27	347.47	1.0082	1.5359
4.00	647.86	0.765	27.31	204.57	348.29	1.0164	1.5350
6.00	687.79	0.770	25.73	206.87	349.10	1.0246	1.5341
8.00	729.51	0.775	24.26	209.19	349.89	1.0327	1.5332
10.00	773.05	0.780	22.88	211.53	350.67	1.0409	1.5323
12.00	818.45	0.785	21.60	213.88	351.44	1.0490	1.5315
14.00	865.77	0.790	20.40	216.24	352.20	1.0572	1.5306
16.00	915.05	0.796	19.27	218.62	352.94	1.0653	1.5298
20.00	1019.7	0.807	17.23	223.42	354.36	1.0815	1.5282
24.00	1132.7	0.819	15.44	228.28	355.72	1.0976	1.5265
28.00	1254.5	0.832	13.85	233.19	357.01	1.1137	1.5249
32.00	1385.6	0.846	12.44	238.16	358.20	1.1297	1.5231
36.00	1526.2	0.860	11.18	243.18	359.30	1.1457	1.5213
40.00	1677.0	0.877	10.05	248.27	360.28	1.1617	1.5194
44.00	1838.3	0.894	9.04	253.43	361.14	1.1776	1.5172
48.00	2010.7	0.914	8.13	258.66	361.95	1.1935	1.5148
52.00	2194.9	0.936	7.30	263.99	362.37	1.2094	1.5120
56.00	2391.5	0.961	6.54	269.44	362.67	1.2255	1.5088
60.00	2601.4	0.990	5.84	275.05	362.70	1.2418	1.5049
65.00	2884.0	1.033	5.04	282.38	362.19	1.2628	1.4988
70.00	3191.7	1.091	4.29	290.31	360.80	1.2851	1.4905
75.00	3528.4	1.175	3.55	299.48	357.79	1.3105	1.4780
80.00	3900.4	1.342	2.71	312.52	350.37	1.3461	1.4533
82.20	4075.0	1.78	1.78	332.0	332.0	1.399	1.399

TABLE B–24 (continued) Saturated R-502 (R-22, 48.8%/R-115, 51.2%) (English units)

Temperature, °F	Pressure psia	Specific volume, ft^3/lbm		Enthalpy, Btu/lbm		Entropy, Btu/lbm · °R	
		v_f	v_g	h_f	h_g	s_f	s_g
−60.00	11.182	0.0107	3.3248	−4.441	70.777	−0.01080	0.17740
−50.00	14.602	0.0108	2.5915	−2.252	71.975	−0.00541	0.17578
−40.00	18.802	0.0109	2.0453	0.000	73.162	0.00000	0.17433
−30.00	23.897	0.0111	1.6328	2.317	74.335	0.00543	0.17304
−20.00	30.006	0.0112	1.3172	4.696	75.491	0.01088	0.17189
−10.00	37.256	0.0114	1.0727	7.138	76.627	0.01633	0.17087
−6.00	40.504	0.0114	0.9907	8.131	77.075	0.01852	0.17049
−2.00	43.964	0.0115	0.9161	9.135	77.520	0.02070	0.17012
0.00	45.776	0.0115	0.8813	9.640	77.741	0.02180	0.16995
2.00	47.644	0.0116	0.8482	10.147	77.960	0.02289	0.16978
4.00	49.569	0.0116	0.8166	10.657	78.179	0.02398	0.16961
6.00	51.552	0.0116	0.7864	11.169	78.397	0.02508	0.16944
8.00	53.594	0.0117	0.7575	11.684	78.613	0.02617	0.16928
10.00	55.697	0.0117	0.7299	12.200	78.828	0.02726	0.16912
20.00	67.155	0.0119	0.6088	14.818	79.887	0.03272	0.16838
30.00	80.287	0.0121	0.5112	17.490	80.913	0.03818	0.16770
40.00	95.229	0.0123	0.4318	20.216	81.903	0.04362	0.16707
50.00	112.12	0.0125	0.3666	22.991	82.852	0.04904	0.16649
60.00	131.10	0.0127	0.3126	25.814	83.755	0.05444	0.16594
70.00	152.32	0.0130	0.2677	28.685	84.606	0.05982	0.16539
80.00	175.92	0.0133	0.2299	31.602	85.397	0.06517	0.16484
90.00	202.06	0.0136	0.1980	34.566	86.118	0.07049	0.16427
100.00	230.89	0.0139	0.1708	37.578	86.758	0.07579	0.16366
110.00	262.61	0.0143	0.1474	40.644	87.298	0.08107	0.16297
120.00	297.41	0.0147	0.1271	43.774	87.716	0.08635	0.16216
130.00	335.54	0.0152	0.1094	46.987	87.977	0.09167	0.16118
140.00	377.30	0.0159	0.0936	50.316	88.024	0.09706	0.15994
150.00	423.06	0.0166	0.0793	53.834	87.763	0.10265	0.15830
160.00	473.39	0.0177	0.0660	57.698	86.997	0.10867	0.15595
165.00	500.50	0.0185	0.0595	59.878	86.293	0.11202	0.15430
170.00	529.11	0.0195	0.0527	62.386	85.198	0.11584	0.15207
179.90	591.0	0.0286	0.0286	74.81	74.81	0.1346	0.1346

TABLE B–25 Heat of formation, Gibbs free energy of formation, and absolute entropy for selected substances and ions at standard conditions of 298 K and 1 atmopshere; $\Delta \bar{h}_f^\circ$ and $\Delta \mathcal{G}$ (kJ/kg mol); $\bar{s}^\circ$ (kJ/kg mol · K)

Substance	Chemical formula	Phase	$\Delta \bar{h}_f^\circ$	$\Delta \mathcal{G}$	$\bar{s}^\circ$
Carbon	C	Solid	0	0	5.74
Hydrogen	H_2	Gas	0	0	130.57
Nitrogen	N_2	Gas	0	0	191.50
Oxygen	O_2	Gas	0	0	205.04
Carbon monoxide	CO	Gas	−110,530	−137,150	197.56
Carbon dioxide	CO_2	Gas	−393,520	−394,380	213.67
Water	H_2O	Gas	−241,820	−228,590	188.72
Water	H_2O	Liquid	−285,830	−237,180	69.95
Hydrogen	H	Gas	218,000	203,920	114.61
Nitrogen	N	Gas	472,680	455,510	153.19
Oxygen	O	Gas	249,170	231,770	160.95
Methane	CH_4	Gas	−74,852	−50,836	186.27
Acetylene	C_2H_2	Gas	226,731	209,200	200.83
Ethylene	C_2H_4	Gas	52,300	68,116	219.45
Ethane	C_2H_6	Gas	−84,684	−32,928	229.49
Propane	C_3H_8	Gas	−103,847	−23,472	269.91
n-Butane	C_4H_{10}	Gas	−126,147	−17,154	310.12
Benzene	C_6H_6	Gas	82,927	129,662	269.20
n-Octane (2,2,4-trimethylpentane)	C_8H_{18}	Gas	−224,137	13,682	423.21
n-Octane (2,2,4-trimethylpentane)	C_8H_{18}	Liquid	−259,282	6,904	328.03
Methyl alcohol	CH_3OH	Liquid	−238,572	−166,230	126.78
Ethyl alcohol	C_2H_5OH	Liquid	−276,981	−174,138	160.67
Cellulose	$C_6H_{10}O_5$	Solid	−961,100		
Sodium chloride	$NaCl$	Aqueous		−393,310	
Sulfuric acid	H_2SO_4	Aqueous	−909,270	−744,630	20.10
Silver ion	Ag^+	Aqueous		77,160	
Sulfate ion	So_4^{2-}	Aqueous		−742,490	
Chlorine ion	Cl^-	Aqueous		−131,260	
Copper ion	Cu^{2+}	Aqueous		65,020	
Hydrogen ion	H^+	Gas		0	
Zinc ion	Zn^{2+}	Aqueous		−147,290	

Source: Data from *JANAF thermochemical tables* (Midland, Mich.: Dow Chemical Co., 1971); D. R. Stull, E. F. Westrum, Jr., and G. C. Sinke, *The chemical thermodynamics of organic compounds* (New York: John Wiley & Sons, Inc., 1969); R. S. Jessup and E. J. Prosen, *J. Res. Nat. Bur. Stand.* 44, 387–93, 1950; F. D. Rossini, D. D. Wagman, W. H. Evan, S. Levine, and I. Jaffe, "Selected Values of Chemical Thermodynamic Properties," Circular of the National Bureau of Standards 500, Washington, D.C., 1952.

TABLE B–25 (continued) Heat of formation, Gibbs free energy, and absolute entropy at 537 R, 1 atm, in Btu/lbm-mol or Btu/lbm-mol-R

Substance	Chemical formula	Phase	$\Delta \bar{h}_f^\circ$	$\Delta \mathcal{G}$	$\bar{s}^\circ$
Carbon	C	Solid	0	0	1.37
Hydrogen	H_2	Gas	0	0	31.21
Nitrogen	N_2	Gas	0	0	45.75
Oxygen	O_2	Gas	0	0	49.01
Carbon monoxide	CO	Gas	−47,551	−59,003	47.22
Carbon dioxide	CO_2	Gas	−169,296	−169,666	51.07
Water	H_2O	Gas	−104,033	−98,342	45.11
Water	H_2O	Liquid	−122,967	−102,037	16.72
Hydrogen	H	Gas	93,786	87,457	27.39
Nitrogen	N	Gas	203,352	195,965	36.61
Oxygen	O	Gas	107,195	99,710	38.47
Methane	CH_4	Gas	−32,202	−21,870	44.52
Acetylene	C_2H_2	Gas	97,542	90,000	48.00
Ethylene	C_2H_4	Gas	22,500	29,304	52.45
Ethane	C_2H_6	Gas	−36,432	−14,166	54.85
Propane	C_3H_8	Gas	−44,676	−10,098	64.51
n-Butane	C_4H_{10}	Gas	−54,270	−7,380	74.12
Benzene	C_6H_6	Gas	35,676	55,782	64.34
n-Octane (2,2,4-trimethylpentane)	C_8H_{18}	Gas	−96,426	5,886	101.15
n-Octane (2,2,4-trimethylpentane)	C_8H_{18}	Liquid	−111,546	2,970	78.40
Methyl alcohol	CH_3OH	Liquid	−102,636	−71.514	30.30
Ethyl alcohol	C_2H_5OH	Liquid	−119,160	−74,916	38.40
Cellulose	$C_6H_{10}O_5$	Solid	−413,475		
Sodium chloride	$NaCl$	Aqueous		−169,206	
Sulfuric acid	H_2SO_4	Aqueous	−391,177	−320,348	4.80
Silver ion	Ag^+	Aqueous		33,195	
Sulfate ion	So_4^{2-}	Aqueous		−320,348	
Chlorine ion	C^-	Aqueous		−56,469	
Copper ion	Cu^{2+}	Aqueous		27,972	
Hydrogen ion	H^+	Gas		0	
Zinc ion	Zn^{2+}	Aqueous		−63,366	

TABLE B–26 Heat of combustion ($-\Delta H°$ at 77°F or 25°C)*

Substance	Symbol	h (h_{fg}) of Vaporization Btu/lbm	HHV, $H_2O(l)$ and $CO_2(g)$		LHV, $H_2O(g)$ and $CO_2(g)$	
			kJ/kg	Btu/lbm	kJ/kg	Btu/lbm
Acetylene	$C_2H_2(g)$		49,916	21,460	48,227	20,734
Benzene	$C_6H_6(g)$	186	42,268	18,172	40,579	17,446
n-Butane	$C_4H_{10}(g)$	156	49,504	21,283	45,717	19,655
Isobutane	$C_4H_{10}(g)$	141	49,360	21,221	45,573	19,593
1-Butene	$C_4H_{10}(g)$	156	48,436	20,824	45,229	19,475
Carbon	C(graphite)		32,764	14,086		
Carbon monoxide	CO(g)		10,103	4,343.6		
n-Decane	$C_{10}H_{22}(g)$	155	48,004	20,638	44,601	19,175
n-Dodecane	$C_{12}H_{26}(g)$	155	47,832	20,564	44,473	19,120
Ethane	$C_2H_6(g)$		51,879	22,304	47,488	20,416
Ethylene	$C_2H_4(g)$		50,300	21,625	47,162	20,276
n-Heptane	$C_7H_{16}(g)$	157	48,439	20,825	44,924	19,314
n-Hexane	$C_6H_{14}(g)$	157	48,679	20,928	45,103	19,391
Hydrogen	$H_2(g)$		141,786	60,957	119,954	51,571
Methane	$CH_4(g)$		55,500	23,861	50,014	21,502
n-Nonane	$C_9H_{20}(g)$	156	48,118	20,687	44,685	19,211
n-Octane	$C_8H_{18}(g)$	156	48,258	20,747	44,789	19,256
n-Pentane	$C_5H_{12}(g)$	157	49,013	21,072	45,355	19,499
Isopentane	$C_5H_{12}(g)$	147	48,904	21,025	45,243	19,451
Propane	$C_3H_8(g)$	147	50,349	21,646	46,355	19,929
Propylene	$C_3H_6(g)$		48,920	21,032	45,783	19,683

Source: E. F. Obert, *Concepts of thermodynamics* (New York: McGraw-Hill Book Company), copyright 1960; by permission of McGraw-Hill Book Company.

* Data from "Selected Values of Properties of Hydrocarbons and Related Compounds," American Petroleum Institute Research Project 44, Thermodynamics Research Center, Texas A & M University, College Station, Texas (Loose Leaf Data Sheets, extant 1972).

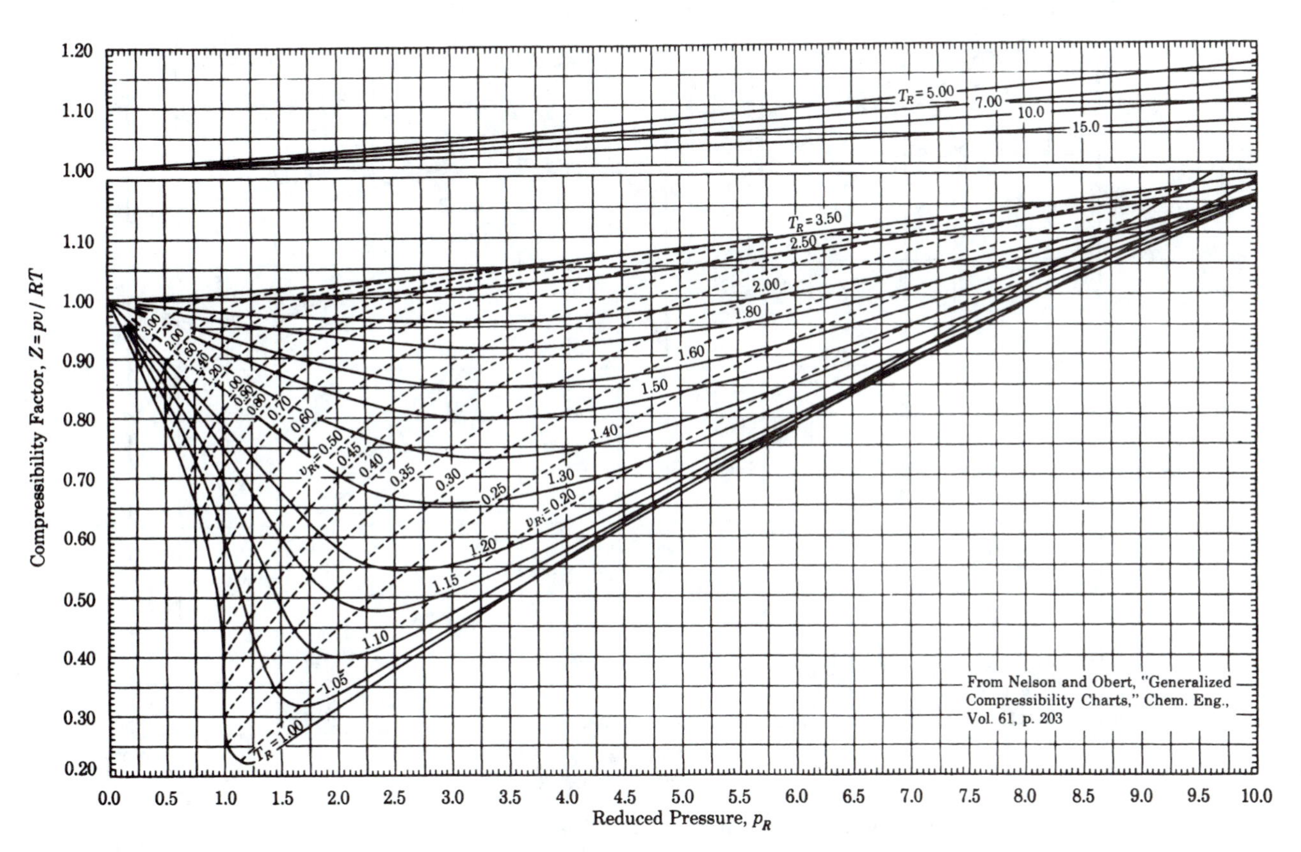

CHART B–1 Generalized compressibility chart for extended pressure range (from L. C. Nelson and E. F. Obert, "Generalized compressibility charts," excerpted by special permission from *Chemical Engineering*, Volume 61, copyright 1954, by McGraw-Hill, Inc., NY 10020)

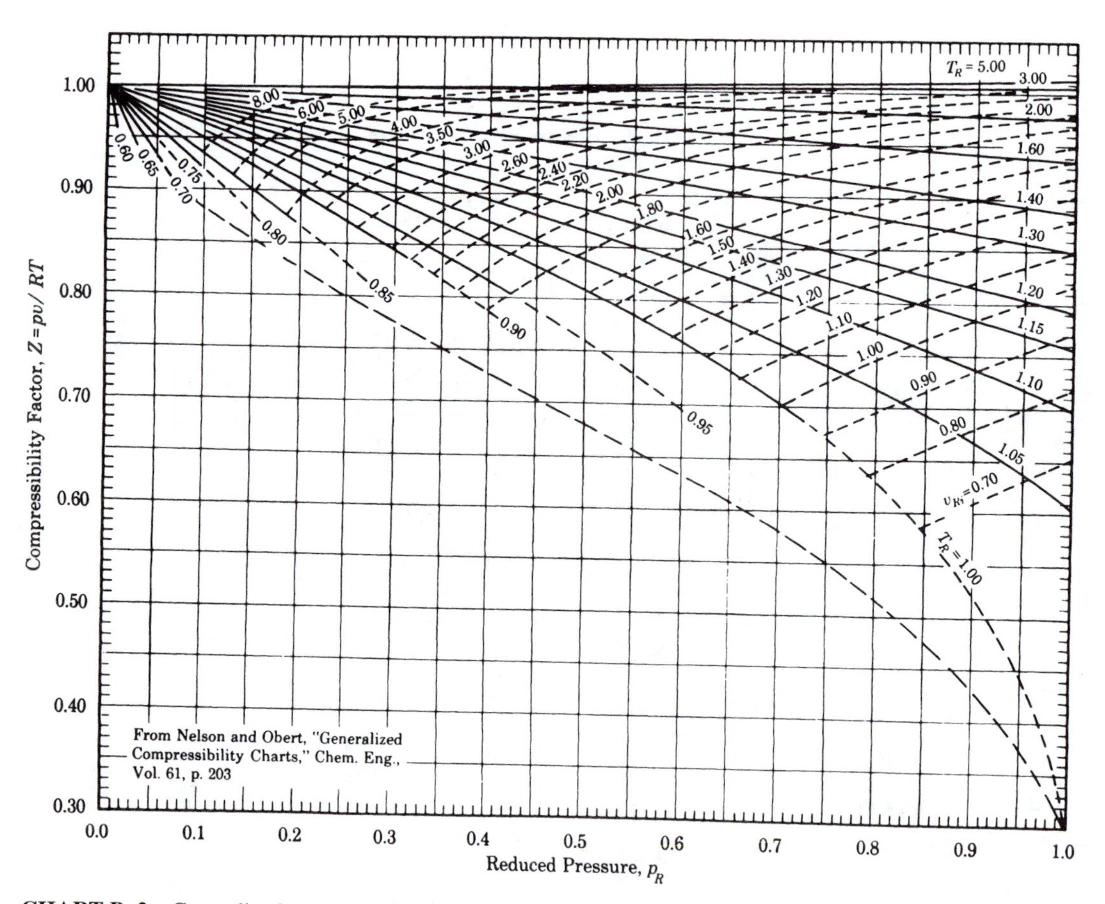

CHART B–2 Generalized compressibility chart (from L. C. Nelson and E. F. Obert, "Generalized compressibility charts," excerpted by special permission from *Chemical Engineering*, Volume 61, copyright 1954, by McGraw-Hill, Inc, NY 10020)

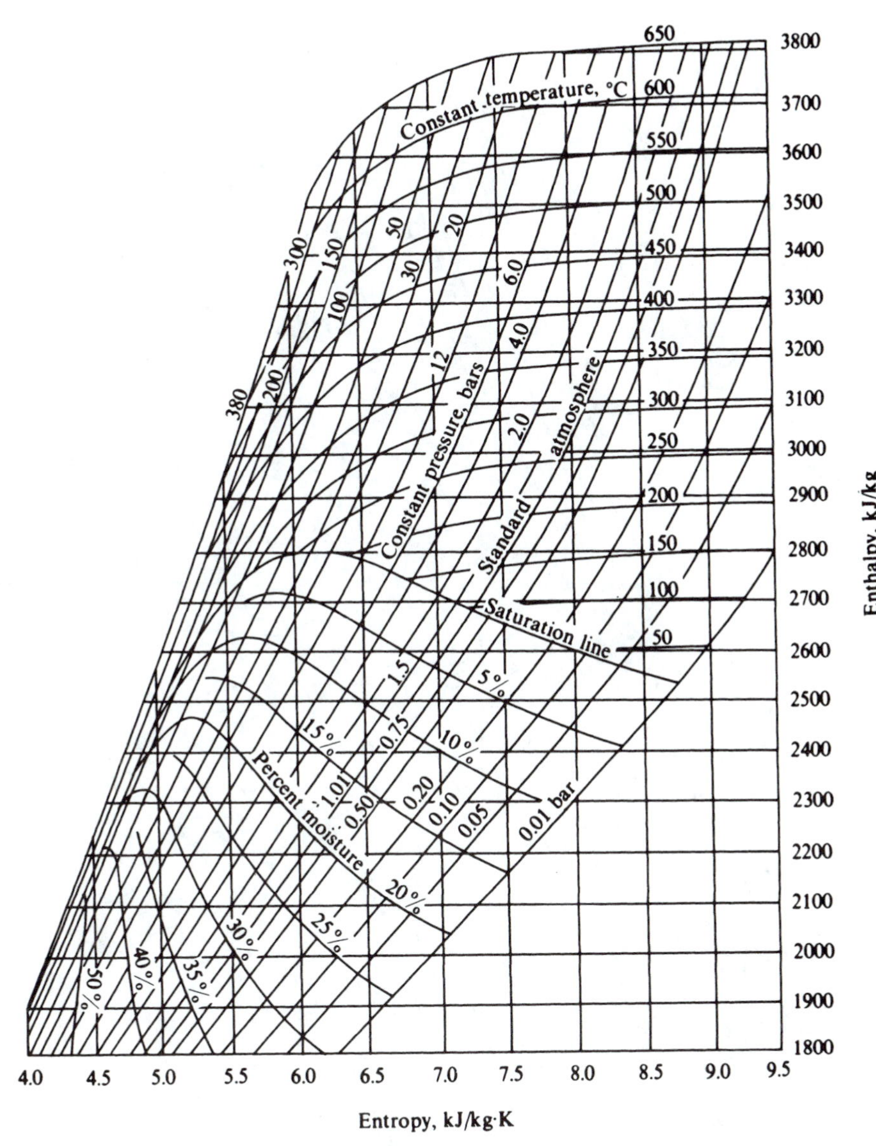

CHART B–3 Mollier diagram, SI units (*note:* 1 bar = 10^5 Pa = 100 kPa) [from Combustion Engineering, Inc., *Steam tables* (Windsor, Conn.: Combustion Engineering, Inc., 1940)]

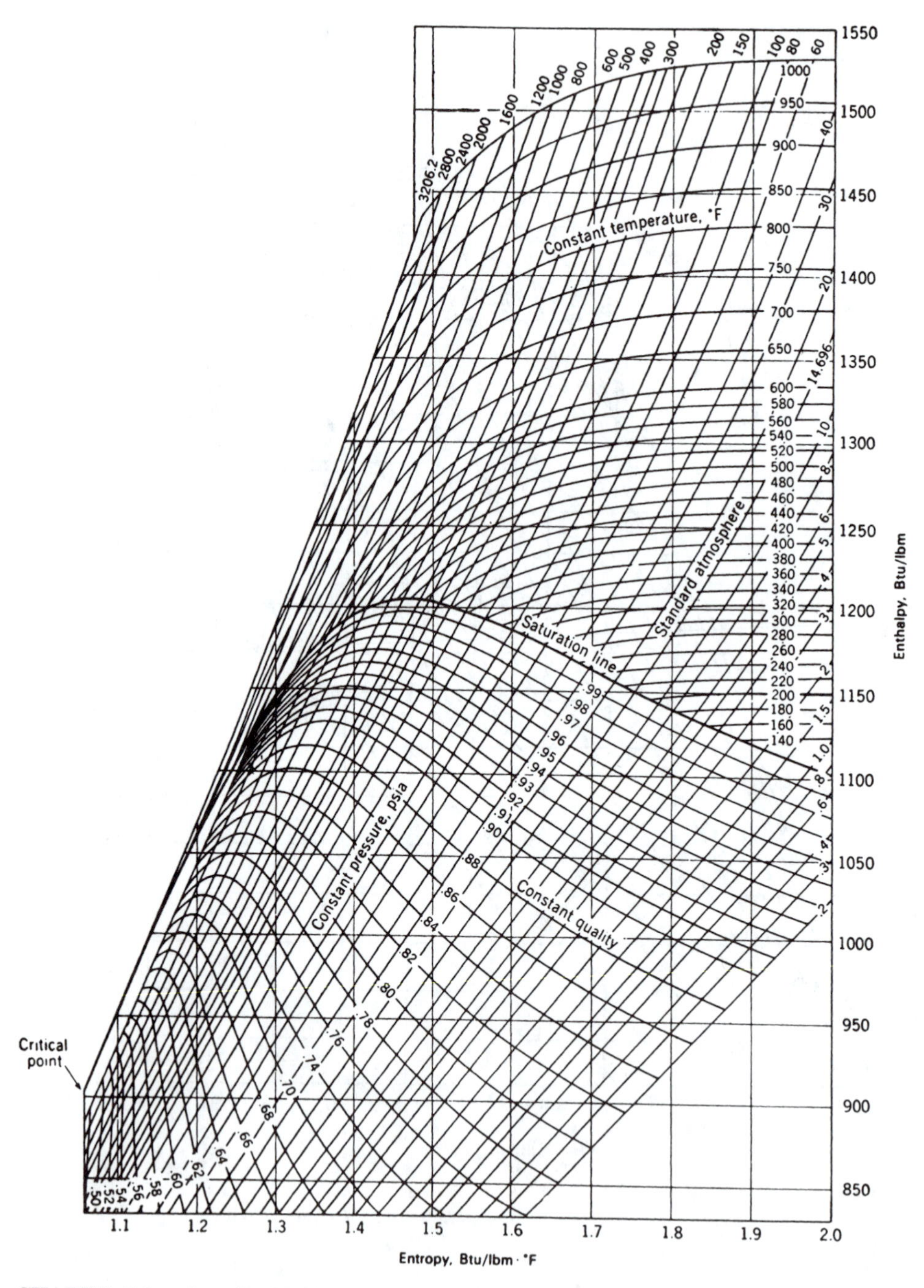

CHART B–3 (continued) Mollier diagram, English units [from Combustion Engineering, Inc., *Steam tables* (Windsor, Conn.: Combustion Engineering, Inc., 1940)]

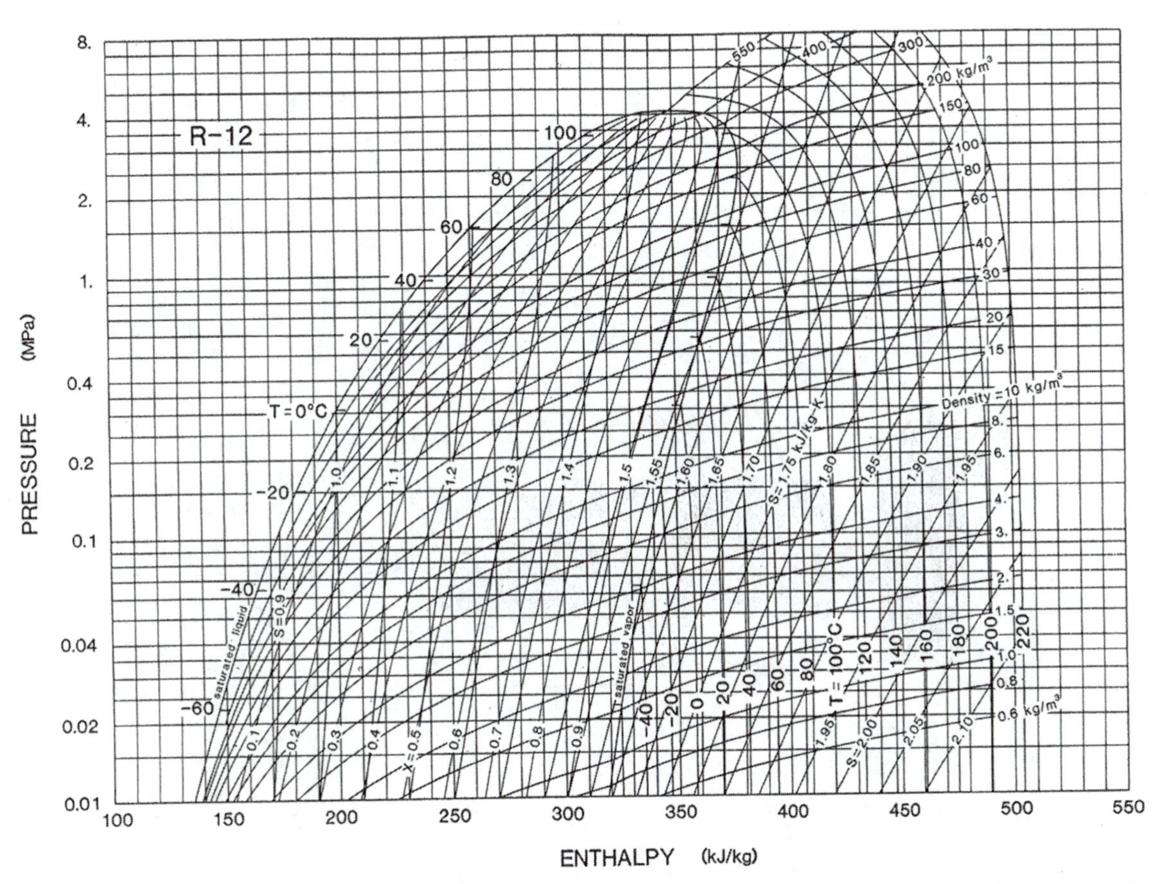

CHART B–4 Pressure–enthalpy diagram for Refrigerant-12 [from the American Society of Heating, Refrigerating, and Air Conditioning Engineers, *ASHRAE handbook, 2001 fundamentals* (Atlanta, Ga: ASHRAE, 2001)]

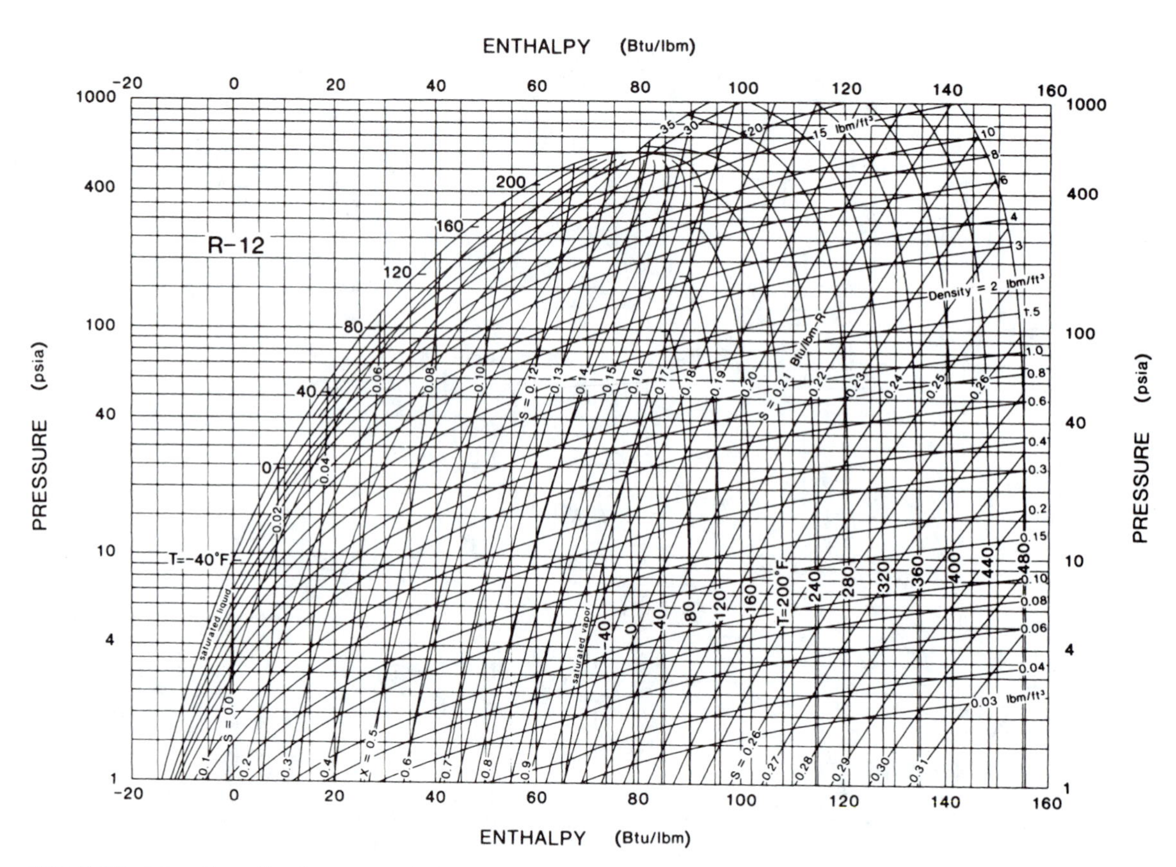

CHART B–4 (continued) Pressure–enthalpy diagram for Refrigerant-12 [from the American Society of Heating, Refrigerating, and Air Conditioning Engineers, *ASHRAE handbook, 1985 fundamentals* (Atlanta, Ga.: ASHRAE, 1985)]

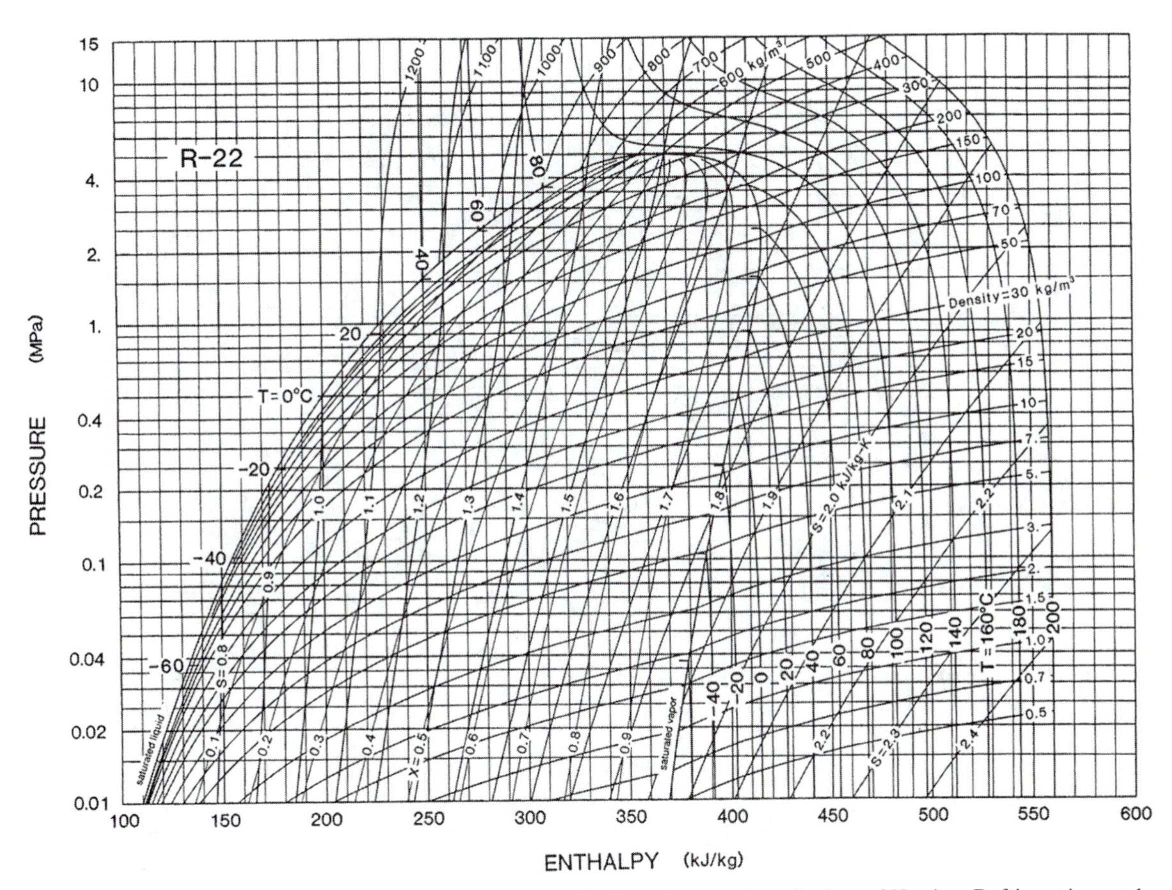

CHART B–5 Pressure–enthalpy diagram for Refrigerant-22 [from the American Society of Heating, Refrigerating, and Air Conditioning Engineers, *ASHRAE handbook, 2001 fundamentals* (Atlanta, Ga.: ASHRAE, 2001)]

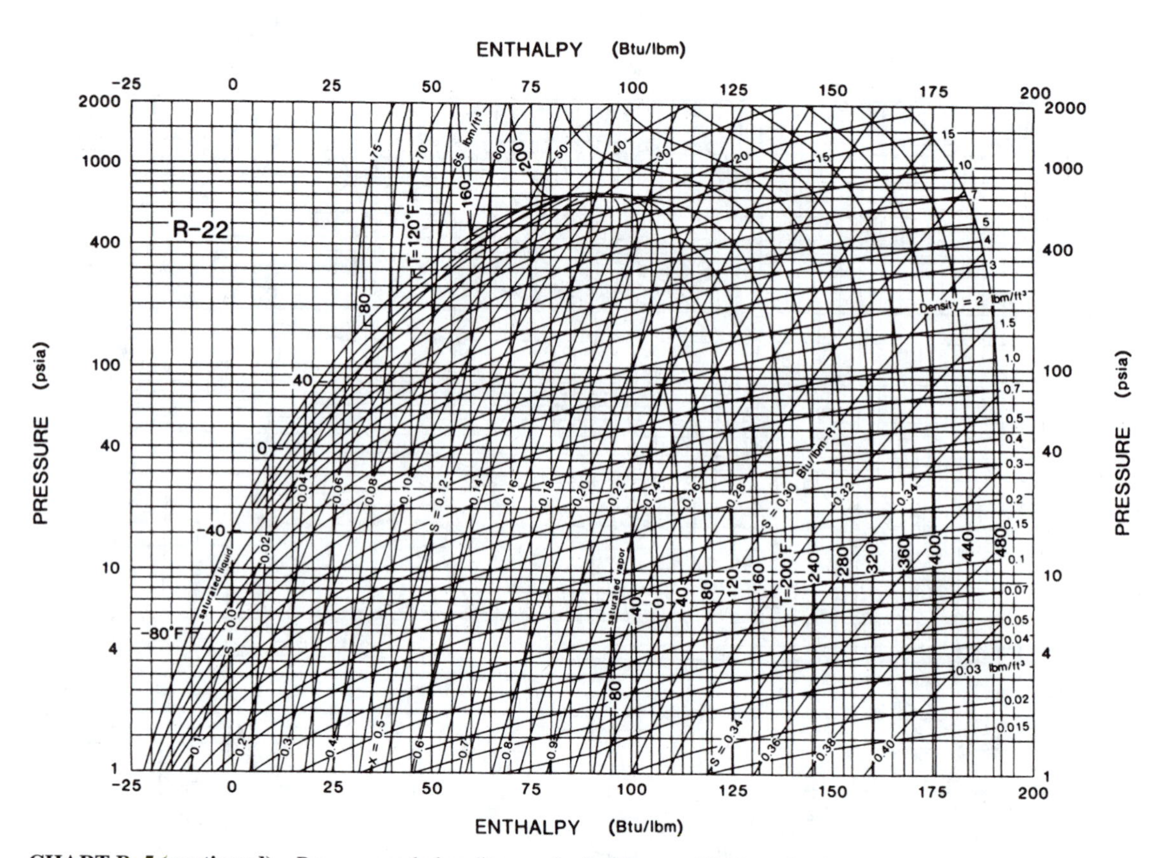

CHART B–5 (continued) Pressure–enthalpy diagram for Refrigerant-22 [from the American Society of Heating, Refrigerating, and Air Conditioning Engineers, *ASHRAE handbook*, *1985 fundamentals* (Atlanta, Ga.: ASHRAE, 1985)]

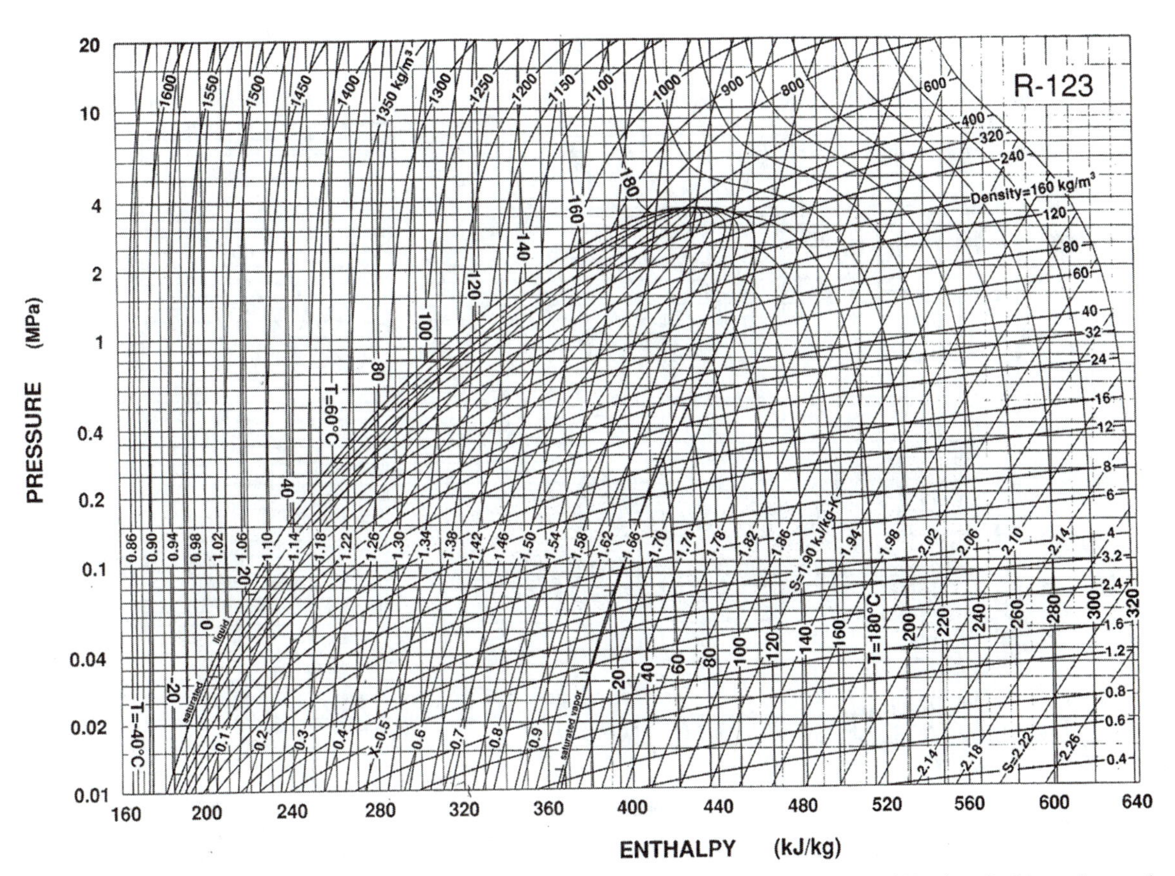

CHART B–6 Pressure–enthalpy diagram for Refrigerant-123 [from the American Society of Heating, Refrigerating, and Air Conditioning Engineers, *ASHRAE handbook, 2001 fundamentals* (Atlanta, Ga.: ASHRAE, 2001)]

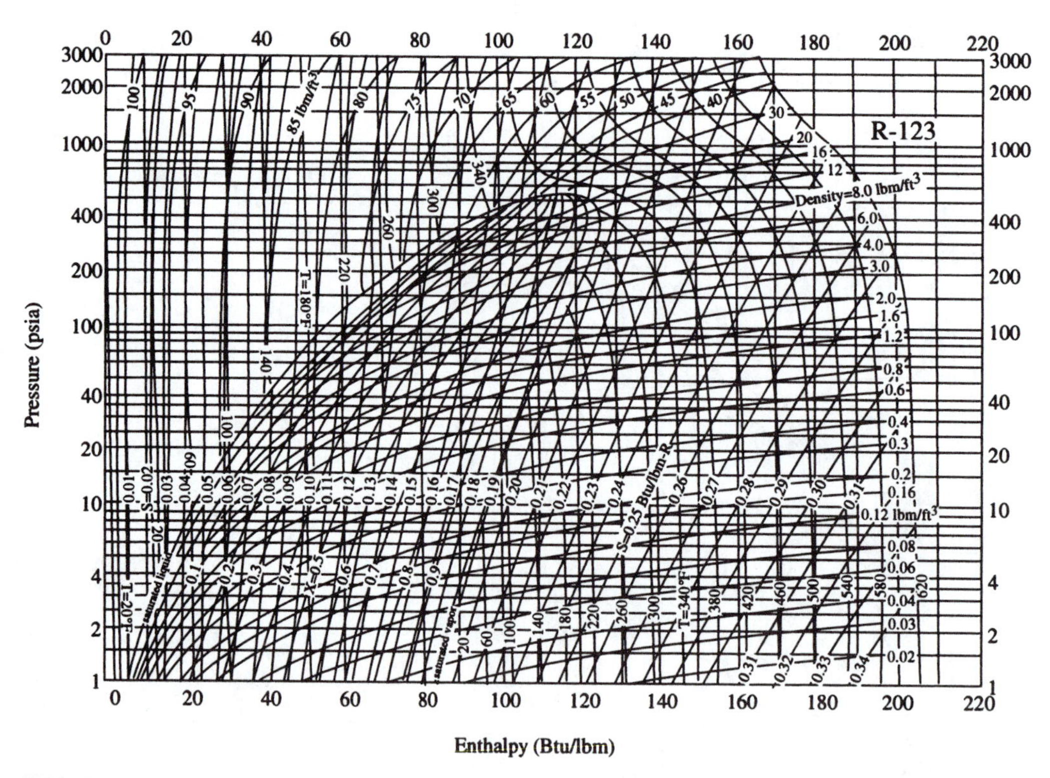

CHART B–6 (continued) Pressure–enthalpy diagram for Refrigerant-123 [reprinted with permission from the American Society of Heating, Refrigerating, and Air Conditioning Engineers, 1997, *ASHRAE handbook, fundamentals* (Atlanta, Ga.: ASHRAE, 1997)]

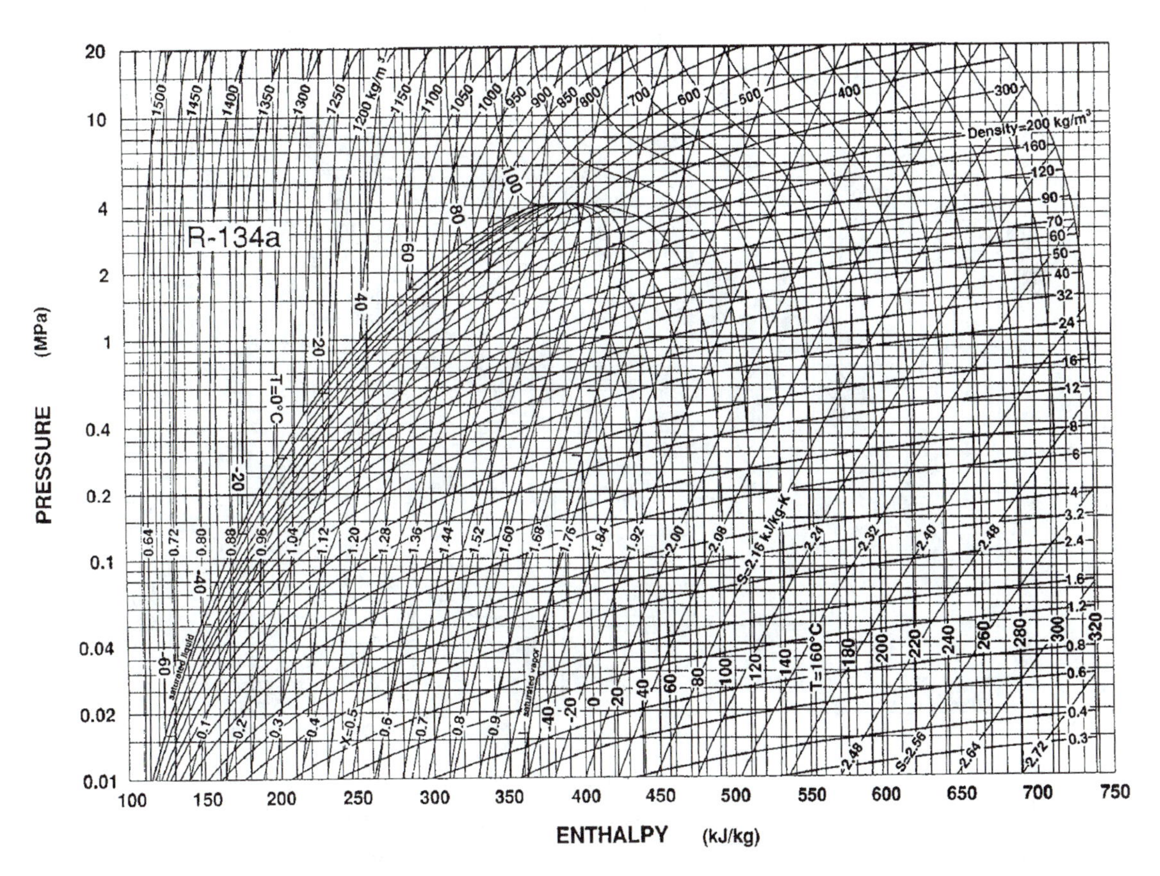

CHART B–7 Pressure–enthalpy diagram for Refrigerant-134a [from the American Society of Heating, Refrigerating, and Air Conditioning Engineers, *ASHRAE handbook, 2001 fundamentals* (Atlanta, Ga.: ASHRAE, 2001)]

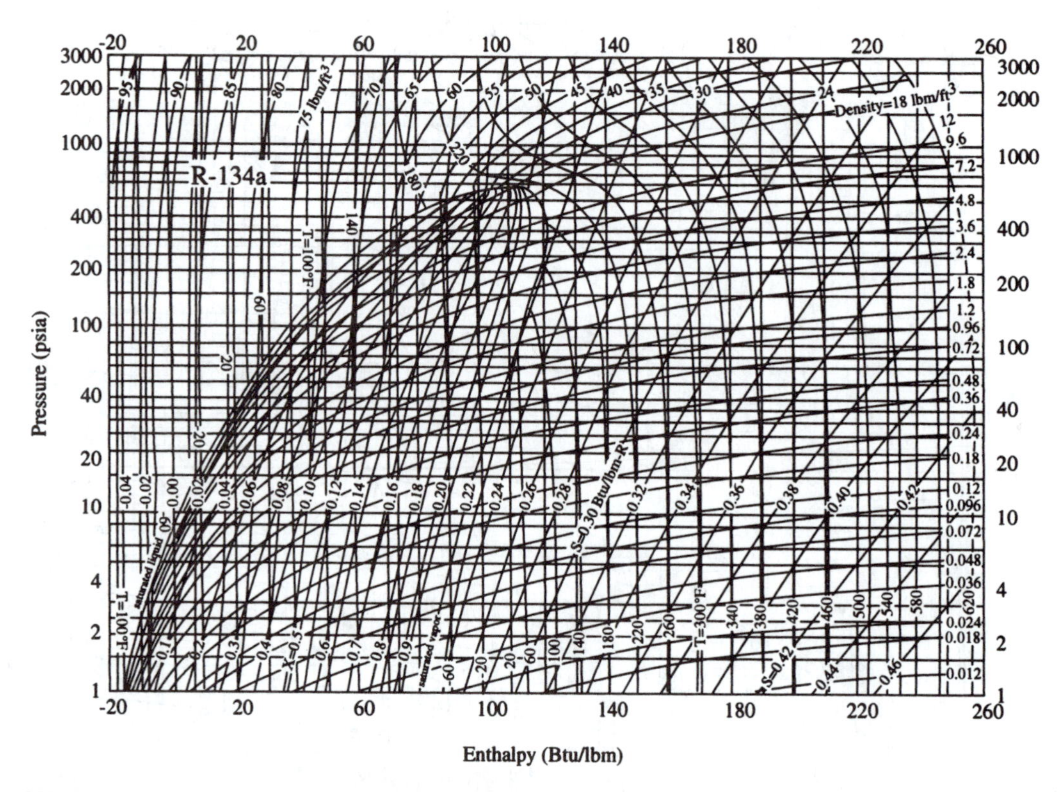

CHART B–7 (continued) Pressure–enthalpy diagram for Refrigerant-134a [reprinted with permission from the American Society of Heating, Refrigerating, and Air Conditioning Engineers, 1997 *ASHRAE handbook, fundamentals* (Atlanta, Ga.: ASHRAE, 1997)]

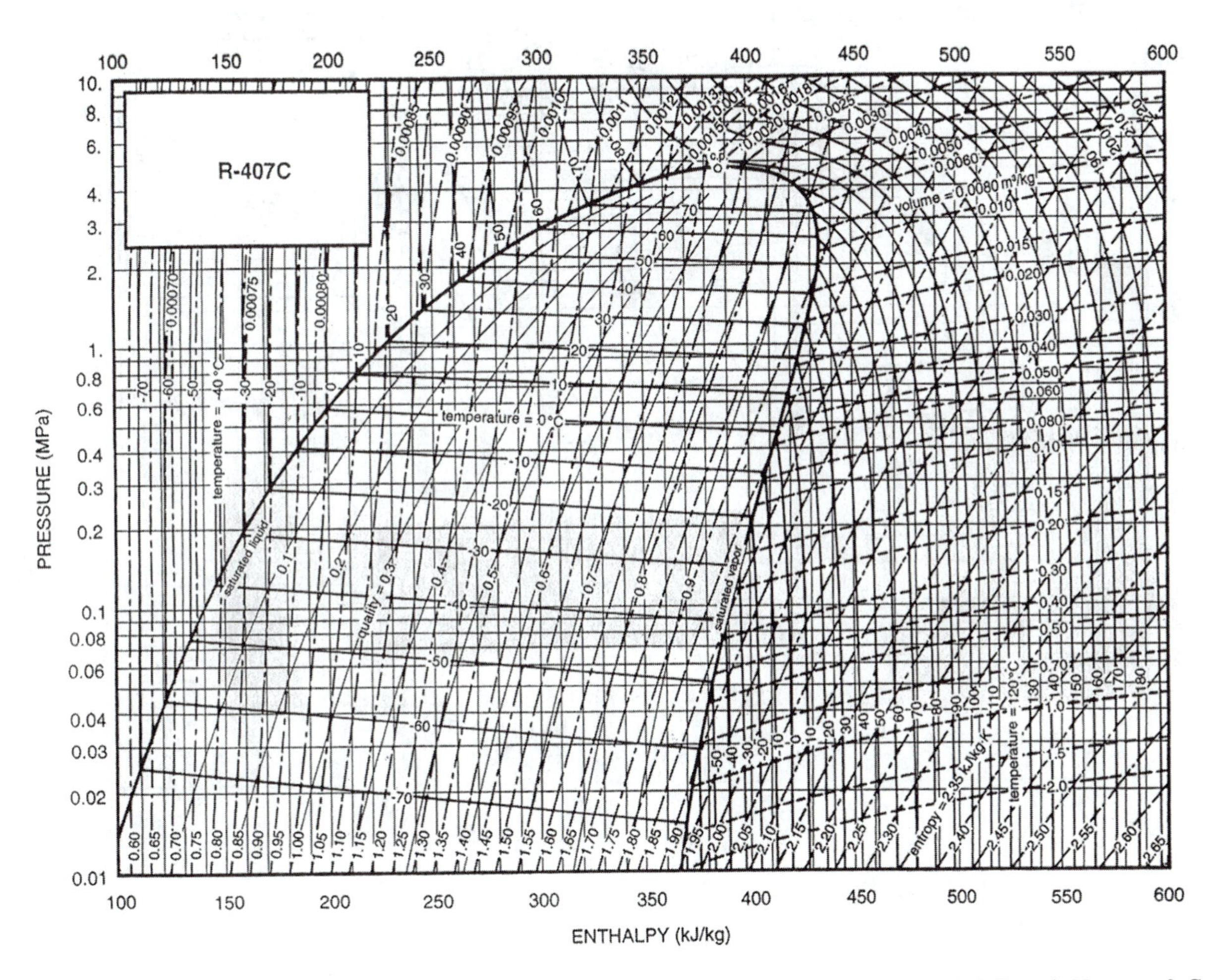

CHART B–8 Pressure–enthalpy diagram for Refrigerant-407c [reprinted with permission from E. I. du Pont de Nemours & Co.]

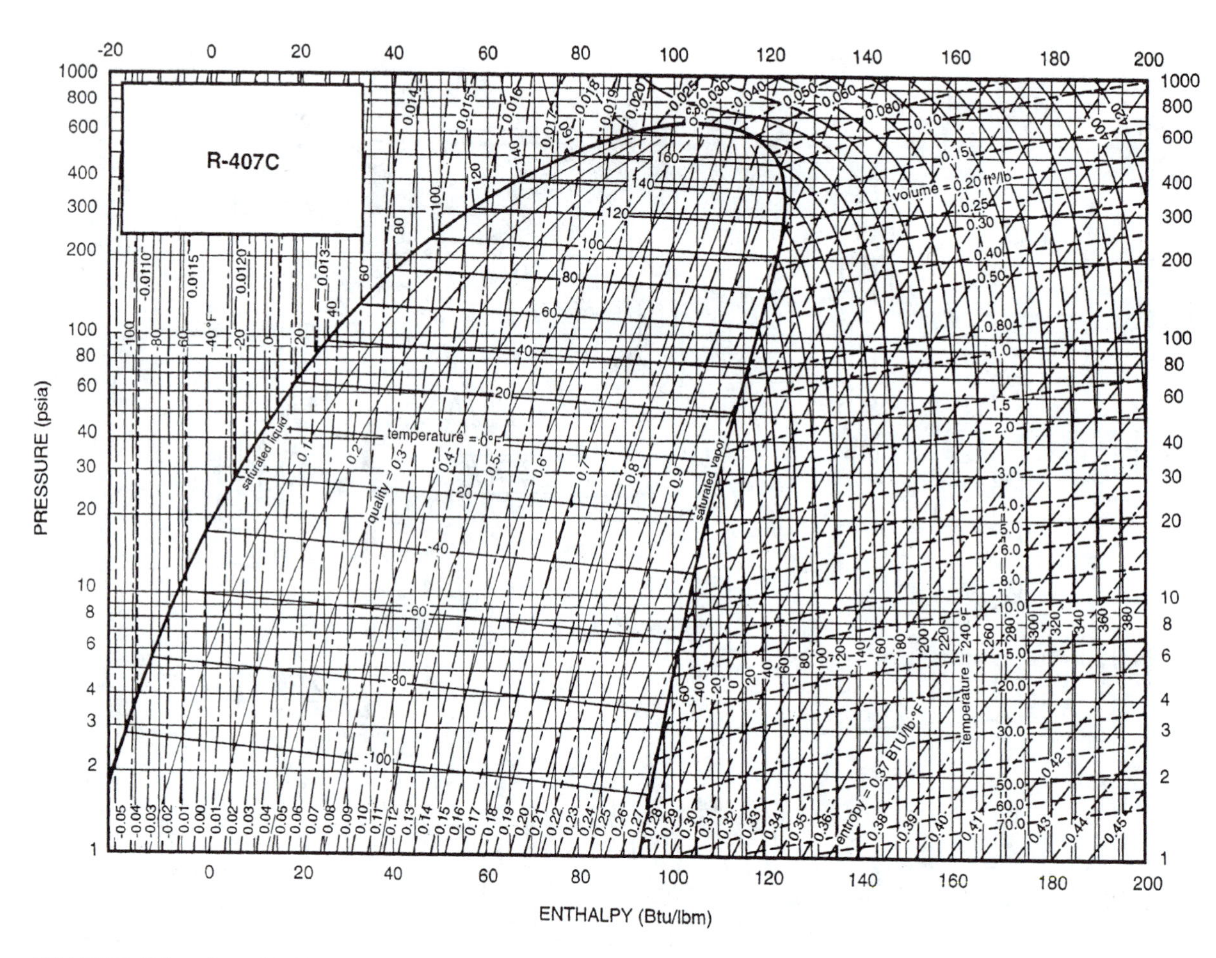

CHART B–8 (continued) Pressure–enthalpy diagram for Refrigerant-407c [reprinted with permission from E. I. du Pont de Nemours & Co.]

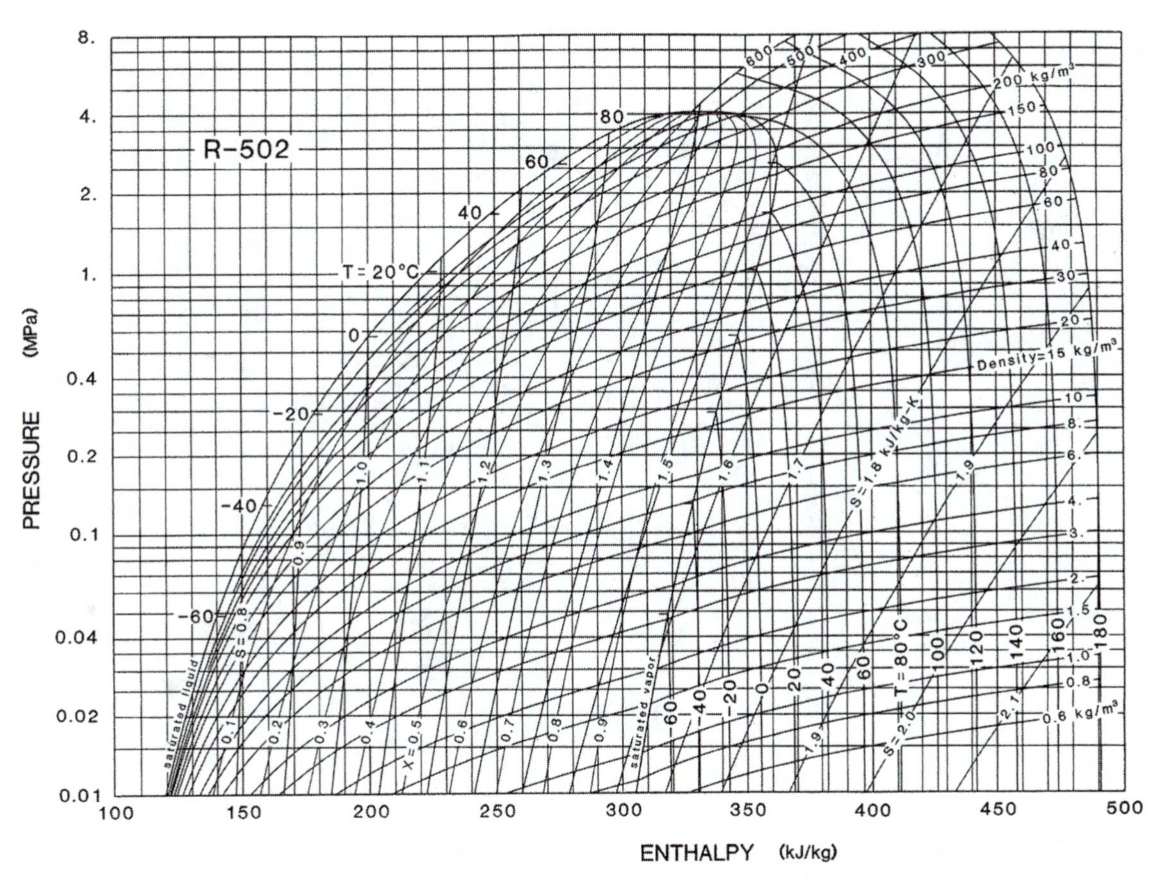

CHART B–9 Pressure–enthalpy diagram for Refrigerant-502 [from the American Society of Heating, Refrigerating, and Air Conditioning Engineers, *ASHRAE handbook, 2001 fundamentals* (Atlanta, Ga.: ASHRAE, 2001)]

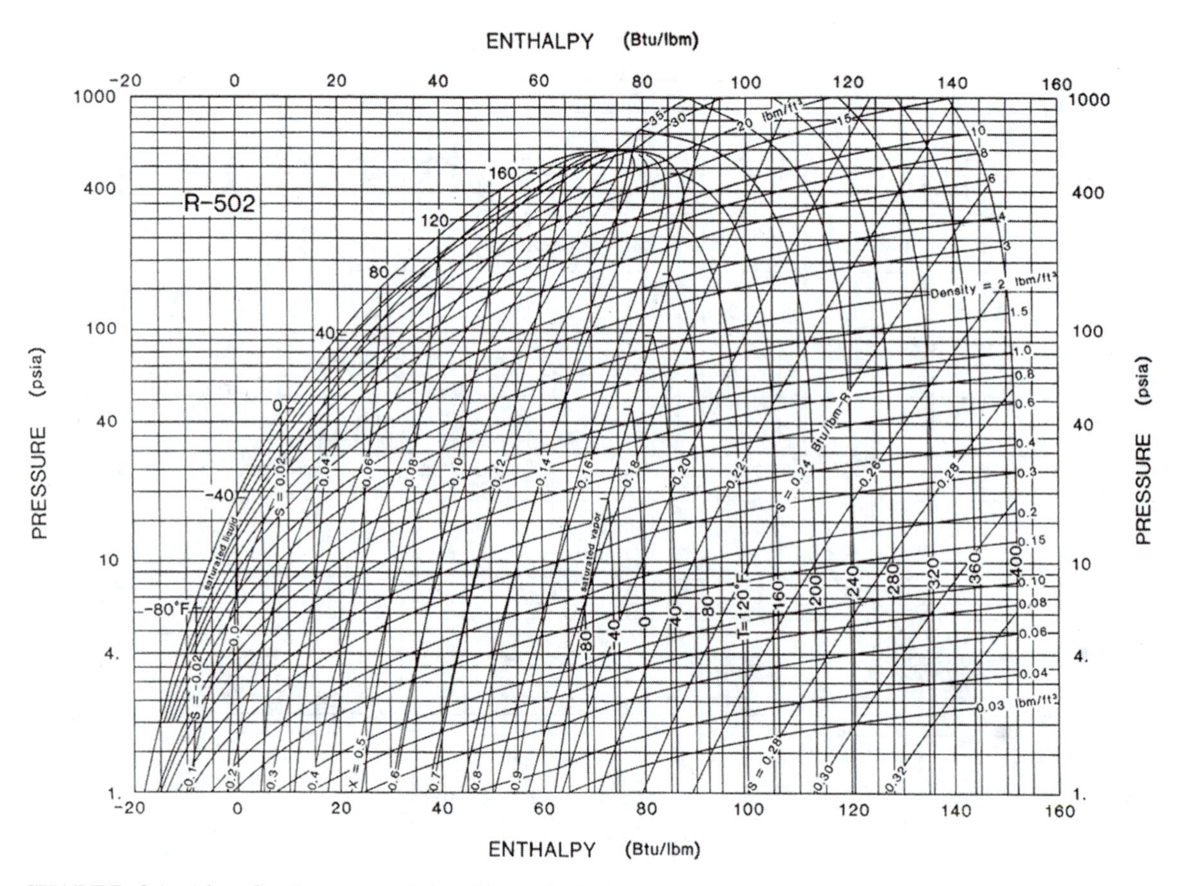

CHART B–9 (continued) Pressure–enthalpy diagram for Refrigerant-502 [from the American Society of Heating, Refrigerating, and Air Conditioning Engineers, *ASHRAE handbook*, *2001 fundamentals* (Atlanta, Ga.: ASHRAE, 2001)]

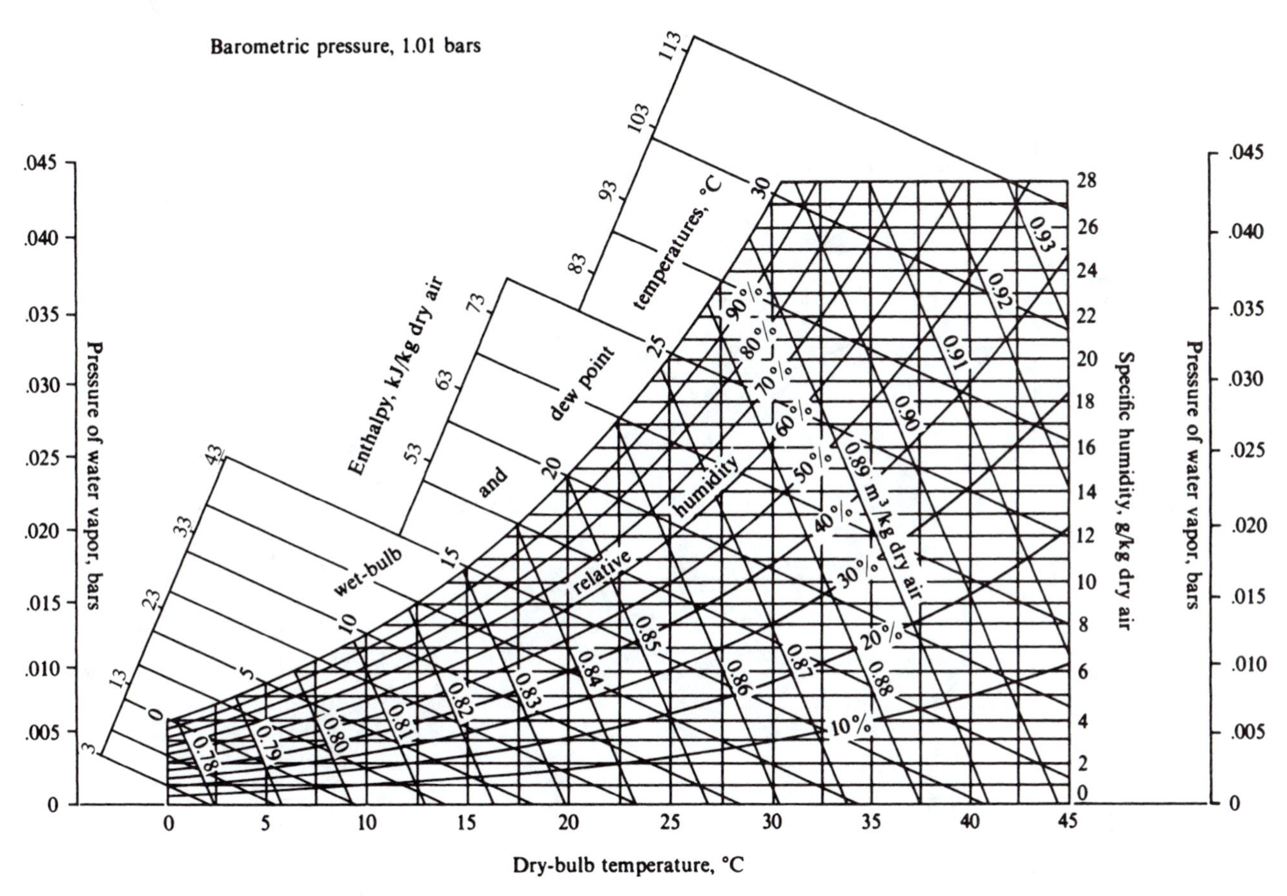

CHART B–10 Psychrometric chart, SI units [from General Electric Company Air Conditioning Division, *Psychrometric chart* [Louisville, Ky.: General Electric Company, 1968)]

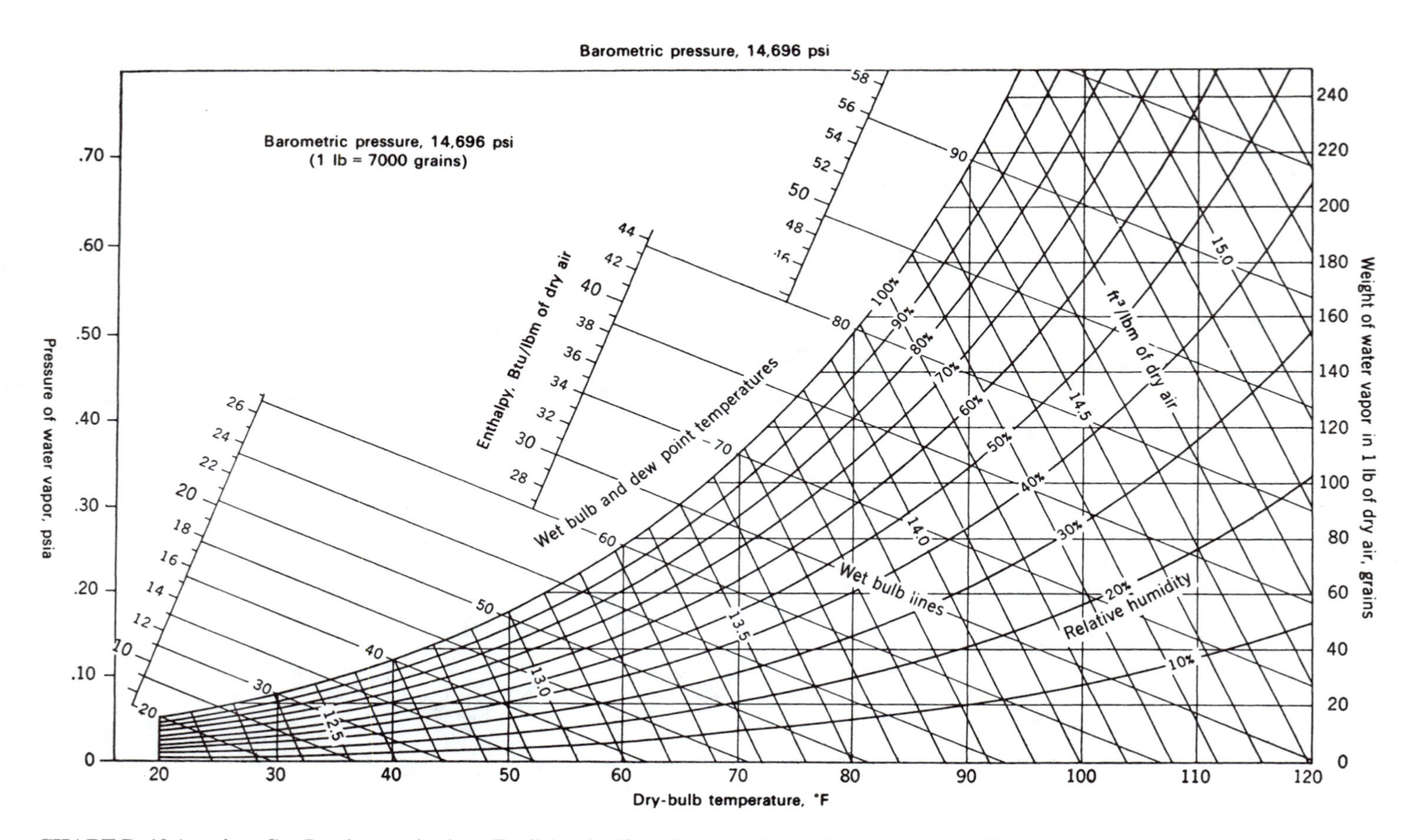

CHART B–10 (continued) Psychrometric chart, English units [from General Electric Company Air Conditioning Division, *Psychrometric chart* (Louisville, Ky.: General Electric Company, 1968)]

SELECTED REFERENCES

The following publications were used directly in the preparation of this book. This list is not intended as a complete bibliography of the literature of thermodynamics but can be used as a source of readings for the interested student.

C–1 General Thermodynamics

Bent, H. *The second law: An introduction to classical and statistical thermodynamics*. New York: Oxford University Press, 1965.

Cengel, Y. A. *Introduction to thermodynamics and heat transfer*. Boston: Irwin McGraw-Hill, 1997.

Faires, V. M. *Thermodynamics*. New York: Macmillan, 1970.

Grigoriev, I. S., and E. Z. Meilikhov. *Handbook of physical quantities*. Boca Raton, FL: CRC Press, 1997.

Hatsopoulos, G. N., and J. H. Keenan. *Principles of general thermodynamics*. New York: Wiley, 1965.

Hawkins, G. A. *Thermodynamics*. New York: Wiley, 1951.

Keenan, J. H. *Thermodynamics*. New York: Wiley, 1951.

Keenan, J. H., and J. Kaye. *Gas tables*. New York: Wiley, 1948.

Keenan, J. H., and F. G. Keyes. *Thermodynamic properties of steam*. New York: Wiley, 1937.

Lee, J. F., and F. W. Sears. *Thermodynamics*. Reading, MA: Addison-Wesley, 1963.

Mooney, D. A. *Mechanical engineering thermodynamics*. Englewood Cliffs, NJ: Prentice-Hall, 1953.

Moran, M. J., and H. N. Shapiro. *Fundamentals of engineering thermodynamics*. 4th ed. New York: Wiley, 2000.

Obert, E. F. *Concepts of thermodynamics*. New York: McGraw-Hill, 1960.

Reynolds, W. C., and H. C. Perkins. *Engineering thermodynamics*. New York: McGraw-Hill, 1977.

Solberg, H. L., O. C. Cromer, and A. R. Spalding. *Elementary heat power*. 2d ed. New York: Wiley, 1956.

Van Wylen, G. J. *Thermodynamics*. New York: Wiley, 1959.

Van Wylen, G. J., and R. E. Sonntag. *Fundamentals of classical thermodynamics*. 3d ed. New York: Wiley, 1985.

Wark, K. *Thermodynamics*. 4th ed. New York: McGraw-Hill, 1983.

Zemansky, M. W. *Heat and thermodynamics: An intermediate textbook*. New York: Wiley, 1957.

C–2 History and General Discussions of Matter and Energy

The way things work: An illustrated encyclopedia of modern technology. New York: Simon and Schuster, 1963.

Toulmin, S., and J. Goodfield. *The architecture of matter: The physics, chemistry, and physiology of matter, both animate and inanimate, as it has evolved since the beginnings of science*. New York: Harper & Row, 1962.

Ubbelhode, A. R. *Man and energy*. Baltimore: Penguin, 1963.

C–3 Internal Combustion Engines

Ferguson, C. R. *Internal combustion engines*. New York: Wiley, 1986.

Lichty, L. C. *Internal combustion engines*. 6th ed. New York: McGraw-Hill, 1951.

Obert, E. F. *Internal combustion engines*. Scranton, PA: International Textbook, 1959.

______. *Internal combustion engines and air pollution*. New York: Harper & Row, 1973.

Stone, Richard. *Introduction to internal combustion engines*. 2d ed. Warrendale, PA: Society of Automotive Engineers (SAE), 1994.

Taylor, C. F. *The internal-combustion engine in theory and practice: Thermodynamics, fluid flow, performance*. Cambridge, MA: MIT Press, 1960.

Taylor, C. F., and E. S. Taylor. *The internal combustion engine*. Scranton, PA: International Textbook, 1948.

C–4 Gas Turbines and Jet Propulsion

Bonney, E. A., M. J. Zucrow, and C. W. Besserer. *Aerodynamics, propulsion, and design practice*. New York: Van Nostrand, 1956.

Harman, R. T. C. *Gas turbine engineering*. New York: Wiley, 1981.

Pratt and Whitney Aircraft. *The aircraft gas turbine engine*. East Hartford, CT., 1970.

Sawyer, R. T. *The modern gas turbine*. Englewood Cliffs, NJ: Prentice-Hall, 1945.

Zucrow, M. J. *Principles of jet propulsion*. New York: Wiley, 1948.

C–5 Steam Turbines and Power Generation

Babcock and Wilcox Co. *Steam, its generation and use*. 38th ed. Lynchburg, VA: Babcock and Wilcox, 1972.

Combustion Engineering. *Combustion engineering*. Windsor, CT.: Combustion Engineering, 1957.

El-Wakil, M. M. *Power plant technology*. New York: McGraw-Hill, 1984.

Fraas, A. P. *Engineering evaluation of energy systems*. New York: McGraw-Hill, 1982.

Morse, F. T. *Power plant engineering and design*. New York: Van Nostrand, 1953.

Termuehlen, Heinz, and W. Emsperger. *Clean and efficient coal fired power plants*. New York: American Society of Mechanical Engineers (ASME), 2003.

Weisman, J., and L. E. Eckart. *Modern power plant engineering*. Englewood Cliffs, N.J.: Prentice-Hall, 1985.

C–6 Steam Engines

Bruce, A. W. *The steam locomotive in America*. New York: Norton, 1952.

Morrison, L. H. *Power*. New York: McGraw-Hill, 1930.

Peabody, C. H. *Thermodynamics of the steam engine*. New York: Wiley, 1909.

Sinclair, A. *Development of the locomotive engine*. Annotated edition prepared by John H. White. Cambridge, MA: MIT Press, 1970.

C–7 Refrigeration and Air Conditioning

American Society of Heating, Refrigerating and Air Conditioning Engineers. *ASHRAE Handbook, Applications*. Atlanta: ASHRAE, 1999.

______. *ASHRAE Handbook, Fundamentals*. Atlanta: ASHRAE, 2001.

______. *ASHRAE Handbook, Systems*. Atlanta: ASHRAE, 1984.

Clifford, G. E. *Heating, ventilating, and air conditioning*. Reston, VA: Reston, 1984.

Jordan, R. C., and G. B. Priester. *Refrigeration*. Englewood Cliffs, NJ: Prentice-Hall, 1948.

McQuiston, F. C., and J. D. Parker. *Heating, ventilating, and air conditioning*. 2d ed. New York: Wiley, 1982.

Sparks, N. R. *Theory of mechanical refrigeration*. New York: McGraw-Hill, 1938.

Stoecker, W. F., and J. W. Jones. *Refrigeration and air conditioning*. 2d ed. New York: McGraw-Hill, 1982.

Threlkeld, J. L. *Thermal environmental engineering*. Englewood Cliffs, NJ: Prentice-Hall, 1970.

C–8 Heat Transfer and Fluid Mechanics

Binder, R. C. *Fluid mechanics*. 4th ed. Englewood Cliffs, NJ: Prentice-Hall, 1962.

Chapman, A. J. *Heat transfer*. New York: Macmillan, 1974.

Fox, R. W., and A. T. McDonald. *Introduction to fluid mechanics*. 3d ed. New York: Wiley, 1985.

Hirschfelder, J. O., C. F. Curtis, and R. B. Bird. *Molecular theory of gases and liquids*. New York: Wiley, 1966.

Incropera, F. P., and D. P. DeWitt. *Fundamentals of heat and mass transfer*. 2d ed. New York: Wiley, 1985.

Jacob, M., and G. A. Hawkins. *Elements of heat transfer*. 3d ed. New York: Wiley, 1957.

MacAdams, W. H. *Heat transmission*. 3d ed. New York: McGraw-Hill, 1954.

Owczarek, J. A. *Fundamentals of gas dynamics*. Scranton, PA: International Textbook, 1964.

Özisik, M. N. *Basic heat transfer*. New York: McGraw-Hill, 1977.

Pao, R. H. F. *Fluid mechanics*. New York: Wiley, 1961.

C–9 Direct Energy Conversion and Solar Energy

Angrist, S. W. *Direct energy conversion*. Boston: Allyn and Bacon, 1965.

Bridgeman, P. W. *The thermodynamics of electrical phenomena in metals*. New York: Dover, 1961.

Daniels, F. *Direct use of the sun's energy*. New Haven, CT: Yale University Press, 1964.

Duffie, J. A., and W. E. Beckman. *Solar engineering of thermal processes*. New York: Wiley, 1980.

Johnson, G. L. *Wind energy systems*. Englewood Cliffs, NJ: Prentice-Hall, 1985.

Shercliff, J. A. *A textbook of magnetohydrodynamics*. Elmsford, NY: Pergamon, 1965.

Sutton, G. W., and A. Sherman. *Engineering magnetohydrodynamics*. New York: McGraw-Hill, 1965.

C–10 Combustion

American Society of Mechanical Engineers. *Combustion fundamentals for waste incineration*. New York: ASME, 1974.

Chigier, N. *Energy, combustion, and environment*. New York: McGraw-Hill, 1981.

C–11 Fuel Cells

Blomen, L.J.M.J., and M.N. Mugerwa. *Fuel cell systems*. New York: Plenum Press, 1993.

Ellis, M.W., *Fuel cells for building applications*. Atlanta, GA: American Society of Heating, Refrigerating, and Air Conditioning Engineers (ASHRAE), 2000.

Hoogers, Gregor. *Fuel cell technology*. Boca Raton, FL: CRC Press, 2003.

Larminie, J., and A. Dicks. *Fuel cell systems explained*. Chichester, UK: John Wiley & Sons, Ltd., 2000.

C–12 Biology, Natural Science

Keeton, W. T. *Biological science*. New York: Norton, 1967.

King, B. G., and M. J. Showers. *Human anatomy and physiology*. 5th ed. Philadelphia: W. B. Saunders, 1963.

Langley, L. L., and E. Cheraskin. *The physiology of man*. 2d ed. New York: McGraw-Hill, 1958.

Tuttle, W. W., and B. A. Schottelius. *Textbook of physiology*. 16th ed. St. Louis: C. V. Mosby, 1969.

THERMODYNAMIC NOTATION AND LIST OF SYMBOLS

D

Symbol	Variable	Units		Chapter Introduced
		English	SI	
a	Acceleration	ft/s^2	m/s^2	1
A	Area	ft^2	m^2	2
AF	Air/fuel ratio	lbm/lbm	kg/kg	11
b	Spring constant	lbf/in	N/cm	3
bhp	Brake horsepower	hp	kW	9
Bi	Biot number	unitless	unitless	15
bmep	Brake mean effective pressure	psi	kPa	9
bsfc	Brake specific fuel consumption	lbm/h · h	kg/kW · h	9
c	Concentration ratio	lbm/lbm	kg/kg	13
C, c	Constants	varied	varied	
c	Specific heat for incompressible substances	Btu/lbm · °R	kJ/kg · K	5
C_p	Total specific heat at constant pressure	Btu/°R	kJ/K	5
c_p	Specific heat at constant pressure	Btu/lbm · °R	kJ/kg · K	5
C_v	Total specific heat at constant volume	Btu/°R	kJ/K	5
c_v	Specific heat at constant volume	Btu/lbm · °R	kJ/kg · K	5
COP	Coefficient of performance	unitless	unitless	7
COR	Coefficient of refrigeration	unitless	unitless	7
D	Diameter	ft	m	3
DD	Degree-day	°F · day	°C · day	16
E	Energy	ft · lbf	kJ	4
e	Specific energy	ft · lbf/lbm	kJ/kg	4
F	Force	lbf	N	1
F_f	Friction force	lbf	N	3
F_I	Thrust	lbf	N	10
F_N	Normal force	lbf	N	3
g	Local gravitational acceleration	ft/s^2	m/s^2	2
g_c	Gravitational constant	$32.17\ ft \cdot lbm/lbf \cdot s^2$		2
Gr	Grashof number	unitless		15
G_u	Universal gravitational constant			2
H	Enthalpy	Btu	kJ	4
h	Specific enthalpy	Btu/lbm	kJ/kg	4
h_g	Enthalpy of saturated vapor	Btu/lbm	kJ/kg	5
h_f	Enthalpy of saturated liquid	Btu/lbm	kJ/kg	5
h_{fg}	Heat of vaporization	Btu/lbm	kJ/kg	5
HHV	Higher heating value	Btu/lbm	kJ/kg	9
imep	Indicated mean effective pressure	psi	kPa	9
ihp	Indicated horsepower	hp	hp	9
I_{sp}	Specific impulse	lbf · s/lbm	N · s/kg	10

Symbol	Variable	Units		Chapter Introduced
		English	SI	
J	Mechanical–thermal conversion factor	778 ft · lbf/Btu		3
k	cp/cv	unitless	unitless	5
K_h	Coefficient of conversion heat transfer	Btu/h · ft^2 · °F	W/m^2 · K	15
KE	Kinetic energy	ft · lbf	kJ	2
ke	Specific kinetic energy	ft · lbf/lbm	kJ/kg	2
L, l	Length	ft	m	2
LHV	Lower heating value	Btu/lbm	kJ/kg	9
m	Mass	lbm	kg	1
$\dot{m}$	Mass flow rate	lbm/s	kg/s	4
mep	Mean effective pressure	psi	kPa	9
MW	Molecular weight	lbm/lbm · mole	kg/kg · mole	5
n	Polytropic exponent	unitless	unitless	3
n	Number of cylinders in an engine			9
N	Angular or rotational speed	rpm	rpm	3
N_m	Number of moles	lbm · moles	kg · moles	5
p_a	Atmospheric pressure	psia	kPa	2
p_c	Critical pressure	psia	kPa	5
p_g	Gage pressure	psig	kPa	2
p_{gv}	Vacuum gage pressure	psiv	kPa	2
p	Pressure	psia	kPa	2
P	Perimeter	ft	m	15
PE	Potential energy	ft · lbf	kJ	2
pe	Specific potential energy	ft · lbf/lbm	kJ/kg	2
p_R	Reduced pressure	unitless	unitless	5
Pr	Prandtl number	unitless		15
p_r	Relative pressure	unitless		9
Q	Heat	Btu	kJ	3
$\dot{Q}$	Heat transfer	Btu/s	kW	3
q	Specific heat transfer	Btu/lbm	kJ/kg	3
Q_c	Heat of combustion	Btu/lbm	kJ/kg	14
$\dot{q}_C$	Heat of combustion	Btu/lbm	kJ/kg	14
$\dot{q}_A$	Heat transfer per unit area	Btu/s · ft^2	kW/m^2	15
$\dot{q}_L$	Heat transfer per unit length	Btu/s · ft	kW/m	15
R	Gas constant	ft · lbf/lbm · °R	N · m/kg · K	2
Re_D	Reynolds number based on diameter	unitless		15
Re_L	Reynolds number based on length	unitless		15
R_u	Universal gas constant	ft · lbf/lbm · mole · °R	N · m/kg · mole · K	5
r	Radius	ft	m	2
r_c	Cutoff ratio	unitless	unitless	9
r_p	Pressure ratio	unitless	unitless	10
r_v	Compression ratio	unitless	unitless	9
R_T	Thermal resistance	h · °F/Btu	K/W	15
R_v	*R*-value	h · ft^2 · °F/Btu		15
S	Entropy	Btu/°R	kJ/K	7
$\dot{S}$	Entropy generation	Btu/s · °R	kW/K	7
s	Specific entropy	Btu/lbm · °R	kJ/kg · K	5
s_f	Entropy of saturated liquid	Btu/lbm	kJ/kg	5
s_g	Entropy of saturated vapor	Btu/lbm	kJ/kg	5
SE	Strain energy	ft · lbf	kJ	6
SG	Specific gravity	unitless	unitless	2
sfc	Specific fuel consumption	lbm/hp · h	kg/kW · h	9

Symbol	Variable	Units		Chapter Introduced
		English	SI	
T	Temperature	°F, °R	°C, K	2
t	Time	second(s)	s	3
t_e	Time constant	s	s	15
T_c	Critical temperature	°R	K	5
T_R	Reduced temperature	unitless	unitless	5
U	Internal energy	Btu	kJ	2
u	Specific internal energy	Btu/lbm	kJ/kg	2
u_f	Internal energy of a saturated liquid	Btu/lbm	kJ/kg	5
u_g	Internal energy of a saturated vapor	Btu/lbm	kJ/kg	5
V	Volume	ft^3	m^3	2
v	Specific volume	ft^3/lbm	m^3/kg	2
v_c	Critical volume	ft^3/lbm	m^3/kg	5
V_D	Displacement	ft^3, in^3	m^3, cm^3	9
v_f	Volume of a saturated liquid	ft^3/lbm	m^3/kg	5
v_g	Volume of a saturated vapor	ft^3/lbm	m^3/kg	5
$\overline{V}$	Velocity	ft/s	m/s	2
$\dot{V}$	Volume flow rate	ft^3/s	m^3/s	4
v_R	Reduced volume	unitless	unitless	5
$v_{R'}$	Pseudoreduced volume	unitless	unitless	5
v_r	Relative volume	unitless		9
W	Weight	lbf	N	2
Wk	Work	ft · lbf	kJ	3
Wk_{cs}	Closed-system work	ft · lbf	kJ	3
Wk_{os}	Open-system work	ft · lbf	kJ	4
Wk_{shaft}	Shaft work	ft · lbf	kJ	3
wk	Work per unit mass	ft · lbf/lbm	kJ/kg	3
$\dot{Wk}$	Power	ft · lbf/s, hp	kW	3
x, y	Length	ft	m	2
Y	Young's modulus	psi	Mpa	6
Z	Compressibility factor	unitless	unitless	5
z	Reference elevation above plane of zero potential energy	ft	m	2
alpha, α	Absorptivity	unitless	unitless	15
alpha, α	Seebeck coefficient	joules/coulomb		17
beta, β	Relative humidity	percent		13
beta, β_p	Volumetric expansion	°R	K^{-1}	15
gamma, γ	Specific weight	lbf/ft^3	N/m^3	2
delta, Δ	Change in a variable			1
delta, δ	Very small change in a variable			3
epsilon, ϵ	Emissivity	unitless		15
eta, η	Efficiency	percent		2
eta, η_{mech}	Mechanical efficiency	percent	percent	9
eta, η_s	Adiabatic efficiency	percent	percent	6
eta, η_T	Thermal efficiency	percent	percent	7
eta, η_v	Volumetric efficiency	lbm/lbm	kg/kg	9
theta, θ	Angular displacement or rotation	radians or degrees	radians or degrees	3
kappa, κ	Thermal conductivity	Btu/h · ft · °F	kW/m · K	15
lambda, λ	Function of availability			8
mu, μ	Chemical potential	Btu/lbm	kJ/kg	13
mu, μ_k	Viscosity	lbm/ft · s	g/m · s	15

Symbol	Variable	Units		Chapter Introduced
		English	SI	
rho, ρ	Density	lbm/ft^3	kg/m^3	2
sigma, σ	Stefan–Boltzmann constant	$Btu/h \cdot ft^2 \cdot °R^4$	$W/m^2 \cdot K^4$	15
tau, T	Torque	ft · lbf	N · m	3
tau, τ	Transmissivity	unitless	unitless	15
phi, Φ	Total availability	Btu or ft · lbf	kJ	8
phi, ϕ	Availability	Btu/lbm or ft · lbf/lbm	kJ/kg	8
phi, ϕ	Encrety	unitless		7
chi, χ	Quality	percent		5
psi, ψ	Mass fraction	lbm/lbm	kg/kg	13
omega, Ω	Mole fraction	moles/mole	moles/mole	13
omega, ω	Specific humidity	$\frac{\text{lbm vapor}}{\text{lbm dry air}}$	$\frac{\text{kg vapor}}{\text{kg dry air}}$	13
xi, ξ	Strain			6
$\mathscr{D}$	Diffusivity	ft^2/s	m^2/s	13
$\mathscr{E}$	Electrical potential	volts		17
$\mathscr{F}$	Faraday's constant	coulombs/g · mole		17
$\mathscr{G}$	Total Gibbs free energy	Btu or ft · lbf	kJ	8
g'	Gibbs free energy	Btu/lbm or ft · lbf/lbm	kJ/kg	8
H'	Total Helmholtz free energy	Btu or ft · lbf	kJ	8
h'	Helmholtz free energy	Btu/lbm or ft · lbf/lbm	kJ/kg	8
$\mathscr{I}$	Electric current	amperes		17
$\mathscr{Q}$	Electric charge	coulombs		17
$\mathscr{R}$	Resistance	ohms		17
$\mathscr{T}$	Thomson coefficient	joules/coulombs · K		17
$\mathscr{V}$	Voltage	volts		17
$\mathscr{J}$	Radiosity	Btu/s	W	15
$\mathscr{U}$	Overall conductance for heat exchangers	$Btu/ft^2 \cdot h \cdot °F$	$kW/m^2 \cdot k$	15

ANSWERS TO SELECTED PROBLEMS

Chapter 1

1–2 476,476
1–4 4.0155...
1–6 −1665
1–8 **(a)** –5.262 kJ **(b)** 8.925 Btu/lbm
1–10 5.4739
2.061×10^{-9}
4.810
1–12 $x = 6.868$ ft
1–14 $T = 95°C$
1–16 $x = 2.3/y^{1.6}$, $y = (2.3/x)^{(1/1.6)}$
1–18 364.5 kJ
1–20 6327.95 in · lbf
1–22 $A = 665$ kJ/kg
1–24 for $n = 10$, $A = 1014.75$
1–26 $A = 76454.44$
1–28 $s = 3.458$, $T = 784.5$

Chapter 2

2–2 $W = 29.37$ N
2–4 $W = 5.44$ lbf
2–6 **(a)** 1.178 m^3 **(b)** 5093 N/m^3 **(c)** 518.6
(d) 0.5186
2–8 **(a)** 102 kPa **(b)** 103.4 kPa
2–10 **(a)** 0.0993lbm/ft^3 **(b)** 0.0991 lbf/ft^3
(c) 0.00159
2–12 28,274 lbf
2–14 **(a)** 29.9389 in Hg **(b)** 61.3 kPa **(c)** 147.3 psi
(d) 344.75 kPa **(e)** 2.9008 psi **(f)** 20 in
WG = 0.722 psig **(g)** 50 cm WG = 4.903 kPa
2–16 20.4 kPa
2–18 9.523 mV
2–20 $T_N = -45.58°N$ at absolute zero
2–22 $T_L = 0.255 + \log T_k$
2–24 **(a)** 484 K **(b)** 273 K **(c)** 702°R **(d)** 621°R
2–26 **(a)** 1.736×10^6 kJ **(b)** 13.2435×10^5 kJ
2–28 **(a)** 392 J **(b)** 588 J
2–30 288 J/kg
2–32 **(a)** 305 kJ **(b)** 270 kJ
2–34 180 Btu
2–36 2100 Btu/lbm
2–38 0.062 ft · lbf/lbm
2–40 12.88×10^6 Btu
2–42 949.45 kW
2–44 23.7%
2–46 65.789 MW

Chapter 3

3–2 5880 J
3–4 3960 ft · lbf
3–6 1.414 cm
3–8 128 N, 640 N, 30.72 J
3–10 0
3–12 0.6 kJ
3–14 14.07 kJ
3–16 366,841.44 ft · lbf
3–18 47,124 N · m
3–20 0.0625 m^3, −207.9 kJ
3–22 $Wk = 255.7$ kJ, $p_2 = 0.0305$ MPa
3–24 2.77 J
3–26 $Wk = 117$ J (approx.)
3–28 $Wk = 312$ J
3–30 −433.3 in · lbf
3–32 −2872.8 kJ
3–34 6.2569×10^7 J for 1 kg mass, 1.2685×10^6 km for 99.5% of work done
3–36 63.24 hp
3–38 123,750,000 ft · lbf
3–40 1,272,727 hp
3–42 0.0545 hp
3–44 4.3982 kW
3–46 5.744 ft · lbf
3–48 15 m/s, 112,500 N
3–50 87.5 s
3–52 29.85 s
3–54 2.16×10^6 Btu (total), 90,000 Btu/h (average)
3–56 0.00555 hp
3–58 4.36 lbf
3–60 210 kW
3–62 0.159 N · m
3–64 1.63 mN

Chapter 4

4–2 1.79 cm
4–4 87.95 J/kg
4–6 16 m/s
4–8 1.9 kg/s
4–10 48.97 ft/min
4–12 **(a)** 2.45 lbm/s **(b)** 1875 ft/s **(c)** 360 ft/s
4–14 10.11 ft/s
4–16 0.32 in
4–18 706.8 ft/s
4–20 15,000 lbm
4–22 30,000 kg
4–24 761.45 gpm
4–26 $t = 35.4$ min, $m = 2000 - 1.879t + 4.41 \times 10^{-4}t^2$
4–28 26.486 . . . s
4–30 200 J
4–32 −3000 kJ/kg
4–34 0
4–36 16.8 Btu
4–38 −4.1 hp (cooling)
4–40 **(a)** −24 kJ **(b)** −24 kJ, 0 **(c)** 0
(d) $wk_{cs} = -8$ kJ/kg, $q = 0$
4–42 21,600 ft · lbf
4–44 69.4 J
4–46 24.04 ft · lbf
4–48 0.177 Btu/lbm
4–50 10.28 hp
4–52 $pv = 272$ kJ/kg, $h = 3111$ kJ/kg, $H = 21{,}777$ kJ
4–54 $pv = 37$ Btu/lbm, $h = 157$ Btu/lbm, $H = 157$ Btu
4–56 14.51 Btu/lbm, 137.44 Btu/lbm, 1374.4 Btu
4–58 3277.1 kJ/kg
4–60 1232 Btu/lbm
4–62 192 ft/s

Chapter 5

5–2 Solid
5–4 Solid
5–6 Liquid
5–8 317.7 kPa
5–10 1.636 m^3/kg
5–12 1.401 kJ/kg · K
5–14 31.9 psig
5–16 203.77 ft^3
5–18 0.2 liters
5–20 Not a perfect gas
5–22 0.590 ft^3/lbm
5–24 363.8°F
5–26 311°R (van der Waal), 313°R (perfect gas)
5–28 $\Delta U = -23.2785$ kJ, $\Delta u = -36.95$ kJ/kg
5–30 618 kJ/kg
5–32 5.865 Btu/lbm
5–34 6083.105 Btu/lbm · mole
5–36 60 kJ/kg
5–38 $cv = 35.3$ kJ/kg · K, $c_p = 49.6$ kJ/kg · K
5–40 −74.46 Btu/lbm
5–42 $c_v = 0.360$ Btu/lbm · °R, $c_p = 0.4608$ Btu/lbm · °R
5–44 −305.912 kJ/kg
5–46 −24 kJ/kg
5–48 298.125 kJ/kg
5–50 $u = 199.8$ kJ/kg, $h = 280.1$ kJ/kg
5–52 $\Delta h = 372.8$ kJ/kg
5–54 $\Delta h = 133.56$ Btu/lbm
5–56 $\Delta h = 99.55$ Btu/lbm
5–58 $p = 140.9$ psia
5–60 $p = 244.83$ kPa
5–62 $v = 0.03883$ m^3/kg, $h = 221.173$ kJ/kg
5–64 $h = 1022.942$ Btu/lbm, $V_g/V = 99.9995\%$
5–66 $u_g = 2792.9$ kJ/kg
5–68 $T = 17.5$°C, $h_f = 72.998$ kJ/kg
5–70 $h = 788.72$ Btu/lbm, $s = 1.2297$ Btu/lbm°R
5–72 $s = 7.224$ kJ/kg-K, $u = 2705$ kJ/kg
5–74 $v = 0.22591$ ft^3/lbm, $u = 103.93$ Btu/lbm
5–76 $\rho = 5.57886$ kg/m^3, $V_f/V = 0.473\%$
5–78 $wk = -2.195$ Btu/lbm, $h = 38.675$ Btu/lbm

Chapter 6

6–2 $v_2 = 1.123$ cm^3, $Wk = -0.13476$ J
6–4 26.292 Btu/lbm
6–6 234.158 kPa, $Wk = 0$
6–8 217.86 psia
6–10 −55,108.5 ft · lbf
6–12 $Q = Wk = 244$ Btu, $\Delta U = 0$
6–14 $Wk = 29.064$ kJ, $\Delta U = -27.1956$ kJ
6–16 $\dot{W}k = 85.8$ kW, $\dot{Q} = -14.55$ kW, $\dot{H} = -100.35$ kW
6–18 $n = 1.143$, 210.28 Btu
6–20 $\dot{W}k = 1076$ hp, $\dot{Q} = -105$ Btu/s
6–22 $T_2 = 75.63$°C, $p_2 = 118.99$ kPa
6–24 $\dot{W}k = 785.1$ kW, $\dot{u} = -452.96$ kW
6–26 19,339 hp
6–28 $p_2 = 2390$ kPa, $\Delta U = 3.04$ kJ, $Wk = 23.04$ kJ
6–32 206.455 kJ/kg, 294.98 K
6–34 $Wk = -307.85$ kJ/kg, $T_2 = 588.7$ K
6–36 $Wk = -222.79$ kJ/kg, $T_2 = 1010.07$ K
6–38 406 K, 8.49 kg
6–40 352.7°R
6–42 766.2°R
6–44 14.21 kJ/kg
6–46 0
6–48 0.654 hp
6–50 −20.295 Btu/s
6–52 108.24 kJ
6–54 0.153 kJ
6–56 376,570 kW
6–58 $\Delta V = 29.3832$ ft^3, $Q = 195.87$ Btu
6–60 −1266.5 kJ/kg
6–62 29.335 psia, −102.585 Btu

6–64 96.1°F, $wk_{cs} = -32.08$ Btu/lbm, $wk_{os} = -41.5$ Btu/lbm
6–66 −2.08 Btu/lbm
6–68 **(a)** −35.4°F, **(b)** 25.3%
6–70 $\Delta V = 7.0035$ ft^3, $Q = 219.7$ Btu
6–72 620 Btu/lbm
6–74 −127.57 Btu/lbm, 174.4°F
6–76 $wk = -17.82$ kJ/kg, $h = 394.82$ kJ/kg
6–78 $T = -20$°F, $x = 66.6\%$
6–80 $wk = 480.555$ Btu/lbm
6–82 $x = 99.7\%$
6–84 $T = -40$°C, $x = 31.49\%$
6–86 $x_2 = 79.9\%$, $wk = -19.6$ Btu/lbm
6–88 $x_2 = 70.0\%$, $wk = 322.9$ kJ/kg
6–90 $T_2 = 357$°F, $p_2 = 506.9$ psia

Chapter 7

7–2 $\Delta U_{12} = 0$, $Wk_{23} = 0.1$ Btu, $\Delta U_{34} = -2$ Btu, $\Delta U_{41} = 2.3$ Btu, $\Delta T_{41} = -375$°F, $\Delta P_{23} = 360$ psi
7–4 $Wk_{12} = -1000$ Btu, $\Delta E_{23} = 4000$ Btu, $Q_{45} = -2001$ Btu, $\Delta E_{34} = -3200$ Btu, $\Delta E_{51} = 301$ Btu, $\Delta E_{45} = -2101$ Btu
7–6 1027°C
7–8 −5600 Btu
7–10 62.84%
7–12 97.9%
7–14 71.9%, 0.281
7–18 **(a)** 974.2 kJ/kg **(b)** −1144.128 kJ/kg **(c)** −169.928 kJ/kg **(d)** COP = 6.733, COR = 5.733
7–20 **(a)** 19.97 hp **(b)** 66.132 Btu/s **(c)** −80.298 Btu/s **(d)** COP = 5.668, COR = 4.668
7–22 yes
7–24 1.296 Btu/°R
7–26 0.0144 Btu/s · °R
7–28 impossible
7–30 0.00936 kW/K
7–32 399.4 K
7–34 0.1194 kJ/kg · K
7–36 $\Delta S = 0.0024$ kJ/K
7–38 $\Delta s = -0.0357$ kJ/kg · K
7–40 $\Delta s = 0.0602$ Btu/lbm · °R
7–42 $\Delta s = -0.03265$ Btu/lbm · °R
7–44 63.47 Btu/°R
7–46 $\Delta s = 0.669$ kJ/kg · K
7–48 608°F
7–50 **(a)** −0.01615 kJ/kg · K **(b)** 0.08685 kJ/kg · K
7–52 $\Delta s = 1.3636$ Btu/lbm · °R
7–54 $wk_{net} = 271.2$ kJ/kg, $q_{add} = 375.55$ kJ/h, $\eta_T = 72.2\%$
7–56 **(a)** 72.19% **(b)** 148,249.55 ft · lbf **(c)** $Q_{add} = 263.9$ Btu
7–60 net work = 1.05 kJ $Q_{rej} = -0.45$ kJ $\eta_T = 70.0\%$

Chapter 8

8–2 202 kJ
8–4 2.1 m
8–6 **(a)** −82.22 Btu/lbm **(b)** −3.93 Btu/lbm
8–8 **(a)** 329.7 Btu **(b)** 388.9 Btu **(c)** 348.6 Btu
8–10 −53.2 kJ
8–12 2784 Btu/s
8–14 **(a)** $g' = -1172$ kJ/kg, $H' = -1440$ kJ/kg **(b)** $g' = -2423$ kJ/kg, $H' = -2697$ kJ/kg
8–16 −314.4 Btu/lbm
8–18 $g' = -286.43$ kJ/kg, $h' = -314.2$ kJ/kg

Chapter 9

9–2 11.057
9–4 **(a)** 172.4 kW **(b)** −172.6 kW
9–6 2.85 in
9–8 **(a)/(b)** $T_3 = 7102$°R, $p_3 = 1398$ psia, $p_4 = 83.3$ psia **(c)** 4103.1 Btu/lbm **(d)** 7380 Btu/cycle, −3276.9 Btu/cycle
9–10 49.6%
9–12 60.19%
9–14 246 psi
9–16 $\eta_{mech} = 82.2\%$, bmep = 510 kPa, bsfc = 0.0263 kg/hp · hr
9–18 $\eta_v = 66.26\%$
9–20 **(a)** 504.8 kW **(b)** 1019.4 kW **(c)** 49.5% **(d)** −514.6 kW
9–22 **(a)** $T_2 = 1142$°R, $p_2 = 198.7$ psia, $T_3 = 4966$°R, $p_3 = 864$ psia, $T_4 = 2872$°R **(b)** 387.16 Btu/lbm **(c)** 4.47 inches **(d)** −5320.86 Btu/min
9–24 **(a)** 78.1 psi **(b)** 0.889 lbm/bhp · hr **(c)** 82.3%
9–26 **(a)** 272.8 kPa **(b)** 6.369 kW **(c)** 47.1%
9–28 **(a)** 188 in^3 **(b)** 458.5 Btu/s **(c)** 46.3% **(d)** 0.04317 Btu/s · °R
9–30 77.9%
9–32 **(a)** 1.508 g/kW · min **(b)** 1077 kPa **(c)** 86.2%
9–34 **(a)** 634.5 kJ/cycle **(b)** −308.6 kJ/cycle **(c)** 325.9 kJ/cycle **(d)** 51.36% **(e)** 968.9 kPa
9–36 3.11
9–38 **(a)** 631.43 Btu/s **(b)** $\eta_T = 56.04\%$ **(c)** $T_{max} = T_4 = 5988.4$°F, $p_{max} = p_3 = p_4 = 1132.8$ psia
9–42 60.9%, 58.0%

Chapter 10

10–2 56.2%
10–4 5772 Btu/s
10–8 754.2 kW
10–10 **(a)** 691 kJ/kg **(b)** −117 kJ/kg **(c)** 4146 kW
10–12 −278 kJ/kg (rev. adiabatic), −229 kJ/kg (polytropic)
10–14 265.6 Btu/lbm, 1155°R
10–16 0.0191 lbm fuel/lbm air
10–18 998 m/s

10–20 337 K
10–22 0.201 lbm/s
10–24 41.6 psia
10–26 1.359 (for 10.20), 0.364 (for 10.21), 0.0969 (for 10.22), 0.135 (for 10.25)
10–28 −419 kJ/kg
10–30 −173 Btu/lbm
10–32 **(a)** 44.8% **(b)** 57,298 kW **(c)** −31,620 kW **(d)** 25,678 kW
10–34 **(a)** 80.2 lbm air/lbm fuel **(b)** 182 Btu/lbm **(c)** −84 Btu/lbm **(d)** 40.8%
10–36 **(a)** 74.1 lbm fuel/lbm air **(b)** 186.8 Btu/lbm **(c)** −84.5 Btu/lbm **(d)** 39.4%
10–38 **(a)** 0.908 lbm/fuel/s, 19.8 lbm air/s **(b)** 32.5%
10–40 $\dot{W}k = 1956.45$ hp, $\eta_T = 42.68\%$ $\dot{S}_{min} = 1.156$ Btu/s · °R
10–42 $\dot{m} = 15.342$ lbm/s, $\dot{Q}_{rej} = 4019.89$ Btu/s, $\eta_T = 50.56\%$
10–44 $\dot{m} = 9.918$ kg/s, $\dot{Q}_{add} = 13{,}785.6$ kW, $\dot{Q}_{rej} = 3785.6$ kW, $\eta_T = 72.54\%$
10–46 **(a)** 70,000,000 Btu/h **(b)** 14052 hp **(c)** 1170°R **(d)** −8772 Btu/s **(e)** 51.1%
10–48 **(a)** 702.8 kJ/kg **(b)** −210.4 kJ/kg **(c)** 492.4 kJ/kg **(d)** −627.3 kJ/kg
10–50 5076.6 ft/s, 4944 lbf, 130.14 lbf · s/lbm
10–52 459.6 lbf · s/lbm

Chapter 11

11–4 **(a)** 28,000 kW **(b)** 196.15°C
11–6 1236.1 Btu/lbm
11–8 363.7 K, 1105.5 kJ/kg
11–10 5086 hp
11–12 **(a)** 500 kW **(b)** 1360 hp
11–14 **(a)** −6.05 kJ/kg **(b)** −1.28 Btu/lbm
11–16 −2159 kJ/kg
11–18 49.1°F
11–20 21,267.9 kg/s
11–22 **(a)** 2580 kJ/kg, 6.72 kJ/kg · K **(b)** 1118 Btu/lbm, 1.629 Btu/lbm · °R
11–24 **(a)** 2960 kJ/kg, 700 kPa **(b)** 1275 Btu/lbm, 165 psia
11–26 38.36%
11–28 **(a)** 7565 kW **(b)** 6658 kW **(c)** 1744 kJ/kg **(d)** −2257.3 kJ/kg **(e)** 1737.2 kJ/kg **(f)** 38.3% **(g)** 2.07 kg/kW · h **(h)** 3.634 kJ/s · K
11–30 **(a)** −0.562 Btu/lbm **(b)** 10,020 kW **(c)** 13,077 lbm/h **(d)** 95,300,000 Btu/hr
11–34 **(a)** 127,000 hp **(b)** 15,490,000 Btu/min **(c)** 4.74 lbm/hp · h **(d)** 34.7%
11–36 **(a)** $m_7 = 0.167$ kg/kg steam **(b)** 1333.8 kJ/kg **(c)** 45.4%
11–40 **(a)** 167,038 hp **(b)** 166,155 hp **(c)** −196,600 hp **(d)** 45.8% **(e)** 4.33 lbm/hp · h

Chapter 12

12–2 11.0
12–4 230.4 kW, −30.4 kW
12–6 −334 Btu/min
12–10 1049.4 kW
12–12 −28.64 kJ/kg, 132.96 kJ/kg, −161.6 kJ/kg
12–14 **(a)** −11.9 Btu/lbm **(b)** 58 Btu/lbm **(c)** −69.9 Btu/lbm
12–16 0.27 kW, 2.37 g/s
12–18 **(a)** 21.2 kg/h **(b)** −2402 kJ/h **(c)** COP = 2.66, COR = 1.66
12–20 0.48 lbm/min, −0.83 hp
12–22 $\dot{m} = 47.6$ lbm/h, $\dot{Q}_{rej} = 2849$ Btu/h, $\dot{W}k = 1349.0$ Btu/h, COP = 2.11, COR = 1.11
12–24 COP = 6.285, COR = 5.285, $\dot{w}k = 348$ hp
12–26 COP = 1.78, COR = 0.78
12–28 COR = 0.828, $\dot{Q}_{rej} = -0.682$ kW, $\dot{m} = 0.0025$ kg/s
12–30 COR = 0.78
12–32 $\dot{m} = 10.1$ kg/s
12–34 COP = 2.86, $\dot{Q}_{add} = 16.0$ tons
12–36 for $\dot{m} = 0.2$ lbm/s, COP = 10.56, for $\dot{m} = 0.17$ lbm/s, COP = 7.41, for $\dot{m} = 0.08$ lbm/s, COP = 5.69
12–38 $\dot{w}k = 4.1$ hp, COP = 6.27
12–40 COP = 7.93
12–42 $\dot{Q}_{rej} = \dot{Q}_{supplied} = 121.7$ kJ/kg

Chapter 13

13–2 mole fractions:

Xenon	0.588
Argon	0.358
Helium	0.054

13–4 Partial pressures (kPa)

Oxygen	14.3
Carbon Dioxide	20.8
Argon	22.9
Helium	142.0

13–6 Partial pressures (psia)

Water Vapor	1.8816
Air	11.6718
Carbon Dioxide	1.1466

13–8 0.46 Btu/lbm · °R
13–10 6.678 Btu/lbm · °R
13–12 **(a)** 0.256 psia **(b)** 47.4 kPa
13–14 **(a)** 1.4 kPa **(b)** 8.6 g/kg **(c)** 58%
13–16 265 kg/h
13–18 27.1 kg/min, 24.55 m^3/min
13–20 $\dot{m}_w = 0.045$ lbm water/s $T_w = 65$°F
13–22 $m_w = 3.9$ g/kg d.a. $\beta_2 = 40\%$
13–24 1.02 Btu/lbm, 0.679 lbm/min
13–26 $m_w = 8.2$ g/kg d.a. $\beta_2 = 58\%$
13–28 74%
13–30 $\dot{Q}_{add} = 43.2$ kW, $\dot{m}_w = 6.875$ kg/min $\dot{Q}_{rej} = -455$ kW

13–32 23°C, 28%
13–34 **(a)** −17.3 kJ/kg **(b)** −1.61 Btu/lbm
13–36 −0.78 Btu/lbm
13–38 0.225×10^{26} g
13–40 0.9946 lbm/lbm
13–42 $h = 295$ kJ/kg

Chapter 14

14–4 mole fractions: Carbon 27%, Oxygen 73%
14–6 mole fractions: CH_3OH 60%, O_2 40%
14–8 13.19
14–10 3.677
14–12 48,154 kJ/kg, 44,684 kJ/kg
14–14 HHV = 17,455 kJ/kg, LHV = 16,098 kJ/kg
14–16 110.18°F
14–18 1.445 lbm · mol water condensed per lbm · mol fuel 21,280 Btu/lbm · mol
14–20 2616.7°C
14–22 2515.4°C
14–26 $\Delta s = 7040.94$ kJ/kg · mol · K
14–28 737.58 Btu/lbm · mol · °R

Chapter 15

15–2 R-value = 58.65
15–4 1.89 Btu/h · ft^2 from wallboard to brick
15–6 R-value = 26.3, $\dot{q} = 1.23$ Btu/h · ft^2
15–8 42.75 W/m
15–10 79.6084°C
15–12 2037 kW/m^2
15–14 565.5 W/m
15–16 0.75124 m^2 · K/W
15–18 7.5124°C
15–20 23.07
15–24 42%, 2553.6 W/m
15–26 91.2 W
15–28 0.45 h
15–30 257,264 Btu/h
15–32 9282 Btu/h
15–34 137.6°F
15–36 44.2 W/m^2 · K, 12.72 W
15–38 4.48 W/m^2 · K
15–40 110,700 m^3/min
15–42 901.8 W/m
15–44 493,264 Btu/h, 305°F
15–46 0.294 kW/m^2
15–48 $F_{12} = F_{21} = 0.09$, $F_{13} = F_{23} = 0.91$
15–50 $F_{12} = 0.1425$, $F_{21} = 0.19$
15–52 20.829×10^6 Btu/h · ft^2
15–54 166.5 Btu/h
15–56 159.52 Btu/h
15–58 $A = 2.1705$ m^2, $\dot{m} = 1.66$ kg/s
15–60 $T_4 = 35$°C, $A = 7.35$ m^2

Chapter 16

16–2 4438°F · days, 2465°C · days, $T_{ave} = 11.9$°C
16–4 6 to 20
16–8 127,300,000,000 kJ
16–10 1066.4 lbm/month
16–12 $\dot{Q}_c = 4{,}956{,}800$ Btu, $\dot{Q}_{a/c} = 6.298$ tons $\dot{m}_w = 725.16$ lbm/day
16–14 767,932 Btu
16–16 2187.5 lbm/h

Chapter 17

17–2 123 volts, 97.6%
17–4 10.2 amps, 99.7%
17–6 72.04 kcal/g · mol, 1.562 volts
17–8 93.94 kcal/g · mol, 2.03 volts
17–10 −685,330 kJ/kg mol, 1.1836 volts
17–12 29.9 ohms
17–14 32.4 amps, 1.62 volts
17–16 63.6 kW
17–18 $\dot{wk} = 1.2$ W
17–20 mep = 960 kPa
17–22 $wk = 0.3$ Btu/cycle, $p_3 = 50.43$ psia
17–24 2.4998 W

INDEX